AQUATIC CHEMISTRY

ENVIRONMENTAL SCIENCE AND TECHNOLOGY

A Wiley-Interscience Series of Texts and Monographs

Edited by JERALD L. SCHNOOR, *University of Iowa*
ALEXANDER ZEHNDER, *Swiss Federal Institute for Water Resources and Water Pollution Control*

A complete list of the titles in this series appears at the end of this volume

AQUATIC CHEMISTRY

Chemical Equilibria and Rates in Natural Waters

Third Edition

WERNER STUMM
EAWAG, Swiss Federal Institute of Technology (ETH), Zurich

JAMES J. MORGAN
Environmental Engineering Science, California Institute of Technology

A WILEY-INTERSCIENCE PUBLICATION
JOHN WILEY & SONS, INC.
New York • Chichester • Brisbane • Toronto • Singapore

This text is printed on acid-free paper.

Copyright © 1996 by John Wiley & Sons, Inc.

All rights reserved. Published simultaneously in Canada.

Reproduction or translation of any part of this work beyond that permitted by Section 107 or 108 of the 1976 United States Copyright Act without the permission of the copyright owner is unlawful. Requests for permission or further information should be addressed to the Permissions Department, John Wiley & Sons, Inc., 605 Third Avenue, New York, NY 10158-0012.

Library of Congress Cataloging in Publication Data:
Stumm, Werner
   Aquatic chemistry: chemical equilibria and rates in natural waters / Werner Stumm, James J. Morgan.—3rd ed.
     p. cm.—(Environmental science and technology)
   "A Wiley-Interscience publication."
   Includes bibliographical references and index.
   ISBN 0-471-51184-6 (acid-free).—ISBN 0-471-51185-4 (pbk.: acid-free)
   1. Water chemistry. I. Morgan, James J. II. Title.
III. Series.
GB855.S78 1995
359.9—dc20   94-48319

Printed in the United States of America

20 19 18 17 16 15 14

SERIES PREFACE
Environmental Science and Technology

We are in the third decade of the Wiley Interscience Series of texts and monographs in Environmental Science and Technology. It has a distinguished record of publishing outstanding reference texts on topics in the environmental sciences and engineering technology. Classic books have been published here, graduate students have benefited from the textbooks in this series, and the series has also provided for monographs on new developments in various environmental areas.

As new editors of this Series, we wish to continue the tradition of excellence and to emphasize the interdisciplinary nature of the field of environmental science. We publish texts and monographs in environmental science and technology as it is broadly defined from basic science (biology, chemistry, physics, toxicology) of the environment (air, water, soil) to engineering technology (water and wastewater treatment, air pollution control, solid, soil, and hazardous wastes). The series is dedicated to a scientific description of environmental processes, the prevention of environmental problems, and to preservation and remediation technology.

There is a new clarion for the environment. No longer are our pollution problems only local. Rather, the scale has grown to the global level. There is no such place as "upwind" any longer; we are all "downwind" from somebody else in the global environment. We must take care to preserve our resources as never before and to learn how to internalize the cost to prevent environmental degradation into the product that we make. A new "industrial ecology" is emerging that will lessen the impact our way of life has on our surroundings.

In the next 50 years, our population will come close to doubling, and if the developing countries are to improve their standard of living as is needed, we will require a gross world product several times what we currently have. This will create new pressures on the environment, both locally and globally. But there are new opportunities also. The world's people are recognizing the need for sustainable development and leaving a legacy of resources for future generations at least equal to what we had. The goal of this series is to help understand the environment, its functioning, and how problems can be overcome; the series will also provide new insights and new sustainable technologies that will allow us to preserve and hand down an intact environment to future generations.

JERALD L. SCHNOOR
ALEXANDER J. B. ZEHNDER

PREFACE

The field of natural water chemistry has continued to grow and develop over the time since the publication of the previous edition of *Aquatic Chemistry*. Our general objective in this substantially revised edition is to draw on basic chemical principles in presenting a quantitative treatment of the processes that determine the composition of natural waters. The concept of chemical equilibrium remains a major theme in our approach, but, as reflected in the new subtitle, rates of processes and chemical reactions receive greater attention than previously, reflecting increased information on these aspects of natural water chemistry acquired over the past decade. Understanding aquatic chemistry calls for both a grasp of key chemical principles and the incorporation of these principles into models that capture the essential aspects of the systems being considered. Numerical examples have been chosen to illustrate methods for attacking the most important aspects of natural water chemistry in a quantitative fashion.

There are several new features of this edition to be noted. A new chapter, Chapter 5, treats atmosphere–water interactions. This chapter illustrates that water, although a minor component of the atmosphere, plays an important role in carrying out major chemical reactions in cloud, fog, and rain—important in linking land, water, and air environments.

There are major revisions in the treatment of solid–water interfaces. Chapter 9 reflects significant progress in concepts and experimental approaches during the last decade. Interactions of solutes with solid surfaces in adsorption are characterized in terms of two basic processes: (1) formation of coordinative bonds (surface complexation) with H^+, OH^- metal ions, and ligands; and (2) hydrophobic adsorption, driven by incompatibility of nonpolar compounds with water. Both of these processes need to be understood in order to explain a variety of processes in natural systems. Surface chemistry is essential for the quantitative treatment of rate laws for geochemical processes in Chapter 13, and for a proper interpretation of the behavior of colloidal systems in particle–particle interactions in Chapter 14.

Important advances in understanding mechanisms of redox processes are treated in Chapter 8, and new interpretations of rates of electron transfer processes are considered in Chapter 11. Chapter 12, on photochemistry, analyzes important light-induced and light-catalyzed processes.

The consideration of metal ions and aqueous coordination chemistry has been updated substantially in Chapter 6. This chapter reflects recent progress made in understanding metal ion speciation and kinetics of complexation. In Chapter 10, particular attention has been directed to the cycling and the biological role of trace metals in nutrition and toxicity in aquatic systems.

Aquatic Chemistry continues to emphasize a teaching approach to the subject. The aim is to enable the reader to learn from the general concepts and methods of problem-solving so that they can then be applied to other aquatic systems of interest. The core chapters, 2 through 9, can be used as a text in an introductory course for advanced undergraduate and beginning graduate students in environmental science and engineering, earth sciences, and oceanography. The later chapters, 10 through 15, are more advanced and detailed. The combination of Chapter 9 (The Solid–Solution Interface), Chapter 13 (Kinetics at the Solid–Water Interface), and Chapter 14 (Particle–Particle Interaction) could serve for a comprehensive treatment of surface chemical principles and applications in the geochemistry of natural waters, in soil and sediment science, and in water technology.

The combination of Chapter 8 (Oxidation and Reduction), Chapter 11 (Kinetics of Redox Processes), and Chapter 12 (Photochemical Processes) introduces the reader to abiotic and biologically mediated redox processes and transformations, emphasizing, in addition to redox energetics, electron transfer mechanisms, linear free energy relationships, and photochemical processes. Chapter 6 (Metal Ions in Aqueous Solutions) and Chapter 10 (Trace Metals) provide a rather complete treatment of coordination chemistry in water and highlight new developments in chemical speciation, bioavailability, and toxicity of metals. The concluding chapter, Chapter 15 (Regulation of the Chemical Composition of Natural Waters), has the aim of acquainting the reader with major factors that regulate the chemical composition of our environment, and to emphasize the great elemental cycles moving through the rocks, water, atmosphere, and biota. We wish to illustrate the concept that pollution is no longer a local and regional problem, and that we humans are able to alter global chemical cycles.

WERNER STUMM
JAMES J. MORGAN

Zurich, Switzerland
Pasadena, California

ACKNOWLEDGMENTS

We thank Drs. Charles O'Melia, Laura Sigg, Bruce James, Dieter Diem, and Stefan Hug, and our student colleagues, Tom Lloyd and Phil Watts, for helpful reviews of individual chapters. We are also grateful to our many colleagues who provided valuable suggestions for corrections and improvements to the third edition. Appreciation is expressed to Mrs. L. Schwarz for typing the manuscript and to Mrs. L. Scott for assistance in revision of some of the chapters.

The authors are pleased to acknowledge the significant contribution that the book *Aquatische Chemie* (Sigg and Stumm, 1994) has made in the writing of this edition of our book.

We are grateful for permission to reproduce Table A6 from Morel and Hering (1993), and the tables on thermodynamic data by Nordstrom et al. and by Byrne et al. in Appendixes 1 and 2 of the book.

The hospitality of EAWAG extended to JJM in 1990 during preparation of the manuscript for this book is appreciated.

W.S.
J.J.M.

CONTENTS

1. Introduction   1

     1.1 Scope of Aquatic Chemistry   1
     1.2 The Solvent Water   6
     1.3 Solute Species   9
     Suggested Readings   11
     Appendix 1.1: Some Useful Quantities, Units, Conversion Factors, Constants, and Relationships   11

2. Chemical Thermodynamics and Kinetics   16

     2.1 Introduction   16
     2.2 Chemical Thermodynamic Principles   20
     2.3 Systems of Variable Composition: Chemical Thermodynamics   29
     2.4 Gibbs Energy and Systems of Variable Chemical Composition   32
     2.5 Chemical Potentials of Pure Phases and Solutions   35
     2.6 Chemical Potentials of Aqueous Electrolytes   38
     2.7 The Equilibrium Constant   41
     2.8 The Gibbs Energy of a System   44
     2.9 Driving Force for Chemical Reactions   49
     2.10 Temperature and Pressure Effects on Equilibrium   52
     2.11 Equilibrium Tools   57
     2.12 Kinetics and Thermodynamics: Time and Reaction Advancement, ξ 58
     2.13 Rate and Mechanism   61
     2.14 Concentration Versus Time   64
     2.15 Theory of Elementary Processes   69
     2.16 Elementary Reactions and ACT   76
     2.17 Equilibrium Versus Steady State in Flow Systems   79
     Suggested Readings   81
     Problems   82
     Answers to Problems   85

3. Acids and Bases — 88

3.1 Introduction — 88
3.2 The Nature of Acids and Bases — 90
3.3 The Strength of an Acid or Base — 92
3.4 Activity and pH Scales — 97
3.5 Equilibrium Calculations — 105
3.6 pH as a Master Variable; Equilibrium Calculations Using a Graphical Approach — 118
3.7 Ionization Fractions of Acids, Bases, and Ampholytes — 127
3.8 Titration of Acids and Bases — 130
3.9 Buffer Intensity and Neutralizing Capacity — 134
3.10 Organic Acids — 140
Suggested Readings — 144
Problems — 144
Answers to Problems — 147

4. Dissolved Carbon Dioxide — 148

4.1 Introduction — 148
4.2 Dissolved Carbonate Equilibria (Closed System) — 150
4.3 Aqueous Carbonate System Open to the Atmosphere — 157
4.4 Alkalinity and Acidity, Neutralizing Capacities — 163
4.5 Alkalinity Changes — 172
4.6 Analytical Considerations: Gran Plots — 179
4.7 Equilibrium with Solid Carbonates — 186
4.8 Kinetic Considerations — 192
4.9 Carbon Isotopes and Isotope Fractionation — 195
Suggested Readings — 202
Problems — 202
Answers to Problems — 204

5. Atmosphere–Water Interactions — 206

5.1 Introduction — 206
5.2 Anthropogenic Generation of Acidity in the Atmosphere — 207
5.3 Gas–Water Partitioning: Henry's Law — 212
5.4 Gas–Water Equilibria in Closed and Open Systems — 216
5.5 Washout of Pollutants from the Atmosphere — 227
5.6 Fog — 229
5.7 Aerosols — 233
5.8 Acid Rain–Acid Lakes — 235
5.9 The Volatility of Organic Substances — 238
5.10 Gas Transfer Across Water–Gas Interface — 241
Suggested Readings — 248
Problems — 249
Answers to Problems — 251

6. Metal Ions in Aqueous Solution: Aspects of Coordination Chemistry 252

   6.1 Introduction 252
   6.2 Protons and Metal Ions 258
   6.3 Hydrolysis of Metal Ions 263
   6.4 Solubility and Hydrolysis: Solid Hydroxides and Metal Oxides 272
   6.5 Chelates 275
   6.6 Metal Ions and Ligands: Classification of Metals 281
   6.7 Speciation in Fresh Waters 289
   6.8 Seawater Speciation 305
   6.9 Kinetics of Complex Formation 311
       Suggested Readings 319
       Problems 320
       Answers to Problems 322
       Appendix 6.1: Stability Constants 325
       Appendix 6.2: The Various Scales for Equilibrium Constants, Activity Coefficients, and pH 335

7. Precipitation and Dissolution 349

   7.1 Introduction 349
   7.2 The Solubility of Oxides and Hydroxides 359
   7.3 Complex Formation and Solubility of (Hydr)oxides 368
   7.4 Carbonates 370
   7.5 The Stability of Hydroxides, Carbonates, and Hydroxide Carbonates 389
   7.6 Sulfides and Phosphates 398
   7.7 The Phase Rule: Components, Phases, and Degrees of Freedom 409
   7.8 Solubility of Fine Particles 413
   7.9 Solid Solutions 416
       Suggested Readings 420
       Problems 420
       Answers to Problems 424

8. Oxidation and Reduction; Equilibria and Microbial Mediation 425

   8.1 Introduction 425
   8.2 Redox Equilibria and the Electron Activity 426
   8.3 The Electrode Potential: The Nernst Equation and the Electrochemical Cell 441
   8.4 pε–pH, Potential–pH Diagrams 455
   8.5 Redox Conditions in Natural Waters 464
   8.6 Effect of Complex Formers on the Redox Potential 489

	8.7	Measuring the Redox Potential in Natural Waters	491
	8.8	The Potentiometric Determination of Individual Solutes	498
		Suggested Readings	506
		Problems	507
		Answers to Problems	512
		Appendix 8.1: Activity Ratio Diagrams for Redox Systems	513

9. The Solid–Solution Interface   516

	9.1	Introduction	516
	9.2	Adsorption	519
	9.3	Adsorption Isotherms	521
	9.4	Hydrous Oxide Surfaces; Reactions with H^+, OH^-, Metal Ions, and Ligands	533
	9.5	Surface Charge and the Electric Double Layer	549
	9.6	Correcting Surface Complex Formation Constants for Surface Charge	568
	9.7	Sorption of Hydrophobic Substances on Organic Carbon-Bearing Particles	575
	9.8	Ion Exchange	586
	9.9	Transport of (Ad)sorbable Constituents in Groundwater and Soil Systems	594
		Suggested Readings	599
		Problems	601
		Appendix 9.1: The Gouy–Chapman Theory	604
		Appendix 9.2: Contact Angle, Adhesion and Cohesion, the Oil–Water Interface	608

10. Trace Metals: Cycling, Regulation, and Biological Role   614

	10.1	Introduction: Global Cycling of Metals	614
	10.2	Analytical Approaches to Chemical Speciation	615
	10.3	Classification of Metal Ions and the Inorganic Chemistry of Life	625
	10.4	Organometallic and Organometalloidal Compounds	628
	10.5	Bioavailability and Toxicity	632
	10.6	Metal Ions as Micronutrients	637
	10.7	The Interaction of Trace Metals with Phytoplankton at the Molecular Level	641
	10.8	Regulation of Trace Elements by the Solid–Water Interface in Surface Waters	648
	10.9	Regulation of Dissolved Heavy Metals in Rivers, Lakes, and Oceans	654
	10.10	Quality Criteria in Fresh Waters: Some Aspects	666
		Suggested Readings	670

11. Kinetics of Redox Processes — 672

11.1 Introduction — 672
11.2 How Good an Oxidant Is O_2? — 672
11.3 Can pε Be Defined for a Nonequilibrium System? — 677
11.4 Kinetics of Redox Processes: Case Studies — 679
11.5 Oxidants Used in Water and Waste Technology: A Few Case Studies — 691
11.6 Linear Free Energy Relations (LFERs) — 702
11.7 The Marcus Theory of Outer-Sphere Electron Transfer: An Introduction — 703
11.8 Nucleophile–Electrophile Interactions and Redox Reactions Involving Organic Substances — 710
11.9 Corrosion of Metals as an Electrochemical Process — 720
Suggested Readings — 725

12. Photochemical Processes — 726

12.1 Introduction — 726
12.2 Absorption of Light — 729
12.3 Photoreactants — 735
12.4 Photoredox Reactions: Photolysis of Transition Metal Complexes — 743
12.5 Photochemical Reactions in Atmospheric Waters: Role of Dissolved Iron Species — 744
12.6 Heterogeneous Photochemistry — 748
12.7 Semiconducting Minerals — 753
Suggested Readings — 759

13. Kinetics at the Solid–Water Interface: Adsorption, Dissolution of Minerals, Nucleation, and Crystal Growth — 760

13.1 Introduction — 760
13.2 Kinetics of Adsorption — 760
13.3 Surface-Controlled Dissolution of Oxide Minerals: An Introduction to Weathering — 771
13.4 Simple Rate Laws in Dissolution — 776
13.5 Rates of $CaCO_3$ Dissolution (and of $CaCO_3$ Crystal Growth) — 788
13.6 Inhibition of Dissolution — 795
13.7 Nucleation and Crystal Growth — 800
Suggested Readings — 816

14. Particle–Particle Interaction: Colloids, Coagulation, and Filtration — 818

14.1 Colloids — 818
14.2 Particle Size Distribution — 826

xvi   Contents

14.3	Surface Charge of Colloids	834
14.4	Colloid Stability: Qualitative Considerations	837
14.5	Effects of Surface Speciation on Colloid Stability	842
14.6	Some Water-Technological Considerations in Coagulation, Filtration, and Flotation	852
14.7	Filtration Compared with Coagulation	857
14.8	Transport in Aggregation and Deposition	858
	Suggested Readings	866
	Appendix 14.1: A Physical Model (DLVO) for Colloid Stability	867

15. Regulation of the Chemical Composition of Natural Waters   872

15.1	Introduction	872
15.2	Weathering and the Proton Balance	875
15.3	Isothermal Evaporation	880
15.4	Buffering	884
15.5	Interactions Between Organisms and Abiotic Environment: Redfield Stoichiometry	886
15.6	The Oceans: Relative Constancy of the Composition and Chemical Equilibria	895
15.7	Constancy of Composition: Steady State	897
15.8	Hydrothermal Vents	901
15.9	The Sediment–Water Interface	903
15.10	Biological Regulation of the Composition	908
15.11	Global Cycling: The Interdependence of Biogeochemical Cycles	914
15.12	The Carbon Cycle	916
15.13	Nitrogen Cycles: Pollution by Nitrogen Compounds	927
15.14	The Sulfur Cycle	932
	Suggested Readings	933

References   935

Appendixes: Thermodynamic Data   976

1	Revised Chemical Equilibrium Data for Major Water-Mineral Reactions	977
2	Thermodynamic Data for Trace Metal Speciation in Seawater	984
3	Thermodynamic Properties	990

Index   1005

1

INTRODUCTION

1.1 SCOPE OF AQUATIC CHEMISTRY

Aquatic chemistry is concerned with the chemical reactions and processes affecting the distribution and circulation of chemical species in natural waters. The objectives include the development of a theoretical basis for the chemical behavior of ocean waters, estuaries, rivers, lakes, groundwaters, and soil water systems, as well as the description of processes involved in water technology. Aquatic chemistry draws primarily on the fundamentals of chemistry, but it is also influenced by other sciences, especially geology and biology.

A theme of this book is that fundamental principles of physical chemistry can be used to identify the pertinent variables that determine the composition of natural water systems. The student of chemistry is perhaps not fully aware that the well known laws of physical chemistry not only apply in the chemical laboratory but also regulate the course of reactions taking place in nature. During the hydrological cycle, water interacts continuously with the earth. Thus a progressive differentiation of geological material is achieved by processes of weathering, soil erosion, and soil and sediment formation. These processes accomplished by nature on a large scale have been likened (Rankama and Sahama, 1950) to the sequence of separations carried out during the course of a chemical analysis. The basic processes—dissolution and precipitation, oxidation and reduction, acid–base and complexation interactions—are the same in nature as in the laboratory. Sillén (1965) likened the evolution of the earth's atmosphere–ocean system to a set of gigantic, coupled acid–base and oxidant–reductant titrations in which volatile acids from the interior of the earth were titrated by the bases of the rocks, and the reduced volatiles were titrated by the oxygen of the evolving atmosphere–biosphere system.

While this book treats several topics similar to those found in an analytical chemistry text, it endeavors to consider the spatial and temporal scales of the reactions in nature as distinctly different from those of the laboratory. For example, in chemical analysis, precipitates (frequently of metastable and active compounds) are formed from strongly oversaturated solutions, whereas in natural water systems, the solid phase is often formed under conditions of slight supersaturation; often crystal growth and aging may continue over geological time spans. Interfacial phenomena are particularly important because chemical processes of significance often occur only at phase discontinuities.

2   Introduction

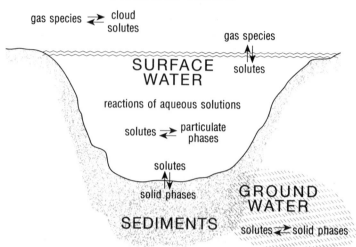

Figure 1.1. Natural water environments of interest in aquatic chemistry. Water links elemental cycles of the atmosphere with those of the sediments. Atmospheric chemistry, water chemistry, sediment geochemistry, soil chemistry, and groundwater chemistry of the elements are needed.

Natural water systems typically consist of numerous mineral assemblages and often include a gas phase in addition to the aqueous phase; they almost always involve a portion of the biosphere. Hence natural aquatic habitats are characterized by a complexity seldom encountered in the laboratory. In order to select the pertinent variables out of a sometimes bewildering number of possible ones, it is advantageous to compare the real systems with their idealized counterparts. Figure 1.1 shows in a very general way the kinds of natural water environments of interest to aquatic chemistry. The cycle of water links the elemental cycles of the atmosphere with those of the sediments. Thus atmospheric chemistry, water chemistry, sediment geochemistry, soil chemistry, and groundwater chemistry of the elements are all connected, on a range of time scales. Chemical models aim to capture the most important variables of complex natural water systems.

Models

To deal with the complexity of natural water systems we employ simplified and workable models to illustrate the principal regulatory factors that control the chemical composition of natural waters. In general, these models must link water composition with that of the atmosphere and the sediments. A model need not be completely realistic in order to be useful. A useful model leads to

1.1 Scope of Aquatic Chemistry

fruitful generalizations and valuable insight into the nature of aquatic chemical processes, and improves our ability to describe and measure natural water systems. Models are simplifications of a more complex reality. In simplifying, we try to be guided by sound chemical concepts. In accord with the aphorism of Albert Einstein: "Everything should be made as simple as possible, but not simpler."

Chemical equilibrium appears to be the most helpful model concept initially to facilitate identification of key variables relevant in determining water–mineral relations and water–atmosphere relations, thereby establishing the chemical boundaries of aquatic environments. Molar Gibbs free energies (chemical potentials) describe the thermodynamically stable state and characterize the direction and extent of processes approaching equilibrium. Discrepancies between predicted equilibrium composition and the data for the actual system provide valuable insight into those cases in which important chemical reactions have not been identified, in which non-equilibrium conditions prevail, or where analytical data for the system are not sufficiently accurate or specific. Such discrepancies are incentive for research and the improvement of existing models.

By comparing the actual composition of seawater (sediments + sea + air) with a model in which the pertinent components (minerals, volatiles) are allowed to reach true equilibrium, Sillén (1961) epitomized the application of equilibrium models for portraying the prominent features of the chemical composition of this system. His analysis indicated that, contrary to the traditional view, the pH of the ocean is not buffered primarily by the carbonate system; his results suggest that heterogeneous equilibria of silicate minerals comprise the principal pH buffer systems in oceanic waters. This approach has provided a more quantitative basis for Forchhammer's suggestion of 100 years ago that the quantity of the different elements in seawater is not proportional to the quantity of elements that river water pours into the sea but is inversely proportional to the facility with which the elements in seawater are made insoluble by general chemical actions in the sea. Although inland waters represent more transitory systems than the sea, equilibrium models are also useful here for interpreting observed facts. We can obtain some limits on the variational trends of chemical composition even in highly dynamic systems, and we can speculate on the type of dissolved species and solid phases one may expect.

Thermodynamic models of various systems within overall natural water systems are illustrated in Figure 1.2. Such models are employed in assessing global, partial, and local equilibrium conditions for water, air, and sediment interactions.

Natural waters indeed are open and dynamic systems with variable inputs and outputs of mass and energy for which the state of equilibrium is a construct. *Steady-state models* reflecting the time-invariant condition of a reaction system may frequently serve as an idealized counterpart of an open natural water system. The concept of free energy is not less important in dynamic systems than in equilibrium systems. The flow of energy from a higher to a lower potential or energy "drives" the hydrological and the geochemical cycles (Ma-

4   Introduction

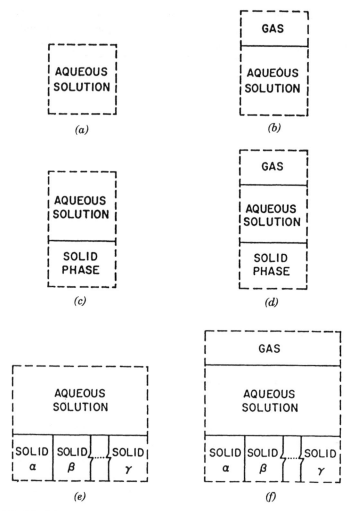

Figure 1.2. Models of various parts comprising overall natural water systems: (a) aqueous solution phase model; (b) aqueous solution and gas phase model; (c) aqueous solution and solid phase model; (d) three-phase aqueous, gas, and solid phase model; (e) aqueous solution plus several solid phases model; and (f) multiphase model for solids, aqueous solution, and a gas phase.

son, 1966; Morowitz, 1968). The ultimate source of the energy flow is the sun's radiation.

Ecosystems

In natural waters, organisms and their abiotic environments are interrelated and interact with each other. Because of the continuous input of solar energy (photosynthesis) necessary to maintain life, ecological systems are never in equilibrium. The ecological system, or ecosystem, may be considered a unit of the environment that contains a biological organization made up of all the organisms interacting reciprocally with the chemical and physical environment. In an ecosystem, the flow of energy and of negative entropy is reflected by the characteristic trophic structure and leads to material cycles within the system (Odum, 1969). In a balanced ecological system a steady state of production and destruction of organic material, as well as of production and consumption of O_2, is maintained.

The distribution of chemical species in waters and sediments is strongly influenced by an interaction of mixing cycles and biological cycles (Buffle and DeVitre, 1994). Radioisotope measurements may often be used to establish the time scale of some of these processes. Similarly, evaluation of the fractionation of stable isotopes aids in the quantitative interpretation of biogeochemical and environmental processes and cycles.

Biogeochemical Cycles It is now widely recognized that the earth is one giant biogeochemical system (Schlesinger, 1991). Within this system the atmosphere, the waters of the earth, and the sedimentary reservoirs are linked to the activity of the biosphere. Global cycles of water, carbon, nitrogen, phosphorus, and sulfur are interconnected with one another and have now been affected to a noticeable degree by human activities. The minor element cycles, for example, those of mercury and lead, have been appreciably perturbed over time. Biogeochemical cycles work at local, regional, and global levels (Bidoglio and Stumm, 1994). The elucidation of the scale of these cycles is facilitated by the appropriate integration of stoichiometric (for complete reactions), equilibrium, steady-state, and kinetic (time-dependent) models.

Kinetics

Our understanding of natural water systems has, until recently, been seriously limited by a lack of kinetic information on critical reactions in water, in sediments, and at interfaces. Earlier in atmospheric chemistry (Seinfeld, 1986) and more recently in aquatic chemistry (Brezonik, 1993), a considerable growth of information on rates and mechanisms for reactions central to environmental chemistry has taken place. As a result, we are now better able to assess the characteristic time scales of chemical reactions in the environment and compare these with, for example, residence times of water in a system of interest. Schematically, as shown here, for chemical vs. fluid time scales

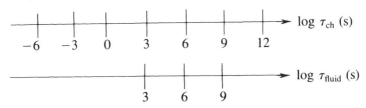

6   Introduction

we would like to distinguish between *slow* reactions and *rapid* reactions, judged against the time scale of fluid flow (e.g. in a lake, river, or groundwater). The fast reactions are well described by equilibrium models (reversible reactions) or stoichiometric models (irreversible reactions). Kinetic (i.e., time-dependent) descriptions are needed when τ_{ch} and τ_{fluid} are comparable.

Water as a Resource and Life Preservation System

Aquatic chemistry is of practical importance because water is an essential resource for humans. We are concerned with the quality of water and its distribution, not with the quantity, because water is an abundant substance on earth. Ecosystem resilience and buffering against change notwithstanding, human activity has become so powerful as to influence global chemical cycles (Schlesinger, 1991) as well as the local and regional chemical and hydrologic cycles. Locally and regionally, groundwater quality has been impaired by release of hazardous chemicals. The restoration of these systems will require the creative integration of chemical, biological, and hydrologic understanding. In the chemical realm, progress in water quality improvement calls for a synthesis of physical, inorganic, organic, and interfacial chemistry.

Conservation of aquatic resources cannot be accomplished by avoiding human influences on the aquatic environment. Control of water pollution and protection of the water resource demand more than waste treatment technology. We need to address questions such as these: To what extent are the oceans able to absorb wastes without harmful effects? Can we improve the fertility of the oceans? How can the ecological balance between photosynthetic and respiratory activities in nutritionally enriched but polluted waters be restored? Answers to such questions call for a greatly improved understanding of the aquatic environment.

Science needs to provide the basis for human harmony with nature, not dominance. "Man masters nature not by force but by understanding" (Bronowski, 1965).

1.2   THE SOLVENT WATER

Water is an unusual liquid. It has a very high boiling point and high heat of vaporization; ice has a very high melting point. The maximum density of liquid water is near 4 °C, not the freezing point, and water thus expands upon freezing. It has a very high surface tension. It is an excellent solvent for salts and polar molecules. It has the greatest dielectric coefficient of any liquid. These unusual properties are a consequence of the dipolar character of the H_2O molecule. Figure 1.3, from Horne (1969), depicts the electron cloud of the angular water molecule, resulting from the hybridization of s and p electrons to yield two bonding orbitals between the O and the two H atoms, and two non-bonding sp^3 orbitals on the oxygen. The molecule thus has high negative charge density

1.2 The Solvent Water 7

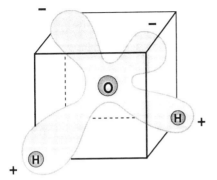

Figure 1.3. Electron cloud depiction for the H_2O molecule (Horne, 1969).

near the oxygen atom and high positive charge density near the protons. It is a dipolar molecule.

Figure 1.4 (Horne, 1969) shows the measured angle of 105° between the hydrogens and the direction of the dipole moment. The measured dipole moment of water is 1.844 debye (a debye unit is 3.336×10^{-30} C m). The dipole moment of water is responsible for its distinctive properties in the liquid state. The O—H bond length *within* the H_2O molecule is 0.96 Å (an angstrom unit, Å, is 10^{-10} m). Dipole–dipole interaction between two water molecules forms a hydrogen bond, which is electrostatic in nature. The lower part of Figure 1.4 (not to the same scale) shows the measured H-bond distance of 2.76 Å, or 0.276 nm.

The hydrogen-bonded structure of ice is shown in Figure 1.5a (Gray, 1973), and one of the several models proposed for the structure of water in the liquid state is shown in Figure 1.5b (Nemethy and Scheraga, 1962). In the open tetrahedral ice structure, each oxygen atom is bound to four nearby oxygen atoms by H bonds. The energy of each H bond is estimated to be about 20 kJ mol^{-1}. (For comparison, covalent bond energies are typically 20 times greater.) H bonds are of low energy, but they are numerous in ice and water.

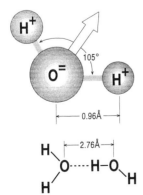

Figure 1.4. Structure of the angular water molecule and the hydrogen bond (Horne, 1969).

8   Introduction

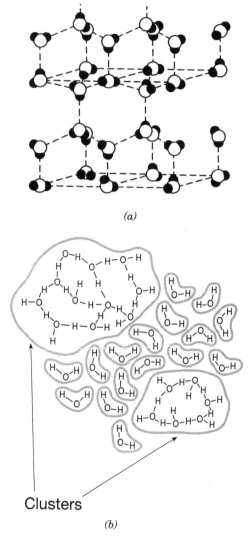

Figure 1.5. (a) Hydrogen-bonded open tetrahedral structure of ice (Gray, 1973). (b) Frank–Wen flickering cluster model of liquid water (Nemethy and Scheraga, 1962).

Upon melting, ice loses its open structure with the "melting" of some fraction of the hydrogen bonds and so the volume of the liquid water decreases, reaching a minimum at 4°C; above this temperature thermal expansion dominates the density.

In Figure 1.5b the Frank–Wen "flickering cluster" model envisions larger clusters of H-bonded water surrounded by noncluster waters, which nonetheless interact with neighbors by dipole–dipole forces. The lifetime of the clusters is estimated at around 100 picoseconds (ps), which is long with respect to the period of a molecular vibration, approximately 0.1 ps. The persistence (or re-formation) of hydrogen bonding in liquid water is a key to understanding the physical properties of water as well as its poorer solvent properties for nonpolar, hydrophobic solutes. The highly structured water linked by H bonds must be disrupted by *any solute* (Tanford, 1980). When the solute is ionic, the attractive interactions between ion and water molecule *favor* dissolution. When the solute is a nonpolar molecule, the structural cost to the hydrogen-bonded water makes dissolution an *unfavorable* process.

1.3 SOLUTE SPECIES

Dissolution of *ionic and ionizable solutes* in water is favored by ion–dipole bonds between ions and water. Figure 1.6 illustrates a hydrated sodium ion, $Na^+(aq)$, for example, from dissolution of NaCl in water, surrounded by six water molecules in octahedral positions. The energy of the ion–dipole bonds depends on the size of the ion and its charge. Higher charge and smaller ionic radii favor the bonding. Further away from the central ion, the water molecules are structured through additional dipole–dipole interactions. In a similar way, the $Cl^-(aq)$ ion interacts with the solvent water to form ion-dipole bonds, with the hydrogen side (local positive charges) of H_2O pointing toward the central ion.

The dissolution of *polar molecules* in water is favored by dipole–dipole interactions. The solvation of the polar molecules stabilizes them in solution. *Nonpolar molecules* are difficultly soluble in water because of the already mentioned unfavorable energy cost of disrupting and re-forming the hydrogen-

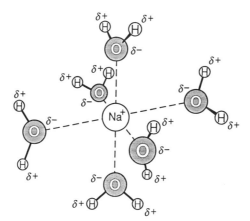

Figure 1.6. Hydrated sodium ion, Na^+, in aqueous solution. H_2O molecules form ion-dipole bonds to the central metal ion. The waters are in octahedral coordination to the sodium ion (Gray, 1973).

bonded water. Complicated molecules or ions with both ionic and nonpolar regions, or polar and nonpolar regions, show more complicated behavior in water (surface active species), but the stabilizing and destabilizing contributions of ion–dipole, dipole–dipole, and H-bond disrupting energies can still be recognized.

Speciation

The particular chemical form in which an element exists in water is its *speciation*. For example, an element can be present as a simple hydrated ion, as a molecule, as a complex with another ion or molecule, and so forth. From what was said previously, *bare* ions or *bare* polar molecules do not exist in water. At the least, they would be solvated species. Species of an element are distinguishable from one another stoichiometrically, structurally, and energetically. In addition to *aqueous* species, one can distinguish elements in different phases, for example, as gaseous species, as solid phases, or in adsorbed states, and on the basis of particle sizes. In the atmosphere, for example, speciation extends over liquid, gas, and aerosol phases (Seinfeld, 1986). The notion of chemical speciation is central to equilibrium and kinetic aspects of aquatic chemistry, as will be evident throughout this book.

Concentrations and Species of Important Elements in Water

Figure 1.7 illustrates, in the form of an abbreviated periodic table, the river water and marine water concentrations, oceanic residence times, and major

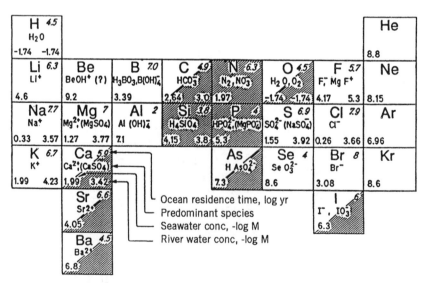

Figure 1.7. Some of the more important elements in natural waters: their concentrations, species, and residence times in river water and seawater (Sigg and Stumm, 1994). Elements whose distribution is significantly affected by biota are shaded.

species for several important elements. For example, consider sulfur: the predominant species present is sulfate, SO_4^{2-}, the river water concentration is given by log molarity = -3.92, or a sulfate concentration of 1.2×10^{-4} M; the total sulfate concentration in seawater is 0.028 M; and the major ion pair species of sulfate in seawater is $NaSO_4^-$(aq); the oceanic residence time is 8 million years. The information presented shows that the elements exist in a variety of oxidation states, protonated versus deprotonated forms, and free (aquated) versus complexed ion forms in water. As will be discussed later, the speciation of an element is influential with respect to its residence time in natural waters.

SUGGESTED READINGS

Berner, E. and Berner, R. A. (1987) *Global Water Cycle, Geochemistry and Environment*, Prentice Hall, Englewood Cliffs, NJ.

Brezonik, P. L. (1994) *Chemical Kinetics and Process Dynamics in Aquatic Systems*, Lewis, Boca Raton, FL.

Harte, J. (1988) *Consider a Spherical Cow*, University Science Books, Mill Valley, CA.

Horne, R. A. (1969) *Marine Chemistry*, Wiley-Interscience, New York.

Morel, F. M. M. and Hering, J. G. (1993) *Principles and Applications of Aquatic Chemistry*, Wiley-Interscience, New York.

Pankow, J. F. (1991) *Aquatic Chemistry Concepts*, Lewis, Chelsea, MI.

Schlesinger, W. H. (1991) *Biogeochemistry: An Analysis of Global Change*, Academic Press, San Diego, CA.

Schwarzenbach, R. P., Gschwend, P. M., and Imboden, D. M. (1993) *Environmental Organic Chemistry*, Wiley-Interscience, New York.

Tanford, C. (1991) *The Hydrophobic Effect*, 2nd ed., Krieger, Melbourne, FL.

APPENDIX 1.1 SOME USEFUL QUANTITIES, UNITS, CONVERSION FACTORS, CONSTANTS, AND RELATIONSHIPS

The Earth–Hydrosphere System (Sigg and Stumm, 1994)

Earth area	5.1×10^{14} m^2
Ocean area	3.6×10^{14} m^2
Land area	1.5×10^{14} m^2
Atmosphere mass	52×10^{17} kg
Ocean mass	$13,700 \times 10^{17}$ kg
Groundwater to 750 m	42×10^{17} kg
Groundwater to 4000 m	95×10^{17} kg
Water in ice	165×10^{17} kg
Water in lakes and rivers	1.3×10^{17} kg
Water in atmosphere	0.105×10^{17} kg

Introduction

Water in biosphere	0.006×10^{17} kg
Total stream discharge	0.32×10^{17} kg yr^{-1}
Precipitation = evaporation	4.5×10^{17} kg yr^{-1}

Some Properties of Water (m kg s units)

Temperature (°C)	$k_B T/10^{-21}$	Density (ρ)	Viscosity/10^{-3} (η)	Surface Tension (γ)	Dielectric Coefficient (ϵ)
5	3.8	999.965	1.5188	0.0749	86.04
20	4.0	998.203	1.0050	0.07275	80.36
25	4.1	997.044	0.8937	0.07197	78.54
30	4.2	995.646	0.8007	0.07118	76.75

The International Units

Physical Quantity	Unit	Symbol
Length	meter	m
Mass	kilogram	kg
Time	second	s
Electric current	ampere	A
Temperature	kelvin	K
Luminous intensity	candela	cd
Amount of material	mole	mol
The main derived units are:		
Force	newton	N = kg m s^{-2}
Energy, work, heat	joule	J = N m
Pressure	pascal	1 Pa = N m^{-2}
Power	watt	W = J s^{-1}
Electric charge	coulomb	C = A s
Electric potential	volt	V = W A^{-1}
Electric capacitance	farad	F = A s V^{-1}
Electric resistance	ohm	Ω = V A^{-1}
Frequency	hertz	Hz = s^{-1}
Conductance	siemens	S = A V^{-1}
Amount of photons	einstein	einstein
Dipole moment	debye	D

Useful Conversion Factors

Energy, Work, Heat

$$1 \text{ joule} = 1 \text{ volt coulomb} = 1 \text{ newton meter}$$
$$= 1 \text{ watt second} = 2.7778 \times 10^{-7} \text{ kilowatt hours}$$
$$= 10^7 \text{ erg}$$

Appendix 1.1 Useful Quantities, Units, and Relationships

$= 9.9 \times 10^{-3}$ liter atmospheres
$= 0.239$ calorie
$= 1.0364 \times 10^{-5}$ volt faraday
$= 6.242 \times 10^{18}$ eV
$= 5.035 \times 10^{22}$ cm^{-1} (wave number)
$= 9.484 \times 10^{-4}$ BTU (British thermal unit)

Power

1 watt $= 1$ kg m^2 s^{-3}
$= 2.39 \times 10^{-4}$ kcal s^{-1} $= 0.860$ kcal h^{-1}

Entropy (S)

1 entropy unit, cal mol^{-1} K^{-1} $= 4.184$ J mol^{-1} K^{-1}

Pressure

1 atm $= 760$ torr $= 760$ mm Hg
$= 1.013 \times 10^5$ N m^{-2} $= 1.013 \times 10^5$ Pa (pascal)
$= 1.013$ bars

Coulombic Force Coulomb's law of electrostatic force is written, in SI units, as

$$F = \frac{q_1 q_2}{4\pi\epsilon\epsilon_0 d^2} \tag{1}$$

The charges q_1 and q_2 are expressed in coulombs (C), the distance d in meters (m), and the force F in newtons (N). The dielectric constant ϵ is dimensionless. The permittivity in vacuum is $\epsilon_0 = 8.854 \times 10^{-12}$ J^{-1} C^2 m^{-1}. Thus to calculate a coulombic energy E, we have

$$E(\text{joules}) = \frac{q_1 q_2}{4\pi\epsilon\epsilon_0 d} \tag{2}$$

Some Important Constants

Avogadro's constant, N	6.022×10^{23} mol^{-1}
Electron charge, e	1.602×10^{-19} C
Faraday, F	96,485 C mol^{-1} (charge of 1 mol of electrons)
Electron mass, m_e	9.109×10^{-31} kg
Atomic mass unit	1.66054×10^{-27} kg
Permittivity of vacuum, ϵ_0	8.854×10^{-12} J^{-1} C^2 m^{-1}
Molecular vibration period, ν	6.2×10^{-12} s

Earth gravitation, g — 9.806 m s^{-2}
Speed of light in a vacuum, c — 2.998 × 10^8 m s^{-1}
Gas constant, R — 8.314 J mol^{-1} K^{-1}
0.082057 liter atm deg^{-1} mol^{-1}
1.987 cal deg^{-1} mol^{-1}
Molar volume (ideal gas, 0°C, 1 atm) — 22.414 × 10^3 cm^3 mol^{-1}
Planck constant, h — 6.626 × 10^{-34} J s
Boltzmann constant, k_B — 1.3805 × 10^{-23} J K^{-1}
Ice point — 273.15 K

Useful Relationships

$\ln x = 2.303 \log x$

$RT_{298.15} \ln x = 5.71 \log x$ (kJ mol^{-1})

$RTF^{-1} = 25.69$ mV at 298.15 K

$\ln 10\, RTF^{-1} = 59.16$ mV at 298.15 K

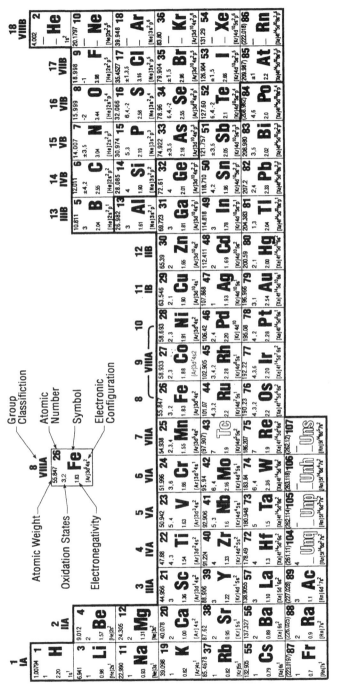

2

CHEMICAL THERMODYNAMICS AND KINETICS

2.1 INTRODUCTION

Natural waters obtain their equilibrium composition through a variety of chemical reactions and physicochemical processes. In this chapter we consider principles and applications of two alternative models for natural water systems: *thermodynamic* models and *kinetic* models. Thermodynamic, or *equilibrium*, models for natural waters have been developed more extensively than kinetic models. They are simpler in that they require less information, but they are nevertheless powerful when applied within their proper limits. Equilibrium models for aquatic systems receive the greater attention in this book. However, kinetic interpretations are needed in description of natural waters when the assumptions of equilibrium models no longer apply. Because rates of different chemical reactions in water and sediments can differ enormously, kinetic *and* equilibrium are often needed in the same system.

Figure 2.1 portrays the agenda of aquatic chemistry in broad outline. The essential connections between natural system observations, models, and experiments are emphasized.

The composition of a natural water, symbolized by the concentration of a constituent A, C_A, results from chemical reactions in the water itself, from processes that transfer constituents between the water and other parts of the system (atmosphere, solid matter in suspended or sedimentary form, the biota, other liquid phases), and from fluxes into and out of the system. Figure 2.2 is a schematic representation of a natural water system model. The concentration $C_A = n_A/V$, where n_A is the mole number and V is the volume of water, can be altered by variations in n_A (i.e., dn_A) brought about by fluxes, transfers, and reactions. The time-invariant, or stationary, state of the chemical composition of the water, C_A, is given by $dC_A/dt = 0$; this state has different origins in models for closed and continuous, open systems.

2.1 Introduction 17

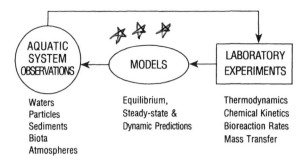

Figure 2.1. Outline of the components of aquatic chemistry. Observations on aquatic systems and laboratory experiments are connected through models describing the composition of the system of interest.

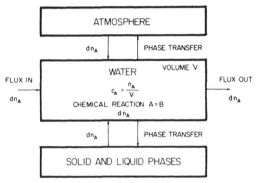

Figure 2.2. General representation of a natural water system treated as a continuous, open system. The system receives fluxes of matter from the surroundings and undergoes chemical changes, symbolized by the reaction A = B. The time-invariant condition is represented by $dC_A/dt = 0$.

Equilibrium and Kinetic Models

The basic differences between thermodynamic models and kinetic models can be exemplified by considering a single hypothetical chemical reaction

$$A = B$$

taking place in the water. We will assume no transfer of A or B between the water and other phases.

A stationary-state *thermodynamic* model for this simple system requires that the system be *closed*, that is, that no matter be exchanged with the surroundings. This means, strictly, that the material flows of A and B between the system and the surroundings must be zero: $\Sigma\, dn_{A,\text{flux}} = 0$, $\Sigma\, dn_{B,\text{flux}} = 0$. Then, application of equilibrium thermodynamic principles, to be developed below, yields an *equilibrium composition* for C_A and C_B in the water. The equilibrium model requires fixed temperature and pressure, homogeneous distribution of A and B throughout the water, a specified total mole number for component A, and the volume of the water. In practical terms, the information needed to find C_A and C_B for the equilibrium state is an effective *equilibrium constant* K_{AB} (a function of pressure, temperature, and composition) and the *total* concentration of A (or B) in the system, $\text{TOTA} = n_{A,0}/V$. The relationships are

$$\text{TOTA} = C_A + C_B \tag{1}$$

$$\frac{C_B}{C_A} = K_{AB} \tag{2}$$

Equations 1 and 2 can be solved to yield the equilibrium values of C_A and C_B. The results are:

$$C_A = \frac{\text{TOTA}}{1 + K_{AB}} \tag{3}$$

$$C_B = \text{TOTA} - C_A \tag{4}$$

A kinetic model for the closed aqueous phase requires kinetic information about the rates of transformation of A and B. Values of the rate constants k_f and k_b for the reversible reaction

$$A \underset{k_b}{\overset{k_f}{\rightleftharpoons}} B \tag{5}$$

can be found from the rate expressions, which are

$$v_f = k_f C_A \tag{6}$$

$$v_b = k_b C_B \tag{7}$$

and the time rates of change of C_A and C_B from the *reaction* are given by

$$\frac{dC_A}{dt} = -k_f C_A + k_b C_B \tag{8a}$$

$$\frac{dC_B}{dt} = k_f C_A - k_b C_B \tag{8b}$$

These expressions, with known initial concentrations of A and B, allow the time course of the approach to equilibrium to be described. (We consider this simple reversible reaction later when we take up rates of reversible reactions in Section 2.13.)

For a completely mixed flow the needed material balance information consists of the steady mole fluxes per unit volume of A and B to the system, given by $r\overline{C}_A$ and $r\overline{C}_B$, where $r = Q/V$ is the fluid transfer rate constant for the system (the rate of flow Q divided by the volume V) and $\overline{C}_A$ and $\overline{C}_B$ are the *inflow* concentrations of A and B, respectively. For the steady state, $dC_A/dt = dC_B/dt = 0$, and assuming inflow and outflow rates to be the same, we have

$$\frac{dC_A}{dt} = r\overline{C}_A - k_f C_A + k_b C_B - rC_A = 0 \tag{9}$$

and a corresponding expression for dC_B/dt.

The stoichiometry of the reaction leads to $C_A + C_B = \overline{C}_A + \overline{C}_B$. These equations can be solved for the steady-state values C_A and C_B. The result for C_A is

$$C_A = \frac{r\overline{C}_A + k_b(\overline{C}_A + \overline{C}_B)}{k_f + k_b + r} \tag{10}$$

Comparison of equations 3 and 10 shows the essential difference between the stationary states of closed and continuous, open systems. For the closed system, *equilibrium* is the time-invariant condition. The total of each independently variable constituent and the equilibrium constant (a function of temperature, pressure, and composition) for each independent reaction (K_{AB} in the example) are required to define the equilibrium composition C_A. For the continuous, open system, the *steady state* is the time-invariant condition. The mass transfer rate constant, the inflow mole number of each independently variable constituent, and the rate constants (functions of temperature, pressure, and composition) for each independent reaction are required to define the steady-state composition C_A. It is clear that open-system models of natural waters require more information than closed-system models to define time-invariant compositions. An equilibrium model can be expected to describe a natural water system well when fluxes are small, that is, when flow time scales are long and chemical reaction time scales are short.

Introducing additional chemical reactions and including transfer processes between the water and atmosphere on the water and solid or liquid phases will increase the mathematical complexity of a closed-system or open-system model. Additional equilibrium constants for chemical reactions and distribution of constituents between phases are required for the closed system; additional rate constants are required for the kinetic processes in the open system, and more

20   Chemical Thermodynamics and Kinetics

material transfer rate constants ($r = Q/V$) may be required for chemical fluxes into other phases. Kinetic expressions of reactions in closed systems will be needed for slow reactions.

In summary, it is proposed that complex natural water systems can be investigated quantitatively by means of idealized models. The models are basically of two kinds:

Continuous, open systems, which exchange material with their surroundings, that is, can vary their mole numbers both by flows and reactions

Closed systems, which do not exchange material with their surroundings, that is, contain a fixed total mass of components but that can vary their mole number by reactions and internal processes.

For each of these idealized models there is a stationary state. For a continuous *open system*, this is the steady state. Rate laws and steady material flows are required to define the steady state. For a *closed system*, equilibrium is the stationary state. Equilibrium may be viewed as simply the limiting case of the stationary state when the flows from the surroundings approach zero. The simplicity of closed-system models at equilibrium is in the rather small body of information required to describe the time-invariant composition. We now turn our attention to the principles of chemical thermodynamics and the development of tools for the description of equilibrium states and energetics of chemical change in closed systems.

2.2 CHEMICAL THERMODYNAMIC PRINCIPLES

Goals

Our principal goals in discussing chemical thermodynamic applications to aquatic chemistry are the following:

1. To compute equilibrium compositions of natural water system models.
2. To find out how far a system is from equilibrium, that is, to compute the driving forces of reactions in a system.
3. To evaluate the minimum external work (energy) required to make a given reaction possible, or to evaluate the maximum external work (energy) available from a given process.
4. To combine partial or local equilibria with rate models of natural water systems.
5. To compute equilibrium constants at different temperatures and pressures from tabulated standard thermodynamic data.

Chemical Amounts and Concentrations   The *amount* of a chemical substance is expressed as the number of moles, denoted by n. Energies, entropies,

2.2 Chemical Thermodynamic Principles

volumes, and other *extensive* thermodynamic quantities are frequently expressed per mole, as in *molar and partial molar* properties (intensive properties): for example, joule mol^{-1} and m^3 mol^{-1}.

Concentrations will be expressed as *mole fraction* of a component or species i, $x_i = n_i / \Sigma\, n_j$; as *molality*, mole per mass of solvent, mol kg^{-1}; or *molarity*, mole per volume of solution. The concentration scale will depend on the properties of the solutes (i.e., ionic, polar, nonpolar, etc.). Pressure, p, and the gas phase partial pressure of species i, p_i, will be expressed in bars (approximately equal to atmospheres).

Thermodynamics and Chemical Thermodynamics

Chemical thermodynamics involves the thermodynamic description of systems subject to *chemical change*, for example, reactions and phase transfers. For systems containing only a single chemical compound (one component), the issue of chemical reaction does not arise, but phase transfer is still possible in heterogeneous (i.e., multiphase) systems. Chemical thermodynamics is a part of the science of thermodynamics. The governing principles of axioms are the same: the first and second laws. In chemical thermodynamics applied to aquatic systems, we deal with *changes in composition*, that is, changes in number of moles of species, dn_i, resulting from chemical reactions and transfers of species between phases, and with *equilibrium compositions* of aqueous phases, sedimentary phases, and atmospheres.

For example, we might wish to find the equilibrium composition of the three-phase system comprising atmospheric carbon dioxide, water, and calcium carbonate. Or, more generally, we might inquire whether a given state of a system with these components is far from or close to equilibrium. A system such as this might be characterized by several alternative constraints: for example, the three-phase system could be *closed* overall, with a fixed content; or the atmosphere could be assumed an infinite reservoir of constant CO$_2$ partial pressure; and so on. Our first step in establishing a framework for such chemical thermodynamic models will be to review some basic principles of thermodynamics. Then we will concentrate on the chemical thermodynamic concepts and tools helpful in setting up and solving models for composition of natural water systems.

The Four Principles of Thermodynamics

0. There is an absolute temperature scale, T.
1. The internal energy, E, is a state function of a system. It is altered by heat transfer to the system, q, and work done by the system, w. Thus $dE = dq - dw$.
2. Entropy, S, is a state function of a system. The entropy change of a system is related to the heat transferred by $dq/T \leq dS$. If we denote the

entropy change of the surroundings of a system by $d_e S$ and the entropy changes interior to the system by $d_i S$, then $dS = d_e S + d_i S$ is an alternative statement of the second law, wherein $d_i S \geq 0$.

3. The entropy of a body is zero when T is zero.

The concepts of internal energy and entropy are viewed as "primitive" concepts *within* thermodynamics, and so the first and second principles of thermodyanmics are primitive, irreducible ideas *within* thermodynamics. Internal energy and the first law are coupled, as are entropy and the second law. Thus *Gibbsian* thermodynamics (J. W. Gibbs, 1876) is based on postulation of the internal energy, the temperature, and the entropy, and all else follows from the "primitive" principles and the calculus. The principles of thermodynamics are based on experience and experiments. [*Note*: It is the province of statistical mechanics to obtain E, V, S as a function of T and P by considering the behavior or molecules in the system (e.g., their translational, vibrational, rotational, and electronic energies). In this context, the notions of *work* and *heat transfer* are given molecular interpretations in terms of energy level of species quantum states and their occupancies by molecules.]

First Law: Heat and Work

As usually presented for equilibrium systems of fixed composition, the first law is

$$dE = dq - dw \qquad (11)$$

(heat transfer *to* the system is considered positive and work done *by* the system is considered negative)

Forms of Work For *PVT* systems, only "pV" work is considered: $dw = -p\,dV$. dE is an exact differential, E being a function of the state of the system only, but dw and dq are both *path-dependent* quantities; that is, their values depend on whether changes in a system take place under quasistatic (reversible) or under irreversible conditions.

When a change in a system takes place quasistatically, small variations are opposed by restoring forces, and the prior condition can be attained by small increases in these forces. For example, in an electrochemical cell, an opposing electromotive force can cause the current (rate of reaction) to be very small and the direction of the reaction reversible. By contrast, an irreversible change proceeds without opposing force. The magnitudes of dq and dw are different under reversible and irreversible conditions, but their *sum* remains the same; that is, only a function of the state of the system (e.g., T and p).

Other forms of work than pV work can be done by a system, for example, gravitational work, electrical work, work of increasing interfacial area, magnetic work. We can represent the total work variation, dw, by $dw = \Sigma\, Y_j\, dX_j$,

2.2 Chemical Thermodynamic Principles

where Y_j is a generalized force (e.g., pressure, electric potential, or interfacial free energy) and dX_j is the conjugate displacement (e.g., dV, de, or dA). Electrical work and interfacial work are encountered in a number of important aqueous chemical processes, for example, galvanic cells, electrolytic cells, and fine particle solubilities. Work other than pV work is referred to as *external work*, $w_{external}$, or w'. Thus $dw = p\,dV + dw'$. For example, we might have a change in the internal energy of a very complex system given by $dE = dq - p\,dV + \mu\,dn - E\,de + \gamma\,dA + mg\,dh$, accounting for mechanical, chemical, electric, interfacial, and gravitational work.

Second Law: Heat and Entropy Changes

The concept of the *system plus its surroundings* is central to thermodynamic models. The *system* is the defined object of study (region, assemblage, object, etc.), separated from the *surroundings* by a *boundary*. See Figure 2.3 for a concise version of the *system + surroundings* concept, and a summary of first and second law relationships for the system and surroundings. This summary draws upon the treatments of Prigogine (1961), Atkins (1990), and Blandamer (1992). (See Figure 1.2 for *model thermodynamic systems* abstracted from natural water environments.) A familiar axiomatic statement of the second law is that, for a *reversible* process, the change in the entropy of the *system* is given by equality 2c of Figure 2.3:

$$dS_{sys} = dq/T \qquad (12)$$

This statement connects entropy with measureable quantities, the heat transferred to the system and the absolute temperature.

Entropy Change of the Universe In a familiar form, for a system plus its surroundings, or "universe," the second law for *reversible* change is

$$dS_{univ} = dS_{sys} + dS_{surr} = 0 \qquad (13)$$

For an *irreversible* change in a system,

$$dq/T < dS_{sys} \qquad (14)$$

and the heat transfer is *less* than the quantity $T\,dS_{sys}$. The general statement for *all* possible changes in the system, reversible and irreversible, is

$$dq \leq T\,dS_{sys} \qquad (15)$$

In terms of system plus surroundings, the universe,

$$dS_{univ} = dS_{sys} + dS_{surr} \geq 0 \qquad (16)$$

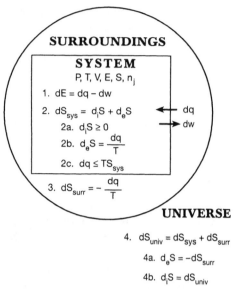

Figure 2.3. System, surroundings, and universe for basic thermodynamic analysis. The system is characterized by the *extensive variables* n_j, V, E, and S, and *intensive variables* P and T. The system receives heat (>0) from the surroundings and does work (>0) on the surroundings. Equations 1 and 2 state the first and second laws, respectively. The entropy change (extensive property) comprises two terms, the entropy change *within* the system, $d_i S$, and the entropy change *from the surroundings*, $d_e S$. The entropy change of the *universe* (system and surroundings, dS_{univ}, is equal to the entropy within the system, $d_i S$. For a *reversible* process, $d_i S = 0$; for a *spontaneous* process, $d_i S > 0$. As discussed in the text, $T\, d_i S = -dG$, the change in the Gibbs energy of the system (Atkins, 1990; Blandamer, 1992; Prigogine, 1961).

Clausius, Equilibrium and Change The *equilibrium* state of the system plus its surroundings is thus one in which S has attained a maximum value. The well-known Clausius formulation of this is: "Die Energie der Welt ist konstant; die Entropie der Welt strebt einem maximum zu." While it would be possible to characterize equilibrium and disequilibrium of chemical systems in terms of these entropies, there are more convenient ways to operate in chemical thermodynamics by concentrating on changes in *the system itself* (Atkins, 1990).

Internal versus Environmental Entropy Change The entropy change of the system can be described as the sum of the entropy change *within* the system, $d_i S$, and the environmental entropy change resulting from heat transfer to the system *from* the surroundings, $d_e S$:

$$dS_{sys} = d_i S + d_e S \tag{17a}$$

2.2 Chemical Thermodynamic Principles

For a finite state change,

$$\Delta S_{sys} = \Delta_i S + \Delta_e S \qquad (17b)$$

which is clearly a restatement of equation 16 focused on the *system* itself. For *all* possible processes in accordance with the second law,

$$d_i S \geq 0 \qquad (18)$$

we see that $d_i S$ and dS_{univ} are equal. An advantage of using $d_i S$ and $\Delta_i S$ is that they are properties of the *system*.

Example 2.1 Consider the mineral transformation

$$CaCO_3 \text{ (aragonite)} \rightarrow CaCO_3 \text{ (calcite)}$$

in a constant-temperature, constant-pressure system at 25°C and 1 atm. The observed total entropy change of the system per mole of reaction is +3.7 J mol^{-1} K^{-1} and the heat transferred from the surroundings to the system is zero. Is the transformation spontaneous under the conditions in the system?

Applying the relationship 17b, with $q = T\Delta_e S = 0$, we find $\Delta S_{sys} = \Delta_i S = +3.7$ J mol^{-1} K^{-1}. The transformation is spontaneous as written. Calcite is the stable form.

First and Second Laws Combined

For *reversible* changes in a system equation the entropy change of the system, $dS_{sys} = dq/T$. Considering first a system doing no external work—that is, only pV work, $dw = p\,dV$—the internal energy change is given by

$$dE = T\,dS_{sys} - p\,dV \qquad (19)$$

At equilibrium for a system with $dV = 0$ and $dS = 0$, dE of the system is a *minimum*, because the *entropy of the universe* is *maximum* with respect to changes in the system.

For a system *not* at equilibrium, that is, one driving (however slowly or rapidly) toward equilibrium by irreversible processes, equations 12 and 15 combined give

$$dE + p\,dV - T\,dS_{sys} < 0 \qquad (20)$$

In a system of constant entropy and volume, that is, $dV = 0$, $dS = 0$ (it is not easy to arrange such constraints experimentally), the internal energy decreases in an irreversible process (e.g., chemical reaction or diffusion). All possible variations in the internal energy of a system are thus described by

$$dE + p\,dV - T\,dS_{sys} \leq 0 \qquad (21)$$

We can describe the possible changes of a system in terms of properties of the system itself (i.e., p, V, T, and S). Again, note that the *driving force* for spontaneous change, and thus the lowering of the internal energy, is an *increase* in the entropy of the universe (the system plus its surroundings). The inequality of equation (21) can be replaced by an *equality* by introducing the internal entropy change, $d_i S$, yielding

$$dE = T\,dS_{sys} - T\,d_i S - p\,dV \qquad (22)$$

Irreversible processes of phase transfer and chemical reaction within a closed system, whether homogeneous (a single phase) or heterogeneous (more than one phase), lead to $T\,d_i S > 0$. At equilibrium, $T\,d_i S = 0$. For fixed S and V constraints, $dE = -T\,d_i S$. A reversible process corresponds to zero internal entropy change and a minimum in dE.

The internal energy, or the "thermodynamic energy" as it is often called, arises in the statement of the first law. It is not the most convenient thermodynamic function for chemical applications. There are other, more useful, state functions available to describe systems of chemical interest, in particular, constant-temperature, constant-pressure systems undergoing chemical reactions.

Auxiliary Thermodynamic Functions

The internal energy, E, is an axiomatic thermodynamic function within the statement of the first law. Its "natural variables," as shown by equation 19, are volume and entropy. Although it is possible to carry out all thermodynamic analyses in terms of $E = E(V, S)$, that choice is not the most convenient one for chemical thermodynamics [apart from first law thermochemistry applications in constant-volume calorimetry, where dE is obtained from dq experimentally, giving the constant-volume heat capacity, $C_V = (dq/dT)_V$]. Chemical reactions usually take place under constant temperature and constant-pressure conditions; sometimes under constant-volume and constant-temperature conditions. For these conditions, other thermodynamic functions are advantageous. Auxiliary thermodynamic state functions are obtained from the internal energy, E, by the method of Legendre transformations (Abbott and VanNess, 1989). The transformed functions are $H = E + pV$, the *enthalpy*; $A = E - TS$, the *Helmholtz free energy*; and $G = E + pV - TS$, the *Gibbs free energy*. The corresponding set of differentials for *constant-composition* systems are then: $dE = T\,dS - p\,dV$; $dH = T\,dS + V\,dp$; $dA = -S\,dT - p\,dV$; and $dG = -S\,dT + V\,dp$. We will be particularly interested in the last of these transformed functions.

Gibbs Free Energy, Enthalpy, and Entropy

From $H = E + pV$ and $G = E + pV - TS$, we obtain $G = H - TS$. At constant T, the differential form is $dG = dH - T\,dS$. For a finite state change at constant pressure and temperature,

$$\Delta G = \Delta H - T\Delta S \tag{23}$$

This is a very useful relationship among three state functions, free energy, enthalpy, and entropy. It is a key tool in the application of thermodynamics to chemical problems. Close examination of equation 23 reveals that, in the form $\Delta S = \Delta H/T - \Delta G/T$, it is a different version of equation 17b; that is, it reexpresses the entropy changes of the second law in terms of state functions of the system itself.

Enthalpy Change and Heat Transferred, q

The enthalpy change, $dH = T\,dS + V\,dp$, can be described as $dH = dq + V\,dp$, and for a *constant-pressure* process, $dp = 0$, we have $dH = dq_p$. For a finite state change at constant pressure, $q_p = \Delta H$; that is, the heat transferred is equal to the enthalpy change of the system. This relation is the basis of constant pressure calorimetry, the constant-pressure heat capacity being $C_p = (dq/dT)_p$. The relationship $q_p = \Delta H$ is valid *only* in the absence of external work, w'. When the system does external work, the first law must include dw'. Then, the heat transferred to the system under constant-pressure conditions is $q_p = \Delta H + w'$. Thus, if a given chemical reaction has an enthalpy change of -50 kJ mol^{-1} and does 100 kJ mol^{-1} of electrical work, the heat transferred to the system is $-50 + 100 = 50$ kJ mol^{-1}.

The Gibbs Free Energy, G

The successive Legendre transformations of E yield a state function, G, for which the natural variables p and T, are both *intensive* properties (independent of the size of the system). Furthermore, for $dp = 0$ and $dT = 0$ (isobaric, isothermal system), the state of the system is characterized by dG. This is clearly convenient for chemical applications under atmospheric pressure, constant-temperature conditions (or at any other isobaric, isothermal conditions). Then, in place of equation (21) for internal energy variation, we state the conditions for irreversible or reversible processes in terms of the Gibbs energy as

$$dG - V\,dp + S\,dT \leq 0 \tag{24}$$

At *fixed T and p*, $dG < 0$ for an *irreversible* change in the system and $dG = 0$ for a *reversible* change. In the laboratory, the conditions for application of the Gibbs energy are met by a reactor with diathermal walls in contact with a heat reservoir, open to the atmosphere (or subjected to higher constant pressures in a suitable apparatus). Introducing $T\,d_iS$, the energy associated with the internal entropy increase, we can replace equation 24 with the equality

$$dG = -S\,dT + p\,dV - T\,d_iS \tag{25a}$$

At fixed T and p,

$$dG = -T\,d_iS = -T\,dS_{\text{univ}} \tag{25b}$$

For any spontaneous processes in a fixed T and p system, $T\,d_iS > 0$, and the free energy change is negative. At equilibrium, d_iS vanishes and dG is zero. Summarizing,

$$dG < 0 \quad \text{for a spontaneous process}$$
$$dG = 0 \quad \text{for a reversible process}$$
$$dG > 0 \quad \text{for an impossible process}$$

Equivalently (Figure 2.2), $dS_{\text{univ}} = 0$ at equilibrium. The Gibbs function thus expresses the second law as $dG \leq 0$ for all possible processes in constant-temperature, constant-pressure systems. (Similarly, the Helmholtz free energy function, A, expresses the second law for isochoric, isothermal systems as $dA \leq 0$ for all possible processes.)

When the System Does External Work

When the system does external work, w' (e.g., electrical work), the expression for the first law must reflect all work done by the system. Inequality 21 becomes

$$dE + p\,dV + dw' - T\,dS_{\text{sys}} \leq 0 \tag{26}$$

For *constant V and* S_{sys}, the external work variation, dw', is $dw' \leq -dE$. For a *reversible* process, the external work is $-dE$, the *maximum* external work available with $dV = 0$ and dS_{sys}. For an *irreversible* process, the available work is less than the maximum. In terms of the internal entropy production, equation 26 becomes

$$dE = -dw' - T\,d_iS \tag{27}$$

and the *available external work* is $dw' = -dE - T\,d_iS$. Creation of entropy within the system decreases the work the system can do, because $T\,d_iS \geq 0$ in accord with the second law.

For constant T and p systems, the available external work is obtained from the change in the Gibbs free energy (the name of the Gibbs function reflects the energy "free" to do external, i.e., non-pV work). The relevant equality

2.3 Systems of Variable Composition: Chemical Thermodynamics

incorporating the internal entropy change is

$$dG = V\,dp - S\,dT + dw' - T\,d_i S \quad (28)$$

At fixed T and p, the *available external work* is

$$dw' = -dG - T\,d_i S \quad (29)$$

In a reversible process, the maximum external work is $dw'_{max} = -dG$. Irreversible change in the system decreases the maximum external work of the system by $T\,d_i S$, the energy associated with creation of entropy in the system. For a finite state change, the available external work is

$$w' = -\Delta G - T\,\Delta_i S \quad (30)$$

Example 2.2

(a) What is the maximum electrical work obtainable from the reaction

$$\tfrac{1}{2} O_2(g) + H_2(g) = H_2O(l)$$

under *standard* conditions (25°C and 1 atm pressure and specified amounts of reactants)?
(b) What is the heat transferred if this maximum work is obtained?

The standard free energy change for the reaction, $\Delta G°$, is -237.18 kJ mol^{-1}. The standard enthalpy change, $\Delta H°$, is -285.85 kJ mol^{-1}.

(a) According to equation 30, the maximum electrical work available under reversible conditions (i.e., when $T\,\Delta_i S = 0$) is $w' = -\Delta G° = 237.18$ kJ mol^{-1}. (Two electrons are transferred per mole of the reaction. Therefore the *standard potential difference* for the electrochemical reaction is: $-(-237,180)\,J/2 \times 96,485\,J/V = 1.23$ V.)
(b) The heat transferred is $q_p = \Delta H° + w' = -285.85 + 237.18 = -48.67$ kJ mol^{-1}. The system releases heat to the surroundings, but much less than under irreversible conditions for this reaction.

(Note that here the *system* is the electrochemical reaction *plus* the external electric circuit required to reversibly extract the free energy released.)

2.3. SYSTEMS OF VARIABLE COMPOSITION: CHEMICAL THERMODYNAMICS

For a closed system of constant composition, that is, the variation in the number of moles for any species i is zero ($dn_i = 0$), and with volume change the only form of work, the fundamental equation is

$$dE = T\,dS - p\,dV \tag{19}$$

The general differential for change in E is

$$dE = \left(\frac{\partial E}{\partial S}\right)_{V,n_j} dS + \left(\frac{\partial E}{\partial V}\right)_{S,n_j} dV \tag{31}$$

The composition of a binary solution can be varied by addition of components. For example, additions of solute (species 2) and water (species 1) change the energy, increasing the ability of the system to do work, and equation 31 must be rewritten to reflect this:

$$dE = \left(\frac{\partial E}{\partial S}\right)_{V,n_j} dS + \left(\frac{\partial E}{\partial V}\right)_{S,n_j} dV + \left(\frac{\partial E}{\partial n_1}\right)_{S,V,n_2} dn_1 + \left(\frac{\partial E}{\partial n_2}\right)_{S,V,n_1} dn_2 \tag{32}$$

The quantities $(\partial E/\partial n_1)_{S,V,n_2}$ and $(\partial E/\partial n_2)_{S,V,n_1}$ are defined as the *chemical potentials*, μ_1 and μ_2, of species 1 and 2, respectively. These chemical potentials depend on the composition of the solution, that is, the mole fractions, x_1 and x_2. Comparing equations 32 and 19, we note that $T = (\partial E/\partial S)_{V,n_j}$ and $p = -(\partial E/\partial V)_{S,n_j}$. The fundamental chemical thermodynamic equation for a binary solution can then be written as

$$dE = T\,dS - p\,dV + \mu_1\,dn_1 + \mu_2\,dn_2 \tag{33}$$

Chemical Reaction

A *chemical reaction* produces a variation in the *mole numbers* of a closed solution, $dn_i = \nu_i\,d\xi$, where the ν_i are the *stoichiometric coefficients* of the species in the reaction (< 0 for reactants, > 0 for products) and ξ is the *advancement variable* (> 0). For example, a weak acid, HA, dissociates in aqueous solution according to the reaction $HA + H_2O = H_3O^+ + A^-$, and the respective stoichiometric coefficents are -1, -1, and $+1$, and $+1$. The fundamental chemical thermodynamic equation for this reaction is then

$$dE = T\,dS - p\,dV - \mu_{HA}\,dn_{HA} - \mu_{H_2O}\,dn_{H_2O}$$
$$+ \mu_{H_3O^+}\,dn_{H_3O^+} + \mu_{A^-}\,dn_{A^-} \tag{34}$$

Variations in mole numbers through additions or reaction in a homogeneous system thus require that the internal energy be represented as

$$E = E(S, V, n_i) \tag{35}$$

2.3 Systems of Variable Composition: Chemical Thermodynamics

and the differential as

$$dE = \left(\frac{\partial E}{\partial S}\right)_{V, n_i} dS + \left(\frac{\partial E}{\partial V}\right)_{S, n_j} dV + \sum_i \left(\frac{\partial E}{\partial n_i}\right)_{S, V, n_j} dn_i \qquad (36)$$

where the chemical potential of species i is $\mu_i = (\partial E/\partial n_i)_{S, V, n_j}$.

The fundamental thermodynamic equation for the solution may be written

$$dE = T\,dS - p\,dV + \sum \mu_i\,dn_i \qquad (37)$$

Comparing equation 37 with 22, we note $\sum \mu_i\,dn_i = -T\,d_iS$. In a homogeneous system the entropy production is a consequence of change in composition. For a fixed S and V system, the condition of equilibrium is: $dE = \sum \mu_i\,dn_i = -T\,d_iS = 0$. For an irreversible change in composition: $dE = \sum \mu_i\,dn_i = -T\,d_iS < 0$.

When the change in the mole numbers, dn_i, is due solely to *chemical reaction* within a single phase, the fundamental equation is then written

$$dE = T\,dS - p\,dV + \sum v_i \mu_i\,d\xi \qquad (38)$$

in which $\sum v_i \mu_i$ is a fundamental quantity in chemical thermodynamics of reactions. The quantity $\sum v_i \mu_i$ describes the energetics of an individual reaction in the system. It is known, following Lewis and Randall (1961), as *the free energy change of the reaction*, ΔG. (However, the quantity $\sum v_i \mu_i$ is the *natural* one for describing the energetics of a chemical reaction in terms of *any* state function, E, H, A, or G. Its use in connection with the Gibbs energy is the most common.) In an alternative notation widely used in European writings on chemical thermodynamics (Prigogine, 1961), $-\sum v_i \mu_i$ is called the *affinity*, A. In terms of the affinity of the reaction, $dE = T\,dS - p\,dV - A\,d\xi$; in terms of the free energy change, $dE = T\,dS - p\,dV + \Delta G\,d\xi$. We will use the latter expression. We note that for a spontaneous reaction at fixed S and V, $A > 0$ or $\Delta G < 0$. At chemical reaction equilibrium the affinity vanishes. We can summarize the relationship between affinity, ΔG, and internal entropy variation, thus: $A = -\Delta G = T\,d_iS\,d\xi$. At equilibrium all are zero.

Heterogeneous Chemical System For a system comprising more than one phase, (e.g., water and a gas phase; or water and a solid phase; or water, solid, and a gas phase, see Figure 1.2), the quantities S and V refer to the *total* system, that is, $S^{\text{total system}}$ and $V^{\text{total system}}$. For a heterogeneous system, the term describing change in composition in equation (37) needs to be modified to account for *both* chemical reaction and transfer between phases. For example, carbon dioxide, $CO_2(g)$, dissolves in water and the dissolved form, $CO_2(aq)$, reacts with water to produce additional species, that is, $H_2CO_3(aq)$, HCO_3^-, and CO_3^{2-} [and thereby to alter $H^+(aq)$ and $OH^-(aq)$).

A double summation is then required, over the species i *and* over the phases.

For several phases, we can label a particular phase p and the total number of phases π; in a phase there are m different species i. The fundamental expression for a *heterogeneous system of variable composition* is (Abbott and VanNess, 1989):

$$dE^{tot} = T\, dS^{tot} - p\, dV^{tot} + \sum_p^\pi \sum_i^m \mu_i^p\, dn_i^p \qquad (39)$$

At complete system equilibrium, dE^{tot} is zero, and E^{tot} is a minimum. For spontaneous processes, $dE^{tot} < 0$.

2.4. GIBBS ENERGY AND SYSTEMS OF VARIABLE CHEMICAL COMPOSITION

Up until now we have employed both the internal (or thermodynamic) energy, E, and the Gibbs energy, G, in describing the *general* critiera for (reversible) processes and spontaneous (irreversible) processes. The use of the internal energy, E, followed naturally from its introduction in the first law of thermodynamics. Form this point on, however, we will rely on the Gibbs energy, G, in considering principles and applications of chemical thermodynamics to systems of fixed temperature and pressure.

The Gibbs free energy of a system depends on P, T, and composition: $G(p, T, n_i)$. The differential is

$$dG = \left(\frac{\partial G}{\partial T}\right)_{p,n_j} dT + \left(\frac{\partial G}{\partial p}\right)_{T,n_j} dp + \sum_i \left(\frac{\partial G}{\partial n_i}\right)_{p,T,n_j} dn_i \qquad (40)$$

The chemical potential of species i is

$$\mu_i = \left(\frac{\partial G}{\partial n_i}\right)_{p,T,n_j} \qquad (41)$$

Equation 41 shows that the chemical potential is a partial molar property. We will need other partial molar quantities (e.g., those for volume, enthalpy, and entropy) in dealing with pressure and temperature effects on energetics of reactions.

Comparing equation 40 with 25a, we note

$$\left(\frac{\partial G}{\partial T}\right)_{p,n_j} = -S \quad \text{and} \quad \left(\frac{\partial G}{\partial p}\right)_{T,n_j} = V \qquad (42)$$

The fundamental chemical thermodynamic equation in terms of the Gibbs energy may now be written compactly as

2.4 Gibbs Energy and Systems of Variable Chemical Composition 33

$$dG = -S\,dT + V\,dp + \Sigma\,\mu_i\,dn_i \tag{43}$$

Again comparing with equation 25a, we note that $\Sigma\,\mu_i\,dn_i = -T\,d_iS$. The internal entropy production is a result of change in the composition of the system. At equilibrium, the dn_i vanish, as does the entropy production within the system. For a fixed temperature and pressure system, $dG = \Sigma\,\mu_i\,dn_i = -T\,d_iS$.

Gibbs Energy Change in Chemical Reactions

A chemical reaction causes a change in the Gibbs energy of a phase according to

$$dG = -S\,dT + V\,dp + \Sigma\,\nu_i\,\mu_i\,d\xi \tag{44}$$

Or, in terms of the free energy change of the reaction, $\Delta G = \Sigma\,\nu_i\,\mu_i$,

$$dG = -S\,dT + V\,dp + \Delta G\,d\xi \tag{45}$$

At fixed T and p,

$$dG = \Delta G\,d\xi \tag{46}$$

Recalling that T and p are fixed, we can write $\Delta G = (\partial G/\partial \xi)_{p,T}$ i.e., the quantity $\Sigma\,\nu_i\,\mu_i$ is the rate of change of the system free energy with respect to the progress of the reaction.

To describe the state of a reaction in a phase, we need to know the stoichiometric coefficients, ν_i, and the chemical potential, μ_i, for each species in the reaction. For reaction equilibrium, the quantity $\Delta G = \Sigma\,\nu_i\,\mu_i = 0$ (as is $T\,d_iS$). For a possible, or spontaneous, reaction, $\Delta G < 0$. For multireaction systems, complete equilibrium corresponds to $dG = 0$ for the system, that is, the Gibbs energy of the phase is a minimum. The total internal entropy production must vanish for the entire system. Similar consideration apply to multiphase systems. An expression analogous to equation 39 for dE^{tot}, but for fixed T and p conditions, is:

$$dG^{\text{tot}} = \sum_p^\pi \sum_i^m \mu_i^p\,dn_i^p \tag{47}$$

At complete multiphase, multireaction equilibrium, dG^{tot} is zero, and the Gibbs energy of the total system is a minimum.

Extensive Properties and Partial Molar Properties of a Phase

The chemical potential of a species in a phase is a partial molar quantity (equation 41). The total Gibbs energy of a phase is $G = \Sigma\,\mu_i\,n_i$. Other partial molar properties are those of enthalpy, volume, entropy, and heat capacity. A partial molar property of a species i, $\overline{X}_i$, is defined by $\overline{X}_i = (\partial X/\partial n_i)_{p,T,n_j}$. The extensive property of the phase is then $X = \sum_1^m \overline{X}_i n_i$. For example, the enthalpy of a phase is calculated by $H = \sum_i n_i \overline{H}_i$.

For systems of more than one phase, the summation must include all phases and all components. For example, the total free energy of a system of π phases is

$$G^{\text{tot}} = \sum_p^\pi \sum_i^m n_i^p \mu_i^p. \qquad (48)$$

Gibbs-Duhem Relationship The partial molar properties of a multicomponent phase cannot be varied independently (the mole fractions, $x_i = n_i/\Sigma\, n_j$, of the components total unity). For example, for the chemical potentials, μ_i, the Gibbs–Duhem relationship is $\Sigma n_i\, d\mu_i = 0$ (for details, see e.g., Atkins, 1990; Blandamer, 1992; Denbigh, 1971). Similar constraints apply to the partial molar volumes, enthalpies, entropies, and heat capacities. For pure substances, the *partial molar* property is equal to the *molar* property. For example, the chemical potential of a pure solid or liquid is its energy per mole. For gaseous, liquid, or solid solutions, $X_i = X_i(n_j)$, that is, the chemical potentials and partial molar volumes of the species depend on the mole fractions.

One approach to computing the equilibrium composition of a system involves minimizing the total Gibbs energy of the system, for example, minimizing an expression such as equation 48, subject to the material balance constraints and others of the system. Thus it is essential in applying this approach to know the bahavior of the chemical potentials as a function of system parameters, namely, p, T, and composition.

Enthalpy, Entropy, and Volume Changes of Reaction

The enthalpy of reaction, ΔH, is evaluated by the expression $\Delta H = \sum_i \nu_i \overline{H}_i$, which is analogous to the expression for the free energy of reaction, $\Delta G = \Sigma\, \nu_i \mu_i$. Expressions for ΔV and ΔS are similar: $\Delta V = \sum_i \nu_i\, \overline{V}_i$ and $\Delta S = \sum_i \nu_i\, \overline{S}_i$. Using these expressions, we can readily calculate the changes in state functions—ΔG, ΔH, ΔS, and ΔV—needed to find the state of a chemical reaction and the influences of temperature and pressure on that reaction.

Chemical Potentials: Temperature and Pressure

Equation 42 gives the partial derviatives of the Gibbs energy with respect to temperature, $(\partial G/\partial T)_{p,n_j} = -S$, and pressure, $(\partial G/\partial p)_{T,n_j} = V$. Derivatives of the chemical potentials with respect to temperature give the following expressions used in predicting the effects of temperature and pressure on chemical equilibrium:

$$\left(\frac{\partial \mu_i}{\partial T}\right)_{p,n_j} = -\overline{S}_i; \quad \left(\frac{\partial \mu_i/T}{\partial T}\right)_{p,n_j} = \frac{-\overline{H}_i}{T^2}; \quad \left(\frac{\partial \mu_i}{\partial p}\right)_{T,n_j} = \overline{V}_i \qquad (49)$$

Further deviatives yield the partial molar compressibility, $\overline{\kappa}_i$, and the partial molar heat capacity, $\overline{c_{p_i}}$. These quantities are required to calculate chemical potentials at higher pressures and over a wider range of temperatures, respectively.

2.5 CHEMICAL POTENTIALS OF PURE PHASES AND SOLUTIONS

The foregoing relationships are necessary to evaluate $\mu_i(p, T)$. In order to achieve our goals of describing equilibrium compositions of natural water systems (aqueous, gaseous, solid phases), predicting the direction of possible changes, and evaluating available work and energy requirements for chemical processes, we need to know the relationship between chemical potentials and composition, that is, $\mu_i(x_1, x_2, \ldots, x_j \ldots x_m)$. With this, we will have $\mu_i(p, T, x_j)$ and the tools for describing equilibrium and disequilibrium in aquatic systems.

The Activity The chemical potential of any species in any phase will be expressed (Lewis and Randall, 1961) as

$$\mu_i = \mu_i^\circ + RT \ln \{i\} \qquad (50)$$

in which $\mu_i^\circ = \mu_i^\circ(p, T)$ and the activity of species i, $\{i\}$, is dimensionless. For pure phases (e.g., solid, liquid, ideal one-component gas) $\{i\} = 1$ by definition. For liquid, gaseous, and solid solutions, the composition needs to be described on a convenient scale. It would be attractive, in thermodynamic principle, to use the mole fraction, x_i, for all phase compositions. For good chemical and practical reasons, that choice is not always convenient. Other scales are used: partial pressure, p_i, for the gas phase; and molality, m_i, for aqueous solutions of electrolytes, inorganic molecules, and lower molecular weight organic solutes.

The Fugacity An alternative definition for the chemical potential (Lewis and Randall, 1961) involves the *fugacity*, an ideal pressure characterizing escaping tendency from a phase:

$$\mu_i = \mu_i^\circ + RT \ln \left(\frac{f_i}{f_i^\circ}\right) \qquad (51)$$

where f_i is the fugacity and f_i° is a standard fugacity, such that $\mu_i = \mu_i^\circ$ when $f_i = f_i^\circ$. Under low pressure conditions, the fugacity can be equated with the partial pressure, p_i. The chemical potential variation and the fugacity variation are obviously related by $d\mu_i = RT\, d \ln f_i$ ($d\mu_i = RT\, d \ln p_i$, for low total pressures). At low concentrations, the limiting behavior of the fugacity is more convenient than that of the chemical potential. For that reason, many workers prefer the fugacity, or partial pressure, as a measure of escaping tendency from a phase (Schwarzenbach et al., 1993). The fugacity is not a natural choice for ionic species.

Concentration and Activity   Activity and actual concentration in any real solution must be related by an *activity coefficient* (Blandamer, 1992; Lewis and Randall, 1961). For example, using mole fractions

$$\{i\} = x_i \gamma_i \tag{52}$$

where γ_i is a dimensionless activity coefficient. The chemical potential of a species in a mixture is then

$$\mu_i = \mu_i^\circ + RT \ln x_i \gamma_i \tag{53}$$

Energy Terms in the Chemical Potential Expression   For solutions, a general form for the chemical potential is

$$\underset{\substack{\text{chemical}\\\text{potential}}}{\mu_i} = \underset{\substack{\text{standard}\\\text{potential}}}{\mu_i^\circ} + \underset{\substack{\text{free energy}\\\text{of mixing}}}{RT \ln x_i} + \underset{\substack{\text{potential}\\\text{free energy}}}{RT \ln \gamma_i}$$

where the free energy of mixing is given by the $\ln x_i$ term and the activity coefficient term accounts for nonideality due to potential energies of species interactions.

Standard State and Reference State   Equation 53 contains two quantities which require precise definition in order to make use of thermodynamic data or render experimental observation into the framework of the equation. We need to specify the condition of a mixture for which $\gamma_i \to 1$. This is the *reference state* for the species in the mixture. We also need to specify the *standard state*. This is the condition of the mixture for which $\gamma_i = 1$ *and* $x_i = 1$. (The standard state of a solution may be either an actual condition or a hypothetical one; it is frequently the latter.)

Concentration Scales   For other, more practical, concentration scales for mixtures, similar considerations arise regarding reference and standard states. For aqueous solutions, both the mole fraction scale and the molal scale are used for thermodynamic interpretations. The solvent water is described on the

2.5 Chemical Potentials of Pure Phases and Solutions

mole fraction scale. The standard state and reference state are the same: for pure solvent, i.e., $x_w = 1$.

Many nonionizable *organic* solutes in water are described thermodynamically on the mole fraction scale, although their solubilities may commonly be reported in practical units, for example, molality. [Refer to Schwarzenbach et al. (1993) and Klotz (1964); for detailed discussion of such aqueous solutions.] Here, the standard state is the pure liquid state of the organic solute, that is, $x_i = 1$. The reference state is $x_i \rightarrow 1$, that is, a solution in which the organic solute molecules interact with one another entirely. Activity coefficients of solute molecules in dilute aqueous solutions are generally much greater than unity for this reference state choice, $x_i \rightarrow 1$. For example, with this reference state, aqueous benzene has an experimental infinitely dilute solution activity coefficient, $\gamma_{\text{benzene}}^\infty$, of 2400; for an infinite dilution reference state, $x_i \rightarrow 0$, the activity coefficient would be approximately 1 (Tanford, 1991).

The Molal Scale The molal scale (mol solute kg^{-1} solvent water) is used for ions, inorganic molecules, low molecular weight organic solutes (e.g., methane and ethane), and ionizable organic molecules (e.g., organic acids and bases). The expression for the chemical potential for the solute species i in water can be written

$$\mu_i = \mu_i^\circ + RT \ln \left(\frac{m_i}{m^\circ}\right) \gamma_i \tag{54}$$

where m_i is the molal concentration; m° is the standard concentration of the scale: $m^\circ = 1$; and γ_i is the molal scale activity coefficient.[†]

Comparing equation 54 with 50, we identify the activity of species i as

$$\{i\} = \left(\frac{m_i}{m^\circ}\right) \gamma_i \tag{55}$$

Reference State. The infinite dilution reference state for the solute, species 2, on the molal scale is $\gamma_2 \rightarrow 1$ as $m_2 \rightarrow 0$ (using the label 1 for the solvent and 2 for the only solute). For a multisolute aqueous solution, one logical reference state is $\gamma_i \rightarrow 1$ as $\Sigma\, m_j \rightarrow 0$; another reference state, the constant ionic medium, is $\gamma_i \rightarrow 1$ as $m_i \rightarrow 0$ in a solution of high univalent–univalent electrolyte (e.g., NaClO$_4$ or KNO$_3$). In the infinite dilution reference state, the solute species experiences only solute–solvent interactions; in the ionic medium reference state, the solute interacts with both solvent and the medium ions.

[†]While we might use different labels for different activity coefficients (e.g., γ_i^*, γ_i'') we prefer to emphasize the *matching* of the scales with the activity coefficient reference states for different species (solvent, ionic solute, etc.). In Chapters 3 and 6, for example, we use the symbol f, or f_i, to discuss activities on a *molar* scale. The usage will be clear in the context.

Standard State. The molal scale standard state is a *hypothetical* one. In the standard state $m_i = 1$ and $\gamma_i = 1$, with the result that $\{i\} = 1$. Note that it is not sufficient that the activity be unity to be in the standard state: the solution must have the properties of the *ideal solution*, with $\gamma_i = 1$. The contributions to the chemical potentials are:

$$\underset{\substack{\text{chemical}\\\text{potential}}}{\mu_i} = \underset{\substack{\text{standard}\\\text{potential}}}{\mu_i^\circ} + \underset{\substack{\text{free energy}\\\text{of mixing}}}{RT \ln (m_i/m^\circ)} + \underset{\substack{\text{potential}\\\text{free energy}}}{RT \ln \gamma_i}$$

Gas Phase At lower pressures (say, less than 50 atm), the partial pressure can be used to calculate the chemical potential of a gaseous species in a mixture. The expression is

$$\mu_i = \mu_i^\circ + RT \ln (p_i/P^\circ) \tag{56}$$

where P° is a standard pressure (e.g., $P^\circ = 1$ atm); $\mu_i^\circ(T)$. Where high pressures are involved, the fugacity is used instead of the partial pressure.

Solid Phases For pure solids, $\mu_i = \mu_i^\circ(p, T)$. For solid solutions, the mole fraction scale is used (e.g., equation 53), $\mu_i = \mu_i^\circ + RT \ln x_i \gamma_i$, with, for example, $\gamma_i \to 1$ as $x_i \to 1$ as a possible reference state. Here, as for organic solutes in water, the activity coefficients must be interpreted in light of the reference state chosen.

In summary, thermodynamic models of natural water systems require manipulation of chemical potential expressions in which three concentration scales may be involved: mole fractions, partial pressures, and molalities. For aqueous solution species, we will use the molal scale for most solutes, with an infinite dilution reference state and a unit molality standard state (of unit activity). For the case of nonpolar organic solutes, the pure liquid reference and standard states are used. Gaseous species will be described on the partial pressure (atm $\simeq$ bar) scale. Solids will be described using the mole fraction scale. Pure solids (and pure liquids) have $x_i = 1$, and hence $\mu_i = \mu_i^\circ$.

2.6. CHEMICAL POTENTIALS OF AQUEOUS ELECTROLYTES

Aqueous electrolyte chemical potentials are described on the molal scale. To illustrate the additional issues that enter into the thermodynamic interpretation of individual ion activity coefficients and chemical potentials and the relation of these to actual electrolyte experimental measurement, we briefly review the properties of the system NaCl(aq), that is, NaCl dissolved in water. [For a detailed discussion, see. e.g., Denbigh, 1971; Harned and Owen, 1959; Klotz, 1964; Robinson and Stokes, 1959).

NaCl is a strong electrolyte in water (i.e., fully dissociated). There are no

2.6 Chemical Potentials of Aqueous Electrolytes

NaCl "molecules" in its aqueous solution. Vapor pressure lowering of dilute NaCl solutions obeys Raoult's law for the solvent water—that is, $p_W = p_W^\circ x_W$—and the vapor pressure corresponds to *two* solute species.

The chemical potential of aqueous NaCl must then be the *sum* of the chemical potentials of the $Na^+(aq)$ and $Cl^-(aq)$ ionic species:

$$\mu_2 \equiv \mu_{NaCl} = \mu_{Na^+} + \mu_{Cl^-} \tag{i}$$

In dilute solution, *theoretical* expressions for the individual ions are

$$\mu_{Na^+} = \mu_{Na^+}^\circ + RT \ln m_{Na^+} \tag{ii}$$

$$\mu_{Cl^-} = \mu_{Cl^-}^\circ + RT \ln m_{Cl^-} \tag{iii}$$

with

$$\mu_{Na^+}^\circ + \mu_{Cl^-}^\circ \equiv \mu_{NaCl(aq)}^\circ \tag{iv}$$

Therefore the *total* chemical potential is of the form

$$\mu_{NaCl(eq)} = \mu_{NaCl(aq)}^\circ + RT \ln m_{Na^+} m_{Cl^-} \tag{v}$$

Adopting the convenient brief notation

$$m_+ = m_{Na^+}$$
$$m_- = m_{Cl^-} \tag{vi}$$

and recognizing that stoichiometry requires

$$m_{NaCl} = m_{Na^+} = m_{Cl^-} \tag{vii}$$

the result is

$$\mu_{NaCl} = \mu_{NaCl}^\circ + RT \ln m_+ m_-$$
$$= \mu_{NaCl}^\circ + RT \ln m_{NaCl}^2 \tag{viii}$$

For a nonideal electrolyte solution, for example, NaCl, the chemical potential on the molal scale must then be expressed as

$$\mu_{NaCl} = \mu_{NaCl}^\circ + RT \ln a_+ a_- \tag{ix}$$

or

$$\mu_{NaCl} = \mu_{NaCl}^\circ + RT \ln \gamma_+ m_+ \gamma_- m_- \tag{x}$$

where a_+ and a_-, and γ_+ and γ_-, represent the molal-scale activities and activity coefficients for Na$^+$(aq) and Cl$^-$(aq), respectively, and $m_+ = m_- = m$. The product $a_+ a_-$ is denoted by a_{NaCl}, the *activity* of the aqueous solute, or $a_2 = a_+ a_-$. The *mean activity* (a geometric mean) is defined by $a_\pm = (a_+ a_-)^{1/2} = a_2^{1/2}$. The quantity that can be determined experimentally is $a_\pm$. The individual activites a_+ and a_- are not independently measurable quantities.

The product $\gamma_+ \gamma_-$ is experimentally measurable. The quantity $(\gamma_+ \gamma_-)^{1/2}$ is referred to as the *mean molal activity coefficient* $\gamma_\pm$. The mean ionic molality $m_\pm$ is defined as $(m_+ m_-)^{1/2}$ and is simply m for a univalent–univalent electrolyte. Summarizing these definitions for a nonideal, *univalent–univalent* solution, where the solute is component 2.

$$a_2 = a_+ a_- \qquad (\text{xi})$$

$$a_\pm = (a_+ a_-)^{1/2} = a_2^{1/2} \qquad (\text{xii})$$

$$a_+ = \gamma_+ m_+ \qquad (\text{xiii})$$

$$a_- = \gamma_- m_- \qquad (\text{xiv})$$

$$\gamma_\pm = (\gamma_+ \gamma_-)^{1/2} \qquad (\text{xv})$$

$$a_\pm = \gamma_\pm m_\pm = \gamma_\pm m \qquad (\text{xvi})$$

Example 2.3. Activity and Activity Coefficient of Aqueous NaCl Water vapor pressures have been measured over NaCl solutions of varying molal concentrations. Results of such measurements allow calculations of relative vapor pressure lowering, a_w, ϕ, and $\gamma_\pm$, for a range of NaCl concentrations in water. Robinson and Stokes (1959) provide such data for concentrations ranging from 0.1 to 6.0 molal. Table 2.1 shows results for three different concentrations. The mole fraction of H$_2$O, X_W, is also included. From such data the activity of aqueous NaCl can be computed. (see also Figure 2.4.)

Table 2.1. Molal Concentrations of Aqueous NaCl, Water Activity, and Mean Molal Activity Coefficient of NaCl in Water

m	a_W	X_W	$\gamma_\pm$
0.1	0.996646	0.99641	0.778
0.3	0.99009	0.98931	0.709
1.0	0.96686	0.96522	0.657

Source: Robinson and Stokes (1959).

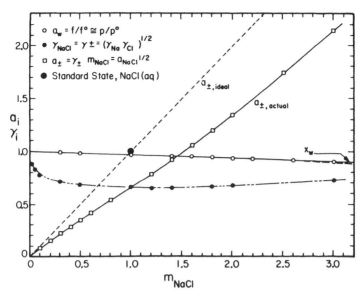

Figure 2.4. Example 2.3. Activities of aqueous NaCl (actual and ideal), activity coefficient for NaCl, and the activity of water at various molal concentrations. The *standard state* ($m = 1$, $a = 1$, $\gamma = 1$) is a hypothetical solution.

2.7 THE EQUILIBRIUM CONSTANT

A reaction involving solutions can be described by an equilibrium constant. The "law of mass action" of Guldberg and Waage in 1867 originally resulted from setting rates of opposing reactions equal and was subsequently brought within the framework of thermodynamics by van't Hoff (Servos, 1990). Reactions involving *only* pure phases do not have equilibrium constants in the ordinary sense (although they can be described in terms of *relative*, nonequilibrium activities). Thermodynamically speaking, reactions involving *only* pure phases proceed until exhaustion of that substance which is present in limiting amount.

For mixtures, the relationship between the Gibbs free energy of a reaction and the composition of the system is obtained by substituting the expression for the chemical potential in terms of the activity of a species i

$$\mu_i = \mu_i^\circ + RT \ln \{i\} \tag{50}$$

into the expression for ΔG for a reaction $\Sigma_i \nu_i M_i = 0$,

$$\Delta G = \sum_i \nu_i \mu_i \tag{45}$$

The direct result is

$$\Delta G = \sum_i \nu_i \mu_i^\circ + RT \sum_i \nu_i \ln \{i\} \qquad (57)$$

or

$$\Delta G = \Delta G^\circ + RT \ln \prod_i \{i\}^{\nu_i} \qquad (58)$$

where

$$\Delta G^\circ = \sum_i \nu_i \mu_i^\circ \qquad (59)$$

is the *standard* Gibbs free energy change of the reaction.

The reaction quotient Q is defined as

$$Q = \prod_i \{i\}^{\nu_i} \qquad (60)$$

So that we have

$$\Delta G = \Delta G^\circ + RT \ln Q \qquad (61)$$

At equilibrium, $\Delta G = 0$ and the numerical value of Q becomes K, the equilibrium constant:

$$K \equiv Q_{eq} = \prod_i \{i\}_{eq}^{\nu_i} \qquad (62)$$

and

$$\Delta G^\circ = -RT \ln K \qquad (63)$$

This is a central relationship in the chemical thermodynamics of mixtures. Under any conditions,

$$\Delta G = RT \ln \frac{Q}{K} \qquad (64)$$

Comparison of Q (actual composition) with the value of K (equilibrium composition) provides a test for equilibrium ($\Delta G = 0$).

The equilibrium constant, K, and the reaction quotient, Q, are expressed fundamentally in terms of activities, $\{i\}$. To describe equilibrium or disequilibrium in terms of concentrations, we must bring in the relationships discussed in Section 2.3, namely, *activity = concentration × activity coefficient.* In each phase and for each type of species, a reference state defines the numerical scale

2.7 The Equilibrium Constant

of activity coefficients and the standard state defines the scale of concentration. For example, for aqueous species expressed on the molal scale, equation (55) gives the relationship between $\{i\}$, m_i, and γ_i: $\{i\} = (m_i/m°)\gamma_i$; the chemical potential equation (54) is $\mu_i = \mu_i° + RT \ln (m_i/m°)\gamma_i$; and the reaction quotient in terms of molalities then takes the form

$$Q = \prod_i \left(\frac{m_i}{m°}\right)^{\nu_i} \prod_i \gamma_i^{\nu_i} \tag{65}$$

For the equilibrium state of the reaction,

$$K = \left(\prod_i \left(\frac{m_i}{m°}\right)^{\nu_i} \prod_i \gamma_i^{\nu_i}\right)_{eq} \tag{66}$$

Similar considerations apply to other choices of scales and reference states in aqueous and solid phases.

A few examples may serve to make the form of molal scale equilibrium constants clear.

(a)   Reaction: $H_2O \longleftrightarrow H^+ + OH^-$ (i)

        Scales: mole fraction   molal   molal

$$\frac{(m_{H^+}/m°)\gamma_{H^+}(m_{OH^-}/m°)\gamma_{OH^-}}{(X_{H_2O}/X°)\gamma_{H_2O}} = K_W \tag{ii}$$

The expression routinely used for *dilute* solutions is

$$m_{H^+} \, m_{OH^-}/X_{H_2O} = K_W(m^2) \tag{iii}$$

which: (1) is based on $m° = 1$ (molal) and $X° = 1$ (dimensionless), hence the numerical value, K_W, is unchanged; (2) reflects the units of K correctly; and (3) takes the *activity coefficient quotient*, $\gamma_{H_2O}/\gamma_{H^+}\gamma_{OH^-}$ to be unity and dimensionless.

(b)   Reaction: $HA \longleftrightarrow + H^+ + A^-$ (iv)

        Scales: molar   molar   molar

This choice is the common one for water chemistry in fresh waters. For dilute solutions containing a weak acid, HA, the practical result is

$$\frac{[H^+][A^-]}{[HA]} = K_a \ (M) \tag{v}$$

[ ]° = 1 M having been chosen implicitly for the standard concentration.

44   Chemical Thermodynamics and Kinetics

For nondilute solutions, the activity coefficient must be included, and the result is

$$\frac{[H^+][A^-]}{[HA]} = K_a\, \gamma_{HA} \div \gamma_{H^+}\gamma_{A^-} \quad (M) \tag{vi}$$

(c)   Reaction: $CO_2(g) \leftrightarrow CO_2(aq)$ (vii)
    Scales: atm   molar

For dilute gas and solution,

$$\frac{[CO_2]}{p_{CO_2}} = K_H \quad (M\ atm^{-1}) \tag{viii}$$

The *general* expression underlying the foregoing expression is

$$\frac{([CO_2(aq)]/[\ ]^\circ)\gamma_{CO_2}}{p_{CO_2}/P^\circ} = K_H \tag{ix}$$

with $[\ ]^\circ = 1$ M and $P^\circ = 1$ atm. The reference state for γ_{CO_2} is infinite dilution for all solutes (the activity coefficient depends on salt concentration, with $\gamma_{CO_2} > 1$, i.e., CO_2 is "salted out") of water.

2.8 THE GIBBS ENERGY OF A SYSTEM

In an earlier section the free energy of a phase and the free energy of a total system were discussed generally in terms of the potentials (e.g., equation 48). With the definition of the chemical potential as a function of activity in hand, we will now consider the Gibbs energy of a system. In a similar fashion, the enthalpy and entropy of a system can be computed using the partial molar quantities and the mole numbers of each phase.

The Gibbs energy of a system can be obtained from

$$G^{tot} = \sum_p^\pi \sum_i^m n_i^p \mu_i^p \tag{48}$$

by substituting the expression for the mole numbers in terms of the advancement of one reaction, ξ,

$$n_i = n_i^\circ + \nu_i \xi \tag{67}$$

and equation 50 for the chemical potential in terms of activity, $\mu_i = \mu_i^\circ + RT$

ln $\{i\}$, giving

$$G = \sum_i (n_i^\circ + \nu_i \xi)(\mu_i^\circ + RT \ln \{i\}) \tag{68}$$

The Gibbs free energy versus ξ for reactions involving solutions and gas mixtures is characterized by a minimum value at the equilibrium composition. We have already seen that the minimum value of G corresponds to $\Delta G = 0$, ΔG being the derivative of G with respect to ξ. The $(n_i^\circ + \nu_i\xi)RT \ln \{i\}$ terms in equation 68 account for the free energy of mixing. For reactions between pure substances, these terms are absent and there is no minimum in the G function. Figure 2.5 depicts generalized relationships between the G function, the Gibbs free energy of reaction, and the extent of reaction for a chemical reaction mixture. Point e corresponds to the equilibrium state for the reaction. Point f represents a nonequilibrium state from which the system may proceed to equilibrium spontaneously, the reaction proceeding from left to right as written. Point r represents a state in which the reaction is spontaneous in the reverse direction (right to left as written).

For the case of a single reaction illustrated in Figure 2.5, the free energy of reaction can readily be calculated from equation 58. The activity, $\{i\}$ can be calculated from the initial mole numbers and the advancement variable. In general, the calculation may require information of activity coefficients of the species, unless we can assume an ideal mixture.

For a reaction involving only pure phases, the shape of the G function and the profile of ΔG is much simpler and more easily calculated. In equation 68, $\{i\} = 1$ for all pure solid or liquid species. Figure 2.6 illustrates G and ΔG versus ξ for reactions involving only pure phases. In type 1, the driving force is zero. The system of phases is stable. In type 2, advancement in the direction chosen as positive is energetically possible. In type 3, the reaction is not spontaneous in the chosen direction.

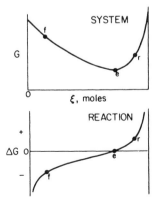

Figure 2.5. Variation of the Gibbs free energy function G and the Gibbs free energy change of the reaction ΔG for a single reaction in a system of variable composition, for example, reactions in solution.

46 Chemical Thermodynamics and Kinetics

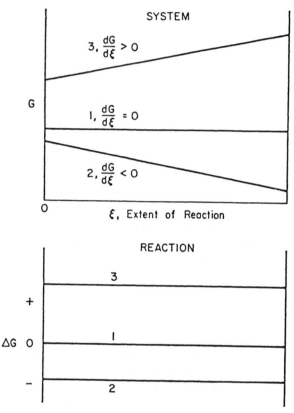

Figure 2.6. Variation of the Gibbs free energy function G and the Gibbs free energy change of the reaction ΔG with extent of reaction for reactions between pure substances, for example, solids or solids and liquids.

Example 2.4. Thermodyanmics of CO_2 Dissolution in Water Describe the variations in the entropy, enthalpy, and Gibbs energy with extent of CO_2 dissolution for a two-phase system comprising a gas phase and an aqueous phase. Find the equilibrium state. Initially, a liter of gas at 1 atm total pressure contains 2×10^{-5} mol of CO_2. It is brought into contact with a liter of pure water. The dissolution process is

$$CO_2(g) = CO_2(aq) \tag{i}$$

The standard-state thermodynamic data at 25°C are as follows:

Species	$\overline{H}_f^\circ$ (kcal mol^{-1})	$\overline{G}_f^\circ$ (kcal mol^{-1})	$\overline{S}^\circ$ (cal K^{-1} mol^{-1})
$CO_2(g)$	−94.05	−94.26	51.06
$CO_2(aq)$	−98.69	−92.31	29.0

2.8 The Gibbs Energy of a System

The Gibbs energy of the two-phase system is given by

$$G = n_{CO_2(g)}\mu_{CO_2(g)} + n_{CO_2(aq)}\mu_{CO_2(aq)} \quad \text{(ii)}$$

The mole fraction of the solvent water is essentially unity, so a term is not included for the water in equation ii. The gas phase and solution phase will be approximately ideal (why?), so the chemical potentials can be well approximated by

$$\mu_{CO_2(g)} = \mu^\circ_{CO_2(g)} + RT \ln p_{CO_2} \quad \text{(iii)}$$

and

$$\mu_{CO_2(aq)} = \mu^\circ_{CO_2(aq)} + RT \ln m_{CO_2} \quad \text{(iv)}$$

with

$$m_{CO_2} \simeq [CO_2(aq)] = n_{CO_2(aq)}$$

for an aqueous volume of 1 liter.

The extent of reaction ξ is numerically equal to $n_{CO_2(aq)}$, and $n_{CO_2(g)} = 2 \times 10^{-5} - \xi$. The partial pressure of $CO_2(g)$ is calculated from the equation of state:

$$p_{CO_2} V_g = n_{CO_2(g)} RT \quad \text{(v)}$$

Thus G can readily be calculated as a function of the extent of reaction, ξ.

In similar fashion, the system enthalpy H for the dilute gas and solution is given by

$$H = n_{CO_2(g)}\overline{H}^\circ_{CO_2(g)} + n_{CO_2(aq)}\overline{H}^\circ_{CO_2(aq)} \quad \text{(vi)}$$

and change in the quantity TS for the system can be obtained from the relationship

$$T \Delta S = \Delta H - \Delta G \quad \text{(vii)}$$

The results for $G - G_0$, $H - H_0$, and $T(S - S_0)$ are shown in Figure 2.7. (G_0, H_0, and S_0 are values at $\xi = 0$). The dissolution of CO_2 under these conditions is favored by a decrease in enthalpy and opposed by a decrease in entropy. The net effect is a decrease in G in proceeding from $\xi = 0$ to the equilibrium value, $\xi = 9.5 \times 10^{-6}$ mol. At equilibrium, $dG/d\xi = 0$. The dissolution of CO_2 in the system is accompanied by the release of 44 mcal of heat to the surroundings. The *decrease* in free energy driving dissolution, 8 mcal, corresponds to an internal entropy *increase* of 0.027 mcal deg^{-1}.

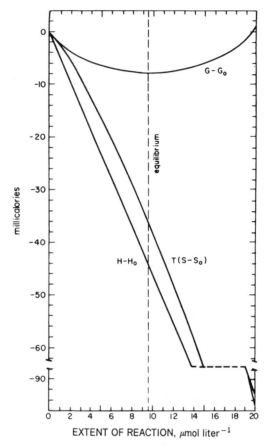

Figure 2.7. Gibbs function, system enthalpy, and system entropy variations with the extent of reaction for the dissolution of gaseous CO_2 in water: $CO_2(g) = CO_2(aq)$ at 25°C in a two-phase system. Total number of moles of CO_2 is 2×10^{-5} mol, the gas-phase volume is 1 liter, and the water volume is 1 liter. The extent of reaction is given by the number of moles of CO_2 dissolved. At equilibrium in the system, $[CO_2(aq)] = 9.5$ μM. The dissolution is favored by $dH/d\xi$ and opposed by $dS/d\xi$.

The standard free energy change $\Delta G°$ for the dissolution process is $\mu°_{CO_2(aq)} - \mu°_{CO_2(g)}$, or 1.95 kcal mol^{-1}. We note that, even though $\Delta G° > 0$, the actual free energy change ΔG is negative for the dissolution process between $\xi = 0$ and $\xi = 9.5 \times 10^{-6}$ mol. The actual free energy change is

$$\Delta G = \Delta G° + RT \ln \frac{[CO_2(aq)]}{p_{CO_2}} \tag{viii}$$

At equilibrium, $\Delta G = 0$, and the equilibrium constant, obtained from $\Delta G° = -RT \ln K$, is $K = 3.72 \times 10^{-2}$ mol liter^{-1} atm^{-1}. At equilibrium,

$$[CO_2(aq)]/p_{CO_2} = 3.72 \times 10^{-2}$$

for the temperature and total pressure considered.

It is of interest to note that CO_2 distributes itself approximately equally between the gas phase and aqueous phase at ordinary temperatures. Here, $[CO_2(aq)]/[CO_2(g)] = 9.5 \times 10^{-6}/1.05 \times 10^{-5} \simeq 1$. ($[CO_2(g)]$ represents mol liter^{-1} of gaseous CO_2.)

2.9 DRIVING FORCE FOR CHEMICAL REACTIONS

The driving force is the Gibbs free energy of reaction, ΔG. The free energy change comprises an enthalpy and an entropy contribution:

$$\Delta G = \Delta H - T\Delta S \tag{23}$$

or, for standard-state conditions,

$$\Delta G° = \Delta H° - T\Delta S° \tag{69}$$

The driving force for a reaction (stability of products with respect to reactants) can be the result of a negative ΔH, a positive ΔS, or both.

The magnitude and direction of the driving force for a reaction depend on the magnitude and sign of the enthalpy change, the magnitude and sign of the entropy change, and the temperature. For $\Delta H < 0$ and $\Delta S < 0$, as well as for $\Delta H > 0$ and $\Delta S > 0$, the possibility of spontaneous reaction depends on the temperature.

The driving force for the ionization of aqueous CO_2 (Example 2.4) was found to be the result of an increase in entropy: $T\Delta S > \Delta H$ at 25°C. Under standard-state conditions at 25°C, the process

$$CO_2(g) = CO_2(aq) \tag{i}$$

is favored by an enthalpy decrease: $\Delta H° = -19.41$ kJ mol^{-1}. However, the decrease in entropy results in $T\Delta S° = -27.57$, and the standard free energy change is $+8.11$ kJ mol^{-1}.

The small stability difference in favor of calcite with respect to aragonite:

$$CaCO_3(\text{aragonite}) = CaCO_3(\text{calcite}) \quad \Delta G° = -1.05 \text{ kJ mol}^{-1} \tag{ii}$$

is associated with an entropy increase of 3.7 J mol^{-1} K^{-1}.

For the precipitation of ferric phosphate,

$$Fe^{3+} + PO_4^{3-} = FePO_4(s) \tag{iii}$$

$$\Delta G^\circ = -102.1 \quad \text{and} \quad \Delta H^\circ = +78.2 \text{ kJ mol}^{-1}$$
$$T \Delta S^\circ = 180.3 \text{ J mol}^{-1} \text{ K}^{-1} \tag{iv}$$

The reaction is endothermic but favored by positive ΔS°.

The precipitation of ferric hydroxide is favored both by $T \Delta S^\circ = 125.9$ kJ mol^{-1} and by $\Delta H^\circ = -86.2$ kJ mol^{-1}, with a resulting ΔG° of -212.1 kJ mol^{-1} (25°C). Large standard entropy changes in the precipitation of a metal ion are associated with an increase in the randomness of water (decreased aquation of metal ions).

Ion association reactions and chelation reactions of aqueous metal ions are generally characterized by significant entropy increases (decreased orientation of solvent molecules and configurational entropy). For example, the ion-pair reaction

$$\text{Co(NH}_3)_5\text{H}_2\text{O}^{3+} + \text{SO}_4^{2-} = \text{Co(NH}_3)_5(\text{H}_2\text{O})\text{SO}_4^+ \tag{v}$$

has $\Delta H^\circ = 0$ and $\Delta S^\circ = +68.6$ J mol^{-1} K^{-1} at 25°C. The association reaction

$$\text{Cu}^{2+} + \text{P}_3\text{O}_{10}^{5-} = \text{CuP}_3\text{O}_{10}^{3-} \tag{vi}$$

has $\Delta H = +20.5$ kJ mol^{-1} and $\Delta S = +247$ J mol^{-1} K^{-1} with a resulting ΔG of -52.2 kJ mol^{-1} (20°C, 0.1 M constant ionic medium).

The chelation reaction of Ca^{2+} with ethylenediaminetetraacetate (Y^{4-}) has the following thermodynamic properties at 25°C:

$$\text{Ca}^{2+} + \text{Y}^{4-} = \text{CaY}^{2-} \tag{vii}$$

$$\Delta H^\circ = -34 \text{ kJ mol}^{-1}, \quad \Delta S^\circ = 92 \text{ J mol}^{-1} \text{ K}^{-1},$$
$$\Delta G^\circ = -59.4 \text{ kJ mol}^{-1} \quad \log K = 10.4 \tag{viii}$$

The binding of aqueous protons to the anions of weak acids is generally favored by sizable entropy increases. For example, the reaction

$$\text{SO}_4^{2-} + \text{H}^+ = \text{HSO}_4^- \tag{ix}$$

has $\Delta S^\circ = 110$ J mol^{-1} K^{-1} and $\Delta H^\circ = 21.8$ kJ mol^{-1}

It is clear that the enthalpy and entropy changes of reactions in aqueous systems vary greatly. The enthalpy change does *not* serve as a criterion of spontaneous reaction. Free energy changes provide the only general description of the driving force of reactions.

Example 2.5. Dissolution of Ammonia in Water From enthalpy and entropy data (Latimer, 1952), compute the equilibrium constant for the solution of

2.9 Driving Force for Chemical Reactions

gaseous ammonia in water at 25°C. The standard molar enthalpies of formation and the standard molar entropies (molal scale for aqueous solution and atm scale for the gas phase) are as follows:

Species	$\overline{H}_f^\circ$ (kJ mol^{-1})	$\overline{S}^\circ$ (J K^{-1} mol^{-1})
NH$_3$(g)	−46.19	192.5
NH$_3$(aq)	−80.83	110.0

For the process

$$NH_3(g) = NH_3(aq) \tag{i}$$

$$\Delta H^\circ = -80.83 - (-46.19) = -34.6 \text{ kJ mol}^{-1} \tag{ii}$$

$$\Delta S^\circ = 110.0 - 192.5 = -82.4 \text{ J mol}^{-1} \text{ deg}^{-1} \tag{iii}$$

$$T\Delta S^\circ = 298.16(-82.4) = -24.6 \text{ kJ mol}^{-1} \tag{iv}$$

The dissolution process is favored by a negative enthalpy change but opposed by a decrease in entropy. Decrease in entropy upon dissolution is characteristic of uncharged solutes. The standard free energy change is

$$\Delta G^\circ = \Delta H^\circ - T\Delta S^\circ = -34.6 - (-24.6) \tag{v}$$
$$= -10.04 \text{ kJ mol}^{-1}$$

The equilibrium constant is computed from

$$-RT \ln K = \Delta G^\circ \tag{vi}$$

or

$$\log K = \frac{-\Delta G^\circ}{5.71} = \frac{10.04}{5.71} = 1.76; \quad K = 57.5 \tag{vii}$$

From this result we can estimate the equilibrium partial pressure of NH$_3$(g) over a solution containing 5×10^{-9} M NH$_3$(aq), as has been found for acid rains with a high NH$_4^+$ concentration. Assuming ideal behavior, and assuming molality ≃ molarity for these conditions,

$$\frac{[NH_3(aq)]}{p_{NH_3}} = K = 57.5 \tag{viii}$$

$$p_{NH_3} = \frac{5 \times 10^{-9}}{56} = 8.7 \times 10^{-11} \text{ atm} \tag{ix}$$

2.10 TEMPERATURE AND PRESSURE EFFECTS ON EQUILIBRIUM

Temperature and the Equilibrium Constant

By applying equation 49 to standard-state partial molar enthalpies of species in a reaction, we obtain the basic expressions for describing the effect of change in temperature on the equilibrium constant of the reaction.

For a chemical reaction $\Sigma_i \nu_i M_i = 0$, $\Delta G° = \Sigma_i \nu_i \mu_i°$ and $\Delta H° = \Sigma_i \nu_i \overline{H}_i°$ is the standard enthalpy change of the reaction. The variation of the standard Gibbs free energy of the reaction is

$$\left(\frac{\partial(\Delta G°/T)}{\partial T}\right)_P = -\frac{\Delta H°}{T^2} \tag{70}$$

The equilibrium constant for the reaction K is related to the standard free energy by $\Delta G° = -RT \ln K$. Therefore, at constant pressure,

$$\frac{d \ln K}{dT} = \frac{\Delta H°}{RT^2} \tag{71}$$

which is the van't Hoff equation.

In general, $\Delta H°$ depends on temperature. At temperatures T_1 and T_2,

$$H_2 - H_1 = \int_{T_1}^{T_2} C_p \, dT \tag{72}$$

Table 2.2 summarizes the pertinent relationships for describing the influence of temperature on the equilibrium constant of a chemical reaction or phase equilibrium. The thermodynamic information of interest includes $\Delta H°$ for the reaction, $\Delta C_p°$ for the reaction, and the variation of $\Delta C_p°$ with temperature.

Equations 4 and 7 of Table 2.2 suggest that a plot of the logarithm of the equilibrium constant (or a representative equilibrium activity) of a reaction versus the reciprocal of absolute temperature can yield information concerning $\Delta H°$. For many reactions $\Delta C_p°$ is close to zero and $\Delta H°$ is essentially independent of temperature, and a linear plot of log K versus $1/T$ is obtained over an appreciable temperature range. The equilibrium constant can then be computed readily by the simple relationship of equation 3 in Table 2.2. When $\Delta C_p°$ is constant over a range of temperature, equation 6 of Table 2.2 can be used to compute the equilibrium constant-temperature coefficient.

Pressure and the Equilibrium Constant

By applying equation 49 to standard-state partial molar volumes of species in a reaction, we obtain the basic expressions for describing the effect of change in pressure on the equilibrium constant of the reaction.

Table 2.2. Influence of Temperature on the Equilibrium Constant

The basic relationships are

$$\frac{d \ln K}{dT} = \frac{\Delta H°}{RT^2} \tag{1}$$

$$\ln \frac{K_2}{K_1} = \int_{T_1}^{T_2} \frac{\Delta H°}{RT^2} dT \tag{2}$$

When $\Delta H°$ is independent of temperature,

$$\ln \frac{K_2}{K_1} = \frac{\Delta H°}{R} \left(\frac{1}{T_1} - \frac{1}{T_2} \right) \tag{3}$$

or

$$\ln K = -\frac{\Delta H°}{RT} + \text{constant} \tag{4}$$

When the heat capacity of the reaction, $\Delta C_p°$, is independent of temperature,

$$\Delta H_2° = \Delta H_1° + \Delta C_p° (T_2 - T_1) \tag{5}$$

Integration of equation 1 then yields

$$\ln \frac{K_2}{K_1} = \frac{\Delta H_1°}{R} \left(\frac{1}{T_1} - \frac{1}{T_2} \right) + \frac{\Delta C_p°}{R} \left(\frac{T_1}{T_2} - 1 - \ln \frac{T_1}{T_2} \right) \tag{6}$$

or

$$\ln K = B - \frac{\Delta H_0}{RT} + \frac{\Delta C_p°}{R} \ln T \tag{7}$$

where ΔH_0 and B are constants.
When $\Delta C_p°$ is a function of temperature, if the heat capacity of each reactant and product is given by an expression of the form

$$C_p° = a_i + b_i T + c_i T^2 \tag{8}$$

then the heat capacity of the reaction is given by

$$\frac{d\Delta H°}{dT} = \Delta C_p° = \Delta a + \Delta b \, T + \Delta c \, T^2 \tag{9}$$

Integration of equations 9 and 1 yields

$$\ln K = B - \frac{\Delta H_0}{RT} + \frac{\Delta a}{R} \ln T + \frac{\Delta b}{2R} T + \frac{\Delta c}{6R} T^2 \tag{10}$$

where ΔH_0 and B are constants and $\Delta a = \Sigma_i \nu_i a_i$, and so on.

For a chemical reaction $\Sigma_i \nu_i M_i = 0$, $\Delta G° = \Sigma_i \nu_i \mu_i°$ and $\Delta V° = \Sigma_i \nu_i \overline{V}_i°$. Using $\Delta G° = -RT \ln K$, where K is the conventional thermodynamic equilibrium constant, the variation of K with pressure is obtained:

$$\left(\frac{\partial \ln K}{\partial P}\right)_T = -\frac{\Delta V°}{RT} \tag{73}$$

Table 2.3 summarizes the essential relationships for pressure effects on chemical equilibrium for the variable-pressure standard-state convention. Note, that these relationships can apply to any consistent choice of standard partial molar volumes, for example, one for which an ionic medium such as seawater is adopted as the solute reference state. For detailed discussion of applications to seawater see, for example, Millero (1969) and Whitfield (1975). A compre-

Table 2.3. Influence of Pressure on Equilibrium

Basic relationships

$$\left(\frac{\partial \mu_i}{\partial P}\right)_T = \overline{V}_i \tag{1a}$$

$$\left(\frac{\partial \mu_i°}{\partial P}\right)_T = \overline{V}_i° \tag{1b}$$

where $\overline{V}_i$ and $\overline{V}_i°$ are the partial molar volumes of i under actual conditions and under defined standard-state conditions, respectively,

$$\overline{k}_i° = -\left(\frac{\partial \overline{V}_i°}{\partial P}\right)_T \tag{2}$$

where $\overline{k}_i°$ is the standard partial molar compressibility, the rate of change of molar volume with pressure,

$$\Delta \overline{V} = \sum \nu_i \overline{V}_i \tag{3a}$$

$$\Delta \overline{V}° = \sum \nu_i \overline{V}_i° \tag{3b}$$

where $\Delta \overline{V}$ and $\Delta \overline{V}°$ are the volume changes of reaction under actual and under defined standard-state conditions, respectively,

$$\left(\frac{\partial \ln K}{\partial P}\right)_T = -\frac{\Delta V°}{RT} \tag{4}$$

where K is the equilibrium constant.

2.10 Temperature and Pressure Effects on Equilibrium

Table 2.3. *(Continued)*

When $\Delta V°$ is independent of pressure,

$$\ln \frac{K_P}{K_1} = -\frac{\Delta V°(P-1)}{RT} \qquad (5)$$

When $\Delta k°$ is independent of pressure

$$\ln \frac{K_P}{K_1} = -\frac{1}{RT}[\Delta V°(P-1) - \tfrac{1}{2}\Delta k°(P-1)^2] \qquad (6)$$

where $\Delta k° = \sum_i \nu_i \bar{k}_i°$

For aqueous solutions specifically,

$$\mu_i = \mu_i° + RT \ln \gamma_i m_i \qquad (7)$$

$$\left(\frac{\partial \ln K}{\partial P}\right)_{T,m} = -\frac{\Delta V°}{RT} \qquad (8)$$

$$\left(\frac{\partial \ln \gamma_i}{\partial P}\right)_{T,m} = \frac{\bar{V}_i - \bar{V}_i°}{RT} \qquad (9)$$

$$\left(\frac{\partial \ln K'}{\partial P}\right)_{T,m} = \frac{\Delta V}{RT} \qquad (10)$$

where $K' = \prod_i m_i^{\nu_i} = K/\prod_i \gamma_i^{\nu_i}$, the concentration product at equilibrium.

---

hensive treatment of the physicochemical effects of pressure has been provided by Hamann (1957). Calculations of the effect of pressure on ionic equilibria in solutions have been detailed by Owen and Brinkley (1941). The reader should consult the references mentioned for details of measurement techniques and for compilations of data.

Pressure and Temperature Influences on the Ionization of Water

Figure 2.8 depicts the dependence of $\Delta G° = -RT \ln K$ on both pressure and temperature. The plot is schematic. The equilibrium constants of aqueous reactions are, in general influenced by both pressure and temperature. The range of temperatures experienced at the surface of the earth is appreciable. Temperature over the range from 5°C to 45°C or greater needs to be considered in using equilibrium constant data for many reactions of interest. The influence

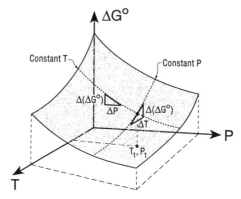

Figure 2.8. The standard free energy change of a reaction depends on the temperature and the pressure. (See Table 2.5 for illustrative data on the standard free energy change for the ionization of water.) $(\partial \Delta G°/\partial T)_P = -\Delta S°$ and $(\partial \Delta G°/\partial P)_T = \Delta V°$ are the thermodynamic relationships governing the influence of temperature and pressure on free energy of a reaction.

of pressure is felt only at great pressures; such pressures are encountered in the deep oceans. Owen and Brinkley (1941) illustrated the effect of 1000-bar (987 atm) pressure on the ionization of water. Their calculations are shown in Table 2.4.

Table 2.5 shows the thermodynamic behavior of the water ionization reaction. The variation of log K_W and $\Delta G_W°$ (molal scale for ions, mole fraction scale for water) with temperature at a fixed pressure of 1 atm and the variation of these quantities with pressure at 25°C are given. These data can be used to obtain the enthalpy change of reaction, $\Delta H_W°$, and the volume change of reaction, $\Delta V_W°$.

Table 2.4. Calculated Ratios of Ion Product of Water at Elevated Pressures

P (bars)[a]	$K_{W,P}/K_{W,1}$		
	5°C	25°C	45°C
1	1	1	1
200	1.24	1.202	1.16
1000	2.8	2.358	2.0

[a]Three temperatures were considered. The increase in ion product is 2.358-fold, at 25°C.
Source: Owen and Brinkley (1941).

Table 2.5. Influence of Pressure and Temperature on the Energetics of Water Ionization[a]

P (bars)	$\log K_W$	ΔG_W° (kJ mol^{-1})
	$T = 298.17\ K$	
1	−14.00	79.94
200	−13.92	79.48
400	−13.84	79.02
600	−13.77	78.63
800	−13.70	78.22
1000	−13.63	77.83
T (°C)	$\log K_W$	ΔG_W° (kJ mol^{-1})
	$P = 1\ bar$	
0	−14.93	85.25
10	−14.53	82.97
20	−14.17	80.91
30	−13.83	78.97
50	−13.26	75.71

[a]Data from Harned and Owen (1958).

2.11 EQUILIBRIUM TOOLS

Some Basic Relationships

Practical calculations pertaining to thermodynamic equilibria draw on a relatively small number of basic relationships. We have considered these throughout this chapter and now collect them together in Table 2.6 for convenience.

Equilibrium Computations for Multicomponent, Multispecies Systems

For finding solutions of a "large," natural water system equilibrium model having many reactions and many species, hand solutions are extremely slow and essentially infeasible. High-speed digital computers have enabled rapid computations of equilibrium composition for mulitcomponent, multiphase, multispecies systems (Ingri et al., 1967; Morel and Morgan, 1972; Schecher and McAvoy, 1994; Westall, 1979, 1980; Westall et al., 1976, Zeleznik and Gordon, 1968).

Two approaches to the treatment of large systems have been developed: the Gibbs energy minimization method and the equilibrium constant approach. Both are based on a knowledge of the chemical potentials of species under standard conditions and under actual conditions of composition of conditions in the chemical system. In the Gibbs energy minimization approach (Clason,

Table 2.6. Basic Relationships for Equilibria: Fixed T and p Systems

1. Chemical potential of a species	$\mu_i = \mu_i^\circ + RT \ln \{i\} = \mu_i$
	$= \mu_i^\circ + RT \ln c_i \gamma_i$
2. Reference states for γ_i	$\gamma_i \to 1$ as $x_i \to 1$ or $x_i \to 0$
	$\gamma_i \to 1$ as $\Sigma m_j \to 0$ or $\gamma_i \to 1$ as $m_i \to 0$
3. Standard states for c_i	$\gamma_i = 1$ and: $x_i = 1$ or $m_i = 1$
4. Reaction $\Sigma \nu_i M_i = 0$	$\Delta H = \sum_i \nu_i \overline{H}_i, \quad \Delta V = \sum_i \nu_i \overline{V}_i,$
	$\Delta S = \sum_i \nu_i \overline{S}_i \quad \Delta G = \sum_i \nu_i \mu_i$
5. State function relationship	$\Delta G = \Delta H - T \Delta S$
6. Equilibrium constant	$K = \prod_i \{i\}_{eq}^{\nu_i}$
7. Reaction quotient	$Q = \prod_i \{i\}^{\nu_i}$
8. Standard free energy and K	ΔG° (kJ mol^{-1}) $= -5.71 \log K$ at 25°C
9. Free energy and Q	$\Delta G = RT \ln \dfrac{Q}{K}$
10. K, m_i, and ν_i	$K = \left(\prod_i \left(\dfrac{m_i}{m^\circ} \right)^{\nu_i} \prod_i \gamma_i^{\nu_i} \right)_{eq}$
11. K and T	$\log \dfrac{K_{T_2}}{K_{T_1}} = \dfrac{\Delta H^\circ}{2.3R} \left(\dfrac{1}{T_1} - \dfrac{1}{T_2} \right)$
12. K and P	$\log \dfrac{K_{P_2}}{K_{P_1}} = -\dfrac{\Delta V^\circ (P_2 - P_1)}{2.3RT}$

1965), one guesses a solution to the equilibrium problem (the free concentrations of species) and proceeds to minimize G^{tot}. In the equilibrium constant approach, one guesses the free component concentrations, solves the mass law nonlinear equations for the species concentrations, and verifies the total component (material expressions). The "QL" computer codes, for example, MINEQL (Schecher and McAvoy, 1994), follow the equilibrium constant method.

In subsequent chapters we will make frequent use of the algebraic language of the equilibrium constant approach: components, species, stoichiometric matrices, and equilibrium constant vectors in describing multispecies systems. This language will be found helpful for system visualization and essential for numerical solution of multispecies problems.

2.12 KINETICS AND THERMODYNAMICS: TIME AND REACTION ADVANCEMENT, ξ

Time

Time has been in the background since Section 2.1, although a sense of time has been implicit in our goal: to apply chemical thermodynamics in describing

2.12 Kinetics and Thermodynamics: Time and Reaction Advancement, ξ 59

the composition of natural water systems. Recall that, in Section 1.1 and again in Section 2.1, the steady-state model (for flow systems) and the equilibrium model (for closed systems) were contrasted, and the applicability of equilibrium models was framed in terms of relative time scales of flows and reactions. Success in applying equilibrium, or thermodynamic, models to aquatic systems depends on the existence of differing time scales between chemical reactions and separations of time scales for reactions versus transport. With respect to appropriate models of natural systems, *slow* reactions and *fast* reactions are treated differently, and the terms *slow* and *fast* imply different *characteristic chemical times* with respect to *physical time scales* (Morgan, 1967; Morgan and Stone, 1985; Stone and Morgan, 1987, 1990). Fast, reversible chemical reactions may appear to be at equilibrium when compared to slower, connected chemical steps within the system. This is the idea of *partial equilibrium* (Pankow and Morgan, 1981). A related notion is local equilibrium: some reactions within a total system may appear to be at equilibrium, while elsewhere in the system, other, connected reactions may be proceeding slowly. This situation (e.g., in multiphase systems) allows some regions to be treated *as if* closed, in an overall open system.

Thermodynamics focuses on the system as a whole. Its power may be said to be in its generality. Chemistry focuses on individual reactions. Its power is in its specificity. Species matter to our understanding of chemistry. As to thermodynamics, the well-known dictum of Lewis and Randall (1923) is worth recalling: "thermodynamics is not compelled to take cognizance of the various molecular species which may exist in a system." To understand time in relation to chemistry in aquatic systems we must take cognizance of species. Indeed, the need to attend to *speciation* (Section 1.3) is, in our present state of understanding, more imperative in relation to rates and mechanisms of reactions than to equilibrium models. As stated by Blandamer (1992), "thermodynamics predicts neither compositions of systems at equilibrium nor rates of change." We must turn elsewhere for the rates of change. [We turned to experiments to obtain chemical potentials (Section 2.5) in order to "predict" the equilibrium compositions.]

Advancement, ξ

For any chemical reaction, written in the general form, $\Sigma \nu_i M_i$, where the ν_i are stochiometric coefficients of the species and the M_i are the molecular symbols of the species (corresponding to the molecular masses), the driving force for advancement of the reaction. (Section 2.3) is $\Delta G = \Sigma \nu_i \mu_i$. Advancement is described by the variable ξ, which is defined by $dn_i = \nu_i \, d\xi$. ξ is positive. A spontaneous reaction, that is, one driving toward equilibrium, is characterized by $\Delta G \, d\xi < 0$. Or, in terms of the entropy creation within the system, $T \, d_i S \, d\xi > 0$. The progress of the chemical reaction is described by the chemical coordinate ξ. The entropy creation, $d_i S$, is positive for the reaction "driving toward equilibrium." If entropy is "time's arrow," how fast does the arrow fly (or drive) toward equilibrium? The answer offers a link between thermodynamics and kinetics of reactions.

The rate of the reaction, R, is

$$R = \frac{1}{\nu_i} \frac{dn_i}{dt} = \frac{d\xi}{dt} \tag{74}$$

If the reaction is homogeneous, the volumetric rate is

$$R^V = \frac{R}{V} = \frac{1}{V} \frac{1}{\nu_i} \frac{dn_i}{dt} = \frac{1}{V} \frac{d\xi}{dt}. \tag{75}$$

Similar expressions can be written for heterogeneous reactions (Morgan and Stone, 1985; Stone and Morgan, 1990). The *rate of entropy production*, θ, is then

$$\theta = \frac{1}{V} \frac{d_i S}{dt} = -\frac{1}{V} \frac{\Delta G}{T} \frac{d\xi}{dt} = -\frac{\Delta G}{T} R^V \tag{76}$$

But, without the reaction rate R^V, we can go no further. We must either find the rate from experiment or, perhaps, predict the rate from a theory of reaction rates. We must turn to the discipline of chemical kinetics in order to find $d\xi/dt$.

Rates and Equilibria

For an elementary, reversible chemical reaction with $\Delta G = \Sigma \nu_i \mu_i$, we know (Section 2.7, equation 64) that

$$\Delta G = RT \ln \frac{Q}{K} \tag{77}$$

For an elementary reaction, we also know the ratio of the forward and backward rates, R_f/R_b, is

$$\frac{R_f}{R_b} = \frac{k_f}{k_b} \frac{1}{Q} = \frac{K}{Q} \tag{78}$$

where k_f and k_b are the elementary reaction rate constants (Morgan and Stone, 1985). Therefore the free energy change of the elementary reaction is related to the rate constants by

$$\Delta G = RT \ln \frac{k_f}{k_b} \frac{1}{Q} \tag{79}$$

But, we obtain a *ratio* of rate coefficients, not either one alone. As a *difference* in rates, the foregoing relationship yields

$$R_f - R_b = R_f \left[1 - \exp\left(\frac{\Delta G}{RT}\right) \right] \tag{80}$$

This appears to be the limit of what can be said from the principle of detailed balancing (microscopic reversibility) about the link between thermodynamics and kinetics (Gardiner, 1969; Hoffmann, 1981; Moore and Pearson, 1981). However, Lasaga (1983) argues that for reactions *close to equilibrium*, the *difference* in the rates,

$$R = R_f - R_b \cong -R_{\text{exchange}} \left(\frac{\Delta G}{RT}\right) \tag{81}$$

so that, the rate is proportional to the driving force of the reaction, $-\Delta G$. We have one more remark to close this limited consideration of thermodynamics and rates. For a series of reversible elementary reactions, the driving force near equilibrium is proportional to $n \, (-\Delta G)$ (Morgan and Stone, 1985).

2.13 RATE AND MECHANISM

There are two central questions in chemical kinetics: (1) How fast can the fastest chemical reactions be? (2) Why are many chemical reactions slow? We will try to provide some elementary insights when answering these questions. Kinetics has several levels. First, there is a level of correct *stoichiometry* for a reaction. Second, there is a level of *energetic characterization* of a reaction, that is, free energy, enthalpy, entropy, and volume changes of reaction (see Section 2.4 and Table 2.6). Third, there is a level of *experimental study of reaction rates* and the formulation of *rate laws* that correctly describe the observed rates. Finally, there is the level of *mechanism*, where *elementary reaction steps* are proposed, verified experimentally, and used to *predict rate expressions*, which are then compared with observation.

The total subject of chemical kinetics is obviously too great to summarize in an introductory overview. There is a remark, perhaps apocryphal, to the effect that one page on kinetics is too much, but a thousand pages are not enough. We will err on the side of brevity. For the reader wishing to delve more deeply into kinetics, we suggest that one of the available treatments of the subject be consulted; for example, Laidler (1987), Moore and Pearson (1981), Gardiner (1969), Brezonik (1994); Lasaga (1983), Burgess (1978), Edwards (1964), and King (1964); review articles by Hoffmann (1981) and Pankow and Morgan (1981) on aquatic kinetic rate laws and data, and chapters by Morgan and Stone (1985), Stone (1987), and Stone and Morgan (1990), on homogeneous and heterogeneous chemical processes in natural waters, provide introductions to kinetic applications in aquatic chemistry. The volume *Aquatic Chemical Kinetics* (Stumm, 1990) deals with kinetics in aquatic systems on the experimental and mechanistic levels. The reader will find that examples of rate laws and mechanisms for homogeneous and heterogeneous chemical reactions are taken up in all subsequent chapters of this book. It is recommended that these examples be consulted in order to see applications of the basic theories outlined here.

Illustrations of Rates and Mechanisms

The *chemical mechanism* of a reaction is a proposed set of *elementary* (molecular) reactions, which provide a sequential path or a number of parallel paths that account for both the stoichiometry and the observed rate law of the overall reaction. If the reaction mechanism is *simple*, it consists of a single elementary step (apart from molecular diffusion of reactants and products, which is always a step in aqueous reactions) capable of accounting for the rate.

Experimental rate laws often point to *complex* mechanisms, that is, a sequence of elementary steps, or two or more such sequences in parallel. Complex mechanisms frequently introduce *intermediate* species, that is, neither reactants nor products. An energetic or equilibrium description of an overall reaction deals only with reactant and product species, whereas a *mechanistic* description of reaction kinetics must recognize, in addition, ground-state catalyst species and intermediate species in the ground state and excited electronic states created by photon absorption.

The reaction of dissolved carbon dioxide, $CO_2(aq)$, with water is complex. It is instructive to consider briefly the presently understood mechanism and the rate expression based on it.

$$CO_2(aq) + 2H_2O = HCO_3^- + H_3O^+ \tag{82}$$

An experimentally derived rate expression for this reaction is

$$R = -\frac{d[CO_2]}{dt} = (k_{f1} + k_{f2}[OH^-])[CO_2]$$

$$- (k_{b1} + k_{b2}[H^+])[HCO_3^-] \tag{83}$$

A mechanism consistent with the rate expression consists of the concurrent reversible reactions

$$CO_2 + H_2O \underset{k_{-1}}{\overset{k_1}{\rightleftharpoons}} H_2CO_3 \tag{84}$$

$$H_2CO_3 + H_2O \underset{k_{-2}}{\overset{k_2}{\rightleftharpoons}} HCO_3^- + H_3O^+ \tag{85}$$

2.13 Rate and Mechanism

$$CO_2 + OH^- \underset{k_{-3}}{\overset{k_3}{\rightleftharpoons}} HCO_3^- \tag{86}$$

The first and third steps are slow at room temperature, while the second is extremely rapid. The relative importance of the third step to the overall hydration rate increases with increasing pH in the range of natural waters.

A second example concerns the oxidation of ferrous iron, Fe(II), by dissolved oxygen. In mildly acidic to near-neutral waters, the reaction is described by the stoichiometry

$$4Fe^{2+} + O_2 + 10H_2O =. 4Fe(OH)_3(s) + 8H^+ \tag{87}$$

The experimental rate law (under low reactant iron concentration conditions when heterogeneous catalysis by product is negligible) has been found to be

$$R = -\frac{1}{4}\frac{d[Fe^{2+}]}{dt} = k[O_2][Fe^{2+}][H^+]^{-2} \tag{88}$$

in which k is a rate constant depending on temperature.

A proposed mechanism, which in part accounts for the rate law for aqueous ferrous iron oxygenation (equation 88), comprises the following elementary steps (Fallab, 1967; Benson, 1968):

(89)

$$Fe^{2+} + H_2O \rightleftharpoons FeOH^+ + H^+ \tag{i}$$

$$FeOH^+ + O_2 \rightleftharpoons Fe(OH)O_2^+ \tag{ii}$$

$$Fe(OH)O_2^+ \rightarrow FeOH^{2+} + O_2^{\pm} \tag{iii}$$

$$O_2^{\pm} + H^+ \rightleftharpoons HO_2^{\cdot} \tag{iv}$$

$$HO_2^{\cdot} + Fe^{2+} + H_2O \rightarrow FeOH^{2+} + H_2O_2 \tag{v}$$

$$H_2O_2 + Fe^{2+} \rightarrow FeOH^{2+} + OH^{\cdot} \tag{vi}$$

$$OH^{\cdot} + Fe^{2+} \rightarrow FeOH^{2+} \tag{vii}$$

Further hydrolysis of $FeOH^{2+}$, followed by nucleation, yields the product $Fe(OH)_3(s)$. Radical species, HO_2 and $OH^{\cdot}$, and hydrogen peroxide, H_2O_2, are reaction intermediates in the proposed mechanism. The importance of speciation in kinetics goes beyond reactants and products in the overall reaction. Some of the features of the experimental rate law are accounted for if steps ii and iii are *slow*. See Wehrli (1990) for a thorough account of the role of reactant speciation in the rates of metal ion oxygenations.

2.14 CONCENTRATION VERSUS TIME

Most elementary reactions involve either one or two reactants. Elementary reactions involving three species are infrequent, because the likelihood of simultaneous three-body encounter is small. In closed, well-mixed chemical systems, the integration of rate equations is straightforward. Results of integration for some important rate laws are listed in Table 2.7, which gives the concentration of reactant A as a function of time. First-order reactions are particularly simple; the rate constant k has units of s^{-1}, and its reciprocal value ($1/k$) provides a measure of a characteristic time for reaction. It is common to speak in terms of the half-life ($t_{1/2}$) for reaction, the time required for 50% of the reactant to be consumed. When

$$[A] = \tfrac{1}{2}[A]_0, \tag{90}$$

$$\tfrac{1}{2} = \exp(-kt_{1/2})$$

$$t_{1/2} = \frac{0.693}{k} \tag{91}$$

Table 2.7. Analytical Solutions to Differential Equations Describing Elementary Reactions

Boundary conditions: at $t = 0$,

$$[A] = [A]_0$$
$$[B] = [B]_0 \tag{1}$$

First-order reaction k (units s^{-1})

$$A \xrightarrow{k} P, \quad \frac{d[A]}{dt} = -k[A], \quad [A] = [A]_0 e^{-kt} \tag{2}$$

Second-order reactions k (units M^{-1} s^{-1})

$$A + A \xrightarrow{k} P, \quad \frac{d[A]}{dt} = -2k[A]^2, \quad [A] = \frac{[A]_0}{1 + 2k[A]_0 t} \tag{3}$$

$$A + B \xrightarrow{k} P, \quad \frac{d[A]}{dt} = -k[A][B], \tag{4}$$

$$[A] = \frac{[A]_0([A]_0 - [B]_0)}{[A]_0 - [B]_0 e^{-k([A]_0 - [B]_0)t}} \tag{5}$$

Third-order reactions k (units M^{-2} s^{-1})

$$A + A + A \xrightarrow{k} P, \quad \frac{d[A]}{dt} = -3k[A]^3, \quad [A] = \frac{[A]_0}{\sqrt{1 + 6kt[A]_0^2 kt}} \tag{6}$$

For first-order reactions in closed vessels, the half-life is independent of the initial reactant concentration. Defining characteristic times for second- and third-order reactions is somewhat complicated in that concentration units appear in the reaction rate constant k. Integrated expressions are available in standard references (e.g., Capellos and Bielski, 1980; Laidler, 1987; Moore and Pearson, 1981).

If the chemical kinetics of interest can be defined by a single elementary reaction, but the reaction occurs in an open system, the differential equation for reaction must be elaborated accordingly. In a later section, we will discuss how both chemical reaction and mass transport can be accounted for in calculating changes in species concentrations as a function of time.

Reaction Order versus Reaction Molecularity

Molecularity is a characteristic of elementary reactions. For example, the elementary reaction $A + B \rightleftharpoons C$ has a molecularity of two in the forward direction and one in the reverse direction; the reaction $2A \rightleftharpoons B + C$ has a molecularity of two in each direction. The rate expression for the latter is $R = k_f[A]^2 - k_b[B][C]$. For elementary reactions, the *order* of the rate expression is the same as the molecularity. For complex reactions, the order of the rate expression reflects the joining of the elementary steps. For a general stoichiometric reaction

$$\nu_A A + \nu_B B + \cdots = \nu_P P + \cdots \tag{92}$$

or

$$\sum_i \nu_i M_i = 0 \tag{93}$$

the rate is

$$R = \frac{1}{\nu_i} \frac{d[M_i]}{dt} \tag{94}$$

If the observed rate law takes the simple form

$$R = k[A]^{\gamma_A}[B]^{\gamma_B} \cdots [P]^{\gamma_P} \cdots [X]^{\gamma_X} \cdots \tag{95}$$

where the exponents are determined by experiment and may be positive or negative integers or fractions, the species X whose concentration appears in the rate law need not be a reactant or product in the overall stoichiometric reaction. The reaction is γ_A-order in [A], γ_B-order in [B], and γ_X-order in [X], and so on. The total order of the rate law is said to be $\gamma = \Sigma_i \gamma_i$, where γ_i is the exponent of species i.

For example, the rapid oxidation of iodide by hypochlorite in alkaline solution,

$$I^- + OCl^- = OI^- + Cl^- \qquad (96)$$

has the rate law

$$R = \frac{d[OI^-]}{dt} = k\frac{[I^-][OCl^-]}{[OH^-]} \qquad (97)$$

The rate is first-order in $[I^-]$, first-order in $[OCl^-]$, and inverse first-order in $[OH^-]$. The inverse dependence on hydroxide suggests a rapid equilibrium preceding a slow step, namely,

$$OCl^- + H_2O \xrightleftharpoons{K} HOCl + OH^- \qquad (98)$$

Complex Reactions

Mechanisms for most chemical processes involve two or more elementary reactions. Our goal is to determine concentrations of reactants, intermediates, and products as a function of time. In order to do this, we must know the rate constants for all pertinent elementary reactions. The principle of mass action is used to write differential equations expressing rates of change for each chemical involved in the process. These differential equations are then integrated with the help of stoichiometric relationships and an appropriate set of boundary conditions (e.g., initial concentrations). For simple cases, analytical solutions are readily obtained. Complex sets of elementary reactions may require numerical solutions.

Reactions in Series Two first-order elementary reactions in series are

$$A \xrightarrow{k_1} B \xrightarrow{k_2} C \qquad (99)$$

From the principle of mass action, rates of the first and second steps are given by

$$r_1 = k_1[A] = -d[A]/dt \qquad (100)$$

$$r_2 = k_2[B] = d[C]/dt \qquad (101)$$

A is consumed by the first elementary reaction. Equation 100 can be integrated directly, giving typical first-order decay in A:

$$[A] = [A]_0 e^{-k_1 t} \qquad (102)$$

2.14 Concentration versus Time

Two processes act on B; it is produced by the first elementary reaction but consumed by the second:

$$d[B]/dt = k_1[A] - k_2[B] \tag{103}$$

Combining equations 102 and 103 yields a differential equation that can readily be integrated:

$$d[B]/dt = k_1[A]_0 e^{-k_1 t} - k_2[B] \tag{104}$$

$$[B] = \frac{k_1[A]_0}{k_2 - k_1}(e^{-k_1 t} - e^{-k_2 t}) \tag{105}$$

As expected, the dynamic behavior of [B] depends on the relative magnitudes of k_1 and k_2. When $k_1 \gg k_2$, the maximum value of [B] will be high; when $k_1 \ll k_2$, the maximum value of [B] will be low.

Only one process acts on C; it is produced by the second elementary reaction. The concentration of C as a function of time is found by inserting equation 105 into equation 101, or by taking advantage of the mass-balance equation:

$$[A]_0 + [B]_0 + [C]_0 = [A] + [B] + [C] \tag{106}$$

$$[C] = [C]_0 + ([A]_0 - [A]) + ([B]_0 - [B]) \tag{107}$$

Thus the concentrations of all reactants, intermediates, and products have been determined as a function of time.

Consider three reactions in series:

$$A \xrightarrow{k_1} B \xrightarrow{k_2} C \xrightarrow{k_3} D \tag{108}$$

$$r_1 = k_1[A] = -d[A]/dt \tag{109}$$

$$r_2 = k_2[B] \tag{110}$$

$$r_3 = k_3[C] = d[D]/dt \tag{111}$$

$$d[B]/dt = r_1 - r_2 = k_1[A] - k_2[B] \tag{112}$$

$$d[C]/dt = r_2 - r_3 = k_2[B] - k_3[C] \tag{113}$$

The equations are considerably more complex than in the preceding case, but an analytical solution can still be found [Capellos and Bielski (1980) provide useful compilations of analytical solutions]. The mass-balance equation and its derivative with respect to time are useful in solving these equations.[†]

[†]Computer codes (see e.g., Braun, et al., 1988) facilitate solution of complex reaction system kinetics.

$$[A]_0 + [B]_0 + [C]_0 + [D]_0 = [A] + [B] + [C] + [D] \tag{114}$$

$$0 = d[A]/dt + d[B]/dt + d[C]/dt + d[D]/dt \tag{115}$$

Species constants and rates of the three contributing elementary reactions are shown in Figure 2.9 for the case when $k_1 = k_2 = k_3 = 0.1$ day^{-1} and $[B]_0 = [C]_0 = [D]_0 = 0$. As the reaction progresses, the predominant species shifts from A to B to C, eventually forming D.

Reversible Reactions Many chemical reactions important in the water environment are reversible. When the reaction of interest is far from equilibrium, the concentration of the product is small, and the rate of the back reaction is low relative to the rate of the forward reaction. Thus, far from equilibrium the back reaction can be ignored, and the reaction can be modeled as an irreversible process. As equilibrium is approached, however, the rate of the back reaction becomes significant and can no longer be ignored.

Consider the following simple reversible reaction:

$$A \underset{k_2}{\overset{k_1}{\rightleftharpoons}} B \tag{116}$$

for which the rates of change can be described by

$$-d[A]/dt = d[B]/dt = k_1[A] - k_2[B] = r_1 - r_2 \tag{117}$$

If we assume that $[B]_0 = 0$ at the onset of reaction, equation 117 can be integrated to obtain [B] as a function of time:

$$[B] = \frac{k_1[A]_0}{k_1 + k_2} (1 - e^{-(k_1 + k_2)t}) \tag{118}$$

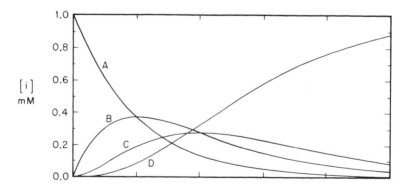

Figure 2.9. Consecutive irreversible reactions. Rate constants for the three elementary reactions are the same ($k_1 = k_2 = k_3 = 0.1$ day^{-1}) and $[B]_0 = [C]_0 = [D]_0 = 0$.

2.15 Theory of Elementary Processes

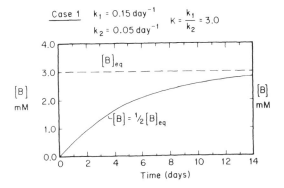

Figure 2.10. Single reversible reaction.

The equilibrium concentration of B can be calculated from equation 118, by letting t approach infinity:

$$[B]_{eq} = \frac{k_1[A]_0}{k_1 + k_2} \tag{119}$$

The characteristic time for approach to equilibrium is given by the reciprocal of the *sum* of the rate constants: $1/(k_1 + k_2)$. This result has important consequences. In Figure 2.10 a case is considered where the equilibrium condition favors product over reactant. Note that the *ratio* of the rate constants gives the position of equilibrium, while their *reciprocal sum* gives the characteristic time for approach to equilibrium. The reader is referred to Stone and Morgan (1990) for further examples of reversible reaction rates.

Table 2.8 (p. 70) shows the influence of K and k_1 on the half-time to equilibrium.

2.15 THEORY OF ELEMENTARY PROCESSES

Fast Reactions are Transport Limited

An elementary bimolecular solution reaction

$$A + B \rightarrow \text{products} \tag{120}$$

in which reactants encounter one another prior to transformation into products,

Table 2.8. Half-Times of Equilibration for the Reaction $A \rightleftharpoons B$ for Different Values of the Forward Rate Constant k_1 and the Equilibrium Constant K

K	k_1 (s^{-1})	$t_{1/2}$
$\frac{1}{10}$	10^{-3}	~1 min
	10^{-5}	~100 min
	10^{-7}	~7 days
1	10^{-3}	~6 min
	10^{-5}	~10 h
	10^{-7}	~40 days
10	10^{-3}	~10 min
	10^{-5}	~17 h
	10^{-7}	~75 days

may be viewed in simplified fashion as a two-step sequence:

$$A + B \xrightarrow{k_E} [A, B] \xrightarrow{k_R} \text{products} \tag{121}$$

where k_E is a rate constant for the *encounter* (collision) step which brings the solute reactants into first contact, say in a "cage" of solvent molecules or as a weak "pair," denoted by [A,B], and k_R is a rate constant for the *chemical* step as such. The rate of encounter is then $k_E[A][B]$ and the rate of the chemical reaction is $k_R[A,B]$. If the intrinsic rate of the chemical step is very *fast*, then the encounter step is rate determining. An upper limit to the "fast" elementary reaction rate is set by molecular diffusion and can be described by the Smoluchowski–Debye theories of solute molecule or ion collision frequency in water (Weston and Schwarz, 1972). The equation for the diffusion-controlled rate constant (M^{-1}s^{-1}) is

$$k_E = \frac{4\pi N}{1000}(D_A + D_B)(r_A + r_B)f \tag{122}$$

where N is Avogadro's constant, D is the diffusion coefficient (cm^{2}s^{-1}), r the solute species radius, and f a factor that accounts for long-range forces, for example, electrostatic, between the approaching reactants. At 25°C, for aqueous neutral species with radii of approximately 0.5 nm, k_E is about 10^{10} M^{-1}s^{-1}; for singly charged ions of opposite charge, k_E is about 10^{11} M^{-1}s^{-1}. Thus the characteristic time scale for "fast" biomolecular solution reactions, at 10^{-5} M concentrations, is measured in μs.

A large number of proton-transfer reactions in dilute aqueous solutions are found to have rate constants close to those predicted by the diffusion model. For example, consider the following proton transfers and the observed rate constants (Laidler, 1965):

Reaction	$k(M^{-1} s^{-1})$
$H_3O^+ + OH^- = 2H_2O$	1.4×10^{11}
$H_3O^+ + SO_4^{2-} = H_2O + HSO_4^-$	1×10^{11}
$OH^- + NH_4^+ = H_2O + NH_3$	3.3×10^{10}
$HCO_3^- + H_3O^+ = H_2CO_3 + H_2O$	4.7×10^{10}

Slow Reactions

If the specific rate of the chemical step is *slow* compared to the encounter rate, that is, $k_R \ll k_E \sim 10^{10}$ $M^{-1}s^{-1}$, then the rate of the elementary reaction, $A + B \rightarrow$ products, is governed by chemical reaction. [Similar considerations enter in the case of heterogeneous reactions, for example, dissolution or adsorption, where, for spherical particles, the quantities of interest are k_s (cm s^{-1}) and D/a (cm s^{-1}), where k_s is a rate constant and a the radius].

Theoretical understanding of slow elementary chemical reactions is a central problem in chemical dynamics. A fundamental quantitative molecular picture of elementary reaction rates in aqueous solutions and at interfaces has proved extremely difficult to construct because of the role of the solvent, that is, the variety of physical and chemical effects resulting from interactions between reactants and solvent water molecules. The presence of ions at significant concentrations, extensive hydration of most species, and participation of water molecules in reactions as catalyst, intermediate, reactant, or product all serve to complicate mechanistic interpretation of slow aqueous reactions. Nonetheless, useful qualitative and quantitative interpretation is possible through considering known effects of hydration on thermodynamic properties.

Activated Complex Theory

The most widely used theoretical framework for elementary solution and interfacial reactions is *activated complex theory* (ACT), also referred to as absolute reaction rate theory, or transition-state theory (TST). The term *activated complex* refers to a high-energy ground-state species formed from reactants, for example,

$$A + B \xrightleftharpoons{K^{\ddagger}} AB^{\ddagger} \qquad (123)$$

and at local equilibrium ($K^{\ddagger}$ is a kind of equilibrium constant) with them. Figure 2.11 illustrates the essential features of a potential energy profile for the elementary reaction. The profile is meant to represent a lowest-energy path on a three-dimensional surface. The potential energy state of the activated complex is the *transition state* (in transition between reactants and products). The ac-

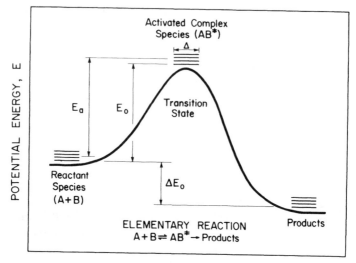

Figure 2.11. Potential energy along a one-dimensional reaction coordinate for an elementary reaction. E_0 is the difference in zero-point energies between reactants and activated complex, $AB^\ddagger$; ΔE_0 is the energy difference between reactants and products. $E_a \simeq E_0$, the activation energy of the elementary process.

tivated complex then falls apart to yield products:

$$AB^\ddagger \rightarrow \text{products} \qquad (124)$$

The TST was developed originally by Eyring and others on the basis of statistical mechanics [see, e.g., Lasaga (1983) or Moore and Pearson (1981)]. The fundamental result is a bimolecular rate constant for an elementary process expressed in terms of (1) the total molecular partition functions per unit volume (q_i) for reactant species and for the activated complex species ($q^\ddagger$), and (2) the difference in zero-point potential energies between the activated complex and reactants (E_0):

$$k = \frac{k_B T}{h} \frac{q^\ddagger}{q_A q_B} \exp\left(-\frac{E_0}{kT}\right) \qquad (125)$$

where k_B is Boltzmann's constant, T is absolute temperature, and h is Planck's constant. If $q^\ddagger$ could be evaluated from molecular properties and E_0 calculated from potential energy surfaces, an elementary rate constant could then be obtained for any bimolecular reaction. Evaluation of total partition functions by calculation on the basis of fundamental translational, vibrational, and rotational properties proves to be too difficult for the complicated solvated species in solution. Activated complex theory (ACT) must then be applied in a different way, using a *thermodynamic* formulation instead of that in equation 125. Briefly, the term $k_B T/h$ is regarded as the frequency of decomposition of the activated complex species, so that the rate of reaction becomes

$$R = \frac{k_B T}{h} [AB^{\ddagger}] = \frac{k_B T}{h} K^{\ddagger}[A][B] \frac{\gamma_A \gamma_B}{\gamma_{\ddagger}} \qquad (126)$$

with $K^{\ddagger}$ viewed as the *thermodynamic* formation constant for $AB^{\ddagger}$ and the γ's are activity coefficients relating activities and concentrations (functions of ionic strength). The elementary rate constant k is then

$$k = \frac{k_B T}{h} K^{\ddagger} \frac{\gamma_A \gamma_B}{\gamma^{\ddagger}} = \frac{k_B T}{h} \frac{\gamma_A \gamma_B}{\gamma_{\ddagger}} \exp\left(-\frac{\Delta G^{\ddagger\circ}}{RT}\right) \qquad (127)$$

where $\Delta G^{\ddagger\circ}$ is a standard free energy of activation.

Because $\Delta G^{\ddagger\circ} = \Delta H^{\ddagger\circ} - T \Delta S^{\ddagger\circ}$,

$$k = \frac{k_B T}{h} \frac{\gamma_A \gamma_B}{\gamma_{\ddagger}} \exp\left(\frac{\Delta S^{\ddagger\circ}}{R}\right) \exp\left(-\frac{\Delta H^{\ddagger\circ}}{RT}\right) \qquad (128)$$

Figure 2.12 illustrates schematically the essential features of the thermodynamic formulation of ACT. If it were possible to evaluate $\Delta S^{\ddagger\circ}$ and $\Delta H^{\ddagger\circ}$ from a knowledge of the properties of aqueous and surface species, the elementary bimolecular rate constant could be calculated. At present, this possibility has been realized for only a limited group of reactions, for example, certain (outer-sphere) electron transfers between ions in solution. The ACT framework finds wide use in interpreting experimental bimolecular rate constants for elementary solution reactions and for *correlating*, and sometimes interpolating, rate constants within families of related reactions. It is noted that a parallel development for *unimolecular* elementary reactions yields an expression for k analogous to equation 128, with appropriate $\Delta S^{\ddagger\circ}$.

It is essential to limit quantitative interpretation of the ACT relationship, for example, equations 127 and 128, to slow *elementary reactions*. Complex reactions, for example, a sequence of elementary steps, need to be first resolved through examination of the rate law and consideration of the mechanism in order to identify possible rate-determining steps and obtain entropies and enthalpies for activated complex formation.

Temperature The effect of temperature on rate constants for elementary reactions will now be examined. To assist in the interpretation of experimental information, Arrhenius (in 1889) postulated the following relationship:

$$k = A e^{-E_a/RT} \qquad (129)$$

74   Chemical Thermodynamics and Kinetics

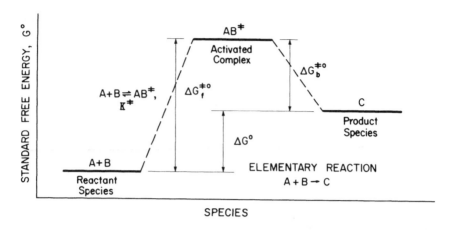

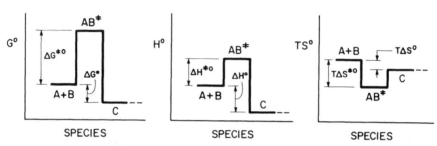

Figure 2.12. Schematic illustration of the essential features of thermodynamics of activated complex theory for an elementary reaction. $\Delta G^{\ddagger\circ} = \Delta H^{\ddagger\circ} - T\Delta S^{\ddagger\circ}$, and $\Delta G^\circ = \Delta G_f^{\ddagger\circ} - \Delta G_b^{\ddagger\circ}$, where f and b denote the forward and back reactions. $K^{\ddagger}$ is the equilibrium constant for activation.

$A = A(T)$ and E_a are referred to as the *Arrhenius parameters*. The logarithmic form of equation 129,

$$\ln k = \ln A - \frac{E_a}{RT} \tag{130}$$

suggests plotting logarithms of experimental rate constants versus reciprocal absolute temperatures ($1/T$) to estimate the preexponential factors A and activation energies E_a. We can relate the Arrhenius parameters to ACT by postulating a Gibbs free energy of activation, $\Delta G^{\circ\ddagger}$, related to $K^{\ddagger}$ in the following manner:

$$\Delta G^{\ddagger\circ} = \Delta H^{\ddagger\circ} - T\Delta S^{\ddagger\circ} = -RT \ln K^{\ddagger} \tag{131}$$

Equation 127 can now be rewritten in terms of $\Delta G^{\ddagger\circ}$, $\Delta H^{\ddagger\circ}$, and $\Delta S^{\ddagger\circ}$:

2.15 Theory of Elementary Processes

$$k = \frac{k_B T}{h} \frac{\gamma_A \gamma_B}{\gamma_\ddagger} e^{-\Delta G^{\ddagger \circ}/RT} = \frac{k_B T}{h} \frac{\gamma_A \gamma_B}{\gamma_\ddagger} e^{\Delta S^{\ddagger \circ}/R} e^{-\Delta H^{\ddagger \circ}/RT} \quad (132)$$

For an elementary reaction, comparison of the Arrhenius equation (129) with the corresponding ACT equation (132) (and with $\gamma_A = \gamma_B = \gamma_\ddagger = 1.0$) yields the following values for the Arrhenius parameters:

$$A = \frac{k_B T}{h} e^{\Delta S^{\ddagger \circ}/R}, \quad E_a = \Delta H^{\ddagger \circ} + RT \quad (133a)$$

The relationship between E_d and $\Delta H^{\ddagger \circ}$ is obtained by differentiating the Arrhenius relationship and substituting (1) of Table 2.2. Thus the Arrhenius equation, predicting a linear relationship between $\ln k$ and $1/T$, is confirmed by the ACT treatment. The expression for k in terms of $\Delta S^{\ddagger \circ}$ and E_a is:

$$k = e \frac{k_B T}{h} \frac{\gamma_A \gamma_B}{\gamma_\ddagger} e^{\Delta S^{\ddagger \circ}/R} e^{-E_a/RT} \quad (133b)$$

Temperature Dependence of Fast Reactions It is to be noted that rate constants for fast (diffusion-controlled) steps are also temperature dependent, since the diffusion coefficient depends on temperature. The usual experimental procedure, suggested by the Arrhenius equation, of plotting $\ln k$ versus $1/T$ will indicate *apparent* activation energies for diffusion control of approximately 12–15 kJ mol^{-1}. For fast heterogeneous chemical reactions in which intrinsic chemical and mass transfer rates are of comparable magnitude, care needs to be taken in interpretation of apparent activation energies for the overall process.

Ionic Strength The effect of ionic strength on rates of elementary reactions readily follows. Using equation 127, we can let k_0 be the value of the second-order rate constant in the reference state, such as an infinitely dilute solution (where all the activity coefficients are unity); k is the rate constant at any specified ionic strength:

$$k = k_0 \frac{\gamma_A \cdot \gamma_B}{\gamma_\ddagger} \quad (134)$$

From the ionic strength, values of γ_A, γ_B, and $\gamma_\ddagger$ can be calculated using the Davies equation (Section 3.4). The charge of the activated complex is known; it is simply the sum of the charges of the two reactants.) Activity coefficients for anions and cations typically decrease as the ionic strength is increased. According to equation 134, increasing the ionic strength: (1) lowers the reaction rate between a cation and anion, (2) raises the reaction rate between like-charged species, and (3) has little or no effect on reaction rate when one or both of the reactants is uncharged.

Pressure Transition-state theory accounts for pressure effects on solution reaction rates through the effect of pressure on the equilibrium constant $K^\ddagger$. At constant temperature and electrolyte concentration (constant ionic strength),

$$\left(\frac{d \ln K^\ddagger}{dP}\right) = -\frac{\Delta V^\ddagger}{RT} \tag{135}$$

where $\Delta V^\ddagger$ is the volume of activation, $\overline{V}^\ddagger - \overline{V}_A - \overline{V}_B$, for $A + B \rightarrow X^\ddagger$. Therefore the rate constant variation with pressure is given by

$$\left(\frac{d \ln k}{dP}\right)_{T,I} = -\frac{\Delta V^\ddagger}{RT} \tag{136}$$

and the integration (for $\Delta V^\ddagger$ assumed independent of pressure) is given by

$$\frac{k_P}{k_1} = \exp\left(-\frac{\Delta V^\ddagger (P - 1)}{RT}\right) \tag{137}$$

The volume of activation can be obtained from a plot of $\log k_P$ versus P. If $\Delta V^\ddagger < 0$, the rate constant increases with pressure, and vice versa. For a $\Delta V^\ddagger$ of, say, -10 cm^3 mol^{-1}, $P = 500$ atm or 507 bars, and $T = 298.15$ K, $k_P/k_1 \simeq 1.25$; a $\Delta V^\ddagger$ of -50 cm^3 mol^{-1} for the same conditions gives $k_P/k_1 \simeq 2.25$. Negative activation volumes, like negative activation entropies, are indicative of "slow" reactions (compared to diffusion-limited, or "fast" reactions).

2.16 ELEMENTARY REACTIONS AND ACT

Why Are Some Reactions Slow?

In the framework of activated complex theory, aqueous solution reactions and heterogeneous reactions are "slow" because they have large $\Delta H^{\ddagger\circ}$ or small $\Delta S^{\ddagger\circ}$, or both. At 25°C, the quantity ek_BT/h is 1.7×10^{13} s^{-1}. An *activation energy* of 30 kJ mol^{-1} reduces the rate constant by a factor of 6.1×10^{-6}; an *activation entropy* of -80 J K^{-1} mol^{-1} reduces the rate constant by a factor of 6.5×10^{-5}. A negative entropy of activation is associated with an activated complex that is more solvated and highly ordered than the reactants. For aqueous reactions, experimentally determined activation entropies cover a range from about -160 to $+160$ J K^{-1} mol^{-1}. (The numerical values of $\Delta S^{\ddagger\circ}$ reflect *units* of the bimolecular or unimolecular rate constant and depend on the concentration scale.) Aqueous reaction activation energies for low-temperature reactions ($E_a = \Delta H^{\ddagger\circ} + RT \simeq \Delta H^{\ddagger\circ} + 2.5$ kJ mol^{-1}) range from about 10–15 kJ mol^{-1} at the lower end up to about 150 kJ mol^{-1} for very slow elementary reactions. For an uncatalyzed unimolecular reaction at 25°C with $\Delta S^{\ddagger\circ} \simeq -50$ J K^{-1} mol^{-1} and $\Delta H^{\ddagger\circ} \sim 100$ kJ mol^{-1}, the time constant is ~ 900 years. Biological catalysis can speed reactions by factors of from 10^7 to 10^{14} (e.g., the urease enzyme, $\sim 10^{13}$), with an associated lowering of $\Delta H^{\ddagger\circ}$ and/or increasing of $\Delta S^{\ddagger\circ}$. Less specific homogeneous or heterogeneous catalysts (e.g., transition metals in solution, acids or bases in solution, metal or ligand sites on solid surfaces) can accelerate slow aqueous reactions significantly (e.g, by 10^3–10^7) by providing *alternate* lowered energy or increased entropy reaction pathways. [In addition to *thermal* reactions, with species in their ground electronic states, *photochemical* reactions may provide the needed activation energy for bond breaking in a reaction by absorption of a photon, either by a reactant itself, for example, an organic molecule or a metal–ligand complex, thereby generating high-energy electronic states, or by other species capable of transferring energy to reactant species (indirect photolysis). The energy per mole of photons ($h\nu$) *absorbed* in the visible range is from 170 to 300 kJ, corresponding to wavelengths from 700 nm to 400 nm.]

Examples

To illustrate some typical values of rate constants for elementary processes in water, we have collected pertinent information on kinetic parameters for a few reactions investigated experimentally. Table 2.9 presents information on rate constants, activation energies, and activation entropies for some aqueous solution reactions. While most of these are thought elementary, some are complex, for example, reactions 5 and 7. The reader should consult original sources for detailed interpretations and for measures of experimental uncertainty. By inspecting Table 2.9 it is seen that both entropic and enthalpic factors strongly influence the rate constants of elementary reactions. The examples of solvent exchange rates for three cations, Cu^{2+}, Al^{3+}, and Cr^{3+}, indicate a small energy barrier for solvent exchange with Cu^{2+}(aq), but large and comparable barriers for both Al^{3+}(aq) and Cr^{3+}(aq). The Al^{3+} water exchange is almost a million times faster than for Cr^{3+}.

Reduction of $FeOH^{2+}$ by Cr^{2+} is considerably faster than for Fe^{3+}; the hydrolyzed oxidant species is associated with a smaller negative activation entropy. Bridging by OH^- groups in the activated complex is probably an important factor, as is also the case in $FeOH^{2+}$ and Fe^{2+} isotopic exchange. Reactions between ions of like charge are generally accompanied by large negative entropies, for example, reactions 6 and 10, which can be interpreted qualitatively as loss of freedom of motion of water molecules in the neighborhood of the activated complex. Entropies of activation are often correlated with volumes of activation, positive values of each viewed as diagnostic of the *dissociative* mechanism.

For reactions of specified activation enthalpies, for example, 40 kJ mol^{-1}, the bimolecular reaction rate depends on the activation entropy. Thus, a *negative* entropy of 60 J K^{-1} mol^{-1} *slows* the reaction 1000-fold, that is, lowers the preexponential Arrhenius factor A from $\sim 10^{13}$ to 10^{10} M^{-1} s^{-1}.

Table 2.9. Kinetic Parameters for Elementary Aqueous Reactions[a]

Reactants	$\log k$[b] $(M^{-1} s^{-1})$	E_a $(kJ\,mol^{-1})$	$\Delta S^{\ddagger\circ}$ $(J\,K^{-1}\,mol^{-1})$	References
Solvent exchange[c]:				
1 $Cu^{2+}(aq) + H_2O \rightarrow$	8.9	26	7	Burgess (1978)
2 $Cr^{2+}(aq) + H_2O \rightarrow$	−7.3	112	−18	Burgess (1978)
3 $Al^{3+}(aq)$	−1.8	116	99	Burgess (1978)
Neutralization:				
4 $CO_2(aq) + OH^- \rightarrow$	4.0	56	12	Moore and Pearson (1980)
Hydration:				
5 $CO_2(aq) + H_2O \rightarrow$	−3.3	63	−107	Edsall (1969)
Substitution:				
6 $Co(NH_3)_5Cl^{2+} + OH^- \rightarrow$	0.2	112	132	Moore and Pearson (1980)
Redox:				
7 $S_2O_3^{2-} + SO_3^{2-} \rightarrow$	−4.3	61	−115	Moore and Pearson (1980)
Substitution:				
8 $CH_3I(aq) + OH^- \rightarrow$	−4.1	93	−22	Moore and Pearson (1980)
Redox:				
9 $Fe^{2+}(aq) + H_2O_2 \rightarrow$	1.8	31	−116	Wells and Salam (1960)
10 $Fe^{2+} + FeOH^{2+} \rightarrow$	3.4 (21.6°C)	31	−75	Burgess (1978)
11 $Fe^{3+} + Cr^{2+} \rightarrow$	3.4	24	−117	Burgess (1978)
12 $FeOH^{2+} + Cr^{2+} \rightarrow$	6.5	22	−54	Burgess (1978)
13 $Co(NH_3)_5Cl^{2+} + Fe^{2+} \rightarrow$	−2.9	55	−126	Burgess (1978)
Dissociation:				
14 $H_2CO_3 \rightarrow CO_2 + H_2O$	1.2 s^{-1}	64	−15	Edsall (1969)

[a] 25°C, except as noted.
[b] $M^{-1}\,s^{-1}$, except for reaction 14, unimolecular, s^{-1}.
[c] Solvent exchange formulated as second order; for first order, $k = k_2(55/6)$, that is, $[H_2O] \simeq 55$ M.

Reactions with larger $\Delta H^{\ddagger\circ}$ values are affected more strongly by temperature variation. For example, for $\Delta H^{\ddagger\circ} = 120$ kJ mol^{-1}, lowering temperature from 30°C to 5°C slows the rate by a factor of 70, whereas for $\Delta H^{\ddagger\circ} = 60$ kJ mol^{-1}, the same temperature change slows the rate by a factor of about 9.

Example 2.6. Hydration of CO_2 in Terms of Transition-State Parameters A reaction of importance in natural waters and biochemistry, the hydration of CO_2,

$$CO_2 + H_2O \rightarrow H_2CO_3$$

is described kinetically by the parameters $E_a = 63.0$, $\Delta H^{\ddagger\circ} = 60.5$ (kJ mol^{-1}), $\Delta S^{\ddagger\circ} = -107$ J K^{-1} mol^{-1}, and log $k_2 = -3.3$ (M^{-1} s^{-1}) at 25°C. The slowness of this reaction ($t_{1/2} \simeq 3$ min at 10°C) is associated with a large energy barrier as well as a highly negative activation entropy. The free energy of activation, $\Delta G^{\ddagger 0}$, is 92 kJ mol^{-1}, giving $K^{\ddagger} \simeq 10^{-16}$ M^{-1}, and indicating that the quasi-equilibrium concentration of the activated complex, $[H_2O, CO_2]^{\ddagger}$, for a 10^{-5} M CO_2 solution would be $\sim 10^{-19}$ M! For the corresponding *dehydration* reaction, $H_2CO_3 \rightarrow CO_2 + H_2O$, experimental values of the *unimolecular* rate constant vary from about 3 to 30 s^{-1} over the temperature range 5–30°C. Interpretation of these data in terms of ACT yields $\Delta H^{\ddagger\circ} = 61.5$ kJ mol^{-1} and $\Delta S^{\ddagger\circ} = -15$ J K^{-1} mol^{-1} for dehydration of H_2CO_3, consistent with $\Delta H^{\circ} = -1$ and $\Delta S^{\circ} = -92$ for the reversible *overall* reaction, and with an equilibrium constant, $K = k_f/k_b$, equal to the ratio of the hydration and dehydration rate constants, a consequence of the principle of detailed balancing (microscopic reversibility) (Moore and Pearson, 1981).

2.17. EQUILIBRIUM VERSUS STEADY STATE IN FLOW SYSTEMS

Open Flow Systems Versus Closed Systems

Most natural water systems are continuous, open systems. Flows of matter and energy occur in the real system. The time-invariant state of a continuous system with flows at the boundaries is the *steady state*.[†] This state may be poorly approximated by the *equilibrium* state of a closed system.[‡] In Figure 2.2 we indicated the important features of an open-system model with material fluxes and chemical reactions. The simple reversible reaction (a "model" reaction),

$$A \underset{k_b}{\overset{k_f}{\rightleftharpoons}} B$$

[†] The term *steady state* as used in this context means the time-invariant state of a flow system with chemical reactions. *Steady state*, with respect to chemical *mechanisms*, means that certain intermediates in a complex reaction are of low concentration, so that $dC/dt \simeq 0$. It is important to keep these usages of "steady state" distinct.
[‡] For *irreversible* reactions in flow systems, see Stone and Morgan (1990).

was introduced to illustrate elementary differences between closed- and open-system models. Solution for the steady-state concentration values of A and B gives the results

$$C_A = \frac{r\overline{C}_A + k_b(\overline{C}_A + \overline{C}_B)}{k_f + k_b + r} \qquad (10)$$

$$C_B = \frac{r\overline{C}_B + k_f(\overline{C}_A + \overline{C}_B)}{k_f + k_b + r} \qquad (138)$$

in which $r = Q/V$, the fluid flow rate constant (time^{-1}), and overbars denote inflowing concentrations.

We examine the *ratio* of C_B and C_A and compare it with the ratio expected for chemical equilibrium. The ratio of C_B/C_A for the steady state is, dividing equation 138 by equation 10,

$$\frac{C_B}{C_A} = \frac{r\overline{C}_B + k_f(\overline{C}_A + \overline{C}_B)}{r\overline{C}_A + k_b(\overline{C}_A + \overline{C}_B)} \qquad (139)$$

Equation (139) shows that C_B/C_A will tend toward k_f/k_b as the material flow to the system becomes small, that is, as $r\overline{C}_B$ and $r\overline{C}_A$ vanish. For $r = 0$, the system is a closed system, and $C_B/C_A = k_f/k_b = K$, the equilibrium constant. The quantity $r = Q/V$ is the reciprocal of the fluid residence time τ_R of the well-mixed system: $r = \tau_R^{-1}$. As τ_R tends to very large values, the steady-state concentration ratio of the system approaches the equilibrium constant.

A simple result is obtained when only A enters the system. Then, $\overline{C}_B = 0$ and equation 139 reduces to

$$\frac{C_B}{C_A} = \frac{k_f\overline{C}_A}{r\overline{C}_A + k_b\overline{C}_A} = \frac{K}{(r/k_b) + 1} \text{ to r.h.s.} \qquad (140)$$

The steady-state concentration ratio thus depends on the chemical rate constants and the flux rate constant. For $r \ll k_b$, $C_B/C_A \simeq k_f/k_b = K$; if $r \gg k_b$, then $C_B/C_A \simeq k_f/r$. In terms of the residence time, τ_R, and the half-time of the backward reaction, τ_b, the steady-state ratio approximates the equilibrium ratio if $\tau_R \gg \tau_b$. (If only B flows in, then $C_B/C_A = K + (r/k_b)$.)

Example 2.7. Steady-State Composition of an Open Completely Mixed System with the Reaction $A \rightleftharpoons B$ as a Function of Residence Time Assume that $\overline{C}_B = 0$, $k_f = 10^{-5}$ s^{-1}, and $k_b = 10^{-6}$ s^{-1}, so that $K = 10$. The residence time, τ_R, is varied over the range from 10^5 to 10^8 s. Table 2.10 shows the relationship of steady-state C_B/C_A to K for a 1000-fold variation in the residence time relative to the characteristic reaction time ($\sim k_b^{-1}$).

The equilibrium assumption is justified for many reactions with short τ_{chem} and long τ_R. Rate data for a large number of first- and second-order aqueous

Table 2.10. Composition^a of a Completely Mixed Flow System at Steady State for A $\underset{k_b}{\overset{k_f}{\rightleftharpoons}}$ B: Only A Enters

τ_R (s)	r (s^{-1})	τ_b/τ_R	$\dfrac{C_B/C_A}{K}$
10^5	10^{-5}	10	0.09
2×10^5	5×10^{-6}	5	0.17
5×10^5	2×10^{-6}	2	0.33
10^6	1×10^{-6}	1	0.50
2×10^6	5×10^{-7}	0.5	0.67
5×10^6	2×10^{-7}	0.2	0.83
10^7	1×10^{-7}	0.1	0.91
10^8	1×10^{-8}	0.01	0.99

$^a(C_B/C_A)/K$ versus τ_R; $k_f = 1 \times 10^{-5}$ s^{-1}; $k_b = 1 \times 10^{-6}$ s^{-1}.

reactions indicate τ_{chem} less than seconds to minutes, and many other aqueous reactions have τ_{chem} less than hours to days (Hoffmann, 1981). The residence times of a number of freshwater systems are greater than these ranges (Imboden and Lerman, 1979). Critical attention needs to be directed to slow chemical reactions for which $\tau_{chem} \gtrsim \tau_R$.

For many systems it is known that there exist regions or environments in which the time-invariant condition closely approaches equilibrium. The concept of local equilibrium is important in examining complex systems. Local equilibrium conditions are expected to develop, for example, for kinetically rapid species and phases at sediment–water interfaces in fresh, estuarine, and marine environments. In contrast, other local environments, such as the photosynthetically active surface regions of nearly all lakes and ocean waters and the biologically active regions of soil–water systems, are clearly far removed from total system equilibrium.

SUGGESTED READINGS

Atkins, P. W. (1990) *Physical Chemistry*, 4th ed., Freeman, New York.

Berner, R. A. (1971) *Principles of Chemical Sedimentology*, McGraw-Hill, New York.

Blandamer, M. J. (1992) *Chemical Equilibria in Solution*, Ellis Horward, New York.

Denbigh, K. G. (1971) *The Principles of Chemical Equilibrium: With Applications in Chemistry and Chemical Engineering*, 3rd ed., Cambridge University Press, Cambridge.

Edsall, J. T. (1969) Carbon Dioxide, Carbonic Acid and Bicarbonate Ion: Physical Properties and Kinetics of Interconversion. In *CO$_2$: Chemical, Biochemical and Physiological Aspects*, Forster, R. E., et al., Eds., NASA SP-188, Washington, DC.

82   Chemical Thermodynamics and Kinetics

Eisenberg, D., and Crothers, D. (1979) *Physical Chemistry*, Benjamin, Menlo Park, CA.

Gardiner, W. C. (1969) *Rates and Mechanisms of Chemical Reactions*, Benjamin, Menlo Park, CA.

Garrels, R. M., and Christ, C. L. (1965) *Solutions, Minerals, and Equilibria*, Harper & Row, New York.

Harned, H. S., and Owen, B. B. (1958) *The Physical Chemistry of Electrolytic Solutions*, 3rd ed., Van Nostrand, Reinhold, New York.

King, E. L. (1964) *How Chemical Reactions Occur*, Benjamin, Menlo Park, CA.

Klotz, I. M. (1964) *Chemical Thermodynamics: Basic Theory and Methods*, revised ed., Benjamin, Menlo Park, CA.

Laidler, K. J. (1987) *Chemical Kinetics*, Harper, New York.

Latimer, W. M. (1952) *The Oxidation States of the Elements and Their Potentials in Aqueous Solutions*, 2nd ed., Prentice-Hall, Englewood Cliffs, NJ.

Lee, T. S. (1959) "Chemical Equilibrium and the Thermodynamics of Reactions." In *Treatise on Analytical Chemistry*, Part I, Vol. 1, I. M. Kolthof and P. J. Elving, Eds., Interscience, New York, pp. 185–275.

Lewis, G. N., and Randall, M. (1961) *Thermodynamics*, 2nd ed., revised by K. S. Pitzer and L. Brewer, McGraw-Hill, New York.

Millero, F. J. (1978) "Thermodynamic Models for the State of Metal Ions in Seawater." In *The Sea*, E. D. Goldberg et al., Eds., Vol. 6, Wiley-Interscience, New York, pp. 653–692.

Moore, J. W., and Pearson, R. G. (1981) *Kinetics and Mechanism*, 3rd ed., Wiley-Interscience, New York.

Prigogine, I., and Defay, R. (1954) *Chemical Thermodynamics*, translated by D. H. Everett, McKay, New York.

Robinson, R. A., and Stokes, R. H. (1959) *Electrolyte Solutions: The Measurement and Interpretation of Conductance, Chemical Potential and Diffusion in Solutions of Simple Electrolytes*, 2nd ed., Butterworths, London.

Wells, C. F., and Salam, M. A. (1967) Complex Formation Between Fe(II) and Inorganic Ions, *Trans. Faraday Soc.* **63**, 620.

Whitfield, M. (1975) "Seawater as an Electrolyte Solution." In *Chemical Oceanography*, J. P. Riley and G. Skirrow, Eds., Vol. 1, 2nd ed., Academic, New York, pp. 44–171.

Wilkins, R. G. (1991) *The Study of Kinetics and Mechanisms of Reactions of Transition Metal Complexes*, 2nd ed., VCH, New York.

PROBLEMS

2.1. For the reaction $A \rightleftharpoons B$ in a closed system, TOTA = 1.0×10^{-3} M. The equilibrium constant for the reaction is 10. The rate constant in the forward direction is 1.0 day^{-1}.

(a) What are the concentrations of species A and B at equilibrium.

(b) If the initial concentration of species A is 1.0×10^{-3} M, what is the initial rate of change in concentration of A in mol per liter per day?

(c) What is the exchange rate (velocity), $v_{exch} = v_f = v_b$, at equilibrium?

2.2. The neutralization reaction $H^+(aq) + OH^-(aq) \rightarrow H_2O(l)$ is carried out irreversibly under 1 atm pressure and under standard solution conditions. The measured heat *released* by the reaction at 25°C is 55,830 J per mole of reaction. When the reaction is allowed to take place quasistatically (reversible conditions), at 25°C the heat *taken up* by the solution is 24,050 J per mole.

(a) What is the enthalpy of the neutralization reaction?
(b) What is the entropy change of the solution?
(c) What is the maximum available work that can be obtained from the reaction?
(d) What is the entropy change of the universe under the irreversible conditions of reaction?

2.3. Consider the following systems at 25°C and 1 atm pressure: (a) equal amounts of the minerals SiO_2 (α-quartz) and SiO_2 (α-tridymite); (b) equal amounts of the minerals FeS_2 (marcasite) and FeS_2 (pyrite); (c) equal amounts of the solid phases Fe_2O_3 and $Fe(OH)_3$ in water; and (d) the four solid phases $PbCO_3$, $PbSO_4$, $CaCO_3$, and $CaSO_4$. Write a reaction for the transformation of phases, for example, $PbCO_3 + CaSO_4 = CaCO_3 + PbSO_4$ in (d), and identify the stable phase or group of phases in each system.

2.4. Hydrofluoric acid, HF, is a *weak* acid in water, while HCl is a *strong* acid. The acidity constant of the former is $\sim 10^{-3}$ m, while that of the latter is $\sim 10^8$ m. Assess the "driving force" for these respective acid strengths using the following thermodynamic data for 25°C (Dasent, 1970): for $HF(aq) = H^+(aq) + F^-(aq)$, $\Delta G° = 16$ kJ mol^{-1}, $\Delta H° = -13$ kJ mol^{-1}; for $HCl(aq) = H^+(aq) + Cl^-(aq)$, $\Delta G° = -48$ kJ mol^{-1}, $\Delta H° = -59$ kJ mol^{-1}.

2.5. A geochemically important reaction is $CaCO_3(s) + CO_2(g) + H_2O(l) = Ca^{2+}(aq) + HCO_3^-(aq)$.

(a) Identify the three concentration scales involved in the standard states for the reaction species.
(b) Calculate the standard free energy change and the enthalpy for the reaction at 25°C and 1 atm.
(c) Calculate the equilibrium constant.
(d) Will an increase of temperature favor a shift toward products or reactants?

84   Chemical Thermodynamics and Kinetics

2.6. The vaporization of liquid water, $H_2O(l) = H_2O(g)$, has a reported vapor pressure at 25°C and 1 atm of 23.756 mm mercury (Hg), or 3167.5 pascal (Pa).

(a) Compare this value with that obtainable from the standard partial molar Gibbs energies for water and its vapor.

(b) Estimate an approximate enthalpy of vaporization (kJ mol^{-1}) from the reported vapor pressure at 20°C, 2337.4 Pa.

2.7. HCl (gas) is equilibrated with water in a two-phase system. The effective equilibrium process is $HCl(g) \rightleftharpoons H^+(aq) + Cl^-(aq)$ [because HCl(aq) is "fully dissociated"; see Problem 2.4]. The concentration of dissolved HCl at equilibrium is 1 m. The mean molal activity coefficient of the HCl solution is 0.809, from experiment. The standard-state chemical potentials are: for HCl(g), -95.3 kJ mol^{-1}; for Cl$^-$(aq), -131.2 kJ mol^{-1}; for H$^+$(aq), zero (by convention). Calculate the equilibrium vapor pressure of HCl(g) over this solution.

2.8. Calculate the maximum energy theoretically available (kJ mol^{-1}) from the oxidation of reduced manganese, Mn^{2+}(aq), by dissolved oxygen in water under the following conditions: $\{Mn^{2+}(aq)\} = 1.0 \times 10^{-5}$; water saturated with oxygen of the atmosphere (0.209 atm); and pH 7.0, that is, $\{H^+(aq)\} = 1.0 \times 10^{-7}$. The activity of the water can be assumed to be unity. The product of the oxidation is manganese dioxide, MnO$_2$(s). The reaction is:

$$Mn^{2+}(aq) + \tfrac{1}{2} O_2(aq) + H_2O(l) = MnO_2(s) + 2 H^+(aq)$$

2.9. The partial molar volumes of the species in the reaction

$$CaCO_3(s) = Ca^{2+}(aq) + CO_3^{2-}(aq)$$

are as follows:

Species	$\bar{V}°$ (cm^3 mol^{-1})
CaCO$_3$(s)	36.9
Ca^{2+}	-17.7
CO$_3^{2-}$	-3.7

(a) Calculate the approximate difference in the chemical potential of calcite, CaCO$_3$(s), under a pressure of 1000 atm as compared to 1 atm (neglecting the compressibility.)

(b) Calculate the solubility equilibrium product of calcite, $\{Ca^{2+}(aq)\}$ $\{CO_3^{2-}(aq)\}$, at a pressure of 1000 atm.

2.10. The reaction for precipitation of Ca^{2+} in water $Ca^{2+} + HCO_3^- = CaCO_3(s) + H^+$ is characterized by the following actual concentration activities: $\{Ca^{2+}\} = 1.0 \times 10^{-3}$, $\{H^+\} = 1.0 \times 10^{-8}$, and $\{HCO_3^-\} = 2.0 \times 10^{-3}$. What is the ratio of the forward and reverse rates for the reaction under these solution conditions?

2.11. Figure 2.10 shows the time course of approach to equilibrium for certain values of the rate constants

(a) Compare the characteristic times for approach to equilibrium for the following conditions: (i) For the same ratio of the rate constants, the magnitude of each rate constant increases by a factor of 10; (ii) for the inverse ratio of the rate constants, their sum remains the same; and (iii) for the inverse ratio of the rate constants, their sum increases by a factor of 10.

(b) For (i) what is the relationship between the characteristic time and the "half-time" for approaching equilibrium?

2.12. Calculate the half-life of an irreversible bimolecular reaction in aqueous solution with the following ACT parameters: $\Delta S^{\ddagger\circ} = 12$ J mol^{-1} K^{-1}, $E_a = 56$ kJ mol^{-1}. Assume 25°C.

2.13. For the conditions assumed in Example 2.7, calculate the composition, i.e. $(C_B/C_A)/K$, versus r for the case in which *only* B enters the system.

ANSWERS TO PROBLEMS

2.1. (a) $C_{A,eq} = 9.09 \times 10^{-5}$ M; $C_{B,eq} = 9.09 \times 10^{-4}$ M
(b) $v_{f,initial} = 1.0 \times 10^{-3}$ M d^{-1}
(c) $v_{exchange} = v_{f,eq} = 9.09 \times 10^{-5}$ M d^{-1}

2.2. (a) Under conditions where no external work is involved $\Delta H° = q_P = -55.83$ kJ mol^{-1}.

(b) Entropy change for the system is q/T under reversible conditions. Therefore, $\Delta S_{sys} = q/T = 24{,}050/298.15 = +80.67$ J K^{-1} mol^{-1}.

(c) Maximum work, w', is obtained under reversible conditions for the entire system (e.g., chemical reaction and an external circuit for electrical work), when the internal entropy change is zero. Then, $w' = -\Delta G° = -(\Delta H° - T\Delta S_{sys}) = 79.88$ kJ mol^{-1}.

(d) Under irreversible conditions equation 25 b gives $\Delta S_{universe} = \Delta_i S = -\Delta G°/T = +267.9$ J K^{-1} mol^{-1}.

2.3. Use the partial molar Gibbs free energies of Appendix 3.

(a) For $SiO_2(\alpha\text{-quartz}) = SiO_2(\alpha\text{-tridymite})$, $\Delta G° = +1.38$ kJ mol^{-1}; quartz is the stable phase.

(b) For $FeS_2(\text{marcasite}) = FeS_2(\text{pyrite})$, $\Delta G° = -1.80$; pyrite is the stable phase.

(c) For $\alpha\text{-Fe}_2O_3 + 3H_2O = 2Fe(OH)_3$, $\Delta G° = +30$ kJ mol^{-1}; $\alpha\text{-Fe}_2O_3$ is the stable phase.

(d) For $PbCO_3 + CaSO_4 = CaCO_3 + PbSO_4$, $\Delta G° = +6.2$ kJ mol^{-1}. Therefore, $PbCO_3$ and $CaSO_4$ are a stable pair of solids.

2.4. For $HF = H^+ + F^-$, $\Delta H° = -13$ and $T\Delta S° = -29$ kJ mol^{-1}. For $HCl = H^+ + Cl^-$, $\Delta H° = -59$ and $T\Delta S° = -11$ kJ mol^{-1}. The "driving force" for dissociation is a favorable enthalpy change in each case, but much more favorable for HCl. For HF, a substantial entropy decrease opposes dissociation.

2.5. (a) The mole fraction scale is used for $CaCO_3(s)$ and $H_2O(l)$; the pressure (atm) scale is used for $CO_2(g)$; the molal scale is used for the aqueous ions.
(b) $\Delta G° = +33.2$ kJ mol^{-1}; $\Delta H° = -40.1$ kJ mol^{-1}.
(c) $\log K = -\Delta G°/2.3RT = -5.8$; $K = 1.6 \times 10^{-6}$ (m^3 atm^{-1})
(d) Applying equation (1) of Table 2.2 we find $d \ln K/dT < 0$; therefore increased temperature favors reactants.

2.6. (a) Applying equations 62 and 63 for the process $H_2O(l) = H_2O(g)$ yields $\Delta G° = 8.61$ kJ mol^{-1}. The equilibrium constant, $K = p_{H_2Oeq}$, is obtained from $\log K = -\Delta G°/2.3RT = -1.508$. We find $K = 0.0310$ atm, or 3141.4 Pa. (A 0.8% discrepancy.)
(b) Applying equation 3 of Table 2.2 and solving for $\Delta H°$:

$$\Delta H° = R \ln \frac{K_2}{K_1} \left(\frac{1}{T_1} - \frac{1}{T_2}\right)^{-1}. \quad \Delta H° = +44.1 \text{ kJ mol}^{-1}.$$

2.7. For $HCl(g) = H^+(aq) + Cl^-(aq)$, $\Delta G° = -35.9$ kJ mol^{-1}. $\log K = -\Delta G°/2.3RT = 6.28$. Solving the equilibrium expression $m_{HCl}^2 \gamma_\pm^2/p_{HCl} = K$ for p_{HCl} gives 3.4×10^{-7} atm. (Note that the equilibrium expression is the form required by Henry's law for a dissociating solute, i.e., $m_{H^+} m_{Cl^-} = m_{HCl}^2$. The mean molal activity coefficient squared, $\gamma_\pm^2$, accounts for the product of $\gamma_+ \gamma_-$ converting ion concentrations to ion activities.)

2.8. The maximum energy available, w' is $-\Delta G$, the *actual* free energy change of the reaction. The standard change, $\Delta G°$, is found to be $+12.08$, using the thermodynamic data in Appendix C. The reaction quotient for the conditions given is $Q = \{MnO_2\}\{H^+\}^2/\{Mn^{2+}\}p_{O_2}^{1/2}\{H_2O;\}$. Substitution of the actual activities gives $Q = 2.2 \times 10^{-9}$. The actual free energy change is $\Delta G = 12.08 + RT \ln (2.2 \times 10^{-9}) = -37.4$ kJ mol^{-1}. w' is 37.4 kJ mol^{-1}. (Note that the available energy results from the variable activity term $\{H^+\}$ and not from the standard term, which is unfavorable.)

2.9. The reaction is $CaCO_3 = Ca^{2+} + CO_3^{2-}$. Equation 4 of Table 2.6 gives $\Delta V° = -17.7 - 3.7 - 36.9 = -58.3$ cm^3 mol^{-1}.

(a) Applying equation 1 of Table 2.3: $\Delta \mu_{CaCO_3} \cong \bar{V}^{\circ}_{CaCO_3} \Delta P = 36.9 \times 999 = 36,860$ atm cm^3, or 3.7 kJ mol^{-1} (corresponding to an activity for CaCO$_3$(s) of 4.5, relative to that at 1 atm).

(b) Applying equation 5 of Table 2.3 gives ln $(K_{1000}/K_1) = 2.38$; $K_{1000}/K_1 = 10.8$. (This calculation does not include compressibilities, which are required for the accurate prediction of pressure effects.)

2.10. For the reaction, $\Delta G^{\circ} = 11.54$, and log $K = -2.02$; $K = 9.5 \times 10^{-3}$. The reaction quotient is $Q = \{CaCO_3\}\{H^+\}/\{Ca^{2+}\}\{HCO_3^-\} = 5 \times 10^{-3}$. Apply equation 78 to obtain $R_f/R_b = K/Q = 1.9$. Energetically, $\Delta G = -1.6$ kJ mol^{-1}. The reaction is nearing equilibrium in this system.

2.11. The characteristic time shown in Figure 2.10 is 5 days $((k_1 + k_2)^{-1})$. The "half time" is ln $2/(k_1 + k_2)$, about 3 1/2 days.
(i) $t_{char} = 0.5$ day
(ii) $t_{char} = 5$ days
(iii) $t_{char} = 0.5$ day
(See, also, other examples in Table 2.8.)

2.12. Applying equation 128 for A + B → products.

$$k_2 = \frac{k_B T}{h} \exp(\Delta S^{\ddagger\circ}/R) \exp(\Delta H^{\ddagger\circ}/RT)$$

with $\Delta H^{\ddagger\circ} = E_a - RT = 56 - 2.5 = 53.5$ kJ mol^{-1}, we find the bimolecular rate constant at 25°C:

$$k_2 = 6.21 \times 10^{12} \exp(12/8.314) \times \exp(-53.5/2.48)$$
$$= 1.12 \times 10^4 \text{ M}^{-1} \text{ s}^{-1}$$

The half-life of the bimolecular reaction depends on the concentration of reactants. For $[A]_0 = [B]_0$ the half-life is given by $t_{1/2} = 1/k_2[A]_0$ (Brezonik, 1993). If the concentration is, for example, 10^{-6} M, $t_{1/2} = 1/(1.12 \times 10^4 \times 10^{-6}) = 89$ s; if reactant concentration is 10^{-9} M, the half-life becomes 89,000 s, about one day.

2.13. When $\bar{C}_A = 0$, equation (139) yields $C_B/C_A = K + (r/k_b)$. Then $(C_B/C_A)/K = 1 + (r/Kk_b) = 1 + (\tau_b/\tau_R K) = 1 + (r/k_f)$. Some results for r versus $(C_B/C_A)/K$ are: 10^5 and 2.0; 10^6 and 1.1; 10^7 and 1.01; 10^8 and 1.001. The steady state is within 1% of reaction *equilibrium* for $\tau_b/\tau_R = 0.1$.

3

ACIDS AND BASES

3.1 INTRODUCTION

pH values for most mineral-bearing waters are known to lie generally within the narrow range 6–9 and to remain very nearly constant for any given water. The composition of natural waters is influenced by the interactions of acids and bases. According to Sillén (1961), one might say that the ocean is the result of a gigantic acid-base titration; acids that have leaked out of the interior of the earth are titrated with bases that have been set free by the weathering of primary rock. [H^+] of natural waters is of great significance in all chemical reactions associated with the formation, alteration, and dissolution of minerals. The pH of the solution will determine the direction of the alteration process.

Biological activities, such as photosynthesis and respiration, physical phenomena, such as natural or induced turbulence with concomitant aeration, and above all processes such as the precipitation and dissolution of $CaCO_3$ and of other minerals influence pH regulation through their respective abilities to decrease and increase the concentration of dissolved carbon dioxide. Besides photosynthesis and respiration, other biologically mediated reactions affect the H^+ ion concentrations of natural waters. Oxygenation reactions often lead to a decrease in pH, whereas processes such as denitrification and sulfate reduction tend to increase pH.

Because of the ubiquitousness of carbonate rocks and the equilibrium reactions of CO_2, bicarbonate and carbonate are present as bases in most natural waters. In addition, small concentrations of the bases borate, phosphate, arsenate, ammonia, and silicate may be present in the solution. Volcanoes and certain hot springs may yield strongly acid water by adding gases like HCl and SO_2.

Of special importance are the acid atmosphere depositions ("acid rain"); the acids (H_2SO_4, HNO_3, HCl, NH_4^+, and organic acids) can potentially disturb the rather delicate proton balance in water and soil systems. (Figure 3.1 gives a simple example of the stoichiometry of the interaction of strong acids and relatively strong bases occurring in the atmosphere.)

Free acids also enter natural water systems as a result of the disposal of industrial waters. From the point of view of interaction with solid and dissolved bases, the most important acidic constituent is CO_2, which forms H_2CO_3 with

3.1 Introduction

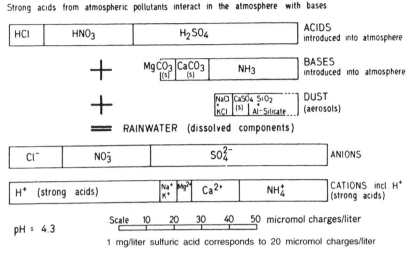

Figure 3.1. Strong acids in rainwater. The acid–base reaction involved in the genesis of a typical acid rainwater. Acids formed from atmospheric pollutants react in the atmosphere with bases and dust particles. The resulting rainwater contains an excess of strong acids. H_2SO_4 originates mostly from S in fossil fuels; after combustion the SO_2 formed is oxidized to SO_3, which gives, with H_2O, H_2SO_4; HNO_3 originates from NO and NO_2. These molecules are formed in the combustion of fossil fuels and to a large extent in the combustion of the automobile engine. For each molecule of NO, one of HNO_3 is formed; for example, $NO + O_3 \rightarrow NO_2 + O_2$; $3NO_2 + H_2O \rightarrow 2HNO_3 + NO$. HCl may largely originate from the combustion of Cl-bearing polymers, for example, polyvinyl chloride, in refuse incinerations. Most bases in the atmosphere are often of natural origin. Atmospheric dust may contain carbonates (calcite and dolomite). NH_3 is released from many soils and liquid manure (e.g., feed).

water. Proton transfer reactions are usually very fast (half-lives less than milliseconds).

Hydrogen ion regulation in natural waters is provided by numerous homogeneous and heterogeneous buffer systems. It is important to distinguish in these systems between intensity factors (pH) and capacity factors (e.g., the total acid- or base-neutralizing capacity). The buffer intensity is found to be an implicit function of both these factors. In this chapter, we discuss acid–base equilibria primarily from a general and didactic point of view. In Chapter 4 we address ourselves more specifically to the dissolved carbonate system.

Equilibria characterizing hydrogen ion transfer reactions are among the simpler types of models. In this chapter we demonstrate the use of numerical and graphical methods and mass law equilibria in order to establish the equilibrium composition. We try to go from the simple to the more complex. Many examples are given and the equilibrium compositions are graphically displayed. Dealing with dilute solutions, we will initially often set concentrations = activities. For didactic reasons we will use Morel's "Tableau method" (see Morel and Hering, 1993) for setting up chemical equilibrium problems and compile systematically in a matrix components, species, formulae and mole balance equations, that define the system.

3.2 THE NATURE OF ACIDS AND BASES

It is well known that a hydrogen ion, that is, a proton, cannot exist as a bare ion in water solution. Theoretical calculations show that a proton would strongly react with a water molecule to form a hydrated proton, a hydronium or a hydroxonium ion (H_3O^+). Actually, the H_3O^+ ion in an aqueous solution is itself associated through hydrogen bonds with a variable number of H_2O molecules: $(H_7O_3)^+$, $(H_9O_4)^+$, and so on. The formula H_3O^+ or H^+ is generally used, however, to denote a hydrated hydrogen ion. Formulas analogous to H_3O^+ for the solvated proton are used for other solvents; for example, NH_4^+ in liquid ammonia or $C_2H_5OH_2^+$ in ethanol. The hydroxide ion is also strongly hydrated in aqueous solutions. Similarly, metal ions do not occur as bare metal ions but as aquo complexes.

The Brønsted Concept The fact that hydrogen ions cannot exist unhydrated in water solution is incompatible with the notion of a simple dissociation of an acid. The ionization of an acid in water may more logically be represented as a reaction of the acid with the water

$$HCl + H_2O = H_3O^+ + Cl^- \tag{1}$$

The function of a base in its reaction with water is the opposite of that of an acid

$$NH_3 + H_2O = NH_4^+ + OH^- \tag{2}$$

This general idea has led to the very broad concept of acids and bases proposed by Brønsted. (The same concept was suggested by Lowry.) Accordingly, an acid is simply defined as any substance that can donate a proton to any other substance, and a base is defined as any substance that accepts a proton from another substance; that is, an acid is a proton donor and a base is a proton acceptor. Thus a proton transfer can occur only if an acid reacts with a base:

$$\begin{array}{c} Acid_1 = Base_1 + proton \\ \underline{proton + Base_2 = Acid_2} \\ Acid_1 + Base_2 = Acid_2 + Base_1 \end{array} \tag{3}$$

Further illustrations of such proton transfers for different solvents are the following:

3.2 The Nature of Acids and Bases

		Acid$_1$(A$_1$)	+ Base$_2$(B$_2$) (solvent)	= Acid$_2$(A$_2$)	+ Base$_1$(B$_1$)	
Perchloric acid		HClO$_4$	+ H$_2$O	= H$_3$O$^+$	+ ClO$_4^-$	(4a)
Carbonic acid		H$_2$CO$_3$	+ H$_2$O	= H$_3$O$^+$	+ HCO$_3^-$	(4b)
Bicarbonate		HCO$_3^-$	+ H$_2$O	= H$_3$O$^+$	+ CO$_3^{2-}$	(4c)
Ammonium		NH$_4^+$	+ H$_2$O	= H$_3$O$^+$	+ NH$_3$	(4d)
Ammonium		NH$_4^+$	+ C$_2$H$_5$OH	= C$_2$H$_5$OH$_2^+$	+ NH$_3$	(4e)
Acetic acid^a		HAc	+ NH$_3$	= NH$_4^+$	+ Ac$^-$	(4f)
Water		H$_2$O	+ H$_2$O	= H$_3$O$^+$	+ OH$^-$	(4g)
Water		H$_2$O	+ NH$_3$	= NH$_4^+$	+ OH$^-$	(4h)

^{a}HAc and Ac$^-$ stand for acetic acid and the acetate ion, respectively.

In the same way the reaction of different solvent bases accepting protons from acids can be illustrated:

		B$_1$	+ A$_2$ (solvent)	= B$_2$	+ A$_1$	
Ammonia^a		NH$_3$	+ H$_2$O	= OH$^-$	+ NH$_4^+$	(5a)
Cyanide		CN$^-$	+ H$_2$O	= OH$^-$	+ HCN	(5b)
Bicarbonate		HCO$_3^-$	+ H$_2$O	= OH$^-$	+ H$_2$CO$_3$	(5c)
Carbonate		CO$_3^{2-}$	+ H$_2$O	= OH$^-$	+ HCO$_3^-$	(5d)
Ammonia		NH$_3$	+ C$_2$H$_5$OH	= C$_2$H$_5$O$^-$	+ NH$_4^+$	(5e)
Amine		RNH$_2$	+ HAc	= Ac$^-$	+ RNH$_3^+$	(5f)
Hydroxide		OH$^-$	+ NH$_3$	= NH$_2^-$	+ H$_2$O	(5g)

^{a}The ammonia molecule is represented by NH$_3$ rather than NH$_4$OH.

Reaction 4g illustrates the self-ionization of the solvent water; that is, water is both a proton donor and a proton acceptor, an acid and a base in the Brønsted sense. Similarly, self-ionization in liquid NH$_3$ is represented by

$$NH_3 + NH_3 = NH_4^+ + NH_2^- \tag{4}$$

and self-ionization in the solvent H$_2$SO$_4$ is represented by

$$H_2SO_4 + H_2SO_4 = H_3SO_4^+ + HSO_4^- \tag{5}$$

Reactions of a salt constituent, cation or anion, with water (equations 4d, 5b, 5c, 5d) have been referred to as *hydrolysis* reactions. Within the framework of the Brønsted theory, the term hydrolysis is no longer necessary, since in principle there is no difference involved in the protolysis of a molecule and that of a cation or anion to water.

Metal ions, like hydrogen ions, exist in aqueous medium as hydrates. Many metal ions coordinate four or six molecules of H_2O per ion. H_2O can act as a weak acid. The acidity of the H_2O molecules in the hydration shell of a metal ion is much larger than that of water. This enhancement of the acidity of the coordinated water may, in a primitive model, be visualized as the result of the repulsion of the protons of H_2O molecules by the positive charge of the metal ion or as a result of the immobilization of the lone electron pair of the hydrate–H_2O molecule. Thus hydrated metal ions are acids:

$$[Al(H_2O)_6]^{3+} + H_2O = H_3O^+ + [Al(OH)(H_2O)_5]^{2+} \qquad (6)$$

To a first approximation their acidity increases with the decrease in the radius and an increase in the charge of the central ion. Similarly, the acidity of boric acid, H_3BO_3 or $B(OH)_3$, can be represented formally as

$$H_3BO_3(H_2O)_x + H_2O = B(OH)_4^-(H_2O)_{x-1} + H_3O^+ \qquad (7)$$

because the borate ion is $B(OH)_4^-$ and not $H_2BO_3^-$. Acidic properties of some substances in aqueous solutions can be interpreted in terms of proton transfer only by assuming hydration of the substance and loss of protons from the primary hydration shell. This is demonstrated in the acidity of metal ions.

In the illustration given above for proton transfer reactions (equations 4 and 5) $Acid_1$ and $Base_1$ or $Acid_2$ and $Base_2$ form *conjugate acid–base pairs*. Thus chloride is the conjugate base of hydrogen chloride; the latter is the conjugate acid of chloride. The conjugate acid of the base water is the hydronium ion, and the conjugate base of the acid water is the hydroxide ion.

Many acids can donate more than one proton. Examples are H_2CO_3, H_3PO_4, and $[Al(H_2O)_6]^{3+}$. These acids are referred to as *polyprotic acids*. Similarly, bases that can accept more than one proton, for example, OH^-, CO_3^{2-}, and NH_2^-, are polyprotic bases. Many important substances, for example, proteins or polyacrylic acids, so-called polyelectrolytic acids or bases, contain a large number of acidic or basic groups.

The *Lewis concept* of acids and bases (G. N. Lewis, 1923) interprets the combination of acids with bases in terms of the formation of a coordinate covalent bond. A Lewis acid can accept and share a lone pair of electrons donated by a Lewis base. Because protons readily attach themselves to lone electron pairs, Lewis bases are also Brønsted bases. Lewis acids, however, include a large number of substances in addition to proton donors: for example, metal ions, acidic oxides, or atoms.

3.3 THE STRENGTH OF AN ACID OR BASE

The strength of an acid or base is measured by its tendency to donate or accept a proton, respectively. Thus a weak acid is one that has a weak proton-donating tendency; a strong base is one that has a strong tendency to accept protons. It is, however, difficult to define the "absolute" strength of an acid or base, since the extent of proton transfer (protolysis) depends not only on the tendency of proton donation by $Acid_1$ but also on the tendency of proton acceptance by $Base_2$. Under these circumstances the *relative* strengths of acids are measured with respect to a standard $Base_2$—usually the solvent. In aqueous solutions, the acid strength of a conjugate acid–base pair, HA–A$^-$, is measured relative to the conjugate acid–base system of H_2O, that is, H_3O^+–H_2O. In a similar way, the relative base strength of a conjugate base–acid pair, B–HB$^+$, is defined in relation to the base–acid system of water, OH$^-$–H_2O.

The rational measure of the strength of the acid HA relative to H_2O as proton acceptor is given by the equilibrium constant for the proton transfer reaction

$$HA + H_2O = H_3O^+ + A^- \qquad K_1 \qquad (8)$$

which may be represented formally by two steps:

$$HA = proton + A^- \qquad K_2 \qquad (9)$$

$$H_2O + proton = H_3O^+ \qquad K_3 \qquad (10)$$

Because the concentration (activity) of water is essentially constant in dilute aqueous solutions (a mole fraction of unity), the hydration of the proton can be ignored in defining acid–base equilibria. Because the equilibrium activity of the proton and of H_3O^+ are not known separately, the thermodynamic convention sets the standard free energy change $\Delta G°$ for reaction 10 equal to zero; that is, $K_3 = 1$. In dealing with dilute solutions we can, because of this convention, represent the aquo hydrogen ion by H$^+$(aq), or more conveniently by H$^+$; that is,

$$[H^+] \equiv [H^+(aq)] = \sum_a [H(H_2O)_a^+(aq)] \qquad (11)$$

and the free energy change ΔG involved in the proton transfer reaction 8 may be expressed in terms of the equilibrium constant of reaction 9, that is, the acidity constant of the acid HA, K_{HA}. Ignoring activity coefficients, we have

$$K_2 = K_1 = K_2 K_3 = K_{HA} = \frac{[H^+][A^-]}{[HA]} \qquad (12)$$

which upon rearrangement gives

$$pH = pK_{HA} + \log \frac{[A^-]}{[HA]} \qquad (13)$$

For concentrations (activities of HA and A$^-$) in a molal $\simeq$ molar scale, pK_{HA} is commonly referred to as pK_a.

Acids and Bases

Strong Acids and Bases In aqueous systems some acids are stronger than H_3O^+ and some bases are stronger than OH^-. Water exerts a leveling influence because of its very high concentration, and pH values much lower than zero or much higher than 14 cannot be achieved in dilute aqueous solutions. In such solutions, acids stronger than H_3O^+ and bases stronger than OH^- are not stable as protonated or deprotonated species, respectively.

Self-ionization of Water In all aqueous solutions the autoprotolysis

$$H_2O + H_2O = H_3O^+ + OH^- \tag{14}$$

has to be considered. In dilute aqueous solutions ($\{H_2O\} = 1$), the equilibrium constant for equation 14, usually called the ion product of water, is

$$K_W = \{OH^-\}\{H_3O^+\} \equiv \{OH^-\}\{H^+\} \tag{15}$$

At 25°C, $K_W = 1.008 \times 10^{-14}$ and the pH = 7.00 corresponds to exact neutrality in pure water ($[H^+] = [OH^-]$). Because K_w changes with temperature (Table 3.1), the pH of neutrality also changes with temperature. (For ion product in seawater see Table A6.4 in Appendix 6.2.)

Equation 15 interrelates the acidity constant of an acid with the basicity constant of its conjugate base; for example, for the acid–base pair HB^+–B, the basicity constant for the reaction

$$B + H_2O = OH^- + BH^+$$

is

$$K_B = \frac{\{HB^+\}\{OH^-\}}{\{B\}} \tag{16}$$

Table 3.1. Ion Product of Water[a]

°C	K_W	pK_W
0	0.12×10^{-14}	14.93
5	0.18×10^{-14}	14.73
10	0.29×10^{-14}	14.53
15	0.45×10^{-14}	14.35
20	0.68×10^{-14}	14.17
25	1.01×10^{-14}	14.00
30	1.47×10^{-14}	13.83
50	5.48×10^{-14}	13.26

[a] $\log K_W = -4470.99/T + 6.0875 - 0.01706T$ (T = absolute temperature).

Source: Harned and Owen (1958). Reproduced with permission from Reinhold Publishing Corporation.

and the acidity constant for

$$HB^+ + H_2O = H_3O^+ + B$$

is

$$K_{HB^+} = \frac{\{H^+\}\{B\}}{\{HB^+\}} \tag{17}$$

Thus

$$K_{HB^+} = \frac{\{H^+\}\{B\}}{\{HB^+\}} = \frac{\{H^+\}\{OH^-\}}{K_B}$$

or

$$K_W = K_{HB^+} \cdot K_B \tag{18}$$

Thus either the acidity or basicity constant describes fully the protolysis properties of an acid–base pair. The stronger the acidity of an acid, the weaker the basicity of its conjugate base, and vice versa. For illustration purposes Table 3.2 lists a series of acids and bases in the order of their relative strength.

Composite Acidity Constants

It is not always possible to specify a protolysis reaction unambiguously in terms of the actual acid or base species. As it is possible to ignore the extent of hydration of the proton in dilute aqueous solutions, the hydration of an acid or base species can be included in a composite acidity constant; for example, it is difficult analytically to distinguish between $CO_2(aq)$ and H_2CO_3. The equilibria are

$$H_2CO_3 = CO_2(aq) + H_2O \qquad K = \frac{\{CO_2(aq)\}}{\{H_2CO_3\}} \tag{19}$$

$$H_2CO_3 = H^+ + HCO_3^- \qquad K_{H_2CO_3} = \frac{\{H^+\}\{HCO_3^-\}}{\{H_2CO_3\}} \tag{20}$$

and a combination of equations 19 and 20 gives

$$\frac{\{H^+\}\{HCO_3^-\}}{(\{H_2CO_3\} + \{CO_2(aq)\})} = \frac{K_{H_2CO_3}}{1 + K} = K_{H_2CO_3^*} \tag{21}$$

where $K_{H_2CO_3^*}$ is the composite acidity constant. Under conditions in which activities can be considered equal to concentration, the sum $\{H_2CO_3\} + \{CO_2\}$

Table 3.2. Acidity and Basicity Constants of Acids and Bases in Aqueous Solutions (25°C)

Acid[a]		−Log Acidity Constant, pK_a (approximate)	Base[b]	−Log Basicity Constant, pK_b (approximate)
$HClO_4$	Perchloric acid	−7	ClO_4^-	21
HCl	Hydrogen chloride	~−3	Cl^-	17
H_2SO_4	Sulfuric acid	~−3	HSO_4^-	17
HNO_3	Nitric acid	−1	NO_3^-	15
H_3O^+	Hydronium ion	−1.74	H_2O	15.74
HSO_4^-	Bisulfate	1.9	SO_4^{2-}	12.1
H_3PO_4	Phosphoric acid	2.1	$H_2P_4^-$	11.9
$[Fe(H_2O)_6]^{3+}$	Aquo ferric ion	2.2	$[Fe(H_2O)_5(OH)]^{2+}$	11.8
CH_3COOH	Acetic acid	4.7	CH_3COO^-	9.3
$[Al(H_2O)_6]^{3+}$	Aquo aluminum ion	4.9	$[Al(H_2O)_5(OH)]^{2+}$	9.1
$H_2CO_3^*$	Carbon dioxide[c]	6.3	HCO_3^-	7.7
H_2S	Hydrogen sulfide	7.1	HS^-	6.9
$H_2PO_4^-$	Dihydrogen phosphate	7.2	HPO_4^{2-}	6.8
HOCl	Hypochlorous acid	7.6	OCl^-	6.4
HCN	Hydrogen cyanide	9.2	CN^-	4.8
H_3CO_3	Boric acid	9.3	$B(OH)_4^-$	4.7
NH_4^+	Ammonium ion	9.3	NH_3	4.7
$Si(OH)_4$	O-Silicic acid	9.5	$SiO(OH)_3^-$	4.5
HCO_3^-	Bicarbonate	10.3	CO_3^{2-}	3.7
H_2O_2	Hydrogen peroxide	11.7	HO_2^-	2.3
$SiO(OH)_3^-$	Silicate	12.6	$SiO_2(OH)_3^{2-}$	1.4
HS^-	Bisulfide	~17[d]	S^{2-}	~−3
H_2O	Water[e]	15.74	OH^-	−1.74
NH_3	Ammonia	~23	NH_2^-	~−9
OH^-	Hydroxide ion	~24	O^{2-}	~−10
CH_4	Methane	~34	CH_3^-	~−20

[a] In order of decreasing acid strength.
[b] In order of increasing base strength.
[c] Total un-ionized CO_2 in water. $[H_2CO_3^*] = [CO_2(aq)] + [H_2CO_3]$.
[d] Value uncertain.
[e] The acidity constants given here are based on the convention $\{H_2O\} = 1$ for the solvent water. H^+(aq), HA, and A^- are on a molal ≈ molar concentration scale. In order to compare the acidity of water with that of the other acids, we must express the Brønsted acid water on a molal (or molar scale); accordingly, the ion product of water has to be divided by $\{H_2O\} = 55.4$, giving $K_{a,w} = 10^{-15.74}$.

96

is approximately equal to the sum of the concentrations, $[H_2CO_3] + [CO_2(aq)]$. $[H_2CO_3^*]$ is defined as the analytic sum of $[CO_2(aq)]$ and $[H_2CO_3]$. The true H_2CO_3 is a much stronger acid ($pK_{H_2CO_3} = 3.8$) than the composite $H_2CO_3^*$ ($pK_{H_2CO_3^*} = 6.3$), because less than 0.3% of the CO_2 is hydrated at 25°C.

An amino acid can lose two protons by two different paths:

$$\begin{array}{c} NH_3^+RCOO^- \\ \nearrow \qquad \searrow \\ ^+NH_3R\,COOH \qquad\qquad\qquad NH_2RCOO^- \\ \searrow \qquad \nearrow \\ NH_2RCOOH \end{array} \qquad (22)$$

Four microscopic constants can be defined, but potentiometrically only two composite (macroscopic) acidity constants can be determined.

3.4 ACTIVITY AND pH SCALES

In dealing with quantitative aspects of chemical equilibrium, we inevitably are faced with the problem either of evaluating or maintaining constant the activities of the ions under consideration. G. N. Lewis (1907) defined the chemical activity of a solute A, $\{A\}$, and its relationship to chemical concentration of that solute, $[A]$, by

$$\mu_A = k_A + RT \ln \{A\} = k_A + RT \ln [A] + RT \ln f_A \qquad (23)$$

where μ_A is the chemical potential of species A and k_A is a constant that identifies the concentration scale adopted (mol liter^{-1}, mol kg^{-1}, or mole fraction) and corresponds to the value of μ at $\{A\} = 1$. It will be recognized that k_A corresponds to the more familiar standard-state chemical potential μ_A°, the standard partial molar Gibbs energy, $\overline{G}_A^\circ$.

Concentration Scales

Any activity can be written as the product of a concentration and activity coefficient. Here we usually express concentration in terms of mol liter^{-1} of solution (molarity, M). Concentration may also be expressed in terms of mol kg^{-1} of solvent (molality, M). The *molal scale* gives concentrations that are independent of temperature and pressure and is used in precise physicochemical calculations.[†] The difference between molarity and molality is small in dilute solutions, especially in comparison to the uncertainties involved in determining

---

[†]In seawater analysis one often uses the unit mol kg^{-1} seawater, which is also pressure and temperature independent.

equilibrium constants or in estimating activity coefficients; for example, a 1.00 molal solution of NaCl is 0.98 M (25°C, 1 atm). Fundamentally the concentration scale most suitable for expressing deviations from ideality is the *mole fraction* scale [$x_i = n_i/(n_i + \Sigma_j n_j)$, i.e., moles of solute per total moles in solution]; but this scale is inconvenient to use experimentally. Molality [weight solute × 1000/(formula weight solute × weight water)] can be converted into molarity [weight solute × 1000/(formula weight solute × volume solution)] by

$$m = M \frac{\text{weight solution}}{\text{weight solution} - \text{weight solutes}} \times \frac{1}{\text{density}} \quad (24)$$

Two activity scales are useful:

The Infinite Dilution Scale This activity convention is defined in such a way that the activity coefficient, $f_A = \{A\}/[A]$, approaches unity as the concentration of all solutes approaches zero; that is,

$$f_A \to 1 \text{ as } \left(c_A + \sum_i c_i\right) \to 0 \quad (25a)$$

where c is the concentration in molar or molal units.

The Ionic Medium Scale This convention can be applied to solutions that contain a "swamping" concentration of inert electrolyte in order to maintain a constant ionic medium. The activity coefficient, $f' = \{A\}/[A]$, becomes unity as the solution approaches the pure ionic medium, that is, when all concentrations other than the medium ions approach zero:

$$f'_A \to 1 \text{ as } c_A \to 0 \text{ in a solution} \quad (25b)$$

where the total concentration is still $\Sigma_i c_i$. If the concentration of the medium electrolyte is more than approximately 10 times the concentration of the species under consideration, activity coefficients remain very close to 1. Thus generally no extrapolation is necessary (Figure 3.2).

Figure 3.2. Activity coefficients depend on the selection of the reference state and standard state. (a) On the *infinite dilution scale* the reference state is an infinitely dilute aqueous solution; the standard state is a hypothetical solution of concentration unity and with properties of an infinitely dilute solution. For example, the activity coefficient of H$^+$ in a HCl solution, $f \pm_{\text{HCl}}$, varies with [HCl] in accordance with a Debye–Hückel equation (dashed line, left ordinate) (see Table 3.3); only at very great dilutions does f become unity. On the *ionic medium scale*, for example, in 1 M KCl, the reference

3.4 Activity and pH Scales 99

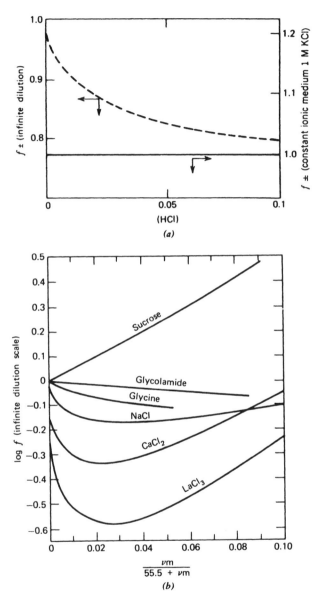

state is the ionic medium (i.e., infinitely diluted with respect to HCl only). In such a medium f_{HCl} (solid line, right ordinate) is very nearly constant, that is, $f_{HCl} = 1$. Both activity coefficients are thermodynamically equally meaningful. (Adapted from P. Schindler.) (b) A comparison of activity coefficients (infinite dilution scale) of electrolytes and nonelectrolytes as a function of concentration (mole fraction of solute) m = moles of solute per kg of solvent (molality); ν = number of moles of ions formed from 1 mol of electrolyte; 1 kg solvent contains 55.5 mol of water. (From Robinson and Stokes, 1959. Reproduced with permission from Butterworths, Inc., London.)

100   Acids and Bases

As equation 23 illustrates, a change in the activity scale convention merely changes k_A. In an ideal constant ionic medium, equation 23 becomes

$$\mu_A = k'_A + RT \ln [A] \tag{26}$$

Both activity scales are thermodynamically equally well defined. In constant ionic medium, activity ($\simeq$ concentration) can frequently be determined by means of emf methods.

Reference and Standard States

Because we cannot define absolute values for the chemical potential, we restrict ourselves to give changes in chemical potential (resulting from a change in a chemical composition or a change in pressure or temperature). By choosing a suitable *reference state*, we are able to define the chemical potential (usually at $T = 298.2$ K and $P = 1$ atm). In equations 23–26, we define the reference state either as infinite dilution or as constant ionic medium.

The *standard state* is given by setting the activity and the activity coefficient of the dissolved species A, $= 1$. For the activity $\{A\}$ the following is valid:

$$\mu_A = \mu_A^\circ + RT \ln (\{A\}/[A_0]) \tag{27}$$

where $[A_0] = 1$ and the chemical potential can be defined as equation 23 (for infinite dilution as a reference condition) or equation 26 (for the constant ionic medium reference condition). Using this convention, any activity under consideration becomes dimensionless, although the concentration scale used (molar or molal) has to be kept in mind. (In this book we will usually use the molar concentration scale, because most equilibrium constants have been determined by this scale.)

For water as solvent, pure liquid water is used as a standard state for the reference condition "infinite dilution"; the activity of water is then defined as the mole fraction of pure water X_{H_2O}:

$$\mu_{H_2O(l)} = \mu_{H_2O(l)}^\circ + RT \ln X_{H_2O} \tag{28}$$

where X_{H_2O} is the mole fraction of water. For dilute solutions $X_{H_2O} = 1$ and the activity of water is set as $\{H_2O\} = 1$. Because the activity of water does not change appreciably in the constant ionic medium scale, one uses $\{H_2O\}$ = constant within this reference condition.

In concentrated salt solutions (e.g., seawater), the activity of water may then be defined as the ratio of the vapor pressure of the salt solution, p_s, to that of pure water, p_{H_2O}, that is, $\{H_2O\} = p_s/p_{H_2O}$, but even in seawater $\{H_2O\}$ does not fall below 0.98.

3.4 Activity and pH Scales

In considering the mass law expression, for example, for an acid–base reaction HA = A$^-$ + H$^+$, the activities of the solutes are defined in reference to the standard state of a 1-molar concentration. These activities are, in principle, ratios to $\{A^-\}_0$, $\{HA\}_0$, and $\{H^+\}_0$ and are entered in molar units. The units are implicitly given for the $\mu°$ given, and the equilibrium constants are therefore dimensionless.

In an ideal solution activities are equal to concentrations. In real solutions, activity coefficients are introduced to correct for the nonideal effects of the different solutes.

Note that capacity factors, like alkalinity, acidity, C_T (= [H$_2$CO$_3^*$] + [HCO$_3^-$] + [CO$_3^{2-}$]), and TOTH, must be given in terms of concentrations (and not activities). These capacity factors are useful mole balances that are independent of pressure and temperature.

pH Conventions The original definition of pH (Sørensen, 1909) was

$$pH = -\log [H^+] \tag{29}$$

Within the infinite dilution concept, pH may be defined in terms of hydrogen ion activity:

$$p^aH = -\log \{H^+\} = -\log [H^+] - \log f_{H^+} \tag{30}$$

Within the ionic medium convention, Sørensen's original definition (equation 29) may be used operationally because one can usually measure $-\log [H^+]$ = $-\log \{H^+\}$ rather accurately. An electrode system is calibrated with solutions of known concentrations of strong acid (e.g., HClO$_4$), which are adjusted with an electrolyte to the appropriate ionic strength,[†] the observed potentiometer (pH meter) reading being compared with [H$^+$].

An operational definition endorsed by the International Union of Pure and Applied Chemistry (IUPAC) and based on the work of Bates determines pH relative to that of a standard buffer (where pH has been estimated in terms of p^{a}H) from measurements on cells with liquid junctions: the NBS (National Bureau of Standards) pH scale. This operational pH is not rigorously identical to p^{a}H defined in equation 30 because liquid junction potentials and single ion activities cannot be evaluated without nonthermodynamic assumptions. In dilute solutions of simple electrolytes (ionic strength, $I < 0.1$) the measured pH corresponds to within ± 0.02 to p^{a}H. Measurement of pH by emf methods is discussed in Chapter 8.

[†]Ionic strength, I, is a measure of the interionic effect resulting primarily from electrical attraction and repulsions between the various ions; it is defined by the equation $I = \frac{1}{2} \Sigma_i c_i z_i^2$. The summation is carried out for all cations and anions in the solution.

Operational Acidity Constants

According to the different activity conventions the following equilibrium expressions may be defined.

1. For the infinite dilution scale,

$$K = \frac{\{H^+\}\{B\}}{\{HB\}} \qquad (31)$$

   In this and subsequent equations charges are omitted: B can be any base of any charge.

2. In a constant ionic medium, the concentration quotient becomes the equilibrium constant

$$^cK = \frac{[H^+][B]}{[HB]} \qquad (32)$$

   More exactly, on the ionic medium scale, the equilibrium constant may be defined as the limiting value (as the solution composition approaches that of the pure ionic medium) for the equilibrium concentration quotient L. In usual measurements at low reactant concentrations, the deviations of L from ^cK are usually smaller than the experimental errors; hence it is preferable to set $L = {^cK}$ rather than to extrapolate.

3. A so-called mixed acidity constant is frequently used:

$$K' = \frac{\{H^+\}[B]}{[HB]} \qquad (33)$$

   This convention is most useful when pH is measured according to the IUPAC convention (pH $\approx$ p^{a}H), but the conjugate acid–base pair is expressed in concentrations.

In a way similar to the expression of activity as a product of concentration and activity coefficient, an acidity constant can be expressed in terms of a product of an equilibrium concentration quotient $L = ([H^+][B])/[HB]$ and an activity coefficient factor. Correspondingly, we can interrelate the various constants defined above in the following way:

$$K = L \frac{f_{H^+} f_B}{f_{HB}} \qquad (34)$$

$$^cK = \lim L \quad \text{(pure ionic medium)} \qquad (35)$$

$$K' = K \frac{f_{HB}}{f_B} \qquad (36)$$

Table 3.3. Individual Ion Activity Coefficients

Approximation	Equation[a]		Approximate Applicability [Ionic Strength (M)]
Debye–Hückel	$\log f = -Az^2 \sqrt{I}$	(1)	$< 10^{-2.3}$
Extended Debye–Hückel	$= -Az^2 \dfrac{\sqrt{I}}{1 + Ba\sqrt{I}}$	(2)	$< 10^{-1}$
Güntelberg	$= -Az^2 \dfrac{\sqrt{I}}{1 + \sqrt{I}}$	(3)	$< 10^{-1}$ useful in solutions of several electrolytes
Davies	$= -Az^2 \cdot \left(\dfrac{\sqrt{I}}{1 + \sqrt{I}} - 0.2I \right)$	(4)[b]	< 0.5

[a] I (ionic strength) $= \frac{1}{2} \Sigma \, C_i z_i^2$; $A = 1.82 \times 10^6 (\epsilon T)^{-3/2}$ (where ϵ = dielectric constant); $A \simeq 0.5$ for water at 25°C; z = charge of ion; $B = 50.3(\epsilon T)^{-1/2}$; $B \simeq 0.33$ in water at 25°C; a = adjustable parameter (angstroms) corresponding to the size of the ion. (See Table 3.4.)
[b] Davies has proposed 0.3 (instead of 0.2) as a coefficient for the last term in parentheses.

Expressions for Activity Coefficients of Ions

The theoretical expressions based on the Debye–Hückel limiting law together with more empirical expressions are given in Table 3.3. In defining the mean activity coefficient $f_\pm$ of a solute, z^2 in the equations of Table 3.3 should be replaced by $z_+ z_-$, where the charges are taken without regard to sign.

Single-ion activity coefficients are constructs; they are not measurable individually; only ratios or products of ionic activity coefficients are measureable. The use of single-ion activity coefficients greatly simplifies calculations.

Table 3.4 lists some activity coefficients as calculated from the extended Debye–Hückel limiting law for various values of I. In dilute solutions, such calculated values agree well with experimental data for mean activity coefficients of simple electrolytes. At higher concentrations, the Davies equation usually represents the experimental data better.

In natural water systems and under many experimental conditions, several electrolytes are present together. The limiting laws for activity coefficients can no longer be applied satisfactorily for electrolyte mixtures of unlike charge types. For estimating unknown activity coefficients, Güntelberg proposed that a in equation 2 of Table 3.3 be taken as 3.0 Å, resulting in a formula containing no adjustable parameter. For aqueous systems containing several electrolytes, equation 3 of Table 3.3 is most useful.

Activity coefficients defined within the infinite dilution activity scale cannot be formulated theoretically for the ionic medium of seawater. Since the oceans contain an ionic medium of practically constant composition, the ionic medium activity scale might be used advantageously in studying acid–base and other equilibria in seawater (see also Appendix 6.2 in Chapter 6).

Table 3.4. Parameter a and Individual Ion Activity Coefficients

Ion Size Parameter a (Å)[a]	Ion	Activity Coefficients Calculated with Equation 2 of Table 3.3 for Ionic Strength				
		10^{-4}	10^{-3}	10^{-2}	0.05	10^{-1}
9	H^+	0.99	0.97	0.91	0.86	0.83
	Al^{3+}, Fe^{3+}, La^{3+}, Ce^{3+}	0.90	0.74	0.44	0.24	0.18
8	Mg^{2+}, Be^{2+}	0.96	0.87	0.69	0.52	0.45
6	Ca^{2+}, Zn^{2+}, Cu^{2+}, Sn^{2+}, Mn^{2+} Fe^{2+}	0.96	0.87	0.68	0.48	0.40
5	Ba^{2+}, Sr^{2+}, Pb^{2+}, CO_3^{2-}	0.96	0.87	0.67	0.46	0.39
4	Na^+, HCO_3^-, $H_2PO_4^-$, CH_3COO^-	0.99	0.96	0.90	0.81	0.77
	SO_4^{2-}, HPO_4^{2-}	0.96	0.87	0.66	0.44	0.36
	PO_4^{3-}	0.90	0.72	0.40	0.16	0.10
3	K^+, Ag^+, NH_4^+, OH^-, Cl^- ClO_4^-, NO_3^-, I^-, HS^-	0.99	0.96	0.90	0.80	0.76

[a] After Kielland (1937). Reproduced with permission from American Chemical Society.

Activity Coefficients of Neutral Species

Neutral molecules have activity coefficients essentially equal to unity in solutions of less than 10 mM ionic strength. At higher salt concentrations, most neutral molecules are increasingly "salted out" of water; that is, the activity coefficient > 1, so that $a_i/c_i > 1$ for molecules in higher ionic strength solutions. In our discussion of dilute aqueous acids and bases, we will assume ideal behavior of the neutral species. The importance of salting out of dissolved CO_2 will be reflected in considering dissolved carbonic species in seawater (Chapter 4).

Example 3.1. Individual Activity Coefficients Estimate p^aH of a 10^{-3} M HCl solution that is 0.05 M in NaCl. The ionic strength is 0.051. Using equation 3 of Table 3.3, $-\log f_{H^+} = 0.092$. Thus

$$p^aH = -\log [H^+] + 0.092 = 3.09_2.$$

Approximately the same result would be obtained from the extended Debye–Hückel limiting law (equation 1, Table 3.3): $p^aH = -\log [H^+] + 0.113 = 3.11_3$.

Salt Effects on Acidity Constants The equilibrium condition of an acid–base reaction is influenced by the ionic strength of the solution. With the help of the formulations of the Debye–Hückel theory or with empirical expressions (Table 3.3), an estimate of the magnitude of the salt effect can be obtained. For example, the mixed acidity constant K' has been related to K (i.e., the

Table 3.5. Nonideality Corrections for Mixed Acidity Constants; Numerical Values of Second Term in Equation 37

Charge Acid, Z_{HB}	Charge Base, Z_B	Ionic Strength, I(M)							
		0.0001	0.0005	0.001	0.002	0.005	0.01	0.05	0.1
+1	0	+0.005	0.01	0.015	0.02	0.03	0.05	0.09	0.12
0	−1	—							
+2	+1	+0.015	0.03	0.05	0.06	0.10	0.14	0.27	0.36
−1	−2	—							
+3	+2	+0.02	0.05	0.08	0.11	0.17	0.23	0.46	0.60
−2	−3	—							

acidity constant valid at infinite dilution) by equation 36. Using the Güntelberg approximation for the single-ion activities, we can write instead of equation 36

$$pK' = pK + \frac{0.5(z_{HB}^2 - z_B^2)\sqrt{I}}{1 + \sqrt{I}} \quad (37)$$

The numerical values for the correction term (i.e., the second term in equation 37) are given in Table 3.5. Since we are frequently interested in the equilibrium concentrations of the various species in a solution of a given pH ($\simeq p^aH$), it is convenient to convert K to K'. With the operational K', calculations can be carried out for all species in terms of concentrations, with the exception of H^+. In equilibrium calculations, concentration conditions and charge balance or proton conditions must be formulated in terms of concentrations.

Example 3.2. Effect of Ionic Strength on pK' Estimate the effect of ionic strength on the successive mixed acidity constants (pK' values) of a dilute ($<10^{-3}$ M), neutralized tribasic acid (e.g., H_3PO_4) in a constant ionic medium of 0.01 M Na_2SO_4 solution. The contribution of the acid and its bases to the ionic strength can be neglected. $I = \frac{1}{2}([Na^+] + 4[SO_4^{2-}]) = 0.03$ M. Using equation 37, we have

$$pK_1' = pK_1 - 0.07$$

$$pK_2' = pK_2 - 0.21$$

$$pK_3' = pK_3 - 0.35$$

3.5 EQUILIBRIUM CALCULATIONS

The quantitative evaluation of the systematic relations that determine equilibrium concentrations (or activities) of a solution constitutes a purely mathematical problem, which is amenable to exact and systematic treatment.

106 Acids and Bases

Any acid–base equilibrium can be described by a system of fundamental equations. The appropriate set of equations comprises the equilibrium constant (or mass law) relationships (which define the acidity constants and the ion product of water) and any two equations describing the constitution of the solution, for example, equations describing a concentration and an electroneutrality or proton condition. Table 3.6 gives the set of equations and their mathematical combination for pure solutions of acids, bases, or ampholytes in monoprotic or diprotic systems.

The principle of such equilibrium calculations is best explained by a series of illustrative examples. In this and subsequent examples, a temperature of 25°C is assumed and conditions such that all activity coefficients are equal to 1.

Example 3.3a. pH and Equilibrium Composition of a Monoprotic Acid: Numerical Calculation
Calculate the pH ($= -\log [H_3O^+]$) of a 5×10^{-4} M aqueous boric acid solution [$B(OH)_3$].

In attacking such a problem, it is convenient to proceed systematically through a number of steps.

1. Establish all the *species present* in the solution: H_3O^+, OH^-, $B(OH)_3$, $B(OH)_4^-$. For brevity we call $B(OH)_3$ = HB and $B(OH)_4^-$ = B^-. Since four chemical species in addition to H_2O are involved in solution, four independent mathematical equations are necessary to interrelate the equilibrium concentrations.

2. The *equilibrium constants* that relate the concentrations of the various species must be found. For the boric acid solution the following equilibrium constants can be used:

$$[H^+][OH^-] = K_W = 10^{-14} \quad \text{(i)}$$

$$\frac{[H^+][B^-]}{[HB]} = K = 7 \times 10^{-10} \quad \text{(ii)}$$

3. A *concentration condition* or a mass balance must be established. Since the analytical concentration C of the HB solution is known, we can write

$$C = 5 \times 10^{-4} \text{ M} = [HB] + [B^-] \quad \text{(iii)}$$

The analytical concentration is the total number of moles of a pure substance that has been added to 1 liter of solution. In our example 5×10^{-4} mol of HB has been added per liter of solution. Some of the HB has protolyzed to form B^-, but the sum of [HB] + [B^-] must equal the number of moles (5×10^{-4}) that was added originally.

So far we have established three independent equations. One additional relation has to be found in order to have as many equations as unknowns.

Table 3.6. [H$^+$] of Pure Aqueous Acids, Bases, or Ampholytes

I. Monoprotic

Species[a]: HA A H$^+$ OH$^-$

Equilibrium constants[b]:
$[H^+][A]/[HA] = K$ (1)
$[H^+][OH^-] = K_w$ (2)

Concentration condition: $[HA] + [A] = C$ (3)

	Acid	Base
Proton condition[c]	$[H^+] = [A] + [OH^-]$ (4)	$[HA] + [H^+] = [OH^-]$ (5)
Numerical solution	$[H^+]^3 + [H^+]^2K - [H^+](CK + K_w) - KK_w = 0$ (6)	$[H^+]^3 + [H^+]^2(C + K) - [H^+]K_w - KK_w = 0$ (7)

II. Diprotic

Species[a]: H$_2$X HX X H$^+$ OH$^-$

Equilibrium constants[b]:
$[H^+][HX]/[H_2X] = K_1$ (8)
$[H^+][X]/[HX] = K_2$ (9)
$[H^+][OH^-] = K_w$ (2)

Concentration condition: $[H_2X] + [HX] + [X] = C$ (10)

Acid (H$_2$X)	Ampholyte (NaHX)	Base (Na$_2$X)
Proton condition[c]: $[H^+] = [HX] + 2[X] + [OH^-]$ (11)	$[H_2X] + [H^+]$ $= [X] + [OH^-]$ (12)	$2[H_2X] + [HX] + [H^+]$ $= [OH^-]$ (13)
Numerical solution: $[H^+]^4 + [H^+]^3 K_1 + [H^+]^2$ $\times (K_1 K_2 - CK_1 - K_w)$ $- [H^+]K_1(2CK_2 + K_w)$ $- K_1 K_2 K_w = 0$ (14)	$[H^+]^4 + [H^+]^3(C + K_1) + [H^+]^2$ $\times (K_1 K_2 - K_w)$ $- [H^+]K_1(CK_2 + K_w)$ $- K_1 K_2 K_w = 0$ (15)	$[H^+]^4 + [H^+]^3(2C + K_1) + [H^+]^2$ $\times (CK_1 + K_1 K_2 - K_w)$ $- K_1 K_w [H^+]$ $- K_1 K_2 K_w = 0$ (16)

[a] Charges are omitted for acid or base species. Equations given are independent of charge type of the acid.
[b] Equilibrium constants are either ^cK or are defined in terms of the constant ionic medium activity scale.
[c] Instead of the proton condition, the electroneutrality equation can be used. Independent of charge type, a combination of electroneutrality and concentration condition gives the proton condition. Na in NaHX or Na$_2$X is used as a symbol of a nonprotolyzable cation (Li$^+$, K$^+$, ...).

108   Acids and Bases

4. This additional equation follows from the fact that the solution must be electrically neutral. This can be expressed in a charge balance or in an *electroneutrality equation*. The total number of positive charges per unit volume must equal the total number of negative charges; that is, the molar concentration of each species is multiplied by its charge:

$$[H^+] = [B^-] + [OH^-] \qquad \text{(iva)}$$

There are alternative algebraic functions. Instead of writing the electroneutrality equation, we can derive a relation called the *proton condition*. If we made our solution from pure H_2O and HB, after equilibrium has been reached the number of excess protons must be equal to the number of proton deficiencies. Excess of deficiency of protons is counted with respect to a "zero level" reference condition representing the species that were added, that is, H_2O and HB. The number of excess protons is equal to $[H^+]$; the number of proton deficiencies must equal $[B^-] + [OH^-]$. This proton condition gives, as in equation iva, $[H^+] = [B^-] + [OH^-]$.

Alternatively, an explicit mole balance for the total sum of (free and bound) protons can be written:

$$\text{TOTH} = [H^+] - [OH^-] + [HB] = 5 \times 10^{-4} \, M \qquad \text{(ivb)}$$

and combination of equations ivb and iii yields, again, equation iva.

5. Four equations (i to iv) for four unknown concentrations must be solved simultaneously. The exact solution, although straightforward, is tedious. Frequently, the problem is readily solved by making suitable approximations. But in order to know what approximations can be made, we must have a qualitative knowledge of the chemical system within the given concentration range.

We first discuss the exact numerical solution; then we illustrate what we mean by making approximations. In the next section we discuss a method of graphical representation that is expedient in making such calculations.

Exact Numerical Solution   A numerical approach may start out by eliminating $[OH^-]$ in equations i and iva. After this substitution we have

$$[H^+] = \frac{K_W}{([H^+] - [B^-])} \qquad \text{(v)}$$

We now solve equation iii for [HB] and substitute in equation ii, eliminating [HB]:

$$[H^+][B^-] = K(C - [B^-]) \qquad \text{(vi)}$$

3.5 Equilibrium Calculations

Now we solve equation v for [B⁻] and substitute the result in equation vi to obtain a single equation in [H⁺]:

$$[H^+]^3 + K[H^+]^2 - [H^+](CK + K_W) - KK_W = 0 \quad \text{(vii)}$$

A value for [H⁺] can be obtained by trial and error from this equation:

$$[H^+]^3 + [H^+]^2(7 \times 10^{-10}) - [H^+](3.6 \times 10^{-13}) - (7 \times 10^{-24}) = 0$$

(viia)

Most numerical methods for solving polynomial equations are based on an initial guess for the answer followed by some iterative procedure for obtaining successively better approximations. A preliminary guess may be obtained by neglecting one or more items in equation viia. A convenient way to obtain a root for this equation is to plot various values of $f([H^+])$ versus $[H^+]$ and to locate $[H^+]$, where $f([H^+])$ crosses the zero axis. Programmable scientific calculators can readily be used to make such trial-and-error calculations; the Newton–Raphson method (Eberhart and Sweet, 1966) as well as computer methods can be used to achieve a rather systematic convergence of the successive approximations to the required root.

The value obtained by trial and error ([H⁺] = 6.1 × 10⁻⁷ M) is substituted into equation i to give [OH⁻] and into equation v to obtain [B⁻]. Then [HB] is calculated from equation iii. The following results obtain:

$$[H^+] = 6.10 \times 10^{-7} \text{ M (pH} = 6.21) \quad [OH^-] = 1.64 \times 10^{-8} \text{ M}$$

$$[B^-] = 5.94 \times 10^{-7} \text{ M} \quad [HB] = 4.99 \times 10^{-4} \text{ M}$$

Approximate Mathematical Expressions We can examine the linear equations for the exact solution for chemically justified approximations, which might simplify the solution. These approximations can then be checked to see if they are acceptable. For example, a solution of boric acid must yield [H⁺] > [OH⁻], and equation iva might be approximated by [H⁺] = [B⁻], and algebraic combination with the remaining equations yields a quadratic expression for [H⁺]

$$[H^+] = \frac{-K + (K^2 + 4KC)^{1/2}}{2}$$

which yields [H⁺] = 5.9 × 10⁻⁷ M. Furthermore, because HB is a rather weak acid (pK_a = 9.2), we might also assume that [HB] > [B⁻] in equation iii, yielding C = [HB] and a resulting simple expression, $[H^+]^2/C = K$. The solution of [H⁺] = 5.9 × 10⁻⁷ M, as previously. The error compared to the exact numerical solution is approximately −3%, or pH 6.23 instead of 6.21, a tolerable difference in most practical situations.

110   Acids and Bases

Tableaux   Examination of the foregoing analysis of the equilibrium composition of a weak acid solution reveals the general structure of all such problems: a set of linear equations describing mass balance and a set of nonlinear equations describing chemical equilibrium. As such problems are usually posed, they involve a description of the construction of the system and a listing of the species in the system at equilibrium. As seen for the example of boric acid, HB, the task is to find a satisfactory mathematical solution to the full set of governing equations. We will adopt a systematic approach to organization of aquatic chemical equilibrium problems using the tableau format advocated by Morel (1983) and Morel and Hering (1993). Their "canonical form" entails (1) a *recipe*, for example, how the system is constructed from reagents, including the molar amounts, the imposition of phases at equilibrium, and the imposition of fixed activities; (2) a list of the *species* at equilibrium; and (3) a list of the *independent* reactions among the species and their associated equilibrium constants. Species are formed from *components*, "a set of chemical entities that permits a complete description of the stoichiometry of the system" (Morel and Hering, 1993). All stoichiometric information on species and components is contained in the *stoichiometric matrix* of a tableau, satisfying conservation and chemical requirements. Each species activity must be expressible in a mass law in terms of the component activities, their stoichiometric coefficients, and a formation equilibrium constant. Let us return to the boric acid system in Example 3.3a, using the tableau methodology and the underlying algebraic formulation of the equilibrium problem. The recipe is $[HB]_T = 5 \times 10^{-4}$ M. The five species are B^-, HB, H_2O, OH^-, and H^+; we first choose as components HB, H^+, and H_2O. (We will always choose H^+ and H_2O among our components.) The species are represented stoichiometrically in terms of the components:

$$(B^-) = (HB)_1(H^+)_{-1} \tag{38a}$$

$$(HB) = (HB)_1 \tag{38b}$$

$$(H_2O) = (H_2O)_1 \tag{38c}$$

$$(OH^-) = (H^+)_{-1}(H_2O)_1 \tag{38d}$$

$$(H^+) = (H^+)_1 \tag{38e}$$

Thus the five species are formed from the three components, according to the reactions: $HB = B^- + H^+$, $HB = HB$, $H_2O = H_2O$, $H_2O - H^+ = OH^-$, and $H^+ = H^+$, with formation equilibrium constants given by log K of -9.2, 0, 0, -14.0, and 0, respectively. Tableau 3.1a presents in a compact way the essential information for the boric acid solution discussed in Example 3.3a.

The tableau comprises a list of components, a list of species, the stoichiometric matrix, and the log K vector for the species formed from the components. The *rows* of the matrix give the stoichiometric coefficients for forming each

3.5 Equilibrium Calculations

Tableau 3.1a. Reflecting the Equilibrium Composition of a 5×10^{-4} M H_3BO_3 (= HB) Solution

Components		HB	H^+	H_2O	log K (25°C)
Species	B^-	1	-1	0	-9.2
	HB	1	0	0	0
	H_2O	0	0	1	0
	OH^-	0	-1	1	-14.0
	H^+	0	1	0	0
	TOT	5×10^{-4} M	0	55.4 M	

species, and these are the exponents of the components in the mass laws. The columns of the matrix are the stoichiometric coefficients of the mole conservation equations of the components. Algebraically, the five species are given by the mass law expressions:

$$[B^-] = [HB]^1[H^+]^{-1}[H_2O]^0 \times 10^{-9.2} \tag{39a}$$

$$[HB] = [HB]^1[H^+]^0[H_2O]^0 \times 1 \tag{39b}$$

$$[H_2O] = [HB]^0[H^+]^0[H_2O]^1 \times 1 \tag{39c}$$

$$[OH^-] = [HB]^0[H^+]^{-1}[H_2O]^1 \times 10^{-14} \tag{39d}$$

$$[H^+] = [HB]^0[H^+]^1[H_2O]^0 \times 1 \tag{39e}$$

in which $[H_2O]$ is expressed on a mole fraction scale. The conservation equations of the three components are

$$\text{TOTHB} = 5 \times 10^{-4} \text{ M} = 1 \times [B^-] + 1 \times [HB] + 0$$
$$\times [H_2O] + 0 \times [OH^-] + 0 \times [H^+] \tag{40a}$$

$$\text{TOTH} = 0 = -1 \times [B^-] + 0 \times [HB] + 0 \times [H_2O] - 1$$
$$\times [OH^-] + 1 \times [H^+] \tag{40b}$$

$$\text{TOTH}_2\text{O} = 55.4 \text{ M} = 0 \times [B^-] + 0 \times [HB] + 1$$
$$\times [H_2O] + 1 \times [OH^-] + 0 \times [H^+] \tag{40c}$$

in which $[H_2O]$ is now expressed on a molar scale.

We see that we can "read" the tableau and immediately translate the entries in it as giving the composition of the species in terms of components, the exponents of the mass law expressions, the coefficients for the mass balance of components, the K vector for the mass laws, and the TOT vector of the components. The tableau is a compact representation of the equilibria and mass

112   Acids and Bases

conservation relationships governing the system. Here, we have included all the zero elements in the matrix and in the algebraic expressions to illustrate the details; henceforth, in reading the tableaux we will ignore the zero elements. We can also express the recipe in terms of the components and the matrix: (HB) = (HB)$_1$ and TOTHB = (HB)$_T$ = 5×10^{-4} M. Such relationships are needed in order to calculate the numerical values of TOTX$_j$ from the recipe of a system.

We note that equation 40b is identical to the proton condition (iva) of Example 3.3a; that is, the choice of HB as a component is equivalent to our earlier notion of HB and H$_2$O as reference level. In examining equations 39 and 40, we see that there are effectively four unknown species activities (B$^-$, HB, OH$^-$, and H$^+$), the activity of H$_2$O being constant for the dilute solution. As in Example 3.3a, there are four independent relationships: 39a, 39d, 40a, and 40b, the remaining expressions being irrelevant for our conditions. Although water is always a component in our tableaux, we will omit it from subsequent tableaux for simplicity, the concentration of H$_2$O being effectively constant for dilute solutions.

In this example there are 5 species, 3 components, and $5 - 3 = 2$ independent equations (mass laws). If equilibrium problems are solved by computer programs, the tableau often corresponds to the computer input (e.g., MICROQL; see Westall, 1986).

A survey of computer programs described in the literature is given by Basset and Melchior (1990).

Example 3.3b. Alternative Tableau   We can represent the same problem (5×10^{-4} M HB) in terms of different components, for example, B$^-$, H$^+$, H$_2$O. See Tableau 3.1b.

The horizontal rows give the following equilibrium expressions:

$$[B^-] = [B^-]$$
$$[HB] = [B^-][H^+]10^{9.2}$$
$$[OH^-] = [H^+]^{-1}10^{-14}$$
$$[H^+] = [H^+]$$

Tableau 3.1b.   Alternative Matrix for the Equilibrium of a 5×10^{-4} M H$_3$BO$_3$ (= HB)

Components		B$^-$	H$^+$	log K (25°C)
Species	B$^-$	1	0	0
	HB	1	1	9.2
	OH$^-$	0	−1	−14.0
	H$^+$	0	1	0
	TOT	5×10^{-4} M	5×10^{-4} M	

The mass balances are

$$\text{TOTB} = [\text{B}^-] + [\text{HB}] = 5 \times 10^{-4} \text{ M}$$

$$\text{TOTH} = [\text{HB}] + [\text{H}^+] - [\text{OH}^-] = 5 \times 10^{-4} \text{ M}$$

Example 3.4. pH of a Strong Acid Compute the equilibrium concentrations of all the species of a 2×10^{-4} M HCl solution (25°C). The acidity constant of HCl as a strong acid is very large, and essentially complete protolysis may be assumed. Nevertheless we would like to show that the problem is essentially analogous to that in Example 3.3a and can be approached the same way:

1. Mass laws:

$$[\text{H}^+][\text{OH}^-] = K_W = 10^{-14} \tag{i}$$

$$\frac{[\text{H}^+][\text{Cl}^-]}{[\text{HCl}]} = K = 10^{+3.0} \tag{ii}$$

2. Concentration condition:

$$[\text{HCl}] + [\text{Cl}^-] = C = 2 \times 10^{-4} \tag{iii}$$

3. Electroneutrality or proton condition:

$$[\text{H}^+] = [\text{Cl}^-] + [\text{OH}^-] \tag{iv}$$

As in Example 3.3a, the exact numerical solution would lead first to

$$[\text{H}^+]^3 + K[\text{H}^+]^2 - (K_W + CK)[\text{H}^+] - K_W K = 0 \tag{v}$$

Solving this by trial and error gives

$$[\text{H}^+] = 2.0 \times 10^{-4} \text{ M} \quad [\text{HCl}] = 4 \times 10^{-11} \text{ M}$$

$$[\text{Cl}^-] = 2.0 \times 10^{-4} \text{ M} \quad [\text{OH}^-] = 5.0 \times 10^{-11} \text{ M}$$

Since HCl has a large protolysis constant, $[\text{H}^+] \gg [\text{OH}^-]$ and correspondingly $[\text{OH}^-] \ll [\text{Cl}^-]$, the electroneutrality condition reduces to $[\text{H}^+] = [\text{Cl}^-]$. Furthermore, $[\text{HCl}] \ll [\text{Cl}^-]$; correspondingly, the reaction $\text{HCl} + \text{H}_2\text{O} = \text{H}_3\text{O}^+ + \text{Cl}^-$ has gone very far to the right. As shown in this example, the ratio of acid to base, that is, $[\text{HCl}]/[\text{Cl}^-]$, is extremely small. Strong acids are virtually completely protolyzed.

The matrix defining this problem is given in Tableau 3.2.

Tableau 3.2. Strong Acid

Components	H$^+$	Cl$^-$	log K
Species H$^+$	1	0	0
OH$^-$	−1	0	−14.0
HCl	1	1	−3
Cl$^-$	0	1	0
TOT	2×10^{-4} M	2×10^{-4} M	

The two mole balance equations are:

$$\text{TOTH} = [\text{H}^+] - [\text{OH}^-] + [\text{HCl}] = 2 \times 10^{-4} \text{ M} \tag{vi}$$

This proton balance is with regard to the reference Cl$^-$, H$_2$O.

$$\text{TOTCl} = [\text{HCl}] + [\text{Cl}^-] = 2 \times 10^{-4} \text{ M} \tag{vii}$$

and the equilibria are given by rows

$$[\text{OH}^-] = [\text{H}^+]^{-1} \, 10^{-14.0} \tag{viii}$$

$$[\text{HCl}] = [\text{H}^+][\text{Cl}^-] \, 10^{-3} \tag{ix}$$

Example 3.5. Weak Base Compute the equilibrium concentrations of a $10^{-4.5}$ M sodium acetate (NaAc) solution (25°C).

1. Species:

$$\text{HAc}, \quad \text{Ac}^-, \quad \text{H}^+, \quad \text{OH}^-, \quad \text{Na}^+$$

2. Constants:

$$\frac{[\text{H}^+][\text{Ac}^-]}{[\text{HAc}]} = K \quad \text{or} \quad \frac{[\text{HAc}][\text{OH}^-]}{[\text{Ac}^-]} = K_B \tag{i}$$

$$p K = 4.70 \qquad p K_B = 9.30$$

3. Concentration condition:

$$[\text{HAc}] + [\text{Ac}^-] = C = [\text{Na}^+] \tag{ii}$$

4. Electroneutrality:

$$[\text{Na}^+] + [\text{H}^+] = [\text{Ac}^-] + [\text{OH}^-]; \text{ since } [\text{Na}^+] = [\text{HAc}] + [\text{Ac}^-] \tag{iii}$$

3.5 Equilibrium Calculations

Tableau 3.3. $10^{-4.5}$ M Na Acetate

Components		H^+	Ac^-	Na^+	log K (25°C)
Species	H^+	1	0	0	0
	Ac^-	0	1	0	0
	HAc	1	1	0	4.7
	OH^-	−1	0	0	−14.0
	Na^+	0	0	1	0
	TOT	0	$10^{-4.5}$ M	$10^{-4.5}$ M	

this becomes

$$[HAc] + [H^+] = [OH^-] \quad \text{(iv)}$$

and is identical to the proton condition.

From Tableau 3.3 we read the following mole balance equations:

$$TOTH = [H^+] + [HAc] - [OH^-] = 0 \quad \text{(v)}$$

(this corresponds to equation iv and

$$[Ac^-] + [HAc] = 10^{-4.5} \text{ M} = [Na^+] \quad \text{(vi)}$$

(this corresponds to equation iv) and

Approximations in the proton condition similar to those used in the previous examples are not permissible in this case. However, an approximation might be possible in the concentration condition. Because Ac^- is a weak base, $[Ac^-] > [HAc]$ and $[Ac^-] \simeq C$. If we combine this approximation with the proton condition and the two equilibrium constants, we obtain

$$[H^+]^2(C + K) = KK_W \quad \text{(vii)}$$

which results in $[H^+] = 10^{-7.2}$. Subsequent calculations give $[HAc] = 10^{-7.01}$ and $[Ac^-] = 10^{-4.51}$, thus confirming that the result is consistent with the approximation made.

If one is not aware of the possibility of an approximation, one can always attempt to solve the exact equation (equation 7, Table 3.6).

Example 3.6. Diprotic System; Ampholyte Calculate the equilibrium pH of a solution prepared by diluting $10^{-3.7}$ mol of sodium hydrogen phthalate to 1 liter with water:

$$C_6H_4\begin{matrix}\diagup COOH \\ \diagdown COONa\end{matrix} = NaHP$$

116 Acids and Bases

The acidity constants at the appropriate temperature (25°C) for H_2P and HP^- are $10^{-2.95}$ and $10^{-5.41}$, respectively.

1. Species:

$$H_2P, \ HP^-, \ P^{2-}, \ H^+, \ OH^-, \ Na^+$$

2. Equilibrium constants:

$$\frac{[H^+][HP^-]}{[H_2P]} = K_1 \qquad \frac{[H^+][P^{2-}]}{[HP^-]} = K_2$$

$$[H^+][OH^-] = K_W \tag{i}$$

3. Concentration condition:

$$P_T = [H_2P] + [HP^-] + [P^{2-}] \tag{ii}$$

4. Proton condition:

$$[H_2P] + [H^+] = [P^{2-}] + [OH^-] \tag{iii}$$

Solution of equation 15 (from Table 3.6) by trial and error gives pH = 4.55, and a closer analysis shows that $[H_2P] < [HP^-] > [P^{2-}]$ and that the proton condition can be approximated by $[H^+] > [H_2P]$ and $[P^{2-}] \gg [OH^-]$.

Example 3.7. Mixture of Acid and Base (Buffer) Calculate the pH of a solution containing 10^{-3} mol of NH_4Cl and 2×10^{-4} mol of NH_3 per liter of aqueous solution.

1. Species:

$$NH_4^+, \ NH_3, \ H^+, \ OH^-, \ HCl, \ Cl^-$$

2. Equilibria, in addition to the ion product of water:

$$\frac{[NH_3][H^+]}{[NH_4^+]} = K = 10^{-9.3} \tag{i}$$

As shows by Example 3.4, the acidity equilibrium for HCl can be ignored because $[HCl] < [Cl^-]$. Similarly, HCl can be neglected in the subsequent concentration and proton conditions.

3. Concentration condition:

$$[NH_4^+] + [NH_3] = C_0(NH_4Cl) + C_0(NH_3) = 1.2 \times 10^{-3} \ M \tag{ii}$$

4. Electroneutrality condition:

$$[NH_4^+] = [Cl^-] + [OH^-] - [H^+] \quad \text{(iii)}$$

Because $[Cl^-] = C_0(NH_4Cl)$ and considering the concentration condition, we can also write

$$[NH_3] = C_0(NH_3) - [OH^-] + [H^+] \quad \text{(iv)}$$

Combining these equations with the acidity equilibrium we obtain

$$[H^+] = K \frac{C_0(NH_4Cl) + [OH^-] - [H^+]}{C_0(NH_3) - [OH^-] + [H^+]} \quad \text{(v)}$$

Neglecting as a justified approximation $[H^+]$ and $[OH^-]$ in the numerator and denominator gives

$$[H^+] = 2.5 \times 10^{-9} \, M \qquad pH = 8.6$$

Example 3.8 Volatile Acid or Base Estimate the pH of an aqueous electrolyte solution exposed to a partial pressure of NH_3 of 10^{-4} atm. Equilibrium constants valid at this temperature are $p^c K_{NH_4^+} = 9.5$; log $K_H = 1.75$ (Henry's law constant $K_H = [NH_3(aq)]/p_{NH_3}$); $p^c K_W = 14.2$. The information given by the equilibrium constants can be rearranged as

$$NH_3(aq) + H^+ = NH_4^+ \qquad -\log {}^c K_{NH^+4} = 9.5 \quad \text{(i)}$$

$$NH_3(g) = NH_3(aq) \qquad \log K_H = 1.75 \quad \text{(ii)}$$

$$H_2O = H^+ + OH^- \qquad \log {}^c K_W = -14.2 \quad \text{(iii)}$$

Summing up the reaction formulas and the log K values, we obtain

$$NH_3(g) + H_2O = NH_4^+ + OH^- \qquad \log K = -2.95 \quad \text{(iv)}$$

that is, $[NH_4^+][OH^-]/p_{NH_3} = 10^{-2.95}$. At equilibrium the proton condition is $[NH_4^+] + [H^+] = [OH^-]$ or $[NH_4^+] \simeq [OH^-]$. Thus $[OH^-]^2/10^{-4} = 10^{-2.95}$, which gives $[OH^-] = 10^{-3.5}$ and pH $\simeq 10.7$.

Tableau 3.4 summarizes the equilibrium condition of a solution in equilibrium with a gas phase with $p_{NH_3} = 10^{-4}$ atm.

$$TOTH = [H^+] - [OH^-] + [NH_4^+] \qquad = 0 \quad \text{(v)}$$

$$TOTNH_3 = [NH_3(g)] + [NH_3(aq)] + [NH_4^+] = ? \quad \text{(vi)}$$

118 Acids and Bases

Tableau 3.4. $NH_3(g)$–Water

Components	$NH_3(g)$	H^+	log K
Species H^+	0	1	
OH^-	0	−1	−14.0
$NH_3(g)$	1	0	0
$NH_3(aq)$	1	0	1.75
NH_4^+	1	1	11.05
	$p_{NH_3} = 10^{-4}$ atm	0	

Equation vi does not apply because the total quantity in the "open" system is not known. The open system is characterized, however, by a given p_{NH_3} (or with a given $[NH_3(g)]$).

3.6 pH AS A MASTER VARIABLE; EQUILIBRIUM CALCULATIONS USING A GRAPHICAL APPROACH

Surveys of the influence of master variables, such as pH, and the rapid solution of even complicated equilibria can be accomplished with relative facility by graphic representation of equilibrium data. The concepts of graphical representation of equilibrium relationships were first introduced by Bjerrum (1914) and have more recently been developed and popularized by Sillén (1959).

As we have seen, a direct numerical approach is often quite difficult because rigorous simultaneous solutions of equilibrium relationships lead to equations of third, fourth, or higher order; these equations are obviously not amenable to convenient numerical resolution.

The simplest example of the application of graphical representation of equilibrium data is that for acid-base equilibria involving a monoprotic acid, such as the acid HA, for which the equilibrium expression for solution in water may be written in terms of a concentration acidity constant, that is, an acidity constant valid at the appropriate temperature and corrected for activity by, for example, the Güntelberg approximation:

$$^cK = \frac{[H^+][A^-]}{[HA]} \qquad (41)$$

For the purpose of illustration it has been assumed that ^cK has the value 10^{-6} ($p^cK = 6$) and that a quantity of HA sufficient to give an exactly 10^{-3} M solution has been added to the water; the total concentration C_T of soluble A-containing species in the water at any position of equilibrium is then 10^{-3} M, or

$$C_T = 10^{-3} \text{ M} = [HA] + [A^-] \qquad (42)$$

3.6 pH as a Master Variable: Equilibrium Calculations

The control variable in any acid–base equilibrium is pH; hence it is desirable to represent graphically the equilibrium relationships of all species as functions of pH. For any value of pH the unknowns in the present example are of course [HA] and [A$^-$], each of which may now be expressed in terms of the known quantities C_T and [H$^+$] by combining equations 41 and 42 as follows:

$$[HA] = \frac{C_T[H^+]}{{}^cK + [H^+]} \tag{43}$$

and

$$[A^-] = \frac{C_T {}^cK}{{}^cK + [H^+]} \tag{44}$$

It is convenient in the construction of the equilibrium diagram to consider first the asymptotes of the individual curves of solute concentration against pH and in this manner to determine the slopes of the separate sections of each curve. After examining one or two examples of this, the method becomes quite obvious and it is not usually necessary to go through any computations; however, for purposes of illustration, the method is discussed one step at a time (see Figure 3.3).

For values of pH less than p^{c}K, the values of ^cK in the denominators of equations 43 and 44 are much smaller than the asymptotic value of [H$^+$] and may be neglected. Taking logarithms, equation 43 then becomes

$$\log[HA] = \log C_T \tag{45}$$

Thus the slope of that portion of the curve of HA against pH in the region pH < p^{c}K is zero.

Similarly, equation 44, relating [A$^-$] to pH, may be written

$$\log[A^-] = \log C_T - p^cK + pH \tag{46}$$

Differentiation of equation 46 with respect to pH yields $d \log[A^-]/d$ pH $= 1$; thus the slope of the part of the curve representing the variation of [A$^-$] with pH in the region pH < p^{c}K is unity.

Consideration of the asymptotes of the sections of the two curves in the region pH > p^{c}K permits us to ignore the quantity [H$^+$] in the denominators of equations 43 and 44. Equation 44 then becomes

$$\log[A^-] = \log C_T \tag{47}$$

hence $d \log[A^-]/d$ pH $= 0$. In a similar way it can be shown that, for pH > p^{c}K, $d \log[HA]/d$ pH $= -1$.

The slopes of the straight-line plots for each solute species against pH have

120 Acids and Bases

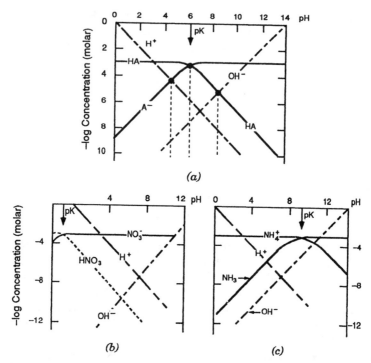

Figure 3.3. (a) Construction of logarithmic diagram and titration points for a monoprotic acid. The method of constructing logarithmic equilibrium diagrams and titration curves by graphical representation of mass law and electroneutrality relations is illustrated. For this example a monoprotic acid (HA) with a p^cK value of 6 and a total concentration of 10^{-3} mole liter^{-1} was used. A 10^{-3} M HA solution is characterized by the condition $[H^+] = [A^-] + [OH^-]$, pH = 4.5 neglect $[OH^-]$); a 10^{-3} M NaA solution is characterized by the condition $[HA] + [H^+] = [OH^-]$, pH = 8.5 (neglect $[H^+]$). (See Figures 3.8 and 3.10 for illustrations of titration curves.) (b, c) Equilibrium composition of 10^{-3} M HNO$_3$ and 10^{-3} M NH$_4$X. X$^-$ is an anion like Cl (where protolysis, under the conditions given, can be neglected).

now been calculated from the asymptotic values for the curves in the regions of pH < p^cK and pH > p^cK. None of these curve asymptotes is rigorous in the immediate region of pH = p^cK. At this point, log[HA] = log[A$^-$] = log(C_T/2). Therefore, at pH = p^cK, the curves must intersect at an ordinate value of log C_T − log 2, or 0.3 unit below the ordinate value of log C_T.

Computing the equilibrium composition of a 10^{-3} M HA solution, we simply have to find where on the graph the appropriate proton condition is fulfilled. The condition to be satisfied is

$$[H^+] = [A^-] + [OH^-] \tag{48}$$

Equation 48 is fulfilled at the intersection of the [H$^+$] line with the [A$^-$] line because obviously at this point [OH$^-$] ≪ [A$^-$]. Equilibrium [H$^+$] and the

3.6 pH as a Master Variable: Equilibrium Calculations 121

concentrations of all other species at this $[H^+]$ can be read directly from the logarithmic concentration diagram:

$$-\log[H^+] = -\log[A^-] = 4.5 \quad -\log[HA] = 3.0$$

If the diagram is drawn on graph paper (where one logarithmic unit corresponds to about 2 cm), the result can be read within an accuracy of better than ± 0.05 logarithmic unit and the relative error $(d\Delta[X]/d[X] = 2.3 \times \log \Delta[X])$ is smaller than 10%. The slight loss of accuracy involved in substituting graphical for numerical procedures is usually not significant. If a very exact answer is necessary, the graphical procedure will immediately show which concentrations can be neglected in the numerical calculations.

The same graph can be used to compute the equilibrium concentrations of a 10^{-3} M solution of NaA. In this case the proton condition is

$$[HA] + [H^+] = [OH^-] \tag{49}$$

This condition is fulfilled at the intersection $[HA] = [OH^-]$; since $[H^+]$ is 1000 times smaller than $[HA]$, it can be neglected in equation 49. This point gives $-\log[H^+] = 8.5$, $-\log[HA] = 5.5$, and $-\log[A^-] = 3.0$.

The proton conditions of equations 48 and 49 correspond to the two equivalence points in acid-base titration systems. The half-titration point is usually (not always) given by pH = pK. Thus the qualitative shape of the titration curve can be sketched readily along these three points (Figure 3.3a).

Example 3.9. Graphical Representation for HNO_3 and NH_4^+ Prepare graphs illustrating the species distribution (a) in a 10^{-3} M nitric acid system and (b) in a 10^{-3} M NH_4^+, NH_3 system.

For comparison, Figure 3.3b and 3.3c give double logarithmic diagrams for 10^{-3} M HNO_3 (pK = -1) and for a 10^{-3} M solution of NH_4^+ (pK = 9.3).

Example 3.10. Ecological or Toxicological Effects Depend on Speciation: NH_3 Toxicity In many circumstances it is necessary to compute the concentration (activity) of an individual species at a given pH. For example, bactericidal effect of Cl(+I) (HOCl and OCl^-) is primarily related to the concentration of HOCl. Similarly, the toxicity to fish of ammonia solutions is related to $[NH_3]$. The total N_T concentration $([NH_4^+] + [NH_3])$, as determined analytically, is 3×10^{-5} M (0.42 mg N/liter^{-1}), pH value is 8.5, and temperature is 15°C (pK at this temperature = 9.57) (see Table 3.7). How much NH_3 is present?

$$N_T = [NH_4^+] + [NH_3] = 3 \times 10^{-5} \text{ M}$$

$$\frac{[NH_3][H^+]}{[NH_4^+]} = 10^{-9.57}$$

$$[NH_3] = \frac{10^{-9.57}}{10^{-9.57} + 10^{-8.5}} \times N_T = 2.4 \times 10^{-6} \text{ M}$$

Table 3.7. Temperature Dependence of the Equilibrium $NH_4^+ + H_2O = NH_3 + H^+$; $pK_{NH_4^+}$

T	$pK_{NH_4^+}$
5°C	9.90
10°C	9.73
15°C	9.57
20°C	9.40
25°C	9.26

Diprotic Acid–Base System

The graphical approach is especially advantageous for more complicated equilibria. Figure 3.4 illustrates a logarithmic pH–concentration diagram for the diprotic acid, 4-hydroxybenzoic acid: total concentration = 2×10^{-3} M. The two acidity constants are $pK = 4.3$ (—COOH group) and 9.4 (—arOH group); the acid is abbreviated with the shorthand notation H_2B.

A combination of the equilibrium expressions for the two acidity constants with the concentration condition ($B_T = [H_2B] + [HB^-] + [B^{2-}]$) gives the equations that define the log concentration–pH dependence of $[H_2B]$, $[HB^-]$, and $[B^{2-}]$:

$$[H_2B] = \frac{B_T}{1 + K_1/[H^+] + K_1K_2/[H^+]^2} \tag{50}$$

$$[HB^-] = \frac{B_T}{[H^+]/K_1 + 1 + K_2/[H^+]} \tag{51}$$

$$[B^{2-}] = \frac{B_T}{[H^+]^2/K_1K_2 + H^+/K_2 + 1} \tag{52}$$

Considering equation 50, it is apparent that the log H_2B–pH line can be constructed as a sequence of three linear asymptotes that prevail in the three pH regions:

I: $pH < pK_1 < pK_2$; $\log [H_2B] = \log B_T$

$$\frac{d\log [H_2B]}{d\,pH} = 0 \tag{53}$$

II: $pK_1 < pH < pK_2$; $\log [H_2B] = pK_1 + \log B_T - pH$

$$\frac{d\log [H_2B]}{d\,pH} = -1 \tag{54}$$

3.6 pH as a Master Variable: Equilibrium Calculations

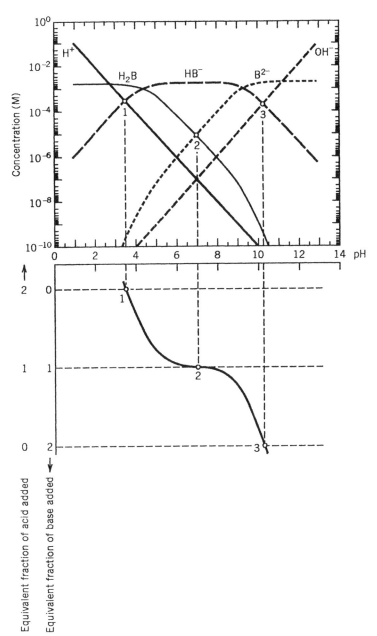

Figure 3.4. (a) Equilibrium diagram for a diprotic acid system (H_2B). Proton conditions: (1) solution of H_2B: $[H^+] = [HB^-] + 2[B^{2-}] + [OH^-]$; (2) solution of NaHB: $[H_2B] + [H^+] = [B^{2-}] + [OH^-]$; and (3) solution of Na_2B: $2[H_2B] + [HB^-] + [H^+] = [OH^-]$. Equilibrium composition: (1) pHB = pH = 3.6; pH_2B = 2.7; pB^{2-} = 9.6; (2) pH = 6.9; pH_2B ≃ pB^{2-} ≃ 5.1; pHB$^-$ ≅ 2.7; (3) pH = 10.4; pH_2B = 10; pHB$^-$ ≅ 3.6; pB^{2-} ≅ 2.7. (b) Sketch of alkalimetric and acidimetric titration curve.

Tableau 3.5. Hydroxybenzoic Acid (H_2B, 2×10^{-3} M)

Components		H_2B	H^+	log K
Species	H_2B	1	0	0
	HB^-	1	-1	-4.3
	B^{2-}	1	-2	-13.7
	OH^-	0	-1	-14.0
	H^+	0	1	0
		2×10^{-3} M	0	

$$\text{TOTB} = [H_2B] + [HB^-] + [B^{2-}] = 2 \times 10^{-3} \text{ M}$$
$$\text{TOTH} = -[HB^-] - 2[B^{2-}] - [OH^-] + [H^+] = 0$$

III: $pK_1 < pK_2 < pH$; $\log[H_2B] = pK_1 + pK_2 + \log B_T - 2pH$

$$\frac{d \log [H_2B]}{d \text{pH}} = -2 \tag{55}$$

These linear portions can readily be constructed; they change their slopes from 0 to -1 and from -1 to -2 at pH = pK_1 and pH = pK_2, respectively.

Similar considerations apply to the plotting of equations 51 and 52. The sections having slopes of -2 or $+2$ are usually unimportant because they occur only at extremely small concentrations. Diagrams of the types given in Figures 3.3 and 3.4 are not only useful in evaluating specific positions of equilibrium, but permit us to survey the entire spectrum of equilibrium conditions as a function of pH as a master variable.

Tableau 3.5 defines the equilibrium conditions of 2×10^{-3} M hydroxybenzoic acid.

Example 3.11. Weak Base Compute the equilibrium concentrations in a $10^{-4.5}$ M sodium acetate (NaAc) solution; pK (25°C) = 4.70 (See Tableau 3.6). Figure 3.5 plots the expressions for the acidity constant and the ion

Tableau 3.6. $10^{-4.5}$ M Na Acetate

Components		Ac^-	H^+	log K
Species	Ac^-	1	0	0
	HAc	1	1	4.7
	OH^-	0	-1	-14.0
	H^+	0	1	0
		$10^{-4.5}$	0	

$$\text{TOTAc} = [Ac^-] + [HAc] = 10^{-4.5} \text{ M}$$
$$\text{TOTH} = [HAc] - [OH^-] + [H^+] = 0$$

3.6 pH as a Master Variable: Equilibrium Calculations

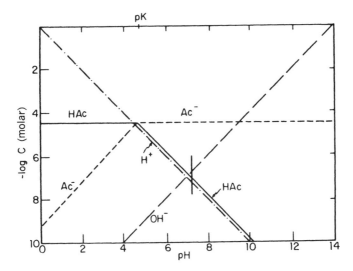

Figure 3.5. Equilibrium composition of $10^{-4.5}$ M NaAc (Example 3.11). Proton condition; $[HAc] + [H^+] = [OH^-]$. $[H^+] = 10^{-7.2}$ M; $[HAc] = 10^{-7.0}$ M $[Ac^-] = 10^{-4.5}$ M.

product of water as well as the concentration condition. For a pure solution of NaAc, the following proton condition is valid: $[HAc] + [H^+] = [OH^-]$.

It is obvious from the graph that no approximation in the proton condition is possible. In order to find the point where the proton condition is fulfilled, we move slightly to the right of the intersection of log[HAc] with log[OH$^-$] and find by trial and error where the proton condition is fulfilled, that is, at pH = 7.2, the condition is $10^{-7.0} + 10^{-7.2} \approx 10^{-6.8}$.

Example 3.12. Diprotic Base Compute the equilibrium composition of a $10^{-1.3}$ M sodium phthalate (Na$_2$P) solution (25°C).

$pK_1 = 2.95$; $pK_2 = 5.41$. Using the Güntelberg approximation, we first convert the pK values into p^{c}K values. In order to calculate the ionic strength, we assume that $[P^{2-}] > [HP^-]$. Thus $I = \frac{1}{2}([Na^+] + 4[P^{2-}]) = 0.15$.

$$p^c K_1 = pK_1 - \frac{\sqrt{I}}{1 + \sqrt{I}} = 2.95 - 0.28 = 2.67$$

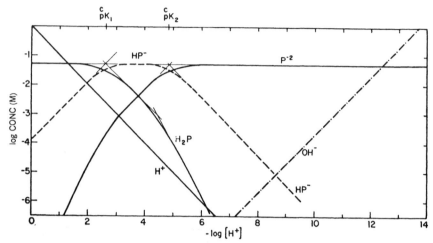

Figure 3.6. Equilibrium composition of diprotic acid (Example 3.12). A 5×10^{-2} M sodium phthalate solution (Na_2P) has a pH of 8.6. Proton condition: $2[H_2P] + [HP^-] + [H^+] = [OH^-]$; $[HP^-] \simeq [OH^-]$.

$$p^cK_2 = pK_2 - \frac{2\sqrt{I}}{1 + \sqrt{I}} = 5.41 - 0.56 = 4.85$$

$$p^cK_W = pK_W - \frac{\sqrt{I}}{1 + \sqrt{I}} = 14.00 - 0.28 = 13.72$$

The logarithmic pH concentration diagram (Figure 3.6) is now constructed using p^cK values. Note that the $\log[OH^-]$ line intersects the $\log C = 0$ at pH = 13.72. The proton condition for a solution of Na_2P is: $2[H_2P] + [HP^-] + [H^+] = [OH^-]$. This proton condition is fulfilled at $-\log[H^+] = 8.60$. The other species are present at the following concentrations:

$$[P^{2-}] = 10^{-1.3} \text{ M} \qquad [HP^-] = 10^{-5.1} \text{ M} \qquad [H_2P] < 10^{-6} \text{ M} \ (10^{-8.85} \text{ M})$$

$$-\log\{H^+\} = -\log[H^+] + \frac{0.5\sqrt{I}}{1 + \sqrt{I}} = 8.76$$

Example 3.13. Volatile Diprotic Acid Estimate the distribution of H_2S, HS^-, and S^{2-} as a function of pH for an aqueous solution (20°C) in equilibrium with an atmosphere containing 0.5% (by volume) of H_2S. Equilibrium constants: $p^cK_{H_2S} = 6.9$; $p^cK_{HS^-} \simeq 17.0$ $\log K'_H = \log([H_2S]/p_{H_2S}) = -1.0$. Since Henry's law is fulfilled over the entire pH range, $[H_2S] = 10^{-3.3}$; $p_{H_2S} = 10^{-2.3}$ (see Figure 3.7 and Tableau 3.7). The line for $[HS^-]$ is defined by $[HS^-] = K_{H_2S}[H_2S]/[H^+]$; ($d \log[HS^-]/d$ pH $= +1$). Similarly, the pH dependence for S^{2-} is given by $[S^{2-}] = K_{H_2S}K_{HS^-}[H_2S]/[H^+]^2$; ($d \log[S^{2-}]/d$ pH $= +2$).

3.7 Ionization Fractions of Acids, Bases, and Ampholytes

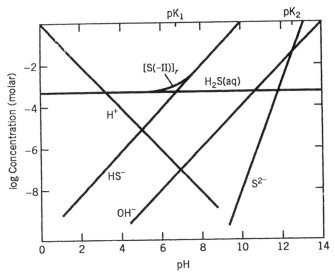

Figure 3.7. Distribution of S(−II) species of a solution in equilibrium with $p_{H_2S} = 5 \times 10^{-3}$ atm (Example 3.13).

Tableau 3.7. S(-II) Species for $p_{H_2S} = 5 \times 10^{-3}$ atm

Components		$H_2S(g)$	H^+	log K
Species	$H_2S(g)$	1	0	0
	$H_2S(aq)$	1	0	−1
	HS^-	1	−1	−7.9
	S^{2-}	1	−2	−24.9
	OH^-	0	−1	−14.0
	H^+	0	1	0
		$p_{H_2S} = 5 \times 10^{-3}$ atm		

The sum of all the S(-II) species ($[S(-II)_T] = [H_2S] + [HS^-] + [S^{2-}]$) is also given in the diagram as a function of pH.

3.7 IONIZATION FRACTIONS OF ACIDS, BASES, AND AMPHOLYTES

The relation between the concentration condition, $C = [HB] + [B]$, and the acidity constant, $K = [H^+][B]/[HB]$, permits us to calculate the relative distribution of acid and conjugate base as a function of pH:

$$\alpha_B = \alpha_1 = \frac{[B]}{C} = \frac{K}{K + [H^+]} \tag{56}$$

$$\alpha_{HB} = \alpha_0 = \frac{[HB]}{C} = \frac{[H^+]}{K + [H^+]} \tag{57}$$

where $\alpha_1 + \alpha_0 = 1$. Historically, α_1 has been called the degree of dissociation; it might be better to call it the ionization fraction or the degree of protolysis. α_0 has been called the degree of formation of the acid, and a plot of α_0 versus pH has been called the formation function, or better the *distribution diagram*. Figure 3.8a gives a schematic distribution diagram. Sometimes it is more convenient to plot the logarithms of the ionization fractions as a function of pH.

Since the ionization fractions are independent of total concentration, their tabulation or graphical representation is very convenient when calculations with the same equilibrium system have to be carried out repeatedly or with more complicated systems. The computation of α values can readily be programmed on programmable scientific calculators. The concentration of the species [HB] and [B] can then always be represented by [HB] = $C\alpha_0$ and [B] = $C\alpha_1$, respectively. The logarithmic equilibrium diagram, discussed in the preceding section, is essentially the additive combination of the line, log C = constant, and the logarithmic distribution diagram (Figure 3.8b). In a diprotic acid–base system, we define similarly

$$[H_2A] = C\alpha_0 \tag{58}$$

$$[HA^-] = C\alpha_1 \tag{59}$$

$$[A^{2-}] = C\alpha_2 \tag{60}$$

where the subscript on α refers to the number of protons lost from the most protonated species. The α values are implicit functions of $[H^+]$:

$$\alpha_0 = \frac{1}{1 + K_1/[H^+] + K_1K_2/[H^+]^2} \tag{61}$$

$$\alpha_1 = \frac{1}{[H^+]/K_1 + 1 + K_2/[H^+]} \tag{62}$$

$$\alpha_2 = \frac{1}{[H^+]^2/K_1K_2 + [H^+]/K_2 + 1} \tag{63}$$

The α values are interrelated by

$$\alpha_0 + \alpha_1 + \alpha_2 = 1 \tag{64}$$

$$\alpha_0 = \frac{[H^+]}{K_1} \alpha_1 \tag{65}$$

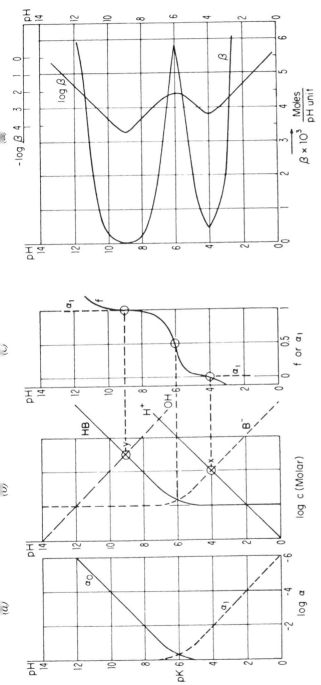

Figure 3.8. The titration curve and the buffer intensity are related to the equilibrium species distribution. For a monoprotic acid–base system (HB–B), (a) gives the logarithmic distribution diagram. α_1 and α_0 are the ionization fractions of B and HB, respectively. (b) The log concentration–pH diagram is the superimposition of the semi-logarithmic distribution diagram (a) and the concentration (log C = constant). Points X and Y correspond to the equivalence points on the alkalimetric or acidimetric titration curve. (c) Plot of the titration curve. The equivalence points (X and Y) and the half-titration point pH = pK are as given in (b). The equivalence fraction of the titrant added, f, shows, over a significant portion of the titration curve (0.1 < f < 0.9), the same dependence on pH as α_1. (d) The buffer intensity, β, corresponding to the inverse slope of the titration curve (dC_B/d pH), can be computed from a log concentration–pH diagram by multiplying by 2.3 the sum of all concentrations represented by a line of slope +1 or −1 at that particular pH in the diagram. (See Section 3.9.)

130 Acids and Bases

$$\alpha_1 = \frac{[H^+]}{K_2}\alpha_2 \tag{66}$$

α values for polyprotic acid–base systems can readily be derived.

3.8 TITRATION OF ACIDS AND BASES

In the titration of an aqueous solution containing C moles per liter of an acid HA with a quantity of strong base (C_B), such as NaOH, the titration curve can readily be deduced because at any point in the titration the following condition of electroneutrality must be fulfilled:

$$[Na^+] + [H^+] = [A^-] + [OH^-]$$

or

$$C_B = [A^-] + [OH^-] - [H^+] = -\text{TOTH} \tag{67}$$

With this equation and the logarithmic concentration–pH diagram, the titration curve relating pH to the quantity of base added can be constructed. In Figure 3.8c, pH is plotted as a function of the equivalent fraction of the titrant (strong base) added:

$$f = \frac{C_B}{C} = \frac{[Na^+]}{C} \tag{68}$$

Equation 67 can be rearranged into

$$C_B = C\alpha_1 + [OH^-] - [H^+] \tag{69}$$

$$f = \alpha_1 + \frac{[OH^-] - [H^+]}{C} \tag{70}$$

where α_1 is the degree of protolysis (ionization fraction) (see Section 3.7); $\alpha_1 = K/(K + [H^+]) = [A^-]/C$.

If we want to consider any dilution resulting from the addition of V milliliters of strong base to V_0 milliliters of solution containing a concentration C_0 of acid HA before dilution, we simply have to introduce a dilution factor and substitute for C in the above equations:

$$C = C_0 \frac{V_0}{V + V_0} \tag{71}$$

For the titration of a C molar solution of the conjugate base (i.e., the salt of a strong base with the weak acid HA, such as KA) with a strong acid (i.e.,

HCl), the curve for variation of pH with quantity of acid (C_A) added can be derived similarly from the electroneutrality condition:

$$[K^+] + [H^+] = [A^-] + [OH^-] + [Cl^-] \tag{72}$$

$$C + [H^+] = [A^-] + [OH^-] + C_A \tag{73}$$

$$C_A = [HA] + [H^+] - [OH^-] \tag{74}$$

$$C_A = C\alpha_0 + [H^+] - [OH^-] \tag{75}$$

where $\alpha_0 = [H^+]/(K + [H^+])$ (see Section 3.7).

The equivalent fraction of the titrant (strong acid) added, $g = C_A/C$, can be given as an implicit function of H^+ by

$$g = \frac{C_A}{C} = \alpha_0 + \frac{[H^+] - [OH^-]}{C} \tag{76}$$

Comparison of equation 76 with equation 70 shows that $g = 1 - f$. Equations 69 and 75 can be generalized into

$$C_B - C_A = C\alpha_1 + [OH^-] - [H^+]$$

or

$$C_A - C_B = C\alpha_0 + [H^+] - [OH^-] \tag{77}$$

Either of equations 77 can be used to evaluate pH changes that result from the addition of strong acid or strong base to a monoprotic weak acid–base system, or to characterize the titration curve of a mixture of a strong acid or base with a weak acid or base. The equivalence points marked X and Y in Figure 3.8b correspond to equilibrium conditions prevailing in pure equimolar solutions of HA ($g = 1$ and $f = 0$) and NaA ($f = 1$ and $g = 0$). In other words, at the equivalence point ($f = 1$) of an alkalimetric titration of HA with NaOH, the solution cannot be distinguished from an equimolar solution of the salt NaA (proton condition: $[HA] + [H^+] = [OH^-]$). Correspondingly, in the acidimetric titration of the salt, NaA, with HCl, the proton condition at the equivalence point ($f = 0$) is identical to that of an equimolar solution of HA ($[H^+] = [A^-] + [OH^-]$).

Example 3.14. Acidimetric Titration of a Strong and a Weak Base Describe the acidimetric titration curve for a solution made up of 2×10^{-3} M NaOCl and 1×10^{-3} M NaOH.

The titration curve can be drawn with the help of equation 76. $C_B = 1.0 \times 10^{-3}$ M. The equivalence points at $C_A = 1.0 \times 10^{-3}$ M and $C_A = 3.0 \times 10^{-3}$ M are at pH values 9.5 and 5.2, respectively, the second equivalence point being somewhat sharper than the first ones (Figure 3.9).

132 Acids and Bases

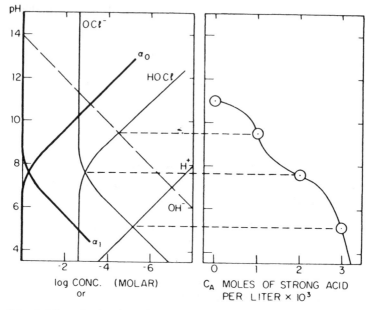

Figure 3.9. Acidimetric titration of a strong base and a weak base (Example 3.14). The solution to be titrated corresponds to "Eau de Javel," made by dissolving Cl_2 in NaOH. $[Na^+] = 3 \times 10^{-3}$ M; $[HOCl] + [OCl^-] = 2 \times 10^{-3}$ M, $[OH^-] = 1 \times 10^{-3}$ M. The first equivalence point (equivalent to the strong base) is at pH $\simeq 9.5$.

Because proton conditions corresponding to equivalent points $f = 0$ and $f = 1$ can readily be identified, titration curves can be sketched expediently with the help of the log concentration diagram.

The principles outlined above can readily be extended to *multiprotic acids*. The alkalimetric titration of an acid H_2L^+ added as the salt $H_2L^+X^-$ (e.g., an amino acid, $RNH_2COOH = HL$) is given by the electroneutrality condition

$$C_B + [H_2L^+] + [H^+] = [L^-] + [OH^-] + [X^-] \tag{78}$$

which can be rearranged with the concentration condition

$$C = [H_2L^+] + [HL] + [L^-] = [X^-] \tag{79}$$

to give

$$C_B = [HL] + 2[L^-] + [OH^-] - [H^+] = -\text{TOTH} \tag{80}$$

The relation is expressed more generally by

$$C_B = C(\alpha_1 + 2\alpha_2) + [OH^-] - [H^+] \tag{81}$$

$$f = \frac{C_B}{C} = \alpha_1 + 2\alpha_2 + \frac{[OH^-] - [H^+]}{C} \tag{82}$$

3.8 Titration of Acids and Bases

where the α values are defined as $\alpha_1 = [HL]/C$, $\alpha_2 = [L^-]/C$, and $\alpha_0 = [H_2L^+]/C$. In such a diprotic system, three equivalence points may be defined. In Figure 3.10a the points x, y, and z correspond to the equivalence points $f = 0$ (proton condition: $[H^+] = [HL] + 2[L^-] + [OH^-]$); $f = 1 ([H_2L^+] + [H^+] = [L^-] + [OH^-]$); and $f = 2$ ($2[H_2L^+] + [HL] + [H^+] = [OH^-]$), respectively.

Similarly, the equation describing the acidimetric titration curve of the base ML (where M^+ does not protolyze) can be derived from the concentration condition 79 and the electroneutrality condition:

$$[M^+] + [H_2L^+] + [H^+] = [L^-] + [OH^-] + C_A \qquad (83)$$

The morphology of the titration curve is given by

$$g = 2 - f = \frac{C_A}{C} = 2\alpha_0 + \alpha_1 + \frac{[H^+] - [OH^-]}{C} \qquad (84a)$$

The alkalimetric or acidimetric titration of the ampholyte HL is described simply by the appropriate portions of equations 82 and 84a, respectively.

For a mixture of protolysis systems, the variation in $[H^+]$ as a result of the

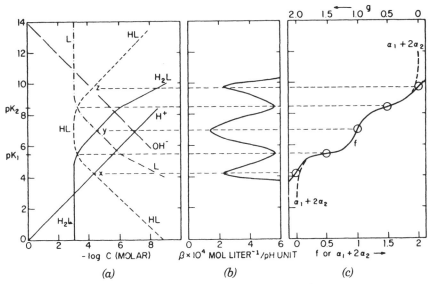

Figure 3.10. Equilibrium composition, buffer intensity, and titration curve of diprotic acid-base system. (a) Species distribution. (b) Buffer intensity. (c) Titration curve. The equivalence points, x, y, and z (a), are representative of the composition of pure solutions of H_2L, NaHL, and Na_2L respectively, and correspond to minima in the buffer intensity. The smaller the buffer intensity, the steeper is the titration curve.

addition of strong acid and/or strong base can be defined by

$$C_B - C_A = {}^IC({}^I\alpha_1 + 2{}^I\alpha_2 + \cdots)$$
$$+ {}^{II}C({}^{II}\alpha_1 + 2{}^{II}\alpha_2 + \cdots) + \cdots + [OH^-] - [H^+] \quad (84b)$$

3.9 BUFFER INTENSITY AND NEUTRALIZING CAPACITY

The slope of a titration curve (pH versus C_B) is related to the tendency of the solution at any point in the titration curve to change pH upon addition of base. The buffer intensity at any point of the titration is inversely proportional to the slope of the titration curve at that point and may be defined as

$$\beta = \frac{dC_B}{d\,\text{pH}} = -\frac{dC_A}{d\,\text{pH}} \quad (85)$$

where dC_B and dC_A are the numbers of mol liter^{-1} of strong acid or strong base required to produce a change in pH of $d\,\text{pH}$. β has also been called the buffer capacity or the buffer index.

Buffer Intensity

Obviously the buffer intensity can be expressed numerically by differentiating the equation defining the titration curve with respect to pH. For a monoprotic acid–base system (see equations 67 and 69).

$$\beta = \frac{dC_B}{d\,\text{pH}} = \frac{d[A^-]}{d\,\text{pH}} + \frac{d[OH^-]}{d\,\text{pH}} - \frac{d[H^+]}{d\,\text{pH}}$$
$$= C\frac{d\alpha_1}{d\,\text{pH}} + \frac{d[OH^-]}{d\,\text{pH}} - \frac{d[H^+]}{d\,\text{pH}} \quad (86)$$

The terms on the right-hand side of equation 86 can be differentiated as follows:

$$-\frac{d[H^+]}{d\,\text{pH}} = \frac{-d[H^+]}{-(1/2.3)\,d\ln[H^+]} = 2.3[H^+] \quad (87)$$

$$\frac{d[OH^-]}{d\,\text{pH}} = \frac{d[OH^+]}{(1/2.3)\,d\ln[OH^-]} = 2.3[OH^-] \quad (88)$$

$$C\frac{d\alpha_1}{d\,\text{pH}} = C\frac{d[H^+]}{d\,\text{pH}}\frac{d\alpha_1}{d[H^+]} = 2.3C\frac{K[H^+]}{(K+[H^+])^2} \quad (89)$$

Because $\alpha_0 = [H^+]/(K + [H^+])$ and $\alpha_1 = K/(K + [H^+])$, the right-hand side of equation 89 can be expressed in terms of ionization fractions or concentrations of [HA] and [A$^-$]:

$$C\frac{d\alpha_1}{d\,\text{pH}} = 2.3\alpha_1\alpha_0 C = 2.3\frac{[\text{HA}][\text{A}^-]}{C} \tag{90}$$

Summing up the individual terms of equation 86 results in

$$\beta = \frac{dC_B}{d\,\text{pH}} = 2.3([\text{H}^+] + [\text{OH}^-] + C\alpha_1\alpha_0)$$

$$= 2.3\left([\text{H}^+] + [\text{OH}^-] + \frac{[\text{HA}][\text{A}^-]}{[\text{HA}] + [\text{A}^-]}\right) \tag{91}$$

The terms on the right-hand side of equation 91 and in the logarithmic concentration–pH diagram, and β can readily be computed (Figure 3.8d). Maximum buffer intensity occurs (inflection point of titration curve) where $d\alpha_1/d(\text{pH})^2 = 0$. This occurs when $\alpha_1 = \alpha_0$ or where [HA] = [A$^-$] and pH = pK. Accordingly, *buffers* usually made by mixing an acid and its conjugate base have their maximum buffer intensity at a pH where [HA] = [A$^-$]. The pH of a solution containing C_{HA} M HA and C_{NaA} M NaA corresponds to a point in the titration curve and can readily be computed by equation 92 (which can be derived with the help of the electroneutrality condition; see Example 3.7):

$$[\text{H}^+] = K\frac{C_{\text{HA}} - [\text{H}^+] + [\text{OH}^-]}{C_{\text{NaA}} + [\text{H}^+] - [\text{OH}^-]} \tag{92}$$

Aqueous solutions are well buffered at either extreme of the pH scale. If in an alkalimetric or acidimetric titration curve the pH at the equivalence point falls into a pH range where the buffer intensity caused by [H$^+$] or [OH$^-$] exceeds that of the other protolytes, obliteration of a pH jump at the equivalence point results (see Figure 3.11). The concept of pH buffers can be extended to ions other than H$^+$. Metal-ion buffers will be discussed in Chapter 6.

If various acid-base pairs—HA, A; HB, B; and so on—are present in the solution, the buffer intensity is given by

$$\beta = 2.3\{[\text{H}^+] + [\text{OH}^-] + C_A\alpha_{\text{HA}}\,\alpha_A + C_B\alpha_{\text{HB}}\,\alpha_B + \cdots\}$$

$$= 2.3\left\{[\text{H}^+] + [\text{OH}^-] + \frac{[\text{HA}][\text{A}^-]}{[\text{HA}] + [\text{A}]} + \frac{[\text{HB}][\text{B}]}{[\text{HB}] + [\text{B}]} + \cdots\right\} \tag{93}$$

Polyprotic Systems In the same fashion as equation 91 has been derived, expressions for the buffer intensity of polyprotic acid–base systems can be developed. In Table 3.8, the buffer intensity of a diprotic acid–base system is derived. A polyprotic acid can be treated the same way as a mixture of indi-

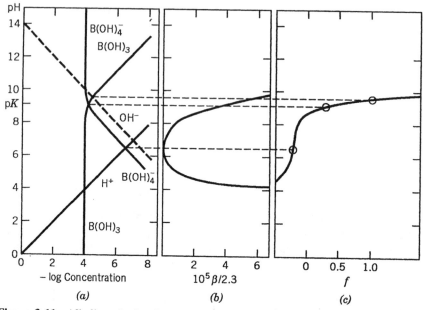

Figure 3.11. Alkalimetric titration of a weak acid (10^{-4} M boric acid). (a) Equilibrium distribution. (b) Buffer intensity. (c) Alkalimetric titration. No pH jump occurs at the equivalence point ($f = 1$) because of buffering by OH^- ions.

vidual monoprotic acids; for example, for the dibasic acid H_2C,

$$\beta \approx 2.3\left\{[H^+] + [OH^-] + \frac{[H_2C][HC^-]}{[H_2C] + [HC^-]} + \frac{[HC^-][C^{2-}]}{[HC^-] + [C^{2-}]}\right\}. \quad (94)$$

This approximate equation holds with an error of less than 5% if $K_1/K_2 > 100$. Note that in the last term of equation 91 and in the two last terms of equations 93 and 94, one of the concentrations in the denominator can usually be neglected: for example, at pH $<$ pK, equation 97 becomes $\beta \approx 2.3([H^+] + [OH^-] + [A^-])$; at pH $>$ pK, it reduces to $\beta \approx 2.3([H^+] + [OH^-] + [HA])$. In accord with this well-justified approximation, the buffer intensity of a solution can be computed from a logarithmic equilibrium diagram by multiplying by 2.3 the sum of all concentrations represented by a line of slope $+1$ or -1 at that particular pH in the diagram.

The concept of buffer intensity considered above may be extended and defined in a generalized way for the incremental addition of a constituent to a closed system at equilibrium. Thus, in addition to the buffer intensity with respect to strong acids or bases, buffer intensities with respect to weak acids

3.9 Buffer Intensity and Neutralizing Capacity

Table 3.8. Titration Curve and Buffer Intensity of a Two-Protic Acid (H_2C)[a]

I. Definitions

$$C = [H_2C] + [HC^-] + [C^{2-}] \qquad \alpha_0 = 1/[1 + K_1/[H^+] + K_1K_2/[H^+]^2]$$

$$[H_2C] = C\alpha_0 \qquad \alpha_1 = 1/[1 + [H^+]/K_1 + K_2/[H^+]]$$

$$[HC^-] = C\alpha_1 \qquad \alpha_2 = 1/[1 + [H^+]/K_2 + [H^+]^2/K_1K_2]$$

$$[C^{2-}] = C\alpha_2$$

II.

$$\frac{d\alpha_1}{d\,\text{pH}} = 2.3\alpha_1(\alpha_0 - \alpha_2); \qquad \frac{d\alpha_2}{d\,\text{pH}} = 2.3\alpha_2(\alpha_1 + 2\alpha_0);$$

$$\frac{d\alpha_0}{d\,\text{pH}} = -2.3\alpha_0(\alpha_1 + 2\alpha_2)$$

III. Titration Curve

$$C_B - C_A = [\text{ANC}] = [HC^-] + 2[C^{2-}] + [OH^-] - [H^+]$$

$$C_B - C_A = 2C - [\text{BNC}] = C(\alpha_1 + 2\alpha_2) + [OH^-] - [H^+]$$

IV. Buffer Intensity

$$\beta = \frac{dC_B}{d\,\text{pH}} = \frac{-dC_A}{d\,\text{pH}} = \frac{d[\text{ANC}]}{d\,\text{pH}} = \frac{-d[\text{BNC}]}{d\,\text{pH}}$$

$$= \frac{d[OH^-]}{d\,\text{pH}} - \frac{d[H^+]}{d\,\text{pH}} + \frac{Cd(\alpha_1 + 2\alpha_2)}{d\,\text{pH}}$$

$$= 2.3\{[H^+] + [OH^-] + C[\alpha_1(\alpha_0 + \alpha_2) + 4\alpha_2\alpha_0]\}$$

$$\beta = 2.3\left\{CK_1[H^+]\frac{[H^+]^2 + 4K_2[H^+] + K_1K_2}{([H^+]^2 + K_1[H^+] + K_1K_2)^2} + [H^+] + [OH^-]\right\}$$

[a]The equations given are rigorous. In very good approximation, a polyprotic acid can be treated the same way as a mixture of individual monoprotic acids (see equation 94).

and bases and for heterogeneous systems may be defined. In general,

$$\beta_{C_j}^{C_i} = \frac{dC_i}{d\,\text{pH}} \tag{95}$$

where $\beta_{C_j}^{C_i}$ is the buffer intensity for adding C_i incrementally to a system of constant C_j; for example, $\beta_{CaCO_3(3)}^{CCO_2}$ measures the tendency of a solution in contact and equilibrium with solid $CaCO_3$ to resist a pH change resulting from the addition or withdrawal of CO_2. In principle, the concept can be extended further to the buffering of metal ions, that is, to the stability of water with respect to the concentration of other ions and parameters such as

$$\beta = \frac{dC}{d\,\mathrm{pCa}} \qquad (96)$$

can be elucidated. The buffer intensity can always be found analytically by differentiating the appropriate function of C_i for the system with respect to the pH or pMe. The buffer intensity is an intrinsic function of the pH or pMe.

Acid- and Base-Neutralizing Capacity

Operationally we might define as a base-neutralizing capacity [BNC] the equivalent sum of all the acids that can be titrated with a strong base to an equivalence point. Similarly, the acid-neutralizing capacity [ANC] can be determined from the titration with strong acid to a preselected equivalence point. At every equivalence point a particular proton condition defines a reference level of protons. Conceptually, [BNC] measures the concentration of all the species containing protons in excess minus the concentration of the species containing protons in deficiency of the proton reference level; that is, it measures the net excess of protons over a reference level of protons. Similarly, [ANC] measures the net deficiency of protons.

In an aqueous monoprotic acid–base system, [ANC] is defined by the right-hand side of equation 67 or 69 in Section 3.8.

$$[\mathrm{ANC}] = [\mathrm{A}^-] + [\mathrm{OH}^-] - [\mathrm{H}^+]$$

or

$$[\mathrm{ANC}] = C\alpha_1 + [\mathrm{OH}^-] - [\mathrm{H}^+] \qquad (97)$$

The reference level is defined by the composition of a pure solution of HA in H_2O ($f = 0$; [ANC] $= 0$), which is defined by the proton condition, $[\mathrm{H}^+] = [\mathrm{A}^-] + [\mathrm{OH}^-]$. (In this and subsequent equations, the charge type of the acid is unimportant; the equation defining the net proton excess or deficiency can always be derived from a combination of the concentration condition and the condition of electroneutrality.) Thus in a solution containing a mixture of HA and NaA, [ANC] is a conservative capacity parameter. It must be expressed in concentrations (and not activities). Addition of HA (a species defining the reference level) does not change the proton deficiency and thus does not affect [ANC].

In the same monoprotic acid–base system, the base-neutralizing capacity with respect to the reference level ($f = 1$) of a NaA solution (proton condition: $[\mathrm{HA}] + [\mathrm{H}^+] = [\mathrm{OH}^-]$) is defined by

$$[\mathrm{BNC}] = [\mathrm{HA}] + [\mathrm{H}^+] - [\mathrm{OH}^-] = \mathrm{TOTH}$$

$$= C\alpha_0 + [\mathrm{H}^+] - [\mathrm{OH}^-] \qquad (98)$$

3.9 Buffer Intensity and Neutralizing Capacity

Example 3.15. [ANC] of Buffer Solutions Compare [ANC] of the following NH_4^+–NH_3 buffer solutions:

(a) $[NH_4^+] + [NH_3] = 5 \times 10^{-3}$ M, pH = 9.3

(b) $[NH_4^+] + [NH_3] = 10^{-2}$ M, pH = 9.0

Despite the lower pH, solution b has a slightly larger acid-neutralizing capacity than solution a. With a pK value of 9.3, the α values are 0.5 and 0.33, and the corresponding [ANC] capacities are 2.5×10^{-3} and 3.3×10^{-3} equivalents per liter, for a and b, respectively.

In a *multiprotic acid–base system* various reference levels ($f = 0, 1, 2, \cdots$) may be defined; for example, in a sulfide-containing solution the acid-neutralizing capacity with reference to the equivalence point defined by the pH of a pure H_2S solution ($f = 0$, $g = 2$) is

$$[ANC]_{f=0} = [HS^-] + 2[S^{2-}] + [OH^-] - [H^+]$$
$$= S_T(\alpha_1 + 2\alpha_2) + [OH^-] - [H^+] \qquad (99)$$

where S_T is the sum of the S(−II) species ($[H_2S] + [HS^-] + [S^{2-}]$).

The base-neutralizing capacity of the phosphoric acid system with reference to the equivalence point, $f = 2$ (solution of Na_2HPO_4 with the proton condition: $2[H_3PO_4] + [H_2PO_4^-] + [H^+] = [PO_4^{3-}] + [OH^-]$), is given by

$$[BNC]_{f=2} = 2[H_3PO_4] + [H_2PO_4^-] + [H^+] - [PO_4^{3-}] - [OH^-]$$
$$= P_T(2\alpha_0 + \alpha_1 - \alpha_3) + [H^+] - [OH^-] \qquad (100)$$

These relations can be generalized into

$$[BNC]_{f=n} = C[n\alpha_0 + (n-1)\alpha_1 + (n-2)\alpha_2 + (n-3)\alpha_3 + \cdots]$$
$$+ [H^+] - [OH^-] \qquad (101)$$

and

$$[ANC]_{f=n} = C[-n\alpha_0 + (1-n)\alpha_1 + (2-n)\alpha_2$$
$$+ (3-n)\alpha_3 + \cdots] - [H^+] + [OH^-] \qquad (102)$$

(As mentioned before, α_x refers to the ionization fraction of the species that has lost x protons from the most protonated acid species; f defines the equivalence points, the point at the lowest pH being $f = 0$.)

The ANC and BNC concept can be extended readily to mixed acid-base systems. For example, a natural carbonate-bearing water containing some

NH_4^+ and borate has an $[ANC]_{f=0}$ (reference, pure CO_2 solution) of the equivalent sum of all the bases that have proton-unpopulated energy levels of HCO_3^- or less, minus the equivalent sum of all the acids of energy levels higher than $H_2CO_3^*$; that is,

$$\begin{aligned}[ANC]_{f=0} &= [HCO_3^-] + 2[CO_3^{2-}] + [NH_3] + [B(OH)_4^-] \\ &\quad + [OH^-] - [H^+] \\ &= C_T(\alpha_1 + 2\alpha_2) + [NH_3] + [B(OH)_4^-] + [OH^-] - [H^+]\end{aligned}$$

(103)

[ANC] and [BNC] are very useful in defining and characterizing an acid–base system. The proton-free energy level, that is, the pH of the system, is independent of the quantity of the solution (*intensity factors*). On the other hand, the number of protons (added coulometrically or with strong acids) required to attain a certain pH represents a *capacity factor* because it is proportional to the quantity of the solution. [ANC] is an integration of the buffer intensity over a pH range:

$$[ANC] = \int_{f=n}^{f=x} \beta \, d\,pH \qquad (104)$$

and gives us a conservative parameter that is not affected by temperature and pressure.

Any acid–base system of unknown distribution can be characterized fully with the help of two parameters. For example, in a solution of phosphates (Na salts), the equilibrium composition with regard to the six species (H_3PO_4, $H_2PO_4^-$, HPO_4^{2-}, PO_4^{3-}, H^+, and OH^-) can be resolved completely if the concentration of at least two of the species or two of certain combinations thereof are evaluated analytically; capacity factors such as [ANC] and [BNC] are especially valuable for defining acid–base systems in terms of conservative parameters. They can be determined frequently with ease and relatively good accuracy; thus the discrepancy between conceptual and operational definition is very small.

In carbonate systems and in natural waters, [ANC] is referred to as *alkalinity*, while [BNC] is called *acidity*. In the context of natural waters, these terms will be discussed in the next chapter.

3.10 ORGANIC ACIDS

Acidity constants of organic compounds depend on the type of functional acid groups present. As Table 3.9 illustrates, there is quite a range of pK_a values

3.10 Organic Acids

Table 3.9. Acidity of Organic Acids

Acid	Formula	pK_a Range
Carboxylic acids	$RC\begin{smallmatrix}\diagup O\\ \diagdown OH\end{smallmatrix}$	1–5
Saturated alcohols	R—OH	>14
Phenols	ar—OH	1–11
Saturated thiols	R—SH	8.5–12.5
Aromatic thiols	ar—SH	3–8
Saturated amines	$R-NH_3^+$	8.5–12.5
Aniline	$C_6H_6NH_3^+$	4.6
m-Nitroaniline	$NO_2C_6H_4NH_3^+$	2.6
Pyridine	C_5H_5NH	5.1

for a given functional group because of the structural characteristics of the remainder of the molecule.

Humic and Fulvic Acids

These are polyelectric acids occurring in soils and natural waters. Humic substances can be extracted from soils with 0.1 M NaOH. Aquatic humic substances are the polymeric acids that are isolated from natural waters with a nonionic XAD-resin[†] or a weakly basic ion exchanger. They are nonvolatile and usually have molecular weights of 500–5000 g. The humic substances are mixtures that are formed from transformations of biogenic organic matter. An idea of the functional groups involved is given by Figure 3.12.

Humic acids are those humic substances that are precipitated at pH = 1. Fulvic acids stay in solution at pH = 1; they contain more —COOH and —OH groups than humic acids. Dissolved organic acids, which are not retained by a nonionic XAD resin, are *hydrophilic acids*; they are not very well defined: They contain, in addition to single aliphatic acids, uronic and polyuronic acids.

In Figure 3.13 the titration of a solution containing equimolar acetic acid (pK = 4.8) and phenol (pK ≃ 10) is compared with that of a humic acid. As can be seen, the humic acid titration curve is flatter than that of the acetic acid–phenol mixture. This is because of the polyfunctionality of the humic acid (the functional —COOH and —OH groups are present in different configurations and thus have somewhat different pK_a values; carboxylic acid groups and phenolic groups have pK values between 4 and 6, and between 9 and 11, respectively). A further effect is due to the polyelectric nature of the humic acid; by titrating with base, the molecule becomes progressively more negatively charged

---

[†]XAD-resins are non-electrolytic macroporous polyacrylacid-esters $CH_3(CH_2)_n$ COOR which adsorb humic and fulvic acids on the basis of their hydrophobic properties. The adsorption of these compounds ocurs only at low pH when the carboxylic groups are protonated.

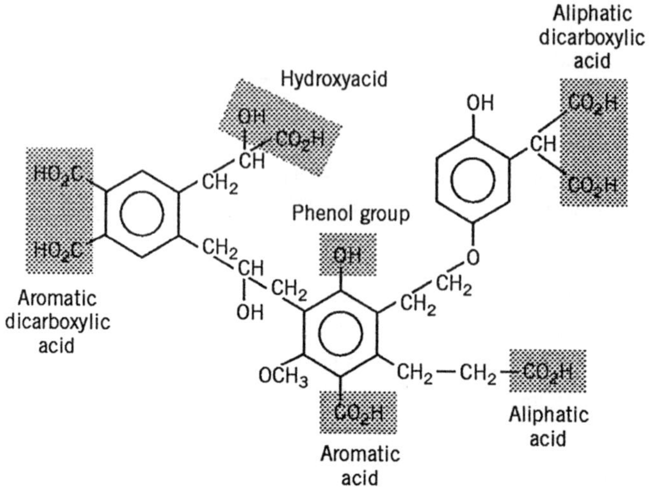

142   Acids and Bases

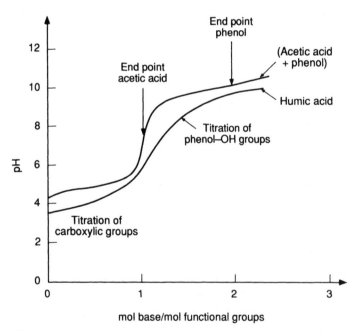

Figure 3.12. Exemplification of different possible —OH and —COOH groups in a hypothetical humic acid polymer. (From Thurman, 1985.)

Figure 3.13. Comparison of alkalimetric titration of an equimolar (10^{-4} M) acetic acid ($pK_a = 4.8$) and phenol ($pK_a = 10$) with titration of humic acid containing $\sim 10^{-4}$ mol carboxylic groups.

Influence of Structural Moieties on Acidity Constants: the Hammet Correlation

In an oxyacid, if we substitute a hydrogen atom with a chlorine atom, that is, HOH versus HOCl, then its acidity is enhanced: pK_a of HOCl = 7.2 versus pK_a of HOH = 15.74. The chlorine atom is more electronegative than hydrogen and induces a shift in electron density away from the O—H bond, which facilitates the loss of the proton. Similar increases in K_a values are observed for chlorine substitution in aliphatic and aromatic carboxylic acids. Similar electron withdrawing effects occur with NO_2, F, Br, phenyl, and NH_3^+ groups. In unsaturated chemicals such as aromatic or olefinic compounds (mobile π electrons), the inductive effect may be felt over longer distances. Furthermore, in aromatic acids, the electron density of the acidic groups is changed not only by induction but also by resonace. Most substituents enter into resonance with the aromatic ring. Such conjugation is largest at the ortho and para positions. Groups may be classed according to whether they withdraw electron density (NO_2) or supply it from their lone pair electrons (F, Cl, NH). Groups in the first class will strengthen acids by a shift in electron density from the acid group to the ring. Groups in the second class increase the electron density or the acidic groups by conjugation through the aromatic ring; resonance has an acid-weakening effect. Meta derivatives cannot conjugate directly with the acidic groups (King, 1965; Schwarzenbach et al., 1993).

Hammet (1940) postulated that the effects of substituents in benzoic acids could be expressed as the sum of the free energy change of the protolysis of the unsubstituted acid, ΔG_H°, and the contributions of the various substituents, ΔG_i°;

$$\Delta G^\circ = \Delta G_H^\circ + \Delta G_i^\circ \tag{105}$$

or

$$-2.3\,RT \log K_a = -2.3\,RT \log K_{a_H} - 2.3\,RT \log \sigma_i \tag{106}$$

or

$$pK_a = pK_{a_H} - \sigma_i \tag{107}$$

In the case of more than one substituent,

$$pK_a = pK_{a_H} - \sum_i \sigma_i \tag{108}$$

For illustrative purposes, Table 3.10 lists a few representative values of Hammet constants for substituents in benzoic acid.

Table 3.10. Hammet Constants for Substituents in Benzoic Acids[a]

Substituent	σ_{meta}	σ_{para}
—CH_3	−0.07	−0.17
—Phenyl	0.06	0.01
—CN	0.56	0.66
—OH	0.1	−0.37
—NH_2	−0.04	−0.66
—NO_2	0.71	0.78
—Cl	0.37	0.23
—Br	0.39	0.23

[a]Data from Lyman et al. (1990).

SUGGESTED READINGS

Jensen, W. B. (1980) *The Lewis Acid–Base Concepts*, Wiley-Interscience, New York.

Bell, R. P. (1969) *The Proton in Chemistry*, Cornell University Press, Ithaca, NY.

Shriver, D. F., Atkins, P. W., and Langford, C. H. (1990) *Inorganic Chemistry*, Oxford University Press, Oxford. Chapters 5 and 6 give a good introduction to Brønsted and Lewis acids and bases.

Sposito, G. (1981) *The Thermodynamics of Soil Solutions*, Clarendon Press, Oxford. Chapters 1 and 2 of this book give a rigorous introduction of the chemical thermodynamics in soil solution.

PROBLEMS

3.1. Two acids, of approximately 10^{-2} M concentration, are titrated separately with a strong base and show the following pH at the end point (equivalence point, $f = 1$):

$$HA: \text{pH} = 9.5$$
$$HB: \text{pH} = 8.5$$

(a) Which one (HA or HB) is the stronger acid?

(b) Which one of the conjugate bases (A^- or B^-) is the stronger base?

(c) Estimate the pK values for the acids HA and HB.

3.2. Hypochlorous acid (HOCl) has an acidity constant $K = 3 \times 10^{-8}$ (25°C). The strength of a sodium hypochlorite solution can be determined by titration with a strong acid.

(a) Sketch a titration curve for a 10^{-3} M solution indicating pH at the beginning and at the end point of the titration.

(b) Because HOCl, different from OCl$^-$, is bactericidal, the disinfection is pH dependent. Sketch a curve representing how the efficiency of disinfection depends on pH. (Cl$_2$, which is added to the water, disproportionates to hypochlorous acid and chloride.)

$$Cl_2 + H_2O = HOCl + H^+ + Cl^-$$

3.3. A 4×10^{-3} M solution of an acid HX has a pH of 2.4. What is the pH of an equimolar solution of the Na$^+$ salt of its conjugate base?

3.4. Calculate $-\log[H^+]$ and $-\log[OH^-]$ of a solution of 0.14 M ammonia:
   (a) at 25°C
   (b) at 100°C
   (^cK_B at 25° and at 100°C = 1.8×10^{-5}).

3.5. What is the acidity produced by the addition of 1 g of pyrite agglomerate (FeS$_2$) to 1 liter of distilled water? Assume that Fe(II) and S_2^{2-} are oxidized to Fe(III) and SO_4^{2-}, respectively.

3.6. Report qualitatively the following titration curves:
   (a) pH versus f, strong acid titrated with strong base.
   (b) [H$^+$] versus f, strong acid titrated with strong base.
   (c) pH versus f, weak base titrated with strong acid.
   (d) pH versus f, weak acid titrated with strong base.

3.7. Arrange the following solution in order of increasing buffer intensity:
   (a) 10^{-3} M NH$_3$–NH$_4^+$, pH = 7.
   (b) 10^{-3} M NH$_3$–NH$_4^+$, pH = 9.2.
   (c) 10^{-3} M H$_2$CO$_3^*$–HCO$_3^-$–CO$_3^{2-}$, pH = 8.2.
   (d) 10^{-3} M H$_2$CO$_3^*$–HCO$_3^-$, pH = 6.3.

3.8. One liter of rainwater contains 10^{-5} mol HCl, 5×10^{-6} mol H$_2$SO$_4$, 10^{-5} mol HNO$_3$, and 10^{-6} mol of a volatile organic acid with a pK_a value of 6.0.
   (a) Calculate the pH value of the rainwater. (Suggestion: Calculate first what the pH would be without the organic acid and then consider the effect of the weak acid).
   (b) How could one distinguish analytically between the mineral acids (the strong acids) and the organic acid?
   (c) What is the mineral acidity of this sample?
   (d) Sketch an appropriate titration curve (titration with dilute NaOH under a N$_2$ atmosphere).

3.9. Benzoic acid, pK_a 4.2, can be used to inhibit biological growth. The following data show the dependence of the minimum required toxic concentration on the pH of the solution.

pH	Minimum Benzoic Acid Required (mM)
3.5	1.2
4.0	1.6
4.5	3.0
5.0	7.3
5.5	21.0
6.0	64.0

Interpret the concentration required for growth inhibition in terms of the acid–base chemistry of benzoic acid. Which is the effective species: benzoic acid or benzoate anion?

3.10. The following data describe rainfall composition averaged over a year at Ithaca, New York:

Ion	mg liter^{-1}
Na^+	0.15
K^+	0.09
Ca^{2+}	0.83
Mg^{2+}	0.08
NH_4^+	0.32
SO_4^{2-}	4.96
NO_3^-	2.88
Cl^-	0.47

The average (geometric) p^{a}H for the rainfall, measured with a glass electrode calibrated with a NBS (activity) buffer, was 4.05.

(a) On the basis of the composition data, estimate the hydrogen-ion concentration of the rainfall, assuming that all important ions have been accounted for.
(b) What is the ionic strength of the rainfall?
(c) Compare the measured p^{a}H with that which you estimate from the composition data.
(d) Discuss the relative importance of these atmospheric components in establishing the acidity of the rainfall: (i) HNO_3, (ii) H_2SO_4, and (iii) CO_2.

3.11. How does the pH change for a 10^{-4} M NH_4Cl solution if one adds 10^{-2} mol Na_2SO_4 per liter? The Güntelberg approximation may be used to estimate the activity coefficients.

ANSWERS TO PROBLEMS

3.1. (a) HB; (b) A$^-$; (c) $pK_{HA} = 7$, $pK_{HB} = 5$.

3.2. (a) $g = 0$, pH = 9.3; $g = 0.5$, pH = 7.5; $g = 1.0$, pH = 5.3.

3.3. pH = 7.

3.4. (a) $-\log[OH^-](25°C) = -\log[OH^-](100°C) = 2.8$.
(b) $-\log[H^+](25°C) = 11.2$; $-\log[H^+](100°C) = 9.2$.
For the calculation $^cK_W = 10^{-14}$ (25°C) and $^cK_W = 10^{-12}$ (100°C) have been used.

3.5. 3.3×10^{-2} eq liter^{-1}.

3.6.

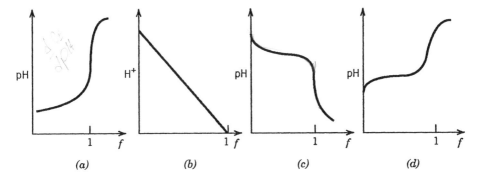

3.7. a < c < d < b.

3.8. (a) pH 4.52.
(b) Organic acid has no effect on pH (pH = 4.519).
(c) pH = 8.

4

DISSOLVED CARBON DIOXIDE

4.1 INTRODUCTION

Inorganic constituents dissolved in fresh water and seawater have their origin in minerals and the atmosphere. Carbon dioxide from the atmosphere provides an acid that reacts with the bases of the rocks. The water may also lose dissolved carbon to the sediments by precipitation reactions. Representative dissolution and precipitation reactions with $CaCO_3(s)$ and a feldspar are

$$\underset{\text{calcite}}{CaCO_3(s)} + CO_2 + H_2O \rightleftharpoons Ca^{2+} + 2HCO_3^- \tag{1}$$

$$\underset{\text{albite}}{NaAlSi_3O_8(s)} + CO_2 + \tfrac{11}{2}H_2O \rightarrow$$

$$Na^+ + HCO_3^- + 2H_4SiO_4 + \tfrac{1}{2}\underset{\text{kaolinite}}{Al_2Si_2O_5(OH)_4(s)} \tag{2}$$

In these and similar reactions, HCO_3^- and CO_3^{2-} (alkalinity) are imparted to or withdrawn from the water. CO_2 is also added to the atmosphere by volcanic activity and through the combustion of fossil fuels. Carbon dioxide is reduced in the course of photosynthesis and set free during respiration and oxidation of organic matter; it thus occupies a unique position in the biochemical exchange between water and biomass. Dissolved carbonate species participate in homogeneous and heterogeneous acid–base and exchange reactions with the lithosphere and the atmosphere. Such reactions are significant in regulating the pH and the composition of natural waters. Table 4.1 gives a survey of the distribution of carbon in its various forms in the atmosphere and biosphere.

In this chapter we describe the distribution of CO_2, H_2CO_3, HCO_3^-, and CO_3^{2-} in natural waters, examine the exchange of CO_2 between atmosphere and waters, evaluate the buffering mechanisms of fresh waters and seawater, and define their capacities for acid and base neutralization.

To make the discussion more pragmatic with regard to application in natural systems, we introduce now the solubility equilibrium with $CaCO_3$, although we will resume this topic in more detail later.

Two idealized equilibrium models—a system closed and a system open to

4.1 Introduction

Table 4.1. Carbon in Sedimentary Rocks, Hydrosphere, Atmosphere, and Biosphere[a]

Location	Total on Earth (10^{18} mol C)	Total on Earth (units of atmospheric CO_2, A_0)
Sediments		
Carbonate	1530	24700
Organic carbon	570	9200
Land		
Organic carbon	0.065	1.05
Ocean		
$CO_2 + H_2CO_3$	0.018	0.3
HCO_3^-	2.6	42
CO_3^{2-}	0.33	5.3
Dead organic	0.23	3.7
Living organic	0.007	0.01
Atmospheric		
$CO_2(A_0)$	0.062	1.0
Various Waters	C_T (M)[b]	Alkalinity (eq liter^{-1})[c]
Seawater	2.3×10^{-3}	$2.3 - 2.6 \times 10^{-3}$
River waters, average	$\sim 10^{-3}$	$\sim 10^{-3}$
River waters, typical range	$10^{-4} - 5 \times 10^{-3}$	$10^{-4} - 5 \times 10^{-3}$
Groundwaters, typical range, United States	$5 \times 10^{-4} - 8 \times 10^{-3}$	$10^{-4} - 5 \times 10^{-3}$
Rainwater, typical range	$10^{-5} - 5 \times 10^{-5}$	$0 - 4 \times 10^{-5d}$
Atmospheric CO_2 = 0.0355% by volume in dry air; (preindustrial $p_{CO_2} = 2.9 \times 10^{-4}$ atm); $p_{CO_2} = 3.55 \times 10^{-4}$ atm		

[a] Compare Figure 15.17.
[b] $C_T = [CO_2(aq)] + [H_2CO_3] + [HCO_3^-] + [CO_3^{2-}]$ (Oceanographers often use the symbol $\Sigma\, CO_2$.)
[c] Alkalinity, the acid-neutralizing capacity, is expressed as moles of protons per liter, or equivalents per liter.
[d] Some rainwaters contain mineral acidity of up to 10^{-3} eq liter^{-1}.

the atmosphere—will be emphasized in order to account for the distribution of the carbonate species. The results will be applied in discussing the major acid-base system of fresh water, seawater, and rainwater.

Of serious concern is human alteration of the natural CO_2 cycle (combustion of fossil fuel and deforestation) and the resulting progressive increase in the CO_2 concentration in the atmosphere. We will evaluate the acidification of surface waters and of the surface layers of the oceans resulting from an increase

150   Dissolved Carbon Dioxide

in atmospheric CO_2. We will consider some rate factors of the carbonate systems, especially the kinetics of CO_2 hydration and dehydration reactions.

4.2 DISSOLVED CARBONATE EQUILIBRIA (CLOSED SYSTEM)

We first consider a system that is closed to the atmosphere; that is, we treat $H_2CO_3^*$ as a nonvolatile acid. See Figure 4.2 for the organization of closed and open systems. For simple aqueous carbonate solutions, the interdependent nature of the equilibrium concentrations of the six solute components—CO_2, H_2CO_3, HCO_3^-, CO_3^{2-}, H^+, and OH^-—can be described completely by a system of six equations. The appropriate set of equations comprise four equilibrium relationships [which define the hydration equilibrium of CO_2 ($H_2CO_3 = CO_2(aq) + H_2O$), the first and second acidity constants of H_2CO_3, and the ion product of water] and any two conservation equations describing a concentration and an electroneutrality or proton condition. As shown earlier, it is convenient to define for the aqueous carbonate system a composite constant for all dissolved CO_2, hydrated or not. For the total analytical concentration of dissolved CO_2 we write $[H_2CO_3^*] = [CO_2(aq)] + [H_2CO_3]$; the number of equations necessary to describe the distribution of solutes reduces to five. These equations together with the relationships that describe the distribution of solutes are given in Table 4.2.

Equilibrium constants valid for different temperatures are given in Table 4.3.

The equilibrium for the reaction

$$H_2O + CO_2(aq) = H_2CO_3(aq)$$

lies rather far to the left, and by far the greater fraction of un-ionized CO_2 is present in the form of $CO_2(aq)$. The commonly used "first acidity constant" K_1 is a composite constant for the protolysis of $H_2CO_3^*$, reflecting both the hydration reaction and the protolysis of true H_2CO_3. The acidity constants of true H_2CO_3, $K_{H_2CO_3}$, and the composite acidity constant of $H_2CO_3^*$, K_1, are interrelated (see Table 4.2) by

$$K_1 = \frac{K_{H_2CO_3}}{1 + K} \qquad (3)$$

where K is the constant describing the hydration equilibrium [equation 1 in Table 4.2]. At 25°C, K is on the order of 650.[†] Thus equation 3 can often be simplified to

$$K_1 \simeq \frac{K_{H_2CO_3}}{K} \qquad (4)$$

[†]At 25°C, values of K range from 350 to 990. The corresponding range in $pK_{H_2CO_3}$ ("true" carbonic acid) is from 3.8 to 3.4.

4.2 Dissolved Carbonate Equilibria (Closed System)

Table 4.2. The Equilibrium Distribution of Solutes in Aqueous Carbonate Solution (System Closed to the Atmosphere)

Species:

$$CO_2(aq), H_2CO_3, HCO_3^-, CO_3^{2-}, H^+, OH^-$$
$$[H_2CO_3^*] = [CO_2 \cdot aq] + [H_2CO_3]$$

Equilibrium constants:[a]

$$[CO_2(aq)]/[H_2CO_3] = K \quad (1)$$

$$[H^+][HCO_3^-]/[H_2CO_3^*] = K_1 \quad (2)$$

$$[H^+][HCO_3^-]/[H_2CO_3] = K_{H_2CO_3} \quad (2a)$$

$$[H^+][CO_3^{2-}]/[HCO_3^-] = K_2 \quad (3)$$

$$[H^+][OH^-] = K_w \quad (4)$$

Concentration condition:

$$C_T = [H_2CO_3^*] + [HCO_3^-] + [CO_3^{2-}] \quad (5)$$

Ionization fractions:[b]

$$[H_2CO_3^*] = C_T\alpha_0 \quad [HCO_3^-] = C_T\alpha_1 \quad [CO_3^{2-}] = C_T\alpha_2$$

$$\alpha_0 = \left(1 + \frac{K_1}{[H^+]} + \frac{K_1 K_2}{[H^+]^2}\right)^{-1} \quad (6)$$

$$\alpha_1 = \left(\frac{[H^+]}{K_1} + 1 + \frac{K_2}{[H^+]}\right)^{-1} \quad (7)$$

$$\alpha_2 = \left(\frac{[H^+]^2}{K_1 K_2} + \frac{[H^+]}{K_2} + 1\right)^{-1} \quad (8)$$

Proton conditions of pure solutions (equivalence points): (e.g., of $H_2CO_3^*$, $NaHCO_3$, Na_2CO_3, respectively):[c] (see also Tableau 4.1)

$$[H^+] = [HCO_3^-] + 2[CO_3^{2-}] + [OH^-] \quad (9)$$

$$[H_2CO_3^*] + [H^+] = [CO_3^{2-}] + [OH^-] \quad (10)$$

$$2[H_2CO_3^*] + [HCO_3^-] + [H^+] = [OH^-] \quad (11)$$

with respective values of $[H^+]$ at equivalence points:[d]

$$[H^+] = (C_T K_1 + K_w)^{0.5} \quad (12)$$

$$[H^+] \simeq [K_1(K_2 + K_w/C_T)]^{0.5} \quad (13)$$

$$[H^+] \simeq K_w/2C_T + [K_w^2/4C_T^2 + K_2 K_w/C_T]^{1/2} \quad (14)$$

As corresponding approximate values:

$$[H^+] \simeq (C_T K_1)^{0.5} \quad (15)$$

Table 4.2. *(Continued)*

As corresponding approximate values (Continued):

$$[H^+] \simeq (K_1 K_2)^{0.5} \tag{16}$$

$$[H^+] \simeq (K_2 K_W / C_T)^{0.5} \tag{17}$$

Titration:
  Alkalimetric

$$f = C_B/C_T = \alpha_1 + 2\alpha_2 + ([OH^-] - [H^+])/C_T \tag{18}$$

  Acidimetric

$$g = C_A/C_T = 2 - f = 2\alpha_0 + \alpha_1 - ([OH^-] - [H^+])/C_T \tag{19}$$

Acid- and base-neutralizing capacity:[e]
  Alkalinity

$$[Alk] = C_T(\alpha_1 + 2\alpha_2) + [OH^-] - [H^+] \tag{20}$$

  Acidity

$$[Acy] = C_T(\alpha_1 + 2\alpha_0) + [H^+] - [OH^-] \tag{21}$$

Buffer intensity:[f]

$$\beta_{C_T}^{C_B} = 2.3\{[H^+] + [OH^-] + C_T[\alpha_1(\alpha_0 + \alpha_2) + 4\alpha_2\alpha_0]\} \tag{22}$$

[a] Equilibrium constants are defined for a constant ionic medium activity scale (Section 3.4).
[b] Equations 6–8 are derived as in Section 3.7.
[c] Na^+ (in $NaHCO_3$ or Na_2CO_3) is used as a symbol of a nonprotolyzable cation (Na^+, Li^+, K^+, . . .). In Figure 4.1 the equivalence points corresponding to equations 9, 10, and 11 are marked x, y, and z, respectively.
[d] The exact numerical solution is of the fourth degree in $[H^+]$ (see Table 3.6). For many practical purposes ($C_T > 10^{-6}$ M), equations 12–14 are sufficiently exact. If $C_T > 10^{-5}$ M, equation 15, and if $C_T \geq 10^{-3}$ M, equations 16 and 17, may be used.
[e] For derivation and further discussion see Sections 3.9 and 4.4.
[f] See Section 3.9.

Table 4.3. Equilibrium Constants for Carbonate Equilibria and the Dissolution of $CaCO_3$ (Calcite)[a]

		\-log K (I = 0)						
Reaction		5°C	10°C	15°C	20°C	25°C	40°C	100°C
$CaCO_3(s)$	$= Ca^{2+} + CO_3^{2-}$	8.35	8.36	8.37	8.39	8.42	8.53	
$CaCO_3(s) + H^+$	$= HCO_3^- + Ca^{2+}$	−2.2	−2.13	−2.06	−1.99	−1.91	−1.69	
$H_2CO_3^*$	$= H^+ + HCO_3^-$	6.52	6.46	6.42	6.38	6.35	6.35	6.45
$CO_2(g) + H_2O$	$= H_2CO_3^*$	1.20	1.27	1.34	1.41	1.47	1.64	1.99
HCO_3^-	$= H^+ + CO_3^{2-}$	10.56	10.49	10.43	10.38	10.33	10.22	10.16
H_2O	$= H^+ + OH^-$	14.73	14.53	14.34	14.16	14.0	13.53	11.27

[a] Reported data for the solubility of $CaCO_3(s)$ (calcite) at 25°C ($I = 0$) vary from $pK_{s0} = 8.34$ to 8.52.

4.2 Dissolved Carbonate Equilibria (Closed System)

Correspondingly, the concentration of $CO_2(aq)$ is nearly identical to the analytical concentration of $H_2CO_3^*$. K_1 is the equilibrium constant known, from direct experimental determination, with a high degree of accuracy.

Representative Equilibrium Diagrams for Fresh Water

Figure 4.1 illustrates the equilibrium distribution of the carbonate solutes as a function of pH (cf. Sections 3.6–3.9). The construction of the double logarithmic diagram has been explained in connection with Figure 3.4. The equations 5, 6, 7, and 8 of Table 4.2 can be drawn graphically as linear asymptotes in different pH ranges. For example, for the equations (see 5 and 6 of Table 4.2)[†]

$$[H_2CO_3^*] = C_T \left(1 + \frac{K_1}{[H^+]} + \frac{K_1 K_2}{[H^+]^2}\right)^{-1} \tag{5}$$

the following relations are valid:

I: $pH < pK_1 < pK_2$; $\log [H_2CO_3^*] = \log C_T$

$d (\log [H_2CO_3^*])/d\, pH = 0$

II: $pK_1 < pH < pK_2$; $\log [H_2CO_3^*] = pK_1 + \log C_T - pH$

$d (\log [H_2CO_3^*])/d\, pH = -1$

III: $pK_1 < pK_2 < pH$; $\log [H_2CO_3^*] = pK_1 + pK_2 + \log C_T - 2\, pH$

$d (\log [H_2CO_3^*])/d\, pH = -2$

These linear asymptotes can be graphed readily; they change their slope from 0 to -1 and from -1 to -2 at the values $pH = pK_1$ and $pH = pK_2$, respectively. Similar considerations apply to the representation of log $[HCO_3^-]$ versus pH (equations 5 and 7, Table 4.2) and of log $[CO_3^{2-}]$ versus pH (equations 5 and 8, Table 4.2).

Furthermore, the α values can be plotted versus pH (Figure 4.2a), and with the help of the α values, the diagrams of $[H_2CO_3^*]$, $[HCO_3^-]$, and $[CO_3^{2-}]$, respectively, versus pH can be sketched readily (Figure 4.2b). In these diagrams the sections having slopes of $+2$ or -2 occur at small concentrations and can often be neglected.

Figure 4.1 illustrates the equilibrium distribution of the carbonate solutes as a function of pH (cf. Sections 3.6–3.9). The pH values of the pure solutions of the acid, the ampholyte, and the base, that is, of pure solutions of

---

[†]Oceanographers usually use ΣCO_2 instead of C_T for the sum of the concentrations of the carbonic species.

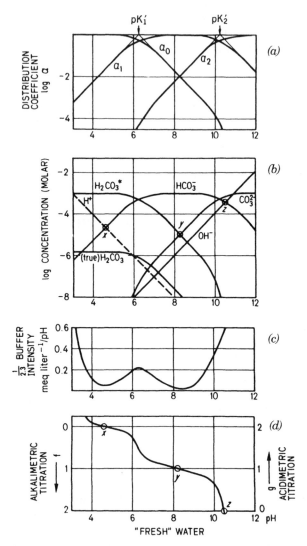

Figure 4.1. Distribution of solute species and buffering in aqueous carbonate systems: fresh water, 25°C. This figure has been constructed under the assumption that $C_T = [H_2CO_3^*] + [HCO_3^-] + [CO_3^{2-}] = $ constant (10^{-3} M). The following equilibrium constants corrected for salt effects have been used: $I = 10^{-3}$, $pK_1' = 6.3$, $pK_2' = 10.25$ (25°C, $p = 1$ atm). The hydration constant K (equation 1 in Table 4.2) was taken to be 630. (a) Ionization fractions as a function of pH. (b) Logarithmic equilibrium diagram for fresh water. Because $[CO_2(aq)] \gg [H_2CO_3]$, $[CO_2\,aq] \simeq [H_2CO_3^*]$. Note that H_2CO_3 is a much stronger acid than $CO_2(aq)$ or $H_2CO_3^*$. Pure H_2CO_3 has a pK value (where $[H_2CO_3] = [HCO_3^-]$) of $pK_{H_2CO_3} \simeq 3.5$. The equivalence points corresponding to pure solutions (C_T molar) of $H_2CO_3^*$, $NaHCO_3$, and Na_2CO_3 are marked x, y, and z, respectively. (c) Buffer intensity is plotted as a function of pH. (d) Alkalimetric or acidimetric titration curve. Note that no pH jump occurs at the equivalence point z, because at this point the buffer intensity caused by high $[OH^-]$ is too large.

4.2 Dissolved Carbonate Equilibria (Closed System)

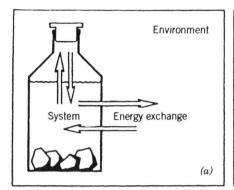
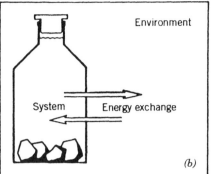
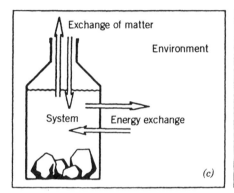
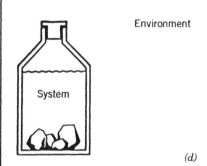

Figure 4.2. Schematic representation of closed (a, b), open (c) and isolated (d) systems. In system (a) a volatile substance can be exchanged between water and the gas phase. The total quantity of matter within the system remains constant. In system (b) the water phase is closed toward the gas phase; no exchange with the gas phase occurs; $H_2CO_3^*$ or NH_3 are treated as nonvolatile species. In the open system (c) exchange of matter with the environment occurs; for example, a water in equilibrium with the atmosphere is characterized by a constant partial pressure of CO_2 (p_{CO_2}). System (d) represents an isolated system. No exchange of matter and energy occurs with the environment. (Metaphorically, the system is like a thermos bottle.)

$H_2CO_3^*$, $NaHCO_3$, and Na_2CO_3, correspond to the equivalence points, x, y, and z in alkalimetric and acidimetric titrations of natural waters; they are defined by the appropriate proton conditions (equations 9–11 of Table 4.2 and equations i–iii of Tableau 4.1a) of $C_T = 10^{-3}$ M ($I = 10^{-3}$, 25°C, $p = 1$ atm) or $C_T = 2.3 \times 10^{-3}$ M (I = seawater conditions, 10°C, $p = 1$ atm) solutions, respectively, of $H_2CO_3^*$, $NaHCO_3$, and Na_2CO_3.

Obviously, natural waters are not closed systems. The idealized model discussed so far is still useful because a natural water sample in the laboratory, for example, during acid–base titration, or waters in groundwater systems or in water supply distribution systems often behave, in first approximation, as in a closed system. Figure 4.2 illustrates metaphoric models for various types of open and closed systems.

Example 4.1. Equivalence Points in Alkalimetric and Acidimetric Titrations: pH of Pure CO_2, $NaHCO_3$, and Na_2CO_3 Solutions Estimate the pH as a function of concentration for pure solutions of CO_2, $NaHCO_3$, and Na_2CO_3, respectively. (Assume a closed system; i.e., treat $H_2CO_3^*$ as a nonvolatile acid.) The answer is given in Figure 4.3 where pH is plotted for the different solutions as a function of log C_T. The numerical calculation of these curves is based on equations 12–14 of Table 4.2, which characterize the appropriate proton conditions (see also equations vii, xiv, and xxi in Tableaux 4.1). The pH values at the equivalence points can readily be estimated from logarithmic equilibrium sketches (Figure 4.1b–d). For example, the proton condition of a pure $NaHCO_3$ solution is given by $[H_2CO_3^*] + [H^+] = [CO_3^{2-}] + [OH^-]$. As Figure 4.3c illustrates, at high concentrations ($C_T > 10^{-3}$ M) the equilibrium is characterized by the simplified condition $[H_2CO_3^*] \simeq [CO_3^{2-}]$. In this concentration range, $NaHCO_3$ solutions maintain a constant pH (ignoring small changes that may result from activity variations). In a more dilute concentration range (10^{-4} M $> C_T > 10^{-7}$ M), the pH of pure $NaHCO_3$ solutions is characterized by the appropriate proton condition $[H_2CO_3^*] \simeq [OH^-]$. As $C_T \to 0$, the electroneutrality condition becomes $[H^+] \simeq [OH^-]$ and neutrality (pH $\simeq$ 7) exists.

Seawater

Figure 4.4 gives an equilibrium diagram for seawater (10°C), which contains, in addition to 2.3 × 10^{-3} M carbonate species (C_T), 4.1 × 10^{-4} M boric acid ($[H_3BO_3] + [B(OH)_4^-]$). The following pK' values (corrected for salt effects) have been used (10°C): $pK_1' = 6.1$, $pK_2' = 9.3$, and $pK_{H_3BO_3}' = 8.8$.

Obviously the diagrams for "fresh water" and for "seawater" are generally similar but differ in certain details, for example, the buffer intensity at high pH, point z. Because of the ionic strength effects, the operational acidity constants are larger for seawater than for fresh waters; that is, the pK' values and thus the pH at the equivalence point—especially at the equivalence point y—are lower for seawater than for fresh water. Seawater contains, in addition to dissolved CO_2, boric acid, H_3BO_3 (representative concentration of total boron = 4.1 × 10^{-4} M). Its presence does not contribute markedly to the buffering of seawater. At the pH of seawater (pH = 8.1), its buffer intensity is near the minimum.

4.3 Aqueous Carbonate System Open to the Atmosphere

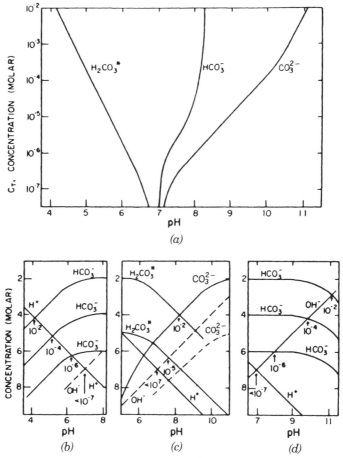

Figure 4.3. pH of pure solutions of $CO_2(H_2CO_3^*)$, $NaHCO_3$, and Na_2CO_3 at various dilutions (see Example 4.1). The curves in (a) can be computed by equations 12–14 of Table 4.2 or with the help of logarithmic equilibrium diagrams. (b–d) sketches for pure CO_2, $NaHCO_3$, and Na_2CO_3 solutions, respectively. For a few concentrations the intersections that characterize the appropriate proton conditions are indicated by arrows.

4.3 AQUEOUS CARBONATE SYSTEM OPEN TO THE ATMOSPHERE

A very simple model showing some of the characteristics of the carbonate system of natural waters is provided by equilibrating pure water with a gas phase (e.g., the atmosphere) containing CO_2 at a constant partial pressure. One may then vary the pH by the addition of strong base or strong acid, thereby keeping the solution in equilibrium with p_{CO_2}. This simple model has its coun-

Tableau 4.1. Closed Carbonate System[a]

a. 10^{-3} M $H_2CO_3^*$ Solution

Components		$H_2CO_3^*$	H^+	log K (25°C) (I = 0)	
Species	$H_2CO_3^*$	1	0	0	(i)
	HCO_3^-	1	−1	−6.35 (log K_1)	(ii)
	CO_3^{2-}	1	−2	−16.68 (log K_1 + log K_2)	(iii)
	OH^-	0	−1	−14.0	(iv)
	H^+	0	1	0	(v)
		$C_T = 10^{-3}$ M	0		(vi)
TOTH = $-[HCO_3^-] - 2[CO_3^{2-}] - [OH^-] + [H^+] = 0$					(vii)

b. 10^{-3} M $NaHCO_3$ Solution

Components		HCO_3^-	H^+	log K	
Species	$H_2CO_3^*$	1	1	6.35 ($-\log K_1$)	(viii)
	HCO_3^-	1	0	0	(ix)
	CO_3^{2-}	1	−1	−10.33 (log K_2)	(x)
	OH^-	0	−1	−14.0	(xi)
	H^+	0	1	0	(xii)
		$C_T = 10^{-3}$ M	0		(xiii)
TOTH = $[H_2CO_3^*] - [CO_3^{2-}] - [OH^-] + [H^+] = 0$					(xiv)

c. 10^{-3} M Na_2CO_3 Solution

Components		CO_3^{2-}	H^+	log K	
Species	$H_2CO_3^*$	1	2	16.68 ($-\log K_1 - \log K_2$)	(xv)
	HCO_3^-	1	1	10.33 ($-\log K_2$)	(xvi)
	CO_3^{2-}	1	0	0	(xvii)
	OH^-	0	−1	−14.0	(xviii)
	H^+	0	1	0	(xix)
		$C_T = 10^{-3}$ M	0		(xx)
TOTH = $2[H_2CO_3^*] + [HCO_3^-] - [OH^+] + [H^+] = 0$					(xxi)

[a] Recall how we read these tableaux. The rows in Tableau 4.1a give the equilibrium expressions as follows:

(ii) $[HCO_3^-] = [H_2CO_3^*][H^+]^{-1} K_1$

(iii) $[CO_3^{2-}] = [H_2CO_3^*][H^+]^{-2} K_1 K_2$

(iv) $[OH^-] = [H^+]^{-1} K_W$

The columns give the mole balances:

(vi) $C_T = [H_2CO_3^*] + [HCO_3^-] + [CO_3^{2-}]$

(vii) TOTH = $-[HCO_3^-] - 2[CO_3^{2-}] - [OH^-] + [H^+] = 0$

4.3 Aqueous Carbonate System Open to the Atmosphere

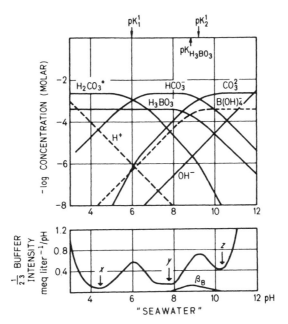

Figure 4.4. (Top) Logarithmic equilibrium diagram for seawater (10°C). Because seawater contains 4.1×10^{-4} M boric acid and borate [H_3BO_3 or $B(OH)_3$ and $B(OH)_4^-$], the distribution of these species is also given. (Bottom) Buffer intensity of seawater. Note that seawater has its minimum buffer intensity in the slightly alkaline pH range (end point y) approximately half a pH unit lower than fresh water. The H_3BO_3–$B(OH)_4^-$ couple does not contribute significantly to the total buffer intensity (β_B for the contribution of aqueous B to the buffer intensity).

terpart in nature when CO_2 reacts with bases of rocks (i.e., with silicates, oxides, or carbonates).

Figure 4.5 shows the distribution of the solute species of such a model. A partial pressure of CO_2 ($p_{CO_2} = 10^{-3.5}$ atm) representative of the atmosphere and equilibrium constants valid at 25°C have been assumed. The equilibrium concentration of the individual carbonate species can be expressed as a function of p_{CO_2} and pH. By combining equation 6, 7, or 8 from Table 4.2 with Henry's law.

As we have seen in Example 3.8, the equilibrium $CO_2(g) + H_2O \rightleftharpoons H_2CO_3^*$ can be expressed by a mass law relationship of the type (Henry's law)

$$[H_2CO_3^*] = K_H \, p_{CO_2} \tag{6}$$

K_H has the units M atm^{-1}; its values for the CO_2 system are given in Table 4.3. The concept of gas–water partitioning equilibria (Henry's law) is discussed more fully in Chapter 5.

One obtains the following relationships that plot linearly in a double loga-

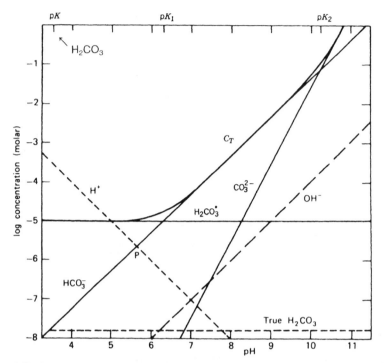

Figure 4.5. Aqueous carbonate equilibrium; constant p_{CO_2}. Water is equilibrated with the atmosphere ($p_{CO_2} = 10^{-3.5}$ atm), and the pH is adjusted with strong base or strong acid. Equations 7–9 with the constants (25°C) $pK_H = 1.5$, $pK_1 = 6.3$, $pK_2 = 10.25$, and pK (hydration of CO_2) $= -2.8$ have been used. The pure CO_2 solution is characterized by the proton condition $[H^+] = [HCO_3^-] + 2[CO_3^{2-}] + [OH^-]$ (see point P) and the equilibrium concentrations $-\log[H^+] = -\log[HCO_3^-] = 5.65$; $\log[CO_2 aq] = -\log[H_2CO_3^*] = 5.0$; $-\log[H_2CO_3] = 7.8$; and $-\log[CO_3^{2-}] = 10.3$.

rithmic diagram. First, the logarithmic expression of equation 6 gives

$$\log [H_2CO_3^*] = \log K_H + \log p_{CO_2} \tag{7a}$$

Using values for K_H (Table 4.3) and the partial pressure of CO_2 in the atmosphere ($\log p_{CO_2} = -3.5$), we find

$$\log [H_2CO_3^*] = -1.5 + (-3.5) = -5.0 \tag{7b}$$

This line can be entered as a horizontal line. At equilibrium with the atmosphere, $[H_2CO_3^*]$ is independent of pH.

Second, $[HCO_3^-]$ is obtained from equation 2 of Table 4.2:

$$[HCO_3^-] = (K_1/[H^+]) [H_2CO_3^*] \tag{8a}$$

4.3 Aqueous Carbonate System Open to the Atmosphere

$$\log [HCO_3^-] = \log K_1 + pH + \log [H_2CO_3^*] \quad (8b)$$

$$= -6.3 + pH - 5.0 = -11.3 + pH \quad (8c)$$

According to equations 8, $d \log [HCO_3^-]/d\, pH = +1$ and $\log [HCO_3^-] = \log [H_2CO_3^*]$ when $pH = pK_1$.

This line in the double logarithmic diagram (equal abscissa and ordinate scales) shows for $\log [HCO_3^-]$ versus pH a slope of $+1$ and the line intercepts at $pH = pK_1$ with $[H_2CO_3^*]$.

Similarly, we can derive for $[CO_3^{2-}]$

$$\log [CO_3^{2-}] = \log (K_2/[H^+]) + \log [HCO_3^-] \quad (9a)$$

$$= \log (K_2 K_1/[H^+]^2) + \log [H_2CO_3^*] \quad (9b)$$

$$= \log K_2 + \log K_1 + 2\, pH + \log [H_2CO_3^*] \quad (9c)$$

$$= -10.3 - 6.3 + 2\, pH - 5.0 \quad (9d)$$

In other words, the line for $\log [CO_3^{2-}]$ versus pH is characterized by a slope of $+2$ and an intersection with the $\log [H_2CO_3^*]$ line when $pK_1 + pK_2 = 2\, pH$ (or $pH = 8.3$).

Figure 4.5 reflects the carbonate species as a function of pH for all natural waters in equilibrium with the atmosphere. There are, of course, some minor variations with temperature and ionic strength.

Example 4.2. "Pristine" Rainwater "Pure" water in equilibrium with the atmosphere is characterized by Tableau 4.2. In Figure 4.5 the proton condition (or the charge balance) is given by equation viii of Tableau 4.2. This condition is fulfilled essentially at point P, where $[H^+] \simeq [HCO_3^-]$, where $pH = 5.65$ (25°C).

Tableau 4.2. Open CO$_2$ System

Components		$CO_2(g)$	H^+	$\log K$ (25°C, $I = 0$)	
Species	$CO_2(g)$	1	0	0	(i)
	$H_2CO_3^*$	1	0	-1.5	(ii)
	HCO_3^-	1	-1	-7.8	(iii)
	CO_3^{2-}	1	-2	-18.1	(iv)
	OH^-	0	-1	-14.0	(v)
	H^+	0	1	0	(vi)
		$p_{CO_2} = 10^{-3.5}$ atm	0		(vii)
TOTH = $[H^+] - [OH^-] - [HCO_3^-] - 2\,[CO_3^{2-}] = 0$					(viii)

162 Dissolved Carbon Dioxide

If the rainwater comes into contact with acids or bases (dissolution of minerals), the pH will vary, and the equilibria given are, of course, valid for the entire pH range. For every pH the composition can be read from the graph. The tableau has to be enlarged with a component for the base (e.g., Na^+ for NaOH, or Ca^{2+} from dissolved $CaCO_3$) or a component for the acid (e.g., Cl^- for HCl, or SO_4^{2-} for H_2SO_4).

The proton (or electroneutrality) condition is then given by

$$TOTH = [H^+] - [OH^-] - [HCO_3^-] - 2[CO_3^{2-}]$$
$$= -C_B + C_A \tag{10}$$

where C_B corresponds to a "base" cation and C_A corresponds to an "acid" anion (anion of a strong acid). (For example, $C_B = [Na^+]$ and $C_A = [Cl^-]$.)

The rows in Tableau 4.2 give the equilibrium expressions as follows:

(ii) $[H_2CO_3^*] = p_{CO_2} K_H$

(iii) $[HCO_3^-] = p_{CO_2} K_H K_1 [H^+]^{-1}$

(iv) $[CO_3^{2-}] = p_{CO_2} K_H K_1 K_2 [H^+]^{-2}$

(v) $[OH^-] = [H^+]^{-1} K_W$

Example 4.3. Seawater in Equilibrium with the Atmosphere ($p_{CO_2} = 10^{-3.5}$ atm) Construct a double logarithmic diagram—similar to that of Figure 4.5 for seawater (35‰ salinity). At 20°C the following constant can be used:

$$^cK_H = [H_2CO_3^*] p_{CO_2}^{-1} = 10^{-1.47} \tag{i}$$

$$K_1' = \{H^+\} [HCO_{3T}^-]/[H_2CO_3^*] = 10^{-6.03} \tag{ii}$$

$$K_2' = \{H^+\} [CO_{3T}^{2-}]/[HCO_{3T}^-] = 10^{-9.18} \tag{iii}$$

The constants given reflect that at high salt concentrations complexes of HCO_3^- and CO_3^{2-} with cations of the medium occur. Thus the concentrations with a subscript T are defined as follows:

$$[HCO_3^-]_T = [HCO_3^-] + [NaHCO_3^0] + [MgHCO_3^+] + [CaHCO_3^+]$$
$$[CO_3^{2-}]_T = [CO_3^{2-}] + [MgCO_3^0] + [CaCO_3^0] + [NaCO_3^-]$$

Figure 4.6 gives the equilibrium diagram.

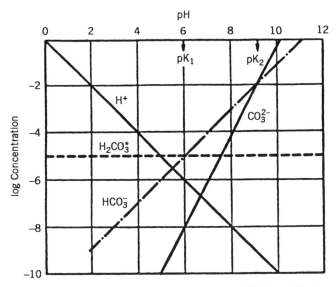

Figure 4.6. Carbonate species of seawater (20°C) in equilibrium with the atmosphere. A comparison with Figure 4.5 shows the influence of the salt concentrations on the equilibrium distribution.

4.4 ALKALINITY AND ACIDITY, NEUTRALIZING CAPACITIES

One has to distinguish between the H^+ ion concentration (or activity) as an intensity factor and the availability of H^+, the H^+ ion reservoir as given by the H-acidity or the deficiency of H^+ ions, or the alkalinity. Alkalinity and acidity are very important concepts; although there are different ways to define these capacity factors, all definitions essentially relate to the proton condition at a given reference level. For the carbonate system, alkalinity [Alk] refers conceptually to the proton condition with reference to $H_2CO_3^*$, H_2O:

$$[Alk] = [HCO_3^-] + 2[CO_3^{2-}] + [OH^-] - [H^+] \tag{11a}$$

$$-TOTH = [HCO_3^-] + 2[CO_3^{2-}] + [OH^-] - [H^+] = [Alk] \tag{11b}$$

(cf. equation vii in Tableau 4.1A and equation viii in Tableau 4.2); that is, it is the equivalent sum of bases that have one or two protons less than the reference species $H_2CO_3^*$ and H_2O minus the H^+ (or H_3O^+) that has one proton more than H_2O. Equations 11 relate also to the operational definition of alkalinity. The acidimetric titration of a carbonate bearing water to the appropriate equivalence point (i.e., the zero proton condition) represents an operational

164   Dissolved Carbon Dioxide

procedure for determining alkalinity, that is, the equivalent sum of the bases that are titratable with strong acid. Thus alkalinity represents the acid-neutralizing capacity, ANC, of an aqueous system. An operational definition can then be given as follows: for solutions that contain no protolysis system other than aqueous carbonate, alkalinity is a measure of the quantity of strong acid required to attain a pH equal to that of a C_T molar solution of $H_2CO_3^*$ (corresponding to point X in Figure 4.1b).

[Alk] or [ANC] is expressed in M (moles of protons per liter) or in equivalents per liter.

Example 4.4. Equilibrium Composition of Open Systems for a Given Alkalinity Characterize the equilibrium composition of a groundwater in equilibrium with a CO_2 partial pressure of 10^{-2} atm having an alkalinity of 6×10^{-3} M (5°C). We take the equilibrium constants from Table 4.3 and correct these constants for an approximate ionic strength of 1.8×10^{-2} M. Tableau 4.3 characterizes the equilibrium information.

Note that the "input" for the [Alk] is given by TOTH = -6×10^{-3} M.

The composition of the groundwater is $pH_2CO_3^* = 3.2$, $pHCO_3^- = 2.2$, $pCO_3^{2-} = 5.2$, $pOH^- = 7.2$, and pH = 7.4.

If the aquatic system contains weak bases (or acids) other than CO_2 and water, we can extend equation 1 to include these other proton acceptors. For example, in the presence of boric acid–borate we can extend our reference level to H_3BO_3 (in addition to $H_2CO_3^*$ and H_2O) and write

$$[Alk] = [HCO_3^-] + 2[CO_3^{2-}] + [B(OH)_4^-] + [OH^-] - [H^+]$$
$$= -TOTH \tag{12}$$

Tableau 4.3. Equilibrium Composition for Given Alkalinity

Components	$CO_2(g)$	H^+	log K (5°C, $I = 1.8 \times 10^{-2}$)
Species p_{CO_2}	1	0	0
$H_2CO_3^*$	1	0	−1.2
HCO_3^-	1	−1	−7.66
CO_3^{2-}	1	−2	−18.05
OH^-	0	−1	−14.61
H^+	0	1	0
	$p_{CO_2} = 10^{-2}$ atm	-6×10^{-3}	

TOTH = $-[HCO_3^-] - 2[CO_3^{2-}] - [OH^-] + [H^+] = -6 \times 10^{-3}$ M

4.4 Alkalinity and Acidity, Neutralizing Capacities

In most natural waters, acid–base systems other than aqueous carbonate can often be neglected; but in special cases one might wish to include them

$$[\text{Alk}] = [\text{HCO}_3^-] + 2[\text{CO}_3^{2-}] + [\text{NH}_3] + [\text{HS}^-] + 2[\text{S}^{2-}]$$
$$+ [\text{H}_3\text{SiO}_4^-] + 2[\text{H}_2\text{SiO}_4^{2-}] + [\text{B(OH)}_4^-] + [\text{Org}^-] + [\text{HPO}_4^{2-}]$$
$$+ 2[\text{PO}_4^{3-}] - [\text{H}_3\text{PO}_4] + [\text{OH}^-] - [\text{H}^+] \qquad (13)$$

The reference (zero proton condition species) includes now, in addition to H_2CO_3^* and H_2O, the species NH_4^+, H_2S, H_4SiO_4, H_3BO_3, Horg (a collective term for organic acids) and H_2PO_4^-. Since these noncarbonate species are usually present in concentrations much smaller than the carbonate species, we should select as reference species those that will not be titrated at the CO_2 (H_2CO_3^*) equivalence point; thus $\text{H}_2\text{PO}_4^{2-}$ (p$K \simeq 7$) is selected rather than H_3PO_4 (p$K \simeq 2$).

Alternative Definition of Alkalinity

Consider a charge balance for a typical natural water ([H$^+$] and [OH$^-$] are negligible):

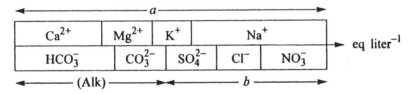

We realize that [Alk] and [H-Acy] also can be expressed by a charge balance—the equivalent sum of conservative cations, less the sum of conservative anions ([Alk] = $a - b$). The conservative cations are the base cations of the strong bases Ca(OH)$_2$, KOH, and the like; the conservative anions are those that are the conjugate bases of strong acids (SO$_4^{2-}$, NO$_3^-$, and Cl$^-$).

$$[\text{Alk}] = [\text{HCO}_3^-] + 2[\text{CO}_3^{2-}] + [\text{OH}^-] - [\text{H}^+]$$
$$= [\text{Na}^+] + [\text{K}^+] + 2[\text{Ca}^{2+}] + 2[\text{Mg}^{2+}] - [\text{Cl}^-]$$
$$- 2[\text{SO}_4^{2-}] - [\text{NO}_3^-] \qquad (14)$$

The [H-Acy] for this particular water, obviously negative, is defined ([H-Acy] = $b - a$) as

$$[\text{H-Acy}] = [\text{Cl}^-] + 2[\text{SO}_4^{2-}] + [\text{NO}_3^-] - [\text{Na}^+]$$
$$- [\text{K}^+] - 2[\text{Ca}^{2+}] - 2[\text{Mg}^{2+}] \qquad (15)$$

166 Dissolved Carbon Dioxide

A simple accounting can be made: every base cationic charge unit (e.g., Ca^{2+} or K^+) that is removed from the water by whatever process is equivalent to a proton added to the water, and every conservative anionic charge unit (from anions of strong acids—NO_3^-, SO_4^{2-}, or Cl^-) removed from the water corresponds to a proton removed from the water, or generally,

$$\Sigma\Delta \text{ [base cations]} - \Sigma\Delta \text{ [conservative anions]} = +\Delta[\text{Alk}] = -\Delta[\text{H-Acy}] \quad (16)$$

In a similar way, [H-Acy], for example, in an acid rain, can be defined as

$$[\text{H-Acy}] = [Cl^-] + 2[SO_4^{2-}] + [NO_3^-] - [Na^+] - [K^+]$$
$$- 2[Ca^{2+}] - 2[Mg^{2+}] \quad (17)$$

H-Acidity If a natural water contains more protons than that given by the zero proton condition (or the H_2CO_3 equivalent point), then this water contains H-acidity (H-Acy) also called mineral acidity:

$$[\text{H-Acy}] = [H^+] - [OH^-] [HCO_3^-] - 2[CO_3^{2-}] \quad (18)$$

Obviously,

$$[\text{H-Acy}] = -[\text{Alk}]$$

and [H-Acy] can be determined by an alkalimetric titration of the (acid) solution to the H_2CO_3 end point. The concentrations in parentheses on the right-hand side of equation 18 can often be neglected (see Figure 4.7). As this figure shows, there are other capacity factors that can be defined with reference to the zero proton conditions defined for a $NaHCO_3$ solution or for a Na_2CO_3 solution (equations xiv and xxi of Tableaux 4.1).

The equations given in Figure 4.7 are of analytical value because they represent rigorous conceptual definitions of the acid-neutralizing and the base-neutralizing capacities of carbonate systems. As discussed in Section 3.9, the definitions of alkalinity and acidity algebraically express the proton excess or proton deficiency of the system with respect to a reference proton level (equivalence point).[†] Alkalimetric and acidimetric titrations to the equivalence points $f = 0$ and $f = 1$ give inherently accurate values of base- and acid-neutralizing capacity. As Figure 4.7 and equations 11–14 of this figure suggest, C_T $(= [H_2CO_3^*] + [HCO_3^-] + [CO_3^{2-}])$ can also be obtained from an acidimetric

---

[†]The formulas for the species may include medium ions. In other words, $[HCO_3^-]$ includes the concentration of free HCO_3^- as well as the concentration of complex-bound HCO_3^-; that is, $[HCO_3^-] = [\text{true } HCO_3^-] + [NaHCO_3] + [MgHCO_3^+] \cdots$. Similarly, $[CO_3^{2-}] = [\text{true } CO_3^{2-}] + [MgCO_3] + [NaCO_3^-]$.

4.4 Alkalinity and Acidity, Neutralizing Capacities

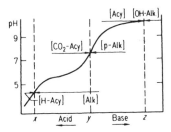

Figure 4.7. Conservative quantities: alkalinity and acidity as acid-neutralizing and base-neutralizing capacity. These parameters can be determined by acidimetric and alkalimetric titration to the appropriate end points. The equations given below define the various capacity factors of an aqueous carbonate system rigorously. If the solution contains protolytic systems other than that of aqueous carbonate, these equations have to be corrected; for example, in the presence of borate one has to add $[B(OH)_4^-]$ to the right-hand side of equation 3 and $[H_3BO_3]$ to the right-hand side of equation 6.

Term	End Point	Definition	
		I. Acid-Neutralizing Capacity (ANC)	
Caustic alkalinity	$f = 2$	$[OH^- \text{-Alk}] = [OH^-] - [HCO_3^-] - 2[H_2CO_3^*] - [H^+]$	(1)
p-Alkalinity	$f = 1$	$[p\text{-Alk}] = [OH^-] + [CO_3^{2-}] - [H_2CO_3^*] - [H^+]$	(2)
Alkalinity	$f = 0$	$[\text{Alk}] = [HCO_3^-] + 2[CO_3^{2-}] + [OH^-] - [H^+]$	(3)
		II. Base-Neutralizing Capacity (BNC)	
Mineral acidity	$f = 0$	$[H\text{-Acy}] = [H^+] - [HCO_3^-] - 2[CO_3^{2-}] - [OH^-]$	(4)
CO_2-acidity	$f = 1$	$[CO_2\text{-Acy}] = [H_2CO_3^*] + [H^+] - [CO_3^{2-}] - [OH^-]$	(5)
Acidity	$f = 2$	$[\text{Acy}] = 2[H_2CO_3^*] + [HCO_3^-] + [H^+] - [OH^-]$	(6)

III. Combinations

$f = 2 - g$	(7)	$[\text{Alk}] + [CO_2\text{-Acy}] = C_T$	(11)
$[\text{Alk}] + [H\text{-Acy}] = 0$	(8)	$[\text{Alk}] + [\text{Acy}] = 2C_T$	(12)
$[\text{Acy}] + [OH\text{-Alk}] = 0$	(9)	$[\text{Alk}] - [p\text{-Alk}] = C_T$	(13)
$[p\text{-Alk}] + [CO_2\text{-Acy}] = 0$	(10)	$[CO_2\text{-Acy}] - [H\text{-Acy}] = C_T$	(14)

or alkalimetric titration (difference between the equivalence points $f = 0$ and $f = 1$).

The pH values at the respective equivalence points (around 4.5 and 10.3) for titrations of alkalinity and acidity represent approximate thresholds beyond which most life processes in natural waters are seriously impaired. Thus alkalinity and acidity are convenient measures for estimating the maximum capacity of a natural water to neutralize acidic and caustic wastes without permitting extreme disturbance of biological activities in the water.

If mineral acid is added to a natural water beyond equivalence point x ($f = 0$) (compare Figures 4.1 and 4.7), the H^+ added will remain as such in solution. Such a water is said to contain *mineral acidity*. Correspondingly, a water with a pH higher than point z ($f = 2$) contains *caustic alkalinity*.

Titration to the intermediate equivalence point y ($f = 1$) is a measure of the CO_2 acidity in an alkalimetric titration and of *p*-alkalinity in an acidimetric titration.

Conservative Properties

Although individual concentrations or activities, such as $[H_2CO_3^*]$, and pH are dependent on pressure and temperature, ANC, BNC, and C_T are conservative properties that are pressure and temperature independent. (Akalinity, acidity, and C_T must be expressed as a concentration, e.g., molarity or molality.)[†]

Furthermore, these conservative quantities remain constant for selected changes in the chemical composition. The case of the addition or removal of dissolved carbon dioxide is of particular interest in natural waters. Any increase in carbon dioxide or, more rigorously, any increase in $[H_2CO_3^*]$, $dC_{H_2CO_3^*}$, increases both the acidity of the system and C_T, the total concentration of dissolved carbonic species. Unlike the case for the addition of strong acid, however, alkalinity remains unaffected by increases or decreases in $[H_2CO_3^*]$. The fact that alkalinity is unaffected by CO_2 can be understood if we consider that alkalinity measures the proton deficiency with respect to the reference proton level $H_2CO_3^*$-H_2O. An analogous argument considers that the addition of CO_2 does not affect the net charge balance (which inherently defines alkalinity) of the solution. It can be shown in a similar way that acidity remains unaffected by the addition or removal of $CaCO_3(s)$ or $Na_2CO_3(s)$. Acidity is thus a valuable capacity parameter for solutions in equilibrium with calcite. C_T, on the other hand, remains unchanged in a closed system upon addition of strong acid or strong base. Table 4.4 summarizes the effects of various chemical

[†]In seawater, these capacity factors are rigorously independent of pressure and temperature if expressed in mol kg^{-1} seawater.

Table 4.4. Change in Capacity Parameters as a Result of Chemical Changes[a]

Capacity	Addition of Molar Increments of				
	C_B	C_A	$H_2CO_3^*$ (CO_2)	$NaHCO_3$	$CaCO_3$ or Na_2CO_3
C_T, M	0	0	+1	+1	+1
[Alk] (eq liter^{-1})	+1	−1	0	+1	+2
[Acy] (eq liter^{-1})	−1	+1	+2	+1	0
[CO_2-Acy]	−1	+1	+1	0	−1
[*p*-Alk]	+1	−1	−1	0	+1
[H-Acy]	−1	+1	0	−1	−2
[OH-Alk]	+1	−1	−2	−1	0

[a]Examples: $dC_B/dC_T = 0$; $d[H_2CO_3^*]/d[Acy] = +2$; $d[NaHCO_3]/d[Alk] = +1$.

4.4 Alkalinity and Acidity, Neutralizing Capacities

changes upon the capacity parameters. For each chemical change, one capacity parameter that remains independent of this change can be found.

Alkalimetric and Acidimetric Titration Curves

The alkalimetric titration of $H_2CO_3^*$ with a strong base, C_B, is given by

$$C_B = [HCO_3^-] + 2[CO_3^{2-}] + [OH^-] - [H^+] = [\text{Alk}]$$
$$= C_T(\alpha_1 + 2\alpha_2) + K_W/[H^+] - [H^+] \quad (19)$$

or the titration of a Na_2CO_3 solution with a strong acid, C_A, is given by

$$C_A = 2[H_2CO_3^*] + [HCO_3^-] + [H^+] - [OH^-] = [\text{Acy}]$$
$$= C_T(2\alpha_0 + \alpha_1) + [H^+] - K_W/[H^+] \quad (20)$$

and can be calculated as a function of

$$f = \frac{C_B}{C_T} \text{ versus pH}$$

or

$$g = (2 - f) = \frac{C_A}{C_T} \text{ versus pH} \quad \text{(see Figure 4.1d).}$$

At any point of the titration, all the species in equilibrium with each other can be computed (Figure 4.8).

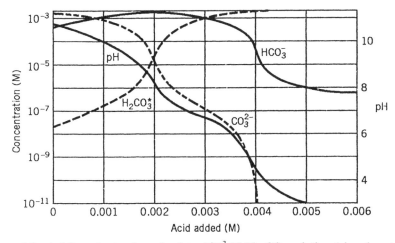

Figure 4.8. Acidimetric titration of a 2×10^{-3} M Na_2CO_3 solution (closed system). Compare Tableau 4.1c.

Example 4.5. pH of Solutions at Given p_{CO_2} Estimate the pH of the following solutions that have been equilibrated with the atmosphere ($p_{CO_2} = 10^{-3.5}$ atm):

10^{-3} M KOH
10^{-3} M NaHCO$_3$
5×10^{-4} M Na$_2$CO$_3$
5×10^{-4} M MgO

All the solutions have the same alkalinity; [Alk] = 10^{-3} eq liter^{-1}. They must all have the same pH at equilibrium with the same p_{CO_2} (apart from essentially negligible differences because of ionic strength). We use the constants given in Table 4.3. From

$$[\text{Alk}] = [\text{HCO}_3^-] + 2[\text{CO}_3^{2-}] + [\text{OH}^-] - [\text{H}^+] \tag{i}$$

$$= C_T(\alpha_1 + 2\alpha_2) + [\text{OH}^-] - [\text{H}^+] \tag{ii}$$

we can express C_T as a function of p_{CO_2} by considering that

$$[\text{H}_2\text{CO}_3^*] = C_T\alpha_0 = K_H p_{CO_2} \tag{iii}$$

or

$$C_T = \frac{K_H p_{CO_2}}{\alpha_0} \tag{iv}$$

Thus

$$\text{Alk} = \frac{K_H p_{CO_2}}{\alpha_0}(\alpha_1 + 2\alpha_2) + K_W/[\text{H}^+] - [\text{H}^+] \tag{v}$$

According to equation v, Alk is implicitly a function of pH. [By having a graph of log α versus pH values (Figure 4.1a), equation v can be solved by trial and error: pH = 8.3.)

We may also compute a general diagram giving H-acidity and alkalinity as a function of pH for a carbonate system in equilibrium with the atmosphere (Figure 4.9).

Example 4.6. Mixing of Waters The effluent of an acid lake [H-Acy] = 10^{-5} M, pH = 5, mixes in equal proportions with a river of pH = 7.5, [Alk] = 2×10^{-4} eq liter^{-1}. The mixed water remains in equilibrium with the atmosphere (10°C). Calculate the pH of the mixed water.

$$[\text{Alk}] = \tfrac{1}{2}(2 \times 10^{-4} - 10^{-5}) = 9.5 \times 10^{-5} \text{ eq liter}^{-1}$$

4.4 Alkalinity and Acidity, Neutralizing Capacities

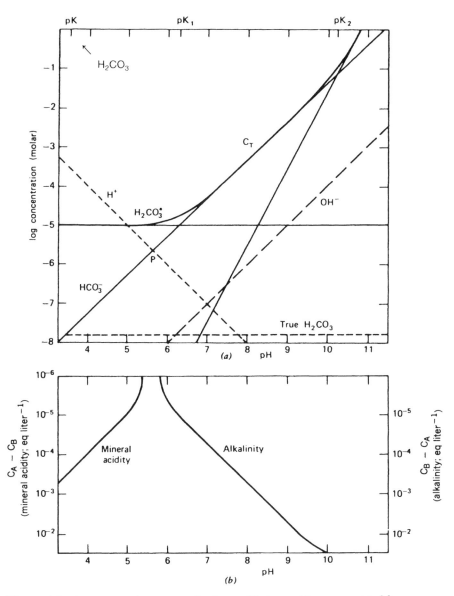

Figure 4.9. Aqueous carbonate species in equilibrium with $p_{CO_2} = 10^{-3.5}$ atm (see Figure 4.5). At pH values different from that of a pure CO_2 solution, the solution contains either alkalinity (Alk) or mineral acidity (H-Acy).

In this pH range, [Alk] ≃ [HCO$_3^-$]. Thus the pH is given by $\{H^+\}$ [HCO$_3^-$]/p_{CO_2} = $10^{-7.71}$ (this constant is valid for 10°C and I = 3 × 10^{-4}). $\{H^+\}$ = 6.5 × 10^{-8}; pH = 7.2.

4.5 ALKALINITY CHANGES

Obviously, alkalinity is also affected by all the processes that yield or consume H^+ or OH^- in stoichiometric equations, for example, the oxygenation of soluble ferrous iron to ferric oxide, $4Fe^{2+} + O_2 + 4H_2O \rightleftharpoons 2Fe_2O_3(s) + 8H^+$, decreases alkalinity, while the reduction of $MnO_2(s)$ by CH_2O, $2MnO_2(s) + CH_2O + 4H^+ = CO_2 + 3H_2O + 2Mn^{2+}$, increases alkalinity. A few other examples are given in Table 4.5. Of particular interest are the alkalinity changes caused by photosynthesis and respiration.

Photosynthesis and Respiration As we have seen, the addition or removal of CO_2 has no effect on alkalinity. This would be true for the photosynthesis process only if it were not accompanied by the assimilation of ions such as NO_3^-, NH_4^+, and HPO_4^{2-}. Since alkalinity is associated with charge balance, such assimilation processes must be accompanied by the uptake of H^+ or OH^- (or release of OH^- or H^+), that is, by alkalinity changes. Thus the photosynthetic assimilation of NH_4^+ causes the uptake of OH^- or the release of H^+ ions (Table 4.5). Similarly, alkalinity increases as a result of photosynthetic NO_3^- assimilation; conversely, the aerobic bacterial decomposition of biota to NO_3^- is accompanied by a decrease in alkalinity. Such processes occurring in land ecosystems are often not without influence on the pH and alkalinity of the adjoining aquatic ecosystems.

Decrease in alkalinity (acidification and cation depletion in soils) occurs whenever the production of organic matter (assimilation of NH_4^+) is larger than the decomposition. This takes place, for example, when peat bogs or forest peats are formed; these systems are very acidic. The harvest of crops on agricultural and forest land often causes discrepancies between production and decomposition.[†]

Example 4.7. Decrease in pH from Aerobic Respiration What is the pH change resulting from the aerobic decomposition of organic matter—6 µg organic carbon—in 1 cm³ of interstitial lake water (10°C), which initially has an alkalinity of 1.2 × 10^{-3} eq liter^{-1}, a pH of 6.90, and an ionic strength of 3 × 10^{-3}? The respiration releases NH_4^+.

Lacking other information, we may assume that the reaction occurs with the stoichiometry given in equation 1c of Table 4.5. The pH change may be estimated in two steps: (1) from the CO_2 increase at constant alkalinity and (2)

[†]For a discussion of alkalinity changes resulting from respiration *and* $CaCO_3$ dissolution in deep oceans, see C. Chen, *Science* (1978), **201**, 735.

Table 4.5. Processes Affecting Alkalinity

Process	Alkalinity Change for Forward Reaction
Photosynthesis and respiration:	
(1a) $nCO_2 + nH_2O \underset{\text{respir.}}{\overset{\text{photos.}}{\rightleftharpoons}} (CH_2O)_n + nO_2$	No change
(1b) $106CO_2 + 16NO_3^- + HPO_4^{2-} + 122H_2O + 18H^+ \underset{\text{respir.}}{\overset{\text{photos.}}{\rightleftharpoons}} \{C_{106}H_{263}O_{110}N_{16}P_1\}_{\text{"algae"}} + 138O_2$	Increase
(1c) $106CO_2 + 16NH_4^+ + HPO_4^{2-} + 108H_2O \underset{\text{respir.}}{\overset{\text{photos.}}{\rightleftharpoons}} \{C_{106}H_{263}O_{110}N_{16}P_1\} + 107O_2 + 14H^+$	Decrease
Nitrification:	
(2) $NH_4^+ + 2O_2 \longrightarrow NO_3^- + H_2O + 2H^+$	Decrease
Denitrification:	
(3) $5CH_2O + 4NO_3^- + 4H^+ \longrightarrow 5CO_2 + 2N_2 + 7H_2O$	Increase
Sulfide oxidation:	
(4a) $HS^- + 2O_2 \longrightarrow SO_4^{2-} + H^+$	Decrease
(4b) $FeS_2(s) + \frac{15}{4}O_2 + 3\frac{1}{2}H_2O \longrightarrow Fe(OH)_3(s) + 4H^+ + 2SO_4^{2-}$ pyrite	Decrease
Sulfate reduction:	
(5) $SO_4^{2-} + 2CH_2O + H^+ \longrightarrow 2CO_2 + HS^- + H_2O$	Increase
CaCO$_3$ dissolution:	
(6) $CaCO_3 + CO_2 + H_2O \rightleftharpoons Ca^{2+} + 2HCO_3^-$	Increase

174   Dissolved Carbon Dioxide

from the alkalinity change (from the uptake of H^+) at constant C_T. We apply the equation $[Alk] = C_T(\alpha_1 + 2\alpha_2) + [OH^-] - [H^+]$, and use the equilibrium constants (10°C) corrected for ionic strength effects, $pK_1' = 6.43$ and $pK_2' = 10.39$. At the initial pH, $\alpha_1 = 0.747$ and α_2 is negligible; thus $C_T = 1.61 \times 10^{-3}$ M. After addition of the CO_2, $C_T = 2.11 \times 10^{-3}$ M, $\alpha_1 = 0.569$, and the new pH = 6.55. Then, at this C_T, alkalinity is changed by $\Delta[Alk] = 5 \times 10^{-4}$ ($\frac{14}{106}$) $= 6.6 \times 10^{-5}$ eq liter^{-1}. Using the same equation again, with $[Alk] = 1.266 \times 10^{-3}$ eq liter^{-1}, $\alpha_1 \simeq [Alk]/C_T = 0.600$, which corresponds to pH = 6.6.

Example 4.8. pH Change Resulting from Photosynthetic CO_2 Assimilation   As a result of photosynthesis with NO_3^- assimilation, a surface water with an alkalinity of 8.5×10^{-4} eq liter^{-1} showed within a 3-h period a pH variation from 9.0 to 9.5. What is the rate of net CO_2 fixation? [Assume a closed aqueous system, that is, no exchange of CO_2 with the atmosphere and no deposition of $CaCO_3$ (25°C), $pK_1 = 6.3$, $pK_2 = 10.2$.]

The pH change results (1) from CO_2 uptake and (2) from an uptake of 18 mol of H^+ per 106 mol of CO_2 assimilated (equation 1b of Table 4.5). Initially, $[Alk]_{pH=9} = C_T(\alpha_1 + 2\alpha_2)_{pH=9} + [OH^-]$, and we obtain $C_T = 7.94 \times 10^{-4}$ M.

For the conditions at pH = 9.5 we have two independent equations:

$$[Alk]_{pH=9.5} = [Alk]_{pH=9} + \tfrac{18}{106}\Delta C_T \tag{i}$$

and

$$[Alk]_{pH=9.5} - [OH^-] = (C_T - \Delta C_T)(\alpha_1 + 2\alpha_2)_{pH=9.5} \tag{ii}$$

Solving, $\Delta C_T = 7.9 \times 10^{-5}$ M and the alkalinity change is $+1.3 \times 10^{-5}$. Its effect is negligible; nearly the same result would have been obtained by considering the CO_2 change only.

In a more careful analysis, it was detected that the alkalinity actually decreased by 3×10^{-5} eq liter^{-1} as a result of $CaCO_3$ precipitation. Considering the equation

$$[Alk]_{final} = [Alk]_{pH=9.5} - 3 \times 10^{-5} \tag{iii}$$

we obtain $[Alk]_{final} = 8.35 \times 10^{-4}$ eq liter^{-1}. At pH = 9.5, this alkalinity corresponds to a $C_{T,final} = 6.9 \times 10^{-4}$ M. Because

$$C_{T,final} = C_{T,pH=9} - \Delta C_{T,photosynthesis} - 1.5 \times 10^{-5} \tag{iv}$$

we calculate $\Delta C_{T,photosynthesis} = 8.9 \times 10^{-5}$ M or an hourly loss of approximately 3×10^{-5} mol CO_2 liter^{-1}.

4.5 Alkalinity Changes

Example 4.9. pH of Seawater Calculate the pH of seawater at 15°C in equilibrium with the atmosphere ($p_{CO_2} = 3.5 \times 10^{-4}$ atm). The seawater has an alkalinity of 2.47×10^{-3} eq liter^{-1} and contains a total boron concentration B_T of 4.1×10^{-4} M. Constants corrected for ionic strength effects of seawater at 15°C are $^cK_H = 4.8 \times 10^{-2}$ M atm^{-1}; $K_1 = 8.8 \times 10^{-7}$, $K_2 = 5.6 \times 10^{-10}$; the acidity constant for H_3BO_3 is $K' = 1.6 \times 10^{-9}$; $K'_W = 2.0 \times 10^{-14}$.

pH for a given alkalinity is defined by

$$[\text{Alk}] = [HCO_3^-] + 2[CO_3^{2-}] + [B(OH)_4^-] + [OH^-] - [H^+] \quad \text{(i)}$$

$$= C_T(\alpha_1 + 2\alpha_2) + B_T\alpha_{B^-} + [OH^-] - [H^+] \quad \text{(ii)}$$

$$= \frac{K_H p_{CO_2}}{\alpha_0}(\alpha_1 + 2\alpha_2) + B_T\alpha_{B^-} + [OH^-] - [H^+] \quad \text{(iii)}$$

where

$$B_T = [H_3BO_3] + [B(OH)_4^-] \quad \text{and} \quad \alpha_{B^-} = \frac{[B(OH)_4^-]}{B_T}$$

One approach would be to solve equation iii by trial and error. For every assumed $[H^+]$ we can obtain the corresponding α values (either from graphs that plot α values versus pH or with the help of a programmable calculator or minicomputer). As a first approximation we will consider the first term in equation iii only, neglecting $B_T\alpha_{B^-}$, $[OH^-]$, and $[H^+]$. We can determine later whether the other terms are indeed negligible.

It is perhaps easier to construct a double logarithmic diagram of the equilibrium distribution of $[H_2CO_3^*]$, $[HCO_3^-]$, $[CO_3^{2-}]$, $[H_3BO_3]$, $[B(OH)_4^-]$, and $[OH^-]$—similar to Figure 4.6—and to see at what pH the condition $[\text{Alk}] = [HCO_3^-] + 2[CO_3^{2-}] + [B(OH)_4^-] = 2.47 \times 10^{-3}$ eq liter^{-1} is fulfilled. Tableau 4.4 characterizes the problem.

A computer calculation for the conditions in Tableau 4.4 gives the results in Table 4.6.

Example 4.10. pH Change Resulting from p_{CO_2} Increase The partial pressure of CO_2 in the atmosphere is increasing, mostly as a result of the release of CO_2 from fossil fuel burning, and will double within the next 40–70 years. What is the pH of surface seawater (as specified in Example 4.9) in equilibrium with a p_{CO_2} of 7.1×10^{-4} atm?

Because the sea is oversaturated with regard to solid $CaCO_3$, it appears justified to disregard any heterogeneous reaction of the seawater with $CaCO_3$; thus we can assume that the alkalinity will remain contant. As before, we can apply equation iii from Example 4.9. A pH = 7.87 is obtained. As Table 4.6 illustrates, the doubling of p_{CO_2} results under equilibrium conditions in a lowering of pH by 0.27 pH units or an increase in $[H^+]$ by a factor of 1.9.

Dissolved Carbon Dioxide

Tableau 4.4. Seawater (15°C)

Components		H_3BO_3	$CO_2(g)$	H^+	log K
Species	$CO_2(g)$	0	1	0	0
	$H_2CO_3^*$	0	1	0	-1.32
	HCO_3^-	0	1	-1	-7.37
	CO_3^{2-}	0	1	-2	-16.62
	H_3BO_3	1	0	0	0
	$B(OH)_4^-$	1	0	-1	-8.8
	OH^-	0	0	-1	-13.7
	H^+	0	0	1	0
		$B_T = 4.1 \times 10^{-4}$	$p_{CO_2} = 3.55 \times 10^{-4}$ atm or $p_{CO_2} = 7.1 \times 10^{-4}$ atm	TOTH $= -2.47 \times 10^{-3}$	

Table 4.6. Change in Seawater Composition Resulting from the Doubling of Atmospheric CO_2 (compare Figure 15.17)

		$p_{CO_2} = 3.55 \times 10^{-4}$ atm	$p_{CO_2} = 7.1 \times 10^{-4}$ atm
$-\log$ conc (M)	$H_2CO_3^*$	4.77	4.47
	HCO_3^-	2.68	2.65
	CO_3^{2-}	3.80	4.04
	H_3BO_3	3.47	3.44
	$B(OH)_4^-$	4.14	4.37
	OH^-	5.62	5.83
	H^+	8.138	7.87

Capacity Diagrams

Graphs using variables with conservative properties (C_T, [Alk], or [Acy]) as coordinates can be used expediently to show contours of pH, [$H_2CO_3^*$], [HCO_3^-], and so on. On such diagrams, as shown by Deffeyes (1965), the addition or removal of base, acid, CO_2, HCO_3^-, or CO_3^{2-} is a vector property. These graphs can be used to facilitate equilibrium calculations.

Such a diagram is given in Figure 4.10. In Figure 4.10 [Alk] is plotted as a function of C_T. The construction of the diagram can be derived readily from

4.5 Alkalinity Changes

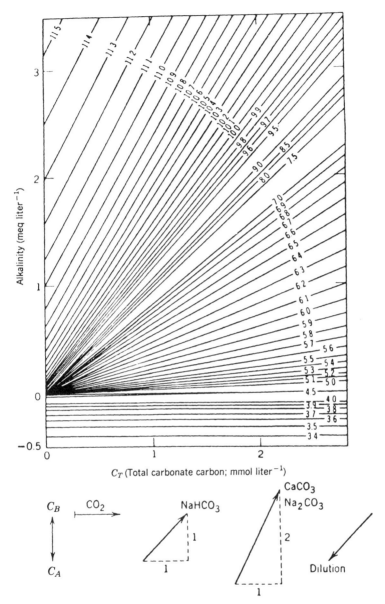

Figure 4.10. pH contours in alkalinity versus C_T diagram. The point defining the solution composition moves as a vector in the diagram as a result of the addition (or removal) of CO_2, $NaHCO_3$, and $CaCO_3(Na_2CO_3)$ or C_B and C_A.

the definition of [Alk]:

$$[Alk] = [HCO_3^-] + 2[CO_3^{2-}] + [OH^-] - [H^+]$$

$$[Alk] = C_T(\alpha_1 + 2\alpha_2) + [OH^-] - [H^+] \qquad (21)$$

It is apparent from equation 21 that, for any $[H^+]$, [Alk] is a linear function of C_T; the system is completely defined by specifying C_T and [Alk]. For each value of C_T and [Alk], pH is fixed.

Most of the usefulness of this diagram stems from the fact that changes in the solution move the point representing the solution composition in definite directions on the diagram. Adding strong acid decreases [Alk] without changing C_T. Vertical lines on Figure 4.10 therefore give alkalimetric or acidimetric titration curves for each C_T value. On the other hand, addition or removal of CO_2 increases or decreases C_T without changing [Alk] and the point moves to the right by the amount of CO_2 added. The change in solution composition (pH) by changes in CO_2, caused, for example, by respiration or photosynthesis, can readily be elucidated because the horizontal line defines a curve of titration with CO_2 for any alkalinity. Figure 4.10 gives the vector nature of the changes in C_T and [Alk] with the addition or removal of various substances. The diagram is less sensitive in the pH regions of low buffer intensity—pH = 8 and pH = 4.5. Note that the line at pH = 4.5 has a small positive slope and does not intercept exactly at [Alk] − 0 (in fact, the intercept is -3×10^{-5} eq liter^{-1}).

A similar diagram can be constructed on the basis of equation 22:

$$[Acy] = 2[H_2CO_3^*] + [HCO_3^-] + [H^+] - [OH^-]$$

$$[Acy] = C_T(2\alpha_0 + \alpha_1) + [H^+] - [OH^-] \qquad (22)$$

In such a diagram the ordinate value ([Acy]) is unaffected by the addition or removal of crystalline carbonates such as $CaCO_3$ or Na_2CO_3, and is of value in evaluating the effect of precipitation or dissolution of $CaCO_3$. Examples 4.3 and 4.4 illustrate the application of such diagrams for equilibrium calculations.

In solutions that contain bases and acids other than H^+, OH^-, and the carbonate species, the contours in the diagrams will become displaced vertically. For example, if [Alk] contains other bases (see equation 12), the intercept of the pH lines with the $C_T = 0$ line will be displaced vertically proportional to the concentration of the noncarbonate base.

Example 4.11. Increase in pH by Addition of Base or Removal of CO_2 The pH of a surface water having an original alkalinity of 1 meq liter^{-1} and a pH of 6.5 (25°C) is to be raised to pH 8.3. The following methods are considered:

  (i) increase in pH with NaOH
  (ii) increase in pH with Na_2CO_3
  (iii) increase in pH by removal of CO_2 (i.e., in a cascade aerator).

For each case the chemical change and the final composition of the solution are given by the following:

Change per Liter of Solution	pH	C_T (mM)	Final Composition Alk (meq liter^{-1})	Acy (meq liter^{-1})
Original solution	6.5	1.7	1.0	2.4
(i) +0.7 mmol NaOH	8.3	1.7	1.7	1.7
(ii) +0.7 mmol Na$_2$CO$_3$	8.3	2.4	2.4	2.4
(iii) −0.7 mmol CO$_2$	8.3	1.0	1.0	1.0

By first using Figure 4.10, we find the intersection of the pH = 6.5 line with the [Alk] = 1 meq liter^{-1} line and see that $C_T = 1.7 \times 10^{-3}$ M. If no CO$_2$ is exchanged with the atmosphere, the quantity of NaOH (C_B) necessary to reach pH = 8.3 can be found directly from the graph (vertical displacement) as ~0.7 meq liter^{-1}.

For the pH increase with Na$_2$CO$_3$, we have to draw a line with slope 2 from the intersection (pH = 6.5; [Alk] = 1 meq liter^{-1}). About 0.7 mM liter^{-1} of Na$_2$CO$_3$ is necessary to attain a water of pH = 8.3.

Finally, a pH of 8.3 can also be attained by decreasing C_T(CO$_2$) and maintaining a constant alkalinity. At the horizontal 1.0 meq liter^{-1} line (Figure 4.10), the difference between the points pH 6.5 and 8.3 corresponds to 0.7 mM liter^{-1} of CO$_2$.

The table summarizing the results of this example illustrates that a given pH change may be attained by different pathways and that the final composition depends on the pathway.

4.6 ANALYTICAL CONSIDERATIONS: GRAN PLOTS

The sharpness of the end point in an alkalimetric or acidimetric titration is related to the slope, $d\,\text{pH}/d\,C_B$ or β^{-1}, of the titration curve at the equivalence point. The relative error involved in the titration is related to the buffer intensity at the end point. There are many situations (e.g., low alkalinities or mineral acidities) where the end point recognition must be more precise than that given by a pH versus acid or base plot. A graphical procedure developed by Gran is based on the principle that added increments of mineral acid linearly increase [H$^+$] or decrease [OH$^-$]. Similarly, increments of strong base linearly decrease [H$^+$] or increase [OH$^-$].

Acidity

In rain and other atmospheric depositions, it is often necessary to determine H-acidity (sum of strong acids) in the presence of weak acids (H$_2$CO$_3^*$, organic acids, NH_4^+). Gran plots are useful for the determination of mineral acidity as well as of *total acidity* (sum of strong and weak acids).[†]

pH Measurements in Rain and Fog

pH and acidity measurements in rain, fog, and cloud waters are, due to the very small concentrations of H^+, the small buffer intensity and low ionic strength not as trivial as one would assume. When the pH measurement is carried out by comparing the response of a glass electrode of the sample with that of a standard buffer solution of given pH, large errors due to the difference in ionic strength between buffer and sample (differences in activity coefficients and change in liquid junction potential) may occur. Calibration using a strong acid adjusted to 0.05 M with respect to KCl is recommended, the observed potentiometer (pH meter) reading being compared with the H^+ concentration. Sample solutions must also be adjusted to 0.05 M with respect to KCl so that calibration and measurement are carried out in the same ionic medium. This calibration technique leads to a measurement of pH in terms of p^cH. For a careful calibration of the glass electrode, known increments of strong acid are added to a solution of inert electrolyte (e.g., 0.05 M KCl) in a thermostated vessel and the response (potential) of the glass electrode is plotted against $[H^+]$.

Alkalimetric Titration of an Acid

Added increments of base cause a linear decrease in $[H^+]$. It is assumed that no other acids dissociate in this first part of the titration plot:

$$v_0 \cdot c_0 = v_1 \cdot c_b^* \tag{23}$$

[†]Exact definitions of acidity and procedures for analytical determination were given by Johnson and Sigg (1983). Reference conditions for H-Acy are H_2O, H_2CO_3, SO_4^{2-}, NO_3^-, Cl^-, NO_2^-, F^-, HSO_3^-, NH_4^+, H_4SiO_4, and $\Sigma H_n Org$ (organic acids with $pK_a < 10$).

H-Acy can then rigorously be defined by

$$[\text{H-Acy}] = [H^+] + [HSO_4^-] + [HNO_2^-] + [HF] + [SO_2 \cdot H_2O]$$
$$- [OH^-] - [HCO_3^-] - 2[CO_3^{2-}] - [NH_3] - [H_3SiO_4^-] - \Sigma n[Org^n] \tag{i}$$

For many rain samples this equation can be simplified to $[\text{H-Acy}] \simeq [H^+]$. In fog and cloud water samples, $[HSO_4^-]$ and $[SO_2 \cdot H_2O]$ are often not negligible. Total acidity $[Acy_T]$ includes all acid with $pK_a \lesssim$ ca. 10. It is defined with regard to the following reference conditions: H_2O, CO_3^{2-}, SO_4^{2-}, NO_2^-, Cl^-, NO_3^-, F^-, SO_3^{2-}, NH_3, $H_3SiO_4^-$, and ΣOrg^{n-}.

$$[Acy_T] = [H^+] + [HSO_4^-] + [HNO_2^-] + [HF] + 2[SO_2 \cdot H_2O] + [HSO_3^-]$$
$$+ 2[H_2CO_3^*] + [HCO_3^-] + [NH_4^+] + [H_4SiO_4]$$
$$+ \Sigma n[H_n Org] - [OH^-] \tag{ii}$$

The difference $[Acy_T] - [\text{H-Acy}]$ gives the sum of the weak acids: NH_4^+ and organic acids are usually most important.

4.6 Analytical Considerations: Gran Plots

where v_0 = initial volume of sample (liters)
c_0 = initial concentration of H-Acy
v_1 = volume of added base (liters)
c_b^* = concentration of base added (M)

For each point on the titration curve, we obtain (considering the volume correction)

$$[H^+] = \frac{c_0 \cdot v_0 - v \cdot c_b^*}{(v_0 + v)} = \frac{v_1 \cdot c_b^* - v \cdot c_b^*}{(v_0 + v)} \quad (24)$$

and

$$10^{-pH}(v_0 + v) = (v_1 - v) \cdot c_b^* \quad (25)$$

where v = volume of added base.

In plotting the left-hand side of equation 25,

$$F_1 = 10^{-pH}(v_0 + v) \quad (26)$$

against v, F_1 becomes 0 when $v_1 = v$. Thus extrapolating F_1 to zero gives the end point. In Figure 4.11b, F_1 is plotted versus $(v_1 - v)\, c_b^*$, the total quantity of base added (μmol liter^{-1}).

Total acidity includes, in addition to $[H^+]$ and strong acids, the weak acids. The latter acids can be titrated with strong base if their pK_a values are less than ~ 10. CO_2 is usually removed from the samples by blowing it out with an inert gas. The end point of total acidity is characterized by the volume v_2 of strong base added. After this end point, further additions of strong base are linearly related to an increase in $[OH^-]$:

$$[OH^-] = \frac{v \cdot c_b^* - v_2 \cdot c_b^*}{(v_0 + v)} \quad (27)$$

with

$$[OH^-] = \frac{K_W}{[H^+]} \quad (28)$$

$$(v_0 + v) \cdot 10^{pH} = \frac{1}{K_W} \cdot c_b^*(v - v_2) \quad (29)$$

In plotting

$$F_2 = 10^{pH}(v_0 + v) \quad (30)$$

versus v, the interaction $F_2 = 0$ is obtained when $v = v_2$ (see Figure 4.11b).

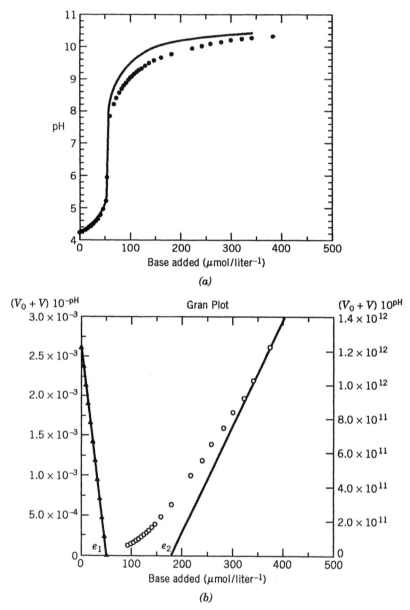

Figure 4.11. Alkalimetric titration of an acid rainwater with a pH = 4.26, consisting of a mineral acidity of [H-Acy] = 50 μeq liter^{-1} and a weak acidity of 90 μeq liter^{-1} ([NH_4^+] = 85 μM and acetic acid [HA] = 5 μM). CO_2 has been expelled with N_2 prior to the titration. In the "conventional" titration curve, the pH jump is equivalent to the mineral acidity [H-Acy]. If only mineral acidity (no weak acids) were present, the titration curve would correspond to the drawn-out line (a). The Gran titration procedure permits one to distinguish between total acidity [Acy$_T$] (end point e_2) = 140 μeq liter^{-1} and mineral acidity, [H-Acy] (end point e_1) = 50 μeq liter^{-1}. (Adapted from Sigg and Stumm, 1994.)

4.6 Analytical Considerations: Gran Plots

Gran Titration in the Carbonate System

Consider Figure 4.12 and the following definitions.

Symbols: v_0, original volume of sample
v, volume of strong acid added
$\bar{c}_A$, molarity of strong acid
H_2C, HC^-, and C^{2-} = $H_2CO_3^*$, HCO_3^-, and CO_3^{2-}, respectively

Capacities:

$$C_T = [H_2C] + [HC^-] + [C^{2-}] \tag{31a}$$

$$[Alk] = [HC^-] + 2[C^{2-}] + [OH^-] - [H^+] \tag{31b}$$

$$[H^+\text{-Acy}] = [H^+] - [HC^-] - 2[C^{2-}] - [OH^-] \tag{32}$$

$$[CO_2\text{-Acy}] = [H_2C] + [H^+] - [C^{2-}] - [OH^-] \tag{33}$$

$$[Acy] = 2[H_2C] + [HC^-] + [H^+] - [OH^-] \tag{34}$$

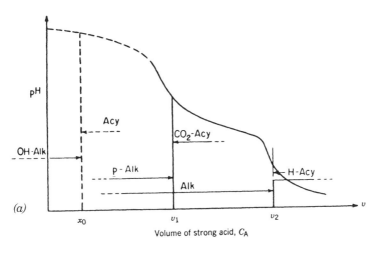

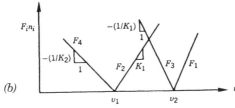

Figure 4.12. Sketch of an acidimetric titration curve (a). In (b) the results of (a) are plotted in terms of Gran functions: F_i is multiplied by scale factors n_i. The F_i values are defined by Equations 48 and 51–54. x_0, v_1, and v_2 are the volumes of strong acid corresponding to the equivalence points $f = 2$, $f = 1$, and $f = 0$, respectively.

$$[p\text{-Alk}] = [C^{2-}] + [OH^-] - [H^+] - [H_2C] \tag{35}$$

$$[OH^-\text{-Alk}] = [OH^-] - [HC^-] - 2[H_2C] - [H^+] \tag{36}$$

At the various equivalence points the following equalities exist:

$$v_0[\text{Alk}] = v_2\bar{c}_A \tag{37}$$

$$v_0[p\text{-Alk}] = v_1\bar{c}_A \tag{38}$$

$$v_0[OH^-\text{-Alk}] = x_0\bar{c}_A \tag{39}$$

$$v_0 C_T = (v_2 - v_1)\bar{c}_A \tag{40}$$

At any point on the titration curve the following equalities exist:

$$(v_0 + v)\,[H^+\text{-Acy}] = (v - v_2)\bar{c}_A \tag{41}$$

$$(v_0 + v)\,[CO_2\text{-Acy}] = (v - v_1)\bar{c}_A \tag{42}$$

$$(v_0 + v)\,[\text{Alk}] = (v_2 - v)\bar{c}_A \tag{43}$$

$$(v_0 + v)\,[\text{Acy}] = (v - x_0)\bar{c}_A \tag{44}$$

$$(v_0 + v)\,[p\text{-Alk}] = (v_1 - v)\bar{c}_A \tag{45}$$

I. Beyond v_2, equation 41 can be simplified to

$$(v_0 + v)\,[H^+] \simeq (v - v_2)\bar{c}_A \tag{46}$$

because for $v > v_2$

$$[H^+] \gg [HC^-] + 2[C^{2-}] + [OH^-] \tag{47}$$

The equivalence point v_2 can be obtained by plotting the left-hand side of equation (46) versus v.

Instead of $[H^+]$ we may write $10^{-\text{"pH"}}$. Independent of any pH or activity conventions, operationally one can define "pH" = pH° $-\log[H^+]$, where pH° is a constant. If the pH meter reading is made in volts, one can also use $[H^+] = 10^{-\text{"pH"}} = 10^{(E-e)/k}$, because $E = e + k \log [H^+]$, where $k = 2.3\,RTF^{-1}$ (where F = faraday). Rewriting equation 46,

$$F_1 = (v_0 + v)10^{-\text{"pH"}} \simeq (v - v_2)\bar{c}_A \tag{48}$$

where $F_1 = 0$ for $v = v_2$.

II. Between v_1 and v_2, equations 42 and 43 can be simplified because in this range $[H_2C] \gg [H^+] - [C^{2-}] - [OH^-]$ and $[HC^-] \gg 2[C^{2-}] + [OH^-]$

$- [H^+]$. Accordingly, instead of equations 42 and 43, respectively,

$$(v_0 + v)[H_2C] \simeq (v - v_1)\bar{c}_A \tag{49}$$

and

$$(v_0 + v)[HC^-] \simeq (v_2 - v)\bar{c}_A \tag{50}$$

Substituting $[H_2C] = (1/K_1)[H^+][HC^-]$ in equation 49 and combining with equation 50 gives

$$F_2 = (v_2 - v)10^{-\text{"pH"}} = (v - v_1)K_1 \tag{51}$$

$F_2 = 0$ for $v = v_1$, the slope of F_2 versus v being K_1.

III. Equation 50 can be rearranged:

$$F_3 = (v - v_1)10^{\text{"pH"}} = (v_2 - v)K_1^{-1} \tag{52}$$

F_3 gives an additional check on v_2, the slope being $-1/K_1$.

IV. Another linear function can be derived, for $x_0 < v < v_1$, where $[C^{2-}] > [OH^-] - [H^+] - [H_2C]$, so that equation 45 can be written approximately as

$$(v_0 + v)[CO_3^{2-}] \simeq (v_1 - v)\bar{c}_A \tag{53}$$

An expression for $[HCO_3^-]$ can be derived from already estimated values of v_1 and v_2, giving

$$(v_0 + v)[HCO_3^-] \simeq (v_2 - 2v_1 + v)\bar{c}_A \tag{54}$$

combining equations 53 and 54 yields

$$\frac{[CO_3^{2-}]}{[HCO_3^-]} = \frac{K_2}{[H^+]} \simeq \frac{v_1 - v}{v_2 - 2v_1 + v} \tag{55}$$

Define

$$F_4 = (v_2 - 2v_1 + v)10^{\text{"pH"}} = \frac{v_1 - v}{K_2} \tag{56}$$

For $F_4 = 0$, $v = v_1$, the slope of F_4 versus v being $-1/K_2$.

Gran plots permit the determination of C_T (see equation 40) from the difference of v_1 and v_2 end points. Dyrssen and Sillén were among the first to characterize seawater by an emf acidimetric titration. Figure 4.13 shows that a simple acidimetric titration can give quite accurate data on [p-Alk], [Alk],

186 Dissolved Carbon Dioxide

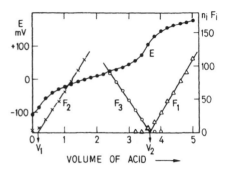

Figure 4.13. Emf titration curve for 154 g of seawater with v milliliters (0.1000 M HCl plus 0.4483 M NaCl); characterization of equivalence points by the Gran method. F_1, F_2, and F_3 are Gran functions, and n_i is an arbitrary scale factor. For finding the equivalence points v_1 and v_2 corresponding to $f = 1$ and $f = 0$, respectively. (From Dryssen and Sillén, 1967.)

and C_T. Carbonate alkalinity (e.g., in seawater), that is, the portion of [Alk] due to $[HCO_3^-] + 2[CO_3^{2-}]$, can only be derived from the measured [Alk] by deducting the contribution of $B(OH)_4^-$ and of $[OH^-]$ and other bases.

$$[\text{Carbonate Alk}] = [\text{Alk}] - B_T \alpha_{B^-} - ([OH^-] - [H^+]) \tag{57}$$

For recent assessments of alkalinity in seawater, see Bradshaw and Brewer (1988) and Butler (1992). Equilibrium constants valid in seawater are given in Appendix 6.2 of Chapter 6.

4.7 EQUILIBRIUM WITH SOLID CARBONATES

The interaction of carbon dioxide with solid carbonates, above all with $CaCO_3$ (calcite), is of general importance in accounting for the composition of most natural waters. Although we deal more extensively with solid phase systems in Chapter 7, it seems expedient to include at this point equilibria with solid carbonates. The *open system* may be characterized by

$$CaCO_3(s) + CO_2(g) + H_2O \rightleftharpoons Ca^{2+} + 2\,HCO_3^- \tag{58}$$

How soluble is calcite in equilibrium with the atmosphere ($p_{CO_2} = 10^{-3.5}$ atm) and what is the composition of the water at equilibrium? The following species are in equilibrium: Ca^{2+}, H^+, $H_2CO_3^*$, HCO_3^-, CO_3^{2-}, OH^-, and $CO_2(g)$.[†] In order to calculate the equilibrium concentrations of the six species in solution, we need to consider, in addition to the equilibria already used in the previous calculations,

$$[H^+][HCO_3^-]/[H_2CO_3^*] = K_1 \tag{59}$$

$$[H^+][CO_3^{2-}]/[HCO_3^-] = K_2 \tag{60}$$

[†]In dilute solutions, ion pairs such as $CaCO_3^0$ and $CaHCO_3^+$ can be neglected.

4.7 Equilibrium with Solid Carbonates

$$[H_2CO_3^*]/p_{CO_2} = K_H \qquad (61)$$

$$[H^+][OH^-] = K_W \qquad (62)$$

the solubility product of $CaCO_3(s)$ and the charge (or proton) balance

$$[Ca^{2+}][CO_3^{2-}] = K_{s0}; \quad K_{s0} = 10^{-8.4} \quad (25°C) \qquad (63)$$

$$2[Ca^{2+}] + [H^+] = [HCO_3^-] + 2[CO_3^{2-}] + [OH^-] \qquad (64)$$

The result obtained with a graphical double log plot (Figure 4.14) or calculated by computer gives

$$pH = 8.3, \quad [H_2CO_3^*] = 10^{-5}\ M, \quad [HCO_3^-] = 10^{-3}\ M$$

$$[Ca^{2+}] = 5 \times 10^{-4}\ M, \quad [CO_3^{2-}] = 1.6 \times 10^{-5}\ M.$$

Figure 4.14 consists of the superposition of Figure 4.5 with the lines of equations 63 and 64. The line for log $[Ca^{2+}]$ intersects with the line of log $[CO_3^{2-}]$, where $[Ca^{2+}] = [CO_3^{2-}] = (K_{s0})^{1/2}$. Because the product of $[Ca^{2+}]$

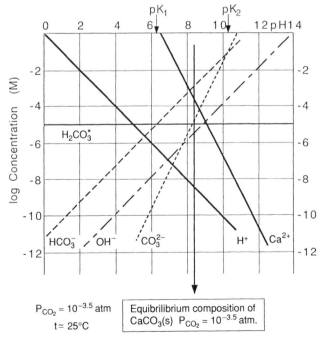

Figure 4.14. Equilibrium composition of a solution in presence of $CaCO_3(s)$ (calcite) at constant partial pressure of CO_2 ($p_{CO_2} = 10^{-3.5}$ atm) at 25°C. If no acid or base is added, the equilibrium composition is indicated by the arrow.

188   Dissolved Carbon Dioxide

with $[CO_3^{2-}]$ must be constant, the line for $\log [Ca^{2+}]$ is a straight line with a slope $d \log [Ca^{2+}]/d\, pH = -2$.

The straight lines representing the mass laws, equations 59–62, give the composition of every water that is in equilibrium with the atmosphere and with solid calcite ($CaCO_3$). The change of pH results from the interaction of the system with acids or bases. The charge balance (equation 64) corresponds to an equilibrium system consisting only of $CO_2(g)$, $CaCO_3(s)$, and H_2O. The simplified charge balance is: $2\,[Ca^{2+}] \simeq [HCO_3^-]$ or $\log [Ca^{2+}] = \log [HCO_3^-] - 0.3$. Figure 4.14 illustrates the strong dependence of soluble $[Ca^{2+}]$ on pH. The equilibrium system is summarized in Tableau 4.5.

Equation iv results from the combination of the following:

$$CaCO_3(s) = Ca^{2+} + CO_3^{2+}; \qquad K_{s0} = 10^{-8.4} \qquad (ix)$$

$$CO_3^{2-}(s) + 2\,H^+ = CO_2(g) + H_2O; \qquad (K_1 K_2 K_H)^{-1} = 10^{18.1} \qquad (x)$$

$$CaCO_3(s) + 2\,H^+ = CO_2(g) + H_1O + Ca^{2+}; \qquad K = 10^{9.7} \qquad (xi)$$

that is,

$$\frac{[Ca^{2+}]\, p_{CO_2}}{[H^+]^2} = 10^{9.7} \text{ or}$$

$$[Ca^{2+}] = 10^{9.7}\,[H^+]^2\,(p_{CO_2})^{-1} \qquad (xii)$$

River Water Composition and Carbonate Equilibrium

In Figure 4.15 HCO_3^- and Ca^{2+} concentrations of different rivers of the world are plotted. The drawn-out lines correspond (1) to the charge balance of the most important species in these rivers,

$$2\,[Ca^{2+}] = [HCO_3^-] \qquad (65)$$

Tableau 4.5. Equilibrium System CO_2, $CaCO_3$ (Calcite), H_2O (Open CO_2 System)

Components		$CO_2(g)$	$CaCO_3(s)$	H^+	$\log K(25°C, I = 0)$	
Species	$H_2CO_3^*$	1	0	0	−1.5	(i)
	HCO_3^-	1	0	−1	−7.8	(ii)
	CO_3^{2-}	1	0	−2	−18.1	(iii)
	Ca^{2+}	−1	1	2	9.7	(iv)
	OH^-	0	0	−1	−14.0	(v)
	H^+	0	0	1	0	(vi)
		$p_{CO_2} = 10^{-3.5}$	$\{CaCO_3\} = 1$	0		(vii)
$TOTH = [H^+] - [HCO_3^-] - 2\,[CO_3^{2-}] - [OH^-] + 2\,[Ca^{2+}] = 0$						(viii)

4.7 Equilibrium with Solid Carbonates

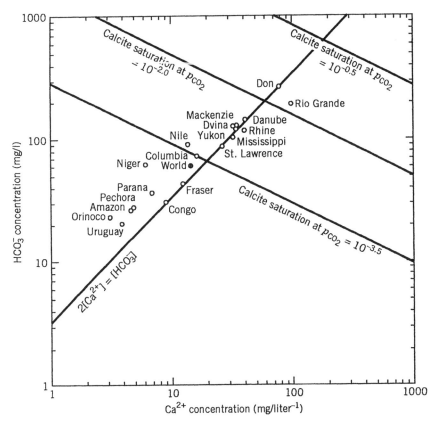

Figure 4.15. The relationship between the bicarbonate and calcium concentrations in some major rivers. The composition of many rivers is characterized by the charge balance $2[Ca^{2+}] \simeq [HCO_3^-]$ and the saturation of $CaCO_3$ $[CaCO_3(s) + CO_2(g) + H_2O = Ca^{2+} + 2\,HCO_3^-]$. (Adapted from Holland, 1978.)

and (2) to the solubility of calcite ($CaCO_3$),

$$CaCO_3(s) + CO_2(g) + H_2O = Ca^{2+} + 2\,HCO_3^- \tag{66}$$

This equilibrium relationship corresponds to the summing up of the following reactions:

$$CaCO_3(s) = Ca^{2+} + CO_3^{2-}; \quad K_{s0} = 10^{-8.4} \quad (25°C) \tag{67}$$

$$CO_3^{2-} + H^+ = HCO_3^-; \quad K_2^{-1} = 10^{10.3} \tag{68}$$

$$CO_2(g) + H_2O = H_2CO_3^*; \quad K_H = 10^{-1.5} \tag{69}$$

$$H_2CO_3^* = H^+ + HCO_3^-; \quad K_1 = 10^{-6.3} \tag{70}$$

190 Dissolved Carbon Dioxide

Thus

$$[Ca^{2+}][HCO_3^-]^2/p_{CO_2} = K_{s0}K_HK_1K_2^{-1} = 10^{-5.9} \qquad (66a)$$

Thus, in a plot of log $[HCO_3^-]$ versus log $[Ca^{2+}]$, we obtain a straight line with a slope of $d \log [HCO_3^-]/d \log [Ca^{2+}] = -0.5$. Apparently many dilute rivers are undersaturated with regard to $CaCO_3$; many other rivers reach a saturation at CO_2 partial pressures between $10^{-3.5}$ and 10^{-2} atm. Because of the organic loading of the waters (respiration of organic material to CO_2) and because of the inflow of groundwaters (characterized by higher p_{CO_2}), many river waters have a higher p_{CO_2} than the atmosphere.

Ca^{2+} in Equilibrium with Aqueous Carbonate in a Closed System

If in a closed system the sum of aqueous carbonate species

$$C_T = \text{TOTC} = [H_2CO_3^*] + [HCO_3^-] + [CO_3^{2-}]$$

remains constant, Ca^{2+} (in saturation equilibrium with calcite, $CaCO_3$) is given by

$$[Ca^{2+}] = \frac{K_{s0}}{[CO_3^{2-}]} = \frac{K_{s0}}{C_T \alpha_2} \qquad (71)$$

In Figure 4.16 the aqueous carbonate species for TOTC = 10^{-3} M[†] and the $[Ca^{2+}]$ corresponding to equation 71 are given.

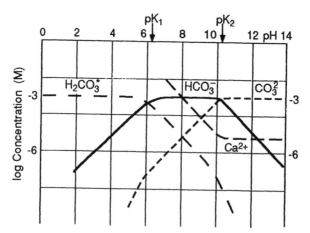

Figure 4.16. $[Ca^{2+}]$ in saturation equilibrium with $CaCO_3$ for TOTC = 10^{-3} M.

[†]Note that TOTC = constant is controlled independent of the dissolution of $CaCO_3(s)$.

4.7 Equilibrium with Solid Carbonates

Groundwater

The uppermost part of the earth's rocks constitutes a porous medium in which water is stored and through which it moves. Up to a certain level, these rocks are saturated with water that is free to flow laterally under the influence of gravity. Subsurface water in this saturated zone is *groundwater*, and the uppermost part of the zone is the water table. The chemistry of the groundwater is influenced by the composition of the aquifer and by the chemical and biological events occurring in the infiltration.

A general relationship between the composition of the water and that of the solid minerals with which the water has come into contact during infiltration and in the aquifer can be expected. Biological activity, especially in the organic layer above the mineral part, has a pronounced effect on the acquisition of solutes. Because of microbial respiration, the CO_2 pressure is increased. (The relative change in CO_2 is much larger than that of O_2 because CO_2 is present in a much smaller concentration in the atmosphere than O_2.) The increased CO_2 pressure tends to increase the alkalinity and the concentration of Ca^{2+} and other solutes.

The $CaCO_3(s)$–CO_2 equilibrium in groundwater carriers is often modeled as a closed system (in the sense of Figure 4.2a) or as an open system.

Example 4.12. Solubility of $CaCO_3$ as a Function of p_{CO_2} How does the solubility of $CaCO_3$ (calcite) increase, for example, in a groundwater carrier with increase in CO_2 pressure? The system is characterized by Tableau 4.5. $[Ca^{2+}]_{eq}$ can be calculated from the equilibrium

$$CaCO_3(s) + CO_2(g) + H_2O = Ca^{2+} + 2\, HCO_3^-; \quad K \quad \text{(i)}$$

The equilibrium constant K was given in equation 66a as

$$\frac{[Ca^{2+}][HCO_3^-]^2}{p_{CO_2}} = K = K_{s0} K_H K_1 K_2^{-1} \quad \text{(ii)}$$

Since at pH values < 10, $2[Ca^{2+}] \simeq [HCO_3^-]$, equation ii can be written

$$\frac{[Ca^{2+}]^3}{p_{CO_2}} \simeq \frac{K}{4}$$

and thus

$$\frac{d\log p_{CO_2}}{d\log [Ca^{2+}]} \simeq 3 \quad \text{and} \quad \frac{d\log p_{CO_2}}{d\log [HCO_3^-]} \simeq 3 \quad \text{(iii)}$$

Thus the equilibrium solubility of $CaCO_3$ as measured by $[Ca^{2+}]$ or $[HCO_3^-]$ increases with $(p_{CO_2})^{1/3}$. For example, an increase in p_{CO_2} from $10^{-3.5}$

to $10^{-0.5}$ atm (as may occur in microbially active subsoil systems) would increase $[Ca^{2+}]$ from $\sim 2 \times 10^{-4}$ M to $\sim 3 \times 10^{-3}$ M. (An exact calculation needs to consider changes in ionic strength.)

4.8 KINETIC CONSIDERATIONS

Many natural waters are not in equilibrium with the atmosphere, partially because of unfavorable mixing conditions but primarily because of the slowness of the gas transfer reaction; this transfer is frequently slower than reactions that produce or consume CO_2 in the aqueous phase [respiration, photosynthesis, precipitation, and dissolution reactions—e.g., $Ca^{2+} + 2HCO_3^- = CaCO_3 + CO_2 + H_2O$—mineral alterations—e.g., $NaAlSi_3O_8(s) + H_2CO_3^* + \frac{9}{2}H_2O = Na^+ + HCO_3^- + 2H_4SiO_4 + \frac{1}{2}Al_2Si_2O_5(OH)_4(s)$].

On the other hand, ionization equilibria in the dissolved carbonate system are established very rapidly. Somewhat slower (seconds), however, is the attainment of equilibrium in the hydration or dehydration reaction of CO_2 (Kern, 1960):

$$CO_2(aq) + H_2O \rightleftharpoons H_2CO_3$$

Kinetics of Hydration of CO_2

The hydration of CO_2 leads to the formation of H_2CO_3, but it may also yield H^+ and HCO_3^-. The reaction scheme may be written

$$\text{(1)} \quad H^+ + HCO_3^- \underset{k_{21}}{\overset{k_{12}}{\rightleftharpoons}} H_2CO_3 \quad \text{(2)}$$

with k_{31}/k_{13} and k_{23}/k_{32} connecting to $CO_2 + H_2O$ (3). (72)

The same reaction scheme applies to the hydration of SO_2. Formally, we may write the rate law for the disappearance of CO_2:

$$-\frac{d[CO_2]}{dt} = (k_{31} + k_{32})[CO_2] - k_{13}[H^+][HCO_3^-] - k_{23}[H_2CO_3] \quad (73)$$

Considering that k_{21}/k_{12} equals the first acidity constant of "true" H_2CO_3, $K_{H_2CO_3}$, and that both k_{21} and k_{12} are far larger than any other of the four other rate constants in equation 72, we can replace $[H^+][HCO_3^-]$ with $K_{H_2CO_3}[H_2CO_3]$ in equation 73 and write

4.8 Kinetic Considerations

$$-\frac{d[CO_2]}{dt} = (k_{31} + k_{32})[CO_2] - (k_{13}K_{H_2CO_3} + k_{23})[H_2CO_3] \quad (74)$$

$$= (k_{31} + k_{32})[CO_2] - [k_{13} + k_{23}(K_{H_2CO_3})^{-1}][H^+][HCO_3^-] \quad (75)$$

Setting

$$k_{CO_2} = k_{31} + k_{32} \quad (76)$$

and

$$k_{H_2CO_3} = k_{13}K_{H_2CO_3} + k_{23} \quad (77)$$

equation 79 can be written

$$-\frac{d[CO_2]}{dt} = k_{CO_2}[CO_2] - k_{H_2CO_3}[H_2CO_3] \quad (78a)$$

$$= k_{CO_2}[CO_2] - \frac{k_{H_2CO_3}}{K_{H_2CO_3}}[H^+][HCO_3^-] \quad (78b)$$

The rate law (equation 78) corresponds to the simplified scheme

$$CO_2 + H_2O \underset{k_{H_2CO_3}}{\overset{k_{CO_2}}{\rightleftharpoons}} H_2CO_3 \overset{\text{very fast}}{\rightleftharpoons} H^+ + HCO_3^- \quad (79)$$

That is, the hydration reaction is first order with respect to dissolved CO_2 and has a rate constant of $k_{CO_2} = 0.025$–0.04 s^{-1} (25°C). The activation energy is approximately 15 kcal mol^{-1}. Similarly, the rate of dehydration has a first-order rate constant $k_{H_2CO_3}$ of 10–20 s^{-1} (20–25°C); its activation energy is ~16 kcal mol^{-1}. Considering the order of magnitude of the reaction rate constants, it is obvious that not more than a few minutes are necessary to establish the hydration equilibrium (Figure 4.17).

The ratio of these velocity constants permits estimation of the equilibrium constant of equation 79 (cf. equations 1 and 2 of Table 4.2):

$$K' = K + 1 = \frac{[CO_2] + [H_2CO_3]}{[H_2CO_3]} = \frac{[H_2CO_3^*]}{[H_2CO_3]} = \frac{K_{H_2CO_3}}{K_1} = \frac{k_{H_2CO_3}}{k_{CO_2}} + 1 \quad (80)$$

K' has reported values of about 350–990 at 25°C.

Superimposed on these first-order reactions are the processes

$$CO_2 + OH^- \underset{k_{41}}{\overset{k_{14}}{\rightleftharpoons}} HCO_3^- \quad (81)$$

194　Dissolved Carbon Dioxide

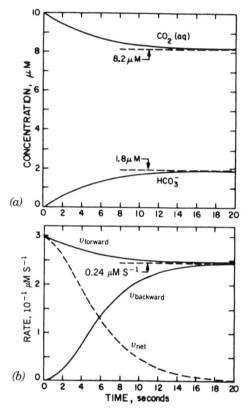

Figure 4.17. Computed concentrations of $CO_2(aq)$ and HCO_3^- as a function of time for the reversible reaction $CO_2(aq) + H_2O = HCO_3^- + H^+$ at 25°C in a closed aqueous system. The total concentration, $C_T = [CO_2(aq)] + [HCO_3^-]$, is 1×10^{-5} M. $CO_2(aq)$ is assumed to be nonvolatile. (b) Computed velocities, $v_{forward}$, $v_{backward}$, and v_{net} in the reaction mixture as a function of time for $CO_2(aq) + H_2O = HCO_3^- + H^+$. At equilibrium, the velocity in both directions is 0.24 μM s^{-1}.

with approximate constants $k_{14} = 8.5 \times 10^3$ M^{-1} s^{-1} and $k_{41} = 2 \times 10^{-4}$ s^{-1} (25°C). Process 81 is kinetically insignificant at pH values below about 8. Above pH = 10, it dominates the hydration reaction.

Catalysis Hydration and dehydration reactions are catalyzed by various bases, by certain metal chelates, and by the enzyme carbonic anhydrase. Natural waters may contain some natural catalysts. Reaction 81 is essentially a catalysis of reaction 79. The conversion of CO_2 into H_2CO_3 or HCO_3^- by both mechanisms is

$$\frac{d[CO_2]}{dt} = -(k_{CO_2} + k_{14}[OH^-])\,[CO_2]$$

$$+ (k_{H_2CO_3}(K_{H_2CO_3})^{-1}\,[H^+] + k_{41})\,[HCO_3^-] \qquad (82)$$

For a treatment on the CO_2 transfer at the gas-water interface, see Section 5.10.

Chemical Enhancement of CO_2 Transfer into Lakes

Lakes often become greatly undersaturated in CO_2 during photosynthesis because CO_2 assimilation is faster than the resupply by gas exchange from the atmosphere. Under such circumstances—high pH, higher film thickness—the invasion rates of CO_2 into the lake are more likely to be enhanced by reactions of CO_2 to form HCO_3^-.

4.9 CARBON ISOTOPES AND ISOTOPE FRACTIONATION

Carbon has two stable isotopes, ^{12}C and ^{13}C, and a radioactive isotope, ^{14}C, with a half-life of 5720 years. These isotopes are present in the following abundances: $^{12}C = 98.89\%$, $^{13}C = 1.11\%$, and $^{14}C \simeq 10^{-10}\%$ of the total.

Carbon-14

The main source of natural ^{14}C is the upper atmosphere, where it is formed by the interaction of cosmic ray neutrons with ^{14}N; as a β emitter it returns to the original form, ^{14}N, after ejection of an electron from its nucleus. About 100 ^{14}C atoms are being generated over each square centimeter of the earth's surface each minute.[†]

Despite its low abundance, ^{14}C is an interesting tracer because of its radioactive decay:

$$-\frac{dN}{dt} = \lambda N \tag{83}$$

where N is the number of ^{14}C atoms (or, generally, of radioactive elements) and λ is the decay constant ($\lambda = 1.2 \times 10^{-4}$ year^{-1} for ^{14}C). This decay constant corresponds to a half-life $t_{1/2}$ (the time necessary to reduce radioactive atoms present at time zero, N_0, to one-half, $\frac{1}{2}N_0$):

$$t_{1/2} = \frac{1}{\lambda} \ln \frac{N_0}{\frac{1}{2}N_0} = \frac{\ln 2}{\lambda} \tag{84}$$

$t_{1/2}$ is 5720 years for ^{14}C. The mean lifetime, τ, that is, the mean life expectancy of a radioactive atom, is given by the sum of the life expectancies of all the

---

[†]In addition, the output of CO_2 from fossil fuels—which contain no ^{14}C—has tended to dilute the natural level [Suess effect (Suess, 1955)], and recent injections of nuclear bomb-produced—artificial—^{14}C have enhanced the ^{14}C level in the opposite direction.

196   Dissolved Carbon Dioxide

atoms divided by the initial number:

$$\tau = -\frac{1}{N_0} \int_{t=0}^{t=\infty} t\, dN = \frac{1}{N_0} \int_0^\infty t\lambda N\, dt$$

$$= \lambda \int_0^\infty t e^{-\lambda t}\, dt = \frac{1}{\lambda} \tag{85}$$

The average ^{14}C atom has a life expectancy of 8300 years; that is, each year 1 atom out of 8300 undergoes radioactive transformation. In *radiocarbon dating* of sea shells, one assumes that the $^{14}C/C$ ratio in seawater has always been the same as it is today. In accordance with equation 83, the age t is given by

$$t = \frac{1}{\lambda} \ln \frac{(^{14}C/C)_{\text{formation}}}{(^{14}C/C)_{\text{today}}} \tag{86}$$

Example 4.13. Carbon-14 as a Tracer for Oceanic Mixing   In a simplified two-box model of the ocean, the warm waters and the cold waters may be subdivided into two well-mixed reservoirs—an upper one a few hundred meters in depth and a lower one of 3200 m depth. The C_T content of the upper and lower reservoirs (corresponding to the Pacific) are, respectively, 1.98×10^{-3} mol liter^{-1} and 2.44×10^{-3} mol liter^{-1}, whereas the $^{14}C/C$ ratios for upper and lower reservoirs are, respectively, 0.92×10^{-12} and 0.77×10^{-12} mol/mol. Estimate from this information the rate of vertical mixing and the residence time of the water in the deep sea (Broecker, 1974).

Figure 4.18 sketches the situation. Water conservation requires that the flux of water downwelling equal the flux of water upwelling ($= v_{\text{mix}}$). The input of carbon to the deep reservoir must be balanced by the output of carbon:

$$v_{\text{mix}} C_{\text{surf}} + B = v_{\text{mix}} C_{\text{deep}} \tag{i}$$

B is the carbon added to the deep reservoir each year by the destruction of biological debris, formed in the surface layer, falling from the surface. From the mass balance,

$$B = v_{\text{mix}}(C_{\text{deep}} - C_{\text{surf}}) \tag{ii}$$

We can now write a mass balance for ^{14}C. The radioactive decay of ^{14}C has to be considered an "output" of the deep reservoir; its rate is given by equation 83:

$$\frac{dN_{^{14}C}}{dt} = V_{\text{deep}}[^{14}C_{\text{deep}}]\lambda \tag{iii}$$

where V_{deep} is the volume of the deep reservoir and $[^{14}C]$ is the concentration of ^{14}C. Because laboratory measurements are typically given as $^{14}C/C$ ratios,

4.9 Carbon Isotopes and Isotope Fractionation

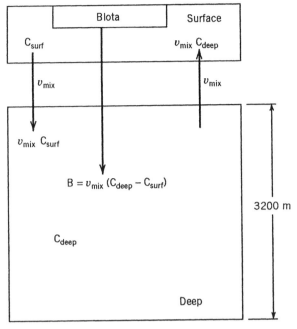

Figure 4.18. Two-box steady-state model for the cycle of water, carbon, and ^{14}C between the surface and deep sea. (See Example 4.13) (After Broecker, 1974).

we may express the mass balance in these ratios:

$$v_{mix} C_{surf} \frac{^{14}C_{surf}}{C_{surf}} + B \frac{^{14}C_{surf}}{C_{surf}} = v_{mix} C_{deep} \frac{^{14}C_{deep}}{C_{deep}} + V_{deep} \frac{^{14}C_{deep}}{C_{deep}} \quad \text{(iv)}$$

Substituting equation ii into equation iv and solving for v_{mix}, one obtains

$$v_{mix} = \lambda V_{deep} \frac{^{14}C_{deep}/C_{deep}}{^{14}C_{surf}/C_{surf} - {}^{14}C_{deep}/C_{deep}} \quad \text{(v)}$$

V_{deep} can be expressed as area times mean depth. Using 3200 m as a mean and using the data given, we obtain

$$v_{mix} = 1.2 \times 10^{-4} \times 3.2 \times 10^3 \times A \frac{9.77}{0.92 - 0.77}$$

$$v_{mix} = 2 \text{ m year}^{-1} \left(= 2 \frac{m^3}{m^2} \text{ yr}^{-1}\right) \quad \text{(vi)}$$

that is, the yearly volume of water exchanged between the surface and the deep ocean is equal in volume to a layer 2 m thick with an area equal to that of the

Isotope Effects

Differences exhibited by isotopic species in coexisting phases are primarily caused by the following:

1. Kinetic differences between isotopic molecules resulting in differences in such properties as diffusional velocity and rate of vaporization. For example, for $^{12}CO_2$ and $^{13}CO_2$, the ratio of diffusional flux would be predicted—since the diffusion coefficient, D, is inversely proportional to the square root of the molecular weight—to be

$$\frac{D_{44}}{D_{45}} = \left(\frac{45}{44}\right)^{1/2} = 1.011$$

The vapor pressure of isotopic compounds decreases with increasing isotopic mass; that is, the tendency to evaporate is larger for the compound with the light isotope.

2. Purely chemical differences among isotopic compounds are primarily caused by differences in the vibrational frequencies of isotopic molecules. The electronic structure of a given element is not changed by isotopic substitution. Thus isotopes of a given element in compounds form the same bonds. Because the energy levels for vibration depend on mass as well as on bond strength, isotopic substitution causes a change in vibration frequency. These differences lead to small fractionations. The bonds formed by light isotopes are more readily broken than bonds involving heavy isotopes. Isotope fractionation measurements taken during irreversible chemical reactions always show a preferential enrichment of the lighter isotope in the products of the reaction (For a comprehensive discussion of isotope effects see Broecker and Oversby, 1971).

Isotopic data are typically recorded as δ values:

$$\delta(‰) = \frac{(^{13}C/^{12}C)_{sample} - (^{13}C/^{12}C)_{standard}}{(^{13}C/^{12}C)_{standard}} \times 1000$$

$$= \left(\frac{(^{13}C/^{12}C)_{sample}}{(^{13}C/^{12}C)_{standard}} - 1\right) 1000 \qquad (87)$$

For example, a sample with a δ ^{13}C [or $\delta(^{13}C/^{12}C)$] $= -15‰$ indicates that the sample is depleted in ^{13}C relative to the standard such that the $^{13}C/^{12}C$ ratio differs by 15‰. There are some worldwide standards in use: for C it is CO_2 prepared from a Cretaceous belemnite from the Peedee formation of South Carolina; for O the usual standard is standard mean ocean water (SMOW).

4.9 Carbon Isotopes and Isotope Fractionation

Isotopic Fractionation During Evaporation and Condensation

Fractionation within the hydrosphere occurs almost exclusively during vapor-to-liquid or vapor-to-solid phase changes. For example, it is evident from the vapor pressure data for water (21.0, 20.82, and 19.51 mm Hg for $H_2^{16}O$, $H_2^{18}O$, and $HD^{16}O$, respectively) that the vapor phase is preferentially enriched in the lighter molecular species, the extent depending on the temperature (Raleigh distillation). The progressive formation and removal of raindrops from a cloud and the formation of crystals from a solution too cool to allow diffusive equilibrium between the crystal interior and the liquid, that is, isotopic reactions carried out in such a way that the products are isolated immediately after formation from the reactants, show a characteristic trend in isotopic composition.

Ocean water of $\delta\ ^{18}O = 0$ stays in equilibrium with a vapor of $\delta\ ^{18}O = -10‰$ and $\delta D = -9‰$ (at about 10°C). If this vapor condenses during cloud formation, the first droplets formed have $\delta\ ^{18}O = 0$. If the liquid phase is always removed from the vapor phase during condensation, as is approximately the case in the atmosphere, the remaining vapor phase will become more and more depleted in ^{18}O as condensation proceeds, so that the δ values of both the vapor and the liquid will decrease.

In the atmosphere–ocean system, the subtropical oceanic areas are the main supply regions for atmospheric water vapor. As this vapor spreads north and over the continents and decreases as a result of precipitation, the δ values decrease until they reach values of -6/mil in fresh water in midlatitudes and about -30/mil in polar precipitations. Since the δ values reflect the fraction of moisture removed and since this fraction is controlled by the atmospheric temperature, the δ values of precipitation are strongly correlated with the atmospheric temperature (Epstein et al., 1977). $^{18}O/^{16}O$ ratios serve as oceanic water mass tracers. There is a linear relationship between ^{18}O and salinity in water samples of different geographic locations in the North Atlantic Ocean. This suggests that all these samples are mixtures of fresh water (salinity $\simeq 0$ and $\delta\ ^{18}O \simeq -20$/mil) and water typical of the open (salinity $= 36$/mil and $\delta\ ^{18}O = 1$/mil) (Broecker, 1974).

Carbon-13

We first consider a typical exchange reaction expressing the distribution of ^{12}C and ^{13}C between two phases:

$$^{13}CO_2(g) + H^{12}CO_3^-(aq) \rightleftharpoons {}^{12}CO_2(g) + H^{13}CO_3^-(aq)$$

which is characterized by the equilibrium constant

$$K = \frac{[^{12}CO_2(g)]\,[H^{13}CO_3^-(aq)]}{[^{13}CO_2(g)]\,[H^{12}CO_3^-(aq)]} \tag{88}$$

Dissolved Carbon Dioxide

One often defines a fractionation factor $\bar{\alpha}$:

$$\bar{\alpha} = \frac{(^{13}C/^{12}C)_{HCO_3^-}}{(^{13}C/^{12}C)_{CO_2(g)}} \tag{89}$$

For equilibrium 88, $K = \bar{\alpha}$. In general, for an isotopic equilibrium reaction,

$$aA_1 + bB_2 \rightleftharpoons aA_2 + bB_1 \tag{90}$$

where A and B are different molecules, both of which contain a common element existing in the light isotopic form (subscript 1) and the heavy isotopic form (subscript 2), the fractionation factor is given by

$$\bar{\alpha} = K^{1/ab} \tag{91}$$

The equilibrium constant K for an isotopic exchange reaction depends on the vibrational frequency shifts produced upon isotopic substitution and can be calculated in principle from knowledge of the fundamental frequencies of the isotopic species. At high temperatures differences in chemical properties cease.

For reaction 88, the distribution factor has been determined experimentally (Deuger and Degens, 1967): $\bar{\alpha} = K = 1.0092$ (0°C) and $= 1.0068$ (30°C). Thus $\Delta_{CO_2-HCO_3^-} = \delta_{HCO_3^-} - \delta_{CO_2(g)} = 9.2‰$ (0°C) or 6.8‰ (30°C). This fractionation occurs predominantly in the hydration stage and not during the passage of the atmospheric carbon dioxide through the air–water interface; that is, the reaction

$$^{13}CO_2(g) + {}^{12}CO_2(aq) \rightleftharpoons {}^{12}CO_2(g) + {}^{13}CO_2(aq) \tag{92}$$

has an equilibrium constant of $K = \alpha = 1.00$.

Example 4.14. δ ^{13}C of Aqueous Carbonate System Estimate the $\delta\ ^{13}C$ of an aqueous carbonate system of pH = 7.2 in equilibrium with a reservoir of a given $p_{CO_2} = 10^{-2}$ atm and a given $\delta\ ^{13}C_{CO_2} = -25‰$ (10°C). (This p_{CO_2} pressure and $\delta\ ^{13}C_{CO_2}$ value may be representative of the gas phase under soil conditions.) At 10°C the following equilibrium constants, using gaseous CO_2 as a reference, interrelating the various carbon species are given (Deines, Langmuir, and Harmon, 1974):

$$K_0 = \frac{(^{13}C/^{12}C)_{H_2CO_3^*}}{(^{13}C/^{12}C)_{CO_2(g)}} = 0.999 \tag{i}$$

$$K_1 = \frac{(^{13}C/^{12}C)_{HCO_3^-}}{(^{13}C/^{12}C)_{CO_2(g)}} = 1.0092 \tag{ii}$$

$$K_2 = \frac{(^{13}C/^{12}C)_{CO_3^{2-}}}{(^{13}C/^{12}C)_{CO_2(g)}} = 1.0075 \tag{iii}$$

4.9 Carbon Isotopes and Isotope Fractionation

The composition of the solution can be computed (cf. equations 5–8) using the following equilibrium constants valid at 10°C and corrected for an ionic strength of 4×10^{-3} M ($-\log K_H = 1.27$, $-\log K_1' = 6.43$, $-\log K_2' = 10.38$) as $C_T = 3.7 \times 10^{-3}$ M, $[H_2CO_3^*] = 5.4 \times 10^{-4}$ M, $[HCO_3^-] = 3.2 \times 10^{-3}$ M, and $[CO_3^{2-}] = 2 \times 10^{-6}$ M.

Combining equation i with the definition of $\delta\ ^{13}C$, we obtain

$$K_0 = \frac{\delta\ ^{13}C_{H_2CO_3^*} + 1000}{\delta\ ^{13}C_{CO_{2(g)}} + 1000} \quad\text{(iv)}$$

or

$$\delta\ ^{13}C_{H_2CO_3^*} = K_0 \delta\ ^{13}C_{CO_{2(g)}} + (K_0 - 1) \times 1000 = -26‰ \quad\text{(v)}$$

and

$$\delta\ ^{13}C_{HCO_3^-} = K_1 \delta\ ^{13}C_{CO_2} + (K_1 - 1) \times 1000 = -16‰ \quad\text{(vi)}$$

and

$$\delta\ ^{13}C_{CO_3^{2-}} = K_2 \delta\ ^{13}C_{CO_2} + (K_2 - 1) \times 1000 = -17.7‰ \quad\text{(vii)}$$

The isotopic composition of the carbon in the solutions is

$$\delta\ ^{13}C_{sol} = ([H_2CO_3^*]\delta\ ^{13}C_{H_2CO_3^*} + [HCO_3^-]\delta\ ^{13}C_{HCO_3^-}$$
$$+ [CO_3^{2-}]\delta\ ^{13}C_{CO_3^{2-}})/C_T \delta\ ^{13}C_{sol} = -17.6‰ \quad\text{(viii)}$$

In this example it was assumed that the carbon gas reservoir of a given $\delta\ ^{13}C$ was large in comparison to the C reservoir of the water and the solid phases and was in continuous isotopic exchange with these phases. If such a water becomes isolated from the gas reservoir and then interacts with calcite or dolomite, then the ^{13}C content of the solution will also depend on that of the dissolving carbonate rock.

Biogenic Organic Matter

During photosynthesis, plants discriminate against ^{13}C in favor of ^{12}C, and as a result the $^{13}C/^{12}C$ ratios of biogenic materials are lower than those of atmospheric CO_2. The photosynthetic fractionation was shown to be comprised of two steps: (1) preferential uptake of ^{12}C from the atmosphere and (2) preferential conversion of ^{12}C-enriched dissolved CO_2 to phosphoglyceric acid, the first product of photosynthesis. The subsequent metabolism of photosynthetic products may also be accompanied by isotopic fractionation, but these fractionations are relatively small.

McKenzie (1985) has exemplified how carbon isotopes can be used to interpret productivity in the lacustrine and marine environments.

SUGGESTED READINGS

Berner, R. A., and Lasaga, A. C. (1989), Modeling the Geochemical Carbon Cycle, *Sci. Am.* **260**, 74–81.

Busenberg, E., and Plummer, L. N. (1986) A Comparative Study of the Dissolution and Crystal Growth Kinetics of Calcite and Aragonite. In *Studies in Diagenesis*, F. A. Mumpton, Ed., *U.S. Geol. Surv. Bull.* **1578**, 139–168.

Butler, J. N. (1982) *CO_2-Equilibria and Their Applications.* Newly printed by Lewis Publishers, Chelsea, MI.

Dickson, A. G. (1984) pH Scales and Proton Transfer Reactions in Saline Media Such as Seawater, *Geochim. Cosmochim. Acta* **48**, 2299–2308.

Hem, J. D. (1985) Study and Interpretation of the Chemical Characteristics of Natural Water, *U.S. Geological Survey Water Supply Paper* **2254**.

Hemond, H. F. (1990) ANC, Alk and the Acid–Base Status of Natural Waters Containing Organic Acids, *Environ. Sci. Technol.* **24**, 1486–1489.

Millero, F. J., Zhang, J. Z., Lee, K., and Campbell, D. M. (1993) Titration Alkalinity of Seawater. *Marine Chem.* **44**, 153–165.

Morse, J. W., and Mackenzie, F. J. (1990) *Geochemistry of Sedimentary Carbonates*, Elsevier, Amsterdam. Comprehensive treatment of carbonate geochemistry, covering the range from electrolyte chemistry of carbon-containing waters to the global cycles of carbon.

Nordstrom, D. K., Plummer, L. N., Langmuir, D., Busenberg, E., May, H. M., Jones, B. F., and Parkhurst, D. L. (1990) Revised Chemical Equilibrium Data for Major Water–Mineral Reactions and Their Limitations. In *Chemical Modeling of Aqueous Systems II*, D. C. Melchior and R. L. Bassett, Eds., ACS Series **416**, 398–413.

Sarmiento, J. L. (1993) Ocean Carbon Cycle: Most of the Carbon Released from Fossil Fuels Will End Up in the Oceans Where a Complex Cycle of Circulation and Other Processes Control Its Fate, *Chem. Eng. News* **72**, No. 22 (May 31), 30–43.

Wollast, R. (1990), Rate and Mechanism of Dissolution of Carbonates in the System $CaCO_3$–$MgCO_3$. In *Aquatic Chemical Kinetics, Reaction Rates of Processes in Natural Waters*, W. Stumm, Ed., Wiley-Interscience, New York, pp. 431–445.

PROBLEMS

4.1. Does the alkalinity of a natural water (isolated from its surroundings) increase, decrease, or stay constant upon addition of small quantities to the following:
   (a) HCl
   (b) NaOH
   (c) Na_2CO_3

(d) NaHCO₃
(e) CO₂
(f) AlCl₃
(g) Na₂SO₄

4.2. If deep ocean water were stored in the laboratory, how would its pH, [Alk], and [CO_3^{2-}] change under the following conditions:
  (a) At 5°C, 1 atm pressure?
  (b) At 20°C, 1 atm pressure?

4.3. The following short method for the determination of alkalinity has been proposed: add a known quantity of standard mineral acid and measure the final pH. (Mineral acid quantity should preferably be such that final pH is between 3 and 4.3.) Present the theory and show how you compute the alkalinity.

4.4. A water containing 1.0×10^{-4} mol CO_2 liter^{-1} and having an alkalinity of 2.5×10^{-4} eq liter^{-1} has a pH of 6.7. The pH is to be raised to pH 8.3 with NaOH.
  (a) How many moles of NaOH per liter of water are needed for this pH adjustment? ($pK_1 = 6.3$ and $pK_2 = 10.3$.)
  (b) How many moles of lime (Ca(OH)₂) would be required?

4.5. An industrial waste from a metals industry contains approximately 5×10^{-3} M H_2SO_4. Before being discharged into the stream, the water is diluted with tap water in order to raise the pH. The tap water has the following composition: pH = 6.5 and alkalinity = 2×10^{-3} eq liter^{-1}. What dilution is necessary to raise the pH to approximately 4.3?

4.6. A surface water has an alkalinity of 2 meq liter^{-1} and a measured pH of 7.8.
  (a) What is the direction of the flux of CO_2 across the air–water interface?
  (b) Calculate the instantaneous flux.
  (c) What is the time scale for reaching 99% of air–water equilibrium with respect to CO_2? The film thickness is estimated at 300 μm. The molecular diffusion coefficient of CO_2 in water is 1.9×10^{-5} cm² s^{-1}. Assume that the partial pressure of CO_2 in the air is 0.00033 atm.

4.7. A cascade aerator operating on well water will reduce the dissolved CO_2 content from 45 to 18 mg liter^{-1} in one pass. The atmospheric saturation value of CO_2 may be assumed to be 0.5 mg liter^{-1}. Laboratory tests indicate that the gas transfer coefficient for H_2S is about 80% that for CO_2. Assume that enough H_2S will be present in the air around the aerator to give a saturation value of 0.1 mg liter^{-1}. Estimate the effluent concentration of H_2S if the well water contains 12 mg liter^{-1} of H_2S.

204   Dissolved Carbon Dioxide

4.8. How is the relative distribution of $H_2CO_3^*$, HCO_3^-, and CO_3^{2-} in a solution with $[Alk] = 2.5 \times 10^{-3}$ eq liter^{-1} affected by a variation in partial pressure of CO_2? Plot distribution versus p_{CO_2}.

4.9. Compute the pH variation resulting from isothermal evaporation (25°C) of an incipiently 10^{-5} M $NaHCO_3$ solution that remains in equilibrium with the partial pressure of CO_2 of 3×10^{-4} atm.

4.10. Estimate the flux of CO_2 carried into a surface water by rainfall. Compare this flux with the exchange rate at the air–water interface. An annual rainfall of 100 cm may be assumed.

4.11. A little pond at noon has a pH = 8.2; in the morning the pH is 7.5. If one blows air through a sample of the pond water, one obtains a pH of 7.7.

(a) Explain these pH differences.

(b) What is the approximate alkalinity of the water?

4.12. We have shown that [Alk] does not change upon addition (or withdrawal) of CO_2. In the literature one finds the plausibility argument: if CO_2 dissolves in a natural water, the CO_2 reacts with CO_3^{2-} ions: $CO_2 + CO_3^{2-} + H_2O = 2\ HCO_3^-$. Thermodynamically there is the following equilibrium:

$$[HCO_3^-]^2 / ([CO_3^{2-}][H_2CO_3^*]) = K$$

How good is the argument? Under what condition are the two equations valid?

ANSWERS TO PROBLEMS

4.1. Decrease for (a) and (f); increase for (b), (c), and (d); no change for (e) and (g).

4.3. $\bar{C}_A[V/(V_0 + V)] - [H^+] \simeq [Alk]$ (V_0 = original volume of sample, V = volume of strong acid added, $\bar{C}_A$ = molarity of strong acid). The strong acid that has been added is equivalent to the alkalinity originally contained in the sample plus the mineral acidity that remains after the acid was added. The mineral acidity can be estimated by measuring pH, because $[H\text{-}Acy] \simeq [H^+]$.

4.4. (a) 1.0×10^{-4}; (b) 0.5×10^{-4}. Note that pH = 8.3 corresponds to the equivalence point ($f = 1$). Compare Figure 4.1 (point y).

4.5. Fivefold dilution. Note that pH = 4.3 corresponds to the equivalence point ($f = 0$, x in Figure 4.1) for the titration of alkalinity. The problem is equivalent to the titration of [Alk] of tap water with the acid of the waste water.

Answers to Problems 205

4.7. 6 mg liter^{-1} H$_2$S. The rate of decrease in the concentration of the gas will be proportional to the oversaturation; that is, $-(C_0 - C_T)/(C_0 - C_3) = \exp(-kt)$.

4.8. Use equation v of Example 4.5 and solve by trial and error. Perhaps more conveniently diagrams such as Figure 4.9 may be constructed in order to compute the equilibrium concentrations.

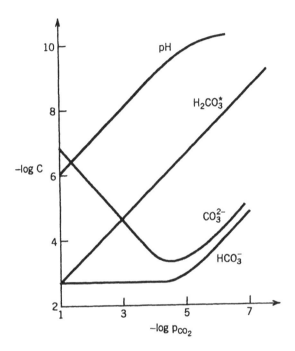

4.9. Use equation v of Example 4.5 or read result from Figure 4.9. For [Na$^+$] = 10^{-5} M, pH = 6.3; for [Na$^+$] = 10^{-3} M, pH = 8.3; for [Na$^+$] = 10^{-2} M, pH = 9.2; for [Na$^+$] = 10^{-1} M, pH = 9.9.

4.11. (b) [Alk] ≃ 2.5 × 10^{-4} M.

5

ATMOSPHERE–WATER INTERACTIONS

5.1 INTRODUCTION

Although water is a very minor component of the atmosphere—less than 10^{-6} vol % of the atmosphere consists of water—many important reactions occur in the water droplets of cloud, fog, and rain. The atmosphere is an oxic environment; in its water phase, gigantic quantities of reductants, such as organic substances, Fe(II), SO_2, CH_3SCH_3, and nitrogen oxides, are oxidized by oxidants such as oxygen, $OH^{\cdot}$ radicals, H_2O_2, and Fe(III).

Hydrogeochemical cycles couple atmosphere, land, and water. The atmosphere is an important conveyor belt for many pollutants. The atmosphere reacts most sensitively to anthropogenic disturbance because it represents a much smaller reservoir than land and water; furthermore, the residence times of many constituents of the atmosphere are smaller than those in the other exchange reservoirs. Water and atmosphere are interdependent systems. Many pollutants, especially precursors of acids and photooxidants, originate directly or indirectly from the combustion of fossil fuels. Hydrocarbons, carbon monoxides, and nitrogen oxides released in thermal power plants and, above all, by automobile engines can produce, under the influence of sunlight, ozone and other photooxidants.

In this chapter we will deal with some important reactions at the gas–water interface and discuss above all the partitioning of molecules between the gas phase and the water phase (Henry's law). We will also explain the processes that influence wet and dry deposition and the composition of atmospheric water droplets (clouds, fog, rain, snow, dew) and illustrate how pollutants released into the atmosphere are transferred back to the land. Attention will be paid to the disturbance of the proton balance by the oxides of C, N, and S, anthropogenically released into the atmosphere, and how this disturbance is transferred from the atmosphere to the terrestrial and aquatic ecosystems.

5.2 ANTHROPOGENIC GENERATION OF ACIDITY IN THE ATMOSPHERE

Acid atmospheric deposition results from the disturbance of cycles that couple atmosphere, land, and water. The surface of our environment is, in a global sense, in a stationary state with regard to a proton and electron balance.

Oxidation and reduction are accompanied by proton release and proton consumption, respectively. (In order to maintain charge balance, the production of e^- will eventually be balanced by the production of H^+.) Furthermore, the dissolution of rocks and the precipitation of minerals are accompanied by H^+ consumption and H^+ release, respectively.

Genesis of Acid Precipitation

The oxidation of carbon, sulfur, and nitrogen, resulting mostly from fossil fuel burning, disturbs redox conditions in the atmosphere.

In oxidation–reduction reactions, electron transfers (e^-) are coupled with the transfer of protons (H^+) to maintain a charge balance. A modification of the redox balance corresponds to a modification of the acid–base balance. The net reactions of the oxidation of C, S, and N exceed reduction reactions in these elemental cycles. A net production of H^+ ions in atmospheric precipitation is a necessary consequence. The disturbance is transferred to the terrestrial and aquatic environments, and it can impair terrestrial and aquatic ecosystems.

Figure 5.1 shows the various processes that involve atmospheric pollutants and natural components in the atmosphere. The following reactions are of particular importance in the formation of acid precipitation: oxidative reactions, either in the gaseous phase or in the aqueous phase, leading to the formation of oxides of C, S, and N (CO_2; SO_2, SO_3, H_2SO_4; NO, NO_2, HNO_2, HNO_3); absorption of gases into water (cloud droplets, falling raindrops, or fog) and interaction of the resulting acids ($SO_2 \cdot H_2O$, H_2SO_4, HNO_3) with ammonia (NH_3) and the carbonates of airborne dust; and the scavenging and partial dissolution of aerosols into water. In this case, aerosols are produced from the interaction of vapors and airborne (maritime and dust) particles; they often contain $(NH_4)_2SO_4$ and NH_4NO_3.

Deposition The products of the various chemical and physical reactions are eventually returned to the earth's surface. Usually, one distinguishes between *wet* and *dry deposition*. Wet deposition (rainout and washout) includes the flux of all those components that are carried to the earth's surface by rain or snow, that is, those dissolved and particulate substances contained in rain or snow. Dry deposition is the flux of particles and gases (especially SO_2, HNO_3, and NH_3) to the receptor surface during the absence of rain or snow. Deposition also can occur through fog aerosols and droplets, which can be

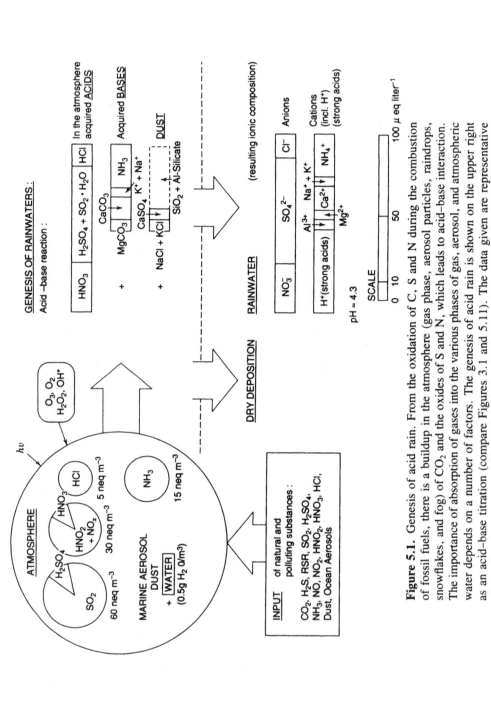

Figure 5.1. Genesis of acid rain. From the oxidation of C, S and N during the combustion of fossil fuels, there is a buildup in the atmosphere (gas phase, aerosol particles, raindrops, snowflakes, and fog) of CO_2 and the oxides of S and N, which leads to acid–base interaction. The importance of absorption of gases into the various phases of gas, aerosol, and atmospheric water depends on a number of factors. The genesis of acid rain is shown on the upper right as an acid–base titration (compare Figures 3.1 and 5.11). The data given are representative of the situation encountered in Zürich, Switzerland. (From Schnoor and Stumm, 1985.)

5.2 Anthropogenic Generation of Acidity in the Atmosphere

deposited on trees, plants, or the ground. With forests, approximately half of the deposition of SO_4^{2-}, NO_3^-, and H^+ occurs as dry deposition (Lindberg et al., 1986) (Table 5.1).

Evapotranspiration of water from the receptor surfaces causes an increase in the concentration of conservative solutes in the soil solution and in water draining from the root zone.

Stoichiometric Model The rainwater shown in Figure 5.1 contains an excess of strong acids, most of which originate from the oxidation of sulfur during fossil fuel combustion and from the fixation of atmospheric nitrogen to NO and NO_2, for example, during combustion of gasoline by motor vehicles. It also should be mentioned that there are natural sources of acidity, resulting from volcanic activity, from H_2S from anaerobic sediments, and from dimethyl sulfide and carbonyl sulfide that originate in the ocean. HCl results from the combustion and decomposition of organochlorine compounds such as polyvinyl chloride. Bases originate in the atmosphere as the carbonate of wind-blown dust and from NH_3, generally of natural origin. The NH_3 comes from NH_4^+ and from the decomposition of urea in soil and agricultural environments.

The reaction rates for oxidation of atmospheric SO_2 (0.05–0.5 day^{-1}) yield a sulfur residence time of several days at the most; this corresponds to a transport distance of several hundred to 1000 km. The formation of HNO_3 by oxidation is more rapid and, compared with H_2SO_4, results in a shorter travel distance from the emission source. H_2SO_4 also can react with NH_3 to form NH_4HSO_4 or $(NH_4)_2SO_4$ aerosols. In addition, the NH_4NO_3 aerosols are in equilibrium with $NH_3(g)$ and $HNO_3(g)$ (Seinfeld, 1986).

The flux of dry deposition is usually assumed to be a product of its concentration adjacent to the surface and the deposition velocity. Deposition velocity depends on the nature of the pollutant (type of gas, particle size), the turbulence of the atmosphere, and the characteristics of the receptor surface (water, ice, snow, vegetation, trees, rocks). Dry deposition is usually collected for study in open buckets during dry periods, but this method underestimates the flux of pollutants, particularly gases (SO_x, NO_x), to a foliar canopy or lake surface. [Fog can be collected with special devices (Zobrist et al., 1993).]

The foliar canopy receives much of its dry deposition in the form of sulfate, nitrate, and hydrogen ion, which occur primarily as SO_2, HNO_3, and NH_3 vapors. Dry deposition of coarse particles has been shown to be an important source of calcium and potassium ion deposition on deciduous forests in the eastern United States (Table 5.1).

In addition to the acid–base components shown in Figure 5.1, various organic acids are often found. Many of these acids are by-products of the atmospheric oxidation of organic matter released into the atmosphere. Of special interest are formic, acetic, oxalic, and benzoic acids, which have been found in rainwater in concentrations occasionally exceeding a few micromoles per liter. Thus they must be included in the calculated acidity of rainwater, and

Table 5.1. Total Annual Atmospheric Deposition of Major Ions to an Oak Forest at Walker Branch Watershed[a]

Process	Atmospheric Deposition (meq m^{-2} yr^{-1})					
	SO_4^{2-}	NO_3^-	H^+	NH_4^+	Ca^{2+}	K^+
Precipitation	70 ± 5	20 ± 2	69 ± 5	12 ± 1	12 ± 2	0.9 ± 0.1
Dry deposition						
Fine particles	7 ± 2	0.1 ± 0.02	2.0 ± 0.9	3.6 ± 1.3	1.0 ± 0.2	0.1 ± 0.05
Coarse particles	19 ± 2	8.3 ± 0.8	0.5 ± 0.2	0.8 ± 0.3	30 ± 3	1.2 ± 0.2
Vapors[b]	62 ± 7	26 ± 4	85 ± 8	1.3	0	0
Total deposition	160 ± 9	54 ± 4	160 ± 9	18 ± 2	43 ± 4	2.2 ± 0.3

[a]Values are means ± standard errors for 2 years of data. Numbers of observations range from 15 (HNO$_3$) to 26 (particles) to 128 (precipitation) to 730(SO$_2$). In comparing these deposition rates it must be recalled that any such estimates are subject to considerable uncertainty. The standard errors given provide only a measure of uncertainty in the calculated sample means relative to the population means; hence additional uncertainties in analytical results, hydrologic measurements, scaling factors, and deposition velocities must be included. The overall uncertainty for wet deposition fluxes is about 20% and that for dry deposition fluxes is approximately 50% for SO_4^{2-}, Ca^{2+}, K^+, NH_4^+ and approximately 75% for NO_3^- and H^+.
[b]Includes SO$_2$, HNO$_3$, and NH$_3$. Complete conversion of deposited SO$_2$ to H$_2$SO$_4$ and of NH$_3$ to NH$_4^+$ was assumed in determining the vapor input of H^+. NH$_3$ deposition was estimated from the literature.

Source: Lindberg et al. (1986).

5.2 Anthropogenic Generation of Acidity in the Atmosphere

their presence in larger concentrations has an influence on the pH of rain and fog samples.

The composition of the rain—an "average" inorganic composition is given in Figures 3.1 and 5.1—reflects the acid–base titration that occurs in the atmosphere. Total concentrations (the sum of cations or anions) typically vary from 20 to 500 μeq liter^{-1} and pH from 3.5 to 6.

Figure 5.2 illustrates some of the interactions that occur at the interface of the atmosphere with a water droplet or at a surface water.

When fog is formed from water-saturated air, water droplets condense on aerosol particles. In addition to components of the aerosols, the fog droplets can absorb such gases as NO_x, SO_2, NH_3, and HCl; they form a favorable milieu for various oxidation processes, especially the formation of H_2SO_4. Fog droplets (10–50 μm in diameter) are much smaller than rain droplets; the liquid water content of fog is often in the range of 1×10^{-4} liter m^{-3} air;

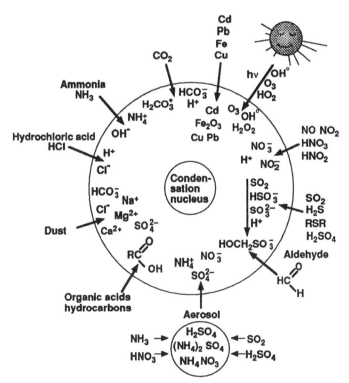

Figure 5.2. Various interactions that determine the composition of a water droplet in the atmosphere (e.g., cloud, fog). Aerosol particles, which to a large extent consist of $(NH_4)_2SO_4$ and NH_4NO_3, can form the nuclei for the condensation of liquid water. Various gases can become absorbed into the aqueous phase. The atmosphere is an oxidative environment; the water phase, often assisted by light, promotes oxidation reactions, for example, the oxidation of SO_2 to H_2SO_4 and of organic matter to CO_2. NH_3 neutralizes mineral acids and buffers the solution phase.

acids in fog are typically 10–50 times more concentrated than those found in rain.

Rain clouds process a considerable volume of air over relatively large distances and thus are able to absorb gases and aerosols from a large region. Because fog is formed in the lower air masses, fog droplets are efficient collectors of pollutants close to the earth's surface. The influence of local emissions (such as NH_3 in agricultural regions or HCl near refuse incinerators) is reflected in the fog composition.

Example 5.1. Composition of Rainwater A rainwater contains the following constituents (in μM): NO_3^-, 60; SO_4^{2-}, 44; Cl^-, 25; Mg^{2+}, 4.5; Ca^{2+}, 16; NH_4^+, 85; K^+, 2; Na^+, 5; H^+, 50; and CH_3COOH, 5.

Sketch the titration curve and estimate the concentrations of acids and bases that participated in establishing the composition of the rainwater. This rainwater contains 50 μM strong acid (mineral acidity) and 90 μM weak acid (NH_4^+ and CH_3COOH). A strong base titrates first the mineral acidity, then the acetic acid, and finally the NH_4^+. In a pH versus strong base plot (see Figure 4.11a) the main pH jump occurs at the end point of the strong acid base titration; the effect of NH_4^+ is only seen in a pH buffering in the pH range 9–10. As we have shown in Chapter 4 (Figure 4.11b), the Gran titration procedure permits us to assess separately mineral acidity (H-Acy) and total acidity (H-Acy + $[NH_4^+]$ + $[CH_3COOH]$).

Although the rainwater consists of ions (and of CH_3COOH molecules), we can think in terms of the approximate stoichiometric mass balance that produced the composition of the rainwater. In 1 liter of water 88 μeq of H_2SO_4 (from SO_2), 60 μmol of HNO_3 (from NO_x), and 25 μmol of HCl have reacted with 85 μmol of NH_3, 41 μeq base particles ($CaCO_3$ and $MgCO_3$), and traces of Na^+ and K^+ aluminum silicates.[†] Note that the charge balance is not exact (173 μeq liter^{-1} of anions versus 183 μeq liter^{-1} of cations). Some of the difference may be due to analytical errors, but, more likely, the difference may be due to analytically unaccounted for organic anions. The acetic acid (pK = 4.7) occurs, at the pH of the rainwater, 70% as free acid and 30% as anion.

5.3 GAS–WATER PARTITIONING: HENRY'S LAW

The distribution of gas molecules between the gas phase and the water phase depends on the Henry's law equilibrium distribution. In the case of CO_2, SO_2, and NH_3, the dissolution equilibrium is pH dependent because the species in the water phase—$CO_2(aq)$, H_2CO_3, $SO_2 \cdot H_2O(aq)$, $NH_3(aq)$—undergo acid–base reactions.

Two varieties of calculation are possible: in an open-system model, a constant partial pressure of the gas component is maintained; in a closed-system model, an initial partial pressure of a component is given, for example, for a

[†]The answer given is a reasonable guess; we could have assumed alternatively that some of the Ca^{2+} and SO_4^{2-} originated from the dissolution of $CaSO_4$ in the dust.

5.3 Gas–Water Partitioning: Henry's Law

cloud before rain droplets are formed or for a package of air before fog droplets condense. In these cases, the system is considered closed: from then on, the total concentration in the gas phase and in the solution phase is constant.

We will briefly review Henry's law, the gas–water equilibrium concept, which we have already applied to heterogeneous equilibria (Chapter 4), and then demonstrate its application in a few examples.

Henry's Law

Henry's law describes the equilibrium distribution of a volatile species between liquid and gaseous phases. In the original form, Henry's law is an observational result for a two-phase equilibrium $A(l) = A(g)$ under dilute solution conditions and for low pressures,

$$p_A = K' C_A \tag{1a}$$

where p_A is the partial pressure and C_A is the liquid-phase concentration for species A. (For example, with p_A in atmospheres and C_A moles per liter, units of K' are atm liter mol^{-1}.) Thermodynamic generality requires that the original law be reexpressed in terms of fugacity and activity:

$$f_A = K a_A \tag{1b}$$

We will assume dilute conditions in the following discussion.

The physicochemical significance of Henry's law is this: there is a linear relationship between the activity of a volatile species in the liquid phase and its activity in the gas phase. (This simple notion is sometimes lost sight of when fundamental gas solubility equilibria are combined with other equilibria, e.g., acid–base, in order to generate overall distribution constants.) Since we may write either $A(aq) = A(g)$ or $A(g) = A(aq)$ for the two-phase distribution, a variety of "Henry's law constants" exist, but that need not concern us. Conversions are straightforward. We make use of two convenient forms in this book, both written in the direction $A(g) = A(aq)$: (1) H (dimensionless) and (2) K_H (M atm^{-1}).

The equilibrium constant H is defined as

$$\frac{[A(aq)]}{[A(g)]} = H \quad \text{(dimensionless)} \tag{2}$$

For example, for $O_2(g) = O_2(aq)$

$$\frac{[O_2(aq)] \; (\text{mol liter}^{-1})}{[O_2(g)] \; (\text{mol liter}^{-1})} = H_{O_2} \tag{3}$$

The equilibrium constant K_H is defined as

$$\frac{[A(aq)]}{p_A} = K_H \quad (\text{M atm}^{-1}) \tag{4}$$

For example, for $O_2(g) = O_2(aq)$,

$$\frac{[O_2(aq)] \text{ (mol liter}^{-1})}{p_{O_2} \text{ (atm)}} = K_H \text{ (M atm}^{-1}) \tag{5}$$

Conversion between H and K_H is straightforward. For $A(g) = A(aq)$

$$K_H = H/RT \tag{6}$$

where R is the gas constant (0.082057 liter atm K^{-1} mol^{-1}) and T is temperature (K).

For example, at 25°C the value of H for oxygen dissolution in water is

Table 5.2. Equilibrium Constants of Importance in Gas–Water Equilibria

Reactions		$K_{25°C}$[a]
1. $CO_2(g) + H_2O(l)$	$= H_2CO_3^*(aq)$	3.39×10^{-2}
2. $H_2CO_3^*$	$= H^+ + HCO_3^-$	4.45×10^{-7}
3. HCO_3^-	$= H^+ + CO_3^{2-}$	4.69×10^{-11}
4. $SO_2(g) + H_2O(l)$	$= SO_2 \cdot H_2O(aq)$	1.25
5. $SO_2 \cdot H_2O$	$= H^+ + HSO_3^-$	1.29×10^{-2}
6. HSO_3^-	$= H^+ + SO_3^{2-}$	6.24×10^{-8}
7. $NH_3(g)$	$= NH_3(aq)$	57
8. $NH_3(aq) + H_2O$	$= NH_4^+ + OH^-$	1.77×10^{-5}
9. $HNO_3(g)$	$= H^+ + NO_3^-$	3.46×10^6
10. $HCl(g)$	$= H^+ + Cl^-$	2.00×10^6
11. $HNO_2(g)$	$= HNO_2(aq)$	49
12. HNO_2	$= H^+ + NO_2^-$	5.13×10^{-4}
13. $H_2S(g)$	$= H_2S(aq)$	1.05×10^{-1}
14. H_2S	$= H^+ + HS^-$	9.77×10^{-8}
15. HS^-	$= H^+ + S^{2-}$	1.00×10^{-19}
16. $NO(g) + NO_2(g) + H_2O(l)$	$= 2 HNO_2(aq)$	1.24×10^2
17. $CH_3COOH(g)$	$= CH_3COOH(aq)$	7.66×10^2
18. CH_3COOH	$= H^+ + CH_3COO^-$	1.75×10^{-5}
19. $CH_2O(g)$	$= CH_2O(aq)$	6.3×10^3
20. $N_2(g)$	$= N_2(aq)$	6.61×10^{-4}
21. $O_2(g)$	$= O_2(aq)$	1.26×10^{-3}
22. $CO(g)$	$= CO(aq)$	9.55×10^{-4}
23. $CH_4(g)$	$= CH_4(aq)$	1.29×10^{-3}
24. $NO_2(g)$	$= NO_2(aq)$	1.00×10^{-2}
25. $NO(g)$	$= NO(aq)$	1.9×10^{-3}
26. $N_2O(g)$	$= N_2O(aq)$	2.57×10^{-2}
27. $H_2O_2(g)$	$= H_2O_2(aq)$	1.0×10^5
28. $O_3(g)$	$= O_3(aq)$	9.4×10^{-3}

[a] Henry constants are K_H values given in M atm^{-1}; Henry constants for organic substances are given in Figures 5.3 and 5.15 and in Table 5.3.

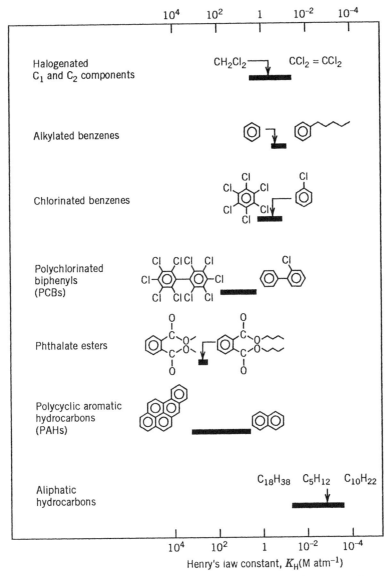

Figure 5.3. Ranges in Henry's law constants (K_H) for some important classes of organic compounds. Compounds with high vapor pressure should partition appreciably from water into air (low K_H values). (Adapted from Schwarzenbach et al., 1993.)

0.0308. Applying equation 6 at $T = 298.15$ K, we obtain $K_H = 0.0308/(0.082057 \times 298.15) = 1.26 \times 10^{-3}$ M atm^{-1}.

Table 5.2 lists some equilibrium constants of importance in atmospheric water interactions. Figure 5.3 gives ranges of Henry's law constants (K_H) for some important classes of organic compounds.

Henry's law constants, as other thermodynamic constants, are valid for ideal solutions; ideally, the expression should be written in terms of activities and fugacities. Since activity coefficients for uncharged species are much nearer unity than those for ions, we can use expressions such as equation 3 for dilute solutions (fresh water) and atmospheric pressures. However, corrections are necessary for seawater and concentrated solutions. Since activity coefficients for molecules in aqueous solution become larger than 1 (salting out effect), the solubility of gases in concentration units is smaller in the salt solution than in the dilute aqueous medium.

For example, the solubility of CO_2 at 20°C in dilute water is 3.91×10^{-2} M atm^{-1} while in seawater of 10‰ and 35‰ salinity, it is 3.73×10^{-2} M atm^{-1} and 3.32×10^{-2} M atm^{-1}, respectively.

5.4 GAS-WATER EQUILIBRIA IN CLOSED AND OPEN SYSTEMS

In dealing with gas-water equilibria, one needs to distinguish between "open" and closed systems (see Figure 4.2).

- In the *open* system, water is in contact with an unlimited amount of a gas; that is, the partial pressure is constant and is not changed by the quantity of the gas that is absorbed in the water phase (cf. Figure 4.2c). This system represents an appropriate model for equilibrium of a surface water with the atmosphere or for rainwater in contact with large quantities of air.
- In the *closed* system, a limited quantity of a gas becomes distributed between the gas and water phase. Equilibrium concentrations always correspond to the Henry constants but the relative proportions in the gas and water phase depend on the ratio of the volumes of water and gas. Such a closed system may, for example, serve as a model for fog droplets, when under stagnant conditions water droplets are in contact with a limited amount of a gas. The assumption of a closed system is often justified in situations where a significant proportion of a volatile substance becomes absorbed in the water phase.

Groundwater systems can often be looked at as a system such as symbolized in Figure 4.2a. Depending on the volume ratio of gas to water, the groundwater system can be treated ideally either as an open system (relative large reservoir of gaseous components, i.e., $p_A \simeq$ constant) or as a closed system [(A)$_{tot} \simeq$ constant].

The gas concentration is given by

$$(A)_g \text{ (mol m}^{-3}) = p_A/RT \qquad (7)$$

5.4 Gas-Water Equilibria in Closed and Open Systems

and the concentration of A in the water phase for the entire system (per m^3 atmosphere), $R = 8.2057 \times 10^{-5}$ m^3 atm K^{-1} mol^{-1}, $(A)_w$, is given by

$$(A)_w = (\text{mol m}^{-3}) = [A(aq)] \times q \tag{8}$$

where $[A(aq)]$ = concentration in water (mol liter^{-1} water)
q = content of water per m^3 of the entire system (liter m^{-3})

Typical water contents are

5×10^{-5}–5×10^{-4} liter m^{-3} for fog
1×10^{-4}–1×10^{-3} liter m^{-3} for clouds

The total concentration is then

$$(A)_{tot} = (A)_g + (A)_w = p_A/RT + [A(aq)]q \; (\text{mol m}^{-3}) \tag{9}$$

or

$$(A)_{tot} = (A)_g + K_H RT \cdot (A)_g \cdot q = (A)_g (1 + K_H \cdot RT \cdot q) \tag{10}$$

Example 5.2. Gas Concentrations The atmosphere contains 500 ppt of COS (carbonyl sulfide). (This gas is to a large extent released—together with dimethylsulfide—by algae and microorganisms in the sea.) What is its partial pressure and its concentration in g m^{-3} at 0°C? In atmospheric science ppm, ppb, and ppt are usually given on a per volume basis. Thus 1 ppm = 10^{-6} atm or 1 ppt = 10^{-12} atm. Thus 500 ppt = 5×10^{-10} atm. Then

$$(COS)_g \; (\text{mol m}^{-3}) = 5 \times 10^{-10} \; \text{atm}/RT \tag{i}$$

RT at 0°C = $8.2057 \times 10^{-5} \times 273.15 = 2.24 \times 10^{-2}$ m^3 atm^{-1} mol^{-1}. Thus $(COS)_g = 2.2 \times 10^{-8}$ mol m^{-3}. The molecular weight of COS is 60. Thus the atmosphere contains about 1.3×10^{-6} g COS per m^3.

Example 5.3a. Closed System: Dissolution of H_2O_2 and Ozone H_2O_2 and O_3 are important oxidants in the atmosphere. Their solubility is pH independent and given at 25°C by (Table 5.2)

$$K_H \; (H_2O_2) = 1.0 \times 10^5 \; \text{M atm}^{-1}$$

$$K_H \; (O_3) = 9.4 \times 10^{-3} \; \text{M atm}^{-1}$$

The fraction of these gases, which will dissolve in a closed system in water, can be calculated as a function of the water content, q (Figure 5.4) (see equations 9 and 10).

Atmosphere-Water Interactions

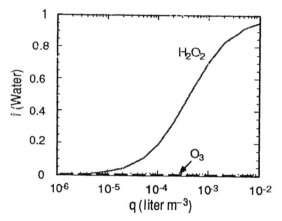

Figure 5.4. Proportion of H_2O_2 and O_3 dissolved in water (closed system) as a function of water content, q (liter water m^{-3}). Because of the large difference in the K_H values of these gases, H_2O_2 is mostly present in the water phase if $q > 10^{-4}$ liter m^{-3}, whereas f_{water} for O_3 is only 2×10^{-8} for $q = 10^{-4}$.

$$\frac{(A)_w}{(A)_{tot}} = \frac{K_H \cdot RT \cdot (A)_g \cdot q}{(A)_g + K_H \cdot RT \cdot (A)_g \cdot q} \qquad \text{(i)}$$

$$f_{water} = \frac{(A)_w}{(A)_{tot}} = \frac{K_H \cdot RT \cdot q}{1 + K_H \cdot RT \cdot q} \qquad \text{(ii)}$$

$$f_{gas} = \frac{(A)_g}{(A)_{tot}} = \frac{1}{1 + K_H \cdot RT \cdot q} \qquad \text{(iii)}$$

Example 5.3b. Distribution Between Water and Atmosphere One liter of water contains initially 1 µg liter^{-1} of the contaminants *n*-octane, toluene, DDT, and Hg. Placed in a closed 10-liter bottle, what fractions of these contaminants are lost at equilibrium? Table 5.3 (see p. 239) gives the necessary data. H values ($H = K_H RT$) for *n*-octane, toluene, DDT, and Hg are, respectively, 7.7×10^{-3}, 3.7, 2.6×10^3, and 2.1. Using equations 9 and 10 or equation iii of Example 5.3a we obtain for the fractions lost from the water: *n*-octane, 0.999; toluene, 0.71; DDT, 0.014; and Hg, 0.81.

Because the equilibrium distribution of these substances between the water and the atmosphere is so much in favor of the latter—except for DDT—their removal by dissolution in rain is not very efficient (compare Section 5.5 and Figure 5.10). Some pollutants, especially those with saturation vapor pressures lower than 10^{-6} atm (e.g., DDT and dieldrin), may be removed by adsorption to atmospheric aerosols and subsequently washed and rained out.

5.4 Gas–Water Equilibria in Closed and Open Systems

Distribution of NH_3 Between Gas Phase and Water

Ammonia is the most important basic component in the atmosphere. The concentration of gaseous NH_3 and its pH-dependent dissolution in the aqueous phase are relevant.

Example 5.4. Ammonia in Open and Closed Systems Compare the solubility of NH_3 in an open and in a closed system. For the open system $p_{NH_3} = 5 \times 10^{-9}$ atm. This corresponds to $NH_3(g) = 2 \times 10^{-7}$ mol m^{-3}. For the closed system $(NH_3)_{tot} = 2 \times 10^{-7}$ mol m^{-3} and the water content is $q = 5 \times 10^{-4}$ liter m^{-3}. Assume 25°C.

The *open system* for NH_3 has already been discussed in Example 3.8. Tableau 3.4 summarized the equilibrium conditions. The solubility of NH_3 is given by (see Figure 5.5)

$$[NH_3]_{aq} = K_H \cdot p_{NH_3} = K_H\, RT \cdot (NH_3)_g \qquad (i)$$

$$[NH_4^+] = [NH_3]_{aq} \cdot [H^+] \cdot K_a^{-1} = K_H K_a^{-1} [H^+] \cdot p_{NH_3}$$

$$= K_H\, RT\, K_a^{-1} [H^+] (NH_3)_g \qquad (ii)$$

For the *closed system*, the mass balance is given by

$$(NH_3)_{tot} = (NH_3)_g + q([NH_3] + [NH_4^+])\ \text{mol m}^{-3} \qquad (iii)$$

$$(NH_3)_{tot} = (NH_3)_g + qK_H\, RT \cdot (NH_3)_g\, (1 + K_a^{-1} [H^+]) \qquad (iv)$$

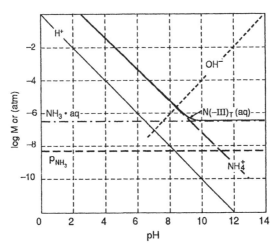

Figure 5.5. NH_3 species in an open system with $p_{NH_3} = $ constant $= 5 \times 10^{-9}$ atm (25°C). The proton condition ($[NH_4^+] + [H^+] = [OH^-]$, or approximately $[NH_4^+] \approx [OH^-]$) corresponds to the condition of a pure ammonia solution (no other acid or base added to the system).

220 Atmosphere-Water Interactions

The fraction in the gas phase is given by

$$\frac{(NH_3)_g}{(NH_3)_{tot}} = \frac{1}{1 + qK_HRT(1 + K_a^{-1}[H^+])} \tag{v}$$

Tableau 5.1 summarizes the equilibrium conditions. Note that the quantity $(NH_3)_{tot}/q$ is formally identical to $TOTNH_3$ (mole per liter of water) used in previous tableaux (e.g., Chapter 3).

Figure 5.6a gives the distribution of the species in the aqueous phase. The graph can be sketched readily by considering the following:

1. The maximum concentration of ammonia in the low pH range (where most ammonia is in the aqueous phase) is given by

$$[NH_4^+(aq)]_{max} = \frac{(NH_3)_{tot}}{q} \text{ (mol liter}^{-1}) = 4 \times 10^{-4} \text{ M} \tag{vi}$$

2. The maximum concentration of ammonia at high pH in the aqueous phase is given by the Henry's law partitioning between the gas phase concentration (2×10^{-7} mol m^{-3}) and the solution

$$[NH_3(aq)]_{max} = (NH_3)_{tot} K_H RT = 2.8 \times 10^{-7} \text{ M} \tag{vii}$$

3. The line for $[NH_4^+]$ and $[NH_3]$ must cross at $pH = pK_{NH_4^+}$. The mass balance equation in Tableau 5.1 is based on equation iii; it might be convenient to express all the quantities on a per liter water basis; that is, equation iii is divided by q:

$$\frac{(NH_3)_{tot}}{q} = \frac{(NH_3)_g}{q} + [NH_3] + [NH_4^+] \text{ (mol liter}^{-1}) \tag{viii}$$

Tableau 5.1. Closed System NH_3-Water

Components		$NH_{3(aq)}$	H^+	log K
Species	$NH_{3(aq)}$	1		0
	NH_4^+	1	1	9.2
	$(NH_3)_g/q$	1		$-0.14 - \log q$
	OH^-		-1	-14
	H^+		1	0
Composition			TOTH = 0	

$$\frac{(NH_3)_{tot}}{q} = 2 \times 10^{-7} \times \frac{1}{q} = 4 \times 10^{-4} \text{ mol liter}^{-1}$$

5.4 Gas–Water Equilibria in Closed and Open Systems

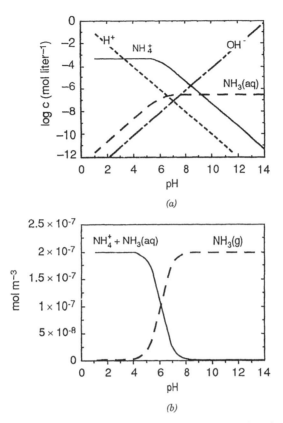

Figure 5.6. Distribution of NH_3 between gas and water phase in closed systems $(NH_3)_{tot}$ = 2×10^{-7} mol m^{-3}; $q = 5 \times 10^{-4}$ liter m^{-3}. (a) Aqueous phase. Without additional acids or bases, the pH and solution composition are given by $[NH_4^+] + [H^+] = [OH^-]$. (b) Concentrations in the gas phase and in the aqueous phase in the closed system. At pH < 5, most ammonia is in the water phase; while at pH > 7, most ammonia is in the gas phase.

Line 3 of Tableau 5.1 gives the first term of the right-hand side of equation viii.

Example 5.5. Solubility of SO_2 in Water Estimate the solubility of SO_2 (a) in an open system ($p_{SO_2} = 2 \times 10^{-8}$ atm) and (b) in a closed system with $(SO_2)_{tot} = 8.2 \times 10^{-7}$ mol m^{-3} (corresponding to 2×10^{-8} atm in the absence of water) and a water content of 5×10^{-4} liter water m^{-3}. (25°C)

(a) *Open System.* Most conveniently the concentrations of the species are represented as a function of p_{SO_2} (compare with open CO_2 system, Section 4.3 and Figure 4.5).

222   Atmosphere–Water Interactions

Tableau 5.2. SO_2–Water: Open System

Components		H^+	$SO_2(g)$	log K (25°C)
Species	H^+	1		
	OH^-	−1		−14
	$SO_2 \cdot H_2O$		1	0.097
	HSO_3^-	−1	1	−1.79
	SO_3^{2-}	−2	1	−9.00
Composition		0	$p_{SO_2} = 2 \times 10^{-8}$ atm	

$$[SO_2 \cdot H_2O] = K_H p_{SO_2} \quad \text{(i)}$$

$$[HSO_3^-] = \frac{K_1}{[H^+]}[SO_2 \cdot H_2O] = \frac{K_1 K_H}{[H^+]} p_{SO_2} \quad \text{(ii)}$$

$$[SO_3^{2-}] = \frac{K_1 K_2}{[H^+]^2}[SO_2 \cdot H_2O] = \frac{K_1 K_2 K_H}{[H^+]^2} p_{SO_2} \quad \text{(iii)}$$

The equilibria involved are summarized in Tableau 5.2. The concentrations are plotted in Figure 5.7.

A pure SO_2-water system at this p_{SO_2} is characterized by pH $\simeq 4.8$, where the proton condition is $[H^+] \simeq [HSO_3^-]$. At higher pH values, that is, upon addition of a base, the solubility becomes very large. At low pH values the solubility is relatively small.

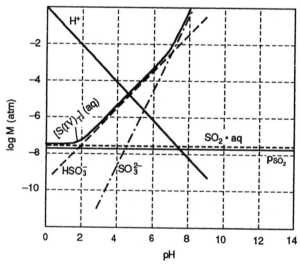

Figure 5.7. S(IV) species in an open SO_2 system; $p_{SO_2} = 2 \times 10^{-8}$ atm.

5.4 Gas–Water Equilibria in Closed and Open Systems

(b) *Closed System.* The mass balance is given by

$$(SO_2)_{tot} = (SO_2)_g + q\,([SO_2 \cdot H_2O] + [HSO_3^-] + [SO_3^{2-}]) \quad \text{(iv)}$$

The concentrations of the species in water depend on pH and the water content of the system:

$$[SO_2 \cdot H_2O] = K_H\,RT \cdot (SO_2)_g \quad \text{(v)}$$

$$[HSO_3^-] = K_H\,RT\,K_1\,[H^+]^{-1} \cdot (SO_2)_g \quad \text{(vi)}$$

$$[SO_3^{2-}] = K_H\,RT\,K_1\,K_2\,[H^+]^{-2} \cdot (SO_2)_g \quad \text{(vii)}$$

Substituting the values in equation iv, one obtains

$$(SO_2)_{tot} = (SO_2)_g + q\,(SO_2)_g\,K_H\,RT\,(1 + K_1\,[H^+]^{-1} + K_1 K_2\,[H^+]^{-2}) \quad \text{(viii)}$$

$$\Sigma\,S(IV)_{aq} = [SO_2 \cdot H_2O] + [HSO_3^-] + [SO_3^{2-}]$$
$$= (SO_2)_g\,K_H\,RT\,(1 + K_1\,[H^+]^{-1} + K_1 K_2\,[H^+]^{-2}) \quad \text{(ix)}$$

and

$$\frac{(SO_2)_g}{(SO_2)_{tot}} = \frac{1}{1 + K_H\,RTq\,(1 + K_1\,[H^+]^{-1} + K_1 K_2\,[H^+]^{-2})} \quad \text{(x)}$$

Figure 5.8a gives the proportions of SO_2 in the gas and aqueous phase as a function of pH. For pH < 5, sulfur dioxide occurs mainly in the gas phase; for pH > 7, it occurs mainly in the solution phase. The fraction of SO_2 in the aqueous phase is given in Figure 5.8b as a function of q (water content) for a few pH values. The double logarithmic graphic representation is particularly convenient to plot the equilibrium distribution of the aqueous species (Figure 5.8c). For a sketch of this diagram it is convenient to recall the following:

1. The maximum concentration in the water phase occurs at high pH and is given by the total dissolution of SO_2

$$(\Sigma\,S(IV)_{aq})\,\text{max} = \frac{(SO_2)_{tot}}{q}$$

2. In the acid pH range, the minimal solubility is given by the Henry's law constant. The lines for HSO_3^- and SO_3^{2-} are sketched initially as in Figure 5.7, but their concentrations are limited by the value $(SO_2)_{tot}/q$.

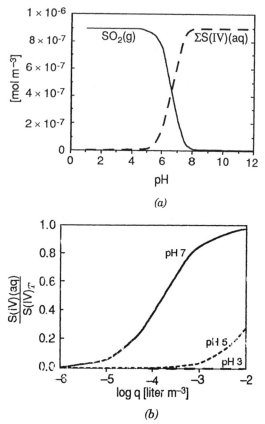

Figure 5.8. (a) Distribution of SO_2 between gas and aqueous phase for the condition $(SO_2)_{tot} = 9 \times 10^{-7}$ mol m^{-3}; water content $q = 5 \times 10^{-4}$ liter m^{-3}. (b) Fraction of S(IV) in the aqueous phase as a function of the water content q of the system for three different pH values. (c) Equilibrium distribution of S(IV) species in the aqueous phase. Same condition as in Figure 5.8a. If no other acid or base is present, the equilibrium composition is given by

$$[H^+] = [HSO_3^-] + 2[SO_3^{2-}] + [OH^-] \quad \text{or} \quad [H^+] \simeq [HSO_3^-]$$

3. Intersections of $SO_2 \cdot H_2O$ and HSO_3^- occur at pH $= pK_1$ and of HSO_3^- and SO_3^{2-} at pH $= pK_2$.

Tableau 5.3 summarizes all the equations needed.

As in the closed NH_3 system, it is important to consider the mass balances correctly. The mass balance of equation viii can be divided by q in order to obtain all concentrations on a per liter water basis:

$$\frac{(SO_2)_{tot}}{q} \text{ (mol liter}^{-1}) = \frac{(SO_2)_g}{q} + [SO_2 \cdot H_2O] + [HSO_3^-] + [SO_3^{2-}] \quad \text{(xii)}$$

5.4 Gas–Water Equilibria in Closed and Open Systems

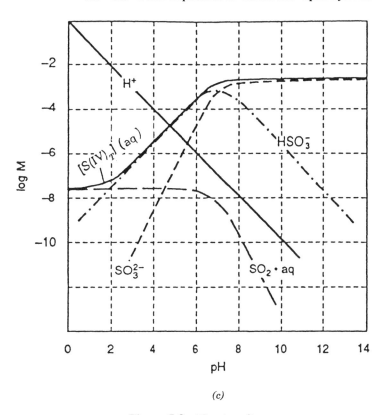

(c)

Figure 5.8. (*Continued*)

Tableau 5.3. SO$_2$–Water: Closed System

Components	SO$_2$ · H$_2$O	H$^+$	log K
Species SO$_2$ · H$_2$O	1		0
HSO$_3^-$	1	−1	−1.89
SO$_3^{2-}$	1	−2	−9.09
(SO$_2$)$_g$/q	1		−1.51 − log q
OH$^-$		−1	−14.0
H$^+$		1	0
Composition		TOTH = 0	

$$\frac{(SO_2)_{tot}}{q} = 9 \times 10^{-7} \times \frac{1}{q} = 1.8 \times 10^{-3} \text{ mol liter}^{-1}$$

In Tableau 5.3 the concentration in the gas phase (fourth row) is given

$$\frac{(SO_2)_g}{q} \text{ (mol liter}^{-1}) = [SO_2 \cdot H_2O] \cdot \frac{1}{K_H \cdot RT} \cdot \frac{1}{q} \quad \text{(xiii)}$$

and can be converted into mol m^{-3}. The equilibrium constant in the tableau is given as log $K = -\log(K_H RT) - \log q$.

Reactions of SO_2 with Aldehydes

Aldehydes—formaldehyde, acetaldehyde, and glyoxal (CHOCHO)—are present in the atmosphere because they are formed in the oxidation of hydrocarbons. These aldehydes are quite soluble. SO_2 reacts with aldehydes, for example,

$$\underset{H}{\overset{H}{\diagdown}}C=O + HSO_3^- \rightleftharpoons H_2\overset{\overset{OH}{|}}{C}-SO_3^- \quad (11)$$

The adducts formed are relatively stable. The kinetics of the formation of these adducts depends on various factors; the rates are often slow in acid solutions. The S(IV) aldehyde compounds, especially hydroxymethanesulfonate ($CH_2OHSO_3^-$), can make up a large fraction of dissolved S(IV) in atmospheric water droplets. Thus the adduct formation may enhance the solubility of S(IV), especially in the low pH range. These adducts are less reactive with regard to oxidation by oxidants.

Example 5.6. SO_2 and Formaldehyde Formaldehyde dissolves in water according to

$$CH_2O(g) \rightleftharpoons CH_2O(aq) \quad \log K_H = 3.8 \quad \text{(i)}$$

$CH_2O(aq)$ consists of two species, free CH_2O and its hydrate $CH_2(OH)_2$:

$$CH_2O + H_2O \rightleftharpoons CH_2(OH)_2 \quad \log K_{Hyd} = 3.26 \quad \text{(ii)}$$

Free CH_2O reacts with HSO_3^- to form hydroxymethanesulfonate:

$$CH_2O + HSO_3^- \rightleftharpoons CH_2OHSO_3^- \quad \log K_{HMSA} = 9.82 \quad \text{(iii)}$$

Combining equilibrium iii with equilibrium ii, and $[CH_2(OH)_2] \approx [CH_2O(aq)]$ one obtains the equilibrium constant

$$\frac{[CH_2OHSO_3^-]}{[CH_2O(aq)][HSO_3^-]} = K'_{HMSA} \quad \log K'_{HMSA} = 6.56 \quad \text{(iv)}$$

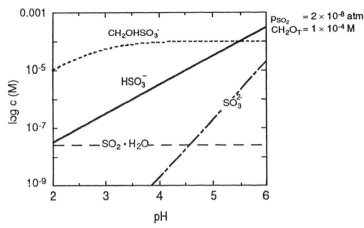

Figure 5.9. Solubility of SO_2 in the water phase (open system) in the presence of formaldehyde. Hydroxymethanesulfonate increases the solubility of SO_2, especially at pH < 5.

Compute the solubility of $SO_2(g)$ in an open system under the following conditions:

$$[CH_2O(aq)_{tot}] = 1 \times 10^{-4} \text{ M}$$

$$p_{SO_2} = 2 \times 10^{-8} \text{ atm}$$

The mass balance for the soluble species is

$$\Sigma \, S(IV) \, (aq) = [SO_2 \cdot H_2O] + [HSO_3^-] + [SO_3^{2-}] + [CH_2OHSO_3^-]$$

Figure 5.9 plots the results.

5.5 WASHOUT OF POLLUTANTS FROM THE ATMOSPHERE

To what extent are atmospheric pollutants washed out by rain? We can try to answer this question by considering the gas-absorption equilibria. Our estimate will be based on the following assumptions.

The height of the air column is 5×10^3 m. This column is "washed out" by a rain of 25 mm (corresponding to 25 liters m^{-2}). In other words,

$$\text{gas volume } V_g = 5 \times 10^3 \text{ m}^3$$

$$\text{water volume } V_w = 0.025 \text{ m}^3$$

$$\frac{V_g}{V_w} = 2 \times 10^5 \tag{12}$$

228   Atmosphere–Water Interactions

The total quantity of a pollutant is

$$A_{tot} = (A)_g \times V_g + (A)_w \times V_w \tag{13}$$

The fraction of pollutants in the water phase, f_{water}, is given by

$$f_{water} = \frac{(A)_w \times V_w}{(A)_g \times V_g + (A)_w V_w} = \frac{1}{\dfrac{(A)_g}{(A)_w} \times \dfrac{V_g}{V_w} + 1}$$

$$= \frac{1}{\dfrac{1}{K_H \times RT} \times \dfrac{V_g}{V_w} + 1} \tag{14}$$

Figure 5.10 illustrates the distribution of different compounds between the water and gas phase in dependence on pH. The following conclusions can be drawn:

1. The distribution of pollutants that undergo proton transfer is strongly pH dependent.

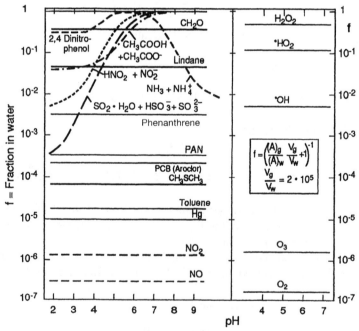

Figure 5.10. Distribution of pollutants between gas (atmosphere) and water (rain) phase as a function of pH. A volume ratio of gas to water of 2×10^5 was assumed. (Adapted from Sigg and Stumm, 1994.)

5.6 FOG

2. The washout of many pollutants by rain is rather small. Some of these pollutants are adsorbed to particles and aerosols and are returned to the surface of the earth by settling and precipitation.
3. The various oxidants in the atmosphere are soluble in the water to a very different extent. Note the small solubility of O_2 and O_3 in comparison to H_2O_2 and $\cdot HO_2$.

5.6 FOG

Fog droplets (10–50 μm diameter) are formed in the water-saturated atmosphere (relative humidity = 100%) by condensation on aerosol particles (see Figure 5.2). The fog droplets absorb gases such as SO_2, NH_3, HCl, and NO_x. The water droplets are a favorable milieu for the oxidation of many reductants, above all, of SO_2 to H_2SO_4. The liquid water content of a typical fog is often on the order of 10^{-4} liter water per m³ air. The concentrations of ions in fog droplets are often 10–50 times larger than those of rain (Figure 5.11). Clouds process substantial volumes of air and transfer gas and aerosols over large distances. On the other hand, fog droplets are important collectors of local pollutants in the proximity of the earth's surface.

Exemplification: Genesis of a Fog Droplet

We will simulate the "synthesis" of a typical fog by considering first a packet of air containing equimolar quantities of NH_3 and SO_2 [$(NH_3)_{tot} = 5 \times 10^{-7}$ mol m⁻³, $(SO_2)_{tot} = 5 \times 10^{-7}$ mol m⁻³]. We add to the packet 10^{-4} liter m⁻³ water and let the water condense to fog. The system is treated as closed with regard to SO_2 and NH_3. The system is also under the influence of CO_2 ($p_{CO_2} = 10^{-3.5}$ atm = constant). (Because of the size of the CO_2 reservoir, we assume the system to be open with regard to CO_2.)

The equilibrium problem is primarily the combination of dissolution of NH_3 and SO_2 (see Figures 5.6 and 5.8). Tableau 5.4 summarizes the problem (compare Tableaux 5.1 and 5.3).

Figure 5.12 gives the heterogeneous equilibrium diagram as a function of pH. It is the "titration curve" of the system. It is evident from Figure 5.12 that the buffering of the heterogeneous gas-water system results primarily from the *components in the gas phase* (NH_3 above pH = 5, SO_2 below pH = 5).

The proton balance TOTH = 0 ($[NH_4^+] \approx [HSO_3^-]$) is characterized by pH = 6.3. If TOTH = 5×10^{-3} M (i.e., by the addition of [HCl(g)] = 5×10^{-7} mol m⁻³], pH = 3.8.

After the absorption of SO_2 into the fog droplets, the SO_2 is oxidized by an oxidant "O" to H_2SO_4

$$SO_2 + \text{``O''} + H_2O = SO_4^{2-} + 2\,H^+ \tag{15}$$

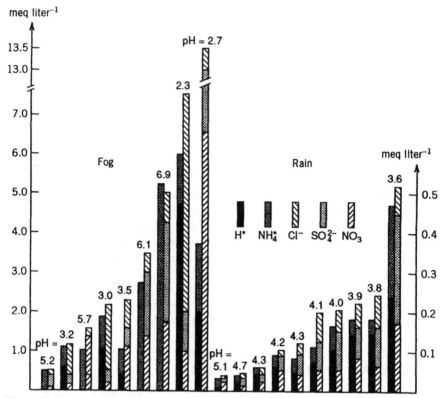

Figure 5.11. Comparison of rain and fog analyses. The data are from samples taken in the proximity of Zürich, Switzerland. Note the difference in ordinate scale. Rain is characterized by 0.05–0.5 meq liter^{-1}, while the concentration of ions in fog (lower liquid water content than with rain) is larger by one to two orders of magnitude. (From Sigg and Stumm, 1994.)

The oxidant "O" can be either O_2, H_2O_2, or O_3. Often the SO_2 in fog is oxidized by O_3. This oxidation reaction is very pH dependent because—as will be discussed more fully in Chapter 11—the oxidation rate is dependent on the concentration of the individual S(IV) species (parallel reaction)

$$-\frac{d[\text{S(IV)}]}{dt} = \frac{d[\text{SO}_4^{2-}]}{dt}$$
$$= (k_0[\text{SO}_2 \cdot \text{H}_2\text{O}] + k_1[\text{HSO}_3^-] + k_2[\text{SO}_3^{2-}]) \cdot [\text{O}_3(\text{aq})] \quad (16)$$

For every SO_2 that is oxidized, two protons are liberated (equation 15). (This is equivalent to the addition of acid.). Thus the equilibrium composition is shifted along the titration curve. The NH_3 is of importance because of the following:

Tableau 5.4. Genesis of a Fog Droplet, Absorption of $SO_2(g)$, $NH_3(g)$, and CO_2, into the Water Phase; Then Strong Acid (C_A) and Strong Base (C_B) Are Added

Components		$SO_2 \cdot H_2O$	$NH_{3(aq)}$	$CO_{2(g)}$	H^+	C_B	C_A	$\log K$
Species	$SO_2 \cdot H_2O$	1						0
	HSO_3^-	1			-1			-1.89
	SO_3^{2-}	1			-2			-9.09
	$(SO_2)_g/q$	1						$1.51 - \log q$
	$NH_3(aq)$		1					0
	NH_4^+		1		1			9.2
	$(NH_3)_g/q$		1					$-0.14 - \log q$
	$H_2CO_3^*$			1				-1.47
	HCO_3^-			1	-1			-7.77
	CO_3^{2-}			1	-2			-18.1
	$CO_2(g)$			1				0
	C_B^+					1		0
	C_A^-						1	0
	OH^-				-1			-14
	H^+				1			0

Composition

$$\frac{(SO_2)_{tot}}{q} = 5 \times 10^{-3} \text{ mol liter}^{-1}$$

$$\frac{(NH_3)_{tot}}{q} = 5 \times 10^{-3} \text{ mol liter}^{-1}$$

$$p_{CO_2} = 10^{-3.5} \text{ atm}$$

(a)	0	0	$= 5 \times 10^{-3}$ mol liter^{-1}
(b)	5×10^{-3} M	0	5×10^{-3} M $= 5 \times 10^{-3}$ mol liter^{-1}

$\text{TOTSO}_2 = [SO_2 \cdot H_2O] + [HSO_3^-] + [SO_3^{2-}] + (SO_2)_g/q$

$\text{TOTNH}_3 = [NH_3(aq)] + [NH_4^+] + (NH_3)_g/q$

$\text{TOTH} = [H^+] + [NH_4^+] - [HSO_3^-] - 2[SO_3^{2-}] - [HCO_3^-] - 2[CO_3^{2-}] - [OH^-] = C_A - C_B$
(a) TOTH $= 0$
(b) TOTH $= 5 \times 10^{-3}$ mol liter^{-1}

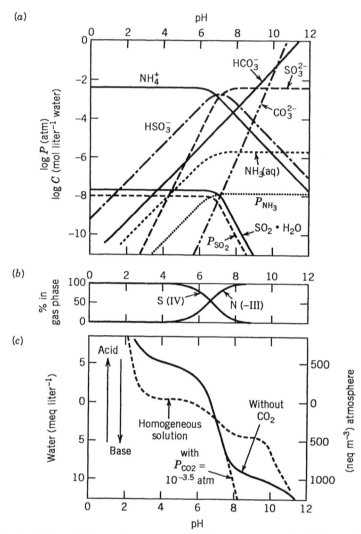

Figure 5.12. (a) Equilibrium diagram of a fog-air system (compare Tableau 5.4). The system is closed with regard to SO_2 and NH_3 (TOTSO$_2$) = 5×10^{-7} mol S(IV) per m^3, (TOTNH$_4$) = 5×10^{-7} mol N(−III) per m^3. Water content $q = 10^{-4}$ liter m^{-3}. The system is open with regard to CO_2 ($p_{CO_2} = 10^{-3.5}$ atm). (b) Fraction of TOTNH$_3$ as NH_3(g) and of TOTSO$_3$ as SO_2(g). (c) Titration curve with strong acid or base. The drawn-out curve corresponds to the equation for TOTH in Tableau 5.4. The dashed curve corresponds to the titration of a homogeneous aqueous NH_4HSO_3 solution (both NH_3 and SO_2 are treated as nonvolatile):

$$\text{TOTH} = C_A - C_B = [SO_2 \cdot H_2O] + [H^+] - [NH_3(aq)] - [OH^-]$$

Note the difference in buffering $dC_{\text{base}}/d\,\text{pH}$ between the heterogeneous and homogeneous system. NH_3(g) is the main buffering component at pH > 5 and SO_2(g) at pH < 5. (From Behra et al., 1989.)

1. It regulates the pH in the aqueous phase.
2. The NH_3 in the gas phase buffers the aqueous solution against a fast pH drop. The lower the pH, the slower is the oxidation rate with O_3; below pH ≃ 5 the oxidation is so slow that it does not occur during the lifetime of the fog.
3. $NH_3(g)$ determines the acid-neutralizing capacity of the system.

Figure 5.13 illustrates the composition of a radiation fog. In this particular case, the effect of HCl (probably from a refuse incineration plant at a distance of 3 km) caused a lowering of pH in the fog water as far down as pH 1.94.

5.7 AEROSOLS

Atmospheric aerosols are important nuclei for the condensation of water droplets (cloud, rain, fog). The dissolution of the water-soluble aerosol components contributes to the composition of the aqueous phase [e.g., NH_4NO_3, $(NH_4)_2SO_4$]. Aerosols may contain, in addition to the absorbed gases, a substantial fraction of atmospheric components that return ultimately to the earth surface by dry or wet deposition. The particle diameter ranges from 0.01 μm up to a few hundred micrometers. Primary atmospheric aerosols consist of dust and smoke particles while secondary aerosols are made up of constituents of the gas phase.

The following reactions in the gas phase can lead to aerosols:

$$H_2SO_4(g) + 2\ NH_3(g) \rightleftharpoons \{(NH_4)_2SO_4\}_{aerosol} \qquad (17)$$

$$H_2SO_4(g) + NH_3(g) \rightleftharpoons \{NH_4HSO_4\}_{aerosol} \qquad (18)$$

$$HNO_3(g) + NH_3(g) \rightleftharpoons \{NH_4NO_3\}_{aerosol} \qquad (19)$$

$$HCl(g) + NH_3(g) \rightleftharpoons \{NH_4Cl\}_{aerosol} \qquad (20)$$

or

$$H_2SO_4(g) + 2\ HNO_3(g) + 4\ NH_3(g) \rightleftharpoons \{(NH_4)_2SO_4 \cdot NH_4NO_3\}_{aerosol} \qquad (21)$$

$$H_2SO_4(g) \rightleftharpoons \{H_2SO_4(l)\}_{aerosol} \qquad (22)$$

Mixed aerosols are also formed:

$$\{(NH_4)_2SO_4 \cdot 2\ NH_4NO_3\}_{aerosol} \rightleftharpoons \{(NH_4)_2SO_4\}_{aerosol}$$
$$+ 2\ HNO_3(g) + 2\ NH_3(g) \qquad (23)$$

These ammonium aerosols (d = 0.3–1 μm) occur at low humidity as solid particles. The reactions given above can be compared with precipitation reac-

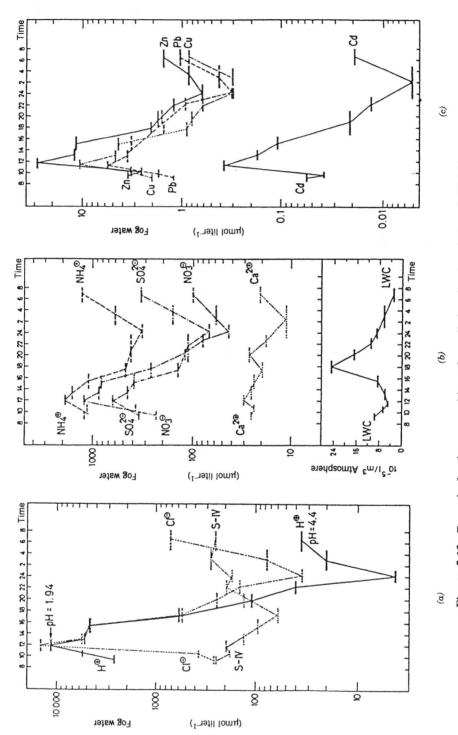

Figure 5.13. Example for the composition of a radiation fog (proximity of Zürich) as a function of time (LWC = liquid water content). (From Sigg et al., 1987.)

tions. Equilibria can be defined as follows:

$$K_p (17) = p_{NH_3}^2 \cdot p_{H_2SO_4} = 2.33 \times 10^{-38} \text{ atm}^3 \quad (25°C) \quad (24)$$

$$K_p (19) = p_{NH_3} \cdot p_{HNO_3} = 3.03 \times 10^{-17} \text{ atm}^2 \quad (25°C) \quad (25)$$

The aerosols are formed, usually rather fast, when the products of partial pressures become exceeded.

In humid air, solid aerosols formed in the reactions 17–21 are converted into liquid droplets (deliquescence); for example above 75% humidity (5°C). The liquid aerosols are very concentrated salt solutions (up to 26 M).

The formation of ammonium sulfate and ammonium nitrate aerosols is an acid–base reaction in the atmosphere. Ammonia neutralizes the acids. Sulfuric acid has a very low vapor pressure ($< 10^{-7}$ atm) and thus exists in the atmosphere as liquid particles, which react with NH_3 and H_2O (reaction 22).

Example 5.7. Dissolution of Aerosols in Fog Water The following aerosol concentrations are measured prior to fog formation

2×10^{-8} mol/m^3 NH_4NO_3
5×10^{-8} mol/m^3 $(NH_4)_2SO_4$

What concentrations result in fog water (liquid water content $q = 10^{-4}$ liter m^{-3}) when 80% of the aerosols become dissolved in the water droplets?

$$[NO_3^-] = \frac{2 \times 10^{-8} \times 0.8}{1 \times 10^{-4}} = 1.6 \times 10^{-4} \text{ M}$$

$$[SO_4^{2-}] = 4 \times 10^{-4} \text{ M}$$

$$[NH_4^+] = 9.6 \times 10^{-4} \text{ M}$$

This simple example illustrates that high concentrations of NH_4^+, NO_3^-, and SO_4^{2-} in fog droplets may result from the dissolution of aerosols.

For a more comprehensive introduction into the chemistry of the aerosols, see Seinfeld (1986).

5.8 ACID RAIN—ACID LAKES

Weathering

The atmospheric constituents react with the crust of the earth through the process of rock weathering. The general term weathering encompasses a variety of processes by which parent rocks are broken down mechanically or are chemically dissolved. In these processes, water occupies a central role, serving both as a reactant and as a transporting agent of suspended and dissolved material. The atmosphere provides a reservoir of weak acids (CO_2 and some organic acids) and oxidants (oxygen, ozone, radicals) and may contain anthropogenic pollutants (strong acids, heavy metals, organic compounds). Mechanical weathering is the fragmentation by primarily physical processes into small grain particles; it includes wind abrasion and rock fragmentation by the freezing of water. During chemical weathering, rocks and primary minerals undergo chemical reactions with water (and its acids and oxidants), thereby being transformed first to solutes and soils and eventually to sediments and sedimentary rocks. Alpine lakes are the first to receive the products of weathering; the ultimate receptacle is the sea. Chemical weathering is an important feature of the global hydrochemical cycle of elements. The CO_2 consumption by silicate weathering (and its temperature dependence) is of global significance as a factor in regulating the CO_2 concentration of the atmosphere.

The atmospheric agents, above all acids formed from carbon dioxide and the oxides of nitrogen and sulfur (carbonic acid, nitric acid, and sulfuric acid), "titrate" the bases of the minerals, that is, the carbonates, the silicates, and the oxides to form solutes. Depending on whether the "end point of the titration" is exceeded or not, the residual waters contain alkalinity (excess of bases such as HCO_3^-) or acidity (excess of acids, especially hydrogen ions). The interaction of acids with minerals can be formulated as follows (we use carbonic acid, H_2CO_3):

Calcium carbonate:

$$CaCO_3 + CO_2 + H_2O = Ca^{2+} + 2\ HCO_3^- \qquad (26)$$

Dolomite:

$$CaMg(CO_3)_2 + 2\ CO_2 + 2\ H_2O = Ca^{2+} + Mg^{2+} + 4\ HCO_3^- \qquad (27)$$

Aluminum silicates (generally):

$$\text{Cation Al silicate} + CO_2 + H_2O$$
$$= \text{Cation} + HCO_3^- + H_4SiO_4 + \text{Al silicates} \qquad (28)$$

Chemical weathering will be discussed in more detail in Chapter 13.

Acid Lakes

Figure 5.14 gives the water composition of four lakes at the top of the Maggia Valley in the Southern Alps of Switzerland. Although these lakes are less than 10 km apart, they differ markedly in their water composition as influenced by different bedrocks in their catchments. These lakes are at an elevation of 2100–

5.8 Acid Rain—Acid Lakes 237

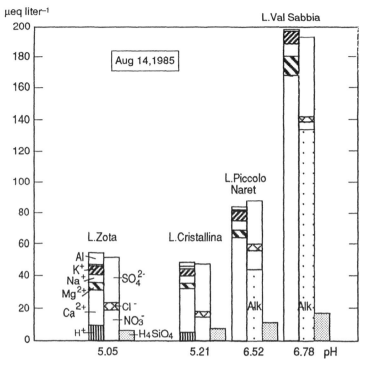

Figure 5.14. Water composition of four lakes in southern Alps of Switzerland. The difference in composition is caused by the geology of the bedrocks in the catchment areas. Lakes Zota and Cristallina are situated within a drainage area of gneissic rocks. The other two lakes are in catchment areas that contain calcite and dolomite.

2550 m. The small catchments are characterized by sparse vegetation (no trees), thin soils, and steep slopes.

The composition of Lakes Cristallina and Zota, which are situated within a drainage area characterized by the preponderance of gneissic rocks and the absence of calcite and dolomite, and that of Lake Val Sabbia, the catchment area of which contains dolomite, are markedly different. The waters of Lake Cristallina and Lake Zota exhibit mineral acidity (i.e., caused by mineral acids and HNO_3), their calcium concentrations are 10–15 μmol liter,$^{-1}$ and their pH is < 5.3. On the other hand, the water of Lake Val Sabbia is characterized by an alkalinity of 130 μmol liter^{-1} and a calcium concentration of 85 μmol liter^{-1}. The water of Lake Piccolo Naret is intermediate; its alkalinity is < 50 μmol liter^{-1} and appears to have been influenced by the presence of some calcite or dolomite in its catchment area.

With regard to acid deposition, *critical loads* were defined in terms of tolerable acidity deposition, for example, in μeq m^{-2} yr^{-1}, that must not be exceeded to avoid harmful effects. The sensitivity of a region is strongly influenced by the minerals of the bedrock. Regions where crystalline rocks prevail are particularly sensitive.

An experimental study on the acidification of a lake (pH was lowered within a period of 8 years from pH 6.8 to 5.1) is especially noteworthy (Schindler et al., 1985).

5.9 THE VOLATILITY OF ORGANIC SUBSTANCES

The tendency of an *organic liquid or solid chemical substance to partition into the atmosphere* is controlled by its vapor pressure. Compounds with high vapor pressures tend to accumulate in the gas phase. [The vapor pressure of a solid or liquid has been likened to a kind of "solubility" of the compound in the atmosphere (Mackay, 1991).] Partitioning *between the gas phase and water* (Section 5.5) is described by Henry's law:

$$\frac{[A(aq)]}{p_A} = K_H \quad \text{or} \quad \frac{[A(aq)]}{(A)_g} = H = K_H RT \qquad (29)$$

A range of Henry's law constants for different classes of organic compounds is given in Figure 5.3.

Henry's law constants can be calculated from experimental measurements of saturation concentrations of a volatile organic compound in the gas phase and in the aqueous phase, that is, from the phase equilibria:

$$A(\text{gas, sat'd}; (A°)_g) = A \text{ (liquid or solid)} = A \text{ (aq, satd; } [A°(aq)]) \qquad (30)$$

The Henry's law constants obtained for saturated conditions are then

$$\frac{[A°(aq)]}{p_A°} = K_H^{sat} \quad \text{or} \quad \frac{[A°(aq)]}{(A°)_g} = H^{sat} = K_H^{sat} RT \qquad (31)$$

Here $[A°(aq)]$ and $p_A°$ are, respectively, the concentration of A (M) in a saturated aqueous solution and the vapor pressure of A (atm) at equilibrium with the pure solid or liquid; K_H^{sat} and H^{sat} are the Henry's law constants computed as in equation 31. What is the relationship between H and H^{sat}? Under ideal solution conditions, they are essentially the same. For extremely soluble organic compounds, the aqueous solutions depart from ideality, and H and H^{sat} are no longer equal. Prausnitz (1969) estimated the upper solubility limit for $H \simeq H^{sat}$ at approximately 1.5 M. For many compounds we can then assume $H \simeq H^{sat}$. Important details of air–water partition coefficients and Henry's law constants for organic substances have been reviewed by Mackay and Shin (1981), Schwarzenbach et al. (1993), and Brezonik (1993).

Figure 5.15 plots double logarithmically the saturation concentration in air

Table 5.3. Vapor Pressure and Henry Constants for Various Compounds at 25°C

Compound A	Molecular Weight	Water Solubility $[A°(aq)]$ (M)	Vapor Pressure $p_A°$ (atm)	Concentration in Gas Phase $(A°)_g = p_A°/RT$ (mol liter^{-1})	$H^{sat} = K_H^{sat} RT = \dfrac{[A°(aq)]}{(A°)_g}$
Alkanes					
n-Octane (C_8H_{18})	114	5.8×10^{-6}	1.8×10^{-2}	7.5×10^{-4}	7.7×10^{-3}
1-Hexane (C_6H_{12})	84	7.0×10^{-4}	2.5×10^{-1}	1.0×10^{-2}	7×10^{-2}
Aromatic substances					
Benzene	78	2.3×10^{-2}	1.2×10^{-1}	5.1×10^{-3}	4.5
Toluene	92	5.6×10^{-3}	3.7×10^{-2}	1.5×10^{-3}	3.7
Naphthalene	128	2.6×10^{-4}	1.0×10^{-4}	4.3×10^{-6}	6.1×10^{1}
Biphenyl	154	4.9×10^{-5}	7.5×10^{-5}	3.1×10^{-6}	1.6×10^{1}
Pesticides					
p,p′-DDT ($C_{14}H_9Cl_5$)	355	1.4×10^{-8}	1.3×10^{-10}	5.4×10^{-12}	2.6×10^{3}
Lindane	291	2.6×10^{-5}	8.3×10^{-8}	3.4×10^{-9}	7.6×10^{3}
Dieldrin	381	5.8×10^{-7}	6.6×10^{-9}	2.7×10^{-10}	2.1×10^{3}
Polychlorinated biphenyls (PCBs)					
2,2′,4,4′-CBP ($C_{12}H_6Cl_4$)	292	2.0×10^{-8}	6.2×10^{-9}	2.5×10^{-10}	79
2,2′,3′,4,4′-CBP ($C_{12}H_4Cl_6$)	361	1.9×10^{-9}	2.2×10^{-10}	9.2×10^{-12}	2.1×10^{2}
Other					
Dimethylsulfide (C_2H_6S)	62	3.5×10^{-1}	6.3×10^{-1}	2.6×10^{-2}	1.4×10^{1}
Mercury (Hg°)	201	1.5×10^{7}	1.7×10^{-6}	7×10^{-8}	2.1
Water	18	(55.5)	3.1×10^{-2}	1.3×10^{-3}	(4.3×10^{4})

[a] Data from various sources. Detailed compilations with references are available from Lyman et al. (1982) and from the Appendix in Schwarzenbach et al. (1993).

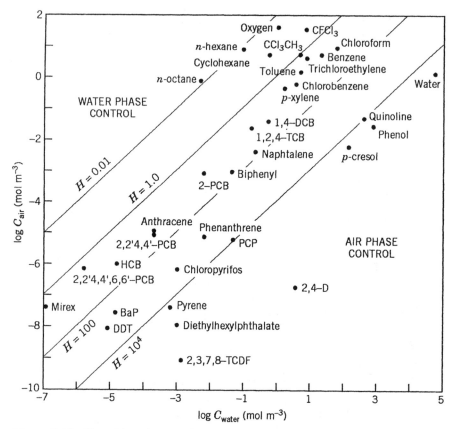

Figure 5.15. Plot of log air saturation concentration $C_{air}^\circ = p^\circ/RT$ versus water solubility C_{water}° for selected chemicals (25°C). Chemicals of equal H lie on the same 45° diagonal. Water phase control and air phase control refer to the transfer kinetics (rate-controlling steps) in the two-film theory. (Adapted from Mackay, 1991.)

p_A°/RT versus the saturation concentration in water [A°(aq)] for 25°C.[†] The volatile compounds ($H < 10$) are in the upper left, while those in the lower right ($H > 100$) are relatively involatile. It is interesting to note that a homologous series of chemicals tend to lie at a line of constant H. Substituting methyl groups or chlorines for hydrogen tends to reduce both vapor pressure and solubility, by a factor of 4–6. Thus K_H or H tends to remain relatively constant (Mackay, 1991). The H value for water in Figure 5.15 can be deduced

---

[†]The temperature dependence of the Henry's law constant can be estimated from the temperature dependence of the vapor pressure: $d \ln p^\circ/dT = \Delta H_{vap}/RT^2$ (ΔH_{vap} = heat of vaporization; always positive) or $\ln p^\circ = -\Delta H_{vap}/RT$ + constant and from the temperature dependence of the aqueous solubility; since the latter is smaller than the former, an increase in temperature reduces K_H—that is, it favors the partition into the gas phase.

from its vapor pressure (3×10^{-2} atm) and the concentration in the water phase (55.4 mol liter^{-1}).

Some individual data are given in Table 5.3. It is seen from these data that at equilibrium the ratio of most contaminants to water in the gas phase is larger than in the liquid phase: $p_A/p_W^\circ > [A(aq)]/55.5$.

If a compound has a higher H than water, it may concentrate in water as a result of faster water evaporation. An analytical chemist often extracts organic chemicals from water samples by bubbling air through the water and stripping the chemical from solution. This is effective when $H \leq 100$. These chemicals are often called "volatiles."

5.10 GAS TRANSFER ACROSS WATER–GAS INTERFACE

The rate of mass transfer of a substance across a water–gas phase boundary has been described in terms of a diffusion film model. In general, it is necessary to consider *two diffusion films*, one in the liquid phase and one in the gas phase. The two bulk phases are well mixed to within a small distance of the interface. From Fick's first law we conclude that the flux through the film of thickness z is given by

$$F = -D\frac{dc}{dz} \tag{32}$$

F, for example, is in mol cm^{-2} s^{-1} if c is in mol cm^{-3}, z in cm, and D in cm^2 s^{-1}. The negative sign corresponds to the convention for the orientation of the z axis. The flux through both boundary layers will attain a steady state

$$F = F_a = F_w \tag{33}$$

that is, the number of molecules passing through each boundary film per square centimeter and second will be the same (see Figure 5.16).

Thus, applying equations 32 and 33, we can quantify the flux as

$$F = -\frac{D_a}{z_a}(c_a - c_{a/w}) = -\frac{D_w}{z_w}(c_{w/a} - c_w) \tag{34}$$

Suffixes a and w refer to the air and water film, respectively, and $c_{a/w}$ and $c_{w/a}$ refer to the concentration in the air film at the air–water interface and the concentration in the water film at the water–air interface.

Equation 34 presumes that the chemical does not undergo a chemical reaction within the layer (i.e., fast in comparison to the transfer process). We then imply that the interface concentrations can be interpreted in terms of the Henry factor, H.

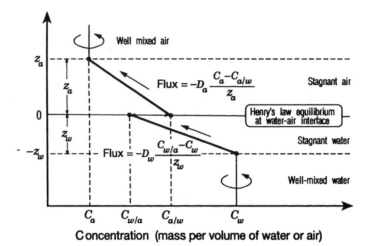

Figure 5.16. Stagnant boundary layer model for air–water exchange of chemicals. Note that the vertical coordinate z is defined as positive upward and that $z = 0$ at the level of air–water contact. (Adapted from Schwarzenbach et al., 1993.)

$$H = \frac{c_{w/a}}{c_{a/w}} \left(\frac{\text{mol (liter water)}^{-1}}{\text{mol (liter air)}^{-1}} \right) = K_H\, RT \tag{35}$$

(We repeat the units because some sources define H by the inverse relationship.)

With the help of equation 35 we can eliminate the interface concentrations in equation 34. First, substituting equation 35 in equation 34 gives

$$F = \frac{D_w}{z_w}(c_w - c_{w/a}) = \frac{D_a}{z_a}\left(\frac{c_{w/a}}{H} - c_a\right) \tag{36}$$

Finally, the flux through the water film can be written

$$F = \frac{1}{V_w^{-1} + V_a^{-1} H}(c_w - c_a H) \tag{37}$$

where

$$V_w = \frac{D_w}{z_w} \quad \text{and} \quad V_a = \frac{D_a}{z_a} \tag{38}$$

are the transfer coefficients (units: cm s^{-1}), that is, the mass transfer velocities, sometimes called "piston velocities," through water and air, respectively.

5.10 Gas Transfer Across Water–Gas Interface

Equation 37 can also be written

$$F = V_{tot}(c_w - c_a H) \tag{39}$$

Comparing equation 37 with equation 39 yields

$$\frac{1}{V_{tot}} = \frac{1}{V_w} + \frac{H}{V_a} \tag{40}$$

In analogy with electrical conductance and as shown by Liss and Slater (1974) and as discussed by Mackay and Leinonen (1975) and Mackay (1991) the reciprocal of the mass transfer velocity is a measure of resistance to the gas transfer and is composed of two resistances in series, the liquid phase resistance, r_w, and the gas phase resistance, r_g:

$$\frac{1}{V_{tot}} = R = r_w + r_g \tag{41}$$

Gas or Liquid Film Controlling Transfer

The primary variable that determines whether the controlling resistance is in the liquid or gas film is the H or Henry constant. As shown in Figure 5.15, and as is apparent from equation 39, for small values of H the water phase film controls the transfer, and for high values of H the transfer is controlled by the air phase film. Gas transfer conditions that are liquid film controlled sometimes are expressed in terms of thickness, z_w, of the water film. As indicated by equation 38, this can be done from a measured value of V_w (or V_{tot}) and the diffusion coefficient of the substance; z_w decreases with the extent of turbulence (current velocity, wind speed, etc.). Typical values for z_w are in the range of micrometers for seawater, a few hundred micrometers in lakes and up to 1 mm in small wind-sheltered water bodies (Brezonik, 1994).

When H is less than about 5, we have water film control; when H is larger than 500, gas film control prevails. For compounds with intermediate values for H, both films contribute to gas transfer resistance. Large molecules or polar compounds like phenols are air–film controlled. Small molecules and nonpolar compounds are water–film controlled.

Inorganic gases (O_2, N_2, CO_2, H_2S, CH_4, NO_x) are—with the exception of HCl, NH_3, SO_2, and SO_3 (which are extremely soluble)—sufficiently volatile that the boundary layer in the gas phase need not be considered. Because the molecular diffusion coefficient of typical inorganic solutes span a relatively narrow range of values (2–5 × 10^{-5} cm^2 s^{-1}), the transfer of inorganic gases is dominated by the hydrodynamic characteristics of the water; it is independent of the nature of the gas.

According to the two-film theory, the transfer rate is proportional to the molecular diffusion coefficient for the compound of interest in the rate-limiting

Atmosphere-Water Interactions

film. Thus the ratio of the mass transfer rate of compound i will be related to the transfer rate of O_2: $F_i/F_{O_2} = D_i/D_{O_2}$. Other tracers such as radon or naturally occurring CH_4 can be used as reference compounds.

Chemical Enhancement of Air-Water Exchange Rates

If the chemical reaction—we consider a water film—is fast compared to the transport time, the conditions used in defining equation 34 are no longer valid. In this case, we may assume immediate equilibrium between the species linked by the fast reaction. The degree of mass transport enhancement caused by chemical reactions may be quantified by adding a chemical enhancement factor, α, to the term of liquid film resistance.

$$\frac{1}{V_{tot}} = \frac{1}{\alpha V_w} + \frac{H}{V_g} \approx \frac{1}{\alpha V_w} \quad (42)$$

or (cf. equations 37 and 38)

$$F = \alpha \frac{D_w}{z_w}(c_w - c_a H) \quad (43)$$

Example 5.8. Flux of Freon into Oceans Estimate the flux of freon 11 (CCl_3F) into the oceans from the following data (Lovelock et al., 1973): mean concentration of CCl_3F in the marine atmosphere over the Atlantic = 50×10^{-6} ppm (by volume); mean surface water concentration = 7.6×10^{-12} cm^3 CCl_3F per cm^3 water. H and D/z_w as given by Liss and Slater (1974) are 0.2 and 11.3 cm h^{-1}.

We may first convert the volumetric units into gram per cm^3:

Air: 50×10^{-6} ppm = 50×10^{-12} cm^3 CCl_3F per cm^3

Air: $\dfrac{50 \times 10^{-12} \text{ cm}^3 \text{ }(CCl_3F) \times 138 \text{ g/mol}}{22.4 \times 10^3 \text{ cm}^3 \text{ air/mol }(CCl_3F)} = 3.1 \times 10^{-13}$ g cm^{-3}

Water: $\dfrac{7.6 \times 10^{-12} \text{ cm}^3 \text{ }(CCl_3F) \times 138 \text{ g/mol}}{22.4 \times 10^3 \text{ cm}^3 \text{ water/mol }(CCl_3F)} = 4.7 \times 10^{-14}$ g cm^{-3}

The concentration gradient within the water film is

$$\Delta C = 0.2 \times 3.1 \times 10^{-13} - 4.7 \times 10^{-14} = 1.5 \times 10^{-14} \text{ g cm}^{-3}$$

The flux

$$F = (D/z_w) \Delta C = 11.3 \text{ cm h}^{-1} \times 1.5 \times 10^{-14} \text{ g cm}^{-3}$$
$$= 1.7 \times 10^{-13} \text{ g cm}^{-2} \text{ h}^{-1} = 1.5 \times 10^{-9} \text{ g cm}^{-2} \text{ y}^{-1}$$

5.10 Gas Transfer Across Water–Gas Interface

The entire flux from the atmosphere to the entire ocean was estimated at that time (1974) as $5.4 \text{ g} \times 10^9 \text{ g yr}^{-1}$, which was estimated to be about 2% of the world production.

Example 5.9a. CO_2 Transfer Water–Air A lake is fed by groundwater infiltration. The composition of the lakewater is characterized by a pH = 6.7 and an alkalinity of 3×10^{-3} M (T = 25°C).

Assuming *steady state*, how fast is the CO_2 transfer from the lake to the atmosphere? A representative water film thickness is 40 μm; $D_{CO_2} = 2 \times 10^{-5}$ cm^2 s^{-1}.

The water film controls the CO_2 transfer. The following conditions apply:

$$[\text{Alk}] \simeq [\text{HCO}_3^-] = \text{constant} \tag{i}$$

$$[\text{H}_2\text{CO}_3^*] \simeq [\text{CO}_2(\text{aq})] \tag{ii}$$

$$[\text{H}_2\text{CO}_3^*] = [\text{CO}_2]_w = [\text{H}^+][\text{HCO}_3^-]/K_1 = 1.2 \times 10^{-3} \text{ M}$$

$$= 1.2 \times 10^{-6} \text{ mol cm}^{-3} \tag{iii}$$

$$[\text{H}_2\text{CO}_3^*]_{w/a} = [\text{CO}_2]_{w/a} = K_H\, p_{CO_2} = 10^{-5} \text{ M} = 10^{-8} \text{ mol cm}^{-3} \tag{iv}$$

The transfer rate is given by

$$F = \frac{D_{CO_2}}{z_w}([\text{CO}_2]_w - [\text{CO}_2]_{w/a}) = \frac{2 \times 10^{-5}}{4 \times 10^{-3}}(1.2 \times 10^{-6} - 10^{-8})$$

$$= 6 \times 10^{-9} \text{ mol cm}^{-2}\text{ s}^{-1} \tag{v}$$

Example 5.9b. CO_2 Transfer Non-steady-state[†] A lake with an average depth of $h = 10$ m has sufficient mixing to ensure uniform concentrations throughout; the initial conditions are as in Example 5.9a.

As in Example 5.9a,

$$[\text{CO}_2]_{w/a} = 10^{-5} \text{ M} = 10^{-8} \text{ mol cm}^{-3}$$

$$[\text{CO}_2]_w = 1.2 \times 10^{-3} \text{ M} = 1.2 \times 10^{-6} \text{ mol cm}^{-3}$$

$$z_w = 4 \times 10^{-3} \text{ cm}$$

$$D_{CO_2} = 2 \times 10^{-5} \text{ cm}^2 \text{ s}^{-1}$$

$$\text{Alk} = [\text{HCO}_3^-] = 3 \times 10^{-3} \text{ M}$$

$$C_T = [\text{H}_2\text{CO}_3^*] + [\text{HCO}_3^-]$$

$$[\text{H}_2\text{CO}_3^*]_{t=0} = 1.2 \times 10^{-3} \text{ M} \quad \text{(see equation iii in Example 5.9a)}$$

[†]Modified from Morel and Hering (1993).

For the mixed water column, the exchange rate J_{CO_2} (mol cm^{-2} s^{-1}) is set equal to the change in concentration of $H_2CO_3^*$

$$J_{CO_2} = \frac{dn_{CO_2}}{dt}\frac{1}{A} = \frac{d[H_2CO_3^*]}{dt}\frac{V}{A} \qquad (i)$$

where n_{CO_2} = number of moles, A = surface, V = volume, and V/A = mean depth = 10^3 cm. Since [Alk] = constant,

$$\frac{d[H_2CO_3^*]}{dt} = \frac{dC_T}{dt} = \frac{A}{V}\frac{D_{CO_2}}{z_w}([H_2CO_3]_{w/a} - [H_2CO_3^*]) \qquad (ii)$$

$$= \frac{1}{10^3 \text{ cm}} \frac{2 \times 10^{-5} \text{ cm}^2 \text{ s}^{-1}}{4 \times 10^{-3} \text{ cm}}$$

$$\cdot \left(\frac{10^{-8} \text{ mol}}{\text{cm}^3} - [H_2CO_3^*]\right) (\text{mol cm}^{-3}\text{ s}^{-1}) \qquad (iii)$$

All concentrations are given in mol cm^{-3}.
In equation iii we set $[H_2CO_3^*] = C_T - [\text{Alk}]$

$$\frac{dC_T}{dt} = 5 \times 10^{-6} \text{ s}^{-1} (10^{-8} + [\text{Alk}] - C_T)$$

$$= 5 \times 10^{-6} (3 \times 10^{-6} - C_T) (\text{mol cm}^{-3}\text{ s}^{-1}) \qquad (iv)$$

The rate coefficient 5×10^{-6} s^{-1} can be converted to 1.8×10^{-2} h^{-1}. The solution of this differential equation gives (t in hours)

$$C_T = 3 \times 10^{-6} + 1.2 \times 10^{-3} \times e^{-0.018t} \qquad (v)$$

The following brief table shows the time dependence of the adjustment in C_T. A few days are necessary to approach equilibrium with the atmosphere.

Time (h)	C_T (M) (Diffusion Boundary Layer = 40 μm)
0	4.2×10^{-3}
20	3.8×10^{-3}
50	3.5×10^{-3}
100	3.2×10^{-3}
∞	3×10^{-3}

Chemical Enhancement of CO_2 Transfer

At higher pH values, the reaction rate of CO_2 to form HCO_3^- increases (compare Section 4.8 and equation 81 of Chapter 4):

5.10 Gas Transfer Across Water–Gas Interface

$$CO_2(aq) + OH^- \rightarrow HCO_3^- \qquad k = 8.5 \times 10^3 \text{ M}^{-1}\text{ s}^{-1}$$

and at high pH values the reaction rate of CO_2 becomes faster than the diffusion process through the water film. In the extreme case, we have to consider for the flux, J_T, all the carbonate species. We can compare the flux, J_T, with that of CO_2:

$$J_T = \sum_i J_i = \frac{D_{CO_2}}{z_w}(C_{T,w/a} - C_T) \qquad (44)$$

and

$$J_{CO_2} = \frac{D_{CO_2}}{z}([H_2CO_3^*]_{w/a} - [H_2CO_3^*]) \qquad (45)$$

and the chemical enhancement factor $\alpha = J_T/J_{CO_2}$ (compare equation 43).

Example 5.10. Enhancement of CO_2 Transfer by Photosynthesis At high photosynthetic intensity, a lake with $[Alk] = 10^{-3.5}$ M attains a pH value of 9.0 (25°C). What is the maximum chemical enhancement factor for the CO_2 uptake from the atmosphere?

We calculate $[H_2CO_3^*]_{w/a}$, $[HCO_3^-]_{w/a}$, $[CO_3^{2-}]_{w/a}$, and $[H^+]_{w/a}$ for the equilibrium with the atmosphere ($10^{-3.5}$ atm). $[H_2CO_3^*]_{w/a} = 10^{-5}$ M, $[HCO_3^-]_{w/a} = 10^{-3.5}$ M, $[CO_3^{2-}]_{w/a} = 10^{-6}$ M, and $[H^+]_{w/a} = 10^{-7.8}$ M.

On the bulk side we have $[H^+] = 10^{-9}$ M and $[Alk] = [HCO_3^-](1 + 2K_2/[H^+]) = 10^{-3.5}$. Thus $[HCO_3^-] = 10^{-3.54}$ M, $[CO_3^{2-}] = 10^{-4.86}$ M, and $[H_2CO_3^*] = 5.7 \times 10^{-7}$ M. We calculate the gradient

$$C_{T,w/a} - C_T = 2.4 \times 10^{-5} \text{ M}$$

and

$$[H_2CO_3^*]_{w/a} - [H_2CO_3^*] = 1 \times 10^{-5} \text{ M}$$

Thus the enhancement factor, α, is ~2.5.

More detailed analysis on the enhancement of CO_2 transfer has been made by Emerson (1975) and Schwartz (1984) for the transfer of reactive gases in atmospheric water droplets.

The Exchange Rate of CO_2 Between the Sea and the Atmosphere

For seawater (20°C), a value for D/z_w of about 1000 m yr^{-1} (~11 cm h^{-1}) (e.g., corresponding to diffusivity of CO_2 of about 5×10^{-2} m^2 yr^{-1} and a z_w of ~50 μm) is representative for the world oceans (Broecker and Peng, 1974).

[$CO_2(aq)$]$_{eq,atm}$ is, in accordance with Henry's law for $p_{CO_2} = 3 \times 10^{-3}$ atm, $\sim 10^{-2}$ mol m^{-3}. By making [$CO_2(aq)$]$_{bulk}$ = 0, we obtain the exchange rate F_0, that is, the flux of CO_2 in either direction:

$$F_0 = \frac{D}{z_w} [CO_2(aq)]_{eq,atm} \tag{46}$$

An exchange rate of ~ 10 mol m^{-2} yr^{-1} results. The residence time of CO_2 in the atmosphere, τ_{CO_2}, is obtained by dividing the total mass of CO_2 in the atmosphere per unit area by the exchange rate:

$$\tau_{CO_2} = \frac{m_{CO_2(atm)}}{F_0} \tag{47}$$

Since there are about 100 mol of CO_2 above each square meter (1 atm $\simeq$ 1000 g air cm^{-2}; and the average molecular weight of air is $\approx$ 29 g; thus there are 100 g/29 g $\simeq$ 34 mol air molecules above 1 cm^2 of which 0.03% are CO_2), a residence time of τ_{CO_2} = 10 years is obtained. Similar residence times have been established on the basis of carbon-14 data. Thus, under these circumstances, the chemical enhancement of the exchange rate does not appear to be very significant. Thus the diffusion of CO_2 is more important than the chemical cycle of HCO_3^- and CO_3^{2-} for the ocean–atmosphere CO_2 exchange.

As shown by Broecker and Peng (1974), the rate-limiting step for removal of anthropogenic CO_2 from the air is vertical mixing within the sea rather than transfer across the air–sea interface.

SUGGESTED READINGS

Charlson, R. J., and Wigley, T. M. L. (1994) Sulfate Aerosol and Climate Change, *Sci. Am.* 270(2), 28-35.

Graedel, T. E., and Crutzen, P. J. (1993) *Atmospheric Change, an Earth System Perspective*, W. H. Freeman, New York.

Hemond, H. F., and Fechner, E. J. (1994) The Atmosphere. (chapter 4) In *Chemical Fate and Transport in the Environment*, Academic Press, San Diego, CA, pp. 227-320.

Jacob, D. J., Munger, J. W., Waldman, J. M., and Hoffmann, M. R. (1986) The H_2SO_4-HNO_3-NH_3 System at High Humidities and in Fogs, *J. Geophys. Res.* **91**/D1, 1073-1088 and 1089-1096.

Lelieveld, J. (1994) Modeling of Heterogeneous Chemistry in the Global Environment. In *Chemistry of Aquatic Systems; Local and Global Perspectives*, ed. G. Bidoglio and W. Stumm, Eds., Kluwer, Dordrecht (NL).

Mackay, D. (1991) *Multimedia Environmental Models; the Fugacity Approach*, Lewis, Chelsea, MI.

Schindler, D. W., Mills, K. H., Mallay, D. F., Findlay, D. L., Shearer, J. A., Davies, I. J., Turner, M. A., Lindsay, G. A., and Cruikshank, D. R. (1985) Long-Term Ecosystem Stress: The Effects of Experimental Acidification on a Small Lake. *Science* **228**, 1395–1401.

Schlesinger, W. H. (1991) The Atmosphere. In *Biogeochemistry, an Analysis of Global Change*, Academic Press, San Diego, pp. 40–71.

PROBLEMS

5.1. What is the pH and the chemical composition of rain droplets, which are in equilibrium with the atmosphere and a partial pressure of NH_3 of $p_{NH_3} = 10^{-8}$ atm? The temperature is 10°C. The available constants are:

$$K_H(CO_2) = 5.37 \times 10^{-2} \text{ M atm}^{-1}$$
$$K_H(NH_3) = 120 \text{ M atm}^{-1}$$
Acidity constants $K_{NH_4^+} = 1.9 \times 10^{-10}$
$$K_{H_2CO_3^*} = 3.5 \times 10^{-7}$$
$$K_{HCO_3^-} = 3.2 \times 10^{-11}$$
$$K_w = 0.4 \times 10^{-14}$$

5.2. A sewer system contains 10 liters anoxic water per m^3. The total concentration of sulfide $S_T = [H_2S] + [HS^-] + [S^{2-}] = 10^{-4}$ mol m^{-3}. Which fraction can be found in the gas phase as a function of pH? (See Table 5.2 for constants.)

5.3. Atmospheric water droplets (10^{-4} liter per m^3 atmosphere) contain a total of 10^{-8} mol m^{-3} NH_4^+ and 5×10^{-9} mol m^{-3} SO_2.
  (a) What is the pH of atmospheric water before and after oxidation of SO_2 by H_2O_2?
  (b) What is the acidity of the water droplets before and after the oxidation of the SO_2? Give reference conditions for acidity.
  (c) After deposition of the water droplets on the soil, NH_4^+ will be oxidized to nitrate (nitrification). What is the total acidity that results from the deposition of the pollutants from 1 m^3 of atmosphere?

5.4. One cubic meter of cloud atmosphere contains, in a closed system, 10^{-7} mol NH_3 m^{-3}, 2×10^{-7} mol SO_2 m^{-3}, and 10 cm^3 water. What is the approximate alkalinity (zero, positive, negative) of the cloud water under toxic conditions (i.e., SO_2 is oxidized to SO_4^{2-})?

5.5 In the figure below, the concentrations of NH_4^+ and of SO_4^{2-} are plotted for different samples of fog (data from Sigg et al., 1987). Which explanations could you give for the observed correlation?

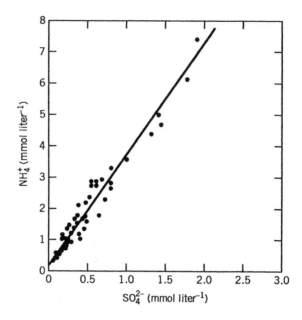

5.6. A surface water has an alkalinity of 2 meq liter^{-1} and a measured pH of 7.8.
  (a) What is the direction of the flux of CO_2 across the air–water interface?
  (b) Calculate the instantaneous flux.
  (c) What is the time scale for reaching 99% of air–water equilibrium with respect to CO_2? The film thickness is estimated at 300 μm. The molecular diffusion coefficient of CO_2 in water is 1.9×10^{-3} cm^2 s^{-1}. Assume that the partial pressure of CO_2 in the air is 0.00033 atm.

5.7. A cascade aerator operating on a well-water will reduce the dissolved CO_2 content from 45 to 18 mg liter^{-1} in one pass. The atmospheric saturation value of CO_2 may be assumed to be 0.5 mg liter^{-1}. Laboratory tests indicate that the gas transfer coefficient for H_2S is about 80% that for CO_2. Assume that enough H_2S will be present in the air around the aerator to give a saturation value of 0.1 mg liter^{-1}. Estimate the effluent concentration of H_2S if the well-water contains 12 mg liter^{-1} of H_2S.

5.8. Estimate the flux of CO_2 carried into a surface water by rainfall. Compare this flux with the exchange rate at the air–water interface. An annual rainfall of 100 cm may be assumed.

5.9. In a closed gas system (25°C) $NH_3(g)$ and $HNO_3(g)$ are present, initially at concentrations given by $p_{NH_3} = 10^{-6}$ atm and $p_{HNO_3} = 5 \times 10^{-7}$ atm. How much NH_3 and HNO_3 remain in the gas phase after aerosols were formed ($K_p = 3 \times 10^{-17}$ atm^2)?

ANSWERS TO PROBLEMS

5.1. pH 7.45,

$$[NH_4^+] = 2.2 \times 10^{-4} \text{ M}$$
$$[NH_3(aq)] = 1.2 \times 10^{-6} \text{ M}$$
$$[H_2CO_3^*] = 1.8 \times 10^{-5} \text{ M}$$
$$[HCO_3^-] = 2.2 \times 10^{-4} \text{ M}$$
$$[CO_3^{2-}] = 1.6 \times 10^{-7} \text{ M}$$

5.2. pH $< pK_{a1}$ 98% $H_2S(g)$

5.3. (a) pH 6.5 prior to oxidation, pH 4.7 after oxidation
(b) [Acy] = 1.44×10^{-5} M before, [Acy] = 1.03×10^{-4} M after
(c) [Acy] = 8.5×10^{-8} mol m^{-3}

5.4. Negative

5.7. 6 mg liter^{-1} H$_2$S. The rate of decrease in the concentration of the gas will be proportional to the oversaturation; that is, $-(C_0 - C_T)/(C_0 - C_3) = \exp(-kt)$.

5.9. $p_{NH_3} = 7.7 \times 10^{-9}$ atm, $p_{HNO_3} = 3.9 \times 10^{-9}$ atm

6

METAL IONS IN AQUEOUS SOLUTION: ASPECTS OF COORDINATION CHEMISTRY

6.1 INTRODUCTION

All chemical reactions have one common denominator: the atoms, molecules, or ions involved tend to improve the stability of their electron configuration, that is, of the electrons in their outer shell. In a broad classification of chemical reactions, we distinguish between two general groups of reactions by which atoms achieve such stabilization. (1) Redox processes, in which the oxidation states of the participating atoms change, and (2) reactions in which the coordinative relationships change. What do we mean by a change in coordinative relations? The coordinative relations change if the coordinative partner changes or if the coordination number[†] of the participating atoms is changed. This may be illustrated by the following examples.

1. If an acid is introduced into water,

$$HClO + H_2O = H_3O^+ + ClO^-$$

   the coordinative partner of the hydrogen ion (which has a coordination number of 1) is changed from ClO^- to H_2O.
2. The precipitation that frequently occurs in the reaction of a metal ion with a base,

$$Mg \cdot aq^{2+} + 2OH^- = Mg(OH)_2(s) + aq$$

   can be interpreted in terms of a reaction in which the coordinative relations are changed, in the sense that a three-dimensional lattice is formed in which each metal ion is surrounded by and coordinatively "saturated" by the appropriate number of bases.

[†]The coordination number is indicative of the structure and specifies the number of nearest neighbors (ligand atoms) of a particular atom.

3. Metal ions can also react with bases without formation of precipitates in reactions such as

$$Cu \cdot aq^{2+} + 4NH_3 = [Cu(NH_3)_4]^{2+} + aq$$

In this simple classification of reactions, no distinction needs to be made between acid–base, precipitation, and complex formation reactions; they are all coordinative reactions, hence phenomenologically and conceptually similar.

Concentrations of Heavy Metals in Natural Waters

Table 6.1 gives some data on heavy metal concentrations in lakes, rivers, and oceans. The concentrations of dissolved metal ions are remarkably low (10^{-7}–10^{-11} M). The data given are based on advanced instrumentation, paying attention to the elimination of contamination during sampling, storage, and analysis. We will return in Chapter 10 to global aspects of metal pollution and the factors that regulate the residual concentration of heavy metals in natural waters.

Definitions

In the following, any combination of cations with molecules or anions containing free pairs of electrons (bases) is called coordination (or complex formation) and can be electrostatic, covalent, or a mixture of both. The metal cation will be called the *central atom*, and the anions or molecules with which it forms a coordination compound will be referred to as *ligands*. If the ligand is composed of several atoms, the one responsible for the basic or nucleophilic nature of the ligand is called the ligand atom. If a base contains more than one ligand atom, and thus can occupy more than one coordination position in the complex, it is referred to as a *multidentate* complex former. Ligands occupying one, two, three, and so on, positions are referred to as unidentate, bidentate, tridentate, and so on. Typical examples are oxalate and ethylenediamine as bidentate ligands, citrate as a tridentate ligand, and ethylenediamine tetraacetate (EDTA) as a hexadentate ligand. Complex formation with multidentate ligands is called *chelation*, and the complexes are called chelates. The most obvious feature of a chelate is the formation of a ring. For example, in the reaction between glycine and $Cu \cdot aq^{2+}$, a chelate with two rings, each of five members (a), is formed. Glycine is a bidentate ligand: O— and N— are the donor atoms. If there is more than one metal atom (central atom) in a complex, we speak about *multi-* or *polynuclear complexes*.

One essential distinction between a proton complex and a metal complex is that the *coordination number* of protons is different from that of metal ions. The coordination number of the proton is 1 (in hydrogen bonding, H^+ can also exhibit a coordination number of 2). Most metal cations exhibit an even coordination number of 2, 4, 6, or occasionally 8. In complexes of coordination number 2, the ligands and the central ion are linearly arranged. If the coordi-

Table 6.1. Representative Concentrations of Some Dissolved Heavy Metals in Natural Waters

Area	Cu (nM)	Zn (nM)	Cd (nM)	Pb (nM)	References
U.S. East Coast rivers	17	13	0.095	0.11	Windom et al. (1991)
Mississippi River	23	3	0.12	—	Shiller and Boyle (1987)
Amazon River	24	0.3–3.8	0.06	—	Shiller and Boyle (1987)
Lake Constance	5–20	15–60	0.05–0.1	0.2–0.5	Sigg et al. (1982)
Lake Michigan	10	9	0.17	0.25	Shafer and Armstrong (1990)
Lago Cristallina[a]	5	30	0.5	3	Sigg (1993, personal communication)
Pacific Ocean	0.5–5	0.1–10	0.01–1	0.005–0.08	Bruland and Franks (1983)
Rainwater[b]	10–300	80–900	0.4–7	10–200	Sigg (1993, personal communication)

[a] Alpine lake at 2200 m altitude in the southern Swiss Alps. Its dissolved Al(III) concentration at pH 6 is 600 nM.
[b] As measured near Zürich, Switzerland.
Source: Sigg (1994).

(a) [structure of Cu complex with ethylenediaminediacetate]

(b) [structure of Co-EDTA complex]

nation number is 4, the ligand atoms surround the central ion either in a square planar or in a tetrahedral configuration. If the coordination number is 6, the ligands occupy the corners of an octahedron, in the center of which stands the central atom. An example is given in (b), where the hexadentate ligand ethylenediaminetetraacetic acid (EDTA),

$$\begin{array}{c} HOOCCH_2 \\ HOOCCH_2 \end{array} N-CH_2-CH_2-N \begin{array}{c} CH_2COOH \\ CH_2COOH \end{array}$$

forms a Co(III) complex.

Ion Pairs and Complexes Two types of complex species can be distinguished.[†]

1. *Ion Pairs.* Ions of opposite charge that approach within a critical distance effectively form an ion pair and are no longer electrostatically effective. The metal ion or the ligand or both retain the coordinated water when the complex compound is formed; that is, the metal ion and the base are separated by one or more water molecules.
2. *Complexes.* Most stable entities that result from the formation of largely covalent bonds between a metal ion and an electron-donating ligand—the interacting ligand is immediately adjacent to the metal cation—are called complexes (inner-sphere complexes).

[†]Our discussion here is based on Bjerrum's *ion-association model*. An alternative treatment of short-range interaction, the *specific interaction model*, will be discussed in Appendix 6.2 to this chapter.

Ion pairs are temporary partnerships that, in general, form between "hard" cations and "hard" anions (see Table 6.3); they are also called outer-sphere complexes. In some cases a distinction between the two types of associations is possible through kinetic or spectrophotometric investigation. Kinetically, when true complexes are being formed, a dehydration step must precede the association reaction. Association accompanied by changes in the absorbance of visible light is indicative of complex formation reactions as such, whereas the formation of ion pairs may be accompanied by changes in the ultraviolet region.

Estimates of stability constants of ion pairs can be made on the basis of simple electrostatic models that consider *coulombic* interactions between the ions. Calculations made in this way indicate the following ranges of stability constants (25°C):

For ion pairs with opposite charge of 1, $\log K \simeq 0$ to 1 ($I = 0$); $\log K = -0.5$ to 0.5 in seawater (SW).

For ion pairs with opposite charge of 2, $\log K \simeq 1.5$ to 2.4 ($I = 0$); $\log K = 0.1$ to 1.2 (SW).

For ion pairs with opposite charge of 3, $\log K \simeq 2.8$ to 4.0 ($I = 0$).

The range for seawater (SW) was estimated on the basis of assumed single-ion activity coefficients for a medium of ionic strength 0.7.

The MgSO$_4$ System An interesting example is given by the interpretation obtained from ultrasonic absorption measurements on 2–2 electrolyte systems. Eigen and Tamm (1962) proposed a three-step process, for example,

$$\text{Mg}^{2+} \cdot \text{aq} + \text{SO}_4^{2-} \cdot \text{aq} \underset{k_{2,1}}{\overset{k_{1,2}}{\rightleftharpoons}} \left[\text{Mg}^{2+} \text{O}_{HH}^{HH} \text{OSO}_4^{2-} \right] \text{aq}$$

State 1 — State 2

$$\left[\text{Mg}^{2+} \text{O}_H^H \text{SO}_4^{2-} \cdot \text{aq} \right] \underset{k_{4,3}}{\overset{k_{3,4}}{\rightleftharpoons}} \left[\text{MgSO}_4 \right] \text{aq}$$

State 3 — State 4

(with $k_{2,3}$, $k_{3,2}$ connecting State 2 and State 3)

Interaction of Metal Ions with Particles

Particles—because of their high surface areas—are scavengers for metal ions and often are reactive elements in their transport from land to rivers and lakes and from continents to the floor of the oceans. Hydrous oxide and aluminum

6.1 Introduction

silicate surfaces, as well as organically coated and organic surfaces, contain functional surface groups (=MOH, =ROH, =R—COOH) that are able to act as coordinating sites of the surface (see Section 9.4.)

The functional groups at the surface undergo acid–base and other coordinative interactions; thus their coordination properties are similar to those of their counterparts in soluble compounds. The concentration of metals is typically much larger in the solid phase than in the solution phase. Thus the buffering of metals is much higher in the presence of particles than in their absence (Sigg et al., 1984).

Adsorption of H^+ and OH^- ions is thus based on protonation and deprotonation:

$$S-OH + H^+ \rightleftharpoons S-OH_2^+$$

$$S-OH\ (+OH^-) \rightleftharpoons S-O^- + H^+\ (+H_2O).$$

Deprotonated surface hydroxyls exhibit Lewis base behavior. Adsorption of metal ions is therefore understood as competitive complex formation involving one or two surface hydroxyl groups:

$$S-OH + M^{z+} \rightleftharpoons S-OM^{(z-1)+} + H^+$$

$$2\ S-OH + M^{z+} \rightleftharpoons (S-O)_2M^{(z-2)+} + 2\ H^+$$

Chemical Speciation

The term *species* refers to the actual form in which a molecule or ion is present in solution. For example, iodine in aqueous solution may conceivably exist as one or more of the species I_2, I^-, I_3^-, HIO, IO^-, IO_3^-, or as an ion pair or complex, or in the form of organic iodo compounds. Figure 6.1 shows the various forms in which metals are thought to occur in natural waters. It is operationally difficult to distinguish between dissolved and colloidally dispersed substances. Colloidal metal-ion precipitates, such as $Fe(OH)_3(s)$ or FeOOH(s) may occasionally have particle sizes smaller than 100 Å—sufficiently small to pass through a membrane filter. Organic substances can assist markedly in the formation of stable colloidal dispersions. Information on the types of species encountered under different chemical conditions (types of complexes, their stabilities, and rates of formation) is a prerequisite to a better understanding of the distribution and functions of trace elements in natural waters.

Baes and Mesmer (1976) provide a critical evaluation of the extensive information on the identity of metal-ion species and their hydrolysis products in solutions, as well as the solid oxides and hydroxides they produce. They provide a critical compilation of hydrolysis equilibrium and oxide and hydroxide solubility constants.

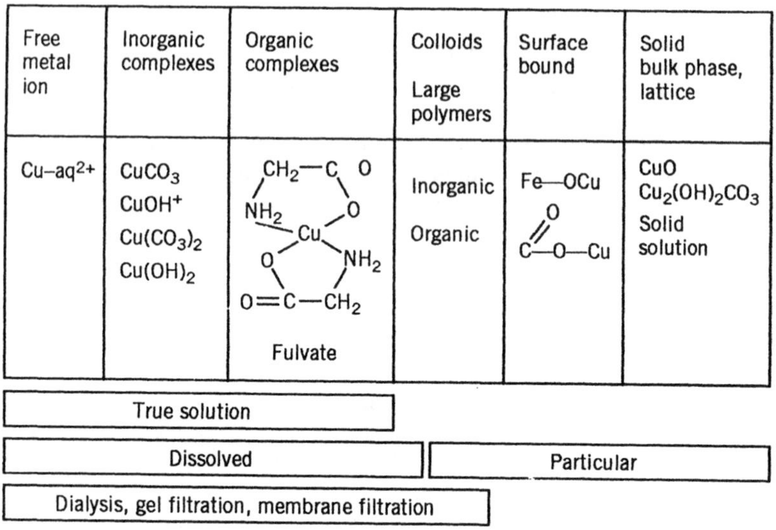

258 Metal Ions in Aqueous Solution: Aspects of Coordination Chemistry

Free metal ion	Inorganic complexes	Organic complexes	Colloids Large polymers	Surface bound	Solid bulk phase, lattice
Cu–aq^{2+}	$CuCO_3$ $CuOH^+$ $Cu(CO_3)_2$ $Cu(OH)_2$	$\begin{array}{c}CH_2-C\diagdown O\\ NH_2\diagdown \quad \diagdown O\\ \quad Cu\\ O\diagup \quad \diagup NH_2\\ O=C-CH_2\end{array}$ Fulvate	Inorganic Organic	Fe—OCu $\diagdown O$ C—O—Cu	CuO $Cu_2(OH)_2CO_3$ Solid solution

True solution
Dissolved		Particular
Dialysis, gel filtration, membrane filtration	

Figure 6.1. Forms of occurrence of metal species.

6.2 PROTONS AND METAL IONS

In all solution environments the bare metal ions are in continuous search of a partner. All metal cations in water are hydrated; that is, they form aquo complexes. The coordination reactions in which metal cations participate in aqueous solutions are exchange reactions with the coordinated water molecules exchanged for some preferred ligands. The barest of the metal cations is the free hydrogen ion, the proton. Hence in some regards there is little difference in principle between a free metal ion and a proton.

Brønsted Acidity and Lewis Acidity In Figure 6.2 alkalimetric titration curves for the reaction of phosphoric acid and $Fe(H_2O)_6^{3+}$, respectively, with a base (OH^- ion) are compared. Millimolar solutions of H_3PO_4 and ferric perchlorate have a similar pH value. Both acids (Fe · aq^{3+} and H_3PO_4) are multiprotic acids; that is, they can transfer more than one proton.

In Figure 6.3 the titration of H_3O^+ with ammonia is compared with the titration of Cu · aq^{2+} with ammonia. pH and pCu ($= -\log[Cu \cdot aq^{2+}]$) are plotted as a function of the base added. In both cases "neutralization curves" are observed. In the case of the H_3O^+–NH_3 reaction a pronounced pH jump occurs at the equivalence point. The pCu jump is less pronounced in the Cu · aq^{2+}–NH_3 reaction because NH_3 is bound to the Cu^{2+} ion in a stepwise consecutive way: ($CuNH_3^{2+}$, $Cu(NH_3)_2^{2+}$, $Cu(NH_3)_3^{2+}$, $Cu(NH_3)_4^{2+}$, $Cu(NH_3)_5^{2+}$) (Figure 6.3d). If, however, four NH_3 molecules are packaged together in one single molecule such as trien (triethylenetetramine, $H_2N-CH_2-CH_2-NH-CH_2-CH_2-NH-CH_2-CH_2-NH_2$), a 1:1 Cu–trien complex is formed and a

6.2 Protons and Metal Ions

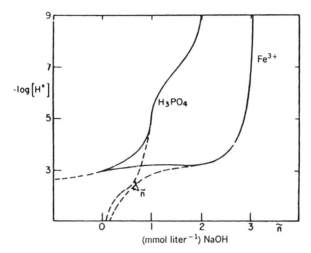

Figure 6.2. Alkalimetric titration of 10^{-3} M H_3PO_4 and 10^{-3} M Fe · aq^{3+}. Both H_3PO_4 and Fe · aq^{3+} are multiprotic Brønsted acids. Millimolar solutions of H_3PO_4 and $Fe(ClO_4)_3$ have similar pH values.

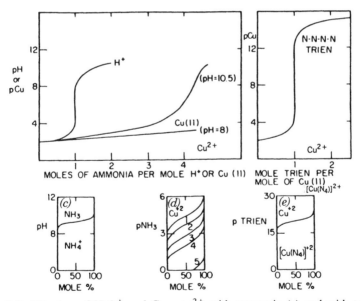

Figure 6.3. Titration of H_3O^+ and Cu · aq^{2+} with ammonia (a) and with tetramine (trien) (b). Equilibrium diagrams for the distribution of NH_3-NH_4^+ (c) of the amino copper(II) complexes (d) and of Cu^{2+}, Cu-trien (e). The similarity of titrating H^+ with a base and titrating a metal ion with a base (Lewis acid–base interaction) is obvious. Both neutralization reactions are used analytically for the determination of acids and metal ions. A pH or pMe indicator electrode (glass electrode for H^+ and copper electrode for Cu^{2+}) can be used for the end point indication.

simple titration curve with a very pronounced pCu jump is observed at the equivalence point (Figure 6.3b). In this case the Cu–trien equilibrium (Figure 6.3e) is as simple as the H^+–NH_3 equilibrium (Figure 6.3c). Such neutralization reactions are exploited analytically for the determination of acids or metal ions; a hydrogen ion electrode (glass electrode) and a metal-ion-sensitive electrode (e.g., a copper electrode for Cu^{2+}), or a pH or pMe indicator, are used as sensors for H^+ and Me^{n+}, respectively. The examples given illustrate the phenomenological similarity between the "neutralization" of H^+ with bases and that of metal ions with complex formers. The bases, molecules or ions, that can neutralize H^+ or metal ions possess free pairs of electrons. Acids are proton donors according to Brønsted. Lewis, on the other hand, has proposed a much more generalized definition of an acid in the sense that he does not attribute acidity to a particular element but to a unique electronic arrangement: the availability of an empty orbital for the acceptance of a pair of electrons. Such acidic or acid analog properties are possessed by H^+, metal ions, and other Lewis acids such as $SOCl_2$, $AlCl_3$, SO_2, and BF_3. In aqueous solutions, protons and metal ions compete with each other for the available bases.

The Acidity of the Metal Ions: Hydrolysis

It is frequently difficult to determine the number of H_2O molecules in the hydration shell, but many metal ions coordinate four or six H_2O molecules per ion. Water is a weak acid. The acidity of the H_2O molecules in the hydration shell of a metal ion is much larger than that of water. As pointed out before (Section 3.2), this enhancement of the acidity of the coordinated water may be interpreted qualitatively as the result of repulsion of the protons of H_2O molecules by the positive charge of the metal ion.

For example, in the case of $Zn \cdot aq^{2+}$:

$$Zn(H_2O)_6^{2+} \rightleftharpoons Zn(H_2O)_5OH^+ + H^+ \qquad *K_1^\dagger$$

$$*K_1 = \frac{[Zn(OH^+)][H^+]}{[Zn^{2+}]} \tag{1a}$$

$$Zn(H_2O)_5OH^+ \rightleftharpoons Zn(H_2O)_4(OH)_2 + H^+ \qquad *K_2$$

$$*K_2 = \frac{[Zn(OH)_2][H^+]}{[ZnOH^+]} \tag{1b}$$

†The stability of the hydrolysis species may also be expressed in terms of the hydroxo complex formation (cf. Table 6.2, where in case of hydrolysis L = OH^- and HL = H_2O).

$$Zn(H_2O)_6^{2+} + OH^- \rightleftharpoons Zn(H_2O)_5OH^+ + H_2O \qquad K_1$$
$$H_2O \rightleftharpoons H^+ + OH^- \qquad K_W$$
$$Zn(H_2O)_6^{2+} \rightleftharpoons Zn(H_2O)_5OH^+ + H^+ \qquad *K_1 = K_1 K_W$$

Thus the K constants and β constants can readily be converted into $*K$ and $*\beta$ constants.

where the cumulative stability constant $*\beta_2$ is given by

$$*\beta_2 = *K_1 \, *K_2$$

$$*\beta_2 = \frac{[\mathrm{Zn(OH)_2}] \, [\mathrm{H^+}]^2}{[\mathrm{Zn^{2+}}]} \tag{1c}$$

Generally, for a species with m hydroxo groups,

$$*\beta_m = \frac{[\mathrm{Me(OH)}_m^{(n-m)+}] \, [\mathrm{H^+}]^m}{[\mathrm{Me^{n+}}]} \tag{2}$$

Hence the acidity of aquo metal ions is expected to increase with a decrease in the radius and an increase in the charge of the central ion. Figure 6.4a attempts to illustrate how the oxidation state of the central atom determines the predominant species (aquo, hydroxo, hydroxo–oxo, and oxo complexes) in the pH range of aqueous solutions. Metal ions with $z = +1$ are generally coordinated with H_2O atoms. Most bivalent metal ions are also coordinated with water up to pH values of 6–12. Most trivalent metal ions are already coordinated with OH^- ions within the pH range of natural waters. For $z = +4$ the aquo ions have become too acidic and are out of the accessible pH range of aqueous solutions with few exceptions, for example, Th(IV). There, O^{2-} already begins to appear as a ligand, for example, for C(IV) where we have oxo–hydroxo complexes, $H_2CO_3 = CO(OH)_2$ or $HCO_3^- = CO_2(OH)^-$, in the pH range 4.5–10; above pH $= 10$, O^{2-} becomes the exclusive ligand (CO_3^{2-}). With even higher oxidation states of the central atom, hydroxo complexes can only occur at very low pH values. The scheme given in Figure 6.4a represents an oversimplification. For every oxidation state, a distribution of acidity according to the ionic radius exists; thus the acidity, as indicated by the pK values given in parentheses, increases in the following series of aquo ions of $z = +2$:

$$\mathrm{Ba^{2+}}(14.0), \quad \mathrm{Ca^{2+}}(13.3), \quad \mathrm{Mg^{2+}}(12.2), \quad \mathrm{Be^{2+}}(5.7)^\dagger$$

The electrostatic rules given are quite useful, but other factors related to the electron distribution are involved. As Figure 6.4b indicates, different cation groups can be distinguished. The cation group furthest to the right contains cations that are most likely to form ionic M—O bonds, whereas the cation groups on the left, which have a stronger tendency to hydrolyze for their size and charge, tend to form bonds of a more covalent character.

Equilibrium constants for hydrolysis are quoted in Appendix 6.1 of this chapter and in Appendixes 1 and 2 at the end of the book. A still reliable source of hydrolysis constants is Baes and Mesmer (1976).

[†]$\mathrm{Be(aq)^{2+}}$ tends to form polynuclear hydrolysis products [e.g., $\mathrm{Be_3(OH)_3^{3+}}$] but on a comparable basis it is more acidic than $\mathrm{Mg(aq)^{2+}}$.

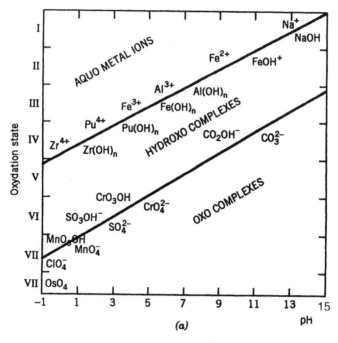
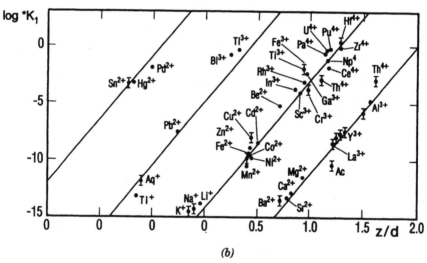
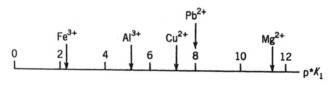

6.3 HYDROLYSIS OF METAL IONS

More than 50 years ago, Brϕnsted postulated that multivalent metal ions participate in a series of consecutive proton transfers:

$$Fe(H_2O)_6^{3+} = Fe(H_2O)_5OH^{2+} + H^+ = Fe(H_2O)_4(OH)_2^+ + 2H^+$$
$$= Fe(OH)_3(H_2O)_3(s) + 3H^+ = Fe(OH)_4(H_2O)_2^- + 4H^+$$

In the case of Fe(III), hydrolysis can go beyond the uncharged species $Fe(OH)_3(H_2O)_3(s)$ to form anions such as the ferrate(III) ion, probably $[Fe(OH)_4 \cdot 2H_2O]^-$. All hydrated ions can in principle donate a larger number of protons than that corresponding to their charge and can form anionic hydroxo metal complexes but, because of the limited pH range of aqueous solutions, not all elements can exist as anionic hydroxo or oxo complexes.

Polynuclear Hydroxo Complexes

The scheme of a consecutive stepwise hydroxide binding is too simple. Although the hydrolysis products listed for hydrolysis of Fe $\cdot$ aq^{3+} are all known and identified, the intermediate steps are frequently complicated. In a few cases, the main products are monomeric. Polymeric hydrolysis species (isopolycations) have been reported for most metal ions. Thus the existence of multinuclear hydrolysis products is a rather general phenomenon especially in solutions of higher metal concentrations. The hydrolyzed species such as $Fe(H_2O)_5OH^{2+}$ can be considered to dimerize by a condensation process. The existence of the dimer has been corroborated experimentally by potentiometric, spectrophotometric, and magnetochemical methods. The dimer may undergo additional hydrolytic reactions that could provide additional hydroxo groups, which then could form more bridges. The terms "ol" and "oxo" are often used in referring to the —OH— and —O— bridges. A sequence of such hydrolytic and condensation reactions, sometimes called olation and oxolation,[†] leads, under condi-

---

[†]Olation may be followed by oxolation, a process in which the bridging OH group is converted to a bridging O group.

Figure 6.4. Hydrolysis of metal ions. (a) Predominant pH range for the occurrence of aquo, hydroxo, hydroxo-oxo, and oxo complexes for various oxidation states. The scheme attempts to show a useful generalization, but many elements cannot be properly placed in this simplified diagram because other factors, such as radius and those related to electron distribution, have to be considered in interpreting the acidity of metal ions. (b) The linear dependence of the $\log_{10}$ of the first hydrolysis constant $*K_1 = \{MOH^{(z-1)+}\}\{H^+\}/\{M^{z+}\}$ on the ratio of the charge to the M—O distance (z/d) for four groups of cations (25°C). (Note changes of abscissa zero for different groups.) (From Baes and Mesmer, 1976.) (c) Hydrolysis constants of some important metal ions.

tions of oversaturation with respect to the (usually very insoluble) metal hydroxide, to the formation of colloidal hydroxo polymers and ultimately to the formation of precipitates. In the pH range lower than the zero point of charge of the metal hydroxide precipitate, positively charged metal hydroxo polymers prevail. In solutions more alkaline than the zero point of charge, anionic hydroxo complexes (isopolyanions) and negatively charged colloids exist. Al-

$$2Fe(H_2O)_5OH^{2+} = [(H_2O)_4Fe\underset{OH}{\overset{OH}{<\!\!\!>}}Fe(H_2O)_4]^{4+} + 2H_2O$$

though multinuclear complexes have been recognized for many years for a few hydrolysis systems such as Cr(III) and Be(II) and for anions of Cr(VI), Si(IV), Mo(VI), and V(V), more recent studies have shown that multinuclear hydrolysis products of metallic cations are of almost universal occurrence in the water solvent system.

The Stability of Hydrolysis Species

The establishment of hydrolysis equilibria is usually very fast, as long as the hydrolysis species are simple. Polynuclear complexes are often formed rather slowly. Many of these polynuclear hydroxo complexes are kinetic intermediates in the slow transition from free metal ions to solid precipitates and are thus thermodynamically unstable. Some metal-ion solutions "age," that is, they change their composition over periods of weeks because of slow structural transformations of the isopoly ions. Such nonequilibrium conditions can frequently be recognized if the properties of metal-ion solutions (electrode potentials, spectra, conductivity, light scattering, coagulation effects, sedimentation rates, etc.) depend on the history of the solution preparation.

Hydrolysis equilibria can be interpreted in a meaningful way if the solutions are not oversaturated with respect to the solid hydroxide or oxide. Occasionally, it is desirable to extend equilibrium calculations into the region of oversaturation; but quantitative interpretations for the species distribution must not be made unless metastable supersaturation can be demonstrated to exist. Most hydrolysis equilibrium constants have been determined in the presence of a swamping "inert" electrolyte of constant ionic strength ($I = 0.1$, 1, or 3 M). As we have seen before, the formation of hydroxo species can be formulated in terms of acid–base equilibria. The formulation of equilibria of hydrolysis reactions is in agreement with that generally used for complex formation equilibria (see Table 6.2).

Table 6.2. Formulation of Stability Constants[a]

I. Mononuclear Complexes

(a) Addition of ligand

$$M \xrightarrow{L}_{K_1} ML \xrightarrow{L}_{K_2} ML_2 \cdots \xrightarrow{L}_{K_i} ML_i \cdots \xrightarrow{L}_{K_n} ML_n$$

$$\longleftarrow \beta_2 \longrightarrow$$
$$\longleftarrow \beta_i \longrightarrow$$
$$\longleftarrow \beta_n \longrightarrow$$

$$K_i = \frac{[ML_i]}{[ML_{(i-1)}][L]} \quad (1)$$

$$\beta_i = \frac{[ML_i]}{[M][L]^i} \quad (2)$$

(b) Addition of protonated ligands

$$M \xrightarrow{HL}_{*K_1} ML \xrightarrow{HL}_{*K_2} ML_2 \cdots \xrightarrow{HL}_{*K_i} ML_i \cdots \xrightarrow{HL}_{*K_n} ML_n$$

$$\longleftarrow *\beta_2 \longrightarrow$$
$$\longleftarrow *\beta_i \longrightarrow$$
$$\longleftarrow *\beta_n \longrightarrow$$

$$*K_i = \frac{[ML_i][H^+]}{[ML_{(i-1)}][HL]} \quad (3)$$

$$*\beta_i = \frac{[ML_i][H^+]^i}{[M][HL]^i} \quad (4)$$

II. Polynuclear Complexes

In β_{nm} and $*\beta_{nm}$ the subscripts n and m denote the composition of the complex M_mL_n formed. [If $m = 1$, the second subscript ($=1$) is omitted.]

$$\beta_{nm} = \frac{[M_mL_n]}{[M]^m[L]^n} \quad (5)$$

$$*\beta_{nm} = \frac{[M_mL_n][H^+]^n}{[M]^m[HL]^n} \quad (6)$$

[a] The same notation as that used in Sillén and Martell (1964, 1971) is used here.

The following rules can be established:

1. The tendency of metal-ion solutions to protolyze (hydrolyze) increases with dilution and with decreasing $[H^+]$.
2. The fraction of polynuclear complexes in a solution decreases on dilution.

The first rule can be illustrated by comparing the equilibria ($I = 0$, 25°C)

$$Mg^{2+} + H_2O = MgOH^+ + H^+ \quad \log {}^*K_1 = -11.4 \quad (3)$$

$$Cu^{2+} + H_2O = CuOH^+ + H^+ \quad \log {}^*K_1 = -8.0 \quad (4)$$

At great dilution (pH → 7), a substantial fraction of the Cu(II) of a pure Cu-salt solution [e.g., $Cu(ClO_4)_2$] will occur as a hydroxo complex

$$\alpha_{CuOH^+} = \frac{[CuOH^+]}{Cu_T} = \left(1 + \frac{[H^+]}{{}^*K_1}\right)^{-1} = 0.091 \quad (5)$$

On the other hand, because of the low acidity of Mg^{2+}, even at infinite dilution, the fraction of hydrolyzed Mg^{2+} ions of a solution of an Mg^{2+} salt is very small:

$$\alpha_{MgOH^+} = \frac{[MgOH^+]}{Mg_T} = \left(1 + \frac{[H^+]}{{}^*K_1}\right)^{-1} = 0.00004 \quad (6)$$

Accordingly, only the salt solutions of sufficiently acid metal ions that fulfill the condition

$$p^*K_1 < \frac{1}{2} pK_W \quad \text{or} \quad p^*\beta_n < \frac{n}{2} pK_W \quad (7)$$

where $^*\beta_n$ is the cumulative acidity constant (see Table 6.2), undergo substantial hydrolysis upon dilution. The progressive hydrolysis upon dilution is the reason that some metal-salt solutions tend to precipitate upon dilution.

Mononuclear Wall If hydrolysis leads to mononuclear and polynuclear hydroxo complexes, it can be shown that mononuclear species prevail beyond a certain dilution. If we consider, for example, the dimerization of $CuOH^+$:

$$2CuOH^+ = Cu_2(OH)_2^{2+} \quad \log {}^*K_{22} = 1.5 \quad (8)$$

it is apparent from the dimensions of the equilibrium constant ($conc^{-1}$) that the dimerization is concentration dependent. Thus for a Cu(II) system where $Cu_T = [Cu^{2+}] + [Cu(OH)^+] + 2[Cu_2(OH)_2^{2+}]$, equilibrium 8 can be formulated as

$$\frac{[Cu_2(OH)_2^{2+}]}{[CuOH^+]^2} = \frac{[Cu_2(OH)_2^{2+}]}{(Cu_T - [Cu^{2+}] - 2[Cu_2(OH)_2^{2+}])^2} = {}^*K_{22} \quad (9)$$

and it becomes obvious that $[Cu_2(OH)_2^{2+}]$ is dependent on Cu_T. With the help of equations 9 and 4 for each pH, the mononuclear wall (e.g., Cu_T for $[Cu_{dimer}]$

= 1/100 [Cu$_{monomer}$]) can be calculated (compare Example 6.1). As pointed out before, for many metals the polynuclear species are formed only under conditions of oversaturation with respect to the metal hydroxide or metal oxide and are thus not stable thermodynamically, for example,

$$Cu_2(OH)_2^{2+} = Cu^{2+} (aq) + Cu(OH)_2(s) \quad \Delta G° = -2.6 \text{ kcal mol}^{-1} \quad (10)$$

Multinuclear hydrolysis species usually are not observed during dissolution of the most stable modification of the solid hydroxide or oxide; they are formed, however, by oversaturating a solution with respect to the solid phase. Such polynuclear species, even if thermodynamically unstable, may be of significance in natural water systems. Many multinuclear hydroxo complexes may persist as metastable species for years.

Quantitative application of known hydrolysis equilibria is illustrated in the next two examples.

Example 6.1. The Hydrolysis of Iron (III) The addition of $Fe(ClO_4)_3$ to H_2O leads to the following dissolved species: Fe^{3+}, $Fe(OH)^{2+}$, $Fe(OH)_2^{1+}$, $Fe(OH)_3$ (aq), $Fe(OH)_4^-$, $Fe_2(OH)_2^{4+}$, and $Fe_3(OH)_4^{5+}$. The following equilibrium constants are available $I = 3$ M (NaClO$_4$) (25°C).

$$Fe^{3+} + H_2O = FeOH^{2+} + H^+ \quad \log {}^*K_1 = -3.05 \quad (i)$$

$$Fe^{3+} + 2H_2O = Fe(OH)_2^+ + 2H^+ \quad \log {}^*\beta_2 = -6.31 \quad (ii)$$

$$2Fe^{3+} + 2H_2O = Fe_2(OH)_2^{4+} + 2H^+ \quad \log {}^*\beta_{22} = -2.91 \quad (iii)$$

$$Fe^{3+} + 3 H_2O = Fe(OH)_3(aq) + 3 H^+ \quad \log {}^*\beta_3 = -13.8 \quad (iv)$$

$$Fe^{3+} + 4 H_2O = Fe(OH)_4^- + 4 H^+ \quad \log \beta_4 = -22.7 \quad (v)$$

$$3 Fe^{3+} + 4 H_2O = Fe_3(OH)_4^{5+} + 4 H^+ \quad \log {}^*\beta_{43} = -5.77 \quad (vi)$$

$$Fe(OH)_3(s) + 3 H^+ = Fe^{3+} + 3 H_2O \quad \log {}^*K_{s0} = 3.96 \quad (vii)$$

The existence of the species $Fe(OH)_3$(aq) and the equilibrium constant of equation iv are uncertain. We also include in the tabulation the solubility product of amorphous $Fe(OH)_3$(s) (equation viii).

(a) Compute the equilibrium composition of a 10^{-9} M solution of Fe(III) as a function of pH.

We assume that the solution is homogeneous, that is, that $Fe(OH)_3$(s) would not precipitate. In the *homogeneous system*, the concentration condition (equation ix or x) must be fulfilled:

Tableau 6.1. Fe(III) Hydrolysis Species in Homogeneous Solution

Components	Fe^{3+}	H$^+$	log K (I = 3 M)
Species Fe^{3+}	1	0	0
FeOH$^+$	1	−1	−3.05
Fe(OH)$_2^+$	1	−2	−6.31
Fe(OH)$_3^0$	1	−3	−13.8
Fe(OH)$_4^-$	1	−4	−22.7
Fe$_2$(OH)$_2^{4+}$	2	−2	−2.91
Fe$_3$(OH)$_4^{5+}$	3	−4	−5.77
H$^+$	0	1	0
	TOTFe(III) = 10^{-9}	pH given	

$$Fe_T = [Fe^{3+}] + [FeOH^{2+}] + [Fe(OH)_2^+] + [Fe(OH)_3(aq)]$$
$$+ [Fe(OH)_4^-] + 2[Fe_2(OH)_2^{4+}] + 3[Fe_3(OH)_4^{5+}] \quad (ix)$$

$$Fe_T = [Fe^{3+}]\left(1 + \frac{*K_1}{[H^+]} + \frac{*\beta_2}{[H^+]^2} + \frac{*\beta_3}{[H^+]^3} + \frac{*\beta_4}{[H^+]^4}\right.$$
$$\left. + \frac{2[Fe^{3+}]*\beta_{22}}{[H^+]^2} + \frac{3[Fe^{3+}]^2*\beta_{43}}{[H^+]^4}\right) \quad (x)$$

Tableau 6.1 summarizes the problem.

Equation ix or x can readily be solved for various [H$^+$] and for TOTFe(III) = 10^{-9} M, especially if one assumes that the dimer and the trimer can be neglected in this dilute solution. Figure 6.5a gives the answer. The multimeric species are present at less than 10^{-15} M and can be neglected. The figure illustrates that Fe^{3+} is essentially a four-protic acid.

(b) Show how the multimeric species Fe$_2$(OH)$_2^{4+}$ and Fe$_3$(OH)$_4^{5+}$ depend on Fe(III)$_T$ and pH.

Figure 6.5. Hydrolysis of Fe(III). (a) Distribution of Fe(III) species in a 10^{-9} M Fe(III) solution as a function of pH. The solution is considered homogeneous [it is, in the neutral pH, range slightly oversaturated with respect to amorphous Fe(OH)$_3$] (cf. Figure 6.8b). The concentrations of the multimeric species Fe$_2$(OH)$_2^{4+}$ and Fe$_3$(OH)$_4^{5+}$ are below 10^{-15} M. (b, c) Distribution diagrams for the various hydrolysis species in a hypothetically homogeneous 10^{-4} M and 10^{-2} M Fe(III) solution, respectively. Formation of the dimer and trimer increases with increasing Fe(III)$_{tot}$. The shaded area indicates the approximate pH range of oversaturation with regard to Fe(OH)$_3$(s). The polynuclear species occur in appreciable concentrations only when the solutions become oversaturated. It is thus plausible that these polynuclear species are intermediates in the formation of the solid phase.

6.3 Hydrolysis of Metal Ions

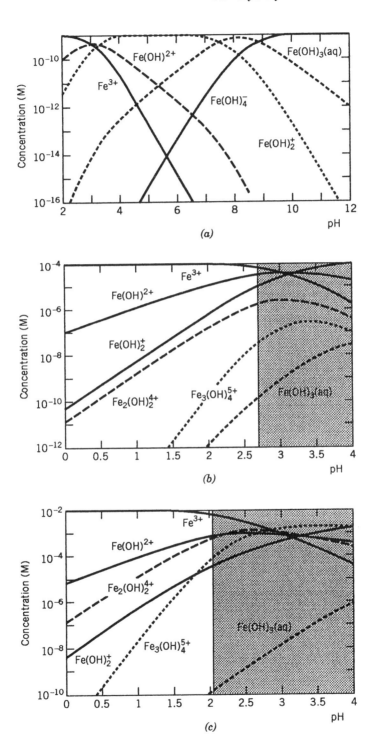

We solve equation ix for $Fe(III)_T = 10^{-4}$ M and for $Fe(III)_T = 10^{-2}$ M for the range of pH values 0–5. The results of our calculation are given in Figure 6.5b,c. The tendency to form multimeric species increases with increasing TOT[Fe(III)] and increasing pH.

Example 6.2. Hydrolysis of Cu^{2+} and Complex Formation with CO_3^{2-} Estimate the equilibrium distribution of Cu(II) species in a carbonate-bearing water of pH = 8. $[Cu(II)]_T = 5 \times 10^{-8}$ M and $C_T = 2 \times 10^{-3}$ M.

Tableau 6.2 summarizes the conditions of the problem and the necessary equilibrium constants.

In this case the concentration of Cu(II) is much smaller than C_T; that is, in the equation for C_T (second column of the tableau)

$$C_T = [H_2CO_3] + [HCO_3^-] + [CO_3^{2-}] + [CuCO_3^0] + 2[Cu(CO_3)_2^{2-}] \quad \text{(i)}$$

the two last terms (the carbonate species) can be neglected. Thus we calculate first, for pH = 8, the CO_3^{2-} concentration:

$$C_T \simeq [CO_3^{2-}] \left(\frac{[H^+]^2}{K_1 K_2} + \frac{[H^+]}{K_2} + 1 \right) \quad \text{(ii)}$$

$$[CO_3^{2-}] = 9.8 \times 10^{-6} \text{ M}$$

Tableau 6.2. Hydrolysis and CO_3^{2-} Complex Formation of Cu^{2+}

Components		Cu^{2+}	CO_3^{2-}	H^+	log K (25°C, I = 0)
Species	Cu^{2+}	1	0	0	0
	$Cu(OH)^+$	1	0	−1	−8.0
	$Cu(OH)_2^0$	1	0	−2	−16.2
	$Cu(OH)_3^-$	1	0	−3	−26.8
	$Cu(OH)_4^{2-}$	1	0	−4	−39.9
	$CuCO_3^0$	1	1	0	6.77
	$Cu(CO_3)_2^{2-}$	1	2	0	10.01
	$H_2CO_3^*$	0	1	2	16.6
	HCO_3^-	0	1	1	10.3
	CO_3^{2-}	0	1	0	0
	OH^-	0	0	−1	−14
	H^+	0	0	1	0
	TOTCu^{2+} = 5×10^{-8} M $\quad C_T = 2 \times 10^{-3}$ M $\quad$ pH given				

6.3 Hydrolysis of Metal Ions

Now we can compute the individual Cu(II) species:

$$[Cu]_T = [Cu^{2+}] + [CuOH^+] + [Cu(OH)_2^0] + [Cu(OH)_3^-]$$
$$+ [Cu(OH)_4^{2-}] + [CuCO_3^0] + [Cu(CO_3)_2^{2-}] \quad \text{(iii)}$$

$$[Cu]_T = [Cu^{2+}] (1 + {}^*\beta_1[H^+]^{-1} + {}^*\beta_2[H^+]^{-2} + {}^*\beta_3[H^+]^{-3} + {}^*\beta_4[H^+]^{-4}$$
$$+ \beta_{1CO_3}[CO_3^{2-}] + \beta_{2CO_3}[CO_3^{2-}]^2) \quad \text{(iv)}$$

We obtain for pH = 8.0 the following distribution:

pH = 8.0	mol liter^{-1}	% Cu$_T$
Cu^{2+}	8.2×10^{-10}	1.6
$CuOH^+$	8.2×10^{-10}	1.6
$Cu(OH)_2^0$	5.2×10^{-10}	1.0
$Cu(OH)_3^-$	1.3×10^{-12}	3×10^{-3}
$Cu(OH)_4^{2-}$	1×10^{-17}	2×10^{-8}
$CuCO_3^0$	4.7×10^{-8}	94.0
$Cu(CO_3)_2^{2-}$	7.8×10^{-10}	1.6

The speciation of Cu (for $C_T = 2 \times 10^{-3}$ M), as a function of pH, is given in Figure 6.6.

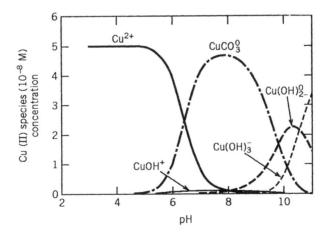

Figure 6.6. Speciation of 5×10^{-8} M Cu(II) in a carbonate-bearing water ($C_T = 2 \times 10^{-3}$ M).

6.4 SOLUBILITY AND HYDROLYSIS: SOLID HYDROXIDES AND METAL OXIDES

It has been pointed out that the formation of a precipitate can often be considered the final stage in the formation of polynuclear complexes. Aggregates of ions that form the building stones in the lattice are produced in the solution, and these aggregates combine with other ions to form neutral compounds.

Plausibly there is a correlation of the solubility of the stable oxide, hydroxide, or oxyhydroxide of a cation with the stability of the first hydrolysis product $MOH^{(z-1)+}$ (Figure 6.7). Many multivalent hydrous oxides are amphoteric because of the acid–base equilibria involved in the hydrolysis reactions of aquo metal ions. Alkalimetric or acidimetric titration curves for hydrous metal oxides provide a quantitative explanation for the manner in which the charge of the hydrous oxide depends on the pH of the medium. The amphoteric behavior of solid metal hydroxides becomes evident from such titration curves. From an operational point of view, such hydrous oxides can be compared with amphoteric polyelectrolytes and can be considered hydrated solid electrolytes, fre-

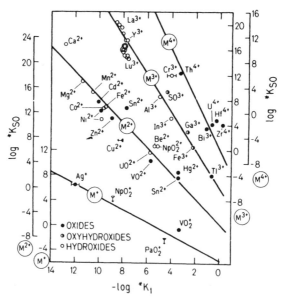

Figure 6.7. Correlation of the solubility product $*K_{s0}$ with the first hydrolysis constant $*K_1$ for M^+, M^{2+}, M^{3+}, and M^{4+} cations. The lines have slopes of -1, -2, -3, and -4 (25°C). This relationship results because the equilibrium constant for the reaction $M(OH)_z(s) + (z-1)M^{z+} = zMOH^{(z-1)+}$, $K = (*K_1)^z K_{s0}$, $K = \{MOH^{(z-1)+}\}^z/\{M^{z+}\}^{(z-1)}$, is often close to $10^{-5.6}$. This relatively low value reflects the general tendency of cations to precipitate shortly after hydrolysis begins unless $[M^z]$ is quite low. The strong tendency for solutions to supersaturate often allows hydrolysis to proceed much further in solution than would be expected from the value of K. (From Baes and Mesmer, 1976.)

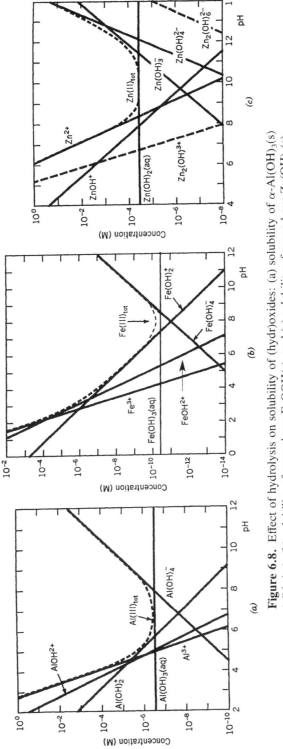

Figure 6.8. Effect of hydrolysis on solubility of (hydr)oxides: (a) solubility of α-Al(OH)$_3$(s) (gibbsite), (b) solubility of amorphous FeOOH(s), and (c) solubility of amorphous Zn(OH)$_2$(s). Data from Baes and Mesmer (1976). Multinuclear species are omitted.

quently possessing a variable space lattice in which the proportion of different ions, cations as well as anions, is variable within the limits of electrical neutrality of the solid. These hydroxides show a strong tendency to interact specifically with anions as well as with cations.

The role of hydrolysis products in influencing the solubility of Al(III), Fe(III), and Zn(II)(hydr)oxides is illustrated in Figures 6.8a, 6.8b, and 6.8c, respectively. Tableaux 6.3a, 6.3b, and 6.3c summarize the equilibrium constants used in the construction of the diagrams (Baes and Mesmer, 1976).

Tableau 6.3a. (Hydr)oxide Solubility as Influenced by Hydrolysis: Gibbsite α-Al(OH)$_3$ (25°C, $I = 0$)

Components		α-Al(OH)$_3$(s)	H$^+$	log K
Species	Al^{3+}	1	3	8.5
	AlOH^{2+}	1	2	3.53
	Al(OH)$_2^+$	1	1	−0.8
	Al(OH)$_3$	1	0	−6.5
	Al(OH)$_4^-$	1	−1	−14.5
	Al$_3$(OH)$_4^{5+}$	3	5	11.6
	Al$_{13}$O$_4$(OH)$_{24}^{7+}$	13	7	11.8
	H$^+$	0	1	0

{Al(OH)$_3$(s)} = 1   pH given

$$\text{Al(III)}_{tot} = [\text{Al}^{3+}] + [\text{AlOH}^{2+}] + [\text{Al(OH)}_2^+] + [\text{Al(OH)}_3]$$
$$+ [\text{Al(OH)}_4^-] + 3\,[\text{Al}_3(\text{OH})_4]^{5+} + 13\,[\text{Al}_{13}\text{O}_4(\text{OH})_{24}^{7+}]$$

Tableau 6.3b. (Hydr)oxide Solubility as Influenced by Hydrolysis: Amorphous FeOOH

Components		FeOOH(s)	H$^+$	log K
Species	Fe^{3+}	1	3	2.5
	FeOH^{2+}	1	2	0.31
	Fe(OH)$_2^+$	1	1	−3.17
	Fe(OH)$_3$	1	0	−10.5
	Fe(OH)$_4^-$	1	−1	−19.1
	Fe$_2$(OH)$_2^{4+}$	2	4	2.05
	Fe$_3$(OH)$_4^{5+}$	3	5	1.2
	H$^+$	0	1	0

{FeOOH(s)} = 1   pH given

$$\text{Fe(III)}_{tot} = [\text{Fe}^{3+}] + [\text{FeOH}^{2+}] + [\text{Fe(OH)}_2^+] + [\text{Fe(OH)}_3]$$
$$+ [\text{Fe(OH)}_4^-] + 2\,[\text{Fe}_2(\text{OH})_2]^{4+} + 3\,[\text{Fe}_3(\text{OH})_4]^{5+}$$

Tableau 6.3c. Amorphous $Zn(OH)_2$ (25°C, $I = 0$)

Components		$Zn(OH)_2(s)$	H^+	log K
Species	Zn^{2+}	1	2	12.45
	$ZnOH^+$	1	1	3.49
	$Zn(OH)_2$	1	0	−4.45
	$Zn(OH)_3^-$	1	−1	−15.95
	$Zn(OH)_4^{2-}$	1	−2	−28.75
	Zn_2OH^{3+}	2	3	15.9
	$Zn_2(OH)_6^{2-}$	2	−2	−32.9
	H^+	0	1	0
		$\{Zn(OH)_2(s)\} = 1$	pH given	

$$Zn(II)_{tot} = [Zn^{2+}] + [ZnOH^+] + [Zn(OH)_2] + [Zn(OH)_3^-]$$
$$+ [Zn(OH)_4^{2-}] + 2[Zn_2OH^{3+}] + 2[Zn_2(OH)_6^{2-}]$$

As before, the log species concentrations and equilibrium constants, defined by the horizontal rows, can be read as follows:

$$\log [Al^{3+}] = 8.5 - 3 \, pH \tag{i}$$

$$\log [AlOH^{2+}] = 3.53 - 2 \, pH \tag{ii}$$

$$\log [Al(OH)_2^+] = -0.8 - pH \tag{iii}$$

$$\log [Al(OH)_3] = -6.5 \tag{iv}$$

$$\log [Al(OH)_4^-] = -14.5 + pH \tag{v}$$

$$\log [Al_3(OH)_4^{5+}] = 11.6 - 5 \, pH \tag{vi}$$

$$\log [Al_3O_4(OH)_{24}^{7+}] = 11.8 - 7 \, pH \tag{vii}$$

In a double logarithmic diagram (e.g., Figure 6.8), the intercepts and slopes are defined by equations i–vii; for example, the log $[Al^{3+}]$ line intercepts with the log concentration $= 0$ line at pH $= \frac{1}{3} \times 8.5$ at pH $= 2.85$.

In Figures 6.8a and 6.8b the polynuclear species, whose concentrations are small relative to the other species, have been omitted. (They can readily be included in the diagram if one needs this information.)[†]

6.5 CHELATES

As with hydroxide, complexes with other ligands are formed stepwise. In a series such as

---

[†]The multinuclear species may be metastable especially if the solubility equilibrium is approached from oversaturation.

Cu^{2+}, $Cu(NH_3)^{2+}$, $Cu(NH_3)_2^{2+}$, $Cu(NH_3)_3^{2+}$, $Cu(NH_3)_4^{2+}$, $Cu(NH_3)_5^{2+}$

the successive stability constants generally decrease; that is, the Cu^{2+} ion takes up one NH_3 molecule after the other. (There are exceptions to this behavior.) As can be seen from Figure 6.3d, relatively high concentrations of NH_3 are necessary to complex copper(II) effectively and to form a tetramine complex. In natural waters the concentrations of the ligands and the affinity of the ligands for the metal ion, with the exception of H_2O and OH^-, are usually sufficiently small so that at best a one-ligand complex may be formed.

The Chelate Effect

Complexes with monodentate ligands are usually less stable than those with multidentate ligands. More important is the fact that the degree of complexation decreases more strongly with dilution for monodentate complexes than for multidentate complexes (chelates). This is illustrated in Figure 6.9, where the degree of complexation is compared as a function of concentration for monodentate, bidentate, and tetradentate copper(II) amine complexes. Free $[Cu \cdot aq^{2+}]$ is plotted as a function of dilution in the left-hand graph, while the quantitative degree of complexation, as measured by $\Delta pCu = \log(Cu_T/[Cu^{2+}])$ is given in the right-hand graph. It is obvious from this figure that the complexing effect of NH_3 on Cu^{2+} becomes negligible at concentrations that might be encountered in natural water systems. Chelates, however, remain remarkably stable even at very dilute concentrations.

The curves drawn in Figure 6.9 have been calculated on the basis of constants taken from Sillén and Martell (1964, 1971). The calculations are essentially the same as those outlined for the hydroxo complex formation; although algebraically simple, they are tedious and time-consuming. In the case of the Cu–NH_3 system, the following species have to be considered:

Cu^{2+}, $Cu(NH_3)^{2+}$, $Cu(NH_3)_2^{2+}$, $Cu(NH_3)_3^{2+}$, $Cu(NH_3)_4^{2+}$, NH_4^+, NH_3

Thus, for every $[H^+]$, seven equations have to be solved simultaneously in order to compute the relative concentrations of each species present. Five mass laws (four stability expressions for the four different amine complexes and the acid–base equilibrium of NH_4^+–NH_3) and two concentration conditions make up the seven equations. As concentration conditions one can formulate equations defining Cu_T and NH_{3T}:

$$Cu_T = [Cu^{2+}] + [Cu(NH_3)^{2+}] + [Cu(NH_3)_2^{2+}] + \cdots \quad (11)$$

$$NH_{3T} = [NH_4^+] + [NH_3] + [Cu(NH_3)^{2+}] + 2[Cu(NH_3)_2^{2+}] + \cdots \quad (12)$$

Guidelines for coping with these and more involved types of calculations have been provided by Ringbom (1963), Schwarzenbach (1959), and others. Computers are of course also very useful.

6.5 Chelates   277

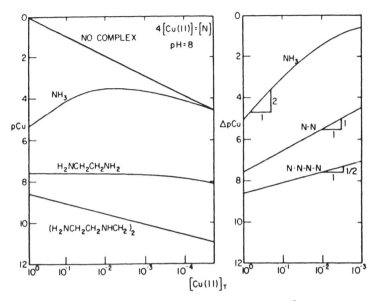

Figure 6.9. The chelate effect on complex formation of Cu · aq^{2+} with monodentate, bidentate, and tetradentate amines. pCu is plotted as a function of concentration in the left-hand diagram. On the right the relative degree of complexation as measured by ΔpCu as a function of concentration is depicted. The extent of complexing is larger with chelate complex formers than with monodentate ligands. Unidentate complexes are dissociated in dilute solutions while chelates remain essentially undissociated at great dilutions.

In Figure 6.9 a concentration of complex former equivalent to that of metal ion was considered. ΔpM of course increases with increasing concentration of the complex former over the metal. Figure 6.10a,b shows the effect of various ligands on complex formation with ferric iron and with Cu(II). Here the concentration of the complexing agent is kept at a constant value and in excess of the metal.

In Figure 6.10a we observe that in going from a monodentate (F$^-$) to a bidentate (salicylate) to a tridentate (citrate) to a hexadentate (EDTA) ligand the relative stability, as measured by Σ FeL/Fe$_{tot}$ and of Σ CuL/Cu$_{tot}$, increases.

Figure 6.10b compares complexation of Cu(II) by citrate, glycine, salicylate, and ammonia as a function of pH. As for Fe(III) complexation, multidentate ligands have relatively greater stability than monodentate (citrate versus ammonia; citrate versus salicylate). The quantitatively different interactions of ligands, metal ions, protons, and hydroxide ions are evident in both Figures 6.10a and 6.10b.

Figure 6.10 illustrates that [H$^+$] and [OH$^-$] influence markedly the degree of complexation. At low pH, H$^+$ competes successfully with the metal ions for the ligand. At high pH, OH$^-$ competes successfully with the ligand for the coordinative positions on the metal ion. Furthermore, at low and high pH,

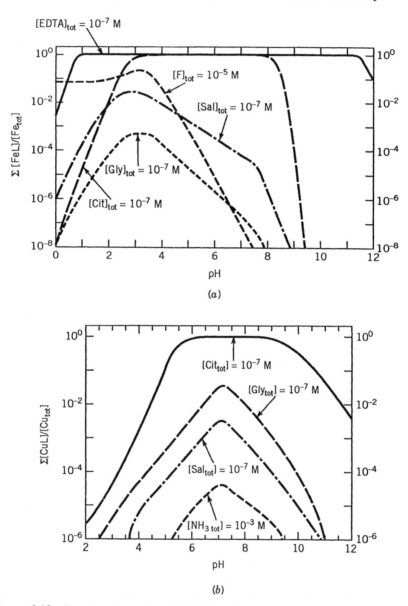

Figure 6.10. Complex formation of Fe(III) and of Cu(II) by various ligands. The stability cannot be predicted alone from complex stability constants but competitive effects of H^+ (with metal ions) and of OH^- (with ligands) need to be considered. Multidentate complex formers form more stable complexes, especially at high dilutions, than monodentate ligands (e.g., F^-, NH_3). To solutions of TOTFe(III) = 10^{-8} M and TOTCu(II) = 10^{-8} M, respectively, complex formers (at the concentrations indicated in the figures) were added (points are calculated). (Cit = citrate, gly = glycinate, sal = salicylate.) The relative extent of complex formation $\Sigma[\text{FeL}]/[\text{Fe}]_{tot}$ or $\Sigma[\text{CuL}]/[\text{Cu}]_{tot}$, respectively, is plotted as a function of pH.

mixed hydrogen–metal and hydroxide–ligand complexes can be formed. [In the case of EDTA (=L), in addition to FeL$^-$, the complexes FeHL, FeOHL^{2-}, and Fe(OH)$_2$L^{3-} have to be considered. Because of the competing influence of H$^+$ or OH$^-$, the complexing effect cannot be estimated solely from the stability constants.]

Increase in Entropy Is Primarily Responsible for the Chelate Effect A chelate ring is more stable than the corresponding complex with unidentate ligands. The enthalpy change resulting from either complex formation, however, is frequently about the same; for example, ΔH for the formation of a diamine complex is approximately equal to ΔH for the formation of an ethylenediamine complex (Schwarzenbach, 1961). Hence essentially the same type of bond occurs in these complexes. That $-\Delta G°$ for the chelate is larger than for the corresponding complex with a unidentate ligand must be accounted for primarily by the fact that the formation of a ring is accompanied by a larger increase in entropy than that encountered in the formation of a nonchelate complex.

Metal-Ion Buffers

The analogy between Me^{z+} and H$^+$ can be extended to the concept of buffers. pH buffers are made by mixing acids and conjugate bases in proper proportions:

$$[H^+] = \frac{K[HA]}{[A^-]} \quad (13)$$

Metal ions can be similarly buffered by adding appropriate ligands to the metal-ion solution:

$$[Me^{2+}] = \frac{K[MeL]}{[L]} \quad (14)$$

Such pMe buffers resist a change in [Me^{z+}]. It is well known that the living cell controls not only pH but also pCa, pCu, pMn, pMg, and so on, and that complex formers are used as the buffering component; pMe buffers are convenient tools for investigating phenomena pertaining to metal ions. It is unwise to prepare a pH 6 solution by diluting a concentrated HCl solution, but this mistake is frequently made with metal-ion solutions. If, for example, we want to study the toxic effect of Cu^{2+} on algae, it might be more appropriate to prepare a suitable pCu buffer. If a copper salt solution is simply diluted, the concentration (or activity) of the free Cu^{2+} may, because of hydrolysis and adsorption and other side reactions, be entirely different from that calculated by considering the dilution only.

Example 6.3. pCa Buffer Calculate pCa of a solution of the following composition: EDTA (ethylenediamine tetraacetate) = $Y_T = 1.95 \times 10^{-2}$ M, Ca$_T$

$= 9.82 \times 10^{-3}$, pH $= 5.13$, $I = 0.1$ (20°C). EDTA is a tetraprotic acid. For the conditions given, the four acidity constants are characterized by $pK_1 = 2.0$, $pK_2 = 2.67$, $pK_3 = 6.16$, and $pK_4 = 10.26$. Two Ca complexes are formed with EDTA, CaHY$^-$ and CaY^{2-}. Hence we need the stability constants

$$\frac{[\text{CaY}^{2-}]}{[\text{Ca}^{2+}][\text{Y}^{4-}]} = K_{\text{CaY}} = 10^{10.6}$$

$$\frac{[\text{CaHY}^-]}{[\text{Ca}^{2+}][\text{HY}^{3-}]} = K_{\text{CaHY}} = 10^{3.5}$$

The computation may start by setting up equations for the concentration conditions (for simplicity charges are omitted):

$$\text{Ca}_T = [\text{Ca}] + [\text{CaY}] + [\text{CaHY}]$$
$$= [\text{Ca}] (1 + [\text{Y}]K_{\text{CaY}} + [\text{Y}][\text{H}]K_4^{-1}K_{\text{CaHY}})$$
$$= [\text{Ca}]\alpha_{\text{Ca}}^{-1} \qquad \text{(i)}$$

and, as before,

$$\alpha_{\text{Ca}} = \frac{[\text{Ca}]}{\text{Ca}_T}$$

$$Y_T = [\text{H}_4\text{Y}] + [\text{H}_3\text{Y}] + [\text{H}_2\text{Y}] + [\text{HY}] + [\text{Y}] + [\text{CaY}] + [\text{CaHY}]$$
$$= [\text{Y}](\alpha_4')^{-1} + [\text{Ca}][\text{Y}]K_{\text{CaY}} + [\text{Ca}][\text{H}][\text{Y}]K_4^{-1}K_{\text{CaHY}} \qquad \text{(ii)}$$

where

$$\alpha_4' = \frac{[\text{Y}]}{\sum_{i=0}^{i=4}[\text{H}_i\text{Y}]} = \left(1 + \frac{[\text{H}]}{K_4} + \frac{[\text{H}]^2}{K_4K_3} + \frac{[\text{H}]^3}{K_4K_3K_2} + \frac{[\text{H}]^4}{K_4K_3K_2K_1}\right)^{-1} \qquad \text{(iii)}$$

α_4' can be plotted most conveniently in a double logarithmic diagram (log α_4' versus $-\log[\text{H}^+]$). Equations i and ii, containing the two unknowns [Y] and [H], are best solved by trial and error. Most conveniently, we may start by calculating α_4' from equation iii or taking its value from a graphical plot. Then [Y] may be estimated from equation iii by assuming that $\Sigma [\text{H}_i\text{Y}] \approx (Y_T - \text{Ca}_T)$.

With this tentative [Y], first values of α_{Ca}^{-1} and [Ca] are estimated, respectively, with the help of equation i. We then may check whether the assumption that $\Sigma [\text{H}_i\text{Y}] = [Y_T - \text{Ca}_T]$ was appropriate. Subsequent reiteration gives the result $[\text{Ca}] = 4.12 \times 10^{-5}$ (pCa $= 4.39$) and $[\text{Y}] = 6.05 \times 10^{-9}$. For illustration the concentrations of all the other species are also given:

$[CaY] = 9.66 \times 10^{-3}$ $[CaHY] = 1.09 \times 10^{-4}$ $[H_4Y] = 2.26 \times 10^{-8}$

$[H_3Y] = 3.07 \times 10^{-5}$ $[H_2Y] = 8.8 \times 10^{-3}$ $[HY] = 8.21 \times 10^{-4}$

Note: If $[Ca^{2+}]$ is lost from the solution, that is, by adsorption at the glass wall, the Ca^{2+} ion buffer described will have a tendency to maintain constant pCa. For example, removal from the solution of 2×10^{-5} M Ca^{2+} per liter ($dCa_T = 2 \times 10^{-5}$ M, corresponding to approximately 50% of the free $[Ca^{2+}]$) will change the free $[Ca^{2+}]$ by approximately 4×10^{-7} M (dpCa = 0.005); that is,

$$\beta_{pH,Y_T}^{Ca_T} = \frac{dCa_T}{d\text{pCa}} \simeq 4 \times 10^{-3} \text{ mol per liter per pCa unit}$$

6.6 METAL IONS AND LIGANDS: CLASSIFICATION OF METALS

The metals and metalloids (semimetals) comprise all of the elements except the noble gases (Group 0) and H, B, C, N, O, F, P, S, Cl, Br, I, and At. The metalloids are Si, Ge, As, Se, Sb, and Te. Hydrogen exhibits metallic properties as the Lewis acid H^+ and nonmetallic properties as the Lewis base H^-. Groups Ia and IIa, the "s block" metals, form monovalent cations (alkali metal cations) and bivalent cations (earth alkali cations), respectively, each of these metal ions achieving the "noble gas structure." Groups IIIb through VIb contain the "p block" metals; among the "p block" metal ions are Al (III oxidation state), Pb (II and IV oxidation states), and Tl (I and II oxidation states). The Al^{3+} ion has the noble gas electron configuration of Ne.

The noble gas configuration (d^0) for metal ions is associated with high spherical symmetry and electron sheaths that are not readily deformed by electric fields; that is, their polarizabilities are low, or, in more descriptive language, they are "hard" spheres.

Those metal cations that have an electron number corresponding to that of Ni^0, Pd^0, and Pt^0 ($g = 10$ or 12) have electron clouds more readily deformable by the electric fields of other species; that is, they have many valence electrons and higher polarizabilities and may be visualized as "soft sphere" ions. In general, higher polarizabilities are found to have increased strength of covalent bonding.

The transition elements have between 0 and 10 d electrons. The three transition series of the Periodic Table, in which the $3d$, $4d$, and $5d$ electronic orbitals are being successively filled, occupy rows 4, 5, and 6 of the Periodic Table. The first transition series runs from Sc (21) through Zn (30); the second from Y (39) through Cd (48); the third from La (57) to Hg (80). The orbital energies and configuration d electrons of the first and subsequent transition series result in a wide variety of oxidation states (except for the Group IIIb metals, which have only the III oxidation state).

Electrons in s, p, or d orbitals may be paired or unpaired, resulting in different energies for the complex electron configuration of the metal. The energy levels of the d orbitals are altered by the electric field of the electron pair donors in ligands, for example, O, N, F, and S. The pairing or unpairing of electrons in octahedral transition metal complexes depends on the number of valence electrons and the electron energies of the ligands in relation to the acceptor metal. Basically, the five d electron orbitals of the central metal are split into two higher-energy orbitals (e_g) and three lower-energy orbitals (t_{2g}), separated by an energy, Δ_0.

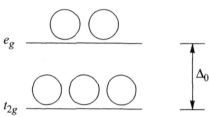

If a ligand is "weak," Δ_0 is small; for "strong" ligands Δ_0 is greater. The resulting set of electron orbitals in an octahedral complex of a metal with d^4, d^5, d^6, or d^7 configurations may have predominantly paired electrons in the orbitals, in a "low-spin" complex; or it may have several unpaired electrons, in a "high-spin" complex.

Whether a complex is high or low spin, depends on the relation of Δ_0 and the energy for electron pairing. The resulting stability of a ML complex, as well as its kinetic behavior, either "labile" or "inert," is thus a direct consequence of the weak or strong interaction of the transition metal d orbitals with ligand orbitals (for more than 3d electrons). For example the valence electron configuration of Mn^{2+}(aq) is $3d^5$. The complex is known to be $Mn(H_2O)_6^{2+}$, and magnetic properties show that all five electrons are unpaired, thus

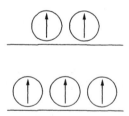

and the complex is high spin. Water is only a moderate-strength ligand. As a result, Mn^{2+}(aq) is labile; that is, it readily releases the H_2O ligands. An example of a low-spin transition metal complex is $Fe(CN)_6^{4-}$, with Fe(II) as the central metal. There are six d electrons. The stable (lowest energy) electron arrangement for large Δ_0 is t_{2g}^6, or t_{2g} ⊖ ⊖ ⊖.

The relative strengths of ligands with respect to energy of d orbital splitting,

6.6 Metal Ions and Ligands: Classification of Metals

Δ_0, is known as the "spectrochemical" series:

$$F^- < H_2O < SCN^- < NH_3 < NO_2^- < CN^-$$

(smallest Δ_0)   (highest Δ_0)

As the first transition series is crossed, the d^5 configuration of Mn^{2+} represents a relatively weak tendency toward covalent bonding, as does Zn^{2+} at the end of the series. In general, covalent character of bonding tends to increase from left to right for the other transition metals.

The lanthanides, elements 58 through 71, constitute a so-called inner transition series, as do the actinides, elements 90 through 103. Scandium (21) and yttrium (39), together with the lanthanides, are traditionally referred to as the rare earth elements. The lanthanides, with 3+ ions and decreasing radii, show strong ionic bonding and weaker covalent bonding characteristics. As discussed below, the lanthanides tend to exhibit "hard sphere" or A-type behavior in their coordination compounds.

A and B Behavior of Metal Cations (Hard and Soft Acid–Base Rules)

Inorganic and organic ligands contain the following possible donor atoms in the fourth, fifth, sixth, and seventh vertical column of the Periodic Table:

C	N	O	F
	P	S	Cl
	As	Se	Br
		Te	I

In water, the halogens are effective complexing agents only as anions, but not if bound to carbon. For special reasons the cyanide ion is a particularly strong complex former. The more important donor atoms include nitrogen, oxygen, and sulfur.

According to the Ahrland et al. (1958) and the Schwarzenbach (1961) analysis of metal–ligand complex stability constants data in aqueous solutions, there is an evident classification of the metal ions into "A," or "hard," and "B," or "soft," type metals (cf. Table 6.3): for example, $Ag^+ + F^- \rightleftharpoons AgF(aq)$ versus $Ag^+ + Cl^- \rightleftharpoons AgCl(aq)$ (giving the difference between Cl^- and F^- as Lewis bases in aqueous solution), or $Mg^{2+} + OH^- \rightleftharpoons MgOH^+(aq)$ versus $Mg^{2+} + F^- \rightleftharpoons MgF^+(aq)$ giving relative strengths of F^- and OH^- as electron donors, or $Cu^+ + Cl^- \rightleftharpoons CuCl(aq)$ versus $Ag^+ + Cl^- \rightleftharpoons AgCl(aq)$ giving electron accepting strengths of Ag^+ and Cu^+ as Lewis acid metal ions.

As Table 6.3 shows, this classification into A- and B-type metal cations is governed by the number of electrons in the outer shell. A-type metal cations having the inert gas type (d^0) electron configuration correspond to those that were classified above as "hard sphere" cations. These ions may be visualized

Table 6.3. Classification of Metal Ions

A-Type Metal Cations	Transition-Metal Cations	B-Type Metal Cations
Electron configuration of inert gas; low polarizability; "hard spheres"; (H^+), Li^+, Na^+, K^+, Be^{2+}, Mg^{2+}, Ca^{2+}, Sr^{2+}, Al^{3+}, Sc^{3+}, La^{3+}, Si^{4+}, Ti^{4+}, Zr^{4+}, Th^{4+}	One to nine outer shell electrons; not spherically symmetric; V^{2+}, Cr^{2+}, Mn^{2+}, Fe^{2+}, Co^{2+}, Ni^{2+}, Cu^{2+}, Ti^{3+}, V^{3+}, Cr^{3+}, Mn^{3+}, Fe^{3+}, Co^{3+}	Electron number corresponds to Ni^0, Pd^0, and Pt^0 (10 or 12 outer shell electrons); low electronegativity; high polarizability; "soft spheres"; Cu^+, Ag^+, Au^+, Tl^+, Ga^+, Zn^{2+}, Cd^{2+}, Hg^{2+}, Pb^{2+}, Sn^{2+}, Tl^{3+}, Au^{3+}, In^{3+}, Bi^{3+}

According to Pearson's (1963) Hard and Soft Acids

Hard Acids	Borderline	Soft Acids
All A-type metal cations plus Cr^{3+}, Mn^{3+}, Fe^{3+}, Co^{3+}, UO^{2+}, VO^{2+}	All bivalent transition-metal cations plus Zn^{2+}, Pb^{2+}, Bi^{3+}, SO_2, NO^+, $B(CH_3)_3$	All B-type metal cations minus Zn^{2+}, Pb^{2+}, Bi^{3+}
Also species such as BF_3, BCl_3, SO_3, RSO_2^+, RPO_2^+, CO_2, RCO^+, R_3C^+		All metal atoms, bulk metals I_2, Br_2, ICN, I^+, Br^+
Preference for ligand atom:		
$N \gg P$		$P \gg N$
$O \gg S$		$S \gg O$
$F \gg Cl$		$I \gg F$
Qualitative generalizations on stability sequence:		
Cations:	Cations:	
Stability ∝ (charge/ radius)	Irving-Williams order: $Mn^{2+} < Fe^{2+} < Co^{2+} < Ni^{2+} < Cu^{2+} > Zn^{2+}$	
Ligands:		Ligands:
$F > O > N = Cl > Br > I > S$		$S > I > Br > Cl = N > O > F$
$OH^- > RO^- > RCO_2^-$		
$CO_3^{2-} \gg NO_3^-$		
$PO_4^{3-} \gg SO_4^{2-} \gg ClO_4^-$		

as being of spherical symmetry; their electron sheaths are not readily deformed under the influence of electronic fields, such as those produced by adjacent charged ions. B-type metal cations have a more readily deformable electron sheath (higher polarizability) than A-type metals and were characterized as "soft sphere" cations.

6.6 Metal Ions and Ligands: Classification of Metals

Metal cations of type A form complexes preferentially with the fluoride ion and ligands having oxygen as the donor atom. Water is more strongly attracted to these metals than are ammonia or cyanide. No sulfides (precipitates or complexes) are formed by these ions in aqueous solution, since OH^- ions readily displace HS^- or S^{2-}. Chloro or iodo complexes are weak and occur most readily in acid solutions, under which conditions competition with OH^- is minimal. The univalent alkali ions form only relatively unstable ion pairs with some anions; some weak complexes of Li^+ and Na^+ with chelating agents, macrocyclic ligands, and polyphosphates are known. Chelating agents containing only nitrogen or sulfur as ligand atoms do not coordinate with A-type cations to form complexes of appreciable stability. A-type metal cations tend to form sparingly soluble precipitates with OH^-, CO_3^{2-}, and PO_4^{3-}; no reaction occurs with sulfur and nitrogen donors (addition of NH_3, alkali sulfides, or alkali cyanides produces solid hydroxides). Some stability sequences are indicated in Table 6.3.

In contrast, B-type metal ions coordinate preferentially with bases containing I, S, or N as donor atoms. Thus metal ions in this class may bind ammonia more strongly than water, CN^- in preference to OH^-, and form more stable I^- or Cl^- complexes than F^- complexes. These metal cations, as well as transition-metal cations, form insoluble sulfides and soluble complexes with S^{2-} and HS^-.

Noncolored components often yield a colored compound (charge transfer bands), thus indicating a significant deformation of the electron orbital overlap. Hence, in addition to coulombic forces, types of interactions other than simple electrostatic forces must be considered. These other types of interactions can be interpreted in terms of quantum mechanics, and in a somewhat oversimplified picture the bond is regarded as resulting from the sharing of an electron pair between the central atom and the ligand (covalent bond). The tendency toward complex formation increases with the capability of the cation to take up electrons (increasing ionization potential of the metal) and with decreasing electronegativity of the ligand, (increasing tendency of the ligand to donate electrons). In the series F, O, N, Cl, Br, I, S, the electronegativity decreases from left to right, whereas the stability of complexes with B-type cations increases. Turner et al. (1981) have applied

$$\Delta \beta = \log \beta_{MF}^0 - \log \beta_{MCl}^0$$

that is, the difference in stability between the fluoro and chloro complexes of a particular element, as a guide to its propensity to form covalent bonds. Elements for which $\Delta \beta > 2$ are A-type cations, and elements for which $\Delta \beta < -2$ are B-type cations and form strong complexes that are largely covalently bound.

For transition metal cations, a reasonably well-established rule on the sequence of complex stability, the Irving-Williams order, is valid. According to this rule, the stability of complexes increases in the series

$$Mn^{2+} < Fe^{2+} < Co^{2+} < Ni^{2+} < Cu^{2+} > Zn^{2+}$$

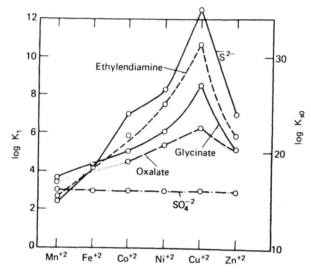

Figure 6.11. Stability constants of 1:1 complexes of transition metals and solubility products of their sulfides (Irving–Williams series).

An example is given in Figure 6.11, giving both complex stabilities and sulfide solubility products.

The following expected gradation of bivalent metals from (a) to (b) character was predicted.

(a) Be > Mg > Ca, Ba, Sr > Sn > Pb > Zn > Cd > Hg (b)

For electron donors, a simple attempt at classifying "hard" and "soft" bases is illustrated in Table 6.3. Such classification schemes reveal, not surprisingly, that "hard" and "soft" are not absolute, but gradually varying qualities. Pearson (1963) has proposed, in generalizing "hardness" and "softness" properties for Lewis acids and bases for many kinds of systems, the so-called "HSAB" (hard and soft acid–base) rules.

- Rule 1: *Equilibrium*. Hard acids prefer to associate with hard bases and soft acids with soft bases.
- Rule 2: *Kinetics*. Hard acids react readily with hard bases and soft acids with soft bases.

Nieboer and Richardson (1980) proposed an index for B-character (i.e., the tendency of a metal ion to form covalent bands). This index, $X_m^2 r$, is based on Pauling's electronegativity, X_m (in simple terms, electron-attracting capability of an atom in a molecule). On the other hand, the ionic potential, Z^2/r, (where Z is the formal charge and r the radius of a metal ion) is an estimate of a metal

6.6 Metal Ions and Ligands: Classification of Metals

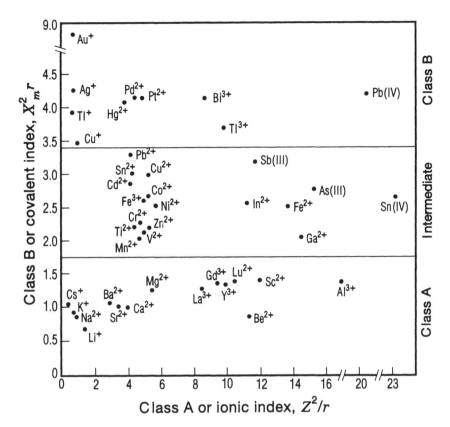

Figure 6.12. A separation of metal ions and metalloid ions [As(III) and Sb(III)] into three categories: class A, borderline, and class B. The class B index $X_m^2 r$ is plotted for each ion against the class A index Z^2/r. In these expressions, X_m is the metal-ion electronegativity, r its ionic radius, and Z its formal charge. (Adapted from Nieboer and Richardson, 1980.)

ion's propensity to form ionic bonds. Figure 6.12 shows the plot of Nieboer and Richardson (1980), which allows us to separate metal ions into the three classes: A, B, and borderline.

Metal–Carbon Compounds

The stability of the metal–carbon bond in aqueous solution decreases from germanium to lead. Besides forming a variety of inorganic complexes, Hg has very high stability constants with organic ligands and can form stable bonds to carbon and therefore true organometallic compounds (Andreae, 1986a).

Metals and metalloids that form alkyl compounds (e.g., methylmercury) deserve special concern, because these compounds are volatile and accumulate in cells; they are poisonous to the central nervous system of higher organisms.

Methylmercury species (or other metal alkyls) may be produced at a rate faster than they are degraded by other organisms. They may accumulate in organisms. Metals and metalloids that have been reported to be biomethylated in minor amounts include Ge, Sn, As, Se, Te, Pd, Pt, Au, Hg, Tl and probably Pb. Anthropogenic pollution by alkylated metal ions is certainly more significant than biomethylation. Methylated tin species have been observed in near-shore polluted waters. For data on As, Sb, Se, and Te see the recent review by Andreae (1986a).

Conditional Constants

As shown in Figure 6.10, the extent of complex formation depends on pH; for a given set of total concentrations, the activities of both the ligand and the (free) metal ion depend on pH because of protolysis. If conditions such as pH, ionic strength, and possibly other variables are kept constant, a constant that is *apparent* or *effective* at these conditions can be defined. One speaks of conditional constants. These depend on the experimental conditions.

In order to clarify the nature of the conditional constants, we consider the reaction between Fe(III) and glycine. This complexation reaction can be characterized by the stability constant

$$Fe^{3+} + Gly^- = FeGly^{2+} \qquad \log K_{FeGly} = 10.8 \qquad (15)$$

In most of the pH range of natural waters, both Fe(III) and glycine undergo acid–base reactions. In order to keep the equilibrium equation in simple form, we consider the side reactions of Fe^{3+} and Gly^- and define a conditional constant

$$K^{cond} = K_{Fe'Gly'} = \frac{[FeGly^{2+}]}{[Fe'][Gly']} \qquad (16)$$

where [Fe'] denotes the sum of all the Fe(III) in solution that has not reacted with glycine, and [Gly'] denotes the sum of all glycine species that are not bound to Fe(III).

$$[Fe'] = [Fe(III)]_T - [FeGly^{2+}] \qquad (17)$$

$$[Gly'] = [Gly]_T - [FeGly^{2+}] \qquad (18)$$

$$[Fe'] = [Fe^{3+}] + [FeOH^{2+}] + [Fe(OH)_2^+] + [Fe(OH)_3] + [Fe(OH)_4^-] \qquad (19a)$$

In the concentration range considered, polynuclear species are negligible.

6.7 Speciation in Fresh Waters

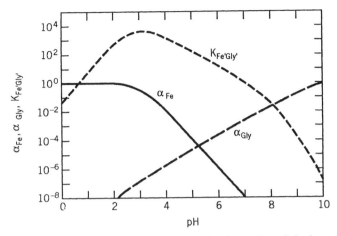

Figure 6.13. The conditional constant for the complex formation of glycine with Fe(III), $K_{Fe'Gly'}$, and α_{Fe} and α_{Gly} values. $K_{Fe'Gly'} = K_{FeGly} \alpha_{Fe} \alpha_{Gly}$.

$$[Fe'] = [Fe^{3+}]\left(1 + \frac{*K_1}{[H^+]} + \frac{*\beta_2}{[H^+]^2} + \frac{*\beta_3}{[H^+]^3} + \frac{*\beta_4}{[H^+]^4}\right) \quad (19b)$$

$$= [Fe^{3+}] \alpha_{Fe}^{-1} \quad (19c)$$

$$[Gly'] = [H_2Gly^+] + [HGly] + [Gly^-] \quad (20a)$$

$$= [Gly^-]\left(1 + \frac{[H^+]}{K_2} + \frac{[H^+]^2}{K_1 K_2}\right) \quad (20b)$$

$$= [Gly^-] \alpha_{Gly}^{-1} \quad (20c)$$

The α values can be calculated as a function of pH and the conditional constant can be written

$$K^{cond} = K_{Fe'Gly'} = \frac{[FeGly^{2+}]}{[Fe'][Gly']} = K_{FeGly} \alpha_{Fe} \alpha_{Gly} \quad (21)$$

α Values and $K_{Fe'Gly'}$ are given in Figure 6.13. It illustrates that the effective tendency to form $FeGly^{2+}$ complexes goes through a maximum around pH 3.5. At pH $= 3.5$, $K_{Fe'Gly'} = 10^{3.6}$.

6.7 SPECIATION IN FRESH WATERS

The concentration range of ligands in natural waters is given in Table 6.4.

While the inorganic ligands are well known, organic ligands are usually only

Table 6.4. Concentration Range of Some Ligands in Natural Waters (log conc. (M))

	Fresh Water	Seawater
HCO_3^-	−4 to −2.3	−2.6
CO_3^{2-}	−6 to −4	−4.5
Cl^-	−5 to −3	−0.26
SO_4^{2-}	−5 to −3	−1.55
F^-	−6 to −4	−4.2
HS^-/S^{2-} (anoxic conditions)	−6 to −3	—
Amino acids	−7 to −5	−7 to −6
Organic acids	−6 to −4	−6 to −5
Particle surface groups	−8 to −4	−9 to −6

known as collective parameters. The information on amino acids and the sum of organic acids is from Buffle (1988). In many waters anthropogenic ligands such as NTA (nitrilotriacetic acid) and EDTA (ethylenediaminetetraacetic acid) occur. For example, in Swiss rivers both NTA and EDTA were found in a concentration range of 10^{-7}-10^{-8} M.

The complexation with sulfate ions is rather unspecific and corresponds mostly to electrostatic interactions; the stability constants for sulfate complexes fall therefore within a narrow range (M + SO_4^{2-} ⇌ MSO_4, log K = 1-3). Much larger differences occur in the hydrolysis constants, in the complexation with chloride (B-cations > A-cations) and with carbonate. HCO_3^- and CO_3^{2-} are of special importance as ligands because they are present in all natural waters at substantial concentrations (HCO_3^- usually in the millimolar range). While alkali and earth alkali ions form primarily ion pairs with carbonate, metal ions such as Cd, Zn, Cu(II), Pb, and Hg(II) form quite stable carbonato complexes. The stability constants and the species composition of carbonato complexes of different elements are still a subject of investigation (Bruno, 1990; Byrne et al., 1988; Ferri et al., 1987a; Fouillac and Criaud, 1984; Millero, 1992). Discrepancies in stability constants of the carbonate complexes of various elements are found in the literature (Fouillac and Criaud, 1984). Recent data on mixed hydroxo–carbonate complexes (Bruno, 1990; Ferri et al., 1987) have indicated that carbonate complexes may be even more important than previously assumed. Sulfide becomes an important ligand under anoxic conditions. The stability constants of the sulfide complexes of many metal ions are still controversial (Dyrssen, 1988; Dyrssen and Kremling, 1990).

Example 6.4. Hydrogen Carbonato and Carbonato Complexes of Mg(II) To what extent are bicarbonato and carbonato Mg complexes species significant in fresh waters? Riesen et al. (1977) give the following equilibrium constants:

$$Mg^{2+} + HCO_3^- = MgHCO_3^+ \qquad \log \beta_1 = 0.69 \quad (25°C) \qquad (i)$$

6.7 Speciation in Fresh Waters

$$Mg^{2+} + 2HCO_3^- = Mg(HCO_3)_2^0 \quad \log \beta_2 = 1.06 \quad (25°C) \quad \text{(ii)}$$

$$Mg^{2+} + CO_3^{2-} = MgCO_3^0 \quad \log K_{1,CO_3^{2-}} = 2.85 \quad (25°C) \quad \text{(iii)}$$

For our calculation we assume a composition $C_T = 4 \times 10^{-3}$ M, $Mg(II)_T = 10^{-3}$ M, $I = 4 \times 10^{-3}$ M, and 25°C and vary the pH between 6 and 9.5.

Using the Davies equation (Table 3.3) to correct for ionic strength, we obtain for the carbonate protolysis system $pK_1' = 6.33$ and $pK_2' = 10.27$; for the complexing with HCO_3^- and CO_3^{2-},

$$\frac{[MgHCO_3^+]}{[Mg^{2+}][HCO_3^-]} = \beta_1' = 4.1 \quad \text{(iv)}$$

$$\frac{[Mg(HCO_3^-)_2^0]}{[Mg^{2+}][HCO_3^-]^2} = \beta_2' = 8.5 \quad \text{(v)}$$

$$\frac{[MgCO_3^0]}{[Mg^{2+}][CO_3^{2-}]} = K_{1,CO_3^{2-}}' = 500 \quad \text{(vi)}$$

We might also have to consider the hydrolysis of Mg(II):

$$Mg^{2+} + H_2O = MgOH^+ + H^+ \quad *K_{1,Mg(II)}$$

$$\frac{[MgOH]\{H^+\}}{[Mg^{2+}]} = *K_{1,Mg(II)}' = 10^{-11.52} \quad \text{(vii)}$$

Equations for total concentrations are

$$[Mg(II)]_T = [Mg^{2+}] + ([MgOH^+] + [MgHCO_3^-]$$
$$+ [Mg(HCO_3^-)_2^0] + [MgCO_3^0]) \quad \text{(viii)}$$

and

$$C_T = [H_2CO_3^*] + [HCO_3^-] + [CO_3^{2-}] + ([MgHCO_3^+]$$
$$+ 2[Mg(HCO_3)_2^0] + [MgCO_3^0]) \quad \text{(ix)}$$

In principle, equations iv–ix and the two protolysis equilibria of $H_2CO_3^*$ have to be solved simultaneously to establish the equilibrium concentration of the eight species [$H_2CO_3^*$, HCO_3^-, CO_3^{2-}, Mg^{2+}, $MgHCO_3^-$, $Mg(HCO_3)_2^0$, $MgCO_3^0$, $MgOH^+$]. Obviously, the complex Mg species are present at low concentrations relative to Mg^{2+} and C_T; we can assume that in equations viii and ix the terms in parentheses on the right side are negligible. We then can compute $[HCO_3^-]$ and $[CO_3^{2-}]$ and estimate the concentration of the complex Mg species from equations iv–vii by setting $[Mg^{2+}] = [Mg(II)_T]$. We can then test whether our assumptions were correct and, if necessary, recalculate $[Mg^{2+}]$ and $C_T = [H_2CO_3] + [HCO_3^-] + [CO_3^{2-}]$; with the recalculated concentrations, equations iv–vii can be solved again, and occasionally more than one iteration is necessary. The result is given in Figure 6.14. $MgHCO_3^+$ is the predominant complex species, but it is present at a concentration of less than 1% of $[Mg^{2+}]_T$. $MgCO_3^0$ becomes important only at high pH, where in our example the solution is already oversaturated with respect to $MgCO_3(s)$. Because of the higher concentration of Mg^{2+} in seawater, extent of HCO_3^- and CO_3^{2-} binding of Mg^{2+} is much greater in seawater.

Representative Fresh Water

We now consider the inorganic chemical speciation for a representative freshwater composition. The speciation calculations are illustrated in detail for two metal ions, Pb(II) and Zn(II), in Example 6.5. For these calculations we assume a fresh water with the following major inorganic analytical component concentrations:

$C_T = 2 \times 10^{-3}$ M
$[SO_4^{2-}]_T = 3 \times 10^{-4}$ M
$[Cl^-]_T = 2.5 \times 10^{-4}$ M
$[Ca^{2+}]_T = 10^{-3}$ M,
$[Mg^{2+}]_T = 3 \times 10^{-4}$ M,
$[Na^+]_T = 2.5 \times 10^{-4}$ M

Table 6.5 gives the predominant inorganic species that have been computed for natural waters (Sigg and Xue, 1994; Turner et al. 1981). The freshwater conditions, in addition to the analytical concentrations mentioned above, were pH = 8, $I = 5 \times 10^{-3}$ M, and O_2 saturation with air. (Major seawater species are presented in Table 6.7.)

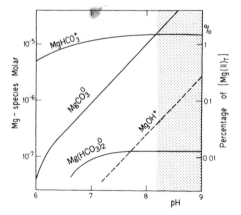

Figure 6.14. Extent of complex (ion-pair) formation by Mg^{2+} with the carbonate and OH^- species under freshwater conditions. $C_T = 4 \times 10^{-3}$ M, $[Mg^{2+}] = 10^{-3}$ M. In most fresh waters the species $Mg(HCO_3)_2$ and $MgCO_3(aq)$ are less than a few percent of the total Mg(II) concentration. Above pH 8.2, the solution is oversaturated with magnesite ($MgCO_3$) (shaded area).

Table 6.5. Major Inorganic Species in Natural Waters

Condition	Element	Major Species	Fresh Water $[M^{n+}]/M_T$	Seawater $[M^{n+}]/M_T$
Hydrolyzed, anionic	B(III)	H_3BO_3, $B(OH)_4^-$		
	V(V)	HVO_4^{2-}, $H_2VO_4^-$		
	Cr(VI)	CrO_4^{2-}		
	As(V)	$HAsO_4^{2-}$		
	Se(VI)	SeO_4^{2-}		
	Mo(VI)	MoO_4^{2-}		
	Si(IV)	$Si(OH)_4$		
Predominantly free aquo ions	Li	Li^+	1.00	1.0^a
	Na	Na^+	1.00	0.98^a
	Mg	Mg^{2+} (Mg^{2+}, $MgSO_4$)	0.94	0.90^a
	K	K^+	1.00	0.98^a
	Ca	Ca^{2+} (Ca^{2+}, $CaSO_4$)	0.94	0.89^a
	Sr	Sr^{2+}	0.94	0.71
	Cs	Cs^+	1.00	0.93
	Ba	Ba^{2+}	0.95	0.86
Complexation with OH^-, CO_3^{2-}, HCO_3^-, Cl^-	Be(II)	$BeOH^+$, $Be(OH)_2^0$	1.5×10^{-3}	1.8×10^{-3}
	Al(III)	$Al(OH)_3(s)$, $Al(OH)_2^+$, $Al(OH)_4^-$	1×10^{-9}	6×10^{-10}
	Ti(IV)	$TiO_2(s)$, $Ti(OH)_4^0$		
	Mn(IV)	$MnO_2(s)$		
	Fe(III)	$Fe(OH)_3(s)$, $Fe(OH)_2^+$, $Fe(OH)_4^-$	2×10^{-11}	1×10^{-12}
	Co(II)	Co^{2+}, $CoCO_3^0$	0.5	0.58
	Ni(II)	Ni^{2+}, $NiCO_3^0$ (Ni^{2+}, $NiCl$)	0.4	0.47
	Cu(II)	$CuCO_3^0$, $Cu(OH)_2^0$	0.01	9.3×10^{-2}
	Zn(II)	Zn^{2+}, $ZnCO_3^0$ (Zn^{2+}, $ZnCl$)	0.4	0.45
	Ag(I)	Ag^+, $AgCl^0$ ($AgCl_2^-$, $AgCl$)	0.6	5.5×10^{-6}
	Cd(II)	Cd^{2+}, $CdCO_3^0$ ($CdCl_2$)	0.5	2.7×10^{-2}
	La(III)[b]	$LaCO_3^+$, $La(CO_3)_2^-$	8×10^{-3}	0.38
	Hg(II)	$Hg(OH)_2^0$ ($HgCl_4^{2-}$)	1×10^{-10}	6×10^{-15}
	Tl(I), (III)	Tl^+, $Tl(OH)_3$, $Tl(OH)_4^-$	2×10^{-21c}	3×10^{-21c}
	Pb(II)	$PbCO_3^0$ ($PbCl^+$, $PbCO_3$)	5×10^{-2}	3×10^{-2}
	Bi(III)	$Bi(OH)_3^0$	7×10^{-16}	1.6×10^{-15}
	Th(IV)	$Th(OH)_4^0$		2×10^{-16}
	U(VI)	$UO_2(CO_3)_2^{2-}$, $UO_2(CO_3)_3^{4-}$	1×10^{-7d}	1×10^{-7d}

[a] See Tableau 6.6.
[b] La(III) is representative of the lanthanides.
[c] Redox state of Tl(I) under natural conditions is uncertain; ratio is for Tl(III).
[d] As UO_2^{2+}.

Note: Freshwater data from Sigg and Xue (1994); most seawater data from Turner et al. (1981). Where major species in seawater deviate from those in fresh waters, the major seawater species are given in parentheses. Fresh water conditions are:

pH = 8   Alk = 2×10^{-3} M
$[SO_4^{2-}]_T = 3 \times 10^{-4}$ M $[Cl^-] = 2.5 \times 10^{-4}$ M
$[Ca^{2+}]_T = 10^{-3}$ M $[Mg^{2+}]_T = 0.3 \times 10^{-3}$ M
$[Na^+]_T = 2.5 \times 10^{-4}$ M
O_2 at saturation with air, $I = 5 \times 10^{-3}$ M

For seawater conditions see Tableau 6.7.

Example 6.5. Speciation of Pb(II) and Zn(II) in Fresh Water The set of equilibrium constants and mole balances are given in Tableaux 6.4a and 6.4b. The mole balance for Pb(II)$_T$ is given by

$$Pb(II)_T = [Pb^{2+}] + \Sigma\,[Pb(OH)_i] + \Sigma\,[Pb(CO_3)_i]$$
$$+ [PbSO_4] + \Sigma\,[Pb(Cl)_i] \tag{i}$$

$$Pb(II)_T = [Pb^{2+}]\,(1 + \Sigma\,\beta_{i,OH}\,[OH^-]^i + \Sigma\,\beta_{i,CO_3}\,[CO_3^{2-}]^i$$
$$+ \beta_{SO_4}\,[SO_4^{2-}] + \Sigma\,\beta_{i,Cl}\,[Cl^-]^i) \tag{ii}$$

Similarly, for each of the ligands, a mole balance can be given:

$$C_T = [CO_3^{2-}] + [HCO_3^-] + [H_2CO_3] + [PbCO_3(aq)]$$
$$+ 2[Pb(CO_3^{2-})_2] \tag{iii}$$

$$[Cl^-]_T = [Cl^-] + [PbCl^+] + 2[Pb(Cl_2)] + 3[Pb(Cl)_3^-]$$
$$+ 4[Pb(Cl)_4^{2-}] \tag{iv}$$

$$[SO_4^{2-}]_T = [SO_4^{2-}] + [PbSO_4(aq)] \tag{v}$$

The results for the major Pb(II) species are given in Figure 6.15a.

Tableau 6.4a. Pb(II) Speciation in Fresh Water

Components	Pb^{2+}	CO_3^{2-}	SO_4^{2-}	Cl^-	H^+	log K ($I = 5 \times 10^{-3}$, 25°C)
Species Pb^{2+}	1	0	0	0	0	0
$PbOH^+$	1	0	0	0	−1	−7.77
$Pb(OH)_2$	1	0	0	0	−2	−17.17
$Pb(OH)_3^-$	1	0	0	0	−3	−28.1
$PbCO_3(aq)$	1	1	0	0	0	6.2
$Pb(CO_3)_2^{2-}$	1	2	0	0	0	9.4
$PbSO_4$	1	0	1	0	0	2.54
$PbCl$	1	0	0	1	0	1.47
$PbCl_2$	1	0	0	2	0	1.60
$PbCl_3$	1	0	0	3	0	1.44
$PbCl_4$	1	0	0	4	0	1.40
HCO_3^-	0	1	0	0	1	10.20
$H_2CO_3^*$	0	1	0	0	2	16.51
CO_3^{2-}	0	1	0	0	0	0
SO_4^{2-}	0	0	1	0	0	0
Cl^-	0	0	0	1	0	0
TOTX	10^{-9}	2×10^{-3}	3×10^{-4}	2.5×10^{-4}	pH given	

Tableau 6.4b. Zn(II) Speciation in Fresh Water

Components		Zn^{2+}	CO_3^{2-}	SO_4^{2-}	Cl^-	H^+	$\log K$ ($I = 5 \times 10^{-3}$, 25°C)
Species	Zn^{2+}	1	0	0	0	0	0
	ZnOH	1	0	0	0	−1	−9.1
	$Zn(OH)_2$	1	0	0	0	−2	−17.0
	$Zn(OH)_3^-$	1	0	0	0	−3	−28.4
	$Zn(OH)_4^{2-}$	1	0	0	0	−4	−42.3
	$ZnCO_3$	1	1	0	0	0	4.52
	$ZnSO_4$	1	0	1	0	0	1.84
	ZnCl	1	0	0	1	0	0.26
	$ZnCl_2$	1	0	0	2	0	0
	HCO_3^-	0	1	0	0	1	10.20
	$H_2CO_3^*$	0	1	0	0	2	16.51
	CO_3^{2-}	0	1	0	0	0	0
	SO_4^{2-}	0	0	1	0	0	0
	Cl^-	0	0	0	1	0	0
TOTX		10^{-8}	2×10^{-3}	3×10^{-4}	2.5×10^{-4}	pH given	

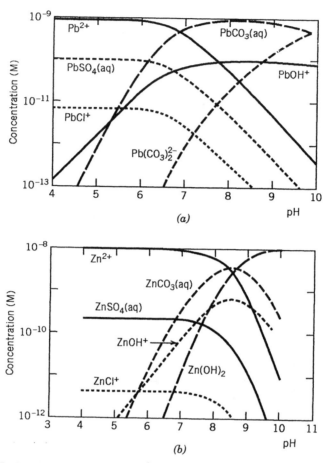

Figure 6.15. Speciation of Pb(II) (10^{-9} M) and Zn(II) (10^{-8} M) under freshwater conditions ($C_T = 2 \times 10^{-3}$ M). (Points are calculated.) For species and equilibrium constants see Tableaux 6.4a and 6.4b.

A similar set of equations can be developed for the Zn(II) species. The results for the most important zinc species are given in Figure 6.15b.

The mole balance equations illustrated for Pb(II) in connection with Tableau 6.5a can be generalized. For each metal ion, M^{n+},

$$[M]_T = [M^{n+}] + \Sigma \, [ML_{\text{inorg}}] \tag{22}$$

$$= [M^{n+}] + \Sigma \, [M(OH)_i] + \Sigma \, M(CO_3)_i + \Sigma \, M(HCO_3)_i + \cdots \tag{23}$$

For each inorganic ligand, L, the mass balance gives

$$L_T = [L] + \Sigma \, [H_m L] + [ML] + [MHL]$$

$$+ \Sigma \, i[ML_i] + \Sigma \, i[M(H_m L)_i] \tag{24}$$

6.7 Speciation in Fresh Waters

For inorganic ligands, the calculations are often simple because the ligands are present in excess of the trace metal ions and the mass balance expressions for the ligands can then be simplified (i.e., $L_T \simeq [L] + \Sigma [H_mL]$). Equation 23 can be expressed in terms of ligand concentrations and the stability constants for the complexes (β):

$$[M]_T = [M^{n+}] (1 + \Sigma \beta_{i\text{OH}} [\text{OH}^-]^i + \Sigma \beta_{i\text{CO}_3} [\text{CO}_3^{2-}]^i$$
$$+ \Sigma \beta_{i\text{HCO}_3} [\text{HCO}_3^-]^i + \Sigma \beta_{i\text{Cl}} [\text{Cl}^-]^i + \Sigma \beta_{i\text{SO}_4} [\text{SO}_4^{2-}]^i) \quad (25)$$

The ratio $[M^{n+}]/M_T$ is determined by the free-ligand concentrations and the stability constants:

$$\frac{[M^{n+}]}{[M]_T}$$
$$= \frac{1}{1 + \Sigma \beta_{i\text{OH}} [\text{OH}^-]^i + \Sigma \beta_{i\text{CO}_3} [\text{CO}_3^{2-}]^i + \Sigma \beta_{i\text{HCO}_3} [\text{HCO}_3^-]^i + \cdots}$$
(26)

The ratio $[M^{n+}]/[M]_T$ calculated in equation 26 and shown in Table 6.5 is independent of total metal concentration.

For most multiligand systems, the calculations can be carried out easily with the help of computer programs, for example, MINEQL (Westall et al., 1976) or MICROQL (Westall, 1986). The total concentrations of the components (e.g., Pb^{2+}, CO_3^{2-}, etc. of Tableau 6.4a) and the stability constants of the complex species (as illustrated by the log K vector in Tableau 6.4a) are the inputs for such programs.

Following the work of Garrels and Thompson (1962), many chemists have made calculations of the inorganic speciation of fresh and marine waters; among these are Stumm and Brauner (1975), Turner et al. (1981), and, more recently, Byrne et al. (1988). The latter workers paid particular attention to the influence of temperature and pH and provided a useful compilation of relevant thermodynamic data (including enthalpies, from which temperature dependences can be estimated). Palmer and van Eldik (1983) have presented a detailed review of metal complexes with carbonic species.

Table 6.5 and Figures 6.6 and 6.15a,b illustrate that the hydroxo and carbonato complexes are predominant under freshwater conditions.[†]

The chloride complexes are not significant, with the exception of Ag^+; this is in contrast to seawater conditions. The complexation with inorganic ligands

---

[†]Note that the inorganic fresh water used in the calculations is representative of the type of fresh waters encountered in $CaCO_3$-bearing regions; that is, it contains relatively high concentrations of carbonate and Ca^{2+}. The competing effects of carbonate and Ca^{2+} (and Mg^{2+}) are smaller in "soft" fresh waters.

298   Metal Ions in Aqueous Solution: Aspects of Coordination Chemistry

keeps the free metal-ion concentration of various elements at low levels in the pH range around 8, for example, for Pb(II), Cu(II), Hg(II), and the fully hydrolyzed metal ions like Al(III) and Fe(III). Since we considered here only inorganic complexation, these ratios represent upper limits of the range in natural waters. Decreasing pH, however, increases the concentration of free aquo ions; in acidified waters, as a consequence of acidic atmospheric inputs, the concentrations of free metal ions increase and contribute to toxic effects.

The redox conditions, of course, have to be considered also. A number of elements with several redox states have a very different speciation under oxic and anoxic conditions. We will discuss inorganic seawater speciation in Section 6.8.

Interaction with Organic Acids

In order to assess the effect of organic acids, we add to our freshwater system a ligand such as the tridentate citrate, which, as we have seen, has a high affinity for Cu(II). We exemplify the effect of citrate on Cu(II) and on Zn(II). To simulate freshwater conditions, we add, in addition to carbonate ($C_T = 2 \times 10^{-3}$ M), 10^{-3} M Ca^{2+}. Although Ca^{2+} usually forms weaker complexes with organic ligands than heavy metal ions, its competing effect is not negligible, because it is typically present at much higher concentrations than the heavy metal. Figure 6.16 gives the results of "titrating" the Cu(II) and Zn(II) system at pH = 8 with citrate. Tableau 6.5 exemplifies the set of conditions necessary for this type of calculation. In the case of Cu(II), 25 equations need to be solved simultaneously. Of course, many species can be neglected. It is expedient to solve such systems by computer (e.g., MICROQL).

Figure 6.16 illustrates a few important points:

1. Relatively large excesses of citrate over $Cu(II)_T$ and $Zn(II)_T$ have to be added to reduce the concentrations of the "free" metal ions $[Cu^{2+}]$ and $[Zn^{2+}]$.
2. Carbonate present at $[CO_3^{2-}] \simeq 10^{-5}$ M ($C_T = 2 \times 10^{-3}$ M) dominates Cu(II) and Zn(II) speciation as long as Cit_{tot} is smaller than 10^{-7} M and 10^{-4} M, respectively.
3. Ca^{2+}, although forming less stable complexes with citrate than Cu^{2+} and Zn^{2+}, outcompetes these ions in forming citrato complexes, because it is present in much higher concentration than Cu(II) and Zn(II).
4. The citrato complex of Cu(II) or Zn(II) prevails only at relatively large excesses of Cit_{tot}.

The competition of Ca^{2+} with Cu^{2+} and Zn^{2+} for complex formation with citrate shall be explained more quantitatively in the case of the Zn(II) system. Comparing the two complex formation constants, we have

6.7 Speciation in Fresh Waters 299

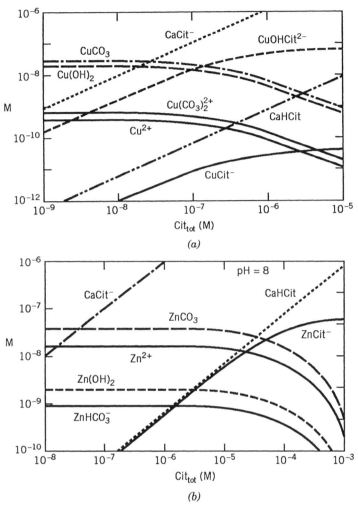

Figure 6.16. Titration of 5×10^{-8} M Cu(II) and 5×10^{-8} M Zn(II) under freshwater conditions ($[Ca^{2+}] = 10^{-3}$ M, $C_T = 2 \times 10^{-3}$ M) at pH = 8 with citrate. Only the major Cu and Zn species and the citrato–Ca species are shown.

$$Zn^{2+} + Cit^{3-} = ZnCit^{-1} \qquad \log K_{ZnCit} = 6.3 \qquad (27)$$

$$Ca^{2+} + Cit^{3-} = CaCit^{-1} \qquad \log K_{CaCit} = 4.7 \qquad (28)$$

$$Zn^{2+} + CaCit^{-1} = ZnCit^{-1} + Ca^{2+} \qquad \log \frac{K_{ZnCit}}{K_{CaCit}} = 1.6 \qquad (29)$$

As shown by equation 29, the ligand form that reacts with Zn(II) is $CaCit^{-1}$ and the equilibrium constant for the exchange reaction is relatively small ($10^{1.6}$).

300   Metal Ions in Aqueous Solution: Aspects of Coordination Chemistry

Tableau 6.5. Titration of Inorganic Cu(II) System with Citrate

Components		Cu^{2+}	Ca^{2+}	Cit^{3-}	CO_3^{2-}	H^+	log K
Species	Cu^{2+}	1	0	0	0	0	0
	CuOH	1	0	0	0	−1	−8
	$Cu(OH)_2$	1	0	0	0	−2	−14.3[a]
	$Cu(OH)_3^-$	1	0	0	0	−3	−26.8
	$Cu(OH)_4^{2-}$	1	0	0	0	−4	−39.9
	$CuCO_3$	1	0	0	1	0	6.77
	$Cu(CO_3)_2$	1	0	0	2	0	10.01
	Ca^{2+}	0	1	0	0	0	0
	$CaCO_3$(aq)	0	1	0	1	0	3.2
	Cit^{3-}	0	0	1	0	0	0
	$HCit^{2-}$	0	0	1	0	1	6.4
	H_2Cit^-	0	0	1	0	2	11.2
	H_3Cit	0	0	1	0	3	14.3
	$CuCit^{-1}$	1	0	1	0	0	7.2
	CuHCit	1	0	1	0	1	17.1
	CuH_2Cit^+	1	0	1	0	2	25
	$CuOHCit^{2-}$	1	0	1	0	−1	2.4
	$Cu_2Cit_2^{2-}$	2	0	2	0	0	16.3
	$CaCit^-$	0	1	1	0	0	4.7
	CaHCit	0	1	1	0	1	9.5
	CaH_2Cit	0	1	1	0	2	12.3
	CO_3^{2-}	0	0	0	1	0	0
	HCO_3^-	0	0	0	1	1	10.2
	$H_2CO_3^*$	0	0	0	1	2	16.51
	H^+	0	0	0	0	1	0
TOTX		5×10^{-8}	10^{-3}	10^{-9} to 10^{-5}	2×10^{-3}	pH given	

[a]There is some uncertainty about this constant; a value of log $\beta = -16.2$ is often used and we used this value in Example 6.2.

The ratio of $[ZnCit^-]$ to $[CaCit^-]$ is given by

$$\frac{[ZnCit^-]}{[CaCit^-]} = \frac{K_{ZnCit}}{K_{CaCit}} \frac{[Zn^{2+}]}{[Ca^{2+}]} \quad (30)$$

$$= 10^{1.6} \frac{[Zn^{2+}]}{[Ca^{2+}]} \quad (31)$$

In accord with equation 30, Figure 6.16 shows that $[ZnCit^-]/[CaCit^-] = 8 \times 10^{-4}$ M and $[Ca^{2+}]/[Zn^{2+}] = 10^{-3}$ M/$[Zn^{2+}] = 5 \times 10^4$ M.

We can define a conditional constant for the reaction of Zn^{2+} with citrate in the presence of 10^{-3} M Ca^{2+}; consider equilibrium 29:

$$\frac{[\text{ZnCit}^-]}{[\text{Zn}^{2+}][\text{CaCit}^-]} = \frac{K_{\text{ZnCit}}}{Z_{\text{CaCit}}} \frac{1}{[\text{Ca}^{2+}]} \tag{32}$$

If $[\text{Ca}^{2+}] = 10^{-3}$, the conditional constant is

$$\text{Zn}^{2+} + \text{Cit} = \text{ZnCit}^{-1} \qquad K^{\text{cond}}_{p\text{Ca}=3,\text{pH}=8} = 4 \times 10^4$$

where Cit represents Cit_{tot} in the presence of 10^{-3} M Ca^{2+} and under the condition of $[\text{Ca}^{2+}] > [\text{Cit}_{\text{tot}}]$.

Complexation by Humic Compounds

Humic substances, humic and fulvic acids, are essentially a mixture of compounds of different molecular weights. The total number of base-titratable groups is in the range of 10–20 meq per gram of carbon. Chelation by neighboring carboxyl and phenolic groups is the major mode of metal complexation. Compounds such as malonic acid, phthalic acid, salicylic acid, and catechol serve as convenient monomeric model compounds for estimating the coordinative properties of humic substances.

Indeed, the metal humic acid complexation constants obtained for trace metals correlate reasonably well with the hydroxide and carbonate stability constants of the metals (Turner et al., 1981). Humic and fulvic acid complexation is therefore most likely to be significant for those cations that are appreciably complexed by CO_3^{2-} and OH^-, for example, with Hg^{2+}, Cu^{2+}, and Pb^{2+}.

Although monomeric model compound ligands may reflect certain features of the metal complex formation by humic substances, they do not provide a reliable representation of the true situation, for the following reasons:

1. Because of the chemical and steric variety of neighboring functional groups on humic and fulvic acids, there may be a range of affinities (binding energies) for metal ions and for protons, that is, a range of complex formation constants and acidity constants.
2. Conformational changes resulting from electrostatic interactions among the various functional groups ($>\text{COO}^-$) on a single molecule can make the coordination properties of the ligand groups ($>\text{COO}^-$ and $>\text{OH}$) highly dependent on the extent of cation binding and on the ionic strength of the water.
3. Electrostatic interactions (repulsion and attraction) between differently charged functional groups influence the acidity and complex formation constants of the humic substances; ionic strength has a marked influence on these electrostatic interactions.
4. It is not yet well established to what extent metal-ion competition, especially between Ca^{2+} or Mg^{2+} and transition metal complexes, for ex-

ample,

$$CuHum + Ca^{2+} = CaHum + Cu^{2+} \tag{33}$$

is operative.

Acid–base titrations of humic substances reflect the nature of the different pK_a values, hence the "smeared out" appearance of these titration curves. While no unique equivalence points are observed, different pK_a regions of carboxylic and phenolic groups can be discerned. Similarly, in metal titrations, metal ions are bound differently by the different ligand groups. The extent of metal-ion binding depends on the ratio of metal ions to humic substances, $[M_T]/[L_T]$. In titrating humic or fulvic acids with metal ions (at fixed pH), the metal is bound first to the highest affinity sites.

The complexing properties of humic and fulvic acids are usually investigated by titrating a known concentration of humic substance with metal ions under suitable conditions (constant pH or varied pH, fixed ionic strength, presence of excess major cations, e.g., Ca^{2+} and Mg^{2+}) and selectively determining the free metal and complex-bound metal concentrations. Figure 6.17 (from Cabiniss and Shuman, 1988) shows titration results for copper–Suwannee River fulvic acid systems. Titrations of the fulvic acid solutions with total copper at fixed pH (Figure 6.17a) and of fulvic acid with base at fixed Cu_T (Figure 6.17b) are shown. pCu increases with increasing fulvic acid concentration at a fixed pH and pCu_T; pCu increases with increasing pH at fixed fulvic acid concentration and fixed pCu_T.

There is presently no unique and completely satisfactory way of quantifying the metal–humic complexation reactions. The approach used depends on the kinds of questions one seeks to answer.

1. Perhaps the simplest approach is to determine, under conditions of interest (pH, ionic strength, pCa, pMg, C_T, and a range of $[M_T]/[L_T]$), a conditional constant for metal complexation. Of course, such conditional constants cannot readily be extrapolated to other conditions.

2. In simple models a combination of different ligands (typically two to five) is used; each ligand is described in terms of a concentration, a metal-ion complexation constant, and an acidity constant. An example of this modeling approach is given in Table 6.6. This model fits best the titration data shown in Figure 6.17.

3. In order to describe the effect of ionic strength and pH, the electrostatic energies resulting from the oligoelectrolytic character of humic and fulvic acids (i.e., intermediate in character between simple ions and true polyelectrolytes) must be assessed. The polyelectrolytic and oligoelectrolytic effect in proton and metal-ion binding has been well studied in physical biochemistry.

A model for the electrostatic effects in fulvic acid–copper ion binding has been proposed recently by Bartschat et al. (1992) (see also Morel and Hering,

6.7 Speciation in Fresh Waters

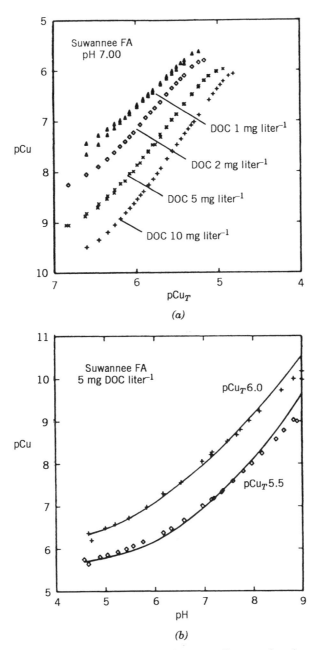

Figure 6.17. Titrations of fulvic acid and Cu. (a) pCu as a function of $[Cu]_T$ for different fulvic concentrations at pH = 7. (b) pCu as a function of pH for two Cu_T concentrations. (From Cabaniss and Shuman, 1988.)

Table 6.6. Fitting Parameters for Five-Site Model

Site	mol (mg C)$^{-1}$	log K	Reaction	Proton Dependence
1 L_a	5×10^{-6}	3.9	$Cu^{2+} + L_a \rightleftharpoons CuL_a$	0
2 L_b	1.9×10^{-7}	1.5	$Cu^{2+} + HL_b \rightleftharpoons CuL_b + H^+$	1
3 L_c	1.1×10^{-6}	-0.36	$Cu^{2+} + HL_c \rightleftharpoons CuL_c + H^+$	1
4 L_d	1.4×10^{-7}	-7.48	$Cu^{2+} + H_2L_d \rightleftharpoons CuL_d + 2\,H^+$	2
5 L_e	9.6×10^{-6}	-10.05	$Cu^{2+} + H_2L_e \rightleftharpoons CuL_e + 2\,H^+$	2

Source: Cabaniss and Shuman (1988).

1993, for a discussion of the application of the model). The model treats fulvic acid as comprising two molecular size fractions (MW $\simeq$ 700 and MW $\simeq$ 5000) and assumes that the concentration of ligand types per weight of carbon is uniform with molecular size. Each of the molecular sizes is treated as an impenetrable sphere of uniform charge density on its surface. This is the same picture of the *physics* as used in the Debye–Hückel theory of ion activity coefficients and in the theory of the electrostatic properties of colloidal spherical particles. The *chemistry* of the fulvic acid is captured in three assumed ligand types: "acetate," or monocarboxylic acid; "malonate," or dicarboxylic acid; and "catechol," with adjacent >OH sites. Bartschat et al. solved the resulting nonlinear Poisson–Boltzmann equation describing the charge–potential relationship for the assumed spherical molecules numerically. Similar approaches have been used by Tipping et al. (1990) and Tipping and Hurly (1992).

4. The spread in binding strength of the functional groups in a humic substance is represented by a continuous distribution of the (log of the) equilibrium constants. Using experimental data, a frequency distribution of affinities (equilibrium constants) is derived, showing the fraction of ligands to be found in a range of stability constants (see, e.g., Perdue and Lytle, 1983, and Buffle and Altman, 1987).

More recently, Westall, et al. (1995) have developed a relatively simple log K spectrum model without inclusion of explicit electrostatic corrections.

All these approaches are not yet fully satisfactory. The Cabaniss–Shuman Five-Site Model and the Barschat et al. (1992) Oligoelectrolyte Model use 15 and 14 parameters, respectively, to fit the titration data; competitive effects of Ca^{2+}, Mg^{2+}, and other metal ions are not yet considered. Nevertheless, these models allow us to gain some chemical insights into effects we might observe and also suggest a need for certain kinds of data sets.

Although humic acids may be important in the regulation of some trace metals in natural waters and in the mobilization of Al(III) and of some radionuclides, they cannot account for the low concentrations of some essential metals like Cu(II) and Zn(II) that are typically found in surface fresh waters and oceans (Bruland, 1989; Sigg and Xue, 1994; Sunda, 1988/89).

Biogenic and Detoxifying Ligands

Cells of phytoplankton and bacteria contain biogenic ligands that assist in buffering intracellular trace metal activity to desirable or tolerable concentrations. Some of these ligands may be remarkably specific and of unusually high affinity to certain metal ions. Some of these biogenic ligands may be released to the water to reduce metal toxicity; others may function as transport ligands to assist the microorganism to acquire essential trace metals. These ligands are usually membrane bound. Better known are chelates that are important in the transport of Fe(III), the so-called *siderophores* that typically contain hydroxamate and catechol moieties that strongly bind to Fe(III). Siderophores are produced under iron-limited growth conditions. Figure 6.18 gives the structure of the well-characterized trihydroxamate siderophore desferriferrioxamine and the complexing functionalities of some biogenic ligands. Phytochelatines are small polypeptides, produced essentially by all plants. *Metallothioneins* are synthesized by bacteria, plants, and animals in response to metal toxicity (Morel and Hering, 1993). Metallothioneins are small proteins (6000–7000 MW) with high cysteine content.

6.8 SEAWATER SPECIATION

Major Inorganic Species

With the help of stability constants valid for seawater, a model for distribution of the most important dissolved species can be calculated. Garrels and Thompson (1962) were the first to establish such a seawater model. Their calculations were based on stability constants (determined in simple electrolyte solutions and corrected or extrapolated to $I = 0$) and estimated activity coefficients of the individual ionic species in seawater. The mean ionic activity coefficients were assumed to be the same as those that would apply to a pure solution of the salt at the same ionic strength as seawater. This assumption is supported phenomenologically.

With the help of operational stability constants valid for seawater (Table 6.7),[†] a direct computation of the major inorganic species (without using individual activity coefficients) was carried out. The calculations consist essentially in solving 20 equations with 20 unknowns simultaneously. Twelve stability constants are available and nine mass balance relations (of the type

$$[SO_{4_T}] = [SO_4^{2-}] + [MgSO_4(aq)] + [CaSO_4(aq)] + [KSO_4^-] + [NaSO_4^-])$$

(34)

[†]The constants in Table 6.7 are slightly different from those given in Appendix 1 of the book.

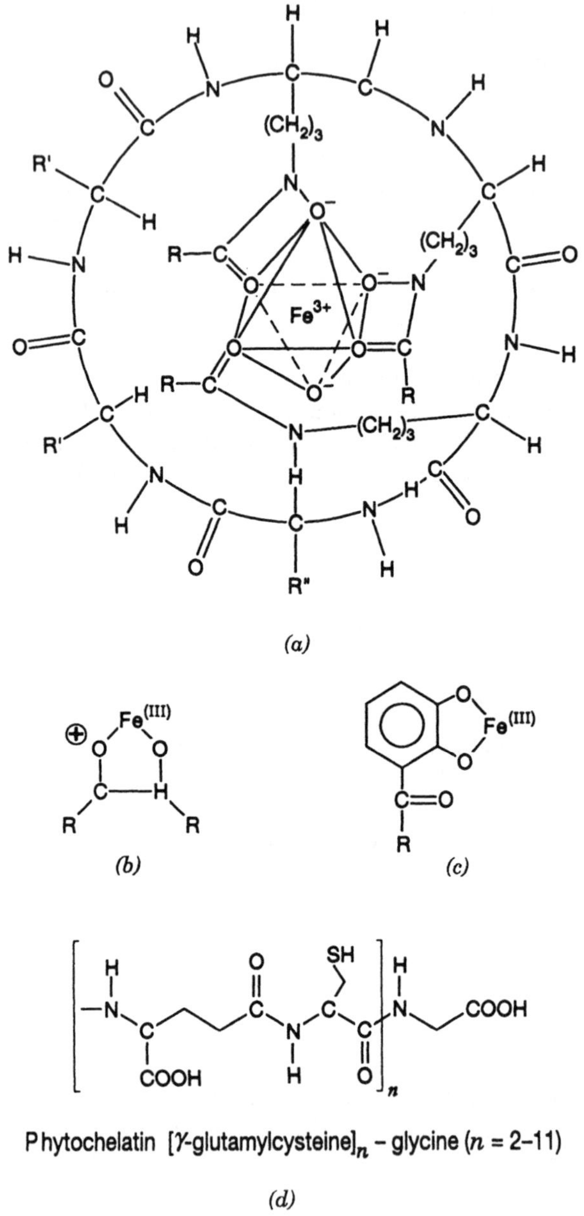

(a)

(b)

(c)

Phytochelatin [γ-glutamylcysteine]$_n$ - glycine (n = 2-11)

(d)

Figure 6.18. Macrocyclic complex formers. (a) Structure of a ferrichrome (desferriferrichrome), one of the strongest complex formers presently known for Fe(III). The iron-binding center is an octahedral arrangement of six oxygen donor atoms of trihydroxamate. Such naturally occurring ferrichromes play an important role in the biosynthetic pathways involving iron. Complexing functionalities of some biogenic ligands: (b) hydroxamate siderophores, (c) catechol siderophores, and (d) phytochelatines. For detailed structures see Neilands (1981).

6.8 Seawater Speciation

Table 6.7. Apparent Stability Constants of Seawater (25°C)
$\log K_1^s = \log([MX]/[M][X])$

Ion	HCO_3^-	CO_3^{2-}	SO_4^{2-}	F
Na^+	-0.55^a	0.97^a	0.34^b	—
K^+	—	—	0.13^b	—
Mg^{2+}	0.21^a	2.20^b	1.01^c	1.3
Ca^{2+}	0.29^a	1.89^b	1.03^b	0.62^d
H^+	$6.035^{e,f}$	$9.09^{e,f}$	1.38^g	2.9^h

[a] Pytkowicz and Hawley (1974).
[b] Pytkowicz and Kester (1969).
[c] Kester and Pytkowicz (1968).
[d] Elgquist (1970).
[e] These constants are defined by

$$K_1' = [H_2CO_{3,T}^*]/\{H^+\}[HCO_{3,T}^-]$$

$$K_2' = [HCO_{3,T}^-]/\{H^+\}[CO_{3,T}^{2-}]$$

where $\{H^+\}$ is H^+ ion activity according to the pH definition of the National Bureau of Standards.
[f] Mehrbach et al. (1973).
[g] Dyrssen et al. (1969).
[h] Srinivasan and Rechnitz (1968).

for Na_T, Mg_T, Ca_T, K_T, Cl_T, SO_{4T}, F_T, HCO_{3T}, and CO_{3T}. The calculation can be simplified if the assumption is made first that the cations are not significantly complexed (e.g., $[Ca^{2+}] = Ca_T$).

Tableau 6.6 shows the equilibria and mole balances considered and gives the results on the inorganic seawater speciation.

The results calculated are in reasonable agreement with the seawater speciation data from Turner et al. (1981) quoted in Table 6.5 and with the qualitative features of the Garrels and Thompson model. Of course, the selection of different suites of stability constants leads to somewhat different speciation pictures. For example, the calculations made by Garrels and Thompson, Dickson et al., and in Tableau 6.6 are based on the assumption that chloride complexes with the major cations are unimportant. This assumption may be wrong and ion pairs with Cl^- may represent nonnegligible fractions of the major cation concentration. Then, of course, a different speciation picture would result; however, the extension of these results to trace metals (see the column for seawater in Table 6.5) would require a reinterpretation of the original experimental coordination data with equilibrium constants with the Cl^- ion pairing model.

What is perhaps more important than the comparison of the various speciation models are the main conclusions that result from all major ion speciation models:

1. The major cations in seawater are predominantly present as free aquo metal ions. This is understandable because they are A-type metals and

Tableau 6.6. Speciation of Inorganic Seawater pH 8.2

	Components									Concentration (mol liter)	log Concentration	Total Concentration	Free Concentration	Species	Metal (%)	Ligand (%)
	Na	K	Mg	Ca	SO$_4$	F	CO$_3$	HCO$_3$	H	log K							
1	1	0	0	0	0	0	0	0	0	0.00	4.568e-1	-0.340	4.680e-1	4.57e-1	Na	98	
2	0	1	0	0	0	0	0	0	0	0.00	1.1005e-2	-1.998	1.020e-2	1.01e-2	K	98	
3	0	0	1	0	0	0	0	0	0	0.00	4.759e-2	-1.322	5.320e-2	4.76e-2	Mg	90	
4	0	0	0	1	0	0	0	0	0	0.00	9.095e-3	-2.041	1.020e-2	9.10e-3	Ca	89	
5	0	0	0	0	0	0	1	0	0	0.00	2.043e-5	-4.690	2.760e-4	2.04e-5	CO$_3$		7
6	0	0	0	0	0	0	0	1	0	0.00	1.739e-3	-2.760	2.140e-3	1.74e-3	HCO$_3$		81
7	0	0	0	0	1	0	0	0	0	0.00	1.086e-2	-1.964	2.820e-2	1.09e-2	SO$_4$		39
8	0	0	0	0	0	1	0	0	0	0.00	3.401e-5	-4.468	6.760e-5	3.40e-5	F		51
9	1	0	0	0	0	0	1	0	0	0.97	8.708e-5	-4.060			NaCO$_3$	0	31
10	1	0	0	0	0	0	0	1	0	-0.55	2.239e-4	-3.650			NaHCO$_3$	0	10
11	0	1	0	0	1	0	0	0	0	0.13	1.472e-4	-3.832			KSO$_4$	1	1
12	0	0	1	0	0	0	1	0	0	2.20	1.541e-4	-3.812			MgCO$_3$	0	56
13	0	0	1	0	0	0	0	1	0	0.21	1.342e-4	-3.872			MgHCO$_3$	0	6
14	0	0	1	0	1	0	0	0	0	1.01	5.287e-3	-2.277			MgSO$_4$	10	19
15	0	0	0	1	0	0	1	0	0	1.89	1.442e-5	-4.841			CaCO$_3$	0	5
16	0	0	0	1	0	0	0	1	0	0.29	3.084e-5	-4.511			CaHCO$_3$	0	1
17	0	0	0	1	1	0	0	0	0	1.03	1.058e-3	-2.975			CaSO$_4$	10	4
18	0	0	0	1	0	1	0	0	0	0.62	1.290e-6	-5.890			CaF	0	2
19	0	0	0	0	0	0	1	0	1	6.03	1.176e-5	-4.930			H$_2$CO$_3^*$	—	—
20	0	0	1	0	0	1	0	0	0	1.30	3.230e-5	-4.491			MgF	0	48
21	1	0	0	0	1	0	0	0	0	0.34	1.085e-2	-1.965			NaSO$_4$	2	38
22	0	0	0	0	0	0	0	0	1	0.00	6.310e-9	-8.200			H	—	—

the concentration of major cations is much greater than the concentration of associating anions.
2. A significant fraction of the anions CO_3^{2-}, SO_4^{2-}, and HCO_3^- are associated with metal ions.

Silicate, Borate, and Phosphate

Borate, $B(OH)_4^-$ and silicate, $H_3SiO_4^-$, form weak complexes with major cations. Because the concentration of major cations is much larger than that of silicate, complexing of cations by silicate is negligible, but complexing of silicate by metal ions may be significant. For the complex formation $M + A \rightleftharpoons MA$, the ratio of complex bound to unbound anion A is

$$\frac{[MA]}{[A]} = K_1[M] \tag{35}$$

Because [M] can be as large as 0.5 or 0.05, even a small K_1 value (e.g., $K_1 = 1$) indicates a marked degree of complexing of A.

The same considerations apply to phosphate, where $[P] \ll [M]$. In seawater of pH = 8, 12% of the inorganic phosphate exists as $PO_{4_T}^{3-}$, 87% as $HPO_{4_T}^{2-}$, and 1% as $H_2PO_{4_T}^-$. Kester and Pytkowicz (1968) estimate that 99.6% of the PO_4^{3-} species and 44% of the HPO_4^{2-} species are complexed with cations other than Na^+ in seawater.

Water as a Ligand As has been mentioned, ions are associated with H_2O. In concentrated solutions such as seawater, the concentration of "free" water is considerably less than that of total water. Christenson and Gieskes (1971) estimate that only about 2.5 mol of H_2O per kg are free. The activity of H_2O, a_{H_2O}, in seawater is given by the ratio of the vapor pressure (fugacity) of seawater, p, to that of pure water, p^0, at the same temperature:

$$a_{H_2O} = \frac{p}{p^0} \tag{36}$$

In seawater of 19.4‰ chlorinity, the activity of H_2O at 25°C is 0.981.

Trace Metals

We can estimate the speciation of trace metals in seawater. The approach is the same as that given for fresh waters in Section 6.7. The concentrations of the free ligands, that is, residual concentrations remaining after complexation with major cations (Tableau 6.6), enter into computation of the complexation equilibria of the trace metals. For inorganic ligands $[L_{free}] \gg [M_{trace}]$ and there is negligible competition in the complexing between individual trace metals. Furthermore, the quantitative degree of complex formation, ΔpM, is independent of the total metal-ion concentration; hence inorganic complex formation systems of the trace metals may be considered individually. Turner et al. (1981) have given a database of stability constants and have calculated the speciation picture in model inorganic seawater. Some of the more important results are incorporated into Table 6.5. Obviously, the speciation pictures of seawater and fresh water are very similar. The most important difference occurs with B-metals such as Cd(II), Hg(II), and Ag(I), that form complexes with Cl^-.

Although calculated equilibrium species distributions remain tentative and are subject to corrections when better data become available, these calculations constitute a useful general framework that permits a better understanding of the observed behavior and functions of trace metals, and of how their chemical forms are influenced by solution variables.

Organic Complexes

Although some exudation products of biota may have special steric arrangements of donor atoms that make them relatively selective toward individual trace metals, most organic matter may not be present in the form of selective complex formers; the organic functional groups compete for association with inorganic cations and protons; cations and protons can satisfy their coordinative requirements with inorganic anions including OH^-, as well as with organic ligands. Most organic ligands in seawater appear to have, at best, complex-forming tendencies similar to those of citrate, amino acids, phthalate, salicylate, carbohydrates, and quinoline-carboxylate. An inspection of "Stability Constants of Metal-Ion Complexes" shows that these classes of compounds are able to form moderately stable complexes with most multivalent cations.

Calculations show that concentrations of amino acids, citrate, or carboxylic acid functional groups as high as 10^{-6} M are not sufficient to appreciably lower free trace metal concentrations under seawater conditions (2 mM Ca^{2+}, 50 mM Mg^{2+}, 2.5 mM C_T, pH = 8). For example, see the multimetal, multiligand seawater computations of Stumm and Brauner (1975). Figure 6.16 exemplifies, in a simplified way, such metal speciation calculations. The figure illustrates that, in the presence of mM Ca^{2+} and mM C_T, citrate can reduce free copper ion concentration, $[Cu^{2+}]$, by a factor of 10 if the total citrate concentration is 3 µM; free zinc concentration can be reduced by a factor of 10 only if the total citrate concentration is at least 6×10^{-4} M. As these calculations show, Cu(II) is more sensitive to complex formation than other trace metals. However, sensitive and reasonably specific analytical measurements (see Section 10.2) document that free concentrations of copper, zinc, and most other essential trace metals are in fact lowered by organic complex formation in aquatic systems (Sunda, 1991). Thus we must conclude (1) that identifiable, low molecular weight organic compounds at their known ambient seawater concentrations cannot affect metal speciation, and (2) that effective complexation of essential trace metals by organics can only occur with naturally produced (probably exuded from biota) ligands exhibiting unusually high specificity and great affinity for trace metals ions.

Humic acid complexation will generally be less significant in seawater than in fresh waters, as more of the organic ligand is bound by Ca^{2+} and Mg^{2+} in seawater. As shown in calculations by Turner et al. (1981), copper(II) and possibly lead are significantly complexed by the humic acid in seawater at pH 8.2. Humic acid complexation may also involve lanthanides and UO_2^{2+}, which form strong carbonate complexes.

But, as has been pointed out already, complexation by humate cannot explain the very low concentrations of nutrient metals (Bruland, 1989; Sunda, 1991) (see Sections 10.2 and 10.5).

Metals, H^+, and ligands form a complicated network of interactions. Because each cation interacts and equilibrates with all ligands and each ligand similarly equilibrates with all cations, the free concentration of metal ions and the distribution of both cations and ligands depend on the total concentrations of all the other constituents of the system. The addition of Fe(III) (or of any other metal) to a water medium, for example, in a productivity experiment, produces significant reverberations in the interdependent "web" of metals and ligands and may lead to a redistribution of all trace metals.

6.9 KINETICS OF COMPLEX FORMATION

So far, we have generally assumed that the complexation reaction is fast and that equilibrium is attained. This is often, but not always, the case. The differentiation of the thermodynamic terms, stable and unstable, from the kinetic terms, labile and inert (or robust), should be made. The classical example of a kinetically inert complex is the hexamine cobalt(III) cation in acid solution:

$$Co(NH_3)_6^{3+} + 6\, H_3O^+ \rightarrow Co(H_2O)_6^{3+} + 6\, NH_4^+ \qquad K \simeq 10^{25} \quad (25°C)$$

(37)

Several days are required to break down the complex, despite the large thermodynamic driving force.

The reaction rate between a metal ion and a ligand usually is partly dependent on the rate at which water leaves and enters the coordination sphere of the ion, the water exchange. For anions and most ligands, this rate is almost invariably fast, on the order of 10^9–10^{10} per second; for cations this rate decreases from this limit for H^+, Na^+, and K^+, down via the large divalent and trivalent ions, Ba^{2+}, Sr^{2+}, Ca^{2+}, and La^{3+} (10^7–10^9 per second), to the smaller divalent ions Mg^{2+} (10^5 per second) to the smaller highly charged ions Al^{3+} (1 per second) and Cr^{3+} (5×10^{-7} per second) (see Table 6.8).

Table 6.8. Rate Constants for Water Exchange[a]

Metal Ion	k_{-w} (s^{-1})
Pb^{2+}	7×10^9
Hg^{2+}	2×10^9
Cu^{2+}	1×10^9
Ca^{2+}	6×10^8
Cd^{2+}	3×10^8
La^{3+}	1×10^8
Zn^{2+}	7×10^7
Mn^{2+}	3×10^7
Fe^{2+}	4×10^6
Co^{2+}	2×10^6
Mg^{2+}	3×10^5
Ni^{2+}	3×10^4
Fe^{3+}	2×10^2
Ga^{3+}	8×10^2
Al^{3+}	1
Cr^{3+}	5×10^{-7}
Rh^{3+}	3×10^{-8}
Fe^{3+}	2×10^2
$FeOH^{2+}$	1×10^5
$Fe(OH)_2^+$	10^7 (est)
$Fe(OH)_4^-$	10^9 (est)

[a] As quoted by Morel and Hering (1993). Most data are from Margerum et al. (1978).

Thus the binding of a ligand to a metal ion can generally be written

$$Me(H_2O)_m^{n+} + L \underset{k_{-1}}{\overset{k_1}{\rightleftharpoons}} Me(H_2O)_m^{n+} \cdot L \qquad (38)^\dagger$$

$$Me(H_2O)_m^{n+} \cdot L \xrightarrow{k_{-w}} Me(H_2O)_{m-1} L^{n+} + H_2O \qquad (39)$$

The rate of the formation of the complex $Me(H_2O)_{m-1} L^{n+}$ is given by (cf. reaction 39)

$$\frac{d[Me(H_2O)_{m-1} L^{n+}]}{dt} = k_{-w} [Me(H_2O)_m^{n+} \cdot L] \qquad (40)$$

The substitution rate of water in the complex $Me(H_2O)_m^{n+} \cdot L$ is similar to that of $Me(H_2O)_m^{n+}$.

†Because the water exchange rate of L is very fast in comparison to that of Me(aq), we can neglect it.

6.9 Kinetics of Complex Formation

The rate of change of $[Me(H_2O)_m^{n+} \cdot L]$, at steady state, is

$$\frac{d[Me(H_2O)_m^{n+} \cdot L]}{dt} = k_1 [Me(H_2O)_m^{n+}] [L]$$
$$- (k_{-1} + k_{-w}) [Me(H_2O)_m^{n+} \cdot L] = 0 \quad (41)$$

where equation 41 under conditions of $k_{-1} \gg k_{-w}$ gives

$$[Me(H_2O)_m^{n+} \cdot L]_{ss} = \frac{k_1}{k_{-1}} [Me(H_2O)_m^{n+}] [L] \quad (42)$$

where $k_1/k_{-1} = K_{OS}$ (K_{OS} = outer sphere or ion pair complex formation equilibrium constant). Equation 40 can now be rewritten:

$$\frac{d[Me(H_2O)_{m-1} L^{n+}]}{dt} = k_{-w} K_{OS} [Me(H_2O)_m^{n+}] [L] \quad (43)$$

or in a general notation (aquo ions omitted):

$$\frac{d[MeL]}{dt} = k [Me] [L] \quad (44)$$

where

$$k = k_{-w} K_{OS} \quad (45)$$

In other words, simple complex formation depends on the energetics of the interaction (ion pair constant, K_{OS}) and, above all, on the water exchange rate of the metal ion.

The dissociation rate of a complex is usually inversely proportional to the stability of the complex, K_{ML}:

$$-\frac{d[MeL]}{dt} = \frac{k}{K_{ML}} [MeL] = \frac{k_{-w} K_{OS}}{K_{ML}} [MeL] \quad (46)$$

Equation 46 derives from detailed balancing of the reaction:

$$Me + L \underset{k_b}{\overset{k}{\rightleftharpoons}} MeL \quad \text{where} \quad K_{ML} = \frac{k}{k_b} \quad (47)$$

Example 6.6. Kinetics of Complex Formation of Co^{2+} with F^- Estimate the rate of inner-spheric complex formation of a 10^{-7} M F^- solution with $Co(H_2O)_n^{2+}$. The following simplifying assumption is made: $[Co(II)] > [F^-]$

(formation of monofluoro complex) and

$$Co(H_2O)_m^{2+} + F^- \rightleftharpoons Co(H_2O)_{m-1} F^+ + H_2O \qquad (i)$$

According to equations 44 and 45 the reaction rate is given by

$$\frac{d[Co(H_2O)_{n-1} F^+]}{dt} = K_{OS}\, k_{-w}\, [Co^{2+}]\, [F^-] = -\frac{d[Co^{2+}]}{dt} \qquad (ii)$$

where $k_{-w} = 2 \times 10^6$ s^{-1} (cf. Table 6.8)
$K_{OS} = 1.6$ [M^{-1}] (for $I = 10^{-3}$ M and 25°C)[†]
$k_{-w} \cdot K_{OS} = 3.2 \times 10^6$ M^{-1} s^{-1}

Equation ii can be written as a first-order reaction:

$$-\frac{d[Co^{2+}]}{dt} = k_{-w}\, K_{OS}\, [F^-]\, [Co^{2+}] = 3.2 \times 10^6 \times 10^{-7}\, [Co^{2+}] \qquad (iii)$$

and the half-life is given by

$$\tau_{1/2} = \ln 2/0.32\ \text{s}^{-1} = 2.2\ \text{s}$$

Metal Exchange Reactions

The overall reaction for metal exchange, M′ + ML → M′L + M, involves three reacting species. The kinetics of such exchange reactions has been investigated by Margerum et al. (1978) and for conditions typically encountered in natural waters by Hering and Morel (1988, 1990). For example, a complex CaY is replaced by CuY. This is a frequently encountered situation in natural waters, where at relatively large concentrations of Ca^{2+}, the Ca complex may prevail. (We omit protonated species and charges.)

$$\text{CaY} + \text{Cu} \rightarrow \text{CuY} + \text{Ca} \qquad (48)$$

Reaction 48 may proceed by two pathways, a disjunctive and an adjunctive mechanism[‡]; both reaction paths may occur in parallel.

The *disjunctive mechanism* starts with a "dissociation" of the complex, which is then followed by the binding to the incoming metal ion.

[†]$K_{OS} = 1.6$ corresponds to a typical ion pair constant for the interaction of a bivalent cation with a monovalent anion (see Section 6.1). K_{OS} can also be estimated from an electrostatic calculation (see Wilkins, 1970).

[‡]One also speaks of dissociative and associative mechanisms.

6.9 Kinetics of Complex Formation

$$CaY \underset{k_-}{\overset{k_+}{\rightleftharpoons}} Ca + Y \tag{49a}$$

$$Cu + Y \xrightarrow{k_{ii}} CuY \tag{49b}$$

Assuming a pseudoequilibrium for reaction 49a,[†]

$$\frac{d[CuY]}{dt} = -\frac{d[CaY]}{dt} = -\frac{d[Cu]}{dt} = k_{ii}\,[Cu]\,[Y] \tag{50a}$$

$$= \frac{k_{ii}\,[Cu]\,K_{CaY}^{-1}\,[CaY]}{[Ca]} \tag{50b}$$

$$= \frac{k_{dis}}{[Ca]}\,[CaY]\,[Cu] \tag{50c}$$

where

$$k_{dis} = k_{ii}\,K_{CaY}^{-1} \tag{50d}$$

According to equations 50, the exchange rate decreases with increased stability of the CaY complex and with increased concentration of Ca^{2+}; it depends on pH.

In the *adjunctive mechanism*, a ternary complex is first formed as an intermediate, which then dissociates:

$$Cu^{2+} + CaY \underset{k_-}{\overset{k_+}{\rightleftharpoons}} CuYCa \tag{55a}$$

$$CuYCa \xrightarrow{k_{iv}} CuY + Ca \tag{55b}$$

[†] A more general derivation of the rate law of reaction 48 can be derived by applying the steady-state approximation to Y:

$$\frac{d[Y]}{dt} = 0 = k_+\,[CaY] - k_-\,[Ca]\,[Y] - k_{ii}\,[Cu]\,[Y] \tag{51}$$

$$[Y]_{ss} = \frac{k_+\,[CaY]}{k_-\,[Ca] + k_{ii}\,[Cu]} \tag{52}$$

$$\frac{d[CuY]}{dt} = \frac{k_{ii}\,k_+\,[Cu]\,[CaY]}{k_-\,[Ca] + k_{ii}\,[Cu]} \tag{53}$$

If $k_{ii} \ll k_-$, then the preequilibrium assumption is fulfilled and equation 52b follows. If $k_{ii} \gg k_-$, then

$$\frac{d[CuY]}{dt} \simeq k_+\,[CaY] \tag{54}$$

and the rate of disjunctive exchange is related to the rate of dissociation of CaY.

The rate law can be derived as follows:

$$\frac{d[CuY]}{dt} = -\frac{d[CaY]}{dt} = -\frac{d[Cu]}{dt} = k_{iv}[CuYCa] \qquad (56)$$

At steady state,

$$[CuYCa]_{ss} = \frac{k_+}{k_- + k_{iv}}[CaY][Cu] \qquad (57)$$

$$\frac{d[CuY]}{dt} = \frac{k_{iv}\,k_+}{k_- + k_w}[CaY][Cu] = k_{adj}[CaY][Cu] \qquad (58)$$

The adjunctive mechanism is very dependent on ligand structure since either formation or dissociation of the intermediate ternary complex, M_1LM_2, can be rate limiting. When the incoming metal, M_1, is a transition metal and the outgoing metal, M_2, is an alkaline earth cation, the formation of the intermediate ternary complex is more likely to be limiting (Hering and Morel; Margerum et al., 1978).

Since disjunctive and adjunctive mechanisms occur as parallel reactions, the overall metal exchange rate is characterized by the sum of the two reactions:

$$\frac{d[CuY]}{dt} = -\frac{d[CaY]}{dt} = -\frac{d[Cu]}{dt} = \left(\frac{k_{dis}}{[Ca]} + k_{adj}\right)[CaY][Cu] \qquad (59)$$

The overall exchange rate is thus given by

$$k_{ex} = \frac{k_{dis}}{[Ca_T]} + k_{adj} \quad (M^{-1}\,s^{-1}) \qquad (60)$$

Figure 6.19 (from Hering and Morel, 1988) illustrates the variation of k_{ex} as a function of $[Ca^{2+}]$ in the exchange reaction Cu(II) + CaEDTA → Cu-EDTA + Ca(II) at pH = 8.2.

Note the kinetic inhibition by excess $[Ca^{2+}]$. At $[Ca^{2+}]_T \simeq 10^{-3}$, there is a transition from the adjunctive to disjunctive exchange pathway.

In the exchange of Ca^{2+} (or Mg^{2+}) for a transition metal ion, the adjunctive exchange pathway usually predominates, perhaps because the stability of the transition metal complex is much larger than that of the alkaline earth complex.

The mechanisms of metal exchange discussed assume competitive binding of the metal ions. In the case of humic substances the competitive situation is not yet clear. There are some indications that Ca^{2+} (and Mg^{2+}) may not bind to the same binding sites of natural humic acids (Cabaniss and Shuman, 1988; Hering and Morel, 1988).

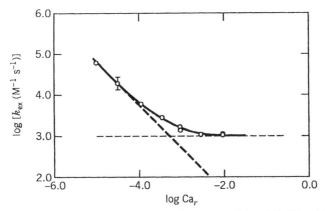

Figure 6.19. Rate constants for the metal-exchange reaction of Cu(II) with CaEDTA at pH = 8.2 as a function of (excess) calcium for rate = $-d[\text{Cu}]/dt = k_{ex}$ [Cu][CaEDTA], where $k_{ex} = (k_{dis}/[\text{Ca}] + k_{adj})$. Data (○) are shown with empirically derived contributions of the adjunctive pathway (----) and disjunctive pathway (-------) to the observed rate constant (——). (Adapted from Hering and Morel, 1989.)

Ligand Exchange Reactions

The pathways for ligand exchange reactions with the overall stoichiometry

$$\text{MX} + \text{Y} \rightarrow \text{MY} + \text{X} \tag{61}$$

are analogous to those described for metal exchange reactions. The disjunctive pathway is given by the sequence

$$\text{MX} \underset{k_-}{\overset{k_+}{\rightleftharpoons}} \text{M} + \text{X} \tag{62a}$$

$$\text{M} + \text{Y} \xrightarrow{k_b} \text{MY} \tag{62b}$$

and the rate law is given by

$$\frac{d[\text{MY}]}{dt} = \frac{k_+ k_b [\text{MX}][\text{Y}]}{k_{-1}[\text{X}] + k_b [\text{Y}]} \tag{63}$$

The adjunctive pathway involves direct attack of Y, the incoming ligand, on the initial complex and formation of the ternary intermediate YMX.

Similar to equation 60, the overall exchange rate under simplifying conditions is given by

$$k_{ex} = \frac{k_{dis}}{[\text{X}]} + k_{adj} \tag{64}$$

Double Exchange Reactions

A double exchange reaction

$$CaX + CuY \rightarrow CuX + CaY \qquad (65)$$

involves four reacting species; the reaction may be either "ligand initiated,"

$$CaX \rightleftharpoons Ca + X$$
$$X + CuY \rightarrow CuX + Y \qquad (66)$$

or "metal initiated,"

$$CaX \rightleftharpoons Ca + X$$
$$Ca + CuY \rightarrow CuX + Y \qquad (67)$$

Double exchange reactions of transition metals often involve coordination chain mechanisms (e.g., further reaction of Y, produced by a ligand-initiated pathway, with the initial complex CaX) (Margerum et al., 1978).

A Case Study: EDTA in Natural Waters

As we have seen, EDTA is present as an anthropogenic pollutant in many natural waters at concentrations above 10^{-8} M. Under many circumstances, EDTA may incipiently occur as Fe(III)EDTA [e.g., in sewage treatment plants where Fe(III) is used to precipitate phosphate]. The question is: How fast is FeEDTA exchanged for a trace metal like Zn^{2+}?

$$Zn^{II} + FeEDTA \rightarrow Fe^{III} + ZnEDTA \qquad (68)$$

where Zn^{II} and Fe^{III} stand for measurable non-EDTA complexed (labile) Zn(II) and for reactive Fe(III); ZnEDTA and FeEDTA, respectively, include all complexed species of Zn and Fe(III) with EDTA (charges are omitted). Stoichiometrically, $\delta[Fe(III)] = -\delta[Zn^{II}]$. Empirically, in accord with equations 59 and 60,

$$-\frac{d[Zn^{II}]}{dt} = \frac{d[ZnEDTA]}{dt} = k_{obs}[FeEDTA][Zn^{II}] \qquad (69)$$

The integrated form of this second-order rate law is

$$\frac{1}{[FeEDTA]_0 - [Zn^{II}]_0} \ln \frac{[Zn^{II}]_0 [FeEDTA]_t}{[FeEDTA]_0 [Zn^{II}]_t} = k_{obs} t \qquad (70)$$

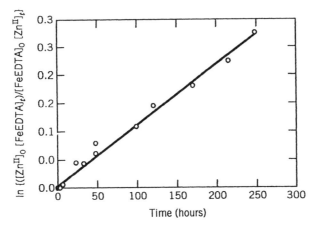

Figure 6.20. Second-order rate function of Zn/FeEDTA exchange in Glatt water. Initial concentrations of Zn^{II} and EDTA: $[Zn^{II}]_0 = 8.23 \times 10^{-8}$ M; $[FeEDTA]_0 = 1 \times 10^{-7}$ M. (Adapted from Xue et al., 1995.)

Figure 6.20 plots results obtained by Xue et al. (1995) according to equation 70 for measurements made in water of the Glatt River (22°C). ($[Ca^{2+}] = 1.8$ mM, $[Mg^{2+}] = 0.71$ mM, pH = 7.9 with initial concentrations of Zn^{II}: $Zn_0^{II} = 8.23 \times 10^{-8}$ M and FeEDTA initially added: $[FeEDTA]_0 = 1 \times 10^{-7}$ M to 4×10^{-7} M.) The evaluation of these data give $k_{obs} = 10.3$ M^{-1} s^{-1}. This constant suggests that, for $[FeEDTA] = 10^{-7}$ M, $-d \ln [Zn^{II}]/dt \simeq 10^{-6}$ s^{-1} with a half-life of $\ln 2/10^{-6}$ $s^{-1} \simeq 8$ days.

These results indicate that the exchange of FeEDTA with Zn(II) under natural water conditions can be rather slow. Why is it so slow? As suggested by Xue et al., the cause is the slow dissociation of FeEDTA (compare equations 46 and 47). For details see Xue and Sigg (1994).

SUGGESTED READINGS

Baes, C. F., and Mesmer, R. E. (1976) *Hydrolysis of Cations,* Wiley, New York.

Burgess, J. (1988) *Ions in Solution; Basic Principles and Interactions,* Ellis-Horwood, Chichester, U.K.

Coale, K. H., and Bruland, K. W. (1988) Copper Complexation in the Northeast Pacific, *Limnol. Oceanogr.* **33,** 1048-1101.

Constable, E. (1990) *Metal Ligand Interactions,* Ellis Horwood, Chichester, U.K.

Dickson, A. G. (1984) pH-Scales and Proton Transfer Reactions in Saline Media Such as Sea-water, *Geochim. Cosmochim. Acta* **48,** 2299-2308.

Dickson, A. G., Friedman, H. L., and Millero, F. J. (1988) Chemical Model of Seawater Systems, a Panel Report, *Appl. Geochem.* **3,** 27-35.

Fraústo da Silva, J. J. R., and Williams, R. J. P. (1991) *The Biological Chemistry of the Elements; The Inorganic Chemistry of Life,* Clarendon Press, Oxford.

Hem, J. D., and Roberson, C. E. (1988) Aluminum Hydrolysis Reactions and Products in Mildly Acidic Aqueous Systems. In *Chemical Modeling of Aqueous Systems,* Vol. II, D. C. Melchior and R. L. Bassett, Eds., ACS Series 416, Washington, DC.

Hering, J. G., and Morel, F. M. M. (1989) Kinetics of Trace Metal Complexation; Role of Alkaline Earth Metals, *Environ. Sci. Technol.* **22,** 1469–1478.

Millero, F. J. (1995) Thermodynamics of the Carbon Dioxide System in the Oceans, *Geochim. Cosmochim Acta,* **59** 661–678.

Öhman, L. O., and Sjöberg, S. (1988) Thermodynamic Calculations with Special Reference to the Aqueous Aluminum System. In *Metal Speciation: Theory Analysis and Application,* J. R. Kramer and H. E. Allen, Eds., Lewis Publishers, Chelsea, MI.

Sigg, L. (1994) Regulation of Trace Elements in Lakes. In *Chemical and Biological Regulation of Aquatic Processes,* J. Buffle and R. R. de Vitre, Eds., Lewis, Chelsea, MI.

Sunda, W. G. (1994) Trace Metal/Photoplankton Interactions in the Sea. In *Chemistry of Aquatic Systems; Local and Global Perspectives,* G. Bidoglio and W. Stumm, Eds., Kluwer Academic, Dordrecht.

Turner, D. R., Whitfield, M., and Dickson, A. G. (1981) The Equilibrium Speciation of Dissolved Components in Freshwater and Sea-water at 25°C and 1 atm Pressure, *Geochim. Cosmochim. Acta* **45,** 855–881.

Xue, H. B., and Sigg, L. (1993) Free Cupric Ion Concentration and Cu(II) Speciation in an Eutrophic Lake, *Limnol Oceanogr.* **38,** 1200–1213.

PROBLEMS

6.1. Make a distribution diagram for the various Cr(VI) species as a function of pH at $Cr_T = 1$, 10^{-2}, and 10^{-4} M. Neglect activity corrections.

Equilibria:
$CrO_4^{2-} + H^+ = HCrO_4^-$ $\log K = 6.5$

$HCrO_4^- + H^+ = H_2CrO_4$ $\log K = -0.8$

$2HCrO_4^- = Cr_2O_7^{2-} + H_2O$ $\log K = 1.52$

$Cr_2O_7^{2-} + H^+ = HCr_2O_7^-$ $\log K = 0.07$

6.2. Find the pH at which "Fe(OH)$_3$" begins to precipitate from a 0.1 M Fe(ClO$_4$)$_3$ solution. Consider the dimerization of FeOH^{2+}.

6.3. (a) In which form does Cd(II) occur mainly in a water of the following composition:

pH = 7.8

$$\text{Alk} = 1.3 \times 10^{-3} \text{ eq liter}^{-1}$$
$$\text{Cd}_T = 1 \times 10^{-9} \text{ M}$$
$$\text{Ca}_T = 1 \times 10^{-3} \text{ M?}$$

(b) Is there a possibility that $CdCO_3(s)$ or $Cd(OH)_2(s)$ precipitates?

(c) How does the speciation of Cd(II) change when 10^{-7} mol EDTA is added to 1 liter of this water?

The following constants are available:

$Cd^{2+} \rightleftharpoons CdOH^+ + H^+$	$\log {^*\beta_1}$	$= -10.1$
$Cd^{2+} + CO_3^{2-} \rightleftharpoons CdCO_3^0$	$\log K$	$= 4.5$
$CdCO_{3(s)} \rightleftharpoons Cd^{2+} + CO_3^{2-}$	$\log K_{s0}$	$= -13.7$
$Cd(OH)_{2(s)} \rightleftharpoons Cd^{2+} + 2OH^-$	$\log K_{s0}$	$= -14.3$
$Cd^{2+} + EDTA^{4-} \rightleftharpoons CdEDTA^{2-}$	$\log K$	$= 16.5$
$Ca^{2+} + EDTA^{4-} \rightleftharpoons CaEDTA^{2-}$	$\log K$	$= 10.7$
$HEDTA^{3-} \rightleftharpoons EDTA^{4-} + H^+$	$\log K$	$= -10.2$

6.4. What equilibria need to be considered in assessing the fish toxicity of a galvanic waste containing Zn(II) and cyanide?

6.5. An organic compound isolated from natural waters is found to contain phenolic and carboxylic functional groups. From an alkalimetric titration curve the intrinsic acidity constants $pK_1 \simeq 4.1$ and $pK_2 \simeq 12.5$ can be estimated. This compound forms colored complexes with Fe(III), the highest color intensity being observed in slightly acidic solutions for a mixture of equimolar concentrations of Fe(III) and the complex former. If to a 10^{-3} M Fe(III) solution at pH $= 3$ this complex former is added to attain a concentration of 5×10^{-2} M, a potentiometric electrode for Fe^{3+} (e.g., a ferro–ferri cell) registers a shift in $[Fe^{3+}]$ corresponding to a $\Delta pFe \simeq 3.5$. Provide a rough estimate for the stability of this complex. Constants for the hydrolysis of Fe(III) are given in Example 6.1.

6.6. In a series of phytoplankton culture experiments the growth rate is found to be affected by the copper chemistry of the medium. The constant pH medium contains a chelating agent, X, so that

$$\text{TOTCu} = [Cu^{2+}] + [CuOH^+] + [CuX]$$

The experimental results show that when the chelating agent concentration is varied at constant TOTCu, the growth rate increases with increasing

TOTX. In the absence of Cu, variation of TOTX produces no effect on growth. Discuss these results with respect to:
(a) Equilibrium properties of the CuX chelate system.
(b) Kinetics of CuX formation and dissociation in comparison to rate of phytoplankton growth.

6.7. Sketch a titration curve for titrating Ca^{2+} with EDTA at a pH of 10. Show the equilibrium composition of CaY^{2-}, Ca^{2+}, and uncomplexed Y as a function of pCa. (H_4Y = EDTA.)

6.8. Is $BaSO_4$ appreciably soluble ($S > 10^{-3}$ M) in a 10^{-1} M solution of citrate of pH = 8? *Information:* Acidity constants of citric acid (H_4L) $pK_1 = 3.0$, $pK_2 = 4.4$, $pK_3 = 6.1$, $pK_4 \approx 16$. BaHL is the only complex to be considered.

$$\frac{[BaHL]}{[Ba^{2+}][HL]} = 10^{2.4}$$

$$[Ba^{2+}][SO_4^{2-}] = 10^{-9}$$

6.9. How much dissolved Hg(II) is in equilibrium with HgS(s) if sulfide complexes are formed in the interstitial water of sediments under the following conditions:

S(−II)total = 10^{-5} M, pH 8

$Hg^{2+} + 2\,HS^- \rightleftharpoons Hg(HS)_2^0$	$\log K = 37.7$
$Hg^{2+} + 2\,HS^- \rightleftharpoons HgHS_2^- + H^+$	$\log K = 31.5$
$Hg^{2+} + 2\,HS^- \rightleftharpoons HgS_2^{2-} + 2\,H^+$	$\log K = 23.2$
$H_2S \rightleftharpoons HS^- + H^+$	$\log K = -7.0$
$HS^- \rightleftharpoons S^{2-} + H^+$	$\log K = -19$
$HgS(s) + H^+ \rightleftharpoons Hg^{2+} + HS^-$	$\log {}^*K_s' = -37.11$

ANSWERS TO PROBLEMS

6.1. The problem may be approached in a fashion similar to that given in Example 6.1. The basic equation to start with is

$$Cr_T = [CrO_4^{2-}] + [HCrO_4^-] + 2[Cr_2O_7^{2-}] + 2[HCr_2O_7^-]$$

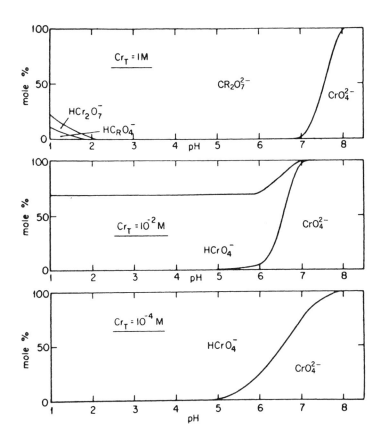

6.2. pH = 1.7.

6.3. (a) $[Cd^{2+}] = 8.8 \times 10^{-10}$ M
$[CdOH^+] = 4.4 \times 10^{-12}$ M
$[CdCO_3] = 1.1 \times 10^{-10}$ M
(b) No saturation
(c) $[Cd^{2+}] = 1.6 \times 10^{-11}$ M

6.4. $Zn^{2+} + H_2O = ZnOH^+ + H^+$; $*K_1$

   $2\,Zn^{2+} + H_2O = Zn_2(OH)^{3+} + H^+$; $*\beta_{12}$

   $Zn^{2+} + CN^- = Zn(CN)^+$; β_1

   $Zn^{2+} + 2\,CN^- = Zn(CN)_2$; β_2

   $Zn^{2+} + 3\,CN^- = Zn(CN)_3^-$; β_3

   $Zn^{2+} + 4\,CN^- = Zn(CN)_4^{2-}$; β_4

   $H^+ + CN^- = HCN$

Chemical species toxic to fish are primarily HCN, Zn^{2+}, and $ZnOH^+$.

$$Zn_T = [Zn^{2+}] + \sum_{n=1}^{4} [Zn(CN)_n^{2-n}] + [ZnOH^+] + 2[Zn_2OH^{3+}]$$

$$CN_T = [HCN] + [CN^-] + \sum_{n=1}^{4} n[Zn(CN)_n^{2-n}]$$

6.8. No. The solubility is increased from $10^{-4.5}$ to $10^{-3.8}$ M. The calculation is made very simple by realizing that [HL] is much larger than any other L-bound species; hence $\alpha_3 \simeq [HL]/L_T \simeq 1.0$, and $\alpha_{Ba(HL)} = [Ba^{2+}]/Ba_T = \{1 + 10^{2.4} [HL]^{-1}\} \simeq 10^{-1.4}$. The conditional solubility product for $BaSO_4$ becomes $P = [Ba_T] [SO_4^{2-}] = K_{s0} \, \alpha_{Ba(HL)}^{-1} = 10^{-7.6}$.

6.9. $[Hg(II)]_T = 3.40 \times 10^{-11}$ M

APPENDIX 6.1 STABILITY CONSTANTS

Table A6.1. Stability Constants[a] for Formation of Complexes and Solids from Metals and Ligands. Compilation by Morel and Hering (1993).

	OH	CO_3^{2-}	SO_4^{2-}	Cl^-	Br^-	F^-	NH_3	$B(OH)_4^-$
H^+	HL·w 14.00	HL 10.33 H_2L 16.68 H_2L·g 18.14	HL 1.99			HL 3.2	HL 9.24 L·g −1.8	HL 9.24 HL_3 10.4 H_2L_3 20.4 H_3L_4 21.0 H_4L_5 38.8
Na^+		NaL 1.27 NaHL 10.08	NaL 1.06					
K^+			KL 0.96					
Ca^{2+}	CaL 1.15 CaL_2·s 5.19	CaL 3.2 CaHL 11.59 CaL·s 8.22 CaL·s 8.35	CaL 2.31 CaL·s 4.62			CaL 1.1 CaL_2·s 10.4		
Mg^{2+}	MgL 2.56 Mg_4L_4 16.28 MgL_2·s 11.16	MgL 3.4 MgHL 11.49 MgL_2·s 4.54 MgL_2·s 7.45	MgL 2.36			MgL 1.8 MgL_2·s 8.2		
Sr^{2+}		SrL·s 9.0	SrL 2.6 SrL·s 6.5			SrL_2·s 8.5		
Ba^{2+}		BaL 2.8 BaL·s 8.3	BaL 2.7 BaL·s 10.0			BaL_2·s 5.8		
Cr^{3+}	CrL 10.0 CrL_2 18.3 CrL_3 24.0 CrL_4 28.6 Cr_3L_4 47.8 CrL_3·s 30.0		CrL 3.0	CrL 0.23		CrL 5.2 CrL_2 9.2 CrL_3 12.0		

Table A6.1. (*Continued*)

		OH		CO_3^{2-}		SO_4^{2-}		Cl^-		Br^-		F^-		NH_3		$B(OH)_4$
Al^{3+}	AlL	9.0									AlL	7.0			
	AlL_2	18.7									AlL_2	12.6			
	AlL_3	27.0									AlL_3	16.7			
	AlL_4	33.0									AlL_4	19.1			
	Al_3L_4	42.1													
	$AlL_3 \cdot s$	33.5													
Fe^{3+}	FeL	11.8			FeL	4.0	FeL	1.5	FeL	0.6	FeL	6.0			
	FeL_2	22.3			FeL_2	5.4	FeL_2	2.1			FeL_2	10.6			
	FeL_4	34.4									FeL_3	13.7			
	Fe_2L_2	25.0													
	$FeL_3 \cdot s$	42.7													
	$FeL_3 \cdot s$	38.8													
Mn^{2+}	MnL	3.4	MnHL	12.1	MnL	2.3	MnL	0.6			MnL	1.3	MnL	1.0	
	MnL_2	5.8	$MnL \cdot s$	10.4									MnL_2	1.5	
	MnL_3	7.2													
	MnL_4	7.7													
	$MnL_2 \cdot s$	12.8													
Fe^{2+}	FeL	4.5	$FeL \cdot s$	10.7	FeL	2.2					FeL	1.4			
	FeL_2	7.4													
	FeL_3	11.0													
	$FeL_2 \cdot s$	15.1													
Co^{2+}	CoL	4.3	$CoL \cdot s$	10.0	CoL	2.4	CoL	0.5			CoL	1.0	CoL	2.0	
	CoL_2	9.2											CoL_2	3.5	
	CoL_3	10.5											CoL_3	4.4	
	$CoL_2 \cdot s$	15.7											CoL_4	5.0	
Ni^{2+}	NiL	4.1	$NiL \cdot s$	6.9	NiL	2.3	NiL	0.6			NiL	1.1	NiL	2.7	
	NiL_2	9.0											NiL_2	4.9	
	NiL_3	12.0											NiL_3	6.6	
	$NiL_2 \cdot s$	17.2											NiL_4	7.7	
													NiL_5	8.3	

Ion	Species	log K	Species	log K	Species	log K	Species	log K	Species	log K	Species	log K
Cu^{2+}	CuL	6.3	CuL	6.7	CuL	2.4	CuL	0.5	CuL	1.5	CuL	4.0
	CuL_2	11.8	CuL_2	10.2	$Cu_4(OH)_6L·s$	68.6					CuL_2	7.5
	CuL_4	16.4	$CuL·S$	9.6							CuL_3	10.3
	Cu_2L_2	17.7	$Cu_2(OH)_2L·s$	33.8							CuL_4	11.8
	$CuL_2·s$	19.3	$Cu_3(OH)_2L_2·s$	46.0								
	$CuL_2·s$	20.4										
Zn^{2+}	ZnL	5.0	ZnL·s	10.0	ZnL	2.1	ZnL	0.4	ZnL	1.2	ZnL	2.2
	ZnL_2	11.1			ZnL_2	3.1	ZnL_2	0.2			ZnL_2	4.5
	ZnL_3	13.6					ZnL_3	0.5			ZnL_3	6.9
	ZnL_4	14.8					$Zn_2(OH)_3L·s$	26.8			ZnL_4	8.9
	$ZnL_2·s$	15.5										
	$ZnL_2·s$	16.8										
Pb^{2+}	PbL	6.3	PbL·s	13.1	PbL	2.8	PbL	1.6	PbL	2.0		
	PbL_2	10.9			$PbL_2·s$	7.8	PbL_2	1.8	PbL_2	3.4		
	PbL_3	13.9					PbL_3	1.7	$PbL_2·s$	7.4		
	$PbL_2·s$	15.3					PbL_4	1.4				
							$PbL_2·s$	4.8				
Hg^{2+}	HgL	10.6	HgL·s	16.1	HgL	2.5	HgL	7.2	HgL	1.6	HgL	8.8
	HgL_2	21.8			HgL_2	3.6	HgL_2	14.0			HgL_2	17.4
	HgL_3	20.9					HgL_3	15.1			HgL_3	18.4
	$HgL_2·s$	25.4					HgL_4	15.4			HgL_4	19.1
							$HgL_2·s$	18.1				
							HgOHL	19.8				
Cd^{2+}	CdL	3.9	CdL·s	13.7	CdL	2.3	CdL	2.0	CdL	1.0	CdL	2.6
	CdL_2	7.6			CdL_2	3.2	CdL_2	2.6	CdL_2	1.4	CdL_2	4.6
	$CdL_2·s$	14.3			CdL_3	2.7	CdL_3	2.4			CdL_3	5.9
							CdL_4	1.7			CdL_4	6.7
Ag^+	AgL	2.0	$Ag_2L·s$	11.1	AgL	1.3	AgL	3.3	AgL	0.4	AgL	3.3
	AgL_2	4.0			$Ag_2L·s$	4.8	AgL_2	5.3			AgL_2	7.2
	$AgL·s$	7.7					AgL_3	6.4			AgL	0.6
							AgL_4	9.7			$AgHL_2·s$	22.9
							AgL	4.7				
							AgL_2	6.9				
							AgL_3	8.7				
							AgL_4	9.0				
							$AgL·s$	13.3				

Table A6.1. (Continued)

	SiO_3^{2-}	S^{2-}	$S_2O_3^{2-}$	PO_4^{3-}	$P_2O_7^{4-}$	$P_3O_{10}^{5-}$	CN^-
H^+	HL 13.1 H₂L 23.0 H₂L₂ 26.6 H₄L₄ 55.9 H₆L₄ 78.2 H₂L·s 25.7	HL 13.9 H₂L 20.9 H₂L·g 21.9	HL 1.6 H₂L 2.2	HL 12.35 H₂L 19.55 H₃L 21.70	HL 9.4 H₂L 16.1 H₃L 18.3 H₄L 19.7	HL 9.3 H₂L 18.8 H₃L 21.3 H₄L 22.3	HL 9.2
Na^+			NaL 0.5	NaHL 13.5	NaL 2.3 Na₂L 4.2 NaHL 10.8	NaL 2.7 NaHL 11.6	
K^+			KL 1.0	KHL 13.4	KL 2.1	KL 2.8	
Ca^{2+}	CaL 4.2 CaHL 14.1 CaH₂L₂ 29.9		CaL 2.0	CaL 6.5 CaHL 15.1 CaH₂L 21.0 CaHL·s 19.0	CaL 6.8 CaHL 13.4 CaOHL 8.9 Ca₃L₂·s 14.7	CaL 8.1 CaHL 14.1 CaOHL 10.4	
Mg^{2+}	MgL 5.3 MgHL 14.3 MgH₂L₂ 30.8		MgL 1.8	MgL 4.8 MgHL 15.3 MgH₂L 20.0 Mg₃L₂·s 25.2 MgHL·s 18.2	MgL 7.2 MgHL 14.1 MgOHL 9.3	MgL 8.6 MgHL 14.5 MgOHL 11.0	
Sr^{2+}			SrL 2.0	SrL 5.5 SrHL 14.5 SrH₂L 20.3 SrHL·s 19.3	SrL 5.4 SrOHL 7.7 Sr₂L·s 12.9	SrL 7.2 SrHL 13.6 SrOHL 9.3	
Ba^{2+}			BaL 2.3 BaL·s 4.8	BaHL·s 19.8		BaL 6.3 BaHL 12.9 Ba₂L·s 16.1	
Cr^{3+}							
Al^{3+}							

Fe^{3+}	FeHL 22.7	FeL 3.3	FeHL 22.5 FeH$_2$L 23.9 FeL·s 26.4			FeL$_6$ 43.6
Mn^{2+}	MnL·s 10.5 13.5			MnL 9.9 MnHL 14.8		
Fe^{2+}	FeL·s 18.1	MnL 2.0	FeHL 16.0 FeH$_2$L 22.3 Fe$_3$L$_2$·s 36.0			FeL$_6$ 35.4
Co^{2+}	CoL·s 21.3 25.6	CoL 2.1	CoHL 15.5	CoL 7.9 CoHL 14.1	CoL 9.7 CoHL 14.8	
Ni^{2+}	NiL·s 19.4 24.9 26.6	NiL 2.1	NiHL 15.4	NiL 7.7 NiHL 14.4	NiL 9.5 NiHL 14.7	NiL 7.3 NiL$_4$ 30.2 NiH$_2$L$_4$ 40.8 NiHL$_4$ 36.1
Cu^{2+}	CuL·s 36.1		CuHL 16.5 CuH$_2$L 21.3	CuL 9.8 CuHL 15.5 CuL$_2$ 12.5 CuH$_2$L 19.2	CuL 11.1 CuHL 15.5	CuL$_2$ 16.3 CuL$_3$ 21.6 CuL$_4$ 23.1
Zn^{2+}	ZnL 16.6 ZnL·s 24.7	ZnL 2.4 ZnL$_2$ 2.5 ZnL$_3$ 3.3 Zn$_2$L$_2$ 7.0	ZnHL 15.7 ZnH$_2$L 21.2 Zn$_3$L$_2$ 35.3	ZnL 8.7 ZnL$_2$ 11.0 ZnOHL 13.1	ZnL 10.3 ZnHL 14.9 ZnOHL 13.6	ZnL 5.7 ZnL$_2$ 11.1 ZnL$_3$ 16.1 ZnL$_4$ 19.6 ZnL$_2$·s 15.9
Pb^{2+}	PbL·s 27.5	PbL 3.0 PbL$_2$ 5.5 PbL$_3$ 6.2 PbL$_4$ 7.3	PbHL 15.5 PbH$_2$L 21.1 Pb$_3$L$_2$·s 43.5 PbHL·s 23.8	PbL 9.5 PbL$_2$ 10.2		
Hg^{2+}	HgL 7.9 HgL$_2$ 14.3 HgOHL 18.5 HgL·s 52.7 HgL·s 53.3	HgL$_2$ 29.2 HgL$_3$ 30.6		HgOHL 18.6		HgL 17.0 HgL$_2$ 32.8 HgL$_3$ 36.3 HgL$_4$ 39.0 HgOHL 29.6

Table A6.1. (*Continued*)

	SiO₃²⁻		S²⁻		S₂O₃²⁻		PO₄³⁻		P₂O₇⁴⁻		P₃O₁₀⁵⁻		CN⁻	
Cd^{2+}			CdL	19.5	CdL	3.9			CdL	8.7	CdL	9.8	CdL	6.0
			CdHL	22.1	CdL₂	6.3			CdOHL	11.8	CdHL	14.6	CdL₂	11.1
			CdH₂L₂	43.2	CdL₃	6.4					CdOHL	12.6	CdL₃	15.7
			CdH₃L₃	59.0	CdL₄	8.2							CdL₄	17.9
			CdH₄L₄	75.1	Cd₂L₂	12.3								
			CdL·s	27.0										
Ag^{+}			AgL	19.2	AgL	8.8	Ag₃L·s	17.6					AgL₂	20.5
			AgHL	27.7	AgL₂	13.7							AgL₃	21.4
			AgHL₂	35.8	AgL₃	14.2							AgOHL	13.2
			AgH₂L₂	45.7	Ag₂L₄	26.3							AgL·s	15.7
			Ag₂L·s	50.1	Ag₃L₅	39.8								
					Ag₆L₈	78.6								

	Ethylene-diamine		NTA		EDTA		CDTA		IDA		Picolinate		Cysteine		Desferri-ferrioxamine B	
H^{+}	HL	9.93	HL	10.33	HL	11.12	HL	13.28	HL	9.73	HL	5.39	HL	10.77	HL	10.1
	H₂L	16.78	H₂L	13.27	H₂L	17.8	H₂L	20.0	H₂L	12.63	H₂L	6.40	H₂L	19.13	H₂L	19.4
			H₃L	14.92	H₃L	21.04	H₃L	23.98	H₃L	14.51			H₃L	20.84	H₃L	27.8
			H₄L	16.02	H₄L	23.76	H₄L	26.62								
					H₅L	24.76	H₅L	28.34								
Na^{+}			NaL	1.9	NaL	2.5			NaL	0.8						
K^{+}					KL	1.7										
Ca^{2+}			CaL	7.6	CaL	12.4	CaL	15.0	CaL	3.5	CaL	2.2			CaL	3.5
					CaHL	16.0					CaL₂	3.8				
Mg^{2+}	MgL	0.4	MgL	6.5	MgL	10.6	MgL	12.8	MgL	3.8	MgL	2.6			MgL	5.2
					MgHL	15.1					MgL₂	4.0				
Sr^{2+}			SrL	6.3	SrL	10.5	SrL	12.4	SrL	3.1	SrL	1.8			SrL	3.1
					SrHL	14.9					SrL₂	3.0				

Metal								
Ba²⁺	BaL 5.9	BaL	BaL 9.6; BaHL 14.6	BaL 10.5; BaHL 17.8	Ba 2.5	BaL 1.6		
Cr³⁺			CrL 26.0; CrHL 28.2; CrOHL 32.2		CrL 12.2; CrL 23.2			
Al³⁺	AlL 13.4; AlOHL 22.1		AlL 18.9; AlHL 21.6; AlOHL 26.6; Al(OH)$_2$L 30.0	AlL 22.1; AlHL 24.3; AlOHL 28.1	AlL 9.9; AlL$_2$ 17.5			
Fe³⁺	FeL 17.9; FeL$_2$ 26.3		FeL 27.7; FeHL 29.2; FeOHL 33.8; Fe(OH)$_2$L 37.7	FeL 32.6; FeOHL 36.5	FeL 12.5	FeL$_2$ 13.9; FeOHL$_2$ 24.9		FeL 31.9; FeHL 32.6
Mn²⁺	MnL 2.8; MnL$_2$ 3.7; MnL$_3$ 5.8	MnL 8.7; MnL$_2$ 11.6	MnL 15.6; MnHL 19.1	MnL 19.2; MnHL 22.4		MnL 4.0; MnL$_2$ 7.1; MnL$_3$ 8.8	MnL 5.6	
Fe²⁺	FeL 4.3; FeL$_2$ 7.7; FeL$_3$ 9.7	FeL 9.6; FeL$_2$ 13.6; FeOHL 12.6	FeL 16.1; FeHL 19.3; FeOHL 20.4; Fe(OH)$_2$L 23.7	FeL 20.8; FeHL 23.9	FeL 6.7; FeL$_2$ 11.0	FeL 5.3; FeL$_2$ 9.7; FeL$_3$ 13.0		FeHL 18.7; FeH$_2$L 21.0
Co²⁺	CoL 6.0; CoL$_2$ 10.8; CoL$_3$ 14.1	CoL 11.7; CoL$_2$ 15.0; CoOHL 14.5	CoL 18.1; CoHL 21.5	CoL 21.4; CoHL 24.7	CoL 7.9; CoL$_2$ 13.2	CoL 6.4; CoL$_2$ 11.3; CoL$_3$ 14.8		CoL 11.2; CoHL 18.0; CoHL 23.6
Ni²⁺	NiL 7.4; NiL$_2$ 13.6; NiL$_3$ 17.9	NiL 12.8; NiL$_2$ 17.0; NiOHL 15.5	NiL 20.4; NiHL 24.0; NiOHL 21.8	NiL 22.1; NiHL 25.4	NiL 9.1; NiL$_2$ 15.7	NiL 7.2; NiL$_2$ 12.5; NiL$_3$ 17.9	NiL 10.7; NiL$_2$ 20.9	NiL 11.8; NiHL 18.3; NiH$_2$L 23.8
Cu²⁺	CuL 10.5; CuL$_2$ 19.6; CuOHL 11.8	CuL 14.2; CuL$_2$ 18.1; CuOHL 18.6	CuL 20.5; CuHL 23.9; CuOHL 22.6	CuL 23.7; CuHL 27.3	CuL 11.5; CuL$_2$ 17.6	CuL 8.4; CuL$_2$ 15.6	Cu(II) → Cu(I)	CuL 15.0; CuHL 24.1; CuH$_2$L 27.0
Zn²⁺	ZnL 5.7; ZnL$_2$ 10.6; ZnOHL 13.9	ZnL 12.0; ZnHL 14.9; ZnOHL 15.5	ZnL 18.3; ZnHL 21.7; ZnOHL 19.9	ZnL 21.1; ZnHL 24.4	ZnL 8.2; ZnL$_2$ 13.5	ZnL 5.7; ZnL$_2$ 10.3; ZnL$_3$ 13.6	ZnL 10.1; ZnL$_2$ 19.1; ZnHL 16.4	ZnL 11.0; ZnHL 17.5; ZnH$_2$L 22.9

Table A6.1. (*Continued*)

	Ethylenediamine		NTA		EDTA		CDTA		IDA		Picolinate		Cysteine		Desferrioxamine B	
Pb^{2+}	PbL	7.0	PbL	12.6	PbL	19.8	PbL	22.1	PbL	8.3	PbL	5.0	PbL	12.5		
	PbL_2	8.5			PbHL	23.0	PbHL	25.3			PbL_2	8.6				
Hg^{2+}	HgL	14.3	HgL	15.9	HgL	23.5	HgL	26.8	HgL	11.7	HgL	8.1	HgL	15.3		
	HgL_2	23.2			HgHL	27.0	HgHL	30.3			HgL_2	16.2				
	HgOHL	24.2			HgOHL	27.7	HgOHL	29.7								
	$HgHL_2$	28.0														
Cd^{2+}	CdL	5.4	CdL	11.1	CdL	18.2	CdL	21.7	CdL	6.6	CdL	5.0			CdL	8.8
	CdL_2	9.9	CdL_2	15.1	CdHL	21.5	CdHL	25.1	CdL_2	11.1	CdL_2	8.3			CdHL	16.2
	CdL_3	11.7	CdOHL	13.4							CdL_3	11.4			CdH_2L	22.7
Ag^+	AgL	4.7	AgL	5.8	AgL	8.2	AgL	9.9			AgL	3.6				
	AgL_2	7.7			AgHL	14.9					AgL_2	6.1				
	AgHL	11.9														

	Glycine		Glutamate		Acetate		Glycolate		Citrate		Malonate		Salicylate		Phthalate	
H^+	HL	9.78	HL	9.95	HL	4.76	HL	3.83	HL	6.40	HL	5.70	HL	13.74	HL	5.51
	H_2L	12.13	H_2L	14.47					H_2L	11.16	H_2L	8.55	H_2L	16.71	H_2L	8.36
			H_3L	16.70					H_3L	14.29						
Na^+									NaL	1.4	NaL	0.7			NaL	0.7
K^+									KL	1.3						
Ca^{2+}	CaL	1.4	CaL	2.1	CaL	1.2	CaL	1.6	CaL	4.7	CaL	2.4	CaL	0.4	CaL	2.4
									CaHL	9.5	CaHL	6.6				
									CaH_2L	12.3						
Mg^{2+}	MgL	2.7	MgL	2.8	MgL	1.3	MgL	1.3	MgL	4.7	MgL	2.9				
									MgHL	9.2	MgHL	7.1				
Sr^{2+}	SrL	0.9	SrL	2.3	SrL	1.1	SrL	1.2	SrL	4.1	SrL	2.1				
											SrHL	6.5				
Ba^{2+}	BaL	0.8	BaL	2.2	BaL	1.1	BaL	1.1	BaL	4.1	BaL	2.1	BaL	0.2	BaL	2.3
									BaHL	9.0						
									BaH_2L	12.4						

Metal	L1	L2	L3	L4	L5	L6	L7
Cr^{2+}					CrL 9.6		
Al^{3+}			CrL 5.4 CrL_2 8.4 CrL_3 11.2 AlL 2.4			AlL 14.2 AlL_2 25.1 AlL_3 31.1	AlL 5.0 AlL_2 8.7
Fe^{3+}	FeL 10.8	FeL 13.8	FeL 4.0 FeL_2 7.6 FeL_3 9.6	FeL 3.7 $FeOHL$ 19.6 $FeOHL_2$ 22.3 $FeOHL_3$ 23.8	FeL 13.5 $Fe_2(OH)_2L_2$ 56.3	FeL 9.3	FeL 17.6 FeL_2 28.6 FeL_3 36.2
Mn^{2+}	MnL 3.2		MnL 1.4	MnL 1.6	MnL 5.5 $MnHL$ 9.4	MnL 3.3	MnL 6.8 MnL_2 10.7
Fe^{2+}	FeL 4.3	FeL 4.6	FeL 1.4	FeL 1.9	FeL 5.7 $FeHL$ 9.9		FeL 7.4 FeL_2 12.1
Co^{2+}	CoL 5.1 CoL_2 9.0 CoL_3 11.6	CoL 5.4 CoL_2 8.7	CoL 1.5	CoL 2.0 CoL_2 3.0	CoL 6.3 $CoHL$ 10.3 CoH_2L 12.9	CoL 3.7 CoL_2 5.1 $CoHL$ 7.0	CoL 7.5 CoL_2 12.3
Ni^{2+}	NiL 6.2 NiL_2 11.1 NiL_3 14.2	NiL 6.5 NiL_2 10.6	NiL 1.4	NiL 2.3 NiL_2 3.4 NiL_3 3.7	NiL 6.7 $NiHL$ 10.5 NiH_2L 12.9	NiL 4.1 NiL_2 5.8 $NiHL$ 7.2	NiL 7.8 NiL_2 12.6
Cu^{2+}	CuL 8.6 CuL_2 15.6	CuL 8.8 CuL_2 15.0	CuL 2.2 CuL_2 3.6	CuL 2.9 CuL_2 4.7 CuL_3 4.7	CuL 7.2 $CuHL$ 10.7 CuH_2L 13.8 $CuOHL$ 16.4 Cu_2L_2 16.3	CuL 5.7 CuL_2 8.2 $CuHL$ 8.3	CuL 11.5 CuL_2 19.3
Zn^{2+}	ZnL 5.4 ZnL_2 9.8 ZnL_3 12.3	ZnL 5.8 ZnL_2 9.5 ZnL_3 9.8	ZnL 1.6 ZnL_2 1.8	ZnL 2.4 ZnL_2 3.6 ZnL_3 3.9	ZnL 6.1 ZnL_2 6.8 $ZnHL$ 10.3 ZnH_2L 13.3	ZnL 3.8 ZnL_2 5.4 $ZnHL$ 7.1	ZnL 7.7
Pb^{2+}	PbL 5.5 PbL_2 8.9		PbL 2.7 PbL_2 4.1	PbL 2.5 PbL_2 3.7 PbL_3 3.6	PbL 5.4 PbL_2 8.1 $PbHL$ 10.2 PbH_2L 13.1	Pb 4.0 PbL_2 4.5	ZnL 2.9 ZnL_2 4.2

Table A6.1. (*Continued*)

	Glycine	Glutamate	Acetate	Glycolate	Citrate	Malonate	Salicylate	Phthalate
Hg^{2+}	HgL 10.9		HgL 6.1		HgL 12.2			
	HgL_2 20.1		HgL_2 10.1					
			HgL_3 14.1					
			HgL_4 17.6					
Cd^{2+}	CdL 4.7	CdL 4.8	CdL 1.9	CdL 1.9	CdL 5.0	CdL 3.2	CdL 6.4	CdL 3.4
	CdL_2 8.4		CdL_2 3.2	CdL_2 2.7	CdL_2 7.2	CdL_2 4.0		
	CdL_3 10.7				CdHL 9.5	CdHL 6.9		
					CdH_2L 12.6			
Ag^+	AgL 3.5		AgL 0.7	AgL 0.4				
	AgL_2 6.9		AgL_2 0.6	AgL_2 0.5				

[a]*Note*: Constants are given as logarithms of the overall formation constants, β, for complexes and as logarithms of the overall precipitation constants for solids, at zero ionic strength and 25°C. The OH^- is a ligand in all oxide and hydroxide species, unlike the IUPAC convention in which H^+ is the computed species. From Smith and Martell (1975, 1976) and Martell and Smith (1974, 1977). Exceptions are major ion interaction constants (Na^+, K^+, Ca^{2+}, Mg^{2+}, CO_3^{2-}, SO_4^{2-}, Cl^-) taken from Whitfield (1974), hydrolysis (OH^-) and some carbonate (CO_3^{2-}) formation constants taken from Baes and Mesmer (1976), Cu^{2+}–CO_3^{2-} complex constants taken from Sunda and Hanson (1979), the ZnS(aq) constant recalculated from the data of Sainte Marie et al. (1964), and the $MnCO_3$(s) constant taken from Morgan (1967). When necessary, constants have been extrapolated to $I = 0$ M using the following values of ($-\log$ of) activity coefficients (applied to all ions including H^+ and tri- and tetravalent ions:

$I \diagdown z$	1	2	3	4
0.1 M	0.11	0.44	0.99	1.76
0.3 M	0.13	0.52	1.17	2.08
0.5 M	0.15	0.60	1.35	2.40
1.0 M	0.14	0.56	1.26	2.24
3.0 M	0.07	0.28	0.63	1.12
4.0 M	0.03	0.12	0.27	0.48

Source: Morel and Hering (1993). Reproduced with permission.

Appendix 6.2 The Various Scales for Equilibrium Constants

APPENDIX 6.2 THE VARIOUS SCALES FOR EQUILIBRIUM CONSTANTS, ACTIVITY COEFFICIENTS, AND pH

In dealing with equilibria in natural waters, we wanted to give above all a feeling of the power of approach. In order not to overwhelm the reader with a large number of intricate details, we attempted to make nonideality corrections for electrolyte solutions simple and effective. The objective of this appendix is to review the various equilibria conventions usable for different natural water media—especially to compare the available conventions for describing (and measuring) pH and ionic equilibria in seawater—and to give an introduction to the ionic interaction theory, which is an expedient alternative and complementary approach to the ion association theory.

In discussing the various refinements in assessing activity coefficients, we should be aware that our insight into the chemical speciation of minor components in natural waters is less hampered by the uncertainty of activity coefficients than by the uncertainty of hydrolysis and complex formation constants.

Procedures for finding the equilibrium distribution of species are based on the principle that at equilibrium the total free energy of the system is at a minimum. This total free energy is the sum of the contributions from each of the constituent chemical species in the system; the contribution of each species depends on its standard free energy of formation, its activity, and the temperature and pressure of the system.

Equilibrium Constants and Activity Conventions

As discussed previously, three types of equilibrium constants are in common use:

1. Constants based on activities (rather than concentrations), the activity scale being based on the infinite dilution reference state.
2. Apparent or stoichiometric equilibrium constants expressed as concentration quotients and valid for a medium of given ionic strength.
3. Conditional constants that hold only under specified experimental conditions (e.g., at a given pH).

Equilibrium constants in the form of concentration quotients are just as thermodynamically valid as the traditional thermodynamic constants, the main difference being the choice of activity scale and reference state.

For the reference state on the infinitely dilute solution scale, the activity coefficient of a species approaches unity as the concentrations of all the solutes approach zero:

$$f_A \to 1 \quad \text{as} \quad \left(c_A + \sum_i c_i\right) \to 0 \tag{1a}$$

An alternative increasingly used convention is defined such that the activity coefficient of a species approaches unity as its concentration approaches zero in the medium of given ionic strength:

$$f'_A \rightarrow 1 \quad \text{as} \quad c_A \rightarrow 0 \tag{1b}$$

in a solution where the total concentration is still $\Sigma_i c_i$ and $\{H_2O\}$ = constant. Activity coefficients of a species on this scale are also close to unity as long as the species concentration is small in comparison to those of the medium ions. Compilations of equilibrium constants usually give data for both thermodynamic equilibrium constants and apparent constants valid for a medium of given ionic strength. Complicated ionic equilibria can only be studied quantitatively in an ionic medium of high total molarity or molality (solutions are adjusted by the addition of an indifferent electrolyte, e.g., $NaClO_4$). By maintaining an ionic medium with a concentration approximately ten times larger than those of the reacting species, usually no correction terms for activity factor changes are necessary. (In addition, constant ionic media "swamp out" changes in the liquid junction potential of galvanic cells.)

For computational approaches, it is possible to use thermodynamic equilibrium constants in conjunction with activities. In order to do this and to relate the constants to concentrations, the values of single-ion activity coefficients must be known. Alternatively, apparent equilibrium constants valid for the medium of particular interest (or a closely similar one) or constants that have been corrected for the medium under consideration can be used in conjunction with concentrations. Nonthermodynamic assumptions are involved in either case.

In our equilibrium calculations so far, we have favored the following approaches:

1. *In dilute solutions* ($I < 10^{-2}$ M), that is, in *fresh waters*, our calculations are usually based on the infinite dilution activity convention and thermodynamic constants. In these dilute electrolyte mixtures, deviations from ideal behavior are primarily caused by *long-range electrostatic interactions*. The Debye–Hückel equation or one of its extended forms (see Table 3.3) is assumed to give an adequate description of these interactions and to define the properties of the ions. Correspondingly, individual ion activities are estimated by means of individual ion activity coefficients calculated with the help of the Güntelberg or Davies (equations 3 and 4 of Table 3.3); or it is often more convenient to calculate, with these activity coefficients, a concentration equilibrium constant valid at a given I,

$$\begin{aligned} K &= \frac{\{AB\}}{\{A\}\{B\}} \\ &= \frac{[AB]}{[A][B]} \frac{f_{AB}}{f_A f_B} \\ &= K^c \frac{f_{AB}}{f_A f_B} \end{aligned} \tag{2}$$

Appendix 6.2 The Various Scales for Equilibrium Constants

2. In more concentrated solutions and *in seawater*, our calculations are usually based on the ionic medium scale; that is, we use for the evaluation of the concentration of coexistent species at equilibrium apparent equilibrium constants (valid for the medium under consideration) expressed in concentration terms; for example, for seawater,

$$K_{s0}^{SW} = [Ca^{2+}][SO_4^{2-}] \tag{3}$$

$$K_{s0}'^{SW} = [Ca_T][SO_{4T}^{2-}] \tag{4}$$

$$K_2'^{SW} = \frac{\{H^-\}[CO_{3T}^{2-}]}{[HCO_{3T}^-]} \tag{5}^{\dagger}$$

The subscript T refers to total (free plus associated) ions, (e.g., $[HCO_{3T}^-]$ = $[HCO_3^-] + [NaHCO_3^0] + [CaHCO_3^-] + [MgHCO_3^-] + 2[Mg(HCO_3^-)_2]$).

By using "medium-bound" constants, we have bypassed activity coefficients in stoichiometric calculations. We can only do so if equilibrium constants valid for seawater or the medium of interest have been determined. If such constants are not known, we are forced to use so-called thermodynamic constants and estimated activity coefficients.[‡]

Activity coefficients in most concentrated solutions reflect deviation from ideal behavior because of (1) the general electric field of the ions, (2) solute–water interactions, and (3) specific ionic interactions (association by ion pair and complex formation). None of the major cations of seawater appears to interact significantly with chloride to form ion pairs; hence activity coefficients in these solutions appear to depend primarily on the ionic strength modified by the extensive hydration of ions. Thus synthetic solutions of these chlorides provide reference solutions in obtaining activity coefficients of the cations. The single activity coefficients for free ions in seawater can be obtained from mean activity coefficient data in chloride solutions at the corresponding ionic strength by various ways.

In the *mean-salt method*,[¶] the behavior of KCl in solution is the standard basis for obtaining individual ion activity coefficients. Various lines of evidence indicate that f_{K^+} and f_{Cl^-} have similar values; that is, as an approximation,

$$f_{\pm(KCl)} = (f_K f_{Cl})^{1/2} = f_{K^+} = f_{Cl^-} \tag{6}$$

[†]We can extend the idea of total concentration also to H^+, where $[H_{SWS}^+] = [H^+] + [HSO_4^-] + [HF]$, so that another apparent ("medium-bound") constant can be defined: $K_2^{*SW} = [H_{SWS}^+][CO_3^{2-}]_T/[HCO_3^-]_T$

[‡]Pytkowicz (1969) writes: "Thermodynamic constants give a superficial appearance of convenience because they do not depend upon the composition of the medium, while apparent constants do. However, in the application of thermodynamic constants to the calculation of concentrations in concentrated multi-electrolyte solutions, activity coefficients have to be determined as a function of composition thereby cancelling what may have seemed to be an advantage."

[¶]Based on the so-called MacInnes convention (MacInnes, 1919).

With the use of equation 6, a table of values for other ions can be built up from the appropriate mean ion activity coefficients; for example, for a monovalent chloride,

$$f_{\pm MCl} = (f_{M^+} f_{Cl^-})^{1/2} = (f_{M^+} f_{\pm KCl})^{1/2} \qquad (7)$$

$$f_{M^+} = \frac{f_{\pm MCl}^2}{f_{\pm KCl}} \qquad (8)$$

For a bivalent chloride,

$$f_{\pm MCl_2} = (f_{M^{2+}} f_{Cl^-}^2)^{1/3} = (f_{M^{2+}} f_{\pm KCl}^2)^{1/3} \qquad (9)$$

$$f_{M^{2+}} = \frac{f_{\pm MCl_2}^3}{f_{\pm KCl}^2}$$

For salts of K^+ other than Cl^- the reverse relation can be used; for example,

$$f_{SO_4^{2-}} = \frac{f_{\pm K_2SO_4}}{f_{\pm KCl}^2} \qquad (11)$$

For a salt like $CuSO_4$, $f_{Cu^{2+}}$ can be obtained, since

$$f_{\pm CuSO_4} = (f_{Cu^{2+}} f_{SO_4^{2-}})^{1/2} \qquad (12)$$

From equations 11 and 12 one obtains

$$f_{Cu^{2+}} = \frac{f_{\pm CuSO_4}^2 f_{\pm KCl}^2}{f_{\pm K_2SO_4}^3} \qquad (13)$$

The MacInnes convention has come under criticism, and other methods have been proposed. Scales of ionic activity at ionic strengths above 0.1 have also been developed based on considerations of the degree of hydration of the ions involved (Bates et al., 1970).

Ion Association and Activity Coefficients: Special Considerations for Seawater

The activity coefficients for single free anions—SO_4^{2-}, HCO_3^-, CO_3^{2-}—cannot be estimated as well as those of the cations[†] because solutions of their salts show association with major seawater cations. In order to understand the prop-

---

[†]As Table 6.5 illustrates, most of the cations in seawater are present as free aquo cations. The concentration of associating SO_4^{2-}, HCO_3^-, and CO_3^{2-} is much smaller than that of major cations.

Appendix 6.2 The Various Scales for Equilibrium Constants

erties of seawater, information on both the nature of the species and the extent of such associations is needed. Many difficulties are involved in evaluation of the degree of association between a cation and a ligand even in electrolyte solutions less complicated than seawater. It is difficult to separate the effects of ionic strength and of ion association. Either the ion association is known from other experiments or the activity coefficient effect is known from other experiments (both involve nonthermodynamic assumptions).

For some strongly associated species, ion association can be measured directly, for example, by spectroscopic means (Byrne and Millero, 1985). Because there is no method by which unequivocal structures of the species present may be obtained, ion association is a phenomenological concept. Ion-pair formation is involved in explaining deviations from normal behavior.[†] However, the impossibility of knowing unambiguously the relevant activity coefficients in seawater implies that the concept of normal behavior is not clearly defined.

Table A6.2 includes historical information on the comparison of ion activity coefficients at the effective ionic strength of seawater. Activity coefficients measured in single salt solutions are compared with those measured in seawater and those calculated from an association model. We have to distinguish between total activity coefficients (cf. equations 3 and 4)

$$^Tf_A = \frac{\{A\}}{[A_T]} \tag{14}$$

and free activity coefficients

$$^Ff_A = \frac{\{A\}}{[A_{\text{free}}]} \tag{15}$$

The relation between ^Tf_A and ^Ff_A is the distribution coefficient

$$\alpha_A = \frac{[A_{\text{free}}]}{[A_T]} = \frac{^Ff_A}{^Tf_A} \tag{16}$$

which can be explained by the association model. For example, α_{Na} in seawater is given by

$$\alpha_{Na} = \frac{[Na^+_{\text{free}}]}{[Na^+]_{\text{free}} + [NaHCO_3] + [NaCO_3^-] + [NaSO_4^-]}$$

$$= (1 + K_{HCO_3}[HCO_3^-] + K_{NaCO_3}[CO_3^{2-}] + K_{NaSO_4}[SO_4^{2-}])^{-1} \tag{17}$$

[†]Nevertheless ion pairs do indeed exist as shown, for example, by Raman spectroscopy (Daly, 1972) and by sound attenuation (e.g., $MgSO_4^0$ in seawater) (Fisher, 1967).

Table A6.2. Comparison of Ion Activity Coefficients at 25°C (1 atm) and Effective Ionic Strength $I = 0.7$ M

Constituent	"Free" Activity Coefficient Measured in Single-Salt Solutions[a,b]	"Total" Activity Coefficient in Seawater Measured in Seawater	Calculated from Association Model	Specific Ion Interaction[c] (1)	(2)	(3)
Na^+	0.71	$0.67^d, 0.70^f$	0.70^e	0.68	0.65	0.64
K^+	—	$0.60^g, 0.61^b$	0.62^e	0.63	0.62	0.61
Mg^{2+}	—	0.26^i	0.25^e	0.23	0.22	0.22
Ca^{2+}	0.26	0.20^j	0.24^e	0.21	0.20	0.21
HCO_3^-	0.68	0.55^j	0.51^e	—	—	—
CO_3^{2-}	0.20	0.021^j	0.021^e	—	—	—
SO_4^{2-}	—	0.11^k	$0.068^e, 0.09^l$	0.11	0.12	0.13

[a] Garrels and Thompson (1962) have used the following activity coefficients of individual species in seawater: Na^+, 0.76; K^+, 0.64; Mg^{2+}, 0.36; Ca^{2+}, 0.28; Cl^-, 0.64; CO_3^{2-}, 0.20; SO_4^{2-}, 0.12; $NaCO_3^- = MgHCO_3^+$, etc., 0.68; $NaHCO_3^0 = MgCO_3^0 = CaCO_3^0 = MgSO_4^0 = CaSO_4^0$, 1.13. Pytkowicz and Kester (1969) have pointed out that uncharged ion pairs are dipolar ions, and they consider an activity coefficient of about 0.8 for $MgSO_4^0$ more appropriate.
[b] Kester and Pytkowicz (1969).
[c] (1) Brønsted-Guggenheim model including interactions between ions of like charge (Leyendekkers, 1975); (2) Brønsted-Guggenheim model (Whitfield, 1973); (3) modified Pitzer equations (Whitfield, 1975a).
[d] Platford (1965).
[e] Berner (1971, p. 48).
[f] Garrels (1967, p. 344).
[g] Garrels and Thompson (1962).
[i] Thompson (1966).
[j] Berner (1965).
[k] Platford and Dafoe (1965).
[l] van Breeman (1972).

The Specific Ionic Interaction Model as an Alternative and Complement to the Ion Association Model

As we have seen, the effect of long-range electrostatic interactions between the ions on the activity coefficients can be predicted fairly accurately in most cases for ionic strengths below 0.01 M. Deviations from the ideal Debye–Hückel theory at higher ionic strengths are attributed mostly to short-range interionic forces. The ion association model (often referred to as the Bjerrum ion association hypothesis) discussed above assumes that the deviations from the Debye–Hückel theory are caused primarily by the relatively strong binding of counterions to form ion pairs. An alternative procedure that can be employed in the evaluation of several thermodynamic properties of aqueous solutions, including the activity coefficients and osmotic coefficients of the individual ions in solution, is the Brønsted–Guggenheim ion interaction model (Guggenheim and Turgeon, 1954). In this model, specific interactions among the various ions are considered and a thermodynamically mutually consistent set of formulas is proposed to deduce—over a broad range of ionic strengths—the thermodynamic properties of mixed electrolytes from those of single electrolytes. These formulas contain parameters called interaction coefficients B_{MX} for each combination of a cation M and an anion X. The value of each interaction coefficient B_{MX} is determined from measurements on a solution containing only the electrolyte whose ions are M and X. The value of each interaction coefficient so determined can then be used for estimating the thermodynamic properties of mixed electrolytes. It is assumed that the multiple interactions with a specific ion are additive. Thermodynamic properties of mixed electrolytes and of seawater can be treated without explicitly considering ion association (ion pairs)[†]; this approach has yielded reasonable results for the activity coefficients and other thermodynamic properties of seawater.

The simplest procedure based on the Brønsted–Guggenheim hypothesis gives rise to an equation for the mean ion activity coefficient that consists only of an electrostatic term and a statistical term linear in the salt concentration (see Whitfield, 1973, 1975b):

$$\log \gamma_{\pm MX} = \log \gamma_{el} + \frac{\nu_M}{\nu_M + \nu_X} \sum_{X^*} B_{MX^*}[X]_T^*$$

$$+ \frac{\nu_X}{\nu_M + \nu_X} \sum_{M^*} B_{M^*X}[M]_T^* \quad (18)$$

where ν_M and ν_X are the number of cations and anions per "molecule" of electrolyte, $[M]_T$ and $[X]_T$ are the total molalities of cations and anions, B_{MX} is a coefficient of interaction between cation M and anion X, and $\log \gamma_{el} =$

[†]"Thermodynamics is not compelled to take cognizance of the various molecular species which may exist in a system, particularly when the existence of such species cannot be absolutely demonstrated (Lewis and Randall, 1961, p. 272).

$-|Z_M Z_X| AI^{1/2}/(1 + I^{1/2})$, where Z_M and Z_X are the charge numbers. The quantities with an asterisk are varied during the summation. Activity coefficients of individual ions in a multicomponent system can be formulated as follows:

$$\log \gamma_X = \left|\frac{Z_X}{Z_M}\right| \log \gamma_{el} + \sum_M B_{M^*X}[M]_T^* \tag{19}$$

$$\log \gamma_M = \left|\frac{Z_M}{Z_X}\right| \log \gamma_{el}^- + \sum_X B_{MX^*}[X]_T^* \tag{20}$$

That is, an interaction term (related to the molality of the solution) is added to the Debye–Hückel function. B values may be calculated from the tabulations in Pitzer and Brewer (1961).

Pitzer and co-workers (1973, 1974) have proposed a more detailed, but at the same time more complex, approach. Whitfield (1973, 1975) has applied these equations to seawater and has shown that this model gives good agreement with available experimental data for the osmotic coefficient and for the mean ion activity coefficient of the major electrolyte components. The results obtained yield numerical results similar to the predictions of the ion association model (see Table A6.2).

One advantage of the ion interaction theory is that it can be applied to solutions of different salinities, that is, to brines, seawaters with different salinities, and estuarine waters. While the ionic medium method provides a very simple solution to many problems, especially for the speciation of constituents in the open ocean, it cannot readily be applied to solutions of different salinity; that is, brines, seawater, and estuarine waters must be treated as separate solvents (Pabalan and Pitzer, 1988).

Extension of the ion interaction model to minor components meets with some difficulty. Many heavy metals form relatively stable ion pairs or inorganic complexes (e.g., $CaHPO_4$, $CuCO_3$, $AgCl_2^-$, $CdCl_2$, $ZnCl^+$) whose existence may be of great biological and geological importance. Whitfield (1975) has extended interaction models for seawater to encompass trace constituents. The two models (ion association and ion interaction) are best seen as providing complementary insights into the chemical nature of seawater, since each treats phenomena not covered by the other. For a detailed treatment of activity coefficients in electrolyte solutions see Pythowicz (1979).

Many models have been developed to estimate the thermodynamic activities of solutes in natural waters (e.g., see Millero, 1984). Of these, the ion pairing and specific interaction theories are the most widely used. A combined model that uses the Pitzer equations to represent specific interactions between ions, together with a thermodynamic description of chemical equilibria, has proved successful in estimating the activities of both major and minor components of seawater (Dickson et al., 1988; Harvie et al., 1984).

Appendix 6.2 The Various Scales for Equilibrium Constants

pH Concepts in Seawater

The difficulties encountered in establishing activity coefficients also relate to the problem of defining pH in seawater. Accurate assessment of the thermodynamic properties of seawater depends on measurements of pH. Essentially, three pH scales are possible.

1. *NBS pH scale.* The NBS pH scale endorsed by IUPAC is based on the infinitely dilute aqueous reference state; the pH is determined relative to that of a standard buffer (whose pH has been estimated in terms of $-\log\{H^+\}$) from measurements in cells with a liquid junction. In dilute solutions ($I < 0.1$ M) the measured pH corresponds to $-\log\{H^+\}$ to within ± 0.02. This is not so in seawater. Because of our ignorance of liquid junction potentials, pH values measured in seawater by the NBS procedure do not approach $-\log\{H^+\}$ on an infinite dilution scale ($\{H^+\}/[H^+] \to 1$ when $[H^+] \to 0$ in pure water). The pH values so measured are on a different activity scale; they may be used to characterize and compare seawater samples and so serve as an index of acid–base balance and speciation.[†]

2. *The H^+ ion concentration.* The hydrogen ion concentration or molality in seawater, m_H, is a clearly defined concept free of the uncertainties attached to the use of single-ion activities (Bates, 1975). But establishing an experimental method for the measurement of m_H in seawater is not without difficulties.

Using seawater as a constant ionic medium we can define pH by the relationship

$$pm_H = -\log m_H \qquad (21)$$

where m_H is the molality of *free* hydrogen ion in the seawater. $\{H^+\}/[H^+] \to 1$ when $[H^+] \to 0$ in the ionic medium. Correspondingly, an electrode system can be calibrated in terms of equation 21 when a salt solution such as NaCl contains a known concentration of strong acid (e.g., HCl), the observed potentiometer reading being compared with m_{H^+}. If we attempt to carry out such a procedure with a seawater medium, let us say with a synthetic seawater free of carbonates and borates, we must consider that some of the H^+ ions become associated with SO_4^{2-}. In other words, we are still faced with the problem of distinguishing between free H^+ and total H^+.

3. *The "total" H^+ ion concentration: the SWS scale.* The total H^+ ion concentration can be defined as

$$-\log[H_T^+] = -\log([H^+] + [HSO_4^-]) \qquad (22)$$

[†]Culberson (1988) writes: "The major problem with the NBS pH scale for equilibrium measurements in sea water is the unreproducibility of its liquid junction potential, not its lack of theoretical meaning."

(Hansson, 1973) or more rigorously as (see Millero, 1995):

$$-\log[H^+_{SWS}] = -\log([H^+] + [HSO_4^-] + [HF^-]) = pH_{SWS} \quad (23)$$

Hansson (1973) introduced the Hansson scale (equation 22) and defined it on the M_W concentration scale (mol kg^{-1} seawater) and determined a new set of acidity constants for carbonic acid and boric acid in seawater.

Obviously, we encounter the same problems in defining $\{OH^-\}$: the total concentration of OH$^-$ in seawater is given by

$$[OH_T^-] = [OH^-] + [MgOH^+] + [CaOH^+] \quad (24)$$

where, for seawater at $-\log[H_T^+] = 8$,

$$p[OH_T^-] = p[OH^-] - 0.45.$$

Thus the formation of HSO_4^- and of $MgOH^+$ ($[CaOH^+]$ and $[HF]$ are nearly negligible) explains the differences in ion products between 0.7 M_W NaCl and standard seawater (25°C):

$$0.7\ M_W\ \text{NaCl}: \quad p[H^+] + p[OH^-] = 13.77$$

$$\text{Standard seawater}: \quad p[H_T^+] + p[OH_T^-] = 13.19$$

For an excellent review on pH concepts in seawater see Dickson (1984).

Obviously, all acid–base equilibrium constants depend on the pH scale used. It is possible to convert approximately an equilibrium constant determined in one scale to that of another scale. The problem of different definitions of equilibrium constants needs attention when applying an infinite dilution scale complex formation constant,—for example, for $CuCO_3$(aq)—in a seawater medium.

Seawater buffers are useful in calibrating glass electrodes and spectroscopic pH indicators. The buffers are prepared in artificial seawater with the protonated and unprotonated form of a given buffer (0.04 M). Dickson (1993) has determined the pK of the buffers of the total proton scale. Suitable buffers are, for example, AP (aminopyridine), $pH_{SWS} = 6.77$; TRIS = 2-amino-2-hydroxymethyl-1.3-propanediol (tris), $pH_{SWS} = 8.07$; and MOR = tetrahydro-1-4-isoxazine (morpholine), $pH_{SWS} = 8.57$.

A Consistent Set of Seawater Equilibrium Constants (Tables A6.3 and A6.4)

Because the relative proportions of the major ions in seawater are constant (see Table 15.2), a consistent set of seawater constants can be given as a function of temperature (0–45°C) and salinity (0–45). Millero (1995) has given sum-

Table A6.3. Summarizing Equations Proposed by Millero (1995)

Equations	References
Ion Product of Water	
$\ln K_W^{SW} = 148.9802 - 13847.26/T - 23.6521 \ln T + (-5.977 + 118.67/T + 1.0495 \ln T)S^{0.5} - 0.01615\, S$	Dickson and Riley (1979)
Acidity Constants of $H_2CO_3^*$	
$\ln K_1^{SW} = 3.17537 - 2329.1378/T - 1.597015 \ln T + (-0.210502 - 5.79495/T)S^{0.5} + 0.0872208\, S - 0.00684651\, S^{1.5}$	Roy et al. (1993)
$\ln K_2^{SW} = -8.19754 - 3403.8782/T - 0.352253 \ln T + (-0.088885 - 25.95316/T)S^{0.5} + 0.1106658\, S - 0.00840155\, S^{1.5}$	Roy et al. (1993)
Solubility of $CaCO_3$ (Calcite and Aragonite)	
$\log K_{Cal}^{SW} = -171.9065 - 0.077993\, T + 2839.319/T + 71.595 \log T + (-0.77712 + 0.0028426\, T + 178.34/T)S^{0.5} - 0.07711\, S + 0.0041249\, S^{1.5}$	Mucci (1983)
$\log K_{Arg}^{SW} = -171.945 - 0.077993\, T + 2903.293/T + 71.595 \log T + (-0.068393 + 0.0017276\, T + 88.135/T)S^{0.5} - 0.10018\, S + 0.0059415\, S^{1.5}$	Mucci (1983) and Weiss (1974)
Henry Constant for CO_2 Exchange (mol $kg^{-1}\, atm^{-1}$)	
$\ln K_0^{SW} = -60.2409 + 93.4517(100/T) + 23.3585 \ln (T/100) + S[0.023517 - 0.023656(T/100) + 0.0047036(T/100)^2]$	

S = salinity, T = absolute temperature (K). For definition, see Table 15.2.

Table A6.4. Seawater Constants[a] on pH_{SWS} Scale and on mol (kg soln)$^{-1}$ Basis (Based on Summarizing Equations in Table A6.3)

Temperature (K)	Henry[b] $-\log K_0^{SW}$	$H_2CO_3^*$ $-\log K_1^{SW}$	$-\log K_2^{SW}$	Calcite $-\log K_{Cal}^{SW}$	Aragonite $-\log K_{Arg}^{SW}$	Ion Product $-\log K_W^{SW}$
273.15	1.2016	6.1004	9.3762	6.3652	6.1113	14.299
278.15	1.2829	6.0454	9.2773	6.3633	6.1105	14.061
283.15	1.3577	5.9926	9.1820	6.3625	6.1118	13.833
288.15	1.4264	5.9419	9.0900	6.3628	6.1153	13.616
293.15	1.4894	5.8930	9.0012	6.3642	6.1209	13.408
298.15	1.5468	5.8460	8.9154	6.3670	6.1289	13.210
303.15	1.5991	5.8008	8.8324	6.3713	6.1391	13.020

[a] Constants given are for $S = 35‰$.
[b] Henry's law constant in mol kg^{-1} atm^{-1}.

Appendix 6.2 The Various Scales for Equilibrium Constants

marizing equations, all valid on the mol per kg solution scale, not mol per kg water, and on the pH$_{SWS}$ scale. The bracketed species refer to stoichiometric "total" concentrations, for example $[HCO_3^-]_T = [HCO_3^-] + [NaHCO_3] + [CaHCO_3^+] + [MgHCO_3^+] + [Ca(HCO_3)_2] + \cdots$.

The constants are defined as follows:

Ion Product of Water

$$K_W^{SW} = [H_{SWS}^+][OH_T^-]$$

where

$$[H_{SWS}^+] = [H^+] + [HSO_4^-] + [HF]$$

and

$$[OH_T^-] = K_W^{SW}/[H_{SWS}^+] = [OH^-] + [Mg(OH)^+] + [Ca(OH)^+]$$

CO$_2$ System

$$K_1^{SW} = [H_{SWS}^+][HCO_3^-]_T/[H_2CO_3^*]$$

$$K_2^{SW} = [H_{SWS}^+][CO_3^{2-}]_T/[HCO_3^-]_T$$

where

$$[H_2CO_3^*] = [CO_2(aq)] + [H_2CO_3]$$

Solubility CaCO$_3$ (Calcite or Aragonite)

$$K_{Cal}^{SW} = [Ca^{2+}]_T[CO_3^{2-}]_T \quad K_{Arg}^{SW} = [Ca^{2+}]_T[CO_3^{2-}]_T$$

Dissolution of CO$_2$ (Henry's Law)

$$K_0^{SW} = [H_2CO_3^*]/p_{CO_2} \quad (\text{mol kg}^{-1} \text{ atm}^{-1})$$

Effect of Pressure

To determine the in situ properties of the carbonate system in the ocean, it is necessary to determine the effect of pressure on the thermodynamic constants. This correction can be made in two ways: (1) using direct measurements of the constants and (2) using partial molal volume and compressibility data (Millero, 1979). The two methods are in good agreement (Millero, 1979) when comparisons are made for the carbonate system. The effect of pressure on the dissociation constants of acids (K_i) can be made from (Millero, 1979) equations

Table A6.5. Coefficients for the Effect of Pressure on the Dissociation Constants of Acids in Seawater

Acid	$-a_0$	a_1	$10^3 a_2$	$-b_0$	b_1
H_2CO_3*	25.50	0.1271	—	3.08	0.0877
HCO_3^-	15.82	-0.0219	—	-1.13	-0.1475
$B(OH)_3$	29.48	-0.1622	2.608	2.84	
H_2O	25.60	0.2324	-3.6246	5.13	0.0794
HSO_4^-	18.03	0.0466	0.316	4.53	0.0900
HF	9.78	-0.0090	-0.942	3.91	0.054
H_2S	14.80	0.0020	-0.400	-2.89	0.054
NH_4^+	26.43	0.0889	-0.905	5.03	0.0814
H_3PO_4	14.51	0.1211	-0.321	2.67	0.0427
$H_2PO_4^-$	23.12	0.1758	-2.647	5.15	0.09
HPO_4^{2-}	26.57	0.2020	-3.042	4.08	0.0714
$CaCO_3(cal)$	48.76	-0.5304		11.76	-0.3692
$CaCO_3(arag)$	35	-0.5304		11.76	-0.3692

of the form

$$\ln(K_i^P/K_i^0) = -(\Delta V_i/RT)P + (0.5\Delta\kappa_i/RT)P^2 \quad (25)$$

where P is the applied pressure in bars, ΔV_i and $\Delta\kappa_i$ are the molal volume and compressibility changes for the association or dissociation reactions ($R = 83.131$ mol bar deg^{-1}). The values of ΔV_i and $\Delta\kappa_i$ for the ionization of acids have been fit to equations of the form for seawater of $S = 35$:

$$\Delta V_i = a_0 + a_1 t + a_2 t^2 \quad (26)$$

$$\Delta\kappa_i = b_0 + b_1 t + b_2 t^2 \quad (27)$$

The coefficients in equations (26) and (27) for the dissociation of a number of acids and the solubility of calcium carbonate are given in Table A6.5 (Millero, 1979, 1995). The results for carbonic and boric acid are taken from the measurements of Culberson and Pytkowicz (1968). The effect of pressure on the solubility of calcite and aragonite has been determined from the measurements of Ingle (1975). The effect of pressure on the dissociation constants of water, hydrogen sulfate, hydrogen sulfide, ammonia, and hydrofluoric and phosphoric acids have been estimated from molal volume and compressibility data.

7

PRECIPITATION AND DISSOLUTION

7.1 INTRODUCTION

The hydrological cycle interacts with the cycle of rocks. Minerals dissolve in or react with the water. Under different physicochemical conditions, minerals are precipitated and accumulate on the ocean floor and in the sediments of rivers and lakes. Dissolution and precipitation reactions impart to the water constituents that modify its chemical properties. Natural waters vary in chemical composition; consideration of solubility relations aids in the understanding of these variations. This chapter sets forth principles concerning reactions between solids and water. Here again the most common basis is a consideration of the equilibrium relations.

Dissolution or precipitation reactions are generally slower than reactions among dissolved species, but it is quite difficult to generalize about rates of precipitation and dissolution. There is a lack of data concerning many geochemically important solid–solution reactions; kinetic factors will be discussed later (Chapter 13). Frequently, the solid phase formed incipiently is metastable with respect to a thermodynamically stable solid phase. Examples are provided by the occurrence under certain conditions of aragonite instead of stable calcite or by the quartz oversaturation of most natural waters. This oversaturation occurs because the rate of attainment of equilibrium between silicic acid and quartz is extremely slow.

The solubilities of many inorganic salts increase with temperature, but a number of compounds of interest in natural waters ($CaCO_3$, $CaSO_4$) decrease in solubility with an increase in temperature. Pressure dependence of solubility is slight but must be considered for the extreme pressures encountered at ocean depths. For example, the solubility product of $CaCO_3$ will increase with increased pressure (by approximately 0.2 logarithmic units for a pressure of 200 atm).†

---

†$(\partial \log K_{s0}/\partial P)_T = (-\Delta V^0/RT) \ln 10$, where ΔV^0, the change in molar volume for the reaction $CaCO_3(s)$ (calcite) $= Ca^{2+} + CO_3^{2-}$ is approximately -58 cm^3 mol^{-1}.

Chemical Weathering

Chemical weathering is one of the major processes controlling the global hydrogeochemical cycle of elements. In this cycle, water operates both as a reactant and as a transporting agent of dissolved and particulate components from land to sea. The atmosphere provides a reservoir for carbon dioxide and for oxidants required in the weathering reactions. The biota assists the weathering processes by providing organic ligands and acids and by supplying locally, upon decomposition, increased CO_2 concentrations. During chemical weathering, rocks and primary minerals become transformed to solutes and soils and eventually to sediments and sedimentary rocks.

Example 7.1. Chemical Erosion Rate The Rhine River above Lake Constance (alpine and prealpine catchment area) averages the following composition:

Ca^{2+} = 43 mg liter^{-1}
Mg^{2+} = 9 mg liter^{-1}
Na^+ = 3.1 mg liter^{-1}
SO_4^{2-} = 53 mg liter^{-1}
Cl^- = 2.8 mg liter^{-1}
HCO_3^- = 115 mg liter^{-1}
H_4SiO_4 = 6.5 mg liter^{-1}

Annual precipitation amounts to 140 cm yr^{-1} of which about 30% is lost due to evaporation and evapotranspiration.

What is the chemical weathering rate in this catchment area? Can it be subdivided to dissolution rates of individual minerals?

The annual runoff approximates 1 m^3 per m^2 geographic area; and 1 m^3 of water contains the concentrations given. This amounts to

1.07 mol of Ca^{2+}
0.4 mol of Mg^{2+}
0.13 mol of Na^+
0.08 mol of Cl^-
1.9 mol of HCO_3^-
0.55 mol of SO_4^{2-}
7 × 10^{-2} mol of H_2SiO_4

per m^3; thus corresponding mole quantities have become dissolved per m^2.

We can try to represent these concentrations in terms of minerals that have dissolved. All the SO_4^{2-}, minus SO_4^{2-} introduced by acid rain (on the average the rain contains 0.05 mol SO_4^{2-} m^{-3}), is "assigned" to the dissolution of $CaSO_4(s)$, all the Mg^{2+} to $MgCO_3(s)$ [component of dolomite, $CaMg(CO_3)_2$], and the remaining Ca^{2+} to $CaCO_3(s)$, Cl^- to NaCl, and the H_4SiO_4 to a feldspar $NaAlSi_3O_8$.

Thus we can estimate the erosion rates:

Mineral	mol m^{-2} yr^{-1}	g m^{-2} yr^{-1}
$CaSO_4$	0.5	68
$MgCO_3$	0.4	33
$CaCO_3$	0.6	60
NaCl	0.08	4.7
Si(s)	0.07	5.5
or $NaAlSi_3O_8$	0.023	

Total chemical erosion: 3.1 eq m^{-2} yr^{-1} or 170 g m^{-2} yr^{-1}.

These estimates are based on some crude assumptions; nevertheless, the chemical erosion rates obtained are believed to be within ±20%.

Note: Mechanical erosion, that is, the production of suspended solids by cracking of rocks, has been estimated for the same catchment area to be 1150 g m^{-2} yr^{-1}.

Heterogeneous Equilibria

The extent of the dissolution or precipitation reaction for systems that attain equilibrium can be estimated by considering the equilibrium constants. A simple example is the solubility of an amorphous silica: that is, $SiO_2(s)$(amorph)

$$SiO_2(s)(amorph) + 2 H_2O \underset{\text{precipitation}}{\overset{\text{dissolution}}{\rightleftharpoons}} H_4SiO_4(aq) \qquad (1)$$

The solubility equilibrium for pure amorphous $SiO_2(s)$ is defined by

$$K_{s0} = \{H_4SiO_4(aq)\} \qquad (2)$$

Equation 2 must be satisfied at equilibrium whether the solution is initially supersaturated or undersaturated with respect to SiO_2, that is, whether $\{H_4SiO_4\}_{initial} > K_{s0}$ or $\{H_4SiO_4\}_{initial} < K_{s0}$. The equilibrium activity in solution is independent of the amount of amorphous $SiO_2(s)$. The activity of pure SiO_2 (i.e., $X_{SiO_2} = 1$) is a constant at fixed temperature and pressure; under standard state conditions, we can choose $\{SiO_2(s)\}$ to be unity.

Comparison with other heterogeneous equilibria is instructive. Consider, for example,

$$H_2O(l) = H_2O(g) \qquad K = p_{H_2O} \qquad (3)$$

$$CaCO_3(s) = CaO(s) + CO_2(g) \qquad K = p_{CO_2} \qquad (4)$$

$$MgCO_3(s) + 3 H_2O(g) = MgCO_3 \cdot 3 H_2O(s) \qquad K = p_{H_2O}^{-3} \qquad (5)$$

for which the respective mass law expressions are

$$\frac{p_{H_2O}}{\{H_2O\}} = K_{eq} \tag{6a}$$

$$\frac{\{CaO\}p_{CO_2}}{\{CaCO_3\}} = K_{eq} \tag{6b}$$

$$\frac{\{MgCO_3 \cdot 3H_2O\}}{\{MgCO_3\}\, p_{H_2O}^3} = K_{eq} \tag{6c}$$

Lowering the water activity by adding appreciable concentrations of solutes (e.g., as in seawater) results in $\{H_2O\} < 1$, and $p_{H_2O} < p_{H_2O}^0$ as a consequence. For pure $CaCO_3(s)$ and pure $CaO(s)$, $p_{CO_2} = K_{eq}$ is the condition for the coexistence of these solid phases. However, should solid solution alter the activity of either solid phase, then the equilibrium p_{CO_2} must change accordingly. Similar considerations dictate that the coexistence of $MgCO_3(s)$ and the trihydrate depends on the activity of water and the activities of the two solids. In an analogous way, the solubility equilibrium of SiO_2 is most generally represented by

$$K = \frac{\{H_4SiO_4\}}{\{SiO_2(s)\}} \tag{7}$$

with the standard and reference states being pure SiO_2.

More complex heterogeneous equilibria involve the formation of two or more aqueous ions, and for such equilibria an ion activity product is required.

In a general way, for an electrolyte that dissolves in water according to the reaction

$$A_mB_n(s) \rightleftharpoons mA^{+n}(aq) + nB^{-m}(aq) \tag{8a}$$

the equilibrium condition can be represented by

$$\{A_mB_n(s)\} = \{A^{+n}(aq)\}^m\{B^{-m}(aq)\}^n \tag{8b}$$

The conventional solubility expression

$$K_{s0} = \{A^{+n}(aq)\}^m\{B^{-m}(aq)\}^n \tag{8c}$$

results if the activity of the pure solid phase is set equal to *unity* and if the common standard-state convention for aqueous solutions is adopted; that is, the solubility product is constant for varying compositions of the liquid phase under constant temperature and pressure and for a chemically pure and compositionally invariant solid phase. Furthermore, the solid should be uniform

7.1 Introduction

and of large grain size (cf. Section 7.8). Only in some cases can the solubility of a salt be calculated from its solubility product alone. Generally, one deals with the solubility of a salt in solutions that contain a common ion, that is, an ion that also exists in the ionic lattice of the solid salt.

Example 7.2. Solubility of Sulfates, Chlorides, Fluorides, and Chromates
Characterize the solubility of the following salts as a function of the concentration of the common anion from the respective solubility products (K_{s0})

$CaSO_4$(pK_{s0} = 4.6), $SrSO_4$(6.2), $BaSO_4$(9.7), $AgCl$(10.0), $PbCl_2$(4.8),

Ag_2CrO_4(12.0), MgF_2(8.1), CaF_2(10.3)

Figure 7.1 is a graphical representation of the solubility product, where the log of the metal-ion concentration is plotted as a function of the $-\log$ of the common anion, for example,

$$\log[Ca^{2+}] = -pK_{s0(CaSO4)} + pSO_4^{2-} \quad \text{(i)}$$

$$\log[Mg^{2+}] = -pK_{s0(MgF2)} + 2\,pF^- \quad \text{(ii)}$$

$$\log[Ag^+] = -\tfrac{1}{2}\,pK_{s0(Ag2CrO4)} + \tfrac{1}{2}\,pCrO_4^{2-} \quad \text{(iii)}$$

Correspondingly, in Figure 7.1 the $\log[Me^{z+}]$ lines have slopes of 1, 2, and $\tfrac{1}{2}$ and intercepts of $-pK_{s0}$, $-pK_{s0}$, and $-\tfrac{1}{2}\,K_{s0}$ for salts of the 1:1, 1:2, and 2:1 type, respectively.

The cations and anions of these salts do not undergo protolysis reactions to any appreciable extent in solutions that are near neutrality. Furthermore, complex formation (or ion-pair binding) between cation and anion may be assumed to be negligible as long as free metal-ion and free anion concentration is small (approximately $< 10^{-1.5}$ M). (Dashed lines in Figure 7.1 indicate where the foregoing assumptions are no longer valid.)

In the absence of a common anion, for example, if the salt is dissolved in pure water, the solubility is given by the electroneutrality requirement. Considering that $[H^+] \simeq [OH^-]$, we have, for example,

$$\text{for } CaSO_4(s): \quad [Ca^{2+}] = [SO_4^{2-}] \quad \text{(iv)}$$

$$\text{for } MgF_2(s): \quad 2\,[Mg^{2+}] = [F^-] \quad \text{(v)}$$

$$\text{for } Ag_2CrO_4(s): \quad [Ag^+] = 2\,[CrO_4^{2-}] \quad \text{(vi)}$$

Hence the solubility of a 1:1 salt in pure water is defined by the intersection of the lines of $\log[Me^{z+}]$ and $\log[\text{anion}]$. For a salt of the 1:2 and 2:1 types, the solubility is slightly displaced from the intersection in such a way that

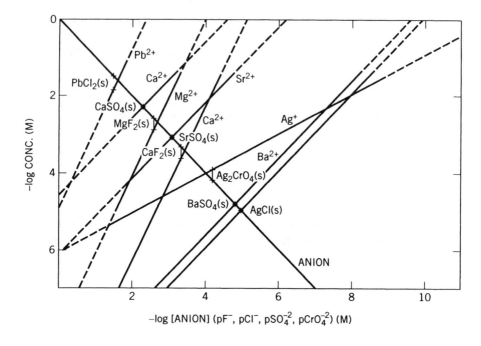

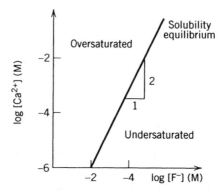

Figure 7.1. Solubility of "simple" salts as a function of the common anion concentration (Example 7.2). The cations and anions of these salts do not protolyze in the neutral pH range. The equilibrium solubility is given by the metal-ion concentration. At high anion or cation concentration, complex formation or ion-pair binding becomes possible (dashed lines). If the salt is dissolved in pure water (or in an inert electrolyte), the solubility is defined by the electroneutrality $z[\text{Me}^{z+}] = n[\text{anion}^{n-}]$. If $z = n$ (e.g., BaSO$_4$), the solubility is given by the intersection (+). If $z \neq n$, the electroneutrality condition is fulfilled at a point slightly displaced from the intersection (‡). The insert exemplifies the solubility equilibrium for CaF$_2$ ($K_{s0} = 10^{-10.3}$) and lists the domains of over- and undersaturation.

7.1 Introduction

$\log[\text{Me}^{z+}] = -0.3 + \log[\text{anion}]$ and $\log[\text{Me}^{z+}] = +0.3 + \log[\text{anion}]$, respectively.

However, the case in which the solubility of a solid can be calculated from the known analytical concentration of added components and from the solubility product alone is very seldom encountered. Ions that have dissolved from a crystalline lattice frequently undergo chemical reactions in solution, and therefore other equilibria in addition to the solubility product have to be considered. The reaction of the salt cation or anion with water to undergo acid–base reactions is very common. Furthermore, complex formation of salt cation and salt anion with each other and with one of the constituents of the solution has to be considered. For example, the solubility of FeS(s) in a sulfide-containing aqueous solution depends on, in addition to the solubility equilibrium, acid–base equilibria of the cation (e.g., $\text{Fe}^{2+} + \text{H}_2\text{O} = \text{FeOH}^+ + \text{H}^+$) and of the anion (e.g., $\text{S}^{2-} + \text{H}_2\text{O} = \text{HS}^- + \text{OH}^-$, and $\text{HS}^- + \text{H}_2\text{O} = \text{H}_2\text{S} + \text{OH}^-$), as well as on equilibria describing complex formation (e.g., formation of FeHS^+ or FeS_2^{2-}).

Data for Solubility Constants

Available compilations of solubility products[†] illustrate that values given by different authors for the same solubility products often differ markedly. Differences of a few orders of magnitude are not uncommon. There are various reasons for such discrepancies:

1. The formation of a sparingly soluble phase and its equilibrium with the solution is a more complicated process than equilibration reactions in a homogeneous solution phase.
2. The composition and properties, that is, reactivity, of the solids vary for different modifications of the same compound or for different active (disordered) forms of the same modification.
3. Species influencing the solubility equilibrium (e.g., species formed by hydrolysis, ion-pair, or complex formation) have been overlooked in defining K_{s0}.

Activity Corrections The solubility equilibrium is influenced by the ionic strength of the solution. Often it is most convenient to use operational equilibrium constants, that is, constants expressed as concentration quotients and valid for a medium of given ionic strength. For freshwater conditions, the Debye-

---

[†]For example, see Sillén and Martell (1964, 1971), Feitknecht and Schindler (1963), Smith and Martell (1976), Martell and Smith (1982, Vol. 5; 1989, Vol. 6), Nordstrom et al. (1990), (see Table 7.1 and Appendix 1 at the end of the book), Robie et al. (1978), Bard et al. (1985), and Baes and Mesmer (1976). Millero (1995) has compiled summarizing equations defining equilibrium constants for seawater as a function of temperature and salinity. They are given in Appendix 6.1, Tables A6.3 and A6.4, of Chapter 6.

Hückel theory [or the Güntelberg or Davies equation (Table 3.3)] may be used to convert the solubility equilibrium constant given at infinite dilution or at a specified I to an operational constant, ^{c}K, valid for the ionic strength of interest. In seawater solubility equilibrium constants, experimentally determined in seawater, may be used. For example, the $CaCO_3$ calcite solubility in seawater of specified salinity may be defined by $K_{s0}^{SW} = [Ca_T^{2+}][CO_{3T}^{2-}]$, where $[Ca_T^{2+}]$ and $[CO_{3T}^{2-}]$ are the total concentrations of calcium and carbonate ions, for example, in mol liter^{-1} seawater.

Precipitation An appreciation of the various types of precipitates that may be formed and an understanding of the changes the precipitates undergo in aging are prerequisites for understanding and interpreting solubility equilibrium constants. An early lucid treatment of these considerations has been given by Feitknecht and Schindler (1963). Figure 7.2 illustrates in a schematic way some of the domains of precipitation and solubility.

An "active" form of the compound, that is, a very fine crystalline precipitate with a disordered lattice, is generally formed incipiently from strongly oversaturated solutions. Such an active precipitate may persist in metastable equilibrium with the solution and may convert ("age") only slowly into a more stable "inactive" form. Measurements of the solubility of active forms give solubility products that are higher than those of the inactive forms. Inactive solid phases with ordered crystals are also formed from solutions that are only slightly oversaturated.

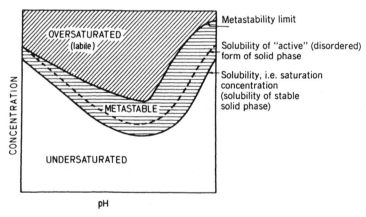

Figure 7.2. Solubility and saturation. A schematic solubility diagram showing concentration ranges versus pH for supersaturated, metastable, saturated, and undersaturated solutions. A supersaturated solution in the labile concentration range forms a precipitate spontaneously; a metastable solution may form no precipitate over a relatively long period. Often an "active" form of the precipitate, usually a very fine crystalline solid phase with a disordered lattice, is formed from oversaturated solutions. Such an active precipitate may persist in metastable equilibrium with the solution; it is more soluble than the stable solid phase and may slowly convert into the stable phase.

7.1 Introduction

Hydroxides and sulfides often occur in amorphous and several crystalline modifications. Amorphous solids may be active or inactive. Initially formed amorphous precipitates or active forms of unstable crystalline modifications may undergo two kinds of changes during aging. Either the active form of the unstable modification becomes inactive or a more stable modification is formed.

In determining solubility equilibrium constants, many investigators have been motivated by a need to gain information that is pertinent primarily for the relatively short-term conditions (minutes to hours) typically encountered in the laboratory or in a water and wastewater treatment system. In operations of analytical chemistry, for example, precipitates are frequently formed from strongly oversaturated solutions; the conditions of precipitation of the incipient active compound rather than the dissolution of the aged inactive solid are often of primary interest. Most solubility products measured in such cases refer to the most active component.[†] On the other hand, in dealing with heterogeneous equilibria of natural water systems, the more stable and inactive solids are frequently more pertinent. Aging often continues for geological time spans. Furthermore, the solid phase has frequently been formed in nature under conditions of slight supersaturation. Solubility constants determined under conditions where the solid has been identified by x-ray diffraction are especially valuable.

Solubility Product and Saturation

In order to test whether a solution, or a natural water, is over- or undersaturated, we inquire whether the free energy of dissolution of the solid phase is positive, negative, or zero; that is, for the solubility equilibrium of calcite.

$$CaCO_3(s) = Ca^{2+} + CO_3^{2-} \qquad K_{s0} \qquad (9)$$

the free energy of dissolution is given by

$$\Delta G = RT \ln \frac{Q}{K_{s0}}$$
$$= RT \ln \frac{\{Ca^{2+}\}_{act} \{CO_3^{2-}\}_{act}}{\{Ca^{2+}\}_{eq} \{CO_3^{2-}\}_{eq}} = RT \ln \frac{IAP}{K_{s0}} \qquad (10)$$

The actual ion activity product, IAP, may be compared with K_{s0}. The state of saturation of a solution with respect to a solid is defined as follows:

$$IAP > K_{s0} \quad \text{(oversaturated)}$$

[†]Strictly speaking, solubility products for active solid compounds are, because of their time dependence, not equilibrium constants; they are of operational value to estimate the conditions (pMe, pH, etc.) under which precipitation occurs.

358   Precipitation and Dissolution

$$IAP = K_{s0} \quad \text{(equilibrium, saturated)}$$

$$IAP < K_{s0} \quad \text{(undersaturated)}$$

By comparing Q with K (equation 10), we can define the state of saturation for all reactions that involve a solid phase. For example, the solubility equilibrium could also be written

$$CaCO_3(s) + H^+ = Ca^{2+} + HCO_3^- \quad *K_s \tag{11}$$

$$\Delta G = RT \ln \frac{Q}{*K_s} = RT \ln \frac{\{Ca^{2+}\}_{act} \{HCO_3^-\}_{act} \{H^+\}_{eq}}{\{H^+\}_{act} \{Ca^{2+}\}_{eq} \{HCO_3^-\}_{eq}}$$

$$= RT \frac{\{Ca^{2+}\}_{act} \{HCO_3^-\}_{act}}{\{H^+\}_{act} *K_s} \tag{12}$$

If in equation 12 $Q/*K_s < 1$ (ΔG is negative), $CaCO_3(s)$ will dissolve; if $Q/*K_s > 1$, $CaCO_3$ will precipitate. The saturation test may often be made by simply comparing the activity (or concentration) of an individual reaction component, for example, H^+ in equation 12, with the activity (or concentration) this component would have if it were in hypothetical solubility equilibrium. Thus the state of saturation of $CaCO_3(s)$ may be interpreted in the following way:

$$pH_{act} > pH_{eq} \quad \text{(oversaturated)}$$

$$pH_{act} = pH_{eq} \quad \text{(saturated)}$$

$$pH_{act} < pH_{eq} \quad \text{(undersaturated)}$$

With regard to $CaCO_3$ saturation, customarily a *saturation index* (SI) has been defined. SI = $pH_{act} - pH_{eq}$. If SI is positive, $CaCO_3$ will deposit. This saturation index can also be determined with relative ease experimentally by measuring the change in pH upon addition of $CaCO_3$.

In a similar way the $CaCO_3(s)$ saturation could be assessed by comparing $[H_2CO_3^*]_{act}$ with $[H_2CO_3^*]_{eq}$ and considering the reaction $CaCO_3(s) + H_2CO_3^* = Ca^{2+} + 2 HCO_3^-$.

Example 7.3. Solubility of Anhydrite and Gypsum Compare the solubility (25°C, 1 atm) of anhydrite [$CaSO_4(s)$, $K_{s0} = 4.2 \times 10^{-5}$] with that of gypsum [$CaSO_4 \cdot 2H_2O(s)$, $K_{s0} = 2.5 \times 10^{-5}$] and determine the state of saturation of seawater of normal salinity ($\{Ca^{2+}\} = 2.4 \times 10^{-3}$ M; $\{SO_4^{2-}\} = 1.9 \times 10^{-3}$ M) with respect to these solid phases.

The IAP of seawater (4.7×10^{-6}) is smaller than either solubility product. Hence seawater is undersaturated approximately five times and approximately nine times, respectively, with regard to gypsum and anhydrite. Since gypsum is less soluble (25°C, 1 atm), it should be more stable in seawater. The equilibrium constant for the conversion of gypsum into anhydrite can be obtained from

$$CaSO_4 \cdot 2H_2O(s) = Ca^{2+} + SO_4^{2-} + 2H_2O$$

$$\log K_{s0\text{gyps}} = -4.60 \qquad (i)$$

$$Ca^{2+} + SO_4^{2-} = CaSO_4(s)$$

$$-\log K_{s0\text{anhyd}} = +4.38 \qquad (ii)$$

$$CaSO_4 \cdot 2H_2O(s) = CaSO_4(s) + 2H_2O$$

$$\log K = -0.22 \qquad (iii)$$

The equilibrium of equation iii (assuming activities of solid phases = 1) is defined by

$$\{H_2O\}^2 = 0.6 \quad \text{or} \quad \{H_2O\} = 0.78 \qquad (iv)$$

As long as the activity of water is larger than 0.78, gypsum at this T and P is more stable than anhydrite. The activity of normal seawater is $\{H_2O\} \approx 0.98$. Upon evaporation of seawater to the point where gypsum becomes saturated, the activity of H_2O ($\{H_2O\} \approx 0.93$) is still much larger than the equilibrium H_2O activity for the conversion. Hence gypsum is more stable than anhydrite even in partially evaporated seawater. Gypsum may, however, become less stable than anhydrite upon burial in sediments (increase in temperature, pressure, and salinity).

7.2 THE SOLUBILITY OF OXIDES AND HYDROXIDES

Many minerals with which water comes into contact are oxides, hydroxides, carbonates, and hydroxide carbonates. The same ligands (hydroxides and carbonates) are dissolved constituents of all natural waters. The solid and solute chemical species under consideration belong to the ternary system $Me^{z+}-H_2O-CO_2$. In connection with hydrolysis (Section 6.4) we have already dealt with solubility equilibria of (hydr)oxides. To be complete, we resume a brief formal treatment.

If a pure solid oxide or hydroxide is in equilibrium with free ions in solution, for example,

$$Me(OH)_2(s) = Me^{2+} + 2OH^-$$

$$MeO(s) + H_2O = Me^{2+} + 2OH^- \qquad (13)$$

the conventional solubility product is given by

$$^cK_{s0} = [Me^{2+}][OH^-]^2 \text{ mol}^3 \text{ liter}^{-3} \quad (14)$$

The subscript zero indicates that the equilibrium of the solid with the simple (uncomplexed) species Me^{2+} and OH^- is considered.[†] Sometimes it is more appropriate to express the solubility in terms of reaction with protons, since the equilibrium concentrations of OH^- ions may be extremely small, for example,

$$Me(OH)_2(s) + 2H^+ = Me^{2+} + 2H_2O \quad (15)$$

$$MeO(s) + 2H^+ = Me^{2+} + H_2O \quad (16)$$

Then the solubility equilibrium can be characterized by

$$^{c*}K_{s0} = \frac{[Me^{2+}]}{[H^+]^2} \text{ mol}^{-1} \text{ liter} \quad (17)$$

or, for a solubility equilibrium with a trivalent metal ion,

$$FeOOH(s) + 3H^+ = Fe^{3+} + 2H_2O$$

by

$$^{c*}K_{s0} = \frac{[Fe^{3+}]}{[H^+]^3} \text{ mol}^{-2} \text{ liter}^2 \quad (18)$$

The definitions for the solubility equilibrium contained in equations 14 and 17 are interrelated. For $MeO_{z/2}$ or $Me(OH)_z$, the following general equation obtains:

$$^{c*}K_{s0} = \frac{^cK_{s0}}{K_W^z} \quad (19)$$

where K_W is the ion product of water and z is the valence. A few representative solubility products are given in Table 7.1.[‡] From these equilibrium constants,

---

[†] The complete dissolution reaction considers the water that participates in the dissolution reaction: $Me(OH)_z(s) + (x + zy)H_2O(l) = Me(H_2O)_x^{z+} + z(OH^-)(H_2O)_y$.

[‡] This table also gives the standard enthalpy change for the reaction, ΔH_r°. With the help of the ΔH_r° values, the equilibrium constant valid for a temperature other than 25°C can be calculated on the basis of the thermodynamic relation

$$\frac{\partial \ln K}{\partial T} = \frac{\Delta H_r^\circ}{RT^2}$$

7.2 The Solubility of Oxides and Hydroxides

$[Me^{z+}]$ in equilibrium with the pure solid phase can be computed readily as a function of pOH or pH. Especially convenient is the graphical representation in a $\log[Me^{z+}]$–pH diagram for which the relationship of equation 20 can be derived:

$$\log[Me^{z+}] = \log {}^{c*}K_{s0} - z\text{pH} \qquad (20)$$
$$\log[Me^{z+}] = \log {}^{c}K_{s0} + zpK_W - z\text{pH}$$

Equation 20 is plotted for a few oxides or hydroxides in Figure 7.3. Obviously, $\log[Me^{z+}]$ plots linearly as a function of pH with a slope of $-z$ ($d \log[Me^{z+}]/d\,\text{pH} = -z$) and an intercept at $\log[Me^{z+}] = 0$, with the value pH $= -(1/z) p^{c*}K_{s0}$.

The relations depicted in Figure 7.3 or characterized by equation 14 do not fully describe the solubility of oxides or hydroxides. We have to consider that the solid can be in equilibrium with hydroxo–metal-ion complexes[¶] $[Me(OH)_n]^{z-n}$.

We already have encountered such hydroxo complexes as conjugate (Brønsted) bases of aquo metal ions, for example,

$$Zn^{2+} + H_2O = ZnOH^+ + H^+ \qquad *K_1 \qquad (21)$$

and

$$FeOH^{2+} + H_2O = Fe(OH)_2^+ + H^+ \qquad *K_2 \qquad (22)$$

We can characterize the solubility of the metal oxide or hydroxide, Me_T, by

$$Me_T = Me^{z+} + \sum_n [Me(OH)_n^{z-n}] \qquad (23)$$

Figures 6.8a, 6.8b, and 6.8c give examples for the solubility of α-Al(OH)$_3$ (gibbsite), amorphous FeOOH, and amorphous Zn(OH)$_2$.

---

which in a first approximation yields

$$\ln \frac{K_{T_2}}{K_{T_1}} = \frac{\Delta H_r^\circ}{R} \left(\frac{1}{T_1} - \frac{1}{T_2} \right)$$

The units kcal mol^{-1} can be converted to kJ mol^{-1} by multiplying by 4.184 ($R = 1.987 \times 10^{-3}$ kcal mol^{-1} K^{-1} or 8.314×10^{-3} kJ mol^{-1} K^{-1}).

[¶]As already discussed in Chapter 6 more generally and more exactly, polynuclear complexes $[Me_m(OH)_n]^{zm-n}$, and complexes with other ligands, L^{y-}, in the solution, $[Me_pL_q]^{(zp-yq)}$, and mixed complexes with OH$^-$ and L^{y-} have to be considered too.

Table 7.1. Thermodynamic Data Related to Solubility (at 25°C and 1 bar) from a Compilation by Nordstrom et al. (1990)

Mineral	Reaction	ΔH_r^0 (kcal mol^{-1})	log K
Portlandite	$Ca(OH)_2 + 2H^+ = Ca^{2+} + 2H_2O$	−31.0	22.8
Brucite	$Mg(OH)_2 + 2H^+ = Mg^{2+} + 2H_2O$	−27.1	16.84
Pyrolusite	$MnO_2 + 4H^+ + 2e^- = Mn^{2+} + 2H_2O$	−65.11	41.38
Hausmanite	$Mn_3O_4 + 8H^+ + 2e^- = 3Mn^{2+} + 4H_2O$	−100.64	61.03
Manganite	$MnOOH + 3H^+ + e^- = Mn^{2+} + 2H_2O$	—	25.34
Pyrochroite	$Mn(OH)_2 + 2H^+ = Mn^{2+} + 2H_2O$	—	15.2
Gibbsite (crystalline)	$Al(OH)_3 + 3H^+ = Al^{3+} + 3H_2O$	−22.8	8.11
Gibbsite (microcrystalline)	$Al(OH)_3 + 3H^+ = Al^{3+} + 3H_2O$	(−24.5)	9.35
$Al(OH)_3$ (amorphous)	$Al(OH)_3 + 3H^+ = Al^{3+} + 3H_2O$	(−26.5)	10.8
Goethite	$FeOOH + 3H^+ = Fe^{3+} + 2H_2O$	—	−1.0
Ferrihydrite (amorphous to microcrystalline)	$Fe(OH)_3 + 3H^+ = Fe^{3+} + 3H_2O$	—	3.0–5.0

Carbonate Species

Reaction	ΔH_r^0 (kcal mol^{-1})	log K	Reaction	ΔH_r^0 (kcal mol^{-1})	log K
$CO_2(g) = CO_2(aq)$[a]	−4.776	−1.468	$Ca^{2+} + CO_3^{2-} = CaCO_3^0$	3.545	3.224
$CO_2(aq) + H_2O = H^+ + HCO_3^-$	2.177	−6.352	$Mg^{2+} + CO_3^{2-} = MgCO_3^0$	2.713	2.98
$HCO_3^- = H^+ + CO_3^{2-}$	3.561	−10.329	$Sr^{2+} + CO_3^{2-} = SrCO_3^0$	5.22	2.81
$Ca^{2+} + HCO_3^- = CaHCO_3^+$	2.69	1.106	$Ba^{2+} + CO_3^{2-} = BaCO_3^0$	3.55	2.71
$Mg^{2+} + HCO_3^- = MgHCO_3^+$	0.79	1.07	$Mn^{2+} + CO_3^{2-} = MnCO_3^0$	—	4.90
$Sr^{2+} + HCO_3^- = SrHCO_3^+$	6.05	1.18	$Fe^{2+} + CO_3^{2-} = FeCO_3^0$	—	4.38
$Ba^{2+} + HCO_3^- = BaHCO_3^+$	5.56	0.982	$Na^+ + CO_3^{2-} = NaCO_3^-$	8.91	1.27
$Mn^{2+} + HCO_3^- = MnHCO_3^+$	—	1.95	$Na^+ + HCO_3^- = NaHCO_3^0$	—	−0.25
$Fe^{2+} + HCO_3^- = FeHCO_3^+$	—	2.0	$Ra^{2+} + CO_3^{2-} = RaCO_3^0$	1.07	2.5

Mineral	Reaction	ΔH_r^0 (kcal mol^{-1})	log K
Calcite	$CaCO_3 = Ca^{2+} + CO_3^{2-}$	−2.297	−8.480
Aragonite	$CaCO_3 = Ca^{2+} + CO_3^{2-}$	−2.589	−8.336
Dolomite (Ordered)	$CaMg(CO_3)_2 = Ca^{2+} + Mg^{2+} + 2CO_3^{2-}$	−9.436	−17.09
Dolomite (Disordered)	$CaMg(CO_3)_2 = Ca^{2+} + Mg^{2+} + 2CO_3^{2-}$	−11.09	−16.54
Strontianite	$SrCO_3 = Sr^{2+} + CO_3^{2-}$	−0.40	−9.271
Siderite (crystalline)	$FeCO_3 = Fe^{2+} + CO_3^{2-}$	−2.48	−10.89
Siderite (precipitated)	$FeCO_3 = Fe^{2+} + CO_3^{2-}$	—	−10.45
Witherite	$BaCO_3 = Ba^{2+} + CO_3^{2-}$	0.703	−8.562
Rhodocrosite (crystalline)	$MnCO_3 = Mn^{2+} + CO_3^{2-}$	−1.43	−11.13
Rhodocrosite (synthetic)	$MnCO_3 = Mn^{2+} + CO_3^{2-}$	—	−10.39

Sulfate Species

Reaction	ΔH_r^0 (kcal mol^{-1})	log K (25°C)	Reaction	ΔH_r^0 (kcal mol^{-1})	log K (25°C)
$H^+ + SO_4^{2-} = HSO_4^-$	3.85	1.988	$Mn^{2+} + SO_4^{2-} = MnSO_4^{2-}$	3.37	2.25
$Li^+ + SO_4^{2-} = LiSO_4^-$	—	0.64	$Fe^{2+} + SO_4^{2-} = FeSO_4^0$	3.23	2.25
$Na^+ + SO_4^{2-} = NaSO_4^-$	1.12	0.70	$Fe^{2+} + HSO_4^- = FeHSO_4^+$	—	1.08
$K^+ + SO_4^{2-} = KSO_4^-$	2.25	0.85	$Fe^{3+} + SO_4^{2-} = FeSO_4^-$	3.91	4.04
$Ca^{2+} + SO_4^{2-} = CaSO_4^0$	1.65	2.30	$Fe^{3+} + 2SO_4^{2-} = Fe(SO_4)_2^-$	4.60	5.38
$Mg^{2+} + SO_4^{2-} = MgSO_4^0$	4.55	2.37	$Fe^{3+} + HSO_4^- = FeHSO_4^{2+}$	—	2.48
$Sr^{2+} + SO_4^{2-} = SrSO_4^0$	2.08	2.29	$Al^{3+} + SO_4^{2-} = AlSO_4^-$	2.15	3.02
$Ba^{2+} + SO_4^{2-} = BaSO_4^0$	—	2.7	$Al^{3+} + 2SO_4^{2-} = Al(SO_4)_2^-$	2.84	4.92
$Ra^{2+} + SO_4^{2-} = RaSO_4^0$	1.3	2.75	$Al^{3+} + HSO_4^- = AlHSO_4^{2-}$	—	0.46

Table 7.1. (Continued)

Mineral	Reaction	ΔH_r^0 (kcal mol^{-1})	log K
Gypsum	$CaSO_4 \cdot 2H_2O = Ca^{2+} + SO_4^{2-} + 2H_2O$	-0.109	-4.58
Anhydrite	$CaSO_4 = Ca^{2+} + SO_4^{2-}$	-1.71	-4.36
Celestite	$SrSO_4 = Sr^{2+} + SO_4^{2-}$	-1.037	-6.63
Barite	$BaSO_4 = Ba^{2+} + SO_4^{2-}$	6.35	-9.97
Radium sulfate	$RaSO_4 = Ra^{2+} + SO_4^{2-}$	9.40	-10.26
Melanterite	$FeSO_4 \cdot 7H_2O = Fe^{2+} + SO_4^{2-} + 7H_2O$	4.91	-2.209
Alunite	$KAl_3(SO_4)_2(OH)_6 + 6H^+ = K^+ + 3Al^{3+} + 2SO_4^{2-} + 6H_2O$	-50.25	-1.4

Silicate Species

Reaction	ΔH_r^0 (kcal mol^{-1})	log K
$Si(OH)_4^0 = SiO(OH)_3^- + H^+$	6.12	-9.83
$Si(OH)_4^0 = SiO_2(OH)_2^{2-} + 2H^+$	17.6	-23.0

Mineral	Reaction	ΔH_r^0 (kcal mol^{-1})	log K
Kaolinite	$Al_2Si_2O_5(OH)_4 + 6H^+ = 2Al^{3+} + 2Si(OH)_4^0 + H_2O$	-35.3	7.435
Chrysotile	$Mg_3Si_2O_5(OH)_4 + 6H^+ = 3Mg^{2+} + 2Si(OH)_4^0 + H_2O$	-46.8	32.20
Sepiolite	$Mg_2Si_3O_{7.5}(OH) \cdot 3H_2O + 4H^+ + 0.5\ H_2O = 2Mg^{2+} + 3Si(OH)_4^0$	-10.7	15.76
Kerolite	$Mg_3Si_4O_{10}(OH)_2 \cdot H_2O + 6H^+ + 3H_2O = 3Mg^{2+} + 4Si(OH)_4^0$	—	25.79
Quartz	$SiO_2 + 2H_2O = Si(OH)_4^0$	5.99	-3.98
Chalcedony	$SiO_2 + 2H_2O = Si(OH)_4^0$	4.72	-3.55
Amorphous silica	$SiO_2 + 2H_2O = Si(OH)_4^0$	3.34	-2.71

A complete set of the thermodynamic data of Nordstrom et al. is given in Appendix 1 at the end of the book.
$^a CO_2(aq) = H_2CO_3^*$.

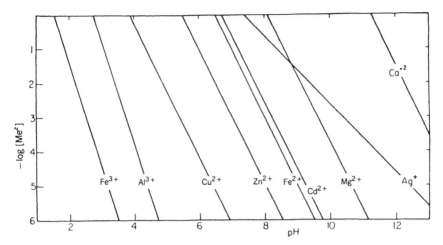

Figure 7.3. Solubility of oxides and hydroxides. Free metal-ion concentration in equilibrium with solid oxides or hydroxides. The occurrence of hydroxo metal complexes must be considered for evaluation of complete solubility.

Example 7.4. Solubility of Hydrous $Cr(OH)_3(s)$ Estimate the solubility of Cr(III) (hydr)oxide as a function of pH. There are thermodynamic data available for the mineral Cr_2O_3 (eskolaite) and for a hydrous $Cr(OH)_3$.

The latter is encountered when Cr(III) precipitates (e.g., by reduction of CrO_4^{2-}). This metastable $Cr(OH)_3(s)$ is more likely to regulate the solubility in wastewater circumstances.

We can use the following equilibrium constants (25°C) (Baes and Messmer, 1976):

Reaction	log K ($I = 0$)	log K ($I = 0.01$)
$Cr(OH)_3(s) = Cr^{3+} + 3\ OH^-$	−30.0	−29.4
$Cr^{3+} + OH^- = CrOH^{2+}$	10	9.8
$Cr^{3+} + 2\ OH^- = Cr(OH)_2^+$	18.3	17.9
$Cr^{3+} + 3\ OH^- = Cr(OH)_3(aq)$	24.0	23.7
$Cr^{3+} + 4\ OH^- = Cr(OH)_4^-$	28.6	28.1
$3\ Cr^{3+} + 4\ OH^- = Cr_3(OH)_4^{5+}$	47.8	47.5
$H^+ + OH^- = H_2O$	14.0	13.91

and establish Tableau 7.1.

Figure 7.4 gives the solubility of amorphous $Cr(OH)_3(s)$ in accordance with the equilibria given in Tableau 7.1. In line with the equations in this tableau, the lines characterizing the logarithmic concentrations of Cr^{3+}, $CrOH^{2+}$, $Cr(OH)_2^+$, $Cr(OH)_3$, $Cr(OH)_4^-$, and $Cr_3(OH)_4^{5+}$, as a function of pH, have slopes

Tableau 7.1. Solubility of $Cr(OH)_3(s)$

Components		$Cr(OH)_3(s)$	H^+	log K ($I = 0.01$)
Species	Cr^{3+}	1	3	12.3 (*K_{s0})
	$CrOH^{2+}$	1	2	8.2 (*K_{s1})
	$Cr(OH)_2^+$	1	1	2.4 (*K_{s2})
	$Cr(OH)_3(aq)$	1	0	−5.7 (*K_{s3})
	$Cr(OH)_4^-$	1	−1	−15.2 (*K_{s4})
	$Cr_3(OH)_4^{5+}$	3	5	28.8 (*K_{s43})
	OH^-	0	−1	−13.9
	H^+	0	1	0
		{$Cr(OH)_3(s)$} = 1	pH given	

$Cr(III)_T = [Cr^{3+}] + [CrOH^{2+}] + [Cr(OH)_2^+] + [Cr(OH)_3] + [Cr(OH)_4^-]$
$+ 3 [Cr_3(OH)_4]^{5+}$

of −3, −2, −1, 0 +1, and −5. The intercepts are also defined readily; for example, for $Cr(OH)_2^+$, log $[Cr(OH)_2^+] = 0$ when pH = log *K_{s2}; for $CrOH^{2+}$, log $[CrOH^{2+}] = 0$ when pH = $\frac{1}{2}$ log *K_{s1}. Summing up all the soluble Cr(III) species indicated by a thick line gives the overall solubility as a function of pH. The solubility of $Cr(OH)_3(s)$ is affected markedly on the acid side by the polynuclear species $Cr_3(OH)_4^{5+}$.

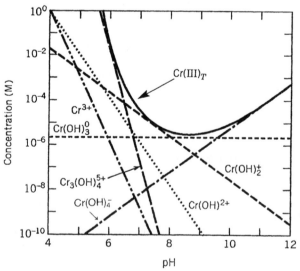

Figure 7.4. Solubility of precipitated $Cr(OH)_3(s)$ (25°C, $I = 10^{-2}$ M) in accord with Tableau 7.1.

7.2 The Solubility of Oxides and Hydroxides

Tableau 7.2. Dissolution of Amorphous SiO₂

Components		SiO_2(s, amorphous)	H^+	log K (25°C, $I = 0.5$)
Species	$Si(OH)_4$	1	0	−2.7
	$SiO(OH)_3^-$	1	−1	−12.16
	$SiO(OH)_2^{2-}$	1	−2	−24.72
	$Si_4O_6(OH)_6^{2-}$	4	−2	−23.37
	H^+	0	1	0
		$\{SiO_2(s)\} = 1$	pH given	

$Si_T = [Si(OH)_4] + [SiO(OH)_3^-] + [SiO(OH)_2^{2-}] + 4\,[Si_4O_6(OH)_6^{2-}]$

Example 7.5. The Chemistry of Aqueous Silica The solubility of SiO_2 can be characterized by the following equilibria:

$$SiO_2(s, \text{quartz}) + 2H_2O = Si(OH)_4 \qquad \log K = -3.7 \qquad \text{(i)}$$

$$SiO_2(s, \text{amorphous}) + 2H_2O = Si(OH)_4 \qquad \log K = -2.7 \qquad \text{(ii)}$$

$$Si(OH)_4 = SiO(OH)_3^- + H^+ \qquad \log K = -9.46 \qquad \text{(iii)}$$

$$SiO(OH)_3^- = SiO_2(OH)_2^{2-} + H^+ \qquad \log K = -12.56 \qquad \text{(iv)}$$

$$4Si(OH)_4 = Si_4O_6(OH)_6^{2-} + 2H^+ + 4H_2O \qquad \log K = -12.57 \qquad \text{(v)}$$

The equilibrium constants given are valid at 25°C. Data are those given by Lagerstrom as valid for 0.5 M NaClO₄.[†]

In these equations silicic acid is written as $Si(OH)_4$ (rather than as H_4SiO_4) in order to emphasize that metalloids (Si, B, Ge) similar to metal ions have a tendency to coordinate with hydroxo and oxo ligands. Like multivalent metal ions, such metalloids tend to form multinuclear species.

The rate of crystallization of quartz is so slow in the low-temperature range that the solubility of amorphous silica represents the upper limit of dissolved aqueous silica (see Tableau 7.2).

Estimate the solubility of amorphous silica.

The equilibrium data of equations ii–v permit computation of the solubility of amorphous SiO_2 for the entire pH range and the relative concentration of the species in equilibrium with amorphous SiO_2 (Figure 7.5). Only $Si(OH)_4$ occurs within neutral and slightly alkaline pH ranges. Under alkaline conditions, the solubility of SiO_2 becomes enhanced because of the formation of

---

[†]Nordstrom et al. (1990) (see Table 7.1) list the following equilibrium constants for $I = 0$, 25°C. Equilibrium i: log $K = -3.98$; ii: log $K = -2.71$; iii: log $K = -9.83$; iv: log $K = -13.17$.

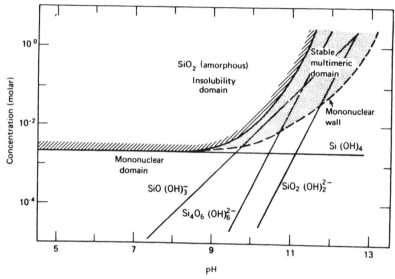

Figure 7.5. Species in equilibrium with amorphous silica. Diagram computed from equilibrium constants (25°C, $I = 0.5$). The line surrounding the shaded area gives the maximum soluble silica. The mononuclear wall represents the lower concentration limit below which multinuclear silica species are not stable. In natural waters the dissolved silica is present as monomeric silicic acid.

monomeric and multimeric silicates. Although there is some uncertainty concerning the exact nature of the multimeric species, the experimental data leave no doubt about their existence. Even if multinuclear species other than $Si_4O_6(OH)_6^{2-}$ were present, the solubility characteristics of SiO_2 would not be changed markedly.

7.3 COMPLEX FORMATION AND SOLUBILITY OF (HYDR)OXIDES

We have already seen that the solubility of solid phases can be enhanced by ligands; for example, for metal oxides, considering OH^- as ligands, only the total solubility as a function of pH is given by

$$Me_T = [Me^{+q}] + \sum [Me(OH)_n^{q-n}] \tag{24}$$

This equation can now be generalized if we consider the possibility of complex formation with any ligand, L, or its protonated form H_jL. Then the total solubility is given by

$$Me_T = [Me]_{free} + \sum Me[OH]_n + m \sum [Me_m H_k L_n[OH]_i] \tag{25}$$

7.3 Complex Formation and Solubility of (Hydr)oxides

Tableau 7.3. Equilibria in the ZnO(s), Oxalate System (Oxalate ≡ Ox)

Components	Ox^{2-}	ZnO(s)	H^+	log K
Species Zn^{2+}	0	1	2	11.2
$ZnOH^+$	0	1	1	2.2
$Zn(OH)_2$	0	1	0	−5.7
$Zn(OH)_3^-$	0	1	−1	−16.9
$Zn(OH)_4^{2-}$	0	1	−2	−29
Ox^{2-}	1	0	0	0
HOx^-	1	0	1	4.2
H_2Ox	1	0	2	5.7
$ZnOx$	1	1	2	16
$ZnOx_2^{2-}$	2	1	2	18.9
H^+	0	0	1	0

$$\{ZnO(s)\} = 1$$

$$TOTOx = [H_2Ox] + [HOx^-] + [Ox^{2-}] + [ZnOx] + 2[ZnOx_2^{2-}]$$
$$= 10^{-4} \text{ M} \tag{i}$$

$$Zn(II)_T = [Zn^{2+}] + [ZnOH^+] + [Zn(OH)_2] + [Zn(OH)_3^-]$$
$$+ [Zn(OH)_4^{2-}] + [ZnOx] + [ZnOx_2^{2-}] \tag{ii}$$

where L represents the different ligand types and where all values of m, n, i, or $k \geq 0$ have to be considered in the summation.

Example 7.6. Effect of 10^{-4} M Oxalate on Solubility of ZnO(s) How does the addition of oxalate affect the solubility of ZnO(s)? The data needed for the calculation are given in Tableau 7.3 and the results are displayed in Figure 7.6. A comparison of Figure 7.6a with Figure 7.6b shows that the solubility is enhanced in the pH region 6–8.

Example 7.7. Effect of EDTA on the Solubility of $Fe(OH)_3(s)$ EDTA (ethylenediaminetetraacetate) is a very powerful complex former, a hexadentate (two nitrogen donor atoms and four oxygen donor atoms from the four carboxyl groups). A comparison of Figure 7.7a with Figure 6.8b[†] shows that EDTA dissolves hydrous ferric oxide, under the conditions given, up to pH = 9. It is important however, to consider that in the presence of an excess of Ca^{2+}, as typically occurs in natural waters, CaEDTA is formed.

$$Ca^{2+} = Y^{4-} = CaY^{2-} \quad \log K = 10.7 \tag{i}$$

[†]Figure 6.8b describes the solubility of amorphous FeOOH(s); its solubility is slightly different from the amorphous $Fe(OH)_3(s)$ used in this example.

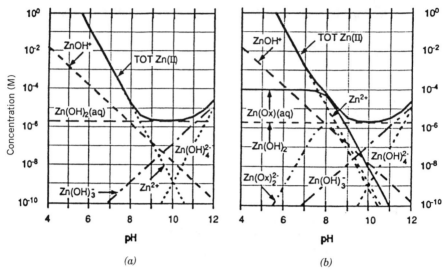

Figure 7.6. Effect of oxalate on the solubility of ZnO(s) (a) solubility of ZnO(s) and (b) solubility of ZnO(s) in presence of 10^{-4} M oxalate.

The stability of the FeY^- complex is much larger:

$$Fe^{3+} + Y^{4-} = FeY^- \qquad \log K = 25.1 \qquad \text{(ii)}$$

Thus the actual ligand is CaY^{2-} rather than Y^{2-}:

$$CaY^{2-} + Fe^{3+} = FeY^- + Ca^{2+} \qquad \log K = \log \frac{K_{ii}}{K_i} = 14.4 \qquad \text{(iii)}$$

As Figure 7.7b, calculated with the help of Tableau 7.4, shows, EDTA in the presence of an excess of Ca^{2+} has little effect in dissolving $Fe(OH)_3(s)$ above pH = 7.

7.4 CARBONATES

In the $Me^{2+}-H_2O-CO_2$ system in the presence of the earth's atmosphere, carbonates are frequently more stable than oxides or hydroxides as solid phases. Thus in natural water systems the concentration of some metal ions is controlled by the solubility of metal carbonates.

We already discussed solubility equilibria with solid carbonates in Section

7.4 Carbonates 371

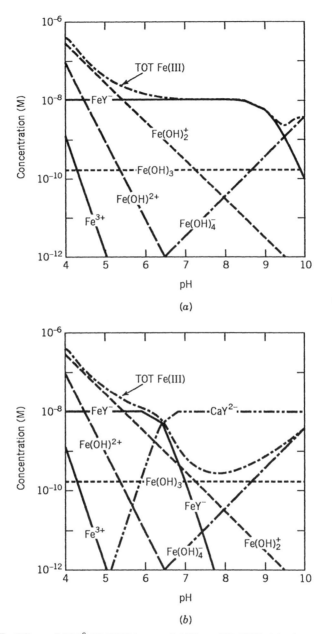

Figure 7.7. Effect of 10^{-8} M EDTA on solubility of $Fe(OH)_3(s)$. As part (b) shows, the effect of EDTA on solubilization of $Fe(OH)_3(s)$ is diminished in presence of 10^{-2} M Ca^{2+}.

Tableau 7.4. Effect of EDTA and Ca^{2+} on the Solubility of $Fe(OH)_3(s)$

Components		Y^{-4}	Ca^{2+}	$Fe(OH)_3(s)$	H^+	log K
Species	Fe^{3+}	0	0	1	3	3.2
	$FeOH^{2+}$	0	0	1	2	1.0
	$Fe(OH)_2^+$	0	0	1	1	−2.5
	$Fe(OH)_3(aq)$	0	0	1	0	−9.8
	$Fe(OH)_4^-$	0	0	1	−1	−18.5
	FeY^-	1	0	1	3	28.3
	Ca^{2+}	0	1	0	0	0
	CaY^{2-}	1	1	0	0	10.7
	HY^{3-}	1	0	0	1	10.26
	H_2Y^{2-}	1	0	0	2	16.42
	H_3Y^-	1	0	0	3	19.09
	H_4Y^a	1	0	0	4	21.08
	H^+	0	0	0	1	0

$$\{Fe(OH)_3(s)\} = 1$$
$$TOTY = [H_4Y] + [H_3Y^+] + [H_2Y^{2-}] + [HY^{3-}] + [Y^{4-}] + [FeY^-]$$
$$+ [CaY^{2-}] = 10^{-8}$$
$$TOTCa = [Ca^{2+}] + [CaY^2] = 10^{-2}\ M$$

$^a H_4Y$ is the abbreviation for the tetraprotic acid H_4EDTA species.

4.7. The equilibrium constants used to characterize solubility equilibria are summarized in Table 7.2 and for the $CaCO_3$ (calcite) system in Table 4.3.)† The various solubility expressions (6–11, Table 7.2) are interrelated and can all be expressed in terms of the conventional solubility product K_{s0}. A listing of the different formulations should indicate merely that the solubility can be characterized by different experimental variables. For example, we can fully define a solubility equilibrium with a solid carbonate by p_{CO_2}, $[Me^{2+}]$, and $[H^+]$ equation 10, Table 7.2; by p_{CO_2}, $[Me^{2+}]$, and $[HCO_3^-]$ (Equation 9, Table 7.2); or by $[H^+]$, $[Me^{2+}]$, and $[HCO_3^-]$ (Equation 7, Table 7.2). Parameters such as these are more accessible to direct analytical determination than $[CO_3^{2-}]$.

Freshwater and Marine Limestones

The simplest process of solid carbonate formation is the precipitation from fresh water. Usually the process is correlated by loss of CO_2 (due to photosynthesis, or escape). Fresh water does not sustain extreme supersaturations

---

†The reader will observe that equilibrium constants for the same reactions may differ depending on the source of the data. (Often, different stability constants for ion pairs are the reason for the difference.) For the solubility of calcite and aragonite in seawater, see Table A6.4.

7.4 Carbonates

Table 7.2. Equilibrium Constants Defining the Solubility of Carbonates

I. Equilibrium Among Solutes and $CO_2(g)$

$H_2O = H^+ + OH^-$	K_W	(1)
$CO_2(g) + H_2O = H_2CO_3^*(aq)$	K_H	(2)
$CO_2(g) + H_2O = HCO_3^- + H^+$	$K_{p1} = K_H K_1$	(3)
$H_2CO_3^* = HCO_3^- + H^+$	K_1	(4)
$HCO_3^- = CO_3^{2-} + H^+$	K_2	(5)

II. Solid–Solutions Equilibria

$MeCO_3(s) = Me^{2+} + CO_3^{2-}$	K_{s0}	(6)
$MeCO_3(s) + H^+ = Me^{2+} + HCO_3^-$	$*K_s = K_{s0} K_2^{-1}$	(7)
$MeCO_3(s) + H_2CO_3^* = Me^{2+} + 2HCO_3^-$	$^+K_{s0} = K_{s0} K_1 K_2^{-1}$	(8)
$MeCO_3(s) + H_2O + CO_2(g) = Me^{2+} + 2HCO_3^-$	$^+K_{ps0} = {^+}K_{s0} K_H$	(9)
$MeCO_3(s) + 2H^+ = Me^{2+} + CO_2(g) + H_2O$	$*K_{ps0} = K_{s0} K_2^{-1} K_1^{-1} K_H^{-1}$	(10)
$MeCO_3(s) + 2H^+ = Me^{2+} + H_2CO_3^*$	$*K_{s0} = K_{s0} K_2^{-1} K_1^{-1}$	(11)

with respect to $CaCO_3$. In most cases, the phase precipitated is calcite, the polymorph of $CaCO_3$ characterized by the lowest solubility product. Carbonate solubility in seawater is governed essentially by the same parameters but kinetically the precipitation of calcium carbonate is complicated by the presence of high concentrations of Mg^{2+}; much of the $CaCO_3$ is in the form of hard parts in calcareous organisms (e.g., foraminifera, cocolithophores). Furthermore, other marine carbonates (aragonite, magnesian calcite) are metastable phases with their origin in the reluctance of the Mg^{2+} to be dehydrated and converted to dolomite, $MgCa(CO_3)_2$, and magnesite, $MgCO_3$.

In treating heterogeneous equilibria of the system Me^{2+}–CO_2–H_2O, it is important to distinguish two cases: (1) systems that are closed to the atmosphere [we consider only the solid phase and the solution phase (Figure 4.2b); i.e., we treat $H_2CO_3^*$ as a nonvolatile acid] and (b) systems that include a (CO_2-containing) gas phase in addition to the solid and solution phase. For each of the two cases a few representative models will be discussed. Because of its significance in natural water systems, calcite will be used as an example of the solid phase.

Model I: System Closed to Atmosphere

$H_2CO_3^*$ is treated as a nonvolatile species (slow gas transfer).

(a) Solubility of $CaCO_3$ for C_T = Constant What is the maximum soluble metal-ion concentration as a function of C_T and pH? This example is analytically quite important. We have a water of a given analytical composition, and

374 Precipitation and Dissolution

we inquire whether the water is oversaturated or undersaturated with respect to a solid metal carbonate. In other words, we compute the equilibrium solubility. Numerically, for any pH and C_T, the equilibrium solubility must be maintained. In the case of calcite,

$$[Ca^{2+}] = \frac{K_{s0}}{[CO_3^{2-}]} = \frac{K_{s0}}{C_T \alpha_2} \tag{26}$$

Because α_2 is known for any pH, equation 26 gives the equilibrium saturation value of Ca^{2+} as a function of C_T and pH. An analogous type of equation can be written for any $[Me^{2+}]$ in equilibrium with $MeCO_3(s)$. Equation 26 is amenable to simple graphical representation in a $\log[Me^{2+}]$ versus pH diagram (Figure 7.8) (See also Figure 4.16.)

The graphical representation consists essentially of a superimposition of the equation for the solubility product upon the carbonate equilibria (cf. Figure 4.1); the product of $[Ca^{2+}]$ and $[CO_3^{2-}]$ must be constant (K_{s0}) (Figure 7.8,

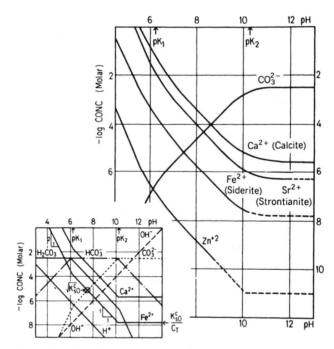

Figure 7.8. Model Ia. Saturation values of Me^{2+} for closed carbonate system with C_T = constant (= 3×10^{-3} M). The diagram gives the maximum soluble $[Me^{2+}]$ as a function of pH for a given C_T. Dashed portions of the curves indicate conditions under which $MeCO_3(s)$ is not thermodynamically stable. The inset gives the essential features for the construction of the diagram.

inset). Thus at high pH (pH > pK_2), where the $\log[CO_3^{2-}]$ line has a slope of zero, the $\log[Ca^{2+}]$ line must also have a slope of zero. Here the saturation concentration $[Ca^{2+}]$ must be equal to $K_{s0}/[CO_3^{2-}]$. In the region where $pK_1 <$ pH $< pK_2$, $\log[CO_3^{2-}]$ has a slope of $+1$; correspondingly, $\log[Ca^{2+}]$ must have a slope of -1. At pH $< pK_1$, $\log[CO_3^{2-}]$ has a slope of $+2$; in order to maintain constancy of the product $[Ca^{2+}][CO_3^{2-}]$, $\log[Ca^{2+}]$ must have a slope of -2. Figure 7.8 shows the equilibrium saturation values of a few metal ions with respect to their carbonates for $C_T = 10^{-2.5}$ M. Such a diagram is well suited for comparing the solubility of various metal carbonates and their pH dependence. Note that C_T is controlled (and kept constant) independent of $MeCO_3$ dissolution.

(b) Dissolution of $CaCO_3(s)$ in Pure Water The following solute species will be encountered: Ca^{2+}, $H_2CO_3^*$, HCO_3^-, CO_3^{2-}, H^+, and OH^-. Since we have six unknowns, at a given pressure and temperature, we need six equations to define the solution composition. Four mass laws interrelate the equilibrium concentrations of the solutes [first and second acidity constant of $H_2CO_3^*$, ion product of water, and solubility product of $CaCO_3(s)$]. An additional equation obtains if one considers that all Ca^{2+} that becomes dissolved must equal in concentration the sum of the dissolved carbonic species

$$[Ca^{2+}] = C_T \tag{27a}$$

Furthermore, the solution must fulfill the condition of electroneutrality:

$$2[Ca^{2+}] + [H^+] = [HCO_3^-] + 2[CO_3^{2-}] + [OH^-] \tag{27b}$$

The set of six equations has to be solved simultaneously. It may be convenient to start with the solubility product

$$[Ca^{2+}] = \frac{K_{s0}}{[CO_3^{2-}]} \tag{28}$$

Using ionization fractions (see Table 4.2), we have

$$[CO_3^{2-}] = C_T \alpha_2 \quad [HCO_3^-] = C_T \alpha_1 \quad [H_2CO_3^*] = C_T \alpha_0$$

Equation 28, rewritten as $[Ca^{2+}] = K_{s0}/C_T \alpha_2$, can be combined with equation 27a to give

$$[Ca^{2+}] = C_T = \left(\frac{K_{s0}}{\alpha_2}\right)^{0.5} \tag{29}$$

376   Precipitation and Dissolution

Tableau 7.5. Solubility of CaCO₃ (Calcite) in Pure Water

Components		CO_3^{2-}	H^+	$CaCO_3(s)$	log K	Equilibrium Concentration (M) (25°C)
Species	Ca^{2+}	-1	0	1	-8.42	1.146×10^{-4}
	CO_3^{2-}	1	0	0	0.00	3.319×10^{-5}
	HCO_3^-	1	1	0	10.30	8.135×10^{-5}
	$H_2CO_3^*$	1	2	0	16.60	1.994×10^{-8}
	OH^-	0	-1	0	-14.00	8.139×10^{-5}
	H^+	0	1	0	0.00	1.229×10^{-10} (pH = 9.91)
Composition		0	0	$\{CaCO_3(s)\} = 1$		

$[Ca^{2+}] = [CO_3^{2-}] + [HCO_3^-] + [H_2CO_3^*]$

$TOTH^+ = [Acy] = 2[H_2CO_3^*] + [HCO_3^-] + [H^+] - [OH^-] = 0$

This can be substituted into the charge condition equation 27b:

$$\left(\frac{K_{s0}}{\alpha_2}\right)^{0.5} (2 - \alpha_1 - 2\alpha_2) + [H^+] - \frac{K_W}{[H^+]} = 0 \tag{30}$$

Equation 30 can be solved by trial and error.

Tableau 7.5 summarizes the equilibrium problem for calcite and gives the calculated values.

(c) CaCO₃ Plus Acid or Base The addition of acid C_A or base C_B will change the pH and the solubility relations. The problem is analogous to the alkalimetric or acidimetric titration of a CaCO₃ suspension. The addition of acid (i.e., $C_A = [HCl]$) or base ($C_B = [NaOH]$) will shift the charge balance (equation 30) to the following:

$$C_A - C_B = \left(\frac{K_{s0}}{\alpha_2}\right)^{0.5} (2 - \alpha_1 - 2\alpha_2) + [H^+] - \frac{K_W}{[H^+]} \tag{31}$$

With the help of this equation, it is possible to compute the quantity of acid or base, C_A or C_B, needed per liter of solution [which remains in equilibrium with CaCO₃(s)] to reach a given pH value (Figure 7.9). For each pH value, the solubility of CaCO₃(s), for example, $[Ca^{2+}]$ or C_T, can be calculated with equation 29, which is still valid under our assumptions. Then the equilibrium concentrations of the other solutes can readily be obtained.

It is convenient to plot equation 29 graphically in a logarithmic concentra-

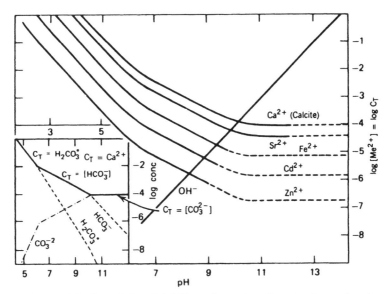

Figure 7.9. Models Ib and Ic. Solubility of metal carbonates in a closed system: [Me^{2+}] = C_T. The inset gives the essential features for the construction of the diagram for CaCO$_3$(s) and equilibrium concentrations of all the carbonate species. A suspension of pure MeCO$_3$(s) ($C_B - C_A = 0$) is characterized by the intersection of [OH$^-$] and [Me^{2+}] = C_T. Dashed portions of the curves indicate conditions under which MeCO$_3$(s) is not thermodynamically stable.

tion–pH diagram. The construction of the plot is facilitated by considering that the following conditions exist in the various pH regions:

For pH > pK_2:

$$d\log\frac{\alpha_2}{d\,\text{pH}} = 0 \quad d\log\frac{C_T}{d\,\text{pH}} = 0 \quad -\log C_T = \frac{1}{2}\text{p}K_{s0} \quad (32)$$

For pK_1 < pH < pK_2:

$$d\log\frac{\alpha_2}{d\,\text{pH}} = +1 \quad d\log\frac{C_T}{d\,\text{pH}} = -\frac{1}{2} \quad (33)$$

For pH < pK_1:

$$d\log\frac{\alpha_2}{d\,\text{pH}} = +2 \quad d\log\frac{C_T}{d\,\text{pH}} = -1 \quad (34)$$

Figure 7.9 gives solubility diagrams for some other metal carbonates.

Model II: Open System in Equilibrium with $CO_2(g)$

(a) *$CaCO_3(s)$–CO_2–H_2O* We open the system previously discussed to the atmosphere. Specifically, we prepare a solution by adding $CaCO_3(s)$ (calcite) to pure H_2O and expose this solution to a gas phase containing CO_2. This model is representative of conditions encountered typically in fresh water. For our example we select a partial pressure of CO_2 corresponding to that of the atmosphere ($-\log p_{CO_2} = 3.5$). We can write, because of the additional phase, in addition to the four mass laws, an independent relationship for the solubility of CO_2 in the aqueous solution (e.g., Henry's law). If we specify the temperature, total pressure, and a partial pressure of CO_2, the system is completely defined. Furthermore, an electroneutrality condition can be formulated. This equilibrium problem has been defined concisely in Tableau 4.5.

Because of the equilibrium with CO_2 in the gas phase, $[Ca^{2+}]$ is no longer equal to C_T, but the same electroneutrality condition (equation 27b) pertains:

$$2[Ca^{2+}] + [H^+] = C_T(\alpha_1 + 2\alpha_2) + [OH^-] \qquad (35)$$

Furthermore, $[Ca^{2+}]$ can be expressed as a function of C_T and $[H^+]$, and C_T is defined by p_{CO_2},

$$[Ca^{2+}] = \frac{K_{s0}}{C_T \alpha_2} \qquad (36)$$

$$C_T = \frac{K_H p_{CO_2}}{\alpha_0} \qquad (37)$$

$$[CO_3^{2-}] = \frac{K_H p_{CO_2} \alpha_2}{\alpha_0} \qquad (38)$$

$$[Ca^{2+}] = \frac{(K_{s0}/K_H \, p_{CO_2})\alpha_0}{\alpha_2} \qquad (39)$$

With the substitution of equations 37 and 39, equation 35 can now be solved for $[H^+]$. It is probably more convenient to plot the various equililbria double logarithmically, as discussed in Sections 4.3 and 4.7. The results can be read from the graph (Figures 4.15 and 7.10). Comparing this result with that of Model 1b (Figure 7.9), we see that the influence of atmospheric CO_2 has depressed the pH markedly and that $[Ca^{2+}]$ and $[Alk]$ have been raised to values very representative of those in natural waters.

(b) *$CaCO_3$–H_2O–CO_2 Plus Acid or Base* pH changes can occur upon addition of acids or bases (e.g., addition of wastes, dissolution of volcanic volatile compounds like HCl, biological reactions) to the model discussed before. The electroneutrality condition can be formulated most generally by considering that the charge balance of equation 35 has been shifted because of the addition

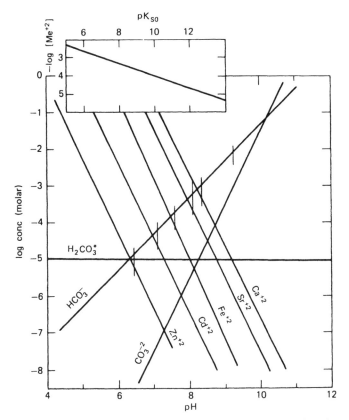

Figure 7.10. Models IIa and IIb. Solubility of MeCO$_3$(s) as a function of pH at constant p_{CO_2}. Here $-\log p_{CO_2} = 3.5$ (corresponding to partial pressure of CO$_2$ in atmosphere). If no excess acid or base is added ($C_B - C_A = 0$), the equilibrium composition of the solution is given by the electroneutrality condition $2[\text{Me}^{2+}] \simeq [\text{HCO}_3^-]$. This condition is indicated by a vertical dash slightly displaced from the intersection of [Me^{2+}] with HCO$_3^-$]. The inset gives $-\log[\text{Me}^{2+}]$ for pure MeCO$_3$(s) suspensions in equilibrium with $p_{CO_2} = 10^{-3.5}$ atm as a function of pK_{s0} (see p. 387).

of C_A (acid) or C_B (base):

$$C_B + 2[\text{Ca}^{2+}] + [\text{H}^+] = C_T(\alpha_1 + 2\alpha_2) + [\text{OH}^-] + C_A \quad (40)$$

Since [Ca^{2+}] and C_T in this equation can again be expressed as a function of p_{CO_2} and [H$^+$], the "titration curve" of a CaCO$_3$ suspension that remains in equilibrium with CO$_2$(g) can be computed.

Equation 39 can be plotted graphically (Figure 7.10). The distribution of the dissolved carbonate species at a given temperature is defined entirely by p_{CO_2}. The log[Me^{2+}] line has a slope of -2 with respect to pH. Differentiating equation 39 with respect to pH and considering that $d \log(\alpha_0/\alpha_2)/d \text{ pH} = -2$ leads to a slope of -2 with respect to pH for the log [Me^{2+}] line in the diagram.

Example 7.8. Calcite in Seawater Compare the composition of a $CaCO_3(s)$ (calcite)–CO_2–H_2O "seawater" model system, made by adding calcite to pure H_2O containing the seawater electrolytes (but incipiently no Ca^{2+} and no carbonates and, for simplicity, no borate) and by equilibrating this solution at 25°C and 1 atm total pressure with the atmosphere ($p_{CO_2} = 3.55 \times 10^{-4}$ atm), with the composition of a real surface seawater whose carbonate alkalinity, Ca(II) concentration, and pH have been determined as 2.4×10^{-3} eq liter^{-1}, 1.06×10^{-2} M, and 8.2, respectively. Estimate the extent of oversaturation of this seawater with respect to calcite. The solubility of calcite at 25°C is taken as $^cK'_{s0} = [Ca_T][CO_{3T}^{2-}] = 5.94 \times 10^{-7}$, where $[Ca_T]$ and $[CO_{3T}^{2-}]$ are the concentration of total soluble Ca(II) ($[Ca^{2+}]$ plus concentration of Ca complexes with medium ions) and of total soluble carbonate ($[CO_3^{2-}]$ and concentration of carbonate complexes with medium ions), respectively. The other constants needed, Henry's law constant and the acidity constant of $H_2CO_3^*$, are taken as[†]

$$p^cK_H = 1.53 \qquad pK'_1 = 6.00 \qquad pK'_2 = 9.11$$

where

$$K'_1 = \frac{\{H^+\}[HCO_{3T}^-]}{[H_2CO_3^*]} \quad \text{and} \quad K'_2 = \frac{\{H^+\}[CO_{3T}^{2-}]}{[HCO_{3T}^-]}$$

Tableau 7.6 summarizes the equilibria and the constants used in this model system (compare Tableau 4.5).

The proton condition corresponds also to the electroneutrality condition

$$TOTH = -[HCO_3^-] - 2[CO_3^{2-}] + 2[Ca^{2+}] + [H^+] - [OH^-] = 0 \quad \text{(i)}$$

Tableau 7.6. Saturation of Seawater with Calcite

Components		$CaCO_3(s)$	$CO_2(g)$	H^+	log K (25°C seawater)
Species	$H_2CO_3^*$	0	1	0	−1.53
	HCO_{3T}^-	0	1	−1	−7.53
	CO_{3T}^{2-}	0	1	−2	−16.64
	Ca_T^{2+}	1	−1	2	10.4
	H^+	0	0	1	0
		$\{CaCO_3(s)\} = 1$	$p_{CO_2} = 3.55 \times 10^{-4}$	0	

[†]Note that the constants given in Appendix 6.2, Table A6.3, are slightly different because a different pH convention (NBS pH scale) is used.

A MICROQL calculation† gives the following results:

pH = 8.34
$[Ca_T^{2+}] = 1.51 \times 10^{-3}$ M
$[HCO_{3T}^-] = 2.77 \times 10^{-3}$ M
$[CO_{3T}^{2-}] = 3.80 \times 10^{-4}$
$[H_2CO_3^*] = 1.05 \times 10^{-5}$ M
[Alk] = 3.03×10^{-3} M

For surface seawater with pH = 8.2 and [Carb-Alk]‡ = 2.4×10^{-3} eq liter^{-1}, we obtain for $[CO_{3T}^{2-}] = C_T \alpha_2 = $ [Carb-Alk]$\alpha_2/(\alpha_1 + 2\alpha_2) = 3.87 \times 10^{-4}$ M. Since $[Ca_T] = 1.06 \times 10^{-2}$ M, we obtain for the ion concentration product

$$[Ca_T]_{act}[CO_{3T}^{2-}]_{act} = 4.1 \times 10^{-6}$$

This may be compared with the equilibrium solubility product $^cK'_{s0} = 5.94 \times 10^{-7}$

$$\frac{[Ca_T]_{act}[CO_{3T}^{2-}]_{act}}{[Ca_T]_{eq}[CO_{3T}^{2-}]_{eq}} = \frac{[Ca_T][CO_{3T}^{2-}]}{^cK'_{s0}} = 6.9 \quad \text{(vi)}$$

Thus the surface seawater, at 25°C (p = 1 atm), is oversaturated by a factor of ~7 with respect to calcite. The model equilibrium system has a slightly higher pH value and a $[Ca_T]$ approximately seven times smaller than that of the surface seawater.

Example 7.9. Effect of Pressure and Temperature on Calcite Solubility in Seawater An enclosed sample of the surface seawater, as discussed in Example 7.8 (25°C; pH = 8.2; $[Ca^{2+}] = 1.06 \times 10^{-2}$ M; [Carb-Alk] = 2.4×10^{-3} eq liter^{-1}), is cooled to 5°C and then subjected to increases in total pressure of up to 1000 atm (equivalent to exposing the sample to increased water depths of approximately 10,000 m). How does the composition, pH, $[CO_3^{2-}]$, $[Ca^{2+}]$, and extent of oversaturation change as a result of the temperature change at 1 atm and as a result of the pressure change at 5°C? The water is incipiently oversaturated with respect to calcite. Assume that CaCO$_3$

---

†Calculation "by hand" can be made by solving equation i rewritten as

$$2[Ca^{2+}] + [H^+] = C_T(\alpha_1 + 2\alpha_2) + [OH^-] \quad \text{(ii)}$$

by trial and error. It is helpful to plot the α values as a function of pH.
‡[Carb-Alk] = $[HCO_3^-] + 2[CO_3^{2-}]$.

does not precipitate or dissolve and that the presence of borate does not affect the calculation significantly.

[Ca_T^{2+}], C_T, and [Carb-Alk] remain constant and independent of pressure and temperature.† At any pressure and temperature the following relationship must hold:

$$[\text{Carb-Alk}] = C_T(\alpha_1 + 2\alpha_2) \qquad (i)$$

α_1 and α_2 are calculated with K_1' and K_2' values valid for seawater and corrected for pressure. The equations for pressure dependence of these constants are

$$\log \frac{(K_i')_P}{(K_i')_{P=1}} = \frac{-\Delta V_i'(P-1)}{2.303RT} \qquad (ii)$$

where $\Delta V_1'(pK_1') = -(24.2 - 0.085t)$ cm^3 mol^{-1} and $\Delta V_2'(pK_2) = -(16.4 - 0.040t)$ cm^3 mol^{-1} and where $t = °C$, $T = K$, and $R = 82.05$ cm^3 atm mol^{-1} deg^{-1}.‡

pH values compatible with a given C_T and [Carb-Alk] can be computed from equation i; then [CO_{3T}^{2-}] = $C_T\alpha_2$ is calculated. The ion product, [Ca_T^{2+}][CO_{3T}^{2-}], can then be compared with $^cK_{s0}'$ of calcite at 25°C and at 5°C; its pressure dependence may be obtained from Gieskes' (1974) summarizing equation. $\Delta V_{\text{calcite}}'$ in equation ii is given as $\Delta V_{\text{calcite}}'(p^cK_{s0}) = -(47.5 - 0.23t)$ cm^3 mol^{-1}. With a value of $^cK_{s0}' = 5.94 \times 10^{-7}$ the results are given in Figure 7.11.

Temperature and pressure have a pronounced effect on pH but cause little variation in [CO_{3T}^{2-}].¶ Since [Ca_T^{2+}] is conservative, the ion product [Ca_T^{2+}][CO_{3T}^{2-}] also does not change appreciably. The extent of oversaturation, however, changes markedly because p$^cK_{s0}'$ (calcite) decreases strongly with both temperature and pressure.

This calculation illustrates that a decrease in temperature and an increase in pressure increase the calcite solubility. In the real ocean the C_T and [Alk] change with depth. Photosynthesis in the upper layers leads to a consumption of CO_2. The organic matter generated by photosynthesis leads, after settling through the water column, to a consumption of O_2 and to an increase in C_T. Some of the $CaCO_3$ formed in the surface waters becomes dissolved, thus increasing C_T and [Alk] in the deeper waters. The extent of over- and under-

---

†This is rigorous only if concentrations are expressed in mol kg^{-1} (i.e., in molal units). The mixed acidity constants K_1' and K_2' are the same whether molar or molal concentration units are used. $^cK_{s0}'$ for calcite is given in molar units. We would correct by considering the density of the seawater at any T and P, but the correction is much smaller ($< 3\%$) than the uncertainty in the $^cK_{s0}'$ value.

‡For recent data on pressure dependence see equation 25 and Table A6.5 in Appendix 6.2.

¶[CO_3^{2-}] is relatively insensitive to pressure and temperature, because at the pH of seawater [Carb-Alk]/$C_T \approx \alpha_1 \approx$ constant. Thus [CO_{3T}^{2-}]/[HCO_{3T}^-] = α_2/α_1 = $K_2'/\{H^+\} \approx$ constant; therefore the effect of P and T upon K' becomes largely reflected in changes in $\{H^+\}$.

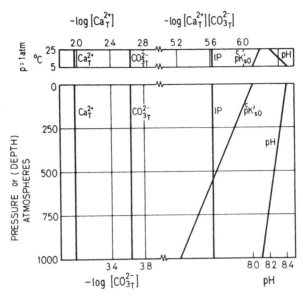

Figure 7.11. Effect of temperature and pressure on composition and extent of calcite oversaturation of an enclosed seawater sample. IP = Ion product ($[Ca_T^{2+}]\ [CO_{3T}^{2-}]$). Initially, the seawater, corresponding to a "typical" surface seawater, has the following composition at 25°C: pH = 8.2, $C_T = 2.18 \times 10^{-3}$ M, [Carb-Alk] = 2.4×10^{-3} eq liter^{-1}, $[Ca_T^{2+}] = 1.06 \times 10^{-2}$ M. See Example 7.9. In real seawater the undersaturation with respect to calcite is due to both a decrease in pK_{s0} caused by increased pressure and a decrease in ion product because of biologically mediated CO_2 production.

saturation is influenced by kinetic factors (inhibition of nucleation and crystal growth of calcite in surface waters and retardation of $CaCO_3$ dissolution in deep waters).

Example 7.10a. Reaction Paths for Calcite Dissolution in Groundwaters
The chemical composition of a newly formed groundwater is initially determined by rainwater (sometimes also by river water) that becomes exposed to increased partial pressure of CO_2 (from the microbially mediated oxidation of organic matter in the soil horizon) after infiltration into the soil. The CO_2-enriched water dissolves minerals such as aluminum silicates, $CaCO_3$, and $CaMg(CO_3)_2$.

How does the groundwater composition change during the dissolution of $CaCO_3$ (calcite)? We assume that $CaCO_3$ is the only mineral being dissolved. The temperature is 10°C.

Two idealized cases may be distinguished (Garrels and Christ, 1965): (i) During the process of $CaCO_3$ dissolution, the water remains in contact and equilibrium with a relatively large reservoir of CO_2 of fixed partial pressure (Figure 4.2a,c). (ii) An initially CO_2-rich water becomes isolated from the

384   Precipitation and Dissolution

$CO_2(g)$ reservoir (Figure 4.2b). We calculate for both cases the reaction progress—the change in composition as a function of the extent of the reaction ($CaCO_3$ dissolution)—for a few selected initial conditions. Equilibrium constants valid at 10°C and corrected for an ionic strength of $I = 4 \times 10^{-3}$ M[†] are: $-\log K_H = 1.27$; $-\log K_1' = 6.43$; $-\log K_2' = 10.38$. $-\log K_{s0(\text{calcite})}' = 7.95$.

(i) Reservoir with Constant p_{CO_2}. During dissolution of $CaCO_3$, conditions characterizing an aqueous carbonate system open to the atmosphere (Figure 4.5) prevail:

$$[HCO_3^-] = \frac{K_1}{[H^+]} K_H p_{CO_2} \tag{i}$$

For every p_{CO_2}, the linear relationship between $[HCO_3^-]$ and pH can be plotted (Figure 7.12). The extent of $CaCO_3$ dissolution is equivalent to the increase in

$$\Delta[Ca^{2+}] = \tfrac{1}{2}\Delta[Alk] \simeq \tfrac{1}{2}\Delta[HCO_3^-] \tag{ii}$$

That is, the solution paths are along the diagonal lines in Figure 7.12. Solubility equilibrium is attained where condition iv is attained:

$$CaCO_3(s) \text{ (calcite)} + H^+ = Ca^{2+} + HCO_3^- \tag{iii}$$

$$*K_s' = \frac{K_{s0}'}{K_2'} = \frac{[Ca^{2+}][HCO_3^-]}{\{H^+\}} \tag{iv}$$

If [Alk] results only from $CaCO_3$ dissolution,

$$2[Ca^{2+}] = [HCO_3^-] \tag{v}$$

Equation iv can be rewritten as

$$[HCO_3^-] = (*K_s' \times 2\{H^+\})^{1/2} \tag{vi}$$

(ii) Enclosed System. When the infiltrated CO_2-enriched water becomes separated from the CO_2 reservoir, the dissolved carbon dioxide, $H_2CO_3^*$, reacts with $CaCO_3(s)$ in a closed system:

$$H_2CO_3^* + CaCO_3(s) = Ca^{2+} + 2HCO_3^- \tag{vii}$$

[†]For more exact calculations, the ionic strength needs to be calculated iteratively for every reaction step.

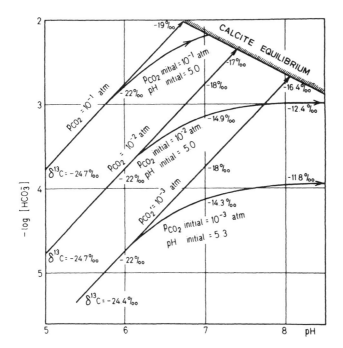

7.12. Dissolution paths of $CaCO_3$ (calcite) (Examples 7.10a, 7.10b). Two idealized cases are distinguished: (i) dissolution of $CaCO_3$ in systems with a reservoir of constant p_{CO_2} (the diagonal straight lines represent dissolution paths) and (ii) dissolution of $CaCO_3$ in a system (of different initial conditions) that becomes enclosed (dissolution paths are given by the curved lines). Calculated δ ^{13}C values of the dissolved carbon (with reference to the BDB standard) are given, assuming δ $^{13}C = -25‰$ for the CO_2 reservoir and δ $^{13}C = 1‰$ for the $CaCO_3(s)$. [A similar figure has been given by Deines et al. (1974).]

The extent of $CaCO_3$ dissolution is given by equation ii. The acidity of the system no longer changes with $CaCO_3(s)$ dissolution (see Section 4.4); thus the acidity, [Acy], acquired at the time of separation from the CO_2 reservoir, remains constant:

$$[Acy] = 2[H_2CO_3^*] + [HCO_3^-] + [H^+] - [OH^-] = \text{constant} \quad \text{(viii)}$$

$$[Acy] \simeq C_T(2\alpha_0 + \alpha_1) \simeq \text{constant} \quad \text{(ix)}$$

We compute [Acy] for the selected initial condition with

$$C_T = \frac{K_H p_{CO_2}}{\alpha_0} \quad \text{(x)}$$

(cf. equation 6 in Chapter 4) and then compute, with the help of the constraint of equation ix for selected pH values, values for C_T and $[HCO_3^-] = C_T\alpha_1$. Results are plotted in Figure 7.12. Reaction progress toward solubility equilibrium follows upward along the curved lines in the $-\log[HCO_3^-]$ versus pH plots. The boundary for $CaCO_3(s)$ (calcite) saturation is given by equation iv.

Example 7.10b. Carbon-13 Isotopes as Indicators of the Existence of a Gas Phase in the Evolution of Groundwaters Deines et al. (1974) have illustrated that the ^{13}C content of dissolved carbonate species can aid our understanding of the evolution of a carbonate groundwater (see Section 4.9). For the two cases discussed above and illustrated in Figure 7.12, the ^{13}C content of the waters can be computed considering appropriate values of $\delta\ ^{13}C$, the enrichment ratio, in the CO_2 reservoir and in calcite. We adopt for the CO_2 reservoir $\delta\ ^{13}C_{CO_2(g)} = -25‰$, and for $CaCO_3(s)$, $\delta\ ^{13}C_{CaCO_3} = +1‰$. The calculation of $\delta\ ^{13}C$ for the open system (reservoir with constant p_{CO_2}) is the same as that described in Example 4.13:

$$\delta\ ^{13}C_{sol} = \frac{([H_2CO_3^*]\delta^{13}C_{H_2CO_3^*} + [HCO_3^-]\delta^{13}C_{HCO_3^-} + [CO_3^{2-}]\delta^{13}C_{CO_3^{2-}})}{C_T} \quad \text{(xi)}$$

where $\delta\ ^{13}C$ values for $H_2CO_3^*$, HCO_3^-, and CO_3^{2-} are computed from the equilibrium constants given in Example 4.14.

For the enclosed system the total ^{13}C content of the solution is composed of the ^{13}C content of the solution at the time of separation from the CO_2 reservoir and the ^{13}C content of the carbon resulting from the $CaCO_3$ dissolution:

$$\delta\ ^{13}C_{sol} = C_{T,\,initial}\delta\ ^{13}C_{initial} + C_{T,\,from\,CaCO_3}\delta\ ^{13}C_{CaCO_3} \quad \text{(xii)}$$

where

$$C_{T,\,from\,CaCO_3} = \Delta[Ca^{2+}] = \tfrac{1}{2}\Delta[Alk] \simeq \tfrac{1}{2}\Delta[HCO_3^-] \quad \text{(xiii)}$$

and where $\delta\ ^{13}C_{initial}$ is calculated using equation xi.

For both cases of $CaCO_3$ dissolution, representative values of $\delta\ ^{13}C_{sol}$ are given in Figure 7.12. A comparison of these figures illustrates that ^{13}C information may aid in deciding whether carbonate rock dissolution occurs mainly under open- or closed-system conditions.

Fresh surface waters typically have $\delta\ ^{13}C$ values between $-5‰$ and $-11‰$, and measurements of ^{13}C in groundwaters can also sometimes be used to evaluate the extent of river water infiltration.

Example 7.11. Solubility of Pure $MeCO_3(s)$ Suspensions Derive an equation that shows how the solubility of various bivalent metal carbonates varies with their solubility product.

7.4 Carbonates

Consider the reaction

$$MeCO_3(s) + CO_2(g) + H_2O = Me^{2+} + 2HCO_3^- \qquad {}^+K_{ps0} \qquad (i)$$

We can represent its equilibrium constant by

$${}^+K_{ps0} = K_{s0} K_1 K_H K_2^{-1} \qquad (ii)$$

This becomes evident from the addition of the following equilibrium reactions:

$$MeCO_3(s) = Me^{2+} + CO_3^{2-} \qquad K_{s0} \qquad (iii)$$

$$CO_2(g) + H_2O = H_2CO_3^* \qquad K_H \qquad (iv)$$

$$H_2CO_3^* = H^+ + HCO_3^- \qquad K_1 \qquad (v)$$

$$CO_3^{2-} + H^+ = HCO_3^- \qquad K_2^{-1} \qquad (vi)$$

The electroneutrality equation can be approximated by

$$2[Me^{2+}] \simeq [HCO_3^-] \qquad (vii)$$

which can now be substituted into the equilibrium expression of equation i:

$$\frac{[Me^{2+}][HCO_3^-]^2}{p_{CO_2}} \simeq \frac{4[Me^{2+}]^3}{p_{CO_2}} \simeq {}^+K_{ps0} \qquad (viii)$$

$$[Me^{2+}] \simeq 0.63 \, {}^+K_{ps0}^{1/3} \, p_{CO_2}^{1/3} \qquad (ix)$$

Hence in a plot of $\log[Me^{2+}]$ versus $\log K_{s0}$ a slope of $\frac{1}{3}$ obtains (see Figure 7.10 inset).

Magnesian Calcite

This mineral, calcite with extensive Mg^{2+} substitution, is a preponderant carbonate phase, mostly of biogenic origin, in seawater. Figure 7.13 gives some data on the solubilities of Mg-calcites[†] as a function of $MgCO_3$ content. It is doubtful that formation and dissolution of any mineral in low-temperature aqueous solutions has been more fully investigated than the magnesian calcite. (For a general review see Morse and Mackenzie, 1990.) There are difficulties with the magnesian calcite system.

One problem is the difficulty to attain reversible equilibrium. Another source of divergence is the use of different models for the aqueous carbonate systems.

[†]Mg-calcite stands for magnesian calcite and not magnesium calcite—the latter could be misunderstood to be $MgCO_3$ (magnesite).

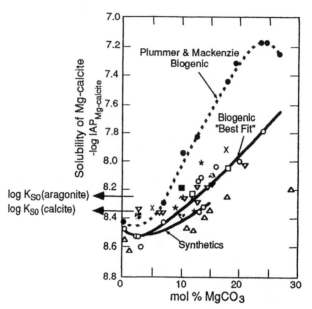

Figure 7.13. Solubilities of the magnesian calcite as a function of $MgCO_3$ content. The solubility is expressed in line with equation 41 as $IAP_{Mg\text{-calcite}} = [Ca^{2+}]^{(1-x)} [Mg^{2+}]^x [CO_3^{2-}]$. The solid curves represent the general trend of results on dissolution of biogenic and synthetic Mg-calcites. The curve fitting the data of Plummer and Mackenzie (1974) is dashed. The various points refer to the results of different researches. (For the origin of the data see Morse and Mackenzie, 1990.) IAP = ion activity product. (Adapted from Morse and Mackenzie, 1990.)

Precipitation and dissolution experiments can be carried out in closed or open systems and various ways of pH adjustments. Surface charge of the solid carbonate depends on the way the pH is adjusted (Stumm, 1992).

Analyses of natural calcites, formed at low temperatures, show $MgCO_3$ contents of up to 30 mol %.

Consider first a formal (equilibrium) approach to the solubility of magnesian calcite and compare its solubility with that of $CaCO_3$ (calcite or aragonite)

$$Ca_{(1-x)} Mg_x CO_3(s) = (1-x)Ca^{2+} + xMg^{2+} + CO_3^{2-} \quad K_{eq(x)} \tag{41}$$

$$Ca^{2+} + CO_3^{2-} = CaCO_3(s) \quad K_{s0(CaCO_3)}^{-1} \tag{42}$$

$$Ca_{(1-x)} Mg_x CO_3(s) + xCa^{2+} = CaCO_3(s) + xMg^{2+} \quad K_{eq(x)} K_{s0(CaCO_3)}^{-1}$$

$$\left(\frac{[Mg^{2+}]}{[Ca^{2+}]}\right)^x = K_{eq(x)} K_{s0(CaCO_3)}^{-1} \tag{43}$$

7.5 The Stability of Hydroxides, Carbonates, and Hydroxide Carbonates

If a magnesian calcite is in contact with a solution whose $([Mg^{2+}]/[Ca^{2+}])^x$ ratio is smaller than $K_{eq(x)}K_{s0(CaCO_3)}^{-1}$, the magnesian calcite is less stable than $CaCO_3(s)$. For example (at 25°C), a 10 mol % magnesian calcite ($-\log K_{(eq41)} \approx 8.0$) is less stable than calcite ($-\log K_{s0} = 8.42$)[†] or aragonite ($-\log K_{s0} = 8.22$).[†] Thus high magnesian calcite should be converted in marine sediments into calcite or aragonite. The conversion to aragonite has been observed. As the magnesian calcite is dissolved, Mg^{2+} becomes enriched in the solution (incongruent dissolution) and a purer $CaCO_3(s)$ is precipitated. As Figure 7.13 suggests, low magnesian calcites ($x = 3$–4 mol %) are probably stable in comparison to calcite. Higher magnesian calcites—although thermodynamically unstable—may persist for considerable time periods.

Some of the differences in solubilities are also related to different disordering of the crystal surface. As shown by Bischoff et al. (1987) with Raman investigations, the biogenic phases are characterized by greater positional disorder than synthetic minerals of the same composition.

The fact that the carbonates of foraminifera buried in marine sediments can be used as a "memory storage" for Cd^{2+} present in the sea when the foraminifera were formed (Delaney and Boyle, 1987) is evidence for the non-reversibility or extremely slow reversibility of biogenic mineral carbonates.

Some discrepancies can be explained by recognizing that *surface processes* attain a degree of metastability rather rapidly, while *phase equilibrium* is not achieved even over quite long periods of time.

Morse and Mackenzie (1990) point out the two fundamental problems:

1. In most experiments to calculate solubilities, the magnesian calcites have been treated as solids of fixed compositions of one component, whereas they are actually a series of at least two-component compounds forming a partial solid solution series.
2. Magnesian calcite phases dissolve incongruently, leading to a formation of a phase different in composition from the original reactant solid.

For a rigorous treatment of the aqueous solubility of magnesian calcite within the framework by chemical thermodynamics, see Lippmann (1991).

7.5 THE STABILITY OF HYDROXIDES, CARBONATES, AND HYDROXIDE CARBONATES

In the previous sections we have applied equilibrium constants for heterogeneous equilibria in a rather formal way. Thermodynamically meaningful conclusions are justified only if, under the specified conditions (concentration, pH, temperature, pressure), the solutes are in equilibrium with the solid phase for which the mass law relationship has been formulated or if under the specified

---

[†]The solubility constants given are those used by Morse and Mackenzie (1990).

Precipitation and Dissolution

conditions the assumed solid phase is really stable or at least metastable. It remains to be illustrated how we can establish which phase predominates under a given set of conditions.

Which Solid Phase Controls the Solubility of Fe(II)?

In order to exemplify how to find out which solid predominates as a stable phase for selected conditions, we may consider the solubility of Fe(II) in a carbonate-bearing water of low redox potential ($P = 1$ atm, $25°C$, $I = 6 \times 10^{-3}$ M). We may first consider the following solubility products of $Fe(OH)_2(s)$ and of $FeCO_3(s)$:

$$K_{s0}(Fe(OH)_2) \simeq 10^{-14.7} \text{ mol}^3 \text{ liter}^{-3}$$

$$K_{s0}(FeCO_3) \simeq 10^{-10.7} \text{ mol}^2 \text{ liter}^{-2}$$

The numerical values of these equilibrium constants cannot be compared directly. Note that the constants have different units. It would be incorrect and misleading to infer from the numerical values of the solubility products that $Fe(OH)_2(s)$ is less soluble than $FeCO_3(s)$. It is appropriate rather to inquire which solid controls the solubility for a given set of conditions (P, T, pH, C_T, [Alk], or p_{CO_2}), that is, gives the smallest concentration of soluble Fe(II).

Example 7.12. Control of Solubility Is it $FeCO_3(s)$ or $Fe(OH)_2(s)$ that controls the solubility of Fe(II) in anoxic water of [Alk] = 10^{-4} eq liter^{-1} and pH = 6.8? We estimate maximum soluble [Fe(II)] by considering the solubility equilibrium ($25°C$) with (a) $FeCO_3(s)$ as well as that with (b) $Fe(OH)_2(s)$.
(a) Assuming solubility equilibrium with $FeCO_3(s)$

$$FeCO_3(s) = Fe^{2+} + CO_3^{2-} \quad \log {}^cK_{s0} \simeq -10.4 \quad \text{(i)}$$

$$H^+ + CO_3^{2-} = HCO_3^- \quad -\log {}^cK_2 \simeq +10.1 \quad \text{(ii)}$$

$$\overline{FeCO_3(s) + H^+ = Fe^{2+} + HCO_3^- \quad \log {}^{c*}K_s \simeq -0.3} \quad \text{(iii)}$$

Correspondingly, $\log[Fe^{2+}] = \log {}^{c*}K_s - \text{pH} - \log[HCO_3^-]$; and since at this pH, $[Fe^{2+}] \simeq [Fe(II)]$ and $[HCO_3^-] \simeq [\text{Alk}]$, we obtain $\log[Fe(II)] = -3.1$.
(b) Assuming solubility equilibrium with $Fe(OH)_2(s)$

$$Fe(OH)_2(s) = Fe^{2+} + 2OH^- \quad \log {}^cK_{s0} = -14.5 \quad \text{(iv)}$$

$$2H^+ + 2OH^- = 2H_2O \quad -2\log {}^cK_W = +27.8 \quad \text{(v)}$$

$$\overline{Fe(OH)_2(s) + 2H^+ = Fe^{2+} + 2H_2O \quad \log {}^{c*}K_{s0} = +13.3} \quad \text{(vi)}$$

Thus $\log[Fe^{2+}] = \log *K_{s0} - 2\text{pH}$; and $\log[Fe(II)] \simeq -0.3$. Because $[Fe^{2+}]$ (or Fe(II)) is smaller for hypothetical equilibrium with $FeCO_3(s)$ than with $Fe(OH)_2$, siderite $[FeCO_3(s)]$ is more stable than $Fe(OH)_2(s)$.

7.5 The Stability of Hydroxides, Carbonates, and Hydroxide Carbonates

Solubility, Predominance, and Activity Ratio Diagrams. From thermodynamic information, diagrams can be constructed that circumscribe the stability boundaries of the solid phases. Depending on the variables used, different kinds of predominance diagrams can be constructed.

For systems closed to the atmosphere, a *solubility diagram* (e.g., $\log[Fe^{2+}]$, $\log[H^+]$, at fixed $\log C_T$) can conveniently illustrate the conditions under which a particular solid phase predominates. Figure 7.14 gives a solubility diagram for Fe(II) considering $FeCO_3(s)$ and $Fe(OH)_2(s)$ as possible solid phases. Construction of the diagram consists essentially of the superposition of a pH-dependent $Fe(OH)_2(s)$ solubility diagram and a pH- and C_T-dependent $FeCO_3(s)$ solubility diagram. For a given $[H^+]$ and C_T the solid compound giving the smaller Fe^{2+} is more stable. Thus, for the conditions in Figure 7.14a, $FeCO_3(s)$ dictates the maximum concentration of Fe(II) below pH values of approximately 10, and $Fe(OH)_2(s)$ limits soluble iron above pH 10.

The same information can be gained from an *activity ratio diagram*. The construction is very simple and is illustrated in Figure 7.14b. We again choose pH as a master variable and make our calculation for a given C_T. In this figure we plot the ratios between the activities of the various soluble and solid species

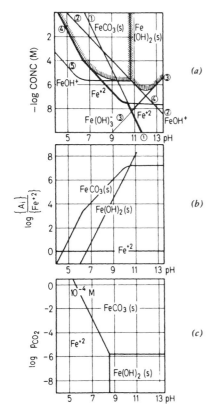

Figure 7.14. Stability of $Fe(OH)_2(s)$ and $FeCO_3(s)$ (siderite) (25°C; $I = 0$). With the help of such diagrams it is possible to evaluate the conditions (pH, C_T, $[Fe^{2+}]$, p_{CO_2}) under which a solid phase [$Fe(OH)_2(s)$ or $FeCO_3(s)$] is stable. (a) Solubility diagram of Fe(II) in a $C_T = 10^{-3}$ M carbonate system. The numbers on the curves refer to equations describing the respective equilibria in this system as follows: (1) $Fe(OH)_2(s) = Fe^{2+} + 2OH^-$, $K_{s0} = 2 \times 10^{-15}$; (2) $Fe(OH)_2(s) = [Fe(OH)]^+ + OH^-$, $K_{s1} = 4 \times 10^{-10}$; (3) $Fe(OH)_2(s) + OH^- = [Fe(OH)_3]^-$, $K_{s3} = 8.3 \times 10^{-6}$; (4) $FeCO_3(s) = Fe^{2+} + CO_3^{2-}$, $K_{s0} = 2.1 \times 10^{-11}$; (5) $FeCO_3(s) + OH^- = [Fe(OH)]^+ + CO_3^{2-}$, $K'_{s1} = 0.1 \times 10^{-5}$. (b) Activity ratio diagram for $C_T = 10^{-3}$ M. The same conclusions as under (a) apply. Above pH $\simeq$ 10, $Fe(OH)_2(s)$ has the highest relative activity; it can precipitate as a pure phase. Below pH $\simeq$ 10, $FeCO_3$ becomes more stable than $Fe(OH)_2(s)$ and controls the solubility of Fe(II). (c) $\log p_{CO_2}$–pH predominance diagram. $FeCO_3(s)$ is the stable phase at p_{CO_2} larger than 10^{-6} atm, that is, in normal atmosphere.

as a function of pH. In our case one of the species $\{Fe^{2+}\}$ (or $[Fe^{2+}]$) is chosen as a reference state. Thus the ordinate values are $\log\{A_i\}/\{Fe^{2+}\}$. Because the diagram gives activities on a relative scale (relative to $[Fe^{2+}]$), we treat the activities of solid phases formally in the same way as the activities (or concentrations) of the solutes. Two equations determine the ratios $\{Fe(OH)_2\}(s)/\{Fe^{2+}\}$ and $\{FeCO_3(s)\}/\{Fe^{2+}\}$. For the solubility of $Fe(OH)_2(s)$ we have

$$\frac{\{Fe^{2+}\}}{\{H^+\}^2\{Fe(OH)_2(s)\}} = *K_{s0Fe(OH)_2} \qquad (44)$$

Thus $\log(\{Fe(OH)_2(s)\}/\{Fe^{2+}\})$ plots as a function of pH as a straight line with a slope of $+2$:

$$\log\frac{\{Fe(OH)_2(s)\}}{\{Fe^{2+}\}} = p*K_{s0Fe(OH)_2} + 2pH \qquad (45)$$

and an intercept $pH = -\frac{1}{2}p*K_{s0Fe(OH)_2}$.

For the solubility of $FeCO_3$,

$$\frac{\{Fe^{2+}\}\{CO_3^{2-}\}}{\{FeCO_3(s)\}} = K_{s0FeCO_3} \qquad (46)$$

Because $\{CO_3^{2-}\} = C_T\alpha_2$, equation 46 can be rearranged to

$$\log\frac{\{FeCO_3(s)\}}{\{Fe^{2+}\}} = pK_{s0FeCO_3} + \log C_T + \log\alpha_2 \qquad (47)$$

Equation 47 can be plotted considering that $\log\alpha_2 = 0$ at $pH > pK_2$ and that $d(\log\alpha_2)/d\,pH$ is $+1$ or $+2$ in the pH regions $pK_1 < pH < pK_2$ and $pH < pK_1$, respectively.

At any pH the ordinate values on the activity ratio diagram give the activities for the various species on a relative scale. Thus, in Figure 7.14b at $pH = 12$. $Fe(OH)_2(s)$ has the highest relative activity. This solid phase will precipitate at this pH; as a pure solid its activity will be unity, and $FeCO_3(s)$ must have an activity of much less than unity and cannot exist as a pure solid phase. Figure 7.14b shows that, for $C_T = 10^{-3}$ M, $FeCO_3(s)$ is stable below approximately $pH = 10$.

For *open systems* it is convenient to select $\log p_{CO_2}$ and $-\log[H^+]$ as variables. An assumption must then be made about the concentrations of the solute. The computation of the straight lines in Figure 7.14c is based on the following equations. For the coexistence of Fe^{2+} and $Fe(OH)_2(s)$,

$$\log *K_{sFe(OH)_2} - 2pH + pFe^{2+} = 0 \qquad (48)$$

7.5 The Stability of Hydroxides, Carbonates, and Hydroxide Carbonates

The thermodynamic coexistence of Fe^{2+} and $FeCO_3(s)$ as a function of pH and p_{CO_2} is expressed, perhaps most conveniently, by considering the solubility equilibrium in the form of the reaction

$$FeCO_3(s) + 2H^+ = Fe^{2+} + CO_2(g) + H_2O \qquad *K_{ps0FeCO_3}$$

$$\log p_{CO_2} = \log *K_{ps0FeCO_3} - 2pH + pFe^{2+} \qquad (49)$$

The coexistence of $FeCO_3(s)$ and $Fe(OH)_2(s)$ is given by the equilibrium

$$FeCO_3(s) + H_2O = Fe(OH)_2(s) + CO_2(g) \qquad (50)$$

where (compare equations 48 and 49)

$$K = \frac{*K_{ps0FeCO_3}}{*K_{s0Fe(OH)_2}} = p_{CO_2} \qquad (51)$$

The equilibrium partial pressure of CO_2 for equation 51 is approximately 10^{-6} atm. At p_{CO_2} higher than 10^{-6} atm, for example, in systems exposed to the atmosphere, $Fe(OH)_2$ is not stable and will be converted to $FeCO_3(s)$. In Figure 7.14a the enhancement of the solubility caused by the formation of $FeOH^+$ ($= 10^{-4}$ M) also has been considered.

The Solubility of Dolomite

The lack of understanding of the dolomite precipitation process is reflected in the discrepancy of solubility products reported by different investigators. Published figures range from $10^{-16.5}$ to $10^{-19.5}$. As mentioned before, solubility equilibrium can be reached (under atmospheric conditions) only from undersaturation. The time of approaching equilibrium is unknown. Thus it is very difficult to ascertain equilibrium in laboratory experiments.

Nature, however, has provided us with a long-term solubility experiment. The well waters of Florida show a constant ratio of magnesium to calcium ($[Mg^{2+}]/[Ca^{2+}] = 0.8 \pm 0.1$). The tendency for subsurface waters to have such a nearly constant magnesium–calcium ratio suggests that waters in porous dolomitic limestones might have equilibrated with both the calcite and dolomite phases [For a general treatment on dolomites in groundwaters, see, for example, Plummer et al. (1990).]

The solubility product of dolomite is given by

$$K_{s0} = \{Ca^{2+}\} \{Mg^{2+}\} \{CO_3^{2-}\}^2 \qquad (52)$$

For the reaction

$$2\,CaCO_3(s)\,(\text{calcite}) + Mg^{2+} = CaMg(CO_3)_2(s) + Ca^{2+} \qquad (53)$$

the equilibrium constant K is defined in terms of the activity ratio of Ca^{2+} and Mg^{2+} and the K_{s0} values of dolomite and calcite:

$$K = \frac{K^2_{s0CaCO_3}}{K_{s0dolomite}} = \frac{\{Ca^{2+}\}}{\{Mg^{2+}\}} \tag{54}$$

For solution in contact and equilibrium with dolomite and calcite, the activity ratio is a constant at any temperature and pressure. In the Florida aquifer waters the concentration ratio remains nearly constant ($f_{Ca^{2+}}/f_{Mg^{2+}} \simeq 1$) even though $[Ca^{2+}]$ and $[Mg^{2+}]$ vary from less than 10^{-3} to 10^{-2} M. With an average value of $[Mg^{2+}]/[Ca^{2+}]$ of 0.78 and $K_{s0calcite} = 5 \times 10^{-9}$ (25°C), a $K_{s0dolomite} = 2.0 \times 10^{-17}$ can be calculated. Nordstrom et al. (1990) list log $K_{s0dolomite} = -17.09$ (25°C).

Dolomites found in nature seldom have exact stoichiometric composition and are frequently structurally rich in calcium (protodolomite). Dolomite, as well as calcite, has a tendency to form solid solutions with many metal ions. Calcite has a tendency to accommodate Mg^{2+} in its structure to form *magnesian calcite*. Kinetically, the deposition of magnesian calcite may be more favorable than the deposition of dolomite.

Carbonates in the System $Mg^{2+}-CO_2-H_2O$

Example 7.13. The Stability of Magnesium Hydroxides, Magnesium Carbonates, and Magnesium Hydroxide Carbonates The following solubility products are available (quoted from Morse and Mackenzie, 1990):

		$-\log K_{s0}$ (25°C)
Magnesite	$MgCO_3$	7.46
Nesquehonite	$MgCO_3 \cdot 3\,H_2O$	4.67
Hydromagnesite	$Mg_4(CO_3)_3(OH)_2 \cdot 3\,H_2O$	36.47
Brucite	$Mg(OH)_2$	11.16

Establish stability domains as a function of the variables C_T, pH, and p_{CO_2}.

In order to obtain a survey of the stability relationships, we construct an activity ratio diagram using Mg^{2+} as a reference state.

For brucite:

$$\log \frac{\{Mg(OH)_2(s)\}}{\{Mg^{2+}\}} = pK_{s0} - 2pK_W + 2pH$$

$$= -16.8 + 2pH \tag{i}$$

7.5 The Stability of Hydroxides, Carbonates, and Hydroxide Carbonates

For magnesite:

$$\log \frac{\{\text{magnesite(s)}\}}{\{\text{Mg}^{2+}\}} = pK_{s0} + \log C_T + \log \alpha_2$$

$$= 7.5 + \log C_T + \log \alpha_2 \quad (ii)$$

For nesquehonite:

$$\log \frac{\{\text{nesquehonite(s)}\}}{\{\text{Mg}^{2+}\}} = pK_{s0} + \log C_T + \log \alpha_2$$

$$= 4.7 + \log C_T + \log \alpha_2 \quad (iii)$$

For hydromagnesite:

$$\log \frac{\{\text{hydromagnesite(s)}\}^{1/4}}{\{\text{Mg}^{2+}\}} = \frac{1}{4} pK_{s0} - \frac{1}{2} pK_W + \frac{3}{4} \log C_T$$

$$+ \frac{3}{4} \log \alpha_2 + \frac{1}{2} \text{pH}$$

$$= 2.1 + \frac{3}{4} \log C_T + \frac{3}{4} \log \alpha_2 + \frac{1}{2} \text{pH} \quad (iv)$$

These equations are plotted in Figure 7.15a for an assumed value of $\log C_T = -2.5$. It is convenient to start to plot these equations at high pH values where $\log \alpha_2 = 0$. In the pH region $pK_1 < \text{pH} < pK_2$, $d \log \alpha_2/d$ pH has a slope of +1. Our activity ratio diagram postulates that, for $\log C_T = -2.5$, magnesite is stable below pH $\simeq 10.7$; nesquehonite is less stable than magnesite; brucite becomes stable above pH $\simeq 10.7$.

In Figure 7.15b a solubility diagram is plotted for $-\log C_T = 2.5$.

For an open system, activity ratio or solubility diagrams for given p_{CO_2} values can be constructed. Figure 7.15c gives $\log p_{CO_2}$–pH predominance diagrams for $p\text{Mg}^{2+} = 2$ and 4.0. The dissolution reaction may be rearranged to give mass law expressions with p_{CO_2} and $[\text{H}^+]$; for example, for magnesite,

$$\text{MgCO}_3(s) = \text{Mg}^{2+} + \text{CO}_3^{2-} \qquad \log K_{s0} = -7.5$$

$$\text{CO}_3^{2-} + 2\text{H}^+ = \text{H}_2\text{CO}_3^* \qquad -\log(K_1 K_2) = 16.6$$

$$\text{H}_2\text{CO}_3^* = \text{CO}_2(g) + \text{H}_2\text{O} \qquad -\log K_H = 1.5$$

$$\overline{\text{MgCO}_3(s) + 2\text{H}^+ = \text{Mg}^{2+} + \text{CO}_2(g) + \text{H}_2\text{O} \qquad \log {}^*K_{ps0} = 10.6}$$

$$(v)$$

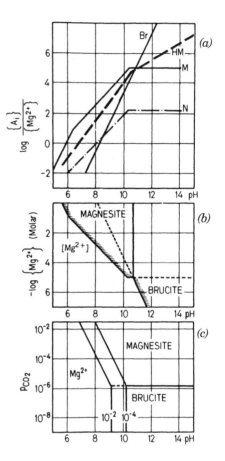

Figure 7.15. Stability in the system Mg^{2+}-CO_2-H_2O ($I = 0$, 25°C) (see Example 7.13). Br, Brucite, $Mg(OH)_2(s)$; HM, hydromagnesite, $Mg_4(CO_3)_3(OH)_2 \cdot 3H_2O(S)$; N, nesquehonite, $MgCO_3 \cdot 3H_2O(s)$: M, magnesite, $MgCO_3(s)$. (a) Activity ratio diagram for $-\log C_T = 2.5$. Equations i to iv define relative activity. Stable phases are M(s): pH < 10.7; Br(s): pH > 10.7. (b) Solubility ($-\log\{Mg^{2+}\}$) versus pH diagram for $-\log C_T = 2.5$. (c) Predominance diagram for $\log\{Mg^{2+}\} = -2$ and -4. Brucite can exist only at low p_{CO_2}.

Similarly, one derives for the conversion of brucite into magnesite,

$$Mg(OH)_2(s) + CO_2(g) = MgCO_3(s) + H_2O \qquad \log K = 6.1 \qquad \text{(vi)}$$

The equilibrium p_{CO_2} for equation vi is $10^{-6.1}$ atm. If p_{CO_2} (actual) is larger than $10^{-6.1}$ atm, brucite is converted to magnesite. On the other hand, hydromagnesite at this temperature and in the pH range of natural water, is unstable (or metastable) with respect to magnesite. As Figure 7.15c suggests, brucite is unstable if brought into contact with humid air.

Example 7.14. Solubility of Cu(II) in Natural Water: Effect of Complexing by Carbonate Estimate the solubility of Cu(II) in carbonate-bearing water of constant $C_T = 10^{-2}$ M (closed system). The pertinent Cu(II) equilibria are given in Table 7.3.

In order to gain insight into the predominant solid phases and soluble species, it is expedient first to construct an activity ratio diagram. Taking $\{Cu^{2+}\}$ as a

7.5 The Stability of Hydroxides, Carbonates, and Hydroxide Carbonates

Table 7.3. Cu(II) Equilibria[a]

Item	Reaction	Log K^a
1	$CuO(s) + 2H^+ = Cu^{2+} + H_2O$ (tenorite)	7.65
2	$Cu_2(OH)_2CO_3(s) + 4H^+ = 2Cu^{2+} + 3H_2O + CO_2(g)$ (malachite)	14.16
3	$Cu_3(OH)_2(CO_3)_2(s) + 6H^+ = 3Cu^{2+} + 4H_2O + 2CO_2(g)$ (azurite)	21.24
4	$Cu^{2+} + H_2O = CuOH^+ + H^+$	-8
5	$2Cu^{2+} + 2H_2O = Cu_2(OH)_2^{2+} + 2H^+$	-10.95
6	$Cu^{2+} + CO_3^{2-} = CuCO_3(aq)$	6.77
7	$Cu^{2+} + 2CO_3^{2-} = Cu(CO_3)_2^{2-}(aq)$	10.01
8	$CO_2(g) + H_2O = HCO_3^- + H^+$	-7.82
9	$Cu^{2+} + 3H_2O = Cu(OH)_3^- + 3H^+$	-26.3
10	$Cu^{2+} + 4H_2O = Cu(OH)_4^{2-} + 4H^+$	-39.4

[a] $I = 0$; 25°C. Given or quoted by Schindler (1967). Schindler assumes that the species $Cu(OH)_2(aq)$ can be neglected.

reference, we obtain the following activity ratios for the equilibria given in Table 7.3:

$$\log \frac{\{CuO(s)\}}{\{Cu^{2+}\}} = -7.65 + 2\,\text{pH} \tag{i}$$

$$\log \frac{\{Cu(OH)(CO_3)_{0.5}(s)\}}{\{Cu^{2+}\}} = -3.17 + 1.5\,\text{pH}$$
$$+ 0.5 \log C_T + 0.5 \log \alpha_1 \tag{ii}$$

Equation ii results from a combination of items 2 and 8 of Table 7.3. Similarly, combining items 3 and 8 gives

$$\log \frac{\{Cu(OH)_{0.67}(CO_3)_{0.67}(s)\}}{\{Cu^{2+}\}} = -1.85 + \text{pH}$$
$$+ 0.67 \log C_T + 0.67 \log \alpha_1 \tag{iii}$$

$$\log \frac{\{CuOH^+\}}{\{Cu^{2+}\}} = -8 + \text{pH} \tag{iv}$$

$$\log \frac{\{CuCO_3(aq)\}}{\{Cu^{2+}\}} = 6.77 + \log C_T + \log \alpha_2 \tag{v}$$

$$\log \frac{\{Cu(CO_3)_2^{2-}(aq)\}}{\{Cu^{2+}\}} = 10.01 + 2 \log C_T + 2 \log \alpha_2 \tag{vi}$$

$$\log \frac{\{Cu(OH)_3^-\}}{\{Cu^{2+}\}} = -26.3 + 3\,pH \tag{vii}$$

$$\log \frac{\{Cu(OH)_4^{2-}\}}{\{Cu^{2+}\}} = -39.4 + 4\,pH \tag{viii}$$

In a double logarithmic diagram, log activity ratio versus pH (equations i–viii) can be plotted in straight-line portions having readily defined slopes. Figure 7.16a shows that under the specified conditions malachite and tenorite qualify as stable solid phases; malachite is stable below pH = 7, while tenorite is more stable in the alkaline region. With respect to dissolved species the activity ratio diagram reveals that under the specified conditions the following species predominate: Cu^{2+} up to pH = 6; $CuCO_3(aq)$ in the pH range 6–9.3; $Cu(CO_3)_2^{2-}(aq)$ in the pH range 9.3–10.7; $Cu(OH)_3^-$ and $Cu(OH)_4^{2-}$ above pH 10.7 and 12.9, respectively. With this information a logarithmic solubility diagram can be sketched (Figure 7.16b).

7.6 SULFIDES AND PHOSPHATES

The principles considered so far for (hydr)oxides and carbonates can of course be applied to salts containing other anions. In the case of sulfides, we also have to consider the formation of complexes with sulfur ligands (SH^-, S^{2-}). This tendency is especially pronounced with B metals. A difficulty is that there is some uncertainty on the value of the second acidity constant of H_2S

$$HS^- = S^{2-} + H^+ \qquad \log K_2 = -13 \text{ to } -19 \tag{55}$$

The values reported vary by six orders of magnitude. The more recently reported spectrophotometrically determined constant gives pK_2 values between 17 and 19. Such values imply that S^{2-}, similar to O^{2-} hardly occurs in aqueous solution. In the past, however, many solubility products have been determined by assuming a $pK_2 \simeq 14$. Furthermore, solid sulfides occur often in different allotropic modifications: for example, FeS occurs as troilite, mackinawite, pyrrhotite, and amorphous FeS; then there is greigite (Fe_3S_4) and different modifications of FeS_2 (pyrite, marcasite). A further complication is that HS^- can form polysulfides in reactions such as $3\,S(s) + HS^- = HS_4^-$.

To avoid the uncertainty of the pK_2 of H_2S, it is expedient to define the solubility of sulfides in terms of the equilibria (see Table 7.4):

$$MeS(s) + H^+ = Me^{2+} + HS^- \qquad {}^*K_s \tag{56}$$

$$M_2S(s) + H^+ = 2\,M^+ + HS^- \qquad {}^*K_s \tag{57}$$

7.6 Sulfides and Phosphates

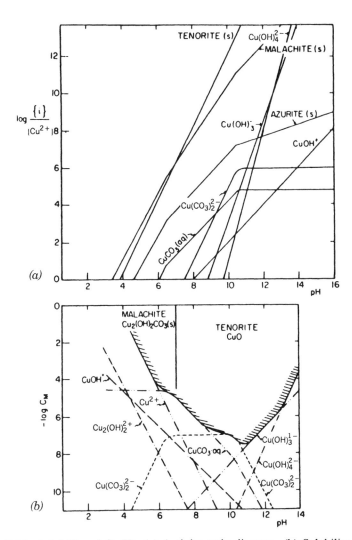

Figure 7.16. Solubility of Cu(II), (a) Activity ratio diagram. (b) Solubility diagram. The solid line surrounding the shaded area gives the total solubility of Cu(II), which up to a pH value of 6.96 is governed by the solubility of malachite [$Cu_2(OH)_2CO_3(s)$]. In the low pH region, azurite [$Cu_3(OH)_2(CO_3)_2(s)$] is metastable but may become stable at higher C_T. Above pH 7, the solubility is controlled by the solubility of CuO (tenorite). The predominant species with increasing pH are Cu^{2+}, $CuCO_3(aq)$, $Cu(CO_3)_2^{2-}$, and hydroxo copper(II) anions. $C_T = 10^{-2}$ M.

Table 7.4. Solubility of Sulfides (25°C, $I = 0$)[a]

$$MeS(s) + H^+ = Me^{2+} + HS^- \qquad {}^*K_s = K_{s0}K_2^{-1}$$

$$M_2S(s) + H^+ = 2\,M^+ + HS^- \qquad {}^*K_s = K_{s0}K_2^{-1}$$

Sulfide		log ${}^*K_s$	Reference
MnS	(green)	0.17	Dyrssen and Kremling (1990)
	(pink)	3.34	Dyrssen and Kremling (1990)
FeS		−4.2	Dyrssen and Kremling (1990)
	(troilite)	−5.25	Davison (1991)
	(mackinawite)	−3.6	Davison (1991)
	(amorphous)	−2.95	Davison (1991)
	(pyrrhotite)	−5.1	Davison (1991)
Fe_3S_4	(greigite)[b]	−4.4	Davison (1991)
FeS_2	(pyrite)[c]	−16.4	Davison (1991)
CoS	(α)	−7.44	Dyrssen and Kremling (1990)
	(β)	−11.07	Dyrssen and Kremling (1990)
NiS	(α)	−5.6	Dyrssen and Kremling (1990)
	(β)	−11.1	Dyrssen and Kremling (1990)
	(γ)	−12.8	Dyrssen and Kremling (1990)
CuS		22.3	Dyrssen and Kremling (1990)
ZnS	(α, sphaelerite)	−10.93	Dyrssen and Kremling (1990)
	(β, wurtzite)	−8.95	Dyrssen and Kremling (1990)
CdS	(greenockite)	−14.36	Daskalakis and Helz (1992)
HgS	(black)	−38.8	Dyrssen and Kremling (1990)
	(red)	−39.5	Dyrssen and Kremling (1990)
SnS		−11.95	Dyrssen and Kremling (1990)
PbS		−13.97	Dyrssen and Kremling (1990)
Cu_2S		−34.65	Dyrssen and Kremling (1990)
Ag_2S		−35.94	Dyrssen and Kremling (1990)
Tl_2S		−7.22	Dyrssen and Kremling (1990)

[a] Where necessary, values of K_{s0} were converted into ${}^*K_s$ by assuming $K_2 = 10^{-13.9}$.
[b] The solubility equilibrium is defined by $Fe_3S_4(s) + 3\,H^+ = 3\,Fe^{2+} + 3\,HS^- + S^0$; $({}^*K_s)^3$.
[c] For the equilibrium $FeS_2(s) + H^+ = Fe^{2+} + HS^- + S^0$.

The ${}^*K_s$ values are related to K_{s0} by

$${}^*K_s = K_{s0}\,K_2^{-1} \qquad (58)$$

Example 7.15. Solubility of FeS at a Given p_{H_2S} Estimate the soluble concentrations of Fe^{2+} at a given partial pressure of H_2S, $p_{H_2S} = 10^{-4}$ atm.
The following equilibrium constants are used:

$$FeS(s) + H^+ = Fe^{2+} + HS^- \qquad \log {}^*K_s = -4.2$$

$$HS^- + H^+ = H_2S(g) \qquad \log K = 7.99$$

$$FeS(s) + 2\,H^+ = Fe^{2+} + H_2S(g) \qquad \log {}^*K_{ps0} = 3.79 \quad (3.88 \text{ for } I = 10^{-2})$$

7.6 Sulfides and Phosphates

Tableau 7.7. Solubility of FeS(s) as a Function of pH at Given p_{H_2S}

Components		$H_2S(g)$	FeS	H^+	log K ($I = 10^{-2}$)
Species	Fe^{2+}	−1	1	2	3.88
	HS^-	1	0	−1	−7.85
	$H_2S(g)$	1	0	0	−0.98
	H^+	0	0	1	0
		$p_{H_2S} = 10^{-4}$ atm	{FeS} = 1		

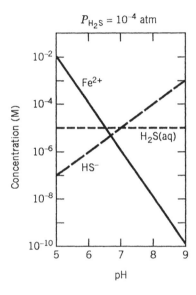

Figure 7.17. Solubility of FeS(s) as a function of pH for $p_{H_2S} = 10^{-4}$ atm in accord with Tableau 7.7. $[Fe^{2+}]$ plots versus pH with a slope of −2.

Thus, at a given $p_{H_2S} = 10^{-4}$ atm, $[Fe^{2+}]$ plots versus $-\log [H^+]$ as a line with slope $d \log [Fe^{2+}]/-d \log [H^+] = -2$ in accord with the equation

$$[Fe^{2+}] p_{H_2S} = 10^{-11.9} [H^+]^2$$

$$\log [Fe^{2+}] = -11.9 - \log p_{H_2S} + 2 \log [H^+]$$

$$= 7.9 - 2 \, p^cH$$

See Tableau 7.7. Figure 7.17 plots the data.

Example 7.16. Stability of FeS(s) and $FeCO_3$ in HS^-- and HCO_3^--Bearing Waters The following conditions are assumed:

$$[Alk] = 5 \times 10^{-3} \, M \quad \text{and} \quad [Fe^{2+}] = 10^{-6} \, M$$

Precipitation and Dissolution

Give the concentration conditions for the existence of FeS(s), FeCO$_3$(s), and Fe(OH)$_2$(s) as a function of S_T (= [H$_2$S] + [HS$^-$] + [S^{2-}]) and pH. (To avoid using $K_{2(H_2S)}$ we can neglect [S^{2-}].) The following equations are needed:

$$\text{FeS(s)} + \text{H}^+ = \text{Fe}^{2+} + \text{HS}^-$$
$$*K_{s(\text{FeS})} = 10^{-4.2} \; (10^{-4.04}) \quad \text{(i)}$$

$$\text{FeCO}_3(\text{s}) + \text{H}^+ = \text{Fe}^{2+} + \text{HCO}_3^-$$
$$*K_{s(\text{FeCO}_3)} = 10^{-0.12} \; (10^{0.05}) \quad \text{(ii)}$$

$$\text{Fe(OH)}_2(\text{s}) + 2\,\text{H}^+ = \text{Fe}^{2+} + 2\,\text{H}_2\text{O}$$
$$*K_{s0} = 10^{12.85} \; (10^{12.92}) \quad \text{(iii)}$$

$$\text{FeS(s)} + \text{HCO}_3^- = \text{FeCO}_3(\text{s}) + \text{HS}^-$$
$$K_{\text{FeS}-\text{FeCO}_3} = *K_{s(\text{FeS})}/*K_{s(\text{FeCO}_3)} \quad \text{(iv)}$$

$$\text{FeS(s)} + 2\text{H}_2\text{O} = \text{Fe(OH)}_2(\text{s}) + \text{HS}^- + \text{H}^+$$
$$K_{\text{FeS}-\text{Fe(OH)}_2} = *K_{s(\text{FeS})}/*K_{s0(\text{Fe(OH)}_2)} \quad \text{(v)}$$

$$\text{H}_2\text{S(aq)} = \text{HS}^- + \text{H}^+$$
$$K_{1(\text{H}_2\text{S})} = 10^{-7.01} \; (10^{-6.92}) \quad \text{(vi)}$$

$$\text{H}_2\text{CO}_3^* = \text{HCO}_3^- + \text{H}^+$$
$$K_1 = 10^{-6.35} \; (10^{-6.27}) \quad \text{(vii)}$$

$$\text{HCO}_3^- = \text{CO}_3^{2-} + \text{H}^+$$
$$K_2 = 10^{-10.33} \; (10^{-10.11}) \quad \text{(viii)}$$

K values given in parentheses are corrected for ionic strength ($I = 10^{-2}$ M). The boundaries between Fe^{2+} (10^{-6} M) and FeS(s) or FeCO$_3$(s) as a function of S_T and pH can be calculated from equations i and ii. [HS$^-$] and [HCO$_3^-$], respectively, in equations i, ii, iv, and v can be replaced by

$$[\text{HS}^-] = S_T \, \alpha_{\text{HS}^-} \quad \text{(ix)}$$

and

$$[\text{HCO}_3^-] = [\text{Alk}] \left(1 + \frac{2 K_2}{[\text{H}^+]}\right)^{-1} \quad \text{(x)}$$

7.6 Sulfides and Phosphates 403

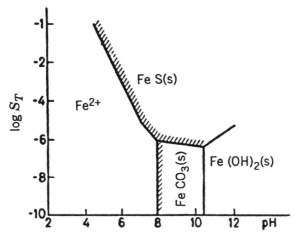

Figure 7.18. Concentration conditions for the existence of the solid phases FeS(s), FeCO$_3$(s), and Fe(OH)$_2$(s). Conditions [Fe^{2+}] = 10^{-6} M, Alk = 5 × 10^{-3} eq liter^{-1}. (Adapted from Sigg and Stumm, 1994.)

where α_{HS^-} is given by

$$\alpha_{HS^-} = \frac{K_{1(H_2S)}}{K_{1(H_2S)} + [H^+]} \tag{xi}$$

Figure 7.18 gives the results. It is based on a calculation where complexes of Fe(II) with S [e.g., Fe(SH)$_n^{2-n}$] are neglected.

Example 7.17. Solubility of CdS Daskalakis and Helz (1992) have carefully determined the solubility of CdS (greenockite). Their constants are as follows:

	log K (I = 0, 25°C)
CdS(s) + H$^+$ = Cd^{2+} + HS$^-$	−14.36
CdS(s) + H$^+$ = CdHS$^+$	≤ −6.7
CdS(s) + H$^+$ + HS$^-$ = Cd(HS)$_2^0$	−1.0
CdS(s) + H$^+$ + 2 HS$^-$ = Cd(HS)$_3^-$	2.08
CdS(s) + H$^+$ + 3 HS$^-$ = Cd(HS)$_4^{2-}$	3.53
CdS(s) + H$_2$O = CdOHS$^-$ + H$^+$	−16.83
H$_2$S(aq) = HS$^-$ + H$^+$	−7.01

Compute CdS(s) (greenockite) solubility as a function of total aqueous sulfide ([$S(-II)_T$]).

Tableau 7.8 can be established for I = 0.01.

The results are plotted for four different pH values in Figure 7.19. Obviously, the solubility depends on total sulfide and on pH. Lowest solubility is

Tableau 7.8. Solubility of Greenockite as a Function of $S(-II)_T$

Components		CdS(s)	HS$^-$	H$^+$	log K (25°C, I = 0.01)
Species	Cd^{2+}	1	−1	1	−14.19
	CdHS$^+$	1	0	1	−6.7
	Cd(HS)$_2^0$	1	1	1	−1.09
	Cd(HS)$_3^-$	1	2	1	1.99
	Cd(HS)$_4^{2-}$	1	3	1	3.51
	CdOHS$^-$	1	0	−1	−16.74
	H$_2$S	0	1	1	6.92
	HS$^-$	0	1	0	0
	H$^+$	0	0	1	0
		{CdS(s)} = 1		pH given	

achieved in mildly acid solution. As Daskalakis and Helz (1992) point out, amorphous CdS may be significantly (two orders of magnitude) more soluble than crystalline greenockite.

Phosphates

The distribution of the several acid and base species of orthophosphates and condensed phosphates in solution is governed by pH. Information on the pH-dependent distribution of the several species is required in interpreting the solubility behavior, complex formation, and sorption processes of phosphorus in water (Table 7.5). The predominant dissolved orthophosphate species over the pH range 5–9 are $H_2PO_4^-$ and HPO_4^{2-}. By employing the solubility equilibrium constants and the acidity constants, it is possible to compute total phosphate solubility (P_T) under specified conditions (pH, calcium concentrations, etc.). For example, we can compute soluble P_T for pure $AlPO_4(s)$ in contact with pure water whose pH is adjusted by addition of acid or base. When $Al(OH)_3(s)$ or $Al_2O_3(s)$ forms, the soluble P_T is then governed by an additional equilibrium condition. Similar considerations apply to $FePO_4(s)$ solubility. Figure 7.20 gives solubility diagrams for a few solid phosphate phases. These diagrams show that $FePO_4(s)$ (strengite) and $AlPO_4(s)$ (variscite) are the stable solid phases if phosphate is precipitated in the low-pH range; the pH of minimum $AlPO_4(s)$ solubility occurs at about 1 pH unit higher than that of $FePO_4(s)$. In the neutral pH range (on the right side of the solubility lines of $FePO_4$ and $AlPO_4$ in Figure 7.20), metastable hydroxophosphate Al(III) or Fe(III) precipitates can be formed.

Example 7.18. Conversion of Calcite into Apatite Under what condition (pH, concentration of inorganic P) can calcite be converted into hydroxylapatite? The simplifying assumptions $P_T \ll [Ca^{2+}]$ and $[Ca^{2+}] = C_T$ may be used.

7.6 Sulfides and Phosphates 405

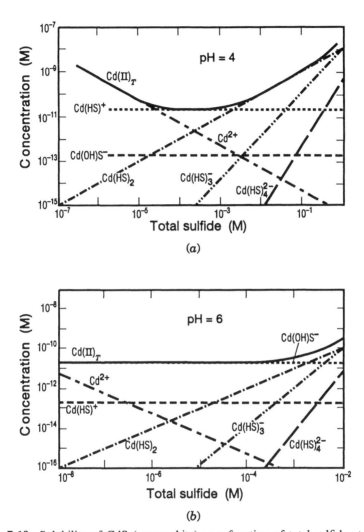

Figure 7.19. Solubility of CdS (greenockite) as a function of total sulfide at various pH values. The solubility is governed in the near neutral and alkaline pH range by Cd complexes with sulfur ligands. Minimum solubility is achieved in mildly acidic solutions. The solubility of precipitated CdS(s) may be larger by one or two orders of magnitude than that of greenockite. In seawater competition of complex formation with Cl^- needs to be considered. The lines are calculated (Tableau 7.8) with the help of stability constants determined by Daskalakis and Helz (1992).

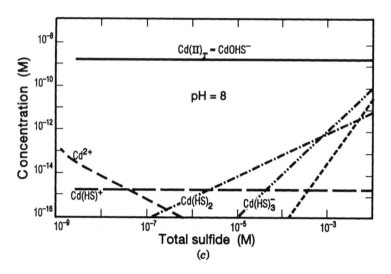

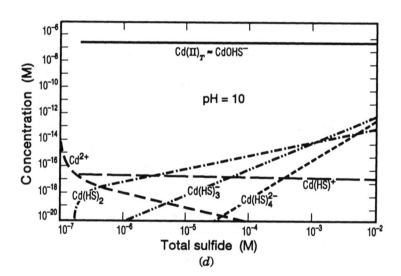

Figure 7.19. (*Continued*)

The reaction may be written

$$10CaCO_3(s) + 2H^+ + 6HPO_4^{2-} + 2H_2O$$
$$= Ca_{10}(PO_4)_6(OH)_2(s) + 10HCO_3^- \quad \text{(i)}$$

The equilibrium constant K for this reaction may be computed from

$$\log K = -\log K_{s0(\text{apatite})} + 10 \log K_{s0(\text{calcite})}$$
$$- 10 \log K_{HCO_3^-} + 6 \log K_{HPO_4^{2-}} + 2 \log K_W \quad \text{(ii)}$$

7.6 Sulfides and Phosphates

Table 7.5. Equilibrium Constants Related to the Solubility of Phosphates of Fe(III), Al(III), Fe(II), and Ca^{2+}

	log K (25°C, $I = 0$)
$FePO_4 \cdot 2H_2O(s)$ (strengite) $= Fe^{+3} + PO_4^{-3} + 2H_2O$	-26
$AlPO_4 \cdot 2H_2O(s)$ (variscite) $= Al^{+3} + PO_4^{-3} + 2H_2O$	-21
$CaHPO_4(s) = Ca^{+2} + HPO_4^{-2}$	-6.6
$Ca_4H(PO_4)_3(s) = 4Ca^{+2} + 3PO_4^{-3} + H^+$	-46.9
$Ca_{10}(PO_4)_6(OH)_2(s) = 10Ca^{+2} + 6PO_4^{-3} + 2OH^-$	-114
$Ca_{10}(PO_4)_6(F)_2(s) = 10Ca^{+2} + 6PO_4^{-3} + 2F^-$	-118
$Ca_{10}(PO_4)_6(OH)_2(s) + 6H_2O = 4[Ca_2(HPO_4)(OH)_2] + 2Ca^{+2} + 2HPO_4^{-2}$	-17
$CaHAl(PO_4)_2(s) = Ca^{+2} + Al^{+3} + H^+ + 2HPO_4^{-3}$	-39
$CaF_2(s) = Ca^{+2} + 2F^-$	-10.4
$MgNH_4PO_4(s) = Mg^{+2} + NH_4^+ + PO_4^{-3}$	-12.6
$FeNH_4PO_4(s) = Fe^{+2} + NH_4^+ + PO_4^{-3}$	~ -13
$Fe_2(PO_4)_2(s) = 3Fe^{+2} + 2PO_4^{-3}$	~ -32

The equilibrium constant for equation i is of course at least as uncertain as the solubility product of apatite. With a value of K_{s0} for $Ca_{10}(PO_4)_6(OH)_2(s)$ of 10^{-114}, an equilibrium constant of $K = 10^{32}$ is obtained. Hence the free energy of this conversion, $\Delta G = RT \ln(Q/K)$, where Q is the quotient of the reactants, may be used to predict under what conditions reaction i is possible. For example, at pH $= 8$ and $[HCO_3^-] = 10^{-3}$ M, a 10^{-4} M solution of HPO_4^{2-} would, thermodynamically speaking, convert calcite into apatite because ΔG is approximately -125 kJ mol^{-1} of apatite formed. Figure 7.21 plots concentrations of soluble phosphate, P_T, necessary to convert $CaCO_3$ into apaptite as a function of pH.

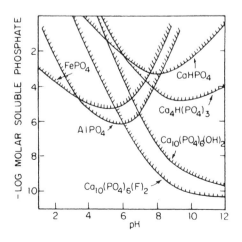

Figure 7.20. Solubility of the metal phosphates. The solubilities of $AlPO_4$ and $FePO_4$ have been calculated on the basis of the equilibria assuming that $FePO_4s(s)$ or $AlPO_4(s)$ can be converted into $Fe(OH)_3(s)$ [or α-FeOOH(s)] or $Al(OH)_3(s)$. The solubility of the calcium phosphate phases has been calculated under the assumption that $[Ca^{2+}] = 10^{-3}$ M and that F^- is regulated by the solubility of $CaF_2(s)$.

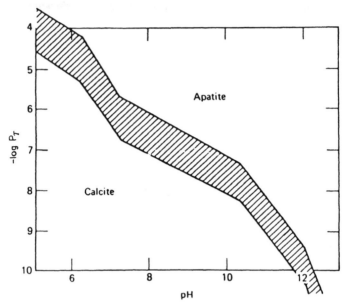

Figure 7.21. Phosphate necessary to convert $CaCO_3(s)$ into $Ca_{10}(PO_4)_6(OH)_2(s)$ (Example 7.18). Calculated by assuming that the solution remains in saturation equilibrium with $CaCO_3(s)$; furthermore, $P_T \ll C_T$ and $[Ca^{2+}] = C_T$. The equilibrium constant for equation i has been taken as 10^{28} (upper line) and 10^{30} (lower line), respectively.

Although equation i represents an oversimplification (apatite is not necessarily formed as a pure solid phase), the tentative result obtained suggests that the phosphorus concentration at the sediment–water interface, especially at higher pH values, is buffered by the presence of hydroxyapatite.

The mineral or chemical composition of phosphorus compounds tends to be different in different environments. Variscite ($AlPO_4 \cdot 2H_2O$) and strengite ($FePO_4 \cdot 2H_2O$) appear to be more common in soils and freshwater sediments, whereas apatite prevails in certain marine sediments. However, the solubility product of apatite does not appear to be exceeded on the deep-ocean floor. In areas of high organic productivity in sediments of eutrophic lakes and of the shallow areas of the ocean, especially the tropical ocean, calcium phosphate minerals (substituted apatites, such as francolite, $Ca_{10}[PO_4CO_3]_6F_2$) are deposited. The main sink of phosphates in the ocean and in many lakes consists of iron(III) oxides on the surface of which phosphates become chemisorbed.

Example 7.19. The Solubility of $MgNH_4PO_4(s)$; Conditional Solubility Product In which pH range is the precipitation of $MgNH_4PO_4$ possible from a water containing $Mg_T(= [Mg^{2+}] + [MgOH^+]) = 10^{-2}$ M, $N_T(= [NH_4^+] + [NH_3]) = 10^{-3}$ M, and $P_T(= [H_3PO_4] + [H_2PO_4^-] + [HPO_4^{2-}] + [PO_4^{3-}]) = 10^{-4}$ M? The conditional solubility product of $MgNH_4PO_4(s)$ is defined by

7.7 The Phase Rule: Components, Phases, and Degrees of Freedom

$$P_s = \frac{K_{s0MgNH_4PO_4}}{\alpha_{Mg}\alpha_N\alpha_P} = Mg_T \times N_T \times P_T \tag{i}$$

The following constants are available (25°C, $I = 0$):

$\log K_{s0MgNH_4PO_4} = -12.6$
$\log {}^*K_{1(\text{hydrolysis of Mg}^{2+})} = -11.4$
$\log K_{1(NH_4^+)} = -9.24$
$\log K_{1(H_3PO_4)} = -2.1$
$\log K_{2(H_2PO_4^-)} = -7.2$
$\log K_{3(HPO_4^{2-})} = -12.3$

We first use these K values without correcting for salinity. (But for a more rigorous answer the K values must be converted into ${}^cK$ values.)

$$\alpha_{Mg} = \left(1 + \frac{{}^*K_1}{[H^+]}\right)^{-1}$$

$$\alpha_N = \left(1 + \frac{K_{1(NH_4^+)}}{[H^+]}\right)^{-1}$$

$$\alpha_P = \left(1 + \frac{[H^+]}{K_3} + \frac{[H^+]^2}{K_2K_3} + \frac{[H^+]^3}{K_1K_2K_3}\right)^{-1}$$

The α values are plotted as a function of pH in Figure 7.22. The conditional solubility product reaches its minimum at pH $= 10.7$ [$= \frac{1}{2}(pK_{3(HPO_4^{2-})} + pK_{(NH_4^+)})$]. Thus precipitation of MgNH$_4$PO$_4$(s) is favored in alkaline solutions. The conditional solubility product can now be compared with the product of the actual concentrations, $Q_{sT} = N_T \times P_T \times Mg_T = 10^{-9}$. As Figure 7.22 shows, only within the pH range 9–12 is $Q_{sT} > P_s$; in principle, a precipitation is possible. However, because the difference between pQ_{sT} and pP_s is quite small (<0.6), no efficient precipitation appears possible. Furthermore, the effect of ionic strength has been ignored. When a Güntelberg approximation and an ionic strength of $I = 0.1$ are used, pK_{s0} (hence p^cP_{sT}) becomes smaller by approximately 1.6 units; the solution actually does not become oversaturated with respect to MgNH$_4$PO$_4$(s).

7.7 THE PHASE RULE: COMPONENTS, PHASES, AND DEGREES OF FREEDOM

In some of the examples discussed so far (e.g., Examples 7.12–7.15), we asked the question: Which solid phase controls the solubility? In addition, we may address the problem of coexistence of phases and ask: How many phases can coexist under the conditions given?

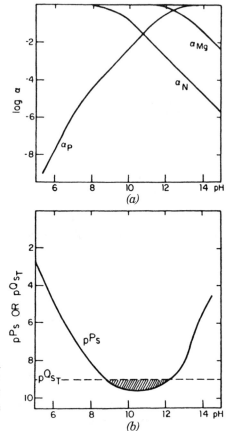

Figure 7.22. Solubility of MgNH$_4$PO$_4$ (Example 7.19). The conditional solubility product $P_S = P_T \times Mg_T \times N_T$ (b) is readily calculated from the α values (a). Minimum solubility at pH $\simeq$ 10.7.

The *phase rule*, derived from thermodynamics by Gibbs, is an important ordering principle, which establishes in equilibrium systems (or models) the relationship between the number of components, the number of phases, and the degrees of freedom:

$$F = C + 2 - P \tag{59}$$

where F = degrees of freedom, that is, independent variables such as concentration conditions, temperature, and pressure.

C = number of components, that is, the minimum number of chemical entities (salts, minerals, molecules, ions, electrons) that are necessary to describe (or duplicate) the system. As we have shown in our tableaux, a system can be described by a variety of choices of components. [Note that in our tableaux, the component H$_2$O(l) with activity = 1 has not been specifically listed.]

7.7 The Phase Rule: Components, Phases, and Degrees of Freedom

P = number of phases. A phase is a domain with uniform composition and properties, for example, a gas, a gaseous mixture, a homogeneous liquid solution, a uniform solid substance, or a solid solution.

In the chapters so far we have considered the phase rule somewhat intuitively; for example, in solving equilibrium problems we used the obvious principle that an equilibrium problem can be solved if for n unknowns (e.g., activities or concentrations of n species) n equations are available. For example, in a closed dissolved carbonate system we need to define the system ($H_2CO_3^*$, HCO_3^-, CO_3^{2-}, H^+, OH^-) and two concentration conditions (e.g., C_T and pH, or [Alk] and $H_2CO_3^*$]), in addition to temperature and pressure, because the five species are interconnected by three mass laws (two acid–base equilibria and the ion product of H_2O). In the example given: $P = 1$ (aqueous solution), $C = 3$ [e.g., HCO_3^-, H^+, $H_2O(l)$], and $F = 4$ (pressure, temperature, and two concentration conditions).

In a similar way, we have implicitly considered the phase rule when we stated, in introducing the tableaux, that the number of components equals the number of species minus the number of independent reactions. An equivalent statement is that in each phase $C - 1$ concentration conditions (including relations for charge balance or proton conditions), at a given temperature and pressure, are necessary in order to describe the system.

The application of the phase rule can be illustrated with Example 7.16. The components of the system are, for example, HCO_3^-, Fe^{2+}, HS^-, H^+, and H_2O. If there is no solid or gas phase present, $F = 5 + 2 - 1 = 6$; that is, we need in addition to T and p, four concentration conditions to describe the system (e.g., S_T, Fe_T, C_T, and pH). If a solid phase, for example, FeS(s), is also present, three concentration conditions (e.g., Fe_T, C_T, and pH) are sufficient to define the system. (As shown in Figure 7.18, for a given C_T and Fe_T, S_T is a function of pH.) If there are three phases [aqueous solution, $FeCO_3$(s), FeS(s)] only two degrees of freedom (in addition to T and p) remain. In Figure 7.18 at the point at which three phases coexist (for the concentration conditions Fe_T and C_T), pH and S_T are given. This simple example illustrates that for a given number of components, every additional phase present in the system reduces the number of degrees of freedom by one, or, in other words, the number of coexisting phases is limited.

Example 7.20. Coexistence of Different Phases Illustrate on the basis of the open $CaCO_3$–CO_2 system (Tableau 7.9) the application of the phase rule.

As shown in Table 7.6 (Type 1), the composition is defined by the proton condition and two independent variables such as temperature and partial pressure of CO_2, p_{CO_2}. When TOTH = 0, this system is constant and independent of concentration (isothermal dilution or evaporation) as long as all three phases remain in equilibrium with each other.

If we add a further phase, for example, $Ca(OH)_2$(s), to the system (maintaining the same components, see Type 2 of Table 7.6) and fix TOTH = 0,

Tableau 7.9. Open CO_2 System with $CaCO_3$ (Calcite)

Components		$CO_2(g)$	$CaCO_3$	H^+	log K ($I = 0$, 25°C)
Species	H_2CO_3	1	0	0	−1.5
	HCO_3^-	1	0	−1	−7.8
	CO_3^{2-}	1	0	−2	−18.1
	Ca^{2+}	−1	1	2	9.8
	OH^-	0	0	−1	−14.0
	H^+	0	0	1	0
		p_{CO_2} given	$\{CaCO_3\} = 1$	0	

only one independent variable remains: if the temperature is fixed, the composition of the system is given; that is, p_{CO_2} is defined. We now have an example of a "manostat," a system of constant composition and fixed p_{CO_2}. In a sense, we have an infinite buffer intensity.

In natural systems, a multitude of phases for a limited number of components enhances the resilience of the equilibrium system toward change.

Table 7.6. System: $CO_2(g)$, $CaCO_3(s)$, $H_2O(l)$, and H^+; Application of Phase Rule

Type	Number of Phases	Number of Components[a]	Degrees of Freedom	Examples of Degrees of Freedom[b,c]
1. Calcite(s) aqueous solution $CO_2(g)$	$P = 3$	$C = 4$ Example: H^+, $CO_2(g)$ $H_2O(l)$ $CaCO_3(s)$	$F = 3$	TOTH and T and p_{CO_2}
2. Calcite(s) $Ca(OH)_2(s)$ aqueous solution $CO_2(g)$	$P = 4$	$C = 4$ Example: H^+, $CO_2(g)$ $H_2O(l)$ $CaCO_3(s)$	$F = 2$	TOTH and T

[a] All species of the system—H^+, HCO_3^-, CO_3^{2-}, $H_2CO_3^*$, OH^-, $CO_2(g)$, $CaCO_3(s)$, $Ca(OH)_2(s)$, $H_2O(l)$—can be composed from these components, for example, $Ca(OH)_2(s) = CaCO_3 - CO_2 - H_2O$.
[b] Proton condition. When TOTH = 0 (TOTH = $[H^+] - [OH^-] - [HCO_3^-] - 2[CO_3^{2-}] + 2[Ca^{2+}] = 0$; see third vertical column in Tableau 7.8), the system corresponds to a $CaCO_3(s) - CO_2(g) - H_2O$ system, to which no acid or base has been added. TOTH = 0 defines also the charge balance.
[c] By defining a partial pressure, a total pressure is also implicitly defined.

7.8 SOLUBILITY OF FINE PARTICLES

Finely divided solids have a greater solubility than large crystals. As a consequence, small crystals are thermodynamically less stable and should recrystallize into large ones. For particles smaller than about 1 μm or of specific surface area greater than a few square meters per gram, surface energy may become sufficiently large to influence surface properties. Similarly, the free energy of a solid may be influenced by lattice defects such as dislocations and other surface heterogeneities.

The change in the free energy ΔG involved in subdividing a coarse solid suspended in aqueous solution into a finely divided one of molar surface S is given by

$$\Delta G = \tfrac{2}{3} \bar{\gamma} S \tag{60}$$

where $\bar{\gamma}$ is the mean free surface energy (interfacial tension) of the solid–liquid interface. [For the derivation of this equation see Schindler (1967)].

Thus

$$\left(\frac{\partial \ln K_{s0}}{\partial S}\right)_T = \frac{2\,\bar{\gamma}}{3\,RT} \tag{61a}$$

or

$$\log K_{s0(\bar{S})} = \log K_{s0(S=0)} + \frac{\tfrac{2}{3}\bar{\gamma}}{2.3\,RT} S \tag{61b}$$

Figure 7.23 illustrates the effect of molar surface (or particle size) on the solubility product of ZnO.

Metastability and Particle Size The particle size may play a significant role in the inversion from one polymorphous form to another. Following arguments presented by Schindler (1967), we may consider the reaction

$$Cu(OH)_2(s) = CuO(s) + H_2O(l) \tag{62}$$

Schindler and his co-workers determined the solubility constants and the influence of molar surface on solubility:

$$Cu(OH)_2: \quad \log {}^*K_{s0} = 8.92 + 4.8 \times 10^{-5} S$$

$$\bar{\gamma} = 410 \pm 130 \times 10^{-7} \text{ J cm}^{-2}$$

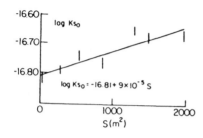

Figure 7.23. Effect of molar surface on solubility product of ZnO (25°C, $I = 0$). (From Schindler, 1967.)

CuO: $\log {}^*K_{s0} = 7.89 + 8 \times 10^{-5} S$

$\bar{\gamma} = 690 \pm 150 \times 10^{-7}$ J cm^{-2}

Thus the solubility of Cu(OH)$_2$ is approximately ten times greater than that of CuO. The inversion of Cu(OH)$_2$ into CuO (reaction (62)) should occur exergonically. However, if CuO is very finely divided ($S_{CuO} > 12{,}000$ m^2), it becomes less stable than coarse Cu(OH)$_2$ ($S_{Cu(OH)_2} = 0$):

$$\Delta G \text{ (cal)} = -1400 + 0.109 S_{CuO} - 0.065 S_{Cu(OH)_2} \qquad (63)$$

Figure 7.24a plots the enhancement of the solubility of CuO and Cu(OH)$_2$ (pH = 7.0) with an increase in molar surface. Recall that the solid phase, which gives the smaller solubility, is the thermodynamically more stable phase. As Figure 7.24 illustrates, at sufficiently large subdivision of the solids ($S_{CuO} = S_{Cu(OH)_2} > 31{,}000$ m^2; $d \simeq 40$ Å) Cu(OH)$_2$(s) may become more stable (less soluble) than CuO(s); this results from the larger interfacial energy of the oxide as compared to the hydroxide. This dependence of stability (solubility) on particle size may be one of the reasons why crystal nuclei of Cu(OH)$_2$ are formed incipiently when Cu(II) is precipitated; upon subsequent crystal growth, and a corresponding decrease in S, CuO becomes more stable and Cu(OH)$_2$ is inverted into CuO(s). These arguments are schematic and simplified. The assumption that $\bar{\gamma}$ remains independent of surface area probably breaks down when we deal with nuclei consisting of a very small number of ions or molecules. Furthermore, the relations given are valid only for particles of uniform size.

Figure 7.24c shows variation of the reaction free energy with particle size for the conversion of hematite to goethite. The stability relation depends on particle size. Figure 7.24b shows the effect of particle size on solubility of calcite.

7.9 SOLID SOLUTIONS

So far we have assumed for the most part that we deal with pure solid phases; their activity is one. The solids occurring in nature are seldom pure solid

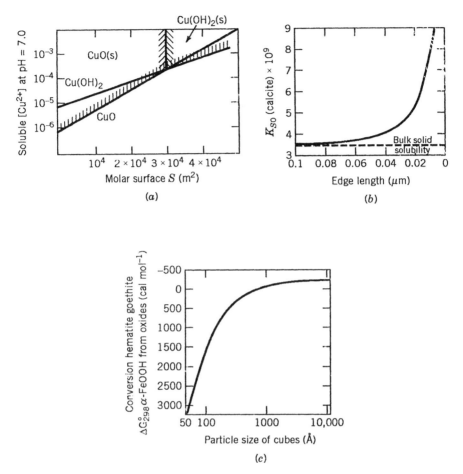

Figure 7.24. (a) Influence of molar surface on solubility of CuO and of $Cu(OH)_2$ at pH = 7.0. [From data on solubility constants and surface tensions by Schindler (1967). The relations depicted have been validated experimentally only for $S < 10^4$ m².] The figure suggests that $Cu(OH)_2(s)$ becomes more stable than CuO(s) for very finely divided CuO crystals ($S > 3 \times 10^4$ m², $d < 40$ Å). Plausibly, in precipitating Cu(II), $Cu(OH)_2(s)$ may be precipitated (d = very small), but CuO(s) becomes more stable than $Cu(OH)_2$ upon growth of the crystals, and an inversion of $Cu(OH)_2$ into the more stable phase becomes possible. (b) Change in calcite solubility with particle size, assuming cubic shape and $\bar{\gamma} = 85$ mJ m^{-2}. (From Morse and Mackenzie, 1990.) (c) $\Delta G°$ for the reaction $\frac{1}{2} \alpha\text{-Fe}_2O_3 + \frac{1}{2} H_2O = \alpha\text{-FeOOH}$ is plotted as a function of particle size assuming equal particle size for goethite and hematite. For equal-sized hematite and goethite crystals, goethite is more stable than hematite when the particle size exceeds 760 Å but less stable than hematite at smaller particle sizes. (From Langmuir and Whittemore, 1971.)

phases. Isomorphous replacement by a foreign constituent in the crystalline lattice is an important factor by which the activity of the solid phase may be decreased. If the solids are homogeneous, that is, contain no concentration gradient, one speaks of homogeneous solid solutions. The thermodynamics of solid solutions has been treated by Vaslow and Boyd (1952).

To express the relationship theoretically, we consider a two-phase system such as $CaCO_3$, $CdCO_3$, where some of the $CdCO_3(s)$ becomes dissolved in $CaCO_3(s)$ (calcite). The reaction may be characterized by the equilibrium

$$CaCO_3(s) + Cd^{2+} \rightleftharpoons CdCO_3 + Ca^{2+} \quad (64)$$

Its equilibrium constant is defined by

$$\frac{\{CdCO_3(s)\}\{Ca^{2+}\}}{\{CaCO_3(s)\}\{Cd^{2+}\}} = \frac{X_{CdCO_3} f_{CdCO_3} \{Ca^{2+}\}}{X_{CaCO_3} f_{CaCO_3} \{Cd^{2+}\}} = \frac{K_{s0(CaCO_3)}}{K_{s0(CdCO_3)}} = D$$

$$\frac{10^{-8.37}}{10^{-11.3}} \simeq 10^3 \quad (65)$$

where the activity ratio of the solids is replaced by the ratio of the mole fractions $[X_{CdCO_3} = n_{CdCO_3}/(n_{CaCO_3} + n_{CdCO_3})]$ multiplied by activity coefficients. According to this equation, the extent of dissolution of Cd^{2+} in solid $CaCO_3$ is a function of the following:

1. the solubility product ratio of $CaCO_3$ and $CdCO_3(s)$,
2. the solution composition, that is, the activity ratio of Cd^{2+} to Ca^{2+}, and
3. a solid solution (nonideality) factor given by the ratio of the activity coefficients of the solid solution components f_{CaCO_3}/f_{CdCO_3}. As a first approximation we will assume that the ratio of activity coefficients is equal to unity.

Consider a solution of 5% $CdCO_3$ in $CaCO_3$ (95%) in equilibrium with Ca^{2+} and Cd^{2+}. The composition of the suspension is:

Aqueous Phase (M)	Solid Phase
$[CO_3^{2-}] = 10^{-5}$	$X_{CaCO_3} = 0.95$
$[Ca^{2+}] = 10^{-3.3}$	$X_{CdCO_3} = 0.05$
$[Cd^{2+}] = 10^{-7.6}$	

The composition of the equilibrium system shows that Cd^{2+} has been enriched significantly in the solid phase in comparison to the aqueous phase. The

7.9 Solid Solutions

solubility of $CdCO_3(s)$ as a minor constituent in the solid solution is reduced in comparison to the solubility of pure $CdCO_3(s)$ (where $[Cd_0^{2+}] = 10^{-6.3}$ M). If one considered the solubility of $CdCO_3$ alone, one might infer by neglecting the solid solution that the solution ($[Cd^{2+}] = 10^{-7.6}$ M $< [Cd_0^{2+}] = 10^{-6.3}$ M) is undersaturated with respect to $CdCO_3(s)$. In fact, it is not.

Although the solubility of the salt that represents the major component of the solid phase is only slightly affected by the formation of solid solutions, the solubility of the minor component is appreciably reduced. The observed occurrence of certain metal ions in sediments formed from solutions that appear to be formally (in the absence of any consideration of solid solution formation) unsaturated with respect to the impurity can, in many cases, be explained by solid solution formation.

Usually, however, the distribution coefficients determined experimentally are not equal to the ratios of the solubility products, because the ratio of the activity coefficients of the constituents in the solid phase cannot be assumed to be equal to 1. Actually, observed D values show that activity coefficients in the solid phase may differ markedly from 1. Let us consider, for example, the coprecipitation of $MnCO_3$ in calcite. Assuming that the ratio of the activity coefficients in the aqueous solution is close to unity, the equilibrium distribution may be formulated as

$$\frac{K_{s0CaCO_3}}{K_{s0MnCO_3}} = D = \frac{[Ca^{2+}]}{[Mn^{2+}]} \frac{X_{MnCO_3}}{X_{CaCO_3}} \frac{f_{MnCO_3}}{f_{CaCO_3}} \tag{66a}$$

$$D = D_{obs} \frac{f_{MnCO_3}}{f_{CaCO_3}} \tag{66b}$$

The solubility product quotient at 25°C ($pK_{s0MnCO_3} = 11.09$, $pK_{s0CaCO_3} = 8.37$) can now be compared with an experimental value of D. The data of Bodine et al. (1965) give $D_{obs} = 17.4$ (25°C). Because D is smaller than the ratio of the K_{s0} values, the solid solution factor acts to lower the solution of $MnCO_3$ in $CaCO_3$ significantly from that expected if an ideal mixture has been formed. If it is assumed that in dilute solid dilutions (X_{MnCO_3} very small) the activity coefficient of the "solvent" is close to unity ($X_{CaCO_3} f_{CaCO_3} \simeq 1$), an activity coefficient for the "solute" is calculated to be $f_{MnCO_3} = 31$. Qualitatively, such a high activity coefficient reflects a condition similar to that of a gas dissolved in a concentrated electrolyte solution, where the gas, also characterized by an activity coefficient larger than unity, is "salted out" from the solution.

Example 7.21. Solid Solution of Sr^{2+} in Calcite It has been proposed that a solid solution of Sr^{2+} in calcite might control the solubility of Sr^{2+} in the ocean.

Estimate the composition of the solid solution phase (X_{SrCO_3}). The following information is available: the solubilities of $CaCO_3(s)$ calcite and $SrCO_3(s)$ in seawater at 25°C are characterized by $p^cK_{s0} = 6.1$ and 6.8, respectively. The equilibrium concentration of CO_3^{2-} is $[CO_3^{2-}] = 10^{-3.6}$ M. The actual concentration of Sr^{2+} in seawater is $[Sr^{2+}] \simeq 10^{-4}$ M. For the distribution, the following equilibrium constant has been found:

$$\left(\frac{[Ca^{2+}]}{[Sr^{2+}]}\right)\left(\frac{X_{SrCO_3}}{X_{CaCO_3}}\right) = 0.14 \quad (25°C) \tag{i}$$

Assuming a saturation equilibrium of seawater with Sr^{2+}-calcite, the equilibrium concentrations would be $[Ca^{2+}] \simeq 10^{-2.5}$ $(= K_{s0}/[CO_3^{2-}])$ and $[Sr^{2+}] = 10^{-4}$ M (= actual concentration). Thus $X_{SrCO_3}/X_{CaCO_3} = 0.004$ and $X_{CaCO_3} = 0.996$.

It may be noted that, since the distribution coefficient is smaller than unity, the solid phase becomes depleted in strontium relative to the concentration in the aqueous solution. The small value of D may be interpreted in terms of a high activity coefficient of strontium in the solid phase, $f_{SrCO_3} \simeq 38$. If the strontium were in equilibrium with strontianite, $[Sr^{2+}] \simeq 10^{-3.2}$ M, its concentration would be more than six times larger than at saturation with $Ca_{0.996}Sr_{0.004}CO_3(s)$. This is an illustration of the consequence of solid solution formation where with $X_{CaCO_3} f_{CaCO_3} \simeq 1$:

$$[Sr^{2+}] = \frac{X_{SrCO_3} f_{SrCO_3} {}^cK_{s0SrCO_3}}{[CO_3^{2-}]} \tag{ii}$$

that is, *the solubility of a constituent is greatly reduced when it becomes a minor constituent of a solid solution phase.*

Heterogeneous Solid Solutions

Besides homogeneous solid solutions, *heterogeneous arrangement* of foreign ions within the lattice is possible. While homogeneous solid solutions represent a state of true thermodynamic equilibrium, heterogeneous solid solutions can persist in metastable equilibrium with the aqueous solution. Heterogeneous solid solutions may form in such a way that each crystal layer as it forms is in distribution equilibrium with the particular concentration of the aqueous solution existing at that time (Doerner and Hoskins, 1925; Gordon et al., 1959). Correspondingly, there will be a concentration gradient in the solid phase from the center to the periphery. Such a gradient results from very slow diffusion within the solid phase. Following the treatment given by Doerner and Hoskins, the distribution equilibrium for the reaction

7.9 Solid Solutions

$$CaCO_3(s) + Mn^{2+} = MnCO_3(s) + Ca^{2+} \qquad (67)$$

is written as in equation (66a), but we consider that the crystal surface is in equilibrium with the solution:

$$\left(\frac{[MnCO_3]}{[CaCO_3]}\right)_{\text{crystal surface}} \times \frac{[Ca^{2+}]}{[Mn^{2+}]} = D' \qquad (68)$$

If $d[MnCO_3]$ and $d[CaCO_3]$, the increments of $MnCO_3$ and $CaCO_3$ deposited in the crystal surface layer, are proportional to their respective solution concentrations, equation (69) is obtained:

$$\frac{d[MnCO_3]}{d[CaCO_3]} = D' \frac{[Mn^{2+}]_0 - [Mn^{2+}]}{[Ca^{2+}]_0 - [Ca^{2+}]} \qquad (69)$$

or, after rearrangement,

$$\frac{d[MnCO_3]}{[Mn^{2+}]_0 - [Mn^{2+}]} = D' \frac{d[CaCO_3]}{[Ca^{2+}]_0 - [Ca^{2+}]} \qquad (70)$$

where $[Ca^{2+}]_0$ and $[Mn^{2+}]_0$ represent initial concentrations in the aqueous solution. Integration of equation 70 leads to

$$\log \frac{[Mn^{2+}]_0}{[Mn^{2+}]_f} = D' \log \frac{[Ca^{2+}]_0}{[Ca^{2+}]_f} \qquad (71)$$

where $[Mn^{2+}]_f$ and $[Ca^{2+}]_f$ represent final concentrations in the aqueous solutions.

Most of the distribution coefficients measured to date for a variety of relatively insoluble solids are characterized by the Doerner–Hoskins relation. This relationship is usually obeyed for crystals that have been precipitated from homogeneous solution (Gordon et al., 1959). If the precipitation occurs in such a way that the aqueous phase remains as homogeneous as possible and the precipitant ion is generated gradually throughout the solution, large, well-formed crystals likened to the structure of an onion are obtained. Each infinitesimal crystal layer is equivalent to a shell of an onion. As each layer is deposited, there is insufficient time for reaction between solution and crystal surface before the solid becomes coated with succeeding layers. Kinetic factors make the metastable persistence of such compounds possible for relatively long—often for geological—time spans.

SUGGESTED READINGS

Berner, R. A. (1980) *Early Diagenesis, a Theoretical Approach*, Princeton University Press, Princeton.

Butler, J. N. (1982) *Carbon Dioxide Equilibria and Their Applications*, Lewis, Chelsea, MI.

Davison, W. (1991) The Solubility of Iron Sulfides in Synthetic and Natural Waters at Ambient Temperatures, *Aquatic Sci.* **53/54**, 309–329.

Dyrssen, D. (1985) Metal Complex Formation in Sulphidic Seawater, *Mar. Chem.* **15**, 285–293.

Nordstrom, D. K., and Munoz, J. L. (1985) *Geochemical Thermo-dynamics*, Benjamin-Cummings, Redwood City, CA.

Siegenthaler, J., and Sarmiento, J. L. (1993) Atmospheric CO_2 and the Ocean, *Nature* **365**, 119–125.

Wollast, R., and Vanderborght, J. P. (1994) Aquatic Carbonate Systems; Chemical Processes in Natural Waters and Global Cycles. In *Chemistry of Aquatic Systems: Local and Global Perspectives*, S. Bidoglio and W. Stumm, Eds., Kluwer, Dordrecht.

PROBLEMS

7.1. (a) How much Fe^{2+} (expressed in mol liter^{-1}) could be present in a 10^{-2} M $NaHCO_3$ solution without causing precipitation of $FeCO_3$? The solubility product of $FeCO_3$ is $K_{s0} = 10^{-10.7}$.

(b) How would this maximum soluble Fe^{2+} concentration change on lowering the pH of that solution by 1 unit?

7.2. Soil chemists have used the following equation for soil water:

$$\log[Ca^{2+}] + 2\text{pH} = \text{const} - \log p_{CO_2}$$

Under what conditions does this equation hold? Express the constant in terms of known equilibrium constants.

7.3. The following concentrations were measured in a 35-m deep lake during summer stagnation:

Depth: 0 m   pH = 8.5
             Ca^{2+} = 1.2×10^{-3} M
             Alk = 3×10^{-3} eq liter^{-1}
             T = 20°C

Depth: 30 m pH = 7.5
Ca^{2+} = 1.6 × 10^{-3} M
Alk = 4 × 10^{-3} eq liter^{-1}
T = 5°C

(a) How can one account qualitatively for the difference between the upper layers and the lower layers?
(b) In these cases is $CaCO_3$ over- or undersaturated?

7.4. If a sample of deep seawater were returned to the laboratory and stored at 20°C and 1-atm pressure, how would its Ca^{2+} and CO_3^{2-} concentration and pH change?

7.5. How much acid or base has to be added to a saturated solution of $CaCO_3$ to adjust the soluble Ca^{2+} level to (a) 10^{-3} M and (b) 5 × 10^{-4} M if the solution is shielded from the atmosphere?

7.6. Tillmans (1907) made investigations on the solubility of marble ($CaCO_3$) in calcium bicarbonate solutions containing different amounts of $H_2CO_3^*$. He shook his solutions for 10 days in closed flasks with marble chips. After this period equilibrium was attained. He determined experimentally the equilibrium concentrations of the $H_2CO_3^*$ and the alkalinity. For all waters that showed pH values below 8.5, after the saturation with $CaCO_3$, he found the following empirical relation: $[H_2CO_3^*]$ = $K[Alk]^3$.
(a) Show that this relationship can be derived by mass law considerations.
(b) Express K of the above equation in terms of K_1 = first acidity constant of $H_2CO_3^*$, K_2 = second acidity constant of $H_2CO_3^*$, K_{s0} = solubility product of $CaCO_3$.

7.7. A groundwater with

pH = 7.5
Ca^{2+} = 2 × 10^{-3} M
Alk = 4 × 10^{-3} M
T = 20°C

is pumped to the surface and there exposed to atmospheric CO_2. How is its composition changed?

7.8. Some solid $CaCO_3$ is in equilibrium with its saturated solution. How will the amount of Ca^{2+} in the solution be affected (increase, decrease, or no effect) by adding small amounts of the following:
(a) KOH
(b) $CaCl_2$
(c) $(NH_4)_2SO_4$
(d) Sodium metaphosphate $(NaPO_3)_n$
(e) $FeCl_3$
(f) Na_2CO_3
(g) H_2O
(h) By increasing pressure

7.9. Consider the system containing phases $Ca_{10}(PO_4)_6(OH)_2(s)$, $CaCO_3(s)$, $CaHPO_4(s)$, aqueous solution, and gas $(CO_2(g))$ in equilibrium at a temperature of 10°C.
(a) Give suitable components and degrees of freedom for this system.
(b) Under conditions of equilibrium does $[Ca^{2+}]$ increase, decrease, or stay constant upon addition of small quantities of the following:
(i) H_3PO_4, (ii) CO_2, (iii) NaOH, (iv) $Ca(OH)_2$, (v) HCl, (vi) H_2O?

7.10. Newer measurements Bruno et al., (1992) on the complex formation of Fe(III) with carbonate have shown that the complexes $Fe^{III}(OH)CO_3$ and $Fe(CO_3)_2^-$ are formed so that, for the dissolution of $Fe(OH)_3(s)$, the following constants can be postulated:

$$Fe(OH)_3(s) + CO_2(g) = Fe(OH)CO_3(aq) + H_2O$$

$$\log K = -4.9$$

$$Fe(OH)_3(s) + 2\ CO_2(g) = Fe(CO_3)_2^-(aq) + H^+ + H_2O$$

$$\log K = -11.4$$

[See Example 7.7 for data on the solubility of $Fe(OH)_3(s)$ and on the hydrolysis of Fe^{3+}.] Do carbonato complexes account for the solubility of $Fe(OH)_3(s)$ in equilibrium with atmospheric $CO_2(g)$ ($p_{CO_2} = 3.5 \times 10^{-4}$ atm)?

7.11. Below are experimentally determined data by various authors (see Davison, 1991) plotted as $-\log [Fe^{2+}]$ versus pH for amorphous iron sulfide at 298 K and 0.1 MPa H_2S ($= 10^{-4}$ atm). The data with numbers refer to various aged (hours) iron sulfide. What solubility constant $*K_s$ for FeS (amorphous) $+ H^+ = Fe^{2+} + HS^-$ can be derived from these data? How does the solubility change with the age of the precipitated FeS?

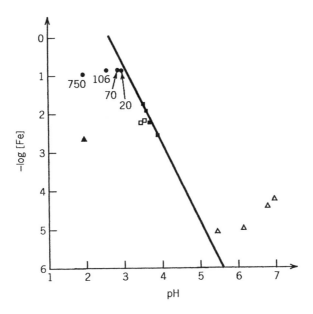

7.12. Explain the following statements (from Morel and Hering, 1993). For example, silica, aluminum oxide (gibbsite), and kaolinite cannot coexist at equilibrium:

$$2SiO_2(s) + Al_2O_3 \cdot 3H_2O(s) \rightarrow Al_2O_3 \cdot 2SiO_2 \cdot 2H_2O + H_2O$$
$$\text{(silica)} \quad\quad \text{(gibbsite)} \quad\quad\quad\quad \text{(kaolinite)}$$

and neither can calcite, kaolinite, and anorthite under a fixed (arbitrary) pressure of CO_2:

$$CaO \cdot Al_2O_3 \cdot 2SiO_2 + CO_2(g) + 2H_2O \rightarrow$$
$$\text{(anorthite)}$$

$$Al_2O_3 \cdot 2SiO_2 \cdot 2H_2O + CaCO_3(s)$$
$$\text{(kaolinite)} \quad\quad\quad \text{(calcite)}$$

7.13. A wastewater contains 10^{-4} M orthophosphate. Fe(III) is added to the system to precipitate phosphate. Is $FePO_4(s)$ or $Fe(OH)_3(s)$ precipitated at pH = 7.5?

$$FePO_4(s) = Fe^{3+} + PO_4^{3-} \quad\quad \log K_{s0} = -26$$

$$Fe(OH)_3(s) = Fe^{3+} + 3\,OH^- \quad\quad \log K_{s0} = -38.7$$

$$H_3PO_4 = H_2PO_4^- + H^+ \quad\quad \log K_1 = -2.1$$

$$H_2PO_4^- = HPO_4^{2-} + H^+ \quad\quad \log K_2 = -7.2$$

$$HPO_4^{2-} = PO_4^{3-} + H^+ \quad\quad \log K_3 = -12.3$$

424   Precipitation and Dissolution

7.14. The figure below (from Rimstidt and Barnes, 1980) shows the energy profiles in the dissolution (forward direction) and precipitation (reverse direction) reactions of amorphous SiO_2 and quartz, α-SiO_2. Use the concept of detailed balancing (microscopic reversibility principle) to explain why the dissolution rate constant is directly proportional to the solubility of the SiO_2 solid phase.

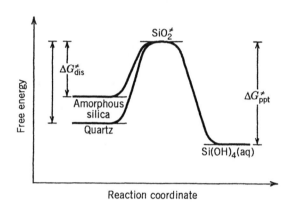

ANSWERS TO PROBLEMS

7.1. (a) $10^{-6.8}$; (b) $10^{-5.8}$.

7.2. Reaction: $CaCO_3(s) + 2H^+ = Ca^{2+} + CO_2(g) + H_2O$. Const = $*K_{ps0}$ = $K_{s0} K_2^{-1} K_H^{-1} K_1^{-1}$.

7.3. Oversaturated with $CaCO_3(s)$ at 0 m and at 30 m.

7.4. $[Ca^{2+}]$ and $[CO_3^{2-}]$ do not change appreciably; the pH increases.

7.5. (a) 1×10^{-3} eq liter^{-1}; (b) 5×10^{-4} eq liter^{-1} acid (by use of equation 30).

7.6. Reaction: $CaCO_3(s) + H_2CO_3^* = Ca^{2+} + 2HCO_3^-$; charge balance: $2[Ca^{2+}] \approx [HCO_3^-]$; [Alk] $\approx [HCO_3^-]$. $[HCO_3^-]^3/[H_2CO_3^*] = 2K_{s0}K_1K_2^{-1} = K$.

7.8. Increase for (h), (c), (e); decrease for (a), (d), (f); no change for (g).

7.9. (a) CO_2, H_2O, P_2O_5, CaO; five phases, hence one degree of freedom, but specification of temperature makes the system nonvariant.
(b) No change for (i), (ii), (iv), and (vi) (because these represent additions of components); decrease for (iii); increase for (v).

7.13. $Fe(OH)_3(s)$.

8

OXIDATION AND REDUCTION; EQUILIBRIA AND MICROBIAL MEDIATION

8.1 INTRODUCTION

In this chapter we stress the stability relations of pertinent redox (oxidation–reduction) components in natural water systems. However, one must be aware that concentrations of oxidizable or reducible species may be far from those predicted thermodynamically, because many redox reactions are slow. In the sea or in a lake there is a marked difference in redox environment between the surface in contact with the oxygen of the atmosphere and the deepest layers of the sediments. In between are numerous localized intermediate zones, resulting from imperfections in mixing or diffusion and from varying biological activities, none of which is truly at equilibrium. The need for biological mediation of most redox processes encountered in natural waters means that approaches to equilibrium depend strongly on the activities of the biota. Moreover, quite different oxidation–reduction levels, different from those prevalent in the overall environment, may be established within biotic microenvironments; diffusion or dispersion of products from the microenvironment into the macroenvironment may give an erroneous view of redox conditions in the latter. Also, because many redox processes do not couple with one another readily, it is possible to have several different apparent oxidation–reduction levels in the same locale. Therefore detailed, quantitative exposition of redox conditions and processes will depend ultimately on understanding the dynamics of aquatic systems—the rates of approach to equilibrium—rather than on describing the total or partial equilibrium compositions. Elementary aspects of biological mediation of some redox reactions will be discussed.

Kinetic considerations of some redox reactions, above all the rate of oxidation by O_2, will be dealt with in Chapter 11, and the influence of light on redox processes (photochemistry) will be discussed in Chapter 12.

Equilibrium considerations can greatly aid attempts to understand in a general way the redox patterns observed or anticipated in natural waters. In all circumstances equilibrium calculations provide boundary conditions toward which the system must be proceeding. Moreover, partial equilibria (those involving some but not all redox couples) are approximated frequently, even though total equilibrium is not reached. In some instances active poising[†] of particular redox couples allows us to predict significant oxidation–reduction levels or to estimate properties and reactions from computed redox levels. Valuable insight is gained even when differences are observed between computations and observations. The lack of equilibrium and the need for additional information or more sophisticated theory are then made clear.

Additional difficulties occur with attempts to measure oxidation–reduction potentials electrochemically in aquatic environments. Values obtained depend on the nature and rates of the reactions at the electrode surface and are seldom meaningfully interpretable. Even when suitable conditions for measurement are obtained, the results are significant only for those components whose behavior is electrochemically reversible at the electrode surface.

8.2 REDOX EQUILIBRIA AND THE ELECTRON ACTIVITY

There is a conceptual analogy between acid–base and reduction–oxidation reactions. In a similar way that acids and bases have been interpreted as proton donors and proton acceptors, reductants and oxidants are defined as electron donors and electron acceptors. e^- stands for the electron. Because there are no free electrons, every oxidation is accompanied by a reduction, and vice versa; or an oxidant is a substance that causes oxidation to occur while being reduced itself.

$$O_2 + 4H^+ + 4e^- = 2H_2O \qquad \text{reduction}$$
$$4Fe^{2+} = 4Fe^{3+} + 4e^- \qquad \text{oxidation}$$
$$O_2 + 4Fe^{2+} + 4H^+ = 4Fe^{3+} + 2H_2O \qquad \text{redox reaction}$$

The Oxidation State

As a result of the electron transfer (mechanistically the transfer may occur as a transfer of a group that carries the electron), there are changes in the oxidation states of reactants and products. Sometimes, especially in dealing with reactions involving covalent bonds, there are uncertainties in the assignment of electron loss or electron gain to a particular element. The oxidation state (or oxidation number) represents a hypothetical charge that an atom would have if the ion or molecule were to dissociate. This hypothetical dissociation, or the assignment of electrons to an atom, is carried out according to rules. The rules and a few examples are given in Table 8.1. In this book roman numerals are used to represent oxidation states, and arabic numbers represent actual electronic charge. The concept of an oxidation state may often have little chemical reality,

---

[†]Buffering with regard to oxidation or reduction.

Table 8.1. Oxidation State

Rules for Assigning Oxidation States:

(1) The oxidation state of a monoatomic substance is equal to its electronic charge.
(2) In a covalent compound, the oxidation state of each atom is the charge remaining on the atom when each shared pair of electrons is assigned completely to the more electronegative of the two atoms sharing them. An electron pair shared by two atoms of the same electronegativity is split between them.
(3) The sum of oxidation states is equal to zero for molecules, and for ions is equal to the formal charge of the ions.

Examples:

Nitrogen Compounds		Sulfur Compounds		Carbon Compounds	
Substance	Oxidation States	Substance	Oxidation States	Substance	Oxidation States
NH_4^+	$N = -III, H = +I$	H_2S	$S = -II, H = +I$	HCO_3^-	$C = +IV$
N_2	$N = 0$	$S_8(s)$	$S = 0$	$HCOOH$	$C = +II$
NO_2^-	$N = +III, O = -II$	SO_3^{2-}	$S = +IV, O = -II$	$C_6H_{12}O_6$	$C = 0$
NO_3^-	$N = +V, O = -II$	SO_4^{2-}	$S = +VI, O = -II$	CH_3OH	$C = -II$
HCN	$N = -III, C = +II, H = +I$	$S_2O_3^{2-}$	$S = +II, O = -II$	CH_4	$C = -IV$
SCN^-	$S = -I, C = +III, N = -III$	$S_4O_6^{2-}$	$S = +2.5, O = -II$	C_6H_5COOH	$C = -2/7$
		$S_2O_6^{2-}$	$S = +V, O = -II$		

but the concept is extremely useful in discussing stoichiometry—as a tool for balancing redox reactions—and in systematic descriptive chemistry.

Sometimes the balancing of redox reactions causes difficulties. One of various approaches is illustrated in the following example.

Example 8.1. Balancing Redox Reactions Balance the following redox reactions: (1) oxidation of Mn^{2+} to MnO_4^- by PbO_2 and (2) oxidation of $S_2O_3^{2-}$ to $S_4O_6^{2-}$ by O_2.

1. Reactants: Mn(II), Pb(IV); Products: Mn(VII), Pb(II)
   Oxidation: Mn(II) = Mn(VII) + 5e$^-$
   Reduction: Pb(IV) + 2e$^-$ = Pb(II)

   Half-reactions:

$$Mn^{2+} = Mn(VII) + 5e^-$$
$$Mn(VII) + 4O(-II) = MnO_4^-$$
$$4H_2O = 4O(-II) + 8H^+$$
$$\overline{}$$
$$Mn^{2+} + 4H_2O = MnO_4^- + 8H^+ + 5e^- \quad (i)$$

$$Pb(IV) + 2e^- = Pb^{2+}$$
$$PbO_2 = Pb(IV) + 2O(-II)$$
$$2O(-II) + 4H^+ = 2H_2O$$
$$\overline{}$$
$$PbO_2 + 4H^+ + 2e^- = Pb^{2+} + 2H_2O \quad (ii)$$

Adding reactions i and ii and eliminating e$^-$:

$$2Mn^{2+} + 5PbO_2 + 4H^+ = 2MnO_4^- + 5Pb^{2+} + 2H_2O$$

2. Reactants: S(II), O(0): Products: S(+2.5), O(−II)
   Oxidation: 2S(II) = 2S(+2.5) + e$^-$
   Reduction: O(0) + 2e$^-$ = O(−II)

   Half reactions:

$$2S(II) = 2S(+2.5) + e^-$$
$$S_2O_3^{2-} = 2S(II) + 3O(-II)$$
$$2S(+2.5) + 3O(-II) = \tfrac{1}{2}S_4O_6^{2-}$$
$$\overline{\phantom{2S(+2.5) + 3O(-II) = \tfrac{1}{2}S_4O_6^{2-}}}$$
$$S_2O_3^{2-} = \tfrac{1}{2}S_4O_6^{2-} + e^- \quad (iii)$$

8.2 Redox Equilibria and the Electron Activity

$$\tfrac{1}{2}O_2 + 2e^- = O(-II)$$

$$O(-II) + 2H^+ = H_2O$$

$$\tfrac{1}{2}O_2 + 2H^+ + 2e^- = H_2O \qquad (iv)$$

Adding half reactions iii and iv:

$$2S_2O_3^{2-} + \tfrac{1}{2}O_2 + 2H^+ = S_4O_6^{2-} + H_2O$$

Electron Activity and pε

Aqueous solutions do not contain free protons and free electrons, but it is nevertheless possible to define relative proton and electron activities. The pH,

$$pH = -\log\{H^+\} \qquad (1)$$

measures the relative tendency of a solution to accept or transfer protons. In an acid solution this tendency is low, and in an alkaline solution it is high. Similarly, we can define an equally convenient parameter for the redox intensity

$$p\varepsilon = -\log\{e^-\} \qquad (2)$$

pε gives the (hypothetical) electron activity at equilibrium and measures the relative tendency of a solution to accept or transfer electrons.[†] In a highly reducing solution the tendency to donate electrons, that is, the hypothetical "electron pressure," or electron activity, is relatively large. Just as the activity of hypothetical hydrogen ions is very low at high pH, the activity of hypothetical electrons is very low at high pε. Thus a high pε indicates a relatively high tendency for oxidation. In equilibrium equations H^+ and e^- are treated in an analogous way. Thus oxidation or reduction equilibrium constants can be defined and treated similarly to acidity constants as shown by the equations in Table 8.2, where the corresponding relations for pH and pε are derived in a stepwise manner. In order to relate pε to redox equilibria, we recall first the relationship derived for pH and acid–base equilibria (left-hand side of Table 8.2). An electron transfer reaction, in analogy with a proton transfer reaction, can be interpreted in terms of two reaction steps (2 and 3 in Table 8.2).

In parallel to the convention of arbitrarily assigning $\Delta G° = 0$ for the hydration of the proton (equation 3a in Table 8.2), we also assign a zero free energy change for the oxidation of $H_2(g)$ (equation 3b in Table 8.2). Equations

---

[†]While H^+ exists as a hydrated species in water, e^- does not. As we shall see, pε is related to the equilibrium redox potential E_H (volts, hydrogen scale). The electron, as discussed here and used as a component in our equilibrium calculations, is different from the solvated electron, which is a transient reactant in photolyzed solutions.

Table 8.2. pH and pε

pH = $-\log\{H^+\}$		pε = $-\log\{e^-\}$	
Acid-base reaction: $HA + H_2O = H_3O^+ + A^-$ $\quad K_1$	(1a)	Redox reaction: $Fe^{3+} + \frac{1}{2}H_2(g) = Fe^{2+} + H^+$ $\quad K_1$	(1b)
Reaction 1a is composed of two steps:		Reaction 1b is composed of two steps:	
$\qquad HA = H^+ + A^- \qquad K_2$	(2a)	$\qquad Fe^{3+} + e^- = Fe^{2+} \qquad K_2$	(2b)
$\qquad H_2O + H^+ = H_3O^+ \qquad K_3$	(3a)	$\qquad \frac{1}{2}H_2(g) = H^+ + e^- \qquad K_3$	(3b)
According to thermodynamic convention: $K_3 = 1$		According to thermodynamic convention: $K_3 = 1$	
Thus		Thus	
$\qquad K_1 = K_2 = K_2 K_3 = \{H^+\}\{A^-\}/\{HA\}$	(4a)	$\qquad K_1 = K_2 = K_2 K_3 = \{Fe^{2+}\}/\{Fe^{3+}\}\{e^-\}$	(4b)
or		or	
$\qquad pH = pK + \log[\{A^-\}/\{HA\}]$	(5a)	$\qquad p\varepsilon = p\varepsilon^\circ + \log[\{Fe^{3+}\}/\{Fe^{2+}\}]$	(5b)
Since $pK = -\log K = \Delta G^\circ/2.3RT$		Since $p\varepsilon^\circ = \log K = -\Delta G^\circ/2.3RT$	
$\qquad pH = \Delta G^\circ/2.3RT + \log[\{A^-\}/\{HA\}]$	(6a)	$\qquad p\varepsilon = -\Delta G^\circ/2.3RT + \log[\{Fe^{3+}\}/\{Fe^{2+}\}]$	(6b)
or for the transfer of 1 mol of H^+ from acid to H_2O:		or for the transfer of 1 mole of e to oxidant from H_2:	
$\qquad \Delta G/2.3RT = \Delta G^\circ/2.3RT + \log[\{A^-\}/\{HA\}]$	(7a)	$\qquad -\Delta G/2.3RT = -\Delta G^\circ/2.3RT + \log[\{Fe^{3+}\}/\{Fe^{2+}\}]$	(7b)
For the general case where n protons are transferred:		For the general case where n electrons are transferred:	
$\qquad H_nB + nH_2O = nH_3O^+ + B^{-n} \quad \beta^*$	(8a)	$\qquad ox + (n/2)H_2 = red + nH^+; \; ox + ne^- = red; \; K^*$	(8b)
$\qquad pH = (1/n)p\beta^* + (1/n)\log[\{B^{-n}\}/\{H_nB\}] \quad \beta$	(9a)	$\qquad p\varepsilon = p\varepsilon^n + (1/n)\log[\{ox\}/\{red\}]$	(9b)
$\qquad pH = \Delta G/n2.3RT$		$\qquad p\varepsilon = -\Delta G^\circ/n2.3RT + (1/n)\log[\{ox\}/\{red\}]$	(10b)
$\qquad = \Delta G^\circ/n2.3RT + (1/n)\log[\{B^{-n}\}/\{H_nB\}]$	(10a)		
$\Delta G = -nFE(E = $ acidity potential$)$	(11a)	$\Delta G = -nFE_H(E_H = $ redox potential$)$	(11b)
$\qquad pH = -E/(2.3RTF^{-1})$		$\qquad p\varepsilon = E_H/(2.3RTF^{-1})$	
$\qquad = -E^\circ/(2.3RTF^{-1}) + (1/n)\log[\{B^{-n}\}/\{H_nB\}]$	(12a)a	$\qquad = E_H^\circ/2.3RTF^{-1} + (1/n)\log[\{ox\}/\{red\}]$	(12b)a
Acidity potential:		Redox potential (Peters–Nernst equation):	
$E = E^\circ + (2.3RT/nF)\log[\{H_nB\}/\{B^{-n}\}]$	(13a)	$E_H = E_H^\circ + (2.3RT/nF)\log[\{ox\}/\{red\}]$	(13b)

a At 25°C, $2.3RTF^{-1} = 0.059$ (V mol^{-1}). From equations 10 and 12: at 25°C, $p\varepsilon = E/0.059$, $p\varepsilon^\circ = E_H^\circ/0.059$. F = faraday

8.2 Redox Equilibria and the Electron Activity

5a and 5b in Table 8.2 show that pH and pε are measures of the free energy involved in the transfer of 1 mol of protons or electrons, respectively.

Any oxidation or reduction can be written as a half-reaction, for example,

$$Fe^{3+} + e^- = Fe^{2+} \quad K_3 \qquad (3)$$

Such a reduction is always accompanied by an oxidation, for example, $I^- = \frac{1}{2} I_2 + e^-$. Even though there are no free electrons in solution, we can formulate an equilibrium expression for the half-reaction (equation 3 in Table 8.2):

$$\frac{\{Fe^{2+}\}}{\{Fe^{3+}\}\{e^-\}} = K_3 \qquad \log K_3 = 13.0 \quad (25°C) \qquad (4)$$

This can be rewritten

$$p\varepsilon = \log K_3 + \log \frac{\{Fe^{3+}\}}{\{Fe^{2+}\}} \qquad (5)$$

Another example—the reduction of IO_3^-—is characterized by the half-reaction

$$IO_3^- + 3H_2O + 6e^- = I^- + 6OH^- \quad K_6 \qquad (6)$$

Such a reduction is also accompanied by an oxidation, for example, $3H_2O = \frac{3}{2} O_2(g) + 6H^+ + 6e^-$. Even though there are no free electrons in solution, we can formulate an equilibrium expression for the half-reaction

$$\frac{\{I^-\}\{OH^-\}^6}{\{IO_3^-\}\{e^-\}^6} = K_6 \qquad \log K_6 = 26.1 \quad (25°C) \qquad (7)$$

or

$$p\varepsilon = \tfrac{1}{6} \log K_6 + \tfrac{1}{6} \log \frac{\{IO_3^-\}}{\{I^-\}\{OH^-\}^6} \qquad (8)$$

Equations 5 and 8 can be generalized as

$$p\varepsilon = p\varepsilon^\circ_{(3)} + \log \frac{\{Fe^{3+}\}}{\{Fe^{2+}\}} \quad \text{or} \quad p\varepsilon = p\varepsilon^\circ + \log \frac{\{ox\}}{\{red\}} \qquad (9)$$

and

$$p\varepsilon = p\varepsilon^\circ_{(6)} + \frac{1}{6} \log \frac{\{IO_3^-\}}{\{I^-\}\{OH^-\}^6} \quad \text{or} \quad p\varepsilon = p\varepsilon^\circ + \frac{1}{n} \log \frac{\Pi_i \{ox\}^{n_i}}{\Pi_j \{red\}^{n_j}} \qquad (10)$$

where

$$p\varepsilon^\circ_{(3)} = \log K_3 \quad \text{and} \quad p\varepsilon^\circ_{(6)} = \tfrac{1}{6} \log K_6$$

or generally

$$p\varepsilon^\circ = \frac{1}{n} \log K \tag{11}$$

where n is the number of electrons involved in the reaction and K is the equilibrium constant for the *reduction* half-reaction. In equation 10, $\Pi_i \{ox\}^{n_i}$ is the product of the activities of the reactants (the oxidants on the left-hand side of the reaction equations) and $\Pi_j \{red\}^{n_j}$ the product of the activities of the products (the reductants on the right-hand side of the reaction equation).

Thus for the reduction

$$SO_4^{2-} + 10\, H^+ + 8\, e^- = H_2S(g) + 4\, H_2O \qquad \log K = 42.0 \tag{12a}$$

$p\varepsilon$ is given by

$$p\varepsilon = p\varepsilon^\circ_{(12)} + \frac{1}{8} \log \frac{\{SO_4^{2-}\}\{H^+\}^{10}}{p_{H_2S}} \qquad p\varepsilon^\circ_{(12)} = 5.25 \tag{12b}$$

Equilibrium Calculations

So far we have treated the electron like a ligand in complex formation reactions. Indeed in *Stability Constants of Metal-Ion Complexes*, the first ligand considered in Section I (inorganic ligands) is the electron.

In subsequent examples we illustrate a few basic equilibrium calculations. For simplicity we assume activities approximately equal concentrations.

Example 8.2. The Formal Computation of $p\varepsilon$ Values Calculate $p\varepsilon$ values for the following equilibrium systems (25°C, $I = 0$):

(a) An acid solution 10^{-5} M in Fe^{3+} and 10^{-3} M in Fe^{2+}.
(b) A natural water at pH = 7.5 in equilibrium with the atmosphere (p_{O_2} = 0.21 atm).
(c) A natural water at pH = 8 containing 10^{-5} M Mn^{2+} in equilibrium with γ-$MnO_2(s)$.

Stability Constants of Metal-Ion Complexes gives the following equilibrium constants:

8.2 Redox Equilibria and the Electron Activity

$$Fe^{3+} + e^- = Fe^{2+} \qquad K = \frac{\{Fe^{2+}\}}{\{Fe^{3+}\}\{e^-\}}$$

$$\log K = 13.0 \qquad \text{(i)}$$

$$\tfrac{1}{2}O_2(g) + 2H^+ + 2e^- = H_2O(l) \qquad K = \frac{1}{p_{O_2}^{1/2}\{H^+\}^2\{e^-\}^2}$$

$$\log K = 41.55 \qquad \text{(ii)}$$

$$\gamma\text{-}MnO_2(s) + 4H^+ + 2e^- = Mn^{2+} + 2H_2O(l) \qquad K = \frac{\{Mn^{2+}\}}{\{H^+\}^4\{e^-\}^2}$$

$$\log K = 40.84 \qquad \text{(iii)}$$

By using equations i, ii, and iii, the following pε values are obtained for the conditions stipulated

(a) $p\varepsilon = 13.0 + \log \dfrac{\{Fe^{3+}\}}{\{Fe^{2+}\}} = 11.0$

(b) $p\varepsilon = 20.78 + \dfrac{1}{2} \log (p_{O_2}^{1/2}\{H^+\}^2) = 13.11$

(c) $p\varepsilon = 20.42 + \dfrac{1}{2} \log \dfrac{\{H^+\}^4}{\{Mn^{2+}\}} = 6.92$

Example 8.3. Equilibrium Composition of Simple Solutions Calculate the equilibrium composition of the following solutions (25°C, $I = 0$) both in equilibrium with the atmosphere ($p_{O_2} = 0.21$ atm):

(a) An acid solution (pH = 2) containing a total concentration of iron, $Fe_T = 10^{-4}$ M.
(b) A natural water (pH = 7) containing Mn^{2+} in equilibrium with γ-$MnO_2(s)$. The equilibrium constants were given in Example 8.2. The redox equilibria are defined by the conditions given (p_{O_2} and pH). Hence pε for both can be calculated with the equation

$$p\varepsilon = 20.78 + \tfrac{1}{2} \log (p_{O_2}^{1/2}\{H^+\}^2)$$

The following values are obtained:

(a) $p\varepsilon = 18.61$
(b) $p\varepsilon = 13.61$

correspondingly, we find

$$p\varepsilon = 13.0 + \log \frac{\{Fe^{3+}\}}{\{Fe^{2+}\}}$$

and

$$p\varepsilon = 20.42 + \frac{1}{2} \log \frac{\{H^+\}^4}{\{Mn^{2+}\}}$$

for solution a, $\{Fe^{3+}\}/\{Fe^{2+}\} = 10^{5.61}$ or $\{Fe^{3+}\} = 10^{-4}$ and $\{Fe^{2+}\} = 10^{-9.61}$; and for solution b, $\{Mn^{2+}\} = 10^{-14.38}$ M.

There are, of course, other approaches in calculating the equilibrium composition; for example, we may first compute the equilibrium constants of the overall redox reactions:

$\frac{1}{2}O_2(g) + 2H^+ + 2e^- = H_2O(l)$	$\log K = 41.55$
$2Fe^{2+} = 2Fe^{3+} + 2e^-$	$\log K = -26.0$
$\frac{1}{2}O_2(g) + 2Fe^{2+} + 2H^+ = 2Fe^{3+} + H_2O(l)$	$\log K = 15.55$
$\frac{1}{2}O_2(g) + 2H^+ + 2e^- = H_2O(l)$	$\log K = 41.55$
$Mn^{2+} + 2H_2O(l) = \gamma\text{-}MnO_2(s) + 4H^+ + 2e$	$\log K = -40.84$
$\frac{1}{2}O_2(g) + Mn^{2+} + H_2O(l) = \gamma\text{-}MnO_2(s) + 2H^+$	$\log K = 0.71$

With the equilibrium constants defined for a given p_{O_2} and pH, the ratio $\{Fe^{3+}\}/\{Fe^{2+}\}$ and the equilibrium activity of Mn^{2+} can be calculated.

pε as a Master Variable The logarithmic equilibrium expressions

$$p\varepsilon = p\varepsilon^\circ + \frac{1}{n} \log \frac{\{ox\}}{\{red\}}$$

lend themselves to graphical presentation in double logarithmic equilibrium diagrams. As we used pH as a master variable in acid–base equilibria, we may use pε as a master variable for the graphical presentation of redox equilibria.

For example, in an acid solution of Fe^{2+} and Fe^{3+} (hydrolysis is neglected), the redox equilibrium

$$\frac{\{Fe^{2+}\}}{\{Fe^{3+}\}\{e^-\}} = K \tag{13}$$

8.2 Redox Equilibria and the Electron Activity

may be combined with the concentration condition

$$[Fe^{2+}] + [Fe^{3+}] = Fe_T \tag{14}$$

to obtain the relations in dilute solution

$$[Fe^{3+}] = \frac{Fe_T K^{-1}}{\{e^-\} + K^{-1}} \tag{15}$$

and

$$[Fe^{2+}] = \frac{Fe_T \{e^-\}}{\{e^-\} + K^{-1}} \tag{16}$$

which, in logarithmic notation, can be formulated in terms of asymptotes. For $\{e^-\} \gg 1/K$ or $p\varepsilon < p\varepsilon°$,

$$\log[Fe^{3+}] = \log Fe_T + p\varepsilon - p\varepsilon° \tag{17}$$

$$\log[Fe^{2+}] = \log Fe_T \tag{18}$$

Similarly, for $\{e^-\} \ll 1/K$ or $p\varepsilon > p\varepsilon°$

$$\log[Fe^{3+}] = \log Fe_T \tag{19}$$

$$\log[Fe^{2+}] = \log Fe_T + p\varepsilon° - p\varepsilon \tag{20}$$

These relations can be conveniently plotted (Figure 8.1). The diagram shows how $p\varepsilon$ changes with the ratio of $\{Fe^{3+}\}$ and $\{Fe^{2+}\}$ for $\log Fe_T = -3$.

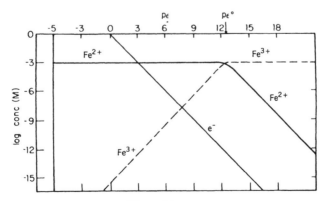

Figure 8.1. Redox equilibrium Fe^{3+}, Fe^{2+}. Equilibrium distribution of a 10^{-3} M solution of aqueous iron as a function of $p\varepsilon$ (acid solution).

Oxidation and Reduction; Equilibria and Microbial Mediation

For aqueous solutions at a given pH, each $p\varepsilon$ value is associated with a partial pressure of H_2 and of O_2:

$$2H^+ + 2e^- = H_2(g) \qquad \log K = 0 \qquad (21)$$

or

$$2H_2O + 2e^- = H_2(g) + 2OH^- \qquad \log K = -28 \qquad (22)$$

and

$$O_2(g) + 4H^+ + 4e^- = 2H_2O \qquad \log K = 83.1 \qquad (23)$$

or

$$O_2(g) + 2H_2O + 4e^- = 4OH^- \qquad \log K = 27.1 \qquad (24)$$

The equilibrium redox equations in logarithmic form are

$$\log p_{H_2} = 0 - 2pH - 2p\varepsilon \qquad (25)$$

$$\log p_{O_2} = -83.1 + 4pH + 4p\varepsilon \qquad (26)$$

Figure 8.2 gives a representation of equilibria 21 and 23 in logarithmic form. Thus, for example, a water of pH = 10 and of $p\varepsilon = 8$ corresponds to $p_{O_2} = 10^{-11}$ atm and $p_{H_2} = 10^{-36}$ atm. Instead of using $p\varepsilon$ as a measure of oxidizing intensity, it is possible to characterize this intensity by specifying pH and p_{O_2} or $p_{H_2}^\dagger$ (Figure 8.2). As Figure 8.2 also illustrates, the $p\varepsilon$ range of natural waters (pH = 4–10) extends from approximately $p\varepsilon = -10$ to 17; beyond these values water is reduced to H_2 or oxidized to O_2, respectively.

Example 8.4. Redox Equilibrium SO_4^{2-}–HS^- Construct a diagram showing the $p\varepsilon$ dependence of a 10^{-4} M SO_4^{2-}–HS^- system at pH = 10 and 25°C. The reaction is

$$SO_4^{2-} + 9H^+ + 8e^- = HS^- + 4H_2O \qquad (i)$$

and the redox equilibrium equation is

$$p\varepsilon = \frac{1}{8}\log K + \frac{1}{8}\log \frac{[SO_4^{2-}][H^+]^9}{[HS^-]} \qquad (ii)$$

---

†The resulting partial pressures are often mere calculation numbers; some correspond to less than one H_2 molecule in a space as large as the solar system. In 1921 Clark proposed a reduction intensity parameter defined by $r_H = -\log p_{H_2}$; fortunately r_H is no longer used.

8.2 Redox Equilibria and the Electron Activity

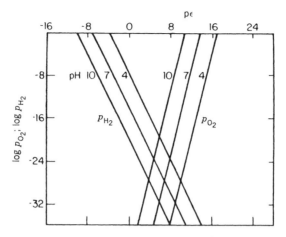

Figure 8.2. The stability of water [$H_2O(l) = H_2(g) + \frac{1}{2}O_2(g)$]. Partial pressure of H_2 and O_2 at various pH values in equilibrium with water.

We may calculate the equilibrium constant from available data on the standard free energy of formation. The National Bureau of Standards gives the following $\bar{G}_f^\circ$ values (kJ mol^{-1}): SO_4^{2-}, -742.0; HS^-, 12.6; $H_2O(l)$, -237.2. $\bar{G}_f^\circ$ for the aqueous proton and the electron are zero. Hence the standard free energy change in equation i is $\Delta G^\circ = -194.2$, and the corresponding equilibrium constant ($K = 10^{-\Delta G^\circ/2.3RT}$) is 10^{34}. Hence, substituting in equation ii, we obtain

$$p\varepsilon = 4.25 - 1.125pH + \tfrac{1}{8}\log[SO_4^{2-}] - \tfrac{1}{8}\log[HS^-] \quad \text{(iii)}$$

Or, for pH = 10,

$$p\varepsilon = -7 + \tfrac{1}{8}\log[SO_4^{2-}] - \tfrac{1}{8}\log[HS^-] \quad \text{(iv)}$$

Equation iv has been plotted in Figure 8.3 for the condition $[SO_4^{2-}] + [HS^-] = 10^{-4}$ M. HS^- is the predominant S($-$II) species at pH = 10. The lines for $[SO_4^{2-}]$ and $[HS^-]$ intersect at $p\varepsilon = -7$; the asymptotes have slopes of 0 and ± 8, respectively. Lines for the equilibrium partial pressures of O_2 and H_2 are also given in the diagram. As the diagram shows, rather high relative electron activities are necessary to reduce SO_4^{2-}. At the pH value selected, the reduction takes place at $p\varepsilon$ values slightly less negative than for the reduction of water. In the presence of oxygen, only sulfate can exist; its reduction is only possible under extremely anoxic conditions ($p\varepsilon < -6$; $p_{O_2} < 10^{-68}$ atm). Note that no consideration was given to solid sulfur (or its complexes, such as HS_4^-) as a possible intermediate state in the reduction of SO_4^{2-} to S($-$II).

438   Oxidation and Reduction; Equilibria and Microbial Mediation

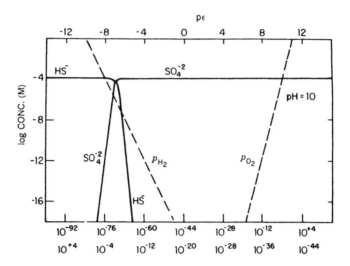

Figure 8.3. Equilibrium distribution of sulfur compounds, as a function of pε at pH = 10 and 25°C. Total concentration of compounds is 10^{-4} M (Example 8.4).

The Electron as a Component

In addition to H_2O and H^+, it is convenient to treat the electron as a component. It is simply interrelated with O_2 and H_2 by the following equilibria:

$$H^+ + e^- = \tfrac{1}{2} H_2(g)$$

$$\tfrac{1}{4} O_2(g) + H^+ + e^- = \tfrac{1}{2} H_2O(l)$$

As will be shown in the following tableaux the electron is a convenient component in an equilibrium matrix.

Example 8.5. Tableaux for Redox Reactions Establishing tableaux for the redox reactions mentioned in Examples 8.2 and 8.4, assuming the pε values to be given.

Tableau 8.1. Fe^{3+}–Fe^{2+} Equilibrium (Example 8.2a, Figure 8.1)

Components		Fe^{2+}	e^-	log K
Species	Fe^{2+}	1	0	0
	Fe^{3+}	1	-1	-13.0
TOTFe^{2+} = [Fe^{3+}] + [Fe^{2+}]		1.0×10^{-3} M	pε given	

8.2 Redox Equilibria and the Electron Activity

Tableau 8.2. O_2–H_2O (Example 8.2b, Figure 8.2)

Components		H^+	e^-	log K
Species	$O_2(g)$	−4	−4	−83.10
	H^+	1	0	0
		pH = 7.5	pε given	

Tableau 8.3. $MnO_2(s)$–Mn^{2+} (Example 8.2c)

Components		−$MnO_2(s)$	H^+	e^-	log K
Species	Mn^{2+}	1	4	2	40.84
	H^+	0	1	0	0
	$MnO_2(s)$	1	0	0	0
		$\{MnO_2(s)\} = 1$	pH = 8	pε given	

Tableau 8.4. SO_4^{2-}, HS^- (pH = 10) (Example 8.4, Figure 8.3)

Components		SO_4^{2-}	H^+	e^-	log K
Species	SO_4^{2-}	1	0	0	0
	HS^-	1	−9	8	34
	H_2S	1	−10	8	41
	S^{2-}	1	−8	8	20
	H^+	0	1	0	0
		$TOTSO_4^{2-} = 10^{-4}$ M	pH = 10	pε given	

Example 8.6. Reduction of NO_3^- to NH_4^+ at pH = 7 The redox equilibrium is given by

$$NO_3^- + 10\,H^+ + 8\,e^- = NH_4^+ + 3\,H_2O \qquad (i)$$

$$[NH_4^+]/([NO_3^-]\,[H^+]^{10}\,\{e^-\}^8) = 10^{119.2} \quad (25°C) \qquad (ii)$$

We can define a *conditional* equilibrium constant valid for pH = 7

$$[NH_4^+]/([NO_3^-]\,\{e^-\}^8) = K_{pH=7} = 10^{119.2}/[H^+]^{10} = 10^{49.2} \qquad (iii)$$

The problem is the same kind as in the previous example, and the double logarithmic diagram (Figure 8.4) can readily be constructed. Because of the dependence on $\{e^-\}^8$ the slopes of $d \log [NO_3^-]/d\,p\varepsilon$ and of $d \log [NH_4^+]/d\,p\varepsilon$ are +8 and −8, respectively. Obviously, $p\varepsilon°(w)$ defines a rather sharp boundary between the predominance of NH_4^+ [$p\varepsilon < p\varepsilon°(w)$] and of NO_3^- [$p\varepsilon < p\varepsilon°(w)$].

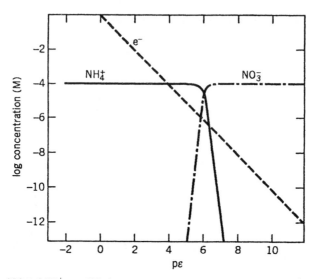

Figure 8.4. NO_3^-–NH_4^+ equilibrium at pH = 7 (Example 8.6). $[NH_4^+] + [NO_3^-] = 10^{-4}$ M. The treatment is simplified in that nitrite, NO_2^-, is not considered as an intermediate (cf. Figures 8.9c and 8.13a). Furthermore, the species NH_4^+ and NO_3^- are treated as species that are metastable with respect to N_2; that is, N_2 is regarded as a redox-inert species (cf. Figure 8.9b).

Example 8.7. Redox Equilibria of Chlorine Chlorine (Cl with an oxidation state of 0), as used in water disinfection, can undergo electron transfer reactions to Cl^I (HOCl) and Cl^{-I} (Cl^-). Formally, the reactions can be written

$\frac{1}{2} Cl_2(aq) + e^- = Cl^-$	$\log K = 23.6$ (25°C)	(i)
$HOCl + H^+ + e^- = \frac{1}{2} Cl_2(aq) + H_2O$	$\log K = 26.9$	(ii)
$HOCl = H^+ + OCl^-$	$\log K = -7.3$	(iii)
$HOCl + H^+ + 2 e^- = Cl^- + H_2O$	$\log K = 50.5$	(iv)

How do $Cl_2(aq)$, HOCl, OCl^-, and Cl^- depend on pε in a solution in which TOTCl = 10^{-5} M? The construction of a diagram of the concentrations of the various species as a function of pε is simplified, if we solve the four equations given above for a few selected pH values, say pH = 2, pH = 5, and pH = 8. The equilibrium condition is summarized in Tableau 8.5.

The stoichiometry of rows 2 and 3 needs some explanation: HOCl = Cl^- + H_2O − H^+ − 2 e^-; and OCl^- = Cl^- + H_2O − 2 H^+ − 2 e^-. The equilibrium constants given can be combined from the mass law equations given in equations i–iv.

8.3 Electrode Potential: Nernst Equation and the Electrochemical Cell

Tableau 8.5. Redox Equilibria of Cl Species

Components		Cl^-	H^+	e^-	log K
(1) Species	$Cl_2(aq)$	2	0	−2	−47.2
(2)	HOCl	1	−1	−2	−50.5
(3)	OCl^-	1	−2	−2	57.8
(4)	Cl^-	1	0	0	0
(5)	H^+	0	1	0	0
		$TOTCl^- = 10^{-5}$ M	pH given	pε given	
(6) $TOTCl^- = 2[Cl_2(aq)] + [HOCl] + [OCl^-] + [Cl^-] = 10^{-5}$ M					

The results are displayed in Figure 8.5. The following points are significant:

1. $Cl_2(aq)$ is *not* a predominant species. Its relative concentration increases with decreasing pH. Obviously, Cl_2 added to the water disproportionates into HOCl or OCl^- and Cl^-:

$$Cl_2(aq) + H_2O = HOCl + H^+ + Cl^- \quad \log K = -3.3 \quad (25°C)$$

This equation results from a combination of equations i and ii. An equilibrium solution of "active chlorine" of Cl_2 can be stored in solution at high pH ("bleach", eau de Javel) because its content of the volatile constituent Cl_2 is minimal.

2. Cl^0 and Cl^I species are stable only at high pε values. A comparison with the equilibrium

$$O_2(g) + 4 H^+ + 4 e^- = 2 H_2O \quad \log K = 10^{-83.1}$$

reveals that at these pε values water should be—thermodynamically speaking—oxidized to O_2; for example, for the conditions of pH = 8 at pε > 13, $p_{O_2} > 1$ (see Figure 8.2). In other words, chlorine solutions (HOCl, OCl^-) are metastable; they should oxidize water.

8.3 THE ELECTRODE POTENTIAL: THE NERNST EQUATION AND THE ELECTROCHEMICAL CELL

The standard electrode potential of an electrode reaction $E°$, for example, of a Zn electrode [$Zn^{2+} + 2 e^- = Zn(s)$], is a standard potential of a cell reaction when that reaction involves the oxidation of molecular H_2 to solvated protons (Parsons, 1985).

$$E° (Zn^{2+}(aq) + H_2(g) \rightarrow 2 H^+(aq) + Zn(s)) \tag{27}$$

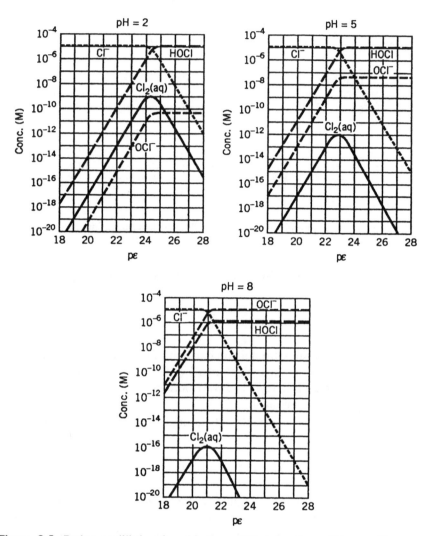

Figure 8.5. Redox equilibria of a chlorine solution for three different pH values. $Cl_2(aq)$ is not a predominant species. Its relative concentration decreases with increasing pH. $TOTCl^- = 2[Cl_2(aq)] + [HOCl] + [OCl^-] + [Cl^-] = 10^{-5}$ M.

which may be abbreviated [assuming that the reaction $H_2(g) \rightarrow 2H^+(aq) + 2e^-$ is always present]

$$E°(Zn^{2+}(aq) + 2e^- \rightarrow Zn(s)) = E°_{Zn^{2+}/Zn} \qquad (28)$$

When all species in a cell reaction are in their standard states (pure solids, unit standard concentrations, etc.) the measured electromotive force (emf) of the

8.3 Electrode Potential: Nernst Equation and the Electrochemical Cell

cell (usually measured in volts) is, to a good approximation, equal to the standard potential of the reaction in the cell, $E°$, which is also given by

$$E_H^\circ = -\Delta G°/nF = (RT/nF) \ln K = \frac{2.3\ RT}{nF} \log K = \frac{2.3\ RT}{F} \text{p}\varepsilon° \quad (29)$$

where $\Delta G°$ is the standard Gibbs free energy change in the cell reaction and K is the thermodynamic equilibrium constant of this reaction; n is the charge number of the reaction, that is, the number of electrons of the reaction as written, or the number of electrons passing around the external circuit, and F is the faraday.

As already given in Table 8.2, pε is related to the electrode potential, E [often also called the redox potential E_H (the suffix H denotes that this potential is on the hydrogen scale, i.e., referred to a standard hydrogen electrode)].

$$\text{p}\varepsilon = \frac{F}{2.3\ RT} E \quad \text{and} \quad \text{p}\varepsilon° = \frac{F}{2.3\ RT} E° \quad (30)$$

The thermodynamic relation of the potential E (or E_H) to the composition of the solution (cf. Table 8.2) is generally known as the Nernst equation:

$$E_H = E_H^\circ + \frac{2.3\ RT}{nF} \log \frac{\Pi_i\ \{\text{ox}\}^{n_i}}{\Pi_j\ \{\text{red}\}^{n_j}} \quad (31)$$

or in pε units

$$\text{p}\varepsilon = \text{p}\varepsilon° + \frac{1}{n} \log \frac{\Pi_i\ \{\text{ox}\}^{n_i}}{\Pi_j\ \{\text{red}\}^{n_j}} \quad (32)$$

For example, for the reaction

$$NO_3^- + 6\ H^+ + 5\ e^- = \frac{1}{2} N_2(g) + 3\ H_2O$$

$$E_H = E_H^\circ + \frac{2.3RT}{5F} \log \frac{\{NO_3^-\}\ \{H^+\}^6}{p_{N_2}^{1/2}} \quad (33)$$

or

$$\text{p}\varepsilon = \text{p}\varepsilon° + \frac{1}{5} \log \frac{\{NO_3^-\}\ \{H^+\}^6}{p_{N_2}^{1/2}} \quad (34)$$

where E_H°, the standard redox potential (i.e., the potential that would be obtained if all substances in the redox reaction were in their standard states of unit activity), is related to the free energy change for the cell reaction, $\Delta G°$, or the equilibrium constant of the reduction reaction (cf. equations 11 and 30):

$$E_H^\circ = -\frac{\Delta G^\circ}{nF} = \frac{RT}{nF} \ln K = \frac{2.3\,RT}{nF} \log K = \frac{2.3\,RT}{F} p\varepsilon^\circ \qquad (35)$$

It is necessary to distinguish between the concept of a potential and the measurement of a potential. Redox or electrode potentials (quoted in tables in *Stability Constants of Metal-Ion Complexes* or by Bard et al., 1985) have been derived from equilibrium data, thermal data, and the chemical behavior of a redox couple with respect to known oxidizing and reducing agents, and from direct measurements of electrochemical cells. Hence there is no a priori reason to identify the thermodynamic redox potentials with measurable electrode potentials.

We can express (relative) electron activity in pε or in volts. The use of pε, which is dimensionless, makes calculations simpler than the use of E_H, because every tenfold change in the activity ratio causes a unit change in pε. Furthermore, because an electron can reduce a proton, the intensity parameter for oxidation might preferably be expressed in units equivalent to pε. Of course, in making a direct electrochemical measurement of the oxidizing intensity, an emf (volts) is being measured, but the same is true of pH measurements, and a half century ago the "acidity potential" was used to characterize the relative H^+ ion activity.

The Electrochemical Series

Table 8.3 lists a few representative standard electrode potentials (or reduction potentials). Figure 8.6 exemplifies the principle of an electrochemical cell. The hydrogen electrode is made up of a Pt-electrode (which does not participate directly in the reaction), which is covered by $H_2(g)$, which acts as a redox partner [$H_2(g) = 2\,H^+ + 2\,e^-$]. Pt acts as a catalyst for the reaction between H^+ and $H_2(g)$ and acquires a potential characteristic of this reaction. The salt bridge between the two cells contains a concentrated solution of salt (such as KCl) and allows ionic species to diffuse into and out of the half-cells; this permits each half-cell to remain electrically neutral.

Liquid junction potentials may arise from the transfer of ionic species through the transition region. The liquid junction potential makes a contribution to the emf of the cell; it increases with increasing difference between the two solutions that form a single junction. The liquid junction potential can often be kept small by using a concentrated salt bridge.

Other electrodes represent redox couples in which oxidized and reduced forms are soluble species in the solution and the electron transfer occurs at the inert electrode (Pt, Au). For example, a platinum electrode immersed in an acid solution of Fe^{2+} and Fe^{3+} (Pt/Fe^{3+}, Fe^{2+}) may under favorable conditions acquire the potential characteristics of a Fe^{3+}/Fe^{2+} couple

$$Fe^{3+} + e^- = Fe^{2+} \qquad (36)$$

8.3 Electrode Potential: Nernst Equation and the Electrochemical Cell

Table 8.3. Equilibrium Constants and Standard Electrode Potentials for Some Reduction Half-Reactions

Reaction	Log K at 25°C	Standard Electrode Potential (V) at 25°C	$p\varepsilon°$
$Na^+ + e^- = Na(s)$	-46	-2.71	-46
$Mg^{2+} + 2e^- = Mg(s)$	-79.7	-2.35	-39.7
$Zn^{2+} + 2e^- = Zn(s)$	-26	-0.76	-13
$Fe^{2+} + 2e^- = Fe(s)$	-14.9	-0.44	-7.45
$Co^{2+} + 2e^- = Co(s)$	-9.5	-0.28	-4.75
$V^{3+} + e^- = V^{2+}$	-4.3	-0.26	-4.30
$2H^+ + 2e^- = H_2(g)$	0.0	0.00	0
$S(s) + 2H^+ + 2e^- = H_2S$	$+4.8$	$+0.14$	2.4
$Cu^{2+} + e^- = Cu^+$	$+2.7$	$+0.16$	2.7
$AgCl(s) + e^- = Ag(s) + Cl^-$	$+3.7$	$+0.22$	3.7
$Cu^{2+} + 2e^- = Cu(s)$	$+11.4$	$+0.34$	5.7
$Cu^+ + e^- = Cu(s)$	$+8.8$	$+0.52$	8.8
$Fe^{3+} + e^- = Fe^{2+}$	$+13.0$	$+0.77$	13.0
$Ag^+ + e^- = Ag(s)$	$+13.5$	$+0.80$	13.5
$Fe(OH)_3(s) + 3H^+ + e^- = Fe^{2+} + 3H_2O$	$+16.0$	$+0.95$	16.0
$IO_3^- + 6H^+ + 5e^- = \frac{1}{2}I_2(s) + 3H_2O$	$+104$	$+1.23$	20.8
$MnO_2(s) + 4H^+ + 2e^- = Mn^{2+} + 2H_2O$	$+43.6$	$+1.29$	21.8
$Cl_2(g) + 2e^- = 2Cl^-$	$+46$	$+1.36$	23
$Co^{3+} + e^- = Co^{2+}$	$+31$	$+1.82$	31

Referring to a list of standard electrode potentials, such as in Table 8.3, one speaks of an *electrochemical series,* and the metals lower down in the series (with positive electrode potentials) are called noble metals. Any combination of half-reactions in an electrochemical cell, which gives a nonzero E value, can be used as a galvanic cell (i.e., a *battery*). If the reaction is driven by an applied external potential, we speak of an *electrolytic cell.* Reduction takes place at the *cathode* and oxidation at the *anode.* The reduction reactions in Table 8.3 are ordered with increasing potential or $p\varepsilon°$ values. The oxidant in reactions with larger $p\varepsilon°$ (or $E°$) can oxidize a reductant at a lower $p\varepsilon°$ (or $E°$) and vice versa; for example, combining half-reactions we obtain an overall redox reaction:

$$Cu^{2+} + 2e^- = Cu°(s) \qquad \log K_1 = 11.4 \qquad p\varepsilon_1° = 5.7$$
$$E_1° = 0.34 \text{ V}$$

$$Fe(s) = Fe^{2+} + 2e^- \qquad \log K_2^{-1} = 14.9 \qquad p\varepsilon_2° = -7.45$$
$$E_2° = -0.44 \text{ V}$$

$$Cu^{2+} + Fe(s) = Fe^{2+} + Cu(s) \qquad \log K = 26.3 \qquad p\varepsilon_1° - p\varepsilon_2° = 13.15$$
$$E_1° - E_2° = 0.78 \text{ V}$$

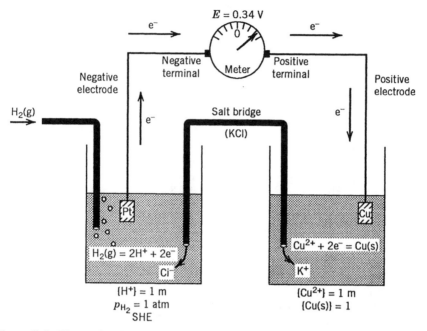

Figure 8.6. Electrochemical cell with a standard hydrogen electrode (SHE) as the left electrode. By convention, the negative terminal of the meter is on the left, and the positive terminal is on the right. Hydrogen gas at 1-atm pressure is being bubbled through the solution near the left electrode (Pt) to keep the solution phase activity of H_2 equivalent to 1 atm. $\{H^+\}$ in the left cell is 1 m. The right electrode is made of Cu(s). $\{Cu^{2+}\}$ in the right cell is 1 m. At the left electrode, H_2 is being oxidized to 2 H^+. At the right electrode, Cu^{2+} is being reduced to Cu(s). Arrows indicate the direction of the flow of electrons. The voltmeter (potentiometer) measures a positive voltage of 0.34 V. (The measurement has to be made when the cell does no work ($i = 0$). (Adapted from Pankow, 1991.)

So far we have seen how to combine log K values or reduction potentials of different half-reactions, thereby obtaining complete redox reactions. Thus

$$\Delta G° = -RT \ln (K_1 - K_2) = -RT n\, (p\varepsilon_1° - p\varepsilon_2°) \qquad (37)$$

The thermodynamic principle here is the addition of $\Delta G°$ values for half-reactions, thereby obtaining standard free energy and potential differences ($\Delta E°$ or $\Delta p\varepsilon°$) for overall redox reactions; for example, $Cu^{2+} + Fe(s) = Fe^{2+} + Cu(s)$, $\Delta p\varepsilon° = 13.15$. It is straightforward to obtain reduction potentials for half-reactions that are linear combinations of other half-reactions following the same approach, that is, algebraic addition of the free energies for the half-reactions. For example, Table 8.3 gives $p\varepsilon°$ values for the following reactions: $Cu^{2+} + e^- = Cu^+$, $Cu^{2+} + 2e^- = Cu(s)$, and $Cu^+ + e^- = Cu(s)$. Only

8.3 Electrode Potential: Nernst Equation and the Electrochemical Cell

two of these reactions are independent, the third being a linear combination of the other two. The standard potential of the first reduction, 0.16 V, is obtained from the second and third reductions via free energies as follows: $(0.34 \times 2 \times F - 0.52 \times 1 \times F)/(1 \times F) = 0.16$ V $= E°$ (or $p\varepsilon° = 2.4$). In general, the resulting $p\varepsilon°$ of combining two independent half-reactions is

$$p\varepsilon° = (n_1 \, p\varepsilon_1° + n_2 \, p\varepsilon_2°)/(n_1 + n_2)$$

where $n > 0$ for reduction and $n < 0$ for oxidation.

Sign Convention As pointed out before (equation 11), all half-reactions are written as reductions with a sign that corresponds to the sign of log K of the reduction reaction.

Reference Electrodes

The standard hydrogen electrode is not very convenient in practical terms. Other reference electrodes, especially those that tend to maintain a constant electrode potential, are usually more convenient. Typical examples are the Ag/AgCl and the Hg/Hg$_2$Cl$_2$ electrode. With these electrodes, the activity of the cations associated with the electrode metal is kept constant by buffering through the solubility product principle. Thus a AgCl/Ag electrode consists in principle of a silver electrode coated with AgCl immersed in a solution of high [Cl$^-$]. Because AgCl(s) is present, the activity of Ag$^+$ is given by

$$\{Ag^+\} = \frac{K_{s0(AgCl)}}{\{Cl^-\}} \tag{38}$$

Furthermore, if some Ag$^+$ is reduced to Ag(s), or some Ag(s) is oxidized to Ag$^+$, the dissolution or precipitation of AgCl, respectively, will keep $\{Ag^+\}$ constant. Hence the electrode potential of a AgCl/Ag electrode remains constant even if some current is flowing through the half-cell (nonpolarizable electrode). Of course, the current must be small enough so that it does not exceed the exchange current of the reaction

$$AgCl(s) + e^- = Ag(s) + Cl^- \tag{39}$$

Such a nonpolarizable electrode is a convenient reference electrode. Another important reference electrode is the calomel electrode with the half-reaction

$$Hg_2Cl_2(s) + 2e^- = 2Hg(l) + 2\,Cl^- \tag{40}$$

Its electrode potential depends on the activity of the chloride ion.

When half-cells are combined in an electrochemical cell, the emf of the cell

Table 8.4. Potentials of Reference Electrodes[a]

Temperature (°C)	Calomel		AgCl/Ag	
	0.1 M KCl	Saturated	0.1 M	Saturated
12	0.3362	0.2528	—	—
20	0.3360	0.2508	—	—
25	0.3356	0.2444	0.2900	0.1988

[a] The values listed (V) included the liquid junction potential. Cell: Pt, H_2/H^+ ($a = 1$) ∥ reference electrode.
Source: Bates (1959).

is given by

$$E_{cell} = E_{right\,cell} - E_{left\,cell} \qquad (41)$$

$$E_{cell} = E_{H\,right} - E_{H\,left} \qquad (42)$$

If the left electrode is a reference electrode, we may also write

$$E_{cell} = E_{H(ox-red)} - E_{ref} \qquad (43)$$

Example 8.8. Electrode Potential of Fe^{3+}-Fe^{2+} System Consider the following cell and compute its emf[†]:

$$Hg|Hg_2Cl_2, KCl_{sat'd} \| HClO_4(1\,M), Fe^{3+}(10^{-3}), Fe^{2+}(10^{-2})|Pt$$

According to Table 8.4, the potential of the saturated calomel electrode (25°C) is 0.244 V. Hence

$$E_{cell} = E_{Fe^{3+},Fe^{2+}} - 0.244$$

and

$$E_{Fe^{3+},Fe^{2+}} = 0.771 + 0.059 \log \frac{10^{-3}}{10^{-2}} = 0.712\,V$$

$$E_{cell} = 0.468\,V$$

Example 8.9a. Solution Composition from Measured emf With an electrode pair consisting of an inert Pt and a saturated calomel reference electrode in a sample of a sediment–water interface of pH = 6.4 (25°C), a potential difference (emf) of 0.47 V is measured. The sediment contains solid amorphous $Fe(OH)_3$, and we assume that the measured potential corresponds to an oxidation–reduction potential of the aquatic environment.

[†] The double vertical line indicates that the liquid junction potential is either ignored or kept small by a suitable salt bridge.

8.3 Electrode Potential: Nernst Equation and the Electrochemical Cell

(a) What is the E_H and pε of the sample?
(b) What is the activity of Fe^{2+}?
(c) Does the redox level found indicate aerobic or anaerobic conditions?

(a) The emf has been measured in a cell

$$Hg, Hg_2Cl_2 | KCl \| ox, red | Pt$$

and

$$E_{cell} = E_{H(ox-red)} - E_{ref}$$

Because $E_{ref} = 0.244$, we obtain $E_{H(ox-red)} = 0.47 + 0.244 = 0.714$ V. This is equivalent to pε = 0.714 V/0.05916 V = 12.1.

(b) At equilibrium $Fe(OH)_3(s)$ is in equilibrium with Fe^{2+}, the redox reaction being

$$Fe(OH)_3(s) + 3H^+ + e^- = Fe^{2+} + 3H_2O \qquad \log K = 16 \qquad (i)$$

Thus $E_{H(ox-red)}$ and pε, respectively, are given by (see equations 11, 29, and 31)

$$E_{H(ox-red)} = E^\circ_{Fe(OH)_3-Fe^{2+}} + 2.3 \frac{RT}{nF} \log \left(\frac{\{H^+\}^3}{\{Fe^{2+}\}} \right) \qquad (ii)$$

$$p\varepsilon = p\varepsilon^\circ_{Fe(OH)_3-Fe^{2+}} + \log \left(\frac{\{H^+\}^3}{\{Fe^{2+}\}} \right) \qquad (iii)$$

The value of $E^\circ_{Fe(OH)_3-Fe^{2+}}$ or of $p\varepsilon^\circ_{Fe(OH)_3-Fe^{2+}}$ can be computed from data on free energy of formation for equation i. The reduction reaction at 25°C has $\Delta G^\circ = -91.4$ kJ, $\log K = 16.0$. From this we calculate

$$E^\circ_{Fe(OH)_3-Fe^{2+}} = \frac{91,400 \text{ J}}{96,485 \text{ J V}^{-1}} = 0.947 \text{ V}$$

or

$$p\varepsilon^\circ_{Fe(OH)_3-Fe^{2+}} = 16.0$$

The activity of Fe^{2+} can be calculated by

$$p\varepsilon = p\varepsilon^\circ + \log \left(\frac{\{H^+\}^3}{\{Fe^{2+}\}} \right)$$

$$12.1 = 16 - 3pH + pFe^{2+}$$

This gives $pFe^{2+} = 15.3$ for pH = 6.4.

(c) In order to decide whether this corresponds to oxic or anoxic conditions, p_{O_2} must be computed. For

$$O_2(g) + 4 H^+ + 4e^- = 2H_2O \qquad p\varepsilon° = 20.8$$

because

$$p\varepsilon = p\varepsilon° + \tfrac{1}{4} \log p_{O_2} \{H^+\}^4$$

$$12.1 = 20.8 - pH + \tfrac{1}{4} \log p_{O_2}$$

$$\log p_{O_2} = 4(12.1 - 20.8 + pH)$$

$$p_{O_2} = 10^{-9.2} \text{ atm}$$

This corresponds to $\sim 10^{-12}$ M dissolved O_2. We may properly speak of anoxic conditions.

Example 8.9b. Standard Potential of the Cl_2/Cl^- Couple Faita et al. (1967) determined the standard potential of the Cl_2/Cl^- electrode. They made emf measurements with the cell

$$Pt|Ag|AgCl|1.75 \text{ M HCl}|Cl_2(= 1 \text{ atm})|Pt-Ir(45\%), Ta|Pt \qquad \text{(i)}$$

The right-hand electrode consisted of a tantalum foil (attached to a Pt wire) coated with a platinum–iridium alloy. This alloy, containing 45% iridium, is used as a Cl_2 electrode because Pt is not fully appropriate since it is subject to corrosive attack by Cl_2 in the presence of HCl. Their results for the cell i are as follows: 25°C: $E_{cell(i)} = 1.13596$ V; 30°C: $E_{cell(i)} = 1.13309$ V; 40°C: $E_{cell(i)} = 1.12711$ V; 50°C: $E_{cell(i)} = 1.12110$ V

(a) What is the chemical reaction taking place in the cell?
(b) What is the standard potential of the Cl_2/Cl^- electrode?
(c) Determine from the experimental data $\Delta G°$, $\Delta H°$, and $\Delta S°$ of the cell reaction of the reduction of Cl_2 to Cl^-, respectively.

The standard Ag/AgCl/Cl$^-$ electrode, $E°_{AgCl/Ag}$, has the following values (volts):

$E°_{AgCl/Ag}$: 25°C, 0.22234; 30°C, 0.21904; 40°C, 0.21208; 50°C, 0.20449

Furthermore, the concentration of HCl in cell i has been chosen such that $\{Cl^-\} = 1.0$.

(a) The chemical reaction taking place in cell i is composed of the half-

8.3 Electrode Potential: Nernst Equation and the Electrochemical Cell

reactions

$$Ag(s) + Cl^- = AgCl(s) + e^-$$

$$\tfrac{1}{2} Cl_2(g) + e^- = Cl^-$$

$$Ag(s) + \tfrac{1}{2} Cl_2(g) = AgCl(s) \tag{ii}$$

The emf of $E_{\text{cell(i)}}$ is a direct measure of the free energy of the reaction, $\Delta G° = -nFE°_{\text{cell}}$. Hence at 25°C, $\Delta G° = -109.612$ kJ.

(b) The standard potential of the Cl_2/Cl^- electrode, $\tfrac{1}{2} Cl_2(g) + e^- = Cl^-$, is given by

$$E_{\text{cell(i)}} = E_{Cl_2/Cl} - E_{AgCl/Cl} \tag{iii}$$

Thus

$$E_{Cl_2/Cl^-} = E_{\text{cell(i)}} + E_{AgCl/Cl} \tag{iv}$$

and, at 25°C, $E°_{Cl_2/Cl^-} = 1.35830$ V. Thus $\Delta G°$ for the reaction $\tfrac{1}{2} Cl_2(g) + e^- = Cl^-$ [which is the same as for the reaction $\tfrac{1}{2} H_2(g) + \tfrac{1}{2} Cl_2(g) = H^+ + Cl^-$] is -131.068 kJ mol^{-1}, corresponding to a log K value of 22.97.

(c) $\Delta S°$ and $\Delta H°$ for the cell reaction $Ag(s) + \tfrac{1}{2}Cl_2(g) = AgCl(s)$ can be calculated by considering that $\Delta S = nF(dE_{\text{cell}}/dT)$ and $\Delta H = -nFE + nFT(dE_{\text{cell}}/dT)$. The data given for $E_{\text{cell(i)}}$ have been fitted as a function of absolute temperature T (least-square procedure) by $E_{\text{cell(i)}} = 1.28958 - (4.31562 \times 10^{-1})T - (2.7922 \times 10^{-7})T^2$. At 25°C the first derivative of this equation is -5.986×10^{-4} (V deg^{-1}). (This temperature dependence may also be obtained approximately by plotting $E_{\text{cell(i)}}$ versus $1/T$.) Hence $\Delta H_{(25°C)} = -126.82$ kJ mol^{-1}. Similarly, $\Delta S°$ and $\Delta H°$ values at 25°C (or at other temperatures) may be obtained from the temperature dependence of $E°_{Cl_2/Cl^-}$. The latter is found (25°C) to be -1.246×10^{-3} V deg^{-1}. Correspondingly, $\Delta S°_{(25°C)} = -120.2$ J deg^{-1} mol^{-1} and $\Delta H_{(25°C)} = -167$ kJ mol^{-1}.

Effect of Ionic Strength and Complex Formation on Electrode Potentials: Formal Potentials

In dealing with redox equilibria, we are also confronted with the problem of evaluating activity corrections or maintaining the activities under consideration as constants. The Nernst equation rigorously applies only if the activities and actual species taking part in the reaction are inserted in the equation. The activity scales discussed before, the infinite dilution scale and the ionic medium scale, may be used. The standard potential or standard pε on the infinite dilution scale is related to the equilibrium constant for $I = 0$ of the reduction reaction by

$$\frac{F}{RT(\ln 10)} E_H^\circ = \frac{1}{n} \log K = p\varepsilon^\circ \qquad (44)$$

and is usually obtained by either extrapolating the measured constant or measured potential to infinite dilution.

In a constant ionic medium the concentration quotient becomes the equilibrium constant ^cK, and correspondingly one might define $^cE_H^\circ$ and $p^c\varepsilon^\circ$.

$$\frac{F}{RT(\ln 10)} {}^cE_H^\circ = \frac{1}{n} \log {}^cK = p^c\varepsilon^\circ \qquad (45)$$

Complex Formation How the standard potential is influenced by complex formation may be illustrated by the dissolution of zinc

$$Zn(s) + 2H^+ = Zn^{2+} + H_2(g) \qquad (46)$$

which is characterized by the half-reaction

$$Zn^{2+} + 2e^- = Zn(s) \qquad \Delta G_1^\circ = -RT \ln K = -2FE^\circ_{Zn^{2+}, Zn(s)} \qquad (47)$$

If the dissolution of Zn occurs in a medium containing ligands that can displace coordinated H_2O from the zinc ions, that is, form complexes, such as in the reaction of Cl^- ions with Zn^{2+} to form $ZnCl_4^{2-}$,

$$ZnCl_4^{2-} = Zn^{2+} + 4\,Cl^- \qquad \Delta G_2^\circ = RT \ln \beta_4 \qquad (48)$$

where β_4 is the formation constant $[\{ZnCl_4^{2-}\}/(\{Zn^{2+}\}\{Cl^-\}^4)]$, then we may characterize the overall half-reaction by

$$ZnCl_4^{2-} + 2e^- = Zn(s) + 4\,Cl^-$$

$$\Delta G_3^\circ = -RT \ln {}^+K = -2FE^\circ_{ZnCl_4^{2-}, Zn(s)} \qquad (49)$$

This is the sum of equations 47 and 48. Correspondingly, we may write any one of the following relations:

$$\Delta G_3^\circ = \Delta G_1^\circ + \Delta G_2^\circ \qquad (50)$$

$$\log {}^+K = \log K - \log \beta_4 \qquad (51)$$

$$p\varepsilon^\circ_{ZnCl_4^{2-}, Zn(s)} = p\varepsilon^\circ_{Zn^{2+}, Zn(s)} - \tfrac{1}{2} \log \beta_4$$

$$E^\circ_{ZnCl_4^{2-}, Zn(s)} = E^\circ_{Zn^{2+}, Zn(s)} - \frac{RT(\ln 10)}{2F} \log \beta_4 \qquad (52)$$

8.3 Electrode Potential: Nernst Equation and the Electrochemical Cell

These equations show that Cl⁻ stabilizes the higher oxidation state, facilitating the dissolution of Zn. The Nernst equation may now be expressed either in terms of free Zn^{2+} or in terms of $ZnCl_4^{2-}$.

$$E = E°_{Zn^{2+},Zn(s)} + \frac{RT(\ln 10)}{2F} \log\{Zn^{2+}\}$$

$$= E°_{ZnCl_4^{2-},Zn(s)} + \frac{RT(\ln 10)}{2F} \log\{ZnCl_4^{2-}\} - \frac{2RT(\ln 10)}{F} \log\{Cl^-\} \quad (53)$$

Formal Potentials As with conditional constants, that is, constants valid under specifically selected conditions, for example, a given pH and a given ionic medium, conditional or formal potentials are of great utility.

$$\frac{F}{RT(\ln 10)} {}^F E°_H = \frac{1}{n} \log P = p^F \varepsilon° \quad (54)$$

The measurement of a formal pε or formal electrode potential consists of measurement of the emf of an electrochemical cell in which, under the specified conditions, the analytical concentration of the two oxidation states is varied. For example, in a 0.1 M H_2SO_4 solution, the formal electrode potential for Fe(III)–Fe(II) is 0.68 V in comparison to 0.77 V for the Fe^{3+}/Fe^{2+} ($I = 0$) system:

$$^F E_{(I=0.1 M H_2SO_4)} = 0.68 + \frac{RT}{F} \ln \frac{Fe(III)_T}{Fe(II)_T} \quad (55)$$

In this case the formal potential includes correction factors for activity coefficients, acid–base phenomena (hydrolysis of Fe^{3+} to $FeOH^{2+}$), complex formation (sulfate complexes), and the liquid junction potential used between the reference electrode and the half-cell in question. Although the correction is strictly valid only at the single concentration at which the potential has been determined, formal potentials may often lead to better predictions than standard potentials because they represent quantities subject to direct experimental measurement.

Example 8.10. Formal Potential of the Fe(III)–Fe(II) System in the Presence of F⁻ Estimate the formal potential of the Fe(III)–Fe(II) couple for solutions of $[H^+] = 10^{-2}$ M and $[F^-] = 10^{-2}$ M, $I = 0.1$ M. The following constants are available:

$$Fe^{3+} + e^- = Fe^{2+} \qquad \log K(I = 0) = 13.0 \quad (i)$$

$$Fe^{3+} + H_2O = FeOH^{2+} + H^+ \qquad \log K_H(I = 0.1) = -2.7 \quad (ii)$$

$$Fe^{3+} + F^- = FeF^{2+} \qquad \log \beta_1(I = 0.1) = 5.2 \quad (iii)$$

$$Fe^{3+} + 2F^- = FeF_2^+ \qquad \log \beta_2(I = 0.1) = 9.2 \qquad \text{(iv)}$$

$$Fe^{3+} + 3F^- = FeF_3 \qquad \log \beta_3(I = 0.1) = 11.9 \qquad \text{(v)}$$

In order to compute the formal potential we first consider the activity correction of the Fe^{3+}–Fe^{2+} electrode. Using the Güntelberg approximation, $K_{(i)}$ is corrected to ${}^cK_{(i)}$:

$$\frac{[Fe^{2+}]}{[Fe^{3+}]\{e\}} = {}^cK = K\frac{f_{Fe^{3+}}}{f_{Fe^{2+}}} = K\frac{0.083}{0.33} = 10^{12.4} \qquad p^c\varepsilon° = 12.4 \qquad \text{(vi)}$$

from which ${}^cE_H°$ is obtained as +0.73 V. It has been assumed that the liquid junction makes a negligible contribution. The same result is obtained if we consider that

$$^cE°_{Fe^{3+},Fe^{2+}} = E°_{Fe^{3+},Fe^{2+}} + \frac{RT}{F}\ln\frac{f_{Fe^{3+}}}{f_{Fe^{2+}}}$$

Next, the correction caused by hydrolysis and complex formation can be taken into account. For the conditions specified, $FeOH^{2+}$ is the only important hydrolysis species:

$$Fe(III)_T = [Fe^{3+}] + [FeOH^{2+}] + [Fe^{2+}] + [FeF^{2}] + [FeF_2^+] + [FeF_3]$$

(vii)

Under the specified conditions, ferrous iron does not form complexes with F^- and OH^-, hence

$$Fe(II)_T = [Fe^{2+}] \qquad \text{(viii)}$$

Equation vii can be rewritten as

$$Fe(III)_T = [Fe^{3+}]\left(1 + \frac{K_H}{[H^+]} + \beta_1[F^-] + \beta_2[F^-]^2 + \beta_3[F^-]^3\right) \qquad \text{(ix)}$$

For the conditions specified,

$$\alpha_{Fe} = \frac{Fe(III)_T}{[Fe^{3+}]} = 9.5 \times 10^5 \qquad \text{(x)}$$

and equilibrium may be formulated as in

$$\frac{Fe(II)_T}{Fe(III)_T\{e^-\}} = \frac{{}^cK}{\alpha_{Fe}} = P \qquad \text{(xi)}$$

and

$$\log P = 6.4 = p^F\varepsilon° \qquad \text{(xii)}$$

8.4 pε–pH, Potential–pH Diagrams

Correspondingly, the formal potential of the Fe(III)–Fe(II) electrode for the given conditions is $^{F}E° = 0.38$ V. The potential of course is the same whether it is expressed in terms of actual concentrations or analytical concentrations:

$$E = 0.73 + \frac{RT}{F} \ln \frac{[Fe^{3+}]}{[Fe^{2+}]} = 0.38 + \frac{RT}{F} \ln \frac{Fe(III)_T}{Fe(II)_T} \quad (xiii)$$

or

$$p\varepsilon = 12.4 + \log \frac{[Fe^{3+}]}{[Fe^{2+}]} = 6.4 + \log \frac{Fe(III)_T}{Fe(II)_T} \quad (xiv)$$

Note that from an electrode kinetic point of view, the Nernst equation does not give any information as to the actual species that establishes the electrode potential. In the case of a fluoride-containing solution, it is very possible that one of the fluoro–iron(III) complexes rather than the Fe^{3+} ion participates in the electron-exchange reaction at the electrode. Complexation usually stabilizes a system against reduction. In the example just considered, because complexation is stronger with Fe(III) than with Fe(II), the tendency for reduction of Fe(III) to Fe(II) is decreased. It is apparent that coordination with a donor group, in general, decreases the redox potential. In the relatively rare instances where the lower oxidation state is favored (e.g., complexation of aqueous iron with phenanthroline), the redox potential is increased as a result of coordination.

Intensity and Capacity pε is an intensity factor; it measures oxidizing intensity. Oxidation or reduction capacity must be expressed in terms of a quantity of system electrons that must be added or removed in order to attain a given pε. This is analogous to the acid- or base-neutralizing capacity with respect to protons; for example, alkalinity and acidity are measured in terms of the proton condition. Thus the oxidative capacity of a system with respect to a given electron energy level will be given by the equivalent sum of all the oxidants below this energy level minus the equivalent sum of all the reductants above it. For example, the oxidative capacity of a solution with respect to an electron level corresponding to Cu(s) is

$$2[Cu^{2+}] + 2[I_3^-] + [Fe^{3+}] + 4[O_2] - 2[H_2] \quad (56)$$

8.4 pε–pH, POTENTIAL–pH DIAGRAMS

An attempt has been made thus far to describe the stability relationships of the distribution of the various soluble and insoluble forms through rather simple graphical representation. Essentially two types of graphical treatments have been used: first, equilibria between chemical species in a particular oxidation state as a function of pH and solution composition; second, equilibria between chemical species at a particular pH as a function of pε (or E_H). Obviously these diagrams can be combined into pε–pH diagrams. Such pε–pH stability field diagrams show in a comprehensive way how protons and electrons simultaneously shift the equilibria under various conditions and can indicate which species predominate under any given condition of pε and pH.

The value of a pε–pH diagram consists primarily in providing an aid for the interpretation of equilibrium constants (free energy data) by permitting the simultaneous representation of many reactions. Of course, such a diagram, like all other equilibrium diagrams, represents only the information used in its construction.

Natural waters are often in a highly dynamic state with regard to oxidation–reduction rather than in or near equilibrium. Most oxidation–reduction reactions have a tendency to be much slower than acid–base reactions, especially in the absence of suitable biochemical catalysis. Nonetheless, equilibrium diagrams can greatly aid attempts to understand the possible redox patterns in natural waters and in water technological systems.

The Construction of pε–pH Diagrams

The construction of a pε–pH diagram may be introduced by considering the redox stability of water. As we have seen (equations 21–24), H_2O can be oxidized to O_2 or reduced to H_2. Equations 25 and 26 can be rewritten as

$$p\varepsilon = 0 - pH - \tfrac{1}{2} \log p_{H_2} \qquad (57)$$

$$p\varepsilon = 20.78 - pH + \tfrac{1}{4} \log p_{O_2} \qquad (58)$$

These equations can be plotted in a pε–pH diagram (Figure 8.9a). The lines of both equations have slopes $d\,p\varepsilon/d\,pH$ of -1, and they intersect the ordinate, at pH = 0, at pε = 20.78 (for $p_{O_2} = 1$) and pε = 0 (for $p_{H_2} = 1$), respectively. Above the upper line, water becomes an effective reductant (producing oxygen); below the lower line, water is an effective oxidant (producing hydrogen) within the H_2O domain, O_2 acts as an oxidant and H_2 as a reductant.

For any partial pressure of O_2, the equilibrium between water and oxygen is characterized by a straight line with a slope $d\,p\varepsilon/d\,pH = -1$; any decrease in p_{O_2} by 10^4 lowers the line by 1 pε unit. (See Figure 8.2).

Example 8.11. pε–pH Diagram for the Sulfur System   Construct a pε–pH diagram for the SO_4^{2-}–S(s)–H_2S(aq) system, assuming that the concentration of soluble S species is 10^{-2} M.

The lines in the pε–pH diagram (Figure 8.7) are characterized by:

8.4 pε–pH, Potential–pH Diagrams

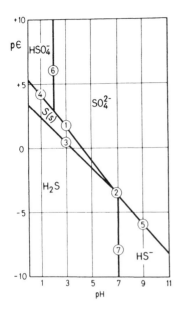

Figure 8.7. pε–pH diagram for the SO_4–S(s)–H_2S system. (The equations for the numbered lines are given in Example 8.11.) Total dissolved S species is 10^{-2} M.

① For the equilibrium

$$SO_4^{2-} + 8H^+ + 6e^- = S(s) + 4H_2O \qquad \log K = 36.2$$

$$p\varepsilon = \frac{36.2}{6} + \frac{1}{6} \log [SO_4^{2-}] - \frac{8}{6} pH \qquad (i)$$

② For the equilibrium

$$SO_4^{2-} + 10H^+ + 8e^- = H_2S(aq) + 4H_2O \qquad \log K = 41.0$$

$$p\varepsilon = \frac{41}{8} + \frac{1}{8} \log \frac{[SO_4^{2-}]}{[H_2S(aq)]} - \frac{10}{8} pH \qquad (ii)$$

③ For the equilibrium

$$S(s) + 2H^+ + 2e^- = H_2S(aq) \qquad \log K = 4.8$$

$$p\varepsilon = \frac{4.8}{2} - pH - \frac{1}{2} \log [H_2S] \qquad (iii)$$

④ For the reaction

$$HSO_4^- + 7H^+ + 6e^- = S(s) + 4H_2O \qquad \log K = 34.2$$

$$p\varepsilon = \frac{34.2}{6} + \frac{1}{6}\log[HSO_4^-] - \frac{7}{6}pH \qquad (iv)$$

⑤ For the reaction

$$SO_4^{2-} + 9H^+ + 8e^- = HS^- + 4H_2O \qquad \log K = 34.0$$

$$p\varepsilon = \frac{34.0}{8} + \frac{1}{8}\log\frac{[SO_4^{2-}]}{[HS^-]} - \frac{9}{8}pH \qquad (v)$$

⑥ For the equilibrium

$$HSO_4^- = SO_4^{2-} + H^+ \qquad \log K = -2.0$$

$$\log\frac{[SO_4^{2-}]}{[HSO_4^-]} - pH = -2.0 \qquad (vi)$$

⑦ For the equilibrium

$$H_2S(aq) = H^+ + HS^- \qquad \log K = -7.0$$

$$\log\frac{[HS^-]}{[H_2S]} - pH = -7.0 \qquad (vii)$$

In the process of reducing SO_4^{2-} or oxidizing sulfide, a number of compounds with intermediate oxidation state can be formed in addition to solid $S^\dagger$ such as SO_3^{2-}, $S_2O_3^{2-}$, and $S_4O_6^{2-}$. The production of these intermediates is usually under biological control. Colloidal sulfur can form polysulfides such as HS_n^- or S_n^{2-} with HS^-.

Example 8.12. pε–pH Diagram for the Fe–CO$_2$–H$_2$O System Construct a pε–pH diagram (25°C) for the Fe–CO$_2$–H$_2$O system; delineate the stability conditions of the solid phases Fe, Fe(OH)$_2$, FeCO$_3$, and amorphous Fe(OH)$_3$; $C_T = 10^{-3}$ M; concentration of soluble Fe species $= 10^{-5}$ M. Pertinent equilibrium equations are given in Table 8.5 and in Tableau 8.6. The results are displayed in Figure 8.8.

---

†Usually orthorhombic sulfur such as $S_8(s)$ on colloidal sulfur.

Table 8.5. Equations Used for the Construction of Figure 8.8

Equations Used for the Construction of the pε–pH Diagram	Functions pε	
$Fe^{3+} + e^- = Fe^{2+}$	$p\varepsilon = 13 + \log [Fe^{3+}]/[Fe^{2+}]$	(1)[a]
$Fe^{2+} + 2e^- = Fe(s)$	$p\varepsilon = -6.9 + \frac{1}{2}\log [Fe^{2+}]$	(2)
$Fe(OH)_3(\text{amorph, s}) + 3H^+ + e^- = Fe^{2+} + 3H_2O$	$p\varepsilon = 16 - \log [Fe^{2+}] - 3\ pH$	(3)
$Fe(OH)_3(\text{amorph, s}) + 2H^+ + HCO_3^- + e^- = FeCO_3(s) + 3\ H_2O$	$p\varepsilon = 16 - 2\ pH + \log [HCO_3^-]$	(4)
	$[HCO_3^-] = C_T\alpha_1$	
$FeCO_3(s) + H^+ + 2e^- = Fe(s) + HCO_3^-$	$p\varepsilon = -7.0 - \frac{1}{2}\ pH - \frac{1}{2}\log [HCO_3^-]$	(5)
$Fe(OH)_2(s) + 2H^+ + 2e^- = Fe(s) + 2H_2O$	$p\varepsilon = -1.1 - pH$	(6)
$Fe(OH)_3(s) + H^+ + e^- = Fe(OH)_2(s) + H_2O$	$p\varepsilon = 4.3 - pH$	(7)
$FeOH^{2+} + H^+ + e^- = Fe^{2+} + H_2O$	$p\varepsilon = 15.2 - pH - \log ([Fe^{2+}]/[FeOH^{2+}])$	(8)
	Functions pH	
$FeCO_3(s) + 2H_2O = Fe(OH)_2(s) + H^+ + HCO_3^-$	$pH = 11.9 + \log [HCO_3^-]$	(a)
$FeCO_3(s) + H^+ = Fe^{2+} + HCO_3^-$	$pH = 0.2 - \log [Fe^{2+}] - \log [HCO_3^-]$	(b)
$FeOH^{2+} + 2H_2O = Fe(OH)_3(s) + 2H^+$	$pH = 0.4 - \frac{1}{2}\log [FeOH^{2+}]$	(c)
$Fe^{3+} + H_2O = FeOH^{2+} + H^+$	$pH = 2.2 - \log ([Fe^{3+}]/[FeOH^{2+}])$	(d)
$Fe(OH)_3(s) + H_2O = Fe(OH)_4^- + H^+$	$pH = 19.2 + \log [Fe(OH)_4^-]$	(e)

[a] The numbers and letters refer to the equations given.

Tableau 8.6. Equilibria in the System Fe–CO$_2$–H$_2$O^a

Components		H$^+$	e$^-$	HCO$_3^-$	Fe^{2+}	log K	Number on Figure 8.8
Species	H$^+$	1					
	OH$^-$	−1				−14.0	
	Fe^{2+}				1		
	Fe^{3+}		−1		1	13.0	①
	Fe^0		2		1	−14.9	②
	FeCO$_3$(s)	−1		1	1	0.2	ⓑ
	Fe(OH)$_2$(s)	−2			1	13.3	
	Fe(OH)$_3$(s)	−3	−1		1	−16.0	③
	FeOH^{2+}	−1	−1		1	−15.2	⑧
	Fe(OH)$_4^-$	−4	−1		1	34.6	
	H$_2$CO$_3^*$	1		1		6.3	
	HCO$_3^-$			1			
	CO$_3^{2-}$	−1		1		−10.3	
Composition	pH fixed	pε fixed	$C_T = 1 \times 10^{-3}$ M	$Fe_T = 1 \times 10^{-5}$ M			

^{a}The other functions are the result of the combination of the equations given in Table 8.5:

$$p\varepsilon = 16.2 - 2\,pH + \log[HCO_3^-] \quad ④$$
$$p\varepsilon = -7.0 - \tfrac{1}{2}pH - \tfrac{1}{2}\log[HCO_3^-] \quad ⑤$$
$$p\varepsilon = -1.1 - pH \quad ⑥$$
$$p\varepsilon = 4.3 - pH \quad ⑦$$
$$pH = 11.9 + \log[HCO_3^-] \quad ⓐ$$

Comparing Various Equilibrium Diagrams Figure 8.9 displays the pε–pH relations for some of the biologically important elements, and Figure 8.10 gives a pε–pH diagram for the Mn–CO$_2$–H$_2$O system. With Figure 8.11 we return to the chlorine equilibria already discussed in Examples 8.7 and Figure 8.5.

The main advantage of a pε–pH diagram is that it provides a good survey and a clear picture of the situation, but it cannot give too much detail, especially about the concentration dependence of the predominance areas. It is possible to draw the boundary lines for various assumed activities and to construct three-dimensional diagrams with activity as one of the axes. There is little limit to the combinations of variables and types of phase diagrams that can be constructed, but we must not forget that the main reason for making a phase diagram is to try to understand or to solve complicated equilibrium problems.

Example 8.13. Chlorine Redox Equilibria Summarize in a pε–pH diagram the information contained in the equilibrium constants, $I = 0$, 25°C, of the following three reactions involving Cl$_2$(aq), Cl$^-$, OCl$^-$, and HOCl. [For convenience, in addition to the equilibrium constant, the standard redox potential,

8.4 pε–pH, Potential–pH Diagrams   461

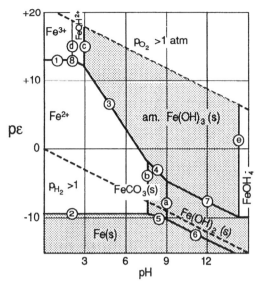

Figure 8.8. Diagram pε versus pH for the system Fe–CO$_2$–H$_2$O. The solid phases are Fe(OH)$_3$ (amorphous), FeCO$_3$ (siderite), Fe(OH)$_2$(s), and Fe(s); $C_T = 10^{-3}$ M. Lines are calculated for Fe(II) and Fe(III) = 10^{-5} M (25°C). The possible conversion of carbonate to methane at low pε values is ignored.

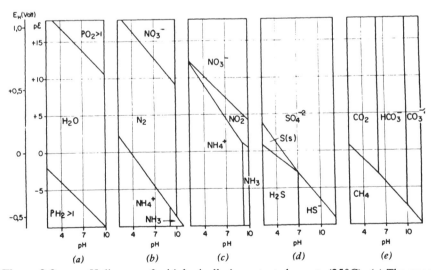

Figure 8.9. pε–pH diagrams for biologically important elements (25°C). (a) The upper and lower lines represent equations 57 and 58 respectively, the oxygen and hydrogen equilibria with water. (b) The nitrogen system, considering only stable equilibria. The only oxidation states involved are (−III), the elemental state, and (V). (c) NH$_4^+$, NH$_3$, NO$_3^-$, and NO$_2^-$ are treated as species metastable with regard to N$_2$; that is, N$_2$ is treated as a redox-inert component. (d) Sulfur species stable for assumed conditions are SO$_4^{2-}$, elemental sulfur, and sulfides (Example 8.11). (e) The thermodynamically possible existence of elemental C (graphite) is ignored.

462   Oxidation and Reduction; Equilibria and Microbial Mediation

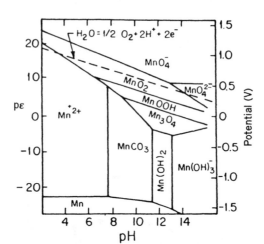

Figure 8.10. pε–pH diagram for the Mn–CO_2–H_2O systems (25°C). The solid phases considered are $Mn(OH)_2(s)$ (pyrochroite), $MnCO_3(s)$ (rhodochrosite), $Mn_3O_4(s)$ (hausmannite), γ-MnOOH(s) (manganite), and γ-MnO_2(s) (nsutite). $C_T = 1 \times 10^{-3}$ M and $Mn_T = 1 \times 10^{-5}$ M.

E_H° (volts), is given.] Total dissolved chlorine is assumed to be $Cl_T = 0.04$ M.

$$HClO + H^+ + e^- = \tfrac{1}{2}Cl_2(aq) + H_2O \qquad \log K = 26.9,\ E_H^\circ = 1.59 \qquad (i)$$

$$\tfrac{1}{2}Cl_2(aq) + e^- = Cl^- \qquad \log K = 23.6,\ E_H^\circ = 1.40 \qquad (ii)$$

$$HClO = H^+ + ClO^- \qquad \log K = -7.3 \qquad (iii)$$

We can write the following equilibrium equations for reactions i and ii:

$$p\varepsilon = 26.9 + \log \frac{[HClO]}{[Cl_2]^{1/2}} - pH \qquad (iv)$$

$$p\varepsilon = 23.6 + \log \frac{[Cl_2]^{1/2}}{[Cl^-]} \qquad (v)$$

Combining reactions i and ii gives the expression for the reduction of HOCl to Cl^-:

$$p\varepsilon = 25.25 + \tfrac{1}{2} \log \frac{[HClO]}{[Cl^-]} - 0.5\,pH \qquad (vi)$$

The reduction of ClO^- to Cl^- is given by

$$p\varepsilon = 28.9 + \tfrac{1}{2} \log \frac{[OCl^-]}{[Cl^-]} - pH \qquad (vii)$$

8.4 pε–pH, Potential–pH Diagrams

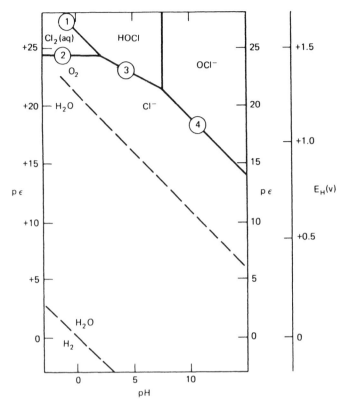

Figure 8.11. pε–pH diagram for the chlorine system. $Cl_T = 0.04$ M; unit atomic ratios for oxidants and reductants at the boundaries. In dilute solutions $Cl_2(aq)$ exists only at low pH. Cl_2, OCl^-, and HOCl are all unstable or metastable in water. Numbers ①–④ refer to the equilibria described by equations iv, v, vi, and vii, respectively, of Example 8.13.

Since $Cl_T = [HOCl] + [ClO^-] + 2[Cl_2] + [Cl^-] = 0.04$ M, and assuming unit atomic ratio for oxidants and reductants, we have the following concentrations: at the Cl_2–HOCl boundary: $[HOCl] = \frac{1}{2}Cl_T = 2 \times 10^{-2}$ M, $[Cl_2] = \frac{1}{4}Cl_T = 10^{-2}$ M; at the Cl_2–Cl^- boundary: $[Cl_2] = Cl_T/4 = 10^{-2}$ M, $[Cl^-] = Cl_T/2 = 2 \times 10^{-2}$ M. Finally, the line separating HOCl from OCl^- is given by (cf. equation iii)

$$\log \frac{[HClO]}{[ClO^-]} + pH = 7.3 \qquad \text{(viii)}$$

With these substitutions the resulting four equations are plotted in Figure 8.11. The line for equation v is pH independent and thus plots as a horizontal line; it intersects with the line representing equation iv, which has $d\,p\varepsilon/d\,pH$ of -1. These lines are discontinued at this intersection. To the right of pH = 1.9

HOCl is a more stable oxidant than Cl_2(aq). (If any doubt should arise about which species predominates thermodynamically, an activity ratio diagram either at a given pε or at a given pH can immediately clarify the stability relations.) Equations vi and vii have slopes of -0.5 and -1, respectively, in the graphical representation; the lines intersect at pH = pK of the hypochlorous acid.

It is convenient to introduce the equations that define the stability limit of H_2O into the same diagram.

Figure 8.11 shows in a different way the same kind of information that was already presented in Figure 8.5.

1. Cl_2(aq) in dilute solutions exists only at rather low pH. Addition of Cl_2 to water is accompanied by disproportionation into HOCl and Cl^-.
2. Cl_2, OCl^-, and $HOCl^-$ are strong oxidants, stronger than O_2. These species are thermodynamically unstable in water; they oxidize H_2O (but the reaction may be slow in the absence of catalysts). Cl^- is the stable species in the pε–pH range of natural waters.

8.5 REDOX CONDITIONS IN NATURAL WATERS

A few elements—C, N, O, S, Fe, Mn—are predominant participants in aquatic redox processes. Tables 8.6a and 8.6b present equilibrium constants for several couples pertinent to consideration of redox relationships in natural waters and their sediments. Data are taken principally from the second edition of *Stability Constants of Metal-Ion Complexes* and *Standard Potentials in Aqueous Solution* (Bard et al., 1985). A subsidiary symbol p$\varepsilon°$(W) is convenient for considering redox situations in natural waters. p$\varepsilon°$(W) is analogous to p$\varepsilon°$ except that $\{H^+\}$ and $\{OH^-\}$ in the redox equilibrium equations are assigned their activities in neutral water. Values for p$\varepsilon°$(W) for 25°C thus apply to unit activities of oxidant and reductant at pH = 7.00. p$\varepsilon°$(W) is defined by

$$p\varepsilon°(W) = p\varepsilon° + \frac{n_H}{2} \log K_W \qquad (59)$$

where n_H is the number of moles of protons exchanged per mole of electrons.

The listing of p$\varepsilon°$(W) values in Tables 8.6a and 8.6b permits an immediate grading of different systems in the order of their oxidizing intensity at pH = 7. Any system in Table 8.6a will tend to oxidize equimolar concentrations of any other system having a lower p$\varepsilon°$(W) value. For example, we see that, at pH = 7, NO_3^- can oxidize HS^- to SO_4^{2-}:

$$\tfrac{1}{8}NO_3^- + \tfrac{5}{4}H^+(W) + e^- = \tfrac{1}{8}NH_4^+ + \tfrac{3}{8}H_2O \quad p\varepsilon°(W) = +6.15,$$

$$\log K(W) = +6.15 \qquad (60)$$

8.5 Redox Conditions in Natural Waters

Table 8.6a. Equilibrium Constants of Redox Processes Pertinent in Aquatic Conditions (25°C)

Reaction		$p\varepsilon°$ (= log K)	$p\varepsilon°$ (W)[a]
(1) $\frac{1}{4}O_2(g) + H^+ + e^-$	$= \frac{1}{2}H_2O$	+20.75	+13.75
(2) $\frac{1}{5}NO_3^- + \frac{6}{5}H^+ + e^-$	$= \frac{1}{10}N_2(g) + \frac{3}{5}H_2O$	+21.05	+12.65
(3) $\frac{1}{2}MnO_2(s) + \frac{1}{2}HCO_3^-(10^{-3}) + \frac{3}{2}H^+ + e^-$	$= \frac{1}{2}MnCO_3(s) + H_2O$	—	+8.9[b,c]
(4) $\frac{1}{2}NO_3^- + H^+ + e^-$	$= \frac{1}{2}NO_2^- + \frac{1}{2}H_2O$	+14.15	+7.15
(5) $\frac{1}{8}NO_3^- + \frac{5}{4}H^+ + e^-$	$= \frac{1}{8}NH_4^+ + \frac{3}{8}H_2O$	+14.90	+6.15
(6) $\frac{1}{6}NO_2^- + \frac{4}{3}H^+ + e^-$	$= \frac{1}{6}NH_4^+ + \frac{1}{3}H_2O$	+15.14	+5.82
(7) $\frac{1}{2}CH_3OH + H^+ + e^-$	$= \frac{1}{2}CH_4(g) + \frac{1}{2}H_2O$	+9.88	+2.88
(8) $\frac{1}{4}CH_2O + H^+ + e^-$	$= \frac{1}{4}CH_4(g) + \frac{1}{4}H_2O$	+6.94	−0.06
(9) $FeOOH(s) + HCO_3^-(10^{-3}) + 2H^+ + e^-$	$= FeCO_3(s) + 2H_2O$	—	−0.8[b,c]
(10) $\frac{1}{2}CH_2O + H^+ + e^-$	$= \frac{1}{2}CH_3OH$	+3.99	−3.01
(11) $\frac{1}{6}SO_4^{2-} + \frac{4}{3}H^+ + e^-$	$= \frac{1}{6}S(s) + \frac{2}{3}H_2O$	+6.03	−3.30
(12) $\frac{1}{8}SO_4^{2-} + \frac{5}{4}H^+ + e^-$	$= \frac{1}{8}H_2S(g) + \frac{1}{2}H_2O$	+5.25	−3.50
(13) $\frac{1}{8}SO_4^{2-} + \frac{9}{8}H^+ + e^-$	$= \frac{1}{8}HS^- + \frac{1}{2}H_2O$	+4.25	−3.75
(14) $\frac{1}{2}S(s) + H^+ + e^-$	$= \frac{1}{2}H_2S(g)$	+2.89	−4.11
(15) $\frac{1}{8}CO_2(g) + H^+ + e^-$	$= \frac{1}{8}CH_4(g) + \frac{1}{4}H_2O$	+2.87	−4.13
(16) $\frac{1}{6}N_2(g) + \frac{4}{3}H^+ + e^-$	$= \frac{1}{3}NH_4^+$	+4.68	−4.68
(17) $H^+ + e^-$	$= \frac{1}{2}H_2(g)$	0.0	−7.00
(18) $\frac{1}{4}CO_2(g) + H^+ + e^-$	$= \frac{1}{24}(\text{glucose}) + \frac{1}{4}H_2O$	−0.20	−7.20
(19) $\frac{1}{2}HCOO^- + \frac{3}{2}H^+ + e^-$	$= \frac{1}{2}CH_2O + \frac{1}{2}H_2O$	+2.82	−7.68
(20) $\frac{1}{4}CO_2(g) + H^+ + e^-$	$= \frac{1}{4}CH_2O + \frac{1}{4}H_2O$	−1.20	−8.20
(21) $\frac{1}{2}CO_2(g) + \frac{1}{2}H^+ + e^-$	$= \frac{1}{2}HCOO^-$	−4.83	−8.33

[a] Values for $p\varepsilon°$ (W) apply to the electron activity for unit activities of oxidant and reductant in neutral water, that is, at pH = 7.0 for 25°C.
[b] These data correspond to $(HCO_3^-) = 10^{-3}$ M rather than unity and so are not exactly $p\varepsilon°$ (W); they represent typical aquatic conditions more nearly than $p\varepsilon°$ (W) values do.
[c] Alternatively one may consider the reaction.
$\frac{1}{2}MnO_2(s) + 2H^+ + e^- = \frac{1}{2}Mn^{2+} (10^{-6} M) + H_2O$; $p\varepsilon°$ (W) = 9.8
[d] $Fe(OH)_3(am) + 3H^+ + e^- = (Fe^{2+}) (10^{-6} M) + H_2O$; $p\varepsilon°$ (W) = 1.0

Table 8.6b. Some Cellular Energy Transfer Reactions[a]

Half Redox Reactions (Reduction)		$p\varepsilon_w°$ (W)
$NAD^+ + 2H^+ + 2e^-$	$= NADH + H^+$	−5.4
$NADP^+ + 2H^+ + 2e^-$	$= NADPH = H^+$	−5.5
2 ferrodoxin(ox) + $2e^-$	= 2 ferrodoxin (red)	−7.1
ubiquinone + $2H^+ + 2e^-$	= ubiquinol	+1.7
2 cytochrome C(ox) + $2e^-$	= 2 cytochrome C (red)	+4.3

[a] *Source*: Partially from Morel and Hering (1993).

Oxidation and Reduction; Equilibria and Microbial Mediation

$$\tfrac{1}{8}HS^- + \tfrac{1}{2}H_2O = \tfrac{1}{8}SO_4^{2-} + \tfrac{9}{8}H^+(W) + e^- \quad p\varepsilon°(W) = -3.75,$$

$$\log K(W) = +3.75 \qquad (61)$$

$$\tfrac{1}{8}NO_3^- + \tfrac{1}{8}HS^- + \tfrac{1}{8}H^+(W) + \tfrac{1}{8}H_2O = \tfrac{1}{8}NH_4^+ + \tfrac{1}{8}SO_4^{2-}$$

$$\log K(W) = +9.9 \qquad (62)$$

or

$$NO_3^- + HS^- + H^+(W) + H_2O = NH_4^+ + SO_4^{2-}$$

$$\log K(W) = 79.2 \qquad (63)$$

[log $K(W)$ is the equilibrium constant for the redox reaction in neutral water, pH = 7.00 at 25°C; note that p$\varepsilon°$, because it is a measure of oxidizing intensity, maintains the same sign independent of the direction in which the reaction is written.] Since $\Delta p\varepsilon°(W)$ or log $K(W)$ is positive, the reaction is thermodynamically possible in neutral aqueous solutions at standard concentrations. Figure 8.9 gives pε–pH diagrams for biologically important elements.

Example 8.14. Oxidation of Organic Matter by SO_4^{2-} Is the oxidation of organic matter, here CH_2O, thermodynamically possible under conditions normally encountered in natural water systems?

The overall process is obtained by combining (12) and (20) in Table 8.6a:

$$\tfrac{1}{8}SO_4^{2-} + \tfrac{5}{4}H^+(W) + e^- = \tfrac{1}{8}H_2S(g) + \tfrac{1}{2}H_2O$$

$$p\varepsilon°(W) = -3.50, \quad \log K(W) = -3.50 \qquad (i)$$

$$\tfrac{1}{4}CH_2O + \tfrac{1}{4}H_2O = \tfrac{1}{4}CO_2(g) + H^+(W) + e^-$$

$$p\varepsilon°(W) = -8.20, \quad \log K(W) = +8.20 \qquad (ii)$$

$$\tfrac{1}{8}SO_4^{2-} + \tfrac{1}{4}CH_2O + \tfrac{1}{4}H^+(W) = \tfrac{1}{8}H_2S(g) + \tfrac{1}{4}CO_2(g) + \tfrac{1}{4}H_2O$$

$$\log K(W) = +4.70 \qquad (iii)$$

This may also be written

$$SO_4^{2-} + 2CH_2O + 2H^+(W) = H_2S(g) + 2CO_2(g) + 2H_2O;$$

$$\log K(W) = 37.6$$

Hence at standard concentrations the reaction at pH = 7 is thermodynamically possible. The same results will also hold for any equal fractions of unit activity

8.5 Redox Conditions in Natural Waters

because the numbers of molecules of sulfur-containing and carbon-containing species do not change as a result of the reaction. Furthermore, for assumed actual conditions a calculation of $\Delta G = RT \ln(Q/K)$ will show whether the oxidation can proceed thermodynamically. Thus, for a set of assumed actual conditions, such as $p_{CO_2} = 10^{-3.5}$ atm, $[CH_2O] = 10^{-6}$ M, $[SO_4^{2-}] = 10^{-3}$ M, and $p_{H_2S} = 10^{-2}$ M, a value of $10^{-31.6}$ is obtained for Q/K; hence ΔG is clearly negative, indicating that SO_4^{2-} can oxidize CH_2O under these conditions.

Redox Intensity and the Biochemical Cycle

The maintenance of life resulting directly or indirectly from a steady impact of solar energy (*photosynthesis*) is the main cause of nonequilibrium conditions (Figure 8.12). Photosynthesis may be conceived as a process producing localized centers of highly negative pε and a reservoir of oxygen. Nonphotosynthetic organisms tend to restore equilibrium by catalytically decomposing the unstable products of photosynthesis through energy-yielding redox reactions. Orga-

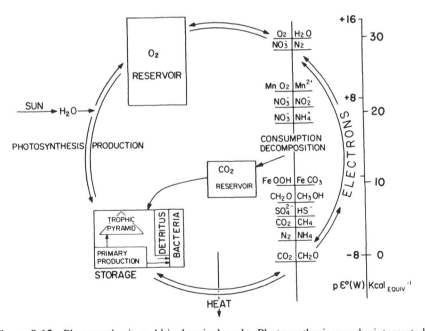

Figure 8.12. Photosynthesis and biochemical cycle. Photosynthesis may be interpreted as a disproportionation into an oxygen reservoir and reduced organic matter (biomass containing high-energy bonds made with hydrogen and C, N, S, and P compounds). The nonphotosynthetic organisms tend to restore equilibrium by catalytically decomposing the unstable products of photosynthesis through energy-yielding redox reactions. The p$\varepsilon°$(W) scale on the right gives the sequence of the redox reactions observed in an aqueous system.

nisms, themselves a product of inorganic matter, are primarily built up from "redox elements," and their relatively constant stoichiometric composition ($C_{106}H_{263}O_{110}N_{16}P$) and the cyclic exchange between chemical elements of the water and the biomass have pronounced effects on the relative concentrations of the elements in the environment. The biologically active elements circulate in a different pattern than water itself or inactive (conservative) solutes.

Photosynthesis, by trapping light energy and converting it to chemical energy, produces reduced states of higher free energy (high-energy chemical bonds) and thus nonequilibrium concentrations of C, N, and S compounds.

As shown in (18) of Table 8.6a, the conversion of CO_2 to glucose at unit activities requires $p\varepsilon°(W) = -7.2$. Although this value may be modified somewhat for actual intracellular activities, it does represent approximately the negative $p\varepsilon$ level that must be reached during photosynthesis.

Of course, the microbially mediated aquatic redox couples must be interrelated to the proper intracellular redox couples. For examples, the NADP system, ubiquitous in living organisms and believed to play a major role in electron transport during photosynthesis, exhibits $p\varepsilon°(W) = +5.5$.[†] Moreover, various ferredoxins, now widely considered to be the primary electron receptors from excited chlorophylls, show $p\varepsilon°(W)$ values in the range -7.0 to -7.5 (Table 8.6b). The coincidence of this range with the $p\varepsilon°(W)$ value for conversion of CO_2 to glucose is suggestive.

In contrast, the respiratory, fermentative, and other nonphotosynthetic processes of organisms tend to restore equilibrium by catalyzing or mediating chemical reactions releasing free energy and thus increasing the mean $p\varepsilon$ level.

Water in solubility equilibrium with atmospheric oxygen has a well-defined $p\varepsilon = 13.6$ (for $p_{O_2} = 0.21$ atm, $E_H = 800$ mV at pH 7 and 25°C). Calculations from the $p\varepsilon°$ values of Table 8.6a show that at this $p\varepsilon$ all the other elements should exist virtually completely in their highest naturally occurring oxidation states: C as CO_2, HCO_3^-, or CO_3^{2-} with reduced organic forms less than 10^{-35} M; N as NO_3^- with NO_2^- less than 10^{-7} M; S as SO_4^{2-} with SO_3^{2-} or HS^- less than 10^{-20} M; Fe as FeOOH or Fe_2O_3 with Fe^{2+} less than 10^{-18} M; and Mn as MnO_2 with Mn^{2+} less than 10^{-10} M. Even the N_2 from the atmosphere should be largely oxidized to NO_3^-.

Since in fact N_2 and organic matter are known to persist in waters containing dissolved oxygen, a total redox equilibrium is not found in natural water systems, even in surface films. At best there are partial equilibria, treatable as approximations to equilibrium either because of slowness of interaction with the other redox couples or because of isolation from the total environment as a result of slowness of diffusional or mixing processes.

The ecological systems of natural waters are thus more adequately represented by dynamic than by equilibrium models. The former are needed to describe the free energy flux absorbed from light and released in subsequent redox processes. Equilibrium models can only depict the thermodynamically

---

[†]Hydrogen resulting from the photolysis of water to be used eventually for the reduction of CO_2 is first bound to the coenzyme nicotinamide adenine dinucleotide phosphate (NADP).

8.5 Redox Conditions in Natural Waters

stable state and describe the direction and extent of processes tending toward it.

When comparisons are made between calculations for an equilibrium redox state and concentrations in the dynamic aquatic environment, the implicit assumptions are that the biological mediations are operating essentially in a reversible manner at each stage of the ongoing processes or that there is a metastable steady state that approximates the partial equilibrium state for the system under consideration.

Microbial Mediation

As already pointed out, nonphotosynthetic organisms tend to restore equilibrium by catalytically decomposing the unstable products of photosynthesis through energy-yielding redox reactions, thereby obtaining a source of energy for their metabolic needs. The organisms use this energy both to synthesize new cells and to maintain the old cells already formed. The energy exploitation is of course not 100% efficient; only a proportion of the free energy released can become available for cell use. It is important to keep in mind that organisms cannot carry out *gross* reactions that are thermodynamically not possible. From a point of view of overall reactions these organisms act only as redox catalysts. Therefore organisms do not oxidize substrates or reduce O_2 or SO_4^{2-}, they only mediate the reaction or, more specifically, the electron transfer, for example, of the specific oxidation of the substrate and of the reduction of O_2 or SO_4^{2-}. Since, for example, SO_4^{2-} can be reduced only at a given pε or redox potential, an equilibrium model characterizes the pε range in which reduction of sulfate is not possible and the pε range in which reduction of sulfate is possible. Thus pε is a parameter that characterizes the ecological milieu restrictively. The pε range in which certain oxidation or reduction reactions are possible can be estimated by calculating equilibrium composition as a function of pε. This has been done for nitrogen, manganese, iron, sulfur, and carbon at a pH of 7. The results are shown in Figure 8.13. There are chemosynthetic bacteria that oxidize reduced sulfur species (sulfate, colloidal sulfur, $S_2O_3^{2-}$) to various intermediate redox states and eventually to SO_4^{2-}.

Nitrogen System Figure 8.13a shows relationships among several oxidation states of nitrogen as a function of pε for a total atomic concentration of nitrogen-containing species equal to 10^{-3} M. The maximum N_2(aq) concentration is therefore 5×10^{-4} M, corresponding to about p_{N_2} of 0.77 atm. For most of the aqueous range of pε, N_2 gas is the most stable species, but at quite negative pε values ammonia becomes predominant and nitrate dominates for pε greater than +12 and pH = 7. The fact that nitrogen gas has not been converted largely into nitrate under prevailing aerobic conditions at land and water surfaces indicates a lack of efficient biological mediation of the reverse reaction also, for the mediating catalysis must operate equally well for reaction in both directions. It appears then that denitrification must occur by an indirect

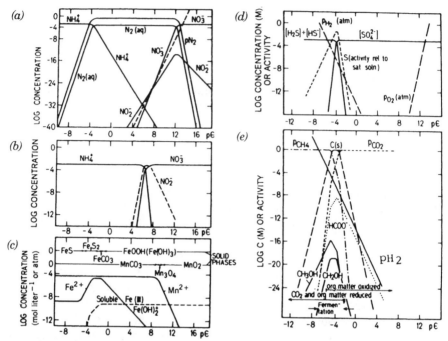

Figure 8.13. Equilibrium concentrations of biochemically important redox components as a function of pε at a pH of 7.0: (a) nitrogen; (b) nitrogen, with elemental nitrogen N_2 ignored; (c) iron and manganese; (d) sulfur; (e) carbon. These equilibrium diagrams have been constructed from equilibrium constants listed in Tables 8.6a and 8.6b for the following concentrations: C_T (total carbonate carbon) = 10^{-3} M; $[H_2S(aq)]$ + $[HS^-]$ + $[SO_4^{2-}]$ = 10^{-3} M; $[NO_3^-]$ + $[NO_2^-]$ + $[NH_4^+]$ = 10^{-3} M; p_{N_2} = 0.78 atm and thus $[N_2(aq)]$ = 0.5×10^{-3} M. For the construction of (b) the species NH_4^+, NO_3^-, and NO_2^- are treated as metastable with regard to N_2.

mechanism such as reduction of NO_3^- to NO_2^- followed by reaction of NO_2^- with NH_4^+ to produce N_2 and H_2O.†

Because reduction of N_2 to NH_4^+ (N_2 *fixation*) at pH 7 can occur to a substantial extent when pε is less than about -4.5, the level of pε required is not as negative as for the reduction of CO_2 to CH_2O. It is not surprising then that blue-green algae are able to mediate this reduction at the negative pε levels produced by photosynthetic light energy. What is perhaps surprising is that nitrogen fixation does not occur more widely among photosynthetic organisms and proceed to a greater extent as compared with CO_2 reduction. Kinetic prob-

---

†Because of the nonreversibility of the biological mediation of the $NO_3^- \rightleftharpoons N_2$ conversion, the NO_3^--N_2 couple cannot be used as a reliable redox indicator. For example, NO_3^- may be reduced to N_2 in an aquatic system even if the bulk phase contains some dissolved oxygen. The reduction may occur in a microenvironment with a pε value lower than that of the bulk, for example, inside a floc or within the sediments; the N_2 released to the aerobic bulk phase is not reoxidized although reoxidation is thermodynamically feasible.

8.5 Redox Conditions in Natural Waters

lems in breaking the strong bonding of the N_2 molecule probably are major factors here.

There are also bacteria that derive their energy from oxidizing NH_4^+ to NO_2^- and NO_3^-. [Nitrosomanas is one type of (aerobic) bacteria that mediates the oxidation (by O_2) to NO_2^- and nitrobacter is another type that catalyzes the oxidation of NO_2^- to NO_3^-.] These bacteria are *autotrophs*; they fix their own organic carbon from CO_2. In some of these transformations N_2O may be formed as an intermediate.

Because of the kinetic hindrance between "bound" nitrogen and N_2, it might also be useful to consider a system in which NO_3^-, NO_2^-, and NH_4^+ are treated as components metastable with respect to gaseous N_2. A diagram for such a system, Figure 8.13b, shows the shifts in relative predominance of the three species all within the rather narrow pε range from 5.8 to 7.2. That each of the species has a dominant zone within this pε range seems to be a factor contributing to the observed highly mobile characteristics of the nitrogen cycle.

Sulfur System The reduction of SO_4^{2-} to H_2S or HS^- provides a good example of the application of equilibrium considerations to aquatic relationships. Figure 8.13d shows relative activities of SO_4^{2-} and H_2S at pH 7 and 25°C as a function of pε when the total concentration of sulfur is 1 mM. It is apparent that a significant reduction of SO_4^{2-} to H_2S at this pH requires pε < -3. The biological enzymes that mediate this reduction with oxidation of organic matter must then operate at or below this pε. Because the system is dynamic rather than static, only an upper bound can be set in this way, for the excess driving force in terms of pε at the mediation site is not indicated by equilibrium computations. Since, however, many biologically mediated reactions seem to operate with relatively high efficiency for utilizing free energy, it appears that the operating pε value is not greatly different from the equilibrium value.

Combining (11) and (14) of Table 8.6a gives

$$SO_4^{2-} + 2H^+(W) + 3H_2S = 4S(s) + 4H_2O \qquad \log K(W) = 4.86 \qquad (64)$$

This equation indicates a possibility of formation of solid elemental sulfur in the reduction of sulfate at pH 7 and standard concentrations. A concentration of 1 M sulfate is unusual, however.

The solubility of $CaSO_4(s)$ is about 0.016 M at 25°C. According to the foregoing rough calculation, sulfur should form in the reduction of SO_4^{2-} in saturated $CaSO_4(s)$ only if the pH is somewhat below 7. There are some indications that this conclusion agrees with the condition of natural sulfur formation. Elemental sulfur may be formed, however, as a kinetic intermediate or as a metastable phase under many natural conditions.

Iron and Manganese In constructing Figure 8.13c, solid FeOOH ($\bar{G}_f^\circ = -462$ kJ mol^{-1}) has been assumed as the stable ferric hydr(oxide). Although thermodynamically possible, magnetite [$Fe_3O_4(s)$] has been ignored as an intermediate in the reduction of ferric oxide to Fe(II). As Figure 8.13 shows, in the presence of O_2, $p\varepsilon > 11$, iron and manganese are stable only as solid oxidized oxides. Soluble forms are present at concentrations less than 10^{-9} M. The concentration of soluble iron and manganese, as Fe^{2+} and Mn^{2+}, increases with decreasing $p\varepsilon$, the highest concentrations being controlled by the solubility of $FeCO_3(s)$ and $MnCO_3(s)$, respectively. ($[HCO_3^-] = 10^{-3}$ M has been assumed for construction of the diagram.)

Carbon System A great number of organic compounds are synthesized, transformed, and decomposed—mostly by microbial catalysis—continually. For operation of the *carbon cycle* degradation is just as important as synthesis. With the exception of CH_4, no organic solutes encountered in natural waters are thermodynamically stable. For example, the disproportionation of acetic acid

$$CH_3COOH = 2H_2O + 2C(s) \qquad \log K = 18$$

$$CH_3COOH = CH_4(g) + CO_2(g) \qquad \log K = 9$$

is thermodynamically favored, but prevented by slow kinetics. Similarly, formaldehyde is unstable with respect to its decomposition into carbon (graphite) and water

$$CH_2O(aq) = C(s) + H_2O \qquad \log K = 18.7$$

but there is no evidence that this reaction occurs.

In redox reactions involving carbon compounds, a great variety of intermediate carbon compounds are formed. Even though reversible equilibria are not attained at low temperatures, it is of considerable interest to compare the equilibrium constants of the various steps in the oxidation of organic matter. The compounds CH_4, CH_3OH, CH_2O, and $HCOO^-$ given in Figure 8.13e represent organic material with formal oxidation states of $-IV$, $-II$, 0, and $+II$, respectively. The diagram has been constructed for the condition $p_{CH_4} + p_{CO_2} = 1$ atm.[†] The major feature of the equilibrium carbon system is simply a conversion of predominant CO_2 to predominant CH_4 with a half-way point at $p\varepsilon = -4.13$. At this $p\varepsilon$ value, where the other oxidation states exhibit maximum relative occurrence, formation of graphite is thermodynamically possible.

Methane fermentation may be considered a reduction of CO_2 to CH_4; this reduction may be accompanied by oxidation of any one of the intermediate oxidation states.[‡] Since all of the latter have $p\varepsilon°(W)$ values less than -6.4

[†]This condition is not fulfilled in the $p\varepsilon$ range where $\{C(s)\} = 1$.
[‡]This statement does not imply a mechanism. Methane may be formed directly, for example, from acetic acid:

$$CH_3COOH \rightarrow CH_4 + CO_2$$

This reaction could be classified as the sum of the reactions $CH_3COOH + 2H_2O = 2CO_2 + 8H^+ + 8e^-$; $CO_2 + 8e^- + 8H^+ = CH_4 + 2H_2O$.

8.5 Redox Conditions in Natural Waters

(this for CH_3OH), each can provide the negative $p\varepsilon$ level required thermodynamically for reduction of CO_2 to CH_4 in its oxidation. Physiologically different organisms may typically be involved in methanogenesis. Certain organisms break down organic materials to organic acids and alcohols, producing eventually acetate, H_2, and CO_2 as intermediates:

$$\text{complex organic material} \longrightarrow \text{organic acids} \begin{array}{l} \nearrow H_2 + CO_2 \longrightarrow CH_4 \\ \searrow CH_3COO^- \longrightarrow CH_4 \end{array}$$

(65)

As indicated in equation 65 H_2—formed by redox disproportionation, for example, by β-oxidation of fatty acids such as $CH_3CH_2CH_2COO^- + H_2O = 2CH_3COO^- + 2H_2(g) + H^+$—acts as a reductant of CO_2:

$$4H_2(g) + CO_2(g) = CH_4 + 2H_2O$$

$$\Delta G° = -31 \text{ kcal} \qquad \log K = 22.9 \; (25°C) \qquad (66)$$

Alcohol fermentation may be exemplified by redox disproportionation of CH_2O (or $C_6H_{12}O_6$):

$$CH_2O + CH_2O + H_2O = CH_3OH + HCOO^- + H^+ \qquad (67)$$

or

$$CH_2O + 2CH_2O + H_2O = 2CH_3OH + CO_2(g) \qquad (68)$$

$$C_6H_{12}O_6 = 2C_2H_5OH + 2CO_2(g) \qquad \Delta G° = -58.3 \text{ kcal} \qquad (69)$$

The reduction of CH_2O to CH_3OH can occur at $p\varepsilon < -3$. Because the concomitant oxidation of CH_2O to CO_2 has $p\varepsilon°(W) = -8.2$, there is no thermodynamic problem.

Microbially Mediated Oxidation and Reduction Reactions

Although, as emphasized, conclusions regarding chemical dynamics may not generally be drawn from thermodynamic considerations, it appears that all the reactions discussed in the previous section, except possibly those involving $N_2(g)$ and $C(s)$, are biologically mediated in the presence of suitable and abundant biota. Table 8.7 and Figure 8.14 survey the oxidation and reduction reactions that may be combined to result in exergonic processes. The possible combinations represent the well-known reactions mediated by heterotrophic and chemoautotrophic organisms. It appears that in natural habitats organisms capable of mediating the pertinent redox reactions are nearly always found.

Table 8.7. Reduction and Oxidation Reactions that May Be Combined to Result in Biologically Mediated Exergonic Processes (pH = 7)

Reduction	$p\varepsilon°(W) = \log K(W)$	Oxidation	$p\varepsilon°(W) = -\log K(W)$
(A) $\frac{1}{4}O_2(g) + H^+(W) + e^- = \frac{1}{2}H_2O$	+13.75	(L) $\frac{1}{4}CH_2O + \frac{1}{4}H_2O = \frac{1}{4}CO_2(g) + H^+(W) + e^-$	−8.20[a]
(B) $\frac{1}{5}NO_3^- + \frac{6}{5}H^+(W) + e^- = \frac{1}{10}N_2(g) + \frac{3}{5}H_2O$	+12.65	(L-1) $\frac{1}{2}HCOO^- = \frac{1}{2}CO_2(g) + \frac{1}{2}H^+(W) + e^-$	−8.73
(C) $\frac{1}{2}MnO_2(s) + \frac{1}{2}HCO_3^-(10^{-3}) + \frac{3}{2}H^+(W) + e^- = \frac{1}{2}MnCO_3(s) + H_2O$	+8.9	(L-2) $\frac{1}{4}CH_2O + \frac{1}{4}H_2O = \frac{1}{2}HCOO^- + \frac{3}{2}H^+(W) + e^-$	−7.68
(D) $\frac{1}{8}NO_3^- + \frac{5}{4}H^+(W) + e^- = \frac{1}{8}NH_4^+ + \frac{3}{8}H_2O$	+6.15	(L-3) $\frac{1}{2}CH_3OH = \frac{1}{2}CH_2O + H^+(W) + e^-$	−3.01
(E) $FeOOH(s) + HCO_3^-(10^{-3}) + 2H^+(W) + e^- = FeCO_3(s) + 2H_2O$	−0.8	(L-4) $\frac{1}{8}CH_4(g) + \frac{1}{4}H_2O = \frac{1}{8}CH_3OH + H^+(W) + e^-$	+2.88
(F) $\frac{1}{2}CH_2O + H^+(W) + e^- = \frac{1}{2}CH_3OH$	−3.01	(M) $\frac{1}{8}HS^- + \frac{1}{2}H_2O = \frac{1}{8}SO_4^{2-} + \frac{9}{8}H^+(W) + e^-$	−3.75
(G) $\frac{1}{8}SO_4^{2-} + \frac{9}{8}H^+(W) + e^- = \frac{1}{8}HS^- + \frac{1}{2}H_2O$	−3.75	(N) $FeCO_3(s) + 2H_2O = FeOOH(s) + HCO_3^-(10^{-3}) + 2H^+(W) + e^-$	−0.8
(H) $\frac{1}{8}CO_2(g) + H^+(W) + e^- = \frac{1}{8}CH_4(g) + \frac{1}{4}H_2O$	−4.13	(O) $\frac{1}{8}NH_4^+ + \frac{3}{8}H_2O = \frac{1}{8}NO_3^- + \frac{5}{4}H^+(W) + e^-$	+6.15
(J) $\frac{1}{6}N_2 + \frac{4}{3}H^+(W) + e^- = \frac{1}{3}NH_4^+$	−4.68	(P) $\frac{1}{2}MnCO_3(s) + H_2O = \frac{1}{2}MnO_2(s) + \frac{1}{2}HCO_3^-(10^{-3}) + \frac{3}{2}H^+(W) + e^-$	+8.9

Examples	Combinations	$\Delta G°(W)$ pH = 7 (kJ eq^{-1})
Aerobic respiration	A + L	−125
Denitrification	B + L	−119
Nitrate reduction	D + L	−82
Fermentation	F + L	−27
Sulfate reduction	G + L	−25
Methane fermentation	H + L	−23
N fixation	J + L	−20
Sulfide oxidation	A + M	−100
Nitrification	A + O	−43
Ferrous oxidation	A + N	−88
Mn(II) oxidation	A + P	−30

[a] CH_2O is used as a formula for an "average" organic substance. The free energy change involved with different specific organic substances may differ from that given for CH_2O. The difference may be large, especially in anoxic processes with substrates whose carbon has a different oxidation state than that in CH_2O.

8.5 Redox Conditions in Natural Waters

The Sequence of Redox Reactions (Figure 8.14) In a closed aqueous system containing organic material—say, CH_2O—oxidation of organic matter is observed to occur first by reduction of O_2 [pε(W) = 13.8]. This will be followed by reduction of NO_3^- and NO_2^-. As seen in Figures 8.12 and 8.13, the succession of these reactions follows the decreasing pε level. Reduction of MnO_2 if present should occur at about the same pε levels as that of nitrate reduction, followed by reduction of $FeOOH(s)$ or $Fe(OH)_3(s)$ to Fe^{2+}. When sufficiently negative pε levels have been reached, fermentation reactions and reduction of SO_4^{2-} and CO_2 may occur almost simultaneously.

The described sequence would be expected if reactions tended to occur in order of their thermodynamic possibility. The reductant (CH_2O) will supply electrons to the lowest unoccupied electron level (O_2); with more electrons available, successive levels—NO_3^-, NO_2^-, $MnO_2(s)$ and so on—will be filled up. The described succession of reactions is mainly reflected in the vertical distribution of components in a nutrient-enriched (eutrophied) lake and in general also in the temporal succession in a closed system containing excess organic matter, such as a batch digester (anaerobic fermentation unit).

Since the reactions considered [with the possible exception of the reduction of $MnO_2(s)$ and $FeOOH(s)$] are biologically mediated, the chemical reaction sequence is paralleled by an *ecological succession* of microorganisms (aerobic heterotrophs, denitrifiers, fermentors, sulfate reducers, and methane bacteria). It is perhaps also of great interest from an evolutionary point of view that there appears to be a tendency for more energy-yielding mediated reactions to take precedence over processes that are less energy-yielding.

The sequence of redox reactions observed in a system with excess CH_2O (or a system, such as a lake or a groundwater, that is "titrated" with organic material such as pollutants or settling algae in a lake) is summarized in Table 8.8. Organic carbon compounds (with the exception of CH_4) are unstable over the entire pε range, but it is frequently assumed that anaerobic conditions are more favorable to the preservation of organic matter than aerobic conditions.

Another type of reaction sequence would be observed in a system of incipient low pε to which O_2 is added. This is the situation commonly encountered after a stream has become polluted with a variety of reducing substances. In such a case it is typically observed that aerobic respiration takes precedence over nitrification, that is, bacterially mediated nitrification is, at least partially, inhibited or represented in the presence of organic material.

It may also be noted (combination J and L in Figure 8.14) that there is a thermodynamic possibility for N_2 reduction accompanied by CH_2O oxidation. This is the gross mechanism mediated by nonphotosynthetic nitrogen-fixing bacteria.

The concern over so-called nonbiodegradable pollutants and the recovery from sediments of organic substances hundreds of millions of years old may serve as a reminder that a state of equilibrium is not often attained, even within geological time spans, and that microorganisms are not "infallible" in catalyzing processes toward the stable state. Equilibrium models can describe the

Table 8.8. Sequence of Progressive Reduction of Redox Intensity by Organic Pollutants

O_2 consumption (respiration)		
$\frac{1}{4}\{CH_2O\} + \frac{1}{4}O_2$	$= \frac{1}{4}CO_2 + \frac{1}{4}H_2O$	(1)
Denitrification		
$\frac{1}{4}\{CH_2O\} + \frac{1}{5}NO_3^- + \frac{1}{5}H^+$	$= \frac{1}{4}CO_2 + \frac{1}{10}N_2 + \frac{1}{2}H_2O$	(2)
Nitrate reduction		
$\frac{1}{4}\{CH_2O\} + \frac{1}{8}NO_3^- + \frac{1}{4}H^+$	$= \frac{1}{4}CO_2 + \frac{1}{8}NH_4^+ + \frac{1}{8}H_2O$	(3)
Production of soluble Mn(II)		
$\frac{1}{4}\{CH_2O\} + \frac{1}{2}MnO_2(s) + H^+$	$= \frac{1}{4}CO_2 + \frac{1}{2}Mn^{2+} + \frac{1}{8}H_2O$	(4)
Fermentation		
$\frac{3}{4}\{CH_2O\} + \frac{1}{4}H_2O$	$= \frac{1}{4}CO_2 + \frac{1}{2}CH_3OH$	(5)
Production of soluble Fe(II)		
$\frac{1}{4}\{CH_2O\} + FeOOH(s) + 2H^+$	$= \frac{1}{4}CO_2 + \frac{7}{4}H_2O + Fe^{+2}$	(6)
Sulfate reduction, production of H_2S		
$\frac{1}{4}\{CH_2O\} + \frac{1}{8}SO_4^{2-} + \frac{1}{8}H^+$	$= \frac{1}{8}HS^- + \frac{1}{4}CO_2 + \frac{1}{4}H_2O$	(7)
Methane fermentation		
$\frac{1}{4}\{CH_2O\}$	$= \frac{1}{8}CH_4 + \frac{1}{8}CO_2$	(8)

conditions of stability of redox components in natural water systems, but more extended quantitative inferences must be made with great caution.

Exemplifications in Soil and Water Systems

In stagnant surface waters, sediment–water, and soil systems, $p\varepsilon$ and pH are especially important master variables that are coupled to one another. An increase in $p\varepsilon$ is accompanied by a decrease in pH. Solid organic matter as well as heterogeneous redox couples involving the solid phases Mn(III,IV)oxides, Fe(II,III)oxides, and FeS and FeS_2 in soils, sediments, and groundwaters provide important $p\varepsilon$ and pH buffering.

The poising (buffering) of a redox system against a $p\varepsilon$ change can be defined—similarly as in acid–base systems—as a redox-buffer or poising intensity

$$\beta_{redox} = \frac{dC_{ox}}{d\,p\varepsilon} \tag{70}$$

where C_{ox} is the concentration of an added oxidant [M].

Figure 8.15 gives representative redox intensity ranges of importance in soils, sediments, and surface and groundwaters. In soils the organic matter (range 2 in Figure 8.15) represents a reservoir of bound H^+ and e^-. When organic matter is mineralized, alkalinity and $[NO_3^-]$ and $[SO_4^{2-}]$ increase and Fe(II) and Mn(II) become mobilized. Phosphate, incipiently bound to Fe(III)(hydr)oxides, is released as a consequence of the partial reductive dissolution of the Fe(III) solid phases. At lower $p\varepsilon$ values (range 3 in Figure 8.15)

8.5 Redox Conditions in Natural Waters

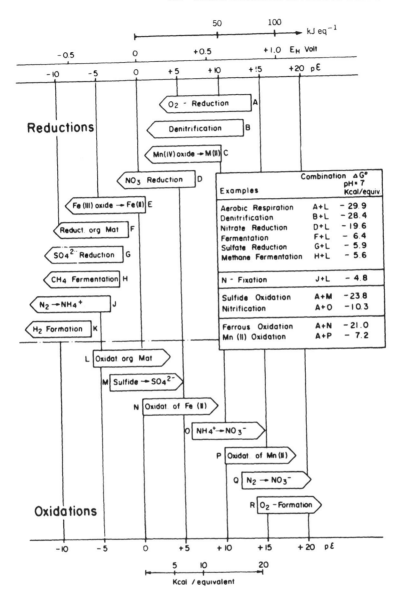

Figure 8.14. Sequence of microbially mediated redox processes. The letters refer to the reactions given in Table 8.7.

the concentrations of Fe(II) and Mn(II) further increase. SO_4^{2-} reduction is accompanied by precipitation of FeS and MnS and by the formation of pyrite.

At the oxic–anoxic boundaries a rapid turnover of iron takes place. This oxic–anoxic boundary may occur in deeper layers of the water column of fresh and marine waters, at the sediment–water interface or within the sediments.

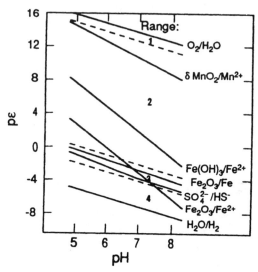

Figure 8.15. Representative ranges of redox intensity in soil and water. Range 1 is for oxygen-bearing waters. The pε range 2 is representative of many ground and soil waters where O_2 has been consumed (by degradation of organic matter), but SO_4^{2-} is not yet reduced. In this range soluble Fe(II) and Mn(II) are present; their concentration is redox-buffered because of the presence of solid Fe(III) and Mn(III,IV) oxides. The pε range 3 is characterized by SO_4^{2-}/HS or SO_4^{2-}/FeS, SO_4^{2-}/FeS$_2$ redox equilibria. Range 4 occurs in anoxic sediments and sludges. (Adapted from Drever, 1988).

Figure 8.16 shows the redox transformations schematically for the water column. The important items are as follows:

1. The principal reductant is the biodegradable biogenic material that settles in the deeper portions of the water column.
2. Electron transfer becomes more readily feasible if, as a consequence of fermentation processes—typically occurring around redox potentials of 0.22 to −0.22 V—molecules with reactive functional groups such as hydroxy and carboxyl groups are formed.
3. Within the depth-dependent redox gradient, concentration peaks of solid Fe(III) and of dissolved Fe(II) develop, the peak of Fe(III) overlying the peak of Fe(II).
4. The Fe(II), forming complexes with these hydroxy and carboxyl ligands, encounter in their upward diffusion the settling of Fe(III)(hydr)oxides and interact with these according to the catalytic mechanism, thereby dissolving rapidly the Fe(III)(hydr)oxides. The sequence of diffusional transport of Fe(II), oxidation to insoluble Fe(OH)$_3$ and subsequent settling and reduction to dissolved Fe(II), typically occurs within a relatively narrow redox cline.

8.5 Redox Conditions in Natural Waters

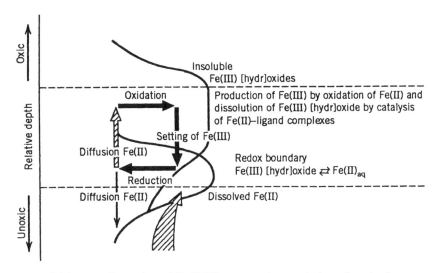

Figure 8.16. Transformation of Fe(II,III) at an oxic–anoxic boundary in the water or sediment column. (Adapted from Davison, 1985.)

Some of the processes mentioned above occur also in soils. Microorganisms and plants produce a larger number of biogenic acids. The downward vertical displacement of Al and Fe observed in the podzolization of soils can be accounted for by considering the effect of pH and of complex formers on both the solubility and the dissolution rates.

Similarly, the redox transformation Mn(III,IV)oxides/Mn^{2+} causes rapid electron cycling at suitable redox intensities. Two differences of Mn and Fe in their redox chemistries are relevant:

1. The reduction of Mn(III,IV) to dissolved Mn(II) occurs at higher redox potentials than does the Fe(III)/Fe(II) reduction.
2. The oxygenation of Mn(II) to Mn(III,IV) oxides, even if catalyzed by surfaces and/or microorganisms, is usually slower than the oxygenation of Fe(II).

The redox cycling of these elements has pronounced effects on the adsorption of trace elements onto oxide surfaces and trace element fluxes under different redox conditions. The hydrous Mn(III,IV) oxides are important mediators in the oxidation of oxidizable trace elements; for example, the oxidation of Cr(III), As(III), and Se(IV) is too slow with O_2; however, these elements subsequent to their relatively rapid adsorption on the Mn(III,IV) oxide are rapidly oxidized by Mn(III,IV).

NO_3^- Reduction Figure 8.17 shows data on groundwater below agricultural areas. The sharp decreases of O_2 and NO_3^- at the redox cline indicate that the

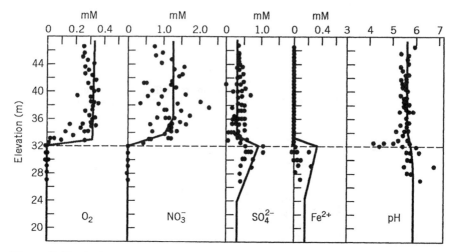

Figure 8.17. Redox components in groundwaters as a function of depth (unconfined sandy aquifer) below agricultural areas for 1988. NO_3^--contaminated groundwaters emanate from the agricultural areas and spread through the aquifer. The redox boundary is very sharp, which suggests that the redox process is fast compared to the rate of downward water transport. The investigators (Postma et al., 1991) suggest that reduction of O_2 and NO_3^- occur by pyrite. (The lines given are based on equilibrium models.) (Adapted from Postma et al., 1991.)

kinetics of the reduction processes are fast compared to the downward water transport rate. Postma et al. (1991) suggest that pyrite, present in small amounts, is the main electron donor for NO_3^- reduction (note the increase of SO_4^{2-} immediately below the oxic–anoxic boundary). Since NO_3^- cannot kinetically interact sufficiently fast with pyrite, a more involved mechanism must mediate the electron transfer. One could postulate a pyrite oxidation by Fe(III) subsequent to the oxidation of the pyrite; the Fe(II) formed is oxidized directly or indirectly (microbial mediation) by NO_3^-.

Example 8.15. Solubility of "Fe(OH)$_3$" from Redox Potential Data in Deep Groundwaters Figure 8.18 gives measured redox potentials and Fe^{2+} data (from Grenthe et al., 1992). These latter data were obtained from analysis of dissolved [Fe(II)] and corrected for complex formation with carbonate (Fe^{2+} + CO_3^{2-} = $FeCO_3$(aq); log K_1 = 5.56, I = 0). Assuming that the measured redox potential refers to the Fe(II)/Fe(III) system, calculate the solubility constant $*K_{s0}$ for the reaction

$$\text{"Fe(OH)}_3\text{"(s)} + 3\,H^+ = Fe^{3+} + 3\,H_2O \qquad *K_{s0} \qquad (i)$$

(The quotation marks signify that direct evidence is not available in that the solid is postulated to form in deep groundwaters.)

8.5 Redox Conditions in Natural Waters

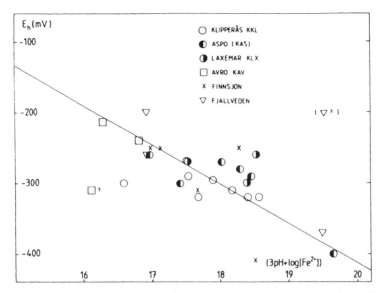

Figure 8.18. Measured redox potentials in a deep groundwater. Experimental values of the measured redox potentials (recalculated to the standard hydrogen electrode scale) versus $(3\text{pH} + \log[\text{Fe}^{2+}])$. The concentration of $[\text{Fe}^{2+}]$ has been obtained from the analytical determinations by correction for the complex formation with carbonate. The notation refers to the different test sites. The *full-drawn line* has been calculated using the selected value of the standard potential E_0^*. The straight line has the theoretical Nernstian slope of $+0.056$ V, at the temperature of measurements. (Adapted from Grenthe et al., 1992.)

The Nernst equation, written in terms of E, is

$$E = 0.771 + \frac{RT}{F} \ln \frac{[\text{Fe}^{3+}]}{[\text{Fe}^{2+}]} \tag{ii}$$

Considering equation i, we obtain

$$E = E_0^* - 2.303\,(RT/F)\,(3\,\text{pH} + \log[\text{Fe}^{2+}]) \tag{iii}$$

where

$$E_0^* = 0.771 + 2.303\,(RT/F)\,\log {}^*K_{s0} \tag{iv}$$

The experimental data given in Figure 8.18 indicate a linear relationship (in accord with equation iii) with a slope of

$$2.3\,(RT/F) = 0.056\text{ V} \tag{va}$$

482 Oxidation and Reduction; Equilibria and Microbial Mediation

and an intercept of

$$E_0^* = 0.707 \text{ V} \quad \text{(vb)}$$

From equation iv we calculate

$$\log {}^*K_{s0} = -1.1$$

or the conventional solubility product

$$\log [Fe^{3+}][OH^-]^3 = \log K_{s0} = -43.1$$

Example 8.16. Groundwater Contamination Variation of Redox Species in the Flow Path of Contaminant Plume Organic wastes from a dump site infiltrate into a groundwater. Sketch the variation in the concentration of the redox species SO_4^{2-}, H_2S, CH_4, Fe, Mn, NO_3^-, O_2, organic carbon in the flow path of the contaminant plume. If we assume a redox sequence as given in Table 8.8 and in Figure 8.14, the following type (Figure 8.19) of concentration profile can be expected. The curve given is of course somewhat speculative, but the sequence given has been observed by many investigators (Bouwer, 1992; von Gunten, and Zobrist, 1993).

Sulfur Transformations

Sulfur undergoes cyclic transformations, which can be viewed at different levels of organization and complexity. Under reducing conditions ($p\varepsilon < -2$) various microorganisms (e.g., Desulfovibrio) mediate the reduction of SO_4^{2-} by organic

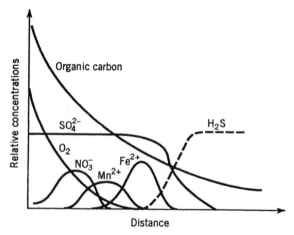

Figure 8.19. Estimation of the variation in concentration of redox species during the flow path of an organic contaminant plume.

matter; as a consequence, some metal sulfides, especially amorphous ferrous sulfide, may be formed, which often gradually crystallize into FeS (mackinawite), which may under suitable condition then form pyrite [e.g., FeS(s) + S(s) → FeS$_2$(s)].

At the oxic–anoxic boundary in the deeper portions of a stagnant water body or a sediment, a redox cycling of sulfur compounds of different oxidation states may occur (Figure 8.20) because reduced S may become reoxidized.

Reactions are schematically given in Figure 8.20. As illustrated schematically, some oxidation reactions are a consequence of photosynthetic sulfur bacteria

$$CO_2 + 2\ H_2S \xrightarrow{h\nu} \{CH_2O\} + 2\ S(s) + H_2O \qquad (71)$$

This reaction may be compared with the simplified equation for algal photosynthesis

$$CO_2 + 2\ H_2O \xrightarrow{h\nu} \{CH_2O\} + O_2 + H_2O \qquad (72)$$

Figure 8.21 illustrates some aspects of S transformations in a eutrophic lake. With progressive lake depth, subsequent to the reduction of O_2, SO_4^{2-} is reduced to sulfide [S(−II)]. This reduction is somewhat simultaneous with CH_4 formation. S(s) is formed by oxidation of sulfide.

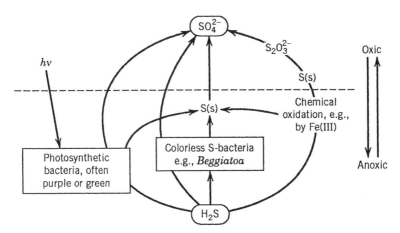

Figure 8.20. Simplified scheme for the oxidation of H_2S by O_2 mediated by a variety of bacteria. The gradient zone between O_2 and H_2S is the environment of many colorless sulfur bacteria, among which the type *Beggiatoa* often reach high population densities and form white mats on the mud or sediment–water interface. If light penetrates at the zonation between O_2 and H_2S, phototrophic, often colorful, sulfur bacteria grow. Reduced sulfur can also be oxidized abiotically, for example, by Fe(III)(hydr)oxides or even by O_2 in the presence of metal-ion catalysts.

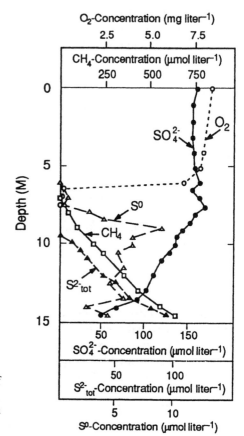

Figure 8.21. Concentration profiles of redox species in the Rotsee, Lucerne, Switzerland, from October 1982. (Data are partially from Kohler et al., 1984.)

Example 8.17. Phosphate Binding in Soils[†] The availability of phosphate for plants depends in various ways on the pε; above all, the interactions of phosphates with Fe(II) and Fe(III) are of importance. The surfaces of Fe(III)(hydr)oxides can adsorb phosphate; this adsorption is pH dependent. Here we consider the pε-dependent solubility of phosphate as strengite, $FePO_4 \cdot 2\ H_2O(s)$, and vivianite, $Fe_3(PO_4)_2 \cdot 8\ H_2O(s)$. The following equilibria are given:

$$3\ \underset{\text{strengite}}{FePO_4 \cdot 2\ H_2O(s)} + e^- = \underset{\text{magnetite}}{Fe_3O_4(s)} + 3\ H_2PO_4^- + 2\ H^+ + 2\ H_2O;$$

$$\log K = -17.1 \tag{i}$$

$$Fe_3O_4(s) + 2\ H_2PO_4^- + 4\ H^+ + 2\ e^- = \underset{\text{vivianite}}{Fe_3(PO_4)_2 \cdot 8\ H_2O(s)};$$

$$\log K = 32.6 \tag{ii}$$

[†]This example has been modified from one given by Sposito (1984).

8.5 Redox Conditions in Natural Waters

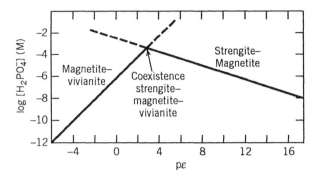

Figure 8.22. Solubility of phosphate (Example 8.17) as regulated by strengite–magnetite and magnetite–vivianite equilibria, respectively. The solubility of phosphate is largest at slightly positive pε values.

We assume slightly acid soil (pH = 5.0) and the presence of soil water and that the constants given are valid for $I = 10^{-3}$ M and a temperature of 10°C.

(a) Calculate $[H_2PO_4^-]$ as a function of pε. At what pε would we have maximum soluble phosphate?

(b) At what pε can the three solid phases coexist?

We can use most conveniently the mass laws of equations i and ii to plot $\log [H_2PO_4^-]$ versus pε:

$$\log [H_2PO_4^-] = -2.37 - \tfrac{1}{3} p\varepsilon \qquad \text{(iii)}$$

and

$$\log [H_2PO_4^-] = -6.3 + p\varepsilon \qquad \text{(iv)}$$

As Figure 8.22 illustrates, the solubility of phosphate at pH = 5 is largest at slightly positive pε values, where $Fe_3O_4(s)$, $FePO_4 \cdot 2 H_2O(s)$, and $Fe_2(PO_4)_3 \cdot 8 H_2O(s)$ can coexist. At higher and lower pε values, respectively, phosphate may be precipitated as strengite and vivianite.

Example 8.18. *"Titrating" a Lake Hypolimnion with Biota* A lake, initially aerobic, is characterized by the following conditions:

$[SO_4^{2-}] = 2 \times 10^{-4}$ M
$C_T = 2 \times 10^{-3}$ M
$[Fe(OH)_3(\text{amorphous})]_{\text{total}} = 10^{-5}$ M
$[NO_3^-] = 1 \times 10^{-4}$ M
$[MnO_2(s)]_{\text{total}} = 10^{-5.5}$ M
pH = 8.0

486 Oxidation and Reduction; Equilibria and Microbial Mediation

Tableau 8.7. Matrix for the "Titration" of an Incipiently Oxic Lake Water with Electrons

Components	NO_3^-	SO_4^{2-}	$CO_2(g)$	$Fe(OH)_3$	MnO_2	H^+	e^-	log K (25°C)
$O_2(g)$	0	0	0	0	0	−4	−4	−83.1
NO_3^-	1	0	0	0	0	0	0	0
NH_4^+	1	0	0	0	0	10	8	119.2
SO_4^{2-}	0	1	0	0	0	0	0	0
HS^-	0	1	0	0	0	9	8	34
Mn^{2+}	0	0	0	0	1	4	2	43.6
Fe^{2+}	0	0	0	1	0	3	1	16
$CH_4(g)$	0	0	1	0	0	8	8	23
$H_2(g)$	0	0	0	0	0	2	2	0
H^+	0	0	0	0	0	1	0	0
e^-	0	0	0	0	0	0	1	0

$$p_{CO_2} = 10^{-3.5} \text{ atm} \qquad \text{pH} = 8$$
$$\text{TOTNO}_3^- = 10^{-4} \text{ M} = [NO_3^-] + [NH_4^+]$$
$$\text{TOTSO}_4^{2-} = 2 \times 10^{-4} \text{ M} = [SO_4^{2-}] + [HS^-]$$
$$[Fe^{2+}] \leq 10^{-5} \text{M} \qquad [Mn^{2+}] \leq 5 \times 10^{-6} \text{ M}$$

The hypolimnion[†] of this lake is "titrated" with biota (sinking phytoplankton). Estimate the concentration of the various species that develop during this "titration" (e.g., progressive time of stagnation during summer stratification versus pε). The "titrant", that is, the sinking biota, can be looked at as an "electron complex"; that is, we can titrate with electrons. Such a calculation is of course only partially correct, but it shows synoptically how anoxia progresses in such a situation. We can use, as a first approximation, the constants valid for 25°C:

$$Fe(OH)_3(s) + 3 H^+ + e^- = Fe^{2+} + 3 H_2O \qquad \log K = 16.0 \qquad \text{(i)}$$

$$MnO_2(s) + 4 H^+ + 2 e^- = Mn^{2+} + 2 H_2O \qquad \log K = 43.6 \qquad \text{(ii)}$$

$$O_2(g) + 4 H^+ + 4 e^- = 2 H_2O \qquad \log K = 83.1 \qquad \text{(iii)}$$

$$NO_3^- + 10 H^+ + 8 e^- = NH_4^+ + 3 H_2O \qquad \log K = 119.2 \qquad \text{(iv)}$$

$$SO_4^{2-} + 9 H^+ + 8 e^- = HS^- + 4 H_2O \qquad \log K = 34 \qquad \text{(v)}$$

$$2 H^+ + 2 e^- = H_2(g) \qquad \log K = 0 \qquad \text{(vi)}$$

$$CO_2(g) + 8 H^+ + e^- = CH_4(g) + 2 H_2O \qquad \log K = 23 \qquad \text{(vii)}$$

[†]Hypolimnion is that portion of the lake that is below the thermocline and is assumed to be below the phototrophic zone.

8.5 Redox Conditions in Natural Waters

The information necessary to make the calculation is contained in Tableau 8.7, and the result of the calculation is given in Figure 8.23; it shows the progressive development of more reductive or less oxic conditions. This succession is also observed at a given time with increasing depth in the lake or in the sediments.

The same principle is illustrated in Figures 8.24 and 8.25, in which the "titration curves" for Lake Bret and a hypothetical groundwater system, respectively, are given.

If the lake hypolimnion is considered a fully closed box, except for the input of organic matter, the length of each plateau region represents the maximum oxidizing capacity of the corresponding compound, that is, its molar concentration multiplied by the number of electrons exchanged during its reduction. The total oxidizing capacity of the lake bottom is the sum of all the individual ones; it is 2.0×10^{-3} M in this particular case. Although such estimations are simplified, since they do not take into account the diffusion of O_2 and SO_4^{2-}

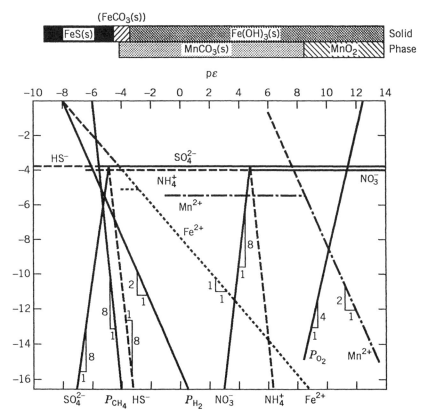

Figure 8.23. "Titrating" a lake with settling biota (see Example 8.18). In the lower part of the lake (summer stagnation), progressively more reducing conditions (decreasing pε) develop. The figure illustrates schematically how the relative concentrations of redox constituents change with decreasing pε.

488   Oxidation and Reduction; Equilibria and Microbial Mediation

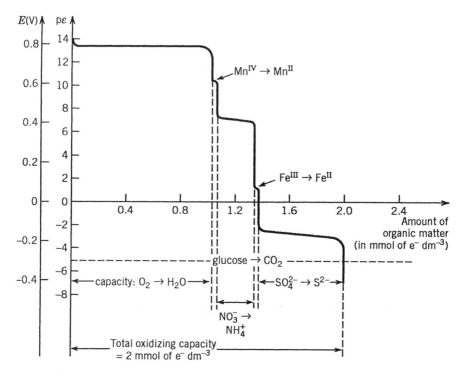

Figure 8.24. Simplified titration curve of the oxidants of Lake Bret (VD, Switzerland). The plateau's lengths correspond to the oxidizing capacity of each oxidant. Computation does not take into account the input of O_2 by diffusion and the oxidizing capacity of $Fe^{III}OOH$ and MnO_2 in sediments. (From Buffle and Stumm, 1994.)

from surface waters and the oxidizing capacity of MnO_2 and $FeOOH$ accumulated in the sediments, Figure 8.24 points out the principle of the sequential reduction process occurring in bottom waters, as well as the relationship between the intensity parameter (vertical scale) and the capacity parameter (horizontal scale).

A further limitation of the above representation is that not only is the lake bottom viewed as a closed box, but it is considered as homogeneously mixed. This, however, is not true. Organic matter degrades only slowly during sedimentation, and oxidant depletion therefore increases with depth. In addition, the horizontal scale of Figure 8.24 is also a time scale corresponding to the time necessary for organic matter production. The net result is a double evolution in redox conditions, both spatially and with time.

Redox titration of a model groundwater is illustrated in Figure 8.25, where $p\varepsilon$ and pH at hypothetical equilibrium are shown versus the amount of organic carbon reacted, $[CH_2O]$. Figure 8.25a shows the $p\varepsilon$ changes for successive titration steps for reduction of O_2, NO_3^-, $MnO_2(s)$, $Fe(OH)_3(s)$, and SO_4^{2-}, and Figure 8.25b shows the corresponding changes in pH for these steps. In the buffered system of the model (alkalinity of 0.83 meq liter^{-1} and pH 7.0 initially) the reduction of O_2 by CH_2O yields additional dissolved CO_2, lowering

8.6 Effect of Complex Formers on the Redox Potential

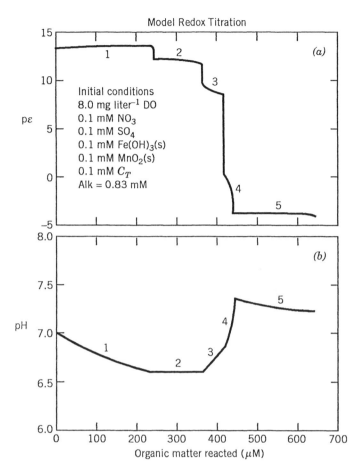

Figure 8.25. Redox titration curve of a model groundwater system: (a) pε response and (b) pH response. Numbered segments correspond to sequential reduction of (1) $O_2(aq)$, (2) $NO_3^-(aq)$, (3) $MnO_2(s)$, (4) $Fe(OH)_3(s)$, and (5) $SO_4^{2-}(aq)$. (From Scott and Morgan, 1990.)

the pH noticeably (and actually causing a slight elevation of pε; compare the relative sensitivities of $p\varepsilon_{O_2}$ to pH and $[O_2(aq)]$). Examination of the changes in TOTH$^+$ and TOTCO$_2$ for each redox couple in relation to ΔCH_2O allows us to rationalize the complete sequence of pH and pε changes. For example, in the reduction of sulfate, the ratio of protons consumed to CO_2 produced is close to unity, and the change in pH consequently is small.

8.6 EFFECT OF COMPLEX FORMERS ON THE REDOX POTENTIAL

We compare the consequences of complexation on pε from a thermodynamic point of view. We use for exemplification Fe(II) and Fe(III) because (1) more

490   Oxidation and Reduction; Equilibria and Microbial Mediation

data are available with this redox pair than with others, and (2) the transformations of iron are especially important in the redox cycling of electrons in natural environments. As Figure 8.26 shows, the Fe(III)/Fe(II) redox couple can be adjusted with appropriate ligands to any redox potential within the stability of water. The principles exemplified here are of course also applicable to other redox systems.

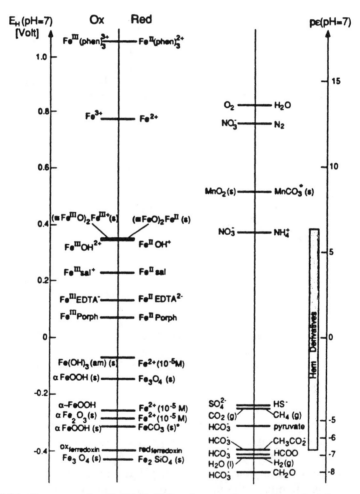

Figure 8.26. Representative Fe(II)/Fe(III) redox couples at pH = 7. (phen = phenanthroline; sal = salicylate; porph = porphyrin; * = valid for [HCO_3^-] = 10^{-3} M.) Complex formation with Fe(II) and Fe(III) both on solid and solute phases has a dramatic effect on the redox potentials; thus electron transfer by the Fe(II),Fe(III) system can occur at pH = 7 over the entire range of the stability of water; E_H (−0.5 to 1.1 V). (= $Fe^{III}O)_2$ Fe^{III} refers to Fe^{3+} adsorbed inner-spherically to a surface of a hydrous ferric oxide. The range of redox potentials for heme derivatives given on the right illustrates the possibilities involved in bioinorganic systems.

8.7 Measuring the Redox Potential in Natural Waters

As illustrated in Figure 8.26, the redox potential at pH = 7, $E_H^°$ (pH = 7), decreases in the presence of most complex formers, especially chelates with oxygen donor atoms, such as a citrate, EDTA, and salicylate, because these ligands form stronger complexes with Fe(III) than with Fe(II). Phenanthroline, which stabilizes Fe(II) more than Fe(III), is an exception, explainable in terms of the electronic configuration of the aromatic N–Fe(II) bond (Luther et al., 1992). But Fe(II) complexes are usually stronger reductants than Fe^{2+}. This stabilization of the Fe(III) oxidation state is also observed with hydroxo complexes and by the binding to O^{-II} in solid phases (see Section 11.4). Thus Fe(II) minerals are, thermodynamically speaking, strong reductants. For example, the couple Fe_2SiO_4 (fayalite)–Fe_3O_4 (magnetite) has an E_H similar to that for the reduction of $H_2O(l)$ to $H_2(g)$ (Baur, 1978) (Figure 8.26). A surface complex of Fe^{2+} adsorbed inner-spherically onto a hydrous oxide surface is more reducing than Fe^{2+}(aq).

Redox reactions are of importance in the dissolution of Fe-bearing minerals. Reductive dissolution of Fe(III)(hydr)oxides can be accomplished with many reductants; especially organic and inorganic reductants, such as ascorbate, phenols, dithionite, and HS^-. Fe(II) in the presence of complex formers can readily dissolve Fe(III)(hydr)oxides. The Fe(II) bound in magnetite and silicate and adsorbed to oxides can reduce O_2 (White, 1990; White and Yee, 1985).

Example 8.19. Effect of Ligand on Reduction of Iodine by Fe(II) Illustrate that Fe^{2+} cannot reduce iodine to iodide but that an Fe(II) complex with a ligand such as citrate or EDTA, respectively, can. Iodine occurs in aqueous solution as I_3^- [I_2(aq) + I^- = I_3^-].

Comparing simple equilibria, we have

$$2\ Fe^{2+} + I_3^- = 2\ Fe^{3+} + 3\ I^- \quad \log K = -16.9$$

$$2\ Fe^{II}L^{2-} + I_3^- = 2\ Fe^{III}L^- + 3\ I^- \quad \log K = 2.1$$

$$L = citrate^{4-}$$

$$2\ Fe^{II}Y^{2-} + I_3^- = 2\ Fe^{III}Y^- + 3\ I^- \quad \log K = 8.6$$

$$Y = EDTA^{4-}$$

Obviously the citrato and the EDTA complexes of Fe(II) are sufficiently strong reductants for iodine.

8.7 MEASURING THE REDOX POTENTIAL IN NATURAL WATERS

It has already been pointed out that it is necessary to distinguish between the concept of a potential and the measurement of a potential. E_H measurements are of great value in systems for which the variables are known or under control. In this section we will discuss the measurement and indirect evaluation of redox potentials. The problems encountered in measuring redox potentials have been reviewed extensively (Grenthe et al., 1992; Lindberg and Runnels, 1984).

Kinetic Considerations Some of the essential principles involved in the measurement of an electrode potential can be described qualitatively by consideration of the behavior of a single electrode (platinum) immersed in an acidified Fe^{2+}-Fe^{3+} solution. To cause the passage of a finite current at this electrode, it is necessary to shift the potential from its equilibrium value. We thus obtain a curve depicting the resulting current as a function of the electrode potential (polarization curve). At the equilibrium potential, that is, at the point of zero applied current, the half-reaction

$$Fe^{3+} + e^- \rightleftharpoons Fe^{2+}$$

is at equilibrium, but the two opposing processes, the reduction of Fe^{3+} and the oxidation of Fe^{2+}, proceed at an equal and finite rate:

$$v_1 \text{ (rate of reduction)} = v_2 \text{ (rate of oxidation)}$$

which is proportional to the rate of passage of electrons in both directions. Although the net rate of passage of electrons is equal in both directions and thus the current is zero, the passage of current in a single direction is not zero and is called the *exchange current* i_0 (see Figure 8.27).

At equilibrium, where no net current flows ($i_a = i_c$), concentrations at the electrode surface equal bulk concentrations, and pε values or potentials measured correspond to equilibrium conditions, yielding

$$\frac{[Fe^{3+}]}{[Fe^{2+}]} = 10^{(p\varepsilon - p\varepsilon^\circ)} = \exp\frac{F}{RT}(E_H - E_H^\circ) \tag{74}$$

or, in a more general sense,

$$\frac{[ox]}{[red]} = 10^{n(p\varepsilon - p\varepsilon^\circ)} = \exp\frac{nF}{RT}(E_H - E_H^\circ) \tag{75}$$

Equations 74 and 75 are identical to the Nernst equation (equation 31).

The net current can be visualized as the summation of two opposing currents (cathodic and anodic). The rate of Fe^{3+} reduction (conventionally expressed as cathodic current) generally increases exponentially with more negative electrode potential values and is, furthermore, a function of the concentration of $[Fe^{3+}]$ and of the effective electrode area. Similar considerations apply to the rate of Fe^{2+} oxidation (anodic current), which is proportional to $[Fe^{2+}]$, electrode area, and the exponential of the potential. It is obvious from the schematic representation that an infinitesimal shift of the electrode potential from its equilibrium value will make the half-reaction proceed in either of the two

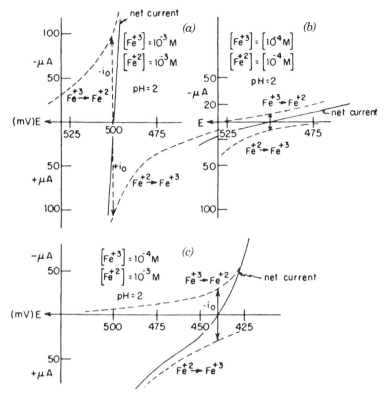

Figure 8.27. Polarization curves for various concentrations of Fe^{2+} and Fe^{3+}. Solid lines are polarization curves (electrode area = 1 cm^2); i_0, exchange current. Dashed lines are hypothetical cathodic ($-i$) and anodic ($+i$) currents. Curves are schematic but based on experimental data at relevant points.

opposing directions, provided the concentration of these ions is sufficiently large. Measurement of the equilibrium electrode potential in such a case is feasible: the amount by which the potential must be shifted to obtain an indication with the measuring instrument, hence the sharpness and reproducibility of the measurement, is determined by the slope of the net current in the vicinity of the balance point. This slope is proportional to the exchange current i_0; i_0 depends upon the concentration of reactants and the electrode area.

Under favorable circumstances with modern instrumentation, for which the current drain is quite low, reliable potential measurements can be made with systems giving i_0 greater than about 10^{-7} A. If tenfold smaller concentrations of both ions are present, i_0 and the slopes are only $\frac{1}{10}$ as great; but with $i_0 = 10^{-5}$ A, reliable measurements can still be made, as shown in Figure 8.27b. If the concentration of only one ion is decreased, the drop in i_0 is not as great, and E_H, the potential corresponding to equal cathodic and anodic currents, is shifted, as shown in Figure 8.27c. If both Fe^{3+} and Fe^{2+} are at 10^{-6} M concentrations (~0.05 mg liter^{-1}), i_0 is 10^{-7} A and measurements are no longer precise. Actually, because of other effects caused by trace impurities, it becomes difficult to obtain measurements in accord with simple Nernst theory when either Fe ion concentration is less than about 10^{-5} M.

Electrodes with large areas are advantageous, but they also tend to magnify the effects of trace impurities or other reactions on the electrode surface itself, such as adsorption of surface-active materials leading to a reduction in the electron-exchange rate or in the effective area A.

"Slow" Electrodes We might contrast such behavior with the conditions we would encounter in attempting the measurement of the electrode potential in water containing O_2. A schematic representation of the polarization curve for this case is given in Figure 8.28. The equilibrium electrode potential again should be located at the point where the net applied current (i.e., the sum of cathodic and anodic currents) is zero. The exact location becomes difficult to determine. Over a considerable span of electrode potentials the net current is virtually zero; similarly, the electron-exchange rate, or the exchange current reflecting the opposing rates of the-half reaction

$$H_2O \rightleftharpoons \tfrac{1}{2}O_2 + 2H^+ + 2e^-$$

is virtually zero. Operationally, a remarkable potential shift must be made to produce a finite net current, and the current drawn in the potentiometric measurement is very large compared with the exchange current. Even with modern instrumentation in which the current drain can be made extremely low, the

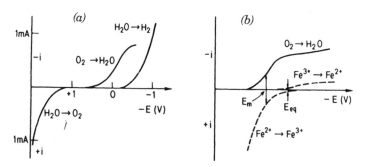

Figure 8.28. Electrode polarization curves for oxygen-containing solutions: (a) in otherwise pure water and (b) in the presence (nonequilibrium) of some Fe^{2+}. Curves are schematic but in accord with available data at significant points. Because the net current (a) is virtually zero over a considerable span of the electrode potentials, the exact location of the redox potential becomes difficult to determine or is determined by insidious redox impurities. A mixed potential (b) may be observed at the point where the anodic and cathodic currents are balanced; but because the various redox partners are not in equilibrium with each other, it is not amenable to quantitative interpretation.

8.7 Measuring the Redox Potential in Natural Waters

experimental location of the equilibrium potential is ambiguous. It has been estimated that for O_2 (1 atm) the specific exchange current i_0/A is 10^{-9} A cm^{-2}, far less than 10^{-7}. The current utilized by impurities may exceed the exchange current. It has been shown that, under conditions of extremely high purity, an oxygen equilibrium potential of 1.23 V can be attained on a Pt electrode (Stumm and Morgan, 1985).

Mixed Potentials Another difficulty arises in E_H measurements. The balancing anodic and cathodic currents at the apparent "equilibrium" potential need not correspond to the same redox process and may be a composite of two or more processes. An example of this is shown in Figure 8.28b for a Fe^{3+}-Fe^{2+} system in the presence of a trace of dissolved oxygen. The measured zero-current potential is the value where the rate of O_2 reduction at the electrode surface is equal to the rate of Fe^{2+} oxidation rather than the value of E_{eq}, since at the latter point simultaneous O_2 reduction produces excess cathodic current. In addition, because the net reaction at E_m converts Fe^{2+} to Fe^{3+}, the measured potential exhibits a slow drift. Such "mixed" potentials are of little worth in determining equilibrium E_H values. Many important redox couples in natural waters are not electroactive. No reversible electrode potentials are established for NO_3^--NO_2^--NH_4^+-H_2S or CH_4-CO_2 systems. Unfortunately, many measurements of E_H (or pε) in natural waters represent mixed potentials not amenable to quantitative interpretation.

In the absence of oxygen, with extreme care, a reasonably reliable E_H can be measured, for example, in groundwater (see Example 8.15 and Figure 8.18). But even in the case mentioned (Grenthe et al., 1992) it took more than 20 days to establish a redox potential of the "Fe(OH)$_3$(s)-Fe(II) system" of that groundwater. For a recent review on redox potential measurements in the laboratory and in the field see Grenthe et al. (1992) and Lindberg and Runnels (1984).

Indirect Evaluation of Redox Potentials Of the standard redox potentials quoted in handbooks or summaries, relatively few have been determined by direct potential measurements; the others have been calculated from a combination of free energy data or from equilibrium constants. It is possible to evaluate the redox level in natural water systems by determining the relative concentrations of the members of one of the redox couples in the system and applying the electrochemical relations in reverse. As Figures 8.3 and 8.23–8.25 suggest, quantitative analytical information on any one of the species O_2, Fe^{2+}, Mn^{2+}, HS^--SO_4^{2-}, and CH_4-CO_2 gives a conceptually defined pε (or E_H), provided the system is in equilibrium. In practice, however, there are limitations even to this basically sound procedure. The system must be in equilibrium or must be in a sufficiently constant metastable state to make the concept of a partial equilibrium meaningful. For example, although most aqueous media are not in equilibrium with regard to processes involving N_2, the inertness of these reactions may allow us to ignore them while treating the

496 Oxidation and Reduction; Equilibria and Microbial Mediation

equilibrium achieved by other species. A singular pε value can be ascribed to a system if equal values of E_H or pε are obtained for each of the redox couples in a multicomponent system; otherwise the couples are not in equilibrium and the concept of a singular redox potential becomes meaningless.

Within these limitations the analytical determination of the oxidized and reduced forms of a redox couple can provide quite precise values for E_H. Depending on the number of electrons transferred in the process, determinations accurate within a factor of 2 will give E_H values within 5–20 mV or pε values within 0.1–0.3 units.

Example 8.20 Computation of pε and E_H from Analytical Information. Estimate pε and E_H for the aquatic habitats characterized by the following analytical information:

(a) Sediment containing FeOOH(s) and $FeCO_3$(s) in contact with 10^{-3} M HCO_3^- at pH = 7.0.
(b) Water from the deeper layers of a lake having a dissolved O_2 concentration of 0.03 mg O_2 liter^{-1} and a pH of 7.0.
(c) An anaerobic digester in which 65% CH_4 and 35% CO_2 are in contact with water of pH = 7.0.
(d) A water sample of pH = 6, containing $\{SO_4^{2-}\} = 10^{-3}$ M and smelling of H_2S.
(e) A sediment–water interface containing $FeCO_3$(s) with a crust of black FeS(s) at pH = 8.

(a) This corresponds to the conditions of equation 9 in Table 8.6a. pε = -0.8 and $E_H = -0.047$ V. This system, in equilibrium, has an infinite poising (redox-buffering) intensity. The redox level can be defined rather precisely.

(b) This corresponds to equation 1 in Table 8.6a and pε = $13.75 + \frac{1}{4} \log p_{O_2}$; 10^{-6} mol O_2 liter^{-1} is equivalent to a partial pressure, $p_{O_2} \approx 6 \times 10^{-4}$, and $\frac{1}{4} \log p_{O_2} = -0.8$. pε for this system is 12.95 and $E_H = +0.77$ V. Note that E_H varies little with p_{O_2} or dissolved O_2 concentration. A reduction in O_2 from 10 (approximately air saturated) to 0.1 mg liter^{-1} will lower E_H by 30 mV.

(c) According to equation 15 in Table 8.6a

$$p\varepsilon = -4.13 + \frac{1}{8} \log \frac{p_{CO_2}}{p_{CH_4}} = -4.16$$

$$E_H = -0.25 \text{ V}$$

(d) The equilibrium is characterized by

$$\tfrac{1}{8}SO_4^{2-} + \tfrac{5}{4}H^+ + e^- = \tfrac{1}{8}H_2S(g) + \tfrac{1}{2}H_2O$$

8.7 Measuring the Redox Potential in Natural Waters

$p\varepsilon°$ can be calculated from the $p\varepsilon°(W)$ value given in Table 8.6a. $p\varepsilon° = p\varepsilon°(W) - (n_H/2)\log K_W$; $p\varepsilon° = -3.5 + 8.75 = 5.25$. Hence

$$p\varepsilon = 5.25 - \tfrac{5}{4}pH + \tfrac{1}{8}\log [SO_4^{2-}] - \tfrac{1}{8}\log pH_2S$$

$$= 2.62 - \tfrac{1}{8}\log p_{H_2S}$$

It is reasonable to set p_{H_2S} between 10^{-2} and 10^{-8} atm. This puts $p\varepsilon$ between -1.6 and -2.4, corresponding to E_H of between -0.09 and -0.14 V.

(e) We do not have all the information necessary, but we can make a guess. A reaction defining the most likely redox poising equilibrium could be formulated as

$$SO_4^{2-} + FeCO_3(s) + 9H^+ + 8e^- = FeS(s) + HCO_3^- + 4H_2O$$

With the use of the free energy values given in Appendix 3 at the end of the book the equilibrium constant of the reaction given above is calculated (25°C) to be $\log K = 38.0$. Correspondingly, $p\varepsilon° = 4.75$, and $p\varepsilon$ can be calculated from

$$p\varepsilon = 4.75 - \tfrac{9}{8}pH + \tfrac{1}{8}(pHCO_3^- - pSO_4^{2-})$$

The term in parentheses has to be estimated; it is very unlikely that it is outside the range -2 to $+2$. This fixes $p\varepsilon$ for this equilibrium system at $pH = 8$ between -4.0 and -4.5. Correspondingly, $-0.27 < E_H < -0.24$ V.

In summary, because many redox reactions in natural waters do not couple with one another readily, different apparent redox levels exist in the same locale; thus an electrode or any other indicator system cannot measure a unique E_H or $p\varepsilon$. If the electrode (or the indicators) reached equilibrium with one of the redox couples, it would indicate the redox intensity of that couple only. A few conditions are necessary to obtain meaningful operational E_H values:

1. The measuring electrode must be inert. As shown by Whitfield (1974) and others, the Pt electrode may form in aerobic milieu PtO and PtO_2 and in sulfide-bearing waters PtS. An oxide-coated electrode is—in the absence of high concentrations of other redox couples—primarily pH responsive.
2. The exchange current should be large in comparison to the drain by the measuring electrode and to "impurity" currents. The measured potential is given by the system with the highest exchange rate. Considering the fact that, even in oxic waters, reductants (CO, H_2, CH_4) other than H_2O (or H_2O_2) are present at a level of 10^{-6} M, the impurity current, that is, the current utilized by the impurities, may exceed by a factor of 100 or

1000 that of the exchange current of the Pt–O_2 system (Bockris and Reddy, 1970).
3. In a system where various redox partners are not in equilibrium with each other, the balancing anodic and cathodic currents at an apparent equilibrium potential need not correspond to the same redox process and may be a composite of two or more processes; thus one may be observing a *mixed potential* not amenable to quantitative interpretation.

8.8 THE POTENTIOMETRIC DETERMINATION OF INDIVIDUAL SOLUTES

Methods that involve the introduction of electrodes into natural media without contamination are most appealing. The glass electrode has proved to be sufficiently specific and sensitive. Other electrodes specific for more than a dozen ions have joined the pH-type glass electrode in commercial production. Many of these electrodes are sufficiently specific and sensitive to permit measurement or monitoring of individual solution components.

An electrochemical cell can be used conveniently to study the properties of a solution quantitatively. For example, it is possible to determine $\{Ag^+\}$ with an Ag electrode, $\{H^+\}$ with a Pt/H_2(g) electrode or with a glass electrode, $\{Cl^-\}$ with an AgCl/Ag electrode, and $\{SO_4^{2-}\}$ with a PbSO$_4$/Pb electrode. Table 8.9 gives a survey of most of the ion-sensitive electrodes employed.

In the case of metal electrodes the potential-determining mechanism is a fast electron exchange, given by the indicator reaction

$$Me^{n+} + ne^- = Me(s) \tag{76}$$

for which the equilibrium potential is

$$E_H = E^\circ_{Me^{n+}, Me(s)} + \frac{RT(\ln 10)}{nF} \log \{Me^{n+}\} \tag{77}$$

If a potential of a given indicator electrode is to be controlled by a specific redox reaction, it is clear that the essential solid phases must be present in adequate amounts and that the electrode reaction must be characterized by a large enough exchange current. The magnitude of the latter determines the concentration at which the indicator solute must be contained in the solution. Obviously, the Nernst equation must not be expected to hold for indefinitively decreasing activity of the potential-determining species. If the exchange current is too small, the electrode response becomes too slow, and reactions with impurities or spurious oxides or oxygen at the electrode surface influence the potential and render it unstable and indefinite. Two of the highest exchange current densities are for H^+ discharge at Pt and for reduction of Hg^{2+} at a Hg surface. It may be noted, however, that metal-ion electrodes often may show

Table 8.9. Ion Selective Electrodes and Optodes

Ion	Electrode Material	Electrode Reaction
I. Metal Electrodes		
H^+	Platinized Pt, H_2	$H^+ + e^- = \frac{1}{2}H_2(g)$
Ag^+, Cu^{2+}, Hg^{2+}	Ag, Cu, Hg	$Me^{n+} + ne^- = Me(s)$
Zn^{2+}, Cd^{2+}	Zn(Hg), Cd(Hg)	$Me^{2+} + 2e^- + Hg(l) = Me(Hg)$
II. Electrodes of the Second Kind		
Cl^-	AgCl/Ag	$AgCl(s) + e^- = Ag(s) + Cl^-$
S^{2-}	Ag_2S/Ag	$Ag_2S(s) + 2e^- = 2Ag(s) + S^{2-}$
SO_4^{2-}	$PbSO_4$/Pb(Hg)	$PbSO_4 + Hg + 2e^- = Pb(Hg) + SO_4^{2-}$
H^+	Sb_2O_3/Sb	$Sb_2O_3 + 6H^+ + 6e^- = 2Sb(s) + 3H_2O$
III. Complex Electrodes		
H^+	Pt, quinhydrone	$C_6H_4O_2 + 2H^+ + 2e^- = C_6H_4(OH)_2$
$EDTA(Y^{4-})$	HgY^{2-}/Hg	$HgY^{2-} + 2e^- = Hg(l) + Y^{6-}$
IV. Glass Electrodes		
H^+	Glass	$Na^+(aq) + H^+_{Membrane} = Na^+_{Membrane} + H^+(aq)$
Na^+, K^+, Ag^+, NH_4^+	Cation-sensitive glass	Ion exchange
V. Solid-State or Precipitate Electrodes		
F^-, Cl^-, Br^-, I^-, S^{2-}, Cu^{2+}, Cd^{2+}, Na^+, Ca^{2+}, Pb^{2+}	Precipitate, impregnated or solid-state electrodes	Ion exchange
VI. Liquid–Polymer–Liquid Membrane Electrodes		
Ca^{2+}, Mg^{2+}, Pb^{2+}, Cu^{2+}, NO_3^-, Cl^-, ClO_4^-	Liquid ion exchange	Ion exchange
VII. Ion Optodes (Ion-Selective Optical Sensors) polymer liquid membranes		

an unusually sensitive response because the electroactive species is actually a complex present at a larger concentration than the free metal ions. For example, if a slower electrode responds properly to $\{Ag^+\}$ in a sulfide-containing medium, the electrode response is caused by the two silver complexes usually present at much higher concentrations than the free Ag^+.

For many metals the system $Me^{n+} + ne^- = Me(s)$ is slow and the equilibrium potential is established too slowly in using such metals as indicator electrodes. Strongly reducing metals cannot be used as indicator electrodes because the potentials established with such metals are mixed potentials. Hence we cannot use an Fe electrode and a Zn electrode to measure $\{Fe^{2+}\}$ and $\{Zn^{2+}\}$, respectively. The Zn electrode is characterized by a mixed potential, because H_2O is reduced at potentials more positive than those at which Zn^{2+} is reduced (corrosion of Zn). If a zinc amalgam electrode is used instead of a zinc electrode, hydrogen discharge occurs at potentials more negative than those at the Zn electrode and the reduction of Zn^{2+} at more positive potentials. The reaction at the electrode is now $Zn^{2+} + Hg(l) + 2e^- = Zn(Hg)$, and the electrode potential is given by

$$E = E_0 + \frac{RT}{nF} \ln \frac{\{Zn^{2+}\}}{\{Zn(Hg)\}} \tag{78}$$

where $\{Zn(Hg)\}$ is the activity of the Zn in the amalgam.

Glass Electrodes

When a thin membrane of glass separates two solutions, an electric potential difference that depends on the ions present in the solutions is established across the glass. Glass electrodes responding chiefly to H^+ ions have become common laboratory tools (Bates 1973). Modification of the glass composition has led to the development of electrodes selective for a variety of cations other than H^+ (Belford and Owen, 1989; Eisenmann, 1967).

The origin of the glass electrode potential is not discussed here, but it may be helpful to mention that the glass membrane functions as a cation exchanger and that a Nernst potential is observed if such a membrane separates two solutions at two different concentrations:

$$E_{cell} = \frac{RT}{F} + \ln \frac{\{^1H^+\}}{\{^2H^+\}} \tag{79}$$

Because the glass electrode contains a solution (acid) of constant $\{^2H^+\}$ inside the glass bulb, the emf measured depends only on $\{^1H^+\}$ in the external solution:

$$E_{cell} = \text{const} + \frac{RT}{F} \ln\{^1H^+\} \tag{80}$$

8.8 The Potentiometric Determination of Individual Solutes

pH Measurement

Presently, most pH values are measured with glass electrodes. Before discussing the cell used for measurement with the glass electrode we discuss the measurement of pH under ideal conditions with a cell such as

$$\text{Pt, H}_2(g)|\text{solution with H}^+, \text{Cl}^-|\text{AgCl(s), Ag} \tag{81}$$

working electrode   sample   reference electrode

The Nernst equation for cell 81 with the cell reaction

$$\text{H}^+ + \text{Ag(s)} + \text{Cl}^- \rightleftharpoons \tfrac{1}{2}\text{H}_2(g) + \text{AgCl(s)}$$

is given, for $p_{H_2} = 1$ atm, by

$$E = E° - k \log (\{\text{H}^+\} \{\text{Cl}^-\}) \tag{82}$$

or

$$E = E° - k \log ([\text{H}^+] [\text{Cl}^-]) - k \log f_{H^+} f_{Cl^-} \tag{83}$$

where $k = RT \ln 10/F = 59.16$ mV at 25°C.

The cell can be calibrated by using a HCl solution of known concentration $C([\text{H}^+] = [\text{Cl}^-] = C)$. Then we have

$$E = E° - 2k \log C - k \log f_{H^+} f_{Cl^-} \tag{84}$$

For a series of solutions of known but variable C, $E°$ and $\log f_{H^+} f_{Cl^-}$ can be determined. (Most conveniently, $E + 2k \log C$ is plotted versus C and extrapolated to $C \to 0$; when $C = 0$, $k \log f_{H^+} f_{Cl^-} \to 0$; then, from E and C, the value of $\log f_{H^+} f_{Cl^-}$ can be calculated.) Using the infinite dilution activity scale, for dilute HCl solutions, $\log f_{H^+} f_{Cl^-}$ can be approximated by the Davies equation (Table 3.3):

$$\log f_{H^+} f_{Cl^-} = -\left(\frac{\sqrt{I}}{1 + \sqrt{I}} - 0.2I\right) \tag{85}$$

where the ionic strength $I = C$.

If we now use a dilute solution of ionic strength I, where $[\text{H}^+] \neq [\text{Cl}^-]$ (e.g., a solution of weak acid or a buffer containing NaCl), we obtain from equation 83

$$-\log \{\text{H}^+\}f_{Cl^-} = \frac{1}{k}(E - E° + k \log [\text{Cl}^-]) \tag{86}$$

Everything on the right-hand side of equation 86 is experimentally accessible. If f_{Cl^-} can be calculated by the Davies equation

$$\log f_{Cl^-} = -0.5\left(\frac{\sqrt{I}}{1 + \sqrt{I}} - 0.2I\right) \tag{87}$$

$-\log\{H^+\}$ can be determined (Bates-Guggenheim convention). The approach outlined above corresponds to the method typically used to measure $-\log\{H^+\}$ of standard buffer solutions.

If the measurements are made in the presence of a swamping inert electrolyte, for example, $NaClO_4$ (constant ionic medium), we would find, instead of equation 85 (compare Figure 3.2),

$$\log \gamma_{H^+}\, \gamma_{Cl^-} \approx 0$$

and, instead of equation 86,

$$-\log [H^+] = \frac{1}{k}(E - E^{\circ\prime} + k \log [Cl^-]) \tag{88}$$

where $E^{\circ\prime} = E^\circ$ + constant. (This correction is due to liquid junction effects.)

pH Measurement with the Glass Electrode Typically, the cell used to measure pH can be represented[†] by

$$\text{Glass electrode} \,\left|\, \begin{array}{c}\text{standard S}\\ \text{or}\\ \text{sample X}\end{array}\right|\, \text{salt bridge (conc. KCl)} | \text{AgCl, Ag} \tag{89}$$

The operational pH definition is

$$\text{pH}(X) = \text{pH}(S) + \frac{1}{k}[E(X) - E(S)] \tag{90}$$

The operation manual of a standard pH meter is based on equation 90. The temperature compensation is based on $k\ (=RT\ \ln 10/F)$. The buffers used in the standardization have pH values measured with cells of type 81 using equation 86 and the convention 87. If the measurement is carried out in the pH range 3-9 and in solutions of $I < 0.1$ M, the measurement represents a very good approximation of the conceptual definition pH = $-\log\{H^+\}$. If $I > 0.1$

[†]Instead of an Ag/AgCl reference electrode, often a calomel electrode, Hg/Hg_2Cl_2, is used.

8.8 The Potentiometric Determination of Individual Solutes

M and pH < 3 and pH > 9, the measured pH cannot be conceptually interpreted.

For measurement of the H^+ *concentration*, $[H^+]$ in a solution of constant ionic medium, the calibration of a pH meter can be carried out with the help of a strong acid solution of known concentration, $[H^+] = C$, to which the electrolyte of the ionic medium has been added. The dial on the pH meter is set in accordance with the known $[H^+]$. For example, seawater has an ionic strength of approximately $I = 0.7$ M; a standard of $-\log[H^+] = 2.00$ may be prepared by adjusting a 1.00×10^{-2} M HCl solution with NaCl to an ionic strength of 0.70 M (composition: $[H^+] = 0.01$ M, $[Na^+] = 0.69$ M, $[Cl^-] = 0.70$ M). In comparing this standard with ocean water, one may find $E(S) = 118$ mV; $E(X) = -249$ mV. Hence $-\log[H^+] = +2.00 + (1/59.2)\,[118 - (-249)] = 8.20$. (See Appendix 6.2 in Chapter 6 for definition of different pH scales for seawater.)

Potentiometric Titration Potentiometry may be used to follow a titration and to determine its end point. The principles have already been discussed in connection with acid–base or complex formation titrations where pH or pMe is used as a variable. Any potentiometric electrode may serve as an indicator electrode, which indicates either a reactant or a reaction product. Usually the measured potential will vary during the course of the reaction and the end point will be characterized by a "jump" in the curve of voltage versus amount of reactant added.

Although the potentiometric measurement permits determination of the concentration (activity) of a species, the potentiometric titration determines the total analytical concentration (capacity); for example, a Ca^{2+} electrode responds to free Ca^{2+} ions only and not to Ca^{2+} complexes present in the solution. Using the Ca^{2+} electrode as an indicator electrode with a strong complex former (forming more stable complexes with Ca^{2+} than those already present in the solution) gives the sum of the concentrations of all Ca^{2+} species. In a similar way, total F^- may be determined by titrating with $La(NO_3)_3$, using a F^- electrode, or total K^+ may be measured from titration with a specific K^+ electrode using Ca tetraphenylborate $Ca[B(C_6H_5)_4]_2$, as a reagent.

The result obtained by a titration is usually more precise than that obtained in a direct potentiometric measurement. It is usually not too difficult to create an end point with a precision of better than 5%. In the direct potentiometric determination the relative error F_{rel} is given by

$$F_{rel} = \frac{\Delta C}{C} = 2.3 \Delta \log C \tag{91}$$

In other words, if the measurement with a glass electrode can be reproduced within 0.04 pH units (2.5 mV), the relative error in $[H^+]$ is 10%. For a similar reproducibility with a Ca^{2+} electrode, $\Delta pCa = 0.08$ pCa units (the Nernst coefficient is half that for the pH electrode), the relative error in $[Ca^{2+}]$ is 20%.

Other Ion-Selective Electrodes

Other types of electrodes are listed in Table 8.9.[†] The glass membrane is replaced by a synthetic single-crystal membrane (solid-state electrodes), by a matrix (e.g., inert silicone rubber) in which precipitated particles are imbedded (precipitate electrodes), or by a liquid ion-exchange layer (liquid–liquid membrane electrodes). The selectivity of these electrodes is determined by the composition of the membrane. All these electrodes show a response in their electrode potentials according to the Nernst equation.

No electrode is fully selective; other species often similar to that to be measured may affect the electrode response. The effect of another (disturbing) ion, S, on the measured electrode potential can typically be expressed by

$$E = E_0 + \frac{2.3\,RT}{nF} \log (\{M^{n+}\} + \Sigma\, K_{M-S}\{S\}^{n_M/n_S}) \qquad (92)$$

where K_{M-S} is a constant representative of the selectivity relation between M and S (e.g., if $K_{M-S} = 10^{-3}$, the electrode responds 1000 times more selectively to M than to S).

The so-called oxygen electrode is based on a different principle: an electrolysis cell is used, and the current measured at the inert cathode is, under standardized conditions, a function of the oxygen concentration (activity). Selectivity is enhanced by covering the cathode with a membrane permeable to molecules only.

Polymer Membrane Ion-Selective Electrodes Similar to Liquid Membranes in Living Cells

In these devices polymer materials containing specific ingredients constitute the backbone of the film covering the electrochemical transducer. Here we deal with a liquid membrane, because the organic solvent provides the medium in which the ions permeate across the membrane. The polymer membrane ion-selective electrodes (ISE) and their ion transport across the membrane function similarly as the ion transport across the membranes of living cells (Figure 8.29). We follow the presentation given by Widmer (1993).

Special proteins, the membrane transport proteins, are responsible for moving the ions across cell membranes. Generally, each protein is designed to transport a particular class of ions. These proteins form a continuous protein pathway across the membrane and therefore allow the ion to migrate across the membrane without coming into direct contact with the hydrophobic interior of the membrane. There are two major classes of membrane transport proteins: the carrier proteins and channel proteins.

The ion transport modes across membranes are visualized in Figure 8.29.

[†]For a recent review, see Widmer (1993).

8.8 The Potentiometric Determination of Individual Solutes

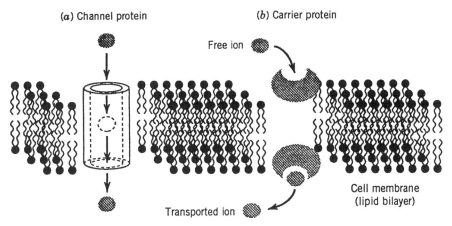

Figure 8.29. Ion transport mechanisms through lipid membranes in living cells. There are principally two kinds of transport protein: (a) channel proteins, that is, a channel-forming ionophore, and (b) carrier proteins, that is, a mobile ion carrier ionophore. The phenomena observed in living cells have much in common with those in artificial polymer membrane ion-selective electrodes. (From Widmer, 1993.)

The phenomena observed in living cells have much in common with those in artificial polymer membrane ISEs. In membrane electrodes, ionophores allow the net movement of ions in a membrane only down their electrochemical gradients. The equilibrium state is reached when the electrochemical gradient becomes zero and the cell potential reaches its final equilibrium value; net transport no longer occurs.

All kinds of chemical substances behave as ionophores, such as chelates, crown ethers, biological agents, enzymes, microbes and specific proteins, and animal and plant whole tissues.

Considerable effort has been given to the synthesis and characterization of ionophores with the goal of improving selectivity and detection limits.

In 1969, Wipf and Simon reported an outstanding potassium ionophore, valinomycin. Valinomycin is an example of a mobile ion carrier. It is a ring-shaped polypeptide that increases the permeability of a membrane to K^+. The ring has a hydrophobic exterior, made up of valine side chains, and a polar interior, where a single K^+ can fit precisely (see Figure 8.30). In the electrode process, valinomycin transports potassium ions across the membrane by picking up K^+ on the solution side of the membrane and releasing it at the transducer surface.

Ion Optodes The principles underlying the sophisticated complex polymer-based matrices for ISEs are the same as those used to design and construct ion-selective optical sensors, the optodes. Ion optodes have been developed for H^+, alkali metal ions, NH_4^+, Ca^{2+}, NO_3^-, and CO_3^{2-}. Numerous sensors for

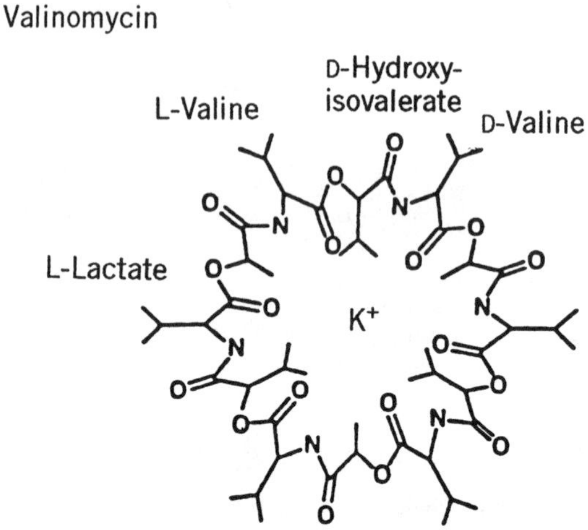

Figure 8.30. Potassium–valinomycin complex. The potassium ion is embedded in the oxygen-rich inner part of the valinomycin moiety, whereas the hydrophobic side chains stretch out into the lipophilic part of the membrane. Valinomycin is composed of three units of the sequence L-lactate, L-valine, D-hydroxyisovalerate, and D-valine.

liquid and gaseous samples with optical signed transduction have been reported (Widmer, 1993).

SUGGESTED READINGS

Appelo, C. A. J., and Postma, D. (1993) *Geochemistry, Groundwater and Pollution,* Balkema, Rotterdam, Chap. 7.

Bard, A. J., Parsons, R., and Jordan, J., Eds. (1985) *Standard Potentials in Aqueous Solution,* IUPAC, Dekker, New York. Contains, in addition to standard potentials, thermodynamic data.

Bidoglio, G. A., Avogadro, A., and De Pleno, A. (1989) Influence of Redox Environment on the Geological Behavior of Radionuclides, *Material Resources Soc. Symp. Proc.* **50,** 709–716.

Grenthe, I., Stumm, W., Laaksuharjn, M., Nilsson, A. C., and Wikberg, P. (1992) Redox Potentials and Redox Reactions in Deep Groundwater Systems, *Chem. Geol.* **98,** 131–150.

Hanselmann, K. (1994) Microbial Activities and New Eco-chemical Influence. In *Chemical and Biological Regulation of Aqueous Systems,* J. Buffle and R. R. de Vitre, Eds., Lewis, Chelsea, MI.

Jacobs, L. A., Von Gunten, H. R., Keil, R., and Kulsys, M. (1988) Geochemical Changes Along a River–Groundwater Infiltration Flow Path: Glattfelden, Switzerland. *Geochim. Cosmochim. Acta* **52,** 2693–2706.

Kempton, J. H., Lindberg, R. D., and Runnels, D. D. (1988) Numerical Modeling of Pt–E_H Measurements by Using Heterogeneous Electron Transfer Kinetics. In *Chem-*

ical Modeling of Aqueous Systems, Volume I, D. C. Melchior and R. L. Bassett, Eds., ACS Series 416, American Chemical Society, Washington, DC.

Kuhn, A., Johnson, C. A., and Sigg, L. (1994) Cycles of Trace Elements in a Lake with a Seasonally Anoxic Hypolimnion. In *Environmental Chemistry of Lakes and Reservoirs*, L. A. Baker, Ed., Advanced Chemical Services, vol. 237, pp. 473–457.

Lindberg, R. D., and Runnels, D. D. (1984) Groundwater Redox Reactions: An Analysis of Equilibrium State Applied to E_H Measurements and Geochemical Modeling, *Science* **228**, 925–927.

Luther, G. W. III (1990) The Frontier Molecular Orbital Theory. In *Aquatic Chemical Kinetics*, W. Stumm, Ed., Wiley, New York, pp. 173–198.

Luther, G. W., Koska, J. E., Church, T. M., Sulzberger, B., and Stumm, W. (1992) Seasonal Iron Cycling in the Salt Marsh Sedimentary Environment; the Importance of Ligand Complexes with Fe(II) and Fe(III) in the Dissolution of Fe(III) Minerals and Pyrite, Respectively. *Mar. Chem.* **40**, 81–103.

Parsons, R. (1985) Standard Electrode Potentials, Units, Conventions and Methods of Deterioration. In *Standard Potentials in Aqueous Solutions*, IUPAC, A. J. Bard, R. Parsons, and J. Jordan, Eds., Dekker, New York.

Scott, M. J., and Morgan, J. J. (1990) Energetics and Conservative Properties of Redox Systems. In *Chemical Modeling of Aqueous Systems II*, Melchior and R. L. Bassett, Eds., ACS Symposium Series 416, Chap. 29, American Chemical Society, Washington, DC.

Stumm, W., and Sulzberger, B. (1992) The Cycling of Iron in Natural Environments; Conditions Based on Laboratory Studies of Heterogeneous Redox Processes, *Geochim. Cosmochim. Acta* **56**, 3233–3257.

Widmer, H. M. (1992) Ion-Selective Electrodes and Ion Optodes, *Anal. Methods Instrum.* **1**, 60–72.

Zehnder, A. J. B., and Stumm, W. (1988) Geochemistry and Biogeochemistry of Anaerobic Habitats. In *Biology of Anaerobic Microorganisms*, A. J. B., Zehnder, Ed., Wiley, New York, pp. 1–38.

PROBLEMS

8.1. (a) Write balanced reactions for the following oxidations and reductions:
  (i) Mn^{2+} to MnO_2 by Cl_2
  (ii) Mn^{2+} to MnO_2 by OCl^-
  (iii) H_2S to SO_4^{2-} by OCl^-
  (iv) MnO_2 to Mn^{2+} by H_2S

(b) Arrange the following in order of decreasing $p\varepsilon$:
  (i) Lake sediment
  (ii) River sediment
  (iii) Seawater
  (iv) Groundwater containing 0.5 mg liter^{-1} Fe^{2+}
  (v) Digester gas (CH_4, CO_2)

508 Oxidation and Reduction; Equilibria and Microbial Mediation

8.2. The II, III, and IV oxidation states of Mn are related thermodynamically by these equilibria in acid solution:

$$\text{Mn(III)} + e^- = \text{Mn(II)} \quad \log K = 25$$
$$\text{Mn(IV)} + 2e^- = \text{Mn(II)} \quad \log K = 40$$

(a) What is the standard electrode potential of the couple

$$\text{Mn(IV)} + e^- = \text{Mn(III)?}$$

(b) Is Mn(III) stable with respect to simultaneous oxidation and reduction to Mn(II) and Mn(IV)? Explain.

8.3. (a) Under what pH conditions can NO_3^- be reduced to NO_2^- by ferrous iron? Assume $Fe(OH)_3(s)$ is product. (Use constants available in Table 8.6a and Table 8.3.)

(b) What is the ratio of the concentrations of NO_3^- to NO_2^- in equilibrium with an aqueous Fe^{2+}-$Fe(OH)_3(s)$ [or $FeOOH(s)$] system that has a pH of 7 and a pFe^{2+} of 4?

8.4. (a) Can the oxidation of NH_4^+ to NO_3^- by SO_4^{2-} be mediated by microorganisms at pH = 7?

(b) Is the oxidation of HS^- to SO_4^{2-} by NO_3^- thermodynamically possible at a pH of 9?

(c) Estimate the pε range in which sulfate-reducing organisms can grow.

8.5. In a closed tank, water and gases have been brought into equilibrium. The water at equilibrium has the following composition: alkalinity = 2 × 10^{-2} eq liter^{-1}; $[Fe^{2+}]$ = 2 × 10^{-5} M; pH = 6.0; dissolved Fe(III) = negligible but ferric hydroxide is precipitated.

(a) What is the partial pressure of CO_2?
(b) What is the pε of the solution?
(c) What partial pressure of O_2 corresponds to this pε?

The following information is available for the appropriate temperature:

$CO_2 + H_2O = H_2CO_3^*$	$\log K = -1.5$
$H_2CO_3^* = H^+ + HCO_3^-$	$\log K = -6.3$
$HCO_3^- = H^+ + CO_3^{2-}$	$\log K = -10.3$
$Fe^{2+} + CO_3^{2-} = FeCO_3(s)$	$\log K = +10.6$
$Fe^{3+} + 3OH^- = Fe(OH)_3(s)$	$\log K = +38.0$
$O_2 + 4e^- + 4H^+ = 2H_2O$	$E° = +1.23$ V
$H^+ + OH^- = H_2O$	$\log K = +14$
$Fe(OH)_3(s) + e^- = Fe^{2+} + 3OH^-$	$E° = -1.41$ V

8.6. What p_{O_2} cannot be exceeded so that a reduction of SO_4^{2-} to HS^- (pH = 7) can take place? (Use Table 8.6a.)

8.7. (a) Can glucose ($C_6H_{12}O_6$) be converted into CH_4 and CO_2 (e.g., in a digester)?
(b) What is the percent composition of the resulting gas?

8.8. Can $Fe_2SiO_4(s)$ in a primitive reducing atmosphere convert $CO_2(g)$ and $N_2(g)$ into alanine ($NH_2CH_3CHCOOH$)? (Compare Baur, 1978.) The following values for standard free energy of formation (kcal mol^{-1}) may be used: $Fe_2SiO_4(s)$, -319; $Fe_3O_4(s)$, -242.4; $H_2O(l)$, -56.690; $NH_2CH_3CHCOOH$, -88.7; $SiO_2(s)$, -204.6; $CO_2(g)$, -94.26.

8.9. Separate each of the following reactions into its half-reactions and in each case write down the schematic representation of a galvanic cell in which the reaction would take place. Wherever possible devise a cell without liquid junction potentials.
(a) $H_2 + PbSO_4(s) = 2H^+ + SO_4^{2-} + Pb$
(b) $AgCl = Ag^+ + Cl^-$
(c) $3Mn^{2+} + 2MnO_4^- + 4OH^- = 5MnO_2 + 2H_2O$
(d) $6Cl^- + IO_3^- + 6H^+ = 3Cl_2 + I^- + 3H_2O$

For each reaction above compute E_{cell} for the corresponding cell and decide the direction in which each reaction would take place spontaneously if all substances were present at unit activity.

8.10. The solubility product of mercurous sulfate is 6.2×10^{-7} at $25°C$, and the cell, $H_2(735 \text{ mm})H_2SO_4(0.001 \text{ M})$, Hg_2SO_4/Hg has an emf of $+0.8630$ V at $25°C$. Compute the standard potential of the half-reaction $Hg_2^{2+} + 2e^- = 2Hg$.

8.11. Potassium iodide (0.05 mol) and iodine (0.025 mol) are dissolved together in a liter of water at $25°C$. By appropriate analysis it is found that only 0.00126 mol of the iodine in the solution has remained in the form of I_2, the remainder having reacted to form I_3^- according to the equation:

$$I_2(aq) + I^- = I_3^-$$

Calculate the equilibrium constant. Check your result on the basis of the following information:

$I_2(s) + 2e^- = 2I^-$	$p\varepsilon° = 9.1$	
$2I^- = I_2(s) + 2e^-$	$E_H° = -0.5355$	
$I_3^- + 2e^- = 3I^-$	$p\varepsilon° = 9.1$	
$3I^- = I_3^- + 2e^-$	$E_H° = -0.536$	
$\Delta G°$ for $I_2(aq) = +3.93$ kcal		

8.12. What is the solubility of Au(s) in seawater? The following equilibrium constants (18–20°C) may be used:

$$AuCl_2^- + e^- = Au(s) + 2Cl^- \qquad \log K = 19.2$$
$$AuCl_4^- + 3e^- = Au(s) + 4Cl^- \qquad \log K = 51.3$$
$$Au(OH)_4^- + 4H^+ + 4Cl^- = AuCl_4^- + 4H_2O \qquad \log K = 29.64$$

8.13. Consider the equilibrium of an aqueous solution containing the solid phases FeOOH(s), $FeCO_3$(s). Compare the reaction

$$FeOOH(s) + HCO_3^- + 2H^+ + e^- = FeCO_3(s) + 2H_2O$$

Does pε increase, decrease, or stay constant upon addition of small quantities of the following: (a) FeOOH(s); (b) CO_2; (c) HCl; (d) EDTA at constant pH; (e) O_2 at constant pH; (f) NaOH?

8.14. The following reaction describes the reduction of *ferrate* iron [Fe^{VI}] to ferric iron:

$$FeO_4^{2-} + 3e^- + 8H^+ = Fe^{3+} + 4H_2O \qquad E_H^\circ = 2.2 \text{ V}$$

(a) What is $p\varepsilon^\circ$?
(b) Plot pε versus pH for the ferrate–ferric equilibrium in a system where $Fe_{T,aq} = 10^{-5}$ M for each aqueous species.
(c) At pH = 7, which is the stronger oxidant: ferrate or HOCl?
(d) What change can you predict for long-term storage of ferrate salts in pure water?

8.15. The oxidation–reduction reaction

$$FeO(s) + \tfrac{1}{4}O_2(g) = \tfrac{1}{2}Fe_2O_3(s)$$

has been suggested as having regulated the oxygen content of the earth's atmosphere in earlier times. You may use Appendix C at the end of the book.

(a) Discuss the stability of FeO(s) with respect to the two other reduced phases of iron, $Fe(OH)_2$(s) and $FeCO_3$ (i.e., ignoring the oxidation–reduction reaction). Is FeO(s) a stable reduced phase in the absence of CO_2? Is $FeCO_3$(s) a stable phase at $p_{CO_2} = 3 \times 10^{-4}$ atm in the absence of oxygen?
(b) Now consider the stability of reduced iron with respect to Fe_2O_3 in the presence of O_2.
   (i) Write a reaction describing the conversion of $FeCO_3$(s) to Fe_2O_3.

(ii) Calculate $\Delta G°$.
(iii) Calculate K_{eq}.
(iv) Discuss the stability relationships at $p_{CO_2} = 3 \times 10^{-4}$ atm and $p_{CO_2} = 0.21$ atm (present conditions).

8.16. Figure P8.1 gives a pε versus pH diagram for the Pb–H$_2$O–CO$_2$ system. Derive the various reactions that are the basis of this diagram and discuss the variables that affect the occurrence of Pb(II) in drinking water.

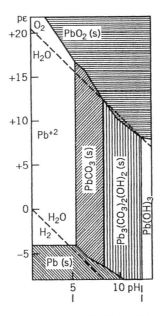

Figure P8.1. pε versus pH diagram for the Pb–H$_2$O–CO$_2$ system. TOTPb = 10^{-4} M, TOTCO$_3$ = 10^{-2} M (25°C).

8.17. (a) Does H$_2$O$_2$ (which may occur as a by-product of photochemical reactions in the surface waters) oxidize Fe(II) to Fe(III)?

(b) Is it plausible, on the basis of thermodynamic arguments, that the oxidation reaction is faster with H$_2$O$_2$ than with O$_2$?

8.18. Figure P8.2 gives a pε–pH diagram for chromium species. Total dissolved chromium, Cr$_T$, is 10^{-6} M and the temperature is 25°C.

(a) From a thermodynamic point of view, what is the predominant form of Cr in (i) oxygen-bearing waters and (ii) an anoxic groundwater?

(b) How could chromium be eluted from a contaminated soil?

(c) How can chromate present in laboratory wastewater be removed safely?

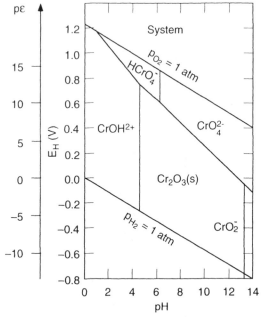

Figure P8.2. pε vs. pH diagram for Cr. [TOTCr] (dissolved) = 10^{-6} M 25°C.

ANSWERS TO PROBLEMS

8.1. (a) $Mn^{2+} + Cl_2 + 2H_2O = MnO_2(s) + 2Cl^-$ to $4H^+$; $Mn^{2+} + ClO^- + H_2O = MnO_2(s) + 2H^+ + Cl^-$; $H_2S + 4OCl^- = SO_4^{2-} + 4Cl^- + 2H^+$; $4MnO_2(s) + H_2S + 6H^+ = SO_4^{2-} + 4Mn^{2+} + 4H_2O$. (b) (iii) ($O_2$-$H_2O$, pε ≥ 12); (ii) ($NH_4^+$-$NO_3^-$, pε ≃ 5–8); (iv) [$10^{-5}$ M Fe^{2+}–FeOOH(s), pε ≃ −2 to 0]; (i) (HS^--SO_4^{2-}, pε ≃ −2 to −4); (v) (CH_4-CH_2O-CO_2, pε ≃ −3 to −5).

8.2. (a) $E_H^\circ = +0.885$ V (pε° = 15); (b) No, 2Mn(III) = Mn(IV) + Mn(II), log K = 10.

8.3. (a) For [Fe^{2+}] ≥ 10^{-4} M and ([NO_2^-]/[NO_3^-]) ≥ 1, pH ≥ 4.3 (consider the reaction $\frac{1}{2}NO_3^- + Fe^{2+} + 2\frac{1}{2}H_2O = \frac{1}{2}NO_2^- + 2H^+ + Fe(OH)_3(s)$; log K ≃ −4.6). (b) ~$10^{-11}$.

8.4. (a) No, log K(W) ≃ −10 for reaction $\frac{1}{8}NH_4^+ = \frac{1}{8}SO_4^- = \frac{1}{8}NO_3^- + \frac{1}{8}H^+ + \frac{1}{8}HS^-$ ([H^+] = 10^{-7}) + $\frac{1}{8}H_2O$. (b) Yes. (c) For pH of natural waters (4.6 < pH < 9.4), 0 > pε > −6.

8.5. (a) 1 atm. (b) 4.8. (c) 10^{-40} atm.

8.6. $p_{O_2} < 10^{-70}$ atm.

8.7. (a) Yes. (b) 50% CO_2, 50% CH_4.

8.8. Yes. $\frac{1}{2}N_2(g) + 3CO_2(g) + 9Fe_2SiO_4(s) + \frac{7}{2}H_2O(l) = NH_2CH_3CH-COOH(aq) + 6Fe_3O_4 + 9SiO_2(s)$, $\Delta G°(25°C) = -32$ kcal. If p_{N_2} is set at 0.01 atm and $p_{CO_2} = 0.03$ atm, alanine would be produced (thermodynamically). Baur shows that the formation of reduced carbon compounds is a thermodynamically spontaneous process in the heterogeneous system comprising CO_2 and N_2.

8.9. (a) $H_2(g) = 2H^+ + 2e^-$; $Pb(s) + SO_4^{2-} = PbSO_4(s) + 2e^-$; $Pb(s)$, $PbSO_4(s)$, H_2SO_4/H_2SO_4, $H_2(Pt)$; $E°_{cell} = +0.35$ V.
(b) $AgCl(s) + e^- = Ag(s) + Cl^-$, $Ag(s) = Ag^+ + e^-$ $Ag(s)$; $AgCl(s)$, $KCl/AgNO_3$, $Ag(s)$, $E°_{cell} = 0.58$ V.
(c) $MnO_4^- + 2H_2O + 3e^- = MnO_2(s) + 4OH^-$; $Mn^{2+} + 4OH^- = MnO_2(s) + 2H_2O + 2e^-$; $(Pt)MnO_2(s)$, Mn^{2+}, $NaOH/MnO_4^-$, $MnO_2(s)(Pt)$; $E°_{cell} = 1.3$ V.
(d) $Cl_2(g) + 2e^- = 2Cl^-$; $IO_3^- + 6H^+ + 6e^- = I^- + 3H_2O$; $(Pt)Cl_2(g)$, $NaCl/NaI$, $NaIO_3(Pt)$; $E°_{cell} = 0.46$ V.

8.10. For $Hg^{2+} + 2e^- = Hg(l)$, $E° = +0.798$ V.

8.11. $K = 10^{2.88}$.

8.12. $\sim 10^{-7.2}$ M.

8.13. No change for (a), (d), and (e); increase for (b) and (c); decrease for (f).

8.17. (a) Yes. (b) Yes.

APPENDIX 8.1 ACTIVITY RATIO DIAGRAMS FOR REDOX SYSTEMS

Activity ratio diagrams provide a simple way to gain a first-hand impression on the meaning of equilibrium data with respect to the relative stability of phases. We illustrate here the approach with examples of stability relations of iron and manganese compounds.

Example A8.1. Stability Relations of Iron and Manganese Compounds at pH = 7 Discuss the stability relations of Fe and Mn compounds with the help of activity ratio diagrams for the conditions pH = 7, $\{HCO_3^-\} = 10^{-3}$ M, and $\{SO_4^{2-}\} = 10^{-3}$ M. The equilibrium constants at 25°C can be obtained from the data on the free energy of formation in Appendix 3 at the end of the book.

For construction of the activity ratio diagram we take $\{Fe^{2+}\}$ as a reference. The following equations can be derived:

$$\log \frac{\{\alpha\text{-}Fe_2O_3(s)\}^{1/2}}{\{Fe^{2+}\}} = -11.1 + 3pH + p\varepsilon \quad \text{(i)}$$

$$\log \frac{\{\alpha\text{-FeOOH(s)}\}}{\{Fe^{2+}\}} = -11.3 + 3\text{pH} + p\varepsilon \tag{ii}$$

$$\log \frac{\{\text{am·Fe(OH)}_3\text{(s)}\}}{\{Fe^{2+}\}} = -17.1 + 3\text{pH} + p\varepsilon \tag{iii}$$

$$\log \frac{\{Fe_3O_4(s)\}^{1/3}}{\{Fe^{2+}\}} = -10.1 + \tfrac{8}{3}\text{pH} + \tfrac{2}{3}p\varepsilon \tag{iv}$$

$$\log \frac{\{FeCO_3(s)\}}{\{Fe^{2+}\}} = -0.2 + \text{pH} + \log\{HCO_3^-\} \tag{v}$$

$$\log \frac{\{Fe(OH)_2(s)\}}{\{Fe^{2+}\}} = -11.7 + 2\text{pH} \tag{vi}$$

$$\log \frac{\{\alpha\text{-FeS(s)}\}}{\{Fe^{2+}\}} = 38 - 8\text{pH} + \log\{SO_4^{2-}\} - 8p\varepsilon \tag{vii}$$

$$\log \frac{\{FeS_2(s)\}}{\{Fe^{2+}\}} = 86.8 - 16\text{pH} + 2\log\{SO_4^{2-}\} - 14p\varepsilon \tag{viii}$$

These equations are plotted for the conditions specified in Figure A8.1. For the Mn species the relevant equations are:

$$\log \frac{\{\gamma\text{-MnO}_2(s)\}}{\{Mn^{2+}\}} = -43.0 + 4\text{pH} + 2p\varepsilon \tag{ix}$$

$$\log \frac{\{\delta\text{-MnO}_2(s)\}}{\{Mn^{2+}\}} = -43.6 + 4\text{pH} + 2p\varepsilon \tag{x}$$

$$\log \frac{\{\gamma\text{-MnOOH(s)}\}}{\{Mn^{2+}\}} = -25.3 + 3\text{pH} + p\varepsilon \tag{xi}$$

$$\log \frac{\{MnS(s)\}}{\{Mn^{2+}\}} = 34.0 - 8\text{pH} + \log\{SO_4^{2-}\} - 8p\varepsilon \tag{xii}$$

$$\log \frac{\{MnO_4^-\}}{\{Mn^{2+}\}} = -124.6 + 8\text{pH} + 5p\varepsilon \tag{xiii}$$

$$\log \frac{\{MnCO_3(s)\}}{\{Mn^{2+}\}} = -0.2 + \text{pH} + \log\{HCO_3^-\} \tag{xiv}$$

$$\log \frac{\{Mn(OH)_2(s)\}}{\{Mn^{2+}\}} = -15 + 2\text{pH} \tag{xv}$$

Because of the uncertainty of free energy data, especially for the various iron oxides, the positions of the lines are not exact. Hematite is more stable than geothite and geothite is more stable than amorphous $Fe(OH)_3$. There is

8.9 Appendix 8.1 Activity Ratio Diagrams for Redox Systems

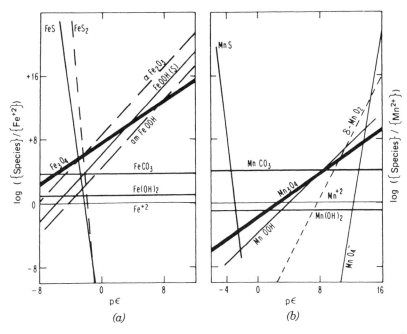

Figure A8.1. Activity ratio diagrams (a,b). Fe and Mn for pH = 7.0, $C_T = 10^{-3}$ M, $\{SO_4^{2-}\} = 10^{-3}$ M.

considerable uncertainty about the pε values at which equilibrium between iron(III) oxide and Fe_3O_4 occurs. $FeCO_3$(s) (siderite) is the most stable solid Fe(II) phase in the oxidative transition of FeS_2 to Fe(III) oxide. In the pε range 8–10, MnO_2, Mn_3O_4, and $MnCO_3$ are of similar relative activity. It appears from the diagram that under the specified conditions Mn_3O_4 may not occur as a stable phase. The diagrams, furthermore, suggest that iron and manganese sulfides start to be formed as one passes to pε values lower than -2 to -3. FeS(s) appears to be metastable with regard to pyrite. Even if sulfides are being formed, the sulfate concentration in many natural waters does not vary appreciably; hence the assumption of a constant sulfate concentration is justified. But if pε values drop further (below p$\varepsilon = -3$ at pH = 7), $\{SO_4^{2-}\}$ does not remain constant but decreases.

9

THE SOLID–SOLUTION INTERFACE

9.1 INTRODUCTION

The various reservoirs of the earth (atmosphere, water, sediments, soils, biota) contain material that is characterized by high area to volume ratios. Even the atmosphere contains solid–water–gas interfaces. There are trillions of square kilometers of surfaces of inorganic, organic, and biological material that cover our sediments and soils and that are dispersed in our waters. Very efficient interface chemistry must occur to maintain appropriate atmospheric chemistry and hydrospheric chemistry. Mineral-based assemblages and humus make up our soil systems that provide the nutrients to support our vegetation. The action of water (and CO_2 and organic matter) on minerals is one of the most important processes that produce extremely high surface areas and reactive and catalytic materials in the surface environments. The geological processes creating topography involve erosion by solution and particle transport. Such processes provide nutrient supply to the biosphere. The mass of material eroded off the continents annually is thus on an order of magnitude similar to that of the rate of crust formation and subduction (Fyfe, 1987). Human activity is greatly increasing erosion, and soil erosion has become a most serious world problem. The oceanic microcosmos of particles—biological particles dominate the detrital phases—plays a vital role in ocean chemistry.

The actual natural systems usually consist of numerous mineral assemblages and often a gas phase in addition to the aqueous phase; they nearly always include a portion of the biosphere. Hence natural systems are characterized by a complexity seldom encountered in the laboratory. In order to understand the pertinent variables out of a bewildering number of possible ones, it is advantageous to compare the real system with idealized counterparts and to abstract from the complexity of nature.

Most chemical reactions that occur in natural waters take place at phase discontinuities, that is, at atmosphere–hydrosphere and lithosphere–hydrosphere interfaces. In this chapter the discussion deals primarily with the solid–solution interface. This is still a very broad topic.

The significance of the solid–solution interface in natural waters becomes apparent when one considers the state of subdivision of the solids typically present in natural waters. The dispersed phase in a natural body of water consists predominantly of inorganic colloids, such as clays, metal oxides, metal hydroxides, and metal carbonates, and of organic colloidal matter of detrital origin, as well as living microorganisms (algae and bacteria). Most of the clay minerals have physical dimensions smaller than 1–2 μm (1 μm = 10^{-6} m). Montomorillonite, for example, has plate diameters of 0.002–0.2 μm and plate thicknesses in the range of 0.002–0.02 μm. Many solid phases have specific *surface energies* on the order of a few tenths J m^{-2}; hence, for substances that have specific surface areas of a few hundred m^2 g^{-1}, the total surface energy is on the order of 100 J g^{-1}.

Forces at Interfaces

Atoms, molecules, and ions exert forces on each other. Molecules at a surface are subject to an inward attraction normal to the surface. This is explained in part by the fact that surface molecules have fewer nearest neighbors and, as a consequence, fewer intermolecular interactions than bulk molecules. Ideally, the energy of interaction at an interface can be interpreted as a composite function resulting from the sum of attractive and repulsive forces, but our insight into intramolecular and interatomic forces is far from satisfactory. At a qualitative—and necessarily introductory—level we briefly enumerate the principal types of forces involved.

Generally speaking, *chemical forces* extend over very short distances; a covalent bond can be formed only by a merging of electron clouds. *Electric forces* extend over longer distances. The electrostatic force of attraction or repulsion between two point charges, q_1 and q_2, separated by distance x is given by Coulomb's law:

$$F = (\text{const}) \frac{1}{\varepsilon} \frac{q_1 q_2}{x^2}$$

where ε is the dielectric constant (relative dielectric permittivity) relative to a vacuum (dimensionless) ($\varepsilon_{H_2O} \simeq 80$).

The potential energy $E(x)$ referred to infinite separation is then given by

$$E(x) = \int_\infty^x F\, dx = -(\text{const}) \frac{1}{\varepsilon} \frac{q_1 q_2}{x} \tag{1a}$$

In the SI system, the charge is measured in coulombs (C), x is in meters (m), F is in newtons (N), and the proportionality constant (or conversion factor), for historical reasons written as

$$(\text{const}) = \frac{1}{4\pi\varepsilon_0} \tag{1b}$$

takes the numerical value 8.99×10^9 N m^2 C^{-2}; ε_0, the permittivity (in vacuum), is 8.854×10^{-12} C^2 N^{-1} or C^2 m^{-1} J^{-1} or C V^{-1} m^{-1}.

There are other forces, principally electrical in nature, present in molecular arrays whose constituents possess a permanent dipole (H$_2$O, NH$_3$). The energy resulting from such dipole–dipole interaction is also called the *orientation energy*. A charged species can also induce a dipole (induction energy).

London (1930) has shown that there is an additional type of force between atoms and molecules, which is always attractive. Its origin can be explained qualitatively by considering that neutral molecules or atoms constitute systems of oscillating charges producing synchronized dipoles that attract each other; hence this force is also, basically, an electric force. It is known as the *dispersion force* or the *London–van der Waals force*. This force is one of the reasons for the departure of real gases from ideal behavior and for the liquefaction of gases. The energy of interaction resulting from these dispersion forces (10–40 kJ mol^{-1}) is small compared to that of a covalent bond or electrostatic (ion-pair) bond ($\gg$ 40 kJ mol^{-1}), but large compared to that of orientation or induction energies ($<$ 10 kJ mol^{-1}). The van der Waals attraction energy between two atoms is inversely proportional to the sixth power of the separating distance over small distances. The total interaction between two semi-infinite flat plates is obtained by summing the pairwise interactions of all the constituent atoms. The resulting attractive energy per unit area between two semi-infinite plates is inversely proportional to the second power of the distance x:

$$V_A = -\frac{A}{12\pi x^2} \tag{2}$$

where A is the Hamaker constant. Its value depends on the density and polarizability of the material and is typically of the order of 10^{-19} to 10^{-20} J in aqueous systems. The Hamaker constant is smaller for organic materials. This equation is valid for distances up to approximately 200 Å; for larger separating distances, correction must be made that make the attractive energy decay as $1/x^3$.

Because of the very small size of the hydrogen ion, its highest coordination number is 2. In a *hydrogen bond*, H$^+$ accommodates two electron pair clouds in order to bind two polar molecules. While van der Waals interactions are principally spherically symmetric, hydrogen bonding occurs at a preferred molecular orientation. Hydrogen bonding, however, is in the same energy range (10–40 kJ mol^{-1}) as van der Waals interactions. The unusually high boiling point of water, for example, in comparison to that of H$_2$S(l), presents evidence for hydrogen bonding in addition to van der Waals association.

Characteristic interactions associated with categories of adsorption have been described by Israelachvili (1991).

9.2 ADSORPTION

Adsorption, the accumulation of matter at the solid–water interface, is the basis of most surface-chemical processes.

1. It influences the *distribution of substances* between the aqueous phase and particulate matter, which, in turn, affects their transport through the various reservoirs of the earth. The affinity of the solutes to the surfaces of the "conveyor belt" of the settling inorganic and biotic particles in the ocean (and in lakes) regulates their (relative) residence time, their residual concentrations, and their ultimate fate. Adsorption has a pronounced effect on the speciation of aquatic constituents.
2. Adsorption affects the *electrostatic properties* of suspended particles and colloids, which, in turn, influences their tendency to aggregate and attach (coagulation, settling, filtration).
3. Adsorption influences the reactivity of surfaces. It has been shown that the rates of processes such as precipitation (heterogeneous nucleation and surface precipitation), dissolution of minerals (of importance in the weathering of rocks, in the formation of soils and sediments, and in the corrosion of structures and metals), and the catalysis and photocatalysis of redox processes are critically dependent on the properties of the surfaces (surface species and their structural identity).

Atoms, molecules, and ions exert forces on each other at the interface. Here, adsorption reactions are discussed primarily in terms of intermolecular interactions between solute and solid phases. This includes:

1. *Surface complexation reactions* (surface hydrolysis, the formation of coordinative bonds at the surface with metals and with ligands).
2. *Electric interactions* at surfaces, extending over longer distances than chemical forces.
3. *Hydrophobic expulsion* of hydrophobic substances (this includes nonpolar organic solutes), which are usually only sparingly soluble in water and tend to reduce the contact in water and seek relatively nonpolar environments, thus accumulating at solid surfaces and becoming absorbed on organic sorbents.
4. *Adsorption of surfactants* (molecules that contain a hydrophobic moiety). Interfacial tension and adsorption are intimately related through the Gibbs adsorption law; its content—expressed in a simple way—is that substances that reduce surface tension become adsorbed at interfaces.

5. *Adsorption of polymers* and of polyelectrolytes—above all humic substances and proteins—is a rather general phenomenon in natural waters and soil systems that has far-reaching consequences for the interaction of particles with each other and on the attachment of colloids (and bacteria) to surfaces.

The process in which chemicals become associated with solid phases is often referred to as *sorption*, especially when one is not sure whether one deals with *ad*sorption (onto a two-dimensional surface) or *ab*sorption into a three-dimensional matrix.

Surface Coordination: The Surface Complex Most naturally occurring solids carry on their surface functional groups such as

$$-\text{OH}, \quad -\text{SH}, \quad \text{and} \quad -\text{C}\underset{\text{OH}}{\overset{\text{O}}{\lesseqgtr}}$$

For example, an oxide such as ferric oxide in the water acquires surface hydroxo groups $\equiv$Fe—OH. These functional groups contain the same donor atoms as in functional groups of solute ligands. We can, for example, compare the following reactions:

$$\text{in solution:} \quad \text{RCOOH} + \text{Cu}^{2+} \rightleftarrows \text{RCOOCu}^+ + \text{H}^+ \qquad (3a)$$

$$\text{on the surface:} \quad \equiv\text{S}-\text{OH} + \text{Cu}^{2+} \rightleftarrows \equiv\text{S}-\text{OCu}^+ + \text{H}^+ \qquad (3b)$$

where $\equiv$S—OH represents a surface functional group.* The complex formation in solution corresponds to the complex formation at the solid–solution interface. The deprotonated ligand RCOO$^-$ and $\equiv$SO$^-$ on the surface behave as Lewis bases. The adsorption of metal ions (and of protons) can be interpreted as competitive complex formation. In a similar way, the adsorption of ligands (anions and weak acids) can be compared with the complex formation in solution:

$$\text{in solution:} \quad \text{Fe(OH)}^{2+} + \text{F}^- \rightleftarrows \text{FeF}^{2+} + \text{OH}^- \qquad (4a)$$

$$\text{on the surface:} \quad \equiv\text{S}-\text{OH} + \text{F}^- \rightleftarrows \equiv\text{SF} + \text{OH}^- \qquad (4b)$$

The central ion on the surface of a mineral, for example, Fe in a Fe(III)(hydr)oxide, behaves (similar to FeOH^{2+} in solution) as a Lewis acid and exchanges the structural OH against ligands (*ligand exchange*).

We will illustrate that surface complex formation equilibria permit us to predict quantitatively the extent of adsorption of H$^+$, OH$^-$ of metal ions and ligands as a function of pH and solution variables and of surface characteristics.

*The various notations $\equiv$S—OH, —SOH, S, SH, $\equiv$MeOH, etc. are used to denote surface sites.

9.3 Adsorption Isotherms 521

Charge development on solid surfaces usually results from coordinative interactions at the solid surface. The surface coordinative model describes quantitatively how surface charge develops and permits us to incorporate the central features of the electric double-layer theory.

The Hydrophobic Effect; Hydrophobic Sorption Hydrophobic ("water-hating") compounds, for example, hydrocarbons and chlorinated hydrocarbons such as the polychlorinated biphenyls are soluble in many nonpolar solvents but not readily soluble in water. Because of the incompatibility of the hydrophobic substance with water, these substances have a tendency to avoid contact with water and seek to associate with nonpolar environments such as the surface of a mineral or an organic particle (Tanford, 1980). The sorption of hydrophobic substances to solid materials (particles, soils, sediments) that contain organic carbon may be compared with the partitioning of a solute between two solvents—water and the organic phase.

Many organic molecules (soaps, detergents, long-chain alcohols) are of a dual nature; they contain a hydrophobic part and a hydrophilic polar or ionic group; they are amphipathic. At an oil–water interface both parts of these molecules can satisfy their compatibility with each medium. Such molecules tend to migrate to the surface or interface of an aqueous solution; they also have a tendency toward self-association (*formation of micelles*). This is considered to be the result of *hydrophobic bonding*. This term may be misleading because the attraction of nonpolar groups to each other arises not primarily from a particular affinity of these groups for each other, but from the strong attractive forces between H_2O molecules, which must be disrupted when any solute is dissolved in water. The hydrophobic effect is perhaps the single most important factor in the organization of the constituent molecules of living matter into complex structural entities such as cell membranes and organelles.

9.3 ADSORPTION ISOTHERMS

Adsorption is often described in terms of isotherms, which show the relationship between the bulk aqueous phase activity (concentration) of adsorbate[†] and the amount adsorbed at constant temperature. Isotherms reflect equilibria. (Kinetic aspects will be discussed later.)

The Langmuir Isotherm

The simplest assumption in adsorption is that the adsorption sites, S, on the surface of a solid (adsorbent) become occupied by adsorbate from the solution, A. Implying a 1:1 stoichiometry, we can derive the Langmuir equation from

---
[†]Adsorbate is the substance in solution to become adsorbed at the adsorbent.

application of the mass law:

$$S + A \rightleftharpoons SA \tag{5}$$

where S = surface site of adsorbent
 A = adsorbate in solution
 SA = adsorbate on surface sites

We imply that the activities of the surface species are proportional to their concentrations. Surface species concentrations can be expressed in moles per liter solution, per g (or kg) solid, per cm² (or m²) of solid surface, or per mole of solid. Applying the mass law of equation 5 we find

$$\frac{[SA]}{[S][A]} = K_{ads} = \exp\left(-\frac{\Delta G°_{ads}}{RT}\right) \tag{6}$$

The maximum concentration of surface sites, S_T, is given by

$$[S_T] = [S] + [SA] \tag{7}$$

Thus from equations (6) and (7)

$$[SA] = [S_T] \frac{K_{ads}[A]}{1 + K_{ads}[A]} \tag{8}$$

If we define the surface concentration,

$$\Gamma = [SA]/\text{mass adsorbent} \tag{9}$$

and

$$\Gamma_{max} = [S_T]/\text{mass adsorbent} \tag{10}$$

we obtain

$$\Gamma = \Gamma_{max} \frac{K_{ads}[A]}{1 + K_{ads}[A]} \tag{11a}$$

Usually the Langmuir equation is known in the form of 11a. One can also write it as

$$\frac{\theta}{1-\theta} = K_{ads}[A] \tag{11b}$$

where

$$\theta = \frac{[SA]}{[S_T]}$$

The conditions for the validity of a Langmuir type adsorption equilibrium are (1) equilibrium up to the formation of a monolayer, $\theta = 1$; and (2) an energy of adsorption that is independent of θ (i.e., equal activity of all surface sites). There is no difference between a surface complex formation constant and a Langmuir adsorption constant (see Example 9.1b).

A Langmuir type adsorption isotherm is shown in Figure 9.1a. As is shown in Figure 9.1b, the evaluation of the equilibrium adsorption constant and of Γ_{max} is readily obtained from experimental data by plotting equation 11a in the

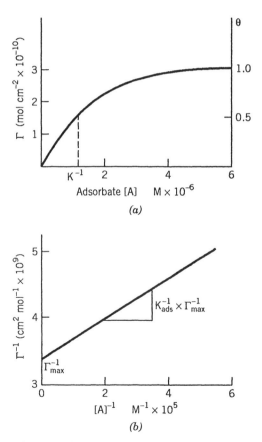

Figure 9.1. Langmuir adsorption isotherm. From the adsorption isotherm (plotted in accordance with equation 11a or equation 11b), the equilibrium constant K_{ads} and the adsorption capacity, Γ_{max}, are obtained by plotting Γ^{-1} versus the reciprocal concentration (activity) of the adsorbate (equation 11c).

reciprocal form

$$\Gamma^{-1} = \Gamma_{max}^{-1} + K_{ads}^{-1} \Gamma_{max}^{-1} [A]^{-1} \qquad (11c)$$

It is often also convenient to plot the Langmuir isotherm in a double logarithmic plot (Figure 9.2).

The fit of experimental data to a Langmuir (or another) adsorption isotherm does not constitute evidence that adsorption satisfies the criteria of the adsorption model. Frequently, adsorption to a surface is followed by additional interactions at the surface; for example, a surfactant undergoes two-dimensional association subsequent to becoming adsorbed; or charged ions tend to repel each other within the adsorbed layer.

If there are two adsorbates, A and B, one can formulate

$$\Gamma_A = \frac{\Gamma_{max} K_A [A]}{1 + K_A [A] + K_B [B]} \qquad (12)$$

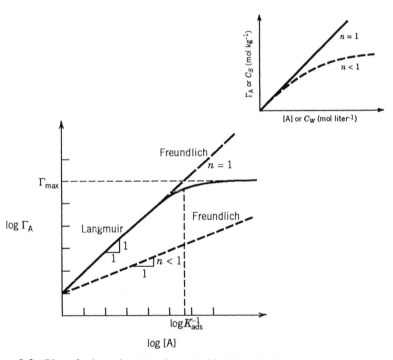

Figure 9.2. Plot of adsorption data in a double logarithmic plot. In a Langmuir isotherm the initial slope is unity. A Freundlich isotherm shows in a double log plot a slope of $n < 1$. Such a Freundlich isotherm is obtained if the adsorbent is heterogeneous (decreasing tendency for adsorption with increasing q). (Adapted from Morel, 1983.) *Inset:* Observed relationship between the concentrations of a chemical in the sorbed state Γ or C_s and the dissolved state [A] or C_w.

where K_A and K_B are the adsorption constants for A and B, respectively. There are further assumptions implied in equation 12, for example, the noninteraction of A and B on the surface.

Alternatively, a two-term series Langmuir equation can be written for two adsorbents (or for an adsorbent with two sites of different affinities):

$$\Gamma_A = \frac{\Gamma_1 K_1 [A]}{1 + K_1 [A]} + \frac{\Gamma_2 K_2 [A]}{1 + K_2 [A]} \tag{13}$$

The Freundlich Isotherm

This equation is very convenient for plotting adsorption data empirically in a log Γ versus log [A] (or log C_s versus log C_w) plot (Figure 9.2):

$$\Gamma = m[A]^n \quad \text{or} \quad C_s = mC_w^n \tag{14}$$

The Freundlich equation is often written in these two (equivalent) forms. Γ or C_s is the quantity of (ad)sorbate associated with the (ad)sorbent, usually expressed as mol kg^{-1} while [A] or C_w is the total chemical concentration of the (ad)sorbate in solution (e.g., in mol liter^{-1}); m is referred to as the Freundlich constant and n is the measure of the nonlinearity involved.

As pointed out by Sposito (1984), this equation initiated the surface chemistry of naturally occurring solids. Maarten van Bemmelen published this equation (now referred to as the Freundlich isotherm) more than 100 years ago and concluded from his results that the adsorptive power of ordinary soils depends on the colloidal silicates, humus, silica, and iron oxides they contain.

The equation applies very well to solids with heterogeneous surface properties and generally for heterogeneous solid surfaces. As has been shown by Sposito (1984), equation 14 can be derived by generalizing equation 13 to an integral over a continuum of Langmuir equations. The van Bemmelen–Freundlich isotherm can be thought of as the result of a log-normal distribution of Langmuir parameters K (i.e., a normal distribution of ln K) in a soil (or on a natural aquatic particle surface) (Sposito, 1984). For an extensive and general review on sorption phenomena see Weber et al. (1991).

The Solid–Water Distribution Ratio

If $n = 1$ in equation 14 (see Figure 9.2 inset), we can define a distribution ratio (one also speaks of a distribution coefficient):

$$K_d = \frac{C_s}{C_w} = \frac{(\text{mol kg}^{-1})}{(\text{mol liter}^{-1})} \; [\text{liter kg}^{-1}] \tag{15}$$

This "linear" isotherm represents the situation where the affinity of the sorbent for the sorbate remains the same for all levels of C_s. (Because of the similarity of this isotherm with the Henry law describing the absorption of gases, it is also referred to in the literature as the Henry equation.) One should, however, realize that over narrow ranges of C_w, especially at low concentrations, most isotherms (Freundlich and Langmuir) appear to be linear (see Figure 9.2). Furthermore, the solid–water distribution ratio is often used "for mathematical convenience" as a conditional constant, for example, valid for a certain concentration range and at a given pH only.

Linear isotherms are typically observed for the (ab)sorption of hydrophobic substances on organic or organically coated particles. In this case the hydrophobic substance becomes absorbed as if the organic matter were the "solvent."

The Frumkin Equation

This equation is also referred to as the Frumkin–Fowler–Guggenheim (FFG) equation and has been developed specifically to take lateral interactions at the surface into account. In the FFG equation, the term $\theta/(1 - \theta)$ in equation 11b is multiplied by the factor $\exp(-2\, a\theta)$, which reflects the extent of lateral interactions:

$$\frac{\theta}{1 - \theta} \exp(-2\, a\theta) = B[A] \qquad (16)$$

This may be compared with the Langmuir equation,

$$\frac{\theta}{1 - \theta} = K_{ads}[A] = B[A] \qquad (11b)$$

where $B = K_{ads}$ is the adsorption constant, [A] is the equilibrium (bulk) concentration (activity) of the adsorbate, and a is the interaction coefficient. If $a = 0$, equation 16 reduces to the Langmuir equation (11b); $a > 0$ indicates attraction, while $a < 0$ means repulsion. The value for the adsorption constant B and for the interaction coefficient a can be determined from the intercept and the slope of the resulting straight line in a plot of $\log[\theta/(1 - \theta)]$ versus θ. Figure 9.3 plots data on the adsorption of caprylic acid at pH = 4 on a hydrophobic surface (liquid mercury). At pH = 4, the caprylic acid is uncharged and lateral interaction (between the hydrocarbon moiety of the fatty acids) occurs ($a > 0$). At higher pH, the species getting into the adsorption layer are anions that repel each other; under such conditions, repulsive lateral interaction is observed ($a < 0$).

Example 9.1a. Adsorption Isotherm from Batch Data The following data for the adsorption of phosphate on α-Al_2O_3 have been reported. All data are at 25°C, at pH = 3.7 and $I = 10^{-2}$ M.

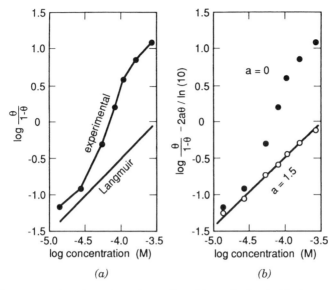

Figure 9.3. Data on the adsorption of caprylic acid on a hydrophobic (mercury) surface in terms of a double logarithmic plot of equation 16. (a) Comparison of the experimental values with a theoretical Langmuir isotherm, using the same values for the adsorption constant B for both curves. (b) The adsorption process can be described by introducing the parameter a, which accounts for lateral interaction in the adsorption layer. Equation 16 postulates a linear relation between the ordinate $[= \log[\theta/1 - \theta)] - 2a\,\theta/(\ln 10)]$ and the abscissa ($\log c$). If the correct value for a is inserted, a straight line results. For caprylic acid at pH 4, a value of $a = 1.5$ gives the best fit. (From Ulrich et al., 1988.)

P adsorbed (μmol/g Al_2O_3)	4.0	10	23	30	34	37	40
Remaining phosphate P_T (μM)	1.0	3.2	5.0	18	33	58	85

(a) Interpret these data in terms of suitable adsorption isotherms.

(b) Is it possible to estimate the area occupied by one phosphate from an estimated maximum adsorption? The specific surface area of the Al_2O_3 used was determined to be 10 m^2 g^{-1}.

It is expedient to plot the data first in terms of a double logarithmic plot (see Figure 9.4a). Obviously, the experimental data are not perfect but a Freundlich isotherm cannot fit because the curve has a trend toward saturation at high concentrations of P_T. A comparison with a logarithmically plotted Langmuir isotherm (Figure 9.2) illustrates that the saturation (capacity) is around $10^{-4.3}$ mol $P_{adsorbed}$ per g Al_2O_3 and that a Langmuir K would be around 10^5 M^{-1}.

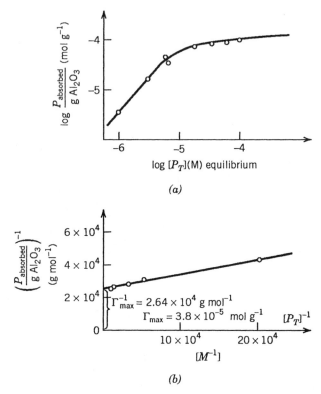

Figure 9.4. Example 9.1a. Adsorption of phosphate on α-Al$_2$O$_3$. The data given are plotted (a) double logarithmically (P$_{ads}$/g Al$_2$O$_3$) versus P_T (residual, M) and at first approximation an interpretation in terms of a Langmuir isotherm has been made (compare with Figure 9.2). (b) The reciprocal of the extent of adsorption is plotted versus the reciprocal of the residual P_T in order to estimate Langmuir parameters (equation 11c). The data are compatible with a Langmuir isotherm; but they do not suffice to postulate a Langmuir type of adsorption (monolayer, energy of adsorption independent of surface coverage).

Plotting the reciprocal of (P$_{adsorbed}$/g Al$_2$O$_3$) versus P_T^{-1} gives us the same information on capacity and also the adsorption equilibrium constant.

Because the experimental data are somewhat uncertain, this information may not be much more reliable than the estimates made before. The plot in Figure 9.4b gives estimates of $\Gamma_{max} = 3.8 \times 10^{-5}$ mol g^{-1} and of $K = 3.2 \times 10^5$ M^{-1}.

Using this estimate of Γ_{max}, we obtain $\sim 4 \times 10^{-5}$ mol P$_{adsorbed}$ per 10 m^2 surface area, or 2.4 P atoms per nm^2, or about 40 Å^2 per P atom. (The area occupied by a freely rotating phosphate ion was estimated to be 23 Å^2.)

Of course, the interpretation of the data made here is based on a "forced fit." A more detailed analysis is needed to investigate whether the requirements

9.3 Adsorption Isotherms

for a Langmuir adsorption (such as adsorption energy independent of θ) are fulfilled.

Multiple Adsorbates Adsorption isotherms are defined for individual adsorbate species. Collective parameters like DOC, phenols, and humic acids can only be used empirically as adsorbate parameters. Adsorption isotherms with collective parameters cannot be used for simple mechanistic interpretation of the data, even if these data can be fitted to such equations (Tomaić and Zutić, 1988).

Radke and Prausnitz (1972) have provided a theoretical treatment on isotherms of multiple adsorbates.

Example 9.1b. Surface Complex Formation and Langmuir Equation Consider a reaction such as

$$SH + Cu^{2+} \rightleftharpoons SCu^+ + H^+ \qquad K^s_{Cu} \qquad (i)$$

where SH is a protonated oxide surface site, for which $SH \rightleftharpoons S^- + H^+$. Illustrate that the mass law of equation i can be converted into a Langmuir-type equation.

The mass law of equation i is

$$\frac{[SCu^+]}{[SH][Cu^{2+}]} = \frac{K^s_{Cu}}{[H^+]} \qquad (ii)$$

Furthermore,

$$[SCu^+] + [SH] + [S^-] = S_T \qquad (iii)$$

Depending on the pH-range, [SH] or $[S^-]$ predominates. For most oxides in the pH range of interest, $[SH] \gg [S^-]$; thus

$$[SCu^+] + [SH] \simeq S_T \qquad (iiia)$$

Combination of equations ii and iiia gives

$$[SCu^+] = \frac{S_T K^s_{Cu} [H^+]^{-1} [Cu^{2+}]}{1 + K^s_{Cu} [H^+]^{-1} [Cu^{2+}]} \qquad (iv)$$

If we define

$$\Gamma_{Cu} = [SCu^+]/\text{mass adsorbent}$$

$$\Gamma_{max} = S_T/\text{mass adsorbent}$$

we obtain

$$\Gamma_{Cu} = \frac{\Gamma_{max} K^s_{Cu} [H^+]^{-1} [Cu^{2+}]}{1 + K^s_{Cu} [H^+]^{-1} [Cu^{2+}]} \quad \text{(v)}$$

where K^s_{Cu} is an *intrinsic* surface complex formation constant and $K^s_{Cu}/[H^+]$ is an intrinsic adsorption constant at a given $[H^+]$; that is,

$$K^s_{Cu}/[H^+] = K_{ads(pH)} \quad \text{(vi)}$$

Thus equation v becomes

$$\Gamma_{Cu} = \frac{\Gamma_{max} K_{ads(pH)} [Cu^{2+}]}{1 + K_{ads(pH)} [Cu^{2+}]} \quad \text{(vii)}$$

As we shall see in Section 9.6, a surface complex formation constant can be corrected for the effects of surface charge:

$$K^s(\text{app}) = K^s(\text{int}) \exp\left(-\frac{\Delta Z F \psi}{RT}\right)$$

In these equations $K(\text{int})$ and $K(\text{app})$ are the intrinsic and apparent equilibrium constants, respectively, F is the Faraday, ψ is the surface potential, and ΔZ is the *change* in the charge of the surface species for the reaction under consideration (as written for the equilibrium reaction for which K is defined).

The intrinsic equilibrium constants are postulated to be independent of the composition of the solid phase. They remain conditional in the sense of the constant ionic medium reference state if interacting ions such as H^+ are expressed as concentrations.

Gibbs Equation on the Relationship Between Interfacial Tension and Adsorption

Molecules in the surface or interfacial region are subject to attractive forces from adjacent molecules, which result in an attraction toward the bulk phase. The attraction tends to reduce the number of molecules in the surface region (increase in intermolecular distance). Hence work must be done to bring molecules from the interior to the interface. The minimum work required to create a differential increment in surface dA is $\gamma \, dA$, where A is the interfacial area and γ is the surface tension[†] or interfacial tension. One also refers to γ as the

---

[†]Strictly speaking, the surface tension of water is the tension of water with respect to vacuum; but one usually refers to the interfacial tension between water and air. As will be discussed in Chapter 13, the interfacial tension between water and solid minerals is of importance in the kinetics of nucleation and precipitation.

9.3 Adsorption Isotherms

interfacial Gibbs free energy for the condition of constant temperature, T, pressure, P, and composition (n = number of moles):

$$\gamma = \left(\frac{\partial G}{\partial A}\right)_{T,P,n} \tag{17}$$

γ is usually expressed in J m^{-2} or in N m^{-1}.

In water, the intermolecular interactions that produce surface tensions are essentially (1) London–van der Waals dispersion interactions, $\gamma_{H_2O(L)}$, and (2) hydrogen bonds, $\gamma_{H_2O(H)}$:

$$\gamma = \gamma_{H_2O(L)} + \gamma_{H_2O(H)} \tag{18}$$

It is estimated that about one-third of the interfacial tension is due to van der Waals attraction, and the remainder is due to hydrogen bonding.

The interfacial tension between two phases is subject to the resultant force field made up of components arising from attractive forces in the bulk of each phase and the forces, usually the London dispersion forces, operating across the interface itself (Adamson, 1990; Fowkes, 1965).

The Gibbs Equation The Gibbs equation,

$$\Gamma_i = -\frac{1}{RT}\left(\frac{\partial \gamma}{\partial \ln a_i}\right)_{T,P} \tag{19}$$

can be derived from equation 17. Here Γ_i is the surface concentration (mol m^{-2}) or more precisely the surface excess[†] with regard to a reference condition (with pure water the adsorption density of $H_2O = 0$).

R = gas constant
T = absolute temperature
γ = surface tension or interfacial tension (J m^{-2})
a_i = activity (or concentration) of species i

The Gibbs equation relates the extent of adsorption at an interface (reversible equilibrium) to the change in interfacial tension; equation 19 predicts that a

---

[†]To define a surface excess concentration rigorously, we must decide whether or not to recognize the finite thickness of surfaces. In view of the difficulty of defining surface thickness, Gibbs defined the surface (for thermodynamic purposes) as a mathematical plane or dividing surface of zero thickness near the physical surface, and surface properties as the *net* positive or negative excess in the vicinity of the surface over the magnitude of the same property in the bulk (Adamson, 1990).

The Solid–Solution Interface

substance that reduces the surface (interfacial) tension $[(\partial \gamma / \partial \ln a_i) < 0]$ will be adsorbed at the surface (interface). Electrolytes have the tendency to increase γ, but most organic molecules, especially surface-active substances (long-chain fatty acids, detergents, surfactants) decrease the surface tension (Figure 9.5). *Amphipathic* molecules (which contain hydrophobic and hydrophilic groups) become oriented at the interface.

Although the Gibbs equation applies also to the solid–liquid interface, direct measurements of the interfacial tension are difficult. In a qualitative sense, it is important to realize that adsorption at a solid surface—of both molecules and ions—reduces interfacial tension. For example, the interfacial tension of an oxide surface (e.g., quartz) is reduced upon adsorption or desorption of (charge-determining) ions. Such an oxide has a relative maximum value at its point of zero charge (corresponding to the capillary maximum of a Hg electrode in water); its interfacial energy decreases at pH values above and below pH_{pzc}. Parks (1984) has elaborated this concept by discussing the surface and interfacial free energies of quartz.

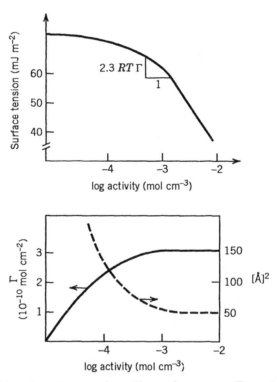

Figure 9.5. Gibbs adsorption equation. The surface excess Γ can be obtained (cf. equation 19) from a plot of surface tension γ versus log activity (concentration) of adsorbate. The area occupied per molecule or ion adsorbed can be calculated.

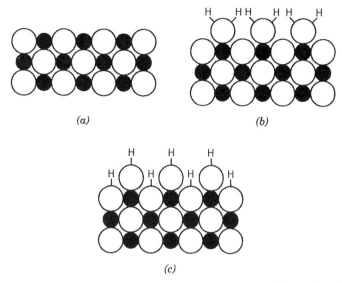

Figure 9.6. Schematic representation of the cross section of the surface layer of a metal oxide. ●, Metal ions; ○, oxide ions. The metal ions in the surface layer (a) have a reduced coordination number. They thus behave as Lewis acids. In the presence of water, the surface metal ions may first tend to coordinate H_2O molecules (b). For most of the oxides, dissociative chemisorption of water molecules (c) seems energetically favored. Geometrical considerations and chemical measurements indicate an average surface density of 5 (typical range 2–12) hydroxyls per nm^2 of an oxide mineral (From Schindler, 1981.)

9.4 HYDROUS OXIDE SURFACES; REACTIONS WITH H^+, OH^-, METAL IONS, AND LIGANDS

Oxides, especially those of Si, Al, and Fe, are abundant components of the earth's crust. Hence a large fraction of the solid phases in natural waters, sediments, and soils contain such oxides or hydroxides. In the presence of water the surfaces of these oxides are generally covered with surface hydroxyl groups (Figure 9.6).

The various surface hydroxyls formed may not be fully structurally and chemically equivalent, but to facilitate the schematic representation of reactions and of equilibria, one usually considers the chemical reaction of "a" surface hydroxyl group, S—OH.[†]

[†]The following surface groups can be envisaged (Schindler, 1985).:

$$\begin{matrix} S\diagdown \\ OH \\ S\diagup \end{matrix} \qquad S-OH \qquad \begin{matrix} \diagup OH_2 \\ S \\ \diagdown OH \end{matrix} \qquad \begin{matrix} \diagup OH \\ S{\leftarrow}OH \\ OH \end{matrix}$$

Table 9.1. Adsorption (Surface Complex Formation Equilibria)

Acid–Base Equilibria

$S-OH + H^+ \rightleftharpoons S-OH_2^+$
$S-OH\ (+\ OH^-) \rightleftharpoons S-O^- + (H_2O)$

Metal Binding

$S-OH + M^{z+} \rightleftharpoons S-OM^{(z-1)+} + H^+$
$2\ S-OH + M^{z+} \rightleftharpoons (S-O)_2 M^{(z-2)+} + 2\ H^+$
$S-OH + M^{z+} + H_2O \rightleftharpoons S-OMOH^{(z-2)+} + 2\ H^+$

Ligand Exchange (L^- = ligand)

$S-OH + L^- \rightleftharpoons S-L + OH^-$
$2\ S-OH + L^- \rightleftharpoons S_2-L^+ + 2\ OH^-$

Ternary Surface Complex Formation

$S-OH + L^- + M^{z+} \rightleftharpoons S-L-M^{z+} + OH^-$
$S-OH + L^- + M^{z+} \rightleftharpoons S-OM-L^{(z-2)+} + H^+$

Source: Adapted from Schindler and Stumm (1987).

As we already discussed, these functional groups contain the same donor atoms as found in functional groups of soluble ligands; that is, the surface hydroxyl group on a hydrous oxide has similar donor properties as the corresponding counterparts in dissolved solutes, such as hydroxides and carboxylates.

Table 9.1 presents the most important adsorption (= surface complex formation) equilibria. The following criteria are characteristic of all surface complexation models (Dzombak and Morel, 1990):

1. Sorption takes place at specific surface coordination sites.
2. Sorption reactions can be described by mass law equations.
3. Surface charge results from the sorption (surface complex formation) reaction itself.
4. The effect of surface charge on sorption (extent of complex formation) can be taken into account by applying a correction factor derived from the electric double-layer theory to the mass law constants for surface reactions.

The Acid–Base Chemistry of Oxides; pH of Point of Zero Charge

Uptake and release of protons can be described by the acidity constants (by assuming a solution of constant ionic strength, we assume that the activity coefficients of the surface species are equal):

$$K_{a1}^s = \frac{\{SOH\}[H^+]}{\{SOH_2^+\}} \text{ mol liter}^{-1} \tag{20}$$

$$K_{a2}^s = \frac{\{SO^-\}[H^+]}{\{SOH\}} \text{ mol liter}^{-1} \tag{21}$$

where { } denotes the concentrations of surface species in moles per kilogram of adsorbing solid and [ ] denotes the concentrations of solutes (M).[†] In many cases, it is more desirable to express the concentrations of the surface species as surface densities (mol m^{-2}) or in the same units as the concentrations of dissolved species (mol liter^{-1}). Then conversion is easily accomplished with the following equations:

$$\langle SOH \rangle = s^{-1} \{SOH\} \text{ mol m}^{-2} \tag{22}$$

$$[SOH] = a\{SOH\} \text{ mol liter}^{-1} \tag{23}$$

where $\langle SOH \rangle$ is the surface concentration in mol m^{-2}, s is the specific surface area of the solid (m^2 kg^{-1}), and a is the quantity of oxide used (kg liter^{-1}).

Example 9.2. Evaluation of Surface Charge from Alkalimetric and Acidimetric Titration Curves and Determination of Surface Acidity Constants We will demonstrate how the surface charge of a hydrous oxide (α-FeOOH) can be calculated from an experimental titration curve[‡] (e.g., Figure 9.7).

In titrating a suspension of α-FeOOH [6 g liter^{-1}, 29 m^2 g^{-1}; 2×10^{-4} mol g^{-1} surface functional groups (TOT≡FeOH)] in an inert electrolyte (10^{-1} M NaClO$_4$) with NaOH or HCl (C_B and C_A are the concentrations of base and acid, respectively, added per liter), we can write for any point on the titration curve

$$C_A - C_B + [OH^-] - [H^+] = [\equiv FeOH_2^+] - [\equiv FeO^-] \tag{i}$$

where [ ] indicates concentrations of solute and surface species per unit volume solution (M). Equation i can also be derived from a charge balance. The right-hand side gives the net number of moles per liter of H$^+$ ions bound to

---

[†]M means mol liter^{-1}. We will frequently use solute concentrations (rather than activities). Often the experiments are done in a constant ionic medium. If the concentrations of the solutes are smaller than the concentrations of the background electrolyte, we are justified in setting the aqueous phase activity coefficients equal to 1 on the scale of the constant *ionic medium reference state*.
[‡][H$^+$] is measured potentiometrically with a glass electrode. Briefly, the method involves the use of a glass electrode and a double-junction calomel reference electrode in the titration cell:

glass electrode	suspension of solid background electrolyte	background electrolyte solution	liquid junctions	calomel electrode

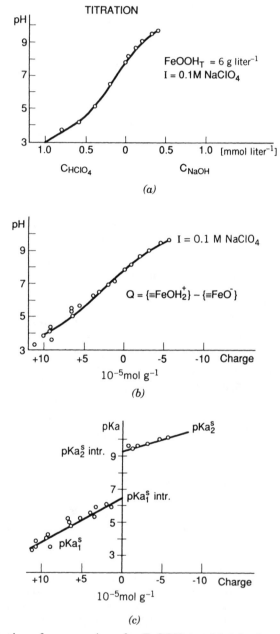

Figure 9.7. Titration of a suspension of α-FeOOH (goethite) in the absence of specifically adsorbable ions. (a) Acidimetric–alkalimetric titration in the presence of an inert electrolyte. (b) Charge calculated from the titration curve (charge balance). (c) Microscopic acidity constants calculated from (a) and (b). Extrapolation to charge zero gives intrinsic pK_{a1}^s and pK_{a2}^s. (Data from Sigg and Stumm, 1981.)

9.4 Hydrous Oxide Surfaces

α-FeOOH. The mean surface charge (i.e., the portion of the charge due to OH^- or H^+) can be calculated as a function of pH from the difference between total added base or acid and the equilibrium OH^- and H^+ ion concentration for a given quantity a (kg liter^{-1}) of oxide used:

$$\frac{C_A - C_B + [OH^-] - [H^+]}{a} = \{\equiv FeOH_2^+\} - \{\equiv FeO^-\} = Q \quad \text{(ii)}$$

where { } indicates the concentration of surface species in mol kg^{-1} (e.g., $[\equiv FeO^-]/a = \{\equiv FeO^-\}$). If the specific surface area s (m^2 kg^{-1}) of the iron oxide used is known, the surface charge σ (C m^{-2}) can be calculated:

$$\sigma = QFs^{-1} = F(\Gamma_{H^+} - \Gamma_{OH^-}) \quad \text{(iii)}$$

where Q is the surface charge in mol kg^{-1}, F is the Faraday constant and Γ_{H^+} and Γ_{OH^-} are the "adsorption" densities of H^+ and OH^- (mol m^{-2}).

The *point of zero charge* pH$_{pzc}$ corresponds to the *zero proton condition* at the surface (for definitions of points of zero charge see Section 9.5):

$$\{\equiv FeOH_2^+\} = \{\equiv FeO^-\}; \quad \Gamma_{H^+} = \Gamma_{OH^-} \quad \text{(iv)}$$

In this case pH$_{pzc}$ = 7.9 (cf. Figure 9.7b).

We can now calculate the surface acidity equilibrium constants (equations 20 and 21). There are five species, $\equiv FeOH_2^+$, $\equiv FeOH$, $\equiv FeO^-$, H^+, OH^-, that are interrelated by the two acidity mass law constants (equations 20 and 21), by the ion product of water ($K'_W = [H^+][OH^-]$) and two mass balance equations:

$$C_A - C_B = [\equiv FeOH_2^+] - [\equiv FeO^-] + [H^+] - [OH^-]$$

(compare equation i) and

$$[TOT \equiv FeOH] = [\equiv FeOH_2^+] + [\equiv FeOH] + [\equiv FeO^-] \quad \text{(v)}$$

The calculation is facilitated by the following initial assumptions:

$$Q \approx \{\equiv FeOH_2^+\} \quad \text{for pH} < \text{pH}_{pzc} \quad\quad Q \approx \{\equiv FeO^-\} \quad \text{for pH} > \text{pH}_{pzc}$$

Then, equation 20 becomes

$$K^s_{a1} = \frac{(\{TOT \equiv FeOH\} - Q)[H^+]}{Q} \quad \text{for pH} < \text{pH}_{pzc}$$

and equation 21 becomes

$$K_{a2}^s = \frac{Q\,[H^+]}{(\{TOT\equiv FeOH\} - Q)} \quad \text{for pH} > \text{pH}_{pzc}$$

The acidity constants calculated from every point in the titration curve (Figure 9.7a,b) are microscopic acidity constants (equations 20 and 21). Each loss of a proton reduces the charge on the surface and thus affects the acidity of the neighboring groups. The free energy of deprotonation consists of the *dissociation* as measured by an intrinsic activity constant, that is, a constant valid for an uncharged surface, and the *removal* of the proton from the site of the dissociation to the bulk of the solution. As shown in Figure 9.7c, the intrinsic values for the acidity constants can be obtained by linear extrapolation of the log K^s versus charge, Q, curve to the zero charge condition. The result shown in the final answer (as we shall show later, this somewhat empirical approach can be justified theoretically). The fits are described by an equation of the form

$$\log K_a = \log K_a^s(\text{int}) - \beta Q \qquad \text{(vi)}$$

where Q is the surface charge in mol kg^{-1}, and β is a coefficient that contains information on the constant capacitance of the oxide–electrolyte interface at high ionic strength. β can be derived (Schindler and Stumm, 1987) as $\beta = (zF)^2/2.3\,RTCs$ where C is the capacitance.

The pH$_{pzc}$ (zero proton condition, point of zero charge) is not affected by the concentration of the inert electrolyte in the absence of a different specific supporting electrolyte ion boundary for cation and anion. The computations of Dzombak and Morel (1990) employ a diffuse layer model coupled with acid–base surface reactions to describe Q versus pH. (This acid–base model incorporates variable capacitance.) As Figure 9.8 shows, there is a common intersection point of the titration curves obtained with different concentrations of inert electrolyte.

Obviously, at the condition where $[\equiv FeOH_2^+] = [\equiv FeO^-]$

$$\text{pH}_{pzc} = 0.5\,[pK_{a1}(\text{int}) + pK_{a2}(\text{int})] \qquad \text{(vii)}$$

Points of Zero Charge Points of zero charge (pzc) are pH values where the net surface charge is zero. We consider here first of all the surface conditions where the charge is established by proton exchange at the surface (Figure 9.9). If the surface charge is established solely by H$^+$ exchange (binding and dissociation of H$^+$), one may also refer to the point of zero net proton charge (pznpc). We can estimate the pH$_{pznpc}$ of metal oxides from electrostatic considerations. Parks (1967) has shown that the pH$_{pznpc}$ of a simple oxide is related to the appropriate cationic charge and radius of the central ion. As shown by Parks, the point of zero charge of a composite oxide is approximately the weighted average of the values of its components. Predictable shifts in points

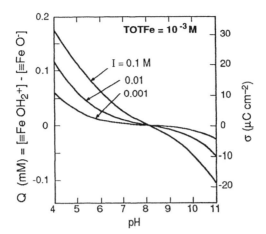

Figure 9.8. Surface charge as a function of pH and ionic strength (1:1 electrolyte) for a 90 mg liter^{-1} (TOTFe = 10^{-3} M) suspension of hydrous ferric oxide. The specific surface area is 600 m^2 g^{-1} and the site concentration is 2×10^{-4} mol sites per liter. (From Dzombak and Morel, 1990.)

of zero charge occur in response to state of hydration, cleavage habit, and crystallinity.

The point of zero charge of salt-type minerals depends on pH and on the concentration (activities) of *all* potential-determining ions. Thus, in the case of calcite, possible potential-determining species, in addition to H$^+$ and OH$^-$, are HCO$_3^-$, CO$_2$ and Ca^{2+}; various mechanisms of charge development are possible. When referring to a point of zero charge of such nonoxides, the solution composition should be specified. In the absence of complications, such

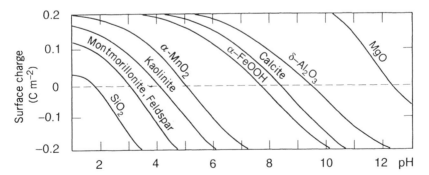

Figure 9.9. Effect of pH on approximate surface charge (C m^{-2}) of a few representative colloids. At pH$_{pzc}$ the surface charge is zero. The measurements of the surface charge depend on the solution composition. The curves given are general trends. The curve for calcite is valid for a suspension of CaCO$_3$ in equilibrium with air (p_{CO_2} = $10^{-3.5}$ atm).

The Solid–Solution Interface

as those caused by structural or adsorbed impurities, the point of zero charge of the solid should correspond to the pH of charge balance (electroneutrality) of potential-determining ions.

Surface Complex Formation with Metal Ions

Surface complex formation of cations by hydrous oxides involves the coordination of the metal ions with the oxygen donor atoms and the release of protons from the surface; for example,

$$S-OH + Cu^{2+} \rightleftharpoons S-OCu^+ + H^+ \quad (24)$$

There is also the possibility of forming bidentate surface complexes:

$$2\ S-OH + Cu^{+2} = (S-O)_2Cu + 2\ H^+ \quad (25a)$$

that is,

$$\begin{matrix}-S-OH \\ -S-OH\end{matrix} + Cu^{2+} \rightleftharpoons \begin{matrix}-S-O \\ -S-O\end{matrix}\!\!>\!Cu + 2\ H^+ \quad (25b)^\dagger$$

Equations 24 and 25 can also be formulated in terms of mass laws; for example

$$K^s_{Cu} = \frac{\{S-OCu^+\}[H^+]}{\{S-OH\}[Cu^{2+}]} \quad (26a)$$

$$\beta^s_{2Cu} = \frac{\{(S-O)_2Cu\}[H^+]^2}{\{(S-OH)_2\}[Cu^{2+}]} \quad (26b)$$

These constants have the rank of conditional stability constants. For exact considerations we need to correct for electrostatic interaction (see Section 6.6).

Inner-Sphere and Outer-Sphere Complexes

As Figure 9.10 illustrates, a cation can associate with a surface as an inner-sphere or outer-sphere complex depending on whether a chemical (i.e., largely

---

†The formula

$$\begin{matrix}-S-OH \\ -S-OH\end{matrix}$$

indicates that we treat the surface —OH groups as a dimer; the line between the two S merely implies that the two S are interconnected, for example, through an oxygen bridge.

9.4 Hydrous Oxide Surfaces 541

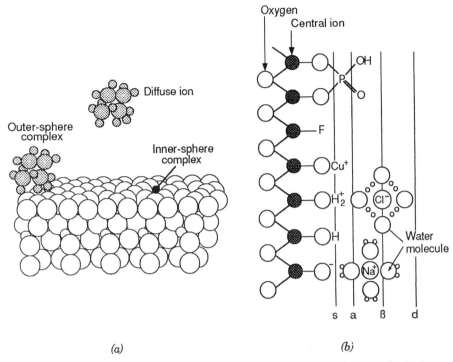

Figure 9.10. (a) Surface complex formation of an ion (e.g., cation) on the hydrous oxide surface. The ion may form an inner-sphere complex ("chemical bond"), an outer-sphere complex (ion pair), or be in the diffuse swarm of the electric double layer. (The inner-sphere complex may still retain some aquo groups toward the solution side.) (From Sposito, 1989.) (b) A schematic portrayal of the hydrous oxide surface, showing planes associated with surface hydroxyl groups ("s"), inner-sphere complexes ("a"), outer-sphere complexes ("β"), and the diffuse ion swarm ("d"). (Adapted from Sposito, 1984.)

covalent) bond between the metal and the electron-donating oxygen ions is formed (as in an inner-sphere type solute complex) or if a cation of opposite charge approaches the surface groups to a critical distance; as with solute ion pairs, the cation and the base are separated by one or more water molecules. Furthermore, ions may be in the diffuse swarm of the double layer.

It is important to distinguish between outer-sphere and inner-sphere complexes. In inner-sphere complexes the surface oxide ions act as σ-donor ligands, which increase the electron density of the coordinated metal ion. Cu(II) bound inner-spherically is a different chemical entity than if it were bound outer-spherically or present in the diffuse part of the double layer; the inner-spheric Cu(II) has chemically different properties; for example, a different redox potential [with regard to Cu(I)], and its equatorial waters are expected to exchange faster than in Cu(II).

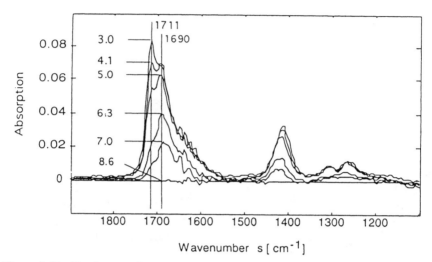

Figure 9.11. Fourier transform infrared (FTIR) spectra of oxalate adsorbed on TiO_2; oxalate solution concentration 10^{-4} M and pH values between 3.0 and 8.6. Surface complex formation starts below pH 8.6. First, a spectrum with a maximum at 1690 cm^{-1} is observed. With decreasing pH, the amplitude increases and an additional peak at 1711 cm^{-1} appears. These bands are assigned to C=O stretching vibrations. The changing spectral shape is an indication of the formation of at least two different inner-sphere complexes (Hug and Sulzberger, 1994.)

Structural Identity

Direct evidence for inner-sphere complexes comes from spectroscopic methods, for example, Fourier transform infrared spectroscopy (Figure 9.11) Biber and Stumm, 1994; Hug and Sulzberger, 1994; Tejedor-Tejedor and Anderson, 1986) and direct in situ x-ray (from synchroton radiation) absorption measurements (EXAFS) (Brown et al., 1989, Charlet and Manceau, 1993; Hayes et al., 1978). For a general review, see Johnston et al. (1993).

A simple method of distinguishing between inner-sphere and outer-sphere complexes is to assess the effect of ionic strength on the surface complex formation equilibria. A strong dependence of ionic strength is typical for an outer-sphere complex. Furthermore, outer-sphere complexes involve electrostatic bonding mechanisms and therefore are less stable than inner-sphere surface complexes, which necessarily involve appreciable covalent bonding along with ionic bonding.

pH Dependence of Surface Binding

As evidenced by the mass laws of equations 24 and 25, the binding of a metal ion by surface ligands—similar to the binding of a metal ion by a solute ligand—is strongly pH dependent (Figure 9.12a). Complex formation is competitive (e.g., metal ion versus H^+ ion or versus another metal ion). Figure 9.12b illustrates the sorption of various metal ions on hydrous ferric oxide. For each

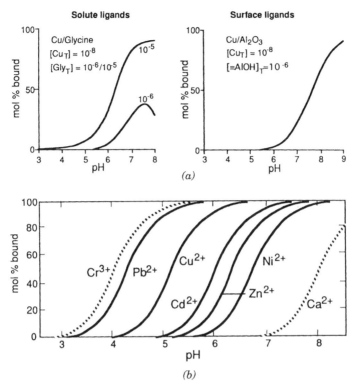

Figure 9.12. pH dependence of the binding of metal ions by solute and surface ligands. (a) Comparison of the complexation of Cu^{2+} by dissolved ligand (glycine) and by surface OH groups of Al_2O_3 as a function of pH. (The curves are calculated on the basis of experimentally determined equilibrium constants.) (b) Extent of surface complex formation on hydrous ferric oxide as a function of pH (measured as mol % of the metal ions in the system adsorbed or surface bound). [TOTFe] = 10^{-3} M (2×10^{-4} mol reactive sites per liter); metal concentration in solution = 5×10^{-7} M; $I = 0.1$ M $NaNO_3$. (The curves are based on data compiled by Dzombak and Morel, 1990.)

metal ion there is a narrow interval of 1–2 pH units where the extent of sorption rises from zero to almost 100%.[†]

Ligand Exchange; Surface Complex Formation of Anions and Weak Acids

The main mechanism of ligand adsorption is ligand exchange; the surface hydroxyl is exchanged by another ligand. This surface complex formation is

---

[†]*Hydrolysis and adsorption.* Some years ago, a theory was advanced that hydrolyzed metal species, rather than free metal ions, are adsorbed to hydrous oxides. Spectroscopic information is in accord with the reaction of (free) metal ions with the surface. There is, however, the possibility, especially with trivalent ions and within given pH ranges, that surface hydroxo species can be formed.

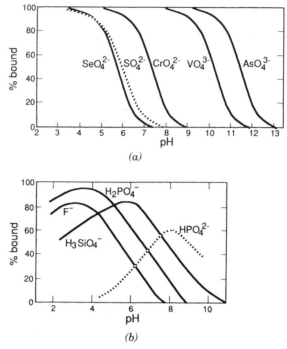

Figure 9.13. Surface complex formation with ligands (anions) as a function of pH. (a) Binding of anions from dilute solutions (5×10^{-7} M) to hydrous ferric oxide [TOTFe = 10^{-3} M]. Based on data from Dzombak and Morel (1990). $I = 0.1$. (b) Binding of phosphate, silicate, and fluoride on goethite (α-FeOOH); the species shown are surface species. (6 g FeOOH per liter, $P_T = 10^{-3}$ M, $Si_T = 8 \times 10^{-4}$ M.) The curves are calculated with the help of experimentally determined equilibrium constants (Sigg and Stumm, 1981).

also competitive; OH^- ions and other ligands compete for the Lewis acid of the central ion[†] of the hydrous oxide [e.g., the Al(III) or the Fe(III) in aluminum or ferric (hydr)oxides]. The extent of surface complex formation (adsorption), as with metal ions, is strongly dependent on pH. Since the adsorption of anions is coupled with a release of OH^- ions, adsorption is favored by lower pH values (Figure 9.13); for example,

$$\equiv AlOH + F^- \rightleftharpoons \equiv AlF + OH^- \tag{27}$$

With bidentate ligands (mononuclear or binuclear), surface chelates are formed.

[†]A Lewis acid site is a surface site capable of receiving a pair of electrons from the adsorbate. A Lewis base is a site having a free pair of electrons (like an oxygen donor atom in a surface OH group) that can be transferred to the adsorbate.

For example, with oxalate

$$\equiv\text{FeOH} + \text{HC}_2\text{O}_4^- \rightleftharpoons \equiv\text{Fe}\begin{pmatrix}\text{O}-\text{C}\diagup^{\text{O}^-}\\|\\ \text{O}-\text{C}_{\diagdown\text{O}}\end{pmatrix} + \text{H}_2\text{O} \quad (28)$$

(oxalate)

or with phosphate

$$\begin{matrix}\equiv\text{FeOH}\\|\\ \equiv\text{FeOH}\end{matrix} + \text{H}_2\text{PO}_4^- \rightleftharpoons \begin{matrix}\equiv\text{Fe}-\text{O}\diagdown_{}\diagup^{\text{O}^-}\\ \text{P}\\ \equiv\text{Fe}-\text{O}\diagup^{}\diagdown_{\text{O}}\end{matrix} + 2\,\text{H}_2\text{O} \quad (29)^\dagger$$

The extent of surface coordination and its pH dependence can again be explained by considering the affinity of the surface sites for metal ion or ligand and the pH dependence of the activity of surface sites and ligands.

Ternary Complexes

Since the coordination sphere of a metal complex on the surface of a hydrous oxide is only partially occupied by the surface ligands, further ligands may be acquired to form a ternary complex (type A) (Schindler, 1990):

$$\text{S}-\text{OH} + \text{Me}^{2+} + \text{L}^- \rightleftharpoons \text{S}-\text{OMe}-\text{L} + \text{H}^+ \quad (30)$$

A (type B) ternary complex can also be formed when it is a polydentate ligand:

$$\text{S}-\text{OH} + \text{L}^- + \text{Me}^{2+} \rightleftharpoons \text{S}-\text{L}-\text{Me}^{2+} + \text{OH}^- \quad (31)$$

Example 9.3. pH Dependence of Surface Complex Formation In order to exemplify simple complex formation equilibria, we calculate the pH dependence of the binding of (a) a metal ion, Me^{2+}, and (b) a ligand, A^-, to a hydrous oxide. The metal oxide, S—OH, is characterized by two surface acidity constants $pK_1^s = 4$ and $pK_2^s = 9$. Its specific surface area is 10 m² g⁻¹. The surface contains 10^{-4} mol surface sites per gram (~6 sites nm⁻²) and we use 1 g liter⁻¹. Thus there are 10^{-4} mol surface sites per liter solution. The ligand A^- is characterized by a pK value of its conjugate acid of $pK_{\text{HA}} = 5$. The total concentration of the adsorbates are (a) TOTMe = 10^{-7} M and (b) TOTHA = 10^{-7} M. We first make the calculation without correction for electrostatic interaction [we will return in Section 9.6 (Example 9.6; see also Figures 9.15 and 9.17) to the same problem and repeat the calculations using charge cor-

---

†See footnote to equation 25b.

rections]. We use the following equilibrium constants:

$$S - OH_2^+ \rightleftharpoons S - OH + H^+ \qquad \log K_1^s = -4 \qquad \text{(i)}$$

$$S - OH \rightleftharpoons S - O^- + H^+ \qquad \log K_2^s = -9 \qquad \text{(ii)}$$

$$S - OH + Me^{2+} \rightleftharpoons S - O Me^+ + H^+ \qquad \log K_M^s = -1 \qquad \text{(iii)}$$

$$S - OH + HA \rightleftharpoons S - A + H_2O \qquad \log K_L^s = 5 \qquad \text{(iv)}$$

$$HA = H^+ + A^- \qquad \log K_{HA} = -5 \qquad \text{(v)}$$

For (a) the following total concentrations can be defined:

$$\text{TOTSOH} = [SOH_2^+] + [SOH] + [SO^-] + [SOMe^+] = 10^{-4} \text{ M} \qquad \text{(via)}$$

$$\text{TOTMe} = [Me^{2+}] + [SOMe^+] = 10^{-7} \text{ M} \qquad \text{(viia)}$$

For (b),

$$\text{TOTSOH} = [SOH_2^+] + [SOH] + [SO^-] + [SA] = 10^{-4} \text{ M} \qquad \text{(vib)}$$

$$\text{TOTHA} = [HA] + [A^-] + [SA] = 10^{-7} \text{ M} \qquad \text{(viib)}$$

This equilibrium information is also summarized in Tableau 9.1.
This example is simple enough to be calculated by hand.
There are various ways to approach solving of the problem. One way is to use a trial and error approach expressing equations via and viia as functions of $[H^+]/[Me^{2+}]$ and [SOH]:

$$[\text{TOTSOH}] = 10^{-4} \text{ M} = [SOH] \left(\frac{[H^+]}{K_1^s} + 1 + \frac{K_2^s}{[H^+]} + \frac{K_M^s [Me^{2+}]}{[H^+]} \right)$$

(viiia)

$$[\text{TOTMe}] = 10^{-7} \text{ M} = [Me^{2+}] \left(1 + \frac{K_M^s}{[H^+]} [SOH] \right) \qquad \text{(ixa)}$$

These two equations contain (for any preselected $[H^+]$) two unknowns ($[Me^{2+}]$ and [SOH]) and can readily be solved simultaneously by trial and error (systematic variations of assumed values of [SOH] and $[Me^{2+}]$) until the left- and right-hand side of these equations are equal; since [SOH] $\gg [Me^{2+}]$, one may start by assuming [SOH] $\approx 10^{-4}$ M.

Results are given in Figure 9.14. Of course, such calculations are more conveniently carried out with a computer program. The solution techniques are described by Dzombak and Morel (1990). A program including adsorption

9.4 Hydrous Oxide Surfaces

Tableau 9.1. Surface Complex Formation of $\equiv$S—OH

Components		$\equiv$S—OH	Me^{2+}	$f(\psi)^a$	H^+	log K
		(a) With Me^{2+}				
Species	$\equiv$S—OH	1	0	0	0	0
	$\equiv$S—OH$_2^+$	1	0	1	1	4
	$\equiv$S—O$^-$	1	0	-1	-1	-9
	$\equiv$S—OMe$^+$	1	1	1	-1	-1
	Me^{2+}	0	1	0	0	0
	H^+	0	0	0	1	0
		10^{-4} M	10^{-7} M		pH given	
Components		$\equiv$S—OH	HA	$f(\psi)^a$	H^+	log K
		(b) With Ligand A^-				
Species	$\equiv$S—OH	1	0	0	0	0
	$\equiv$S—OH$_2^+$	1	0	1	1	4
	$\equiv$S—O$^-$	1	0	-1	-1	-9
	$\equiv$S—A	1	1	0	0	5
	HA	0	1	0	0	0
	A^-	0	1	0	-1	-5
	H^+	0	0	0	1	0
		10^{-4} M	10^{-7} M		pH given	

^{a}Relates to the surface charge of the species $f(\psi) = e^{-F\psi/RT}$, where ψ is the surface potential. In a first approximation (Figures 9.14 and 9.16) we assume that surface charge is negligible. The effect of surface charge will be dealt with in Example 9.6. The results are given in Figures 9.15 and 9.17.

on charged surfaces, developed by Westall (1979) (MICROQL-II), has been adapted to personal computers, e.g., MacμQL by B. Müller, Eawag, 1993.

The same approach can be used to calculate adsorption equilibrium of the ligand A^-. The two equations, expressing total concentrations, are

$$[\text{TOTSOH}] = 10^{-4} \text{ M} = [\text{SOH}] \left(\frac{[H^+]}{K_1^s} + 1 + \frac{K_2^s}{[H^+]} + K_L^s [\text{HA}] \right) \quad \text{(viiib)}$$

$$[\text{TOTHA}] = 10^{-7} \text{ M} = [\text{HA}] \left(1 + \frac{K_{\text{HA}}}{[H^+]} + K_L^s [\text{SOH}] \right) \quad \text{(ixb)}$$

The results are displayed in Figures 9.16. Note that so far no corrections have been made for electrostatic effects. The influence of electrostatic correc-

548 The Solid–Solution Interface

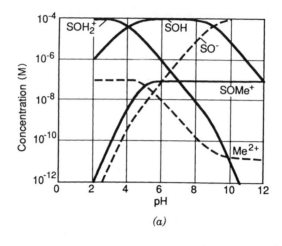

(a)

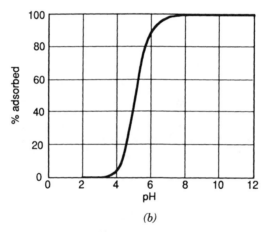

(b)

Figure 9.14. *Metal binding* by a hydrous oxide from a 10^{-7} M solution (SOH + Me^{2+} ⇌ SOMe$^+$ + H$^+$) for a set of selected equilibrium constants (see Example 9.3, equations i–iii) and concentration conditions (see text). No corrections have been made for electrostatic interactions (cf. Figure 9.15). The pH edge reflects a narrow pH range in which the metal ion is adsorbed. Hydrolysis of Me^{2+} is neglected.

tions, depicted in Figures 9.15 and 9.17, will be described in detail in Example 9.6.

Figures 9.14 and 9.16 exemplify the typical pH dependence for cation and ligand adsorption. As Figure 9.16 illustrates, the adsorption of the ligand A (or HA) goes through a maximum at a pH value that is near the pK value of HA. The pH dependence of this maximum is a consequence of the mass laws involved in the surface and solution relations.

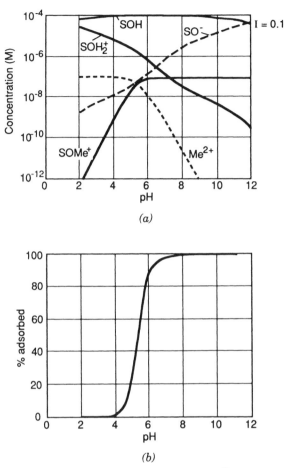

Figure 9.15. *Metal binding* by a hydrous oxide from a 10^{-7} M solution (SOH + Me^{2+} ⇌ SOMe$^+$ + H$^+$) for the same conditions as in Figure 9.14 but corrected for electrostatic interaction by the diffuse double-layer model (Gouy–Chapman) for $I = 0.1$ (see Example 9.6).

9.5 SURFACE CHARGE AND THE ELECTRIC DOUBLE LAYER

Acquiring Surface Charge

Figure 9.9 showed that many oxides, carbonates, and silicates encountered in waters, sediments, and soils have a surface charge and that this charge may be strongly affected by pH. The pH$_{pznpc}$ defines the pH at which the H$^+$ ion-dependent surface charge is zero. Figure 9.18 illustrates that many organic and biotic particles, at the pH range of natural waters, are negatively charged.

Solid particle surfaces can develop electric charge in three principal ways:

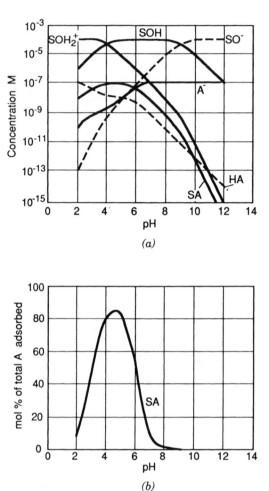

Figure 9.16. *Ligand binding* by a hydrous oxide from a 10^{-7} M solution (SOH + HA $\rightleftharpoons$ SA + H$_2$O) for a set of selected equilibrium constants (see equations i, ii, iv, and v) and concentration conditions (see text). No corrections have been made for electrostatic interactions (cf. Figure 9.17). Part (b) illustrates that a maximum of adsorption of the ligand occurs near the pK value of its conjugate acid.

1. The charge may arise from *chemical reactions* at the surface. Many solid surfaces contain ionizable functional groups: —OH, —COOH, —OPO$_3$H$_2$, and —SH. The charge of these particles becomes dependent on the degree of ionization (proton transfer) and consequently on the pH of the medium. As we have seen, the electric charge of a hydrous oxide can be explained by the acid–base behavior of the surface hydroxyl groups S—OH:

$$\text{S—OH}_2^+ \underset{}{\overset{K_1^s}{\rightleftharpoons}} \text{S—OH} \underset{}{\overset{K_2^s}{\rightleftharpoons}} \text{S—O}^-$$

9.5 Surface Charge and the Electric Double Layer

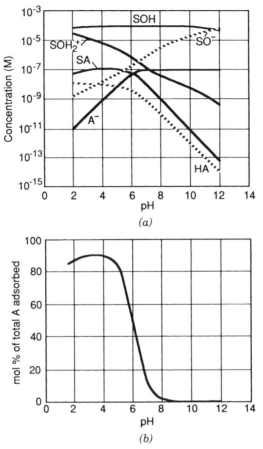

Figure 9.17. *Ligand binding* by a hydrous oxide from a 10^{-7} M solution; same conditions as in Figure 9.16 but corrected for electrostatic interactions by the diffuse double-layer model (Gouy–Chapman) for $I = 0.1$ (see Example 9.6).

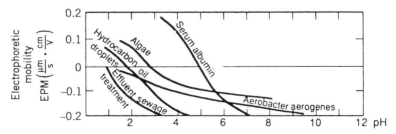

Figure 9.18. Effect of pH on charge and electrophoretic mobility. In the neutral pH range, most organic or biogenic suspended solids typically encountered in natural waters are negatively charged. These simplified curves are based on results by different investigators whose experimental procedures are not comparable and may depend on solution variables other than pH. The curves are meant to exemplify trends and are meaningful in a semiquantitative way only.

Most oxides and hydroxides exhibit such amphoteric behavior; thus the charge is strongly pH dependent, being positive at low pH values. Similarly, for an organic surface, for example, that of a bacterium or of biological debris, one may visualize the charge as resulting from protolysis of functional amino and carboxyl groups, for example,

$$R\begin{matrix}COOH\\NH_3^+\end{matrix} \xrightleftharpoons{K_1} R\begin{matrix}COO^-\\NH_3^+\end{matrix} \xrightleftharpoons{K_2} R\begin{matrix}COO^-\\NH_2\end{matrix}$$

At low pH a positively charged surface prevails; at high pH, a negatively charged surface is formed. At some intermediate pH the net surface charge will be zero.

Charge can also originate by processes in which solutes become coordinatively bound to solid surfaces, for example,

$$\equiv Fe-OH + Cu^{2+} \rightleftharpoons \equiv FeOCu^+ + H^+$$

$$\equiv Fe-OH + HPO_4^{2-} \rightleftharpoons \equiv Fe-OPO_3^{2-} + H_2O$$

$$\equiv S + HS^- \rightleftharpoons \equiv S-SH^-$$

$$\equiv AgBr + Br^- \rightleftharpoons \equiv AgBr_2^-$$

$$\equiv RCOOH + Ca^{2+} \rightleftharpoons \equiv RCOOCa^+ + H^+$$

2. Surface charge at the phase boundary may be caused by lattice imperfections at the solid surface and by *isomorphous replacements* within the lattice. For example, if in any array of solid SiO_2 tetrahedra an Si atom is replaced by an Al atom (Al has one electron less than Si), a negatively charged framework is established:

$$\begin{bmatrix} HO & O & O & O & OH \\ \diagdown Si \diagup \diagdown Si \diagup \diagdown Al \diagup \diagdown Si \diagup \\ HO & O & O & O & OH \end{bmatrix}^-$$

Similarly, isomorphous replacement of the Al atom by Mg atoms in networks of aluminum oxide octahedra leads to a negatively charged lattice.[†] Clays are representative examples of atomic substitution, which causes the charge at the phase boundary. Sparingly soluble salts also carry a surface charge because of lattice imperfections.

Thus the net surface charge density of a hydrous oxide is determined by the proton transfer and reactions with other cations or anions. In general, the net

---

[†]In a similar way the formation of solid solutions, where an ion replaces an ion of different charge in the crystalline lattice, can lead to surface charge.

9.5 Surface Charge and the Electric Double Layer

surface charge density of a hydrous oxide is given by

$$\sigma_P = F[\Gamma_H - \Gamma_{OH} + \Sigma(Z_M \Gamma_M) + \Sigma(Z_A \Gamma_A)] \tag{32}$$

where σ_P is the net surface charge density in C m^{-2}, F is the Faraday constant (96490 C mol^{-1}), Z is the valency of the sorbing ion, and Γ_H, Γ_{OH}, Γ_M, and Γ_A, respectively are the sorption densities (mol m^{-2}) of H$^+$, OH$^-$, metal ions, and anions.

The net proton charge (in C m^{-2}), the charge due to the binding of protons or H$^+$ ions—one also speaks of the surface protonation—is given by

$$\sigma_H = F(\Gamma_H - \Gamma_{OH})$$

The surface charges in mol kg^{-1}, Q_H and Q_{OH}, are obtained as $Q_H = \Gamma_H s$ and $Q_{OH} = \Gamma_{OH} s$, where s is the specific surface area of the solid (m^2 kg^{-1}). The net surface charge, Q_P, is experimentally accessible (by measuring cations and H$^+$ and OH$^-$ and anions that have been bound to the surface), for example, in case of adsorption of a metal, M^{2+}, or a ligand, A^{2-}:

$$Q_P = \{S-OH_2^+\} - \{S-O^-\} + \{S-OM^+\} \tag{33a}$$

$$Q_P = \{S-OH_2^+\} - \{S-O^-\} - \{SA^-\} \tag{33b}$$

where Q_P is the surface charge accumulated at the interface in mol kg^{-1}. Q_P can be converted into σ_P (C m^{-2}): $\sigma_P = Q_P F/s$, where s is the specific surface area in m^2 kg^{-1}.

Although aquatic particles bear electric charge, this charge is balanced by the charges in the diffuse swarm, which move about freely in solution while remaining near enough to colloid surfaces to create the effective (counter) charge σ_D that balances σ_P:

$$\sigma_P + \sigma_D = 0 \tag{34}$$

The following *points of zero charge* (pzc) can be distinguished:

$$\text{pzc:} \quad \sigma_P = 0 \tag{35}$$

This is often referred to as the isoelectric point. It is the condition where particles do not move in an applied electric field. If one wants to specify that the pzc is established solely due to binding of H$^+$ or OH$^-$, one may specify *point of zero net proton charge* (or *condition*) (pznpc). Furthermore we can define a *point of zero salt effect* (pzse)

$$\text{pznpc:} \quad \sigma_H = 0 \tag{36}$$

$$\text{pzse:} \quad \delta\sigma_H/\delta I = 0 \tag{37}$$

where I is ionic strength. At the pzse, the surface charge is not affected by a change in concentration of an "inert" background electrolyte.

Each *diffuse swarm ion* contributes to σ_D. The effective surface charge of an individual ion i can be apportioned according to

$$\sigma_{D_i} = \frac{Z_i}{m_s} \int_V [c_i(x) - c_{0,i}] dV \qquad (38)$$

where Z_i is the valence of the ion, $c_i(x)$ is its concentration at point x in the solution, and $c_{0,i}$ is its concentration in the solution far enough from any particle surface to avoid adsorption in the diffuse ion swarm (Sposito, 1989). The integral in equation 38 is over the entire volume V of aqueous solution contacting the mass m_s of solid adsorbent. Thus this equation represents the excess charge of ion i in aqueous solution: if $c_i(x) = c_{0,i}$ uniformly, there could be no contribution of ion i to σ_D. Note that equation 38 applies to all ions in the solution including H^+ and OH^- and that σ_D is the sum of all σ_{D_i}.

3. A surface charge may be established by *adsorption* of a *hydrophobic species* or a surfactant ion. Preferential adsorption of a "surface-active" ion can arise from so-called hydrophobic bonding or from bonding via hydrogen bonds or from London–van der Waals interactions. The mechanism of sorption of some ions (e.g., fulvates or humates) is not certain. Ionic species carrying a hydrophobic moiety may bind inner-spherically or outer-spherically depending on whether the surface-coordinative or the hydrophobic interaction prevails.

The Net Total Particle Charge; Surface Potential

Thus different types of surface charge density contribute to the *net total particle charge* on a colloid, denoted σ_P.

$$\sigma_P = \sigma_0 + \sigma_H + \sigma_{IS} + \sigma_{OS} \qquad (39)$$

where σ_P = total net surface charge.
σ_0 = permanent structural charge (usually for a mineral) caused by isomorphic substitutions (or the formation of solid solutions) in minerals; significant charge is produced primarily in the 2:1 phyllosilicates.
σ_H = net proton charge, that is, the charge due to the binding of protons or the binding of OH^- ions (equivalent to the dissociation of H^+). Protons in the diffuse layer are not included in σ_H.
σ_{IS} = inner-sphere complex charge.
σ_{OS} = outer-sphere complex charge.[†]

[†]The sum of $\sigma_{IS} + \sigma_{OS}$ is sometimes referred to as the Stern layer surface charge density $\sigma_S = \sigma_{IS} + \sigma_{OS}$. The sum of σ_0 and σ_H is often referred to as intrinsic surface charge density $\sigma_{in} = \sigma_0 + \sigma_H$. Thus $\sigma_P = \sigma_{in} + \sigma_S$.

9.5 Surface Charge and the Electric Double Layer

The unit of σ is usually C m^{-2} (1 mol of charge units equals 1 faraday or 96485 coulombs).

As we have seen, the electric state of a surface depends on the spatial distribution of free (electronic or ionic) charges in its neighborhood. The distribution is usually idealized as an *electric double layer*; one layer is envisaged as a fixed charge or surface charge attached to the particle or solid surface while the other is distributed more or less diffusively in the liquid in contact (Gouy–Chapman diffuse layer model, Figure 9.19). A balance between electrostatic and thermal forces is attained.

According to the Gouy–Chapman theory, the surface charge density σ_P (C m^{-2}) is related to the potential at the surface ψ_0 (volt) (Equation vi in Figure 9.19):

$$\sigma_P = (8\, RT\varepsilon\varepsilon_0 c \times 10^3)^{1/2} \sinh(Z\psi_0 F/2\, RT) \tag{40a}$$

where R is the molar gas constant (8.314 J mol^{-1} K^{-1}), T the absolute temperature (I), ε the relative dielectric constant of water ($\varepsilon = 78.5$ at 25°C), ε_0 the permittivity of free space (8.854 × 10^{-12} C V^{-1} m^{-1} or 8.854 × 10^{-12} C^2 J^{-1} m^{-1}), and c the molar electrolyte concentration (M). Equation 40a is valid for a symmetrical electrolyte (Z = ionic charge). At low potential equation 40a can be linearized as

$$\sigma_P = \varepsilon\varepsilon_0\, \kappa\psi_0 \tag{40b}$$

See Appendix 9.1.

The Debye parameter, κ (compare equation iii in Figure 9.19), is defined by

$$\kappa = \left(\frac{2\, F^2\, I \times 10^3}{\varepsilon\varepsilon_0\, RT}\right)^{1/2} \tag{40c}\dagger$$

where I is the ionic strength (M). The double-layer thickness, κ^{-1} (in meters), is the reciprocal of the Debye parameter.

Equations 40a and 40b can be written for 25°C, where ε, the dielectric constant of water, is 78.5, as

$$\sigma_P = 0.1174\, c^{1/2} \sinh(Z\psi_0 \times 19.46) \tag{40d}$$

and

$$\sigma_P \simeq 2.3\, I^{1/2}\, \psi_0 \tag{40e}$$

where c is the electrolyte concentration (M) and σ_P has the units C m^{-2}.

---

†The simplified equation $\kappa = 3.29 \times 10^9\, I^{1/2}$ (m^{-1}), valid for 25°C, is useful. Thus κ^{-1} for a 10^{-3} M NaCl solution is ~10 nm; for seawater $\kappa^{-1} \simeq 0.4$ nm.

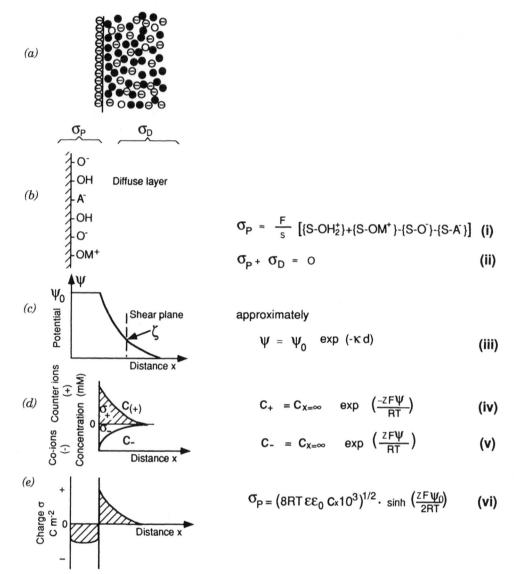

Figure 9.19. The diffuse double layer. (a) Diffuseness results from thermal motion in solution. (b) Schematic representation of ion binding on an oxide surface on the basis of the surface complexation model. s is the specific surface area (m^2 kg^{-1}). Braces refer to concentrations in mol kg^{-1}. (c) The electric surface potential, ψ, falls off (simplified model) with distance from the surface. The decrease with distance is exponential when $\psi < 25$ mV. At a distance κ^{-1} the potential has dropped by a factor of $1/e$. This distance can be used as a measure of the extension (thickness) of the double layer (see equation 40c). At the plane of shear (moving particle) a zeta potential can be established with the help of electrophoretic mobility measurements. (d) Variation of charge distribution (concentration of positive and negative ions) with distance from the surface (Z is the charge of the ion). (e) The net excess charge.

9.5 Surface Charge and the Electric Double Layer

The Stern Layer The Gouy–Chapman treatment runs into difficulties at small κx values when the surface potential is high. The local concentrations of ions near the surface (Equations 3 and 4 in Table A9.1 in Appendix 9.1) become far too large; this is because of the assumption of point charges and neglect of ionic diameter.

Stern (1923) suggested that the interface be divided into two regions, the first a compact layer of ions adsorbed at the surface and the second being the diffuse layer. The compact region could include ions with a specific adsorption potential. The compact region was visualized as having a site density for ions [in the spirit of the Langmuir isotherm (1917)]. Stern had the mercury–water interface in mind. Grahame (1947) elaborated on the Stern picture of the interface by imagining two parts of the compact region: an inner layer of specifically adsorbed ions (e.g., Cl^- on Hg) and an outer layer of fully solvated ions (e.g., F^- on Hg). These two layers are known as the inner Helmholtz plane (IHP) of ions and the outer Helmholtz plane (OHP) of ions, respectively. The oxide–water interface (see Figure 9.10) requires its own model of the interface structure in order to describe electrostatic energies of adsorption. Westall and Hohl (1980) summarized several proposed models of the oxide–water interface, ranging from the simple Helmholtz model to the rather complex Stern–Grahame models, for example, the "triple layer model." The central issue is how to relate *chemical* concepts of the oxide–water interface (inner sphere versus outer sphere complex species) to mean-field electrostatic concepts (charge density and electric potential). Figure 9.10 shows specifically bound species (e.g., protons and phosphate ions) at the surface (inner sphere species) and nonspecifically adsorbed ions (chloride and sodium ions) separated from the surface by one or more water molecules, with the diffuse layer extending outward from the outer sphere ions.

As illustrated in Figure 9.19b, a simple model of the chemistry and electrostatics of the oxide–water interface is obtained by including *all* surface species as part of σ_P; that is, the particle charge includes structural charge, proton charge, and inner- and outer-sphere charge (equation 39). This model requires one surface potential, $\psi_0 = \psi_D$, and one charge density, $\sigma_P = -\sigma_D$. The chemical inputs are total surface site density and the mass law constants for specifically bound ions. This "diffuse layer model" (which is actually a surface complex formation plus diffuse layer model) is employed by Dzombak and Morel (1990) in fitting adsorption data for hydrous ferric oxide. The more elaborate chemical–electrostatic models are obtained if we translate the picture in Figure 9.10b into a greater variety of mean planes of charge. For example, a simple Stern model would place specifically bound ions at the surface and outer-sphere complex ions at the β planes of Figure 9.10b, with the region between these two planes described by a Stern layer capacitance. In this book we rely solely on the surface complex formation/diffuse layer model (SCF/DLM) or the constant capacitance model (CCM) to understand the adsorptive properties of hydrous oxides.

In the surface complex formation model, the amount of surface charge that can be developed on an oxide surface is restricted by the number of surface sites. (This limitation is inherently not a part of the Gouy–Chapman theory.)

The Triple-Layer Model (TLM) This model, first developed by Yates et al. (1974) and Davis (1978), is essentially an expanded Stern–Grahame model: the specifically adsorbed ions are placed as partially solvated ions at a plane of closest approach. Additional capacitances are introduced. Subsequently, Hayes (1987) had to modify the earlier TLM by moving specifically adsorbed ions to the mean plane of the surface.

Zeta Potential

The electrokinetic potential (zeta potential, ζ) is the potential drop across the mobile part of the double layer (Figure 9.19c) that is responsible for electrokinetic phenomena, for example, electrophoresis (motion of colloidal particles in an electric field). It is assumed that the liquid adhering to the solid (particle) surface and the mobile liquid are separated by a shear plane (slipping plane). The electrokinetic charge is the charge on the shear plane.

The surface potential is not accessible by direct experimental measurement; it can be calculated from the experimentally determined surface charge (equations 32–34) by equations 40a and 40b. The *zeta potential*, ζ, calculated from *electrophoretic* measurements, is typically lower than the surface potential, ψ_0, calculated from diffuse double-layer theory. The zeta potential reflects the potential difference between the plane of shear and the bulk phase. The distance between the surface and the shear plane cannot be defined rigorously.

Electrophoresis refers to the movement of charged particles relative to a stationary solution in an applied potential gradient, whereas in *electroosmosis* the migration of solvent with respect to a stationary charged surface is caused by an imposed electric field. The *streaming potential* is the opposite of electroosmosis and arises from an imposed movement of solvent through capillaries; conversely, a *sedimentation potential* arises from an imposed movement of charged particles through a solution. Operationally the zeta potential can be computed from electrophoretic mobility and other electrokinetic measurements. For example, for nonconducting particles whose radii are large when compared with their double-layer thicknesses, the zeta potential, ζ, is related to the electrophoretic mobility, m_e (velocity per unit of electric field).[†] Many corrections that are difficult to evaluate must be considered in the computation of ζ. The measurement of electrophoretic mobility is treated by Hunter (1989); the measurement in natural waters is discussed by Neihof and Loeb (1972).

Example 9.4. Zn(II) Adsorbed on Hydrous Iron(III) Oxide A hydrous iron(III) oxide suspension (10^{-3} mol liter^{-1}) at pH = 6.0 has adsorbed 20% of Zn^{2+} from a solution that contained incipiently 10^{-4} mol liter^{-1} of Zn^{2+}

[†]The relationship between electrophoretic mobility and zeta potential depends on the model of the electrochemical double-layer structure used.

9.5 Surface Charge and the Electric Double Layer

and an inert electrolyte ($I = 10^{-2}$ M) (25°C). The hydrous iron oxide has been characterized to have a specific surface of 600 m² g⁻¹ and 0.2 mol of active sites per mole of Fe(OH)₃. From an alkalimetric–acidimetric titration curve, we know that at pH = 6 the 10^{-3} M Fe(OH)₃ suspension contains $[\equiv\text{Fe–OH}_2^+] - [\equiv\text{Fe–O}^-] = 3 \times 10^{-5}$ M charge units.

Calculate (a) the surface charge σ (C m⁻²) (b) the surface potential, and (c) the "thickness" of the double layer, κ^{-1}.

(a) The surface speciation in mol liter⁻¹ is given by $[\equiv\text{Fe–OZn}^+] = 2 \times 10^{-5}$ M and by $([\equiv\text{Fe–OH}_2^+] - [\equiv\text{Fe–O}^-]) = 3 \times 10^{-5}$. (Since the concentration of active sites in the suspension is $[\equiv\text{Fe–OH}_{\text{tot}}] = 2 \times 10^{-4}$ M and $[\equiv\text{FeOH}_2^+] > [\equiv\text{FeO}^-]$, the molar concentration of $[\equiv\text{Fe–OH}] \simeq 1.5 \times 10^{-4}$ M.) Thus the hydrous ferric oxide suspension carries on its surface a total of 5×10^{-5} mol of positive charge units per liter (cf. equation 33a); that is, ~25% of the total sites are positively charged and nearly 75% are uncharged. On a per surface area basis (m²), we get

$$\sigma_P = \frac{5 \times 10^{-5} \text{ mol charge unit liter}^{-1}}{10^{-3} \text{ mol Fe(OH)}_3\text{(s) liter}^{-1}} \frac{1 \text{ g Fe(OH)}_3}{600 \text{ m}^2} \frac{\text{mol Fe(OH)}_3}{107 \text{ g}} \frac{96485 \text{ C}}{\text{mol}}$$

$$\sigma_P = 7.5 \times 10^{-2} \text{ C m}^{-2} \quad \text{or} \quad 7.5 \ \mu\text{C cm}^{-2} \tag{i}$$

(b) The surface potential is obtained from equation 40d:

$$\sinh(\psi_0 \times 19.46) = \frac{7.5 \times 10^{-2}}{0.1174 \times 0.1} = 6.4$$

$$\psi_0 \times 19.46 = 2.554$$

$$\psi_0 = \frac{2.554}{19.46} = 0.13 \text{ V}$$

(c) κ can be calculated according to equation 40c for $I = 10^{-2}$ M:

$$\kappa = \left(\frac{2 \times (96485)^2 \ \frac{\text{C}^2}{\text{mol}^2} \times 10^{-2} \times 10^3 \text{ mol}}{8.854 \times 10^{-12} \ \frac{\text{C}}{\text{mJ}} \times 78.5 \times 2.478 \times 10^3 \ \frac{\text{J}}{\text{mol}}} \right)^{1/2}$$

$$= 3.29 \times 10^8 \text{ m}^{-1}$$

Thus the thickness of the double layer, as characterized by κ^{-1}, is 3 nm.

The Relation Between pH, Surface Charge, and Surface Potential

Dzombak and Morel (1990) have illustratively and compactly summarized (Figure 9.20) the interdependence of the Coulombic interaction energy with pH

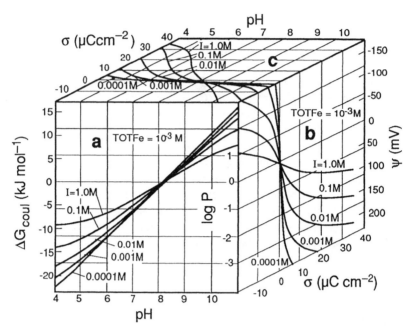

Figure 9.20. Relationship between pH, surface potential, ψ (or Coulombic term, log P, or Coulombic free energy, ΔG_{coul}), and surface charge density, σ (or surface protonation), for various ionic strengths of a 1:1 electrolyte for a hydrous ferric oxide surface [$P = \exp(-F\psi/RT)$]. (a) Dependence of the Coulombic term and surface potential on solution pH; note the near-Nernstian behavior at low ionic strength. (b) ψ versus σ; these curves correspond to the Gouy–Chapman theory. (c) σ versus pH; these are the curves obtained experimentally. (From Dzombak and Morel, 1990.)

and surface charge density at various ionic strengths for hydrous ferric oxide suspensions in which H^+ is the only potential-determining ion. In explaining this figure, we follow largely their explanation. The influence of pH and ionic strength on the Coulombic interaction energy and on the Coulombic correction factor [$\exp(-F\psi_0/RT)$] is calculated according to the diffuse double-layer model. The only experimental basis for the relationships depicted are the pH versus σ curves (panel c), as calculated from the surface protonation as measured from alkalimetric and/or acidimetric titration. The surface potential, ψ, cannot be measured. The ψ versus σ graphs (panel b) are obtained strictly from Gouy–Chapman theory. This relationship is predicted to be linear at low potentials and exponential at higher potentials. Greater surface charges are developed at higher ionic strengths and greater surface potentials at lower ionic strengths. The Coulombic factor $P = [= \exp(-F\psi/RT)]$ varies by approximately seven orders of magnitude between pH $= 4$ and pH $= 7$. The Coulombic effect can be expressed in conventional energy units (kJ mol^{-1}) ($\Delta G = -F\psi$). As shown, Coulombic effects can contribute up to ~ 20 kJ mol^{-1} (corresponding to 200 mV) to surface reactions.

9.5 Surface Charge and the Electric Double Layer

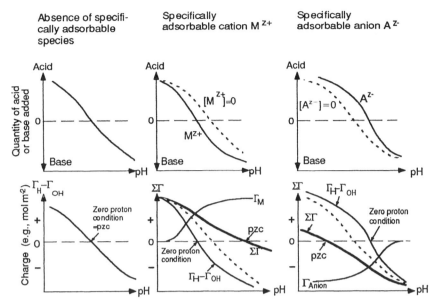

Figure 9.21. The net charge at the hydrous oxide surface is established by the proton balance (adsorption of H$^+$ or OH$^-$ and their complexes) at the interface *and* specifically bound cations or anions. This charge can be determined from an alkalimetric–acidimetric titration curve and from a measurement of the extent of adsorption of specifically adsorbed ions. Specifically adsorbed cations (anions) increase (decrease) the pH of the point of zero charge (pzc) or the isoelectric point but lower (raise) the pH of the zero net proton condition (pznpc). Addition of a ligand, at constant pH, increases surface protonation while the addition of a metal ion (i.e., specifically adsorbed) lowers surface protonation. (Adapted from Hohl et al., 1980.)

Effect of Metals and Ligands on Surface Charge

As we have seen, the net surface charge of a hydrous oxide surface is established by proton transfer reactions and the surface complexation (specific sorption) of metal ions and ligands. As Figure 9.21 illustrates, the titration curve for a hydrous oxide dispersion in the presence of a coordinatable cation is shifted toward lower pH values (because protons are released as a consequence of metal ion binding, S—OH + Me^{2+} $\rightleftharpoons$ SOMe$^+$ + H$^+$) in such a way as to lower the pH of the zero proton condition at the surface.

At this point (pH$_{pznpc}$) the portion of the charge due to H$^+$ and OH$^-$ or their complexes[†] becomes zero. Because of the binding of M^{z+} to the surface ($\Gamma_{M^{z+}}$), the fixed surface charge increases or becomes less negative[‡] and, at the

---

[†] If a hydrolyzed metal ion is adsorbed, its OH$^-$ will be included in the proton balance; similarly, in case of adsorption of protonated anions, their H$^+$ will be included in the proton balance.

[‡] Some colloid chemists often place these specifically bound cations and anions in the Stern layer. From a coordination chemistry point of view, it does not appear very meaningful to assign a surface-coordinating ion to a layer different from H or OH in a ≡MeOH group.

pH where the fixed surface charge becomes zero, the point of zero charge (pzc) is shifted to higher pH values. Correspondingly, specifically adsorbable anions increase the pH of the zero proton condition but lower the pH of the pzc (Figure 9.21).

A Simplified Double-Layer Model (Constant Capacitance)

In Example 9.2, we saw (Figure 9.7c) that there was a linear relationship between the oxide surface charge and the electrostatic contribution to proton binding, that is, ΔpK_{app}, implying a *constant capacitance* with changing surface potential. Under certain conditions of surface charge (potential) and ionic strength, we expect the relationship between surface charge density and surface potential to take a particularly simple form,

$$\sigma_0 = C \Psi_0 \tag{41a}$$

where C is the constant, integral capacitance of the oxide–aqueous electrolyte interface usually expressed as Farad m^{-2}. According to the simplest Stern picture of the interface, the integral capacitance has two components: a Gouy (diffuse layer) component, C_G, and the Helmholtz (compact layer) component, C_H, related by

$$C^{-1} = C_G^{-1} + C_H^{-1} \tag{41b}$$

The Helmholtz capacitance may be described by $C_H = \varepsilon' \varepsilon_0/d$, where ε' is the local value of the dielectric coefficient and d is the Stern layer thickness. Under conditions where ε' is independent of surface charge density, we can identify two domains of nearly constant capacitance under varying surface charge densities (i.e., pH). When $C_G \gg C_H$ (at high ionic strength), $C \simeq C_H$; when $C_G \ll C_H$ (lower ionic strength), $C \simeq C_G$. When the surface potential is small, for example, less than 25 mV, then $C_G \simeq \varepsilon\varepsilon_0\kappa = 2.3\ I^{1/2}$ (25°C) (equation 40e, or equation 14b, Table A9.1). Thus there are conditions under which we expect the charge/potential relationship of the interface to be approximately linear. Example 9.2 is a case of constant capacitance because of high ionic strength.

Surface Charge on Carbonates, Silicates, Sulfides, and Phosphates

Although we have often used for exemplification the surfaces of hydrous oxides, the concepts given apply to all surfaces. As has been pointed out, most hydrous surfaces are characterized by functional groups that acquire charge by chemical interaction with H^+, OH^-, metal ions, and ligands. (For the moment we ignore redox reactions.)

There are various possibilities for functional groups on the surface of carbonates, sulfides, phosphates, and so on. Using a very simple approach similar

9.5 Surface Charge and the Electric Double Layer

to the one in Figure 9.6 for hydrous oxides, one could postulate surface groups for carbonates (e.g., $FeCO_3$) and sulfides (e.g., ZnS); for example, for carbonates,

$$\begin{array}{cccccccc}
\text{H} & \text{OH} & \text{H} & \text{OH} & \text{H} & \text{OH} & \text{Water} \uparrow \\
\hline
CO_3 & \text{Fe} & CO_3 & \text{Fe} & CO_3 & \text{Fe} & \textbf{Solid} \downarrow \\
\text{Fe} & CO_3 & \text{Fe} & CO_3 & \text{Fe} & CO_3 & \\
CO_3 & \text{Fe} & CO_3 & \text{Fe} & CO_3 & \text{Fe} &
\end{array}$$

(42)

Surface complex formation models for carbonates and sulfides, respectively, have been proposed by van Capellen et al. (1993) and by Rönngren et al. (1991).

Surface Charge on Silicates

In addition to the inorganic hydroxyl groups that are exposed on many mineral surfaces (metal oxides, phyllosilicates, and amorphous silicate minerals), we need to consider the particular features relating to charge on the silica surfaces of *layer silicates*.

The tetrahedral silica sheet in a layer silicate (e.g., see the schematic representation of the kaolinite structure in Figures 9.22 and 9.23c,d) is called a *siloxane surface*. Sposito (1984) has described the nature of the *siloxane ditrigonal cavity* (see Figure 9.23a–d) in the following way: "The silica plane is characterized by a distorted hexagonal (i.e., trigonal) symmetry among its constituent oxygen atoms, that is produced when the underlying tetrahedra rotate to fit their apexes to contact points on the octahedral sheet. Further accommodation of the tetrahedra to the octahedral sheet is achieved through the tilting of their bases so that the silicon–oxygen bonds are directed towards the contact points instead of laying normal to the basal plane of the mineral.

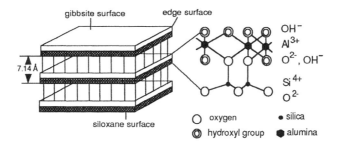

Figure 9.22. Schematic representation of the kaolinite structure. It reveals the 1 : 1 structure due to the alternation of silica-type (black) and gibbsite-type layers (white). Furthermore, the edge surface exposes aluminol and silanol groups.

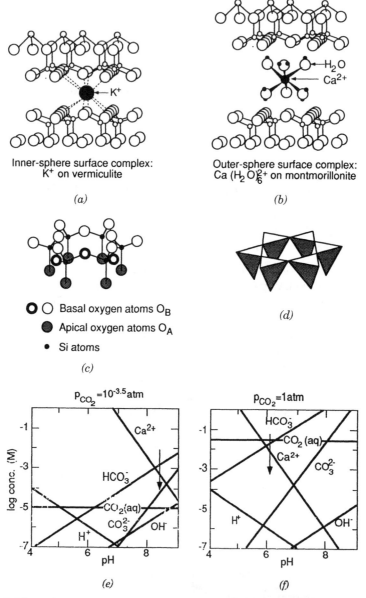

Figure 9.23 (a–d) Surface complexes between metal cations and siloxane ditrigonal cavities on 2:1 phyllosilicates, shown in exploded view: (a, b) linked SiO_4 groups in one siloxane Si_6O_6 ring (from Sposito, 1984); (c) as a "ball and spike" model; and (d) as linked tetrahedra. (e, f) Equilibrium diagrams of $CaCO_3$ solubility as a function of pH at constant partial pressure of CO_2: (e) $p_{CO_2} = 3.2 \times 10^{-4}$ atm and (f) $p_{CO_2} = 1$ atm. The equilibrium composition reflected by the charge balance $2\,[Ca^{2+}] \simeq [HCO_3^-]$ ($\log[Ca^{2+}] + 0.3 = \log[HCO_3^-]$) corresponds to the vertical arrow. The point of zero charge (arrow) is at pH = 8.2 (e) and 6.45 (f).

9.5 Surface Charge and the Electric Double Layer

As a result of this adjustment one of the basal oxygen atoms in each tetrahedron is raised about 0.02 nm above the other two and the siloxane surface becomes corrugated."

The ditrigonal cavity formed by six corner-sharing silica tetrahedra (Figure 9.23) has a diameter of 0.26 nm and is bordered by six sets of lone-pair electron orbitals emanating from the surrounding ring of oxygen atoms. These structural features—as pointed out by Sposito (1984)—qualifies the ditrigonal cavity as a soft Lewis base capable of complexing water molecules (and possibly other neutral dipolar molecules).

Two cases of isomorphic substitution can be distinguished: in the tetrahedral sheet or in the octahedral sheet (Sposito, 1984).

1. If isomorphic substitution of Si(IV) by Al(III) occurs in the tetrahedral sheet, the resulting negative charge can distribute itself over the three oxygen atoms of the tetrahedron (in which the Si has been substituted); the charge is localized and relatively strong inner-sphere surface complexes (Figure 9.23a) can be formed.
2. If, on the other hand, isomorphic substitution occurs in the octahedral sheet [substitution of Al(III) by Fe(II) or Mg(II)], the resulting negative charge distributes itself over the ten surface oxygen atoms of the four silicon tetrahedra that are associated through their apexes with a single octahedron in the layer. This distribution of negative charge enhances the Lewis base character of the ditrigonal cavity and makes it possible to form complexes with cations as well as with dipolar molecules. An outer-sphere surface complex of this type of a Ca^{2+} cation is illustrated (Sposito, 1984) in Figure 9.23b.

Successive surface protonation at the gibbsite plate and the edge surface can account for the pH-dependent surface charge of kaolinite (Wieland and Stumm, 1992).

Surface Charge of $CaCO_3$

Many processes involving carbonates—ubiquitous minerals in natural systems—are controlled by their surface properties. In particular, flotation studies on calcite have revealed the presence of a pH-variable charge and of a point of zero charge (Somasundaran and Agar, 1967). Furthermore, electrokinetic measurements have shown that Ca^{2+} is a charge (potential)-determining cation of calcite (Thompson and Pownall, 1989).

It is reasonable to assume that H^+, OH^-, HCO_3^-, and $CO_2(aq)$ are able to interact as "potential-determining" species (adsorption or desorption) with $CaCO_3(s)$ and affect its surface charge (equation 44). In a system [$CaCO_3(s)$, CO_2, H_2O] where $CaCO_3(s)$ (calcite) is equilibrated with p_{CO_2} = constant, the following equilibrium is valid:

$$CaCO_3(s) + 2\,H^+ \rightleftharpoons Ca^{2+} + CO_2(g) + H_2O \qquad K = 10^{9.7}\;(25°C,\,I = 0)^\dagger$$

The equilibrium concentration (activity) of the species H^+, OH^-, Ca^{2+}, HCO_3^-, CO_3^{2-}, and $CO_2(aq)$ are known for a given p_{CO_2}. In addition to the four equilibria (footnote), the electroneutrality (charge balance) condition

$$2\,[Ca^{2+}] + [H^+] = [HCO_3^-] + 2\,[CO_3^{2-}] + [OH^-] \qquad (43a)$$

or approximately,

$$2\,[Ca^{2+}] \simeq [HCO_3^-] \qquad (43b)$$

is needed to compute the composition. Figure 9.23e,f plots these equilibria in a double logarithmic plot. The pH of a $CaCO_3$ (calcite) suspension in equilibrium with $p_{CO_2} = 10^{-3.5}$ atm and 1 atm is (25°C, $I = 0$) pH = 8.2 and pH = 6.45, respectively.

The pH value, in accordance with equations 43a or 43b, at least in a first approximation, corresponds to *the point of zero charge*, because at this pH_{pzc} the surface species do not contribute to the overall electroneutrality of the aqueous suspension. Thus pH_{pzc} depends on p_{CO_2}; the higher p_{CO_2} the lower is the pH of the suspension and the pH_{pzc}. The pH_{pzc} of different metal carbonates decreases with decreasing solubility (Stumm and Morgan, 1970).

The addition of acid (C_A) or base (C_B) to a $CaCO_3$ system (while p_{CO_2} = constant) will change the alkalinity in solution and produce (1) a shift in the HCO_3^-, CO_3^{2-}, Ca^{2+} equilibrium (and in pH), (2) an adsorption of potential-determining ions on the $CaCO_3$ surface, and (3) a dissolution or precipitation of $CaCO_3$.

$$C_B + 2\,[Ca^{2+}] + [H^+] = [HCO_3^-] + 2\,[CO_3^{2-}] + [OH^-] + C_A$$

Upon addition of acid the charge-determining positively charged species will increase at the expense of the negatively charged charge-determining species:

$$2\,[Ca^{2+}] + [H^+] > [HCO_3^-] + 2\,[CO_3^{2-}] + [OH^-]$$

and upon addition of base

$$2\,[Ca^{2+}] + [H^+] < [HCO_3^-] + 2\,[CO_3^{2-}] + [OH^-]$$

†The equilibrium constant can be obtained by combining the following equilibria:

$$CaCO_3 = Ca^{2+} + CO_3^{2-} \qquad K_{s0} = 10^{-8.3}$$
$$CO_3^{2-} + H^+ = HCO_3^- \qquad K_2^{-1} = 10^{10.2}$$
$$HCO_3^- + H^+ = CO_2(aq) + H_2O \qquad K_1^{-1} = 10^{6.3}$$
$$CO_2(aq) = CO_2(g) \qquad K_H^{-1} = 10^{+1.5}$$
$$CaCO_3(s) + 2\,H^+ = Ca^{2+} + CO_2(g) + H_2O \qquad K_{s0}(K_1 K_2 K_H)^{-1} = 10^{9.7}$$

9.5 Surface Charge and the Electric Double Layer

Thus the surface charge of the $CaCO_3$ will, respectively, increase or decrease with addition of acid or base.

In principle, the surface charge on solid carbonates can be determined—as with hydrous oxides—from alkalimetric or acidimetric titration curves; but the procedure is more involved because in addition to the sorption of charge-determining ions (the extent of which can be assessed from alkalimetric or acidimetric titration) there is also the additional effect of dissolution or precipitation of the carbonates. It was shown in case of $MnCO_3(s)$ (rhodochrosite) and $FeCO_3(s)$ (siderite) that the precipitation or dissolution of the carbonates was slow in comparison to the adsorption of the charge-determining ions (Charlet et al., 1990) and thus a rapid titration procedure permits a surface charge determination. In the case of $CaCO_3$, precipitation and dissolution were too fast, relatively, and thus an experimental surface charge determination was not possible. But the same concept should also be valid for $CaCO_3$ and other carbonates.

We do not know in detail at the atomic scale how the surface charge development on a carbonate mineral is established, but formally and schematically one could visualize the following type of charging reactions for a hydrous surface carbonate ($MeCO_3$) such as $\equiv CO_3H^0$ or (Stipp and Hochella, 1991; van Capellen et al., 1993)

$$C \underset{O}{\overset{OH^0}{\lessgtr}} O$$

$$\equiv CO_3^- \underset{}{\overset{H^+}{\rightleftarrows}} \equiv CO_3H^0 \xrightarrow{Me^{2+}} \equiv CO_3Me^+$$

$$\equiv MeO^- \underset{}{\overset{H^+}{\rightleftarrows}} \equiv MeOH^0 \underset{}{\overset{H^+}{\rightleftarrows}} \equiv MeOH_2^+ \qquad (44)$$

$$\equiv MeCO_3^- \rightleftarrows \equiv MeHCO_3^0$$

(with HCO_3^- and H_2CO_3 indicated)

The charges assigned to the surface species in reaction 44 indicate *relative* values. Surface equilibrium constants need to be established in order to estimate the species distribution as a function of $[H^+]$ and other potential-determining species. An aqueous suspension of $CaCO_3$ crystals in the presence of CO_2 (p_{CO_2} = constant) at the point of zero charge (pH_{pzc}) at equilibrium—in line with the scheme of 44—contains in addition to $\equiv$CaOH and $\equiv$CO$_3$H an equivalent number of positively and negatively charged surface species. If the pH is adjusted, for example, with a base (at constant p_{CO_2}), the surface charge of the CaCO$_3$ becomes negative. It is thus not surprising that under these conditions CaCO$_3$ surfaces can adsorb specifically metal cations as a function of pH (Zachara et al., 1988).

9.6 CORRECTING SURFACE COMPLEX FORMATION CONSTANTS FOR SURFACE CHARGE

We return to the complex formation equilibria described in equations 3, 4, 20, 21, and 26. The equilibrium constants as given in these equations are essentially intrinsic constants valid for a (hypothetically) uncharged surface. In many cases we can use these constants as apparent constants to illustrate some of the principal features of the interdependent variables that affect adsorption. Although it is impossible to separate the chemical and electrical contributions to the total energy of interaction with a surface without extra thermodynamic assumptions, it is useful to operationally break down the interaction energy into a chemical and a Coulombic part:

$$\Delta G_{tot} = \Delta G^\circ_{chem} + \Delta G^\circ_{coul}$$

where ΔG°_{chem} is the "intrinsic" free energy term and ΔG°_{coul} is the variable electrostatic or Coulombic term:

$$\Delta G^\circ_{coul} = \Delta Z F \psi_0$$

Theoretically, this term reflects the electrostatic work in transporting ions through the interfacial potential gradient (Morel, 1983; Stumm et al., 1970) Since

$$\Delta G_{tot} = -RT \ln K$$

we can write, for example, for the proton transfer between $\equiv$FeOH and $\equiv$FeO$^-$,

$$\frac{\{\equiv FeO^-\}[H^+]}{\{\equiv FeOH\}} = K^s_{a2}(app) = K^s_{a2}(int) \exp\left(\frac{F\psi_0}{RT}\right) \quad (45a)$$

that is, the overall proton transfer is interpreted as a two-step reversible process, the proton dissociation from the $\equiv$FeOH given by K^s(int) and by the transfer of the proton from the surface to bulk solution given by $\exp(-F\psi/RT)$

Most generally, for a surface complex formation reaction,

$$K^s(app) = K^s(int) \exp\left(-\frac{\Delta Z F \psi_0}{RT}\right) \quad (45b)$$

9.6 Correcting Surface Complex Formation Constants for Surface Charge

In these equations K(int) and K(app) are the intrinsic and apparent equilibrium constants, respectively, F is the Faraday constant, ψ_0 is the surface potential, and ΔZ is the *change* in the charge of the surface species for the reaction under consideration (as written for the equilibrium reaction for which K is defined).

The intrinsic equilibrium constants are postulated to be independent of the composition of the solid phase. They remain conditional in the sense of the constant ionic medium reference state if interacting ions such as H^+ are expressed as concentrations.

There is no experimental way to measure ψ_0. (As mentioned before, the zeta potential as obtained, for example, from electrophoretic measurements is smaller than ψ.) But as discussed, we can obtain the surface charge (equation 33) and then compute the surface potential ψ on the basis of the diffuse double-layer model with equation 40a: equation 40a in simplified form for 25°C is

$$\sigma = 0.1174 \, c^{1/2} \sinh(Z\psi_0 \times 19.46) \tag{46}$$

where Z is the valence of ions in the symmetrical background electrolyte.

Example 9.5. Adsorption of Zn(II) on Hydrous Ferric Oxide From alkalimetric–acidimetric titration curves on hydrous ferric oxide, the following intrinsic acidity constants have been obtained ($I = 0.1$ M, 25°C):

$$\log K^s_{a1}(\text{int}) = -7.18$$

$$\log K^s_{a2}(\text{int}) = -8.82$$

Furthermore, the surface complex formation with Zn(II) has been determined from adsorption studies in 10^{-3} M suspensions of hydrous ferric oxide with dilute (10^{-7} M) Zn(II) solution. It can be described by the reaction

$$\equiv\text{FeOH} + \text{Zn}^{2+} \rightleftharpoons \equiv\text{FeOZn}^+ + \text{H}^+ \quad K^s_1$$

for which the intrinsic constant is

$$\log K^s_1 (\text{int}) = 0.66$$

The hydrous iron oxide has been characterized to have a specific surface area of 600 m² g⁻¹ and 0.2 mol of active sites per mole of oxide. Then the concentration of the active sites is

$$\text{TOT}\equiv\text{FeOH} = \frac{10^{-3} \text{ mol Fe(OH)}_3}{\text{liter}} \frac{0.2 \text{ mol sites}}{\text{mol Fe}} = \frac{2 \times 10^{-4} \text{ mol sites}}{\text{liter}}$$

$$\text{TOT}\equiv\text{FeOH} = [\equiv\text{FeOH}] + [\equiv\text{FeOH}_2^+] + [\equiv\text{FeO}^-] + [\equiv\text{FeOZn}^+]$$

$$= 10^{-3.7} \text{ M}$$

Calculate the binding [adsorption of Zn(II)] as a function of pH.

The Solid–Solution Interface

The species are, H^+, OH^-, Zn^{2+}, $\equiv FeOH_2^+$, $\equiv FeOH$, $\equiv FeO^-$, and $\equiv FeOZn^+$.

For a given pH, six equations are needed to calculate the speciation. They are four mass laws [in the subsequent equations $P = \exp(-F\psi/RT)$],

$$[OH^-] = [H^+]^{-1} \qquad 10^{-13.8} \qquad \text{(i)}$$

$$[\equiv FeOH_2^+] = [H^+][\equiv FeOH] \qquad P \times 10^{7.18} \qquad \text{(ii)}$$

$$[\equiv FeO^-] = [H^+]^{-1}[\equiv FeOH] \qquad P^{-1} \times 10^{-8.82} \qquad \text{(iii)}$$

$$[\equiv FeOZn^+] = [H^+]^{-1}[Zn^{2+}][\equiv FeOH]\, P \times 10^{0.66} \qquad \text{(iv)}$$

and two mole balance equations,

$$TOT\equiv FeOH = [\equiv FeOH_2^+] + [\equiv FeOH]$$
$$+ [\equiv FeO^-] + [\equiv FeOZn^+] \qquad \text{(v)}$$

$$TOTZn = [Zn^{2+}] + [\equiv FeOZn^+] \qquad \text{(vi)}$$

Furthermore, the surface charge σ is given by

$$\sigma = (F/as)\,([\equiv FeOH_2^+] - [\equiv FeO^-] + [\equiv FeOZn^+]) \qquad \text{(vii)}$$

and the charge potential relationship is

$$\sigma = 0.1174\, c^{1/2}\, \sinh(Z\psi_0 \times 19.46) \qquad \text{(viii)}$$

[As before, a is the concentration of solids (kg liter^{-1}) and s is the specific area (m^2 kg^{-1}).]

It is of course convenient to make this calculation with the help of a computer program (such a program is described in Dzombak and Morel, 1990). But a manual solution is possible; it is simplified if we realize that $[\equiv FeOZn^+]$ is negligible in equations v and vii. Thus the surface species, the surface charge, and the Coulombic interaction depend on pH and ionic strength. The result is given in Figures 9.24 and 9.8 (or Figure 9.20c).

Now the interaction of Zn(II) with the hydrous iron oxide surface can readily be computed because the surface charge is hardly affected by $[\equiv FeOZn^+]$. Thus, for example, at pH = 6, the surface charge $[\equiv FeOH_2^+] - [\equiv FeO^-]$ = 0.07 mM, or $\sigma = 11\ \mu C\ cm^{-2}$. Correspondingly, $\exp(-F\psi/RT) = 2.7 \times 10^{-2}$ and $\log P = -1.57$. This corresponds to a surface potential of +92 mV or to a Coulombic energy $F\psi \approx 9$ kJ mol^{-1}, which the Zn^{2+} has to overcome to be adsorbed at the positively charged hydrous ferric oxide (because ΔZ is +1).

Now we can compute the concentration of surface bound Zn(II), that is, $\equiv FeOZn^+$, by considering equation iv and vi. Considering that at pH = 6

9.6 Correcting Surface Complex Formation Constants for Surface Charge

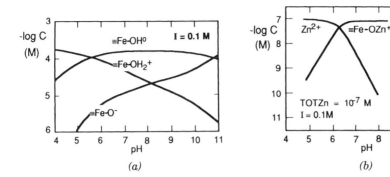

Figure 9.24. (a) Calculated surface speciation as a function of pH at ionic strength 0.1 (1:1 electrolyte) for a 10^{-3} M hydrous ferric oxide suspension. (b) Calculated equilibrium speciation as a function of pH for zinc in a 10^{-3} M suspension of hydrous ferric oxide: TOTZn = 10^{-3} M, I = 0.1 M. (Adapted from Dzombak and Morel, 1990.)

$[\equiv\text{FeOH}] = 5 \times 10^{-5}$ M, one calculates from equation iv that

$$\frac{[\equiv\text{FeOZn}^+]}{[\text{Zn}^{2+}]} = \frac{[\equiv\text{FeOH}]}{[\text{H}^+]} P\, 10^{-0.66} = 0.29$$

Since $[\equiv\text{FeOZn}] + [\text{Zn}^{2+}] = 10^{-7}$ M, $[\equiv\text{FeOZn}^+] = 2.9 \times 10^{-8}$ M, or 29% of the Zn(II) is adsorbed at pH = 6.

The speciation as a function of pH is shown in Figure 9.24.

"High-Affinity" and *"Low-Affinity"* **Surface Sites** Often, in cation adsorption on hydrous ferric oxides two site types (high-affinity and low-affinity) are required for modeling the equilibrium. This is a result of heterogeneity of the hydrous oxide surfaces (see Dzombak and Morel, 1990).

Example of 9.6. pH Dependence of Surface Complex Formation We resume the problem discussed in Example 9.3 and solve the same problem, but now we correct for electrostatic effects[†] (see Tableau 9.1). Summarizing the

---

[†]The set of equilibrium constants are (as in Example 9.3):

$$\text{S—OH}_2^+ \rightleftharpoons \text{SOH} + \text{H}^+ \qquad \log K_1^s = -4 \qquad \text{(i)}$$

$$\text{S—OH} \rightleftharpoons \text{SO}^- + \text{H}^+ \qquad \log K_2^s = -9 \qquad \text{(ii)}$$

$$\text{S—OH} + \text{Me}^{2+} \rightleftharpoons \text{SOMe}^+ + \text{H}^+ \qquad \log K_M^s = -1 \qquad \text{(iii)}$$

$$\text{S—OH} + \text{HA} \rightleftharpoons \text{SA} + \text{H}_2\text{O} \qquad \log K_L^s = 5 \qquad \text{(iv)}$$

$$\text{HA} \rightleftharpoons \text{H}^+ + \text{A}^- \qquad \log K_{\text{HA}} = -5 \qquad \text{(v)}$$

The constants for equations i–iv are intrinsic constants (see Tableau 9.1).

problem, we want to calculate the pH dependence of the binding of (a) a metal ion Me^{2+} and (b) a ligand A^- to a hydrous oxide, SOH, and account for the effect of a charged surface at an ionic strength $I = 0.1$. We have a specific surface area of 10 m² g⁻¹, 10^{-4} mol surface sites per gram (~6 sites nm⁻²), and the concentration used is 1 g liter⁻¹ (10^{-4} mol surface sites per liter solution). As before (Example 9.3), the surface complex formation constants are $\log K_M^s = -1$ and $\log K_L^s = 5$, respectively.

The diffuse double-layer model is used to correct for Coulombic effects. If a constant capacitance model is used, the model requires the input of a capacitance; but the result is not very different.

The results are given in Figures 9.15 and 9.17, respectively. They should be compared with Figures 9.14 and 9.16.

A comparison of Figures 9.14 and 9.16 with Figures 9.15 and 9.17 shows that the Coulombic effects change the relative concentrations of SOH_2^+ and of SO^- in comparison to the charge-uncorrected conditions. But it is of interest to note that the influence of the charge correction is not very significant for the extent of adsorption of Me^{2+} or A^- as a function of pH. In the case of the ligand, there is a less pronounced maximum but the location of the pH edge is similar. The influence of charge on adsorption equilibrium decreases with increasing ionic strength.

Example 9.7. Adsorption of Pb(II) onto a Hematite Surface We have 10^{-8} mol of Pb(II) added to a dispersion of 8.6 mg Fe_2O_3 in 1 liter of water. The following reactions are considered:

$$\equiv FeOH_2^+ = \equiv FeOH + H^+ \qquad K_{a1}^s(\text{int}) = 10^{-7.25}$$

$$\equiv FeOH = \equiv FeO^- + H^+ \qquad K_{a2}^s(\text{int}) = 10^{-9.75}$$

$$\equiv FeOH + Pb^{2+} = \equiv FeOPb^+ + H^+ \qquad K_1^s(\text{int}) = 10^{4.0}$$

The α-Fe_2O_3 is characterized as follows:

Specific surface area, $s = 4 \times 10^4$ m² kg⁻¹
Functional groups per kg $= 3.2 \times 10^{-1}$ mol kg⁻¹
Concentration of hematite, $a = 8.6 \times 10^{-6}$ kg liter⁻¹
Ionic strength, $I = 5 \times 10^{-3}$ M

Compute the equilibrium composition, and the extent of Pb(II) sorption to α-Fe_2O_3, as a function of pH, using the Gouy–Chapman diffuse double-layer model.

Tableau 9.2 summarizes the information necessary for the calculation. These and similar computations discussed here are facilitated by use of a program such as MICROQL (Westall, 1980).

The results are displayed in Figure 9.25.

9.6 Correcting Surface Complex Formation Constants for Surface Charge

Tableau 9.2. Adsorption of Pb^{2+} on $\alpha\text{-}Fe_2O_3$

Components		$\equiv$FeOH	Pb^{2+}	$f(\psi)$	H^+	log K
1 Species	$\equiv$FeOH	1				0
2	$\equiv$FeOH$_2^+$	1		1	1	7.25
3	$\equiv$FeO$^-$	1		-1	-1	-9.75
4	$\equiv$FeOPb$^+$	1	1	1	-1	4.0
5	Pb^{2+}		1			0
6	PbOH$^+$		1		-1	7.7
7	H^+				1	0
		2.8×10^{-6}	10^{-7}		pH given	

$f(\psi) = e^{-F\psi/RT}$
Surface charge = σ_{0T} = $\langle\equiv$FeOH$_2^+\rangle$ + $\langle\equiv$FeOPb$^+\rangle$ - $\langle\equiv$FeO$^-\rangle$ (in mol m^{-2})

Example 9.8. Adsorption of F^- onto a Hematite Surface (Ligand Exchange) What is the extent of F^- adsorption to a hematite (α-Fe$_2$O$_3$) surface as a function of pH? The same surface characteristics for α-Fe$_2$O$_3$ as that given in the previous example can be considered. Charge effects are corrected with the Gouy–Chapman diffuse double-layer model. F_{tot} is 10^{-3} M and ionic strength = 5×10^{-3} M.

Tableau 9.3 summarizes the information needed for the calculation.
The results, calculated with the help of the MICROQL program, are given in Figure 9.26. Note the large effect of F^- adsorption on the concentration of $\equiv$FeOH$_2^+$.

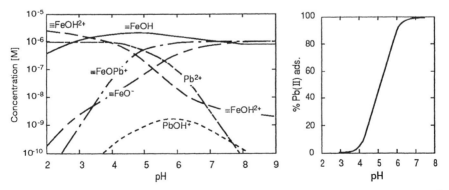

Figure 9.25. Adsorption of Pb(II) on hematite (α-Fe$_2$O$_3$): $\equiv$FeOH$_T$ = 2.7×10^{-6} M; Pb$_T$ = 10^{-6} M; I = 5×10^{-3} M. Electrostatic effects are corrected with the Gouy–Chapman model. (Surface characteristics of α-Fe$_2$O$_3$ are given in Example 9.7.)

Tableau 9.3. Adsorption of F^- on $\alpha\text{-Fe}_2O_3$

Components		$\equiv$FeOH	HF	$f(\psi)$	H^+	log K
1 Species	$\equiv$FeOH	1	0	0	0	0
2	$\equiv$FeOH$_2^+$	1	0	1	1	7.25
2	$\equiv$FeO$^-$	1	0	-1	-1	-9.75
4	$\equiv$FeF	1	1	0	0	5.5
5	HF	0	1	0	0	0
6	F^-	0	1	0	-1	-3.7
7	H^+	0	0	0	1	0
		2.8×10^{-6}	10^{-3}		pH given	

$f(\psi) = e^{-F\psi/RT}$
Surface charge $= \sigma_{0T} = \langle\equiv\text{FeOH}_2^+\rangle - \langle\equiv\text{FeO}^-\rangle$ (in mol m^{-2})

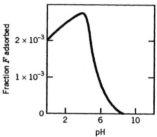

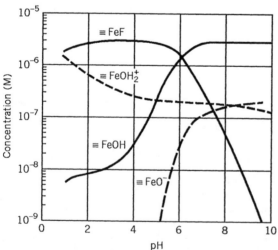

Figure 9.26. Adsorption of fluoride on a hematite ($\alpha\text{-Fe}_2O_3$) surface. $\equiv$FeOH$_T = 2.7 \times 10^{-6}$ M, $F_{\text{tot}} = $ [HF] + [F$^-$] + [$\equiv$FeF] 10^{-3} M; $I = 5 \times 10^{-3}$ M. Surface characteristics of $\alpha\text{-Fe}_2O_3$ are as in Example 9.8.

9.7 SORPTION OF HYDROPHOBIC SUBSTANCES ON ORGANIC CARBON-BEARING PARTICLES

As mentioned in Section 9.1, hydrophobic substances are soluble in nonpolar solvents, while their solubility in water is very limited. Many of these substances are also soluble in fats and lipids and are also called *lipophile* compounds. Such substances have a tendency to avoid contact with water and to associate with a nonpolar, nonaqueous environment, such as a surface, for example, an organic particle or a particle containing organic material or the lipid-containing biomass of an organism.

The solubility of hydrophobic substances on suspended particles, on sediments, on biota, or on soil particles can be related to the solubility of these substances in organic solvents. The solvent *n*-octanol, $CH_3(CH_2)_7OH$, is a kind of surrogate for many kinds of environmental and physiological organic substances and has become a reference phase for organic phase water partitioning of organic solutes. This organic solvent is, because of the OH groups, only partially nonpolar and can dissolve, in addition to nonpolar substances, also partially polar compounds containing O and N functional groups.

The distribution equilibrium of a compound between water and *n*-octanol, K_{OW}, can be determined experimentally:

$$K_{OW} = \frac{[A(oct)]}{[A(aq)]} \left(\frac{\text{mol (liter octanol)}^{-1}}{\text{mol (liter water)}^{-1}} \right) \qquad (47)$$

For extensive compilations see Schwarzenbach et al. (1993, Appendix).

Figure 9.27 compares data on ranges of water solubilities and octanol water partition coefficients for some common xenobiotic compound classes. The juxtaposition of data on the water solubility and K_{OW} partition coefficients in Figure 9.27 illustrates that usually K_{OW} coefficients for compound classes are inversely proportional to water solubility:

$$K_{OW} = \propto \frac{1}{C_w^{sat}}$$

The extent of sorption of hydrophobic solutes at the solid–water interface depends on the organic carbon content of the sorbent. In other words, the organic material in the (porous) soils behaves like octanol; and the hydrophobic material is *ab*sorbed into the organic matrix.

The sorption coefficient, K_p—essentially a distribution coefficient (mol kg^{-1} solid phase)/mol (liter water)$^{-1}$—can be expressed as a function of organic carbon content, f_{OC} (fraction by weight), and the octanol–water partition coefficient

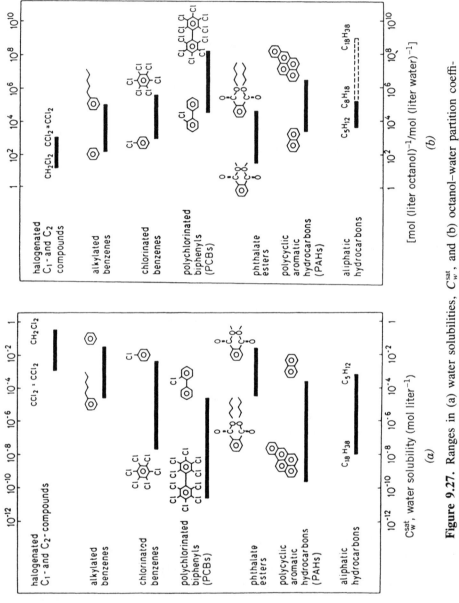

Figure 9.27. Ranges in (a) water solubilities, C_w^{sat}, and (b) octanol–water partition coefficients, K_{OW}, for some important classes of organic compounds. (From Schwarzenbach et al., 1993.)

9.7 Sorption of Hydrophobic Substances

$$K_p = bf_{OC}(K_{OW})^a \left(\frac{\text{mol kg}^{-1}\text{ (solid)}}{\text{mol liter}^{-1}\text{ (water)}}\right) \quad (48)$$

where a and b are constants.

Figure 9.28 illustrates the distribution of nonpolar organic solutes between natural particles and water.

Sorption to Organic Colloids Colloidal organic material (e.g., humic colloids) absorbs nonpolar organic chemicals. Operationally, an increase of "ap-

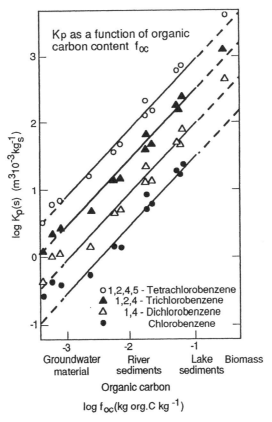

Figure 9.28. The distribution of nonpolar organic substances between aquatic solids and water (as given by the distribution coefficient K_p) is dependent upon the lipophilicity of the compound and the organic C content of the adsorbing material (f_{OC} = weight fraction). The solid phases considered here are coastal sea and lake sediments, river sediments, solids from aquifers and biomass (activated sludge). The octanol/water distribution coefficients are, respectively 500, 2400, 11,200 and 52,000 for chlorobenzene, 1,4-dichlorobenzene, 1,2,4-trichlorobenzene and 1,2,4,5-tetrachlorobenzene. (Modified from Schwarzenbach and Westall, 1980).

parent solubility" over that in pure water may be observed. But the true aqueous solubility is not affected; the additional nonpolar chemicals are associated with the humic colloids.

Ionizable Organic Chemicals Adsorption isotherms and distribution coefficients depend on solution speciation, for example, the sorption of a substance that can be protonated or deprotonated is different for the acid and its conjugate base. The solubility of a neutral molecule in octanol (or in the organic matrix of a particle) will be much larger than its charged conjugate base or acid. In a first approximation, the solubility of the ionic species in the organic solvent can be neglected. For a detailed treatment see Schwarzenbach et al. (1993).

Bioaccumulation The accumulation of xenobiotic substances in organisms and the food chain is important in the assessment of the harmfulness of a substance in the environment. The accumulation of organic substances in organisms occurs often in accordance with their lipophilicity (Figure 9.29); K_{ow} often serves as a measure of lipophilicity and as a predictive parameter for bioaccumulation in the food chain. One also speaks of *biomagnification* to describe progressive accumulation of xenobiotic substances in the food chain.

Bioconcentration Factor

Partitioning models are based on the assumption that chemicals of concern partition in a more or less passive fashion between the aquatic environment

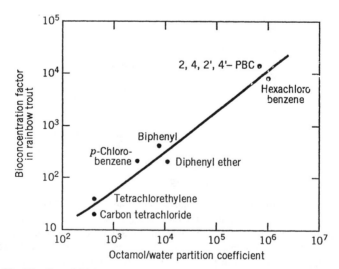

Figure 9.29. The lipophilicity of a substance, as measured by the octanol–water distribution coefficient, is an essential parameter for predicting biomagnification (biostorage and accumulation in the food chain) (Calculated from data by Chiou et al., 1977.)

and the organisms living in that environment. Such assumptions are most justifiable in the case of hydrophobic substances. Empirical equations are often of the form

$$\log \text{BCF} = y \log K_{OW} - x$$

where BCF is the bioconcentration factor and y and x are constants ($y \leq 1$) Lyman et al., 1990). Partitioning models do not account for increases in chemical concentrations as one moves up a food chain. In kinetic models, consideration is given to the dynamics of ingestion, internal transport, storage, metabolic transformation, and excretion processes in each type of organism.

Adsorption of Amphipathic Substances and Surfactants

The Gibbs equation (see equation 19) predicts that a substance that reduces the surface (interfacial tension) will be the adsorbed at the surface (interface). Surface-active substances (especially long-chain fatty acids, detergents, and surfactants) decrease the surface (interfacial) tension. *Amphipathic* molecules (which contain hydrophilic and hydrophobic groups) become oriented at the interface. At solid–water interfaces, the orientation depends on the relative affinities of the adsorbate for water and the solid surface. The hydrophilic groups (sulfate carboxylate, hydroxyl, etc.) may—if the hydrophobic tendency is relatively small—interact coordinatively with the functional groups of the solid surface (Ochs et al., 1994; Ulrich et al., 1988).

Example 9.9. The Adsorption of Sodium Dodecyl Sulfate (SDS) on Al_2O_3 The adsorption isotherm of sodium dodecyl sulfate (SDS) on alumina at pH = 6.5 in 0.1 M NaCl (Figure 9.30a) is characteristic of anionic surfactant adsorption onto a positively charged oxide. As shown by Somasundaran and Fuerstenau (1966) and by Chandar et al. (1987), the isotherm can be divided into four regions. These authors give the following explanation for the adsorption mechanism:

1. The slope of unity in region I indicates that the anionic surfactant adsorbs as individual ions through electrostatic interaction with the positively charged surface.
2. The sharp increase in adsorption in region II marks the onset of surfactant association at the surface through lateral interaction of the hydrocarbon chains.
3. The decreasing slope in region III can be attributed to an increasing electrostatic hindrance to the surfactant association process following interfacial charge reversal.
4. Plateau adsorption could correspond to either complete surface coverage or a value limited by constant surfactant monomer activity in solution as a result of micellization in the aqueous solution bulk phase.

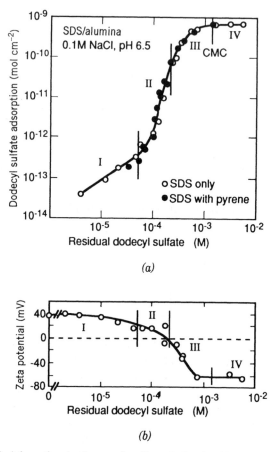

Figure 9.30. (a) Adsorption isotherm of sodium dodecyl sulfate (SDS) on alumina at pH 6.5 in 0.1 M NaCl. (b) Zeta potential of alumina as a function of equilibrium concentration of SDS (designation of regions based on isotherm shape). (From Chandar et al., 1987; based on the data of Somasundaran and Fuerstenau, 1966.) Chandar et al. (1987) have shown with the aid of fluorescent probe studies that in region II and above adsorption occurs through the formation of surfactant aggregates of limited size.

The corresponding zeta potential behavior of alumina particles is found to correlate with the amount of adsorbed SDS, as shown in Figure 9.30b. In region I, the potential is relatively constant and equal to that of alumina particles at pH 6.5 in the absence of SDS (+40 mV). A significant decrease in the positive potential is observed beyond the transition from region I to II, which correlates with the marked increase in anionic surfactant adsorption. The isoelectric point corresponding to the neutralization of the positively charged surface by the adsorbed anionic surfactant appears to coincide with the transition from region II to III. Region III is therefore characterized by SDS adsorption resulting in a net negative potential at the surface. Constant negative potentials

(-65 mV) are measured when the adsorption reaches its plateau level in region IV.

The fluorescent probe studies (Chandar et al., 1987) have shown that in region II and above adsorption occurs through the formation of surfactant aggregates of limited size. As these authors point out, in region II, where the surface is essentially bare and a sufficient number of positive sites are available (zeta potential is positive), an increase in adsorption is achieved mostly by increasing the number of aggregates; a constant aggregation number implies that the number density of aggregates increases with an increase in adsorption. In region II relatively uniform-sized aggregates (aggregation number 120–130) are measured on the surface.

The transition from region II to III marks the point at which most of the positive sites are filled since the zeta potential reverses its sign at this point. Adsorption in region III is therefore likely to occur through the growth of existing aggregates rather than through the formation of new ones. The growth of these aggregates can be expected to be electrostatically hindered.

Humic Substances

Humic and fulvic acids contain various types of phenolic and carboxylic functional (hydrophilic) groups as well as aromatic and aliphatic moieties; the latter import hydrophobic properties to these substances. We refer to Thurman (1985) and Aiken et al. (1985) for a description of the various properties of humic and fulvic acids in soils and waters and to Buffle (1988) for the coordinating properties of humus and humic acids.

Because of hydrophobic interaction, humic and fulvic acids tend to accumulate at the solid–water interface. At the same time, the adsorption is influenced by coordinative interaction; for example, schematically,

$$\equiv Al-OH + RC\underset{OH}{\overset{O}{\diagup\!\!\!\diagdown}} \rightleftharpoons \equiv Al\,O-\overset{O}{\overset{\|}{C}}-R + H_2O$$

This adsorption reaction is reflected in the pH dependence of the adsorption. The adsorption Γ versus pH curve is very similar to that of an organic acid with a pK value in the range 3–5 (Figure 9.31).

For a general review on the adsorption of humic acids with inorganic and organic surfaces see Tipping (1990).

The Sorption of Polymers

Polymers may adsorb readily on solid surfaces. Somasundaran et al. (1964) have investigated the effect of chain length on the adsorption of alkylammonium cations and of surfactant anions on oxide surfaces. These authors consider that the adsorption is influenced by the van der Waals energy of interaction or

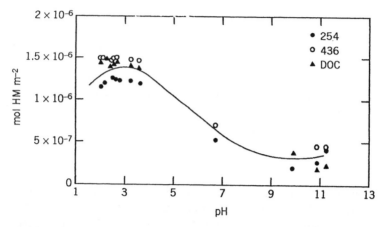

Figure 9.31. Adsorption isotherm of (Aldrich) humic acid (HM) on δ-Al$_2$O$_3$ (mol functional groups adsorbed per m^2 as a function of pH). Extent of adsorption was determined both by measurements of light absorption at 254 and 436 nm and by measurements of dissolved organic carbon (DOC) of the residual HM in solution (original concentration = 25 mg liter^{-1}). (From Ochs et al., 1994.)

hydrophobic bonding between CH$_2$ groups of adjacent adsorbed molecules. The van der Waals interactive energy per CH$_2$ group is found to be approximately 1 RT (2.5 kJ mol^{-1}). For a 12-carbon alkylamine, the van der Waals cohesive energy is therefore on the order of 30 kJ mol^{-1} and exceeds, in this particular case, the electrostatic contribution. Each polymer molecule or polymeric ion can have many groups or segments that can potentially be adsorbed; these groups are often relatively free of mutual interaction. Usually the extent of adsorption, but not necessarily its rate, increases with molecular weight and is affected by the number and type of functional groups in the polymer molecule (see Table 9.2). For example, on a negatively charged silica surface, polystyrene sulfonate is adsorbed readily while monomeric *p*-toluene sulfonate (CH$_3$—C$_6$H$_4$—SO$_3^-$) does not adsorb from 10^{-4} M solutions (Overbeek, 1976). Hydroxyl, phosphoryl, and carboxyl groups can be particularly effective in causing adsorption; the significance of such functional groups directs attention to the specificity of the chemical interactions involved in the adsorption process. Typical natural polymers are, for example, starch, cellulose, tannins, and humic acids.

Because many segments of the polymer can be in contact with the surface, a low bonding energy per segment may suffice to render the affinity of several segments together so high that their adsorption is virtually irreversible.

Lyklema (1985) provides the following semiquantitative thermodynamic argument. Suppose that a given polymer molecule loses a conformational entropy[†] of 100 k_B (where k_B is Boltzmann's constant) upon sorption in a certain

---

[†]Polymer adsorption results in a reduction in the number of conformations that a coil can assume. Thus, for reasons of entropy alone, adsorption would be unfavorable.

9.7 Sorption of Hydrophobic Substances

Table 9.2. Types of Polyelectrolytes

Nonionic	Anionic	Cationic
Polyvinyl alcohol $\left[-\mathrm{CH-CH_2-}\atop\;\;\;\mid\atop\;\;\;\mathrm{OH}\right]_n$	Polystyrene sulfonate $\left[-\mathrm{CH-CH_2-}\atop\text{(phenyl-}SO_3^-\text{)}\right]_n$	Polyvinyl pyridinium $\left[-\mathrm{CH-CH_2-}\atop\text{(pyridinium-}H^+\text{)}\right]_n$
Polyacrylamide $\left[-\mathrm{CH-CH_2-}\atop\;\;\;\mid\atop\;\;\;\mathrm{CONH_2}\right]_n$ $\xrightarrow{\text{hydrolysis}}$	Partially hydrolyzed Polyacrylamid $\left[-\mathrm{CH_2-CH-CH_2-CH-}\atop\;\;\;\;\;\;\;\;\mid\;\;\;\;\;\;\;\;\mid\atop\;\;\;\;\;\;\;\;\mathrm{CONH_2}\;\;\;\mathrm{COO^-}\right]$	Polyethylene imine $\left[\mathrm{CH_2-CH_2-NH_2^+-}\right]$
Polyethylene oxide $\left[-\mathrm{CH_2-CH_2-O-}\right]_n$	Polyacrylate $\left[-\mathrm{CH-CH_2-}\atop\;\;\;\mid\atop\;\;\;\mathrm{COO^-}\right]$	

conformation; that is, $T\Delta_{\text{ads}} S = -100\ k_B T$. Such a molecule would not adsorb in that conformation unless the loss is overcompensated by a free energy gain from another source. This source is the interaction free energy due to contacts between segments and the adsorbent. (This interaction is of an energetic nature in the case where binding takes place through van der Waals forces or hydrogen bridges, but it can have entropic terms, for example, in the case of hydrophobic bonding. Note that these entropic terms are of a nonconfigurational nature: they are proportional to the number of segments adsorbed, independent of the conformation of the polymer as a whole.)

Lyklema (1985) also provides "for the sake of argument" a simplified numerical example. Let us assume that in the adsorbed conformation 100 segments are in contact with the surface. If the adsorption free energy per segment were $-0.9\ k_B T$, the contribution of segment binding to $\Delta_{\text{ads}} G$ would be $-90\ k_B T$ and no adsorption would ensue, but if it is $-1.1\ k_B T$ per segment, the entropy loss would be overcompensated and strong adsorption takes place. What happens in the latter case is that the conformation in the adsorbed state adjusts itself until the overall free energy change $\Delta_{\text{ads}} G = 0$, as required for equilibrium. This simple reasoning makes the finding of polymer adsorption theory plausible; slight changes in the adsorption free energy per segment mark the transition between "nonadsorbing" and "strongly adsorbing." That a change of only a few tenths of a $k_B T$ unit per segment has such dramatic consequences is of course due to the large number of segments involved.

584 The Solid–Solution Interface

In modern statistical theories (Fleer and Lyklema, 1983; Scheutjens and Fleer, 1980) it is found that there does exist a certain critical adsorption free energy, customarily called x_s, above which the molecule adsorbs strongly and below which it does not adsorb. Typical values of x_s (crit) range from $0.2\ k_BT$ to $0.4\ k_BT$.

For an extensive treatment see Lyklema (1985).

Polyelectrolytes The most striking feature of polyelectrolytes is that, due to the electrostatic repulsion between the segments, the formation of thick adsorbed layers is prevented. Polyelectrolytes tend to adsorb in rather flat conformations. If adsorbent and polyelectrolyte bear opposite charges, this attraction can be of an electric (Coulombic) nature; if the charges have the same sign, adsorption takes place only if the nonelectrostatic attraction outweighs the electrostatic repulsion (Lyklema, 1985).

Electric interactions are screened by electrolytes. Hence, by adding electrolytes, the adsorption behavior is made to resemble that of uncharged macromolecules (see Figure 9.33).

For a quantitative theory see Van der Schee and Lyklema (1984).

Figure 9.32 provides an illustration of the adsorption of a neutral polymer, polyvinyl alcohol, on a polar surface, and the resulting effects on the double-layer properties. Adsorption of anionic polymers on negative surfaces—espe-

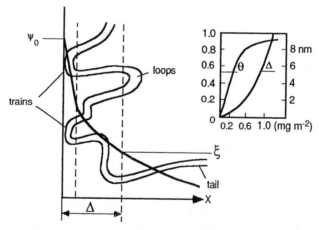

Figure 9.32. Effect of adsorbed polymer on the double layer. Because of the presence of adsorbed train segments, the double layer is modified. The zeta potential, ζ, is displaced because the adsorbed polymer displaces the plane of shear. The parameters for describing adsorbed polymers are the fraction of the first layer covered by segments, θ, and the effective thickness, Δ, of the polymer layer. The inset gives the distribution of segments over trains and loops for polyvinyl alcohol adsorbed on silver iodide. Results were obtained from double-layer and electrophoresis measurements. (Adapted from Lyklema, 1978.)

9.7 Sorption of Hydrophobic Substances

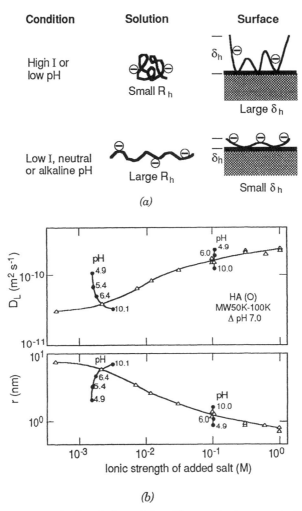

Figure 9.33. (a) Schematic description of the effects of ionic strength (I) and pH on the conformations of a humic molecule in solution and at a surface. R_h denotes the hydrodynamic radius of the molecule in solution and δ_h denotes the hydrodynamic thickness of the adsorbed anionic polyelectrolyte. (Adapted from Yokoyama et al., 1989; and O'Melia, 1991). (b) The influence of ionic strength of pH on diffusion coefficient, D_L, and on Stokes–Einstein radius of a humic acid fraction of 50,000–100,000 Dalton. (From Cornel et al., 1986).

cially in the presence of Ca^{2+} or Mg^{2+}, which may act as coordinating links between the surface and functional groups of the polymer—is not uncommon (Tipping and Cooke, 1982).

As will be discussed in Chapter 14, the adsorption of polymers may either increase or decrease colloid stability.

Humic substances are anionic polyelectrolytes of low to moderate molecular weight (500–20,000); their charge is due primarily to partially deprotonated carboxylic and phenolic groups.

A schematic representation of the effects of pH and ionic strength on the configuration of anionic polyelectrolytes, such as humic substances, is presented in Figure 9.33. In fresh waters at neutral and alkaline pH values, charged macromolecules assume extended shapes (large hydrodynamic radius, R_h) as a result of intramolecular electrostatic repulsive interactions. When adsorbed at interfaces under these conditions, they assume flat configurations (small hydrodynamic thickness, δ_h). At high ionic strength or at low pH, the polyelectrolytes have a coiled configuration in solution (small R_h) and extend further form the solid surface when adsorbed (large δ_h).

9.8 ION EXCHANGE

Ion Exchange Capacity There are different meanings of "ion exchange." In a *most general sense*, any replacement of an ion in a solid phase in contact with a solution by another ion can be called ion exchange. This includes reactions such as

$$CaCO_3(s) + Sr^{2+} \rightleftharpoons SrCO_3(s) + Ca^{2+}$$

$$NaAlSi_3O_8(s) + K^+(aq) \rightleftharpoons KAlSi_3O_8(s) + Na^+$$

$$Fe(OH)_3(s) + HPO_4^{2-} + 2\ H^+ \rightleftharpoons FePO_4(s) + 3\ H_2O$$

In a *more restrictive sense*, the term "ion exchange" is used to characterize the replacement of one *adsorbed, readily exchangeable* ion by another. This circumspection, used in soil science (Sposito, 1989b), implies a surface phenomenon involving charged species in *outer-sphere complexes* or in the *diffuse ion swarm*. It is not possible to adhere rigorously to this conceptualization because the distinction between inner-sphere and outer-sphere complexation is characterized by a continuous transition (e.g., H^+ binding to humus).

A further difficulty is in the distinction between a concept and an operation, for example, in the *definition of ion exchange capacity*. Operationally, "the ion exchange capacity of a soil (or of soil-minerals in waters or sediments) is the number of moles of adsorbed ion charge that can be desorbed from unit mass of soil, under given conditions of temperature, pressure, soil solution composition, and soil–solution mass ratio" (Sposito, 1989b). The measurement of an ion exchange capacity usually involves the replacement of (native) readily exchangeable ions by a "standard" cation or anion.

The cation exchange capacity of a negative double layer may be defined as the excess of counterions that can be exchanged for other cations. This ion exchange capacity corresponds to the area marked σ_+ in Figure 9.19d. It can be shown that σ_+/σ_- (where σ_- is the charge due to the deficiency of anions) remains independent of electrolyte concentration only for constant potential surfaces. However, for constant charge surfaces, such as double layers on clays, σ_-/σ_+ increases with increasing electrolyte concentration. Hence the cation exchange capacity (as defined by σ_+) increases with dilution and becomes equal to the total surface charge at great dilutions.

Since cations are adsorbed electrostatically not only due to the permanent structural charge, σ_0 (caused by isomorphic substitution), but also due to the proton charge, σ_H (the charge established because of binding or dissociating protons), the ion exchange capacity is pH dependent (it increases with pH). Furthermore, the experimentally determined capacity may include inner-spherically bound cations.

Standard cations used for measuring cation exchange capacity are Na^+, NH_4^+, and Ba^{2+}. NH_4^+ is often used but it may form inner-sphere complexes with 2:1 layer clays and may substitute for cations in easily weathered primary soil minerals. In other words, one has to adhere to detailed operational laboratory procedures; these need to be known to interpret the data; and it is difficult to come up with an operationally determined "ion exchange capacity" that can readily be conceptualized unequivocally.

Simple Models The surface chemical properties of clay minerals may often be interpreted in terms of the surface chemistry of the structural components, that is, sheets of tetrahedral silica, octahedral aluminum oxide (gibbsite), or magnesium hydroxide (brucite). In the discrete site model, the cation exchange framework, held together by lattice or interlayer attraction forces, exposes fixed charges as anionic sites.

In clays such as montmorillonites, because of the difference in osmotic pressure between solution and interlayer space, water penetrates into the interlayer space. Depending on the hydration tendency of the counterions and the interlayer forces, different interlayer spacings may be observed. A composite balance among electrostatic, covalent, van der Waals, and osmotic forces influences the swelling pressure in the ion exchange phase and in turn also the equilibrium position of the ion exchange equilibrium. One often distinguishes between inner-crystalline swelling (which is nearly independent of ionic strength) and osmotic swelling (which is strongly dependent on ionic strength) (Weiss, 1958). The extent of swelling increases with the extent of hydration of the counterion. Furthermore, the swelling is much less for a bivalent cation than for a monovalent counterion. Because only half as many Me^{2+} are needed as Me^+ to neutralize the charge of the ion exchanger, the osmotic pressure difference between the solution and the ion exchange framework becomes smaller. Thus, if strongly hydrated Na^+ replaces less hydrated Ca^{2+} and Mg^{2+} in soils, the resulting swelling adversely affects the permeability of soils. Similarly, in waters of high relative $[Na^+]$, the bottom sediments are less permeable to water. Surface complex formation between the counterions and the anionic sites (oxo groups) also reduces the swelling pressure.

From a geometric point of view, clays can be packed rather closely. Muds containing clays, however, have a higher porosity than sand. The higher porosity of the clays is caused in part by the high water content (swelling), which in turn is related to the ion exchange properties.

Ion Exchange Equilibria The double-layer theory predicts qualitatively correctly that the affinity of the exchanger (the ion-exchanging solid—clay, humus, ion exchange resin, zeolite) for bivalent ions is larger than that for monovalent ions and that this selectivity for ions of higher valency decreases with increasing ionic strength of the solution. However, according to the Gouy theory, there should be no ionic selectivity of the exchanger between different equally charged ions.

Relative affinity may be defined quantitatively by *formally* applying the mass law to exchange reactions:

$$\{Na^+ R^-\} + K^+ = \{K^+ R^-\} + Na^+ \tag{49}$$

$$2\,\{Na^+ R^-\} + Ca^{2+} = \{Ca^{2+} R_2^{2-}\} + 2\,Na^+ \tag{50}\dagger$$

where R^- symbolizes the negatively charged network of the cation exchanger. A selectivity coefficient, Q, can then be defined by

$$Q_{(NaR \to KR)} = \frac{X_{KR}\,[Na^+]}{X_{NaR}\,[K^+]} \tag{51}$$

$$Q_{(NaR \to CaR)} = \frac{X_{CaR}\,[Na^+]^2}{X_{NaR}^2\,[Ca^{2+}]} \tag{52}$$

where X represents the equivalent fraction of the counterion on the exchanger (e.g., $X_{CaR} = 2\,[CaR_2]/(2\,[CaR_2] + [NaR])$. The selectivity coefficients may be treated as mass law constants for describing, at least in a semiquantitative way, equilibria for the interchange of ions, but these coefficients are neither constants nor are they thermodynamically well defined. Because the activities of the ions within the lattice structure are not known and vary depending on the composition of the ion exchanger phase, the coefficients tend to deviate from constancy. Nevertheless, it is expedient to use equation 52 for illustrating the concentration dependence of the selectivity for more highly charged ions. A theory on the thermodynamics of ion exchange was given by Sposito (1984). The calculations for the distribution of ions between the ion-exchanging material and the solution involve the same procedures as in other calculations or equilibrium speciation. In addition to the mass law expression, one needs the corresponding mass balance equations for [TOTR], [TOTNa], and [TOTCa].

In Figure 9.34 the mole fraction of Ca^{2+} on the exchanger is plotted as a function of the mole fraction of Ca^{2+} in the solution. For a hypothetical ex-

---

†This reaction could also be formulated as $Na_2R + Ca^{2+} = CaR + 2\,Na^+$.

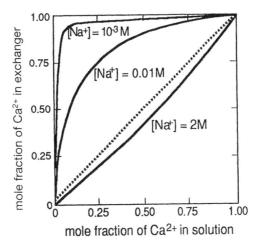

Figure 9.34. Typical exchange isotherms calculated for the reactions $Ca^{2+} + 2\{Na^+R^-\}$ do not separate $R^-\} \rightleftharpoons \{Ca^{2+} \; R_2^{2-}\} + 2\;Na^+$. In dilute solutions the exchanger shows a strong preference for Ca^{2+} over Na^+. This selectivity decreases with increasing ion concentration. The 45° line represents the isotherm with no selectivity.

change with no selectivity, the exchange isotherm is represented by the dashed line. In such a case the ratio of the counterions is the same for the exchanger phase as in the solution. The selectivity of the exchanger for Ca^{2+} increases markedly with increased dilution of the solution; in solutions of high concentration the exchanger loses its selectivity. The representation in this figure makes it understandable why a given exchanger may contain predominantly Ca^{2+} in equilibrium with a fresh water; in seawater, however, the counterions on the exchanger are predominantly Na^+. Figure 9.34 illustrates why, in the technological application of synthetic ion exchangers such as water softeners. Ca^{2+} can be selectively removed from dilute water solutions, whereas an exhausted exchanger in the Ca^{2+} form can be reconverted into a $\{NaR\}$ exchanger with a concentrated brine solution or with undiluted seawater.

Table 9.3 gives the experimentally determined distribution of Ca^{2+} and K^+ for three clay minerals at various equivalent concentrations of Ca^{2+} and K^+. The results demonstrate the concentration dependence of the selectivity; they also show that marked differences exist among various clays.

With the help of selectivity coefficients, such as in equations 51 and 52, a general order of affinity can be given. For most clays the Hofmeister series, which also applies for the affinity of ions to oxide surfaces,

$$Cs^+ > K^+ > Na^+ > Li^+$$
$$Ba^{2+} > Sr^{2+} > Ca^{2+} > Mg^{2+} \tag{53}$$

590   The Solid–Solution Interface

Table 9.3. Ion Exchange of Clays with Solutions of CaCl$_2$ and KCl of Equal Equivalent Concentration

		Ca^{2+}/K$^+$ Ratios on Clay			
	Exchange Capacity (meq g^{-1})	Concentration of Solution $2[Ca^{2+}] + [K^+]$ (meq liter^{-1})			
Clay		100	10	1	0.1
Kaolinite	0.023	—	1.8	5.0	11.1
Illite	0.162	1.1	3.4	8.1	12.3
Montmorillonite	0.810	1.5	—	22.1	38.8

is observed; that is, the affinity increases with the (nonhydrated) radius of the ions. In other words, the ion with the larger hydrated radius tends to be displaced by the ion of smaller hydrated radius. This is an indication that innersphere interaction occurs in addition to Coulombic interaction (Coulombic interaction alone would give preference to the smaller unhydrated ion). For some zeolites and some glasses, a reversed selectivity (rather than that of equation 53) may be observed.

The dependence of selectivity on electrolyte concentration has important implications for analytical procedures for determining ion exchange capacity and the composition of interstitial waters. The rinsing of marine samples with distilled water shifts the exchange equilibria away from true seawater conditions; the influence of rinsing increases the bivalent/monovalent ratio, especially the Mg^{2+}/Na$^+$ ratio. Sayles and Mangelsdorf (1979) have shown that the net reaction of fluvial clays and seawater is primarily an exchange of seawater Na$^+$ for bound Ca^{2+}. This process is of importance in the geochemical budget of Na$^+$.

Example 9.10. Ion Exchange Selectivity: Clays in Fresh Water or Seawater Sayles and Mangelsdorf (1977) have equilibrated clays first with river water (concentrations as $-\log$ M: Na$^+$, 3.6; K$^+$, 4.2; Ca^{2+}, 3.4; Mg^{2+}, 3.8) and then with seawater (concentrations as $-\log$ M: Na$^+$, 0.3; K$^+$, 2.0; Ca^{2+}, 2.0; Mg^{2+}, 1.3).

An example of their data for kaolinite (ion exchange capacity = 16.2 meq per 100 g) is as follows:

Exchange Composition (Clays)/(equiv. fraction)	X_{Na}	X_{Mg}	X_{Ca}	X_K
Kaolinite equilibrated with river water	0.06	0.18	0.51	0.08
Kaolinite equilibrated with seawater	0.38	0.32	0.24	0.06

(a) Interpret the data *qualitatively*, especially from a point of view of (i) Na$^+$/K$^+$ ratios and (ii) the ratio of bivalent to monovalent ions.

(b) Why should clay samples not be rinsed with distilled water prior to analysis?

(a) (i) The preference of clays for K^+ over Na^+ provides a ready explanation why $[Na^+]/[K^+]$ ratios in natural waters are larger than unity although K^+ is only slightly less abundant than Na^+ in igneous rocks. Ion exchange processes continuously remove K^+ from solution and return it to the solid phase.

(a) (ii) Because of high electrolyte concentration of seawater, the clay is no longer selective for bivalent ions; for example,

$$(X_{Na}/X_{Ca})_{\text{seawater solution}} \approx 25 \qquad (X_{Na}/X_{Ca})_{\text{kaolinite in seawater}} \approx 4.3$$

Such processes may be important in the geochemical budget of Na^+. A computation of the selectivity coefficients helps to quantify the answer.

(b) The answer is obvious: distilled water tends to dilute the electrolyte concentration and the ion exchange selectivity increases with dilution; that is, the clay will change its composition.

Sodium Adsorption Ratio Because of the swelling effects of Na^+, the relative amount of sodium (sodicity) in a water, especially in irrigation water quality, is an important measurement in soil science. Decreased permeability can interfere with the drainage required for normal salinity control and with the normal water supply and aeration required for plant growth. The relative sodium status of irrigation waters and soil solutions is often expressed by the *sodium adsorption ratio* (*SAR*):

$$\text{SAR} = \frac{[Na^+]}{([Ca^{2+}]/2 + [Mg^{2+}]/2)^{1/2}} \tag{54}$$

Combining $[Ca^{2+}]$ and $[Mg^{2+}]$ in this expression, derived from ion exchange equilibrium considerations, is not strictly valid but causes little deviation from more exact formulations and is justified because these two bivalent cations behave similarly during cation exchange.

The Sorption of Cations to Layer Silicates; Metal-Ion Binding by Ion Exchange and Surface Complex Formation

Finely dispersed clay minerals are present in soils and sediments as well as in mylonites of crystalline rocks. Thus these materials make up some of the colloids found in the waters draining from these areas. Montmorillonite and illite are relatively abundant.

These three-layer silicates are characterized by permanent surface charges (due to isomorphic substitutions). Therefore the binding of cations is assumed to be caused by stoichiometric ion exchange of interlayer ions. These concepts hold well for alkaline and earth-alkaline cations; their adsorption and their ionic strength dependence can be characterized by distribution coefficients derived from ion exchange theory. It has been known for some time, however, that the binding of heavy metals (and of heavy metal radionuclides) on clays cannot be fully accounted for by an ion exchange mechanism; for example, little dependence of sorption on ionic strength is observed and the distribution often does not reflect the charge of the sorbed ion.

To mitigate *radioactive contamination*, it is important to understand the processes and mechanisms of interactions between radionuclides and the solid material of aquifers. Cationic radionuclides may be sorbed by processes such as ion exchange or surface complex formation, thus retarding their transport by groundwater.

Sorption Depends on Sorption Sites Three-layer silicates contain on the crystal edges (broken bonds) end-standing OH groups, which, similar to the OH groups on hydrous oxide surfaces, can interact with H^+, OH^-, and metal ions (surface complex formation).

Although no detailed theory on the sorption mechanism of heavy metals on three-layer clays is available, many inferences imply the specific coordinative interaction of end-standing hydroxo groups with nonhydrolyzed metal ions.

The sorption of alkaline and earth-alkaline cations on expandable three-layer clays—smectites (montmorillonites)—can usually be interpreted as stoichiometric exchange of *interlayer* ions. Heavy metals, however, are sorbed by surface complex formation to the OH functional groups of the outer surface (the so-called broken bonds). The nonswellable three-layer silicates, micas such as illite, usually can not exchange their interlayer ions; but the outside of these minerals and the weathered crystal edges ("frayed edges") participate in ion exchange reactions.

To understand binding of heavy metals on clays, one needs to consider—in addition to ion exchange—the surface complex formation on end-standing functional OH groups. Furthermore, the speciation of the sorbate ion (free hydroxo complex, carbonato or organic complex) and its pH dependence have to be known.

The surface characteristics of *kaolinite* were discussed in Section 9.5 and in Figure 9.22. The siloxane layer may—to a limited extent—participate in ion exchange reactions. The functional OH groups at the gibbsite and edge surfaces are able to surface complex heavy metal ions (Schindler et al., 1987).

As Figure 9.35 shows, the exchange of Sr^{2+} on Na^+-montmorillonite fits the ion exchange theory very well. But the adsorption of heavy metals cannot be accounted for by this theory. Co(II) behaves as if it were monovalent; K_D for americium is independent of $[Na^+]$ (americium occurs at pH = 6.5 as a hydroxo complex).

Adsorption of Cs^+ on Clays Cs^+, an ion with a simple solution chemistry (no hydrolysis, no complex formation), can be remarkably complex. Grütter et al. (1990) have studied adsorption and desorption of Cs^+ on glaciofluvial deposits and have shown that the isotherms for sorption and exchange on these materials are nonlinear. Part of this nonlinearity can be accounted for by the

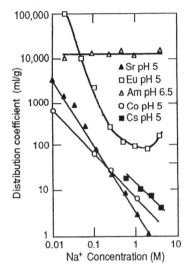

Figure 9.35. Sorption of various radionuclides on Na^+-montmorillonites. (From Shiao, 1979.) (The relationship plotted can be derived from equations 51 and 52; e.g., for the exchange of Sr^{2+} on Na^+-montmorillonites,

$$K_p = \frac{X_{SrR}}{[Sr^{2+}]} = Q \frac{X_{NaR}^2}{[Na^+]^2}, \text{ or}$$

$$\log K_p = \log Q + 2 \log (X_{NaR} - [Na^+])$$

Since X_{NaR} remains relatively constant, $\log K_p$ vs $\log [Na^+]$ should plot with a slope of -2.

collapse of the c-spacing of certain clays (vermiculite, chlorite). As illustrated in Figure 9.36, the Cs^+ sorption on illite and chlorite is characterized by nonlinearity.

For discussions on modeling cation exchange see Fletcher and Sposito (1989), Schindler et al. (1987), Charlet et al. (1993), and Bradbury and Baeyens (1994). Dzombak and Hudson (1995)

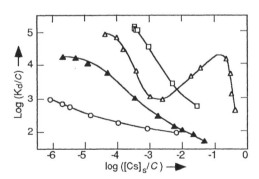

Figure 9.36. Sorption of cesium in synthetic groundwater on clay minerals:

○ Montmorillonite   □ Illite
△ <40 μm Chlorite   ▲ <2-μm Chlorite

The data are "normalized" with regard to the ion exchange capacity C of the sorbents. The sorption curves of the illite and of the <40-μm chlorite are strongly nonlinear, whereas that of the montmorillonite approaches linearity. (From Grütter et al., 1990.)

9.9 TRANSPORT OF (AD)SORBABLE CONSTITUENTS IN GROUNDWATER AND SOIL SYSTEMS

The subsurface environment can be regarded as a giant chromatography system with soil and aquifer solids as the solid phase and the groundwater as the carrier. The concepts for describing the transport of chemicals accompanied by concomitant adsorption and desorption can be borrowed from chromatography theory. However, there are a few differences of importance:

1. Particles of the soil matrix are usually very polydisperse and heterogeneously packed.
2. One has to distinguish between transport in groundwater aquifers, which are saturated with water, and transport in the unsaturated subsurface zone, where pores contain water and air (this might lead to a very heterogeneous flow structure).
3. Cracks and root zones may lead to preferential flow paths resulting in channeling (Borkovec et al., 1991).

In groundwater, environment changes in concentration can occur because of (1) chemical reactions within the aqueous phase[†] and (2) transfer of solute to (or from) solid surfaces in the porous medium or to (or from) the gas phase in the unsaturated zone.

We will concentrate on adsorption–desorption. The one-dimensional form of the advection–dispersion equation for a homogeneous saturated medium can then be written

$$-u\frac{\partial c_i}{\partial x} + D\frac{\partial^2 c_i}{\partial x^2} - \frac{\rho}{\theta}\frac{\partial S_i}{\partial t} = \frac{\partial c_i}{\partial t} \qquad (55)$$

Advection   Dispersion   Sorption

where u = linear velocity (cm s^{-1})
D = dispersion coefficient (cm^2 s^{-1})
c_i = concentration of species i (mol cm^{-3})
x = distance (in direction of flow) (cm)
S_i = species i sorbed (mol kg^{-1})
ρ = bulk density (g cm^{-3})[‡]
θ = porosity (volume of voids/volume total)

[†]Relevant homogeneous reactions include acid–base, complexation, solution precipitation, redox reactions, biodegradation, radioactive decay, and hydrolysis.
[‡]The bulk density, ρ, refers to the density of the entire assemblage of solids plus voids, that is, mass of solids per volume of solids plus voids. [If we refer to the density of the solid material itself, that if, of the "rock:" ρ' = (mass of rock)/(volume of rock), we can replace ρ/θ in equation 55 by $\rho'(1 - \theta)/\theta$.] As with the transport equation itself, these relationships assume negligible "immobile" water within the solid or "rock" phase.

9.9 Transport of (Ad)Sorbable Constituents

$$\frac{\rho}{\theta}\frac{\partial S_i}{\partial t} = \frac{\rho}{\theta}\frac{\partial S_i}{\partial c_i}\frac{\partial c_i}{\partial t} \tag{56}$$

where $(\partial S_i/\partial c_i)$ can be interpreted in terms of a linear sorption constant

$$\frac{\partial S_i}{\partial c_i} = K_p \text{ (liter kg}^{-1}\text{)} \tag{57}†$$

If we set the dispersion (second term in equation 55) equal to zero,

$$D\frac{\partial^2 c_i}{\partial x^2} = 0 \tag{58}$$

we can rewrite the transport–sorption equation 55 (using equations 56 and 57) as the *retardation equation*:

$$-u\frac{\partial c_i}{\partial x} = \frac{\partial c_i}{\partial t}\left(1 + \frac{\rho}{\theta}K_p\right) \tag{59}$$

where the expression in parentheses is the retardation factor, R:

$$R = 1 + \frac{\rho}{\theta}K_p = u_R^{-1} = \frac{\bar{u}}{\bar{u}_i} \tag{60}$$

where $\bar{u}$ and $\bar{u}_i$ are, respectively, the average linear velocity of the groundwater and of the retarded constituent. Both velocities can be measured where $c/c_0 = 0.5$ in the spatial concentration profile (see Figure 9.37a).

Typical values of ρ are 1.6–2.1, kg liter^{-1} (corresponding to a density of the solid material itself of 2.65 and a total porosity, θ, in the range of 0.2–0.4). Thus representative values of ρ/θ are 4–10 kg liter^{-1} and we can rewrite the retardation factor (equation 60)

$$\frac{\bar{u}}{\bar{u}_i} = (1 + 4\ K_p) \text{ to } (1 + 10\ K_p) \tag{61}$$

Figure 9.37b shows how, in addition to (ad)sorption, biodegradation can affect the movement of organic compounds. Note that the abscissa in Figure 9.37b is time, whereas that in Figure 9.37a is distance. While a conservative contaminant would behave just as the water in which it is contained, an ad-

---

†Equation 57 is typically observed for partitioning processes, for example, for the (ab)sorption of nonpolar organic substances in a solid matrix containing organic (humus-like) material. Equation 57 corresponds also to the initial linear portion of a Langmuir isotherm or to a Freundlich equation $S = K_p c^b$, where $b \approx 1$. Note that colloid-facilitated transport is excluded in this treatment.

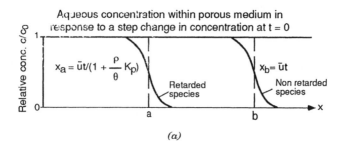

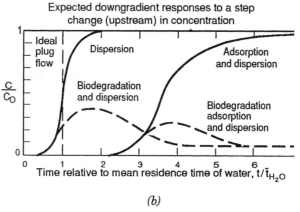

Figure 9.37. (a) Advances of adsorbed and nonadsorbed solutes through a packed column (porous medium). Partitionong of species between solid material and water is described by K_p (equation 50). The relative velocity of sorbing to nonsorbing solutes is given by

$$\frac{\bar{u}_i}{\bar{u}} = 1/[1 + (\rho/\theta)K_p].$$

Solute inputs are at concentration c_0 at time $t > 0$. (Modified from Freeze and Cherry, 1979.) (b) Illustration of the effects of dispersion, sorption, and biodecomposition on the time change in concentration of an organic compound at an aquifer observation well following the initiation of water injection into the aquifer at some distance away from the observation well. c represents the observed concentration and c_0 the concentrations in the injection water. (From McCarty et al., 1980.)

sorbable solute (as in Figure 9.37a) is retarded and arrives at the observation well later than the conservative contaminant. Biodegradation would act to destroy the organic contaminant, so its concentration at the observation well would never reach C_o; its concentration would initially be governed by dispersion alone, until sufficient time had passed to establish an acclimated bacterial population with sufficient mass to change the organic concentration (McCarty et al., 1981).

9.9 Transport of (Ad)Sorbable Constituents

As we have seen in Figure 9.28, the distribution (or partition) coefficient for the (ab)sorption of a nonpolar organic solute onto a solid phase can be estimated from the octanol–water coefficient K_{OW} (see equation 48):

$$K_p = bf_{OC} (K_{OW})^a \tag{48}$$

where f_{OC} is the organic carbon (weight fraction) constant of the solid material and a and b are constant.

The conclusions of retardation are applicable to a wide range of sorption and transport problems, including artificial groundwater recharge and leaching of pollutants from landfills.

The physics of groundwater movement and the dispersion of substances is described in an easily understandable way by Hemond and Fechner (1994).

Surface Complex Formation, Ion Exchange, and Transport in Groundwater and Soil Systems The retardation equation can also be applied to inorganic soluble substances (ions, radionuclides, metals). But here we have to consider, in addition to the sorption or ion exchange process, that the speciation of metal ions or ligands in a multicomponent system influences the specific sorption process and varies during the pollutant transport in the groundwater; chemistry then becomes an important part of the transport.

What does the breakthrough of a solute that interacts by surface complex formation (or an equivalent Langmuir isotherm) with the surfaces look like? The response of the column is no longer linear and will depend on the concentration of the solute. In line with the surface complex formation equilibrium (Γ_M versus [M] shows saturation), the distribution coefficient decreases with increasing residual concentration, thus leading to a decreasing retardation factor. This leads to a self-sharpening front, a traveling nonlinear wave; that is, in a c/c_o versus distance plot (cf. Figure 9.37a) a sharp abrupt change of c/c_o occurs (instead of the smooth symmetrical curve for linear adsorption). Furthermore, the higher the concentration, the earlier the breakthrough (Borkovec et al., 1991).

Mass transport models for multicomponent systems have been developed where the equilibrium interaction chemistry is solved independently of the mass transport equations, which leads to a set of algebraic equations for the chemistry coupled to a set of differential equations for the mass transport (Cederberg et al., 1985).

Example 9.11. Retention of Hydrophobic Organic Pollutants in Groundwater Aquifer Because of a leak in a waste container, traces of tetrachlorethylene and 1,2,4,5-tetrachlorobenzene escape in a groundwater. Provide an estimate of how long it takes to find these pollutants at a distance (direction of flow) of 25 m. The flow velocity of the groundwater is ~2 m day^{-1}. The following characteristics of the aquifer are available:

Density $\rho = 2.0$ kg liter^{-1}
Porosity $\theta = 0.2$
Fraction of organic carbon $f_{OC} = 0.0015$
Octanol–water distribution coefficients
 Tetrachlorethylene $K_{OW} = 400$
 Tetrachlorobenzene $K_{OW} = 50,000$

There is no biodegradation. The partition coefficient K_p can be calculated from equation 48, $K_p = bf_{OC} (K_{OW})^a$. The constants a and b were determined in column experiments with media from the aquifer as $a = 0.7$ and $b = 2.0$. Thus we obtain the following K_p values:

Tetrachlorethylene $K_p = 0.2$ liter kg^{-1}
Tetrachlorobenzene $K_p = 5.8$ liter kg^{-1}

In equation 60, $\rho/\theta = 10$ kg liter^{-1}. The retardation factors (R) are

For tetrachlorethylene $R = 3$
For tetrachlorobenzene $R = 59$

Thus one can estimate that it takes a conservative tracer ~12.5 days, tetrachlorethylene ~37 days, and tetrachlorobenzene ~2 years to reach the point of observation.

Of course, these are crude estimates based on numerous simplifying assumptions.

Determination of Nonlinear Adsorption Isotherms from Column Experiments; an Alternative to Batch Studies

Measurements of isotherms are commonly obtained in the laboratory by performing batch experiments. As has been pointed out recently by Bürgisser et al. (1993), another alternative to the classical batch experiment that allows simple and rapid measurements of an entire possibly nonlinear adsorption isotherm relies on column experiments and the use of techniques borrowed from nonlinear chromatography (Burgisser et al., 1993; Schweich et al., 1983). Experimentally, the column experiment (column is packed with the material in question) is carried out near the relevant solid/solution ratio, and the percolating solution may easily be adjusted to the composition of interest (e.g., composition of groundwater).

In the case of a linear adsorption isotherm, the *slope* of the breakthrough curve is the same as the breakthrough of a conservative tracer. The breakthrough is just delayed by the retention factor $R = 1 + (\rho/\theta) K_p$.

This type of behavior is illustrated in Figure 9.38a with dispersion effects neglected. In the case of a nonlinear isotherm, the response depends on the concentration. If the isotherm is convex (e.g., Freundlich isotherm with $n <$

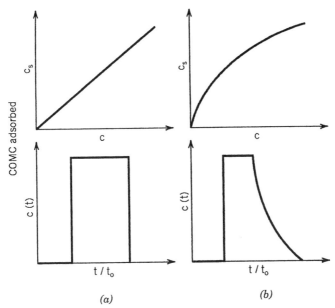

Figure 9.38. Schematic representation of the response of the chromatographic column for different adsorption isotherms (top). The column breakthrough of a step concentration change without disperson effects is shown (below) for different isotherms: (a) linear and (b) convex. The abscissa t/t_0 is equivalent to the number of pore volumes eluted. (Adapted from Bürgisser et al., 1993.)

1) (Figure 9.2), a step concentration increases as the column input produces a self-sharpening front, and in the case of a step decrease, a diffuse front will be formed (Figure 9.38b).

This behavior is caused by the fact that the retention of the chemical decreases with increasing concentration in solution. In the case of the adsorption front, the retention decreases with increasing concentrations, and this effect leads to an instability and to the development of a narrow, self-sharpening front. In the case of the desorption front, the retention increases in time and leads to a broad, diffuse front (Bürgisser et al., 1993).

SUGGESTED READINGS

Adamson, A. W. (1990) *Physical Chemistry of Surfaces,* 5th ed. Wiley-Interscience, New York.

Bolt, G. H., and van Riemsdijk, W. H. (1987) Surface Chemical Processes in Soil, in W. Stumm, Ed., *Aquatic Surface Chemistry,* Wiley-Interscience, New York, pp. 127–164.

Cederberg, G. A., Street, R. L., and Leckie, J. O. (1985) A Groundwater Mass

Transport and Equilibrium Chemistry Model for Multicomponent Systems, *Water Res. Res.* **21**, 1095-1104.

Charlet, L., and Manceau, A. (1993) Structure, Formation and Reactivity of Hydrous Oxide Particles; Insights from X-ray Absorption Spectroscopy. In *Environmental Particles*, Vol. 2, J. Buffle and H. P. Van Leeuwen, Editors. Lewis, Boca Raton, FL.

Ćosović, B. (1990) Adsorption Kinetics of the Complex Mixture of Organic Solutes at Model and Natural Phase Boundaries, in W. Stumm, Eds., *Aquatic Chemical Kinetics*, Wiley-Interscience, New York, pp. 291-310.

Dzombak, D. and R. J. Hudson (1995) Ion Exchange: The Contribution of Diffuse Layer Sorption and Surface Complexation, in **Aquatic Chemistry,** C. P. Huang et al., Eds. ACS Washington D.C.

Dzombak, D. A., and Morel, F. M. M. (1990) *Surface Complexation Modeling; Hydrous Ferric Oxide,* Wiley-Interscience, New York. (This book addresses general issues related to surface complexation and its modeling, using the results obtained for hydrous ferric oxide as a basis for discussion.)

Haderlein, S. B., and Schwarzenbach, R. P. (1993) Adsorption of Substituted Nitrobenzenes and Nitrophenols to Mineral Surfaces, *Environ. Sci. Technol.* **27**, 316-326.

Hiemstra, T., van Riemsdijk, W. H., and Bolt, H. G. (1989) Multisite Proton Adsorption Modeling at the Solid/Solution Interface of (Hydr)Oxides, I. Model Description and Intrinsic Reaction Constants, *J. Colloid Interface Sci.* **133**, 91-104.

Hiemstra, T., de Wit, J. C. M., and van Riemsdijk, W. O. (1989) Multisite Proton Adsorption Modeling at the Solid/Solution Interface of (Hydr)Oxides, II. Application to Various Important (Hydr)Oxides, *J. Colloid Interface Sci.* **133**, 105-117.

Huang, C. P., and Rhoads, E. A. (1989) Adsorption of Zn(II) onto Hydrous Aluminosilicates, *J. Colloid Interface Sci.* **131**, 289-306.

Lyklema, J. (1985) How Polymers Adsorb and Affect Colloid Stability, Flocculation, Sedimentation, and Consolidation. In Proceedings of the Engineering Foundation Conference, Sea Island, GA, B. M. Moudgil and P. Somasundaran, Eds., pp. 3-21.

Lyklema, J. (1991) *Fundamentals of Interface and Colloid Science, Volume I: Fundamentals,* Academic Press, London. (This book treats the most important interfacial and colloidal phenomena starting from basic principles of physics and chemistry.)

Maity, N., and Payne, G. F. (1991) Adsorption from Aqueous Solutions Based on a Combination of Hydrogen Bonding and Hydrophobic Interaction, *Langmuir* **7**, 1247-1254.

McCarty, P. L., Reinhard, M., and Rittmann, B. E. (1981) Trace Organics in Groundwater, *Environ. Sci. Technol.* **15**(1), 40-51.

Parks, G. A. (1990) Surface Energy and Adsorption at Mineral/Water Interfaces: An Introduction, in M. F. Hochella Jr. and A. F. White, Eds., *Mineral-Water Interface Geochemistry,* Mineralogical Society of America, Washington, DC, pp. 133-175.

Schindler, P. W., and Stumm, W. (1987), The Surface Chemistry of Oxides, Hydroxides, and Oxide Minerals, in W. Stumm, Ed., *Aquatic Surface Chemistry*, Wiley-Interscience, New York, pp. 83-110.

Schwarzenbach, R., Imboden, D., and Gschwend, Ph. M. (1992) *Environmental Or-

ganic Chemistry, Wiley-Interscience, New York. (Theory and interpretation of surface phenomena involving xenobiotic organic chemicals.)

Sigg, L., and Stumm, W. (1981) The Interaction of Anions and Weak Acids with the Hydrous Goethite (α-FeOOH) Surface, *Colloids Surfaces* **2**, 101–117.

Somorjai, G. A. (1994) *Chemistry in Two Dimensions: Surfaces*, Cornell University Press, Ithaca, NY.

Sposito, G. (1984) *The Surface Chemistry of Soils*, Oxford University Press, New York. (This monograph gives a comprehensive and didactically valuable interpretation of surface phenomena in soils from the point of view of coordination chemistry.)

Sposito, G. (1990) Molecular Models of Ion Adsorption on Mineral Surfaces, in M. F. Hochella Jr. and A. F. White, Eds., *Mineral–Water Interface Geochemistry*, Mineralogical Society of America, Washington, DC, pp. 261–279.

Sposito, G. (1992) Characterization of Particle Surface Charge. In *Environmental Particles*, Vol. 1, J. Buffle and H. P. van Leeuwen, Eds., Lewis, Chelsea, MI.

Sposito, G. (1994) *Chemical Equilibria and Kinetics in Soils*, Oxford University Press, Oxford.

Sposito, G. (1995) Adsorption as a Problem in Coordination Chemistry: The Concept of the Surface Complex. In *Aquatic Chemistry*, C. P. Huang et al., Eds., ACS, Washington, DC.

Stone, A. T., Torrents, A., Smolen, J., Vasudevan, D., and Hadley, J. (1993) Adsorption of Organic Compounds Possessing Ligand Donor Groups at the Oxide/Water Interface, *Environ. Sci. Technol.* **27**(5), 895–909.

Stumm, W. (1992) *Chemistry of the Solid–Water Interface; Processes at the Mineral–Water and Particle–Water Interface*, Wiley-Interscience, New York. (An introduction into aquatic surface and colloid chemistry.)

Stumm, W. (Ed.) (1987) *Aquatic Surface Chemistry, Chemical Processes at the Particle–Water Interface*, Wiley-Interscience, New York.

Tessier, A. (1993) Sorption of Trace Elements on Neutral Particles in Oxic Environments. In *Environmental Particles*, Vol. 2, J. Buffle and H. P. Van Leeuwen, Eds., Lewis, Boca Raton, FL.

Weber, W. J. Jr., McGinley, P. M., and Katz, L. E. (1991) Sorption Phenomena in Subsurface Systems: Concepts, Models, and Effects on Contaminant Fate and Transport, *Water Res.* **25**, 499–528.

Westall, J. C. (1987) Adsorption Mechanisms in Aquatic Surface Chemistry. In *Aquatic Surface Chemistry*, W. Stumm, Ed., Wiley-Interscience, New York, pp. 3–32.

Zutić, V., and Tomaić, J. (1988) On the Formation of Organic Coatings on Marine Particles: Interactions of Organic Matter at Hydrous Alumina/Seawater Interfaces, *Mar. Chem.* **23**, 51–67.

PROBLEMS

9.1. (a) Explain qualitatively the pH dependence of cation and anion adsorption, respectively.

(b) Why does surface complex formation with a weak acid lead to a relative maximum in the extent of surface complex formation (adsorption) at a pH that is usually near the value of $-\log$(acidity constant) of the weak acid (pK_{HA})?

9.2. A sample of goethite is characterized by the following reactions:

$$\equiv FeOH_2^+ = H^+ + \equiv FeOH \qquad pK_1^s = 6$$

$$\equiv FeOH = H^+ + \equiv FeO^- \qquad pK_2^s = 8.8$$

$$\equiv FeOH + Cu^{2+} = \equiv FeOCu^+ + H^+ \qquad pK^s = -8$$

Electrostatic effects are considered negligible.
(a) Calculate an adsorption isotherm for Cu^{2+} from a dilute $Cu(NO_3)_2$ solution at pH = 7 and $\equiv$FeOHTOT = 10^{-6} M.
(b) What is the qualitative effect on the extent of adsorption of the following factors: (i) presence of HCO_3^- in solution; (ii) increase in temperature of the solution; and (iii) addition of 10^{-3} M Ca^{2+}?

9.3. Discuss the surface binding of metal ions and of ligands in terms of the Lewis acid–base theory.

9.4. Compare the alkalimetric titration of a polyprotic acid (e.g., polyaspartic acid) with that of an Al_2O_3 dispersion; show in either case the effect of the presence of a metal ion (e.g., Cu^{2+}) on the titration curve.

9.5. Describe semiquantitatively the effect of increased pH (at constant alkalinity) or of increased alkalinity (at constant pH) on the binding of Cu(II) to soil particles. (Consider that Cu^{2+} forms soluble carbonato complexes.)

9.6. Does the addition of small quantities of the following solutes to a suspension of α-Fe_2O_3 increase, decrease, or cause no effect on pH_{pznpc}?
(a) NaCl
(b) KF
(c) NaH_2PO_4
(d) $PbCl_2$
(e) $Na_2C_2O_4$ (oxalate)
(f) Humic acid

9.7. The wall of a glass beaker contains $\equiv$SiOH groups. Why can dilute solutions of metal ions (pH > 7) not be stored in glass vessels?

9.8. Check the validity of the following statement: If a suspension of a hydrous oxide does not change its pH upon addition of $NaNO_3$, then this pH value is the pH_{pzc}.

9.9. The surface potential, ψ_0, cannot be measured directly. It can be estimated however, with the help of equation 40a from the surface charge density. Discuss the assumptions involved in applying such a calculation.

9.10. In the figure below (from Stumm et al., 1976), microscopic acidity constants K_{a1}^s of the reaction $\equiv AlOH_2^+ \rightleftharpoons \equiv AlOH + H^+$ for γ-Al_2O_3 are plotted as a function of $\{\equiv AlOH_2^+\}$. (Extrapolation to zero charge gives intrinsic acidity constants.) The data are for 0.1 M $NaClO_4$. The temperature is 25°C. The specific surface area is 120 m² g⁻¹. This figure illustrates (within experimental precision) the conformity of the proton titration data to the constant capacitance model. The relevant expressions are equation vi of Example 9.2, equation 41a, and equation 45b. Calculate the capacitance.

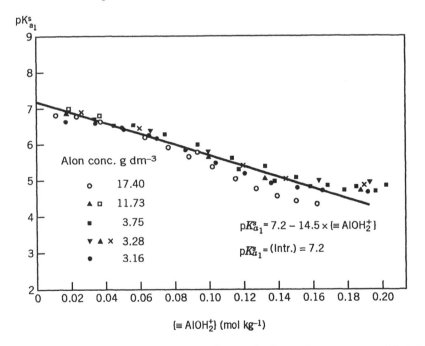

9.11. Some researchers suggest to estimate ψ_0 from the zeta potential ζ by assuming $\psi_0 \simeq \zeta$. Discuss the validity of this approximation.

9.12. Consider the results given below on the zeta potential of Al_2O_3 (corundum) in solutions in various electrolytes by Modi and Fuerstenau (1957). Explain the various potential-increasing and decreasing effects; identify the ions that are specifically adsorbed.

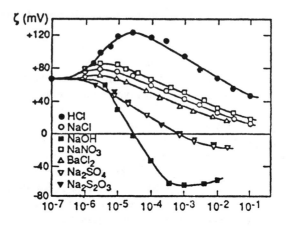

Concentration of electrolyte (eq liter^{-1})

APPENDIX 9.1 THE GOUY–CHAPMAN THEORY

Table A9.1 gives a brief derivation of the equations. The theory is based on the validity of the Poisson equation for distances measured over molecular dimensions. Furthermore, the theory depends on the following assumptions (Sposito, 1984):

1. The surface from which x is measured is a uniform infinite plane of charge.
2. The charged species in the solution are point ions; these ions interact with themselves and with the surface only by Coulomb force.
3. The water in the solution is a uniform continuum characterized by the dielectric constant.
4. The (inner) potential, ψ, at a distance x is proportional to the average energy, $W_i(x)$, required to bring an ion i from infinity to the point x in the solution.
5. We consider *dilute* solutions and equate concentration and activity:

$$W_i(x) = z_i F \psi(x)$$

The limitations imposed on this theory have been discussed (e.g., see Sposito, 1984).

Table A9.1. Gouy–Chapman Theory of Single Flat Double Layer[a]

I. Variation of Charge Density in Solution

Equality of electrochemical potential, $\bar{\mu}\ (=\mu + zF\Psi)$ of every ion, regardless of position

Electrochemical potential:

$$\bar{\mu}_{+(x)} = \bar{\mu}_{+(x=\infty)}, \qquad \bar{\mu}_{-(x)} = \bar{\mu}_{-(x=\infty)} \qquad (1)$$

$$zF(\Psi_{(x)} - \Psi_{(x=\infty)}) = -RT \ln \frac{n_{+(x)}}{n_{+(x=\infty)}} = RT \ln \frac{n_{-(x)}}{n_{-(x=\infty)}} \qquad (2)$$

Cations:

$$n_+ = n_{+(x=\infty)} \exp\left(\frac{-zF\Psi_{(x)}}{RT}\right) \qquad (3)$$

Anions:

$$n_- = n_{-(x=\infty)} \exp\left(\frac{zF\Psi_{(x)}}{RT}\right) \qquad (4)$$

Space charge density:
(if $z_+ = z_-$)

$$q = zF(n_+ - n_-) \qquad (5)$$

II. Local Charge Density and Local Potential

Ψ and q are related by Poisson's equation:

Poisson's equation

$$\frac{d^2\Psi}{dx^2} = -\frac{q}{\varepsilon_0} \qquad (6)$$

Combining equations 3, 4, and 5 with 6 and considering that $\sinh x = (e^x - e^{-x})/2$ gives the

Double-layer equation:

$$\frac{d^2\Psi}{dx^2} = \frac{\kappa^2 \sinh(zF\Psi/RT)}{(2F/RT)} \qquad (7)$$

where κ is the reciprocal thickness of the double layer (the reciprocal Debye length)

$$\kappa = \left(\frac{e^2 \sum_i n_i z_i^2}{\varepsilon_0 k_B T}\right)^{1/2} \qquad (8)$$

Table A9.1. *(Continued)*

For convenience the following substitutions can be made:

$$y = \frac{zF\Psi}{RT} \qquad \bar{z} = \frac{zF\Psi_d}{RT} \qquad \zeta = \kappa x \tag{9}$$

Considering equation 9, equation 8 becomes the

Substituted double-layer equation:

$$\frac{d^2y}{d\zeta^2} = \sinh y \tag{10}$$

For boundary conditions, if $\zeta = \infty$, $dy/d\zeta = 0$ and $y = 0$

First integration:

$$dy/d\zeta = -2 \sinh(y/2), \text{ or} \tag{11}$$

$$d\Psi/dx = -\frac{RT}{zF} 2\kappa \sinh(y/2) \tag{11a}$$

and for boundary conditions, if $\zeta = 0$, $\Psi = \Psi_d$ or $y = \bar{z}$

Second integration:

$$e^{y/2} = \frac{e^{\bar{z}/2} + 1 + (e^{\bar{z}/2} - 1)e^{-\zeta}}{e^{\bar{z}/2} + 1 - (e^{\bar{z}/2} - 1)e^{-\zeta}} \tag{12}$$

Simplified equations for $\Psi_d \ll 25$ mV:

Instead of equation 7

$$\frac{d^2\Psi}{dx^2} = \kappa^2 \Psi \tag{7a}$$

Instead of equation 12

$$\Psi = \Psi_d \exp(-\kappa x) \tag{12a}$$

III. Diffuse Double-Layer Charge and Ψ_d

Surface charge density:

$$\sigma = -\int_0^\infty q \, dx = \varepsilon\varepsilon_0 \int_0^\infty \left(\frac{d^2\Psi}{dx^2}\right) dx \tag{13}$$

$$= -\varepsilon\varepsilon_0 \left[\frac{d\Psi}{dx}\right]_{x=0}$$

Inserting equation 11a,

$$\sigma = (8\varepsilon\varepsilon_0 n_s k_B T)^{1/2} \sinh\left(\frac{zF\Psi_d}{2RT}\right) \tag{14}$$

$$= 0.1174\, c_s^{1/2} \sinh\left(\frac{zF\Psi_d}{2RT}\right) \text{ (C m}^{-2}) \text{ (at 25°C)} \tag{14a}$$

If $\Psi_0 \ll 25$ mV, $\sigma \simeq \varepsilon\varepsilon_0\, \kappa\Psi_d$, or $\sigma = 2.3\, I^{1/2}\, \Psi_d$ \hfill (14b)

^aWe have the following definitions:

μ = chemical potential
$\bar{\mu}$ = electrochemical potential
Ψ = local potential (V)
Ψ_d = diffuse double-layer potential (V)
q = (volumetric) charge density (C cm^{-3})
σ = diffuse surface charge (C m^{-2})
x = distance from surface
n_+ = local cation concentration (mol cm^{-3})
n_- = local anion concentration (mol cm^{-3})
$n_{(x=\infty)}$ = bulk ion concentration (mol cm^{-3})
n_i = number of ions i per cm^3 [$Nn_{(x=\infty)}$] (cm^{-3})
n_s = number of ion pairs (cm^{-3})
c_s = salt concentration (mol liter^{-1} or M)
N = Avogadro's number, 6.02×10^{23} mol^{-1}
I = ionic strength (mol liter^{-1})
z = valence of ion
κ = reciprocal thickness of double layer
e = charge of electron (elementary charge), 1.6×10^{-19} C
k_B = Boltzmann constant, 1.38×10^{-23} J K^{-1}
$k_B T$ = Boltzmann constant times absolute temperature, 0.41×10^{-20} C at 20°C
RT = $N \times k_B T = 2.46 \times 10^3$ V C mol^{-1} at 20°C
ε = relative dielectric permittivity (dimensionless) (ε = 78.5 for water at 25°C)
ε_0 = permittivity in vacuum, 8.854×10^{-14} C V^{-1} cm^{-1}
F = Faraday = $6.02 \times 10^{23} \times e = 96,485$ C mol^{-1}
$F\Psi/RT$ = $e\Psi/k_B T = 1$ for $\Psi = 25$ mV at 20°C

APPENDIX 9.2 CONTACT ANGLE, ADHESION AND COHESION, THE OIL-WATER INTERFACE

In this chapter we introduced the interfacial tension (equivalent to interfacial energy) as the minimum work required to create a differential increment in surface area. The interfacial energy, equally applicable to solids and liquids, was referred to as the interfacial Gibbs free energy at constant temperature, pressure, and composition.

$$\gamma = \left(\frac{\partial G}{\partial A}\right)_{T,p,n} \quad (A.1)$$

where γ is the interfacial tension usually expressed in J m^{-2} or N cm^{-1} (N = newton).

The Contact Angle

Three phases are in contact when a drop of liquid (e.g., water) is placed on a perfectly smooth solid surface and all three phases are allowed to come to equilibrium (Figure A9.1).

The change in surface free energy, ΔG^s, accompanying a small displacement of the liquid such that the change in area of solid covered is ΔA, is

$$\Delta G^s = \Delta A\,(\gamma_{SL} - \gamma_{SV}) + \Delta A\,\gamma_{LV}\cos(\theta - \Delta\theta) \quad (A.2)$$

At equilibrium

$$\lim_{\Delta A \to 0}(\Delta G^s/\Delta A) = 0$$

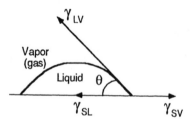

Figure A9.1. Components of interfacial tension (energy) for the equilibrium of a liquid drop on a smooth surface in contact with air or the vapor phase. The liquid, in most instances, will not wet the surface but remains as a drop having a definite *angle of contact* between the liquid and solid phase.

Appendix 9.2 Contact Angle, Adhesion and Cohesion, Oil–Water Interface

and

$$\gamma_{LV} \cos \theta = \gamma_{SV} - \gamma_{SL} \tag{A.3}†$$

This equation, called the *Young equation*, is in accord with the concept that the various surface forces can be represented by surface tensions acting in the direction of the surfaces. Equation A.3 results from equating the horizontal components of these tensions.

For practical purposes, if the contact angle is greater than 90°, the liquid is said not to wet the solid (if the liquid is water one speaks of a hydrophobic surface); in such a case, drops of liquid tend to move about easily and not to enter capillary pores. If $\theta = 0$ (ideal perfect wettability), equation A.3 no longer holds and a *spreading coefficient*, $S_{LS(V)}$, reflects the imbalance of surface free energies:

$$S_{LS(V)} = \gamma_{SV} - (\gamma_{LV} + \gamma_{SL}) \tag{A.4}$$

Young's equation is a plausible, widely used result, but experimental verification is often difficult; for example, the two terms that involve the interface between the solid and the two other phases cannot be measured independently.‡ Furthermore, many complications can arise with contact angle measurements; γ_S values of ionic solids based on contact angle measurements (Table A9.2) are different from those estimated from solubility (see Section 13.7).

The measurement of contact angles for a sessile drop or bubble resting on or against a plane solid surface can be accomplished by direct microscopic examination.

Another method involves adjusting the angle of a plate immersed in the liquid so that the liquid surface remains perfectly flat right up to the solid surface.

---

†γ_{SV} (the surface tension of the solid in equilibrium with the vapor of the liquid) is often replaced by γ_S, the surface energy of a solid (hypothetically in equilibrium with vacuum or its own vapor). The difference between γ_{SV} and γ_S is the equilibrium film pressure, $\pi = \gamma_{SV} - \gamma_S$, which depends on the adsorption of the vapor of the liquid on the solid–gas interface, which lowers surface tension; cf. equation A.3.

‡As shown by Fowkes (1968), the interfacial energy between two phases (whose surface tensions—with respect to vacuum—are γ_1 and γ_2) is subject to the resultant force field made up of components arising from attractive forces in the bulk of each phase and the forces, usually the London dispersion forces, operating across the interface itself. Then the interfacial tension (energy) between two phases γ_{12} is given by

$$\gamma_{12} = \gamma_1 + \gamma_2 - 2 (\gamma_{1(L)} \gamma_{2(L)})^{1/2} \tag{A.5}$$

The geometric mean of the dispersion force components $\gamma_{1(L)}$ and $\gamma_{2(L)}$ may be interpreted as a measure of the interfacial attraction resulting from dispersion forces between adjacent dissimilar phases.

Table A9.2. Attractive Forces at Interfaces: Surface Energy, γ, and London–van der Waals Dispersion Force Component of Surface Energy, $\gamma_{(L)}$[a]

Material	γ (mJ m^{-2})	$\gamma_{(L)}$ (mJ m^{-2})
Liquids[b]		
Water	72.8	21.8
Mercury	484.0	200.0
n-Hexane	18.4	18.4
n-Decane	23.9	23.9
Carbon tetrachloride	26.9	—
Benzene	28.9	—
Nitrobenzene	43.9	—
Glycerol	63.4	37.0
Solids[c]		
Paraffin wax	—	25.5
Polyethylene	—	35.0
Polystyrene	—	44.0
Silver	—	74.0
Lead	—	99.0
Anatase (TiO$_2$)	—	91.0
Rutile (TiO$_2$)	—	143.0
Ferric oxide	—	107.0
Silica	—	123.0 (78)
Graphite	—	110.0

[a] Based on information provided by Fowkes (1968) (20°C). A dash indicates that no value is available.

[b] $\gamma_{(L)}$ values for water and mercury have been determined by measuring the interfacial tension of these liquids with a number of liquid-saturated hydrocarbons. The intermolecular attraction in the liquid hydrocarbons is entirely due to London–van der Waals dispersion forces for all practical purposes. $\gamma_{(L)}$ was derived from contact angle measurements.

[c] $\gamma_{(L)}$ of solids were derived from contact angle measurements or from measurements of equilibrium film pressures of adsorbed vapor on the solid surface.

Adhesion and Cohesion

Processes related to γ and θ are adhesion, cohesion, and spreading. We consider two phases A and B without specifying their physical state; their common interface is AB. We can distinguish the following processes as they affect a unit area using a notation given by Hiemenz (1986).

1. Cohesion:

$$\text{no surface} \rightarrow 2 \text{ A (or B) surfaces} \tag{A.6}$$

Appendix 9.2 Contact Angle, Adhesion and Cohesion, Oil–Water Interface

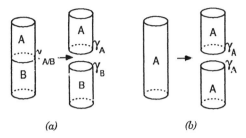

Figure A9.2. (a) Work of adhesion and (b) work of cohesion.

2. Adhesion:

$$1 \text{ AB surface} \rightarrow 1 \text{ A} + 1 \text{ B surface} \quad (A.7)$$

Figure A9.2 gives a schematic illustration.

The work of *adhesion*, W_{AB}, is the work to separate 1 m² of AB interface into two separate A and B interfaces and is given by

$$\Delta G = W_{AB} = \gamma_{\text{final}} - \gamma_{\text{initial}} = \gamma_A + \gamma_B - \gamma_{AB} \quad (A.8)$$

(If AB is a solid–liquid interface, A and B are the A and B interfaces with vapor and, in an exact sense, γ_A and γ_B are γ_{AV} and γ_{BV}, respectively.)

The work of *cohesion*, for example, of a pure liquid, consists of producing two new interfaces, each of 1 m²; it measures the attraction between the molecules of this phase:

$$\Delta G = W_{AA} = 2\gamma_A \quad (A.9)$$

Wetting, Water Repellency, and Detergency

Obviously, the spreading of water on a hydrophobic solid is favored by adding a surfactant. γ_{LV} and γ_{SL} are thereby reduced, and, on both accounts, $\cos \theta$ (equation A.3) increases and θ decreases.

Water repellency is achieved by making the surface hydrophobic; that is, γ_{SV} has to be reduced as much as possible. If $\gamma_{SV} - \gamma_{SL}$ is negative, $\theta > 90°$. Water-repelling materials include waxes, petroleum residues, and silicones; for example, if $\equiv$SiOH groups on glass or SiO_2 surfaces have reacted with silanes to form $\equiv$SiO—Si—Alkyl groups, these surfaces have a hydrocarbon type of surface. If θ is greater than 90°, the water will tend not to penetrate into the hollows or pores in the solid; under these conditions, gas bubbles will attach to the surfaces. The contact angle plays an important role in flotation and in detergency. Table A9.3 gives some data on contact angles. In the latter the

The Solid–Solution Interface

Table A9.3. Some Contact Angle Values for Solid–Liquid–Air Interfaces[a] (25°C)

Solid	Liquid	Contact Angle
Glass	Water	~0
TiO_2	Water	~0
Graphite	Water	86
Paraffin	Water	106
Polytetrafluorethylene	Water	108
Polyethylene	Water	94

[a] Data given are quoted from Adamson (1990) were references are given.

Figure A9.3. The adhesion of solid particles to solid surfaces (or to other particles) can be reduced by surfactants. The work of adhesion is given (cf. equation A.8) by $W_{SP} = \gamma_{PW} + \gamma_{SW} - \gamma_{SP}$. Thus γ_{PW} and γ_{SW} have to be lowered to reduce the adhesion.

action of the detergent is to lower the adhesion between a dirt particle and a solid surface (Figure A9.3). While we employ the arguments used in problems of detergency, the same considerations apply to illustrate that surfactants can reduce the adhesion of particles to solid surfaces or to other particles.

The Oil–Water Interface

Although the oil–water interface may not strictly belong in the realm of this book, we would like to illustrate briefly that the elementary concepts discussed here are directly applicable to this interface and can be used to understand some of the phenomena associated with oil spills.

In Figure A9.4, adhesion and cohesion are illustrated for the oil–water interface.

The work of adhesion is influenced by the orientation of the molecules at the interface. For example, with the help of Table A9.2 and equation A.8, the work of adhesion of n-decane–water (corresponding to a paraffinic oil–water system) and of glycerol–water can be computed to be 40×10^{-3} J m^{-2} and 56×10^{-3} J m^{-2}, respectively. It requires more work to separate the polar glycerol molecules (oriented with the OH groups toward the water) from the water phase than the nonpolar hydrocarbon molecules. For paraffinic oils, W_{OO} is about 44 mJ m^{-2}, for water W_{WW} is 144 mJ m^{-2}, and for glycerol W_{OO} is 127 mJ m^{-2}.

Appendix 9.2 Contact Angle, Adhesion and Cohesion, Oil–Water Interface

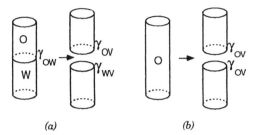

Figure A9.4. (a) Work of adhesion and (b) work of cohesion for the oil–water interface.

$$\text{Adhesion:} \quad W_{OW} = \gamma_{OV} + \gamma_{WV} - \gamma_{OW} \tag{i}$$

$$\text{Cohesion for oil and water:} \quad W_{OO} = 2\gamma_{OV}; \quad W_{WW} = 2\gamma_{WV} \tag{ii}$$

$$\text{Spreading:} \quad S_{O/W} = \gamma_{WV} - \gamma_{OV} - \gamma_{OW} \tag{iii}$$

Spreading (equation iii of Figure A9.4) occurs when the oil adheres to the water more strongly than it coheres to itself; this is generally the case when a liquid of low surface tension is placed on one of high surface tension. Thus mineral oil spreads on water, but water cannot spread on this oil. The initial spreading coefficient does not consider that the two liquids will, after contact, become mutually saturated. The addition of surfactants, which lower γ_{OW} and γ_{SW} (cf. Figure A9.3), causes the dispersion of the oil into droplets.

10

TRACE METALS: CYCLING, REGULATION, AND BIOLOGICAL ROLE

10.1 INTRODUCTION: GLOBAL CYCLING OF METALS

The anthropogenic dispersion of metals into the atmosphere appears to rival, and sometimes exceed, natural mobilizations. Metals are released into the atmosphere, both as particles and as vapors, as a result not only of fossil fuel (coal, oil, natural gas) combustion but also of cement production and extractive metallurgy. Elements are termed atmophile when their mass transport through the atmosphere is greater than that in streams. Many atmophile elements are volatile and have metal oxides of relatively low boiling points. It is also known that some of these metals, especially the B-type metals—Hg, As, Se, Sn, and Pb—can be methylated and/or released into the atmosphere as vapors and that Hg and probably As and Se are released as inorganic vapor from the burning of coal. In contrast, the A-type metals—Al, Ti, Sn, Mn, Co, Cr, V, and Ni—are termed lithophile because their mass transport to the oceans by streams exceeds their transport through the atmosphere (Tables 10.1a and 10.1b).

The concentrations of trace metals in continental waters are controlled by atmospheric precipitation and the weathering processes on soils and bedrocks. Because these pathways and processes have been altered significantly by humankind, the flux and distribution of trace metals in a large fraction of all freshwater resources have increased.

Domestic and industrial wastewaters, sewage discharges, and urban runoff also contribute large quantities of metal pollution to the aquatic environment. These discharges often occur at point sources and can lead to excessive local metal burdens in water.

The increase in loading of heavy metals as a consequence of civilization can be estimated from the memory record of sediments (Figure 10.1).

The atmosphere, in particular, has become a key medium in the transfer of pollutant trace metals to remote aquatic ecosystems. On a global scale, this pathway annually supplies more than 70% of the lead and vanadium, about 30% of the mercury, and about 20% of the cadmium flux into aquatic ecosystems. In many rural and remote regions, the atmosphere actually supplies most

Table 10.1a. Worldwide Emissions (10^3 tonnes per year) of Trace Metals from Natural Sources to the Atmosphere

Element	Windborne Soil Particles	Sea Salt Spray	Volcanoes	Forest Fires	Biogenic Sources	Total[a]
Antimony	0.78	0.56	0.71	0.22	0.29	2.6
Arsenic	2.6	1.7	3.8	0.19	3.9	12
Cadmium	0.21	0.06	0.82	0.11	0.24	1.4
Chromium	27	0.07	15	0.09	1.1	43
Cobalt	4.1	0.07	0.96	0.31	0.66	6.1
Copper	8.0	3.6	9.4	3.8	3.3	28
Lead	3.9	1.4	3.3	1.9	1.7	12
Manganese	221	0.86	42	23	30	317
Mercury	0.05	0.02	1.0	0.02	1.4	2.5
Molybdenum	1.3	0.22	0.40	0.57	0.54	3.0
Nickel	11	1.3	14	2.3	0.73	29
Selenium	0.18	0.55	0.95	0.26	8.4	10
Vanadium	16	3.1	5.6	1.8	1.2	28
Zinc	19	0.44	9.6	7.6	8.1	45

[a]Totals are rounded.
Source: Nriagu (1989).

of the trace metal budgets of aquatic ecosystems. For example, more than 50% of all the trace metal getting into the Great Lakes is transported via the atmosphere (Nriagu, 1986).

Another evaluation of metal transfer rates was made by Sposito (1986) on the basis of the data compiled by Buat-Menard (1985). Despite many uncertainties, the conclusion emerges from Sposito's review (1986) that the trace metals Cu, Zn, Ag, Sb, Sn, Hg, and Pb are the most potentially hazardous on a global or regional scale. Lead is of acute concern on the global scale, because of its prominent showing in all the enrichment factors and transfer rates considered. These conclusions are in accordance with those of Andreae et al. (1986a) in a Dahlem report.

Thus the B-type metals (i.e., the soft Lewis acids) are not only enriched in the natural environment, but they are also, because of their toxic effect (i.e., their tendency to react with soft bases, e.g., with SH and NH groups in enzymes), potentially hazardous to ecology and human health.

10.2 ANALYTICAL APPROACHES TO CHEMICAL SPECIATION

Our ability to accurately determine trace metal concentrations, and thus our knowledge of the chemistry of these elements in marine and freshwater systems, has undergone a revolution (Bruland et al., 1991). Equally or even more important than the development of highly sensitive and partially specific analytical

Table 10.1b. Worldwide Atmospheric Emissions (10^3 tonnes per year) of Trace Metals from Anthropogenic Sources

Element	Energy Production	Mining	Smelting and Refining	Manufacturing Processes	Commercial Uses[a]	Waste Incineration	Transportation	Total[b]
Antimony	1.30	0.10	1.42			0.67		3.5
Arsenic	2.22	0.06	12.3	1.95	2.02	0.31		19
Cadmium	0.79		5.43	0.60		0.75		7.6
Chromium	12.7			17.0		0.84		31
Copper	8.04	0.42	23.2	2.01		1.58		35
Lead	12.7	2.55	46.5	15.7	4.50	2.37	248	332
Manganese	12.1	0.62	2.55	14.7		8.26		38
Mercury	2.26		0.13			1.16		3.6
Nickel	42.0	0.80	3.99	4.47		0.35		52
Selenium	3.85	0.16	2.18			0.11		6.3
Thalium	1.13			4.01				5.1
Tin	3.27		1.06			0.81		5.1
Vanadium	84.0		0.06	0.74		1.15		86
Zinc	16.8	0.46	72.0	33.4	3.25	5.90		132

[a]Including agricultural uses.
[b]Totals are rounded.
Source: Nriagu and Pacyna (1988).

10.2 Analytical Approaches to Chemical Speciation

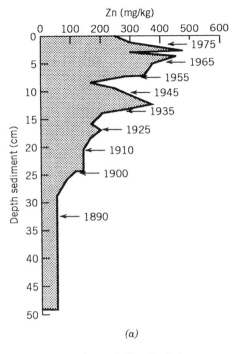

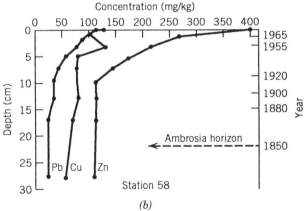

Figure 10.1. Memory records of sediments. (a) Zn loading in sediments of Lake Zurich. Prior to urbanization of the catchment area of the lake, Zn concentrations of the sediments reflect natural levels. After World War II, increased industrial activity is reflected in the large increase in the Zn concentrations of the sediments. In the last years the Zn load has decreased, largely because of more stringent air pollution legislation and because of improved waste treatment. (Adapted from Beer and Sturm, 1992. (b) Depth–concentration profiles in sediment cores of Pb, Cu, and Zn in Lake Erie. (From Nriagu et al., 1979.)

618 Trace Metals: Cycling, Regulation, and Biological Role

techniques were the advances made in using clean techniques in water sample collection and in sample processing during analysis. As a consequence of clean techniques, the concentrations of many of the bioactive trace metals were found to be orders of magnitudes lower than previously thought. Table 10.2 summarizes some data on the distribution of dissolved trace metals in marine surface waters. Refer to Table 6.1 for a comparison of trace metals in fresh waters. The assessment of the extremely low concentrations of these bioactive trace metals in oceanic and fresh waters has prompted a reevaluation of their role with respect to phytoplankton productivity.

In Figure 6.1 we illustrated the various forms of the occurrence of Cu(II) in natural waters, and in Table 6.5 the major inorganic species in fresh water and in seawater were listed. One would wish to have analytical possibilities to distinguish between the various solute and "adsorbed" species or to identify the solid or surface sites [organic surface, iron(III) oxide, aluminum silicates] in which the metal ion is present or bound to. Usually, the evidence for a

Table 10.2. Distribution of Dissolved Trace Metals in Surface Marine Waters

Metal	Environment	Salinity (‰)	Concentration (nmol liter^{-1})	Reference[a]
Iron	Pacific Ocean	35	0.05–0.5	1–4
	Coastal North Carolina	36	4–23	5
	Newport River Estuary, NC	20–33	60–250	5
	San Francisco Bay	12	6000	1
Manganese	Pacific Ocean	34–36	0.3–4	2–4, 6
	Antarctic Ocean		0.08	7
	Sargasso Sea	36	2–5	8
	Coastal Massachusetts	30	21	8
	Newport River Estuary, NC	24–31	16–400	5
Zinc	North Pacific Ocean	32–35.6	0.06–0.24	4, 9
	Sargasso Sea	36	0.06 ± 0.02	8
	Coastal Massachusetts	30	2.4	8
	Beaulieu Estuary	5–28	60–600[b]	10
Copper	North Pacific Ocean	32–35.6	0.4–1.4	4, 9
	Sargasso Sea	36	1.2	8
	Coastal Massachusetts	30	4.1	8
	Narragansett Bay	26–32	3.7–260[b]	11
Molybdenum	Central Pacific and Atlantic Oceans		107	12
Cobalt	Northeast Pacific Ocean	33–34.5	0.004–0.05	3, 4
	Outside San Francisco Bay	33–28	0.25–1.5	13

[a](1) Gordon et al. (1982); (2) Landing and Bruland (1987); (3) Martin and Gordon (1988); (4) Martin et al. (1989); (5) Evans (1977); (6) Klinkhammer and Bender (1980); (7) Martin et al. (1990): (8) Bruland and Franks (1983); (9) Bruland (1980); (10) Holliday and Liss (1976); (11) Mills and Quinn (1984); (12) Collier (1985); (13) Knauer et al. (1982).
[b]Extremely high concentrations probably result from pollution.
Source: Adapted from Sunda (1988/89). See original paper for references.

10.2 Analytical Approaches to Chemical Speciation

particular form of occurrence is circumstantial and is based on complementary evaluations together with kinetic and thermodynamic considerations.

To a first approximation, we can consider—as we shall see in more detail—that the uptake of a metal ion in an aqueous medium by an organic surface (roots in soil or phytoplankton in surface waters) is proportional to the concentration (activity) of free ion (Fraústo da Silva and Williams, 1991).

Analytical Difficulties Although we can make many chemical equilibrium models that predict the existence of complexes in natural waters, analytically one encounters difficulties in identifying unequivocally the various solute species and in distinguishing between dissolved and particulate concentrations. The analytical task is rendered very difficult because the individual chemical species are often present at nano- and picomolar concentrations. The ion-selective electrode (ISE), if it were sufficiently sensitive, would permit the measurement of free metal-ion activity.

The sensitivity of ion-selective electrodes can be excellent in metal-ion-buffered solutions, that is, at fairly high total metal-ion concentrations. Under these conditions, commercial solid-state membrane electrodes were shown to follow the Nernst equation in fresh waters down to pCu 11–12 and after preconditioning in a metal-ion buffer to pCu = 13 (Sigg and Xue, 1994; Sunda and Hanson, 1979; Xue and Sigg, 1990). Some ion-selective electrodes, especially the Cu electrode, show interference in high chloride media (seawater).

Otherwise no single simple method permits unequivocal identification of species. In voltammetric methods, the current resulting from the oxidation or reduction of the metal ion at an electrode is measured; the measured current signal is proportional to the electrochemically *labile* metal concentration, that is, the concentration of the metal species available to the electrode within the time scale of the electrochemical method. The labile metal-ion concentration may include inorganic as well as weak organic complexes. Since the direct electrochemical methods (polarography) are not sufficiently sensitive, the voltammetric techniques used usually involve a preconcentration step. For example, in anodic stripping voltammetry (ASV) the metal is first cathodically reduced at a suitable deposition potential on a hanging mercury drop or film electrode and accumulated on it as mercury amalgam. A subsequent oxidative stripping step (reverse polarography) permits one to measure the anodic current during the oxidative step; its peak current (or the integral of current versus time) relates to the metal concentration. The sensitivity is good (down to about nanomolar concentrations) but, as mentioned above, this method does not determine directly the free metal ions but only the labile species (free aquo ions and inorganic complexes and readily dissociable organic complexes). Therefore the concentration of labile metals as measured by ASV or by the even more sensitive DPASV (differential pulse anodic stripping voltammetry) cannot be directly related to bioavailability. Comprehensive references for electroanalytical methods in trace metal analysis in natural waters are Whitfield and Jagner (1981) and Buffle (1988).

ASV and DPASV are still useful speciation methods because of their sensitivity and their ability to distinguish labile metal species from strong (nonlabile) organic complexes. Furthermore, these techniques are very useful to distinguish dissolved species from species bound to colloidal or particulate phases (Gonçalves et al., 1985). For recent applications see Bruland (1989, 1992), Donat and Bruland (1990), and Muller and Kester (1990, 1991).

Titrating Waters with Metal Ions The application of these techniques often involves titration of the natural water with metal ions, from which the conditional stability constant of metal–organic complexes and the ligand concentrations can be obtained (Coale and Bruland, 1988). These parameters are significant in evaluating the complexation of metals by natural organic ligands; they are used to estimate by extrapolation the free metal-ion concentration at the ambient total concentrations when sensitivities are insufficient. The extrapolation provides only an upper limit.

Figure 10.2 illustrates such a titration curve. The underlying approach for

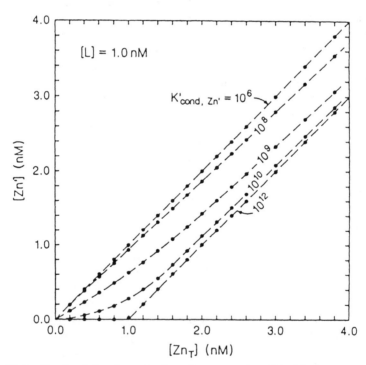

Figure 10.2. Simulated titrations of a ligand concentration of 1 nM, demonstrating the effect of varying the conditional stability constant, $K'_{\text{cond Zn}'}$. A titration curve with a $K'_{\text{cond Zn}'} < 10^6$ ($K'_{\text{cond Zn}'} [L] < 10^{-3}$) is indistinguishable from that of a blank solution containing no ligand. (From Bruland, 1989.)

10.2 Analytical Approaches to Chemical Speciation

this figure is as follows.[†] An electroinactive complex, ZnL, is formed when a dissolved ligand (or class of ligands), L, complexes with electroactive zinc (forming a 1:1 complex):

$$Zn' + L' = ZnL \qquad K'_{\text{cond Zn}'} = [ZnL]/[Zn'][L'] \qquad (1)$$

where [Zn'] is the total concentration of dissolved inorganic zinc species (or, more exactly, the labile zinc) present, [L'] is the concentration of organic ligands that have the potential to strongly bind zinc, and [ZnL] is the concentration of the zinc–organic ligand complex. $K'_{\text{cond Zn}'}$ is the conditional stability constant under specific conditions (pH, ionic strength, concentration of competing cations). See the discussion on conditional constants in Section 6.6. [Zn'] is related to the free aquo zinc ion, [Zn^{2+}], by an inorganic side reaction coefficient $\alpha_{\text{Zn}'}$. As indicated in Table 6.5, theoretical estimates of $\alpha_{\text{Zn}'}$ in seawater are about 2.1; that is, ~47% of the dissolved inorganic Zn exists as free aquo Zn^{2+} ion. Either the zinc species constituting Zn' are directly reduced, or the dissociation rates of the complexes (such as ZnCl$^+$) are kinetically fast. [L'] includes the concentrations of free dissolved ligand, L^{n-}, protonated forms, and ligands that may be complexed with other cations (e.g., CaL).

Figure 10.3a illustrates the titration of a seawater sample. In Figure 10.3b the data of Figure 10.3a are linearly transformed into

$$\frac{[Zn']}{[ZnL]} = \frac{[Zn']}{[L]} + \frac{1}{K'_{\text{cond Zn}'}[L]} \qquad (2)$$

The ratio [Zn']/[ZnL] plotted versus [Zn'] for each titration point yields a straight line with a slope of [L]$^{-1}$ and intercept of $(K'_{\text{cond Zn}'}[L])^{-1}$. Equation 2 can be derived as follows (Ruzic, 1982): [ZnL] is defined by equation 1.

$$[ZnL] = \frac{[L]}{1 + 1/K'_{\text{cond Zn}'}[Zn']} \qquad (3)$$

where [L] is the concentration of dissolved ligand

$$[L] = [ZnL] + [L']$$

With the help of equation 3, one obtains for [Zn']/[ZnL]

$$\frac{[Zn']}{[ZnL]} = \frac{[Zn']}{[L]}\left(1 + \frac{1}{K'_{\text{cond Zn}'}[Zn']}\right) \qquad (4)$$

[†]We summarize some of the arguments presented by Bruland (1989).

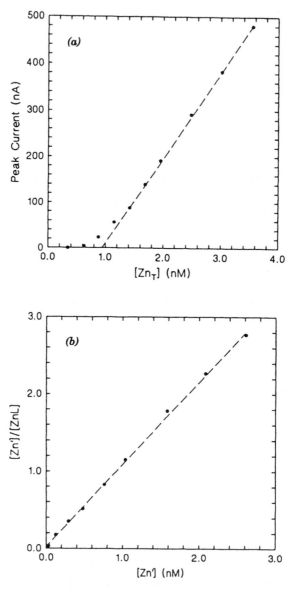

Figure 10.3. (a) Zinc titration of a typical sample from the central North Pacific 90 m (VERTEX-IV). The slope of the line is the sensitivity of the electrode. (b) Linear transformation of the 90-m titration data indicates a ligand concentration of 0.94 nM and $\log K'_{\text{cond Zn'}} = 10.6$. (From Bruland, 1989.)

$$\frac{[Zn']}{[ZnL]} = \frac{[Zn']}{[L]} + \frac{1}{K'_{\text{cond Zn'}}[L]} \tag{5}$$

which is the same as equation 2.

Since free metal-ion activity relates to bioavailability (or toxicity to organisms), *bioassays* instead of chemical analysis have been used successfully to determine free $[Cu^{2+}]$ in fresh and seawater.

Ligand Exchange Some indirect methods are based on exchange with known ligands. This technique is based on the competition for metals between natural organic ligands and added known ligands, and the subsequent specific and sensitive measurement of the complexes with these known ligands. The free metal-ion concentrations are then determined by equilibrium calculations, on the basis of the formation of the complexes formed with the known ligands and of the stability constants. In other words, a variety of organic ligands, such as EDTA, catechol, and hydroxyquinoline, and sensitive analytical techniques such as DPASV and DPCSV (*differential pulse anodic or cathodic stripping voltammetry*) (Van den Berg et al., 1987, 1990) and chemiluminescence (Sunda and Huntsman, 1991) are combined to assess free ion activity.

To illustrate the principle of this method we follow the description of Xue and Sigg (1994) for the determination of free Cu^{2+} in fresh water. In this method, ligand exchange with catechol and determination of the copper–catechol complexes by *cathodic stripping voltammetry* (*CSV*) were used for the speciation of Cu, according to the method developed by Van den Berg (1984). Catechol is added to a natural water sample and forms complexes with copper in competition with the natural ligands; the Cu–catechol complexes are measured specifically by CSV. After addition of catechol to a sample of natural water and equilibration of the ligand exchange between catechol and natural organic ligands, the dissolved copper is distributed as follows:

$$[Cu]_T = [Cu^{2+}] + [Cu]_{\text{inorg}} + \Sigma[CuL_i] + \Sigma[Cu(cat)_i] \tag{6}$$

or

$$[Cu]_T = [Cu^{2+}]\left(1 + \Sigma\beta_i[L_{i\,\text{inorg}}] + \Sigma\beta_{i\,\text{org}}[L_{i\,\text{org}}] + \alpha_{\text{cat}}\right) \tag{7}$$

where $\Sigma[Cu(cat)_i]$ is the concentration of copper–catechol complexes that is specifically determined by CSV; $[Cu]_{\text{inorg}}$ is the concentration of inorganic complexes, and $\Sigma[CuL_i]$ is the concentration of complexes with natural organic ligands. $\alpha_{\text{cat}} = \beta_{1\,\text{cat}}[\text{cat}^{2-}] + \beta_{2\,\text{cat}}[\text{cat}^{2-}]^2$. $\beta_{1\,\text{cat}}$ and $\beta_{2\,\text{cat}}$, respectively, represent the stability constants of $Cu(cat)$ and $Cu(cat)_2^{2-}$ complexes.

The free copper-ion concentration is calculated on the basis of the equilibrium with the free catechol present in solution:

$$[Cu^{2+}] = \Sigma[Cu(cat)_i]/\alpha_{cat} \tag{8}$$

$\Sigma \beta_{i\,org}[L_{i\,org}]$ and $[Cu^{2+}]$ in the absence of catechol can then be calculated from equation 7. The factor $\Sigma \beta_i[L_{i\,inorg}]$ can be calculated from the major ion composition of the water sample, so that $\Sigma \beta_{i\,org}[L_{i\,org}]$ for the organic complexation is obtained. The directly obtained results are thus values for $[Cu^{2+}]$ and for $\Sigma \beta_{i\,org}[L_{i\,org}]$. Models have to be applied in order to separate the term $\Sigma \beta_{i\,org}[L_{i\,org}]$ into various stability constants and ligand concentrations.

Similar calculations can be applied to other ligand exchange reactions.

This method was applied to the speciation of Cu in a eutrophic lake (Lake Greifen, Switzerland), with the objective of relating the speciation to the chemical and biological processes in the lake. Total concentrations of dissolved Cu in the water column of Lake Greifen were 5–20 nM. The speciation of Cu obtained by the ligand exchange method indicated very strong complexation of Cu; pCu was in the range 14–16 (Figure 10.4). These results indicated the presence of very strong ligands for copper, at concentrations that are in excess of the total dissolved Cu. The free Cu^{2+} concentrations were six to seven orders

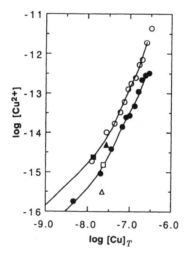

Figure 10.4. $[Cu^{2+}]$ in a eutrophic lake. Titration curves of Lake Greifen water with Cu(II) are in terms of $\log[Cu^{2+}]$ versus $\log[Cu_T]$. The samples shown were collected at 5 m depth on April 23 (●) and May 21 (○), 1990. The curves fitted to the water data were computed from the conditional stability constants and ligand concentrations for strong and weak organic ligands. The following concentrations and conditional stability constants were able to fit the data: L_1 = 40–90 nM, $\log K_1$ = 14.3; and L_2 = 250–550 nM, $\log K_2$ = 12.3. The values obtained from a simplified two-ligand model may represent an average of different ligands with different stability constants. (From Xue and Sigg, 1993.)

of magnitude lower than the concentrations of total dissolved copper; >99% of the total dissolved copper must thus be present in the form of organic complexes.

The free [Cu^{2+}] measured in Lake Greifen is very low, and relatively high concentrations of strong organic ligands are present. This is probably related to the high organic productivity. Seawater values for [Cu^{2+}] are typically two to three orders of magnitude larger.

The high-affinity chelators for Cu(II) in seawater are present at concentrations of 1–5 nM and have conditional stability constants of 10^{12}–10^{13}.

The extremely low concentrations of Cu^{2+} (and of other bioactive trace elements) cannot be explained by the EDTA present in the lake at 10^{-8} M or by humic–fulvic acids. The natural organic ligands that strongly complex Cu are probably biogenic ligands, exuded by phytoplankton. The very low [Cu^{2+}] is related to high productivity of phytoplankton, and possibly to the exudation of complex formers (biochelators) by the algae (Xue and Sigg, 1993).

10.3 CLASSIFICATION OF METAL IONS AND THE INORGANIC CHEMISTRY OF LIFE

Before discussing questions of bioavailability, uptake of metals by biota, and toxicity, we would like to draw a few generalizations on the basis of the chemical properties of metals (Fraústo da Silva and Williams, 1991; Nieboer and Richardson, 1980). We refer to our previous discussion on class A and class B metals (Section 6.6) and recall that class A ions seek out oxygen binding sites while class B ions prefer nitrogen and/or sulfur centers. As pointed out by Nieboer and Richardson, all macronutrients belong to class A and the micronutrient metals belong to the borderline group. The preferences of these metal-ion groups for ligands are summarized in Table 10.3.

The most common class A ions essential for biological processes are K^+, Na^+, Mg^{2+}, and Ca^{2+}. Ca^{2+}, in contrast to Mg^{2+}, is found principally in extracellular locations. Ca^{2+} is bound in proteins via a complex involving some combinations of backbone carbonyls, carboxylate, and alcohol functions. In model systems such as EDTA or NTA, Ca^{2+} is believed to bind weakly to nitrogen only because it is brought into close proximity when bound to the oxygen centers in these ligands (Nieboer and Richardson, 1980). Mg^{2+} plays a major role in living organisms because it stabilizes soft structures, especially the macromolecules RNA and DNA. This is achieved by Mg^{2+} ion acting as a counterion for the negatively charged phosphate moieties in these molecules. Mg^{2+} is more B-type than Ca^{2+} (Figure 6.12). Mg^{2+} is surrounded in chlorophyll by four nitrogen atoms and a single oxygen atom; on the other hand, Mg^{2+} in the chlorophyll complex is relatively labile; it can be displaced by H^+ from weak acids.

The most important role of the borderline trace metals in algal physiology is that of *essential* micronutrients; these serve as cofactors in a number of

Table 10.3. Metal-Ion Binding[a]

(a) Ligands Encountered in Biological Systems

I. Ligands Preferred by Class A Metal Ions	II. Other Important Ligands	III. Ligands Preferred by Class B Metal Ions
F^-, O^{2-}, OH^-, H_2O	Cl^-, Br^-, N_3^-, NO_2^-	H^-, I^-, R^-, CN^-
CO_3^{2-}, SO_4^{2-}, $ROSO_3^-$, NO_3^-	SO_3^{2-}, NH_3, N_2, RNH_2	CO, S^{2-}, RS^-, R_2S, R_3As
HPO_4^{2-}, $-O-\overset{O}{\underset{O^-}{P}}-O-$ etc.	R_2NH, R_3N, $=N-$, $-CO-N-R$	
ROH, RCO^-, $-\overset{O}{\underset{}{C}}-$, ROR	O_2, O_2^-, O_2^{2-}	

(b) Metal-Ion Binding Sites in Proteins Based on Crystallographic Studies

Functional Groups Sought by Class A Metal Ions	Functional Groups Sought by Class B Metal Ions
Carboxylate: $R-\overset{O}{\underset{}{C}}-O^-$	Sulfhydryl: $-SH$
Carbonyl: $R-\overset{O}{\underset{}{C}}-OR$, $R-NH\overset{O}{\underset{}{C}}-R$	Disulfide: $-S-S-$
Alcohol: $R-\overset{}{\underset{}{C}}-OH$	Thioether: $-SR$
Phosphate: $R-OPO_3^{2-}$	Amino: $-NH_2$
Phosphodiester: $R-O-\overset{O}{\underset{O^-}{P}}-O-R$	Heterocyclic nitrogen: imidazole of histidine, nucleotide bases

[a] The symbol R represents an alkyl radical such as CH_3-, CH_3CH_2-, and so on. The RNH_2 could represent an amine such as CH_3NH_2. In a few cases R could also be an aromatic moiety such as the phenyl ring. Class A metal ions have an absolute preference in aqueous solution for the types of ligands in column I, all of which bind through oxygen. Class B metal ions exhibit a high affinity for the ligand types in column III but are also able to form strong complexes in aqueous solutions with the ligands in column II. Borderline metal ions can interact with ligands in all three columns but may exhibit preferences.

Source: Nieboer and Richardson (1980).

enzymes (Table 10.4). All but zinc and nickel can exist in more than one oxidation state in cells, and catalysis of redox reactions and electron transport are major functions of trace-metal-containing enzymes. For example, the borderline character of iron manifests itself in the great variety of ligands that iron can bind and the spectacular range of redox potential Fe(II)/Fe(III) systems can achieve (see Section 8.6, Figure 8.26). Fe is present in the active center of cytochromes and iron–sulfur proteins (e.g., ferredoxin), which are important components of the photosynthetic and respiratory electron transport chains. Manganese serves as the primary electron acceptor in the oxidation of water in photosystem (II). Molybdenum (together with Fe) is essential for nitrogen fixation. Cu serves in electron transfer and catalysis of redox reactions with a

Table 10.4. Some Enzymes and Redox Proteins Containing Trace Metal Cofactors

Metal	Enzyme	Function
Fe	Cytochrome f	Photosynthetic electron transport
	Cytochromes b and c	Electron transport in respiration and photosynthesis
	Ferredoxin	Electron transport in photosynthesis and nitrogen fixation
	Iron–sulfur proteins	Photosynthetic and respiratory electron transport
	Catalase	H_2O_2 breakdown to H_2O and O_2
	Peroxidases	H_2O_2 reduction to H_2O
	Chelatase	Porphyrin and phycobiliprotein synthesis
Fe and Mo	Nitrogenase	Nitrogen fixation
	Nitrate and nitrite reductases	Nitrate reduction to ammonia
Mn or Fe	Superoxide dismutase	Disproportionation of O_2^- radicals to O_2 and H_2O_2
Mn	O_2 evolving enzyme	Oxidation of water to O_2 in photosynthesis
Zn[a]	DNA and RNA polymerases	Nucleic acid replication and transcription
	Carbonic anhydrase	Hydration and dehydration of CO_2
	Alkaline phosphatase	Hydrolysis of phosphate esters
Cu	Plastocyanin	Photosynthetic electron transport
	Cytochrome c oxidase	Mitochondrial electron transport
	Ascorbate oxidase	Ascorbic acid oxidation and reduction
Co	Vitamin B_{12}[b]	Carbon and hydrogen transfer reactions
Ni	Urease	Hydrolysis of urea

[a]Zn may be replaced by Cd in many if not most of these enzymes based on the findings of Price and Morel (1990).
[b]Cofactor in a number of enzymes.
Source: Adapted from Sunda (1988/89).

number of organic molecules. Zn is, among other things, involved in the hydrolysis of phosphate esters and the hydration and dehydration of CO_2 (Sunda, 1991).

Class B ions are more toxic than borderline ions, which are more toxic than class A ions. Class B ions are most effective at binding to SH groups (of cysteine) and to nitrogen centers (e.g., lysine, histidine).

10.4 ORGANOMETALLIC AND ORGANOMETALLOIDAL COMPOUNDS

Several B metals, the borderline metal Co, and all metalloid metals can form element-carbon bonds that are stable in water. Metallorganic compounds of A ions and of borderline ions (exception Co) hydrolyze in contact with water. Methylation and alkylation are common reactions in biological systems; they may also proceed abiotically and generate methyl- or alkyl-element compounds. Examples include selenide, selenoamino acids, methylarsenic acid (MMAA), and dimethylarsenic acid (DMAA):

$$\underset{\text{MMAA}}{\overset{\text{CH}_3}{\underset{\text{OH}}{\overset{|}{\text{O}=\text{As}-\text{OH}}}}} \qquad \underset{\text{DMAA}}{\overset{\text{CH}_3}{\underset{\text{CH}_3}{\overset{|}{\text{O}=\text{As}-\text{OH}}}}}$$

The arsenic cycle shown below (Irgolic and Martell, 1985) may serve to convey an idea about the diversity of compounds formed by arsenic.

Andreae (1986b) has written a review on organoarsenic compounds in the environment. A survey on the various forms of methyltransfer mechanisms of environmental significance has been given by Rapsomanikis (1986) and a general review on organometallic compounds in the environment has been edited by Craig (1986).

Metals and metalloids that form alkyl compounds (e.g., methylmercury) deserve special concern because these compounds are volatile and accumulate in cells; they are poisonous to the central nervous system of higher organisms. Because methylmercury (or other metal alkyls) may be produced at a rate faster than it is degraded by other organisms, it may accumulate in higher organisms such as fish. It is one of the few examples that demonstrate an increase in biological availability of a nonessential element in toxic concentrations through chemical transformation by the system.

Methylmercury Species Because of the role of metal and metalloid alkyls in the toxicity of certain elements and the importance of such forms in the movement, bioaccumulation, and transformation of these elements, we briefly

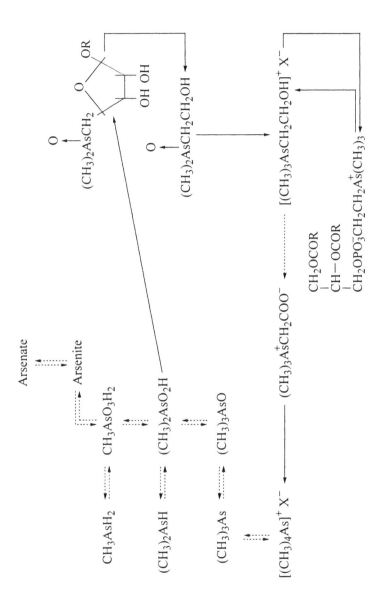

630 Trace Metals: Cycling, Regulation, and Biological Role

discuss here some properties of CH_3Hg^+ and its complexes. CH_3Hg^+ exists in aqueous solutions as an aquo complex $CH_3-Hg-OH_2^+$ with a covalent bond between Hg and O. The cation behaves as a soft acid and has a strong preference for the addition of only one ligand (Table 10.5a). CH_3Hg^+ undergoes rapid coordination reactions with S, P, O, N, halogen, and C; the rate of the formation of Cl^-, Br^-, and OH^- complexes is extremely fast (diffusion controlled). The CH_3Hg^+ unit itself, however, is kinetically remarkably inert toward decomposition. Therefore methylmercury compounds once formed—usually by biologically mediated methylation—are not readily demethylated. The neutral CH_3Hg^+ species are hydrophilic, lipophilic, and volatile; thus they can readily pass through boundaries. This, together with their broad tendency to form stable complexes quickly and the robustness of the CH_3Hg^+ unit, characterizes some of the far-reaching toxicological properties of methylmercury. See also Mason et al. (1993).

Example 10.1. Thermodynamic Stability of CH_3Hg^+ Species Interpret the equilibrium constants given in Table 10.5a to predict under what conditions various CH_3Hg^+ species are formed (thermodynamically) and which species predominate in natural waters.

A concentration ratio diagram may be constructed both for seawater ($[Cl^-] = 0.6$ M) and for fresh water ($[Cl^-] = 2 \times 10^{-4}$ M). A concentration of $[CH_4(aq)] = 10^{-4}$ M is assumed. The resulting diagrams (Figure 10.5) allow the following conclusions: (1) CH_3HgOH is the stable methylmercury species in fresh water. (2) Methylmercury compounds can—thermodynamically speaking—be formed in natural waters containing CH_4; these compounds should

Table 10.5a. Stability of Methylmercury and Its Complexes[a]

Reaction	Log K
(1) $Hg^{2+} + CH_3^- = CH_3Hg^+$	~50
(2) $CH_4(aq) = CH_3^- + H^+$	~−47
(3) $Hg^{2+} + CH_4(aq) = CH_3Hg^+ + H^+$	~3
(4) $CH_3Hg^+ + CH_3^- = (CH_3)_2Hg$	~37
(5) $CH_4(aq) + HgCl_2 = CH_3HgCl + H^+ + Cl^-$	−5.2
(6) $CH_3Hg^+ + Cl^- = CH_3HgCl$	5.25
(7) $CH_3Hg^+ + H_2O = CH_3HgOH + H^+$	−4.63
(8) $CH_3Hg^+ + CO_3^{2-} = CH_3HgCO_3^-$	6.1
(9) $CH_3Hg^+ + SO_4^{2-} = CH_3HgSO_4^-$	0.94
(10) $CH_3Hg^+ + S^{2-} = CH_3HgS^-$	21.02
(11) $CH_3Hg^+ + CH_3HgOH = (CH_3Hg)_2OH$	6.1
(12) $CH_3Hg^+ + CH_3HgS^- = (CH_3Hg)_2S$	16.34
(13) $HgS(s) + CH_4 = CH_3HgS^- + H^+$	~−26

[a] The more important inorganic Hg species are characterized by the following equilibrium constants (25°C, $I = 1$): $Hg(OH)_n^{2-n}$; log $\beta_1 = 10.1$, log $\beta_2 = 21.1$; $HgCl_n^{2-n}$: log $\beta_1 = 6.72$, log $\beta_2 = 13.23$, log $\beta_3 = 14.2$, log $\beta_4 = 15.3$; log $*\beta_{HgOHCl}$ ($Hg^{2+} + H_2O + Cl^- = HgOHCl$) = 3.67.

10.4 Organometallic and Organometalloidal Compounds

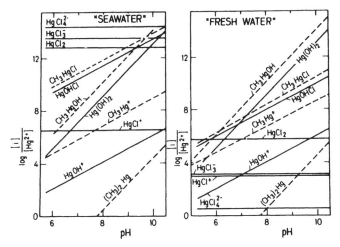

Figure 10.5. Relative thermodynamic stability of CH_3Hg^+ species. With the help of the equilibrium constants given in Table 10.5a, concentration ratio diagrams have been constructed for the following conditions: seawater $[Cl^-] = 0.6$ M, $[CH_4(aq)] = 10^{-4}$ M; fresh water $[Cl^-] = 2 \times 10^{-4}$ M, $[CH_4(aq)] = 10^{-4}$ M (see Example 10.1). CH_3Hg species once formed (usually by biomethylation) are inert with regard to demethylation; their coordination reactions, however, occur fast and in agreement with those predicted by equilibrium constants. Thus, while equilibria among the various CH_3Hg species and equilibria among the inorganic Hg species depicted prevail, the CH_3Hg species are most likely not in equilibrium with the inorganic Hg species.

(thermodynamically) decompose at low pH:

$$CH_3HgOH + 2H^+ = CH_4 + Hg^{2+} + H_2O \qquad \log K \simeq 1.6$$

$$CH_3HgCl + H^+ + Cl^- = CH_4 + HgCl_2 \qquad \log K \simeq 5.2$$

Because of the inertness of the CH_3Hg^+ units, this decomposition occurs only very slowly. In other words, we can treat the CH_3Hg^+ species in their coordination reactions (reactions 5–12 of Table 10.5a) as metastable. Because CH_3Hg^+ species can be formed in a microenvironment (i.e., inside a cell), they can be present in the bulk phase in concentrations that seem thermodynamically incompatible.

Trialkyltin compounds, R_3Sn^+, are ecologically problematic substances because of their high toxicity toward aquatic organisms. As antifouling agents, they are directly introduced into the water. Transport, distribution, bioaccumulation, and bioavailability of trialkyltin compounds depend on speciation (Table 10.5b). The neutral species, R_3SnOH and R_3SnCl, are more readily partitioned into lipophilic material.

Trialkyltin compounds are predominant as cations at low pH and at pH > 6.6 as R_3SnOH. In seawater a small fraction ($\sim 3\%$) is present as R_3SnCl.

Table 10.5b. Equilibria of Trialkyltin Compounds

Reaction[a]	log K (25°C)
$Me_3Sn^+ + H_2O = Me_3SnOH + H^+$	-6.60
$Bu_3Sn^+ + H_2O = Bu_3SnOH + H^+$	-6.2
$Me_3Sn^+ + Cl^- = Me_3SnCl$	-0.17
$Bu_3Sn^+ + Cl^- = Bu_3SnCl$	-0.37[a]

[a] Me = methyl, Bu = butyl.
Source: Weidenhaupt (1994).

The sorption of trialkyltin compounds on organic material or organic carbon containing material can be interpreted as distribution of R_3SnX between water and the organic "phase" (similar to the distribution between water and octanol). R_3Sn^+ behaves also as a "hard" cation, undergoes cation exchange at low pH with clay minerals, and forms, in the pH range of natural waters, surface complexes with oxide surfaces, for example, with the basal gibbsite layers of kaolinite or montmorillonite (Weidenhaupt, 1994).

$$\equiv Al-OH + Bu_3Sn^+ = \equiv AlOBu_3Sn + H^+ \qquad \log K = -3.07$$

10.5 BIOAVAILABILITY AND TOXICITY

Metals are partitioned in the biota by different acid–base affinities, by their kinetics, by spatial partitioning (e.g., by membranes and compartments), and by temporal partitioning (Fraústo da Silva and Williams, 1991). One aspect of metal toxicity, but by no means the only one, is the chemical combination of metals and ligands (Lewis acids and bases) in organisms. The cellular bases are almost exclusively sulfur, nitrogen, and oxygen donor groups (including H_2O and solute bases, e.g., HCO_3^-, HPO_4^{2-}, and OH^-). The acids are H^+; the essential metal cations (e.g., Na, K, Mg, Ca, Cr, Mn, Fe, Co, Ni, Cu, Zn, and Mo) and the potentially hazardous metals, if present (e.g., Hg^{2+}, CH_3Hg^+, Pb^{2+}, Cd^{2+}, Cr). Among the sulfur-seeking ("soft-soft") B-type metals are, of course, the toxicants Hg, Pb, and Tl, as well as the essential protein and enzyme metals Fe, Cu, and Zn. The proton, H^+, has high affinity for all donors, S, N, and O—thus the key role of pH in metal binding in organisms. Mg, Ca, Be, Al, Sn, Ge, and the lanthanides are oxygen-seeking ("hard–hard").

The Relevance of Free Metal-Ion Activity

The chemical behavior of any element in the environment depends on the nature of its compounds and species. Physiological, ecological, and toxicological effects of a metal are usually strongly structure specific; that is, they depend on the species. Thus, for example, the effect of Cu on the growth of algae depends on whether Cu(II) is present as free Cu^{2+} ion or is present as a carbonato or an organic complex. This has been exemplified in some studies on the relationship of Cu^{2+} ion activity and the toxicity of copper to phytoplankton. In these experiments cupric ion activity was altered independently of total Cu(II) concentration by varying the complex-forming concentration and the pH—that is, by using pCu buffers (Jackson and Morgan, 1978; Morel, 1983; Morel and Hudson, 1985; Sunda, 1988/89; Sunda and Guillard, 1976).

Our discussion concentrates primarily on single-cell organisms. They are special in the sense that they are directly in contact with the aqueous medium, being separated from it only by a membrane.

Figure 10.6 illustrates in a simplified way the physiological dependence (growth or biomass production) on the concentration of an essential element. The lowest concentration range limits growth; in the next concentration range an optimum in growth is obtained; and at higher concentrations toxic effects (reduction in growth) are observed. As this figure shows, nonessential elements (such as Cd, Pb) cause a negative physiological effect with increasing concentration.

To demonstrate that the free ion activity of a metal is a key parameter that determines its biological effects, it is necessary to measure the same physiological responses for the same free metal-ion activities, obtained by several

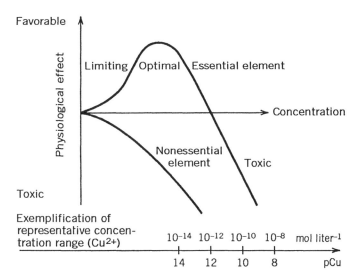

Figure 10.6. Relation between the concentration of an element and its physiological effect. In order to give an idea of the "reactive" concentration range for the interaction with algae, a plausible scale for pCu (= $-\log$ [Cu]) is given. Note, however, that the concentration range depends on the type of algae and on the presence of other (competing) metal ions and differs for different metal ions.

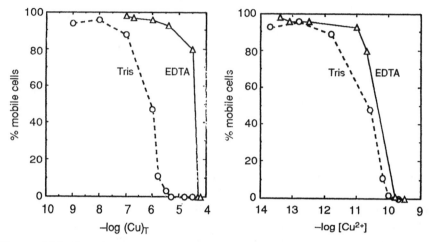

Figure 10.7. Experiments with different chelators and a wide range of trace metal concentrations demonstrate that trace metal toxicity and deficiency are determined by metal-ion activities and not total concentrations. Motility data of the dinoflagellate *Gonyaulax tamarensis* as a function of total copper $[Cu]_T$ and cupric ion activity $[Cu^{2+}]$ for two chelators, Tris and EDTA. (Adapted from Anderson and Morel, 1978).

combinations of chelating agents and ligand concentrations. For example, Figure 10.7 shows how different combinations of complex formers and total metals in algal culture media result in a unique physiological response to the activity of Cu^{2+}. Thus, in designing experiments to study the biological effects of trace metals, one needs first to calculate the trace metal speciation in the medium. This often involves equilibrium calculations for multimetal, multiligand systems.

These concepts, with certain limitations, can also be extended to higher organisms. Andrews et al. (1977) have demonstrated that the toxicity of Cu(II) to a freshwater crustacean is directly connected with free cupric ion activity, but not with the activities of Cu complexes or particulate forms of copper.

Sanders et al. (1983) have also shown that the effects of Cu(II) on the growth of crab larvae and on their metallothionein with copper chelate buffer systems must be interpreted on the basis of free Cu^{2+} ion activity. The data obtained reveal predictable relations between $[Cu^{2+}]$ in seawater and processes at the cellular and organismic levels. Similarly, the uptake of metal ions by plants (e.g., of aluminum) is usually related to free metal-ion activity.[†] Others have shown that the chelation of a variety of metals reduces the toxicity of metals to organisms: for example, a reduction in the uptake of mercury by fish in the presence of EDTA and cysteine; a reduction in copper and/or zinc toxicity to

---

[†] It is often difficult to distinguish experimentally between the free metal ion and its labile inorganic complexes (they are interdependent by equilibria). The free metal-ion activity may be replaced by the total inorganic metal concentration.

fish by NTA, glycine, and humic acids; and a reduction in copper toxicity to clams by EDTA.

The dependence of metal toxicity or availability to aquatic organisms on free metal-ion concentrations in solution may in fact be a rather widespread phenomenon. The existence of such a dependence should allow one to predict changes in the response of an organism to a particular metal through knowledge of the variations in the aqueous chemistry of the metal. Variables such as the total concentrations of the metal in question, pH, alkalinity, the concentration of natural chelators, the concentration of competing metals, and the presence of adsorptive surfaces all can affect the concentration of free metal ions and thus affect the response of an organism to that particular metal.

Metal Uptake by Algae

The uptake of metal ions by phytoplankton appears in most cases to be a two-stage process involving the binding of a pool of metal on the outside surface of the cell either by biologically released ligands or by surface functional ligand groups. Subsequent to this surface complex formation, the metals are carried through the membrane—usually by porter molecules—to the inside of the cell. The uptake model depicted here (Figure 10.8) is much influenced by Fraústo da Silva and Williams (1991) and their research on structural aspects of metal toxicity and on the discussions by Morel and Hudson (1985), Wood and Wang (1985), and Sigg (1987). On the water side, an equilibrium exists between the various complexes of solutes and surface ligands with the free metal ion. A competition between the different ligands for H^+ and different metal ions establishes multidimensional equilibria. Since these equilibria are interdependent, no element is free from the label "toxic" at certain dose levels (Williams, 1981). The kinetic aspects include the rates of the ligand exchange reactions, the rate of reaction and release of the carrier ligand L_1, and the rate of transport of ML_1 across the membrane (usually by active transport). If the transport into the cell is slow in comparison to the pre-equilibration processes on the solution side, then the uptake of the metal ion by the cell depends on the free metal-ion concentration. The production and release of L_1, L_2, and so on are related to the growth rate of the algae. The selectivity of certain metal ions is given by the selectivity of the ligand L_1. Carrier molecules are often proteins, and the stability of their complexes with different metal ions corresponds to the Irwing–Williams stability order. Steric factors are also involved in the selectivity. The concentration inside the cell depends on the availability of the ligand L_1, on the rate of transport across the membrane, and possibly on the rate of transport out of the cell.

The central importance of free metal-ion activities in controlling the biological effects does not necessarily mean that the free aquo metal ions are actually the chemical species taken up by aquatic organisms. It reflects the fact that the chemical reactivity of a metal is measured by the free metal-ion activity and that the physiological effects of a metal are mediated by chemical reactions between the metal and the various cellular ligands (Morel and Hering, 1993).

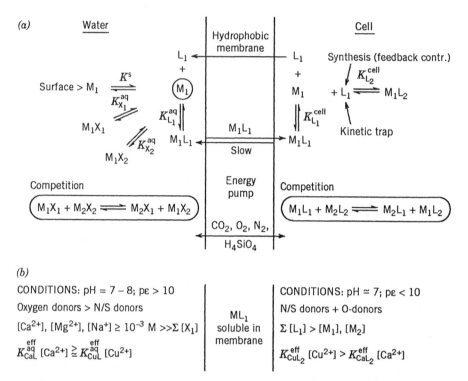

Figure 10.8. Schematic model of metal ion uptake through a membrane of a phytoplankton cell. (a) The metal ion is bound to the outside surface of the cell either by biologically released ligands or by surface functional ligand groups subsequently to the surface complex formation. The metals are carried—usually by porter molecules—to the inside of the cell. If the transport into the cell is slow in comparison to the preequilibration process on the solution side, then the uptake of the metal ion of the cell depends on the free metal ion activity. (b) Solution variables outside and inside the cell.

The metal ions in the cell are used in biochemical processes or become trapped in inactive forms (e.g., as complexes of metallothionein) as a detoxification mechanism. Toxic effects are observed in algae when the cellular concentration of toxic metal ions reaches some critical level approaching the minimum cellular concentration of essential trace metals. The system becomes "overflooded" with toxic metal ions that then *react with critical enzymes.*

Toxic metals are generally transported into cells by nutrient metal uptake systems. Metabolic binding sites are never completely specific for intended nutrient metals, and thus nutrient metals can be competitively displaced from those sites by inhibitory or toxic metals. The toxic metal gains access to interior metabolic sites while the uptake of the nutrient metal is competitively inhibited (Sunda, 1991).

In Figure 10.8b the solution variables outside and inside the cell (pH, redox potential, pε, and types of ligands and cations) are compared. In the surrounding water oxygen donors (hard bases such as OH^-, HCO_3^-, carboxylates) prevail over soft bases (NH_3, HS^-). The hard earth alkali and alkali cations typically exceed the concentrations of trace metals and ligands; thus the tendency to form Ca and Mg complexes exceeds the tendency to form trace metal complexes. On the other hand, on the inside of the cell, reducing conditions prevail, and therefore the concentrations of soft bases (NH_3, HS^-) become preponderant in addition to oxygen donor ligands. The concentrations of the ligands may now exceed the concentrations of the trace metals, and the tendency to form complexes with Cu^{2+} exceeds that of Ca^{2+}.

Methylation (or alkylation) of metal ions may increase the bioavailability of nonessential elements. For example, the methylation or alkylation of Hg(II) results in an increase of toxicity. Hg–alkyl compounds—typically methylmercury is present in seawater as the noncharged species CH_3HgCl and in fresh water as CH_3HgOH—are taken up faster and permeate the membranes more easily than charged inorganic Hg(II) species[†], because they are more lipid soluble. On the other hand, organoarsenic species are claimed to be less toxic in the food chain and for humans than inorganic arsenics.

10.6 METAL IONS AS MICRONUTRIENTS

As we have seen, trace metals are involved as cofactors of metalloenzymes and proteins, in all general metabolic processes of phytoplankton, including photosynthesis and respiration, and in assimilation of macronutrients. The vertical profiles of trace metal concentrations in open oceans (Figure 10.9) are like those of macronutrients; that is, they show surface depletion resulting from algal uptake and partial release at greater depth due to mineralization.

Similar types of depth profiles can also be observed in lakes (Figure 10.10), although the vertical concentration difference is often not as pronounced in lakes as in the ocean. The reason for this is that the water column in lakes is much shorter, photosynthesis and respiration overlap vertically, and mixing and stagnation are now more dynamic than in oceans. Continuous inputs of metal ions to the upper layer make it more difficult to observe a depletion in the epilimnion (upper layer of the lake).

Redfield Stoichiometry

The interaction between trace metals and phytoplankton is reciprocal; trace metals affect the algal community, but this community in turn has a major impact on trace metal chemistry. Phytoplankton assist in regulating the concentrations of many trace metal nutrients. As we have seen in Section 10.5, feedback mechanisms include the exudation of high-affinity, relatively specific complexing ligands. Furthermore, trace metals can have a major impact on the composition (biological species) of the biological community. Algal uptake of

---

[†]$HgCl_2(aq)$ can also permeate passively into the interior of the cell.

638 Trace Metals: Cycling, Regulation, and Biological Role

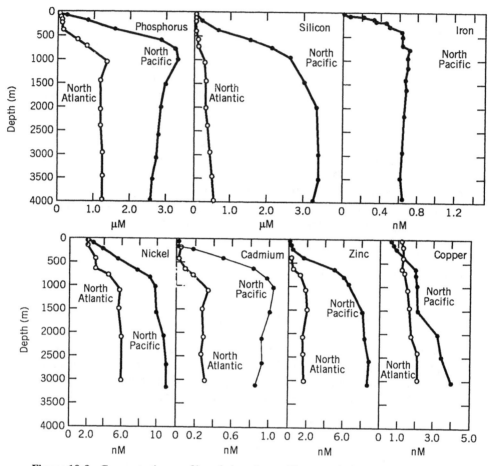

Figure 10.9. Concentration profiles of phosphate, silicate, and dissolved trace elements in open oceans. Fe, Ni, Cd, Zn, and Cu show characteristic surface depletion resulting from uptake by algae. (Cd is nonessential but may be taken up the same way as nutrients.) (Data on iron are from Martin et al., 1989; others from Bruland and Franks, 1983.) (Adapted from Morel and Hering, 1993.)

macronutrients and trace element nutrients in the surface layers is followed by regeneration of these nutrients back into solution in deeper waters with the microbial degradation of the sinking biogenic particles. This cycle depletes the concentrations of nutrients in surface waters in the ratios that they occur in phytoplankton and enriches them in the deeper waters by the same ratios. These ratios are referred to as Redfield ratios (cf. Section 15.5). The observed correlations between the concentrations of macronutrient and general trace metals (Figures 10.9 and 10.10) have invited comparison to the Redfield ratio for major nutrients (Boyle, 1976; Bruland et al., 1978; Morel and Hudson, 1985; Sigg, 1985). This concept suggests that trace metals are present in phytoplank-

10.6 Metal Ions as Micronutrients

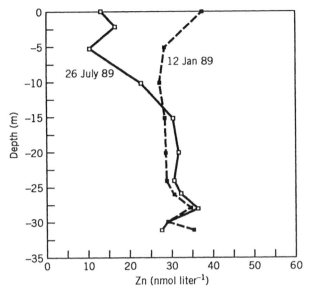

Figure 10.10. Zn in an eutrophic lake (Lake Greifen). Depth-concentration profile is given for Zn in the water column (in January the water is mixed because of winter circulation). (Adapted from Kuhn et al. 1993.)

ton in relatively constant proportions; some of these trace metals could be colimiting the growth of aquatic biota. Morel et al. (1994) have shown that, for certain marine algae, the bicarbonate uptake may be limited by zinc as HCO_3^- transport appears to involve the zinc metalloenzyme carbonic anhydrase and the concentration of Zn in seawater is low enough to limit the growth of certain phytoplankton. The Redfield model for the biological control of major algal nutrients should be extended to many trace elements using the stoichiometric formula

$$\{C_{106}H_{263}O_{110}N_{16}P_1Si_xFe_aMn_bZn_cCu_dCd_eNi_f, \text{etc}\} \tag{9}$$

The mole ratios a, b, c, and so on can be obtained either from the metal/PO_4 or metal/C slope of the trace metals and phosphate and HCO_3^- in the nutricline, as well as from the metal/PO_4 or metal/C in plankton.

Analysis of the composition of algae is difficult because it is not easily possible to separate the algae from other colloidal or suspended material. A further complication is that metal ions in the water column are complexed to various degrees by natural organic ligands. More data are needed to draw conclusions whether, and under what conditions, Redfield ratios are maintained.

Table 10.6 gives data on the stoichiometry of plankton. The information in Table 10.6b may not be true Redfield ratios.

Table 10.6. Composition of Phytoplankton

(a) Elemental composition of plankton[a]

Element	Martin et al. (1976) Sta. 54-88 ($n = 9$)	Martin and Knauer (1973) Monterey Bay ($n = 4$)	Collier and Edmond (1983) MANOP C ($n = 2$)
P	260 ± 80	250	280 ± 30
Ca	480 ± 200	160 ± 10	1400 ± 100
Si	1600 ± 2900	1000 ± 400	470 ± 130
Al	2.0 ± 0.9	1.2 ± 0.7	0.83 ± 0.16
Fe	1.3 ± 0.8	1.3 ± 1.7	1.3 ± 0.1
Zn	0.47 ± 0.30	0.21 ± 0.16	0.83 ± 0.27
Mn	0.10 ± 0.03	0.097 ± 0.052	0.096 ± 0.001
Cu	0.10 ± 0.03	0.05 ± 0.03	0.15 ± 0.03
Ni	0.09 ± 0.04	0.05 ± 0.04	0.24 ± 0.02
Cd	0.12 ± 0.05	0.017 ± 0.004	0.15 ± 0.01

(b) Ratios of Zn and Cu to Phosphate and Organic Carbon in Settling Particles from Different Lakes and in Algae[b]

Lake/Algae	Zn/P (mol/mol)	Zn/C (mol/mol)	Cu/P (mol/mol)	Cu/C (mol/mol)	Reference
Lakes, settling particles					
L. Greifen	0.01–0.06	1–6 × 10^{-4}	2–15 × 10^{-3}	3–15 × 10^{-5}	Sigg et al. (1994)
L. Zurich	0.02–0.07	2–7 × 10^{-4}	4–20 × 10^{-3}	4–20 × 10^{-5}	Sigg et al. (1987)
L. Michigan		5 × 10^{-4}		15 × 10^{-5}	Shafer et al. (1991)
Windermere		5 × 10^{-4}		43 × 10^{-5}	Hamilton-Taylor et al. (1984)
Mountain L		2 × 10^{-4}		8 × 10^{-5}	Nriagu and Wong (1989)
Algae					
Marine algae		0.3–1 × 10^{-4}			Sunda and Huntsman (1992)
Marine plankton	2–5 × 10^{-3}		5–6 × 10^{-4}		Collier and Edmond (1984)
Cyclotella cryptica	1–2 × 10^{-3}		7 × 10^{-5}		Kiefer (1994)

[a] Data from North Pacific samples with Al content < 100 μg g^{-1} dry wt. Values are in units of μmol g^{-1} dry wt. From Bruland et al. (1991).
[b] Adapted from Sigg et al. (1994).

10.7 THE INTERACTION OF TRACE METALS WITH PHYTOPLANKTON AT THE MOLECULAR LEVEL

For trace metals to be taken up, they first must bind (form complexes with) ligands present at the cell surface that are able to transport the metal complex to the membrane (Figure 10.8) (Morel et al., 1991; Sunda, 1988/89).

Kinetically, the simplest scheme representing this uptake is

$$M + L \underset{k_d}{\overset{k_f}{\rightleftarrows}} ML \tag{10a}$$

$$ML \xrightarrow{k_{in}} M_{cell} \tag{10b}$$

As in Michaelis–Menten enzyme kinetics, the steady-state assumption of ML is given by

$$\frac{d[ML]}{dt} = k_f [M][L] - (k_d + k_{in})[ML] = 0 \tag{11}$$

$$[ML]_{ss} = \frac{k_f}{k_d + k_{in}} [M][L] = K_s^{-1} [M][L] \tag{12}$$

Since

$$L_T = [ML] + [L] \tag{13}$$

$$[ML]_{ss} = \frac{L_T [M]}{K_s + [M]} \tag{14}$$

The uptake rate V is given by

$$V = k_{in} [ML]_{ss} \tag{15}$$

When $[M] > K_s$, $V = V_{max}$; that is,

$$V_{max} = k_{in} L_T \tag{16}$$

and the uptake rate of metal ion, M, into the cell, V, is given by

$$V = \frac{V_{max} [M]}{K_s + [M]} \tag{17}$$

where V_{max} is the maximum rate achieved when the transport ligands are fully saturated (equation 16) and [M] is the concentration of the metal ion in the medium to be taken up by cells via the cellular ligand L.[†] K_s is defined by equation 12; it is the half-saturation constant, equal to the metal ion concentration at which half the transport molecules are bound; k_f, k_d, and k_{in} are rate constants defined in the reactions 10a and 10b. In particular, k_{in} is the rate constant for the transport of the metal across the membrane and subsequent transfer to the cytoplasm.

Under usual circumstances, trace metal uptake ligands are far from saturated in natural waters; most of the ligands are free and available to react with the essential metals in the water ([M] < K_s).

Equation 17 simplifies to

$$V = \frac{V_{max}\ [M]}{K_s} \qquad (18)$$

and the uptake rate is then a function of the metal concentration.

Two limiting cases are important:

1. If $k_d \gg k_{in}$, then an equilibrium between the metal bound to the metal transport ligand and that in the water can be inferred ($K_s \simeq k_d/k_f$); the uptake rate, determined by the free metal-ion concentration, is slow compared to the establishment of all other complex formation equilibria.

2. If, on the other hand, $k_{in} \gg k_d$ ($K_s \simeq k_f/k_{in}$), the transport is then kinetically controlled by the rate of the formation of the surface complex ML, which, in turn, is controlled by the dissociation of a complex species, MX, in the water (a dissolved complex or a complex in or on a solid phase). The importance of the free metal-ion concentration results from the assumption that the exchange of the metal between the chelator X and the uptake ligand L proceeds via dissociation of the complex MX.[‡] Then it is the steady-state

---

[†] M may represent either the free metal ion or the labile dissolved metal species, often denoted M' (cf. Section 10.2).

[‡] Consider the reactions

$$MX \underset{k_{-1}}{\overset{k_1}{\rightleftharpoons}} M + X \qquad (i)$$

$$M + L \underset{k_d}{\overset{k_f}{\rightleftharpoons}} ML \qquad (ii)$$

$$\frac{d[M]}{dt} = 0 = k_1\ [MX] - [M]\ (k_{-1}[X] + k_f[L]) \qquad (iii)$$

$$[M]_{ss} = \frac{k_1\ [MX]}{k_{-1}\ [X] + k_f\ [L]} \qquad (iv)$$

If $k_f[L] < k_{-1}[X]$,

$$[M]_{ss} \simeq \frac{k_1\ [MX]}{k_{-1}\ [X]_T} \qquad (v)$$

and

$$V \propto [M]_{ss} \qquad (vi)$$

10.7 The Interaction of Trace Metals with Phytoplankton

concentration of the fast-reacting metal species that determines the reaction rate with the transport ligands, and hence the uptake rate ("kinetic control"). Typically, all the inorganic metal species may be considered to react rapidly and the relevant concentration is then their sum, denoted [M']. (But at constant pH, [M'] is proportional to [M].) Thus one cannot easily distinguish between "thermodynamic" control and "kinetic" control of the uptake (Morel and Hering, 1993; Morel et al., 1991; Sunda, 1988/89).

As noted above, the uptake of metals occurs via binding to specific membrane transport ligands and is related to the external concentration of free metal ions or dissolved inorganic metal species by the saturation kinetics equation (equation 17). In the case of Fe and Mn (the two metal ions studied best to date), the half-saturation constant K_s is fixed for each metal and algal species and does not vary with the availability (concentration) of the metal in solution. On the other hand, V_{max} is not fixed but increases markedly with decreasing external metal concentration. This increase in V_{max} appears to be under negative feedback control by one or more cellular metal pools (Figure 10.8), allowing the cell to regulate its internal iron and manganese concentrations at optimum levels for growth, despite significant changes in the free metal-ion concentration in the environment (Harrison and Morel, 1986; Sunda and Huntsman, 1985). Figure 10.11 gives data from Sunda and Huntsman (1986) on cellular manganese concentrations as a function of pMn^{2+}.

Example of Feedback Interaction Between Zinc and Phytoplankton in Seawater

In the last few years a large amount of data in the field and in the laboratory has established mounting evidence that essential trace elements in the nutricline

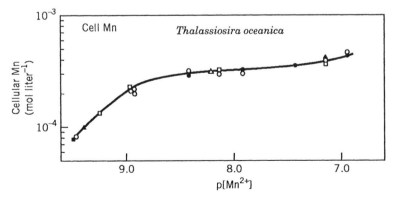

Figure 10.11. Regulation of cellular manganese concentration as a function of pMn^{2+} in *Thalassiosira oceanica*. Uptake follows Michaelis–Menten kinetics. With decreasing $[Mn^{2+}]$, increases in V_{max}, at constant K_s, allow cells to maintain relatively constant cellular Mn concentrations. (Data from Sunda and Huntsman, 1985.)

are controlled by biological uptake and regeneration as occurs for major nutrients. In reporting here some experiments by Sunda and Huntsman (1992) on the biological uptake of Zn and its regulation, we would like to give an excerpt of some of this research and to illustrate the type of experiments that are carried out and give some insights into the argumentation presented. While it is not possible to cover complete details and present all the circumstantial evidence, we hope that the illustration given will encourage the reader to study some of the original papers mentioned in the Suggested Readings at the end of this chapter.

The laboratory experiments were carried out in Zn ion buffers made in natural seawater medium (pH = 8.2) with EDTA. The extent of metal complexation in the medium was computed in the equilibrium calculations, as discussed in Chapter 6. For Zn in EDTA medium, the concentration of inorganic Zn' was computed with an experimentally measured conditional stability constant of 6.4×10^7 M^{-1}. This constant was defined by the equation

$$\frac{[Zn\ EDTA^{2-}]}{[Zn'][EDTA']} = K'_c \tag{19}$$

where [Zn'] is the total concentration of inorganic Zn species (free Zn^{2+} and inorganic complexes) and [EDTA'] is the concentration of EDTA complexes with Ca and Mg ions. Figure 10.12a gives two growth curves as a function of $[Zn^{2+}]$ (out of many more reported by Sunda and Huntsman, 1992)—one for a coastal species and one for an oceanic species. The growth rate of the coastal species was limited at $[Zn^{2+}]$ below 10^{-7} M and was inhibited at the highest Zn level. The oceanic species showed no limitation at the lowest Zn ion concentration ($10^{-12.3}$ M) and was unaffected at $[Zn^{2+}]$ = 10^{-7} M. The higher Zn toxicity of the coastal species was associated with a marked increase in cellular Zn/C ratio (Figure 10.12b), indicating that it could be caused by higher Zn accumulation. Figure 10.12b plots cellular Zn/C ratio versus $[Zn^{2+}]$. These curves have sigmoidal shapes with a plateau (minimal slope) in the pZn^{2+} range 8–10.5. The curves are very similar for all the isolates (including four isolates not shown here). The pZn range for the plateau is characteristic of that for unpolluted marine waters (Bruland, 1989). The establishment of a plateau is indicative of a negative feedback regulation. The laboratory data (Figure 10.12b) were compared in Figure 10.12c with Zn/C mole ratios calculated from field data of three stations in the eastern North Pacific. (These ratios were computed from the slopes of ΔZn versus ΔPO_4 in the nutriclines of the water column; the Zn/PO_4 ratios were converted to Zn/C ratios, using the Redfield C/P ratio for plankton of 106:1.) The agreement between this modeled relationship of the field data with those from the laboratory provides strong evidence that Zn concentrations in the nutricline are controlled by biological uptake and regeneration occurs as for major nutrients (Sunda and Huntsman, 1992).

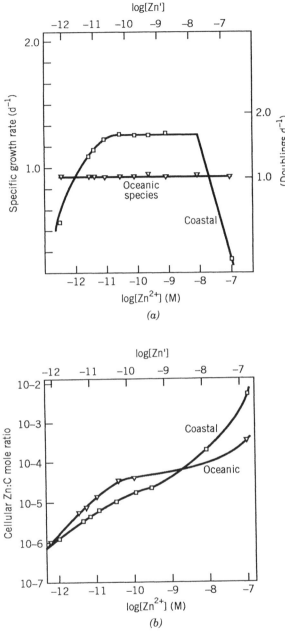

Figure 10.12. Biological uptake of Zn and its regulation. (Selection of data adapted from Sunda and Huntsman, 1992.) (a) Specific growth rates of coastal (*Thalassiosira weissflogii*, □) and oceanic (*Thalassisira oceanica*, ▽) algal species as functions of $\log[Zn^{2+}]$ and $\log[Zn']$. ($[Zn^{2+}]$ and $[Zn']$ are the concentrations of free Zn ions and inorganic Zn species.) (b) Cellular Zn/C mole ratio as functions of $[Zn^{2+}]$ and $[Zn']$ (symbols as in part a). (c) Lines are the same as in part (b). Points represent computed relationships in the nutricline of three stations in the North Pacific; these relationships were computed from dissolved Zn versus PO_4 concentrations in the respective water columns.

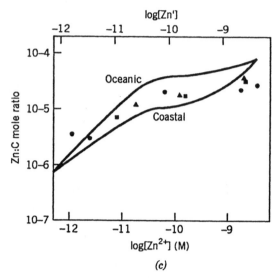

Figure 10.12. (*Continued*)

Example of Reciprocal Interaction Between Cu(II) and Phytoplankton in an Eutrophic Lake

Figure 10.13 plots seasonal variations in Lake Greifen of (1) assimilated ^{14}C and chlorophyll and (2) pCu and $\log[Cu]_T/[Cu^{2+}]$. (Analysis for pCu is by ligand exchange; see Section 10.3 and Figure 10.4.)

Algal blooms occurred in March–April and August–October. Measured pCu and $\log[Cu]_T/[Cu^{2+}]$ exhibit the same pattern as the variation of chlorophyll over time. Both have peaks in March–April, minima in May and low values in fall–winter. A side-by-side comparison of titration curves for $\log[Cu^{2+}]$ versus $\log[Cu]_T$ (Figure 10.4) also shows stronger complexation in the April sample, compared to the May sample, as indicated by the shift in the curve along the x axis. The data shown provide some evidence that biologically produced organic ligands play an important role in Cu complexation; in turn, these interactions have important effects on the growth and physiology of the organisms (Xue and Sigg, 1993).

The Iron Hypothesis

In certain areas of the ocean, especially the Southern Ocean, the subarctic Pacific, and the equatorial Pacific, phytoplankton do not exhaust phosphate and nitrate in the surface waters. In a 1990 issue of *Nature*, Martin et al. published a paper on iron in Antarctic waters and reported exacting measurements of extremely low iron concentrations in these regions and mentioned in passing that "oceanic iron fertilization aimed at the enhancement of phytoplankton production may turn out to be the most feasible method of stimulating the

10.7 The Interaction of Trace Metals with Phytoplankton 647

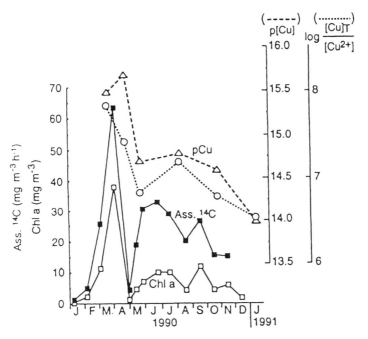

Figure 10.13. Reciprocal interaction of phytoplankton and Cu in eutrophic lake (Lake Greifen). Variations are plotted of chlorophyll, assimilated ^{14}C, p[Cu], and log($[Cu]_T$/$[Cu^{2+}]$) over time in 1990. Chlorophyll and assimilated ^{14}C represent averages from the values of 0–5 m depth; p[Cu] and log($[Cu]_T$/$[Cu^{2+}]$) are the measured values at 5 m depth. (From Xue and Sigg, 1993.)

active removal of greenhouse gas CO_2 from the atmosphere if the need arises.'' A week later the press started a public controversy with the idea that "the greenhouse effect could be ironed out" by dumping thousands of tons of iron into the ocean to create giant blooms of algae that could soak up some of the excess of carbon dioxide. A valuable by-product of the media coverage was a more profound consideration of the scientific issues at hand and the critical role that nutrient-rich seas play, in our understanding of the coupling of the oceanic carbon budget with that of the atmosphere. The American Society of Limnology and Oceanography held a special symposium (February 1991) and a special issue of some of the papers presented at this symposium was published (Chisholm and Morel, 1991).

The symposium participants resolved to urge all governments to regard the role of iron in marine production as an area of further research and not to consider iron fertilization as a policy option that significantly changes the need to reduce emissions of carbon dioxide.

Uptake of Iron Fe is perhaps the most important of all the bioactive trace elements. However, its marine chemistry and inorganic speciation are rather complex and still not adequately understood. Most of the iron comes into the seawater through the atmosphere, which, in turn, receives it from dust. In atmospheric water droplets, a significant fraction of the iron(III) becomes solubilized and reduced to soluble Fe(II) (Behra and Sigg, 1990; Zhuang et al., 1990; Zuo and Hoigné, 1992). In seawater the thermodynamically stable oxidation state is Fe(III).

Table 6.5 lists $[Fe^{3+}]/Fe(III)_T = 10^{-12}$ (cf. Byrne et al., 1988, or Turner et al., 1981) and Figure 6.8b gives the Fe species distribution in equilibrium with solid amorphous FeOOH. The thermodynamic data suggest that hydrolyzed Fe(III) species, especially $Fe(OH)_2^+$, $Fe(OH)_3^0$, and $Fe(OH)_4^-$, are dominant dissolved Fe(III) species. The solid Fe(III)(hydr)oxide present in seawater may be more amorphous, more labile, and more soluble than well-crystallized forms of FeOOH or Fe_2O_3, because in the photic zone a continuous redox cycling of Fe(III)-Fe(II) occurs as a photochemical process. The photochemical Fe(II) formation proceeds probably through different pathways: (1) through the photochemical reductive dissolution of Fe(III)(hydr)oxides, and (2) through photolysis of dissolved iron(III) organic complexes (e.g., with fulvic or humic acids).

The iron(II) formed photochemically is reoxidized by O_2. This oxidation is enhanced if the ferrous iron is adsorbed at a mineral (or biological) surface. The reoxidation of Fe(II) produces a "$Fe(OH)_3$"(s) as a colloid or on the surface that is less polymeric and less crystalline than aged Fe(III)(hydr)oxides, and thus more soluble and in faster equilibrium with monomeric (inorganic) Fe(III) species, which may control iron uptake by phytoplankton (Bruland et al., 1991; Rich and Morel, 1990).

It is suspected that the Fe transport molecules are membrane bound siderophores (Morel et al., 1991) (cf. Figure 6.7).

A similar photochemically catalyzed redox cycle most likely occurs also for manganese. Since Mn(IV) apparently does not occur as a soluble species, the uptake of manganese occurs as Mn^{2+} (Sunda and Kieber, 1994).

10.8 REGULATION OF TRACE ELEMENTS BY THE SOLID-WATER INTERFACE IN SURFACE WATERS

The solid-water interface, mostly established by the particles in natural waters and soils, plays a commanding role in regulating the concentrations of most dissolved reactive trace elements in soil and natural water systems and in the coupling of various hydrogeochemical cycles. Usually the concentrations of most trace elements (M or mol kg^{-1}) are much larger in solid or surface phases than in the water phase. Thus the capacity of particles to bind trace elements (ion exchange, adsorption) must be considered in addition to the effect of solute complex formers in influencing the speciation of the trace metals.

The particles in natural systems are characterized by a great diversity (e.g., minerals, including clays and organic particles; organisms, biological debris,

10.8 Regulation of Trace Elements by the Solid-Water Interface

humus, macromolecules; and inorganic particles coated with organic matter).

The Particle Surface as a Carrier of Functional Groups[†]

Abstracting from the complexity of the real systems, there is one common property of all natural particles. Their surfaces contain functional groups that can interact with H^+, OH^-, metal ions, and—if Lewis acid sites (e.g., $\equiv Al$ and $\equiv Fe$) are available on the surface—with ligands. Many inorganic solids (oxides and silicates) contain hydroxo groups; carbonates and sulfides expose $-C-OH$, $-C\begin{smallmatrix}\diagup O\\ \diagdown OH\end{smallmatrix}$, MeOH, and $-SH$ groups, respectively. While the interaction of alkaline and earth-alkaline ions with clays occurs mostly through ion exchange processes, the adsorption of heavy metals on clays is often dominated by surface complex formation with the single coordinated OH groups on the edge surfaces (aluminol and silanol groups). The surfaces of humic substances are characterized mostly by carboxyl and phenolic $-OH$ groups (some imino and amino as well as some $-SH$ groups may also be present). Biological surfaces contain $-COOH$, $-NH_2$, and $-OH$ groups. Despite the diversity of these functional groups, they all have the characteristics of surface ligands able to bind protons and metal ions.

In order to relate field studies to model studies, distribution coefficients of elements between the dissolved and solid phases are useful. These distribution coefficients are of the following form:

$$K_D = \frac{c_s}{c_w} \text{ (liter kg}^{-1}\text{)} \qquad (20)$$

where c_s is the concentration in the solid particles (mol kg^{-1}) and c_w is the concentration in water (mol liter^{-1}).

The partition coefficients are conditional "constants"; they depend on pH and on the grain size of the particles. Such distribution coefficients can be predicted on the basis of the equilibrium constants defining the complexation of metals by surfaces and their complexation by solutes (Table 10.7).

Distribution coefficients based on adsorption equilibria are independent of the total concentrations of metal ions and suspended solids, as long as the metal concentrations are small compared with the concentration of surface groups. Examples of the K_D obtained from calculations for model surfaces are presented in Figure 10.14. A strong pH dependence of these K_D values is observed. The pH range of natural lake and river waters (7–8.5) is in a favorable range for the adsorption of metal ions on hydrous oxides and organic particles.

Increasing concentrations of a soluble ligand cause a decrease in K_D for the

---

[†]In writing this section we largely depended on Chapter 11 (written by Sigg) in Stumm (1992).

Table 10.7. Determination of the Distribution Coefficient K_D from Surface Complex Formation

Species at surface: $\equiv$S—O—M$^+$, ($\equiv$SO)$_2$M
Species in solution: M^{2+}, MOH$^-$, M(OH)$^{(2-m)+}$, ML$_1$, ML$_2$, where L$_1$ and L$_2$ are known soluble ligands.

$$K_D = \frac{\{\equiv S-O-M^-\} + \{\equiv S-O)_2M\}}{[M^{2+}] + [MOH^-] + [M(OH)_m] + [ML_1] + [ML_2]} \left(\frac{\text{mol kg}^{-1}}{\text{mol liter}^{-1}}\right)$$

$$K_D = \frac{K^s_M\{\equiv SOH\}/[H^+] + \beta^s_M\{\equiv SOH\}^2/[H^+]^2}{1 + K_{OH}[OH^-] + \beta_{OH_m}[OH^-]^m + K_1[L] + \beta_2[L]^2} \text{ (liter kg}^{-1}\text{)}$$

where { } denotes concentration in mol kg^{-1} of solid phase.

$$K^s_M = \{\equiv S-O-M^+\}[H^+]/\{\equiv SOH\}[M^{2+}]$$

$$\beta^s_M = \{\equiv S-O)_2M\}[H^+]/\{\equiv SOH\}^2[M^{2+}]$$

The K^s values can be corrected for electrostatic effects.

$K_{OH} = [MOH^+]/[M^{2+}][OH^-]$ $\qquad \beta_{OH_m} = [M(OH)_m]/[M^{2+}][OH^-]^m$

$K_1 = [ML]/[M^{2+}][L]$ $\qquad \beta_2 = [ML_2]/[M^{2+}][L]^2$

K_D depends on:
- pH
- kind of surface; number of OH groups per surface
- complexation in solution

Source: Schindler (1984) and Sigg (1987).

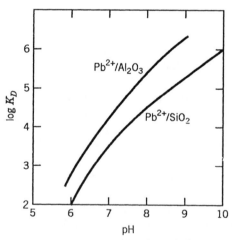

Figure 10.14. Distribution coefficients (liter kg^{-1}) calculated for surface complexation with $\equiv$AlOH and $\equiv$SiOH surface groups. The following species were taken into account for Pb: Pb^{2+}, $PbOH^+$, $PbCO_3^0$, $Pb(CO_3)_2^{2-}$, $\equiv Al-O-Pb^+$, $(\equiv AlO)_2Pb^0$; and with SiO$_2$ ($\equiv$SiO)$_2$Pb^0 and $\equiv$SiO—Pb$^+$. $\{\equiv AlOH\} = 0.25$ mol kg^{-1} and $\{\equiv SiOH\} = 1.5$ mol kg^{-1}.

10.8 Regulation of Trace Elements by the Solid-Water Interface

simple case in which the complexations in solution and at the surface are competing with each other (see Example 10.1 and Figure 10.6).

Natural Particles in Comparison to Oxides

Many naturally occurring particles—even organic particles—and surfaces of bacteria and algae interact with metal ions in a similar way as oxides. In Figure 10.15, the adsorption of Pb(II) on natural particles (isolated from a small river) is compared with that on goethite surfaces. Adsorption isotherms, at a given pH, are compared (data from Müller and Sigg, 1990). The approximate fit of the data of both adsorbents to a Langmuir isotherm (equivalent to a surface complex formation model) is evident from a plot of the reciprocal Langmuir equation.

To normalize the data in reference to the surface parameter, the quantity of mol Pb(II) sorbed is plotted per mol of functional groups present. These were determined experimentally from the adsorption capacity of Pb(II). Perhaps surprising is that the binding of Pb by natural particles at pH = 8 is not significantly different from that by goethite.

Particles as Ligands

The extent of adsorption of reactive elements (e.g., metal ions) to particles can readily be determined by titrating a particle suspension or a sample of natural water containing particles with a metal ion or to inverse the titration, that is,

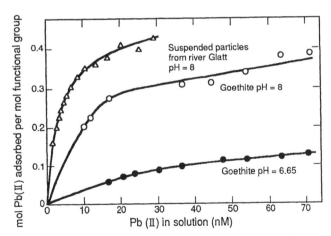

Figure 10.15. Adsorption of Pb(II) ions to the surface of goethite and of natural particles (data from Müller and Sigg, 1990). Surface complex formation of Pb^{2+} on goethite and the surface of natural particles (Glatt River). The data can be fitted by Langmuir adsorption isotherms (or surface complex formation constants). The comparison between goethite and natural particles at pH = 8 shows a slightly larger tendency of the natural particles to bind Pb^{2+} than of goethite.

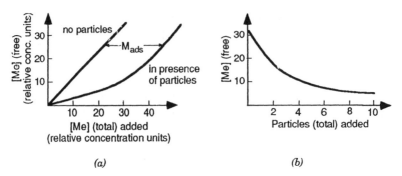

Figure 10.16. (a) Schematic representation of the titration of a solution with a standard metal solution at a constant pH. The displacement of the curve (in comparison to a particle-free solution) by the particles reflects the adsorption isotherm and permits the determination of the extent of adsorption. (b) A given metal-ion concentration is titrated with a particle suspension. The slope of this curve also reflects the adsorption isotherm and permits one to determine the extent of adsorption of metal ions to the particles.

to titrate a dilute standard metal solution with particles (Figure 10.16) (Müller and Sigg, 1990).

Analytical–Operational Difficulties In order to work close to the conditions in natural waters, very low concentrations of metal ions (in the nanomolar range) and of particles, as well as pH values in the neutral range, have to be used. Analytical difficulties occur because of undesired adsorption of metal ions to the experimental devices (walls of beakers, glass filtration devices, etc.)[†] and of insufficient separation of the particulate and dissolved phase (particles in the colloidal size range).

Example 10.2. Effect of Soluble Ligands and of Particles on the Distribution of Zn(II) Between Particulate and Soluble Phases In order to evaluate the two important variables that—in addition to pH—affect the residual concentrations of a metal ion, we use a simple equilibrium approach to assess the effect of these two variables. We assume a constant pH and characterize the effect of particle ligands, $\equiv L$, by the surface complex formation equilibrium:

$$K_{\equiv LZn} = [\equiv LZn]/[Zn^{2+}] \, [\equiv L] = 10^{8.2} \tag{i}$$

and we characterize the equilibrium with the soluble complex former X as

$$K_{ZnX} = [ZnX]/[Zn^{2+}] \, [X] = 10^7 \tag{ii}$$

[†]Such metal-ion adsorption effects become relatively significant—especially in very dilute solutions at pH values above 7.

10.8 Regulation of Trace Elements by the Solid-Water Interface

For the calculation, we assume the following total concentrations:

$$[Zn_T] = [\equiv LZn] + [Zn^{2+}] + [ZnX] = 10^{-7} \text{ M} \tag{iii}$$

$$[\equiv L_T] = [\equiv L] + [\equiv LZn] \qquad = 10^{-9}\text{--}10^{-6} \text{ M} \tag{iv}$$

$$[X_T] = [X] + [ZnX] \qquad = 10^{-9}\text{--}10^{-6} \text{ M} \tag{v}$$

In order to demonstrate the effect, we first "titrate" a 10^{-7} M Zn(II) solution in the presence of $[X] = 5 \times 10^{-8}$ M (= constant) with particles. The results are given in Figure 10.17a. Then we "titrate" the solution in the presence of a small (constant) concentration of particles with a soluble complex former X (Figure 10.17b).

As a consequence of the addition of particles, soluble [Zn(II)] and free $[Zn^{2+}]$ decrease while the particle-borne Zn, $[\equiv LZn]$, increases. The addition

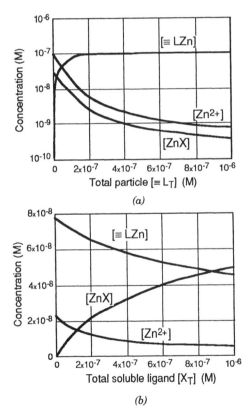

Figure 10.17. Role of particles in regulating metal ions in rivers. (a) In a simple calculation (see Example 10.1), a 10^{-7} M Zn solution is "titrated" with particles $[X_T] = 5 \times 10^{-8}$ M. (b) A 10^{-7} M Zn solution in the presence of a small particle concentration $[\equiv L_T] = 10^{-7}$ M is "titrated" with a soluble complex former X.

of a soluble complex former increases soluble [Zn(II)] but decreases free [Zn^{2+}]; the particle-borne Zn, [≡LZn], decreases with increasing [X_T].

Titration data (particles with metals or vice versa) can often be interpreted in terms of a simple conditional complex formation constant (equation i) and binding capacity L_T (equation iv) by an equation that is analogous to equation 5 given in Section 10.3:

$$\frac{[Zn']}{\{\equiv LZn\}} = \frac{[Zn']}{\{\equiv L_T\}} + \frac{1}{K_L\{\equiv L_T\}} \quad \text{(vi)}$$

where [Zn'] is the analytically detectable soluble Zn species, the conditional constant $K_L = [\equiv LZn]/[Zn'][\equiv L]$, $\{\equiv L_T\}$ is the surface binding capacity in mol kg^{-1}, and $\{\equiv LZn\}$ is the Zn bound to the surface in mol kg^{-1}.

10.9 REGULATION OF DISSOLVED HEAVY METALS IN RIVERS, LAKES, AND OCEANS

Rivers

Table 10.8 gives some representative examples of dissolved trace metal concentrations in rivers, especially in large unpolluted rivers.

The data show that rivers contain remarkably low concentrations (in some cases concentrations are as low as 10^{-11} M) of dissolved metal ions. (Much of the data reported in the literature have been based on total particulate *and* dissolved concentrations; furthermore, analytical procedures have often not been able to discriminate against contamination during sampling and sample processing.) The metal concentrations in rivers are a consequence of (1) the

Table 10.8. Examples of Dissolved Trace Metal Concentrations in Rivers[a]

River	Concentration (nM)				Reference
	Cd	Cu	Zn	Pb	
U.S. East Coast rivers	0.095	17	13.0	0.11	b
Mississippi	0.12	23	3.0		c
Yangtze	<0.01	18–21	0.6–1.2		d
Amazon	0.06	24	0.3–3.8		d
Orinoco	0.035	19	2.0		d

[a]These data are based on advanced instrumentation and sampling methodology, paying attention to the elimination of contamination during sampling, storage, and analysis. See article by Windom et al. (1991) from which the data for this table are taken.
[b]Investigated mean of data from two sampling campaigns by Windom et al. (1991).
[c]Shiller and Boyle (1987).
[d]Data from Shiller and Boyle (1985), Boyle et al. (1982), and Edmond et al. (1985).

10.9 Regulation of Dissolved Heavy Metals in Rivers, Lakes, and Oceans

geochemistry of the rocks in the catchment area (metals released into the water by weathering), (2) anthropogenic pollution (by waste inputs and atmospheric deposition), and (3) river chemistry (adsorption of metal ions to particles and other surfaces and particle deposition into the sediments).

Windom et al. (1991) report that on the average 62%, 40%, 90%, and 80% of the Cd, Cu, Pb, and Zn, respectively, carried by U.S. East Coast rivers is on particles. Similar conclusions were reached in an earlier study by Martin and Whitfield (1983). These high proportions of particulate metal ions are representative of large rivers that are often relatively unpolluted and that are characterized by high loads of turbidity (low ionic strength). In many small rivers with anthropogenic metal pollution and of low turbidities (calcareous drainage area), a significant fraction of metal ions may be present in dissolved form (cf. Figure 10.19). The effect of particles on the residual concentrations is also apparent from the pH dependence. Figure 10.18 shows an example of this pH dependence. (Similar data on the pH dependence of Cd and Pb were reported by Windom et al., 1991.) The experimental results show a decrease in [Zn] with increasing pH. Such a dependence is in a first approximation compatible with a reaction of the type

$$S-OH + Me^{2+} \rightleftharpoons S-OMe^+ + H^+ \qquad (21)$$

Figure 10.19 gives examples of the speciation of Pb(II) in a local river (Glatt, Switzerland) in the presence of concentrations of EDTA and NTA (typically encountered in this river). EDTA (ethylene diaminetetraacetate) and NTA (nitrilo triacetate) are strong complex formers used in industry and households; they are typically found in receiving waters as pollutants. The data of Müller and Sigg (1990) for natural particles are very similar to those determined for marine sediment particles by Ballistrieri and Murray (1983). Johnson (1986) has demonstrated heavy metal regulation in a heavily polluted acid-mine drainage river. The concentration dependence could be accounted for by surface complexation with hydrous iron(III) oxide particles.

Regulation of Trace Elements in Lakes

When trace elements are introduced into a lake by riverine and atmospheric input, they interact (1) with solutes (complex formation) and (2) with inorganic and organic (phytoplankton) particles (adsorption and assimilation). The affinity of the reactive elements for the particles, which settle through the water column, determines essentially the relative residence time of these elements and their residual concentrations and ultimate fate (Figure 10.20).

As shown in Table 6.1, the concentrations of trace elements in the water column is—despite anthropogenic pollution—extremely small (10^{-11}–10^{-7} M), illustrating the remarkable efficiency of the continuous "conveyor belt" of the settling, adsorbing, scavenging, and assimilating particles. Scavenging of met-

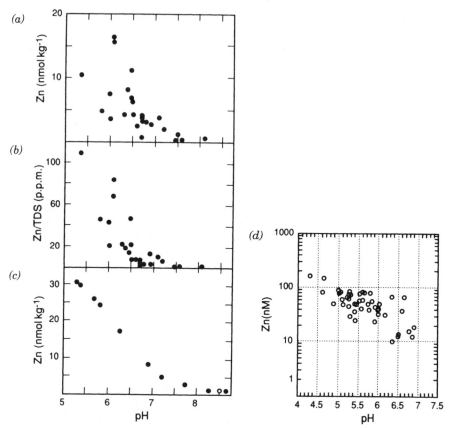

Figure 10.18. Effect of pH on residual metal concentration in fresh waters. Dissolved zinc is plotted against pH. (a) Zinc in relatively undisturbed major rivers including the Yangtze (Chiang Jiang) and tributaries of the Amazon and Orinoco. (b) Zinc normalized to total dissolved solids for the same set of major rivers. (c) Zinc in pH-adjusted aliquots of Mississippi River water (April 1984, 103 mg liter^{-1} suspended load, pH 7.7); the adjusted aliquots were allowed to equilibrate overnight before filtration and analysis. (From Shiller and Boyle, 1985.) (d) Zinc in different mountain lakes in the southern parts of the Swiss Alps. These lakes are less than 10 km apart, so that the atmospheric inputs can be considered to be uniform over this scale, but their water composition (pH) is influenced by different bedrocks in their catchments. A similar dependence on pH has also been observed for Cd and Pb; but this dependence is less pronounced with Cu(II) when solute complex formation counteracts adsorption (data 1983–1992). (From Sigg et al., 1995, in press.)

als like Zn(II), Pb(II), and Cd(II) is pH dependent (see Figure 10.20); thus the residual concentrations of these metals tend to decrease with increasing pH (Figure 10.18d). The *sedimentary record* reflects the accumulation of trace elements in sediments and a profile of concentration versus sediment depth (or age) gives a "memory record" on the loading in the past (see Figure 10.1).

10.9 Regulation of Dissolved Heavy Metals in Rivers, Lakes, and Oceans

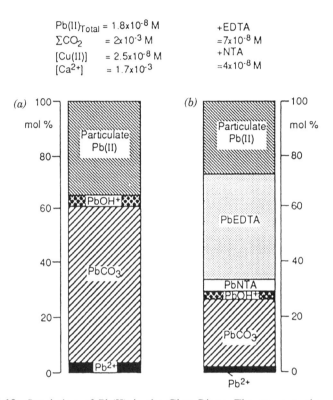

Figure 10.19. Speciation of Pb(II) in the Glatt River. The concentrations given for CO_2, Pb(II), Cu(II), and $[Ca^{2+}]$ as well as for the pollutants EDTA and NTA are representative of concentrations encountered in this river. The speciation is calculated from the surface complex formation constants determined with the particles of the river and the stability constants of the hydroxo, carbonato, NTA, and EDTA complexes. The presence of $[Ca^{2+}]$ and $[Cu^{2+}]$ is considered. (From Müller and Sigg, 1990.)

Role of Settling Particles

Both biogenic organic particles (algae, biological debris) and inorganic particles (e.g., manganese and iron oxides) (cf. Table 10.9) contribute to the binding, assimilation, and transport of reactive elements. The photosynthetic production of algae and their sedimentation are dominant processes, especially in eutrophic lakes. Near the sediment–water interface, anoxic conditions may occur, under which iron and manganese oxides undergo reduction and dissolution. As discussed before, trace elements are affected by these processes in different ways

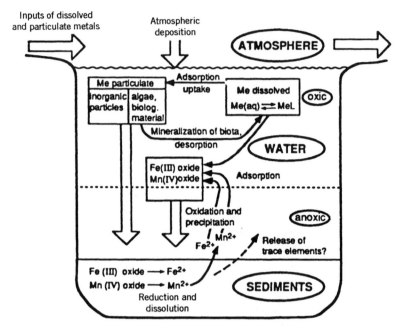

Figure 10.20. Schematic representation of the cycling of trace elements in a lake. Trace elements are removed to the sediments together with settling material, which consists in large part of biological material. (From Sigg and Stumm, 1994.)

(interaction with iron and manganese oxides, precipitation as and complexation with sulfides).

Sedimentation rates of 0.1–2 g m^{-2} d^{-1} are typically observed in lakes; still higher values are found in very eutrophic lakes. The settling material can be collected in sediment traps; it can then be characterized in terms of chemical composition, morphology, and size distribution of the particles. The composition is subject to seasonal variations caused primarily by different biological activities in the various seasons (example in Figure 10.10).

Steady-State Models

Simple steady-state models may be used to relate quantitatively the mean concentration in the lake water column and the residence time of metal ions to the removal rate by sedimentation (for a detailed treatment of lake models see Imboden and Schwarzenbach, 1985). In a simple steady-state model, the inputs to the lake equal the removal by sedimentation and by outflow; the water column is considered as fully mixed; mean concentrations and residence times in the water column can be derived from the measured sedimentation fluxes. The binding of metals to the particles is fast in comparison to the settling.

The quantitative relationships and an example are summarized in Table

Table 10.9. The Role of Settling Particles in Regulating Trace Elements in Lakes

Components of Settling Particles	Characteristics
Phytoplankton and biological debris	Surfaces of organisms have strong affinity for heavy metals such as Cu(II), Pb(II), Zn(II), Cd(II), and Ni(II) (surface complex formation). Organisms also absorb (assimilate) nutrients (P, N, Si, S, etc.) and nutrient metal ions [e.g., Cu(II), Zn(II), Co(II)] and metal ions that are mistaken (by the organisms) as nutrients [e.g., Cd(II), As(V)]. Phytoplankton is mineralized in water column and sediments.
CaCO₃ (usually precipitated within the lake)	Heavy metals and phosphate are adsorbed and may become incorporated; $CaCO_3$ crystals are usually large ($d > 5\ \mu m$); because of the small specific surface area (in comparison to other settling material), the effect of $CaCO_3$ on overall removal of trace elements is small.
Fe(III)(hydr)oxides introduced into the lake and formed within the lake	Strong affinity (surface complex formation) for heavy metals, phosphates, silicates, and oxyanions of As and Se; Fe(III) oxides, even if present in small proportions, can exert significant removal of trace elements. At the oxic–anoxic boundary of a lake, Fe(III) oxides may represent a large part of settling particles. Internal cycling of Fe by reductive dissolution and by oxidation–precipitation is coupled to the cycling of metal ions.
Mn(III, IV) oxides mostly formed within the lake	High affinity for heavy metals and high specific surface area. Redox cycling [$MnO_x(s) \rightleftharpoons Mn^{2+}(aq)$] (Section 9.6) is usually important in regulating trace element concentrations and transformations in lower portion of lake and sediments.
Aluminum silicates clays, oxides	Ion exchange; binding of phosphates and metal ions; unless present in large concentrations, overall effect on trace element removal is small.

10.10. Under the assumption of steady-state conditions, the residence time of an element (τ_M) that is removed from the water column by sedimentation and by outflow is given by

$$\tau_M^{-1} = \tau_w^{-1} + \tau_s^{-1}$$

where τ_w is the residence time of water in the lake and τ_s is the residence time of an element with respect to sedimentation. The removal rate by sedimentation, τ_s^{-1}, can be expressed as $\tau_s^{-1} = f_p \times k_s$, where f_p is the fraction of the element in particulate form and k_s (d^{-1}) is the rate constant characterizing sedimentation. The fraction of an element in the particulate phase relates to its tendency to bind to particles and depends on the partition coefficient of the element between particulate phase and solution (dependent on the chemical interactions with the particles and in solution), and on the concentrations of particles in the water column. The removal rate by sedimentation can be calculated from the flux of an element to the sediments and its total amount in the water column (Table 10.10). The removal rate by sedimentation is large if

Table 10.10. Removal of Metal Ions from Lake Water Column by Sedimentation

Residence time of element M: $\tau_M^{-1} = \tau_w^{-1} + \tau_s^{-1}$ (i)

where τ_M = residence time of element M (days)
τ_w = residence time of water (days)
τ_s = residence time of element with respect to sedimentation (days)

Removal rate by sedimentation (τ_s^{-1}): $\tau_s^{-1} = f_p \times k_s = \dfrac{F_M}{h \times [M]_w}$ (ii)

where f_p = fraction of element bound to particles
F_M = sedimentation rate of element M (mol m^{-2}d^{-1})
$[M]_w$ = concentration of element M in water (mol m^{-3})
k_s = removal rate of particles (d^{-1})
h = mean depth of water column (m)

Example: *Pb in Lake Zurich, summer*
Sedimentation rate of Pb: $F_{M(Pb)} = 1$ μmol m^{-2}d^{-1}
Mean concentration in water column: $[Pb]_w = 8 \cdot 10^{-4}$ μM $= 0.8$ μmol m^{-3}
Removal rate: $\dfrac{F_M}{h \times [Pb]_w} = 2.5 \times 10^{-2}$ d^{-1}
Residence time of water: $\tau_w = 400$ days
With $\tau_w^{-1} = 2.5 \times 10^{-3}$ d^{-1}:
Residence time of Pb: $\tau_M = 36$ days

Steady-state concentration: $[Pb]_w = \dfrac{\tau_w^{-1}}{\tau_w^{-1} + \tau_s^{-1}} \times c_{in} = 0.09$ c_{in} (iii)

where c_{in} = input concentration

Source: Sigg (1992).

10.9 Regulation of Dissolved Heavy Metals in Rivers, Lakes, and Oceans

both the fraction of an element bound to particles and the sedimentation rate are significant, and the residence time of an element in the water column is much smaller than the water residence time. The mean concentration in the water column can be shown to depend on the removal rate by sedimentation and the water residence time according to equation iii in Table 10.10. This means that if the removal rate by sedimentation is high, the mean concentration in the water column turns out to be much lower than the input concentration. Low concentrations of metal ions in the water column can thus be expected to occur if a metal ion binds to a significant extent to particles and if the sedimentation rate is high. The chemical factors that determine the binding to particles (pH, surface ligands, ligands in solution), the tendency of phytoplankton to uptake trace elements, and the chemical factors that influence coagulation and, in turn, sedimentation rates affect the distribution of a metal ion between particulate phase and solution and thus the fraction in particulate form f_p. For a single element under similar chemical conditions in different lakes, the residence times and mean concentrations depend on the sedimentation rates.

Simple steady-state models can only predict mean concentrations. Seasonal variations and concentration–depth profiles in the water columns of lakes give further insight into the mechanisms governing the removal of metal ions. For a more detailed treatment of modeling concepts in lakes, see Chapter 15 of Schwarzenbach et al. (1993).

Oceans

The same processes discussed for lakes are operative in oceans. Autochthonous particles, above all phytoplankton and other organisms, as well as eolian dust make up the continuous conveyor belt of settling materials. The interaction with biota is reciprocal. Nutrient metal ions coregulate the growth of phytoplankton, but the phytoplankton community has a pronounced effect on the metal ions and their speciation. The bulk of the trace elements, adsorbed most efficiently to the biological surfaces in the surface layers, is carried into the deep sea by settling; as the particles fall through the water column, they provide a food source for successive populations of filter feeders so that the material is then repackaged many times en route to the sediment (Whitfield and Turner, 1987).

As we have seen, large spatial and temporal differences in both trace metal concentrations and chemical speciation in the sea have led to wide variations in biological availability of metals and their effects on phytoplankton.

In Figure 10.21 the depth profile of the Pb concentrations of the Central Pacific is compared with that of Lake Constance. In each case, the Pb concentrations of the surface waters are higher than in the deep water; in both cases atmospheric transport plays a significant role in supplying Pb to the surface water. The decrease in the concentration of Pb with depth occurs by particles that scavenge Pb(II) most efficiently. Patterson (e.g., Settle and Patterson, 1980) has used data on the memory record of sediments to compare

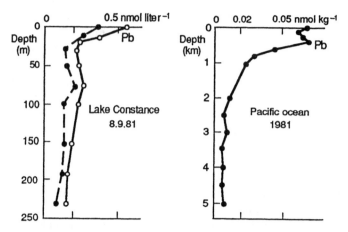

Figure 10.21. Lead profiles in Lake Constance (Summer 1981 data: Sigg, 1985) and in the Pacific Ocean (1981 data: Schaule and Patterson, 1981). The similar shapes of these profiles, despite the difference in length scales (kilometers for the ocean and hectometers for the lake), illustrate the influence of the atmospheric deposition on the upper layers and the scavenging of Pb(II) by the settling particles. (Adapted from Sigg, 1985.)

prehistoric and present-day eolian inputs. These data suggest that the present Pb(II) input is two orders of magnitude larger than that of prehistoric time.

As in lakes, other potential scavenging and metal regeneration cycles operate near the sediment–water interface. Subsequent to early epidiagenesis in the partially anoxic sediments, iron(II), manganese(II), and other elements depending on redox conditions are released by diffusion from the sediments to the overlying water, where iron and manganese are oxidized to insoluble iron(III) and manganese(III, IV) oxides. These oxides are also important conveyors of heavy metals near the sediment surface.

Whitfield (1979) and Whitfield and Turner (1987) have shown that the elements in the ocean can be classified according to their oceanic residence times, τ_i:

$$\tau_i = \frac{\text{total number of moles of } i \text{ in ocean}}{\text{rate of addition or removal (mol time}^{-1})}$$

which are, in turn, a measure of the intensity of their particle–water interaction. Thus the elements that show the strongest interactions with the particulate phase have very short residence times; those elements that interact little with particles are characterized by long residence times (Figure 10.22).

Reactive trace metals exhibit a range of biogeochemical behavior that can be characterized by two endmembers—nutrient type and scavenged type (Figure 10.22c,d). Nutrient-type metals, best exemplified by Zn and Cd, are primarily removed from surface waters by biogenic particles and then remineralized at

10.9 Regulation of Dissolved Heavy Metals in Rivers, Lakes, and Oceans

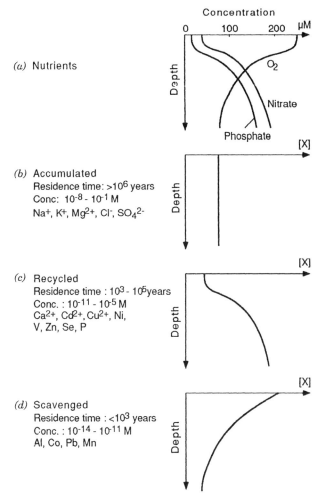

Figure 10.22. Schematic depth ocean profiles for elements. This figure is based on a classification of elements according to their oceanic profiles given by Whitfield and Turner (1987). Uptake of some of the elements, especially the recycled ones, occurs somewhat analogously to that of nutrients. There are some elements such as Cd that are nonessential but may be taken up (perhaps because they mimic essential elements) the same way as nutrients. The concentration ranges given show significant overlap, since the concentrations of the elements also depend on crustal abundance.

depth. See Section 15.10 on an idealized kinetic model for the marine cycle of biologically fixed elements. Internal biogeochemical cycles together with physical mixing and circulation patterns control the distributions of nutrient-type metals. Scavenged-type metals, best exemplified by Al or Pb, are removed onto particles in intermediate and deep waters as well as at the surface. External

664   Trace Metals: Cycling, Regulation, and Biological Role

inputs such as the deposition of eolian dust, control the concentration and distribution of scavenged-type metals. Other metals, such as Fe, exhibit a mixture of the characteristics of these two endmembers (Bruland et al., 1994).

Interbasin Fractionation   Bruland et al. (1994) have analyzed the effect of chemical and biological behavior together with global circulation patterns on interbasin fractionation of nutrient-type and scavenged-type metals. The North Atlantic is a relatively small basin with its surface waters influenced by large atmospheric and river inputs. In contrast, the North Pacific basin is much larger. It is less influenced by atmospheric and riverine inputs, and as a result of global circulation patterns (for a model on the interbasin circulation between Atlantic and Pacific Oceans see Figure 15.18), its deep waters are characterized as old and nutrient-rich. Dissolved Al concentrations are approximately 50-fold higher in the deep waters of the North Atlantic relative to the North Pacific. This is consistent with elevated external sources of Al plus its short oceanic residence time. North Atlantic deep waters are formed from North Atlantic surface waters, which are enriched in dissolved Al due to large external sources of Al to the basin.

On the other hand, a nutrient-type trace metal like Zn attains a concentration of dissolved Zn that is approximately five times greater in the old, nutrient-rich deep waters of the North Pacific than they are in the young, nutrient-poor North Atlantic deep waters. Its distribution in both ocean basins is similar to that of silicic acid. The efficiency with which Zn is recycled in the ocean leads to its relatively long oceanic residence time.

The Biogeochemical Cycling of Elemental Mercury

As shown in Example 10.3 (below), elemental Hg^0 becomes an important dissolved inorganic Hg species below the $p\varepsilon$ range 4–6. [The redox potential for Hg(II) reduction to Hg^0 (0.85 V) is similar to that for the Fe^{3+}/Fe^{2+} couple (0.77 V)]. The water solubility of elemental Hg^0 is about 3×10^{-7} M.

Example 10.3. Elemental Aqueous Mercury, $Hg^0(aq)$: Equilibria

(a) Estimate the water solubility of $Hg^0(aq)$ from the following information (25°C):

$$Hg^{2+} + 2e^- = Hg^0(aq) \quad E_H^0 = 0.659 \quad p\varepsilon^0 = 11.15 \quad \text{(i)}$$

$$Hg^{2+} + 2e^- = Hg(l) \quad E_H^0 = 0.854 \quad p\varepsilon^0 = 14.4 \quad \text{(ii)}$$

For the reaction

$$Hg(l) = Hg^0(aq) \quad \text{(iii)}$$

the log equilibrium constant is given by $\log K_i - \log K_{ii}$ (where $\log K = np\varepsilon^0$).

10.9 Regulation of Dissolved Heavy Metals in Rivers, Lakes, and Oceans

Hence log $K_{iii} = -6.5$. The solubility of aqueous mercury is $10^{-6.5}$ M (3×10^{-7} M, 0.06 mg Hg liter^{-1}).

(b) How does the relative equilibrium contribution of elemental aqueous mercury to the total soluble inorganic mercury in fresh water and seawater and the volatility of Hg^0(aq) depend on pϵ? We assume for fresh water pH = 8, [Cl$^-$] = 10^{-3} M, [SO$_4^{2-}$] = 10^{-3} M; and for seawater, pH = 8, [Cl$^-$] = 0.6 M, [SO$_4^{2-}$] = 10^{-2} M.

From Figure 10.5 we know that the major soluble Hg(II) species are Hg(OH)$_2$(aq) in fresh water and HgCl$_4^{2-}$ in seawater. We may establish diagrams in which relative activities of Hg^0(aq), Hg^0(g), Hg^{2+}, HgCl$_2^0$(aq), HgCl$_4^{2-}$, Hg(OH)$_2$(aq), and HgS(s) are plotted as a function of pϵ for both fresh water and seawater (Figure 10.23). The construction of activity ratio

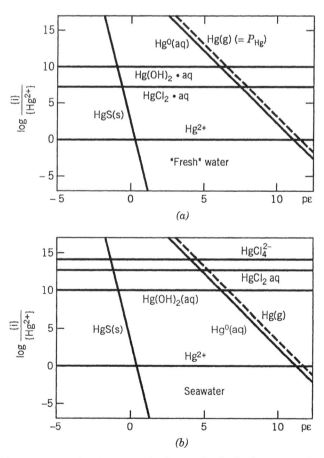

Figure 10.23. Activity ratio diagrams: Hg(II) species in fresh water and seawater. (a) pH = 8, [Cl$^-$] = 10^{-3} M, [SO$_4^{2-}$] = 10^{-3} M. (b) pH = 8, [Cl$^-$] = 0.6 M, [SO$_4^{2-}$] = 10^{-2} M.

diagrams is explained in Appendix 8.1 (Example A8.1).

$$Hg^{2+} + 2e^- = Hg^0(aq) \qquad \log K = 22.3 \qquad \text{(iv)}$$
$$Hg^{2+} + 2Cl^- = HgCl_2(aq) \qquad \log K = 13.2 \qquad \text{(v)}$$
$$Hg^{2+} + 4Cl^- = HgCl_4^{2-} \qquad \log K = 15.1 \qquad \text{(vi)}$$
$$Hg^{2+} + 2OH^- = Hg(OH)_2(aq) \qquad \log K = 21.9 \qquad \text{(vii)}$$
$$Hg^{2+} + SO_4^{2-} + 8H^+ + 8e^- = HgS(s) + 4H_2O \qquad \log K = 70.0 \qquad \text{(viii)}$$
$$Hg^0(aq) = Hg^0(g) \qquad \log K = 0.93 \qquad \text{(ix)}$$

Such activity ratio diagrams illustrate that $Hg^0(aq)$ becomes the major dissolved inorganic Hg species below the $p\epsilon$ range 4–6 for both seawater and fresh water; above this $p\epsilon$ range, $Hg(OH)_2(aq)$ and $HgCl_4^{2-}$ are the preponderant species in fresh water and seawater, respectively.

We may also note from equilibrium iii that the water solubility of Hg^0 (elemental mercury) is substantial and that the volatility of $Hg^0(aq)$ as characterized by equilibrium ix (Henry's law constant = K_H = 1.2×10^{-1} M atm^{-1}) is relatively large; at normal temperatures the un-ionized mercury solute, especially in the not fully aerobic $p\epsilon$ range, is readily lost by volatilization from its water solutions at normal temperatures.

The Global Mercury Cycle A recent review (Mason et al., 1994) on the global Hg cycling is diagrammed in Figure 10.24. The evasion of Hg^0 from the ocean is balanced by the total oceanic deposition of Hg(II) from the atmosphere. The mechanisms, whereby reactive Hg species are reduced to volatile Hg^0 in the oceans, are poorly known, but reduction appears to be biologically mediated. Deposition on land is the dominant sink for atmospheric Hg. Mason et al. (1994) estimate that over the last century anthropogenic emissions have tripled the concentration of Hg in the atmosphere and in the surface ocean.

10.10 QUALITY CRITERIA IN FRESH WATERS: SOME ASPECTS

Aquatic organisms, especially those at the top of the food chain, such as fish-eating birds, are much more sensitive to toxic metals in their habitat than are terrestrial biota. There is some circumstantial evidence to suggest that the "no-effect" thresholds for some metals is fairly close to their current levels in natural waters (see Table 6.1).

Evaluating Toxicity: Some Generalizations

Aquatic toxicologists are still struggling to develop proper test methods to determine ecotoxicological threshold concentrations. The harmfulness of a sub-

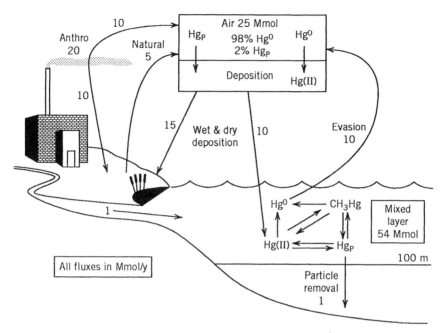

Figure 10.24. The current global mercury cycle (1 Mmol = 10^6 mol). (Adapted from Mason et al., 1994.)

stance depends on its interaction with organisms or entire communities. The intensity of this interaction depends on the specific structure and on the activity of the substance under consideration; but other factors such as temperature, turbulence, and the presence of other substances are also important.

In evaluating toxicity, we need to distinguish between (1) substances that endanger organisms, that is, impair their health, or poison aquatic organisms [one often speaks of acute toxicity, especially if effects are observed within short time periods ($\leq$ days)]; and (2) substances that affect primarily the organization and structure of aquatic ecosystems. In this interaction, contaminants may impair the self-regulatory functions of the system or interfere with food chains.

While there is some knowledge about the impact of chemical substances on individual organisms, less is known about their impact on ecosystems. The natural distribution of organisms depends on their ability to compete under given conditions, not merely their ability to survive the physical and chemical environment; a population will be eliminated when its competitive power is so much reduced that it can be replaced by another species. The competitive abilities of an organism are an interplay of its reproductive rate, which is related to food and physiological potential, and its mortality rate from all sources, including predation and imposed toxicity. Thus, in an ecosystem, a population may be eliminated even at apparently trivial toxicity levels if its competitive ability is marginal or if it is the most sensitive of the competitors. Similarly, a population may be eliminated if its food source is affected even if the population of interest is completely resistant to the toxicant. It is thus often not possible to predict the ecological effects of substances from bioassays with individual organisms. Even if no acutely damaging effects on individual organisms are observed, the ecological consequences of such impairment may, over long time spans, often be more detrimental to aquatic ecosystems than acute toxic effects.

It has been postulated, for example, that altering a phytoplankton community may profoundly affect the healthy distribution and abundance of many animal populations higher in the food web. Thus, in eutrophic environments, alterations in algal communities can further reduce an already decreased species diversity, aggravating problems of algal blooms and contributing to general degradation of the ecosystem.

Chemical Speciation

The speciation is of utmost importance in evaluating ecological effects. The response of organisms is specific to substrates. As shown in Section 10.6, the free metal-ion activity is a key parameter that usually determines its biological effects on phytoplankton. These concepts can, with certain limitations, also be extended to higher organisms and plant roots.

In evaluating ecotoxicological quality criteria, tests or experiments should be carried out in such a way that the free metal-ion concentration (activity) is known or can be estimated from the composition of the solution. In some cases, metal-ion buffers should be used. A pragmatic difficulty arises, because quality criteria have to be given in terms of routinely measurable parameters such as *dissolved* total metal-ion concentration (the determination of free metal-ion activity can be very demanding). But then one has to realize that the biological effect for a given total concentration can be different depending on pH (hydrolysis), alkalinity (carbonate complex formation), organic substances present (organic complex formation), and so on. For example, in the case of *cadmium* and *zinc*, the tendency to form inorganic complexes (see Table 6.5) and organic complexes is relatively small so that $[Cd^{2+}]/[Cd(II)]_{dissolved}$ and $[Zn^{2+}]/[Zn(II)]_{dissolved} \simeq 0.5$ to 0.03.

Lead has a strong tendency to be adsorbed on particle surfaces. Usually only very small concentrations of dissolved Pb^{2+} are found in rivers and groundwaters (see Table 6.1). The tendency for inorganic complex formation (formation of $PbCO_3$, see Table 6.5) and organic complex formation is somewhat greater, but still relatively small so that $[Pb^{2+}]/[Pb(II)]_{dissolved} \simeq 0.05$ to 10^{-4}.

For *nickel and cobalt*, the tendency to form complexes is relatively small, so these cations occur to a significant extent as free Ni^{2+} and Co^{2+}. $[Ni^{2+}]/[Ni(II)]_{dissolved} \simeq [Co^{2+}]/[Co(II)]_{dissolved} \simeq 0.5$.

Copper has a marked tendency to form inorganic complexes (see Table 6.5) ($[Cu^{2+}]/[Cu(II)]_{inorganic} \simeq 0.01$) and a strong tendency to form organic complexes (Irving–Williams series). In presence of 2–5 mg liter^{-1} DOC, $[Cu^{2+}]/[Cu(II)]_{dissolved}$ is on the order of 10^{-5} to 10^{-7} (Behra et al., 1993).

10.10 Quality Criteria in Fresh Waters: Some Aspects

Chromium(III) is mostly present in particular form [as Cr(III)(hydr)oxide] and adsorbed to particles. Cr(VI) occurs in dissolved form as CrO_4^{2-}.

Inorganic *arsenic* occurs mostly as dissolved As(III) ($H_nAsO_3^{3-n}$) and As(V) ($H_nAsO_4^{3-n}$). Different organic As compounds exist.

From a toxicological point of view, alkyl-Hg, especially methyl mercury compounds, are usually more toxic than inorganic species ($HgCl_2$, Hg^0, Hg^{2+}). The same is true for *tin*.

Silver complex formation with Cl^- and S(-II) is important. In oxic fresh waters $[Ag^+]/[Ag(I)_{dissolved}] \approx 0.6$. Table 10.11 lists recommended water quality criteria (concentrations that should not be exceeded to avoid ecologically harmful effects in fresh water) for Germany, the Netherlands, Canada, and Switzerland.

Table 10.11. Exemplification of Recommended Water Quality Criteria[a]

Element	Germany (μg/liter) (diss)		Netherlands (μg/liter) (diss)	Canada (μg/liter) (TOT)		Switzerland (μg/liter) (diss)
As	—		12.5	50		10
Pb(II)	0.1	soft[b]	1.3	1	soft	1
	0.4	middle		2	middle	
	1	hard		4	hard	
Cd(II)	0.01	soft	0.025	0.2	soft	0.05
	0.02	middle		0.8	middle	
	0.05	hard		1.3	hard	
Cr(III) + (VI)	?		2.5	20		2
Co(II)	—		—	—		10
Cu	0.1	soft	1.3	2	soft	2
	0.3	middle		2	middle	
	0.5	hard		4	hard	
Ni	0.2	soft	7.5	25	soft	5
	0.5	middle		65	middle	
	2	hard		110	hard	
Hg(inorg)	0.005	soft				
	0.01	middle				
	0.03	hard				
Hg(org)	—		0.005	0.1 (inorg + org)		0.01 (inorg + org)
Ag(I)	—		—	0.1		0.1
Zn(II)	1	soft	6.5	30		5
	2	middle				
	4	hard				
Sn(org)	—		0.003–0.01	—		0.001 TBT[c]
						0.001 TPT

[a] Concentrations given in μg/liter are either dissolved (diss) or total (TOT).
[b] The classification is as follows:
soft: Alk < 2 mM; middle: Alk 2–5 mM; hard: Alk > 5 mM.
[c] TBT = tributyltin; TPT = triphenyltin.

Source: Adapted from Behra et al. (1993).

Sediments

In soils and sediments, the speciation in the dissolved aqueous phase is primarily important to assess bioavailability and toxicity. The total concentration (adsorbed and solid phase) gives an idea of the capacity of a reactive element, which could under certain circumstances (e.g., acidification, complex formation) be mobilized.

An informative paper by Di Toro et al. (1991), on predicting the acute toxicity of Cd and Ni in sediments by assessing the acid volatile suflide (AVS), illustrates that the sediment properties that determine the concentration (activity) of the sediment in the interstitial water determine the fraction of the metal that is bioavailable and potentially toxic. The study of Di Toro et al. is based on measurements of acute toxicity to benthic organisms (amphipodes, oligochaetes, and snails). It is shown that this toxicity is essentially related to the free metal ions in solubility equilibrium with the solid metal sulfides present.

SUGGESTED READINGS

Bioavailability and Inorganic Chemistry of Life

Frausto da Silva, J. J. R., and Williams, R. J. P. (1991) *The Biological Chemistry of the Elements; the Inorganic Chemistry of Life,* Clarendon Press, Oxford.

Nieboer, E., and Richardson, H. S. (1980) The Replacement of the Nondescriptive Term Heavy Metals by a Biologically and Chemically Significant Classification of Metal Ions, *Environ. Pollution (B)* **1,** 3–26.

Craig, P. J. (Ed.) (1986) *Organometallic Compounds in the Environment; Principles and Reactions,* Longmans, Harlow.

Analytical Approach to Metal Speciation and Metal Analysis

Bruland, K. W. (1989) Oceanic Zinc Speciation Complexation of Zinc by Natural Organic Ligands in the Central North Pacific, *Limnol. Oceanogr.* **34,** 267–283.

Donat, J. R., and Van den Berg, C. M. G. (1992) A New Cathodic Stripping Voltammetric Method for Determining Organic Copper Complexation in Seawater, *Mar. Chem.* **38,** 69–90.

Hudson, R. J. M., Covault, D. T., and Morel, F. M. M. (1992) Investigations of Iron Coordination and Redox Reactions in Seawater Using ^{59}Fe Radiometry and Ion-Pair Solvent Extraction of Amphiphilic Iron Complexes, *Mar. Chem.* **38,** 209–235.

Shiller, A. M., and Boyle, E. A. (1985) Dissolved Zinc in Rivers, *Nature* **317,** 49–52.

Shiller, A. M., and Boyle, E. A. (1987) Variability of Dissolved Trace Metals in the Mississippi River, *Geochim. Cosmochim. Acta 51,* 3272–3277.

Sigg, L., and Xue, H. B. (1994) Metal Speciation, Concepts, Analysis and Effects. In *Chemistry of Aquatic Systems,* G. Bidoglio and W. Stumm, Eds., Kluwer Academic, Norwell, MA.

Sunda, W. G., and Hanson, A. K. (1987) Measurement of Free Cupric Ion Concentration in Seawater by a Ligand Competition Technique Involving Copper Sorption onto C18 SEP-PAK Cartridges, *Limnol. Oceanogr.* **32,** 357–551.

Windom, H. L., Byrd, T., Smith, R. G., and Huan, F. (1991) Inadequacy of NAS-QUAN Data for Assessing Metal Trends in the Nation's Rivers, *Environ. Sci. Technol.* **25,** 1137–1142.

Metal Ions as Micronutrients in Oceans and Lakes

Bruland, K. W., Donat, J. R., and Hutchins, D. A. (1991) Interactive Influence of Trace Metals on Biological Production in Oceanic Waters, *Limnol. Oceanogr.* **36,** 1555–1578.

Morel, F. M. M., Hudson, R. J. M., and Price, N. L. (1991) Limitations of Productivity by Trace Metals in the Sea, *Limnol. Oceanogr.* **36,** 1742–1755.

Xue, H. B., and Sigg, L. (1993) Free Cupric Ion Concentration and Cu(II) Speciation in an Eutrophic Lake, *Limnol. Oceanogr.* **38,** 1200–1213.

Sunda, W. G. (1991) Trace Metal Interaction with Marine Phytoplankton, *Biol. Oceanogr.* **6,** 411–442.

Whitfield, M., and Turner, D. R. (1987) The Role of Particles in Regulating the Composition of Seawater. In *Aquatic Surface Chemistry,* W. Stumm, Ed., Wiley-Interscience, New York.

Broecker, W. S., and Peng, T. H. (1982) Reactive Metals and the Great Particular Sweep (Chapter 4). In *Tracers in the Sea,* Eldigio Press, Palisades New York.

11

KINETICS OF REDOX PROCESSES

11.1 INTRODUCTION

Many redox equilibria are established only slowly. We will discuss in this chapter many nonequilibrium aspects of redox chemistry. The oxidation of transition metal ions Fe(II), Mn(II), and Cu(I), of reduced sulfur compounds (H_2S, HS^-, SO_2, HSO_3^-, SO_3^{2-}), and of organic substances with oxygen (O_2) is of great interest, and the kinetics of these reactions will be treated here in an introductory way. The question will be asked: How good an oxidant is oxygen? We will illustrate some reaction mechanisms that can account for the fact that intermediates in the oxygen reduction can be better oxidants than oxygen. We will sketch a "bridge" between thermodynamics and kinetics and illustrate that in some cases so-called free energy relationships can be established between thermodynamics and kinetics.

As we have shown, the fixation of solar energy by photosynthesis causes a redox disproportionation. Abiotic photochemical processes can similarly induce redox disproportionation. Photochemical reactions produce highly reactive radicals and unstable redox species, which are important in the euphotic zone of natural waters. This will be discussed more in Chapter 12.

Nucleophilic–electrophilic mechanisms of organic transformations are considered together with redox reactions of organic compounds in order to illustrate common chemical properties of these reactions.

Finally, we illustrate the redox processes involved in the corrosion of metals and analyze the corrosion reaction as an electrochemical process.

11.2 HOW GOOD AN OXIDANT IS O_2?

Reaction Steps in the Reduction of O_2

The overall reaction

$$O_2(g) + 4\,H^+ + 4\,e^- \rightleftharpoons 2\,H_2O$$

$$\log K = 83.1,\ E_H^\circ = 1.229;\ p\varepsilon^\circ = 20.78 \qquad (1)$$

11.2 How Good an Oxidant Is O_2?

$$p\varepsilon + pH = 20.78 + \tfrac{1}{4} \log p_{O_2} \tag{1a}$$

can be subdivided into two two-electron sequences:

$$O_2(g) + 2 H^+ + 2 e^- \rightleftharpoons H_2O_2$$
$$\log K = 23.5, \ E_H^\circ = 0.69; \ p\varepsilon^\circ = 11.75 \tag{2}$$

and

$$H_2O_2 + 2 H^+ + 2 e^- \rightleftharpoons 2 H_2O$$
$$\log K = 59.6, \ E_H^\circ = 1.76; \ p\varepsilon^\circ = 29.8 \tag{3}$$

Oxygen, from a thermodynamic point of view, is a strong oxidant if the reduction occurs in a more-or-less synchronous four-electron step ($\log K = 83.1$ for equation 1). However, O_2 is a much weaker oxidant if the first two-electron reduction sequence becomes operative only; if the second reduction sequence ($H_2O_2 \to H_2O$) is much slower (presumably because of the cleavage of the O—O bond) than the first one, the reaction $O_2 + 2 H^+ + 2 e^- \rightleftharpoons H_2O_2$ with a standard potential of only 0.69 V ($\log K = 23.5$) determines the oxidizing ability of O_2. This seems to be the case for the electronation of oxygen in many electrode systems.

On the other hand, the rate of oxygenation of many transition metals [Fe(II), Mn(II), Ti(III), V(IV), U(IV)] is dependent on p_{O_2}; as we shall see, the reaction kinetics correspond to a mechanism in which the first step, that is, the reaction with the O_2 molecule itself, is rate determining, and O_2 often appears to be reduced all the way to H_2O.

Comparing E_H° or $p\varepsilon^\circ$ values for the reduction of $O_2(g)$ and H_2O_2, respectively, to water, it is obvious that H_2O_2 is, thermodynamically speaking, a stronger oxidant than O_2. This appears strange, because H_2O_2 is an intermediate in the reduction of $O_2(g)$.

H_2O_2 is unstable and disproportions into $O_2(g)$ and H_2O:

$$H_2O_2 = H_2O + \tfrac{1}{2} O_2(g) \qquad \Delta G^\circ = -103 \text{ kJ mol}^{-1} \tag{4}$$

In the $p\varepsilon$ versus pH diagram of Figure 11.1, the line for the two-electron reduction of O_2 to H_2O_2 is drawn. Strong oxidizing substances such as Br_2 will be reduced by H_2O_2, but H_2O_2 will be able to oxidize Br^-.

$$Br_2 + H_2O_2 \rightleftharpoons O_2 + 2 H^+ + 2 Br^-$$
$$2 Br^- + H_2O_2 + 2 H^+ \rightleftharpoons Br_2 + 2 H_2O$$
$$2 H_2O_2 \rightleftharpoons O_2 + 2 H_2O$$

The net result is the catalytic decomposition of the hydrogen peroxide.

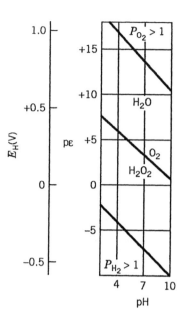

Figure 11.1. Stability of water and O_2–H_2O_2 between oxidation of water [$O_2(g) + 4\,H^+ + 4\,e^- = 2\,H_2O$; $p\varepsilon + pH = 20.78$] and reduction of water [$2\,H^+ + 2\,e^- = H_2(g)$; $p\varepsilon + pH = 0$]. The line for $O_2(g) + 2\,H^+ + 2\,e^- = H_2O_2$; $p\varepsilon^\circ = 11.7$ [$p\varepsilon^\circ = 4.7$ (pH = 7)] is drawn in the diagram. H_2O_2 is not stable in the presence of O_2 and H_2O. In the $p\varepsilon$ domain above the line O_2–H_2O_2, H_2O_2 is a better oxidant than O_2; and below this line, O_2 is a better oxidant than H_2O_2 (compare Figure 8.9a).

The complete reduction of O_2 to water provides the major oxidizing reaction in natural waters, and the O_2–H_2O couple determines effectively "the $p\varepsilon$" of oxic waters, even if in many situations, the acquisition of the last two electrons, from H_2O_2 to H_2O, is relatively slow (Stumm, 1978).

Aerobic life is surprisingly fast; O_2 reduction most likely does not take place through slow one-electron steps. Macromolecular electron transfer catalysts (enzymes) with fixed steric positions are presumably able to catalyze the more-or-less synchronous four-electron reduction of O_2. Thus, for most biologically mediated redox reactions, the O_2–H_2O system appears often to be the operative redox couple.

One-Electron Steps in the Reduction of O_2

In a simplified way, we can look at the one-electron steps in the reduction of O_2:

$$
\begin{aligned}
O_2 + e^- &\longrightarrow O_2^- \qquad &&\text{superoxide} \\
O_2^- + H^+ &\longrightarrow HO_2^\cdot \qquad &&\text{hydroperoxyl radical} \\
HO_2^\cdot + e^- &\longrightarrow HO_2^- \qquad &&\text{base of hydrogen peroxide} \\
HO_2^- + H^+ &\longrightarrow H_2O_2 \qquad &&\text{hydrogen peroxide} \\
H_2O_2 + e^- &\longrightarrow OH^\cdot + OH^- \qquad &&\text{hydroxyl radical} \\
OH^- + H^+ &\longrightarrow H_2O \qquad &&\text{water} \\
OH^\cdot + e^- &\longrightarrow OH^- \qquad &&\text{hydroxide} \\
OH^- + H^+ &\longrightarrow H_2O \qquad &&\text{water}
\end{aligned}
\qquad (5)
$$

11.2 How Good an Oxidant Is O_2?

These intermediates can also occur by electronic excitation, for example, by absorption of photons in photochemical processes. Also, atomic oxygen and singlet oxygen can be produced in these processes.

We first look at the thermodynamics of the one-electron reduction steps.

The O Species from a Thermodynamic Point of View

$\Delta G°$ values and equilibrium constants are available for the one-electron reduction steps (see Appendix 3 on the free energy of formation) (see Table 11.1).

Superoxide O_2^- The proton transfer of O_2^-(aq) is given by

$$HO_2^{\cdot} = H^+ + O_2^- \qquad \log K = -4.8 \tag{6}$$

One should note that the one-electron reduction of molecular O_2 is endergonic. (See Figure 11.2.)

This positive $\Delta G°$ value explains, in part, the relative kinetic inertness of O_2 to react with many reductants, especially with many organic substances (most of which persist in aerobic environments unless light, microorganisms, or other catalysts "activate" the oxygen).

O_2^- is both a one-electron oxidant and a one-electron reductant. Like all oxygen reduction intermediates, O_2^- is unstable and disproportionates:

$$2\ O_2^- + 2\ H^+ = O_2 + H_2O_2 \qquad K = 6.2 \times 10^{31} \tag{7}$$

O_2^- can both oxidize and reduce transition elements; for example,

$$O_2^- + Fe(II) + 2\ H^+ \longrightarrow Fe(III) + H_2O_2 \tag{8}$$

$$O_2^- + Fe(III) \longrightarrow Fe(II) + O_2 \tag{9}$$

Obviously Fe(II)/Fe(III) catalyzes the disproportionation; or O_2^- induces the

Table 11.1. $\Delta G°$ and $\log K$ Values for One-Electron Steps in the Reduction of O_2

Reaction	$\Delta G°$ (kJ mol^{-1})	$\log K$
A[a] O_2(aq) + e$^-$ → O_2^-(aq)	15.5	−2.7
B O_2^-(aq) + e$^-$ + 2H$^+$ → H_2O_2(aq)	−165.9	29.1
C H_2O_2(aq) + e$^-$ + H$^+$ → OH$^{\cdot}$(aq) + H_2O	−95.3	16.7
D OH$^{\cdot}$(aq) + e$^-$ + H$^+$ → H_2O	−244.9	42.9

[a] If the reactant is O_2(g) rather than O_2(aq), the values for $\Delta G°$ and $\log K$ are 31.8 kJ mol^{-1} and −5.58, respectively.

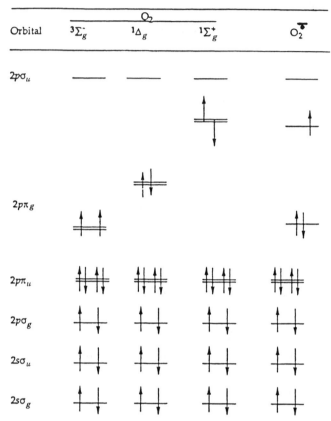

Figure 11.2. Molecular orbital diagrams for O_2 and O_2^-. The ground state "normal" O_2 (dioxygen) ($^3\Sigma_g^-$ = triplet ground state) has two unpaired π-electrons (it is a biradical) and is paramagnetic. Two excited forms of *singlet oxygen* $^1O_2$ ($^1\Delta_g$ and $^1\Sigma_g^+$) can be generated (often by light irradiation in the presence of a sensitizer). The first product in a one-electron reduction of oxygen is the *superoxide* (O_2^-) anion. (Adapted from Sawyer, 1991.)

cycling of Fe:

$$O_2^- \rightarrow \quad Fe(III) \leftarrow \quad \rightarrow H_2O_2$$
$$O_2 \leftarrow \quad \rightarrow Fe(II) \quad \leftarrow O_2^- \qquad (10)$$

This type of reaction plays an important role in sunlit surface waters and in atmospheric water droplets.

The Hydroxyl Radical OH· This radical is (thermodynamically and kinetically) a powerful oxidant. It can react with a large number of organic com-

11.3 Can pϵ Be Defined for a Nonequilibrium System?

pounds. It is formed by various mechanisms including the photolysis of NO_3^- and the reaction of Fe(II) with H_2O_2.

Ozone From a thermodynamic point of view ozone is a strong oxidant in water, producing oxygen as the reduced species:

$$\tfrac{1}{2}O_3(g) + H^+ + e^- = \tfrac{1}{2}O_2(g) + \tfrac{1}{2}H_2O \qquad p\epsilon° = 35.1 \qquad (11)$$

The reaction occurs outside the stability domain of water; that is, ozone can oxidize H_2O:

$$O_3(g) + H_2O = 2\,O_2 + 2\,H^+ + 2e^- \qquad (12)$$

Ozone decomposes in a rather complicated set of reactions. (For an overview see Hoigné, 1988.) OH· radicals are the main secondary oxidants produced from decomposed ozone. Thus, as we shall discuss in Section 11.5, ozone reacts with solutes directly as well as indirectly through OH·.

Example 11.1. pϵ° Values for the Reduction of O_3 and O_2 and their Reduction Intermediates Sketch a figure for $p\epsilon°_{pH=7}$ (or $E°_{pH=7}$) values for ozone, oxygen, and their reduction intermediates. Distinguish between four-electron, two-electron, and one-electron redox pairs.

The data may be calculated from the data on free energy of formation in Appendix 3. H_2O_2, O_2^-, OH·, and O_3 are not stable, but their redox "equilibria" can be treated as if they were metastable (see Figure 11.3).

11.3 CAN pϵ BE DEFINED FOR A NONEQUILIBRIUM SYSTEM?

Obviously, a conceptually meaningful pϵ cannot be defined for a nonequilibrium—that is, nonstable or nonmetastable—system. Based on some of the observed activities of redox components of a seawater system (atmosphere, hydrosphere, and sediments), different pϵ values can be calculated. The examples in Figure 11.4 illustrate immediately that the various redox components are not in equilibrium with each other and that the real system cannot be characterized by a unique pϵ.

An Equilibrium Model for the Sea Abstracting from the complexity of nature, an idealized counterpart of the oxic ocean (atmosphere, water, sediment) may be visualized. Oxygen obviously is the atmospheric oxidant that is most influential in regulating (with its redox partner, water) the redox level of oxic water. It is more abundant—within the time span of its atmospheric residence time—in the atmosphere than in the other accessible exchange reservoirs. It is chemically and biologically reactive; its redox processes (photosynthesis

Kinetics of Redox Processes

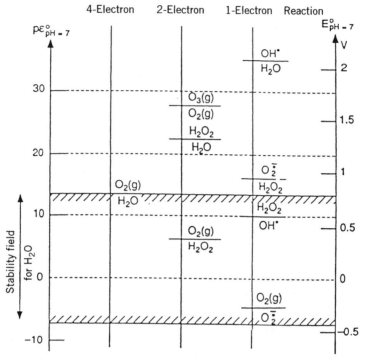

Figure 11.3. Sketch of $p\epsilon^\circ_{pH=7}$ and $E^\circ_{pH=7}$ values for four-electron, two-electron, and one-electron redox processes of oxygen compounds. H_2O_2, O_2^-, $OH^\cdot$, and O_3 are treated as if they were metastable. Thermodynamically, the $OH^\cdot$ radical (reduction to H_2O) and O_3 (reduction to O_2) and H_2O_2 (reduction to H_2O) are the "strongest" oxidants; oxygen with regard to its reduction to H_2O_2 and with regard to its reduction to O_2^- is a "weak" oxidant.

and respiration) occur with a mean flux of ~40 mol of electrons per m² of surface per year.

The fact that sunlit surface water layers contain substantial amounts of H_2O_2 (steady-state concentrations are on the order of 50 nM) reflects a disturbance of equilibrium by solar light. In a similar way, photosynthesis continuously takes the system away from equilibrium.

Another major abundant potential oxidant is N_2; because of its relative inertness in redox reactions (low electron flux), it does not appear to be a suitable candidate for controlling the redox composition of the atmosphere–hydrosphere interface. Thus the oxygen–water couple appears to be the predominant redox buffer and represents the equilibrium model that determines the redox level of other less abundant redox couples: $O_2(g) + 4 H^+ + 4 e^- \rightleftharpoons 2 H_2O(l)$, log K = 83.1 (25°C).

This equilibrium system acquires, at pH = 8 for $p_{O_2} = 0.2$ atm, a $p\epsilon$ of 12.6. As we have seen, at this $p\epsilon$ all biochemically important elements would

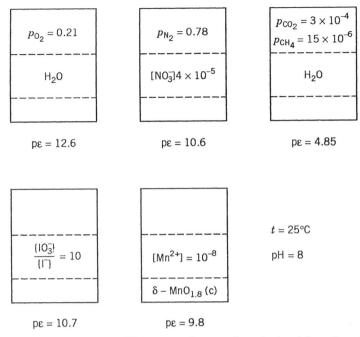

Figure 11.4. pε of seawater. Different pε values can be calculated for a few observed activities of redox components of an oxic seawater system (atmosphere, hydrosphere, and sediments). It is difficult to characterize the real system by a unique pε.

exist virtually completely in their highest naturally occurring oxidation states, if equilibrium prevailed.

11.4 KINETICS OF REDOX PROCESSES: CASE STUDIES

Within the last decades, much progress has been made in understanding the mechanisms of redox reactions, especially how, during the reaction, the changes in the state of coordination are coupled to the changes in oxidation state. Many, if not most, redox reactions involve substitutional changes as an integral part of the overall process.

That the addition of an electron can cause a dramatic change in structural geometry and lability may be exemplified by the reduction of $Cr(H_2O)_6^{3+}$ to $Cr(H_2O)_6^{2+}$. $Cr(H_2O)_6^{3+}$ exists as a symmetric octahedral structure of high stability. The half-time for water exchange between $Cr(H_2O)_6^{3+}$ and the solvent is rather large ($\sim 10^6$ s). $Cr(H_2O)_6^{2+}$ is still octahedral, but its structure is distorted, that is, elongated along one axis; the complex is extremely labile, the half-time for water exchange between aquo complex and solvent being less than 10^{-9} s.[†] Thus the change in oxidation state changes the rate of substitution by a factor of more than 10^{15}. On the other hand, the oxidation of Cr(III) to Cr(VI) changes the coordination number to 4 (CrO_4^{2-} or $Cr_2O_7^{2-}$).

A clearer understanding of the redox kinetics of processes in aquatic systems depends on a better appreciation of the relationship between structural and dynamic aspects of chemical behavior.

We illustrate a few case studies on the kinetics of a few well understood inorganic redox reactions.

Example 11.2. Oxidation of Sulfite by Ozone, an Example for Parallel Oxidation Pathways The oxidation of SO_2 (and its conjugate bases) by various oxidants (O_2, H_2O_2, O_3)

$$SO_2 + \text{``O''} + H_2O = SO_4^{2-} + H_2O \quad \text{where ``O'' is the oxidant} \quad \text{(i)}$$

is of importance in the treatment of certain wastes, but above all in the atmosphere where enormous quantities of SO_2 are oxidized in atmospheric water droplets (clouds, fog, rain).

Here we discuss the oxidation of SO_2 by ozone. $SO_2(g)$ dissolves in water, depending on pH (see Figure 5.7), as $SO_2 \cdot H_2O$, HSO_3^-, and SO_3^{2-}. These species are oxidized (nucleophilic attack by ozone) with different reaction rates. The oxidation reactions occur as parallel reaction steps:

$$-\frac{d [SO_2 \cdot H_2O]}{dt} = k_0 [SO_2 \cdot H_2O] [O_3(aq)] \quad \text{(ii)}$$

$$-\frac{d[HSO_3^-]}{dt} = k_1 [HSO_3^-] [O_3(aq)] \quad \text{(iii)}$$

$$-\frac{d [SO_3^{2-}]}{dt} = k_2 [SO_3^{2-}] [O_3(aq)] \quad \text{(iv)}$$

Thus

$$-\frac{d S(IV)}{dt} = (k_0[SO_2 \cdot H_2O] + k_1[HSO_3^-] + k_2[SO_3^{2-}]) [O_3(aq)] \quad \text{(v)}$$

Since

$$[S(IV)] = [SO_2 \cdot H_2O] + [HSO_3^-] + [SO_3^{2-}] \quad \text{(via)}$$

[†]The splitting of the d orbitals in an octahedral field permits one to rationalize some aspects of stereochemistry and lability. In the +3 oxidation state of chromium, $3s^2p^6d^3$, the three d electrons half fill the threefold degenerate t_{2g} orbitals. This symmetric electronic arrangement gives rise to a regular octahedral shape. In $Cr(H_2O)_6^{2+}$, the four d electrons have parallel spins, three occupying t_{2g} orbitals singly and the fourth an e_g orbital. The nonsymmetric filling of the e_g orbitals gives rise to a pronounced distortion of the regular octahedral structure.

11.4 Kinetics of Redox Processes: Case Studies

and

$$[SO_2 \cdot H_2O] = \alpha_0 [S(IV)], \quad [HSO_3^-] = \alpha_1 [S(IV)], \quad [SO_3^{2-}] = \alpha_2 [S(IV)]$$

where

$$\alpha_0 = (1 + K_1[H^+]^{-1} + K_1K_2 [H^+]^{-2})^{-1} \tag{vii}$$

$$\alpha_1 = ([H^+] K_1^{-1} + 1 + K_2 [H^+]^{-1})^{-1} \tag{viii}$$

$$\alpha_2 = ([H^+]^2 K_1^{-1}K_2^{-1} + [H^+] K_2^{-1} + 1)^{-1} \tag{ix}$$

Equation v can be rewritten as

$$-\frac{d[S(IV)]}{dt} = k_{ozone} [S(IV)] [O_3(aq)] \tag{x}$$

where

$$k_{ozone} = k_0\alpha_0 + k_1\alpha_1 + k_2\alpha_2 \tag{xi}$$

Hoigné et al. (1985) have derived the rate laws given and have determined the individual k values: $k_0 = 2 \times 10^4 \ M^{-1} \ s^{-1}$; $k_1 = 3.2 \times 10^5 \ M^{-1} \ s^{-1}$; and $k_2 = 1 \times 10^9 \ M^{-1} \ s^{-1}$ (25°C).

Obviously, k_{ozone} is highly pH dependent. Using the following K_1 and K_2 values (25°C):

$$SO_2 \cdot H_2O = HSO_3^- + H^+ \quad \log K_1 = -1.86 \ (I = 10^{-3})$$

$$HSO_3^- = SO_3^{2-} + H^+ \quad \log K_2 = -7.15 \ (I = 10^{-3})$$

we can calculate k_{ozone} versus pH (equation xi) and the half-life of S(IV) at a steady-state concentration of $[O_3(aq)] = 10^{-9}$ M (Figure 11.5).

Example 11.3. Oxidation of S(IV) by H_2O_2 The oxidation by H_2O_2 is less pH dependent than that by O_3. The following has been proposed (Hoffmann and Calvert, 1985):

$$HSO_3^- + H_2O_2 \underset{k_{1b}}{\overset{k_{1f}}{\rightleftharpoons}} SO_2OOH^- + H_2O \tag{i}$$

peroxomonosulfurous acid

$$SO_2OOH^- + H^+ \overset{k_2}{\rightleftharpoons} SO_4^{2-} + 2 H^+ \tag{ii}$$

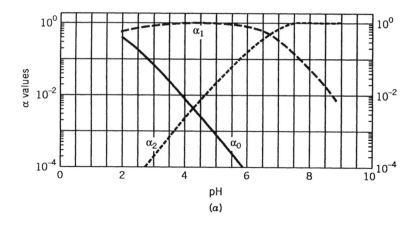

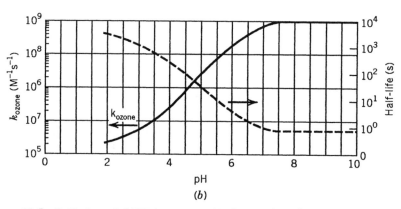

Figure 11.5. Oxidation of S(IV) by ozone: (a) the α values for the $SO_2(aq)$ system and (b) the k_{ozone} (equation xi) and half-life for an assumed steady-state concentration of $[O_3(aq)] = 10^{-9}$ M. Above pH = 5, the oxidation of S(IV) is sufficiently fast that the resupply of O_3 into the water droplets (from the atmosphere) may not be sufficiently fast to maintain a steady-state concentration of $O_3(aq)$.

The rate is given by

$$\frac{d\,[SO_4^{2-}]}{dt} = k_2\,[SO_2OOH^-]\,[H^+] \tag{iii}$$

Assuming a steady state for SO_2OOH^-, we have

$$\frac{d\,[SO_2OOH^-]}{dt} = k_{1f}[H_2O_2]\,[HSO_3^-] - (k_{1b} + k_2[H^+])\,[SO_2OOH^-]$$

$$= 0 \tag{v}$$

11.4 Kinetics of Redox Processes: Case Studies

The steady-state concentration is given by

$$[SO_2OOH^-]_{ss} = \frac{k_{1f}[H_2O_2][HSO_3^-]}{k_{1b} + k_2[H^+]} \quad \text{(v)}$$

Using this concentration in equation iii, we obtain

$$\frac{d[SO_4^{2-}]}{dt} = \frac{k_2 k_{1f}[H_2O_2][HSO_3^-][H^+]}{k_{1b} + k_2[H^+]} \quad \text{(vi)}$$

and specifically

$$-\frac{d[S(IV)]}{dt} = \frac{k[H^+][H_2O_2][S(IV)]\alpha_1}{1 + K[H^+]} \quad \text{(vii)}$$

where α_1 is as given in equation viii of Example 11.2, $k = 7.45 \times 10^7$ M^{-2} s^{-1}, and $K = 13$ M^{-1} at 25°C (Hoffmann and Calvert, 1985).

For other oxidation pathways of S(IV), see Seinfeld (1986). S(IV) can also be oxidized by oxygen in the presence of Fe as a catalyst (Hoffmann and Calvert, 1985; Kotronarou and Sigg, 1993; Sedlak and Hoigné, 1994).

Oxidation of Fe(II) and Mn(II) The rate of oxygenation of Fe(II) in solutions of pH ≥ 5 was found to be first-order with respect to the concentrations of both Fe(II) and O_2 and second-order with respect to the OH$^-$ ion. Thus a 100-fold increase in the rate of reaction occurs for a unit increase in pH. The results of representative kinetic experiments are shown in Figure 11.6. Figures 11.6a and 11.6b show the course of Fe(II) and Mn(II) disappearance from the solution at different pH values. It is evident that the reaction rates are strongly pH dependent. Oxidation and Fe(II) is very slow below pH 6. Figure 11.6e shows rate constants for Fe(II) oxidation by oxygen over the pH range 1-6. Catalysts (especially Cu^{2+} and Co^{2+}) in trace quantities, as well as anions that form complexes with Fe(III) (e.g., HPO_4^{2-}), increase the reaction rate significantly. The oxygenation kinetics follow the rate law (Millero et al., 1987; Singer and Stumm, 1970; Stumm and Lee, 1961):

$$\frac{-d[Fe(II)]}{dt} = k[Fe(II)][OH^-]^2 p_{O_2} \quad (13)$$

where $k = 8.0 \ (\pm 2.5) \times 10^{13}$ min^{-1} atm^{-1} mol^{-2} liter^2 at 20°C. Frequently, it is more convenient to use the rate law in the form

$$\frac{-d[Fe(II)]}{dt} = \frac{k_H[O_2(aq)]}{[H^+]^2}[Fe(II)] \quad (14)$$

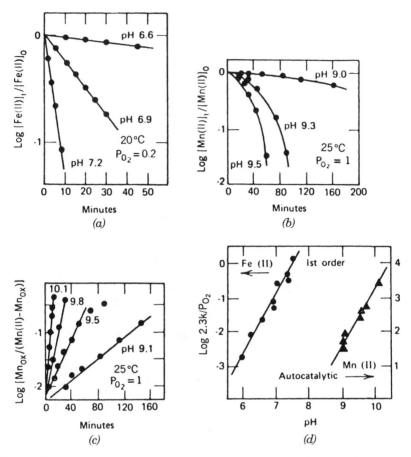

Figure 11.6. Oxidation of Fe(II) and Mn(II) by oxygen. All experiments were conducted with dissolved Fe(II) or Mn(II) concentrations of less than 5×10^{-4} M. In each series of experiments, the pH was controlled by continuously bubbling CO_2- and O_2-containing gas mixtures through HCO_3^- solutions of known alkalinity. (a) Oxygenation of Fe(II) in bicarbonate solutions. (b) Removal of Mn(II) by oxygenation in bicarbonate solutions. (c) Oxidation of Mn(II) in HCO_3^- solutions; autocatalytic plot. (d) Effect of pH on oxygenation rates. (e) Oxidation rate of ferrous iron as a function of pH. At low pH the oxidation rate is independent of pH, while in the higher pH range equation 13 (second-order dependence on [OH$^-$]) is fulfilled. (From Singer and Stumm, 1970. More recent data by Millero et al., 1987.)

where, at 20°C, $k_H = 3 \times 10^{-12}$ min^{-1} mol^1 liter^{-1}. For a given pH, the rate increases about tenfold for a 15°C temperature increase (activation energy $\simeq 23$ kcal mol^{-1} at constant [H$^+$]). Kester et al. (1975) found the same rate law based on measurements in natural seawater from Narragansett Bay (salinity = 31.2‰) and in surface Sargasso seawater (36.0‰). They report half-times

11.4 Kinetics of Redox Processes: Case Studies

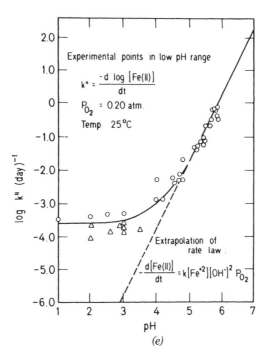

Figure 11.6. (*Continued*)

nearly 100 times larger than those observed in fresh water at the same pH.[†]

By comparing Figures 11.6a and 11.6b, it is obvious that the Mn(II) oxygenation does not follow the same rate law as Fe(II) oxygenation. The manner of the decrease in the Mn(II) concentration with time suggests an autocatalytic reaction (Morgan and Stumm, 1964). The rate expression

$$\frac{-d\,[\mathrm{Mn(II)}]}{dt} = k_0\,[\mathrm{Mn(II)}] + k\,[\mathrm{Mn(II)}]\,[\mathrm{MnO_2}] \tag{15}$$

was found to fit the experimental data well (Figure 11.6c), thus lending support to an autocatalytic model.

The reaction might be visualized as proceeding according to the following pattern (reactions are not balanced with respect to water and protons):

$$\mathrm{Mn(II)} + \mathrm{O_2} \xrightarrow{\text{slow}} \mathrm{MnO_2(s)} \tag{16a}$$

[†]An explanation for the kinetics of Fe(II) oxidation in estuarine and seawater has been suggested by the results of Sung and Morgan (1980), who reported strong effects of ionic strength and medium anions (Cl^-, SO_4^{2-}) in slowing the reaction.

$$\text{Mn(II)} + \text{MnO}_2(s) \xrightarrow{\text{fast}} \text{Mn(II)} \cdot \text{MnO}_2(s) \quad (16b)$$

$$\text{Mn(II)} \cdot \text{MnO}_2(s) + \text{O}_2 \xrightarrow{\text{slow}} 2\text{MnO}_2(s) \quad (16c)$$

Although other interpretations of the autocatalytic nature of the reaction are possible, the following experimental findings are in accord with such a reaction scheme. (1) The extent of Mn(II) removal during the oxygenation reaction is not acccounted for by the stoichiometry of the oxidation reaction alone; that is, not all the Mn(II) removed from the solution [as determined by specific analysis for Mn(II)] is oxidized (as determined by measurement of the total oxidizing equivalents of the suspension). (2) As pointed out before, the products of Mn(II) oxygenation are nonstoichiometric, showing various average degrees of oxidation ranging from approximately $MnO_{1.3}$ to $MnO_{1.9}$ (30–90% oxidation to MnO_2) under varying alkaline conditions. (3) The higher-valence manganese oxide suspensions show large sorption capacities for Mn^{2+} in slightly alkaline solutions. The relative proportions of Mn(IV) and Mn(II) in the solid phase depend strongly on pH and other variables.

The data given in Figure 11.6e can be explained by a rate law, originally proposed by Millero (1985):

$$-\frac{d\,[\text{Fe(II)}]}{dt} = (k_0[\text{Fe}^{2+}] + k_1\,[\text{Fe(OH)}^+] + k_2\,[\text{Fe(OH)}_2(\text{aq})])\,[\text{O}_2] \quad (17)$$

This rate law is in accord with that given in equation 13 for the pH range of natural waters, because the first two terms on the right-hand side of equation 17 can be neglected for pH > 5. Fe(II) and the other transition elements—Mn(II), VO^{2+}, and Cu(I)—may more readily associate (probably outer-spherically) with O_2 if they are present in the form of hydroxo complexes (i.e., hydrolysis species) or as complexes with hydroxo surface groups of hydrous oxides (Luther, 1990; Wehrli, 1990; Wehrli and Stumm, 1989). As explained by Luther (1990) (Figure 11.7), the OH^- ligands donate electron density to the reduced metal ion through both the σ and π systems, which results in metal basicity and increases reducing power; for example, the Fe(III) oxidation state is stabilized by the OH^- ligands (Figure 8.26). Complex formation with other ligands containing oxygen donor atoms, including hydrous oxide surfaces to which Fe(II) becomes adsorbed, usually enhance the oxidation rate by O_2.

Oxidation Mechanism Which reaction mechanisms can account for the rate law (equation 17)? Two factors have to be explained:

1. The pH dependence of the oxidation rate (Figure 11.6e).
2. The first-order dependence on O_2.

The dependence of the oxidation rate on pH can be accounted for by assuming that hydrolyzed Fe(II) reacts faster with oxygen than nonhydrolyzed Fe(II).

11.4 Kinetics of Redox Processes: Case Studies

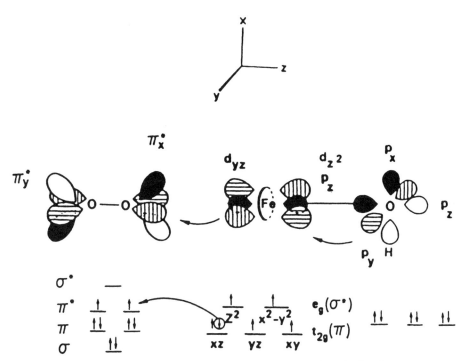

Figure 11.7. Molecular orbital diagram for the oxidation of Fe(II) by O_2. An outer-sphere electron-transfer process from $\pi(d_{yz})$ to π_y^* [or $\pi(d_{xz})$ to π_x^*] is possible. OH^- bound to Fe(II) can enhance the oxidation by transfer of electron density through both the σ and π systems. This stabilizes the Fe(III) formed on oxidation. (From Luther, 1990.)

Furthermore, we expect (similar to the oxidation of the various S species) that the Fe(II) species [Fe^{2+}, $FeOH^+$, $Fe(OH)_2(aq)$] are oxidized in parallel by O_2. These reactions are given in Table 11.2.

Obviously, all the oxidation steps are endergonic, but $\Delta G°$ decreases in going from free Fe^{2+} to FeOH to $Fe(OH)_2(aq)$. These oxidation steps (Table 11.2) can now be combined with the various reduction steps of the oxygen

Table 11.2. $\Delta G°$ and log K_{ox} Values for the Oxidation of Fe(II)[a]

Reaction	$\Delta G°$ (kJ mol^{-1})	log K_{ox}
0 $Fe^{2+} \rightarrow Fe^{3+} + e^-$	74.2	−13.0
1 $FeOH^+ \rightarrow FeOH^{2+} + e^-$	48.0	−8.41[b]
2 $Fe(OH)_2 \rightarrow Fe(OH)_2^+ + e^-$	3.0	−0.53

[a]The data have been calculated with the information on free energy of formation of the species involved, given in Appendix 3.
[b]Based on data by Bard et al., (1985) $\Delta G° = 38.6$ and log $K_{ox} = -6.76$.

Table 11.3. Free Energy of Oxygenation of Fe(II): Combination of Oxidation of Fe(II) (Table 11.2) with Reduction Steps of O_2 (Table 11.1)

Combination	pH = 2 $\Delta G°_{pH=2}$ (kJ mol^{-1})	Combination	pH = 5^a $\Delta G°_{pH=5}$ (kJ mol^{-1})	Combination	pH = 7 $\Delta G°_{pH=7}$ (kJ mol^{-1})
0 + A	89.8	1 + A	63.5	2 + A	18.5
0 + B	−68.8	1 + B	−60.9	2 + B	−83.0
0 + C	−9.7	1 + C	−18.8	2 + C	−52.4
0 + D	−759.2	1 + D	−168.3	2 + D	−202.0

^{a}Using a free energy of formation of −268 kJmol^{-1} for FeOH$^+$ (Bard, et al, 1985) gives the following values for this column: 54.1, −70.2, −28.1, and −177.7, respectively.

species. These reduction steps were already discussed in Section 11.2 and listed in Table 11.1. A combination of oxidation steps and reduction steps is given in Table 11.3.

We can conclude from Table 11.3 that in each combination of the sequential reaction, the first step, involving the reduction $O_2 + e^- = O_2^-$, is endergonic and that in all parallel reactions, the oxidation of Fe(OH)$_2$(aq) is most exergonic (Figure 11.8a,b). If we were to use a thermodynamic criterion for a kinetic argument, we would stipulate (see Wehrli, 1990) that the first step in each sequence is the slowest step and rate determining and that the oxidation of Fe(OH)$_2$(aq) produces the fastest oxygenation sequence. Thus, in the pH range of natural waters, at pH > 5, the overall oxygenation reaction of Fe(II) can be described by

$$\text{Fe(OH)}_2(\text{aq}) + O_2 \xrightarrow{\text{slow}} \text{Fe(OH)}_2^+ + O_2^- \quad (2, A)$$

$$\text{Fe(OH)}_2(\text{aq}) + O_2^- + 2\,H^+ \longrightarrow \text{Fe(OH)}_2^+ + H_2O_2 \quad (2, B)$$

$$\text{Fe(OH)}_2(\text{aq}) + H_2O_2 + H^+ \longrightarrow \text{Fe(OH)}_2^+ + OH^· + H_2O \quad (2, C) \quad (18)$$

$$\text{Fe(OH)}_2(\text{aq}) + OH^· + H^+ \longrightarrow \text{Fe(OH)}_2^+ + H_2O \quad (2, D)$$

$$4\,\text{Fe(OH)}_2(\text{aq}) + O_2 + 4\,H^+ = 4\,\text{Fe(OH)}_2^+ + 2\,H_2O$$

Wehrli has shown (Figure 11.8b) a linear free energy relationship between the free energy of the reaction (log K) and the rate (log k) of the reaction for the oxygenation of Fe^{2+}, FeOH$^+$, and Fe(OH)$_2$(aq).

Fenton Reagent The interaction of H_2O_2 with Fe(II) yields OH$^·$ radicals stoichiometrically:

$$\text{Fe(II)} + H_2O_2 + H^+ \longrightarrow \text{Fe(III)} + OH^· + H_2O$$

Fe(III) then acts as a catalyst to decompose the H_2O_2 to O_2 and H_2O during which a steady-state concentration of Fe(II) as a source of OH$^·$ (as given above)

11.4 Kinetics of Redox Processes: Case Studies

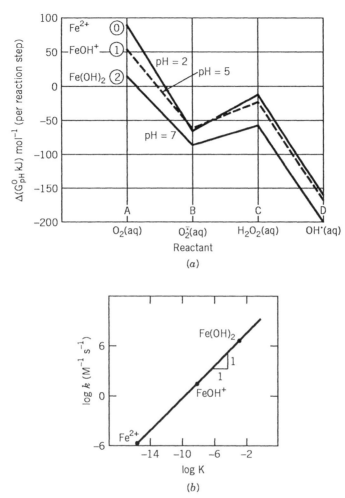

Figure 11.8. Oxidation of Fe(II) by oxygen. (a) Free energy of the redox reaction steps of the oxidation of Fe(II) to Fe(III) coupled with the one-electron reduction of O_2(aq), O_2^-(aq), H_2O_2(aq), and $OH^{\cdot}$(aq), respectively. At pH = 2, the oxidation is $Fe^{2+} \rightarrow Fe^{3+}$; at pH = 5, $Fe(OH)^+ \rightarrow Fe(OH)^{2+}$; and at pH = 7, $Fe(OH)_2 \rightarrow Fe(OH)_2^+$ (cf. Table 11.3). (b) Plot of reaction rates log k (from Figure 11.6e) for the oxidation of different Fe(II) species with O_2:

$$Fe^{II}(OH)_i^{(2-i)} + O_2 \longrightarrow Fe^{III}(OH)_i^{(3-i)} + O_2^-$$

and log equilibrium constant of this reaction (Table 11.3). As will be discussed in Section 11.6, the slope of $d \log k / d \log K = 1$ corresponds to that predicted by the Marcus theory. (Adapted from Wehrli, 1990.)

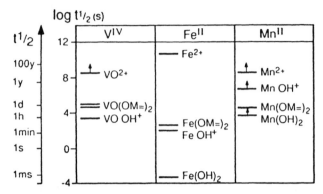

Figure 11.9. Effects of hydrolysis and adsorption on the oxygenation of transition-metal ions. Arrows indicate lower limit. Data represent order of magnitude because surface area concentrations were not determined. (From Wehrli and Stumm, 1989.)

is generated:

$$\text{Fe(III)} + \text{H}_2\text{O}_2 \xrightleftharpoons{-\text{H}^+} \text{Fe} - \text{O}_2\text{H}^{2+} \rightleftharpoons \text{Fe}^{2+} + \text{HO}_2^{\bullet}$$

$$\text{Fe(III)} + \text{HO}_2^{\bullet} \longrightarrow \text{Fe(II)} + \text{H}^+ + \text{O}_2$$

Surface Binding Enhances Oxidation of Transition Ions

The oxidation of Fe(II), VO^{2+}, Mn^{2+}, and Cu^+ by oxygen is favored thermodynamically and kinetically not only by hydrolysis but also by specific adsorption to hydrous oxide surfaces. Adsorption with hydroxo surface groups of hydrous oxides has a similar effect as hydrolysis (association with OH^-) (Luther, 1990; Wehrli et al., 1990). Figure 11.9 compares the catalytic effects of hydrolysis and adsorption for VO(II), Fe(II), and Mn(II). Fe(II) bound to silicates is also more readily oxidized by O_2 than Fe^{2+} (White, 1990).

The Oxidation of Pyrite

The sulfur-bearing minerals that predominate in coal seams are the iron sulfide ores pyrite and marcasite. Both have the same ratio of sulfur to iron, but their crystallographic properties are quite different. Marcasite has an orthorhombic structure, while pyrite is isometric. Marcasite is less stable and more easily decomposed than pyrite. The latter is the most widespread of all sulfide minerals and, as a result of its greater abundance in the eastern United States, pyrite is recognized as the major source of acid mine drainage. $FeS_2(s)$ is used here as a symbolic representation of the crystalline pyritic agglomerates found in coal mines.

When pyrite is exposed to air and water, the following overall stoichiometric reactions may characterize the oxidation of pyrite:

11.5 Oxidants Used in Water and Waste Technology: A Few Case Studies

$$FeS_2(s) + \tfrac{7}{2}O_2 + H_2O = Fe^{2+} + 2SO_4^{2-} + 2H^+ \quad (19a)$$

$$Fe^{2+} + \tfrac{1}{4}O_2 + H^+ = Fe^{3+} + \tfrac{1}{2}H_2O \quad (19b)$$

$$Fe^{3+} + 3H_2O = Fe(OH)_3(s) + 3H^+ \quad (19c)$$

$$FeS_2(s) + 14Fe^{3+} + 8H_2O = 15Fe^{2+} + 2SO_4^{2-} + 16H^+ \quad (19d)$$

The oxidation of the sulfide of the pyrite to sulfate (equation 19a) releases dissolved ferrous iron and acidity into the water. Subsequently, the dissolved ferrous iron undergoes oxygenation to ferric iron, which then hydrolyzes to form insoluble "ferric hydroxide," releasing more acidity to the stream and coating the streambed. Ferric iron can also be reduced by pyrite itself, as in equation 19d, where sulfide is again oxidized and acidity is released along with additional ferrous iron, which may reenter the reaction cycle via reaction 19b.

The oxidation of pyrite by O_2 is mediated by the Fe(II)–Fe(III) system; pyrite is oxidized by Fe(III), which forms a surface complex with pyrite (Luther, 1990).

The rate-determining step at the relatively low pH values often encountered under conditions of pyrite dissolution is the oxygenation of Fe(II) to Fe(III), usually catalyzed by autotrophic bacteria (Singer and Stumm, 1970). Thus the overall rate of pyrite dissolution is insensitive to the mineral surface area concentration. Microbially catalyzed oxidation of Fe(II) to Fe(III) by oxygen could also be of some significance for oxidative silicate dissolution in certain acid environments.

11.5 OXIDANTS USED IN WATER AND WASTE TECHNOLOGY: A FEW CASE STUDIES

Ozone-Initiated Oxidation of Solutes

The primary oxidations initiated by ozone in water can be described by the following reaction sequences (Hoigné and Bader, 1978):

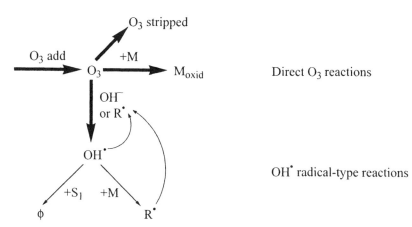

On the one hand, part of the ozone (O_3) dissolved in water reacts directly with the solutes M. Such direct reactions are highly selective and often rather slow (minutes). On the other hand, part of the ozone added decomposes before it reacts with solutes; this leads to free radicals. Among these, the OH˙ radicals belong to the most reactive oxidants known to occur in water. OH˙ can easily oxidize all types of organic contaminants and many inorganic solutes (radical-type reactions). They are therefore consumed in fast reactions (microseconds) and exhibit little substrate selectivity. Only a few of their reactions are of specific interest in water treatment processes. Measured oxidations in model solutions indicate up to 0.5 mol OH˙ formed per mole of ozone decomposed. The higher the pH, the faster the decomposition of ozone, which is catalyzed by hydroxide ions (OH^-). The decomposition is additionally accelerated by an autocatalyzed sequence of reactions in which radicals formed from decomposed ozone act as chain carriers. Some types of solutes react with OH˙ radicals and form secondary radicals (R˙), which still act as chain carriers. Others, for instance, bicarbonate ions, transform primary radicals to inefficient species (ϕ) and thereby act as inhibitors of the chain reaction. Therefore the rate of the decomposition of ozone depends on the pH of the water as well as on the solutes present. The overall effect is a superposition of the direct reaction and the radical-type reaction. For a review, see Hoigné (1988).

Kinetic Formulation Following Hoigné's treatment, we may write the direct reaction of molecular ozone with a solute (M) as follows:

$$O_3 + \eta M \xrightarrow{k} \eta M_{oxid} \tag{20a}$$

The rates of these reactions are first-order with respect to ozone and, as a rule, nearly first-order with respect to solute concentration. Therefore the rate at which a solute is oxidized becomes

$$\frac{-d[M]}{dt} = -\eta \frac{d[O_3]}{dt} = \eta k [O_3][M] \tag{20b}$$

and the elimination of M is given by

$$-\ln \frac{[M]_e}{[M]_0} = \eta k [\bar{O}_3] t \tag{20c}$$

where $[M]_0$ and $[M]_e$ are, respectively, the initial and final concentrations of M: η is the yield factor of M elimination per ozone used (mol mol^{-1}); k is the rate constant of reaction of ozone with M; and $[\bar{O}_3]$ is the mean O_3 concentration during reaction period t. The relative solute elimination by this direct reaction depends only on the mean concentration of ozone, the time the ozonation lasts,

11.5 Oxidants Used in Water and Waste Technology: A Few Case Studies

and the rate constant k. Some experimentally determined data for rate constants are presented in Figures 11.10a and 11.10b for illustration.

Oxidation by OH• Radicals The amount of OH• radicals available for oxidation of a solute M depends on the amount of OH• formed and the relative rate with which it reacts with M when compared with the rate at which it is consumed by all other solutes, in accordance with the following scheme:

Reaction: *Rate of OH• consumption:*

$$\Delta O_3 \to \eta' \text{ OH}^\bullet \begin{cases} + M \longrightarrow & k'_M[M][\text{OH}^\bullet] \\ + \Sigma S_i \longrightarrow & \Sigma(k'_i[S_i])[\text{OH}^\bullet] \end{cases} \qquad (21)$$

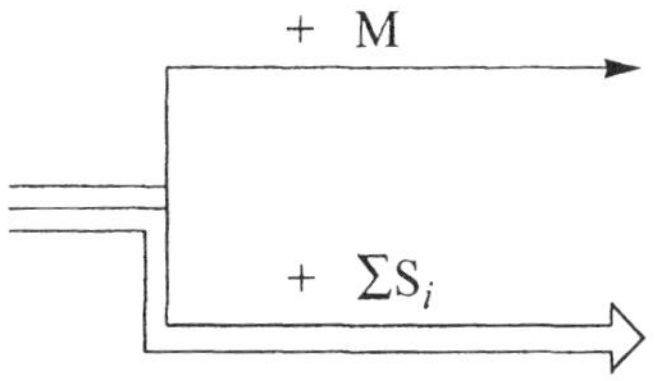

Thereby, ΔO_3 is the amount of O_3 decomposed during the process; η' is the yield for OH• formation from ΔO_3; k' is the second-order reaction rate constant for OH•; and $\Sigma\ (k'_i[S_i])$ is the rate of OH• scavenging by all solutes present including O_3 and M. The reactivity of OH• radicals is very high toward all organic solutes in water. Even free ammonia (NH_3), hydrogen peroxide, and ozone may interfere with OH• (compare Figure 11.10b). Only a fraction (η'') of the OH• radicals reacting with solutes will result in a solute elimination. If this yield factor η'' is also taken into account, then the rate of solute oxidations in the presence of competing scavengers becomes

$$\frac{-d[M]}{dt} = \eta'\eta'' \frac{d\Delta O_3}{dt} = k'_M[M] \left[\Sigma\ (k'_i[S_i]) \right]^{-1} \qquad (22)$$

Within a limited range of oxidation, where $\Sigma\ (k'_i[S_i])$ depends on neither the degree of ozonation nor the ozone concentration, integration of equation 22 yields

$$-\ln\left(\frac{[M]_e}{[M]_0}\right) = \eta'\eta'' \Delta O_3\ k'_M \left[\Sigma\ (k'_i[S_i]) \right]^{-1} \qquad (23a)$$

when

$$k'_M[M] \ll \Sigma\ (k'_i[S_i]) \qquad (23b)$$

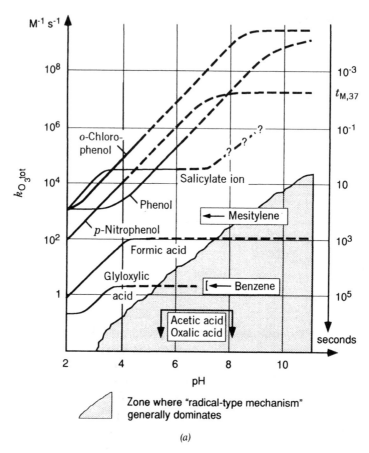

Figure 11.10. Rate constants for reactions of ozone and OH· radicals with solutes. (a, b) Examples of rate constants for direct reactions of ozone with organic and inorganic solutes versus pH (data selected). $k_{O_3 tot}$ is related to k by the yield factor. (c) Rate constants for reactions of OH· radicals with different solutes. $(\Delta O_3)_{37}$ is the required amount of decomposed ozone, which results in the elimination of the quoted substrate to 37% of the initial value (batch-type or plug-flow reactor). This scale is calibrated for eutrophic lakewater (Lac de Bret, DOC = 4 mg liter^{-1}, [HCO$_3^-$] = 1.6 mM, pH = 8.3. The latter changes proportionally to the DOC of water). (From Hoigné, 1988.)

The elimination of M is first-order with respect to the M concentration. The amount of ozone that must be decomposed in order to give a certain elimination factor is expected to increase linearly with the rate at which OH· radicals are consumed by the sum of all radical scavengers present. The amount of ozone required to be decomposed decreases linearly with the rate constant with which M itself reacts with the oxidant.

Rate constants for OH· radical reactions with hundreds of aqueous solutes are known. Figure 11.10c gives an approximation of the magnitudes of such constants.

11.5 Oxidants Used in Water and Waste Technology: A Few Case Studies

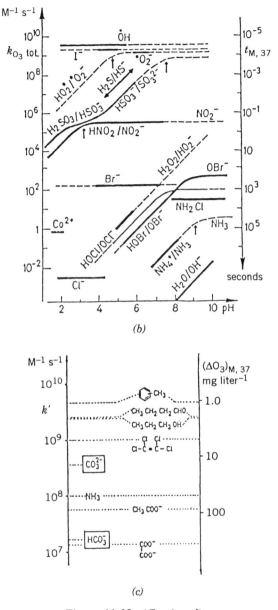

Figure 11.10. (*Continued*)

Oxidation by ClO_2

Chlorine dioxide is a relatively stable free radical; it is paramagnetic because it contains an unpaired electron. It can be prepared by oxidizing chlorite, ClO_2^-, for example, by peroxosulfate or Cl_2. The reaction with Cl_2 is

$$2\ ClO_2^- + Cl_2 = 2\ ClO_2 + 2\ Cl^- \tag{24}$$

Chlorine dioxide is a highly selective oxidant. It reacts very fast with compounds that can easily donate an electron. From a thermodynamic point of view,

$$ClO_2(aq) + e^- = ClO_2^- \quad \Delta G° = -100.4 \text{ kJ mol}^{-1} \quad p\varepsilon° = 17.6 \ (25°C) \tag{25}$$

ClO_2 can oxidize H_2O at pH values above about 4.

For the direct reaction of ClO_2 with P, a substance to be oxidized (a pollutant), Hoigné and Bader (1994) found for many systems that in the rate law

$$-\frac{d[ClO_2]}{dt} = k[ClO_2]_0^n [P]_0^m \tag{26}$$

n and m were equal to unity (first-order with regard to reactants). These authors also showed that ClO_2 reacted more rapidly with deprotonated species. For example, the reaction with a phenolate is orders of magnitude faster than with phenol ($k_{A^-} \gg k_{HA}$):

$$k_{tot} = (1 - \alpha_1) k_{HA} + \alpha_1 k_{A^-} \tag{27}$$

where

$$\alpha_1 = \frac{[A^-]}{A_{tot}} = \frac{1}{1 + 10^{(pK_a - pH)}} \tag{28}$$

Figures 11.11a and 11.11b give a summary of some results obtained by Hoigné and Bader (1994).

Ozonation of Bromide-Containing Waters

The bromide concentration in untreated waters has been of some concern because chlorination and ozonation of Br^- produce hypobromite (OBr^-), which, in the presence of organic matter, yields chlorobromoforms. Furthermore, bromate, which also has been classified as a carcinogen, is formed by ozonation. Kinetic studies by various authors have shown that the formation of bromate strongly depends on the pH of the solution and that the presence of ammonia leads to the formation of monobromamine, which is subsequently oxidized to NO_3^- and Br^-:

$$O_3 + NH_2Br \longrightarrow 2 H^+ + NO_3^- + Br^- + 3 O_2 \tag{29}$$

A schematic overview of this reaction is also given in Figure 11.12. Even relatively small Br^- concentrations can lead to critical amounts of BrO_3^- upon treatment with O_3.

11.5 Oxidants Used in Water and Waste Technology: A Few Case Studies

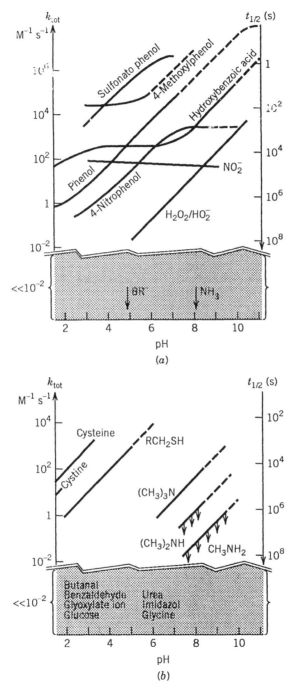

Figure 11.11. Summary of reaction rate constants and half-time values of different compounds in the presence of chlorine dioxide versus pH. Assumptions for $t_{1/2}$ scale: $[ClO_2] = 1\ \mu M$; $[P] \ll [ClO_2]_{const}$. (From Hoigné and Bader, 1994.)

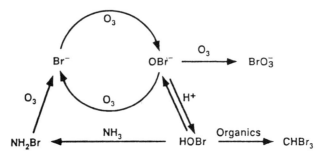

Figure 11.12. Scheme of reactions of O_3 with Br^-, OBr^-, and NH_2Br in aqueous solutions, showing the fate of bromine species. (From Haag and Hoigné, 1983.)

The chemical reaction system can be characterized by the following steps (von Gunten and Hoigné, 1992):

Number	Reaction	k or pK_a (20°C)
1	$O_3 + Br^- \rightarrow O_2 + OBr^-$	160 $M^{-1} s^{-1}$
2	$O_3 + OBr^- \rightarrow 2O_2 + Br^-$	330 $M^{-1} s^{-1}$
3a	$O_3 + OBr^- \rightarrow BrO_2^- + O_2$	100 $M^{-1} s^{-1}$
3b	$O_3 + HOBr \rightarrow BrO_2^- + O_2 + H^+$	≤ 0.013 $M^{-1} s^{-1}$
4	$BrO_2^- + O_3 \rightarrow BrO_3^- + O_2$	$> 10^5$ $M^{-1} s^{-1}$
5	$HOBr \rightleftharpoons H^+ + OBr^-$	9 (8.8)
6	$HOBr + NH_3 \rightarrow NH_2Br + H_2O$	8×10^7 $M^{-1} s^{-1}$
7	$O_3 + NH_2Br \rightarrow Y^a$	40 $M^{-1} s^{-1}$
8	$Y + 2O_3 \rightarrow 2H^+ + NO_3^- + Br^- + 3O_2$	$k_8 \gg k_7$
9	$NH_4^+ \rightleftharpoons H^+ + NH_3$	9.3

^{a}Y are unknown products that react in later reactions.

All the reaction constants are given for 20°C. In a solution without ammonia, reactions 1–3a and equilibrium 5; must be considered; reaction 3b can be neglected.

Based on these kinetic data, von Gunten and Hoigné (1992) have modeled the time course of the ozonation process to gain an understanding of the main parameters controlling the formation of BrO_3^-. Bromine species were computed as a function of $[O_3] \times t$, the product of ozone concentration and time. The model was verified by von Gunten and Hoigné (1994).

Chlorination

The redox chemistry of chlorine (hydrolysis and disproportionation of Cl_2 and acid–base properties of HOCl) was reviewed in Section 8.2 (see Figure 8.5).

11.5 Oxidants Used in Water and Waste Technology: A Few Case Studies

Chlorine and HOCl and its base are used in municipal water treatment for disinfection and oxidative removal of Fe(II), Mn(II), S(−II), and organic compounds.

The hydrolysis of Cl_2,

$$Cl_2 + H_2O = HOCl + H^+ + Cl^- \tag{30}$$

is very fast; the half-life of Cl_2 is on the order of only 0.06 s (Aieta and Roberts, 1985).

HOCl is highly effective as an oxidant and as a disinfectant (rapid rate of microbial kill). Disadvantages arise from the nonselectivity. Organic matter is oxidized and HOCl participates in substitution reactions yielding *organo chlorine* compounds such as chlorophenols and trihalomethanes (e.g., chloroform, $CHCl_3$). In order to circumvent this problem partially, ammonium may be added, to form chloramines ("combined chlorine"), which are less reactive.

Definitions One speaks of

Free chlorine $= (2\ [Cl_2]) + [HOCl] + [OCl^-]$
Combined chlorine $= [NH_2Cl] + 2\ [NHCl_2] + (3\ [NCl_3])$
Chlorine residual $=$ free and combined chlorine

Terms in parentheses are usually negligible.

Chloramines Water chlorination in the presence of NH_4^+ leads to the formation of chloramines (for reviews, see Isaac and Morris, 1983, and Brezonik, 1994). Reaction rates and product distribution depend on pH, relative concentration of NH_4^+ and HOCl, time, and temperature. The pH dependence of NH_2Cl formation is consistent with a rate law involving the neutral species HOCl and NH_3 as reactants:

$$HOCl + NH_3 \longrightarrow NH_2Cl + H_2O \quad k = 4.2 \times 10^6\ M^{-1}\ s^{-1}\ (25°C) \tag{31}$$

The reaction is quite rapid being more than 99% complete in much less than 1 min under typical conditions for water chlorination.

The kinetics of formation of dichloramine is more complicated. Morris and Isaac (1983) report for

$$NH_2Cl + HOCl \longrightarrow NHCl_2 + H_2O \quad k = 3.5 \times 10^2\ M^{-1}\ s^{-1}\ (25°C) \tag{32}$$

This reaction is subject to specific and general acid catalysis. Because the reactions producing NH_2Cl and $NHCl_2$ have different pH dependences, the distribution of chloramines depends on pH. For equimolar concentrations of chlorine and ammonium at 25°C, the following distributions occur according to Jolley and Carpenter (1983).

pH	% NH_2Cl	% $NHCl_2$
5	13	87
6	57	43
7	88	12
8	97	3

Breakpoint Chlorination

Dichloramine is an unstable compound and decomposes by several mechanisms. Most important is an oxidation mechanism producing Cl^- and N_2 resulting in a loss of N_2. Under neutral and basic conditions, decomposition of $NHCl_2$ is thought to proceed by the coupled reactions (Brezonik, 1994) (B is a base)

$$NHCl_2 + HOCl + B \longrightarrow NCl_3 + BH^+ + OH^- \tag{33a}$$

$$NHCl_2 + NCl_3 + 2OH^- \longrightarrow N_2 + 2 HOCl + 3 Cl^- + H^+ \tag{33b}$$

A simplified scheme of breakpoint chlorination is given in Figure 11.13.

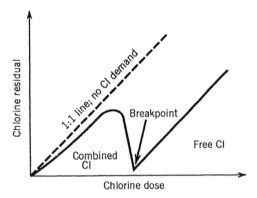

Figure 11.13. General scheme of breakpoint chlorination: difference between total residual Cl and chlorine dose reflects chlorine demand, primarily from ammonium and amines. Before breakpoint, most Cl is in combined forms, primarily mono- and dichloramine; after the breakpoint, the combined residual consists of slow-reacting organic chloramines. Added Cl remains in free form after the breakpoint. Sharpness of breakpoint and minimum observed Cl concentration depend on pH, temperature, and time of reaction. Loss of residual Cl at breakpoint is caused by oxidation of di- and trichloramines to N_2 according to reactions 33a and 33b and other reactions. (Adapted from Brezonik, 1994.)

11.5 Oxidants Used in Water and Waste Technology: A Few Case Studies

Chlorination of Phenols

Phenols react rapidly (electrophilic attack of HOCl on phenoxide anions) and this can lead to taste and odor in water supplies. The reaction can be described by a second-order rate law:

$$\frac{d[Cl]_T}{dt} = k_{obs}[Cl]_T[Ph]_T \tag{34}$$

The subscript refers to the total concentration of free chlorine (Cl) or phenolic compound (Ph) in protonated and deprotonated form. The k_{obs} (see Figure 11.14) shows a pH dependence that can be explained by assuming a specific

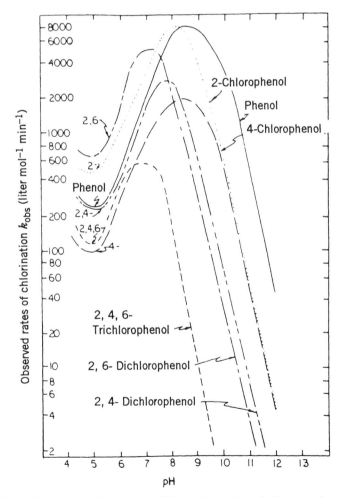

Figure 11.14. Variation in k_{obs} versus pH for chlorination of phenol and various chlorophenol intermediates. (From Lee, 1967.)

rate law:

$$\frac{d[\text{Cl}]_T}{dt} = k_2 [\text{HOCl}][\text{PhO}^-] \qquad (35)$$

(k_2 being nearly constant between pH 6 and 12).

11.6 LINEAR FREE ENERGY RELATIONS (LFERs)

The transition state theory gives us a framework to relate the kinetics of a reaction with the thermodynamic properties of the activated complex (Brezonik, 1990). In kinetics, one attempts to interpret the stoichiometric reaction in terms of elementary reaction steps and their free energies, to assess breaking and formation of new bonds, and to evaluate the characteristics of activated complexes. If, in a series of related reactions, we know the rate-determining elementary reaction steps, a relationship between the rate constant of the reaction, k (or of the free energies of activation, $\Delta G^\ddagger$), and the equilibrium constant of the reaction step, K (or the free energy, ΔG°), can often be obtained. For two related reactions,

$$\frac{-\Delta G_2^\ddagger + \Delta G_1^\ddagger}{RT} = \alpha \frac{-\Delta G_2^\circ + \Delta G_1^\circ}{RT} \qquad (36)$$

or

$$\ln k_2 - \ln k_1 = \alpha (\ln K_2 - \ln K_1) \qquad (37)$$

For a series of i reactants, one obtains

$$\Delta G_i^\ddagger = \alpha \, \Delta G_i^\circ + C \qquad (38a)$$

or

$$\ln k_i = \alpha \ln K_i + C \qquad (38b)$$

Empirically, one plots $\log k_i$ versus $\log K_i$ (or ΔG_i°) and determines α and C from slope and intercept. One simple relationship of this nature has already been given in Figure 11.8b.

Before we generalize, we should discuss the question: Can we estimate redox reactivity with the help of thermodynamics? A simple affirmative answer would be wrong. But it is often useful to consider thermodynamics, in order to gain some insight into reaction mechanisms and kinetics. Most redox reactions are, from a stoichiometric point of view, processes where more than one electron is transferred. But most of these processes occur in a series of one-electron

11.7 The Marcus Theory of Outer-Sphere Electron Transfer

steps. For example, oxidative or reductive transformations of an organic compound require two electrons to yield a stable product. In the majority of abiotic redox reactions, the two electrons are transferred in sequential steps. With the transfer of the first electron, a radical species is formed, which is, in general, more reactive than the parent compound. Thus the rate-determining step very often will be the transfer of the first electron (Eberson, 1987).

In an example that we have already discussed (Section 11.4), the oxidation of Fe(II) by oxygen is initiated by a one-electron transfer to O_2 from a Fe(II) species; thus the radical O_2^- is formed:

$$\text{Fe(II)} + O_2 \longrightarrow \text{Fe(III)} + O_2^- \tag{39}$$

As we have seen, this reaction step is determining for the rate of the overall reaction

$$4 \text{ Fe(II)} + O_2 + 4 \text{ H}^+ = 4 \text{ Fe(III)} + 2 \text{ H}_2\text{O} \tag{40}$$

To investigate or explore LFERs between thermodynamic and kinetic data of redox processes, one should (1) try to interpret the process in terms of elementary (one-electron) reaction steps, and (2) try to identify the rate-determining step.

One can then attempt to relate the free energy of the reaction (or the equilibrium constant or the redox potential) of the one-electron step to the rate of the reaction. Extensive tabulations of one-electron redox potentials have recently become available (e.g., Wardman, 1989). Often it is possible to relate rate constants to free energy parameters ($\Delta G°$, K, p$\varepsilon°$) in a series of related redox reactions (e.g., oxidation of ions of transition elements with O_2, H_2O_2, MnO_2, Fe_2O_3, etc.) or redox reactions involving organic compounds with various substituents or congeners.

Figure 11.15 (from Stone, 1987) illustrates that the rate of oxidation of various substituted phenols by MnO_2 can be correlated with the half-wave potentials, $E_{1/2}$, of these phenols. The half-wave potentials, $E_{1/2}$, measure the tendency of the anode to oxidize the phenols. The half-wave potential corresponds in first approximation to the redox potential of the phenol and its one-electron oxidation product. Figure 11.15 implies that the "thermodynamic" tendency of an electrode to oxidize a certain phenol relates to the "kinetic" tendency of MnO_2 to oxidize this phenol.

11.7 THE MARCUS THEORY OF OUTER-SPHERE ELECTRON TRANSFER: AN INTRODUCTION

Redox reactions involving metal ions occur by two types of mechanisms: inner-sphere and outer-sphere electron transfer. In inner-sphere mechanisms, the oxidant and reductant approach intimately and share a common primary hy-

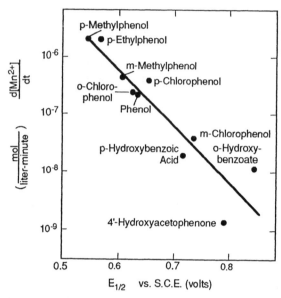

Figure 11.15. Oxidation of substituted phenols by Mn(III,IV) oxides. The oxidation rate increases with the half-wave potentials ($E_{1/2}$) (for the oxidation of various phenols) on a graphite electrode. In a first approximation, $E_{1/2}$ corresponds to the redox potential of the phenol and its one-electron oxidation product. (Modified from Stone, 1987.)

dration sphere. For example, in electron transfer between two metal ions, the activated complex has a bridging ligand between (M—L—M'). Inner-sphere redox reactions involve bond-forming and -breaking processes like other group transfer and substitution reactions, and transition state theory applies directly to them (Brezonik, 1990).

In outer-sphere electron transfer, the primary hydration spheres remain intact; the metal ions are separated by at least two water molecules (or other ligands), and only the electron moves between them. No breaking or formation of chemical bonds occurs, such mechanisms are explained by the *Franck–Condon* principle. This principle states that, because nuclei are more massive than electrons, an electronic transition takes place, while the nuclei in a molecule are effectively stationary. The principle governs the probabilities of transitions between the vibrational levels of different molecular electronic states.

We will not derive or treat the Marcus theory but only illustrate some underlying arguments of this theory. We refer to Eberson (1987), Sutin (1986), and Marcus (1975).

To understand how an outer-sphere reaction occurs, we shall first consider an exchange reaction between normal and isotopically labeled (*Fe) centers:

$$\text{Fe(aq)}^{2+} + {}^*\text{Fe(aq)}^{3+} \longrightarrow \text{Fe(aq)}^{3+} + {}^*\text{Fe(aq)}^{2+} \tag{41}$$

11.7 The Marcus Theory of Outer-Sphere Electron Transfer

The second-order rate constant is 3 M^{-1} s^{-1} at 25°C and the activation energy is 32 kJ mol^{-1}. Bond lengths in Fe^{2+} are greater than those in Fe^{3+}. Hence:

1. A part of the activation energy will arise from the adjustment to a common value in both complexes (inner-sphere rearrangement energy).
2. The solvation shell must be reorganized (outher-sphere reorganization energy).
3. There is an electrostatic energy between the two reactants.

The reaction rate constant, k, can be obtained if we can estimate these various contributions to the activation energy, $\Delta G^{\ddagger}$:

$$k = Z \exp\left(-\frac{\Delta G^{\ddagger}}{RT}\right) \qquad (42)$$

where Z is the collision frequency.

The Potential Energy Curves for Reaction Following the presentation by Shriver et al. (1990), the reactions initially have their normal bond lengths for Fe(II) and *Fe(III), respectively, and the reaction corresponds to motion in which the Fe(II) bonds shorten and the *Fe(III) bonds simultaneously lengthen (Figure 11.16). The potential energy curve for the products of this symmetrical reaction is the same as for the reactants, the only difference being the interchange of the roles of the two Fe atoms. We have assumed that the metal-ligand deformations resemble a harmonic oscillation and have therefore drawn them as parabolas.

The activated complex is located at the intersection of the two curves. However, the noncrossing rule states that molecular potential energy curves of states of the same symmetry do not cross but instead split into an upper and a lower curve. The noncrossing rule implies that if the reactants in their ground states

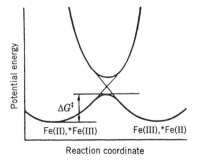

Figure 11.16. A simplified reaction profile for electron exchange in a symmetrical reaction. On the left of the graph, the nuclear coordinates correspond to Fe(II) and *Fe(III); on the right, the ligands and solvent molecules have adjusted locations and the nuclear coordinates correspond to Fe(III) and *Fe(II), where the * denotes the isotope label. (Adapted from Shriver et al., 1990.)

slowly distort, they follow the path of minimum energy and transform into products in their ground states.

More general redox reactions correspond to a nonzero reaction Gibbs energy, so the parabolas representing the reactants and the products lie at different heights. If the potential surface is higher (Figure 11.17a), the crossing point moves up and the reaction has a higher activation energy. Conversely, moving the product curve down (Figure 11.17c) leads to lower crossing points and lower activation energy, and is zero in Figure 11.17d. At the extreme of exergonic reaction (cf. Figure 11.17d), the crossing point rises and rates may become slower again. (We will not discuss this last point further.)

The diagrams in Figure 11.17 suggest that there are two factors that determine the rate of electron transfer. The first is the shape of the potential curves. If the parabolas rise steeply, indicating a rapid increase in energy with distortion, their crossing points will be high and activation energies will be high too. Shallow potential curves, in contrast, imply low activation energies. Similarly,

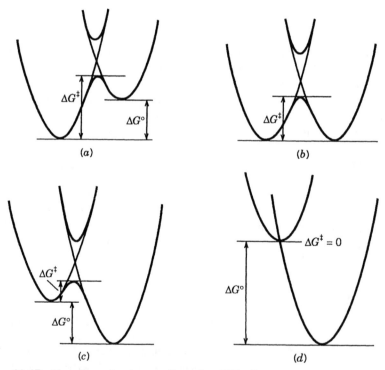

Figure 11.17. The effect of a change of reaction Gibbs free energy for electron transfer on the activation energy when the shape of the potential surfaces remains constant (corresponding to equal self-exchange rates). Note that in going from an endergonic reaction (a) to one with $\Delta G° = 0$ (b) and to a progressively more exergonic reaction (c, d), the activation energy decreases (or the reaction rate increases). (Adapted from Shriver et al., 1990.)

11.7 The Marcus Theory of Outer-Sphere Electron Transfer

large changes in the equilibrium internuclear distance mean that the equilibrium points are far apart and the crossing point will not be reached without large distortions. The second factor is the Gibbs free energy, and the more favorable it is, the lower the activation energy of the reaction.

These considerations were expressed quantitatively by Marcus. He derived an equation for predicting the rate constant for an outer-sphere reaction from the exchange rate constants for each of the redox couples involved and the equilibrium constant for the overall reaction. The *Marcus equation* for the rate constant k_{12} is

$$k_{12} \simeq (k_{11} k_{22} K_{12} f_{12})^{1/2} \qquad (43)$$

with

$$\log f_{12} = \frac{(\log K_{12})^2}{4 \log (k_{11} k_{22}/Z^2)} \qquad (44)$$

where k_{11} and k_{22} are the rate constants for the two exchange reactions and K is the equilibrium constant for the overall reaction. The factor f is a complex parameter composed of the rate constants and the encounter rate. It may be taken as near unity for approximate calculations. The idea of a weighted average of the two rates of self-exchange is emphasized by the name *Marcus cross-relation*, which is sometimes used instead. The encounter rate constant, Z, is taken as 10^{11} M^{-1} s^{-1}.

Example 11.4. Marcus Cross-relation[†] Calculate the rate constant at 0°C for the reduction of $[Co(bipy)_3]^{3+}$ by $[Co(terpy)_2]^{2+}$.

The required exchange reactions are

$$[Co(bipy)_3]^{2+} + [*Co(bipy)_3]^{3+} \xrightarrow{k_{11}} [Co(bipy)_3]^{3+} + [*Co(bipy)_3]^{2+}$$

$$[Co(terpy)_2]^{2+} + [*Co(terpy)_2]^{3+} \xrightarrow{k_{22}} [Co(terpy)_2]^{3+} + [*Co(terpy)_2]^{2+}$$

where $k_{11} = 9.0$ M^{-1} s^{-1} (at 0°C)
$k_{22} = 48$ M^{-1} s^{-1} (at 0°C)
$K_{12} = 3.57$ (at 0°C)

Setting $f = 1$ gives

$$k_{12} = (9.0 \times 48 \times 3.57)^{1/2} \text{ M}^{-1} \text{ s}^{-1} = 39 \text{ M}^{-1} \text{ s}^{-1}$$

This result compares reasonably well with the experimental value, which is 64 M^{-1} s^{-1}.

[†]This example is taken from Shriver et al. (1990).

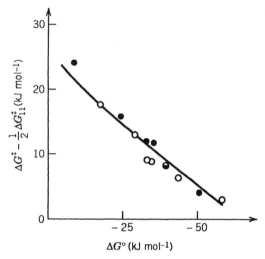

Figure 11.18. The correlation between rates and overall free energy change for the oxidation of a series of phenanthroline complexes of Fe(III) by Ce(IV) in sulfuric acid (open circles) and Ce(IV) in perchloric acid (closed circles). (Adapted from Shriver et al., 1990.)

The Marcus cross-relation can be expressed as a LFER since the logarithms of the rate constants are proportional to the free energies of activation. Thus

$$2 \ln k_{12} \simeq \ln k_{11} + \ln k_{22} + \ln K_{12} \tag{45a}$$

Equation 45a implies that

$$2 \Delta G^{\ddagger} \simeq \Delta G^{\ddagger}_{11} + \Delta G^{\ddagger}_{22} + \Delta G^{\circ}_{12} \tag{45b}$$

Figure 11.18 shows the correlation between the rates and overall free energy change for the oxidation for a series of substituted phenanthroline complexes by Ce(IV).[†]

The Parabolic Relationship Between Rate Constant and Equilibrium Constant A more general expression of the quadratic structure–activity relationship can be given by (Eberson, 1987)

$$k = \frac{k_d}{1 + \dfrac{k_d}{K_d Z} \exp\left\{[W + \tfrac{1}{4}\lambda\,(1 + \Delta G^{\circ}/\lambda)^2]/RT\right\}} \tag{46}$$

[†]Other examples are summarized by Pennington (1978).

11.7 The Marcus Theory of Outer-Sphere Electron Transfer

This expression relates the second-order rate constant, k, for an outer-sphere electron transfer reaction to the free energy of reaction, $\Delta G°$, with one adjustable parameter, λ, known as the reorganization energy. W is the electrostatic work term for the coulombic interaction of the two reactants, which can be calculated from the collision distance, the dielectric constant, and a factor describing the influence of ionic strength. If one of the reactants is uncharged, W is zero. In exact calculations, $\Delta G°$ should be corrected for electrostatic work. The other terms in equation 46 can be treated as constants (Eberson, 1987): the diffusion-limited reaction rate constant, k_d, can be taken to be 10^{10} M^{-1} s^{-1}; K_d is the equilibrium constant for precursor complex formation and Z is the universal collision frequency factor (see Eberson, 1987).

Figure 11.19 illustrates the relationship for the rate of reaction of ClO_2 with phenolic compounds (deprotonated forms) as a function of $\Delta G°$.

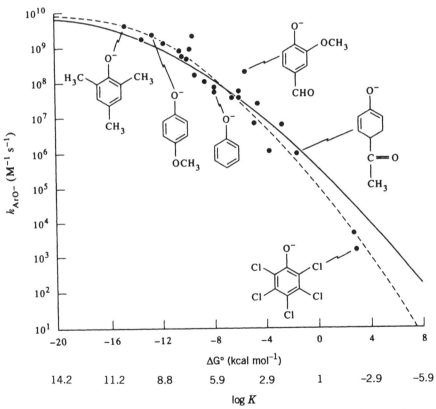

Figure 11.19. Correlation between the second-order rate constants for reactions of substituted phenoxide anions with chlorine dioxide and estimated values of $\Delta G°$. The circles are experimental data and the curves represent fits of the data to the Marcus equation. The solid curve corresponds to equation 46 (λ = 30.1 kcal mol^{-1}). $\Delta G°$ (and log K) values were calculated from electrode half-wave potentials. (Adapted from Tratnyek and Hoigné, 1994.)

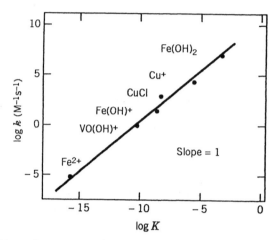

Figure 11.20. Linear free energy relation for the oxygenation of metal ions. The slope of unity is predicted by Marcus theory for endergonic outer-sphere electron transfer steps. (From Wehrli, 1990.)

For the calculation of the curve, the self-organization energy, λ, was taken as 30.1 kcal mol^{-1}; the combined term k_d/K_dZ was assigned a value of 0.1.

The parabolic shape of the curve, going from a slope of $d \log k/d \log K$ of 0 for very exergonic redox reactions to $-\frac{1}{2}$ for reactions close to $\Delta G° = 0$ and to -1 for rather endergonic reactions, is typical for the Marcus relationship (equation 46).

The relationship and trend between $\Delta G^{\ddagger}$ (or log k) and $\Delta G°$ (or log K) of the Marcus relationship can be appreciated by looking at the simplified scheme of Figure 11.17d. Figure 11.17d illustrates that a small activation energy ($\Delta G^{\ddagger} \simeq 0$) corresponds to a very exergonic reaction ($\Delta G° \ll 0$), approaching $d \log k/d \log K \simeq 0$. On the other hand, an endergonic reaction ($\Delta G° > 0$) corresponds (see Figure 11.17a) to a large $\Delta G^{\ddagger}$, approaching a limit where $d \log k/d \log K \simeq 1$. The intermediate situation is characterized by Figure 11.17b, where $\Delta G° \simeq 0$; here the situation corresponds to $d \log k/d \log K \simeq 0.5$.

We discussed earlier the LFER for the oxygenation of Fe(II); the plot of log k versus log K given in Figure 11.8b has a slope of unity. This is in accord with the Marcus prediction ($\Delta G > 0$). The data given in Figure 11.8b can be extended to the oxygenation of other transition metal ions (Wehrli, 1990) (Figure 11.20).

11.8 NUCLEOPHILE–ELECTROPHILE INTERACTIONS AND REDOX REACTIONS INVOLVING ORGANIC SUBSTANCES

As we have discussed before, Lewis defined acids as compounds prepared to accept electron pairs and bases as substances that could provide such pairs. In organic reactions one speaks of *electrophiles* and *nucleophiles*.

11.8 Nucleophile–Electrophile Interactions and Redox Reactions

Electrophiles and nucleophiles can be looked on as acceptors and donors, respectively, of electron pairs from and to other atoms—most frequently carbon.

At first sight, it may appear that a discussion of nucleophile–electrophile interactions does not fit into a chapter on redox processes, but the transfer of electron pairs from a nucleophile to an electrophile, such as in hydrolysis or in chlorination of a phenol, may be looked at as a reaction between an oxidant and reductant, although we are aware that mechanistically, in many "real" redox processes involving organic substances, the electron transfers often occur in one-electron steps.

Thus nucleophiles include negatively charged ions, molecules processing atoms with unshared pairs of electrons, and molecules that contain highly polarized or polarizable bonds. Simple nucleophiles in the aquatic environment are (ordered approximately with increasing nucleophilicity):

Nucleophiles: H_2O, NO_3^-, F^-, SO_4^{2-}, CH_3COO^-, Cl^-, HCO_3^-, HPO_4^{2-}, OH^-, I^-, CN^-, HS^-

Electrophiles can be positively charged ions or molecules containing atoms without full octets; a few examples are:

Electrophiles: H^+, H_3O^+, NO_2, NO, R_3C, SO_3, CO_2, $AlCl_3$, Br_2, O_3, Cl_2, $HOCl$, Cl^+

Because of the large abundance of nucleophiles in the environment, reactive electrophiles are often very short-lived. Because of its great abundance, water plays an important role among the nucleophiles in the environment. A reaction in which a water molecule or an OH^- substitutes for another atom or group of atoms in an organic molecule is called a *hydrolysis* reaction.

We will merely exemplify a few reactions to illustrate that many organic processes can be understood mechanistically, that rate laws can be formulated, and that compilations on rate constants are available; for example, Larson and Weber (1994), Lyman et al. (1982), Schwarzenbach et al. (1993), and Sykes (1986).

Hydrolysis Reactions: Nucleophilic Substitution at a Saturated Carbon Atom

Hydrolysis reactions are very prevalent in the aquatic environment. The products of hydrolysis are often ecologically or toxicologically less harmful than the unhydrolyzed reactants.

Typical examples for reactions of organic electrophiles with nucleophiles (H_2O, OH^- or HS^-) are given in Table 11.4 (Schwarzenbach and Gschwend, 1990).

Relation of Kinetics to Mechanisms The activation barrier in a substitution reaction is determined by (1) the energy needed to break the old bond with the

Table 11.4. Reactions of Organic Electrophiles with Water, OH⁻ or HS⁻

1. $R-X(X=Cl, Br, I) + H_2O \rightarrow R-OH + X^- + H^+$
2. $RCH_2-X(X=Cl, Br, I) + HS^- \rightarrow RCH_2-SH + X^-$
3. $R_1COOR_2 + H_2O/OH^- \rightarrow R_1COO(H) + HOR_2$
4. $R_1R_2NCOOR_3 + H_2O/OH^- \rightarrow R_1R_2NH(H) + CO_2 + HOR_3$
5. $(R_1O)_2\overset{\overset{O}{\|}}{P}-OR_2 + H_2O/OH^- \rightarrow (R_1O)_2\overset{\overset{O}{\|}}{P}-O(H) + HOR_2$ and
   $(R_1O)R_2O)\overset{\overset{O}{\|}}{P}-O(H) + HOR_1$

departing group, and (2) the energy released as the new bond with the entering group is formed.

The latter is given by the relative nucleophilicity of the nucleophile, which can be quantified. The former contribution to $\Delta G^\ddagger$ is more difficult to quantify.

If the reaction occurs by such a two-step reaction (break an old bond and make a new one) (Figure 11.21), one speaks of an S_N2 *reaction* (substitution, nucleophilic bimolecular) and it is given by a second-order kinetic rate law:

$$\text{Rate} = k_{Nu} [Nu^{\nu-}] [-\overset{|}{\underset{|}{C}}-X]$$

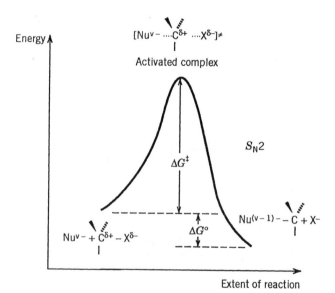

Figure 11.21. Activation energy relationship for an S_N2 process.

11.8 Nucleophile-Electrophile Interactions and Redox Reactions

where Nu^{v-} is the nucleophilic reagent that interacts to form first the activated complex, which then dissociates into the substituted product.

In the S_N1 mechanism, the first (and rate-determining) step consists of the dissociation of the leaving group, which then reacts with the nucleophile:

$$\overset{|}{\underset{|}{C}}{}^{\delta+}\!\!-\!X^{\delta-} \xrightarrow{\text{slow}} \left[\overset{|}{\underset{|}{C}}{}^{+}\quad X^-\right] \xrightarrow{+Nu^{v-}} \overset{|}{\underset{|}{-C}}\!-\!Nu^{(v-1)}$$

$$\text{Intermediate}$$

The reaction rate will be first order:

$$\text{Rate} = -kZ[-\overset{|}{\underset{|}{C}}-X]$$

and is independent of the concentration of the nucleophile.

An illustration of the quantitative information on rates and mechanisms of such substitution reactions available is shown in Table 11.5 (from Schwarzenbach and Gschwend, 1990).

As pointed out by Schwarzenbach and Gschwend (1990), it can be seen from this table that the carbon-iodine bond hydrolyzes much faster than the carbon-chlorine bond. Generally, the resilience of bonds decreases in the order $C-F > -Cl > -Br > -I$. Furthermore, the reaction rates increase dramatically when going from primary $(-CH_2-X)$ to secondary $(> CH-X)$ to tertiary $(-\overset{|}{\underset{|}{C}}-Cl)$ carbon-halogen bonds.

Many hydrolysis reactions are *pH dependent* with base catalysis predominant. The acid-base catalysis is especially significant with carboxylic acid esters. The pseudo-first-order hydrolysis rate constant $k_h(\text{time}^{-1})$ can be expressed as

$$k_h = k_A [H^+] + k_N + k_B [OH^-]$$

For an example see Figure 11.22.

Metal Complexes

The same concept applies to substitution reactions (e.g., ligand exchange) in metal complexes, for example, trans $[PtCl_2(py)_2] + 2NH_3 \rightarrow$ trans $[PtCl_2(NH_3)_2] + 2$ py. The same two substitution mechanisms are operative (see Section 6.9). Coordination chemists speak of associative (or adjunctive) mechanisms and of dissociative (or disjunctive) mechanisms.

Table 11.5. Postulated Reaction Mechanisms and Hydrolysis Half-Lives at 25°C of Some Monohalogenated Hydrocarbons at Neutral pH

Compound	$t_{1/2}$ (Hydrolysis)				Dominant Mechanism in Nucleophilic Substitution Reactions
	X=F	Cl	Br	I	
R—CH$_2$—X	~30 yr[a]	340 days[a]	20–40 days[b]	50–110 days[c]	S_N2
CH$_3$\\CH—X/CH$_3$		38 days	2 days	3 days	$S_N2 \cdots S_N1$
CH$_3$—C(CH$_3$)(CH$_3$)—X	50 days	23 s			S_N1
CH$_2$=CH—CH$_2$—X		69 days	0.5 days	2 days	$(S_N2) \cdots S_N1$
C$_6$H$_5$—CH$_2$—X		15 h	0.4 h		S_N1

[a]R=H.
[b]R=H, C_1 to C_5-n-alkyl.
[c]R=H, CH$_3$.
Source: Data from Mabey and Mill (1978).

Sulfur Nucleophiles

The lower oxidation states of sulfur (−II sulfide to +IV sulfite) are electron rich and act as nucleophiles. These species react with alkylhalides to displace the more electronegative halide. As pointed out by Brezonik (1994), such reactions may be significant in anoxic aquatic environments such as sediments and groundwater under landfills. Such reactions (see reaction 2 in Table 11.4) are analogous to hydrolysis and produce thiols. Barbash and Reinhard (1989) have estimated rate constants for such reactions.

The oxidation of sulfite by oxidizing agents has already been discussed in Section 11.4. The electrophilic oxidant (e.g., O_3) interacts with the nucleophilic S(IV) species in a second-order rate reaction.

An interesting case history was reported by Schwarzenbach et al. (1985) on a groundwater contamination by primary and secondary alkyl bromides. In this case, a series of short-chain alkyl bromides (Figure 11.23) were introduced continuously into the ground by wastewater also containing high concentrations of sulfate (SO_4^{2-}). Due to the activity of sulfate-reducing bacteria, hydrogen sulfide (H_2S/HS^-) was formed, which, in turn, reacted with the alkyl bromides to yield alkyl mercaptans or thiols (Figure 11.23). The mercaptans (RSH/RS$^-$), which are even better nucleophiles than H_2S/HS^-, then reacted further with

11.8 Nucleophile-Electrophile Interactions and Redox Reactions

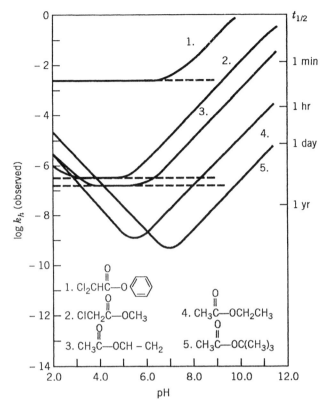

Figure 11.22. Observed hydrolysis constants (log k_h) versus pH for some esters. (Data from Mabey and Mill, 1978; adapted from Brezonik, 1994.)

other alkyl bromide molecules, resulting in the formation of a whole series of dialkyl sulfides and other hazardous products (for more details see Schwarzenbach et al., 1985). Of interest to us here is the fact that all possible dialkyl sulfides exhibiting at least one primary alkyl group were found, but that no compounds with two secondary alkyl groups could be detected. These results suggest that the secondary alkyl bromides were reacting chiefly via an S_N1 mechanism, thereby yielding secondary alcohols, and it was not until the primary alkyl mercaptanates, which are particularly strong nucleophiles, appeared that the secondary bromides also became involved in a more S_N2-like reaction.

Electrophilic Substitution: Chlorination Example

Three types of electrophilic substitution reactions involving chlorine species are of interest in aquatic systems:

1. Substitution into aromatic compounds.
2. Reaction with nitrogenous compounds, especially amines to form N-chloroorganic reactions.

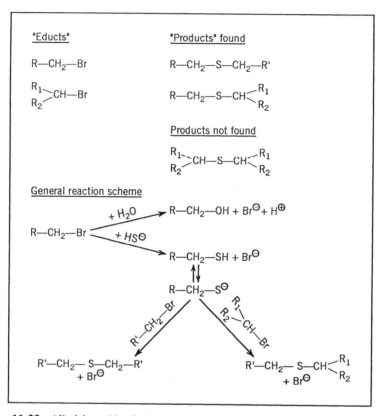

Figure 11.23. Alkyl bromides leaked into groundwater and thioethers found several years later; the reaction scheme shown can account for the products seen. (From Schwarzenbach et al., 1985.)

3. Reaction with natural dissolved organic substances, including aquatic humus, to form chloroform and other trihalomethanes.

The free halogens (Cl_2, Br_2) are unable to substitute benzene itself; a Lewis acid catalyst such as $AlCl_3$ is requires to assist in polarizing the attacking halogen molecule, thereby providing it with an electrophilic end. An "activated" nucleus like phenol can, however, easily be attacked by halogens. The electrophilic agent in natural waters is HOCl (near neutral pH). The Cl atom in HOCl behaves like Cl^+, a strong electrophile, and combines with a pair of electrons of the nucleophile with which it reacts (Brezonik, 1994).

The formation of chlorophenols in water supplies can create severe odor and taste problems. The chlorination of phenols has already been described in Section 11.5. As has been shown in Figure 11.14, the reactants are HOCl (more electrophile than OCl^-) and phenolate (PhO^-) (a better nucleophile than phenol (PhOH)).

Multiple activating groups on the aromatic ring accelerate halogenation re-

11.8 Nucleophile–Electrophile Interactions and Redox Reactions

actions; for example, resorcinol and *m*-dihydroxobenzene react rapidly with HOCl, leading to the formation of chloroform.

Oxidative Transformations The oxidative transformation of organic chemicals is primarily limited to those chemicals containing hetero atoms with lone pairs of electrons, such as phenols, aromatic amines, and sulfides. Oxidation of these classes of chemicals in aquatic ecosystems often results in the formation of polymers or adducts with organic matter that are not well defined.

Quite frequently oxidative transformations occur with reducible hydrous oxides—Fe(III)(hydr)oxide, Fe_3O_4, and Mn(III, IV)(hydr)oxides. Redox reactive metals also include $Pb^{IV}O_2(s)$, $Cr^{VI}O_4^{2-}$, $Hg^{II}(OH)_2(aq)$, and $Co^{III}OOH(s)$ (Stone et al., 1994). The source and reactivity of organic reductants interacting with higher valence metals have been reviewed by Stone et al. (1994).

Organic Reductants and Oxidants

Relative to hydrolysis, our current understanding of the reductive and oxidative transformation of organic chemicals in aquatic ecosystems is very limited. Although the functional groups that are susceptible to redox reactions in environmental systems have been identified, we often do not yet have the ability to predict accurate reaction rates for these chemicals in aquatic ecosystems (Weber, 1994). This is due primarily to the complexity of enviromental systems, which makes it very difficult to identify the electron donors and acceptors in aquatic ecosystems. For example, reduced transition metals [particularly iron(II) in various complexes], iron–sulfur proteins, sulfides, humic materials, and polyphenols have all been suggested as potential reductants. In recent years, a significant increase has occurred in the amount of research focusing on redox transformations. Much of this interest arises from the observation that reductive transformations can result in the formation of reaction products that may be more detrimental to aquatic ecosystems than the parent compound. Furthermore, organic chemicals that had previously been thought to have stability in aquatic ecosystems because they did not contain hydrolyzable functional groups (e.g., nitroaromatics and aromatic azo compounds) have been shown in laboratory studies to undergo facile reduction in anoxic systems (Macalady et al., 1986).

Fe(II)/Fe(III) as a Mediator in Electron Transfer The Fe(III)–Fe(II) system often acts as an electron carrier. A possible schematic example is given by

Electron transfer mediator

Kinetics of Redox Processes

This reaction occurs abiotically. But it has been shown that Fe(III) can act as an electron acceptor in microbially mediated reactions (Lovely et al., 1994).

Table 11.6 gives examples of redox transformations that are likely to occur in aquatic ecosystems (Weber, 1994). We also refer to the growing literature in this field: Sykes (1986), Eberson (1987), Schwarzenbach et al. (1993), Weber (1994), Stone et al. (1994), and Larson and Weber (1994).

The equations given in Table 11.6 symbolize the redox reactions in terms of two-electron transfers. As we pointed out earlier in this chapter, the accepted theory is that electrons are transferred sequentially one by one (Eberson, 1987).

Table 11.6. Redox Transformations Likely to Occur in Aquatic Ecosystems

1. Reductive dehalogenation
   Hydrogenolysis

$$R-X + 2e^- + H^+ \rightarrow R-H + X^-$$

   Vicinal dehalogenation

$$\underset{\underset{X}{|}}{-C}-\underset{\underset{X}{|}}{C}- + 2e^- \longrightarrow \;\;>\!\!C\!\!=\!\!C\!\!<\; + 2X^-$$

2. Nitroaromatic reduction

$$Ar-NO_2 + 6e^- + 6H^+ \rightarrow Ar-NH_2 + 2H_2O$$

3. Aromatic azo reduction

$$Ar-N=N-Ar' + 4e^- + 4H^+ \rightarrow ArNH_2 + H_2NAr'$$

4. Sulfoxide reduction

$$R_1-\overset{\overset{O}{\|}}{S}-R_2 + 2e^- + 2H^+ \longrightarrow R_1-S-R_2 + H_2O$$

5. N-Nitrosoamine reduction

$$\underset{R_1}{}\overset{N=O}{\underset{|}{N}}\underset{R_2}{} + 2e^- + 2H^+ \longrightarrow \underset{R_1}{}\overset{H}{\underset{|}{N}}\underset{R_2}{} + HNO$$

6. Quinone reduction

$$O\!=\!\!\!\bigcirc\!\!\!=\!O + 2e^- + 2H^+ \rightleftharpoons HO-\!\!\bigcirc\!\!-OH$$

7. Reductive dealkylation

$$R_1-X-R_2 + 2e^- + 2H^+ \rightarrow R_1-XH + R_2H$$

Source: Weber (1994).

11.8 Nucleophile–Electrophile Interactions and Redox Reactions

The general mechanism for one-electron transfer as described by Eberson (1987) (see Schwarzenbach and Gschwend, 1990) is

$$P + R \rightleftharpoons (PR) \longrightarrow [PR \longleftrightarrow P^{\bullet-}R^{\bullet+}]^{\neq} \longrightarrow (P^{\bullet-}R^{\bullet+})$$

Educts   Precursor complex   Activated complex   Successor complex

$$\updownarrow$$

$$P^{\bullet-} + R^{\bullet+}$$
Products

where P is an organic pollutant (the electron acceptor) and R is a reductant (the electron donor). The precursor complex, (PR), describes the electron coupling prior to the electron transfer, which occurs in the transition state. Then a successor complex is proposed that subsequently dissociates to provide the radical ions. The concept of the one-electron scheme provides a common feature in the seemingly unrelated redox transformations presented earlier in Table 11.6. Each of these processes occurs initially by transfer of a single electron in the rate-determining step. In each case, a radical ion is formed that is more susceptible to further reaction than is the parent compound.

Reduction of nitroaromatics is a well-studied reductive transformation process. Major steps in the reduction are important naturally occurring organic reductants:

Nitrobenzene $\xrightarrow[2H^+]{2e^-}$ Nitrosobenzene $\xrightarrow[2H^+]{2e^-}$ Hydroxylamine $\xrightarrow[2H^+]{2e^-}$ Aniline

In reducing environments, such as anoxic sediments, the reduction of nitroaromatics can occur with half-lives on the order of minutes to hours.

Kinetic mechanisms of the dissolution of higher valence hydrous oxides by organic reductants have been extensively investigated (Hering and Stumm, 1990; Stone et al., 1994; Stumm, 1992) (Table 11.7) and are discussed in Section 13.3.

Typically, electron transfer is preceded by the formation of an encounter complex. In the case of dissolution of solid, higher valence oxides, usually by organic reductants, this is usually a surface complex (see Figure 13.10). Here we briefly illustrate that the reduction of Cr(VI) by alcohols, α-hydroxycarboxylates, α-carbonyl carboxylates, and mercaptans occurs through formation of a Cr(VI) reductant adduct. A mechanism proposed by Stone (1994) for

Table 11.7. Laboratory Studies of Nonphotochemical Reduction of Mn(III,IV)(hydr)oxides and Fe(II,III)(hydr)oxides by Organics

Metal Oxidants	Organic Reductants Employed
MnO_2, MnOOH	Mono-, di-, trihydroxobenzenes, oxalate pyruvate, phenol-substituted monophenols, hydroquinone, catechol, anilines, fulvic acid, humic acid, ascorbate, thiosalicylate, acetoin
Fe_3O_4, Fe_2O_3, and FeOOH	Ascorbate, mercaptoacetic acid, mono-, di-, and trihydroxobenzenes, hydroquinone, catechol, fulvic acid
CoOOH	Hydroquinone, oxalate, pyruvate, acetoin

chromium(VI) reduction by glycolic acid assumes a product formation by a ligand exchange at the Cr(VI) center:

$$HCrO_4^- + HO-CH_2COOH \longrightarrow O_3Cr^{VI}-O-CH_2-COOH + H^+$$

which is then followed by electron transfer:

$$O_3Cr^{VI}-O-CH_2-COOH \longrightarrow Cr^{IV} + HCOCOO^-$$

In subsequent fast steps, the Cr(IV) is reduced to Cr(III).

11.9 CORROSION OF METALS AS AN ELECTROCHEMICAL PROCESS

Here we wish to exemplify how metal corrosion can be interpreted from both a thermodynamic and electrochemical kinetic point of view. This simple introduction may serve to direct readers to some of the more detailed literature on the chemistry of corrosion.

Thermodynamic Aspects Most metals are, with regard to their conversion to oxides, unstable. The reaction $xM + (y/2) O_2 = M_xO_y$ is characterized by the following free energy values:

Oxide	$\Delta G°$ (kJ mol^{-1})	Oxide	$\Delta G°$ (kJ mol^{-1})
Fe_2O_3	-742.3	CuO	-129.7
Al_2O_3	-1582.4	NiO	-211.7
Cr_2O_3	-1058.2	ZnO	-318.4
MgO	-569.4	SnO_2	-519.7

The pϵ versus pH diagram for the various metals gives the predominance areas where corrosion, from a thermodynamic point of view, is possible. In the case of Fe, we may consult Figure 8.8. The nonshaded area of predominance of Fe^{2+}, Fe^{3+}, and $FeOH^{2+}$ is the thermodynamic range where corrosion occurs. If the pϵ of iron can be kept below -10, for example, by applying some external voltage, it will not corrode. In the shaded predominance area of solid precipitates [$Fe(OH)_2(s)$, $FeCO_3(s)$, and $Fe(OH)_3(s)$], corrosion is still possible but the precipitate may, under suitable conditions, form a partially or fully protective coating that retards corrosion; such a *passive film* of iron(III) oxide may be formed at high pϵ values (anodic polarization or in the presence of suitable oxidants such as chromate).

The electrochemical series (Table 8.3) gives thermodynamic information on the so-called "nobility" of various metals; the higher the standard electrode potential, the more noble is the metal, silver being more noble than Cu and Cu being more noble than Zn.

We may compare the following (hypothetical) cells:

$H_2(g)|H^+\|Cu^{2+}|Cu(s)$ Potential difference 0.34 V

$Fe(s)|Fe^{2+}\|H^+|H_2(g)$ Potential difference 0.44 V

(These cells are written in such a way that electrons in the external circuit flow from left to right.)

$Mg(s)|Mg^{2+}\|Fe^{2+}|Fe(s)$ Potential difference 1.91 V

$Fe(s)|Fe^{2+}\|Cu^{2+}|Cu(s)$ Potential difference 0.78 V

If an iron metal is connected (external circuit) to a Mg metal, the latter becomes an anode and Mg^{2+} goes into solution; the iron becomes the cathode and is thus protected by the electrons delivered to the iron from the Mg electrode. One speaks of *cathodic protection*. A similar situation arises with a Zn-coated iron metal. As long as the Zn acts as an anode (dissolves), the iron is cathodically protected. Cathodic protection can also be achieved by applying some voltage between a counterelectrode and the iron. Alternatively, the connection of Cu with Fe makes the copper the cathode and the iron the anode; the dissolution (corrosion) of Fe is enhanced. A similar phenomenon occurs if the water flowing through the iron pipe contains Cu(II). According to the electrochemical series, Cu(II) will be deposited on the iron metal surface; the resulting "local battery" causes iron to corrode. In this case often a pitting corrosion is observed.

Electrochemical Aspects The half-reactions responsible for the corrosion of iron in water are, for *oxidation*,

$$Fe = Fe^{2+} + 2\ e^-$$

and, for *reduction*, either

$$O_2(g) + 4\ H^+ + 4\ e^- = 4\ H_2O \quad \text{or} \quad 2\ H^+ + 2\ e^- = H_2(g)$$

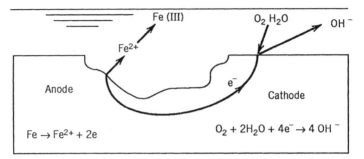

Figure 11.24. Scheme of anodic and cathodic reactions on corroding iron surface in the presence of O_2. The iron metal is the conductor of electrons between local anodes and cathodes; the electrolyte is the ionic conductor. In cathodic areas, the cathodic reduction of O_2 consumes H^+ (or produces OH^-). Of course, the rate of electron production equals the rate of electron consumption.

The Fe(II) formed can become oxidized to Fe(III). Figure 11.24 illustrates schematically how anodic and cathodic processes occur simultaneously on a corroding iron surface.

The Polarization Curve The electrochemical mechanism of corrosion permits us to express corrosion rate in terms of current density. If the corroding iron is used as an electrode and a voltage is applied with the help of an inert auxiliary electrode, the electrode potential can be plotted as a function of the measured current, i_t (Figure 11.25). The resulting polarization curve reveals important characteristics of the corrosion processes occurring. The measured current density, i_t is the sum of the anodic and cathodic partial current densities, i_a and i_c (i_c is counted as a negative current density). Knowing the anodic and cathodic reactions, the resulting corrosion velocity, i_{corr}, can be determined (See Figure 11.25a). At the corrosion potential (open circuit electrode potential of the Fe electrode, $i_t = 0$), $i_c = i_a = i_{corr}$. As a first approximation, the corrosion current can be estimated from the slope of the i_t versus E curve, because $(di_t/dE)_{i_t=0}$ is proportional to i_{corr}.

From the slope of the polarization curve and its variation with time (exposure time of the iron electrode), information on the kind of inhibition can be gained. An inhibition of anodic processes decreases the i_a versus E current density and increases the corrosion potential; correspondingly, an increase in cathodic inhibition causes a decrease in the i_c and lowers the corrosion potential.

Fully Passive Oxide Films

The corrosive behavior of a few metals is essentially determined by the kinetics of the dissolution of the corrosion products. This seems to be the case for Zn in HCO_3^- solutions, for passive iron in acids, and for passive Al in alkaline solutions. The mechanism of the dissolution of iron and of the passivation of

11.9 Corrosion of Metals as an Electrochemical Process

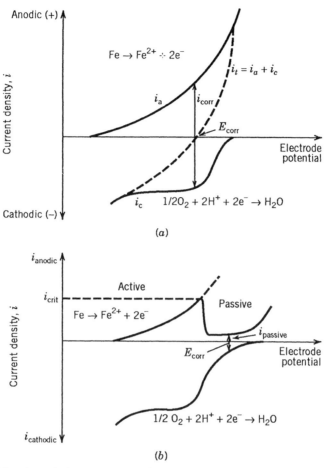

Figure 11.25. The polarization curve. The corroding iron is used as an electrode, a variable voltage is applied (against an auxiliary electrode), and the current density is measured as a function of the electrode potential. The measured current density is composed of the cathodic and anodic current density, $i_t = i_a + i_c$. At the open circuit, the electrode potential is the corrosion potential and the anodic current is the corrosion current ($i_a = i_{corr}$ when $i_t = 0$). (a) Corroding iron. (b) Iron that becomes passive when the corrosion potential becomes sufficiently large. $i_{passive}$ is the current density in the presence of a passive oxide. In a first approximation, a steady state between the corrosion rate and the oxide dissolution rate may be assumed. Steels alloyed with elements such as Cr and Mo are more readily amenable to passivation ("stainless steel").

this dissolution is extremely complex. We may not know exactly the composition of the passive film; but it has been suggested that it consists of an oxide of $Fe_{3-x}O_4$ with a spinel structure. The passive layer seems to vary in composition from Fe_3O_4 (magnetite), in oxygen-free solutions, to $Fe_{2.67}O_4$ in the presence of oxygen. Figure 11.26 represents a schematic model of the hydrated

724   Kinetics of Redox Processes

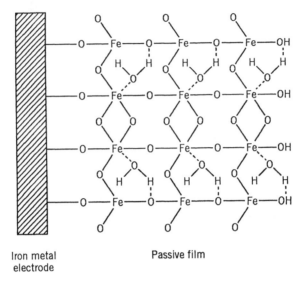

Figure 11.26. Schematic representation of the hydrated passive film on iron. (From Pou et al., 1984.)

passive film on iron as proposed by Bockris and collaborators (Pou et al., 1984). Obviously, the hydrated passive film on iron displays the coordinative properties of the surface hydroxyl groups.

Obviously such oxides may be attacked by acids, by reductants such as H_2S, and certain ligands. We will return to this question in Section 13.4 when we discuss the kinetics of dissolution of oxides.

The $CaCO_3$-Fe(III)(Hydr)oxide Coating in Water Supply Distribution Systems

Fully passive oxide films cannot be formed in natural waters. But often the $CaCO_3$ formed on the surface of the corroding iron walls, together with some corrosion products, may provide some protection and retard corrosion. We have discussed (in Sections 4.7 and 7.4) under what conditions $CaCO_3$ is deposited. The problem, however, is more complicated because the question of $CaCO_3$ deposition is not only one of equilibrium (saturation index). The cathodic reaction, the reduction of O_2 or H^+, on a corroding iron surface produces a pH increase in the proximity of the corroding metal. Although the maintenance of a saturation index (SI) near zero or slightly positive has often been expedient to reduce corrosion and to prevent a clogging of the pipe by $CaCO_3$, the problem of the mitigation of corrosion and "red water" cannot be solved by this simple recipe alone.

SUGGESTED READINGS

Eberson, L. (1987) *Electron Transfer Reactions in Organic Chemistry*, Springer, Berlin.

Hoffmann, M. R. (1990) Catalysis in Aquatic Environments. In *Aquatic Chemical Kinetics*, W. Stumm, Ed., Wiley-Interscience, New York, pp. 71–107.

Hoigné, J. (1990) Formulation and Calibration of Environmental Reaction Kinetics; Oxidations by Aqueous Photooxidants as an Example. In *Aquatic Chemical Kinetics*, W. Stumm, Ed., Wiley-Interscience, New York.

Lappin, A. G., 1994 Redox Mechanisms in Inorganic Chemistry, Ellis Horwood Chichester England.

Luther, G. M. (1990) The Frontier-Molecular-Orbital Theory Approach in Geochemical Processes. In *Aquatic Chemical Kinetics*, W. Stumm, Ed., Wiley-Interscience, New York, pp. 173–198.

Scully, J. C. (1990) *The Fundamentals of Corrosion*, 3rd ed., Pergamon, Oxford.

Weber, E. J. (1994) Abiotic Pathways of Organic Chemicals in Aquatic Ecosystems. In *Chemistry of Aquatic Systems; Local and Global Perspectives*, G. Bidoglio and W. Stumm, Eds., Kluwer Academic, Norwell, MA.

12

PHOTOCHEMICAL PROCESSES

12.1 INTRODUCTION

The most important photochemical process in the biosphere is photosynthesis, leading to conversion of light energy into chemical energy. As we have seen in Chapter 8, typical photosynthesis produces biomass (primary production) and a reservoir of oxygen. Solar energy also generates strictly nonbiotic photochemical processes. In aqueous systems most photochemical processes are redox reactions. As a consequence of photolysis, a redox disproportionation (as in photosynthesis) typically occurs and highly reactive radicals and unstable redox species are produced.

Photochemical processes play an important role in the atmosphere and in atmospheric waters. Sunlight-mediated processes modify both directly and indirectly the chemical transformations of dissolved organic matter and influence the geochemical cycles of iron and other transition elements by affecting redox processes. By creating highly reactive oxygen species, such as $^1O_2$, $OH^{\cdot}$, H_2O_2, and organic peroxides, photochemical reactions in oxygenated surface waters ultimately enhance oxidation reactions; even many refractory compounds are degraded. Thus photoreactions are also increasingly used in water treatment to remediate contaminated water supplies. Many light-induced processes are modified by adsorption of suitable solutes on surfaces, especially on semiconductor surfaces, and are relevant in reactions such as the dissolution of oxides; substances adsorbed to these surfaces (ligands, humic substances) are catalytically transformed (e.g., surface-mediated oxidation of SO_2 and the nonbiotic degradation of organic molecules).

Many atmospheric problems (smog formation, tropospheric ozone formation, and stratospheric ozone depletion) are closely linked to photoreactions of atmospheric constituents and pollutants.

Direct and Indirect Photochemical Transformations

In direct photochemical transformations the light-absorbing substance is changed. In an indirect photochemical transformation, the products of photolysis are often short-lived reactants (transient intermediates):

12.1 Introduction

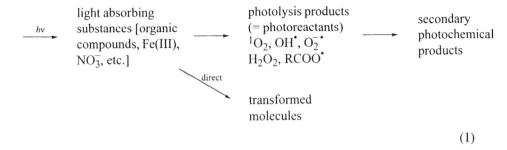

(1)

An example of a direct photolysis is the conversion of a (refractory) chloroorganic substance into a (less refractory) phenolic substance.

(2)

In this chapter we will discuss the principles of light absorption by water and by solutes and the basic steps involved in photoactivation. Special attention will be paid to photolytic reactions producing photoreactants such as singlet oxygen, $^1O_2$, hydroxyl radicals, $OH^{\bullet}$, and hydrogen peroxide, H_2O_2; and we will discuss how steady-state concentrations of these photoreactants can be estimated. It is shown that the photoredox reactions of transition metal complexes are important in surface and atmospheric waters. Finally, we will discuss heterogeneous photochemistry, most importantly the functioning of semiconductor surfaces.

Photons and Photooxidants as Environmental Factors

During a cloudless summer noon hour, surface waters receive approximately 1 kW m^{-2} of sunlight, or about 2 mol of photons per square meter within the wavelength region of 300–500 nm of interest for photochemical reactions (Figure 12.1). Within 1 year about 1300 times this dose is accumulated (Haag and Hoigné, 1986). This irradiation can be considered as a primary environmental constant. A large portion of these photons is absorbed by dissolved organic material (DOM) present in natural water. In addition, a portion can interact with organic and inorganic surfaces of particles and a small fraction of short-wavelength light with nitrate or nitrite or even with "micropollutants" and, mainly in atmospheric waters, also with Fe(III) species.

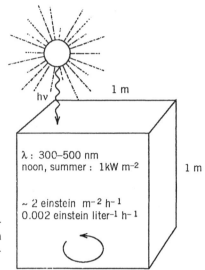

Figure 12.1. Solar radiation. Mean dose intensity in a mixed 1-m water column in which all light is absorbed. (1 einstein = 1 mol photons). (Adapted from Hoigné et al., 1989).

The resulting rate of interactions between photons and absorbers is very high: assuming that most of the photons of all wavelength ranges are absorbed in a well-mixed 1-m water column, about 2 mmol liter^{-1} h^{-1} of interactions occur between photons and absorbing substrates. In cases where DOM is the main absorber, and assuming an average chromophore unit weight of 120 g mol^{-1} in water containing 4 mg of dissolved organic carbon (DOC) per liter, the concentration of absorber unit is 0.033 mM. Thus each chromophore of DOM is excited about once a minute. Even if only a few of these interactions lead to a chemical reaction, high rates of transformations of chemicals are to be expected. Compared with this, chromophores of other compounds may exhibit a lower overlap with the spectrum of the solar light and absorb a correspondingly smaller fraction of the spectrum and therefore exhibit a correspondingly lower rate of excitations.

As pointed out by Hoigné (1990), some of the excitations of the light absorbers will lead to direct photolysis. In addition, as shown in equation 1, dissolved organic material, nitrate (or nitrite), Fe(III) species, and some minerals act as sensitizers or precursors for the production of highly reactive intermediates (so-called photoreactants) such as singlet oxygen ($^1O_2$), OH radicals (OH$^{\cdot}$), DOM-derived peroxy radicals (ROO$^{\cdot}$), solvated electrons (e_{aq}^-), superoxide anions ($O_2^{-\cdot}$), triplet states of humic compounds, electron–hole pairs on semiconducting surfaces, reduced transition metal species, and hydrogen peroxide. In chlorine-containing waters even Cl$^{\cdot}$ and OH$^{\cdot}$ would have to be considered as photoreactants. Of these photoreactants only reduced transition metal species and H_2O_2 accumulate (hours), because of their relatively low reactivity. All other species are highly reactive and short-lived (microsecond range).

12.2 ABSORPTION OF LIGHT

The light absorption by water as a medium can be characterized by equation 3:

$$I = I_0 \times 10^{-\alpha l} \qquad (3)$$

while the absorption by the dissolved substance in the water is given by

$$I = I_0 \times 10^{-(\alpha + \varepsilon c)l} \qquad (4a)$$

or if $\alpha \ll \varepsilon c$

$$I = I_0 \times 10^{-\varepsilon c l} \qquad (4b)$$

The light absorbed, $I_a = I_0 - I$, is given by

$$I_a = I_0 (1 - 10^{-\varepsilon c l}) \qquad (4c)$$

These equations are valid only *for a given wavelength*; I and I_0 are the light intensity emerging from the solution and the incident light, respectively (e.g., in mol photon cm^{-2} s^{-1}). α (cm^{-1}) is the absorption coefficient of water; ε is the molar exinction coefficient (liter mol^{-1} cm^{-1}); c is the concentration (mol liter^{-1}); and l is the length of the light path. The extent of the light absorption is usually expressed as in equation 5:

$$A = \log \frac{I_0}{I} = \varepsilon c l \qquad (5)$$

Equation 5 corresponds to the *Beer–Lambert law*, which implies that the fraction of radiation absorbed is independent of the intensity of the radiation employed, and that the amount of radiation absorbed by the system is proportional to the number of molecules absorbing the radiation.

The primary process in all photochemical reactions is the absorption of a photon of light energy. The energy of light depends on the wavelength. For one photon,

$$E = h\nu = \frac{hc}{\lambda} \qquad (6)$$

where h is Planck's constant (6.63×10^{-34} J s), c is the velocity of light (3×10^8 m s^{-1}), and ν is the frequency (number of cycles in 1 s). Thus, at 400 nm, 1 mol photon (1 einstein) is equivalent to 300 kJ[†]; at 300 nm, 1 mol photon is equivalent to 400 kJ. This energy can be compared with the bond energy (enthalpy) of a chemical bond (e.g., C—H, 415 kJ mol^{-1}; C—C, 350 kJ mol^{-1}; O=O, 500 kJ mol^{-1}; N≡N, 950 kJ mol^{-1}; C=C, 600 kJ mol^{-1}.

Only light that is absorbed can affect chemical change. The *Stark–Einstein law* states that one photon is absorbed by each molecule responsible for the primary photochemical process.

Quantum Yield

The quantum yield is, at a given wavelength, the number of moles of the light-absorbing substance that react for each mole of photons (einstein) absorbed

$$\phi_\lambda = \frac{\text{number of moles reacting}}{\text{number of einsteins absorbed}} \qquad (7)^{\ddagger}$$

Since the energy of the einstein depends on the wavelength of the radiation, in laboratory photochemical work monochromatic light (i.e., of a definite wavelength) is often used. The energy of the absorbed light is determined by an actinometer, a device using a photochemical reaction to estimate the absorbed energy. A solution of iron(III) oxalate is frequently used. When exposed to light of λ = 450 nm, the oxalate is decomposed. From the amount of oxalate decomposed, the amount of energy absorbed can be evaluated.

Photoactivation

As we have seen, a necessary (but not sufficient) condition for a photochemical reaction is that a reactant, S, must be able to absorb light in the ultraviolet (UV)–visible range. Photon absorption produces a more reactive species, S*.

An electron is raised from its ground state into a higher-energy state (Figure 12.2a)

$$S_0 + h\nu \longrightarrow S^* \qquad (8)$$

The subsequent changes may involve chemical reactions and/or physical

---

[†]Energy of 1 mol photons is given by

$$E = \frac{6.02 \times 10^{23}}{\text{mol}} \frac{6.63 \times 10^{-34} \text{ J s} \times 3 \times 10^8 \text{ m s}^{-1}}{400 \times 10^{-9} \text{ m}} = 300 \text{ kJ mol}^{-1}$$

or 300 kJ einstein^{-1}.

[‡]An excited species may undergo several physical and chemical processes. Thus a quantum yield, ϕ_i, for each individual process can be defined.

reactions (thermal equilibration, fluorescence, phosphorescence) (Figure 12.2a):

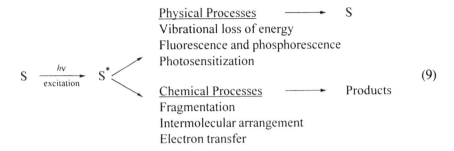
(9)

Most ground-state molecules have an even number of electrons. The electron spins of the ground state (molecular O_2 is unusual as we shall discuss later) are usually paired, which is thus a singlet state.[†] Absorption of a photon promotes an electron to an excited singlet state

$$^1S_0 + h\nu \longrightarrow {}^1S_1^* \text{ or } {}^1S_2^* \qquad (10)$$

Excited singlet states generally are too short-lived ($\tau < 10^{-9}$ s) to react with solutes in natural waters. The process of intersystem crossing can unpair two electron spins and yield a triplet state.[‡] S* states have much longer lifetimes than do singlet states. Thus many photochemical reactions involve excited triplet states (Figure 12.2b).

Molecular O_2 is unusual in that its ground state is characterized by two unpaired electrons, that is, a triplet state.

As illustrated by schematic electron configuration in Figures 12.2 and 12.5, in singlet oxygen, $^1O_2$ ($^1\Delta_g$) is only about 92 kJ mol^{-1} above the ground state and is readily formed. This is because $^1O_2$ has an unfilled π^* orbital and can accept a pair of electrons and thus has a special affinity to interact with electron-rich species such as the olefins.

Rate of Photolysis

The direct photolysis rate, ν, in accord with equation 7, is the number of photons absorbed; that is,

$$\nu_\lambda = I_{a\lambda} \phi_\lambda \qquad (11)$$

[†]It is called singlet state because there is only *one* single arrangement in space for such a pair of spins.

[‡]A triplet state is one in which two electrons in different orbitals have parallel spins. The name "triplet" reflects the quantum mechanical fact that two parallel spins can adopt only three orientations with respect to an external magnetic field (↑↑, ⇉, ↓↓). The ground state of O_2 is a triplet state.

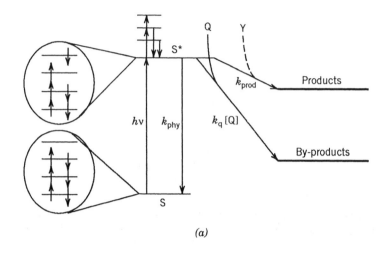

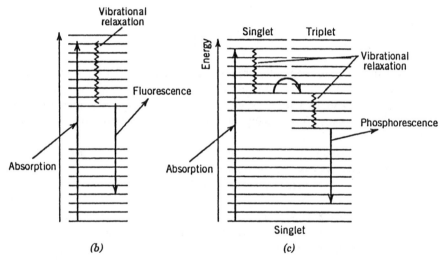

Figure 12.2. (a) Simplified scheme of photoexcitation with reaction and deactivation pathways. Absorption of a photon affects the transition of one electron from a lower electronic level to a higher electronic level, leaving an electron hole at the lower level and creating an excess electron in the higher level. For this reason, excited states are generally better oxidants and better reductants than the corresponding ground states. When the promoted electron is not rapidly transferred to another molecule (or to another part of the same molecule), or the missing electron is not replaced, the promoted electron will return to its initial electronic level, with no net reaction occurring. The energy gained by absorption of a photon is represented by $h\nu$ (h = Planck's constant, ν = light frequency); k_{phy} is the sum of all unimolecular photophysical deactivation rate constants (emission, internal conversion); and k_{prod} and Σk_q are rate constants of product and by-product formation. (Adapted from Sulzberger and Hug, 1994.) (b) A schematic portrayal of molecular electronic and vibrational energy levels. The initial absorption takes the molecule to an excited electronic state. The excited molecule is subject to collisions with the surrounding molecules, and as it gives up energy it steps

where $I_{a\lambda}$ is the light absorbed and ϕ_λ the quantum yield of the photochemical reaction, both at a given wavelength.

According to the Beer-Lambert law (equation 4c),

$$I_{a\lambda} = I'_{0\lambda}(1 - 10^{-\varepsilon_\lambda[S]l}) = I'_{0\lambda}(1 - e^{-2.3\varepsilon_\lambda[S]l}) \quad (4c)$$

where the light intensities I_λ and $I'_{0\lambda}$ are expressed in mol photons per liter per time per nm; $I'_{0\lambda}$ is the volume-averaged light intensity that is available to the light-absorbing substance (chromophore); l is the light pathlength (in cm); ε_λ is the molar extinction coefficient of the chromophore at a given wave length (in M^{-1} cm^{-1}); and [S] is the concentration of chromophore (in M).

Equation 4c can be simplified if $2.3\,\varepsilon_\lambda[S]\,l \ll 1$ to

$$I_{a\lambda} \simeq I'_{0\lambda}(2.3\,\varepsilon_\lambda[S]\,l) \quad (12)$$

For a sufficiently small concentration of the chromophore (and/or the extinction coefficient), the rate of the photochemical reaction, ν_λ, (in M time^{-1}) is

$$\nu_\lambda = -\left(\frac{d[S]}{dt}\right)_\lambda \simeq I'_{0\lambda}\,2.3\,\varepsilon_\lambda l \phi_\lambda[S] \simeq k_\lambda[S] \quad (13)$$

where the experimental rate constant, k_λ, is

$$k_\lambda = I'_{0\lambda}\,2.3\,\varepsilon l\,\phi_\lambda \quad (14)$$

Table 12.1 summarizes some data on direct photolysis rates and quantum yields for some organic contaminants. Rates of direct photochemical reactions of aqueous species, S, as measured in thin-layer samples, correspond to rates that would occur in the top few centimeters of a water column. Corrections have to be considered for mixed water columns at greater depths (see Zafiriou, 1973, and Haag and Hoigné, 1986).

The total rate of photochemical transformation is obtained by integrating over the range in which light absorbs

$$-\frac{d[S]}{dt} = \int_\lambda k_\lambda[S] \quad (15)$$

---

down the energy of vibrational levels. The molecule may emit the remaining excess energy, generating a photon. (c) The intersystem crossing, where the spin of an electron is inverted. Such triplet states have longer half-times than do singlet states. Most photochemical reactions involve excited triplet states. Emission of a photon by a triplet state returning the molecule to the ground singlet state is called phosphorescence. Alternatively, S* may undergo chemical reaction and form other compounds by a variety of bond rearrangements. (Adapted from Atkins, 1992.)

Table 12.1. Direct Photolysis Rates and Quantum Yields for Some Organic Contaminants[a]

Compound	ϕ_d	k_s	$t_{1/2}$	Reference
Anthracene	0.003	8.5	0.081	Zepp, 1980
Benz[a]anthracene	0.0033	10.5	0.066	Zepp and Baughman, 1978
Benzo[a]pyrene	0.032	12.7	0.054	Mill et al., 1981
	0.00089	11.7	0.059	Zepp and Baughman, 1978
Benzo[f]quinoline	0.00089	13.0	0.053	Mill et al., 1981
	0.014	11.5	0.06	Zepp and Baughman, 1978
	0.014	16.0	0.043	
Benzo[b]thiophene	0.1	0.049	14.1	Mill et al., 1981
Carbaryl	0.006	0.105	6.6	Mill et al., 1981
9H-Carbazole	0.0076	5.75	0.12	Zepp and Baughman, 1978
		6.63	0.10	Zepp and Baughman, 1978
7H-Dibenzo[c,g]carbazole	0.0028	16.7	0.041	Mill et al., 1981
Dibenzo[e,g]carbazole	0.0033	20.9	0.033	Mill et al., 1981
Dibenzothiophene	0.0005	0.12	5.7	Zepp and Baughman, 1978
	0.0005	0.13	5.3	Zepp and Baughman, 1978
3,3'-Dichlorobenzidine	—	25.6	0.0027	Mill et al., 1981
DDE	0.3	0.75	0.92	Zepp and Baughman, 1978
2,4-D-Butoxyethyl ester	0.17	0.058	12.0	Zepp and Baughman, 1978
DMDE	0.3	7.3	0.094	Zepp and Baughman, 1978
Diphenyl mercury	0.056	0.86	0.8	Zepp and Baughman, 1978
Methyl parathion	0.00017	0.077	9.0	Zepp and Baughman, 1978
Naphthacene	0.013	230.0	0.003	Zepp, 1980
Naphthalene	0.0015	0.09	7.7	Zepp, 1980
N-Nitrosoatrazine	0.3	75.0	0.0092	Zepp and Baughman, 1978
Parathion	0.00015	0.069	10.0	Zepp and Baughman, 1978
Phenyl mercuric acetate	0.25	2.76	0.25	Zepp and Baughman, 1978
Pyrene	0.0022	9.3	0.074	Zepp, 1980
Quinoline	0.00033	0.028	25.0	Zepp and Baughman, 1978
	0.00033	0.031	22.3	Mill et al., 1981
Trifluralin	0.002	17.7	0.039	Zepp and Baughman, 1978

[a] ϕ_d measured at 313 nm or 366 nm; k_s in day^{-1} (average 24-h rate for midsummer 40° near-surface conditions); $t_{1/2}$ in days for same conditions.

Source: Adapted from Brezonik (1994).

Determination of Quantum Yield: Chemical Actinometry

Equations 13 and 14 can be used to determine, with the help of actinometers, the quantum yield. The light intensity may be determined by exposing a chemical actinometer to the light source in the same way and at the same time that the component of interest is exposed. From measurements of the concentration of the compound S, the first-order photolysis rate constant, k_λ, can be determined from the slope of $\ln([S]/[S_0])$ versus time. The absorbance has to be kept small so that the simplification introduced into equation 12 holds. Then ϕ_λ can be estimated by

$$\phi_\lambda = \frac{k_\lambda}{2.3\, I'_{0\lambda}\varepsilon l} \tag{16}$$

12.3 PHOTOREACTANTS

Table 12.2 gives a qualitative survey of photolysis products that are important reactants in surface and atmospheric waters. Some of the important physicochemical properties of these photoreactants have already been discussed in Chapter 11. As the table illustrates, the formation of singlet oxygen, of OH· radicals, of $O_2^{-\cdot}/HO_2^\cdot$, and of organic peroxides is especially important. Some reaction schemes are given in Figure 12.3.

Table 12.2. Photochemically Produced Reactive Species in Natural Waters

Products		Possible Production Processes
Singlet oxygen	$^1O_2$	Sensitized by light-absorbing dissolved organic matter (humic acids)
Superoxide anion	$O_2^{-\cdot}$	Photolysis of Fe(III) complexes; deprotonation of $HO_2^\cdot$
Hydroperoxyl	$HO_2^\cdot$	Uptake from atmosphere, protonation of $O_2^{-\cdot}$
Hydrogen peroxide	H_2O_2	Photolysis of Fe(III) complexes; disproportionation of superoxide anion[a]; exchange with atmosphere
Ozone	O_3	Formed in atmosphere, uptake from atmosphere
Hydroxyl radical	OH·	Photolysis of hydroxo or other Fe(III) complexes; of NO_3^-, NO_2^-; decomposition of O_3, photolysis of H_2O_2
Organic peroxy radicals	ROO·	Photolysis of dissolved organic material
Polar oxidation products of organic compounds		Photochemical oxidation of dissolved or adsorbed organic material
Aquated electron	e_{aq}^-	Photolysis of dissolved organic matter

[a] $2\,O_2^{-\cdot} + 2\,H^+ = H_2O_2 + O_2$.

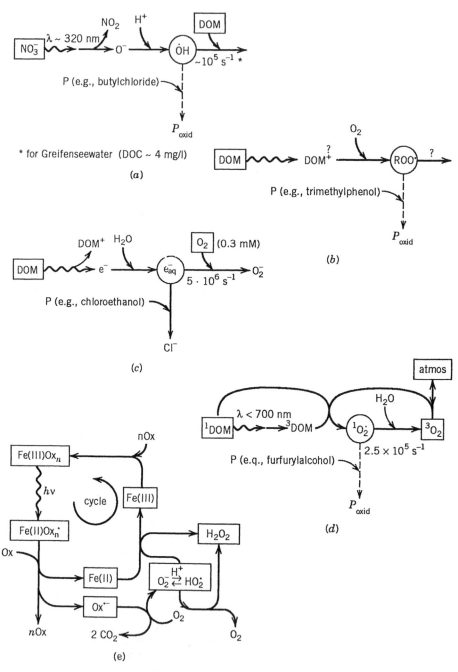

Figure 12.3. (a–d) Schemes of photolytic reactions producing photoreactants. P is a probe molecule (or a micropollutant), which is used to assess the steady-state concentration of the photoreactant. (From Hoigné, 1990.) (e) Scheme for the photochemical/chemical cycling of iron and the formation of H_2O_2. (From Zuo and Hoigné, 1992.)

OH• Radical Production

Hydroxyl radicals are produced in the aqueous environment by photolysis of nitrate (Figure 12.3a), nitrite, and aqueous iron complexes, or, in water treatment, from photolysis of HOCl and catalytic decay of aqueous O_3. OH• radicals formed in natural waters react principally with dissolved organic matter (see Figure 11.10c).

Aqueous OH• is highly nonselective and can abstract an H atom from many sites of an organic molecule or add to any C—C double bond or accept an electron (Hoigné, 1990). A wide spectrum of different products is formed.

In the *troposphere* gaseous OH• is the dominant photooxidant. Tropospheric OH• is produced predominantly from photolyzed O_3, followed by O reacting with H_2O. It is scavenged by CH_4, the main organic material in the atmosphere, and by carbon monoxide.

Photolysis of NO_3^- The irradiation of NO_3^- in water at its long wave absorption band (maximum 302 nm) results in two primary photochemical processes (Figure 12.3a):

$$NO_3^- \xrightarrow{h\nu} NO_3^{-*} \begin{cases} \rightarrow NO_2^- + O \text{ (atomic oxygen)} \\ \rightarrow NO_2 + O^{-\bullet} \xrightarrow{H_2O} OH^{\bullet} + OH^- \end{cases} \quad (17)$$

Atomic oxygen (O) will most likely react with O_2 to form O_3, which is then rapidly consumed. The presence of NO_3^- in fresh surface waters can, by forming OH•, induce the oxidation of DOM.

General Kinetic Approach   In a natural water situation, various competitive effects need to be taken into account. In fresh waters OH• is consumed within a few microseconds by the DOC and carbonate and in seawater predominantly by Br^-. Thus the situation occurring for the oxidation of trace compounds in natural waters is rather different from that occurring when probe molecules are used in high concentrations or when experiments are performed using distilled water.

Example 12.1. Estimation of Steady-State Concentrations of a Photoreactant in a Natural Water   We follow the theoretical treatment by Haag and Hoigné (1985) (photosensitized oxidation in natural waters via OH• radicals.) The scheme is given by reactions ia and ib:

$$A \xrightarrow[r_{\text{formation}}]{-\eta \frac{d[A]}{dt}} {}^{\bullet}OH \begin{cases} \xrightarrow{r = k_M [M][{}^{\bullet}OH]} M_{\text{oxid}} & \text{(ia)} \\ \xrightarrow[r_{\text{consumption}}]{r = k_i [S_i][{}^{\bullet}OH]} S_{\text{oxid}} & \text{(ib)} \end{cases}$$

Here k_M and k_i are the second-order rate constants for the parallel reactions:

$$OH^{\bullet} + M \xrightarrow{k_M} M_{oxidized} \tag{iia}$$

$$OH^{\bullet} + S_i \xrightarrow{k_i} S_{i\,oxidized} \tag{iib}$$

S_i represents the scavenger substance present in the natural water (it may be more appropriate to represent it by $\Sigma[S_i]$; then k_i becomes Σk_i) and M is a microprobe molecule (or a specified micropollutant) added (or present) in the natural water. A is some precursor such as NO_3^- or H_2O_2, which upon photolysis yields $OH^{\bullet}$ radicals, whereby $-\eta(\Delta A) = (\Delta OH^{\bullet})$.

If

$$k_M[M] \ll \Sigma\, k_i[S_i] \tag{iii}$$

(a prerequisite for M to serve as microprobe molecule), the steady-state concentration of $OH^{\bullet}$ can be obtained from the condition that the rate of formation of $OH^{\bullet}$

$$\frac{d[OH^{\bullet}]}{dt} = -\frac{\eta d[A]}{dt} \tag{iva}$$

equals the rate of consumption of $OH^{\bullet}$

$$-\frac{d[OH^{\bullet}]}{dt} = \Sigma k_i[S_i]\,[OH^{\bullet}] \tag{ivb}$$

The combination of equations iva and ivb gives

$$-\frac{\eta\, d[A]}{dt} = \Sigma\, k_i[S_i]\,[OH^{\bullet}] \tag{va}$$

or

$$[OH^{\bullet}]_{ss} = -\frac{\eta\, d[A]}{dt} \times \frac{1}{\Sigma\, k_i[S_i]} \tag{vb}$$

It is this steady-state concentration that can be determined with a probe M (assuming, in line with equation iii, that this probe has a negligible effect on $[OH^{\bullet}]_{ss}$).

For this probe molecule

$$-\frac{d[M]}{dt} = k_M\,[M]\,[OH^{\bullet}]_{ss} = k_{exp}\,[M] \tag{via}$$

12.3 Photoreactants

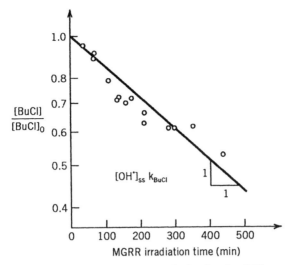

Figure 12.4. Semi-log plot of butyl chloride elimination by OH˙ upon irradiation in the Merry-go-round Reactor (MGRR) in Greifensee water. (Adapted from Haag and Hoigné, 1985.)

or

$$-\ln \frac{[M]}{[M_0]} = k_M [OH˙]_{ss} t = k_{exp} t \quad \text{(vib)}$$

if k_M is known (from experiments in synthetic solution), $[OH˙]_{ss}$ can be calculated

$$[OH˙]_{ss} = \frac{k_{exp}}{k_M} \quad \text{(vic)}^\dagger$$

Figure 12.4 exemplifies such a measurement.

From such types of experiment, steady-state concentrations of various photooxidants have been determined for Lake Greifen by Hoigné and collaborators

---

†Upon integration of equation via and substituting $[OH˙]_{ss}$ from equation vb and considering that $-\eta(\Delta A) = (\Delta OH˙)$, we obtain

$$-\ln \frac{[M]}{[M_0]} = [\Delta OH˙] \frac{k_M}{\Sigma k_i[S_i]} = k_{exp} t \quad \text{(vii)}$$

Here $(\Delta OH˙)$ is the total amount of OH˙ formed from photolyzed A per unit volume during the reaction time t. A condition for this integration is that $\Sigma k_i[S_i]$ remains constant; this assumption is generally valid. Equation vii states that a semi-log plot of residual [M] should decline linearly with the amount of OH˙ formed $(\Delta OH˙)$, that is, with $\eta(\Delta A)$, the slope being inversely proportional to the amount of scavenger present in the water.

Table 12.3. Photoreactants as Environmental Factors E_j in a Surface Water

(a) Dependencies of E_j on Water Composition and Depth

E_j	Functionalities of E_j	
	At Surface	1-m Layer[a]
$[^1O_2]_{ss}$	$\propto$ [DOM]	Independent[b,c]
$[OH^\cdot]_{ss}$	$\propto [NO_3^-]/[DOM]$	$\propto [NO_3^-]/[DOM]^2$
$[ROO^\cdot]_{ss}$	$\propto$ [DOM]	Independent[b,c]
$[e^-_{aq}]_{ss}$	$\propto [DOM]/[O_2]$	$\propto 1/[O_2]$

(b) Numerical Examples for Lake Greifensee in June, Noon Sunlight (1 kW m^{-2})

E_j	Probe or Reference, M	k_M (M^{-1} s^{-1})	$k_M [E_j]_{ss}^{0m}$ (% h^{-1})	$[E_j]_{ss}^{0m}$ (M)	$[E_j]_{ss}^{1m}$ (M)	
$[^1O_2]_{ss}$	Furfuryl alcohol	1.2×10^8	3	8×10^{-14}	5×10^{-14}	(e)
$[OH^\cdot]_{ss}$	Butyl chloride	3×10^9	0.2	2×10^{-16}	4×10^{-17}	(f)
$[ROO^\cdot]_{ss}$	Trimethylphenol	(?)[d]	15	(?)[d]	(?)[d]	(g)
$[e^-_{aq}]_{ss}$	CCl$_4$	3×10^{10}	0.13	1.2×10^{-17}	5.2×10^{-18}	(h)

[a]Light screening by suspended particles is neglected.
[b]Independent of [DOM].
[c]For DOC < 5 mg liter^{-1}.
[d]Assumed relevant wavelength region = 355 nm for corresponding screening factor.
[e]Data from Haag and Hoigné (1986).
[f]Data from Zepp et al. (1987a).
[g]Data from Faust and Hoigné (1987).
[h]Data from Zepp et al. (1987a).
Source: From Hoigné (1990).

(Table 12.3b). Table 12.3a gives the most important factors that influence the occurrence of these photooxidants.

Despite the fact that OH$^\cdot$ is a very reactive radical toward many organic pollutants, its low steady-state concentration makes it a nondominant oxidant in natural waters. As Schwarzenbach et al. (1993) point out, the second-order rate constant for reaction with OH$^\cdot$ (see Figure 11.10c) is on the order of 6×10^9 M^{-1} s^{-1}. Multiplying this rate constant with $[OH^\cdot]_{ss} \simeq 10^{-17}$ gives a first-order rate constant of $k \simeq 6 \times 10^{-8}$ s^{-1}, corresponding to a half-time of ~130 days.

Singlet Oxygen

Several studies have demonstrated that photolysis rates of certain organic chemicals are enhanced in the presence of humic substances. Humic substances (DOM) can act as sensitizers to excite ground-state oxygen molecules ($^3O_2$) to short-lived singlet oxygen ($^1O_2$). $^1O_2$ has a chemical reactivity quite different from that of ground-state oxygen (Figure 12.5). DOM is the dominant dissolved

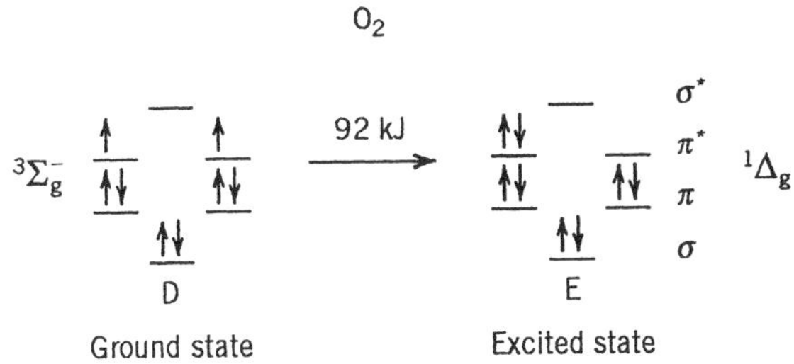

12.3 Photoreactants

Figure 12.5. Idealized molecular orbital diagrams for the excitation of ground-state oxygen to singlet oxygen.

absorber of light (light with wavelength $\lambda \leq 700$ nm; $\lambda = 700$ nm corresponds to an energy of $E = 170$ kJ einstein^{-1}—cf. equation 3—92 kJ mol^{-1} are needed to excite $^3O_2$ to $^1O_2$). The overall reaction can be formulated as in Figure 12.3d or 12.5.

Light absorption promotes the humic "molecule" to its first excited singlet state ^{1}DOM*. Excited molecules in singlet states are very short-lived; they decay, in part, by undergoing intersystem crossing to excited triplet states ^{3}DOM*, which are considerably longer-lived than excited singlet states (see Figure 12.2b). Every transfer to O_2 is very efficient.

Only a fraction of the photoproduced $^1O_2$ reacts chemically; most of it is quenched[†] to ground-state O_2 by water (half-life of ~3 μs). Nevertheless, a significant steady-state concentration can be achieved during sunshine in natural waters (Table 12.3b). The $[^1O_2]_{ss}$ concentrations (Table 12.3, Figure 12.6) were determined by using furfuryl alcohol,

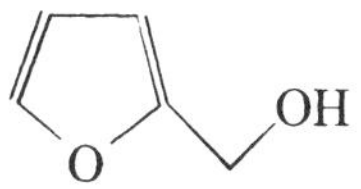

as a probe.

Laboratory experiments have shown that the steady-state concentration of $^1O_2$ increases proportionally to the amount of light absorbed by the DOM.

The examples of reaction rates of $^1O_2$ in Figure 12.7 show that $^1O_2$ reacts only with deprotonated species (e.g., phenolate anions); that is, the apparent rate constants decrease in the pH region below the pK_a of the chemical. Singlet oxygen is selective; it is an electrophile that reacts only with particular functional chemical structures such as are present in 1,3 dienes (see chemical structure of furfuryl alcohol) or polycondensed aromatic hydrocarbons (with delocalized π electron bonds) or in sulfides or mercaptans (Hoigné, 1990).

[†]The chemical species that efficiently accept the electronic energy (of excitement) are called quenchers or acceptors.

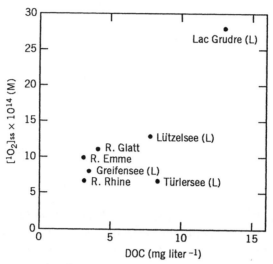

Figure 12.6. Observed $[^1O_2]_{ss}^0$ in water samples from some Swiss rivers (R) and lakes (L) as a function of the dissolved organic carbon (DOC) concentrations of these waters. The results apply for noontime light intensity on a clear summer day at 47.5° N. (Data from Haag and Hoigné, 1986.)

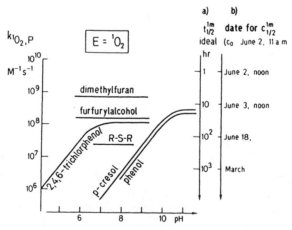

Figure 12.7. Comparison of rate constants for reactions of $^1O_2$ (left scale) and sunlight irradiation times required for solute eliminations (right scales, 1-m average depth) versus pH. (Data are from Scully and Hoigné, 1987.) (a) Half-life of selected pollutants in Greifensee water during exposure to June midday sunlight (1 kW m^{-2}) yielding $[^1O_2]_{ss}^{1m} = 4 \times 10^{-14}$ M (Haag and Hoigné, 1986). (b) Scale of dates when the concentration of the chemicals becomes reduced to 50% of its initial value. These dates are an estimate based on the real sum curve of measured solar irradiations in Dübendorf, Switzerland, starting on a clear summer day. (Adapted from Hoigné, 1990.)

Solvated Electron

A hypothesis describing the source and sink of photolytically produced aqueous electrons is given in Figure 12.3c. As Table 12.3b shows, the steady-state concentration of e^-(aq) in a fresh water is relatively low (10^{-17} M). Reactions with O_2 (to $O_2^{-\cdot}$) constitute the major sink. Because the yield of solvated electrons in natural waters is so low, its reactions can be of only minor importance in producing photooxidants.

Organic Peroxy Radicals

Compounds that are classified as antioxidants (such as alkylphenols, aromatic amines, thiophenols, and imines) generally exhibit moderately high reactivity toward organic peroxy radicals (ROO$^\cdot$). A plausible reaction sequence for the formation of ROO$^\cdot$ is given in Figure 12.3b.

As probe molecule, Hoigné et al. used trimethylphenol

$$\text{(CH}_3\text{)}_3\text{C}_6\text{H}_2\text{-OH} + \text{ROO}^\cdot \longrightarrow \text{(CH}_3\text{)}_3\text{C}_6\text{H}_2\text{-O}^\cdot + \text{ROOH}$$

(18)

DOM-derived ROO$^\cdot$ may well inhibit a wide range of reactivities. This variation introduces considerable uncertainty in the determination of ROO$^\cdot$ steady-state concentration and production rate (Table 12.3b).

Photooxidants in the Atmosphere

Important photooxidants that are formed in the atmosphere may become absorbed into atmospheric water droplets (clouds, fog, dew) and in surface waters. Most important emissions are $HO_2^\cdot$ and O_3. Assuming a dry deposition of 0.1 mg O_3 m^{-2} h^{-1}, a water film of 0.1-mm depth would receive a mean dose rate of up to 1 mg ozone liter^{-1} per hour (20 μM h^{-1}) (Hoigné, 1990).

12.4 PHOTOREDOX REACTIONS: PHOTOLYSIS OF TRANSITION METAL COMPLEXES

A photoredox reaction is a redox reaction that occurs after electronic excitation of one or several reaction partners. A representative example is given by the charge transfer process that occurs as a consequence of light absorption by a metal–ligand complex. The light absorption promotes an electron from an occupied orbital of the ligand to an unoccupied orbital of the metal ion. Photoredox reactions occur either via intermolecular or intramolecular processes. Iron is an ideal metal for charge transfer reactions in natural waters, because it is fairly abundant, has two oxidation states, forms strong complexes [especially Fe(III)], and has a charge transfer absorption region at near-UV wavelength radiation that penetrates water reasonably well. Cu(II)–Cu(I), Co(III)–Co(II), and Mn(III,IV)–Mn(II) also have suitable properties to catalyze decarboxylation reactions by ligand to metal charge transfer (LMCT) mechanisms. A simple example is given by the oxalato iron(III) complex (Zuo and Hoigne, 1992).

$$\left[Fe^{III} \begin{array}{c} O-C=O \\ | \\ O-C=O \end{array} \right]^+ \xrightarrow{h\nu} Fe(II) + C_2O_4^{-\bullet} \qquad (19)$$

$$C_2O_4^{-\bullet} + O_2 \longrightarrow O_2^{-\bullet} + 2\,CO_2 \qquad (20)$$

$$2\,O_2^{-\bullet} + 2\,H^+ \longrightarrow H_2O_2 + O_2 \qquad (21)$$

The same reaction can be formulated for a trioxalato Fe(III) complex (see Figure 12.3e). Similar photoredox reactions occur with other carboxylato complexes such as citrate and humic or fulvic acids. These ligands shift the spectrum of light absorption to longer wavelengths. An important example in surface waters is the photolysis of the Fe^{III}–EDTA complex by the LMCT mechanism. Fe(III)–NTA complexes are similarly photolyzed, but since EDTA—unlike NTA—is not biodegraded, the photodecomposition is the key reaction for its elimination from natural waters. Fe(III) and Mn(III,IV) probably play a significant role in the photodegradation of aquatic humics (T. Voelker-Barschat and B. Sulzberger, *personal communication,* 1994; Brezonik, 1994).

Such processes play an important role not only in surface waters but also in atmospheric water droplets. Both Fe and oxalate are present in cloud and fog water; oxalate is an intermediate in the oxidation of atmospheric organic pollutants. Iron is introduced into the atmosphere from dust; it is present in these mostly slightly acid water droplets as in dissolved or colloidal form as Fe(II) and Fe(III) (Behra and Sigg, 1990). H_2O_2 formed by reactions such as 19–21 can oxidize SO_2 in the aqueous atmosphere (see Example 9.3) (Faust, 1994; Hoigné et al., 1994; Kotronarou and Sigg, 1993; Sedlak and Hoigné, 1993, 1994).

12.5 PHOTOCHEMICAL REACTIONS IN ATMOSPHERIC WATERS: ROLE OF DISSOLVED IRON SPECIES

Importance of Atmospheric Water

Hoigné et al. (1994) point out that in most regions of the globe, about 15% of the volume of the lower troposphere is composed of clouds. A parcel of air

12.5 Photochemical Reactions in Atmospheric Waters

therefore spends about one-seventh of its time in a cloud system. In addition, the distribution patterns of clouds cause air to remain typically for less than 12h between clouds. Within clouds, the high dispersion of the liquid phase, forming a few hundred droplets per milliliter, allows for an efficient mass transfer between the gaseous phase and the droplets. Therefore, for many atmospheric compounds, the liquid phase acts as a very efficient reactor despite its limited volume, only accounting for a fraction of about 50 ppbv of the lower troposphere. Comprehensive critical reviews describing and interpreting the role of cloud systems have been published by Lelieveld and Crutzen (1991), Graedel and Crutzen (1993), and Lelieveld (1994).

Solar Flux and H_2SO_4 Production

Figure 12.8 shows a correlation between the solar flux and H_2SO_4 deposition (Faust, 1994). The correlation observed is consistent with the hypothesis that the deposition of sulfuric acid over the northeastern United States is limited by

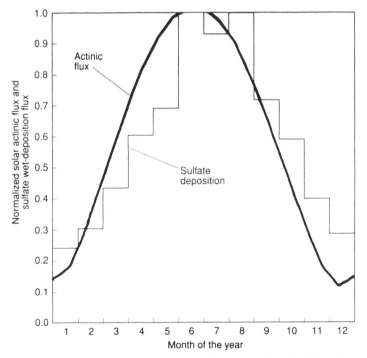

Figure 12.8. Normalized monthly regional wet-deposition flux of sulfate for the northeastern United States, 1976–1979, and normalized spherically integrated solar irradiance. The figure demonstrates that the wet deposition of sulfate on a regional scale (the northeastern United States) is correlated with solar actinic flux, which initiates the formation of all photooxidants that are responsible for the conversion of SO_2 to sulfate. The lifetime of SO_2 in the troposphere is on the order of several days. (Adapted from Faust, 1994.)

the availability (within the time allowed for by transport of the SO_2) of one or more photooxidants, which control the oxidation of SO_2 to H_2SO_4 in the troposphere. Thus photoredox processes responsible for the oxidation of SO_2 to H_2SO_4 can affect the amount of sulfate that is ultimately present in the atmosphere on local, regional, and global scales (Faust, 1994).

The role of iron in atmospheric waters has been reviewed by Hoigné et al. (1994), by Faust (1994), by Sedlak and Hoigné (1993, 1994).

The effectiveness of the atmospheric aqueous phase reactor is due to the following:

1. Some reactants, such as H_2O_2 and $HO_2^{\cdot}/{}^{\cdot}O_2^-$, reach in the equilibrated aqueous phase concentrations that are nearly 1 million times higher in concentration than in the gaseous phase.
2. Reactants of high water solubility (e.g., $HO_2^{\cdot}$) become separated from those exhibiting low solubility (e.g., NO).
3. Compounds dissolved in water may form new species by dissolution, hydration, dissociation, and complex formation.
4. Within water droplets, solar radiation is enhanced because of light scattering and in-droplet reflections.

Photochemical Oxidants in the Atmosphere

It is beyond the scope of this book to review gas phase photochemistry in any detail. We restrict ourselves to a limited discussion of the origin of some of the reactants of relevance in atmospheric water.

NO_x and the Production of Ozone NO and NO_2 are important air pollutants originating mainly in combustion processes. Most of the NO_x formed in combustion is NO but some NO can become oxidized in the combustion process to NO_2. Even small amounts of NO_2 are sufficient to cause a complex series of reactions involving organics that lead to *photochemical smog*. The following is a simplified interpretation of the photochemical cycle of NO_2, NO, and O_3. (For details see Seinfeld, 1986, and Finlayson-Pitts and Pitts, 1986.)

When NO and NO_2 are present in sunlight, O_3 formation occurs as a result of the photolysis of NO_2

$$NO_2 + h\nu \xrightarrow{i} NO + O \tag{22}$$

$$O + O_2 + M \xrightarrow{ii} O_3 + M \tag{23}$$

where M represents a molecule like N_2 or O_2 (a third molecule that absorbs the excess of vibrational energy and thereby stabilizes the O_3 molecule formed). As Seinfeld points out, there are no significant sources of ozone in the atmosphere other than reaction 23. Once formed, O_3 reacts with NO to regenerate NO_2:

12.5 Photochemical Reactions in Atmospheric Waters

$$O_3 + NO \xrightarrow{iii} NO_2 + O_2 \quad (24)$$

To simplify, Seinfeld suggests the following "Gedankenexperiment." We place known initial concentrations of NO and NO_2—$[NO]_0$ and $[NO_2]_0$—in air in a reactor (V_{const}, T_{const}) and irradiate.

The rate of change in the reactor after beginning irradiation is

$$\frac{d[NO_2]}{dt} = -k_i[NO_2] + k_{iii}[NO][O_3] \quad (25)$$

We can formulate similar rate laws for the other species in the reactor; for example, assuming atomic oxygen to be constant

$$\frac{d[O]}{dt} = -k_{ii}[O][O_2][M] + k_i[NO_2] \quad (26)$$

In dealing with highly reactive species such as the oxygen atom, it is customary to make a steady-state approximation and thereby assume that the rate of formation (reaction 22) is equal to the rate of disappearance (reaction 23). Thus

$$[O]_{ss} = \frac{k_i[NO]}{k_{ii}[O_2][M]} \quad (27)$$

The steady-state ozone concentration can be approximated from equations 22 and 24:

$$\frac{d[NO_2]}{dt} \simeq -k_{iii}[O_3][NO] + k_i[NO_2] = 0$$

$$[O_3]_{ss} \simeq \frac{k_i[NO_2]}{k_{iii}[NO]} \quad (28)$$

This expression has been called the *photostationary state relation*.

Graedel and Crutzen (1993) point out that it is important to realize that the potential for ozone formation in the troposphere is large and is limited only by the availability of NO and NO_2 as catalysts. Ozone is a phytotoxic and poisonous gas, and if its concentrations in the troposphere were to grow there would be serious environmental consequences.

OH Radicals

The presence of O_3 in the troposphere is also very important because it generates $OH^\cdot$ radicals:

$$O_3 + h\nu(\leq 310 \text{ nM}) \longrightarrow O + O_2 \qquad (29)$$

$$O + H_2O \longrightarrow 2\ OH^{\bullet} \qquad (30)$$

The central role played by $OH^{\bullet}$ is due to its reactivity; by abstracting a hydrogen atom from a nearby molecule, $OH^{\bullet}$ reverts to H_2O:

$$OH^{\bullet} + RH \longrightarrow R^{\bullet} + H_2O \qquad (31)$$

where R is an organic fragment like C_2H_5 and $R^{\bullet}$ its radical $C_2H_5^{\bullet}$. Without the hydroxyl radical the composition of the atmosphere could be completely different because $OH^{\bullet}$ is the most important agent in cleansing pollutant trace gases (including CO, CH_4, and SO_2). Its concentration is $1-4 \times 10^{-14}$ atm ($0.25-1 \times 10^6$ radicals cm^{-3}); its steady-state concentration and its atmospheric lifetime (~ 0.35) are controlled primarily by its reaction with CH_4.

Atmospheric Water Droplets

Compounds that are strongly water soluble, such as H_2O_2, $HO_2^{\bullet}$, and to some extent OH radicals together with HCHO (formaldehyde) and certain constituents to be oxidized such as SO_2 and carboxylic acids, are incorporated into the liquid phase. This process leaves behind in the gas phase the less-soluble components such as NO, CO, and CH_4. $HO_2^{\bullet}$ participates in the liquid phase in an O_3 destruction cycle, leading to the formation of formic acid, HCOOH. In the water phase $OH^{\bullet}$ reacts with CH_2O and HCOOH to form $HO_2^{\bullet}/O_2^{-\bullet}$. These radicals are especially important in the aquatic atmosphere because they interact with Fe(II), Fe(III), and their complexes to produce H_2O_2, which, in turn, oxidizes aqueous SO_2 (see also Figure 12.3e):

$$\begin{array}{c} O_2^{-\bullet} \\ \\ O_2 \end{array} \rightleftarrows \begin{array}{c} Fe(III) \\ \\ Fe(II) \end{array} \rightleftarrows \begin{array}{c} H_2O_2 + 2\ OH^- \\ \\ 2\ H_2O + O_2^{-\bullet} \end{array} \qquad (32)$$

(See also Figure 12.8.)

12.6 HETEROGENEOUS PHOTOCHEMISTRY

Photochemical processes in aquatic systems are not limited to substances in solution. Photochemical reactions may involve particles. One can distinguish two mechanisms:

1. Organic or inorganic solutes, adsorbed (or surface complexed) to particle surfaces, may act as chromophores and absorb light energy and transfer

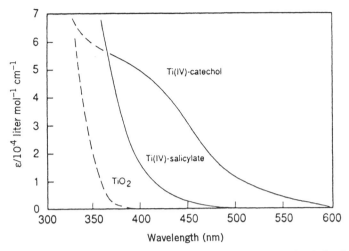

Figure 12.9. UV-visible absorption spectra of TiO_2 transparent sols (0.5 g liter^{-1}, pH 3.6). Addition of salicylic acid and catechol (2×10^{-4} M) produces a red shift of the absorption onset to 500 and 600 nm, respectively, and enhances the electron transfer to the electron acceptors in solution. (From Moser et al., 1991.)

an electron to the lattice metal ion (ligand to metal charge transfer). Thus multivalent metal oxides such as those of iron or manganese are reductively dissolved (Sulzberger et al., 1989; Waite et al., 1988).

2. Many metal oxides and metal sulfide particles behave as semiconductors. If these semiconductor surfaces absorb light energy directly, the resulting excited states—as will be explained more fully—produce a charge separation (electrons, e^-, and holes, h^+), which form reducing and oxidizing sites at the particle surface. Often it is not easy to establish whether the particle (semiconductor) or the adsorbed substance is the chromophore.

Figure 12.9 shows surface complex formation of bidentate ligands to the surface of TiO_2 transparent sols (Moser et al., 1991). Surface complexation causes a "red" shift in the light absorption and, as shown by Moser et al., an enhancement of electron transfer from the conduction band to electron acceptors in solution.

Exemplification: The Photoreductive Dissolution of α-Fe_2O_3

The various elementary steps involved in the surface photoredox reaction, leading to dissolution of hematite in the presence of oxalate, are outlined in Figure 12.10. The two-dimensional structure of the surface of an iron(III) hydroxide given in this figure is highly schematic. The charges indicated correspond to relative charges. An important step is the formation of a hypothetical bidentate, mononuclear surface complex. With pressure jump relaxation technique, it has

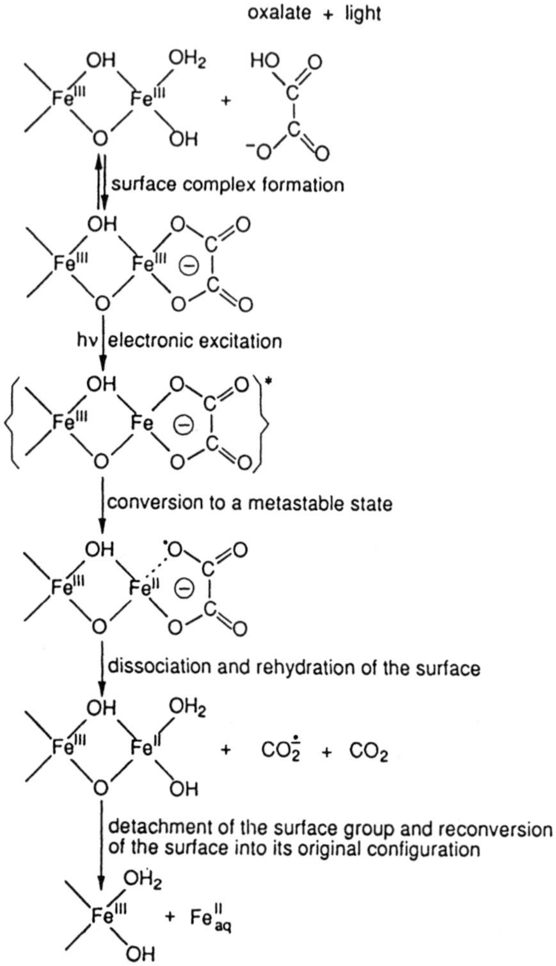

Figure 12.10. Schematic representation of the various steps involved in the light-induced reductive dissolution of hematite in the presence of oxalate. (Adapted from Sulzberger et al., 1989.)

been shown that the adsorption equilibria at the mineral–water interface are generally established fast (Hayes and Leckie, 1986). Election transfer occurs via an electronically excited state (indicated by an asterisk) which is either a ligand-to-metal charge transfer transition of the surface complex and/or a $Fe^{III} \leftarrow O^{-II}$ charge transfer transition of hematite. The oxalate radical undergoes a fast decarboxylation reaction yielding CO_2 and the $CO_2^{\cdot -}$ radical, which is a strong reductant that can reduce a second surface iron(III) in a thermal reaction. Thus two surface iron(II) and two CO_2 may theoretically be formed per absorbed photon. We assume that detachment of Fe^{II} from the crystal lattice is the rate-limiting step of the overall reaction. After detachment of the surface group, the surface of hematite is reconverted into its original configuration; hence the surface concentration of active sites and of adsorbed oxalate does not change throughout the experiment.

This reaction has been shown to be further catalyzed by the adsorption of the Fe(II) released in the dissolution reaction to the surface-bound oxalate (Suter et al., 1988).

The (Photo)redox Cycle of Iron

The iron cycle shown in Figure 12.11 illustrates some of the redox processes typically observed in soils, sediments, waters, and atmospheric water droplets, especially at oxic–anoxic boundaries. The cycle includes the reductive dissolution of iron(III) (hydr)oxides by organic ligands, which may also be photocatalyzed in surface waters, and the oxidation of Fe(II) by oxygen, which is catalyzed by surfaces. The oxidation of Fe(II) to Fe(III) (hydr)oxides is accompanied by the binding of reactive compounds (heavy metals, phosphate, or organic compounds) to the surface, and the reduction of the ferric (hydr)oxides is accompanied by the release of these substances into the water column.

On overall balance, the cycle shown in Figure 12.11 represents a mediation (by iron) of the oxidation of organic matter by oxygen. This oxidation may be important in both the degradation and polymerization of organic matter in soils and waters. The interdependence of the iron cycle with that of other redox cycles is obvious; for example, Fe(II) can reduce Mn(III,IV) oxides, and HS^- is an efficient reductant of hydrous Fe(III) oxides. Many of the processes mentioned above also occur directly in soils or are mediated indirectly by microorganisms. Microorganisms and plants produce a large number of biogenic acids that are effective in ligand-promoted dissolution of Fe(III) and other

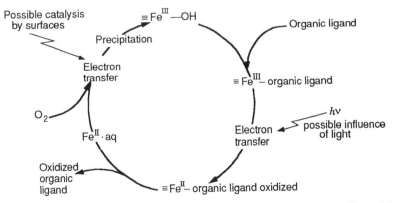

Figure 12.11. Schematic representation of the aquatic (photo)redox cycling of iron. The symbol ≡ denotes the lattice surface of an iron(III) (hydr)oxide.

(hydr)oxides. Oxalic, maleic, acetic, succinic, tartaric, ketogluconic, and *p*-hydroxybenzoic acids have been found in top soils, with oxalic acid, the most abundant, in concentrations as high as 10^{-5}–10^{-4} M (in soil water). The downward vertical displacement of Al and Fe observed during podzolization of soils can be explained by considering the effect of pH and of complex formers on the solubility of iron and aluminum (hydr)oxides and on their dissolution rates. The same biogenic acids are found typically in waters and sediments and are also produced by fermentation reactions under anoxic conditions. The reductive dissolution of Fe(III) (hydr)oxides is also of importance for iron uptake by higher plants.

The light-induced oxidation of humic or fulvic acid by MnO_2 [and by Fe(III) (hydr)oxides] produces, in addition to Mn(II) [and Fe(II)], low molecular weight organic molecules that can serve as substrates for microorganisms (Sunda, 1994).

Photochemical Reduction of Fe(III) in Acid Lakes

In acid surface waters, photochemical reduction of colloidal Fe(III) (hydr)oxides and a diel variation of dissolved Fe(II) can be observed (Colienne, 1983; McKnight et al., 1988; Sulzberger et al., 1990). Figure 12.12 shows such a diel variation in the concentration of dissolved Fe(II) in a slightly acidic alpine lake (Lake Cristallina) in Switzerland.

The net concentration of Fe(II) at any time of day reflects the balance of the reductive dissolution and the oxidation/precipitation reactions and parallels

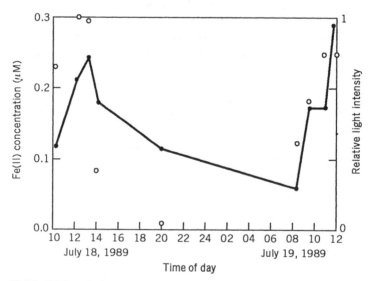

Figure 12.12. Diel variation of the concentration of dissolved Fe(II) (●) and of the incident light intensity (○) in Lake Cristallina. (The maximum measured light intensity is set to 1 on a relative scale.) (From Sulzberger et al., 1990.)

the light intensity. It is believed that the photochemical reductive dissolution of Fe(III) (hydr)oxides is also of importance for the formation of iron(II) in marine systems; however, around pH = 8 the rate of reductive dissolution is likely to be much smaller than at lower pH values. Also, in the photic zone of oceanic waters the concentration of H_2O_2 may be sufficiently high that oxidation of Fe(II) by H_2O_2 may compete with oxygenation, which would increase the overall Fe(II) oxidation rate. H_2O_2 formation occurs through photochemical processes, for example, involving dissolved organic carbon (Cooper et al., 1988; Zuo and Hoigné, 1992).

Even at higher pH values (pH $\simeq$ 8, seawater) a photochemically induced iron cycle may occur even if little or no accumulation of Fe(II) is observed. As a consequence of such a cycle, a very amorphous and highly reactive Fe(III) (hydr)oxide can be formed, which may play a role in the uptake of Fe by phytoplankton [higher solubility and faster equilibration with monomeric Fe(III)].

Reaction Rates

The reaction rates observed in surface-mediated reactions are proportional to the surface concentration of the reactant on the surface. Typically this surface concentration can be expressed as a function of its solute concentration by the Langmuir equation; that is,

$$\langle \equiv S-X \rangle = \langle \equiv SX \rangle_{max} \frac{K_{ads}\ [X]}{1\ +\ K_{ads}\ [X]} \quad (33)$$

The rate law is then

$$-\frac{d[X]}{dt} = k_{a\lambda}\ \phi\ \langle \equiv S-X \rangle_{max} \frac{K_{ads}\ [X]}{1\ +\ K_{ads}\ [X]} \quad (34)$$

This concentration dependence is illustrated for the photochemical oxidation of 4-chlorophenol on TiO_2 particles (Figure 12.13).

12.7 SEMICONDUCTING MINERALS

The electronic properties of solids can be described by various theories that complement each other. For example, band theory is suited for the analysis of the effect of a crystal lattice on the energy of the electrons. When the isolated atoms, which are characterized by filled or vacant orbitals, are assembled into a lattice containing $\sim 5 \times 10^{22}$ atoms cm^{-3}, new molecular orbitals form (Bard and Faulkner, 1980). These orbitals are so closely spaced that they form essentially continuous bands: the filled bonding orbitals form the *valence band* (vb) and the vacant antibonding orbitals form the *conduction band* (cb) (Figure

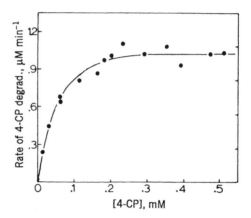

Figure 12.13. Photochemical oxidation of 4-chlorophenol on TiO_2 particles. The oxidation rate as a function of dissolved 4-chlorophenol follows a Langmuir adsorption isotherm and reflects the surface concentrations. (From Al-Ekabi et al., 1989.)

12.14). These bands are separated by a forbidden region or band gap of energy E_g.

An *intrinsic semiconductor* is characterized by an equal density of positive and negative charge carriers, produced by thermal excitation; that is, the density of electrons in the conduction band, n_i, and of holes in the valence band, p_i, are equal

$$n_i = p_i \propto \exp\left(\frac{-E_g}{2 k_B T}\right) \qquad (35)$$

Extrinsic semiconductors are materials that contain donor or acceptor species (called doping substances) that provide electrons to the conduction band or holes to the valence band. If donor impurities (donating electrons) are present in minerals, the conduction is mainly by way of electrons, and the material is called an *n-type semiconductor*. If acceptors are the major impurities present, conduction is mainly by way of holes and the material is called a *p-type semiconductor*. For instance, in a silicon semiconductor elements from the vertical row to the right of Si of the Periodic Table (e.g., As) behave as electron donors (As → As^+ + e^-) while elements from the vertical row to the left of Si (e.g., Ga) behave as hole donors (Ga + e^- → Ga^-); that is, in the latter case, electrons are excited from the vb into the acceptor sites, leaving behind mobile holes in the vb with the formation of isolated negatively charged acceptor sites.

If a semiconductor is brought into contact with an electrolyte containing one or more redox couples, charge transfer between the two phases occurs until electrostatic equilibrium (equality of the free energies of the electron in both phases) is attained.

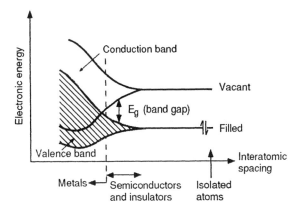

Figure 12.14. Formation of bands in solids by assembly of isolated atoms into a lattice. When the band gap $E_g \ll k_B T$ or when the conduction and valence bands overlap, the material is a good conductor of electricity (metals). Under these circumstances, there exist in the solid filled and vacant electronic energy levels at virtually the same energy, so that an electron can move from one level to another with only a small energy of activation. For larger values of E_g, thermal excitation or excitation by absorption of light may transfer an electron from the valence band to the conduction band. There, the electron is capable of moving freely to vacant levels. The electron in the conduction band leaves behind a hole in the valence band. (Adapted from Bard and Faulkner, 1980.)

Light-Induced Redox Processes at the Semiconductor–Electrolyte Interface Upon absorption of light with energy equal to or higher than the band gap energy, a band-to-band transition occurs; that is, an electron from the filled valence band is raised to the vacant conduction band, leaving behind an electron vacancy, a hole, h^+, in the valence band. As a consequence of electronic excitation by light, a strongly reducing electron (photoelectron) and a strongly oxidizing hole (photohole) are formed. The electrons and holes move in the semiconductor in a manner analogous to the movement of ions in solution, but their mobilities are orders of magnitude larger than those of ions in solution. The charge carriers that reach the surface can undergo redox reactions with adsorbed species at the solid–liquid interface. The redox reaction at the solid–liquid interface is in competition with the recombination of the photogenerated electron–hole pair; its efficiency depends on how rapidly the minority carriers—the photoholes in an n-type semiconductor and the photoelectrons in a p-type semiconductor—reach the surface of the solid and how rapidly they are captured through interfacial electron transfer from or to a thermodynamically suitable reductant or oxidant of the electrolyte solution. Plausibly, the redox reaction at the semiconductor–electrolyte interface is facilitated if the reductant and/or oxidant is adsorbed (by inner- or outer-spheric coordination) at the semiconductor surface (Figure 12.15).

Furthermore, one needs to consider that an inner-spherically adsorbed redox

756 Photochemical Processes

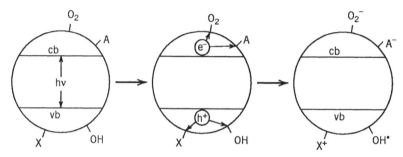

Figure 12.15. Simplified scheme of light absorption by a semiconductor particle and subsequent redox reaction on the surface. vb, Valence band; cb, conduction band; X and A, adsorbed species. Note that in this scheme, the y axis is at the same time energy and distance. Valence band holes and conduction band electrons can oxidize and/or reduce, respectively, molecules adsorbed on the surface directly:

$$h^+ + \equiv S-X \longrightarrow \equiv S-X^+ \longrightarrow \equiv S + X^+$$

or form OH· that can oxidize adsorbed compounds in a secondary reaction:

$$h^+ + \equiv SOH_2 \longrightarrow \equiv S-OH^{\cdot} + H^+$$

$$\equiv S-OH^{\cdot} + \equiv S-X \longrightarrow 2\equiv S + X^+ + OH^-$$

Whether or not h^+ and e^- lead to oxidation and reduction once they reach the surface depends on the redox potentials of (h^+) and (e^-) (which depend on the semiconductor material) and of the compound on the surface. If h^+ and e^- are not transferred to compounds on the surface or are trapped by impurities in the lattice, they recombine before participating in chemical reactions. (Adapted from Sulzberger and Hug, 1994.)

partner is characterized by a different redox potential than a solute redox partner. For theoretical background see Bard and Faulkner (1980); Bard (1988); and Grätzel (1989).

Of major interest in geochemistry and in natural water systems are semiconducting minerals for which the absorption of light occurs in the near-UV or visible spectral region and as a result of which redox processes at the mineral–water interface are induced or enhanced. Table 12.4 gives band gap energies of a variety of semiconductors.

Most of the semiconductors potentially can catalyze a wide variety of redox reactions. Semiconductors that are not naturally occurring offer potential in water technology to treat organic contaminated wastes. Photocatalytic degradation of many organochlorine compounds on illuminated TiO_2 has been demonstrated (see Bahnemann et al., 1994). One drawback to the use of TiO_2 as a photocatalyst is its large band gap energy. That means that it is photoactivated primarily by UV light.

We now return to the light-catalyzed reductive dissolution of Fe(III) (hydr)oxides, which was discussed before (see Figure 12.10). The surface

12.7 Semiconducting Minerals

Table 12.4. Band Gap Energies and Corresponding Wavelengths of Light for a Variety of Semiconductors

Semiconductor	Band Gap (eV)	Equivalent Wavelength (nm)	Semiconductor	Band Gap (eV)	Equivalent Wavelength (nm)
ZrO_2	5.0	248	CdS	2.4	516
Ta_2O_5	4.0	310	$\alpha\text{-}Fe_2O_3$	2.34	530
SnO_2	3.5	354	ZnTe	2.3	539
$KTaO_3$	3.5	354	$PbFe_{12}O_{19}$	2.3	539
$SrTiO_3$	3.4	365	GaP	2.3	539
Nb_2O_5	3.4	365	$CdFe_2O_4$	2.3	539
ZnO	3.35	370	CdO	2.2	563
$BaTiO_3$	3.3	376	$Hg_2Nb_2O_7$	1.8	689
TiO_2	3.0–3.3	376–413	$Hg_2Ta_2O_7$	1.8	689
SiC	3.0	413	CuO	1.7	729
V_2O_5	2.8	443	PbO_2	1.7	729
Bi_2O_3	2.8	443	CdTe	1.4	885
$FeTi_3$	2.8	443	GaAs	1.4	885
PbO	2.76	449	InP	1.3	954
WO_3	2.7	459	Si	1.1	1127
$WO_{3-x}Fe_x$	2.7	459	$\beta\text{-}HgS$	0.54	2296
$YFeO_3$	2.6	476	$\beta\text{-}MnO_2$	0.26	4768
$Pb_2Ti_{1.5}W_{0.5}O_{6.5}$	2.4	516			

Source: Waite (1986).

complex of oxalate adsorbed on an Fe(III) hydroxide surface is metastable in the absence of light. The absorption of photons leads typically to a charge separation and in turn to a redox disproportionation, that is, either (1) in the semiconductor or (2) the surface complex $\equiv Fe^{III}\text{—}Ox$ becomes excited as a consequence of photon absorption.

In the first case, the disproportionation leads to e^- and h^+. The electrons in the conduction band are reducing [formation of Fe(II)]; the holes are oxidizing and oxidize oxalate directly or indirectly to a radical. See the scheme below.

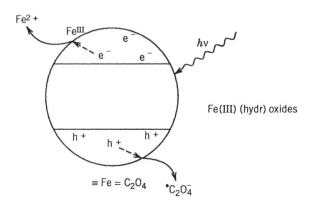

In the second case, an excited surface complex (ligand to metal charge transfer) is formed, which then disproportionates:

$$2 \ (\equiv Fe^{III} - C_2O_4^-) \xrightarrow{h\nu} 2 \ (\equiv Fe^{III} - C_2O_4)^* \longrightarrow 2 \ (\equiv \ldots) \\ + 2 \ Fe^{2+}(aq) + C_2O_4^{2-} + 2 \ CO_2 \quad (36)$$

Example 12.2. Steady-State Concentration of Fe^{2+} in an Acidic Lake On the basis of a mass balance for dissolved Fe(II) derive a simplified equation that illustrates the factors that influence the steady-state concentration of Fe^{2+}.

In addition to inflow and outflow of Fe^{2+}, we need expressions for the rate of reduction of Fe(III) (hydr)oxides and for the reoxidation of Fe^{2+} by oxygen. The rate of photochemical Fe(II) formation depends on the light intensity I and the surface concentration of the specifically adsorbed ligand (electron donor):

$$\nu_\nu = kI \ \{\equiv FeL\} \quad (i)$$

The oxidation rate by O_2 has been given as

$$\nu_0 = k_{ox} \ [O_2(aq)] \ [OH^-]^2 \ [Fe^{2+}] \quad (ii)$$

Then the mass balance for dissolved Fe^{2+} is given by

$$V \frac{d[Fe^{2+}]}{dt} = Q \ [Fe^{2+}]_{in} - Q \ [Fe^{2+}] - V k_{ox} \ [Fe^{2+}] \ [O_2(aq)] \ [OH^-]^2 \\ + VkI \ \{\equiv FeL\} \quad (iii)$$

where

V = water volume
Q = volumetric flow rate
$[Fe^{2+}]_{in}$ = inflow concentration from seepage and/or streams
k_{ox} = oxidation rate constant
k = rate constant for light-induced reductive dissolution
$\{\equiv Fe^{III}L\}$ = surface concentration of the specifically adsorbed electron donor

Assuming a steady state for Fe^{2+}, equation iii can be transformed as follows:

$$[Fe^{2+}]_{ss} = \frac{[Fe^{2+}]_{in} + (V/Q) \ kI \ \{\equiv FeL\}}{1 + (V/Q) \ k_{ox} \ [O_2(aq)] \ [OH^-]^2} \quad (iv)$$

This simple model assumes that the reduction of particulate iron(III) in the photic zone occurs primarily via a light-induced reductive dissolution and that the main oxidation pathway of iron(II) is the homogeneous oxidation by molecular O_2; it ignores nonbiotic reductive dissolution and oxidation of Fe(II) by H_2O_2 (formed photochemically). Since the reoxidation is strongly pH dependent (see Chapter 11), the steady-state concentration will decrease strongly with increasing pH and will depend on the constant and type of dissolved organic matter (see Barry et al., 1994, and Sulzberger, 1990).

SUGGESTED READINGS

Haag, W. R., and Hoigné, J. (1986) Singlet Oxygen in Surface Waters. 3. Photochemical Formation and Steady State Concentration in Various Types of Waters, *Environ. Sci. Technol.* **20**, 341–348.

Helz, G. R., Zepp, R. G., and Crosby, D. G. (Eds.) (1994) *Aquatic and Surface Photochemistry*, Lewis, Boca Raton, FL.

Hoffmann, M. R. (1990) Catalysis in Aquatic Environments. In *Aquatic Chemical Kinetics*, W. Summ, Ed., Wiley-Interscience, New York.

Sulzberger, B. (1990) Photoredox Reactions at Hydrous Metal Oxide Surfaces; A Surface Coordination Chemistry Approach. In *Aquatic Chemical Kinetics*, W. Stumm, Ed., Wiley-Interscience, New York, pp. 401–429.

Waite, T. D. (1990) Photo-Redox Processes at the Mineral–Water Interface. In *Mineral–Water Interface Geochemistry*, M. F. Hochella, Jr. and A. F. White, Eds., Mineralogical Society of America, Washington, DC, pp. 559–603.

Zafiriou, O. C. (1983) Natural Water Photochemistry. In *Chemical Oceanography*, Vol. 8, Academic Press, London, pp. 339–379.

Zepp, R. G., Faust, B. C., and Hoigné, J. (1992) Hydroxyl Radical Formation in Aqueous Reactions of Fe(II) with H_2O_2: The Photo Fenton Reaction, *Environ. Sci. Technol.* **26**, 313–319.

Zika, R. R., and Cooper, W. J. (1989) *Photochemistry of Environment Aquatic Chemistry*, ACS Series **327**, American Chemical Society, Washington, DC.

Zuo, Y., and Hoigné J. (1993) Evidence for Photochemical Formation of H_2O_2 and Oxidation of SO_2 in Authentic Fog Water, *Science* **260**, 71–73.

13

KINETICS AT THE SOLID–WATER INTERFACE: ADSORPTION, DISSOLUTION OF MINERALS, NUCLEATION, AND CRYSTAL GROWTH

13.1 INTRODUCTION

In Section 9.9, in discussing the transport of (ad)sorbable chemicals in groundwater, we exemplified briefly how the spatial distribution of chemicals (e.g., pollutants) depends on the coupling of transport and chemical interfacial processes.

In this chapter we discuss the rates of adsorption, paying special attention to those few cases where information on the rate of specific adsorption (reaction of an adsorbate in the adsorption layer) is available. Furthermore, we elaborate on the chemical processes involved in the dissolution of minerals and concentrate on the dissolution of oxides, silicates, and carbonates, which play an enormous role in the chemical weathering and erosion. We try to demonstrate that in most cases the rate-determining step in the dissolution is a chemical reaction at the surface of the mineral. Thus we have here an excellent example of the relationship between surface structure and reactivity. Surface chemistry plays an equally important role in the formation of the solid phase (precipitation, nucleation, and crystal growth). Nature's selectivity is reflected in the creation of a crystal and its growth.

For the coupling of transport and interfacial processes the reader is referred to Ball and Roberts (1991), Bidoglio (1994), Hemond and Fechner (1994), and Weber et al. (1991).

13.2 KINETICS OF ADSORPTION

Various steps are involved in the transfer of an adsorbate to the adsorption layer, including transport to the surface by convection and/or molecular diffusion, and the attachment to the surface. The latter may include steps like formation of a bond, dehydration, surface diffusion, and association processes of the adsorbed constituents. In a similar way, desorption consists of various steps.

In a simplified way we may distinguish (1) transport of a solute to the interface and (2) the transfer at the interface of the adsorbate from the solution to the adsorption layer (*intrinsic adsorption*).

Effective rates of sorption, especially in subsurface systems, are frequently controlled by rates of solute transport rather than by intrinsic sorption reactions *per se*. In general, mass transport and transfer processes operative in subsurface environments may be categorized as either "macroscopic" or "microscopic." Macroscopic transport refers to movement of solute controlled by movement of bulk solvent, either by advection or hydrodynamic (mechanical) dispersion. For distinction, microscopic mass transfer refers to movement of solute under the influence of its own molecular or mass distribution (Weber et al., 1991).

One of the fundamental steps involved in characterizing and modeling microscopic mass transfer is appropriate representation of associated resistances or impedances, including relevant distances over which solute is transferred and relevant properties of the medium through which transfer occurs. The nature and characteristics of such resistances vary with local conditions associated with particular combinations of sorbent, solute, fluid, and system configuration. Differences in local conditions and associated transport phenomena are typified in subsurface systems by differences between solute transport through the interstitial cracks and crevices of rocks or soil particles, through organic polymer matrices associated with soils, and through internal fluid regions of soil aggregates.

For the mathematical descriptions of microscopic interdependences and mass transfer processes within fluid and sorbing phases see, for example, Roberts et al. (1987), Ball and Roberts (1991), Weber et al. (1991), Hemond and Fechner (1994), and Paces (1994).

Coupled processes in reaction, flow, and transport of contaminants have been reviewed by Bidoglio (1994). Scale effects in the transport of contaminants in natural media have been discussed by Behra (1994).

Semi-infinite Linear Diffusion

The intrinsic adsorption is often much faster than the transport to the interface. For many adsorbates, especially organic substances, the concept of semi-infinite linear diffusion can give us an idea of the time necessary for an adsorbate to be adsorbed. The number of moles of adsorbate, n, diffusing to a unit area of a surface, A, per second is proportional to the bulk concentration of adsorbate, c:

$$\frac{1}{A}\frac{dn}{dt} = c\sqrt{\frac{D}{\pi t}} \tag{1}$$

Kinetics at the Solid–Water Interface

In a similar way, desorption consists of various steps: those that make up an intrinsic desorption and those that are the transport from the surface to the bulk solution.

The following assumptions are made:

1. The bulk concentration c remains constant.
2. Transport by diffusion is rate controlling; that is, the adsorbate arriving at the interface is adsorbed fast (intrinsic adsorption). This intrinsic adsorption, that is, the transfer from the solution to the adsorption layer, is not rate determining or in other words, the concentration of the adsorbate at the interface is zero.
3. Furthermore, the radius of the adsorbing particle is relatively large (no spherical diffusion).

Integration of equation 1 gives

$$\Gamma = \frac{n}{A} = 2c\sqrt{\frac{Dt}{\pi}} \qquad (2)$$

Since $\Gamma_t = (n/A)_t$ is the surface concentration (mol m^{-2}) at a specific time, the time, δ, necessary to cover the surface with a monomolecular layer is given by

$$\delta = \frac{\pi}{4}\frac{\Gamma_{max}^2}{Dc^2} \qquad (3)$$

Equation 3 can be extended to obtain a fractional surface coverage, $\theta = \Gamma/\Gamma_{max}$, at time t

$$t = \frac{\pi}{4}\frac{\theta^2 \Gamma_{max}^2}{Dc^2} \qquad (4)$$

Thus the time necessary to attain a certain coverage, θ, or the time necessary to cover the surface completely ($\theta = 1$) is inversely proportional to the square of the bulk concentration (cf. Figure 13.1). Assuming molecular diffusion only, δ is on the order of 2 min for a concentration of 10^{-5} M adsorbate when the diffusion coefficient D is 10^{-5} cm^2 s^{-1} and $\Gamma_{max} = 4 \times 10^{-10}$ mol cm^{-2}.[†] Considering that transport to the surface is usually by turbulent diffusion up to the boundary layer of the surface, such a calculation illustrates that the

[†]Compare equation 3:

$$\delta = \frac{\pi}{4}\frac{(4 \times 10^{-10})^2}{10^{-5}\text{ cm}^2\text{ s}^{-1}}\left(\frac{\text{mol}^2}{\text{cm}^4}\right)\frac{\text{liter}^2 \times 10^6 \text{ cm}^6}{10^{-10}\text{ mol}^2\text{ liter}^2} = 125 \text{ s}$$

13.2 Kinetics of Adsorption

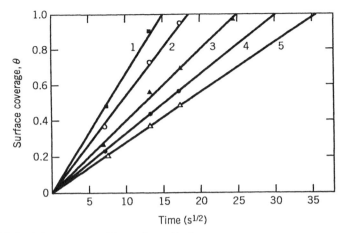

Figure 13.1. Adsorption on Hg surface. Surface coverage versus square root of accumulation time for adsorption of Triton X-100 at Hg surface. Concentration of Triton X-100 (Triton X-100 = tert. octylphenol ethoxylate with 9–10 ethoxy groups) in 0.55 mol liter^{-1} NaCl: (1) 1.25, (2) 0.94, (3) 0.73, (4) 0.63, (5) 0.52 mg liter^{-1}. (From Batina et al., 1985.)

formation of an adsorption layer is relatively rapid at concentrations above 10^{-6} M. But it can become slow at concentrations lower than 10^{-6} M.

Subsequent to the adsorption onto a surface, surfactants, especially long-chain fatty acids and alcohols, tend to undergo alterations such as two-dimensional associations in the adsorbed layer, presumably at rates kinetically independent of preliminary steps. These intralayer reactions have been shown to be very slow.

The applicability of equations 1–4 is somewhat restricted because the bulk concentration is assumed to be constant either because its depletion is negligible (adsorbed quantity $\ll$ quantity present in the system) or because it is kept constant by a steady-state mechanism. Analytical expressions of Γ as a function of time for situations where the approach to adsorption equilibrium is accompanied by a corresponding adjustment of c are available only for a few relatively simple cases.

A rigorous treatment of diffusion to or from a flat surface has been given by Lyklema (1991). Van Leeuwen (1992) has pointed out that in analyzing experimental adsorption data, which are always confined to a limited time window, it is tempting to fit the data to the sum of two or three exponential functions with different arguments. Although such fits are often apparently successful, the merit of the fit is purely mathematical; a mechanistic interpretation in terms of a first-order dependence is usually not justified. With porous materials, diffusion into the pores renders the adsorption process very slow. If one does not wait sufficiently long, one often gains the impression that the process is irreversible (e.g., Figure 13.2).

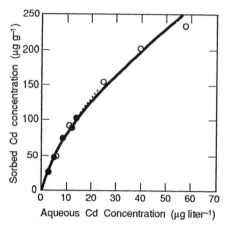

Figure 13.2. Evidence for reversibility for sorption on illite. Adsorption–desorption equilibrium for Cd(II) on illite after 54 days of equilibration. The solution contains HCO_3^-, 2×10^{-3} M Ca^{2+}, and has a pH = 7.8. Freundlich isotherms are based on separate adsorption (●) and desorption (○). Adsorption approaches equilibrium faster than desorption. The data do not allow a conclusion on the specific absorption mechanism. (Data are from Comans, 1987.)

Adsorption of Metal Ions to a Hydrous Oxide Surface

We have argued that (inner-sphere) surface complex formation of a metal ion to the oxygen donor atoms of the functional groups of a hydrous oxide is, in principle, similar to complex formation in homogeneous solution, and we have used the same type of equilibrium constants. We can apply similar concepts in kinetics.

Hachiya et al. (1984) and Hayes and Leckie (1986) used the pressure-jump relaxation method to study the adsorption kinetics of metal ions to oxide minerals. Their results support the same mechanism for adsorption as for homogeneous complex formation.

$$\equiv Al-OH + Me(H_2O)_n^{2+} \underset{}{\overset{K_{OS}}{\rightleftharpoons}} \equiv Al-OH \cdots Me(H_2O)_n^{2+} \quad (5)$$

$$\equiv Al-OH \cdots Me(H_2O)_n^{2+} \xrightarrow{k_{ads}} \equiv Al-O\underset{Me(H_2O)_{n-1}^{2+}}{\overset{H}{\diagup}} + H_2O \quad (6)$$

$$\equiv Al-O\underset{Me(H_2O)_{n-1}^{2+}}{\overset{H}{\diagup}} \xrightarrow[\text{fast}]{k} \equiv Al-O-Me(H_2O)_{n-1}^+ + H^+ \quad (7)$$

The dashed line in the complex in 5 and 6 indicates an outer-sphere (OS) surface complex; K_{OS} stands for the outer-sphere complex formation constant and k_{ads} ($M^{-1} s^{-1}$) refers to the intrinsic adsorption rate constant at zero surface charge (Wehrli et al., 1990). K_{OS} can be calculated with the help of a relation describing charge versus potential for constant capacitance interfaces (Chapter 9, Example 9.2).

$$K_{OS} = \exp\left(-\frac{ZF\psi}{RT}\right) \simeq \exp\left(-Z\frac{F^2}{CsRT}Q\right) \tag{8}$$

where Z is the charge of adsorbate, F is the Faraday constant, C is the integral capacitance of a flat double layer (farad m^{-2}), s is the specific surface area ($m^2 kg^{-1}$), and Q refers to the experimentally accessible surface charge (mol kg^{-1}). The overall rate for the formation of the surface complex is given, in analogy to the homogeneous case, by

$$\frac{d[\equiv AlOMe^+]}{dt} = K_{OS}k_{ads}[M^{2+}][\equiv Al-OH] \tag{9}$$

The term $F^2/CsRT$ is obtained from the constant capacitance model (9.6). Figure 13.3 gives a plot of the linear free energy relation between the rate constants for water exchange and the intrinsic adsorption rate constant, k_{ads}.

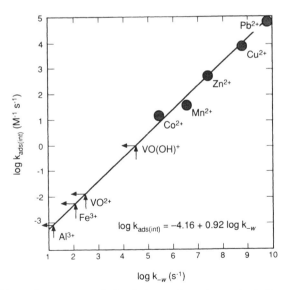

Figure 13.3. Linear free energy relation between the rate constants for water exchange k_{-w} (s^{-1}) and the intrinsic adsorption rate constants $k_{ads(int)}$ ($M^{-1} s^{-1}$) from the pressure-jump experiments of Hachiya et al. The intrinsic constants refer to an uncharged surface. The linear free energy relations based on the experimental points are extended to some ions with lower H_2O exchange rate in order to predict absorption rates. (Adapted from Wehrli et al., 1990.)

The mechanism given is in accord with the existence of inner-sphere surface complexes; it illustrates that one of the water molecules coordinated to the metal ion has to dissociate in order to form an inner-sphere complex; if this H_2O loss is slow, then the adsorption (i.e., the binding of the metal ion to the surface ligands) is slow.

The circles refer to the pressure-jump experiments of Hachiya et al. The relationship is expressed by an almost linear correlation (slope 0.92):

$$\log k_{ads} = -4.16 + 0.92 \log k_{-w} \tag{10a}$$

where as before k_{ads} refers to the intrinsic adsorption rate constant [M^{-1} s^{-1}] and k_{-w} refers to the constant for water exchange (s^{-1}). [This constant can be converted into k_{H_2O} (M^{-1} s^{-1}) by dividing k_{-w} by the molar concentration of water (~ 55.4 M).] Then equation 10a becomes

$$\log k_{ads} = -2.42 + 0.92 \log \left(\frac{k_{-w}}{55.4}\right) \tag{10b}$$

The heterogeneous adsorption process is slower by a factor of ~ 250 than an equivalent homogeneous process.[†]

Most metal ions have water exchange rates larger than 10^5 and then their adsorption rates are characterized by $k_{ads} > 10^{0.5}$ M^{-1} s^{-1}. Thus, with 10^{-3} M surface ligands, the half-time for adsorption, $t'_{1/2}$, to an uncharged surface would be $t'_{1/2} < 4$ min. However, Cr^{3+}, VO^{2+}, Co^{3+}, Fe^{3+}, and Al^{3+} have lower exchange rates (see Figure 13.3; the water exchange rates for Co^{3+} and Cr^{3+}, which are not given in the figure, are $k_{-w} = 10^{-1}$ and $10^{-5.6}$ s^{-1}, respectively). These ions, if nonhydrolyzed, are probably adsorbed inner-spherically rather slowly. But we might recall that hydroxo species can exchange water by two or three orders of magnitude faster than "free" aquo ions. Experiments by Wehrli et al. (1990) have shown that adsorption rates of V(IV) increase with pH (in the pH range 3–5). This increase can be accounted for by assuming that $VO(OH)^+$ is the reacting adsorbate. The half-time for the adsorption of $VO(OH)^+$ onto 10^{-3} M surface ligands obtained from experimental data is on the order of 4 h. This is in good agreement with the rate predicted on the basis of the dissociative mechanism. Cr(III) was observed to adsorb to $Al_2O_3(s)$ above pH = 5. Under these conditions, dihydroxo and possibly multimeric hydroxo species are formed that are adsorbed faster.

Example 13.1. Adsorption Rate of Co^{2+} on Al_2O_3 Calculate the half-time for the adsorption of Co^{2+} on Al_2O_3 at pH = 7. We use for Al_2O_3 the same equilibrium constants as those given before and $I = 0.1$:

[†]The kinetics given imply that intraparticle mass transfer is fast and not rate controlling.

13.2 Kinetics of Adsorption

$$\equiv AlOH_2^+ \rightleftharpoons \equiv AlOH + H^+ \quad \log K = -7.4 \quad \text{(i)}$$

$$\equiv AlOH \rightleftharpoons \equiv AlO^- + H^+ \quad \log K = -10.0 \quad \text{(ii)}$$

and assume a specific surface area for the aluminum oxide of $s = 10^4$ m^2 kg^{-1} and 10^{-1} mol surface sites kg^{-1} and use a suspension $a = 10^{-3}$ kg/liter^{-1}; that is, there are 10^{-4} mol surface sites per liter. At pH = 7, the surface charge is approximately $[\equiv AlOH_2^+] = 1 \times 10^{-5}$ M; k_{ads} for Co^{2+} = 1.2×10^1 M^{-1} s^{-1} according to Figure 13.3.

We can now calculate the outer-sphere complex formation constant according to equation 8:

$$K_{OS} = \exp\left(-\frac{ZF\psi}{RT}\right) \quad \text{(iii)}$$

We calculate ψ for $I = 0.1$ according to the diffuse double-layer theory (equation 40c in Chapter 9). σ_P is calculated from $[\equiv AlOH_2^+]$ as follows:

$$\sigma_P = \frac{[\equiv AlOH_2^+]F}{a \times s}$$

$$= \frac{1 \times 10^{-5} \text{ mol liter}^{-1} \times 96485 \text{ C mol}^{-1}}{10^{-3} \text{ kg liter}^{-1} \; 10^4 \text{ m}^2 \text{ kg}^{-1}} \quad \text{(iv)}$$

$$= 0.095 \text{ C m}^{-2}$$

From this value we compute ψ by equation 40c in Chapter 9:

$$\sinh(z\psi \times 19.46) = \frac{\sigma_P}{0.1174 \sqrt{I}} = \frac{0.095}{0.1174 \times 3.16 \times 10^{-1}} = 2.56$$

$$z\psi \times 19.46 = 1.67 \quad \text{(v)}$$

$$\psi = 0.086 \text{ V}$$

K_{OS} can then be calculated according to equation iii, where $RT = 2.48 \times 10^3$ V C mol^{-1} (at 298.15 K):

$$K_{OS} = \exp\left(-\frac{2 \times 96485 \text{ C mol}^{-1} \times 0.086 \text{ V}}{2480 \text{ V C mol}^{-1}}\right) = 1.25 \times 10^{-3}$$

Accordingly, in line with equation 9, the rate is given by

$$-\frac{d[Co^{2+}]}{dt} = K_{OS} k_{ads}[Co^{2+}][\equiv Al-OH]$$

$$= 1.25 \times 10^{-3} \times 1.2 \times 10^{1} \times 10^{-4}[Co^{2+}] \quad \text{(vi)}$$

$$-\frac{1}{[Co^{2+}]}\frac{d[Co^{2+}]}{dt} = 1.5 \times 10^{-6} \text{ s}^{-1}$$

$$t_{1/2} = \ln 2/1.5 \times 10^{-6} \text{ s}^{-1} = 4.6 \times 10^{5} \text{ s}$$

$$= 128 \text{ h} \quad \text{(vii)}$$

This result illustrates that the rate of adsorption is decreased (by nearly three orders of magnitude in comparison to an uncharged surface) by the opposing charge of adsorbate and adsorbent.

Kinetics of Anion Adsorption A similar mechanism can be postulated for the kinetics of ligand exchange. Since the ligand exchanges with the central Me ion (Lewis acid) of the metal oxide, its water exchange rate may be rate determining. But the central ion exists already as an oxo, hydroxo species and thus the water exchange rate can be expected to be significantly larger than that of a nonhydrolyzed metal ion. But in this context it is interesting to note that the rate of OH^- exchange on α-Cr_2O_3 is known to be orders of magnitudes lower than that on goethite (α-FeOOH) (Yates and Healy, 1965). The extent and rate of NO_3^- adsorption by these two minerals are the same with equilibrium attained in minutes (outer-sphere adsorption). By contrast, the rate of phosphate adsorption is much lower with goethite (inner-sphere ligand exchange retarded because of slow H_2O exchange) than with α-Cr_2O_3, although the extent of adsorption is larger on goethite than on α-Cr_2O_3.

Reversibility of Adsorption

Although adsorption and desorption together establish an adsorption equilibrium, desorption has received relatively little experimental attention.

If adsorption occurs as a bimolecular reaction, as suggested by the Langmuir equation,

$$S + A \underset{k_b}{\overset{k_f}{\rightleftharpoons}} SA \quad (11)$$

One can postulate

$$\frac{d[SA]}{dt} = k_f[S][A] - k_b[SA] \quad (12)$$

and, at equilibrium, $d[SA]/dt = 0$,

13.2 Kinetics of Adsorption

$$\frac{[SA]}{[S][A]} = \frac{k_f}{k_b} = K = K_{ads} \qquad (13)$$

If one of the kinetic constants has been determined and K_{ads} is known, the other rate constant can be calculated.

The principle we have applied here is known as *microscopic reversibility* or the principle of detailed balancing. It shows that there is a link between kinetic rate constants and thermodynamic equilibrium constants. Obviously, equilibrium is not characterized by the cessation of processes; at equilibrium the rates of forward and reverse *microscopic* processes are equal for every *elementary* reaction step. The microscopic reversibility (which is routinely used in homogeneous solution kinetics) applies also to heterogeneous reactions (adsorption, desorption; dissolution, precipitation).

In the application of the principle of microscopic reversibility, we have to be careful. *We cannot apply this concept to overall reactions.* Even equations 11–13 cannot be applied unless we know that other reaction steps (e.g., diffusional transport) are not rate controlling. In a given chemical system there are many elementary reactions going on simultaneously. Rate constants are path dependent (which is not the case for equilibrium constants) and may be changed by catalysts. For equilibrium to be reached, *all elementary* processes must have equal forward and reverse rates and *all* species, not just reactive intermediates, must be at steady state (Lasaga, 1981; Lasaga et al., 1984).

Exemplification: Surface Complex Formation

Hayes and Leckie (1986) postulate the following mechanism on the basis of their pressure-jump relaxation experiments on the adsorption–desorption of Pb^{2+} at the goethite–water interface:

$$SOH + Pb^{2+} \underset{k_{-1}}{\overset{k_1}{\rightleftharpoons}} SOPb^+ + H^+ \qquad (14)$$

where k_1 and k_{-1} represent the forward and reverse rate constants. Defining intrinsic rate constants as rate constants that would be observed in the absence of an electric field, the equilibrium is represented by

$$K_1(int) = \frac{k_1(int)}{k_{-1}(int)} = \frac{k_1}{k_{-1}} \exp\left(\frac{F\psi_0}{RT}\right) \qquad (15)$$

The principle of the pressure-jump method is based on the pressure dependence of the equilibrium constant; that is,

$$\left(\frac{\partial \ln K}{\partial P}\right)_T = -\frac{\Delta V}{RT} \qquad (16)$$

where ΔV is the standard molar volume change of the reaction, and P the pressure.

A pressure perturbation results in the shifting of the equilibrium; the return of the system to the original equilibrium state (i.e., the relaxation) is related to the rates of all elementary reaction steps. The relaxation time constant associated with the relaxation can be used to evaluate the mechanism of the reaction. During the shift in equilibrium (due to pressure-jump and relaxation) the composition of the solution changes and this change can be monitored, for example, by conductivity. A description of the pressure-jump apparatus with conductivity detection and the method of data evaluation is given by Hayes and Leckie (1986).

A representative result of their data on Pb adsorption on goethite is given in Table 13.1.

The Elovich Equation

This equation has been used, especially in soil science, to describe the kinetics of adsorption and desorption on soils and soil minerals:

$$\frac{d\Gamma}{dt} = k_1 \exp(-k_2\Gamma) \tag{17}$$

The solution of equation 17 (Sposito, 1984) is

$$\Gamma(t) = \frac{1}{k_2} \ln(k_1 k_2 t_0) + \frac{1}{k_2} \ln\left(1 + \frac{t}{t_0}\right) \quad (t \geq t_c) \tag{18}$$

where

$$t_0 = \frac{\exp(k_2 \Gamma_c)}{k_1 k_2} - t_c \quad (t_c \geq 0)$$

Table 13.1. Intrinsic Rate Constants for the Reaction

$$\equiv\text{FeOH} + \text{Pb}^{2+} \underset{k_{-1}}{\overset{k_1}{\rightleftharpoons}} \equiv\text{FeOPb}^+ + \text{H}^+$$

ΔP (atm)	k_1 (int) ($M^{-1} s^{-1}$)	k_{-1} (int) ($M^{-1} s^{-1}$)	K_1 (int)
140	1.7×10^5	4.2×10^2	4.0×10^2
100	2.4×10^5	6.1×10^2	4.0×10^2
70	3.6×10^5	8.9×10^2	4.0×10^2

Source: Hayes and Leckie (1986).

13.3 Surface-Controlled Dissolution of Oxide Minerals

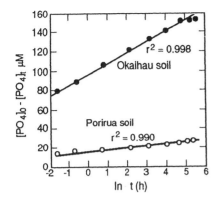

Figure 13.4. Adsorption kinetics of phosphate on soils or soil minerals: plot of Elovich equation. (From Chien and Clayton, 1980.)

Γ_c is the value of Γ at time t_c, the time at which the rate of sorption begins to be described by equation 17.

Although equation 18 can be derived as a general rate law expression under the assumption of an exponential decrease in number of available sorption sites with Γ and/or a linear increase in activation energy of sorption with Γ, the Elovich equation is perhaps best regarded as an empirical one for the characterization of rate data (Sposito, 1984).

As an example, equation 18 appears to describe phosphate sorption by soils quite well (see Figure 13.4).

13.3 SURFACE-CONTROLLED DISSOLUTION OF OXIDE MINERALS: AN INTRODUCTION TO WEATHERING

Chemical Weathering[†]

Chemical weathering is one of the major processes controlling the global hydrogeochemical cycle of elements. In this cycle, water operates as both a reactant and a transporting agent of dissolved and particulate components from land to sea. The atmosphere provides a reservoir for carbon dioxide and for oxidants required in the weathering reactions. The biota assists the weathering processes by providing organic ligands and acids and by supplying increased CO_2 concentrations, locally through decomposition.

During chemical weathering, rocks and primary minerals become transformed to solutes and soils and eventually to sediments and sedimentary rocks.

When a mineral dissolves, several successive elementary steps may be involved:

1. Mass transport of dissolved reactants from bulk solution to the mineral surface.

[†]In writing this section, we depended largely on a review by Stumm and Wollast (1990) and on Stumm (1992).

2. Adsorption of solutes.
3. Interlattice transfer of reacting species.
4. Chemical reactions.
5. Detachment of reactants from the surface.
6. Mass transport into the bulk of the solution.

Under natural conditions the rates of dissolution of most minerals are too slow to depend on mass transfer of the reactants or products in the aqueous phase. One can thus restrict the discussion to weathering reactions where the rate-controlling mechanism is the mass transfer of reactants and products in the solid phase or to reactions controlled by a surface process and the related detachment process of reactants.

Calcareous minerals and evaporite minerals (halides, gypsum) are very soluble and dissolve rapidly and, in general, congruently (i.e., yielding upon dissolution the same stoichiometric proportions in the solution as the proportions in the dissolving mineral and without forming new solid phases). Their contribution to the total dissolved load in rivers can be estimated by considering the mean composition of river water and the relative importance of various rocks to weathering (Garrels and Mackenzie, 1971). Recent estimates (Holland, 1978; Meybeck, 1979; Wollast and Mackenzie, 1983) indicate that evaporites and carbonates contribute approximately 17% and 38%, respectively, of the total dissolved load in the world's rivers. The remaining 45% is due to the weathering of silicates, underlining the significant role of these minerals in the overall chemical denudation of the earth's surface.

There are no unequivocal weathering reactions for the silicate minerals. Depending on the nature of parent rocks and hydraulic regimes, various secondary minerals like gibbsite, kaolinite, smectites, and illites are formed as reaction products. Some important dissolution processes of silicates are given, for example, by the following reactions:

$$\underset{\text{anorthite}}{CaAl_2Si_2O_8} + 2\ CO_2 + 3\ H_2O = \underset{\text{kaolinite}}{Al_2Si_2O_5(OH)_4} + Ca^{2+} + 2\ HCO_3^- \tag{19a}$$

$$7\ \underset{\text{albite}}{NaAlSi_3O_8} + 6\ CO_2 + 26\ H_2O = 3\ \underset{\text{montmorillonite}}{Na_{1/3}AL_{7/3}Si_{11/3}O_{10}(OH)_2} + 6\ Na^+ + 6\ HCO_3^-$$

$$+ 10\ H_4SiO_4 \tag{19b}$$

$$\underset{\text{biotite}}{KMgFe_2AlSi_3O_{10}(OH)_2} + \tfrac{1}{2} O_2 + 3\ CO_2 + 11\ H_2O = \underset{\text{gibbsite}}{Al(OH)_3}$$

$$+ 2\ \underset{\text{Fe(III) (hydr)oxide}}{Fe(OH)_3} + K^+ + Mg^{2+} + 3\ HCO_3^- + 3\ H_4SiO_4 \tag{19c}$$

In all cases, water and carbonic acid, which is the source of protons, are the main reactants. The net result of the reaction is the release of cations (Ca^{2+},

Mg^{2+}, K^+, Na^+) and the production of alkalinity via HCO_3^-. When ferrous iron is present in the lattice, as in biotite, oxygen consumption may become an important factor affecting the weathering mechanism and the rate of dissolution.

These reactions, however, are complex and generally proceed through a series of reaction steps. The rate of weathering of silicates may vary considerably depending on the arrangement of the silicon tetrahedra in the mineral and on the nature of the cations.

The Role of Weathering in Geochemical Processes in Oceanic and Global Systems As indicated in the stoichiometric equations given above, the rate of chemical weathering is important in determining the rate of CO_2 consumption. Furthermore, the global weathering rate is most likely influenced by temperature and is proportional to total continental land area and the extent of its coverage with vegetation; the latter dependence results from CO_2 production in soils, which is a consequence of plant respiration and the decay of organic matter as well as the release of complex-forming substances such as dicarboxylic acids, hydroxycarboxylic acids, and phenols, that is, anions that form soluble complexes with cations that originate from the lattice or form surface complexes with the surface of oxide minerals. Regionally, the extent of weathering is influenced by acid rain (Schnoor, 1990; Schnoor and Stumm, 1985; Warfinge and Sverdrup, 1992).

It has been shown by Berner et al. (1983) and by Berner and Lasaga (1989) that silicate weathering is more important than carbonate mineral weathering as a long-term control on atmospheric CO_2. The HCO_3^- and Ca^{2+} ions produced by weathering of $CaCO_3$,

$$CaCO_3 + CO_2 + H_2O = Ca^{2+} + 2\ HCO_3^- \tag{20}$$

precipitate in the ocean (through incorporation by marine organisms) as $CaCO_3$. The CO_2 consumed in carbonate weathering is released again upon formation of $CaCO_3$ in the ocean (reversal of the reaction given above). Thus, globally, carbonate weathering results in no net loss of atmospheric CO_2. The weathering of calcium silicates, for example, equation 19a or, in a simplified way,

$$CaSiO_3 + 2\ CO_2 + 3\ H_2O = Ca^{2+} + 2\ HCO_3^- + H_4SiO_4 \tag{21}$$

also produces Ca^{2+} and HCO_3^-, which form $CaCO_3$ in the sea, but only half of the CO_2 consumed in the weathering is released and returned to the atmosphere upon $CaCO_3$ formation. Thus silicate weathering results in a net loss of atmospheric CO_2. Of course, ultimately, the cycle is completed by metamorphic and magmatic breakdown, deep in the earth, of $CaCO_3$ with the help of SiO_2, a reaction that may be represented in a simplified way as $CaCO_3 + SiO_2 = CaSiO_3 + CO_2$. Knowledge of the rate of dissolution of minerals is also necessary for the quantitative evaluation of geochemical processes in the oceanic system.

Diffusion-Controlled Versus Surface-Controlled Mechanisms

Among the theories proposed, essentially two main mechanisms can be distinguished: these are either that the rate-determining step is a transport step (e.g., a transport of a reactant or a weathering product through a layer of the surface of the mineral) or that the dissolution reaction is controlled by a surface reaction. The rate equation corresponding to a transport-controlled reaction is known as the parabolic rate law when

$$r = \frac{dC}{dt} = k_p t^{-1/2} \; (\text{M s}^{-1}) \tag{22}$$

where k_p is the reaction rate constant (M s$^{-1/2}$). By integration the concentration in solution, C (M), increases with the square root of time:

$$C = C_0 + 2 k_p t^{1/2} \tag{23}$$

The Surface-Controlled Rate of Dissolution

Alternatively, if the reactions at the surface are slow in comparison with diffusion or other reaction steps, the dissolution processes are controlled by the processes at the surface. In this case, the concentrations of solutes adjacent to the surface will be the same as in the bulk solution. The dissolution kinetics follow a zero-order rate law if the steady-state conditions at the surface prevail:

$$r = \frac{dC}{dt} = kA \; (\text{M s}^{-1}) \tag{24}$$

where the dissolution rate r (M s^{-1}) is proportional to the surface area of the mineral, A (m^2); k is the reaction rate constant (M m^{-2} s^{-1}). Figure 13.5 compares the two control mechanisms. Since most important dissolution reactions are surface controlled, we will concentrate on the kinetics of surface-controlled reactions.

Figure 13.6a shows examples of the results obtained on the dissolution of δ-Al$_2$O$_3$. In batch experiments where pH is kept constant with an automatic titrator, the concentration of Al(III)(aq) (resulting from the dissolution) is plotted as a function of time. The linear dissolution kinetics observed for every pH are compatible with a process whose rate is controlled by a surface reaction. The rate of dissolution is obtained from the slope of the plots.

Figure 13.6b gives results by Schott and Berner (1985) on the dissolution rate of iron-free pyroxenes and olivines, as measured by the silica release.

Figure 13.7 shows the reactivity of the surface; that is its tendency to dissolve depends on the type of surface species present. Surface protonation tends to increase the dissolution rate, because it leads to highly polarized interatomic bonds in the immediate proximity of the surface central ions and thus facilitates the detachment of a cationic surface group into the solution. On the other hand,

13.3 Surface-Controlled Dissolution of Oxide Minerals

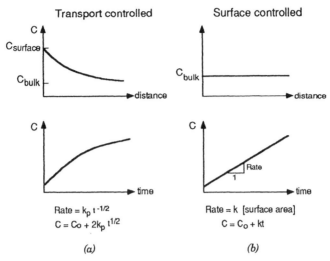

Figure 13.5. Transport vs surface controlled dissolution. Schematic representation of concentration in solution, C, as a function of distance from the surface of the dissolving mineral. In the lower part of the figure, the change in concentration (e.g., in a batch dissolution experiment) is given as a function of time. (a) Transport controlled dissolution. The concentration immediately adjacent to the mineral reflects the solubility equilibrium. Dissolution is then limited by the rate at which dissolved dissolution products are transported (diffusion, advection) to the bulk of the solution. Faster dissolution results from increased flow velocities or increased stirring. The supply of a reactant to the surface may also control the dissolution rate. (b) Pure *surface controlled dissolution* results when detachment from the mineral surface via surface reactions is so slow that concentrations adjacent to the surface build up to values essentially the same as in the surrounding bulk solution. Dissolution is not affected by increased flow velocities or stirring. A situation, intermediate between (a) and (b)—a mixed transport-surface reaction controlled kinetics—may develop.

a surface coordinated metal ion (e.g., Cu^{2+} or Al^{3+}) may block a surface group and thus retard dissolution. An outer-sphere surface complex has little effect on the dissolution rate. Changes in the oxidation state of surface central ions have a pronounced effect on the dissolution rate.

An inner-sphere surface complex with a ligand such as that shown for oxalate

$$\begin{array}{c}\diagdown\\ Me\\ \diagup\end{array}\!\!\begin{array}{c}OH_2\\ \\ OH\end{array} + C_2O_4^{2-} + H^+ \rightleftharpoons \begin{array}{c}\diagdown\\ Me\\ \diagup\end{array}\!\!\begin{array}{c}O-C\!=\!O\\ |\\ O-C\!=\!O\end{array} + 2\,H_2O$$

or other dicarboxylates, dihydroxides, or hydroxycarboxylic acids,

$$R\!\!\begin{array}{c}\diagup COOH\\ \diagdown COOH\end{array}\!,\quad R\!\!\begin{array}{c}\diagup OH\\ \diagdown OH\end{array}\!,\quad \text{or}\quad R\!\!\begin{array}{c}\diagup COOH\\ \diagdown OH\end{array}$$

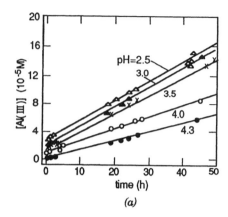

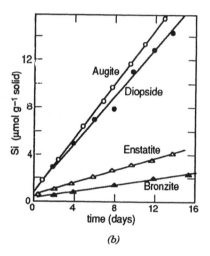

Figure 13.6. (a) Linear dissolution kinetics observed for the dissolution of δ-Al$_2$O$_3$, representative of processes whose rates are controlled by a surface reaction and not by a transport step. (Data from Furrer and Stumm, (1986).) (b) Linear dissolution kinetics of frame silicates. Minerals used were pyroxenes and olivines; their essential structural feature is the linkage of SiO$_4$ tetrahedra, laterally linked by bivalent cations (Mg^{2+}, Fe^{2+}, Ca^{2+}). Plotted are amounts of silica released versus time for the dissolution of etched enstatite, bronzite (p_{O_2} = 0), diopside, and augite at pH 6 (T = 20°C for bronzite; T = 50°C for the other minerals). (From Schott and Berner, 1985.)

facilitates the detachment of a central metal ion and enhances the dissolution. This is readily understandable, because the ligands shift electron density toward the central metal ion at the surface and bring negative charge into the coordination sphere of the Lewis acid center and enhance simultaneously the surface protonation and can labilize the critical Me–oxygen lattice bonds, thus enabling the detachment of the central metal ion into solution.

13.4 SIMPLE RATE LAWS IN DISSOLUTION

In the dissolution reaction of an oxide mineral, the coordinative environment of the metal changes; for example, in dissolving an aluminum oxide layer, the Al^{3+} in the crystalline lattice exchanges its O^{2-} ligand for H$_2$O or another ligand L. In line with Figure 13.7, the most important reactants participating

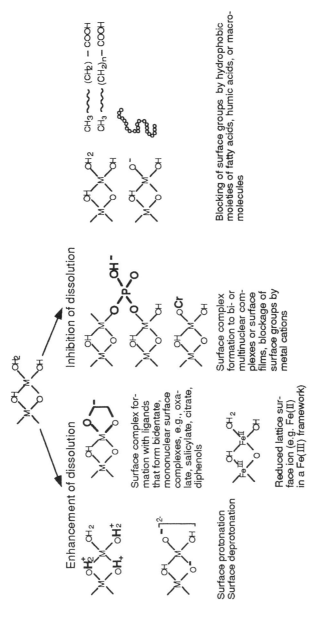

Figure 13.7. Effect of protonation, complex formation with ligands and metal ions and reduction on dissolution rate. The structures given here are schematic short hand notations to illustrate the principal features (they do not reveal the structural properties nor the coordination numbers of the oxides under consideration; charges given are relative).

778   Kinetics at the Solid–Water Interface

in the dissolution of a solid mineral are H_2O, H^+, OH^-, ligands (surface complex building), and reductants and oxidants (in the case of reducible or oxidizable minerals). The following simple arguments on the kinetics of the dissolution reaction depend on the model assumptions given in Table 13.2.

Thus the reaction occurs schematically in two sequences:

$$\text{surface sites} + \text{reactants } (H^+, OH^-, \text{ or ligands}) \xrightarrow{\text{fast}} \text{surface species} \quad (25)$$

$$\text{surface species} \xrightarrow[\text{detachment of Me}]{\text{slow}} \text{Me(aq)} \quad (26)$$

where Me stands for metal. Although each sequence may consist of a series of smaller reaction steps, the rate law of surface-controlled dissolution is based on the idea (1) that the attachment of reactants to the surfaces sites is fast and (2) that the subsequent detachment of the metal species from the surface of the crystalline lattice into the solution is slow and thus rate limiting.

In the first sequence, the dissolution reaction is initiated by the surface coordination with H^+, OH^-, and ligands that polarize, weaken, and tend to break the metal–oxygen bonds in the lattice of the surface. Since reaction 26 is rate limiting, using a steady-state approach, the rate law on the dissolution reaction will show a dependence on the concentration (activity) of the particular

Table 13.2. Model Assumptions

1. *Dissolution of slightly soluble hydrous oxides:*
   Surface process is rate-limiting
   Back reactions can be neglected if far away from equilibrium
2. *The hydrous oxide surface, as a first approximation, is treated like a cross-linked polyhydroxo–oxo acid:*
   All functional groups are treated as if they were identical (mean field statistics)
3. *Steady state of surface phase:*
   Constancy of surface area
   Regeneration of active surface sites
4. *Surface defects, such as steps, kinks, and pits, establish surface sites of different activation energy, with different rates of reaction:*

$$\text{active sites} \xrightarrow{\text{faster}} \text{Me(aq)} \quad (a)$$

$$\text{less active sites} \xrightarrow{\text{slower}} \text{Me(aq)} \quad (b)$$

   Overall rate is given by reaction (a) steady-state condition can be maintained if a constant mole fraction, χ_a, of active sites to total (active and less active) sites is maintained, that is, if active sites are continuously regenerated
5. *Precursor of activated complex:*
   Metal centers bound to surface chelate or surrounded by n protonated functional groups

$$(C_H^s/s) \ll 1; \, (C_L^s/s) \ll 1 \, (s = \text{number of sites})$$

surface species, C_j (mol m^{-2}):

$$\text{dissolution rate} \propto \langle \text{surface species} \rangle \quad (27a)$$

We reach the same conclusion (equation 27a) if we treat the reaction sequence according to the activated complex theory (ACT), often also called the transition state theory. The particular surface species that has formed from the interaction of H$^+$, OH$^-$, or ligands with surface sites is the *precursor* of the activated complex (Figure 13.8):

$$\text{dissolution rate} \propto \langle \text{precursor of the activated complex} \rangle \quad (27b)$$

$$R = kC_j \quad (27c)$$

where R is the dissolution rate (e.g., in mol m^{-2} h^{-1}) and C_j is the surface concentration of the precursor (mol m^{-2}).

The surface concentration of the particular surface species, C_j, corresponds to the concentration of the precursor of the activated complex. Note that we use braces $\langle \ \rangle$ and brackets [ ] to indicate surface concentrations (mol m^{-2}) and solute concentrations (M), respectively.

Thus the following simple *rate laws* describe the dissolution of oxides and silicates.

Proton promoted Dissolution:

$$R_H = k_H \langle \equiv \text{MeOH}_2^+ \rangle^j \quad (28)$$

Basic Dissolution:

$$R_{OH} = k_{OH} \langle \equiv \text{MeO}^- \rangle^i \quad (29)$$

Ligand-Promoted Dissolution:

$$R_L = k_L \langle \equiv \text{MeL} \rangle \quad (30)$$

Reductive Dissolution:

$$R_R = k_R \langle \equiv \text{MeR} \rangle \quad (31)$$

where R is the dissolution rate (in mol m^{-2} time^{-1}) and the k_i are the rate constants. $\langle \equiv \text{MeOH}_2^+ \rangle$, $\langle \equiv \text{MeO}^- \rangle$, $\langle \text{MeL} \rangle$, and $\langle \equiv \text{MeR} \rangle$ are the surface concentrations of, respectively, protons, hydroxide ions (deprotonated surface groups), ligands, and reductants (in mol m^{-2}). $\langle \equiv \text{MeOH}_2^+ \rangle$ and $\langle \equiv \text{MeO}^- \rangle$ correspond to surface protonation and surface deprotonation; that is, the "adsorbed" protons or "desorbed" protons, j and i are exponents where j in the ideal case corresponds to the charge of the central metal ion [i.e., 3 for Al$_2$O$_3$

and Fe(OH)$_3$, 2 for CuO and BeO]. The surface concentrations in equations 27–31 can be determined experimentally; surface protonation and surface deprotonation can be determined by alkalimetric or acidimetric titration of a suspension of the solid phase; the surface concentration of ligands and reductants can be determined analytically (difference of the concentrations before and after adsorption).

Mean Field Statistics and Steady State

However, we need to reflect on one of our model assumptions (Table 13.2). It is certainly not justified to assume a completely uniform oxide surface. The dissolution is favored at a few localized (active) sites where the reactions have lower activation energy. The overall reaction rate is the sum of the rates of the various types of sites. The reactions occurring at differently active sites are parallel reaction steps occurring at different rates (Table 13.2). In parallel reactions, the fast reaction is rate determining. We can assume that the ratio (mole fraction, χ_a) of active sites to total (active plus less active) sites remains constant during the dissolution; that is, the active sites are continuously regenerated after Al(III) detachment and thus steady-state conditions are maintained. Thus a mean field rate law can generalize the dissolution rate. The reaction constant k_L in equation 30 includes χ_a, which is a function of the particular material used (see remark 4 in Table 13.2). In the activated complex theory the surface complex is the precursor of the activated complex (Figure 13.8) and is in local equilibrium with it. The detachment corresponds to the desorption of the activated surface complex.

The postulate of steady state during dissolution reaction (Table 13.2) implies

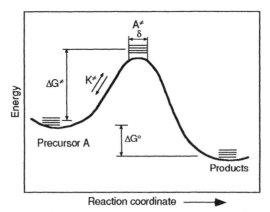

Figure 13.8. Activated complex theory for the surface-controlled dissolution of a mineral far from equilibrium. A is the precursor, that is, a surface site that can be activated to $A^\ddagger$. The latter is in equilibrium with the precursor. The activation free energy for the conversion of the precursor into the product is given by $\Delta G^\ddagger$.

a continuous reconstitution of the surface with the maintenance of a constant distribution of the various surface sites and the steady-state concentration of the surface complexes.

The Overall Rate of Dissolution

The surface charge of the mineral is an important factor in the polarization of the lattice bonds on the surface. Thus one may generalize that the dissolution rate is related to the surface charge imparted to the surface by H^+ and/or OH^-; the rate increases both with increasing positive surface charge with decreasing pH values of the solution and with increasing negative surface charge with increasing pH values. The minimum dissolution rates are observed at the pH_{pzc}.

The overall rate of dissolution is given by the sum of the individual reaction rates, assuming that the dissolution occurs in parallel at different metal centers (Furrer and Stumm, 1986).

Dissolution of Silicates

The mechanism of the dissolution of silicates is similar to that of oxides. The surface protonation of O and OH lattice sites accelerates the dissolution with decreasing pH. Figure 13.9 gives a brief survey on the dissolution rates of some minerals. A pH effect is observed in all cases. There are dramatic differences in the dissolution rates (at pH = 6, four to five orders of magnitude) between carbonates and silicates. Similarly, the *reductive dissolution* of Fe(III) (hydr)oxides (e.g., by H_2S) is faster than the nonreductive dissolution by orders of magnitudes (Biber et al., 1994).

With feldspars and layer silicates (Brady and Walther, 1992) the points of attack of the protons are the O atoms that interlink the Al-oxide groups with the Si-oxide structures. The protonation causes a slow detachment of Al from the crystal lattice, which is coupled with the subsequent detachment of $Si(OH)_4$ species. Oxalates, diphenols, and citric acid—similar substances that occur in soils as by-products of biological decomposition and root exudates—can also accelerate the dissolution of Al silicates.

Note that quartz shows a pH dependence in its dissolution rate although the overall reaction for the dissolution of $SiO_2(s)$,

$$SiO_2(s) + 2\ H_2O = Si(OH)_4(aq)$$

does not involve H^+ or OH^- ions (Figure 13.9).

The pH_{pznpc} is around pH 2–3. Both the positive surface charge, due to bound protons, and the negative surface charge, due to deprotonation (equivalent to bound OH^-), enhance the dissolution rate. The same kind of pH dependence is observed also with silicates.

For more data and interpretation see Brady and Walther (1992). For interesting generalizations on the dissolution of silicates, see Casey and Westrich

782   Kinetics at the Solid-Water Interface

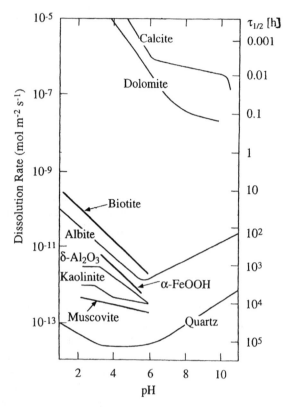

Figure 13.9. Dissolution rates of different minerals as a function of pH (25°C). The general patterns shown are in accord with typical experimental observations. Details of different investigations of course vary. For data see:

Calcite and Dolomite:	Fluidized-bed reactor L. Chou, R. M. Garrels, and R. Wollast, *Chem. Geol.* **78,** 269 (1989);
Biotite:	F. C. Lin and C. Clemency, *Clay Clay Minerals* **29,** 101 (1981);
α-FeOOH:	B. Zinder et al., *Geochim. Cosmochim. Acta* **50,** 1861 (1986);
δ-Al$_2$O$_3$:	(I = 0.1 M NaNO$_3$) G. Furrer and W. Stumm, *Geochim. Cosmochim. Acta* **50,** 1847 (1986);
Kaolinite, Muscovite:	W. Stumm and E. Wieland in *Aquatic Chemical Kinetics,* Wiley, New York (1990);
Quartz:	P. Brady and J. Walther, *Chem. Geol.* **82,** 253 (1990);
Albite ("low albite"):	L. Chou and R. Wollast, *Amer. J. Science* **285,** 963 (1985).

For the calculation of the half-life (right scale) 10 surface groups per nm^2 were assumed. (Modified from Sigg and Stumm, 1994.)

(1992), Sverdrup (1990), Sverdrup and Warfvinge (1993), and Lasaga et al. (1994). Casey et al. (1993) discuss the relevant question: "What do dissolution experiments tell us about natural weathering?" Xiao and Lasaga (1994) have carried out ab-initio quantum mechanical studies of the kinetics and mechanisms of silicate dissolution. Their studies indicate that the attack of H^+ on the bridging oxygen site significantly weakens the Si—O—Si and Si—O—Al bridging bond.

Figure 13.10 gives a general scheme explaining in a simplified way the various mechanisms of the dissolution reaction of an oxide. Figure 13.11 shows a set of representative experimental data and how these data are interpreted. A

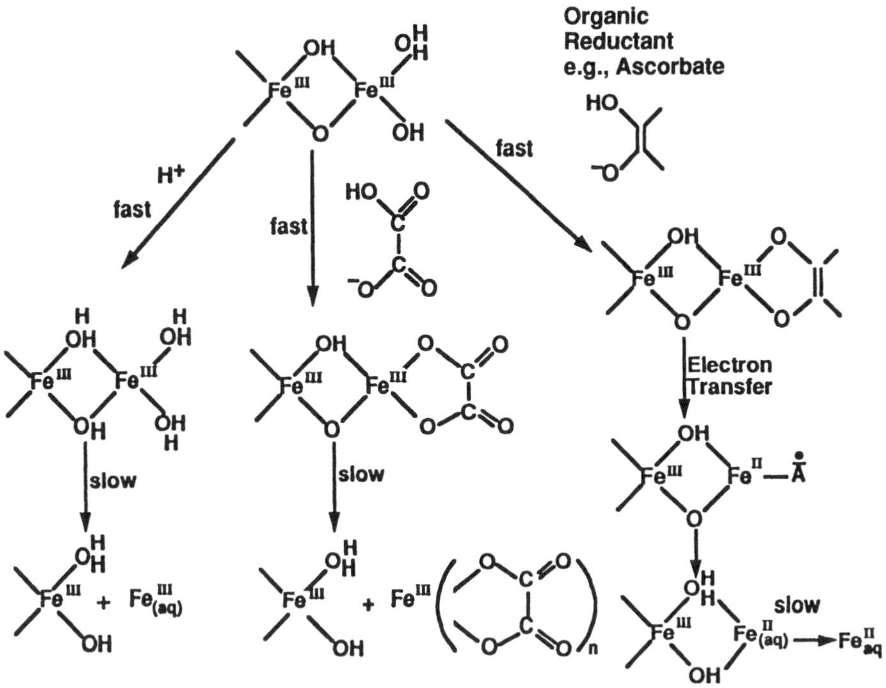

Figure 13.10. Schematic representation of the oxide dissolution processes [exemplified for Fe(III) (hydr)oxides] by acids (H^+ ions), ligands (example oxalate), and reductants (example ascorbate). In each case a surface complex (proton complex, oxalato and ascorbato surface complex) is formed, which influences the bonds of the central Fe ions to O and OH on the surface of the crystalline lattice, in such a way that a slow detachment of a Fe(III) aquo or a ligand complex [in case of reduction an Fe(II) complex] becomes possible. In each case the original surface structure is reconstituted, so that the dissolution continues (steady-state condition). In the redox reaction with Fe(III), the ascorbate is oxidized to the ascorbate radical $A^{\cdot-}$. The principle of proton-promoted and ligand-promoted dissolution is also valid for the dissolution (weathering) of Al-silicate minerals. The structural formulas given are schematic and simplified; they should indicate that Fe(III) in the solid phase can be bridged by O and OH.

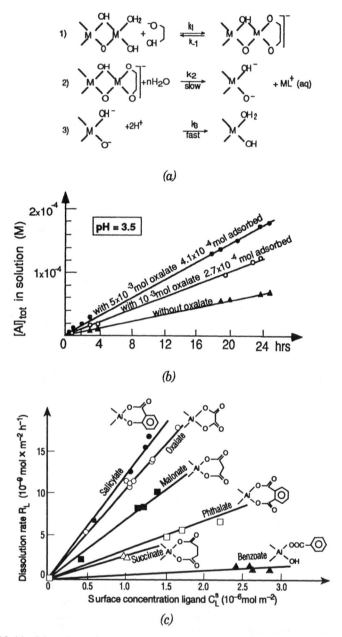

Figure 13.11. Ligand- and proton-promoted dissolution of Al_2O_3. (a) The ligand-catalyzed dissolution of a trivalent metal (hydr)oxide. (b) Measurement of $Al(III)(aq)$ as a function of time at constant pH at various oxalate concentrations. The dissolution kinetics are given by a reaction of zero order. The dissolution rate, R_L, is given by the slope of the $[Al(III)(aq)]$ versus time curve. (c) Dissolution rate as a function of the surface ligand concentration for various ligands. The dissolution is proportional to the surface concentration of the ligand, $\langle \equiv MeL \rangle$ or C_L^s ($R_L = k_L C_L^s$). (d) Proton-promoted

comprehensive treatment of the dissolution of metal oxides has been given by Blesa et al. (1994).

Table 13.3 gives a few data on laboratory dissolution rates and compares these data with information on field measurements on solute export in silicate terrain.

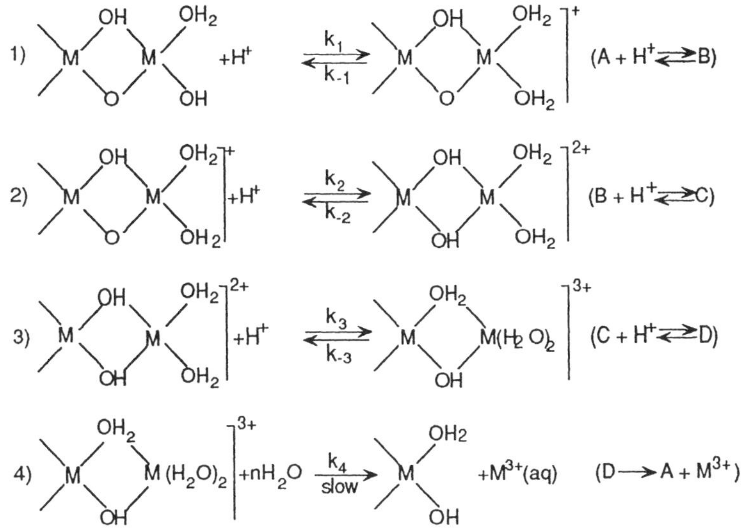

(d)

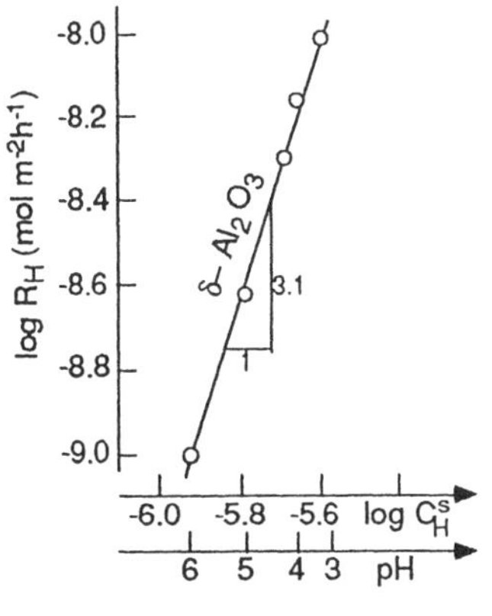

(e)

dissolution of an oxide. The fast surface protonation leads to a polarization of the lattice sites in the proximity of the surface metal center. The slow detachment is the rate-determining step. (e) The dissolution rate $d[\text{Al(III)(aq)}]/dt$ is constant for each pH. The dissolution rate, R_H (mol m^{-2} h^{-1}), is in this case proportional to the surface protonation to the third power; that is, $R_H = k_H \langle \equiv\text{AlOH}_2^+ \rangle^3 = k_H (C_H^s)^3$.

Table 13.3. Laboratory and Field Weathering Rates

Mineral	Laboratory Data Dissolution Rate pH = 4[a] (mol Si m^{-2} h^{-1})		Reference
Plagioclase	1.1×10^{-7}		Busenberg and Clemency, 1976
Olivine	1.5×10^{-7}		Busenberg and Clemency, 1976
Biotite	7×10^{-8}		Grandstaff, 1986
Kaolinite	1×10^{-8}		Stumm and Wieland, 1990
Muscovite	5×10^{-9}		Stumm and Wieland, 1990
Quartz	2×10^{-9}		Brady and Walther, 1990
δ-Al$_2$O$_3$	1×10^{-8}	mol Al(III)	Stumm and Wieland, 1990
CaCO$_3$ (calcite)	1×10^{-1}	mol Ca^{2+}	Chou, Garrels, and Wollast, 1989
Geographic Area	Representative Field Measurements on Cation Export from Silicate Terrain (eq cation m^{-2} (catchment) h^{-1})		Reference
Trnavka River Basin (CZ)	2.3×10^{-6}		Paces, 1983
Coweeta Watershed NC (USA)	4×10^{-6}		Velbel, 1985
Filson Creek MN (USA)	$1.7-3.4 \times 10^{-6}$		Siegel and Pfannkuch, 1984
Cristallina (CH)	$2.3-3.4 \times 10^{-6}$		Giovanoli et al., 1989
Bear Brooks ME (USA)	1.1×10^{-5}		Schnoor, 1990
Swiss Alps	$4.6-11.4 \times 10^{-5}$		Zobrist and Stumm, 1979

[a]Indicative of orders of magnitude, data by different authors vary.

It may be noted that the field data on a per m^2 catchment area basis from different geographical areas are remarkably similar ($10^{-2}-10^{-1}$ eq m^{-2} yr^{-1}). In order to estimate from these rates actual rates of dissolution of rocks on a per m^2 mineral surface area basis, we have to know how many m^2 of effective (active) mineral surface is available per m^2 geographic area. This is not known, but the estimate of 10^5 m^2 surface area of mineral grains active in weathering available per m^2 of geographic area was made (Schnoor, 1990).[†] If we assume such a conversion factor (10^5 m^2 per m^2 geographic area), weathering rates in nature would be significantly slower than predicted from laboratory studies. There are various reasons for this apparent discrepancy. In addition to the possibility that our conversion factor is off—possibly by one or two orders of magnitude—the following reasons could be given: (1) unsaturated flow through soil macropores limits the amount of weatherable minerals exposed to water and (2) dissolved Al(III) has been shown to inhibit the dissolution rates in soil macropores (hydrological control due to macropore flow through soils).

[†]Based on the following assumptions: 50 cm of saturated regolith (mantle rock), 0.5 m^2 g^{-1} surface area of mineral grains, 40% of mineral grains active in weathering, density of minerals ≈ 2.5 g cm^{-3}.

13.4 Simple Rate Laws in Dissolution

Example 13.2. How Long Does It Take to Dissolve One "Monomolecular" Layer? Estimate, on the basis of results on experimental weathering rates and the number of functional groups (or central ions) per unit area, how long it takes to dissolve one "monomolecular" layer of a representative silicate mineral.

There are about 10 functional groups per nm^2. That corresponds to 10^{19} groups or $S_T = 1.7 \times 10^{-5}$ mol groups per m^2. A representative weathering rate is 10^{-3}–10^{-4} mol m^{-2} yr^{-1}. Converting this in a pseudo-first-order rate law, we have

$$R'_H = \frac{\text{weathering rate (mol m}^{-2}\text{ yr}^{-1})}{\text{density of surface groups (mol m}^{-2})} \simeq 6\text{–}60 \text{ yr}^{-1} \qquad (i)^\dagger$$

This rough calculation shows that only a few monolayers are dissolved per year.

(The idea for this example comes from an article by B. Wehrli in *EAWAG News* 28/29, 1990.)

Experimental Apparatus

Various devices can be used to determine the kinetics and rates of chemical weathering. In addition to the batch pH-stats, flow-through columns, fluidized bed reactors, and recirculating columns have been used (Schnoor, 1990). Figure 13.12a illustrates the fluidized bed reactor pioneered by Chou and Wollast (1984) and further developed by Mast and Drever (1987). The principle is to achieve a steady-state solute concentration in the reactor (unlike the batch pH-stat, where solute concentrations gradually build up). Recycle is necessary to achieve the flow rate to suspend the bed and to allow solute concentrations to build to a steady state. With the fluidized bed apparatus, Chou andWollast (1984) could control the Al(III) concentration (which can inhibit the dissolution rate) to a low level at steady state by withdrawing the sample at a high rate.

Figure 13.12b shows a thin-film continuous flow reactor used by Bruno et al. (1991) for determining the dissolution rate of UO_2 under reducing conditions. A known weight of $UO_2(s)$ was enclosed in the reactor between two membrane filters (0.22 μm). The reducing conditions of the feed solution were obtained by bubbling $H_2(g)$ in the presence of a palladium catalyst. The dissolution rates determined using continuous flow reactors are based on the U(IV) concentration of the effluent at steady state. The amount of U(IV) dissolved depends on the reaction time, which is related to the residence time of the test solution in the reactor given by

$$t = V/Q$$

---

†The half-lifes of the functional groups in Figure 13.9 have been calculated by this approach.

788 Kinetics at the Solid–Water Interface

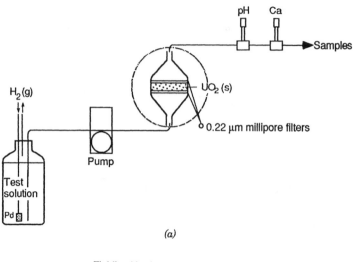

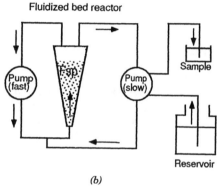

Figure 13.12. Continuous flow-type reactors to measure dissolution rates. (a) Experimental scheme of the thin-film continuous flow reactor used for example by Bruno et al. (1991) to determine dissolution rate of UO_2 under reducing conditions. (b) Schematic diagram of the fluidized-bed reactor by Chou and Wollast (1984), and developed further by Mast and Drever (1987).

where V is the volume of solution in contact with the solid phase, and Q is the flow rate. The dissolution rate values are calculated using the equation

$$r_{\text{diss}} = Q(\text{liter s}^{-1}) \times [\text{U(IV)}](M) = \text{mol s}^{-1}$$

13.5 RATES OF $CaCO_3$ DISSOLUTION (AND OF $CaCO_3$ CRYSTAL GROWTH)

Dissolution of carbonates can only occur if the solution is undersaturated with regard to the solid carbonate. The solubility equilibrium of carbonates and especially of calcite has been discussed extensively in Chapters 4 and 7.

13.5 Rates of CaCO₃ Dissolution (and of CaCO₃ Crystal Growth)

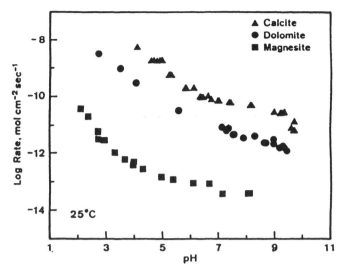

Figure 13.13. Influence of pH on the rate of dissolution of carbonates in the system CaCO$_3$–MgCO$_3$. These experiments were carried out with a fluidized-bed reactor with carbonate species controlled by the steady-state condition reactor in open systems with controlled CO$_2$. (Adapted from Chou et al., 1989.)

At very low pH, the rate of dissolution is so fast that the rate is limited by the transport of the reacting species between the bulk of the solution and the surface of the mineral (Berner and Morse, 1974). The rate can then be described in terms of transport (molecular or turbulent diffusion) of the reactants and products through a stagnant boundary layer. The thickness of this layer depends on the stirring and the local turbulence.

But within the pH range of natural waters, the dissolution (and precipitation) of carbonate minerals is surface controlled; that is, the rate of dissolution is the rate determined by a chemical reaction at the water–mineral interface. Figure 13.13 gives the data on the dissolution rates of various carbonate minerals in aqueous solutions obtained in careful studies by Chou et al. (1989).

The dissolution rates for calcite and aragonite have been described in terms of the following rate law (Busenberg and Plummer, 1986; Chou et al.,[†] 1989, Plummer et al., 1978):

$$\text{rate (mol cm}^{-2}\text{ s}^{-1}) = k_1[\text{H}^+] + k_2[\text{H}_2\text{CO}_3^*] + k_3 \qquad (32)$$

Figure 13.14 illustrates the pH–p_{CO_2} domains where the [H$^+$] dependence (first term on the right-hand side of equation 32, the p_{CO_2} dependence (second term), and the third term (H$_2$O) are primarily operative in the CaCO$_3$ dissolution.

The rate law of equation 32 has been interpreted in terms of the following

---

[†]These authors used activities instead of concentrations, but at constant ionic strength (constant ionic medium scale), the rate law can be written in terms of concentrations.

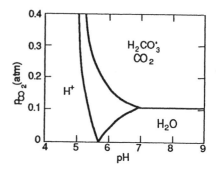

Figure 13.14. Predominance areas where the first, second or third term of equation (32) is important. In the triangle region all three terms have to be considered.

dissolution reactions, which occur as parallel reactions:

$$CaCO_3(s) + H^+ \underset{k_{-1}}{\overset{k_1}{\rightleftharpoons}} Ca^{2+} + HCO_3^- \qquad (33)$$

$$CaCO_3(s) + H_2CO_3^* \underset{k_{-2}}{\overset{k_2}{\rightleftharpoons}} Ca^{2+} + 2\,HCO_3^- \qquad (34)$$

$$CaCO_3(s) \underset{k_{-3}}{\overset{k_3}{\rightleftharpoons}} Ca^{2+} + CO_3^{2-} \qquad (35)$$

In the rate law of equation 32, the back reactions of 33–35 have been neglected because usually the system is sufficiently far from equilibrium.

Surface Reactions As we have seen from the dissolution of oxides, the surface-controlled dissolution mechanism would have to be interpreted in terms of surface reactions; one can reasonably expect that these reactions constitute the elementary steps in the formation of the surface-activated complex.

As we have seen with oxides and silicates, charge- (or potential) determining ions have great influence on the dissolution kinetics because—as their definition implies—they interact chemically with the surface. Charge-determining species for $CaCO_3(s)$ (cf. Section 9.5 and equation 44) are H^+, Ca^{2+}, HCO_3^-, and H_2CO_3.

Recent studies with the Atomic Force Microscope (Gratz et al., 1991) revealed interesting insight into calcite dissolution reactions at the near-atomic scale. Generally, surface dissolution was observed only at steps (both at pits and to a lesser extent at monomolecular ledges) and not in terraces. For salient features of surfaces (steps, kinks, edges), see Figure 13.27d. Dissolution occurred by removal of molecules from preexisting steps. These authors also observed simple growth sequence wherein $CaCO_3$ monolayers were deposited on the calcite surface by uniformly advancing ledges.

In performing experiments with $CaCO_3$, be it on dissolution or on crystal growth, it is essential that we know the variables or keep as many of them as

13.5 Rates of CaCO₃ Dissolution (and of CaCO₃ Crystal Growth)

constant as possible. Since these rates are dependent on pH and on surface charge, it is essential that we realize that the $CaCO_3$ in equilibrium with the solution at a given pH has different concentrations of H^+, Ca^{2+}, HCO_3^-, CO_3^{2-}, and $CO_2(aq)$ and different surface charge characteristics depending on how this pH has been adjusted. Much of the discrepancy in the literature can be accounted for by not considering whether the pH has been adjusted (1) in a closed system, (2) in an open system and constant p_{CO_2} by adding acid or base, or (3) by simply changing in a $CaCO_3$–H_2O–CO_2 system the p_{CO_2}. For example, in case (3) the surface charge, independent of pH, remains nearly zero (see Section 9.5 and Figure 9.23e,f), while in case (1) a decrease in $[H^+]$ will result in a concomitant increase in $[CO_3^{2-}]$ and decrease in $[Ca^{2+}]$. Thus, in the latter case, there are some compensating effects, in the sense that the two oppositely charged bivalent charge-determining ions change with pH according to

$$\frac{d[Ca^{2+}]}{d\,pH} = -\frac{d[CO_3^{2-}]}{d\,pH} = 1 \quad \text{while} \quad \frac{d[HCO_3^-]}{d\,pH} = 0 \qquad (36)$$

In the case of dolomite and magnesite dissolution, a dependence on a fractional order of $[H^+]$ has been observed, indicating that adsorption of charge-determining ions is involved.

Walter and Morse (1984) were able to document the relative importance of *microstructure* for the dissolution of biogenic carbonates. Biogenic magnesian calcites are structurally disordered and chemically heterogeneous. Both these factors play a role in the reactivity of these minerals in natural systems.

***The Crystal Growth of a CaCO₃* (Calcite)** The crystal growth of calcite has been studied by Plummer and et al. (1978), by Kunz and Stumm (1984), and by Chou et al. (1989) to correspond to the reverse of the rate of dissolution (equation 32):

$$R_{tot} = \frac{d[CaCO_3]}{dt} = k_{-3}[Ca^{2+}][CO_3^{2-}] + k_{-1}[Ca^{2+}][HCO_3^-] + (k_{-2}[Ca^{2+}][HCO_3^-]^2) \qquad (37)$$

Since most natural waters are near saturation, this reversal of the rate law would appear to be in line with the concept of microscopic reversibility of the forward and backward reaction steps. Some data are given in Figure 13.15.

In fresh waters, $CaCO_3$ is precipitated chemically. In seawater, the chemical precipitation is inhibited by Mg^{2+}. In seawater, $CaCO_3$ is formed biogenically as (hard) parts of organisms (Figure 13.16).

The crystal growth has to be preceded by nucleation. In fresh waters, this nucleation occurs heterogeneously on particle surfaces. In seawater, nucleation occurs primarily on the templates of calcareous organisms.

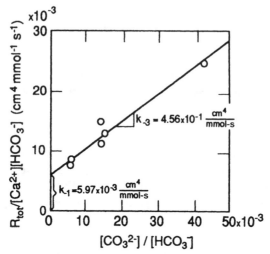

Figure 13.15. Data on the $CaCO_3$ crystal growth (20°C) carried out under conditions of constant p_{CO_2}. The data are plotted in line with Eq. (37) as (the last term on the right hand side can be neglected)

$$\frac{R_{tot}}{[Ca^{2+}][HCO_3^-]} = k_{-1} + k_{-3}\frac{[CO_3^{2-}]}{[HCO_3^-]}$$

where k_{-1} and k_3 are obtained from intercept and slope. (From Kunz and Stumm, 1984.)

Saturation State of Lake Water and Seawater with Respect to $CaCO_3$

Photosynthesis occurring in the upper layers of the oceans and of lakes removes CO_2 from the water and raises the saturation ratio of $CaCO_3$. In a simplified way:

$$CO_2 + H_2O \underset{R}{\overset{P}{\rightleftharpoons}} CH_2O + O_2$$

$$Ca^{2+} + 2\,HCO_3^- \rightleftharpoons CaCO_3 + CO_2 + H_2O$$

$$Ca^{2+} + 2\,HCO_3^- \underset{R}{\overset{P}{\rightleftharpoons}} CaCO_3 + CH_2O + O_2$$

As a consequence of photosynthesis (P), $CaCO_3$ is precipitated in the top layers. Respiration (R) of the phytoplankton and its debris in the deeper water layers causes the dissolution of $CaCO_3$. Figure 13.16 illustrates the type of carbonates formed and Figure 13.17 gives for Lake Constance saturation ratios ($\Omega = IAP/K_{s0}$) and the corresponding sedimentation fluxes of $CaCO_3$ into the

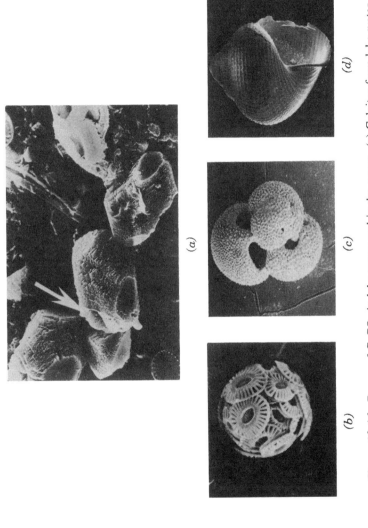

Figure 13.16. Forms of $CaCO_3$ in lake water and in the ocean. (a) Calcites from lake water interconnected into aggregates. Diatoms (arrow) are overgrown with calcite. (b) Calcareous pelagic plants (cocolithophores) (diameter ≈ 10 μm). (c) Animals (foraminifera, 100-μm diameter). (d) Pelagic pteropod (1000-μm diameter). (Parts b–d from Morse and Mackenzie, 1990.)

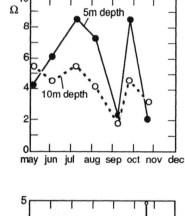
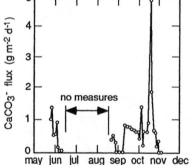

Figure 13.17. Saturation ratios of CaCO$_3$ in Lake Constance and the corresponding sedimentation load. (From Sigg et al., 1982.)

sediments. Obviously the settling of CaCO$_3$ represents a significant sedimentation load ("conveyor belt").

The saturation ratio, Ω, as a function of depth is given for the Atlantic, Indian, and Pacific Oceans in Figure 13.18. The equilibrium saturation is also given in this figure for aragonite. Because of the greater solubility of aragonite, the water column becomes undersaturated with regard to aragonite at smaller depths than for calcite. The Mediterranean Sea is supersaturated everywhere with respect to calcite as well as to aragonite.

Planktonic foraminifera and cocolithophores are composed of low magnesian calcite (<1 mol % MgCO$_3$). Benthic foraminifera are formed of either aragonite or high magnesian calcite. Pteropods are the most abundant aragonite organisms.

Growth of Concretions In sedimentary rocks we commonly find concretions, that is, material formed by deposition of a precipitate, such as calcite or siderite, around a nucleus of some particular mineral grain or fossil. The origin of most concretions is not known but their often-spherical shape with concentric internal structure suggests diffusion as an important factor affecting growth. The rate of growth, if diffusion controlled, is readily amenable to mathematical

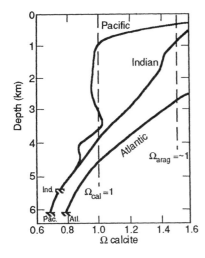

Figure 13.18. Saturation profiles for the northern Atlantic and Pacific oceans, and the central Indian Ocean (GEOSECS stations 31, 221, 450). (From Morse and Mackenzie, 1990.)

treatment. Berner (1968) has provided some idea of the time scale involved in the growth of postdepositional concretions. His calculations illustrate, for example, that for a typical slowly flowing groundwater, with a supersaturation in $CaCO_3$ of 10^{-4} M (assumed to be constant), the time of growth of calcite concretions ranges from 2500 years for concretions of 1-cm radius to 212,000 years for those of 5-cm radius. Hence concretion growth, if diffusion and convection are the rate-controlling steps, is relatively rapid when considered on the scale of geological time.

Kinetics of Carbonate Coprecipitation Reactions and Solid Solution Formation The uptake of a cation into a carbonate has been studied by Davis et al. (1987), by Wersin et al. (1989), and by Stipp and Hochella (1991). Morse and Mackenzie (1990) have reviewed extensively the geochemistry of dolomites and magnesian calcites.

13.6 INHIBITION OF DISSOLUTION

It is important to understand the factors that retard dissolution. The same question is especially relevant in technical systems and in the corrosion of metals and building materials. Passivity is imparted to many metals by overlying oxides; the inhibition of the dissolution of these "passive" layers protects the underlying material.

Figure 13.7 lists some of the factors that promote and inhibit dissolution. Obviously, substances that "block" surface functional groups or prevent the approach of dissolution promoting H^+, OH^-, ligands, and reductants to the functional groups inhibit the dissolution. A very small concentration of inhibitor can often be effective because it may suffice to block the functional groups of the solution active sites such as the ledge sites or the kink sites. Often competitive equilibria exist between the surface and solutes, which block surface sites, and, on the other hand, solutes, which promote dissolution (e.g., H^+, ligands). The main role of any inhibitor is to prevent access of corrosion-promoting agents to the surface.

Surface Protonation

The adsorption of a charged species, at constant pH, may change the surface protonation (Blesa et al., 1994; Sigg and Stumm, 1981). The formation of a negatively charged surface complex (e.g., by adsorption of a multivalent ligand) is accompanied by an increase in surface protonation; the formation of a positively charged surface complex (e.g., due to the adsorption of a cation) is accompanied by a decrease in surface protonation. In doing so, it may cause synergistic or antagonistic effects with respect to the relative acceleration or inhibition of dissolution. For more detailed arguments and experimental data see Biber et al. (1994).

Binuclear Surface Complexes

As suggested in Figure 13.7, binuclear or multinuclear surface complexes tend to block surface sites. A much higher activation energy is involved in detaching simultaneously two metal centers from the surface; hence dissolution is retarded by binuclear surface species. This retardation is especially pronounced when the effect of surface coordination with an inhibitor upon surface protonation is not unfavorable. For example, in the reaction of an anionic inhibitor, L^{2-},

$$(\equiv FeOH)_2 + L^{2-} \rightleftharpoons (\equiv Fe)_2L + 2\, OH^-$$

an uncharged surface complex is formed and the surface protonation is not changed; or the adsorption of chromium(III) to an oxide in acid solution as a binuclear surface complex is accompanied by a significant decrease in surface protonation. Thus the inhibition by blocking surface groups is synergistically enhanced by the decrease in surface protonation.

Exemplification: Inhibition by Oxoanions

To exemplify inhibition effects, we choose a few case studies with Fe(III)(hydr)oxides because these oxides are readily dissolved with protons, ligands, and reductants and are of great importance in the iron cycles in natural waters. The reductive dissolution of Fe(III) minerals by a reductant such as H_2S is much faster than ligand- or proton-promoted dissolution. The dissolution reaction, as shown by Dos Santos-Afonso and Stumm (1992), is initiated by the formation of $\equiv FeS$ and $\equiv FeSH$ surface complexes; the subsequent electron transfer within the complex leads to the formation of Fe(II) centers in the

13.6 Inhibition of Dissolution

surface lattice. The Fe(II)—O bond is characterized by a smaller Madelung energy than the Fe(III)—O bond [larger radius of Fe(II) than Fe(III)]; therefore the Fe(II) is more readily detached from the surface into the solution.

As shown in Figure 13.19a, phosphate and borate inhibit the dissolution of goethite by H_2S. Similarly, the dissolution of lepidocrocite (γ-FeOOH) by EDTA (Y^{4-}) is inhibited by phosphate and arsenate (Figure 13.19b). Both in the reductive dissolution (by H_2S) and the ligand-promoted dissolution (by

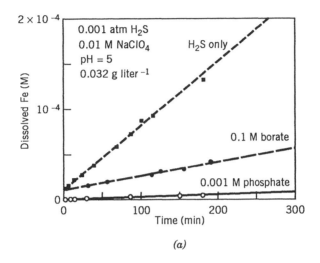

(a)

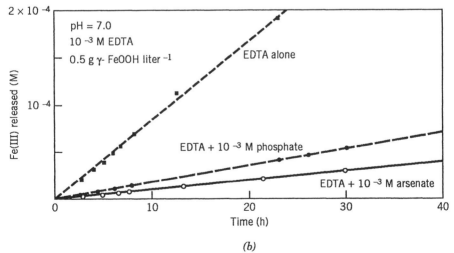

(b)

Figure 13.19. (a) Effect of phosphate and borate on progress of reductive dissolution by H_2S. (10^{-2} M $NaClO_4$; pH = 5.0. Partial pressure of H_2S = 10^{-3} atm; 0.032 g liter^{-1} goethite.) (b) Effect of phosphate and arsenate on retarding the EDTA (10^{-3} M) promoted nonreductive dissolution rate of lepidocrocite (0.5 g liter^{-1}) at pH = 7.0 (0.01 M $NaClO_4$). (Adapted from Biber et al., 1994.)

EDTA) the inhibition effects can be explained by ligand exchange reactions of the type (the definive assignment of the protonation of surface species has not yet been established)

$$2\equiv\text{FeS}^- + \text{H}_2\text{PO}_4^- + 3\text{H}^+ = (\equiv\text{Fe})_2\text{HPO}_4 + 2\text{H}_2\text{S} \tag{38}$$

$$2\equiv\text{FeH}_2\text{Y}^- + \text{H}_2\text{PO}_4^- + \text{H}^+ = (\equiv\text{Fe})_2\text{HPO}_4 + 2\text{H}_3\text{Y}^- \tag{39}$$

The fact that oxoanions can effectively inhibit reductive and nonreductive dissolution with regard to a reference system supports and the concept of competition between dissolution-promoting and dissolution-inhibiting ligands. As Figure 13.20 shows with the help of MICROQL calculations (Westall, 1979; based on complex formation constants fitted to actual data by Sigg and Stumm, 1981, on one hand and by Dos Santos-Afonso and Stumm, 1992, on the other hand), the dissolution system is very sensitive to phosphate.

Exemplification: Inhibition in Acid Solutions by Cr(III)

Figure 13.21 shows that Cr(III) is a very efficient inhibitor of hematite (α-Fe$_2$O$_3$) dissolution. The adsorption of Cr(III) to goethite surfaces has been investigated, using XAS, by Charlet aand Manceau (1993). Scanning tunneling microscopy was used by Eggleston to study the adsorption of Cr(III) to hematite (001) (Eggleston and Stumm, 1993). These studies confirm inner-sphere adsorption and suggest that the most stable sites for Cr(III) on the surfaces are

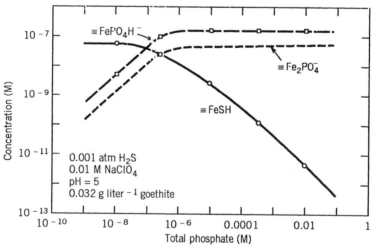

Figure 13.20. Surface equilibrium competition between reductant and inhibitor. Phosphate above a total concentration of 10^{-6} M replaces —SH surface group. (0.001 atm H$_2$S, 0.01 M NaClO$_4$, 0.03 g liter^{-1} goethite.) (Adapted from Biber et al., 1994.)

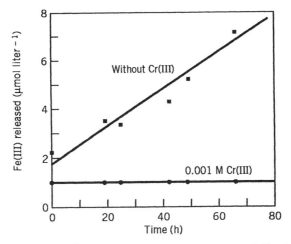

Figure 13.21. Effect of 10^{-3} M Cr(III) on the proton-promoted dissolution (pH = 3) of α-FeOOH (0.5 g liter^{-1}) in 0.1 M KNO$_3$. (From Bondietti et al., 1993.)

associated with octahedral vacancies in the underlying oxide and, depending on surface density, show a preponderance of adsorbate clustering.

Corrosion: Passive Films

To what extent is our knowledge of the reactivity of oxides useful for the understanding of corrosion reactions and passivity? The corrosive behavior of a few metals is essentially determined by the kinetics of the dissolution of the corrosion products. This seems to be the case for Zn in HCO$_3^-$ solutions, for passive iron in acids, and for passive Al in alkaline solutions (Grauer and Stumm, 1982). The mechanism of the dissolution of iron and of the passivation of this dissolution is extremely complex. We may not know exactly the composition of the passive film; but it has been suggested that it consists of an oxide of Fe$_{3-x}$O$_4$ with a spinel structure. The passive layer seems to vary in composition from Fe$_3$O$_4$ (magnetite), in oxygen-free solutions, to Fe$_{2.67}$O$_4$ in the presence of oxygen. It may also consist of a duplex layer consisting of an inner layer of Fe$_3$O$_4$ and an outer layer of γ-Fe$_2$O$_3$. The coulometric reduction of the passive layer gives two waves, which are interpreted either by the reduction of two different layers, Fe$_2$O$_3$ and Fe$_3$O$_4$, or by successive reduction of Fe$_3$O$_4$ to lower valence oxides and its further reduction to metallic iron. Figure 11.26 represents a schematic model of the hydrated passive film on iron as proposed by Bockris and collaborators (Pou et al., 1984). Obviously, the hydrated passive film on iron displays the coordinative properties of the surface hydroxyl groups.

Since the dissolution rate of passive metals is often related to the dissolution rate of the passive film, some of our information on the effect of solution variables on the dissolution reactivity of such types of oxides appears applicable to the interpretation of some of the factors that enhance or reduce passivity.

Some of the passive films have been characterized as semiconductors; in this case, corrosion of these oxides may imply transfer of holes (h^+) from the valence band to the reductant and of electrons (e^-) from the conduction band, in the case of iron(III) oxides as Fe(II).

13.7 NUCLEATION AND CRYSTAL GROWTH

The birth of a crystal and its growth provide an impressive example of nature's selectivity. In qualitative analytical chemistry inorganic solutes are distinguished from each other by a separation scheme based on the selectivity of precipitation reactions. In natural waters certain minerals are being dissolved, while others are being formed. Under suitable conditions a cluster of ions or molecules selects from a great variety of species the appropriate constituents required to form particular crystals.

Minerals formed in natural waters and in sediments provide a record of the physicochemical processes operating during the period of their formation; they also give us information on the environmental factors that regulate the composition of natural waters and on the processes by which elements are removed from the water. The memory record of the sediments allows us to reconstruct the environmental history of the processes that led to the deposition of minerals in the past.

Pronounced discrepancies between observed composition and the calculated equilibrium composition illustrate that the formation of the solid phase, for example, the nucleation of dolomite and calcite in seawater, is often kinetically inhibited, and the formation of phosphates, hydrated clay, and pyrite is kinetically controlled.

The Role of Organisms in the Precipitation of Inorganic Constituents

Organisms produce significant chemical differentiation in the formation of solid phases. The precipitation of carbonates, of opal, and of some phosphatic minerals by aquatic organisms has long been acknowledged. Within the last two decades, many different kinds of additional biological precipitates have been found. Life has succeeded in largely substituting for, or displacing to a varying extent, inorganic precipitation processes in the sea in the course of the last 6 $\times 10^8$ years. It appears that metastable mineral phases and, more commonly, amorphous hydrous phases are the initial nucleation products of crystalline compounds in biological mineral precipitates. Amorphous hydrous substances have been shown to persist in the mature, mineralized hard parts of many aquatic organisms (Mann et al., 1989).

13.7 Nucleation and Crystal Growth

The Processes Involved in Nucleation and Crystal Growth

Various processes are involved in the formation of a solid phase from an oversaturated solution (Figure 13.22). Usually three steps can be distinguished:

1. The interaction between ions or molecules leads to the formation of a critical cluster or nucleus:

$$X + X \rightleftharpoons X_2$$
$$X_2 + X \rightleftharpoons X_3$$
$$X_{j-1} + X \rightleftharpoons X_j \quad \text{(critical cluster)}$$
$$\text{Nucleation:} \; X_j + X \longrightarrow X_{j+1} \tag{40a}$$

Nucleation corresponds to the formation of the new centers from which *spontaneous* growth can occur. The nucleation process determines the size and distribution of crystals produced.

2. Subsequently, material is deposited on these nuclei,

$$X_{j+1} + X \longrightarrow \text{crystal growth} \tag{40b}$$

and crystallites are formed (crystal growth).

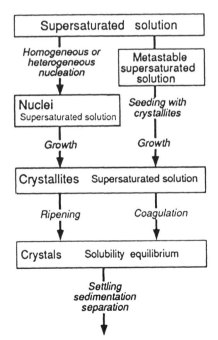

Figure 13.22. Simplified scheme of processes involved in nucleation and crystal growth. (From Nancollas and Reddy, 1974.)

3. Large crystals may eventually be formed from fine crystallites by a process called ripening.

A better insight into the mechanisms of the individual steps in the formation of crystals would be of great help in explaining the creation and transformation of sedimentary deposits and biological precipitates (see Fig. 13.23). Valuable reviews are available on the principles of nucleation of crystals and the kinetics of precipitation and crystal growth (Steefel and Van Cappellen, 1990; Van Cappellen, 1991; Zhang and Nancollas, 1990). Only a few important considerations are summarized here to illustrate the wide scope of questions to be answered in order to predict rates and mechanisms of precipitation in natural systems.

Homogeneous Nucleation: A Large Supersaturation Must Be Exceeded Before Homogeneous Nucleation Can Occur

If one gradually increases the concentration of a solution, exceeding the solubility product with respect to a solid phase, the new phase will not be formed within a specific amount of time until a certain degree of supersaturation has been achieved. Stable nuclei can only be formed after an activation energy barrier has been surmounted.

We first review the classical theory. The free energy of the formation of a nucleus, ΔG_j, consists essentially of energy gained (volume free energy) from making bonds and of work required to create a surface:

$$\Delta G_j = \Delta G_{\text{bulk}} + \Delta G_{\text{surf}} \tag{41a}$$

For a nucleus, the first quantity (always negative for a supersaturated solution) can be expressed as

$$\Delta G_{\text{bulk}} = -jkT \ln \frac{a}{a_0} = -jkT \ln S \tag{41b}$$

where j is the number of molecular units ("monomers") in the nucleus or, expressed in terms of volume for a spherical nucleus, $j = 4\pi r^3/3\,V$, where V is the "molecular" volume; a is the actual concentration (activity) and a_0 is the concentration at solubility equilibrium of the solutes that characterize the solubility. a/a_0 is the saturation ratio S:

$$S = \frac{a}{a_0} = \left(\frac{\text{IAP}_0}{K_{s0}}\right)^{1/\eta} \tag{42}$$

where IAP_0 is the ion activity product (of the actual activities in the oversaturated solution) and K_{s0} is the solubility product; η is the number of ions in

13.7 Nucleation and Crystal Growth

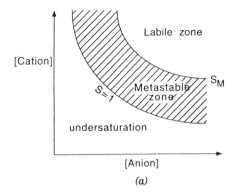

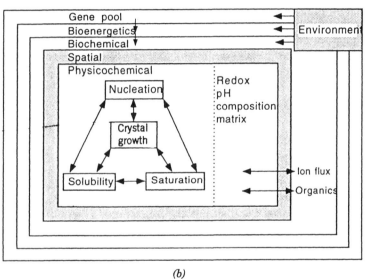

Figure 13.23. (a) Schematic solubility isotherm of a solid electrolyte. Below a certain supersaturation, S_M, the nucleation rate is virtually zero and the solution under these conditions can be stable for long periods without precipitation. The range $1 < S < S_M$ is the metastable zone, within which crystal growth can be achieved without the complication of concomitant nucleation if the solution is seeded with crystallites. Foreign surfaces may also induce nucleation in the metastable zone (heterogeneous nucleation). (Adapted from Zhang and Nancollas, 1990.) (b) The control processes in biomineralization. In organisms, there are several different interconnecting levels that regulate the physicochemical properties of mineralization (solubility, supersaturation, nucleation, and crystal growth). An essential condition for controlled mineralization is spatial localization arising from the compartmentalization of biological space. This permits direct regulation of physicochemical and biochemical properties in the mineralization zone. Nucleation, in particular, can be mediated by organic polymeric substrates in or on the spatial boundary. At a higher level of organization, mineralization is under biochemical and bioenergetic constraints and, ultimately, under control at the gene level. The interplay of these control processes with the external environment is also of fundamental importance. (From Mann, 1988.)

the formula unit of a mineral $A_\alpha B_\beta$ (i.e., $\eta = \alpha + \beta$). Because of the "normalization" by η, the saturation ratio S is independent of the way the formula is written, for example, $Ca_5(PO_4)_3OH$ or $Ca_{10}(PO_4)_6(OH)_2$.

The second quantity in equation 41a is given (spherical nucleus) by

$$\Delta G_{\text{surf}} = 4\pi r^2 \bar{\gamma} \qquad (43)$$

where $\bar{\gamma}$ is the interfacial energy (assumed to be independent of cluster size). Hence the free energy of the formation of a spherical cluster can be written as

$$\Delta G_j = -\frac{4\pi r^3}{3V} kT \ln S + 4\pi r^2 \bar{\gamma} \qquad (44)$$

In Figure 13.24a, ΔG_j is plotted as a function of j for a few values of the saturation ratio $a/a_0 = S$. Obviously the activation energy ΔG^* decreases with increasing saturation ratio, as does the size of the critical nucleus.

The rate at which nuclei form, J, may be represented according to conven-

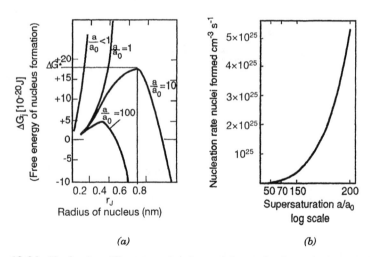

(a)   (b)

Figure 13.24. Nucleation. The energy barrier and the nucleation rate depend critically on the supersaturation. (a) Free energy of formation of a spherical nucleus as a function of its size, calculated for different saturation ratios (a/a_0). The height of the maximum, ΔG^* is the activation barrier to the nucleation process of nucleus of radius r_j. (b) Double logarithmic plot of nucleation rate versus saturation ratios (a/a_0) calculated with equations reflecting that the nucleation rate is proportional to $\exp(-\Delta G^*/kT)$. The curves have been calculated for the following assumptions: $\bar{\gamma} = 100$ mJ m^{-2}; "molecular" volume $V = 3 \times 10^{-23}$ cm^3; collision frequency efficiency $\bar{A} = 10^{30}$ cm^{-3} s^{-1} (equation 45). With increasing size of the cluster the first term on the right hand side of equation (44) (increasing with r^3) outweighs the second term (increasing with r^2); for large crystals the second term becomes negligible.

tional rate theory as

$$J = \overline{A} \exp\left(\frac{-\Delta G^*}{kT}\right) \qquad (45)$$

where $\overline{A}$ is a factor related to the efficiency of collisions of ions or molecules. ΔG^* can be calculated on the basis of the information already given in equations 40–44 (Stumm, 1992). Accordingly, the rate of nucleation is controlled by the interfacial energy, the supersaturation, the collision frequency efficiency, and the temperature.

In the case of heterogeneous nucleation, the interfacial energy needs some redefinition because the nucleus is now formed in part in contact with the solution and in part in contact with the surface of the solid substrate:

Homogeneous Nucleation:

$$\Delta G_{\text{surf}} = \overline{\gamma}_{\text{CW}} A \qquad (46)$$

Heterogeneous Nucleation:

$$\Delta G_{\text{interf}} = \overline{\gamma}_{\text{CW}} A_{\text{CW}} + (\overline{\gamma}_{\text{CS}} - \overline{\gamma}_{\text{SW}}) A_{\text{CS}} \qquad (47)^{\dagger}$$

The suffixes CW, CS, and SW refer to cluster–water, cluster–substrate, and substrate–water, respectively.

A surface-catalytic effect is observed, as mentioned above, when the surface of the solid substrate "matches well" with the crystal to be formed, that is, when

$$\overline{\gamma}_{\text{CS}} < \overline{\gamma}_{\text{CW}} \qquad (48a)$$

In ideal cases (seed crystals), $\overline{\gamma}_{\text{CS}}$ becomes very small or zero

$$\overline{\gamma}_{\text{CS}} \to 0 \qquad (48b)$$

and the solid–solution interfacial energy of the substrate is similar to that of the cluster

$$\overline{\gamma}_{\text{SW}} = \overline{\gamma}_{\text{CW}} \qquad (48c)$$

As a consequence, for a "good" substrate:

$$\Delta G_{\text{interf}} = \gamma_{\text{CW}}(A_{\text{CW}} - A_{\text{CS}}) \qquad (49)$$

[†]The overall equation for the free energy of the heterogeneous nucleation is (cf. equation 41) $\Delta G = -RT \ln S + \overline{\gamma}_{\text{CW}} A_{\text{CW}} + (\overline{\gamma}_{\text{CS}} - \overline{\gamma}_{\text{SW}}) A_{\text{CS}}$.

When the attachment of the substrate to the precipitate to be formed is strong, the clusters tend to spread themselves out on the substrate and form thin surface islands. A special limiting case is the formation of a surface nucleus on a seed crystal of the same mineral (as in surface nucleation crystal growth). As the cohesive bonding within the cluster becomes stronger relative to the bonding between the cluster and the substrate, the cluster will tend to grow three-dimensionally (Steefel and Van Cappellen, 1990).

On the other hand, if $\bar{\gamma}_{SW} \gg \bar{\gamma}_{CW}$, the precipitate tends to form a structurally continuous coating on the substrate grain. The interfacial energy (equation 47) may even become negative and the activation barrier vanishes. An example reflecting this condition is the growth of amorphous silica on the surface of quartz (Wollast, 1974).

As Figure 13.24b suggests, homogeneous nucleation should not occur even within very long time spans for a supersaturation $S \le 10$. At a high degree of supersaturation, the nucleation rate is so high that the precipitate formed consists mostly of extremely small crystallites. Incipiently formed crystallites might be of a different polymorphous form than the final crystals. If the nucleus is smaller than a one-unit cell, the growing crystallite produced initially is most likely to be amorphous; substances with a large unit cell tend to precipitate initially as an amorphous phase ("gels").

Heterogeneous Nucleation

Heterogeneous nucleation, however, is in many cases the predominant formation process for crystals in natural waters. In a similar way as catalysts reduce the activation energy of chemical reaction, foreign solids may catalyze the nucleation process by reducing the energy barrier. Qualitatively, if the surface of the solid substrate matches well with the crystal, the interfacial energy between the two solids is smaller than the interfacial energy between the crystal and the solution, and nucleation may take place at a lower saturation ratio on a solid substrate surface than in solution (Figure 13.25).

Phase changes in natural waters are almost invariably initiated by heterogeneous solid substrates. Inorganic crystals, skeletal particles, clays, sand, and biological surfaces can serve as suitable substrate.

The Interfacial Energy and the Ostwald Step Rule

The above considerations show that the interfacial energy is of the utmost importance in determining the thermodynamics and kinetics of the nucleation process. Unfortunately, however, there are considerable uncertainties on the values of interfacial free energies. Values determined from contact angle measurements are significantly lower than those determined from the dependence of solubility on molar surface of the crystallites. Furthermore, reliable data on $\bar{\gamma}_{CS}$ are lacking.

It is useful to know that for a given type of crystal (oxides, sulfates, car-

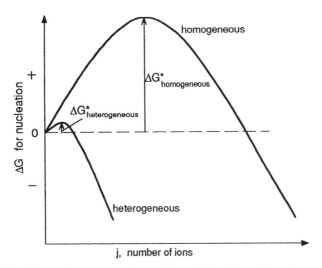

Figure 13.25. Schematic representation of the ability of a solid substrate to catalyze (for a given saturation ratio S) the nucleation. It is assumed that $\bar{\gamma}_{CS} < \bar{\gamma}_{CW}$. The exact curves depend also on the geometry of the crystals formed.

bonates), the interfacial mineral–aqueous solution free energy, $\bar{\gamma}$ (or $\bar{\gamma}_{CW}$), increases with decreasing solubility (Schindler, 1967). Nielsen (1986) cites the following empirical relationship:

$$\frac{4\bar{r}^2\bar{\gamma}}{kT} = 4.70 - 0.272 \ln C_{sat} \tag{50}$$

where C_{sat} is the solubility in moles of formula units of mineral per liter and $\bar{r}$ is the mean ionic radius (m); the units of $\bar{\gamma}$ are J m^{-2}. Table 13.4 lists interfacial energies: the solubilities (for oxides and hydroxides) and C_{sat} values were calculated for pH = 7.

The *Ostwald step rule*, or the rule of stages, postulates that the precipitate with the highest solubility (i.e., the least stable solid phase) will form first in a consecutive precipitation reaction. This rule is very well documented; mineral formation via precursors and intermediates can be explained by the kinetics of the nucleation process. The precipitation sequence results because the nucleation of a more soluble phase is kinetically favored over that of a less soluble phase because the more soluble phase has the lower solid–solution interfacial tension ($\bar{\gamma}_{CW}$) than the less soluble phase (equation 50). In other words, a supersaturated solution will nucleate first the least stable phase (often an amorphous solid phase) because its nucleation rate is larger than that of the more stable phase (Figure 13.26). While the Ostwald step rule can be explained on the basis of nucleation kinetics, there is no thermodynamic contradiction in the initial formation of a finely divided precursor.

Table 13.4. Surface Free Energies, $\bar{\gamma}$

Mineral	Formula	$\bar{\gamma}(mJ/m^2)$	C_{sat} (M) (pH = 7)
Fluorite	CaF_2	120[b]	2×10^{-4}
Calcite	$CaCO_3$	94[b]	6×10^{-5}
Witherite	$BaCO_3$	115[c]	1×10^{-4}
Cerussite	$PbCO_3$	125[c]	4×10^{-6}
Gypsum	$CaSO_4 \cdot 2\,H_2O$	26[d]	1.5×10^{-2}
Celestite	$SrSO_4$	85[c]	4×10^{-4}
Barite	$BaSO_4$	135[c]	1×10^{-5}
F-apatite	$Ca_5(PO_4)_3F$	289[e]	6×10^{-9}
OH-apatite	$Ca_5(PO_4)_3OH$	87[f]	7×10^{-6}
OCP_p[a]	$Ca_4H(PO_4)_3 \cdot H_2O$	26[e]	2×10^{-4}
Portlandite	$Ca(OH)_2$	66[c]	6×10^{-5}
Brucite	$Mg(OH)_2$	123[c]	1.5×10^{-4}
Goethite	$FeOOH$	1600[g]	1×10^{-12}
Hematite	Fe_2O_3	1200[g]	1×10^{-12}
Zincite	ZnO	770[h]	2×10^{-5}
Tenorite	CuO	690[h]	3×10^{-7}
Gibbsite (001)	$Al(OH)_3$	140[i]	7×10^{-8}
Gibbsite (100)		483[i]	
Quartz	SiO_2	350[j]	1×10^{-4}
Amorphous silica	SiO_2	46[k]	2×10^{-3}
Kaolinite	$Al_2Si_2O_5(OH)_4$	200[l]	1×10^{-6}

[a] OCP_p, octacalcium phosphate (subscript p = precursor).
[b] Christoffersen et al. (1988).
[c] Nielsen and Söhnel (1971).
[d] Chiang et al. (1988).
[e] Van Cappellen (1991).
[f] Arends et al. (1987).
[g] Berner (1980).
[h] Schindler (1967).
[i] Smith and Hem (1972).
[j] Parks (1984).
[k] Alexander et al. (1954).
[l] Steefel and Van Cappellen (1990).

Source: Adapted from Van Cappellen (1991). Solubilities (pH = 7) are calculated from a variety of sources compiled by Van Cappellen (1991).

In Section 7.8 we showed that the size dependence of the solubility has also a thermodynamic base. In Figure 7.24a we demonstrated the influence of molar surface on solubility of CuO and $Cu(OH)_2$ at pH = 7. This figure suggested that $Cu(OH)_2(s)$ becomes more stable than CuO(s) for very finely divided CuO crystals. In precipitating Cu(II), $Cu(OH)_2$ may be precipitated incipiently (*d* is very small) but CuO becomes more stable than $Cu(OH)_2(s)$ upon growth of the crystals, and an inversion of $Cu(OH)_2(s)$ into the more stable phase becomes possible (Schindler, 1967).

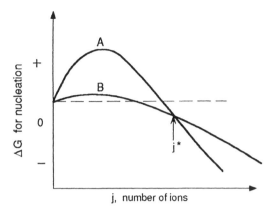

Figure 13.26. Schematic plot of free energy of formation of clusters from solution as a function of size (number of ions in the cluster). Curve A corresponds to the (thermodynamically) ultimately formed more stable phase, while curve B corresponds to the precursor phase. The particle size, j^*, is critical for the inversion from one polymorphous form to another. (Adapted from Van Cappellen, 1991.)

The Precursor as a Substrate (Template) for the Formation of the More Stable Phase

As the precursor (e.g., an amorphous phase) precipitates and brings down the supersaturation of the solution, the more stable phase to be precipitated is using the precursor phase as a substrate for its own precipitation (Steefel and Van Cappellen, 1990). A classical example that documents this principle is the precipitation of calcium phosphates, where a metastable calcium phosphate precursor phase is nucleated initially and is then replaced, in some instances via an intermediate phase, by apatite. (Nancollas, 1989; Steefel and Van Cappellen, 1990).

Biomineralization

In biomineralization, inorganic elements are extracted from the environment and selectively precipitated by organisms. Usually, templates consisting of suitable macromolecules serve as a substrate for the heterogeneous nucleation of bulk mineralized structures such as bone, teeth, and shells. Biological control mechanisms are reflected not only in the type of mineral phase formed but also in its morphology and crystallographic orientation (Lowenstamm and Weiner, 1989; Mann et al., 1989). Two examples (perhaps oversimplified) may illustrate the principle (Ochial, 1991):

1. *Silica Gel Formation in Diatoms.* The proteins in the cell walls of several diatom species contain a relatively large proportion of serine [$H_2N-CH(CH_2OH)COOH$], which contains aliphatic hydroxo groups that can

undergo condensation or ligand exchange reactions with silicic acid

$$\begin{array}{c} \text{HO} \quad \text{OH} \quad \text{HO} \quad \text{OH} \\ \diagdown \diagup \qquad \diagdown \diagup \\ \text{Si} \qquad\qquad \text{Si} \\ \diagup \quad \diagdown \qquad \diagup \quad \diagdown \\ \text{O} \qquad\qquad \text{O} \qquad\qquad \text{O} \\ | \qquad\qquad | \qquad\qquad | \\ -\text{Ser} \cdots \text{Ser} \cdots \text{Ser}- \end{array}$$

The small aggregates of SiO_n so formed are considered to be the template for the nucleation and growth of silica.

2. *$CaCO_3$ Shells in Molluscs*. The major component of the organic matrix protein is a glycoprotein with a predominance of aspartic acid [$H_2NCH(CH_2COOH)COOH$] and glycine (H_2NCH_2COOH). The sequence aspartate–X–aspartate (where X is mostly glycine) in the protein is thought to nucleate $CaCO_3$.

$$\begin{array}{c} \text{CO}_3 \qquad \text{CO}_3 \\ | \qquad\qquad | \\ \text{Ca} \qquad\qquad \text{Ca} \\ \diagup \quad \diagdown \quad \diagup \quad \diagdown \\ \text{O} \qquad \text{O} \qquad \text{O} \\ | \qquad | \qquad | \\ \text{C}{=}\text{O} \quad \text{C}{=}\text{O} \quad \text{C}{=}\text{O} \\ | \qquad\quad | \qquad\quad | \\ -\text{Asp}-\text{X}-\text{Asp}-\text{X}-\text{Asp}-\text{X} \end{array}$$

There are various other factors that determine the crystal structure (calcite versus aragonite).

Several crystals, such as vaterite and calcite, forms of $CaCO_3$, or α-glycine, have been nucleated (*induced oriented crystallization*) at the water surface covered with a monolayer film of carboxylic acids or aliphatic alcohols (compressed to "suitable" distances of the hydrophilic groups with a Langmuir balance) (Mann, 1988).

Molecular Recognition at Crystal Interfaces As we have seen, the key concept for understanding the different processes of nucleation and crystal growth is "molecular" recognition at the interface; the surfaces of templates, nuclei, and crystals can be thought of as being composed of "active sites" that interact stereospecifically with ions or molecules in solution, in a manner similar to the interactions of enzymes and substrates or antibodies and antigens. Crystals can be engineered with desired morphologies (Weissbuch et al., 1991).

The processes controlling biomineralization are summarized in Figure 13.23b. Organized biopolymers at the sites of mineralization are essential to these processes. In unicellular organisms these macromolecules act primarily as spatial boundaries through which ions are selectively transported to produce localized supersaturation within discrete cellular compartments. In many instances, particularly in organisms such as the diatoms that deposit shells of amorphous silica, the final shape of the mineral appears to be dictated by the ultrastructure of the membrane-bound compartment. Thus a diversity of mineral shapes can be biologically molded by constraining the space available to the growing mineral. In many multicellular organisms crystallographic properties are related to the surface structure of the macromolecules; for example, bone is made up of microscopic plate-like calcium phosphate crystals formed within and between fibrils of collagen (Mann, 1988).

"Biomimetic" Processing

Surface functionalization routes have been developed in technology by the mimicking of schemes used by organisms to produce complex ceramic composites such as teeth, bones, and shells. High-quality dense polycrystalline films of oxides, hydroxides, and sulfides have now been prepared by "biomimetic" synthesis techniques (Bunker et al., 1994).

Enhancement of Heterogeneous Nucleation by Specific Adsorption of Mineral Constituents

The "classical" theory of nucleation concentrates primarily on calculating the nucleation free energy barrier, ΔG^*. Chemical interactions are included under the form of thermodynamic quantities, such as the surface tension. A link with chemistry is made by relating the surface tension to the solubility, which provides a kinetic explanation of the Ostwald step rule and the often observed disequilibrium conditions in natural systems. Can the chemical model be complemented and expanded by considering a specific chemical interactions (surface complex formation) of the components of the cluster with the surface?

In addition to the matching of the structures of the surfaces of the mineral to be nucleated and the substrate, adsorption or chemical bonding of nucleus constituents to the surface of the substrate can be expected to enhance the nucleation. Surface complex formation and ligand exchange of crystal-forming ions with the surface sites of the substrate, their partial or full dehydration and structural realignment, and perhaps the formation of ternary surface complexes are essential and at least partially rate-determining steps in the heterogeneous nucleation process. One might argue that a critical surface concentration of surface constituent ions (i.e., a two-dimensional solubility product) must be exceeded before a nucleus is being formed.

As a simple example we illustrate the nucleation of $CaCO_3$ on a metal oxide surface (Kunz and Stumm, 1984):

$$\equiv Me-OH + Ca^{2+} \rightleftarrows \; \equiv Me-OCa^+ + H^+ \tag{51}$$

$$\equiv Me-OH + HCO_3^- \rightleftarrows \; \equiv Me-CO_3^- + H_2O \tag{52}$$

In equation 51 an inorganic surface can also be replaced by an organic surface, $\equiv R-OH$. Hohl et al. (1982) (results are given in Stumm et al., 1983) have shown that the nucleation of CaF_2 can be enhanced by $CeO_2(g)$. CeO_2 and CaF_2 have the same crystalline structure and the same lattice distances. The heterogeneous nucleation rate can be retarded by cations and ligands that are competitively bound to the oxide surface.

Crystal Growth

The overall *kinetics* of crystal precipitation has to consider that the process consists of a series of consecutive processes; in simple cases, the slowest is the rate-determining step. Assuming the volume diffusion is not the rate-determining step, we still have at least the following reaction sequences:

1. *Adsorption:* adsorption of constituent ions onto the substrate.
2. *Surface nucleation:* diffusion of adsorbed ions; partial dehydration; formation of two-dimensional nucleus; growth to three-dimensional nucleus.
3. *Crystal growth.*

Each one of these sequential processes consists of more than one reaction step.

The classical crystal growth theory goes back to Burton, Cabrera, and Frank (BCF) (1951). The BCF theory presents a physical picture of the interface (Figure 13.27c), where at kinks on a surface step—at the outcrop of a screw dislocation—adsorbed crystal constituents are sequentially incorporated into the growing lattice.

Different rate laws for crystal growth have been proposed. The empirical law often used is

$$V = k(S - 1)^n \tag{53}$$

where V is the linear growth rate (length time^{-1}).

Often this law fits the experimental data well, especially at high degrees of supersaturation. Often an exponent $n = 2$ is found. Nielsen (1981) has explained this observation by assuming that for these solids the rate-determining step is the integration of ions at kink sites of surface spirals (see Figure 13.27).

Surface Precipitation

In surface precipitation, cations (or anions), which adsorb to the surface of a mineral, may form a precipitate of the cation (anion) with the constituent ions of the mineral at high surface coverage. Figure 13.28 shows schematically the surface precipitation of a cation Me^{2+} to hydrous ferric oxide. This model, suggested by Farley et al. (1985), allows for a continuum between surface complex formation and bulk solution precipitation of the sorbing ion; that is,

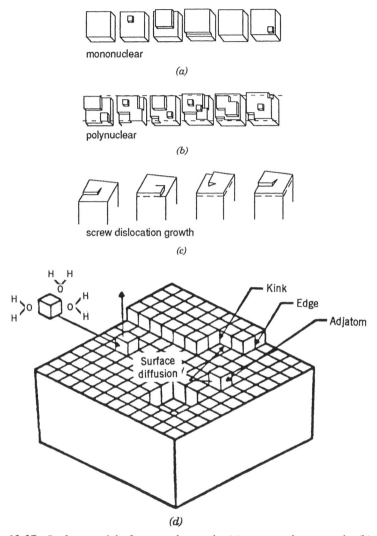

Figure 13.27. Surface models for crystal growth: (a) mononuclear growth, (b) polynuclear growth, and (c) screw dislocation growth. Along the step a kink site is shown. Adsorbed ions diffuse along the surface and become preferentially incorporated into the crystal lattice at kink sites. As growth proceeds, the surface step winds up in a surface spiral. Often the growth reaction observed occurs in the sequence c, a, b. (From Nielsen, 1964.) (d) Salient features and elementary processes at surfaces.

as the cation is complexed at the surface, a new hydroxide surface is formed. In the model, cations at the solid (oxide)–water interface are treated as surface species, while those not in contact with the solution phase are treated as solid species forming a *solid solution*. The formation of a solid solution implies isomorphic substitution. At low sorbate cation concentrations, surface com-

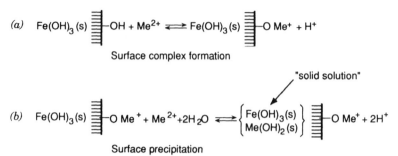

Figure 13.28. Schematic representation of surface precipitation on hydrous ferric oxide, $Fe(OH)_3(s)$. (a) At low surface coverage with Me^{2+}, surface complex formation dominates. Instead of the usual short-hand notation ($\equiv Fe-OH + Me^{2+} \rightleftharpoons \equiv FeOMe^+ + H^+$), we use one that shows the presence of $Fe(OH)_3(s)$. (b) With progressive surface coverage, surface precipitation may occur. The surface precipitate is looked at as a solid solution of $Fe(OH)_3(s)$ and $Me(OH)_2(s)$; some isomorphic substitution of Me(II) for Fe(III) occurs. This model has been proposed by Farley et al. (1985).

plexation is the dominant mechanism. As the sorbate concentration increases, the surface complex concentration and the mole fraction of the surface precipitate both increase until the surface sites become saturated. Surface precipitation then becomes the dominant "sorption" (metal ion incorporation) mechanism. As bulk solution precipitation is approached, the mole fraction of the surface precipitate approaches unity.

Figure 13.29 shows idealized isotherms (at constant pH) for cation binding to an oxide surface. In the case of cation binding onto a solid hydrous oxide, a metal hydroxide may precipitate and may form at the surface prior to its formation in bulk solution and thus contribute to the total apparent "sorption." The contribution of surface precipitation to the overall sorption increases as the sorbate/sorbent ratio is increased. At very high ratios, surface precipitation may become the dominant "apparent" sorption mechanism. Isotherms showing reversals as shown by curve e have been observed in studies of phosphate sorption by calcite (Freeman and Rowell, 1981).

Does Surface Precipitation Occur at Concentrations Lower Than Those Calculated from the Solubility Product? As the theory of solid solutions explains, the solubility of a constituent is greatly reduced when it becomes a minor constituent of a solid solution phase (curve b in Figure 13.29). Thus a solid species [such as $M(OH)_2$] can precipitate at lower pH values in the presence of a hydrous oxide (as a solid solvent) than in its absence.

Adsorption and Precipitation versus Heterogeneous Nucleation and Surface Precipitation There is not only a continuum between surface complexation (adsorption) and precipitation, but there is also obviously a continuum from heterogeneous nucleation to surface precipitation. The two models are

13.7 Nucleation and Crystal Growth

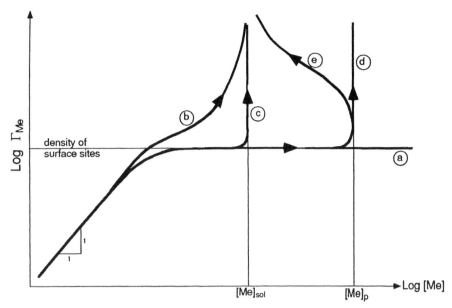

Figure 13.29. Schematic sorption isotherms of a metal ion (Me) on an oxide (XO_n) at constant pH: (a) adsorption only; (b) adsorption and surface precipitation via ideal solid solution; (c) adsorption and heterogeneous nucleation in the absence of a free energy nucleation barrier ($\Delta G^* \rightarrow 0$); (d) adsorption and heterogeneous nucleation of a metastable precursor; (e) same as in (d) but with transformation of the precursor into the stable phase. The arrows show the isotherm evolution for continual addition of dissolved Me. The initial isotherm with the slope of 1 (in the double logarithmic plot) corresponds to a Langmuir isotherm (surface complex formation equilibrium). $[Me]_{sol}$ = solubility concentration of Me for the stable metal oxide; $[Me]_p$ = solubility concentration of Me for a metastable precursor (e.g., a hydrated Me oxide phase). (From Van Cappellen, 1991.)

two limiting cases for the initiation of precipitation. In the heterogeneous nucleation model, the interface is fixed and no mixing of ions occurs across the interface. As a consequence, precipitation of the new solid phase does not occur until the solution becomes supersaturated. In the surface precipitation model, the interface is a mixing zone for the ions of the new solid phase and those of the substrate. The surface phase is treated as an ideal solid solution. This allows precipitation to start from solutions undersaturated with the pure phase. Furthermore, the composition of the surface phase can vary continuously from that of the pure substrate to that of the new phase. Whether in an actual case precipitation of a new phase approaches one or the other limiting model will depend, as has been pointed out by Van Cappellen (*personal communication*, 1991), on the mixing energies of the pure new phase and the substrate. If the mixing enthalpy is small ($\Delta H_{mix} = 0$: ideal solid solution), solid solution formation should be favored; when it is large, nucleation should be favored.

In other words, it should be possible to correlate the occurrence of one or the other mechanism (or some combination of both) with the mismatch strain energy between the lattices of the substrate and the new phase, and with "chemical" differences in electronegativity and polarization between the substituting ions.

For some case studies on surface precipitation and coprecipitation, see Dzombak and Morel (1990), Sposito (1986), and Wersin et al. (1989).

Growth Inhibitors Because the rate-determining step is frequently controlled at the interface, small amounts of soluble foreign constituents may alter markedly the growth rate of crystals and their morphology. The retarding effect of substances that become adsorbed may be explained as being due primarily to the obstruction by adsorbed molecules to the deposition of lattice ions. In some cases it has been shown that the rate constant for crystal growth is reduced by an amount reflecting the extent of adsorption.

Trace concentrations of dissolved organic matter and orthophosphates and polyphosphates act as "crystal poisons" on the nucleation and growth of calcite probably by inhibiting the spread of mononuclear steps on the crystal surface by becoming adsorbed on active growth sites such as kinks.

It is interesting to note that some crystal poisons may not only interfere with the growth of crystals but may also retard their *dissolution*. Insidious trace quantities of organic matter and phosphate inhibit the dissolution of calcite in undersaturated waters. Apparently precipitation and dissolution of ionic solids proceed by the attachment or detachment of ions at kinks in monomolecular steps on the crystalline surface. The detachment of kinks—points of excess surface energy and preferred sites of chemisorption of crystal poisons—is rendered more difficult by the adsorption of organic solutes or phosphates, thus giving rise to kink immobilization and retardation of dissolution. Berner and Morse (1974) have shown that the critical undersaturation of calcite in seawater increases with increasing orthophosphate concentration in solution. Their results provide an explanation for the observed variations of $CaCO_3$ dissolution in the deeper portion of the ocean and for the *lysocline*. The lysocline is a region where the rate of dissolution with depth radically increases (see Figure 13.18).

SUGGESTED READINGS

Benett, P. C. and W. H. Casey (1994) Chemistry and Mechanism of Dissolution of Silicates by Organic Acids in *Organic Acids in Geological Processes*, E. D. Pittman and M. D. Lewan, Eds., Springer, Berlin.

Casey, W. H., and Sposito, G. (1990) On the Temperature Dependence of Mineral Dissolution Rates, *Geochim. Cosmochim. Acta* **56,** 3825–3830.

Lasaga, A. C. (1990) Atomic Treatment of Mineral–Water Surface Reactions. In *Mineral–Water Interface Geochemistry*, M. Hochella and A. F. White Eds., *Rev. Mineralogy* **23,** 17–80.

Lasaga, A. C., Soler, J. M., Ganor, J., Burch, T. E., and Nagy, K. L. (1994) Chemical Weathering Rate Laws and Global Geochemical Cycles, *Geochim. Cosmochim. Acta* **58,** 2361–2386.

Schnoor, J. L. (1990) Kinetics of Chemical Weathering: A Comparison of Laboratory and Field Weathering Rates. In *Aquatic Chemical Kinetics*, W. Stumm, Ed., Wiley-Interscience, New York.

Shiraki, R. and Brantley, S. L. (1995) Kinetics of Near-Equilibrium Calcite Precipitation at 100°C: An Evaluation of Elementary Reaction-Based and Affinity-Based Rate Laws, *Geochim. Cosmochim. Acta* **59**(8), 1457–1471.

Sposito, G. (1994) *Chemical Equilibria and Kinetics in Soils*, Chapter 4: Surface Reactions, pp. 138–176, Oxford University Press, New York.

Sverdrup, H. A. (1990) *The Kinetics of Base Cation Release Due to Chemical Weathering*, Lund University Press, Lund.

Wieland, E., Wehrli, B., and Stumm, W. (1988) The Coordination Chemistry of Weathering: III. A Generalization on the Dissolution Rates of Minerals, *Geochim. Cosmochim Acta* **52,** 1969–1981.

14

PARTICLE–PARTICLE INTERACTION: COLLOIDS, COAGULATION, AND FILTRATION

14.1 COLLOIDS

Colloids are ubiquitous in natural waters; they are present in relatively large concentrations ($>10^6$ cm^{-3}) in fresh surface waters, in groundwaters, in oceans, and in interstitial soil and sediment waters. The solid–water interface established by these particles plays a commanding role in regulating the concentrations of most reactive elements and of many pollutants in soil and natural water systems and in the coupling of various hydrochemical cycles. Wells and Goldberg (1994) estimate that the total surface area of the small colloidal (5–200 nm) fraction alone is >18 m^2 per m^3 of seawater in the upper water column. Processes with colloids are also of importance in technical systems, above all in water technology.

Aquatic suspended particles are usually characterized by a continuous particle size distribution. The distinction between particulate and dissolved compounds, conventionally made in the past by membrane filtration, does not consider organic and inorganic colloids appropriately. Colloids of iron(III) and manganese(III,IV) oxides, sulfur, and sulfides are often present as submicron particles that may not be retained by membrane filters (e.g., Buffle et al., 1992). Recent measurements in the ocean led to the conclusion that a significant portion of the operationally defined "dissolved" organic carbon may in fact be present in the form of colloid particles.

The few measurements of colloidal mass that are available indicate that it may be as large if not larger than the sum of particles of micron size or larger. Furthermore, Honeyman and Santschi (1992) state that in the oceans this pool of "ill-defined" material has a fate quite different from material in the large particle pool (e.g., their residence times are estimated to be orders of magnitude different).

Definition of Colloids

Colloids are usually defined on the basis of size, for example, entities having in at least one direction a dimension between 1 nm and 1 μm. An operational distinction on the basis of size (membrane filtration, centrifugation, diffusion) although useful for many operational questions, is not fully satisfactory. In order to be in agreement with the thermodynamic concept of speciation, the connotation "dissolved" should be used for those species for which a chemical potential can be defined. Colloids are dynamic particles; they are continuously generated (by physical fragmentation and erosion, by precipitation and nucleation from oversaturated solutions), undergo compositional changes, and are continuously removed from the water (by coagulation, by attachment and settling, and by dissolution). Some of the reactions of colloids are reviewed in Table 14.1. Table 14.2 gives the types of colloids encountered in aquatic systems.

Table 14.1. The Colloidal Particle as a Reactant (Various Combinations of Reactions Are Possible)

Type of Property/ Reaction	Property or Reaction
Physical	Collector of other particles (aggregation of colloids, coagulation)
	Conveyor of chemicals (metals, pollutants, nutrients)
Chemical	Collector of hydrophobic solutes that accumulate at the surface because of expulsion from the water
	Organic or inorganic surface ligands (Lewis bases) that interact with protons or metal ions
	Lewis acids that bind ligands (anions and weak acids) (ligand exchange)
	Charged surface (mostly resulting incipiently from the adsorption of metal ions, H^+, and ligands) interacting with charged and polar surfaces
	Redox catalyst sorbing oxidants and reductants and mediating their interaction
	Electron acceptors or donors oxidizing or reducing solutes [Fe(III) oxides, Mn(III, IV) oxides, FeS_2 and sulfides, biogenic organic particles]
	Chromophore absorbing light to induce heterogeneous redox processes (including reductive dissolution of higher-valent oxides) (semiconductors)
Chemical–biological	Biological particle biochemically processing carbon and other nutrients, by generating or destroying alkalinity
	Extracellular enzymes hydrolyzing, oxidizing, or reducing solutes

Source: Adapted from Stumm (1993).

Table 14.2. Type of Colloids Present in Natural Systems

1. *River-borne Particles*

    - Products of weathering and soil colloids (e.g., aluminum silicates, kaolinite, gibbsite, SiO_2)
    - Iron(III) and manganese(III,IV) oxides
    - Phytoplankton, biological debris, humus colloids (colloidal humic acid), fibrils[a]
    - So-called dissolved iron(III) consists mainly of colloidal Fe(III) oxides stabilized by humic or fulvic acids

2. *Soil Colloids*

    - Kaolinite particles. Typically about 50 unit layers of hexagonal plates are stacked irregularly and interconnected through H-bonding between the OH groups of the octahedral sheet and the oxygens of the tetrahedral sheet (Figure 9.22) (Sposito, 1989)
    - Illite and other 2:1 layer type clay minerals. Plate-like particles stacked irregularly
    - Smectites and vermiculites have a lesser tendency to agglomerate
    - Humus, colloidal humic acids, fibrils
    - Iron hydrous oxides
    - Polymeric coatings of soil particles by humus, hydrous iron(III) oxides, and hydroxo-Al(III) compounds

3. *Sediment Colloids*
    In addition to the colloids listed above:

    - Sulfide and polysulfide colloids in anoxic sediments

4. *Organic and Biological Colloids*

    - Microorganisms, viruses, biocolloids, fibrils; aggregation of exudates and macromolecular organic matter

[a]Fibrils are elongated organic colloids with a diameter of 2–10 nm and composed in part of polysaccharide.

The colloids adsorb heavy metal ions and waterborne pollutants: hence the fates of reactive elements and of many pollutants in the environment depend to a large extent on the movement of colloids in aqueous systems. The colloids in natural waters are characterized by an extreme complexity and diversity: organisms, biological debris, organic macromolecules, various minerals, clays, and oxides, partially coated with organic matter.

Figure 14.1 gives a size spectrum of water borne particles and filter pores. Colloids usually remain suspended because their gravitational settling is less than 10^{-2} cm s^{-1}. Under simplifying conditions (spherical particles, low Reynolds numbers), Stokes' law gives for the settling velocity, v_s,

$$v_s = \frac{g}{18} \frac{\rho_s - \rho}{\eta} d^2 \tag{1}$$

14.1 Colloids

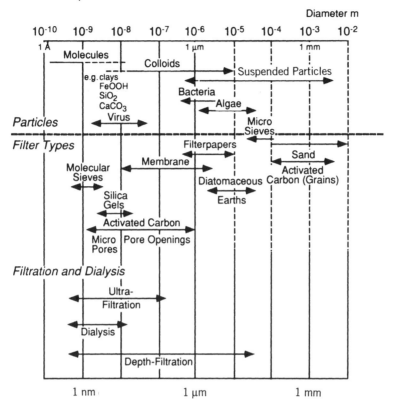

Figure 14.1. Suspended particles or colloids in natural and wastewaters vary in diameter from 0.001 to about 100 μm (1×10^{-9} to 10^{-4} m). For particles smaller than 10 μm, terminal gravitational settling will be less than about 10^{-2} cm s^{-1}. Filter pores of sand filters, on the other hand, are typically larger than 500 μm. The smaller particles (colloids) can become separated either by settling if they aggregate or by filtration if they attach to filter grains. Particle separation is of importance in the following processes: aggregation of suspended particles (clays, hydrous oxides, phytoplankton, biological debris) in natural waters; coagulation (and flocculation) in water supply and wastewater treatment; bioflocculation (aggregation of bacteria and other suspended solids) in biological treatment progresses; sludge conditioning (dewatering, filtration); and filtration, groundwater infiltration, and removal of precipitates.

where g is the gravity acceleration; ρ_s and ρ are the mass density of the particle and of water, respectively; η is the absolute viscosity (at 20°C, 0.001005 kg m^{-1} s^{-1}); and d is the diameter of the particle. Note that v_s is proportional to the square of the particle diameter. Equation 1 applies also to *flotation* (gravitational rising of suspended particles that are lighter than water; when $\rho_s < \rho$, $v_s < 0$).

Colloids are removed from the water either by settling if they aggregate or by filtration if they attach to the grains of the medium through which the solution passes (soils, groundwater carriers, technological filters). *Aggregation of particles* (clays, hydrous oxides, humus, microorganisms, phytoplankton) refers in a general sense to the agglomeration of particles to larger aggregates. The process by which a colloidal suspension becomes unstable and undergoes gravitational settling is called *coagulation*. Sometimes the term *flocculation* is used to describe aggregation of colloids by bridging polymers, but all these terms are often used interchangeably.

Aggregation of particles is important in natural water and soil systems, in groundwater infiltration and groundwater transport, and in water technology: for example, coagulation and flocculation in water supply and wastewater treatment, bioflocculation (aggregation of microorganisms and other suspended solids) in biological treatment processes, and sludge conditioning (dewatering, filtration). Flotation is used both in water technology and in the separation of a specific mineral component from a mixture. Oceans and lakes are settling basins for particles; coagulation in these basins can be sufficiently rapid and extensive to affect suspended particle concentrations and sedimenting fluxes significantly. A significant fraction of river-borne colloids and suspended matter is coagulated and settled in the estuaries.

The role of settling particles in regulating the concentrations of heavy metals in rivers, lakes, and oceans has been discussed in Chapter 10.

Stability of Colloids In dealing with colloids, the term *stability* has an entirely different meaning than in thermodynamics. A system containing colloidal particles is said to be stable if during the period of observation it is slow in changing its state of dispersion. The times for which sols[†] are stable may be years or fractions of a second. The large interface present in these systems represents a substantial free energy, which by recrystallization or agglomeration tends to reach a lower value; hence, thermodynamically, the lowest energy state is attained when the sol particles have been united into aggregates. The term *stability* is also used for particles having sizes larger than those of colloids; thus the stability of sols and suspensions can often be interpreted by the same concepts.

Historically, two classes of colloidal systems have been recognized; hydrophobic and hydrophilic colloids. In colloids of the second kind there is a strong affinity between the particles and water; in colloids of the first kind this affinity is negligible. There exists a gradual transition between hydrophilic and hydrophobic colloids. Gold sols, silver halogenides, and nonhydrated metal oxides are typical hydrophobic colloid systems. Gelatin, starch, gums and proteins, as well as most biocolloids (viruses, bacteria), are hydrophilic. Hydrophobic and hydrophilic colloids have a different stability in the same electrolyte solution. Macromolecular colloids and many biocolloids are often quite stable.

[†]A colloid dispersed in a fluid is known as a sol.

Many colloid surfaces relevant in water systems contain bound H_2O molecules at their surface. Adsorption of suitable polymers may impart stability for steric reasons.

As we shall see, colloid stability can be affected by electrolytes and by adsorbates that affect the surface charge of the colloids and by polymers that can affect particle interaction by forming bridges between them or by sterically stabilizing them.

Actinide Colloids Actinide cations undergo hydrolysis in water. Hydrolysis is a step to polynucleation and thus to the generation of actinide colloids; the polynuclear hydrolysis species become readily adsorbed to the surface of natural colloids. This also applies to ^{234}Th, daughter of the primordial radionuclide ^{238}U. ^{234}Th is partially polynuclear and colloidal and thus highly surface reactive and is removed from the water column through sorption and coagulation onto sinking particles. The power of using radioactive disequilibrium between parent–daughter nuclides of the U–Th decay series is that it adds an inherent kinetic element to scavenging models and the use of in situ radiochemical clocks (Honeyman and Santschi, 1992). A great part of M(III), M(IV), and M(VI) present in groundwater is colloid bound. Because of the possibility that such colloids migrate in an aquifer system, actinide colloids in groundwaters are presently a subject of various investigations (see Kim, 1991). Size distribution has been estimated by ultrafiltration, scanning electron microscopy, and photocorrelation spectroscopy. Ultrafiltration facilitates the characterization of a number of size groups, down to ~ 1-nm diameter (Buffle et al., 1992; Leppard, 1992), while scanning electron microscopy determines the particle number down to ~ 50 nm. The number counting of colloids can also be performed by photoacoustic detection of light scattering. The presence of actinide colloids can be verified by laser-induced photoacoustic spectroscopy (Kim, 1991). The use of field flow fractionation techniques to characterize aquatic particles is described by Beckett and Hart (1993).

Figure 14.1 lists size ranges for particles to become separated in dialysis, ultrafiltration, and membrane filtration, and we give some attention to the use of these procedures based on the experience of Buffle et al. (1992). For more details, their paper should be consulted.

The Use of (Membrane) Filtration to Separate "Particulate" from "Dissolved" Matter

In natural waters and soil and sediment systems, one needs to distinguish analytically between dissolved and particulate material. Figure 14.1 classifies various types of particulate and dissolved materials. Obviously, distinguishing between "dissolved" and "particulate" matter by filtration alone is often not successful, because size distributions of aquatic components vary in a continuous manner from angstroms to micrometers.

The use of filters and membranes of different pore size to accomplish a sequential size fractionation is in principle, and under certain circumstances, possible; it was proposed (for literature see Buffle, 1988, and Buffle et al., 1992) to estimate the size of the various colloids and macromolecules, and to determine to what extent trace elements are associated with various size fractions of colloids and macromolecules. Such sequential size fractionation techniques need to be applied with extreme caution; we list some of the reasons why these techniques may yield erroneous results:

1. Although most particles, larger than a given pore size, are normally retained, many smaller particles (sometimes 10–1000 times smaller than the pore size) may also be retained. If filtration occurs above and within a filter pore, colloidal particles smaller than the pore size become attached to the larger particles. The pore size distribution of membrane filters is often broad.
2. Coagulation may occur in the bulk sample and in the filter. Because of the long times involved in the filtration through <1-μm pores, coagulation in the bulk sample (and in the suspension above the filter) occurs. For the particle concentrations typically encountered in many natural waters (10^5–10^9 particles cm^{-3}), coagulation of half of the particles occurs over a period of hours or days, depending on chemical conditions. Filtration must be done as quickly as possible after sampling. Coagulation on and in the filter can occur, the extent increasing with the flow rate used for filtration.
3. Interaction of solutes can occur with the filter material and the retained particles. The problem is especially serious with heavy metals that adsorb especially at pH values of 7 and above on filter walls and on the filter materials (glass, acrylic copolymers, cellulose esters, polycarbonates, etc.). One also needs to consider that:
    (a) trace metal concentrations are often below 10^{-8} M (Figure 14.2 illustrates some of the problems mentioned for iron(III) hydroxyphosphate particles); and
    (b) the retained particles and the filter material are charged, the concentrations of solutes and colloids change during filtration, and adsorption occurs as a consequence of double-layer properties and of chemical interaction with functional groups.

There is no perfect way to distinguish between what is dissolved and what is nondissolved. Ion-selective electrodes respond selectively to solute ions but are often not sufficiently selective. One other possibility is to determine solute species by voltammetric techniques on Hg (or other) electrodes in the presence of colloids without prior centrifugation or filtration (Gonçalves et al., 1985, 1987; Müller and Sigg, 1990).

If the analytically determined "dissolved" phase contains colloids, artifacts in distribution coefficients, such as a so-called particle concentration effect

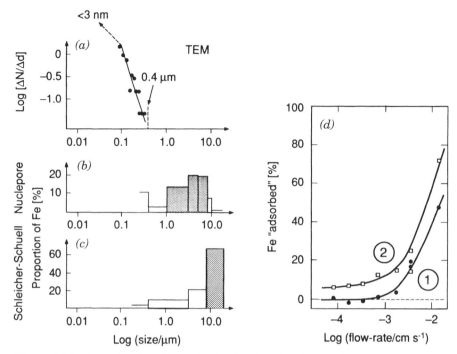

Figure 14.2. Operational problems in size fractionation by membrane filters. (a, b, c) Size distributions of iron oxyhydroxyphosphate particles obtained by transmission electron microscopy (true distribution) and syringe filtration on Nucleopore polycarbonate filters and Schleicher–Schuell cellulose ester depth filters. Iron particles formed at the oxic–anoxic interface of eutrophic Lake Bret (Switzerland). (d) Fraction of iron particles retained on 3.0-μm membranes, as a function of flow rate: ① Nucleopore polycarbonate, and ② Schleicher–Schuell cellulose nitrate. In the absence of coagulation or adsorption, no particle should be retained. (From Buffle et al., 1992.)

(e.g., K_D decreases with the concentration of particles), may be observed. (For details see Morel and Gschwend, 1987).

The arbitrary limit of a pore size of 0.2–0.5 μm in many sampling procedures has—besides all disadvantages—some advantages: particles not retained by such filters do not settle down in natural waters within days and "move" with the solutes; most bacteria and other organisms (except viruses) are retained by these filters and thus the filtered sample is often nearly sterile and less subject to microbially mediated changes. On the other hand, one needs to realize that the thermodynamic basis of all solution and heterogeneous equilibria refers to the conceptually defined solutes.

Interstitial Water The differentiation between solutes and particles is of great importance in the sampling of interstitial water. Often so-called peepers are used. These consist usually of plexiglass plates in which small compartments (0.5 cm deep and 0.5–1 cm high) are separated from the sediments by a dialysis membrane. The compartments are initially filled with degassed distilled water. After 1–2 weeks for equilibration subsequent to the retrieval of the peeper, the "dissolved" components are measured in each compartment. For this type of application the pore size does not seem to be very critical; colloids do not seem to accumulate in the compartments (because of low diffusion coefficients).

Davison et al. (1991) have proposed using devices with probing compartments filled with gels. Very high resolutions on vertical concentration gradients can be obtained this way. Davison and Zhang (1994) used ion permeable gel membranes for in situ speciation measurements of trace components in natural waters.

14.2 PARTICLE SIZE DISTRIBUTION

Particles may be sorted into size fractions (diameter, volume, mass, or number). The analysis of agglomeration, sediment transport, and dissolution processes of particles depends on information about their particle size distribution. Junge (1964) and Friedlander (1960) have advanced the understanding of atmospheric sciences by characterizing and explaining particle size distributions in atmospheric aerosols. In aquatic sciences similar developments should be forthcoming. Perhaps the simplest type of graphical representation is a bar histogram in which the particle number concentrations Δ (number) (vol^{-1}), found in each class interval, Δ (volume) or Δ (diameter), are plotted (discrete particle size distribution, Figure 14.3b). By measuring a large number of particles and making the class intervals approach zero, the smooth curve resulting represents the *continuous particle size distribution*. Usually the particle size distributions of natural materials (suspended, sedimentary, or airborne particles) are not normally distributed; they are often very asymmetric. (The normal frequency distribution as defined by Gauss is produced when an infinite number of factors cause independent variations of equal magnitude.)

Often it is convenient to plot the results of experimental measurements as a *cumulative particle size distribution* (Figure 14.3a).

Here N (number cm^{-3}) denotes the total concentration of particles with a volume equal to or less than v (μm^3); the total concentration of all particles is given by N_∞. The slope of this curve, $\Delta N/\Delta v$ or dN/dv, is called a *particle size distribution* and is represented as $n(v)$. In this case, $n(v)$ has units of number cm^{-3} μm^{-3}.

Particle volume is one of three common measures of particle size. Surface area(s) and diameter (d_p) are also used, so that three particle size distribution functions can be defined:

$$\frac{\Delta N}{\Delta v} = \frac{dN}{dv} = n(v) \quad \text{(number cm}^{-3}\ \mu\text{m}^{-3}\text{)} \tag{2a}$$

14.2 Particle Size Distribution 827

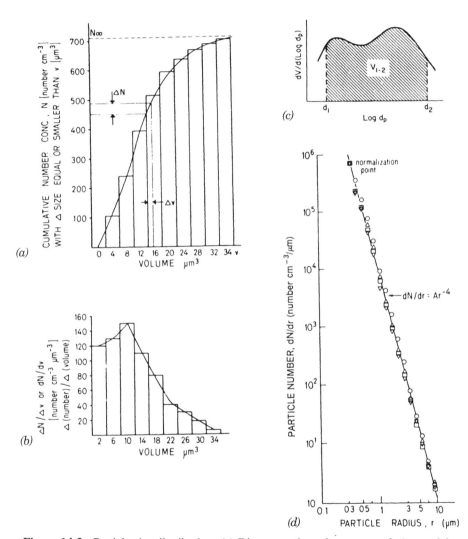

Figure 14.3. Particle size distribution. (a) Discrete and continuous *cumulative* particle size distribution. (b) Discrete and continuous particle size distribution. (c) Volume distribution plotted in accordance with equation 3. (d) Particle size distributions at four depths in a calcareous sediment from west equatorial Pacific Ocean, 1°6.0'S, 161°36.6'E, box core No. 136, water depth 3848 m. (From Lerman, 1979.)

$$\frac{\Delta N}{\Delta s} = \frac{dN}{ds} = n(s) \quad (\text{number cm}^{-3}\ \mu\text{m}^{-2}) \tag{2b}$$

$$\frac{\Delta N}{\Delta d_p} = \frac{dN}{d(d_p)} = n(d_p) \quad (\text{number cm}^{-3}\ \mu\text{m}^{-1}) \tag{2c}$$

These functions can be measured and used in both conceptual and empirical studies on coagulation and other particle transport processes.

The volume concentration of all particles in the interval between size 1 and size 2 may be written as

$$V_{1-2} = \int_1^2 dV$$

Dividing and multiplying the right-hand side by $d(\log d_p)$ yields

$$V_{1-2} = \int_1^2 \frac{dV}{d(\log d_p)} d(\log d_p) \tag{3}$$

A plot of $dV/d(\log d_p)$ versus $\log d_p$ is illustrated in Figure 14.3c. The total volume concentration provided by particles in the size interval from d_1 to d_2 is equal to the integrated (shaded) area in Figure 14.3c. In preparing such plots from field measurements, it is frequently assumed that the particles are spherical. If this assumption is not true, the area is proportional (not equal) to the volume concentration.

For a plot of $dS/d(\log d_p)$ versus $\log d_p$ that is similar to Figure 14.3c, the integrated area under the resulting curve from d_1 to d_2 would represent the total concentration of surface area in a suspension provided by particles in the size interval from d_1 to d_2. A plot of $dN/d(\log d_p)$ versus $\log d_p$ would provide similar information about the total number concentration.

The size distribution of atmospheric aerosols and of aquatic suspensions is often found to follow a power law of the form

$$n(d_p) = \frac{dN}{d(d_p)} = A\, d_p^{-\beta} \tag{4}$$

or

$$\log \frac{dN}{d(d_p)} = \log A - \beta \log d_p$$

in which A is a coefficient related to the total concentration of particulate matter in the system. As noted below, the exponent β has been determined experimentally and has also been shown on theoretical grounds to result from the interaction of various physical processes such as coagulation and settling.[†]

An integrated colloid concentration can be determined from a minimum value of $(d_p)_m$ to a maximum size $(d_p)_M$:

$$N = \frac{A}{\beta - 1} [(d_p)_m^{1-\beta} - (d_p)_M^{1-\beta}]$$

Measurements indicate that in natural waters values of β range from 2 to 5. Lerman (1979) reports measurements of size distributions at four locations in the North Atlantic. Fifty-three size distributions derived from samples taken at depths ranging from 30 to 5100 m yielded a mean value of $\beta = 4.01 \pm 0.28$. Fillela and Buffle (1993) report on size distributions based on particle number for different natural aquatic systems.

Friedlander (1960), Hunt (1980), Filella and Buffle (1993), and others have analyzed the effect of colloid agglomeration by coagulation and particle removal by settling on the shape of the particle size distribution function as expressed by equation 4. The predictions of model calculations are often consistent with the range of values of β observed in aquatic systems.

Figure 14.4a illustrates a simulation model by Filella and Buffle (1993) on the temporal evolution of a colloid system characterized initially by a continuous particle size distribution with $\beta = 4$[‡] and a size range from 1 nm to 100 μm.

Three different coagulation-sedimentation regimes can be identified:

1. Initially small particles that coagulate rapidly to form more stable agglomerates of about 100 nm (peak position depends on initial particle concentration).
2. Particles with intermediate initial size that coagulate slowly to agglomerates hardly larger than the initial particles.
3. Initially larger particles that are mainly affected by sedimentation.

The particle behavior predicted by the model agrees reasonably well with natural distributions.

[†]Equation 4 has some interesting characteristics. A particle size distribution that follows an empirical power law with $\beta = 1$ has an equal number of particles in each logarithmic size interval. Similarly, for $\beta = 3$, the concentration of surface is uniformly distributed in each logarithmic size interval. Finally, for $\beta = 4$, the volume of solids is equally distributed in each logarithmic size interval, while the surface area and number concentrations are primarily in the smaller sizes.

Some pollutants can be characterized in terms of mass or volume concentrations. Examples include oil, suspended solids, and certain precipitates. Other pollutants are concentrated at surfaces. Examples include DDT adsorbed on detritus and trace metals adsorbed on clays. For materials such as these, the surface concentration of the particulate phase is of interest. Still other pollutants (e.g., pathogenic organisms) are best considered in terms of their number concentration.
[‡]It has been suggested by Lerman (1979) that $\beta = 4$ when colloids are produced by erosion; when $\beta < 4$, aggregation may have occurred.

830   Particle–Particle Interaction: Colloids, Coagulation, and Filtration

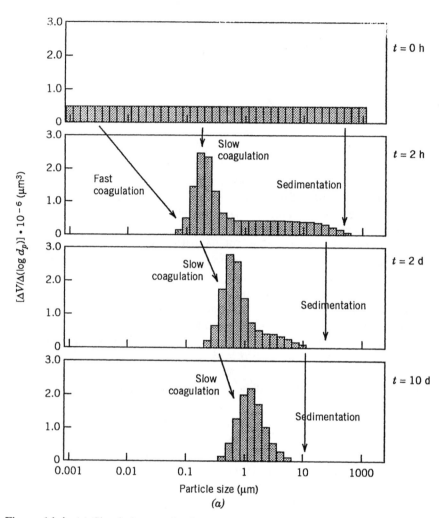

Figure 14.4. (a) Simulation results for a continuous initial particle size distribution, with $\beta = 4$ and size range from 1 nm to 100 μm: initial total mass concentration, 3 mg liter^{-1}: $\rho_s = 1.5$ g cm^{-3}; temperature = 15°C; coagulation with a velocity gradient $G = 10$ s^{-1} and a collision efficiency factor $\alpha = 0.05$ (see Sections 14.4 and 14.8 for the definition of these terms). (b, c) Simulation results for a continuous initial particle size distribution with $\beta = 4$ and size ranging from 1 nm to 100 μm after 2 days with different initial mass concentrations ranging from 0.01 to 10 mg liter^{-1} ($\rho_s = 2.0$ g cm^{-3}; temperature = 15°C; $G = 0.5$; $\alpha = 0.05$). (b) Evolution of particle size versus concentration with ordinates expressed as percentage of initial value for each size class. (c) Evolution of mean size value with concentration. (Adapted from Filella and Buffle, 1993.)

14.2 Particle Size Distribution

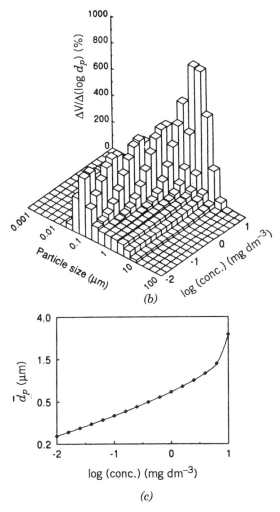

Figure 14.4. (*Continued*)

Figure 14.4b shows predictions on the distributions obtained after 2 days for samples having an initial continuous particle size distribution with $\beta = 4$ and size ranging from 1 nm to 100 μm where the initial mass concentration varies from 0.01 to 10 mg liter^{-1}. Figure 14.4c shows the change in the mean size with concentrations under the same conditions. As can be seen, the behavior of small particles (< 100 nm) is independent of particle concentration because they disappear instantaneously under all conditions. Intermediate and large particles are sensitive to changes in concentration. Obviously, open oceans, quiet rivers, and lakes, where particle concentrations are less than 5–10 mg liter^{-1} will show relatively stable size distribution patterns (Filella and

Buffle, 1993) and the distributions that occur in natural waters may reflect physical and chemical processes that have occurred in these systems.

Water Treatment Very significant effects of particle size distributions on the performance of water treatment processes have been proposed by Lawler et al. (1980). Particle mass or other parameters related to particle mass (turbidity) are important means to characterize the performance of solids separation processes. If particulate volume and particulate density are known, the relation between particle size and particle mass can be established. Assuming spherical particulates and a constant density, this relation can easily be derived.

Significant changes of particulate concentration and size distribution are observed along the different treatment steps of a conventional treatment plant. In Figure 14.5, the relative cumulative solids mass concentrations in a raw wastewater, in a settled wastewater, in an activated sludge mixed liquor, in a secondary effluent, and in a filtered effluent are shown as a function of particle size. If settled or unsettled, the colloidal particulates less than 10 μm dominate the mass distribution in raw wastewaters. These fine particulates are removed in the activated sludge stage by flocculation and hydrolysis and are, from a mass point of view, negligible compared to the large biological flocs of 100–500-μm diameter. The absence of large amounts of colloids in the secondary effluent accounts for the usually clear appearance of the water, which may be intensified by polishing the effluent with tertiary filters. In that last stage,

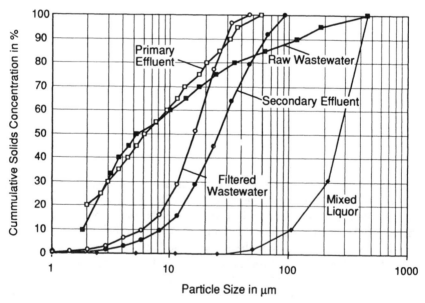

Figure 14.5. Examples of particle size distributions in raw sewage, primary effluent, mixed liquor, and secondary and tertiary effluents. (From Boller, 1992.)

especially particulates above 10 μm are removed nearly quantitatively. However, they account for the major part of the residual solids mass.

Particles in Lakes

With hydraulic residence times ranging from months to years, lakes are efficient settling basins for particles. Lacustrine sediments are sinks for nutrients and for pollutants such as heavy metals and synthetic organic compounds that associate with settling particles. Natural aggregation increases particle sizes and thus particle settling velocities and accelerates particle removal to the bottom sediments and decreases particle concentrations in the water column.

In field studies on coagulation and sedimentation in lakes (Weilenmann et al., 1989), the particle size distribution and concentrations at several depths in the water column of Lake Zurich were measured (Figure 14.6).

In this figure particle concentrations are expressed as the incremental change in particle volume concentration (ΔV) that is observed in an incremental logarithmic change in particle diameter ($\Delta \log d_p$). Plots such as in Figure 14.6 have the useful characteristic that the area under a curve between any two particle sizes is related to the total volume concentration of particles between these two sizes (cf. Figure 14.3c).

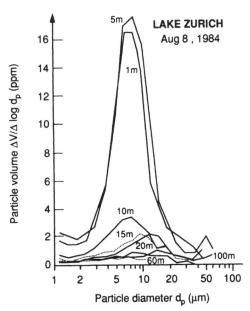

Figure 14.6. Observed particle volume concentration distributions in the water column of Lake Zurich. Samples obtained on August 8, 1984 at the depths indicated. (ppm = μl liter^{-1}.) (From Weilenmann et al., 1989.)

14.3 SURFACE CHARGE OF COLLOIDS

As we have seen in Section 9.5, solid particle surfaces can develop an electric surface charge in two principal ways: either from isomorphic substitutions or from the adsorption of charged solutes above all the interactions of surface functional groups with ions in aqueous solutions.

For the reader's convenience we summarize in Table 14.3 the most important different components of surface charge and the definitions related to them. This information has been described in detail in Section 9.5.

The surface charge is accessible from the quantities of H^+ and OH^- cations and anions that are bound to the surface; for example, in the case of the adsorption of M^{2+} to an oxide surface,

$$Q_P = \{S-OH_2^+\} + \{S-OM^+\} - \{S-O^-\} \tag{5}$$

where Q_P is the surface charge (mol kg^{-1}) accumulated at the interface and braces indicate concentrations in mol kg^{-1}; Q_P can be converted into the net total particle charge density σ_P (C m^{-2}) [$\sigma_P = Q_P F/s$, where F is the Faraday constant and s is the specific surface area (m^2 kg^{-1})]. If a sufficient amount of M^{2+} sorbs, the surface will have a more positive charge than if proton exchange reactions alone were governing the surface charge. Correspondingly, surface-bound ligands tend to decrease the net surface charge.

The stability of colloids is dependent on the surface charge density of the colloidal particles, which, in turn, depends on chemical surface speciation.

How do we determine the net particle surface charge density? Historically, surface charge has often been inferred from electrophoretic mobility (u) (see the discussion on zeta potential in Section 9.5). Semiquantitatively, this is very convenient because of the availability of expedient instrumentation. As shown in Figure 14.7, a qualitative connection between u and σ_P can be inferred from the variation of u and σ_P with pH. Furthermore, when $u = 0$, $\sigma_P = 0$ and $\psi = 0$. But the quantitative relationship between u and σ_P is highly model dependent.

The conversion of u into the zeta potential depends on the molecular structure of the double layer of the colloid surface. The zeta potential is smaller than the surface potential because it reflects the inner potential difference across that portion of the electrochemical double layer which does not migrate in an applied field.

In principle, the surface charge of a colloidal particle is assessable when we can determine the various components (equation i of Table 14.3) that make up the total net surface charge.

The *intrinsic surface charge density* ($\sigma_{in} = \sigma_0 + \sigma_H$) can be determined by the Schoffield method. In this method particles are reacted with an electrolyte (e.g., NaCl) at a given pH value and ionic strength; the specific surface excess of the cation and anion adsorbed from the electrolyte is determined (see Sposito, 1989, 1992).

14.3 Surface Charge of Colloids

Table 14.3. Summary of Components of Surface Charge of Colloidal Particles[a]

1. *Total net surface charge density* σ_P (e.g., in C m^{-2})

$$\sigma_P = \sigma_0 + \sigma_H + \sigma_{IS} + \sigma_{OS} \quad \text{(i)}[b]$$

where σ_0 = surface charge (permanent) density developed from isomorphic substitution
σ_H = net proton charge density due to binding of protons and OH$^-$ ions
σ_{IS} = charge density due to inner-sphere surface complexes
σ_{OS} = charge density due to outer-sphere surface complexes

2. *Diffuse layer charge*

$$\sigma_P + \sigma_D = 0 \quad \text{(ii)}$$

where σ_D = charge density due to ions in diffuse layer

3. *Points of zero charge*

Symbol	Name	Defining condition	
p.z.c.	point of zero charge	$\sigma_p = 0$	(iii)
p.z.n.p.c.	point of zero net proton charge	$\sigma_H = 0$	(iv)

4. *Surface charge density and surface potential,* ψ (V)

$$\sigma_P = (8 \, RT\epsilon\epsilon_0 c \times 10^3)^{1/2} \sinh(Z\psi F/2 \, RT) \quad \text{(v)}$$

where R is the molar gas constant (8.314 J mol^{-1} K^{-1}), T the absolute temperature (K), ϵ the dielectric constant of water (ϵ = 78.5 at 25°C), ϵ_0 the permittivity of free space (8.854 × 10^{-12} C V^{-1} m^{-1} or 8.854 × 10^{-12} C^2 J^{-1} m^{-1}), and c the molar electrolyte concentration (M). Valid for a symmetrical electrolyte (Z = charge number). For 25°C this equation can be written as

$$\sigma_P = 0.1174 \, c^{1/2} \sinh(Z\psi \times 19.46) \quad \text{(va)}$$

and

$$\sigma_P \simeq 2.3 \, I^{1/2} \, \psi \quad \text{(vb)}$$

where c = the electrolyte concentration in M
σ_P = has the units C m^{-2}

[a] Compare Section 9.5.
[b] One refers to the intrinsic surface charge density as $\sigma_{in} = \sigma_0 + \sigma_H$ and sometimes to the Stern surface charge density as $\sigma_S = \sigma_{IS} + \sigma_{OS}$.

The *net proton surface charge density*, σ_H, can be determined from alkalimetric/acidimetric titration, as described in Example 9.2.

The *inner-sphere and outer-sphere surface charge* can be determined analytically, for example, by measuring the concentration of the metal ions or ligands on the surface of the particles; this is usually done by determining the

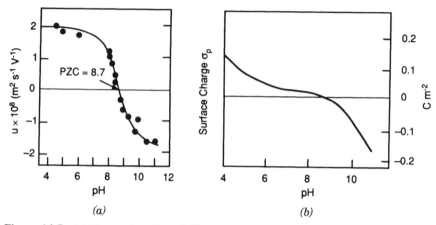

Figure 14.7. (a) Electrophoretic mobility and (b) surface charge of γ-Al$_2$O$_3$ in presence of 10^{-3} M NaCl. Surface charge data from alkalimetric titration curve. (Data from Stumm et al., 1970.)

concentrations of the adsorbates in solution in the presence and absence of the colloids. Sometimes, it is more convenient to measure directly the metal ions and ligands bound to the surface, for example, subsequent to elution from the solid phase.

Determination of Surface Charge from Surface Complex Formation Equilibria (or from Adsorption Isotherms)

Mass law considerations allow the calculation of the surface-bound species and thus of the surface charge. For example, for the interaction of M^{2+} with the surface

$$S\text{—}OH + M^{2+} \rightleftharpoons S\text{—}OM^+ + H^+ \tag{6}$$

the following equilibrium expression is valid:

$$\{S\text{—}OM^+\}\,[H^+]/\{S\text{—}OH\}\,[M^{2+}] = K_M^s = K_M^s(\text{intr}) \exp(-\Delta Z F \psi / RT)$$

K_M^s, the apparent constant, is the product of an intrinsic constant (a constant valid for a hypothetical uncharged surface) and a Boltzmann factor. ψ is the surface potential, F the Faraday constant, and ΔZ the change in the charge number of the surface species of the reaction for which the equilibrium constant is defined (in this case $\Delta Z = +1$). The intrinsic constant is experimentally accessible by extrapolating experimental data to the surface charge where $\sigma_P = 0$ and where $\psi = 0$. The correction, as given above, assumes the classical diffuse double-layer model (a planar surface and a diffuse layer of counterions).

The value of ψ can be estimated from σ_P on the basis of the Gouy–Chapman model (see equation v in Table 14.3).

As we have seen, computer programs are available to make equilibrium calculations where the corrections for charge effects are iteratively considered. Figure 14.8 shows mass law calculations of surface site density and net surface charge for hematite colloids that have interacted at various pH values with different ligands or with Cu^{2+} or Cd^{2+}. Equilibrium constants were corrected for electrostatic effects by the Gouy–Chapman model ($I = 10^{-2}$ M). These calculated curves may be compared with the proton-dependent charge (H^+/OH^- only) obtained for the hematite surface in the absence of specific adsorbates. Obviously ligands decrease surface charge and lower the pH_{pzc}, while metal ions increase surface charge and raise pH_{pzc}. As shown in Figure 14.8b some of the charge versus pH curves may go through minima. This is simply a consequence of competitive proton–metal ion or ligand–hydroxide ion equilibria that are pH dependent.

These calculations reflect equation i of Table 14.3, which shows that different types of surface charges contribute to the net total particle charge density on a colloid.

Since the complex formation model can estimate the net surface charge, σ_P, and thus in turn the surface potential, ψ, (equation v, Table 14.3), the colloid stability and pH_{pzc} can, at least in principle, be predicted (see Figure 14.8b).

Where to Obtain Surface Complex Formation Constants

There are an increasing number of data in the literature. A compilation for surface complexation constants for inorganic surface species on hydrous ferric oxide is available (Dzombak and Morel, 1990). Often correlations from solute complex formation constants with those of surface complex formation are possible. The good correlation obtained in this and similar linear free energy relationships (LFERs) is exemplified in Figure 14.9.

The reasonably good correlation indicates that the same chemical mode of interaction occurs in solution and at the surface and that the available sorption data are consistent with one another, and that therefore such LFERs may be used to predict (intrinsic) sorption constants from solute complex formation constants and vice versa.

14.4 COLLOID STABILITY: QUALITATIVE CONSIDERATIONS

In a qualitative way, colloids are stable when they are electrically charged (we will not consider here the stability of hydrophilic colloids—gelatin, starch, proteins, macromolecules, biocolloids—where stability may be enhanced by steric arrangements and the affinity of organic functional groups to water). In a physical model of colloid stability, particle repulsion due to electrostatic interaction is counteracted by attraction due to van der Waals interaction. The

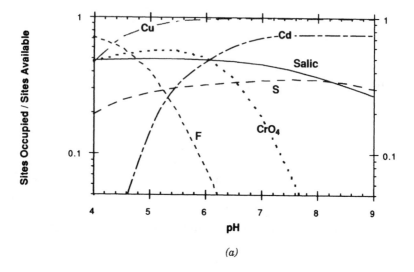

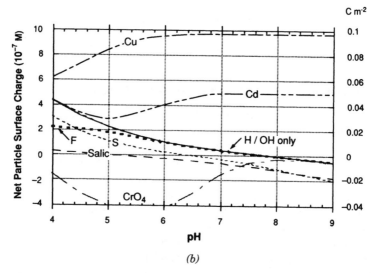

Figure 14.8. (a) Surface site density and (b) net surface charge for surface complexes of some metal ions and ligands in comparison with the charge due to H^+ and OH^- ions only, exemplifying some equilibrium calculations with the surface complex formation model (in each case total concentration of metal ion and of ligand is $[A]_{tot} = 10^{-4}$ M, $[\equiv FeOH]_{tot} = 10^{-6}$ M, and $I = 10^{-2}$ M). 3 mg Fe_2O_3 liter^{-1}, specific surface area = 40 m^2 per g Fe_2O_3, 5 sites per nm^2. The charges of the species were omitted for simplicity: for example, the surfaces are $\equiv FeOCu^+$, $\equiv FeOCd^+$, $\equiv Fe(salicylate)^{-1}$, $\equiv FeS^-$, $\equiv FeCrO_4^-$, and $\equiv FeF$.

14.4 Colloid Stability: Qualitative Considerations

repulsion energy depends on the surface potential and its decrease in the diffuse part of the double layer; the decay of the potential with distance is a function of the ionic strength (see Figures 9.19 and 9.20). The van der Waals attraction energy, to a first approximation, is inversely proportional to the square of the intercolloid distance.

Figure 14.10 illustrates that the repulsion energy is affected by ionic strength while attraction energy is not. At small separations attraction outweighs repulsion and at intermediate separations repulsion predominates. With increasing ionic strength, the attraction is preponderant over larger interparticle distances. As Figure 14.10 shows, a shallow so-called secondary minimum in the net interaction energy may lead to particle attraction at larger distances between the particles.

As this figure illustrates, increasing ionic strength (addition of salts) compresses the electric double layer and decreases the colloid stability. Thus colloids tend to be less stable in estuarine and ocean waters.

But often more important than the effects of the electrolytes that influence the thickness of the electric double layer are many solutes that, upon adsorption onto the colloid surface, reduce or modify the surface charge. The specific adsorption of H^+, OH^-, metal ions, and ligands (as well as the attachment of polymers) to the colloid surface affects the surface charge and the surface potential and, in turn, the colloid stability.

Colloid Stability and Kinetics

Aggregation of colloids is primarily a kinetic phenomenon and colloid stability can be characterized quantitatively by measuring the rate of aggregation. In a simple way the rate of aggregation of a suspension can be written as follows (O'Melia and Tiller, 1993):

$$\frac{dN}{dt} = -k_a N^2 \qquad (7)$$

where N is the number concentration of particles in suspension at time t and k_a is a second-order rate constant that is a function of physical and chemical properties of the system.

k_a can be characterized in terms of

$$k_a = \alpha_a \beta \qquad (8)$$

where β is a mass transport coefficient with dimensions such as $m^3 \, s^{-1}$; it is largely physical and often evaluated theoretically (see also Figure 14.21); and α_a is a dimensionless sticking coefficient that is primarily chemical and often determined experimentally.

α_a can be visualized as a collision efficiency factor; that is, it gives the fraction of collisions that are successful (e.g., $\alpha = 10^{-4}$ means that one out

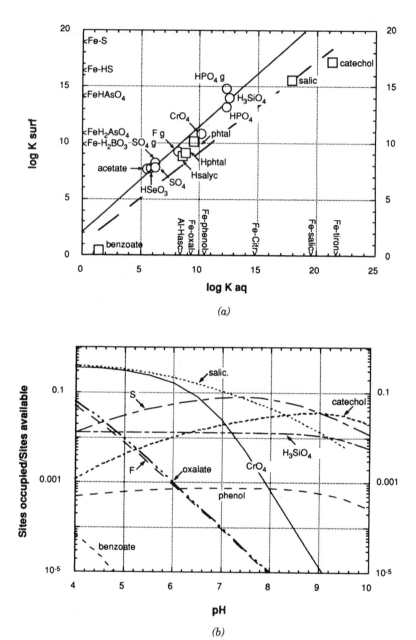

Figure 14.9. (a) Linear free energy relation between the tendency to form solute complexes of Fe(III)aq and Al(III)(aq)

$$\text{MeOH}^{2+} + \text{H}^+ + \text{A}^- = \text{MeA}^{2+} + \text{H}_2\text{O} \qquad K_1(\text{aq})$$

and the tendency to form surface complexes (intrinsic equilibrium constant) on $\gamma\text{-Al}_2\text{O}_3$ and hydrous ferric oxide or goethite surfaces

$$\equiv\text{MeOH} + \text{H}^+ + \text{A}^- = \equiv\text{MeA} + \text{H}_2\text{O} \qquad K^s(\text{surf})$$

14.4 Colloid Stability: Qualitative Considerations

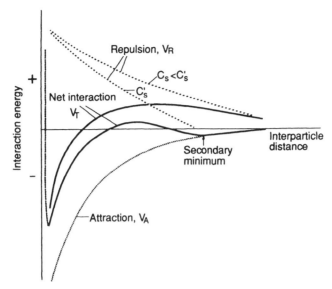

Figure 14.10. Physical model for colloid stability. Schematic forms of the curves of interaction energies [electrostatic repulsion V_R, van der Waals attraction V_A, and total (net) interaction V_T] as a function of the distance of surface separation. Summing up repulsive (conventionally considered positive) and attractive energies (considered negative) gives the total energy of interaction. Electrolyte concentration C_s is smaller than C_s'. At very small distances a repulsion between the atomic electronic clouds (Born repulsion) becomes effective. Thus at the distance of closest approach, a deep potential energy minimum reflecting particle aggregation occurs. A shallow so-called secondary minimum may cause a kind of aggregation that is easily counteracted by stirring.

(A is the actual species that forms the complex; for example, $A^- = H_3SiO_4^-$ and $\equiv FeA = \equiv FeH_3SiO_4$.) Equilibrium constants in solution ($I = 0$) are from Smith and Martell (1976) (constants given for Fe^{3+} were converted into constants valid for $FeOH^{2+}$ by $\log K = -2.2$ for the reaction $Fe^{3+} + H_2O = FeOH^{2+} + H^+$). Data for surface complex formation on hydrous ferric oxide (○) are from Dzombak and Morel (1990), for goethite (marked with a "g") from Sigg and Stumm (1981), and for $\gamma\text{-}Al_2O_3$ (□) from Kummert and Stumm (1980). These data are intrinsic equilibrium constants, that is, extrapolated to zero surface charge. At the ordinate and abscissa a few relevant surface complex formation constants and solute equilibrium constants, respectively, are listed for which the constants in solution or at the surface are not known; they may be used to estimate the corresponding unknown constant. (b) Fractional surface coverage of $\equiv Fe(III)$ surface complexes as a function of pH. The calculation is based on the condition

$$[\equiv FeOH]_{tot} = [A]_{tot} = 10^{-6} \, M \quad I = 10^{-2}$$

Electrostatic correction was made with the diffuse (Gouy–Chapman) double-layer model. The figure shows the effect of pH on the relative extent of surface complexation.

842 Particle–Particle Interaction: Colloids, Coagulation, and Filtration

of 10^4 collisions leads to an agglomeration) or, for given physical circumstances (fluid motion), α_a can be defined as

$$\alpha_a = \frac{\text{rate at which particles attach}}{\text{rate at which particles collide}} \qquad (9)$$

Colloid stability can be measured experimentally by determining α_a (under conditions of constant β) by comparing the effective rate of agglomeration with the theoretical one.

The *stability ratio* is defined as $W = 1/\alpha_a$.

14.5 EFFECTS OF SURFACE SPECIATION ON COLLOID STABILITY

In order to illustrate how surface charge is affected by chemical interactions, we first describe the surface of an oxide colloid and characterize quantitatively the interaction of H^+ and OH^- ions and ligands with this oxide surface. We then adsorb ligands such as humic acids and "coat" the oxide with organic matter.

Example 14.1. Surface Charge as a Function of pH for Hematite We use as an introductory example the pH dependence of the surface charge of hematite (a case already encountered in Chapter 9; see Figure 9.20) and its effect on the coagulation rate (Liang and Morgan, 1990). In Figure 14.11 the potentiometrically determined (alkalimetric–acidimetric titration) surface charge is given as a function of pH for different ionic strengths. The surface charge may be compared with the measured electrophoretic mobility (Figure 14.11b). (From the latter, the zeta potential, ζ, can be calculated, which is, however—with the exception of pH_{pzc} where $\zeta = \psi = 0$—smaller than ψ.) The colloid stability in Figure 14.11c is expressed as the stability ratio, W (equal to α_a^{-1}), and is measured experimentally by comparing the actual coagulation rate with that predicted theoretically. W is the ratio of the "fast" coagulation rate to the actual rate. Relative colloid stability is obtained from measurements of coagulation rates. In these the particle concentrations are measured as a function of time and the rate is calculated from fitting equation 7. The concentration of the particles as a function of time can, under suitable conditions, be determined spectrophotometrically. The theoretical basis of such measurements go back to Troelstra and Kruyt (1943). At $pH = 8.5$, the pH_{pzc} of the colloidal hematite particles, the stability ratio is unity for all ionic strengths. At $pH \simeq 5$ and low ionic strength, coagulation is slowed down a thousand-fold relative to that of pH_{pzc}, a consequence of the high positive charge density from $\equiv FeOH_2^+$.

Calculate the concentration of surface species and the surface charge as a function of pH for a hematite suspension, which has the same characteristic as that used in the experiments of Liang and Morgan (1990).

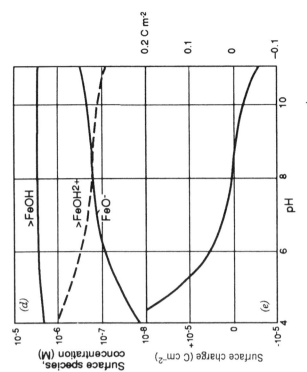
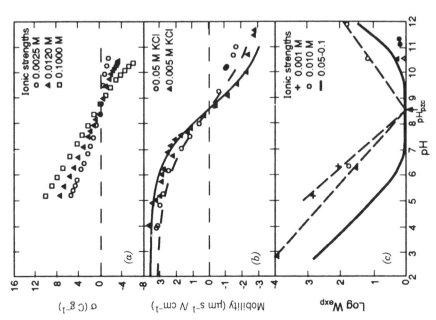

Figure 14.11. Comparison of hematite (a) surface charge (C g^{-1}), (b) electrophoretic mobility, and (c) stability ratio, W_{exp}, as a function of pH. Note that at pH$_{pzc}$ the net surface charge and mobility are both zero, and the stability is a minimum. The experimental stability ratio (W_{exp}), the potentiometrically determined surface charge, and the electrokinetic mobility of 70-nm particles over the pH range from 3 to 11 are shown. The solid line in (c) summarizes experiments obtained with $I = 0.05$–0.1. (Adapted from Liang and Morgan, 1990.) (d, e) Results of simple equilibrium calculations that have been made with equilibrium constants and surface characteristics of α-Fe$_2$O$_3$ given by Liang and Morgan (see Example 14.1).

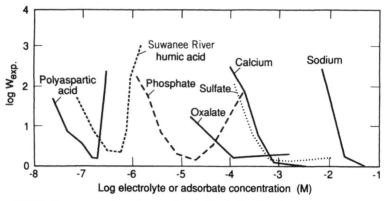

Figure 14.12. Summary plot of experimentally derived stability ratios, W_{exp}, of hematite suspensions, as a function of added electrolyte or adsorbate concentration (in case of polymers, monomer units) at pH around 6.5 (pH = 10.5 for Ca^{2+} and Na^+). Hematite concentration is about 10–20 mg liter^{-1}. The stability ratio, W_{exp}, was determined from measurements on the coagulation rate; it is the reciprocal of the experimentally determined collision efficiency factor, α_p. (From Liang and Morgan, 1990.)

Specific surface area is 40 m^2 g^{-1}; acidity constants of $\equiv FeOH_2^+$ are pK_{a1}^s (int) = 7.25 and K_{a2}^s (int) = 9.75; site density = 4.8 nm^{-2}; hematite concentration = 10 mg liter^{-1}. Ionic strength = 0.005 M. For the calculation, the diffuse double-layer model shall be used.

The results are given in Figures 14.11d and 14.11e. The semiquantitative agreement between experimental data and calculated data is obvious. The surface charge estimated can be converted into a surface potential on the basis of the diffuse double-layer model from which a stability could be calculated. Alternatively, a Stern model approach may be used, incorporating a distance of closest approach of outer-sphere ions (Section 9.5).

Figure 14.12 exemplifies the effect of various solutes on the colloid stability of hematite at pH around 6.5 (or pH = 10.5 for Ca^{2+} and Na^+) (Liang and Morgan, 1990).

Simple electrolyte ions like Cl^-, Na^+, SO_4^{2-}, Mg^{2+}, and Ca^{2+} destabilize the iron(III) oxide colloids by compressing the electric double layer, that is, by balancing the surface charge of the hematite with "counterions" in the diffuse part of the double layer. Usually these "simple" electrolyte ions are not—to any large extent—specifically bound[†] to the oxide surface and are thus not able to cause a charge reversal. In contrast, as Figure 14.12 illustrates, oxalate, phosphate, are humic and fulvic acids, as well as polyelectrolytes such as polyaspartic acid, are specifically bound (ligand exchange) and are able, at

---

[†]Specifically adsorbed species are those that are bound by interactions other than electrostatic ones. To what extent SO_4^{2-} and Ca^{2+} can form inner-sphere complexes is not yet well established. SO_4^{2-} is able to shift the point of zero proton condition of several oxides.

14.5 Effects of Surface Speciation on Colloid Stability

higher concentrations, to cause a charge reversal; that is, the surface complex bound inorganic and organic species alter the metal oxide surface charge (charge or potential-determining ions). The surface complex formation constants of different ligands can be used to predict the surface speciation and the surface charge from which the surface potential can be calculated. The effectiveness of ions in altering the surface charge of iron(III) oxide surfaces and in coagulating or stabilizing colloidal Fe_2O_3 under comparable pH conditions with respect to pH_{pzc} (+ or −) is given by the decreasing sequence of the surface complex-forming constants, for example, phosphate > oxalate > phthalate.

Similar information in a more general way is given in Figure 14.13, where schematic curves of residual turbidity as a function of coagulant dosage for a

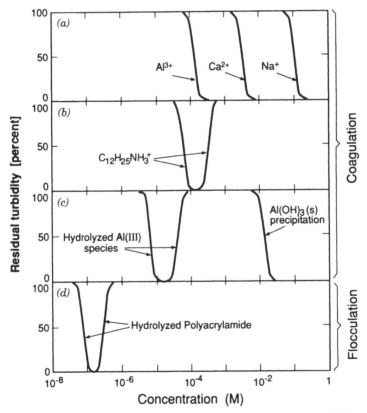

Figure 14.13. Schematic agglomeration curves for several different destabilizing agents: (a) "compaction" of the double layer by counterions in accordance with the Schulze–Hardy rule; (b) coagulation by specifically adsorbable organic cation; (c) coagulation and restabilization by Al-hydroxo polymers; at higher dosage $Al(OH)_3$ precipitates and enmeshes dispersed colloids; and (d) destabilization by strongly sorbable polymers of charge equal to that of the colloid. Coagulation results from "bridging" of the colloids by polymer chains (concentration is given in monomer units).

natural water treated with Na^+, Ca^{2+}, and Al^{3+}, with hydrolyzed metal ions, or with polymeric species are given. It is evident that there are dramatic differences in the coagulating abilities of simple ions (Na^+, Ca^{2+}, Al^{3+}), hydrolyzed metal ions (charged multimeric hydroxo metal species), and other species that interact chemically with the colloid, including species of large ionic or molecular size.

Obviously, *two different mechanisms* can govern coagulation.

1. Inert electrolytes, that is, ions that are not specifically adsorbed, compress the double layer and thus reduce the stability of the colloids (Figure 14.10). A critical coagulation concentration, C_s or ccc, can be defined that is independent of the concentration of the colloids (Schulze–Hardy rule).[†]

2. Specifically adsorbable species (e.g., species that form surface complexes) affect the surface charge of the colloids and therefore their surface potentials, thereby affecting colloid stability. The critical coagulation concentration, ccc, for a given colloid surface area concentration, decreases with increasing affinity of the sorbable species to the colloid. The ccc is no longer independent of the colloid concentration; there is a "stoichiometry" between the concentration necessary to just coagulate and the concentration of the colloids. This is illustrated in the simple example of Figure 14.14. The charge reversal of Fe_2O_3 colloids by phosphate is of importance in the phosphate removal by Fe(III) from sewage and wastes (Figure 14.15).

In *seawater*, the "thickness" of the double layer as given by κ^{-1} (equation 9.40c) is a few angstroms, equal approximately to a hydrated ion. In other words, the double layer is compressed and hydrophobic colloids, unless stabilized by specific adsorption or by polymers, should coagulate. Coagulation of this kind is observed in the estuaries where river water becomes progressively enriched with electrolytes (Figure 14.19a). That these colloids exist in seawater for reasonable time periods may be caused by (1) the small concentrations of colloids (relatively small contact opportunities), (2) the stabilization of colloids by specific adsorption (e.g., of organic polymers), and (3) the presence of at least partially hydrophobic biological particles that dominate the detrital phases.

Specifically sorbable species able to coagulate colloids at low concentrations may also restabilize these dispersions at higher concentrations. When the destabilization agent and the colloid are of opposite charge, this restabilization is accompanied by a reversal of the charge of the colloidal particles. Purely Coulombic attraction would not permit an attraction of counterions in excess of the original surface charge of the colloid.

[†]The Schulze–Hardy rule stipulates that the ccc decreases with the charge of the counterion of mono, di, and trivalent ions in the ratio of $(1/z)^6$. (But the Schulze–Hardy rule is usually used in a qualitative sense to indicate the significance of the valence of the counterion to bring about destabilization.)

14.5 Effects of Surface Speciation on Colloid Stability

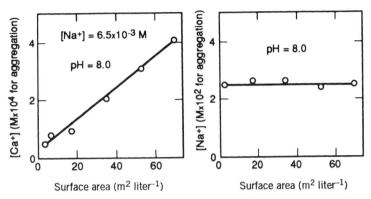

Figure 14.14. Relationship between MnO_2 colloid surface area concentration and ccc of Ca^{2+}. A stoichiometric relationship exists between ccc and the surface area concentration; in the case of Na^+, however, this interaction is weaker, so that primarily compaction of the diffuse part of the double layer causes destabilization. (From Stumm et al., 1970.)

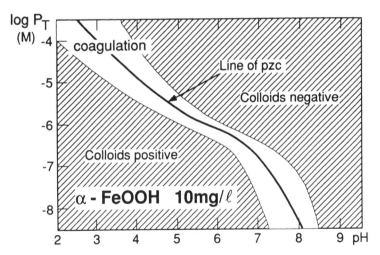

Figure 14.15. Colloid stability in the Fe(III) (hydr)oxide–phosphate system. Surface complexation equilibria were used to calculate the concentration domains of positively charged and negatively charged colloids and of nearly uncharged phosphate surface complexes on FeOOH. (From Stumm and Sigg, 1979.)

Example 14.2. Reversal of Surface Charge of Hematite by the Interaction with a Ligand Estimate the variation of surface charge of a hematite suspension (same characteristics as that used in Example 14.1) to which various concentrations of a ligand H_2U are added that forms bidentate surface complexes with the Fe(III) surface groups $\equiv FeU^-$; such a ligand could be oxalate, phthalate, or salicylate or could serve as a simplified model for a humic acid; we assume acidity constants and surface complex formation constants representative for such ligands. Assume pH = 6.5 and $I = 0.005$ M.

The following equilibrium constants characterize the system:

$$H_2U = H^+ + HU^- \qquad \log K_{a1} = -5 \qquad (i)$$

$$HU^- = H^+ + U^{2-} \qquad \log K_{a2} = -9 \qquad (ii)$$

$$\equiv FeOH_2^+ = \equiv FeOH + H^+ \qquad \log K_1^s = -7.25 \qquad (iii)$$

$$\equiv FeOH = \equiv FeO^- + H^+ \qquad \log K_2^s = -9.75 \qquad (iv)$$

$$\equiv FeOH + H_2U = \equiv FeU^- + H^+ + H_2O \qquad \log K^s = 2 \qquad (v)$$

$$H_2U_T = [H_2U] + [HU^-] + [U^{2-}] + [\equiv FeU] \qquad (vi)$$

$$[\equiv FeOH_T] = [\equiv FeOH_2^+] + [\equiv FeOH]$$
$$+ [\equiv FeO^-] + [\equiv FeU^-] \qquad (vii)$$

The diffuse double-layer model is used to correct for electrostatic effects. The calculation is made with a MICROQL program incorporating surface chemical interactions and adapted for a personal computer. Figure 14.16 gives the calculations for increasing concentrations of H_2U_T. The surface charge is reversed at a concentration of $\sim 2 \times 10^{-6}$ M. This is in qualitative agreement with the experimentally obtained curve for humic acids, given in Figure 14.12. With humic acids the experimentally determined concentration for charge reversal is somewhat smaller than that calculated in Figure 14.16; this can perhaps be accounted for by inferring a polymeric effect (association of humate anions at the surface).

Adsorption of and Coagulation by Fatty Acids

Figure 14.17 gives experimental results on coagulation of hematite suspension by fatty acids of different chain lengths. As concentration is increased, an influence on hematite coagulation rate becomes noticeable. Increase in fatty acid concentration first makes hematite less stable, and a minimum stability ratio is reached. The stability ratio then increases sharply as the fatty acid concentration is increased beyond the critical coagulation concentration. Plots of relative coagulation rate versus fatty acid concentration for C_8, C_{10}, and C_{12} are similar to one another, but successive critical coagulation concentrations differ by a factor of about 10. For C_3, the stability ratio merely flattens out as the concentration is increased beyond a critical coagulation concentration. That a charge reversal occurs for C_8, C_{10}, and C_{12} acids can only be accounted for by two-dimensional association of the adsorbed anions of the fatty acids (hemicelle formation). The situation is comparable to that described for the adsorption of sodium dodecyl sulfate on alumina, described in Figure 9.30.

14.5 Effects of Surface Speciation on Colloid Stability

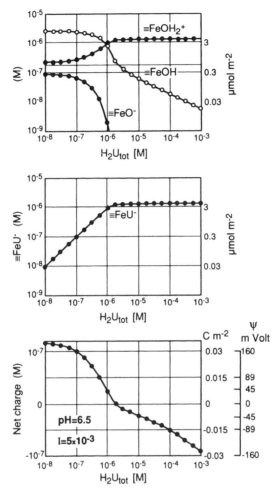

Figure 14.16. Interaction of hematite with a bidentate ligand H_2U. The relative concentrations of surface species, expressed as M, are given as a function of H_2U_T (added to the system). Coagulation is expected to occur at concentrations near the charge reversal. Conditions are given in Example 14.2 (pH = 6.5, $I = 5 \times 10^{-3}$). Individual points refer to computed data.

Polymers

The adsorption of polymers on colloids can (1) enhance colloid stability or (2) induce flocculation.[†]

In steric stabilization the colloids are covered with a polymer sheath, stabilizing the sol against coagulation by electrolytes (Figure 9.32). In sensitiza-

---

[†]Some authors distinguish between coagulation and flocculation and use the latter term for colloid agglomeration by bridging of polymer chains.

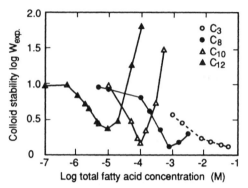

Figure 14.17. Experimentally derived stability ratio, W_{exp}, of hematite suspensions, plotted as a function of fatty acid concentration at pH 5.2. The ionic strength is 50 mM NaCl and hematite concentration is 34.0 mg liter^{-1}. Lauric acid is denoted by C_{12}, capric acid by C_{10}, caprylic acid by C_8, and propionic acid by C_3. (From Liang and Morgan, 1990.)

tion or adsorption flocculation, the addition of very small concentrations of polymers or polyelectrolytes leads to destabilization (Lyklema, 1985).

As shown in Figure 14.13d, polymers can destabilize colloids even if they are of the same charge as the colloids. In polymer adsorption (cf. Figure 9.32) chemical adsorption interaction may outweigh electrostatic repulsion. Coagulation is then achieved by *bridging* of the polymers attached to the particles. LaMer and co-workers have developed a chemical bridging theory that proposes that the extended segments attached to one of the particles can interact with vacant sites on another colloidal particle.

A schematic representation of the effect of ionic strength and pH on the configuration of anionic polyelectrolytes, such as humic substances, is presented in Figure 9.33 (Yokoyama et al., 1988). In fresh waters at neutral and alkaline pH, charged macromolecules assume extended shapes (large hydrodynamic radii) as a result of intramolecular electrostatic repulsive interactions. When adsorbed at interfaces under these conditions, they assume flat configurations (small hydrodynamic thickness). At high ionic strength or at low pH, the polyelectrolytes have a coiled configuration in solution (small hydrodynamic radii) and extend further from the solid surface when adsorbed (with large hydrodynamic thickness).

Tiller and O'Melia (1993) compared their polymer adsorption statistics calculations with data from laboratory experiments on the interaction of polyacrylic and polyaspartic acid as well as of humic acids with hematite. Their conclusion is that, at low ionic strength, anionic polyelectrolytes affect the coagulation of positively charged particles by altering the net surface charge in a way similar to that of multivalent, monomeric anions. Steric repulsion, at low ionic strength (i.e., in fresh water), plays little or no role in the stabilization of hematite colloids by the organic macromolecules used in their work; calcium

14.5 Effects of Surface Speciation on Colloid Stability

attenuates the stabilizing effect. But at high ionic strength (e.g., seawater), the conformational characteristics of adsorbed polyelectrolytes can directly affect coagulation, because under these conditions the adsorbed layer thickness can be of the same order as the Debye length, δ_D (Figure 9.19c, equation 9.40c). The latter is given, in the absence of an organic layer, approximately by

$$\delta_D = \kappa^{-1} = 0.28/I^{0.5} \quad (nm)$$

Thus, for fresh waters ($I \approx 10^{-3}$), the Debye length is on the order of 10 nm, while in seawater, $\delta_D \approx 0.4$ nm.

According to the argument presented, the adsorbed organic layer is characterized in fresh water by a hydrodynamic thickness that is small in comparison with the Debye length; colloid stability is, under these conditions, ruled by double-layer repulsion. In seawater, however, the hydrodynamic thickness of the adsorbed layer is of a similar order of magnitude or larger than the Debye length, so that steric repulsion becomes relatively more important in stabilizing the colloids. Organic colloids can be stable in seawater because they contain sterically stabilizing macromolecular segments; inorganic colloids can be stable because adsorbed polymers such as humic substances provide stabilizing segments.

Coagulation in Lakes

The observation of Weilenmann et al. (1989) (see Figure 14.6) on some Swiss hard water lakes shows that two important factors affecting the stability of particles in lakes are (1) [Ca^{2+}] and (2) organic matter (like humic substances). Ca^{2+} destabilizes and organic matter stabilizes colloidal particles in lakes.

The enhanced stability of particles in waters containing natural organic matter is a consistent observation without a clear cause (O'Melia and Tiller, 1993). Some speculation is presented here. Humic substances comprise the principal fraction of dissolved organic matter in most natural waters. These molecules can be considered as flexible polyelectrolytes with anionic functional groups; they also have hydrophobic components. In fresh waters such molecules assume extended shapes due to intramolecular electrostatic repulsive interactions. When adsorbed at interfaces at low ionic strength, they assume flat configurations. Adsorption on inorganic surfaces such as metal oxides (e.g., iron oxides) could result from the ligand exchange of functional groups on the humic substances (carboxylic, phenolic) with surface hydroxyl groups on the metal oxides, supplemented by a hydrophobic interaction involving nonpolar components of the humic molecules. The result would be an accumulation of negative charge on the surface of the oxide as a result of the adsorbed organic substances.

Humic substances in lakes result from autochthonous biological processes within the lake and from allochthonous inputs from terrestrial sources. The macromolecular biological debris produced by biological wastewater treatment plants can have chemical and physical characteristics similar to natural organic

852 Particle–Particle Interaction: Colloids, Coagulation, and Filtration

substances and might provide a source of stabilizing organic matter when discharged into lakes.

An interesting corollary to these arguments is that extensive sewage treatment not only removes organic C and phosphate (which in turn causes a decrease in DOC) but may affect the entire lake metabolism by indirectly enhancing the coagulation and in turn increasing sediment fluxes and suspended particle concentrations significantly. In a lake of low humic acids (DOC) and high $[Ca^{2+}]$, photosynthetically produced biomass and other particles are transported more rapidly into the hypolimnion.

14.6 SOME WATER-TECHNOLOGICAL CONSIDERATIONS IN COAGULATION, FILTRATION, AND FLOTATION

Coagulation is a unit process in water supply and treatment. Coagulants are used to remove color (humic and fulvic acids) and turbidity (light-scattering particles) from the raw water. Coagulation techniques have also been used in waste treatment and sludge conditioning. Often hydrolyzing metal ions, Al(III) or Fe(III), are used as coagulants (Figures 14.13 and 14.18). The Al(III) salt

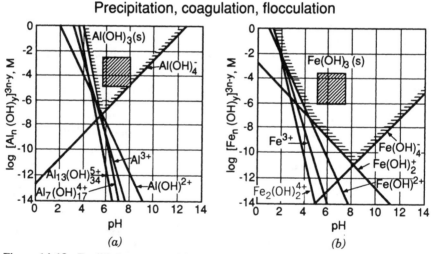

Figure 14.18. Equilibrium composition of solutions in contact with freshly precipitated Al(OH)$_3$ and Fe(OH)$_3$, calculated using representative values for the equilibrium constants for solubility and hydrolysis equilibria. Shaded areas are approximate operating regions in water treatment practice; coagulation in these systems occurs under conditions of oversaturation with respect to the metal hydroxide. Oversaturation is induced by adjusting the pH of incipiently acidic Fe(III) and Al(III) solutions. After initiation of the oversaturation, the hydrolysis of Al(III) and Fe(III) progresses and charged multimeric hydroxo Al(III) and Fe(III) species, the actual coagulants, are formed as intermediates; ultimately, these intermediates polymerize to solid Al(III) and Fe(III) (hydr)oxides, respectively. (From Stumm and O'Melia, 1968.)

14.6 Some Water-Technological Considerations

[or Fe(III) salt] added undergoes hydrolysis and forms polymeric or oligomeric charged hydroxo complexes of various structures such as, for example,

$$\left(\begin{array}{c} HO \\ H_2O \end{array} Al \begin{array}{c} O \\ OH \end{array} Al-OH-Al- \right)_n^{m+}$$

Such species adsorb specifically and modify the surface charge of the colloids. As shown in Figure 14.13 these hydrolysis products have a different effect from Al^{3+} (at low pH); they are able to reverse the surface charge. While nonhydrolyzed Al^{3+} (at low pH) compresses the double layer, hydrolyzed Al(III) coagulants interact with the colloids through specific adsorption; the dose necessary for coagulation depends on the concentration of the colloids (actually their surface area concentration) (Figure 14.19). Stoichiometry prevails also in the removal of color. The Al added reacts largely by ligand exchange with the hydroxo groups of the coagulant hydroxo Al species.

Al(III) and Fe(III) salts are acids. The addition of these chemicals to water is similar to an acidimetric titration of the water. As a result, the pH of the system after the addition of these coagulants will depend on the coagulant dosage and the alkalinity of the water or wastewater. Coagulation by Fe(III) and Al(III) polymers would be expected to exhibit restabilization (overdosing) and stoichiometry, if adsorption is important.

Schematic curves of residual turbidity as a function of coagulant dosage at constant pH for waters treated with Al(III) or Fe(III) salts are presented in

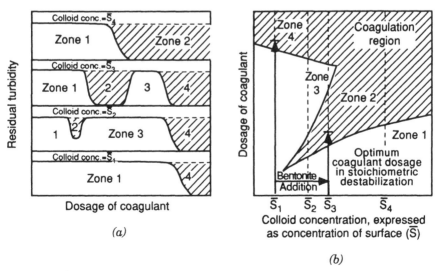

Figure 14.19. Schematic representation of coagulation observed in jar tests using aluminum(III) or iron(III) salts at constant pH. (From Stumm and O'Melia, 1968.)

Figure 14.19. Four curves are presented, each for a water containing a different concentration of colloidal material. The colloid concentration is represented by the concentration of colloidal surface per unit volume of suspension, $\bar{S}$ (e.g., as m^2 $liter^{-1}$). Each of these schematic coagulation curves is subdivided into four zones. In zone 1, corresponding to low dosage, insufficient coagulant has been added to bring about destabilization. Increasing the coagulant dosage produces destabilization and permits rapid aggregation (zone 2). A further increase in dosage can restabilize the dispersions at some pH level (zone 3).

In zone 4, a sufficient degree of oversaturation occurs to produce a rapid precipitation of a large quantity of aluminum of ferric hydroxide, enmeshing the colloidal particles in what has been termed a "sweep floc." Figure 14.19b is a schematic representation of the interrelationships between coagulant dosage and colloid concentration (m^2 $liter^{-1}$) at constant pH. At low colloid concentrations ($\bar{S}_1$ in Figure 14.19b), coagulation requires the production of a large excess of amorphous hydroxide precipitate. For systems containing a low concentration of colloid it has been proposed that insufficient contact opportunities exist to produce aggregates of even completely destabilized particles in a reasonable detention time. Such conditions may prevail in many water treatment plants when the turbidity of the raw water is low.

At higher concentrations of colloid ($\bar{S}_2$ and $\bar{S}_3$ in Figure 14.19b), a smaller coagulant dosage is required than for coagulation that involves the precipitation of the metal hydroxide. In this region, increasing colloid concentration requires an increasing coagulant dosage; that is, a stoichiometry in coagulation is observed. The destabilization zone (zone 2) is observed to widen with increasing colloid concentration (curves for $\bar{S}_2$ and $\bar{S}_3$ in Figure 14.19a). At very high colloid concentrations of the order encountered in sludge conditioning in wastewater treatment plants, a high coagulant dosage is required to destabilize the colloid ($\bar{S}_4$). Under these circumstances the coagulant (metal polymer) dosage required for destabilization may equal or exceed that required for the precipitative coagulation of dilute suspension.

Figure 14.20 exemplifies a coagulation curve; the particles are SiO_2 and the coagulant is hydrolyzed Fe(III). An understanding of these properties of hydrolyzing metal salts can be useful in coagulation practice. Consider first that Al(III) salts can be effective as coagulants in present practice in two ways: by adsorption to produce charge neutralization and by enmeshment in a precipitate. The chemical dosage that is required depends on how destabilization is achieved. High dosages of coagulant to produce a gelatinous metal hydroxide precipitate can be effective in these and other situations where low but objectionable concentrations of colloidal materials are present. Al(III) coagulants are also used to remove "color" from the water. The color is usually due to the presence of humic or fulvic acids. The removal of color is due to the formation of mixed Al(III) hydroxo-, humato aggregates. At a given pH, the Al(III) needed is proportional to the color constituents of the waters. In waters that are both turbid and colored, the required total alum dosage is usually dictated by the concentration of color constituents.

14.6 Some Water-Technological Considerations

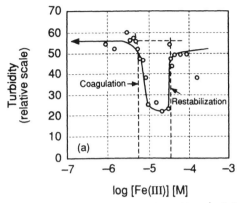

Figure 14.20. Coagulation of a SiO_2 suspension: 0.8 g liter^{-1}; 6.6 m^2 liter^{-1} by Fe(III) at pH = 5. (From data by O'Melia and Stumm, 1967.)

An additional method exists for the effective coagulation of low-turbidity waters, that is, the addition of more particles in the form of a coagulant aid. Bentonite (a finely divided clay) and some forms of activated silica serve this function.

Flotation

Flotation is a solid–liquid separation process that transfers solids to the liquid surface through attachment of gas bubbles to solid particles. Flotation processes are used in the processing of crushed ores, whereby a desired mineral is separated from the gangue or non-mineral-containing material. Various applications in solid separation processes are also in use in waste and water supply treatment. Algae have been removed successfully from surface waters by flotation.

In flotation, all three states of matter—solid, liquid, and gas—are involved and each of these involves surface chemistry. The various process steps that are involved include:

1. The generation of gas (air) bubbles and the selective attachment of these bubbles to the particles to be removed.
2. The addition of chemicals (additives) that adsorb (selectively) on the particles to render the attachment of air bubbles possible.
3. The actual flotation process, the production of froth, and the continuous operation in a flow-through reactor.

Collectors are additives that adsorb on the particle (mineral) surface and prepare the surface for attachment to an air bubble so that it will float to the surface. Collectors must adsorb *selectively* to render a fractionation of different solids possible.

Suitable collectors can render hydrophilic minerals such as silicas or hydroxides hydrophobic. An ideal collector is a substance that attaches with the help of a functional group to the solid (mineral) surface often by ligand exchange or electrostatic interaction and exposes hydrophobic groups toward the water. Thus amphipathic substances, such as alkyl compounds with C_8 to C_{18} chains, are widely used with carboxylates or amine polar heads. Surfactants that form hemicelles on the surface are also suitable. For sulfide minerals, mercaptans, monothiocarbonates, and dithiophosphates are used as collectors. Xanthates or their oxidation products, dixanthogen

$$(R-O-\underset{\underset{S}{\|}}{C}-S-)_2$$

are used as collectors for many ores. The S group can specifically sorb (ligand exchange), for example, to lead and copper ores in reactions such as

$$\equiv Pb-OH + S(CSOR)_2 \longrightarrow \equiv Pb-S(CSOR)_2^+ + OH^-$$

Frothing agents are intended to stabilize the particle (mineral)–air mixture (foam) at the surface of the flotation tank. Alkyl or amyl alcohols in the C_5 to C_{12} range are typical frothers. They lower γ_{SV}, which is beneficial to the stability of the foam; to some extent collectors and frothers may counteract each other in their effects so that compromise conditions must be selected.

Bacterial Adhesion: Hydrophobic and Electrostatic Parameters

In many natural and technical systems, metabolically active bacteria are found to be associated with interfaces. Bacterial adhesion is of importance in the formation of biofilms. Interaction of bacteria in soils is of great importance. *Bioflocculation,* the aggregation of microorganisms in biological waste treatment, is essential for the functioning of biological waste treatment. It has been proposed by many investigators (e.g., Busch and Stumm, 1968) that polymers, which are either excreted by microorganisms or exposed at the surface of cells, are responsible for the aggregation of microorganisms. Such polymers were shown to accumulate under conditions of declining bacterial growth. These polymers were also shown to destabilize hydrous oxide colloids. (Refer to section on polymers, above.)

Van Loosdrecht et al. (1990) have investigated systematically the adhesion of microorganisms to solid surfaces in aquatic environments and describe the initial adhesion process in terms of a colloid-chemical theory. These workers focus on the interplay of (1) electrostatic interactions between bacterial surfaces and solid surfaces and (2) on the hydrophobic and steric energies of these surfaces. Adsorbed polymers can influence bacteria–surface interactions in several ways: (1) electrostatic repulsions, (2) steric hindrance caused by monolayer saturations, and (3) bridging between surfaces at lower polymer coverages. Thus polymer charge, polymer molecular weight, and adsorbent properties of solid substrate and cell surface all play consequential roles in bacterial adhesion.

14.7 FILTRATION COMPARED WITH COAGULATION

Moving particles can come into contact with each other; when they attach to each other, the process is termed aggregation or coagulation. Suspended particles may also come into contact with stationary ones or with other fixed solid boundaries; when they attach, the process is termed deposition or filtration.

It has been proved valuable to consider the processes of physicochemical aggregation and deposition as comprising two separate and sequential steps, a transport step followed by an attachment step.

Thus filtration in many regards is analogous to coagulation. Juxtaposing the simple equation for aggregation (which we already discussed) with that of deposition,

$$\frac{dN}{dt} = -k_a N^2 \tag{7}$$

$$\frac{dN}{dL} = -k_d N \tag{10}$$

where L is the distance along the length of the porous medium and k_d is a pseudo-first-order rate constant that also depends on physical and chemical properties of the system.

Deposition or filtration involves collisions between suspended particles (n) and stationary collectors in the porous medium; the number of these stationary collectors is included in the distance L, whereas we expressed k_a as a product of α_a for aggregation and β, $k_a = \alpha_a \beta$. We can write for deposition or filtration

$$k_d \propto \alpha_d \eta \tag{11}$$

where η is a dimensionless mass transport coefficient that, like β, is primarily physical and frequently determined theoretically; and α_a and α_d are dimensionless sticking coefficients or attachment probabilities that are typically chemical and measured experimentally.

Alpha in Deposition The sticking coefficient or attachment probability, α_d, of particles depositing in porous media is defined as follows:

$$\alpha_d = \frac{\text{rate at which particles attach to the collector}}{\text{rate at which particles approach the collector}} \tag{12}$$

14.8 TRANSPORT IN AGGREGATION AND DEPOSITION

Mass transport coefficients for aggregation (coagulation) (β) and deposition (η) (equations 8 and 11),

$$k_a = \alpha_a \beta \qquad (8)$$

$$k_d \propto \alpha_d \eta \qquad (11)$$

can be derived from physical considerations of the transport mechanisms of suspended particles and for deposition in porous media.

As Figure 14.21 illustrates, three processes of mass transport are considered:

1. *Brownian* or molecular *diffusion* by which random motion of small particles is brought about by thermal effects.
2. *Fluid flow,* either laminar or turbulent. Velocity differences or gradients occur in all real flowing fluids. Particles that follow the motion of the suspending fluid will then travel at different velocities. These fluid and particle velocity differences or gradients can produce interparticle contacts among particles suspended in the fluid (Figure 14.21). Particle transport in this case depends on the mean velocity gradient in the fluid, G

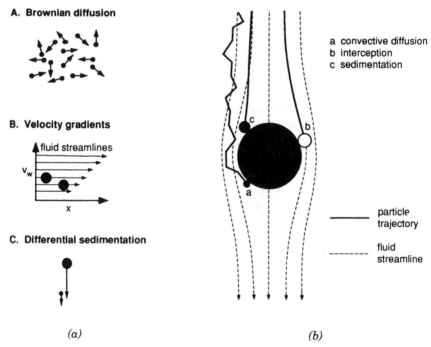

Figure 14.21. Transport mechanisms. (Adapted from O'Melia and Tiller, 1993.)

14.8 Transport in Aggregation and Deposition

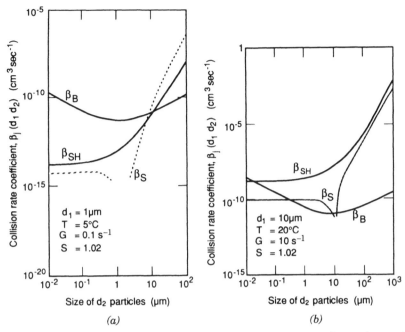

Figure 14.22. Effects of particle size on collision rate constants for agglomeration: (a) $d_1 = 1$ μm and (b) $d_1 = 10$ μm.

(s^{-1}).† When one of the particles is stationary, as in a porous medium, and the other is transported by a flowing fluid, contact is said to be by interception (Figure 14.21).

3. *Gravity*, which produces vertical transport of particles and which depends on the buoyant weight of the particles. Large, dense suspended particles can contact smaller or less dense ones in a process termed differential sedimentation (Figure 14.21a). If one of the particles is stationary, as in a packed bed filter, contacts of suspended particles with the fixed particle can be said to occur by convective sedimentation (Figure 14.21b).

These rate constants are compared for two cases in Figure 14.22. It follows that heterogeneity in particle size can significantly increase agglomeration rates.

Example 14.3. Estimating Agglomeration Rates (a) Estimate the time necessary to halve the particle concentration of a turbid water of 10^6 uniform ($d_1 = d_2$) particles per cm³ that are completely destabilized ($\alpha_a = 1$). Assume that the Brownian motion alone ($d \ll 1$ μm) is responsible for the collision. According to equation 4 in Table 14.4, β_p for 20°C is of the order of 5×10^{-12} cm³ s⁻¹.

---
†The velocity gradient G can be expressed as $G = dv_w/dx$ (cm s⁻¹/cm).

Applying equation 7 with this rate constant, we find

$$-\frac{dN}{dt} = \beta_p N^2 \qquad \text{(i)}$$

$$\frac{1}{N} - \frac{1}{N_0} = \beta_p t \qquad \text{(ii)}$$

with $\beta_p = 5 \times 10^{-12}$ cm^3 s^{-1} and $N = \frac{1}{2}N_0$, we find $t_{1/2} = 2 \times 10^5$ s or ~ 2.3 days.

(b) Estimate the time necessary to halve the particle concentration of a turbid water of 10^6 uniform particles ($d_1 = d_2$) per cm^3 that are brought into contact by shear (velocity gradient, $G = 5$ s^{-1})†; the initial diameter of the particles is $d = 1$ μm.

Using equation 5 of Table 14.4, we calculate for a value of $\beta_0 = 3 \times 10^{-12}$ cm^3 s^{-1}, which gives an estimate for $t_{1/2}$ of 3.3×10^5 s or ~ 3.8 days.

Example 14.4. Effects of Particle Size on Agglomeration Rate Compare the agglomeration rate of an aqueous suspension containing 10^4 virus particles per cubic centimeter ($d = 0.01$ μm) with that of a suspension containing, in addition to the virus particles, 10 mg liter^{-1} bentonite (number conc. $= 7.35 \times 10^6$ cm^{-3}; $d = 1$ μm). The mixture is stirred, $G = 10$ s^{-1}, and the temperature is 25°C. Complete destabilization, $\alpha = 1$, may be assumed. (This example is from O'Melia, 1978.)

Let us neglect bentonite–bentonite particle interactions. We calculate from the equations given in Table 14.4 the following rate constants:

$$\beta_p = 2 \times 10^{-12} \text{ cm}^3 \text{ s}^{-1} \qquad \text{(i)}$$

$$\beta_b = 3.1 \times 10^{-10} \text{ cm}^3 \text{ s}^{-1} \qquad \text{(ii)}$$

$$\beta_{sh} = 1.7 \times 10^{-12} \text{ cm}^3 \text{ s}^{-1} \qquad \text{(iii)}$$

$$\beta_s = 7.8 \times 10^{-13} \text{ cm}^3 \text{ s}^{-1} \qquad \text{(iv)}$$

According to equation 4 (Table 14.4), the time required to halve the concentration of the virus particles in the suspension containing the virus particles alone would be almost 200 days. In the presence of bentonite ($\beta_b = 3.1 \times 10^{-10}$ cm^3 s^{-1} and $N_{d_2} = 7.35 \times 10^6$ cm^{-3}), we find after integrating that the free virus concentration after 1 h of contact is only 2.6 particles cm^{-3}.

This example illustrates that the presence of larger particles may aid significantly in the removal of smaller ones, even when Brownian diffusion is the predominant transport mechanism (cf. Figure 14.4).

†$G = 5$ s^{-1} corresponds to slow stirring in a beaker.

14.8 Transport in Aggregation and Deposition

Table 14.4. Agglomeration Kinetics of Colloidal Suspensions[a]

Transport Mechanism	Rate Constant for Heterodisperse Suspensions		Rate Constant if $d_1 = d_2$[b]	
Brownian diffusion	$\beta_b = \dfrac{2}{3}\dfrac{k_B T}{\eta}\dfrac{(d_1 + d_2)^2}{d_1 d_2}$	(1)	$\beta_p = \dfrac{4 k_B T}{3\eta}$	(4)
Laminar shear	$\beta_{sh} = \dfrac{(d_1 + d_2)^3}{6} G$	(2)	$\beta_0 = \tfrac{2}{3} d_p^3 G$	(5)
Differential settling	$\beta_s = \dfrac{\pi g(\rho_s - \rho)}{72\eta}(d_1 + d_2)^3 (d_1 - d_2)$	(3)	$\beta_s = 0$	(6)

[a] The rate at which particles of sizes d_1 and d_2 come into contact by the jth transport mechanism is given by $F_j = \beta_j N_{d_1} N_{d_2}$.

d = particle diameter
F_j = collision rate in collisions per unit volume
β_j = bimolecular rate constant for the jth mechanism
N_{d_1}, N_{d_2} = number concentrations of particles of size d_1 and d_2, respectively
k_B = Boltzmann constant
η = absolute viscosity (kg m^{-1} s^{-1})
ρ_s, ρ = mass density of the solids and water
g = gravity acceleration
G = mean velocity gradient (s^{-1})
T = absolute temperature

[b] A factor of 2 is applied so that the collisions are not counted twice.

Figure 14.23a illustrates how coagulation in natural systems and in water and waste treatment systems depends on comparable variables. In natural waters, long detention times may provide sufficient contact opportunities despite very small collision frequencies (small G and small N). In fresh water, the collision efficiency is usually also low ($\alpha \simeq 10^{-3}$ to 10^{-6}; that is, only 1 out of 10^3–10^6 collisions leads to a successful agglomeration). In seawater, colloids are less stable ($\alpha \simeq 0.1$–1) because of the high salinity. κ^{-1} for seawater is about 0.36 nm; hence seawater double layers are nondiffuse. Estuaries with their salinity gradients and tidal movements represent gigantic natural coagulation reactors where much of the dispersed colloidal matter of rivers settles. In water and waste treatment systems, we can reduce detection time (volume of the tank) by adding coagulants at a proper dosage ($\alpha \to 1$) and by adjusting the power input (G). If the concentration ϕ is too small, it can be increased by adding additional colloids as so-called coagulation aids. Similar trade-offs in operation and design exist for optimizing particle removal in filtration.

Figure 14.23b shows how, in the progression from very slow filtration in groundwater percolation to ultrahigh-rate contact filtration, a relatively constant efficiency in particle removal (constant product $\alpha \eta L/d$) is maintained despite a dramatic increase in filtration rate. This is achieved by counterbalancing decreased contact opportunities (decreasing η and L and increasing d) by im-

862 Particle–Particle Interaction: Colloids, Coagulation, and Filtration

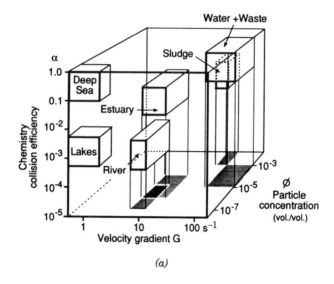

(a)

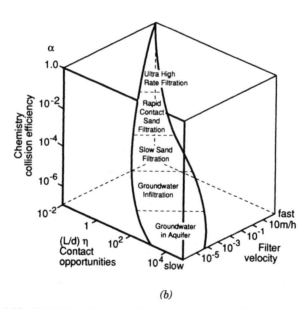

(b)

Figure 14.23. Variables that typically determine the efficiency of coagulation and filtration in natural waters and in water and waste treatment systems. (a) How the variables determine the coagulation efficiency. (b) Marked increase in filtration rate can be achieved by counterbalancing a reduction in contact opportunities by chemically improving the contact efficiency, with similar efficiency in particle removal.

proved effectiveness of particle attachment (through natural release or addition of suitable chemical destabilizing agents that increase α to 1).

Contact Filtration

As is evident from equations 11 and 12, the efficiency of filtration can be improved by improving α_d, the attachment of the particles to the filter grains. The considerations applied to the destabilization of colloids can also be used in filtration. A practical example is given in Figure 14.24 for phosphate elimination in sewage treatment.

Colloid Transport in Aquifers and Packed Bed Filtration

Substantial concentrations of colloids have been found in aquifers. To what extent are such colloids transported in packed bed filtration and in aquifers and to what extent are such colloids generated within the filtration media? Do colloids, capable of adsorbing substantial amounts of contaminants, contribute to the migration of organic pollutants, heavy metals, and radionuclides in groundwater? As shown in Figure 14.22, for colloids, the major transport mechanism is convective diffusion and the capture efficiency should increase with decreasing diameter, whereas for particles larger than ~ 1 μm interception and sedimentation are dominant. The mechanisms by which attached particles can detach from filter grains are not very well understood. In very porous filter grains or in systems with cracks, colloids may become chromatographically

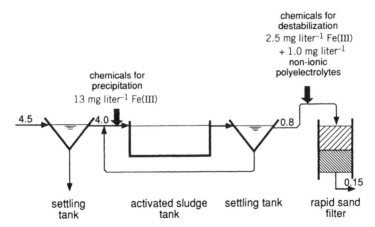

Figure 14.24. Combination of processes to eliminate phosphate in waste treatment. The numbers in the figure give typical concentrations of P (mg P per liter). The arrows after the primary settling tank and after the final settling tank indicate the addition of phosphate precipitants and flocculants. Fe(III) or Al(III) can be used for the precipitation of phosphate in the activated sludge tank and as a destabilizer in the final sand filter. (From Boller, EAWAG, 1993.)

separated from ions by *size exclusion*. [While ions diffuse readily into the pores of the filter grains, colloids cannot enter into the porous structure and thus colloid retardation (with regard to water) may under certain circumstances be smaller than that of solutes.] For a discussion on colloid transport in aquifers, see O'Melia (1990), Gschwend and Reynolds (1987), Ryan and Gschwend (1990), MacDowell-Boyer (1992), and Hahn (1994).

Figure 14.25 gives the travel distance required to deposit 99% of the particles in a suspension as a function of the size (radius) of these particles for two chemical conditions ($\alpha_d = 1$ and $\alpha_d = 0.001$). Theoretical background for the results in Figure 14.25 are given by Tobiason and O'Melia (1988) and experimental corroboration was made by Martin et al. (1992). For questions of colloid transport in porous media see also McCarthy and Zachara (1989), McCarthy and Degueldre (1993), and Geschwend and Reynolds (1987).

To predict the transport and fate of colloids in the subsurface, it is important to understand both the mechanism of particle deposition and that of remobilization in porous media. Experiments by MacDowell-Boyer (1992) and Monte Carlo simulations of Brownian particles near the surfaces of the media indicate that the secondary stability minimum (see physical model on colloid stability, Figure 14.10) can play an important role in the deposition and reentrainment of submicron particles at ionic strengths relevant to groundwater.

The secondary well is established by the surface and solution chemistry of

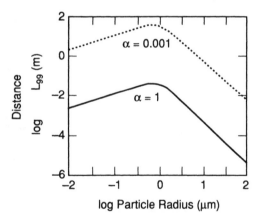

Figure 14.25. Elimination of particles in a porous filter medium. Ordinate gives the length of a filter that is necessary to remove 99% of the particles. The figure illustrates the importance of the attachability (α) and of the radius. α can be influenced by the addition of suitable chemicals. Model calculation by Tobiason and O'Melia (1988) for the following conditions:

$$\text{Linear flow velocity} = 0.1 \text{ m d}^{-1}$$
$$\text{Diameter grain of filter medium} = 0.025 \text{ cm}$$
$$\text{Density of particles} = 1.05 \text{ g cm}^{-3}$$
$$\text{Porosity} = 0.4$$

14.8 Transport in Aggregation and Deposition   865

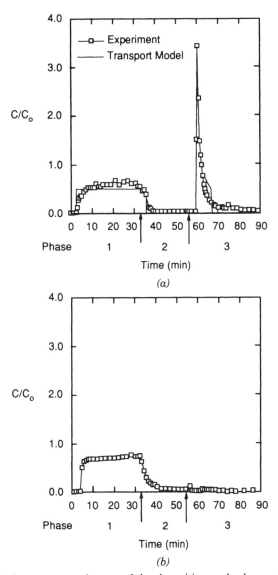

Figure 14.26. Laboratory experiments of the deposition and release of latex particles (0.308-μm diameter) in 20-cm columns of glass beads. (0.4-mm diameter) at a flow rate of 0.136 cm s^{-1} and a temperature of 22°C. (From Hahn and O'Melia, personal communication, 1994.) (a) Typical breakthrough curve for deposition and reentrainment. Particles are deposited at $I = 0.1$ M KCl in phase 1 (0–34 min), eluted with 0.1 M KCl in phase 2 (34–58 min), and eluted with deionized water in phase 3 (58–90 min). Particles deposited in the secondary well are released at low ionic strength. (b) Breakthrough curve for deposition at very high divalent salt concentration (0.2 M CaCl$_2$) in phase 1 (0–32 min), elution at the same salt concentration in phase 2 (32–56 min), and elution with deionized water in phase 3 (56–90 min). Particles deposited in the primary well are not released at low ionic strength.

the system. Particles are deposited in the secondary well, a rate controlled by mass transport, and can escape back into the bulk solution by Brownian motion if the well is sufficiently shallow ($<$ a few $k_B T$). Experimental results by Hahn (1994) are presented in Figure 14.26. Consequently, according to this theory, transport of particles and particle-reactive contaminants in porous media can be very different from that predicted with the classical model that involves irreversible deposition. Multiple events of deposition and release may occur, and small particles could be transported long distances in the subsurface. The longest travel distances would be observed for very small particles because the depth of the secondary minimum is small for these particles.

SUGGESTED READINGS

Filella, M., and Buffle, J. (1993) Factors Controlling the Stability of Submicron Colloids in Natural Waters, *Colloids Surfaces* **A73**, 255–273.

Gschwend, Ph. M., and Reynolds, M. D. (1987) Monodisperse Ferrous Phosphate Colloids in an Anoxic Groundwater Plume, *J. Contam. Hydrol.* **1/3**, 309–327.

Jackson, G. A., and Lochmann, S. (1994) Modelling Coagulation of Algae in Marine Ecosystems. In *Environmental Particles,* Vol. 2, J. Buffle and H. P. van Leeuwen, Eds., Lewis, Chelsea, MI.

Kim, J. I. (1991) Actinide Colloid Generation in Groundwater, *Radiochim. Acta* **52/53**, 71–81.

Liang, L., and Morgan, J. J. (1990) Chemical Aspects of Iron Oxide Coagulation in Water: Laboratory Studies and Implications for Natural Systems, *Aquatic Sci.* **52/1**, 32–55.

MacDowell-Boyer, L. M. (1992) Chemical Mobilization of Micron-Sized Particles in Saturated Porous Media Under Steady Flow Conditions, *Environ. Sci. Technol.* **26**, 586–593.

Morel, F. M. M., and Gschwend, P. M. (1987) The Role of Colloids in the Partitioning of Solutes in Natural Waters. In *Aquatic Surface Chemistry,* W. Stumm, Ed., Wiley-Interscience, New York, pp. 405–422.

O'Melia, C. R. (1990) Kinetics of Colloid Chemical Processes in Aquatic Systems. In *Aquatic Chemical Kinetics, Reaction Rates of Processes in Natural Waters,* W. Stumm, Ed., Wiley-Interscience, New York, pp. 447–474.

O'Melia, C. R., and Tiller, C. L. (1993) Physicochemical Aggregation and Deposition in Aquatic Environments. In *Environmental Particles,* Vol. 2, J. Buffle and H. P. van Leeuwen, Eds., Lewis, Chelsea, MI.

Schnoor, J. L. (1990) Kinetics of Chemical Weathering: A Comparison of Laboratory and Field Weathering Rates. In *Aquatic Chemical Kinetics,* W. Stumm, Ed., Wiley-Interscience, New York, pp. 475–504.

Sposito, G. (1992) Characterization of Particle Surface Charge. In *Environmental Particles,* Vol. 1, J. Buffle and H. P. van Leeuwen, Eds., Lewis, Chelsea, MI.

Stumm, W., and O'Melia, C. R. (1968) Stoichiometry of Coagulation, *J. Am. Water Works Assoc.* **60**, 514–539.

Weilenmann, U., O'Melia, C. R., and Stumm, W. (1989) Particle Transport in Lakes: Models and Measurements, *Limnol. Oceanogr.* **34**, 1–18.

APPENDIX 14.1: A PHYSICAL MODEL (DLVO) FOR COLLOID STABILITY

To what extent can theory predict the collision efficiency factor? Two groups of researchers, *D*erjaguin and *L*andau, and *V*erwey and *O*verbeek, independently of each other, developed a theory (The DLVO theory, 1948) to quantitatively evaluate the balance of repulsive and attractive forces between interacting particles. The theory has proved an effective tool in the interpretation of many empirical facts in colloid chemistry.

The DLVO theory considers van der Waals attraction and diffuse double-layer repulsion as the sole operative factors. It calculates the interaction energy (as a function of interparticle distance) as the reversible isothermal work (i.e., Gibbs free energy) required to bring two particles from distance ∞ to distance d. Physically, the requirement is that at any instant during interaction the two double layers are fully equilibrated (Lyklema, 1978). The mathematics of the interaction are different for the interaction of constant-potential surfaces than for the interaction of constant-charge surfaces. As long as the interaction is not very strong, that is, as long as the surfaces do not come too close, it does not make a great difference (Lyklema, 1978). Table A14.1 gives the approximate equations for the constant-potential case. As an illustration, we calculate the interaction energy of two flat plates in example A14.1.

Example A14.1. Calculating Interaction Energy Calculate the interaction energy of two flat plates with Gouy layers of 25-mV surface potential (assumed to be constant), at 25°C in a 10^{-3} M NaCl solution (1.0 mol m^{-3}), for a separation distance of 10 nm. A Hamaker constant $A_{11(2)} = 10^{-19}$ J will be used.

Using equations 1–3 of Table A14.1, and considering that, for a 10^{-3} M NaCl solution (equation 9.40c), $\kappa^{-1} = 9.5 \times 10^{-9}$ m, we calculate

$$V_R = \frac{64 \times 1.0 \,\text{mol m}^{-3} \times 6.02 \times 10^{23} \times 1.38 \times 10^{-23} \,\text{J K}^{-1} \times 298.13 \,\text{K}}{1.05 \times 10^8 \,\text{m}^{-1}}$$

$$\times \left[\tanh\left(\frac{1.6 \times 10^{-19} \,\text{C} \times 25 \times 10^{-3} \,\text{V}}{4 \times 1.38 \times 10^{-23} \,\text{J K}^{-1} \times 298.13 \,\text{K}} \right) \right]^2$$

$$\times \exp[-(1.05 \times 10^8 \,\text{m}^{-1} \times 10^{-8} \,\text{m})]$$

$$= 1.5 \times 10^{-3} \,\text{J m}^{-2} \times 5.7 \times 10^{-2} \times 0.35$$

$$= 3.0 \times 10^{-5} \,\text{J m}^{-2}$$

$$V_A = \frac{10^{-19} \,\text{J}}{12 \times 3.14 \times 10^{-16} \,\text{m}^2}$$

$$= -2.65 \times 10^{-5} \,\text{J m}^{-2}$$

$$V_T = 3.5 \times 10^{-6} \,\text{J m}^{-2}$$

Table A14.1. Colloid Stability as Calculated from van der Waals Attraction and Electrostatic Diffuse Double-Layer Repulsion[a,b]

Additive interactions of repulsive interaction energy V_R and attraction energy V_A:

$$V_T = V_R + V_A \tag{1}$$

Repulsive interaction per unit area between flat plates:

$$V_R = \frac{64 n_s k_B T}{\kappa} \left[\tanh\left(\frac{ze\Psi_d}{4k_B T}\right)\right]^2 e^{-\kappa d} \tag{2a}$$

for spherical particles:

$$V_R = \frac{64\pi n_s k_B T}{\kappa^2} \frac{(a+\delta)^2}{R} \left[\tanh\left(\frac{ze\Psi_d}{4k_B T}\right)\right]^2 e^{-\kappa(H-2\delta)} \tag{2b}$$

Van der Waals attraction per unit area[c] for flat plates:

$$V_A = \frac{A_{11(2)}}{12\pi d^2} \tag{3a}$$

for spherical particles:

$$V_A = -\frac{A_{11(2)}}{6}\left(\frac{2}{s^2-4} + \frac{2}{s^2} + \ln\frac{s^2-4}{s^2}\right) \tag{3b}$$

where

$$s = \frac{R}{a}$$

For very short particle distances, equation 3b may be replaced by

$$V_A = \frac{A_{11(2)} a}{12 H} \tag{3c}$$

Electrolyte concentration required to just coagulate colloids with flat surfaces (25°C):[d]

$$c_s = 8 \times 10^{-36} \frac{[\tanh(ze\psi/4k_B T)]^4}{A_{11(2)}^2 z^6} \tag{4}$$

Valence effect on stability for small ψ:

$$c_s = \frac{3.125 \times 10^{-38}}{A_{11(2)}^2 z^2} \left(\frac{e\psi}{k_B T}\right)^4 \tag{5}$$

Stability ratio for spherical particles:

$$W = 2a \int_{2a}^{\infty} \exp\left(\frac{V_T}{k_B T}\right) R^{-2}\, dR \tag{6}$$

[a]We have the following definitions:

- a = particle radius
- A = Hamaker constant (J)
- c_s = concentration of salt (M)
- d = interaction distance between two surfaces
- e = elementary charge, 1.6×10^{-19} C

Appendix 14.1 A Physical Model (DLVO) for Colloid Stability

H = shortest interaction distance between two spherical particles
k_B = Boltzmann constant
n_s = number of "molecules" or ion pairs per volume
R = distance between centers of two spheres
W = stability ratio ($1/\alpha$) (in the kinetic equations, k_j can be replaced by k_j/W)
T = absolute temperature
V = interaction energy
z = charge number
κ = reciprocal thickness of double layer
Ψ = potential (V) at the plane where the diffuse double layer begins

[b]Compare Lyklema (1978). Most equations below stem from DLVO theory. Since exact solutions do not exist, recourse must be made to approximations.
[c]$A_{11(2)}$ is the Hamaker constant [dimension (energy)] that applies for the interaction between particles 1 in medium 2. This quantity can be related to the corresponding constants for attraction in vacuum between particles 1 or between particles 2 by $A_{11(2)} = (\sqrt{A_{11}} - \sqrt{A_{22}})^2$.
[d]Calculated for the condition that $V_T \leq 0$ and $\partial V_T/\partial d = 0$.

In Figure 14.10 the energies of interaction (double-layer repulsion, V_R, van der Waals attraction, V_A, and net total interaction, V_T) were plotted as a function of separation distance of the surfaces. As equation 2a of Table A14.1 shows, V_R decreases in an exponential fashion with increasing separation. V_R increases roughly in proportion to ψ_d^2 (for small ψ_d, $\tanh u \approx u$). The distance characterizing the repulsive interaction is similar in magnitude to the thickness of a single double layer (κ^{-1}). Thus the range of repulsion depends primarily on the ionic strength. The energy of attraction due to van der Waals dispersion forces was plotted in the lower part of Figure A14.1 as a function of separation. This curve varies little for a given value of the Hamaker constant A, which depends on the density and polarizability of the dispersed phase but is essentially independent of the ionic makeup of the solution. Hamaker constants for different materials range from approximately 10^{-20} to 10^{-19} (Hamaker constants for organic colloids are smaller than 10^{-20}). Summing up repulsive and attractive energies gives the total energy of interaction. Conventionally, the repulsive potential is considered positive and the attractive potential negative. At small separations, attraction outweighs repulsion, and at intermediate separations, repulsion predominates. This energy barrier is usually characterized by the maximum (net repulsion energy) of the total potential energy curve, $V_{\max}$ (Figure A14.1). The potential energy curve shows, under certain conditions, a secondary minimum at larger interparticle distances ($d \approx 10^{-6}$ cm). This secondary minimum depends directly on the value of the Hamaker constant and on the diameters of the particles involved; it is seldom deep enough to cause strong instability but might help in explaining weaker forms of adhesion or agglomeration.

The Stability Ratio

We have already defined the stability ratio operationally. A stable colloidal dispersion is characterized by a high-energy barrier, that is, by a net repulsive

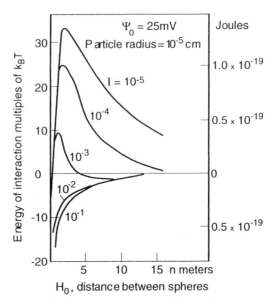

Figure A14.1. Physical model for colloid stability. Net energy of interaction for spheres of constant potential surface for various ionic strengths (1:1 electrolyte). A Hamaker constant ($A = 10^{-19}$ J) was used in the model calculations (Verwey and Overbeek, 1948).

interaction energy. Fuchs has defined a stability ratio W that is related to the area enclosed by the resultant curve of the energy of interaction versus separation distance. W is the factor by which agglomeration is slower than in the absence of an energy barrier. The conceptually defined W should correspond to the operationally determined α ($W = \alpha^{-1}$). To a first approximation the stability ratio W is related to the height of the potential energy barrier V_{max}. The latter is conveniently expressed in units of $k_B T$. If V_{max} exceeds the value of a few $k_B T$, relatively stable colloids will be found. For example, if $V_{max} \simeq 15\ k_B T$, only 1 out of 10^6 collisions will be successful ($\alpha = 10^{-6}$). Figure A14.1 shows how V_{max} typically decreases with an increase in electrolyte concentration.

Some of the pertinent interactions that affect colloid stability are readily apparent from Figures 14.7 and A14.1. The main effect of electrolytes is a more rapid decay of the repulsion energy with distance and to compact the double layer (reducing κ^{-1}). Experimentally, it is known that the charge number of the counterion plays an important role. The critical electrolyte concentration required just to agglomerate the colloids is proportional to $z^{-6} A^{-2}$ for high surface potential, and to $z^{-2} A_{11(2)}^{-2}$ at low potentials (equations 4 and 5 in Table A14.1). This is the theoretical basis for the empirical *valency rule of Schulze and Hardy*.

Limitations of the DLVO Theory

The DLVO theory is a theoretical construct that has been able to explain many experimental data in at least a semiquantitative manner; it illustrates plausibly that at least two types of interactions (attraction and repulsion) are needed to account for the overall interaction energy as a function of distance between the particles.

Measurements on the interaction force between two mica plates (Israelachvili and Adams, 1978) have shown good agreement with the theory in dilute KNO_3 solution at distances larger than ~ 5 nm. Similar results have recently been obtained with measurements between SiO_2 with the help of atomic force microscopic measurements (Ducker et al., 1991). In solutions containing bivalent ions the agreement between theory and measurement is less good for various reasons, but also because the description of the diffuse double-layer theory fails at short distances from the surface (Sposito, 1984). Apparently non-DLVO forces such as solvation, hydration, and capillary effects may become operative at separations below 5 nm. Early validations of the theory were based on coagulation studies of monodisperse suspensions.

Critical Coagulation Concentration (ccc) in Coagulation

Under simplifying conditions, it is possible to derive the relation that ccc values of mono-, di-, and trivalent ions vary in the ratio $(1/z)^6$ at high surface potentials or $100:1.6:0.13$ or at low potentials in the ratio $(1/z)^2$ or $100:25:11$.

Liang and Morgan (1990), for example, found the following critical coagulation concentrations (ccc) for hematite (diameter = 70 nm; $pH_{pznpc} = 8.5$) at pH = 6.5: Cl^-, 80 mM; SO_4^{2-}, 0.45 mM; HPO_4^{2-}, 0.016 mM. Such results are not in accord with the Schulze–Hardy rule. [The experimental ccc values are in the ratio 100 (Cl^-): 0.56 (SO_4^{2-}): 0.02 (HPO_4^{2-}).] The reason is the chemical interaction of the coagulant ions with the colloids; that is, SO_4^{2-} forms moderately stable and the HPO_4^{2-} very stable surface complexes with the hematite. As we have seen, such chemical interaction causes a stoichiometric relationship between the ccc and the (incipient) colloid surface area concentration. The addition of an excess of an adsorbing coagulant can cause a charge reversal and a restabilization of the colloid (Stumm and O'Melia, 1968).

The presence of polymers or polyelectrolytes has important effects on the van der Waals interaction and on the electrostatic interaction (see Figure 14.10).

Particle Size The DLVO theory predicts an increase in the total interaction energy with an increase in particle size. This effect has not been verified in coagulation studies.

15

REGULATION OF THE CHEMICAL COMPOSITION OF NATURAL WATERS

15.1 INTRODUCTION

Natural waters acquire their chemical characteristics by dissolution and by chemical reactions with solids, liquids, and gases with which they have come into contact during the various parts of the hydrological cycle. Waters vary in their chemical composition, but these variations are at least partially understandable if the environmental history of the water and the chemical reactions of the rock–water–atmosphere systems are considered. Dissolved mineral matter originates in the crustal materials of the earth; water disintegrates and dissolves mineral rocks by weathering. Gases and volatile substances participate in these processes. As a first approximation, seawater may be interpreted as the result of a gigantic acid–base titration—acid of volcanoes versus bases of rocks (oxides, carbonates, silicates) Sillén, 1961). The composition of fresh water similarly may be represented as resulting from the interaction of the CO_2 of the atmosphere with mineral rocks.

An early survey of the frequency distribution of various constituents of terrestrial waters by Davies and DeWiest (1966) (see Figure 15.1) makes it apparent that many of the aquatic constituents show little natural variation in their concentrations. For example, 80% of the water analyses for dissolved silica show concentrations between $10^{-3.8}$ and $10^{-3.2}$ M. The range of [H^+] of naturally occurring bodies of water is generally $10^{-6.5}$ to $10^{-8.5}$ M. Bluth and Kump (1994) have undertaken a study of small drainage basins composed of single lithology to develop relationships among rock type, climate, and chemical denudation.

The Circulation of Rocks, Water, Atmosphere, and Biota

Figure 15.2 gives a very simplified picture of some of the more important global biogeochemical transformations. Metaphorically, there are a multitude

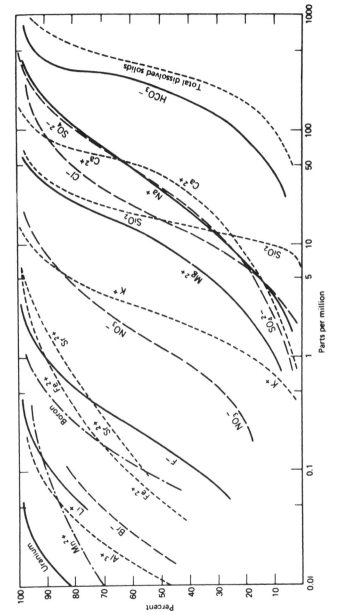

Figure 15.1. Cumulative curves showing the frequency distribution of various constituents in terrestrial water. Data are mostly from the United States from various sources. (Adapted from Davies and DeWiest, 1966.)

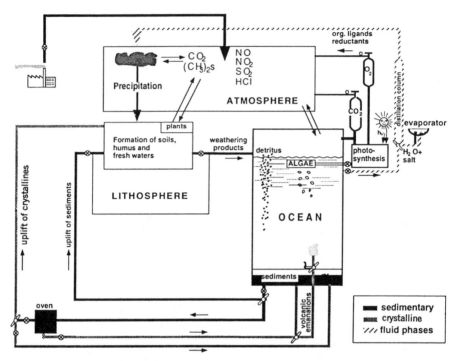

Figure 15.2. Steady state model for the earth's surface geochemical system. The interaction of water with rocks in the presence of photosynthesized organic matter continuously produces reactive material of high surface area. This process provides nutrient supply to the biosphere and along with biota forms the array of small particles (soils). Weathering imparts solutes to the water and erosion brings particles into surface waters and oceans. (Inspired by Siever, 1968.)

of valves and switches that control the system. Of particular concern are the gas regulators of the CO_2 and O_2 tanks. Both O_2 and CO_2 regulators are governed by photosynthesis and by weathering.

The classical geochemical material balance (Goldschmidt, 1933) assumes that the H^+ balance in our environment has been established globally by the interaction of primary (igneous) rocks with volatile substances:

Igneous rocks	+	Volatile substances	$\rightleftharpoons$	Seawater	+	Atmosphere	+	Sediments	(1)
silicates		CO_2		pH = 8		$p_{O_2} = 0.2$		carbonates	
carbonates		H_2O		pε = 12		$p_{CO_2} = 0.0003$		silicates	
oxides		SO_2							
		HCl							

In an oversimplified way, one can say that acids of the volcanoes have reacted with the bases of the rocks; the compositions of the ocean (which is at the first end point (pH = 8) of the titration of a strong acid with a carbonate) and the atmosphere (which with its $p_{CO_2} = 10^{-3.5}$ atm is nearly in equilibrium with the ocean) reflect the proton balance of reaction 1. Oxidation and reduction are accompanied by proton release and proton consumption, respectively. (In order to maintain charge balance, the production of e^- must eventually be balanced by the production of H^+.) Furthermore, the dissolution of rocks and the precipitation of minerals are accompanied by H^+ consumption and H^+ release, respectively. Thus, as shown by Broecker (1971), the pε and pH of the surface of our global environment reflect the levels where the oxidation states and the H^+ ion reservoirs of the weathering sources equal those of the sedimentary products.

15.2 WEATHERING AND THE PROTON BALANCE

In order to understand pH regulation one needs to consider the major H^+-yielding and H^+-consuming processes. Weathering is a major H^+-consuming process and pH-buffering mechanism globally and regionally; it also plays a major role in local watersheds in soil processes, in nutrient uptake by plants, and in epidiagenetic reactions in sediments.

Acid atmospheric deposition causes acidification of waters and soils if the neutralization of the acids by weathering is too slow. Biologically mediated redox processes are important in affecting the H^+ balance. Among the redox processes that have a major impact on H^+ production and consumption are the synthesis and mineralization of biomass. Any uncoupling of linkages between photosynthesis and respiration affects acidity and alkalinity in terrestrial and aquatic ecosystems (Table 15.1).

As Figure 15.3 illustrates, aggrading vegetation (forests and intensive crop production) produces acidity because as more cations are taken up by the plants (trees) H^+ is released through the roots. The protons released react with the weatherable minerals to produce some of the cations needed by the plants. The H^+ balance in soils (production through the roots versus consumption by weathering) is delicate and can be disturbed by acid deposition.

If the weathering rate equals or exceeds the rate of H^+ release by the biota, such as would be the case in a calcareous soil, the soil will maintain a buffer in base cations and residual alkalinity. On the other hand, in noncalcareous "acid" soils, the rate of H^+ release by the biomass may exceed the rate of H^+ consumption by weathering and cause a progressive acidification of the soil. In some instances, the acidic atmospheric deposition may be sufficient to disturb an existing H^+ balance between aggrading vegetation and weathering reactions.

Critical Load

The concept of critical load is based on the dose–response relationship, the critical load being exceeded when the load causes harmful effects to the recep-

Table 15.1. Processes that Modify the H^+ Balance in Waters

Processes[a]	Changes in Alkalinity $\Delta[Alk] = -\Delta[H\text{-}Acy]$ (equivalents per mole reacted)
1. Weathering reactions:	
$\underline{CaCO_3(s)} + 2H^+ \rightleftharpoons Ca^{2+} + CO_2 + H_2O$	+2
$\underline{CaAl_2Si_2O_8(s)} + 2H^+ \rightleftharpoons Ca^{2+} + H_2O + Al_2Si_2O_5(OH)_4(s)$	+2
$\underline{KAlSi_3O_8(s)} + H^+ + 4\tfrac{1}{2}H_2O \rightleftharpoons K^+ + 2H_4SiO_4 + \tfrac{1}{2}Al_2Si_2O_5(OH)_4(s)$	+1
$\underline{Al_2O_3 \cdot 3H_2O} + 6H^+ \rightleftharpoons 2Al^{3+} + 6H_2O$	+6
2. Ion exchange:	
$2ROH + \underline{SO_4^{2-}} \rightleftharpoons R_2SO_4 + 2OH^-$	+2
$NaR + \underline{H^+} \rightleftharpoons HR + Na^+$	+1
3. Redox processes (microbial mediation):	
Nitrification	
$\underline{NH_4^+} + 2O_2 \rightleftharpoons NO_3^- + H_2O + 2H^+$	−2
Denitrification	
$1\tfrac{1}{4}CH_2O + \underline{NO_3^-} + H^+ \rightarrow 1\tfrac{1}{4}CO_2 + \tfrac{1}{2}N_2 + 1\tfrac{3}{4}H_2O$	+1

Oxidation of H$_2$S

$\underline{H_2S}$ + 2O$_2$ → SO$_4^{2-}$ + 2H$^+$ −2

SO$_4^{2-}$ reduction

$\underline{SO_4^{2-}}$ + 2CH$_2$O + 2H$^+$ → 2CO$_2$ + H$_2$S + H$_2$O   +2

Pyrite oxidation

$\underline{FeS_2}$(s) + 3$\tfrac{3}{4}$O$_2$ + 3$\tfrac{1}{2}$H$_2$O → Fe(OH)$_3$
   + 2SO$_4^{2-}$ + 4H$^+$ −4

4. In the *buildup (or breakdown) of biomass*, the uptake (release) of each equivalent of conservative anion causes an equivalent increase (decrease) in alkalinity, and each equivalent of base cations that is taken up (released) results in an equivalent decrease (increase) of alkalinity. The reduction of Fe(III) (hydr)oxides and of Mn(III,IV) (hydr)oxides by biota is particularly efficient in generating alkalinity in the hypolimnion of lakes.

{(CH$_2$O)$_{106}$(NH$_3$)$_{16}$(H$_3$PO$_4$)}$_{Biota}$ + 424 $\underline{Fe(OH)_3}$ + 862 H$^+$
  → 424 Fe^{2+} + 16 NH$_4^+$ + 106 CO$_2$
  + HPO$_4^{2-}$ + 1166 H$_2$O   (a)

{(CH$_2$O)$_{106}$(NH$_3$)$_{16}$(H$_3$PO$_4$)} + 212 $\underline{MnO_2}$ + 398 H$^+$
  → 212 Mn^{2+} + 16 NH$_4^+$ + 106 CO$_2$ + HPO$_4^{2-}$ + 298 H$_2$O   (b)

---

^{a}Reactant is underlined.

878 Regulation of the Chemical Composition of Natural Waters

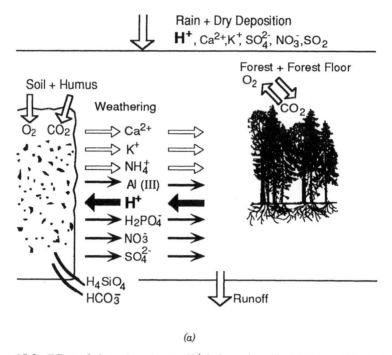

(a)

Figure 15.3. Effect of plants (trees) upon H^+-balance in soils. (a) Competition between H^+ release by the roots of growing trees (aggrading forest) and H^+ ion consumption by chemical weathering of soil minerals. The delicate proton balance can be disturbed by acid deposition. (b) Processes affecting the acid-neutralizing capacity of soils (including the exchangeable bases, cation exchange and mineral bases). H^+ ions from acid precipitation and from the release by the roots react by weathering carbonates, aluminum silicates and oxides and by surface complexation and ion exchange on clays and humus. Mechanical weathering resupplies weatherable minerals. Lines clays and humus. Mechanical weathering resupplies weatherable minerals. Lines drawn out indicate flux of protons; dashed lines show flux of base cations (alkalinity). The trees (plants) act like a base pump. (From Schnoor and Stumm, 1986.)

tor. With regard to acid deposition, the critical load could be defined in terms of tolerable acidity deposition (e.g., in μeq m^{-2} yr^{-1}) that must not be exceeded to avoid harmful effects. There have been various methods to establish critical loads, which we will not discuss here; instead, we refer the reader to two publications (Bricker and Rice, 1989, and Kămări et al., 1992). The sensitivity of a region is strongly influenced by the minerals of the bedrock. As shown in Figure 13.9, the dissolution rate (neutralization rate of acidity) is very fast in carbonate terrain and very slow in regions where crystalline rocks prevail.

Simple steady-state models have been used to determine critical loads for lakes and forest soils. The basic principle is that primary mineral weathering in the watershed is the ultimate supplier of base cations, which are required elements for vegetation and lake water to ensure adequate acid-neutralizing

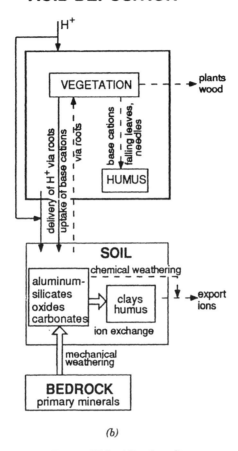

Figure 15.3. (*Continued*)

capacity (see previous definitions in Section 4.4 of ANC as the sum of base cations minus acid anions). If more acid is deposited in a watershed than chemical weathering can neutralize, acidification of the soils and water will eventually occur. It may not happen immediately; there could be a delayed response because ion exchange reactions can supply base cations to vegetation and runoff water for a period, but eventually the exchange capacity of the soil will become depleted and acidification will result. The delay period can be estimated as the base cation exchange capacity (exchangeable Ca, Mg, Na, and K ions) of the soil divided by the cation denudation rate, typically on the order of a few decades in poor, acidic soils. The exchanger (cation exchange complex in soil) is "set-up" by the mineralogy and lithology of the parent material, and this determines the base exchange capacity and even the chemistry of the soil.

Sverdrup et al. (1992) have developed the PROFILE model, which is based on the principle of continuity of alkalinity or ANC in soil. The critical load is defined as the allowable acid loading that will not acidify forest soils and cause the release of aluminum and hydrogen ions to soil solution:

$$CL = BC_w - Alk_L$$

where CL = critical load, meq m^{-2} yr^{-1}
BC_w = weathering rate, meq m^{-2} yr^{-1}
Alk_L = alkalinity leaching, meq m^{-2} yr^{-1}

The amount of alkalinity that is necessary to leach from the soil can be estimated from the ratio of base cations to aluminum, (Ca + Mg)/Al, assuming a critical threshold value for biological effects at 1:1. (Above a ratio of (Ca + Mg)/Al = 1 roots are assumed to be protected from aluminum toxicity and Mg deficiency.) Then, based on gibbsite solubility and a minimum required base cation concentration for forest nutrition of 5 meq m^{-2} yr^{-1}, it is possible to estimate the required alkalinity leaching for healthy soils. Furrer et al. (1989) developed a more detailed steady-state model. It includes weathering, ion exchange, base cation uptake by trees, nitrogen transformation reactions, and equilibrium soil chemistry under steady-state conditions. The model has not been used for critical load assessments, but it represents a more comprehensive approach.

Methods to calculate critical loads from steady-state models do not take time scales for acidification and recovery into account. Detailed non-steady-state models have been used to estimate target loads for three watersheds in Norway and Finland (Warfvinge et al., 1992). For reviews on acidification of soils see Reuss and Johnson (1986) and Stumm and Schnoor (in press). Paces (1994) illustrates the modeling of hydrological and biogeochemical responses of a catchment area to anthropogenic inputs.

15.3 ISOTHERMAL EVAPORATION

Water composition is also affected by concentration resulting from evaporation (and evapotranspiration). Example 15.1 will illustrate the principles, procedures, and calculations of the effect of concentrating natural waters by isothermal evaporation for a few simplified systems. These calculations illustrate how the reaction path of natural waters during evaporation depends on the Ca^{2+}/HCO_3^- ratio. In a *reaction progress model* the effects of initial reactions on a reaction path, for example, on the appearance of a solid stable phase and on the redistribution of aqueous species, are described. Reaction progress models are usually based on the concept of *partial equilibrium*. Partial equilibrium describes a state in which a system is in equilibrium with respect to one reaction, but out of equilibrium with respect to others.

15.3 Isothermal Evaporation

Example 15.1. Isothermal Evaporation of a Natural Water Leading to CaCO$_3$ Precipitation Compute pH and solution composition during isothermal evaporation (25°C) for three types of calcium bicarbonate-containing waters (a) $2[Ca^{2+}] = [Alk]$; (b) $2[Ca^{2+}] < [Alk]$; and (c) $2[Ca^{2+}] > [Alk]$. All waters are assumed to remain in equilibrium with atmospheric CO$_2$ (log p_{CO_2} = -3.5 atm).

During evaporation the ionic strength continuously changes, and corrections for ionic strength should be made. For simplicity we make our calculations for $I = 5 \times 10^{-2}$. The values obtained without further correction must be regarded as approximations that may contain uncertainties of up to 0.3 logarithmic units for a concentration factor of 100. (An iterative procedure may be used to obtain more exact solutions: one first obtains approximate pH values and molarities by using constants valid at $I = 0$; with these tentative values I can be estimated as a function of the concentration factor in order to correct the equilibrium constants used.) At $I = 5 \times 10^{-2}$ and 25°C equilibrium constants (corrected with the Davies equation) with the following numerical values are used: $pK'_H = 1.47$, $pK'_1 = 6.27$, $pK'_2 = 10.16$, $pK'_{s0} = 7.72$. We choose the following initial conditions:

(a) $[Ca^{2+}]_0 = 5 \times 10^{-6}$ M $[Alk]_0 = 10^{-5}$ eq liter^{-1}

(b) $[Ca^{2+}]_0 = 2 \times 10^{-4}$ M $[Alk]_0 = 6 \times 10^{-4}$ eq liter^{-1}

(c) $[Ca^{2+}]_0 = 3 \times 10^{-4}$ M $[Alk]_0 = 4 \times 10^{-4}$ eq liter^{-1}

Initially, all three waters are undersaturated with respect to calcite, and the alkalinity is present as HCO$_3^-$. The solution pH is given by the equilibrium

$$CO_2(g) + H_2O = HCO_3^- + H^+ \quad K_H K_1$$

that is,

$$[H^+] = \frac{K_H K_1 p_{CO_2}}{[HCO_3^-]} \tag{i}$$

At high pH values (above pH ~9) CO$_3^{2-}$ may become a more important species; instead of equation i we may then write

$$[Alk] = \left(\frac{K_H p_{CO_2}}{\alpha_0}\right)(\alpha_1 + 2\alpha_2) + \frac{K_W}{[H^+]} - [H^+] \tag{ii}$$

The relationship between [Alk] and pH has been calculated before for Figure 4.9.

We may now proceed in steps and concentrate the solution successively by various factors (concentration factor = R).

We show here how the calculations can be made without a computer. With the help of a computer program like MICROQL (cf. Tableau 4.2) these calculations can be done expediently.

In waters of type (a), $[Ca^{2+}]$ and [Alk] (or $[HCO_3^-]$) will increase with R: $[Ca^{2+}] = R[Ca^{2+}]_0$ and $Alk = R[Alk]_0$ until the $CaCO_3$ solubility equilibrium is reached. Saturation with calcite and the atmosphere may be characterized by the equilibrium $CaCO_3(s) + CO_2(g) + H_2O = Ca^{2+} + 2HCO_3^-$; $^+K_{ps0}$ (where $^+K_{ps0} = K_{s0}K_1K_HK_2^{-1}$); that is,

$$[Ca^{2+}] = \frac{^+K_{ps0}\, p_{CO_2}}{[HCO_3^-]^2} \qquad (iii)$$

or, more generally (valid also at high pH),

$$[Ca^{2+}] = \frac{K_{s0}\, \alpha_0}{K_H \alpha_2 p_{CO_2}} \qquad (iv)$$

$[Ca^{2+}]$ can be calculated from equation iii as a function of pH; this has already been done in Figure 4.5. As soon as solubility equilibrium is attained, the composition of the solution is constant (Figure 15.4a).

Solution of type (b) is characterized by the electroneutrality condition (Figure 15.4b):

$$[X] = Alk - 2[Ca^{2+}] \qquad (v)$$

where X may be thought of as any (nonprecipitating) cation. During evaporation, the left-hand side is given by $R[X]_0$; the terms on the right-hand side can be expressed as a function of pH (equations i or ii and iii or iv). These calculations are most conveniently carried out with a programmable calculator or a computer; otherwise, approximate results may be read from the diagrams in Figure 4.5. Most conveniently, for a given pH, [Alk] and $[Ca^{2+}]$ are first calculated; then R is obtained from X in equation v. The result is given in Figure 15.4b. At a concentration factor of ~4, calcite starts to precipitate. During the bulk precipitation H^+, Ca^{2+}, and Alk are somewhat buffered. After most calcite has been precipitated, $[Ca^{2+}]$ decreases while pH, [Alk], and $[CO_3^{2-}]$ increase, and eventually the solution composition approaches that of a solution of soda.

Waters of type (c) are characterized by the electroneutrality condition

$$[Y] = 2[Ca^{2+}] - [Alk] \qquad (vi)$$

where [Y] may be thought of as any (nonprecipitating) anion. The extent of concentration can be measured by $R[Y]_0$, and the terms on the right-hand side of equation vi can be expressed as before. As Figure 15.4c shows, pH starts to decrease after precipitation of the bulk of $CaCO_3$.

Garrels and Mackenzie (1967) have calculated the change in composition

15.3 Isothermal Evaporation 883

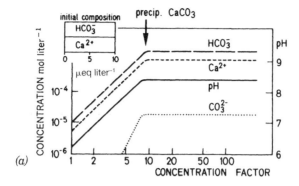

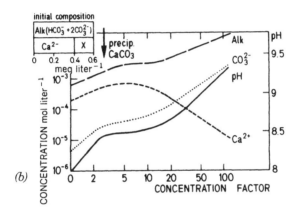

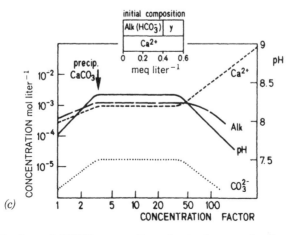

Figure 15.4. Isothermal (25°C) evaporation of natural waters leading to $CaCO_3$ precipitation. (a) $2[Ca^{2+}] = [Alk]$, (b) $2[Ca^{2+}] < [Alk]$, and (c) $2[Ca^{2+}] > [Alk]$ ($p_{CO_2} = 10^{-3.5}$ atm).

resulting from the evaporation of typical Sierra Nevada spring water coupled with sequential precipitation of calcite sepiolite and gypsum. In the early stage of evaporative concentration, the relatively insoluble carbonates, sulfates, and silicates are precipitated. In the later stages of the evolution of saline lakes and brines, the very soluble saline minerals may also be precipitated.

For a review of literature on evaporative concentration and brine evolution see Eugster and Hardie (1978).

As shown in Example 15.1 and in Figure 15.4, the ratio [Alk]/[Ca^{2+}] is of great significance in the genesis of natural waters and evolutionary paths during evaporative concentration. Waters in which 2[Ca^{2+}] ≃ [HCO_3^-] do not change their [Ca]/[HCO_3^-] ratio markedly upon evaporative concentration. Many river waters of the world are along the 2[Ca^{2+}] = [HCO_3^-] line in Figure 4.15. Upon extensive evaporation such waters may precipitate magnesite.

Weathering of silicate minerals usually supplies cations in addition to Ca^{2+} and Mg^{2+}. Such waters with [HCO_3^-] > 2[Ca^{2+}] (residual alkalinity) tend upon evaporation to increase their pH values and concentrations of HCO_3^- and CO_3^{2-} and to decrease [Ca^{2+}]. Upon extensive evaporation, such waters acquire a composition similar to that found in natural soda lakes; eventually alkaline brines of the $Na-CO_3-SO_4-Cl$ type may be formed.

Waters with a negative residual alkalinity (2[Ca^{2+}] > [HCO_3^-]; see Figure 15.4c), as a consequence of evaporation, tend to increase their calcium hardness, while their alkalinity and pH decrease.† Eventually, upon extensive evaporation, they may reach saturation with respect to gypsum and become Ca-Na-SO_4-Cl brines.‡

15.4 BUFFERING

Buffering of pH of natural waters is not caused solely by the $CO_2-HCO_3^--CO_3^{2-}$ equilibrium. Heterogeneous equilibria are the most efficient buffer systems of natural waters. In Section 3.9 the pH buffer intensity, $\beta_{C_j}^{C_i}$, was defined for the incremental addition of C_i to a closed system of constant C_j at equilibrium

$$\beta_{C_j}^{C_i} = \frac{dC_i}{d\,\mathrm{pH}} \qquad (2)$$

We can now compare, for example, the buffer intensities, with respect to strong acid, of the following systems (Figure 15.5): (1) a carbonate solution of constant C_T, β_{C_T}; (2) an aqueous solution in equilibrium with calcite,

---

†The conclusions of Figure 15.4b,c, that residual alkalinity increases upon evaporation (concomitant with pH increase) and that negative residual alkalinity decreases (concomitant with pH decrease), apply generally even if alkalinity is made up of anions other than HCO_3^- and CO_3^{2-}.
‡See also Al-Droubi et al. (1980).

15.4 Buffering

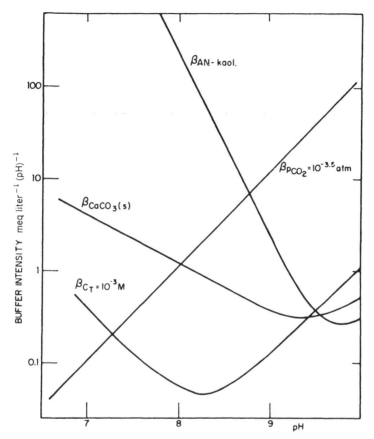

Figure 15.5. Buffer intensity versus pH for some heterogeneous systems and for the homogeneous dissolved carbonate system. Buffer intensities: β_{C_T} (dissolved carbonate, $C_T = 10^{-3}$ M) β_{CaCO_3} (carbonate solution in equilibrium with calcite), $\beta_{p_{CO_2}}$ (carbonate solution in equilibrium with $p_{CO_2} = 10^{-3.5}$ atm), $\beta_{\text{an-kaol}}$ (solution in equilibrium with anorthite and kaolinite).

β_{CaCO_3}; (3) a carbonate solution in equilibrium with a gas phase of constant p_{CO_2}, $\beta_{p_{CO_2}}$; and (4) a solution in equilibrium with both kaolinite and anorthite, $\beta_{\text{an-kaol}}$. As explained earlier (Section 3.9), the buffer intensity is found analytically by differentiating the appropriate function of C_i for the system with respect to pH.

The buffer intensity of a homogeneous carbonate system has been derived before. For heterogeneous systems the buffer intensities are derived in a similar way. (For a detailed derivation refer to Stumm and Morgan, 1981.) As Figure 15.5 shows, the homogeneous buffer intensity β_{C_T} is relatively small. For freshwater systems, β_{CaCO_3} and $\beta_{p_{CO_2}}$ are of considerable practical interest.

$CaCO_3(s)$ is an efficient buffer in the neutral and acid pH range. If, for example, a large quantity of acid is discharged into a natural water system containing $CaCO_3(s)$, an initially large decrease in the pH of that system occurs. The extent of this decrease depends largely on the magnitude of the fraction of total buffer intensity attributable to dissolved buffer components. Ultimately, however, the decrease in pH resulting from addition of the acid leads to dissolution of solid calcium carbonate and the establishment of a new position of equilibrium. Thus the final change in pH is much less than the initial decrease, which is resisted only by the intensity contribution of dissolved buffer components. The addition of a strong base in large quantity, conversely, leads to a deposition of calcium carbonate, thus reducing the pH shift that would occur in the absence of dissolved calcium.

The aluminum silicates provide considerably more resistance toward pH changes. The equilibrium system anorthite–kaolinite at pH = 8 has a buffer intensity a thousand times higher than that of a 10^{-3} M carbonate solution. As has been shown, equilibrium systems consisting of a sufficient number of coexisting phases attain, in principle, an infinite buffer intensity.

What actually protects water from pH changes is also dependent on the kinetics of the heterogeneous reactions. Reactions with solid carbonates and ion exchange processes are faster than alteration reactions of solid silicates. Investigations on the rate of the buffering reactions are in need.

From a kinetic point of view we must also consider that biochemical processes affect pH regulation and buffer action in natural water systems. Photosynthetic activities decrease CO_2, whereas respiratory activities contribute CO_2.

For fresh waters there is a further restraint on pH rise: the CO_2 reservoir of the atmosphere. For a given p_{CO_2} the pH is a function of alkalinity. In order to raise the pH of a water in equilibrium with the atmosphere from 8 to 9, alkalinity must increase by nearly 5 meq liter^{-1} (either by base addition or by evaporation). Hence only soda lakes, that is, lakes containing substantial amounts of soluble carbonates and bicarbonates, can attain high pH values; for example, Sierra Nevada spring waters discharged to the east of the Sierra and evaporated in a plaza of the California desert.

15.5 INTERACTIONS BETWEEN ORGANISMS AND ABIOTIC ENVIRONMENT: REDFIELD STOICHIOMETRY

Aquatic organisms influence the concentration of many substances directly by metabolic uptake, transformation, storage, and release. In order to understand the chemistry of an aquatic habitat, the causal and reciprocal relationship between organisms and their environment must be taken into consideration.

Photosynthesis and Respiration

Energy-rich bonds are produced as a result of photosynthesis, thus distorting the thermodynamic equilibrium. Bacteria and other respiring organisms catalyze the redox processes that tend to restore chemical equilibrium. In a simplified way, we may consider a stationary state between photosynthetic production $P = dp/dt$ (rate of production of organic material; p = algal biomass) and heterotrophic respiration R (rate of destruction of organic material) (Figure 15.6) and chemically characterize this steady state by a simple stoichiometry (Redfield, et al., 1963).

$$106\ CO_2 + 16\ NO_3^- + HPO_4^{2-} + 122\ H_2O + 18\ H^+ (+ \text{trace elements and energy})$$

$$\underset{P}{\overset{R}{\rightleftharpoons}}$$

$$\underset{\text{algal protoplasm}}{\{C_{106}H_{263}O_{110}N_{16}P_1\}} + 138\ O_2 \qquad (3)$$

Alternatively, algal protoplasm may conveniently be expressed as

$$\{(CH_2O)_{106}(NH_3)_{16}(H_3PO_4)\}$$

The flux of energy through the system is accompanied by cycles of nutrients and other elements. Like every ecosystem, lakes and their surroundings and oceans include a biological community (primary producers, various trophic levels of decomposers, and consumers) in which the flow of energy is reflected in the trophic structure and in material cycles.

Although the stoichiometry of equation 3 may vary from one aquatic habitat to another, it is remarkable that the summation of the complicated processes of the P-R dynamics, in which so many organisms participate, results in such simple relations. To a first approximation parcels of water differ in their P, N, and C concentrations to the extent that organisms have removed or added elements in the fixed biomass ratio (equation 3).

A vertical segregation of nutritional elements occurs in lakes (during the stagnation period) and in the ocean (Figure 15.6b). During photosynthesis nitrogen (NO_3^- or NH_4^+) and phosphate are taken up together with carbon (CO_2 or HCO_3^-) in the proportion C/N/P $\simeq$ 106:16:1. As a consequence of respiration (oxidation) of these organism-produced particles after settling, these elements are released again in approximately the same proportions. Respiration is accompanied by a respiration quotient $\Delta O_2/\Delta C \simeq -1.3$ (or $\Delta O_2/\Delta N \simeq -9$).

As shown in Figure 15.7 and 15.8, simple correlations in ΔN, ΔP, and ΔO_2 are typically observed in lakes and in the ocean. The stoichiometric formulation of equation 3 reflects in a simple way *Liebig's law of the minimum*. It follows from Figure 15.7a,b that seawater becomes exhausted simultaneously in dissolved phosphorus and nitrogen as a result of photosynthetic assimilation. We infer that nitrogen and phosphorus together determine the extent of organic

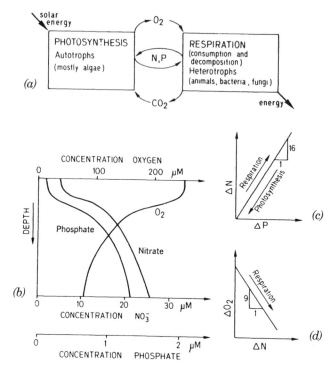

Figure 15.6. Photosynthesis and respiration. (a) A well-balanced ecosystem may be characterized by a stationary state between photosynthetic production, P (rate of production of organic material) and heterotrophic respiration, R (rate of destruction of organic matter). Photosynthetic functions and respiratory functions may become vertically segregated in a lake or in the sea. In the surface waters the nutrients become exhausted by photosynthesis. (b) The subsequent destruction (respiration) of organism-produced particles after settling leads to enrichment of the deeper water layers with these nutrient elements and a depletion of dissolved oxygen. The relative compositional constancy of the aquatic biomass and the uptake (P) and release (R) of nutritional elements in relatively constant proportions (see equation 3) are responsible for a covariance of carbon, nitrate, and phosphate in lakes (during stagnation period) and in the ocean; an increase in the concentration of these elements is accompanied by a decrease in dissolved oxygen. (c, d) The constant proportions $\Delta C/\Delta N/\Delta P/\Delta O_2$ typically observed in these waters are caused by the stoichiometry of the P–R processes.

production if temporary and local deviations are not considered. We might consider the possibility that originally phosphorus (e.g., from apatite) was the sole minimum nutrient, but that the concentration of nitrogen has been adjusted in the course of evolution as a result of nitrogen fixation and denitrification to the ratio presently found. Alternatively, we could also argue that the stoichiometric composition of the organisms as a result of evolution has become the same as that in the sea.

15.5 Interaction Between Organisms and Abiotic Environment

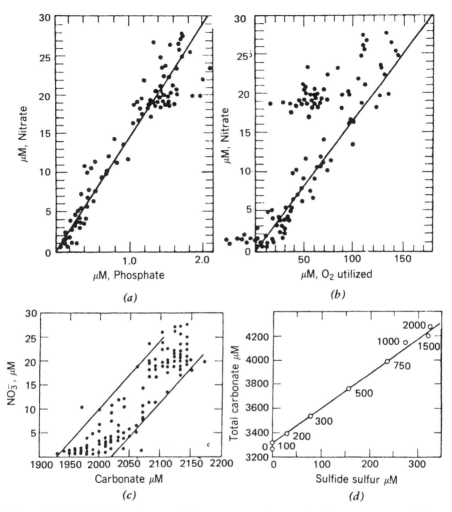

Figure 15.7. Stoichiometric correlations among nitrate, phosphate, oxygen, sulfide, and carbon. The correlations can be explained by the stoichiometry of reactions such as equation 3; concentrations are in micromolar. (a) Correlation between nitrate nitrogen and phosphate phosphorus corrected for salt error in waters of the western Atlantic. (b) Correlation between nitrate nitrogen and apparent oxygen utilization in same samples. The points falling off the line are for data from samples above 1000 m (Redfield, 1934, p. 177). (c) Correlation between nitrate nitrogen and carbonate carbon in waters of the western Atlantic. (d) Relation of sulfide sulfur and total carbonate carbon in waters of the Black Sea. Numbers indicate depth of samples. Slope of line corresponds to $\Delta S^{2-}/\Delta C = 0.36$. (From data of Stopintsev et al., 1958, as quoted in Redfield, et al., 1963.) (e) Correlation of the concentration of nitrogen to phosphate in the Atlantic Ocean (GEOSECS data). The slope through all the data yields an N/P ratio close to 16.

890 Regulation of the Chemical Composition of Natural Waters

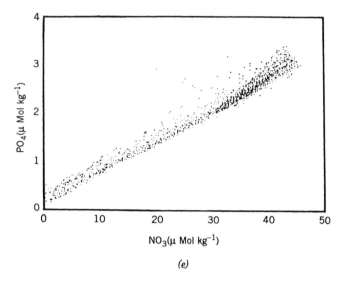

(e)

Figure 15.7. (*Continued*)

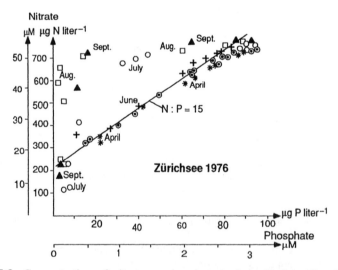

Figure 15.8. Concentration of nitrate vs phosphate in Lake Zurich. The drawn out line has a slope of $\Delta P/\Delta N \approx 15$. The deviations in July–September are related to the development of a metalimnetic oxygen minimum at ca. 20 m depth during this period. (Modified from Sigg and Stumm, 1994.)

15.5 Interaction Between Organisms and Abiotic Environment

There is also a horizontal segregation of biologically utilized elements in the deep Atlantic and the deep Pacific. The water circulation pattern in the ocean is responsible for the enrichment of biologically active elements in the deep Pacific relative to the deep Atlantic. Such effects are lucidly explained by Broecker and Peng (1982).

Data from GEOSECS and TTO Expeditions The GEOSECS and TTO (Transient Tracers in the Oceans) programs have provided a large source of data in the Atlantic (see Broecker 1985, and Takahashi et al., 1985). The interpretation of the chemical data from isopycnal surfaces yields a Redfield ratio of $P/N/C/O_2$ of $1:16:103:172$. The ratio of $P/CaCO_3$ dissolution has been estimated to be about $1:12$. This indicates that the ratio of CO_2 produced by the oxidation of organic carbon to that derived from the dissolution of $CaCO_3$ is about $10:1$.

All the GEOSECS data on the correlation of the concentration of nitrogen to phosphate in the Atlantic Ocean are given in Figure 15.7e.

As we have shown in Chapter 10, the Redfield concept might be extended to nutrient trace metals.

Lakes

Lakes, of course, are more transient in their chemical and biological characteristics than oceans. A stoichiometry of the nutrients can be observed in the elemental composition of the algae as well as in the concentration depth profiles (see Figure 15.8).

Limiting Nutrients For most inland waters phosphorus is the limiting nutrient in determining productivity. In some estuaries and in many marine coastal waters, nitrogen appears to be more limiting to algal growth than phosphorus. Deficiency in trace elements occurs usually only as a temporal or spatial transient.

Because of the complex functional interactions in lake ecosystems, the limiting factor concept needs to be applied with caution. We should distinguish between rate-determining factors (an individual nutrient, temperature, light, etc.) that determine the rate of biomass production and a limiting factor where a nutrient determines in a stoichiometric sense (equation 3) the maximum possible biomass standing crop.

Schindler (1977) used evidence from whole-lake experiments to show convincingly the phosphorus limitation in lakes; he also demonstrated that natural mechanisms compensated for nitrogen and carbon deficiencies in eutrophied lakes. In experimentally fertilized lakes, the invasion of atmospheric carbon dioxide supplied enough carbon to support and maintain phytoplankton populations proportional to P concentrations over a wide range of values. There was a strong tendency in every case for lakes to correct carbon deficiencies—obviously the rate of CO_2 supply from the atmosphere was sufficiently fast—

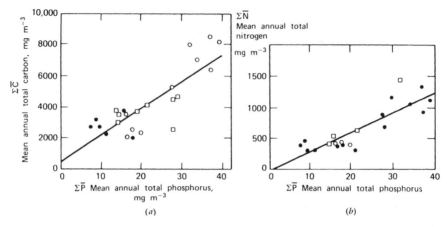

Figure 15.9. Phosphorus limitations in lakes. (a, b) In experimentally fertilized lakes of the experimental lakes area of the Freshwater Institute in Winnipeg (Environment Canada), ratios of mean annual concentrations C/P and N/P tend to become constant. (From Schindler, 1977.) In (a) we see that the C content increases as a consequence of P addition to the lakes, while (b) illustrates that the N content of a lake increases when the P input is increased, even when little or no nitrogen is added with fertilizer. Each point represents the results of a different lake.

maintaining concentrations of both chlorophyll and carbon that were proportional to the P concentration (Figure 15.9).

Schindler (1977) also demonstrated that biological mechanisms were in many cases capable of correcting algal nitrogen deficiencies. While a sudden increase in the P input may cause algae to exhibit symptoms of limitation by either N or C or both, there are long-term processes at work that appear to correct the deficiencies eventually, once again leaving phytoplankton growth proportional to the P concentration. As Schindler points out, this "evolution" of appropriate nutrient ratios in fresh waters involves a complex series of interrelated biological, geological, and physical processes, including photosynthesis, the selection of species of algae that can fix atmospheric nitrogen, alkalinity, nutrient supplies and concentrations, rates of water renewal, and turbulence. Various authors have observed shifts in algal species with changing N/P ratios; low N/P ratios appear to favor N-fixing blue-green algae, whereas high N/P ratios, achieved by controlling P input by extensive waste treatment, cause a shift from a "water bloom" consisting of blue-green algae to one containing forms that are less objectionable.

Typically, N/P ratios change in passing from the land to the sea. If one compares field data with the average ratio in phytoplankton, one sees that fresh waters in most instances receive an excess of N over that needed. Agricultural drainage contains relatively large concentrations of bound nitrogen, because nitrogen is washed out more readily from fertilized soil than phosphorus. In estuaries denitrification is frequently encountered, because NO_3-bearing waters may come into contact with organically enriched water layers; the N/P ratio can also be shifted by differences in the circulation rate of these two nutrients.

For reviews in biotic feedbacks in lake phosphorus cycles and the Redfield concept see Carpenter et al. (1992).

Example 15.2. O_2 Consumption Resulting from Increased Productivity
Estimate the P loading of a lake that may be tolerated without causing anaerobic conditions as a function of the depth of the hypolimnion (water layer below the thermocline).

As is the case in the sea, the deeper portions of a lake receive P in two forms: (1) preformed P, that is, P that enters the lake as such (or adsorbed on clays), and (2) P in the form of biogenic debris. Most of the latter is oxidized to form phosphates of oxidative origin, P_{ox}. For every P atom of oxidative origin found, a stoichiometric equivalent of oxygen atoms—342 oxygen atoms per P atom, or 140 g oxygen per 1 g P (cf. equation 3)—have been consumed. Accordingly, a flux of P_{ox} is paralleled by a flux of O_2 utilization, as indicated schematically in Figure 15.10. The tolerable phosphorus loading of a lake may be related to the hypolimnetic oxygen consumption. The annual P loading per lake surface, L_t (mg P m^{-2} yr^{-1}), causes (under the simplifying assumption that all L_t becomes phosphorus of oxidative origin, P_{ox}) during the stagnation period, T_{st} (days), an approximate oxygen consumption, $\Delta[O_2]$ (mg m^{-3}) of the hypolimnion assumed to be homogeneously mixed of depth $z_H(m)$ that is

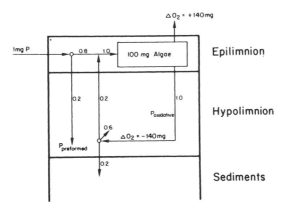

Figure 15.10. Simplified scheme of typical transformation of phosphorus in a stratified lake. One milligram of phosphorus introduced into a lake during the stagnation period may lead to the synthesis of 100 mg algae (dry mass), which upon mineralization cause an oxygen consumption in the hypolimnion of 140 mg O_2; from the organic P mineralized in the hypolimnion 0.6 mg are assumed to accumulate during the stagnation period, while 0.2 mg are assumed to be adsorbed [e.g., on iron(III) oxide] and transferred into the sediments; another 0.2 mg reaches the surface waters as phosphate by eddy diffusion.

given by

$$\Delta[O_2] = 140 \frac{T_{st} L_t}{365 z_H} \quad (i)$$

Correspondingly, a maximum P loading, L_{max}, could be estimated for a tolerable oxygen consumption $[O_2]_{max}$:

$$L_{max} = \Delta[O_2]_{max} \times 7 \times 10^{-3} \times \frac{365}{T_{st}} z_H \quad (ii)$$

Many complicating factors, however, must not be overlooked. The simple stoichiometric relations may be too simplified, and they may change from lake to lake.

As shown schematically in Figure 15.10, 1 mg of P can synthesize (assuming P to be the limiting factor) approximately 0.1 g of algal biomass (dry weight) in a single cycle of the limnological transformation. After settling to the deeper layers, this biomass exerts a biochemical oxygen demand of approximately 40 mg for its mineralization.

From this simple calculation it is obvious that the organic material introduced into the lake by domestic wastes (20–100 mg organic matter liter^{-1}) is small in comparison to the organic material that can be biosynthesized from introduced fertilizing constituents (3–8 mg P liter^{-1}, which can yield 300–800 mg organic matter per liter). Aerobic biological waste treatment with a heterotrophic enrichment culture mineralizes substantial fractions of bacterially oxidizable organic substances but is not capable of eliminating more than 20–50% of nitrogen and phosphate constituents.

Anoxic Conditions The usual sequence of various redox reactions with organic matter is observed where the accumulation of organic matter is great. As shown in Table 8.8, oxygen dissolved in water first becomes exhausted, and then the oxidation of organic matter continues with nitrate serving as the oxidant; subsequently, organic fermentation reactions and redox reactions with SO_4^{2-} and CO_2 as electron acceptors occur. The reduced products of the oxidation of organic matter accumulate in the water in addition to the products of the oxidation of organic matter. The formal equations given in Table 8.6 can be modified to develop stoichiometric models that predict how the components of the anoxic water will change as a result of the mineralization of settled plankton. Such stoichiometric models have been developed by Richards (1965), who corroborated the validity of these models for a large number of anoxic basins and fjords. For example, under conditions of sufficiently low pϵ, planktonic material becomes oxidized by SO_4^{2-}, which is reduced to S(-II). For the oxidation of one $\{C_{106}H_{263}O_{110}N_{16}P_1\}$, approximately 424 electrons are necessary. (Note that the carbon in the plankton formula has a formal oxidation state of approximately 0.) In the reduction of SO_4^{2-} to S(-II), 53 SO_4^{2-} ions

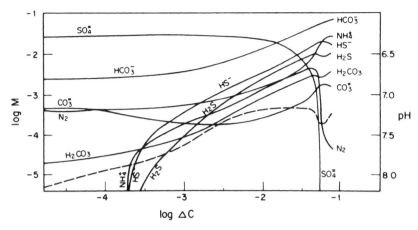

Figure 15.11. Distribution of predominant dissolved species during the decomposition of alanine ($C_3H_7O_2N$) in seawater at 25°C. The molar concentrations of dissolved species are given as a function of ΔC, the number of moles of organic carbon reacted per liter of solution. The dashed line represents pH. (From Thorstenson, 1970.)

can accept the 424 electrons necessary; hence one would expect a ratio $\Delta S(-II)/\Delta C = 0.5$. As Figure 15.7d shows, this ratio is somewhat less for the waters of the Black Sea.

Using the concept of partial equilibria Thorstenson (1970) has predicted in a reaction path calculation the compositional changes in the aqueous phase as a function of the amount of organic matter decomposed. Figure 15.11 gives the distribution of predominant dissolved species at hypothetical equilibrium as a function of increments of organic matter decomposed, represented in this example by $C_3H_7O_2N$ (= alanine).

15.6 THE OCEANS: RELATIVE CONSTANCY OF THE COMPOSITION AND CHEMICAL EQUILIBRIA

The salt dissolved in seawater has remarkably constant major constituents (Table 15.2). Cl^-, SO_4^{2-}, Mg^{2+}, K^+, Ca^{2+}, and Na^+ dominate sea salt. Their ratios one to another are very constant. This constancy does not extend to all the trace components (see Table 6.1), especially not to the biolimiting elements that are removed from the surface seawater by organisms. For nonbiolimited elements, the ratio of the element to the total salt (e.g., chlorinity) in both surface and deep seawater samples is unchanged.

The relative proportions of the major ions in seawater and average river water are quite different. Concentration of river water does not produce ocean water. If average river water were concentrated by evaporation, a variety of minerals would precipitate and the ratio among the elements would change. As

Table 15.2. Major Composition of Seawater

Constituent	Seawater at $S = 35‰$ (g kg^{-1})[a]	(g kg^{-1}) ÷ Chlorinity[b]	(mol kg^{-1}) ÷ Chlorinity	Residence Time in Oceans, log τ (years)[d]
Na$^+$	10.77	0.556	0.0242	7.7
Mg^{2+}	1.29	0.068	0.0027	7.0
Ca^{2+}	0.4121	0.02125	0.000530	5.9
K$^+$	0.399	0.0206	0.000527	6.7
Sr^{2+}	0.0079	0.00041	0.0000047	6.6
Cl$^-$	19.354	0.9989	0.0282	7.9
SO$_4^{2-}$	2.712	0.1400	0.0146	6.9
HCO$_3^-$ [c]	0.1424	0.00735	0.00012	4.9
Br$^-$	0.0673	0.00348	0.000044	8
F$^-$	0.0013	0.000067	0.0000035	5.7
B	0.0045	0.000232	0.0000213	7.0
	$\Sigma = 35$	$\Sigma = 1.82$	$\Sigma = 0.058$	

[a]Salinity, $S(‰)$, is defined as the weight in grams of the dissolved inorganic matter in 1 kg of seawater after all Br$^-$ and I$^-$ have been replaced by the equivalent quantity of Cl$^-$ and all HCO$_3^-$ and CO$_3^{2-}$ are converted to oxide. In over 97% of the seawater in the world, the salinity S is between 33‰ and 37‰. The total number of grams of major constituents (sea salt), g_T, for 1 kg of solution is related to the chlorinity by $g_T = 1.81578$ Cl (‰). Salinity $S(‰)$, is defined as $S(‰) = 1.80655$ (Cl‰); thus $g_T = 1.00511$ $S(‰)$.
[b]The chlorinity, $Cl(‰)$, is determined by the titration of seawater with AgNO$_3$. It was defined as the chlorine equivalent of the total halide concentration in g kg^{-1} seawater; it is now defined as the mass in grams of Ag necessary to precipitate the halogens (Cl$^-$ and Br$^-$) in 328.5233 g of seawater. It has been adequately demonstrated that the relative composition of the major (greater than 1 mg kg^{-1} seawater) components of seawater is nearly constant. By measuring one constituent of seawater, the composition of other components can be characterized. The constituent normally selected is the chlorinity, $Cl(‰)$.
[c]The results given for HCO$_3^-$ are actually values of the carbonate alkalinity expressed as though it were all HCO$_3^-$.
[d]Residence times were computed by $\tau = M/Q$, where M for a particular constituent is equal to its concentration in seawater times the mass of the oceans and Q is equal to the concentration of the constituent in average river water times the annual flux of river water to the ocean.

we have seen in Example 15.1 and Figure 15.4b, the composition of evaporated river water would resemble the composition of carbonate-bearing nonmarine evaporite brines much more than the composition of seawater. Furthermore, the dissolved material present in the ocean is only a small fraction of that delivered to the ocean by the rivers over geological time. Obviously, ions must be removed from the ocean approximately as fast as they are supplied by rivers. Thus the removal processes exert the major control on the chemical composition of seawater.

Essentially two major concepts regarding control of the composition of seawater complement each other: (1) control by chemical equilibria between seawater and oceanic sediments, and (2) kinetic regulation by the rate of supply of individual components and the interaction between biological cycles and mixing cycles.

Obviously the sea is an open, dynamic system with variable inputs and outputs of mass and energy for which the state of equilibrium is a construct. As we have seen, the concept of free energy, however, is no less important in dynamic systems. In considering equilibria and kinetics in ocean systems, it is useful to recall that different time scales need to be identified for the various processes. When a particular reaction of a phase or species has—within the time scale of consideration—a negligible rate, it is permissive to define a metastable equilibrium state. Similarly, in a flow system the time-invariant condition of a well-mixed volume approaches chemical equilibrium when the residence time is sufficiently large relative to the appropriate time scale of the reaction.

15.7 CONSTANCY OF COMPOSITION: STEADY STATE

Much of the chemistry of the oceans and of freshwater systems depends on the kinetics of various physical and chemical processes and on biochemical reactions rather than on equilibrium conditions. The simplest model describing systems open to their environment is the time-invariant steady-state model. Because the sea has remained constant for the recent geological past, it may be well justified to interpret the ocean in terms of a steady-state model.

Input is balanced by output in a steady-state system. The concentration of an element in seawater remains constant if it is added to the sea at the same rate that it is removed from the ocean water by sedimentation. Input into the oceans consists primarily of (1) dissolved and particulate matter carried by streams, (2) volcanic hot spring and basalt material introduced directly, and (3) atmospheric inputs. Often the latter two processes can be neglected in the mass balance. Output is primarily by sedimentation; occasionally, emission into the atmosphere may have to be considered. Note that the system considered is a single box model of the sea, that is, an ocean of constant volume, constant temperature and pressure, and uniform composition.

The concept of the *residence time* (or the passage time) of an element in the sea, τ, is defined as

$$\tau = \frac{\text{amount in the sea}}{\text{amount supplied per unit time or amount removed per unit time}} \tag{4}$$

or in shorter notation

$$\tau = \frac{M}{\Sigma J_i} \quad \text{(time)} \tag{5}$$

where the amount M may be given in the same units (e.g., moles) as the input or removal fluxes; J_i are in the same units per unit of time (e.g., mol yr^{-1}).

Each of the input or removal fluxes corresponds to a fractional mean residence time, τ, defined with respect to the particular process:

$$\tau_i = \frac{M}{J_i} \quad \text{(time)} \tag{6}$$

The mean residence time is then given by

$$\frac{1}{\tau} = \frac{1}{\tau_1} + \frac{1}{\tau_2} + \cdots + \frac{1}{\tau_n} \tag{7}$$

Equation 7 shows that the stronger fluxes (smaller fractional residence times) make the contributions of the weaker fluxes insignificant.

If J_{in} and J_{out} represent the input rate and removal rate (fluxes) of an element in the sea, respectively, then

$$\frac{dM}{dt} = J_{\text{in}} - J_{\text{out}} \tag{8}$$

If one assumes as a first approximation that the removal rate is proportional to the total amount of the element in the sea, that is, $J_{\text{out}} = kM$, where k is the rate constant, then at steady state equation 8 becomes

$$0 = J_{\text{in}} - kM$$

or

$$k^{-1} = MJ_{\text{in}}^{-1} = \tau \tag{9}$$

The inverse of the residence time is equal to the removal rate constant of an element in the sea.

The rate of sedimentation is controlled largely by the rate at which an element is converted (uptake by organisms, precipitation, coprecipitation, ion exchange) into an insoluble, settleable form. Hence the reactivity of the elements influences the time that elements spend, on the average, as constituents of the seawater.[†] For most elements residence times have been determined on the basis of estimates of the input by runoff from the land or from calculations of sedimentation times. Remarkably similar results are obtained by these two

---

[†]Whitfield (1979) has shown that the mean oceanic residence time can be related to an ocean–rock partition coefficient, which in turn is related to the electronegativities of the elements on the assumption that seawater composition is controlled by general adsorption–desorption reactions at surfaces having oxygen donor groups. Lasaga (1980) provides a systematic approach to the kinetic treatment of geochemical cycles by extending the concepts of residence times and response times. This extension is particularly useful when complex cycles are involved.

15.7 Constancy of Composition: Steady State

Table 15.3. Major Ion Composition of Average River[a] and Seawater[b]

Ion	Average River (mM)	Average Seawater (mM)
HCO_3^-	0.86	2.38
SO_4^{2-}	0.069	28.2
Cl^-	0.16	545.0
Ca^{2+}	0.33	10.2
Mg^{2+}	0.15	53.2
Na^+	0.23	468.0
K^+	0.03	10.2

[a]Data from Berner and Berner (1987). Note that the reported concentrations exclude pollution.
[b]Data from Holland (1978).
Source: Morel and Hering (1993).

methods. Residence times calculated from the river input[†] are given in Table 15.2. Elements that are highly oversaturated (e.g., Al, Fe) have short residence times, that is, times that are smaller than those necessary for ocean mixing. On the other hand, elements with low reactivity such as Na and Li have very long residence times. Compare Section 10.10 and Figure 10.22.

Equilibrium considerations pertaining to seawater have been made in Section 6.8.

In comparing the major ion composition of river water to that of seawater (Table 15.3), we are again reminded that seawater is not simply evaporated river water.

Table 15.4 gives a rough mass balance of riverine inputs and the various outputs by atmospheric cycling, ion exchange processes, deposition in sediments, and hydrothermal activities. One important part of the mass balance is the deposition of $CaCO_3$. Biological processes also cause the removal of SO_4^{2-}, which is consumed in biologically mediated sulfate reduction and deposited as pyrite.

Difficulties arise in explaining the removal of Mg^{2+} or K^+ from the annual river flow to the sea. Reverse weathering processes were postulated for a time but little experimental evidence could be found for such processes.

But some processes occurring in *hydrothermal vents* (springs coming from the sea floor in areas of active volcanism) may account for the exchange of Mg^{2+}. High-temperature seawater–basalt reactions occur at active spreading centers at ridge crests; basalt reacts with seawater at lower temperatures on the ridge flanks. Such reactions are also the major sink for magnesium in the ocean. The hot water from the hydrothermal "vents" is effectively stripped of mag-

---

[†]Ions such as Cl^- and Na^+ are cycled through the atmosphere; maritime aerosols from wind spray and bursting seawater bubbles are transported to the continents and returned via rivers. If corrections are made for atmospheric cycling, residence times for Na^+ and Cl^- are obtained that are several times greater than the values given in Table 15.2.

Table 15.4. Long-Term Input-Output Balance for Major Seawater Ions and Alkalinity[a]

Ion	Ocean Inventory (10^{18} mol)	River Input[b]	Atmospheric-Evaporite Cycling[c]	Ion Exchange[d]	Hydrothermal Activity[e]	Pyrite Burial or Other[f]	Carbonate Deposition[g]
Cl^-	710	+6.1	(−6.1)				
Na^+	608	+8.5	−5.7	−1.9	(−0.9)		
Mg^{2+}	69	+5.2	−0.3	−1.2	(−3.1)		
SO_4^{2-}	37	+3.2	−0.3		(−1.7)	−1.2	
K^+	13	+1.2	−0.06	−0.4	(−0.6)	−0.1	
Ca^{2+}	13	+12.5	−0.06	+2.6	(+2.0)		−17
Alk	3.1	+31.9		+0.5	−0.4	+2.4	−35

[a]Data from McDuff and Morel (1980) and Berner and Berner (1987).
[b]Concentrations from Berner and Berner (1987); discharge, 3.74×10^{16} liters yr^{-1} (world average); annual fluxes in 10^{12} mol yr^{-1}.
[c]Includes pore water burial.
[d]Values from McDuff and Morel (1980).
[e]Includes both high- and low-temperature basalt alternation.
[f]For K, fixation on deltaic clays.
[g]Values from Berner and Berner (1987).
Source: Morel and Hering (1993).

nesium. (The results for Na^+ and Cl^- are variable.) For potassium, the major removal mechanisms have not been established.

15.8 HYDROTHERMAL VENTS

Deep-sea hydrothermal vents were discovered in the 1970s after an extensive search along the Galápagos Rift, a part of the globe-encircling system of seafloor spreading axes. During the past 7 years, more hydrothermal vent fields have been located along the East Pacific Rise. They fall into two main groups: (1) warm vent fields with maximum exit temperatures of 5–23°C and flow rates of 0.5–2 cm s^{-1} and (2) hot vent fields with maximum exit temperatures of 270–380°C and flow rates of 1–2 m s^{-1}. Hot vent fields commonly include warm- and intermediate-temperature vents (≤ 300°C) ("white smokers") as well as high-temperature vents (350 ± 2°C) ("black smokers"). A highly efficient microbial utilization of geothermoal energy is apparent at these sites: rich animal populations were found to be clustered around these vents in the virtual absence of a direct photosynthetic food source.

Figure 15.12 summarizes the inorganic processes occurring at warm- and hot-water vent sites. As pointed out by Jannasch and Mottl 1985, during the cycling of seawater through the earth's crust along the mid-ocean ridge system, geothermal energy is transferred into chemical energy in the form of reduced inorganic compounds. These compounds are derived from the reaction of seawater with crustal rocks at high temperatures and are emitted from warm (≤ 25°C) and hot (~ 350°C) submarine vents at depths of 2000–3000 m. Chemolithotrophic bacteria use these reduced chemical species as sources of energy for the reduction of carbon dioxide (assimilation) to organic carbon (Table 15.5):

$$CO_2 + H_2S + O_2 + H_2O \longrightarrow [CH_2O] + H_2SO_4$$

(aerobic chemoautolithotrophy, bacteria) (10)

$$2CO_2 + 6H_2 \longrightarrow [CH_2O] + CH_4 + 3H_2O$$

(anaerobic chemoautolithotrophy, bacteria) (11)

These bacteria form the base of the food chain, which permits copious populations of certain specifically adapted invertebrates to grow in the immediate vicinity of the vents. Such highly prolific, although narrowly localized, deep-sea communities are thus maintained primarily by terrestrial rather than by solar energy. Reduced sulfur compounds appear to represent the major electron donors for aerobic microbial metabolism, but methane-, hydrogen-, iron-, and

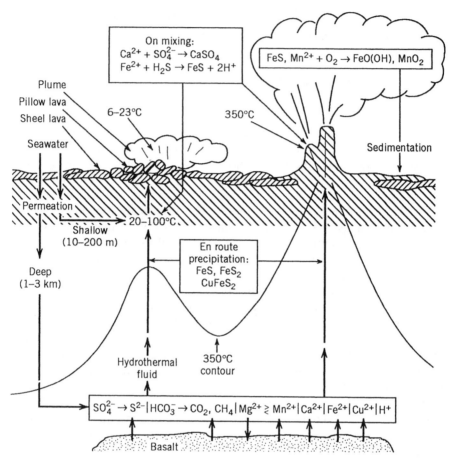

Figure 15.12. Schematic diagram showing inorganic chemical processes occurring at warm- and hot-water vent sites. Deeply circulating seawater is heated to 350°–400°C and reacts with crustal basalts, leaching various species into solution. The hot water rises, reaching the sea floor directly in some places and mixing first with cold, downwelling seawater in others. On mixing, iron–copper–zinc sulfide minerals and anhydrite precipitate. (From Jannasch and Mottl, 1985.)

manganese-oxidizing bacteria have also been found. Methanogenic, sulfur-respiring, and extremely thermophilic isolates carry out anaerobic chemosynthesis. Bacteria grow most abundantly in the shallow crust where upwelling hot, reducing hydrothermal fluid mixes with downwelling cold, oxygenated seawater. The predominant production of biomass, however, is the result of symbiotic associations between chemolithotrophic bacteria and certain invertebrates, which have also been found as fossils in Cretaceous sulfide ores of ophiolite deposits (Jannasch and Mottl, 1985).

Table 15.5. Electron Sources and Types of Bacteria Occurring at Hydrothermal Vents

Electron Donor	Electron Acceptor	Organisms
S^{2-}, S^0	O_2	Sulfur-oxidizing bacteria
S^{2-}, S^0	NO_3^-	Denitrifying and sulfur-oxidizing bacteria
H_2	O_2	Hydrogen-oxidizing bacteria
H_2	NO_3^-	Denitrifying hydrogen bacteria
H_2	S^0, SO_4^{2-}	Sulfur- and sulfate-reducing bacteria
H_2	CO_2	Methanogenic and acetogenic bacteria
NH_4^+, NO_2^-	O_2	Nitrifying bacteria
Fe^{2+}, (Mn^{2+})	O_2	Iron- and manganese-oxidizing bacteria
CH_4, CO	O_2	Methylotrophic and carbon monoxide-oxidizing bacteria

Source: Jannasch and Mottl (1985).

15.9 THE SEDIMENT–WATER INTERFACE

The sediments are not just depositories for the material removed from the ocean or lake water. The flux of constituents from the sediments into the water, and vice versa, is important in controlling the composition of oceans and lakes.

The diagenetic† chemical reactions occurring within the sediments consist of abiotic and biogenic reactions. Because of these reactions, sediments exert a significant effect on the overlying ocean and lake waters (Figure 15.13). Most biogenic reactions depend on the decomposition of organic matter. The sequence of redox reactions occurring in the sediments is the same as that already discussed (see Section 8.5) for the interaction of excess organic matter with O_2, NO_3^-, SO_4^{2-}, and HCO_3^-: that is, the removal of dissolved O_2; the reduction of NO_3^-, SO_4^{2-}, and HCO_3^-; and the production of CO_2, NH_4^+, phosphate, HS^-, and CH_4 (Figure 15.13a).

The products of these reactions may in turn bring further changes in sediment chemistry such as the solubilization of iron and manganese after O_2 has been removed. HS^- resulting from SO_4^{2-} reduction, on the other hand, may react with detrital iron minerals to form iron sulfides. Excess HCO_3^- is produced by SO_4^{2-} reduction and NH_4^+ formation; eventually $CaCO_3$ may be precipitated. Buildup of dissolved phosphate may under suitable conditions bring about the precipitation of apatite. Mg^{2+} may be precipitated as a result of the removal of clay minerals of iron, which reacts with HS^- to form iron sulfides. Reactions that are not biogenically controlled include the dissolution of opaline silica, $CaCO_3$, and feldspars, and various ion-substitution reactions in sediment minerals and ion exchange processes on clay minerals.

†Diagenesis refers to changes that take place within a sediment during and after burial.

904 Regulation of the Chemical Composition of Natural Waters

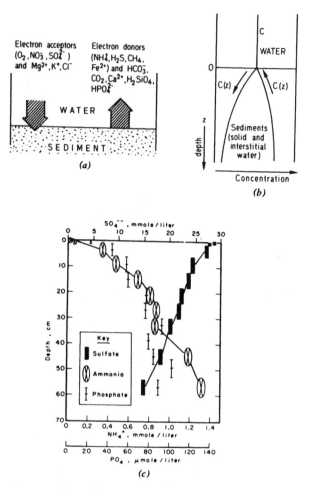

Figure 15.13. The sediment–water interface. (a) Direction of fluxes expected for dissolved constituents between sediment pore waters and the overlying waters (oceans and lakes). (b) For sediments and pore water, the one-dimensional distribution of concentrations is time and depth dependent. Arrows indicate fluxes at the sediment–water interface depending on the concentration gradient in pore water. The overlying water (ocean or lakes) is assumed to be well mixed. (c) Sulfate, phosphate, and ammonia versus depth in pore waters from Santa Barbara Basin, California. (From Sholkovitz, 1973.)

Fluxes of Solids, Water, and Solutes

Sedimentation of solid particles and entrapment of water in the pore spaces are two major fluxes of materials across the sediment–water interface.

Further processes responsible for transport across the sediment–water interface are upward flow of (pore) water caused by hydrostatic pressure gradients

15.9 The Sediment-Water Interface

of groundwater in aquifers or land; molecular diffusional fluxes in pore water; and mixing of sediment and water at the interface (bioturbation and water turbulence). The rates of sediment deposition vary from mm per 1000 years in the pelagic ocean up to cm yr^{-1} in lakes and near-shore oceanic areas. A sediment can be thought of as made up of equal volumes of water and solid particles (sediment volume fraction = 0.5) of density 2.5 g cm^{-3} (Lerman, 1979).

Since ocean water contains ~35 g liter^{-1} and fresh water ~0.01 g liter^{-1} dissolved material, the mean fluxes to the sediments are in the following ranges: solids = 6 × 10^{-4} to 6 × 10^{-1} g cm^{-2} yr^{-1} and dissolved materials = 0.1 × 10^{-4} to 0.1 × 10^{-1} cm^{-2} yr^{-1} (Lerman, 1979). In freshwater lakes dissolved material fluxes are closer to the lower value.

The net flux of a chemical species across the sediment-water interface, $F_{x=0}$, is due to (molecular) diffusion in pore water, $F_d = -\phi D(dC/dz)$, the flux due to advection in pore water, $F_a = \phi UC$, and the flux due to deposition of solid particles, $F_s = \phi U_s C_s$:

$$F_{z=0} = \phi\left(-D\frac{dC}{dz} + UC + U_s C_s\right) \qquad (12)$$

where C and C_s are the concentrations in solution and in solids, respectively (units of mass per unit of pore water), z is the depth (positive and increasing downward from the sediment-water interface), and U and U_s are the rates of pore water advection and of sedimentation, respectively (cm yr^{-1}). U is positive when the sediment and water flow are downward relative to $z = 0$; ϕ is the porosity (volume fraction of sediment occupied by water).[†]

Two simplifying assumptions can be introduced into equation 12. (1) If the sediment porosity ϕ changes little with depth, then the sediment particles and pore water do not move relative to each other in a growing sediment column and we can set U equal to the sedimentation rate $U = U_s$. (2) A relationship between the concentration in pore water, C, and the concentration in the solids, C_s, can be established. Under simplifying assumptions a linear relationship $C_s = KC$ may be assumed, where K is a (dimensionless) distribution coefficient. Thus in noncompacting sediments, the general flux equation may be written

$$F_{z=0} = \phi\left[-D\frac{dC}{dz} + U(K + 1)C\right]_{z=0} \qquad (13)$$

The larger K is, the more important the flux of a chemical species or settling particles, UKC, in comparison to the flux in pore water, UC. In compacting

---

[†]An observer balanced on the sediment-water interface ($z = 0$) as sediment particles continue to arrive from above and pile up will see the particles and pore water flow by in the downward direction. In this sense one can always speak of the fluxes of solids, waters, and solutes as moving up or down (Lerman, 1979).

sediments the total flux becomes (Berner, 1976)

$$F_{z=0} = \phi\left[-D\frac{dC}{dz} + UC\left(\frac{K\phi(1-\phi_\infty)}{\phi_\infty(1-\phi)} + 1\right)\right] \quad (14)$$

where ϕ_∞ is the porosity at a given depth when it has attained a steady value smaller than ϕ, the porosity at the interface.

For the calculation of the fluxes the following parameters are needed: porosity, sedimentation rate, diffusion coefficients, distribution constants, and the concentration gradients for each species $(dC_i/dz)_{z=0}$. In assessing diffusion coefficients, coupling-effects due to electroneutrality between coions and counterions may have to be considered. Serious errors may be introduced by using wrong values of dC_i/dz, because it is very difficult to collect undisturbed sediment.

Redox Transformations in Lake Sediments

Figure 15.14 gives some pore water profiles from Lake Greifen. The idealized redox sequence leads to a picture of vertically separated processes (see Section 8.6).

Remineralization of organic N and P results in increases in ammonium and phosphate below the sediment–water interface. The increase in alkalinity is also a result of the decomposition of organic matter either directly (as organic N is remineralized to NH_4^+) or indirectly (as calcite dissolves in response to the release of CO_2 associated with remineralization of organic C). In this sediment, the profiles of the redox-active species are not clearly separated.

Depletion of both NO_3^- and SO_4^{2-} and release of Mn^{2+} occurs very near the sediment–water interface; $MnO_2(s)$ is found only in the top centimeter of the sediments. The peak in sulfide, however, is slightly lower in the sediment profile than the peak in dissolved Mn^{2+}; the successive appearance of these species is in accord with the corresponding redox potentials. The decrease in the dissolved concentrations of Mn(II) and S(−II) with depth are attributable to precipitation of rhodochrosite [$MnCO_3(s)$] and iron sulfides. Precipitation of phosphate-containing minerals may account for the decrease in pore water phosphate. Vivianite, $Fe_3(PO_4)_2 \cdot 8H_2O$, was calculated to be supersaturated in the pore water, although this mineral could not be identified in the sediments.

Sediments as Memory Records Figure 15.15 gives the Mn_T/Fe_T ratio in dry sediments versus depth and age. This ratio permits some insight into the redox potential of the sediment–water interface at the time of deposition.

Before 1930 the sediment–water interface at the time of deposition was permanently oxic. The change in Mn_T/Fe_T ratio at this depth can be explained

15.9 The Sediment-Water Interface 907

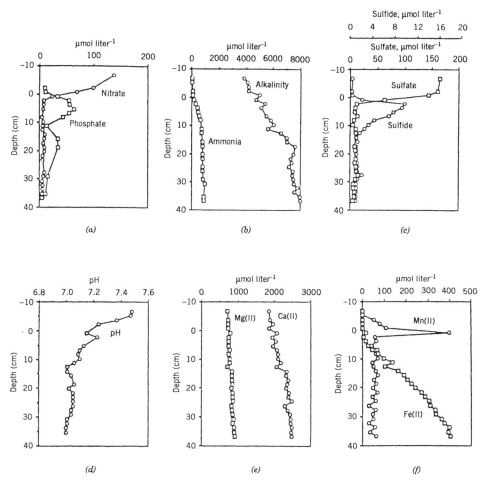

Figure 15.14. Interstitial water concentrations of major compounds, alkalinity, and pH versus depth: (a) nitrate and phosphate, (b) ammonia and alkalinity, (c) sulfate and sulfide, (d) pH, (e) magnesium(II) and calcium(II), and (f) manganese(II) and iron(II). (From Wersin et al., 1991.)

by the different behavior of manganese in oxidizing and reducing environments. With the development of the anoxic sediment–water interface, a significant fraction of deposited Mn(III, IV) oxides was reduced and released to the water column as Mn^{2+}. In contrast, the reduction of the Fe(III) oxides and the release of Fe^{2+} are not markedly enhanced under anoxic conditions. Thus the Mn_T/Fe_T ratio is significantly higher in the light grey zone, deposited before 1930, when the sediment–water interface was permanently oxic, than in the younger dark grey sediments, deposited under partially reducing conditions.

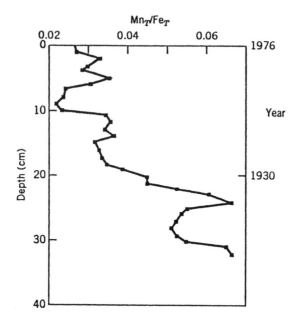

Figure 15.15. Mn_T/Fe_T ratio in dry sediments versus depth and age (data taken from core in 1976 analyzed by G. Schaer, personal communication). Prior to 1930 the sediment–water interface was permanently aerobic. Under anaerobic conditions after 1930 the tendency of the Mn deposition decreased because Mn(II) [as reduced from initially deposited Mn(III,IV) (hydr)oxides] is mobile and released to the water column. (From Wersin et al., 1991.)

15.10 BIOLOGICAL REGULATION OF THE COMPOSITION

We have already shown in the previous section that life—by cycling and transporting chemical elements—plays an important role in maintaining the earth in a homeostatic[†] state. Lotka (1924) wrote more than 70 years ago:

> If we are satisfied to omit innumerable details, we can trace, for each of the most important chemical elements concerned, the broad outline of its cycle in nature. The elements and simple compounds principally concerned are carbon (CO_2), oxygen (O_2), nitrogen (N_2, NH_3, NO_2^-, NO_3^-), water (H_2O), phosphorous (PO_4^{3-} etc.) ... For the drama of life is like a puppet show in which stage, scenery, actors and all are made of the same stuff. The players, indeed, "have their exits and their entrances," but the exit is by way of translation into the substance of the stage; and each entrance is a transformation scene. So stage and players are bound together in the close partnership of an intimate comedy: and if we would catch the spirit of the piece, our attention must not all be absorbed in the characters alone, but must be extended also to the scene, of which they

---

[†]The term homeostatic has been used to indicate constancy maintained by negative feedback.

15.10 Biological Regulation of the Composition

are born, on which they play their part, and with which, in a little while, they merge again.

A Kinetic Model for the Chemical Composition of Seawater

Obviously the composition of natural waters is markedly influenced by the growth, distribution, and decay of phytoplankton and other organisms. The dominant role of organisms in regulating the oceanic composition and its variation with depth of some of the important sea salt components (i.e., C, N, P, and Si) will be illustrated here by introducing certain aspects of Broecker's kinetic model for the chemical composition of seawater (Broecker and Peng, 1982). We summarize Broecker's line of arguments.

The one process that yields geographic and depth variation in the chemical composition of sea salt components is uptake of dissolved constituents by organisms. The remains of these organisms sink under the influence of gravity and are gradually destroyed by oxidation. The superposition of this particular cycle upon the ordinary mixing cycle in the sea accounts for the present distribution of chemical properties.

Broecker's model is shown in Figure 15.16. The ocean is divided into two boxes—an upper one of a few hundred meters depth and a lower one of

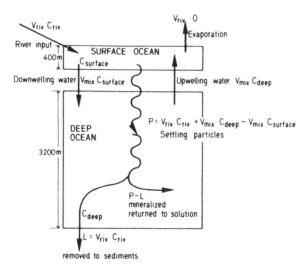

Figure 15.16. Broecker's (1974) idealized kinetic model for the marine cycles of biologically fixed elements. V_{riv}, volume of river water entering the ocean per year expressed as volume per unit sea area (m^3 m^{-2} yr^{-1} or m yr^{-1}) = 0.1 m yr^{-1}; C_{riv}, concentration of an element in average river water (mol m^{-3}); $V_{riv} \times C_{riv}$, input flux (mol m^{-2} yr^{-1}); V_{mix}, volume of water sinking into deep water box = volume of water rising to surface water box (volume m^{-2} yr^{-1}) = 200 m yr^{-1}; $C_{surface}$, C_{deep}, concentration in surface ocean, in deep ocean (e.g., mol m^{-3}); $V_{mix} \times C_{surface}$, flux (mol m^{-2} yr^{-1}) of an element that sinks from the ocean surface into the deep ocean (mol m^{-2}).

3200-m depth. The zone separating the upper warmer water body from the lower cooler water body is a density gradient (the thermocline), which provides an obstruction to mixing. It is furthermore assumed that the only way an element is added is by river runoff from the continents and the only way an element is removed from the ocean is by the fall of organism-produced particles to the sea floor, which are then permanently buried in the sediments.

The ocean and its two compartments are assumed to be at steady state. The warm surface ocean receives its supply of any given element (1) from the water entering from the rivers and (2) from the deep ocean, which is steadily being exchanged with surface water. Downwelling and particulate settling match these two inputs, so that the concentration of any element remains constant in the surface ocean. The amount of a given element entering the surface reservoir each year is then $V_{mix} C_{deep} + V_{riv} C_{riv}$. The return flow to the deep sea carries away an amount of the element equal to $V_{mix} C_{surf}$. Material balance requires that the remainder, $P = V_{mix} C_{deep} + V_{riv} C_{riv} - V_{mix} C_{surf}$, be carried by falling particles (see Figure 15.16).

The yearly amount of runoff from the continents, V_{riv}, is equal in volume to a layer 10 cm thick over the entire ocean surface. The mean residence time of water in the deep ocean—as calculated from the distribution of ^{14}C—is on the order of 1600 years. Since the depth of the deep ocean is ~3200 m, the upwelling rate (which must equal to the downwelling rate) is ~2 m yr^{-1}. Thus $V_{mix}/V_{riv} \simeq 20$ (for an explanation of symbols see Figure 15.16).

Defining the fraction of an element removed from the surface ocean by these particles, g,

$$g = \frac{\text{particle flux of an element into deep ocean}}{\Sigma \text{ fluxes of this element into surface ocean}}$$

we have

$$g = \frac{V_{riv} C_{riv} + V_{mix} C_{deep} - V_{mix} C_{surf}}{V_{riv} C_{riv} + V_{mix} C_{deep}}$$

$$= 1 - \frac{V_{mix} C_{surf}/V_{riv} C_{riv}}{1 + V_{mix} C_{deep}/V_{riv} C_{riv}} \qquad (15)$$

For an element such as phosphorus, C_{deep}/C_{riv} is about 5 and C_{surf}/C_{riv} is 0.25. The corresponding value of g is 0.95; that is, 95% of the phosphate introduced into surface ocean is carried away by falling particles.

The fraction f of a given element carried to the deep sea by the particulate flux P surviving destruction,

$$f = \frac{\text{export } (L) \text{ into sediments}}{\text{import of particles } (P) \text{ into deep sea}}$$

must be equal to the river input $fP = V_{riv}C_{riv}$:

$$f = \frac{L}{P} = \frac{V_{riv}C_{riv}}{V_{riv}C_{riv} + V_{mix}C_{deep} - V_{mix}C_{surf}}$$

$$= \frac{1}{1 + V_{mix}/V_{riv}(C_{deep}/C_{riv} - C_{surf}/C_{riv})} \quad (16)$$

For phosphorus, for example, $f = 0.01$; that is, about 99% of the particulate phosphorus reaching the deep sea is oxidized to phosphate and is recycled; 1% is buried in the sediments.

fg gives the fraction of an element that is removed per oceanic mixing cycle.

$$fg = \frac{\text{export }(L)\text{ into the sediments}}{\Sigma \text{ fluxes into surface ocean}}$$

$$fg = \frac{1}{1 + V_{mix}C_{deep}/V_{riv}C_{riv}} \quad (17)$$

For phosphorus, $fg \simeq 0.01$. Thus only about 1% of the phosphorus entering the surface of the sea is removed to the sediments.

τ is the residence time of an element; that is, the time required to remove an amount of an element equal to that stored in the sea today, while $T_{mix} = (V_{mix}/\text{depth of the deep sea}) \simeq 1600$ years. Thus

$$\tau = \frac{T_{mix}}{fg} \quad (18)$$

Since phosphate is removed at the rate of 1% during every mixing cycle ($fg = 0.01$), the average lifetime of a P atom in the ocean must be about 1600 years. Broecker (1974) describes this in the following way:

> In words, a typical P atom, upon release from some sedimentary rock by erosion, is carried by rivers to the sea. It then goes through an average of 100 oceanic mixing cycles: 100 times it is fixed by an organism in surface water and becomes part of a particle that sinks and is destroyed in the deep sea. Each time, it waits in the dark abyss about 1600 years before being sent back to the surface. On the average, during the hundredth cycle, the particle bearing the P atom survives destruction and is trapped in the sediment. So our P atom makes 100 round trips of 1600 years each during it stay in the ocean. It then becomes part of the sediment, where it remains for several hundred million years until it is uplifted and exposed again to continental erosion. The life of a typical P atom is indeed bleak. It spends 99.9 percent of its time trapped in the sedimentary rocks of the earth; that is, out of every 200,000,000 years it has only one 160,000 year stint in the ocean. Since the warm layer of surface water is very thin and the time required for particulate loss is small, the P atom spends most of its time in the

Table 15.6. Calculation Summary for the Elements Phosphorus, Silicon, Barium, Calcium, Sulfur, and Sodium

Category	Element	$\dfrac{C_{\text{surface}}}{C_{\text{river}}}$	$\dfrac{C_{\text{deep}}}{C_{\text{river}}}$	g	f	$f \times g$	τ (years)
Biolimiting	P	0.25	5	0.95	0.01	$\dfrac{1}{100}$	2×10^5
	Si	0.05	1.6	0.97	0.03	$\dfrac{1}{300}$	6×10^5
Biointermediate	Ba	0.20	0.60	0.70	0.11	$\dfrac{1}{13}$	3×10^4
	Ca	30.0	30.3	0.01	0.16	$\dfrac{1}{500}$	1×10^6
Biounlimited	S	5000	5000	—	—	$\dfrac{1}{10{,}000}$	2×10^7
	Na	50,000	50,000	—	—	$\dfrac{1}{100{,}000}$	2×10^8

Source: Broecker (1974).

ocean in the cold dark abyss; for every 1600-year mixing cycle, it spends about four years in the surface water!

Table 15.6 gives a summary of calculations for some examples of biolimiting, biointermediate, and biounlimited elements. For biolimiting elements g is close to unity and $f < 0.1$; for biounlimiting elements g must be very small ($g < 0.01$).

The mass balance considerations and calculations illustrate how concentrations of these elements are controlled. Rearranging equation 16, we have

$$\frac{C_{\text{deep}}}{C_{\text{riv}}} = \frac{1 - fg}{fg} \frac{V_{\text{riv}}}{V_{\text{mix}}} \tag{19}$$

or, for a biolimiting element,

$$\frac{C_{\text{deep}}}{C_{\text{riv}}} \simeq \frac{1}{f} \frac{V_{\text{riv}}}{V_{\text{mix}}} \simeq \frac{1}{20f} \tag{20}$$

and, for a biounlimiting element,

$$\frac{C_{\text{deep}}}{C_{\text{riv}}} \simeq \frac{1}{fg} \frac{V_{\text{riv}}}{V_{\text{mix}}} \tag{21}$$

and

$$\frac{C_{\text{surf}}}{C_{\text{deep}}} \simeq 1 \tag{22}$$

As shown by equation 20, the phosphorus concentration dissolved in the sea is controlled by (1) the upwelling rate of deep seawater, (2) the fraction of particles falling to the deep sea that survive oxidation, (3) the phosphorus content of average river water, and (4) the rate of continental runoff.

CaCO₃ Precipitation and Dissolution Ca may serve as an example for the application of the Broecker model of a biointermediate element. The warm surface seawater is several times oversaturated with respect to calcite and aragonite. Spontaneous nucleation does not occur under these conditions in the sea (e.g., inhibition by Mg^{2+}).

Unlike that in lakes, removal is entirely by organisms. Particulate $CaCO_3$ becomes dissolved in the deep ocean because of the CO_2 released by the oxidation of biogenic debris. The ratio of $[Ca^{2+}]$ in surface seawater to that in the deep sea is ~0.99. Table 15.6 shows the values for f and g. The average Ca atom resides in the sea for 500 mixing cycles or 8×10^5 years. Oceanic sediments found at depths below 4500 m are nearly free of $CaCO_3$, while those

914 Regulation of the Chemical Composition of Natural Waters

at depths less than 3000 m are usually more than 70% by weight $CaCO_3$. Although there is some kinetic inhibition of $CaCO_3$ dissolution in waters undersaturated with respect to $CaCO_3$, this difference can be explained plausibly. The $CaCO_3$ fragments falling into the deepest part redissolve, while those falling into shallower parts are preserved.

15.11 GLOBAL CYCLING: THE INTERDEPENDENCE OF BIOGEOCHEMICAL CYCLES

Figure 15.17 gives an impression of the complicated net of the geochemical cycles of C, N, and S, which in part regulate our environment. Although the steady-state model of this figure goes back to 1974 (Garrels and Perry, 1974), the essential features on the interdependence of the various cycles are still valid. The figure illustrates the most important geochemical reservoirs that have been important in interconnecting these reservoirs over the last few million years.

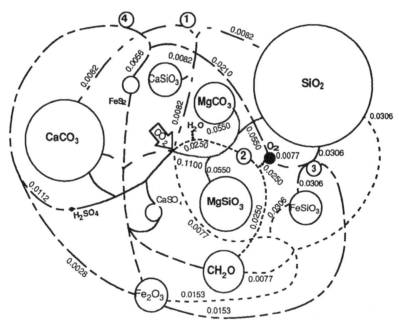

Figure 15.17. The interdependence of the geochemical cycles of C, N, and O. The interlocking of the global chemical cycles illustrates the stationary state that prevailed in regulating our environment for the last 600 million years. Reservoir areas are proportional to the size of the inventories (the number of moles contained); for example, $SiO_2 = 220 \times 10^{20}$ mol, $CaCO_3 = 50 \times 10^{20}$ mol, and $O_2 = 0.38 \times 10^{20}$ mol. The circled numbers refer to the processes given in equation 23-26. The numbers of the interconnecting branches refer to steady-state material fluxes in 10^{14} mol^{-1}. (Adapted from Garrels and Perry, 1974.)

15.11 Global Cycling: The Interdependence of Biogeochemical Cycles

Reservoir areas in Figure 15.17 are proportional to the number of moles contained. Processes that are occurring in the deep earth crust are neglected. The most important interconnections between the reservoirs are the interactions of the sediments with carbon dioxide (CO_2), water (H_2O), and oxygen (O_2). They can be described by the following simplified stoichiometric reactions:

$$\text{①} \quad CaSiO_3 + CO_2 \rightleftharpoons SiO_2 + CaCO_3 \qquad (23)$$

$$\text{②} \quad CO_2 + H_2O \rightleftharpoons \{CH_2O\} + O_2 \qquad (24)$$

$$\text{③} \quad O_2 + 4\, FeSiO_3 \rightleftharpoons 2\, Fe_2O_3 + 4\, SiO_2 \qquad (25)$$

$$\text{④} \quad \tfrac{15}{8} O_2 + \tfrac{1}{2} FeS_2 + H_2O \rightleftharpoons H_2SO_4 + \tfrac{1}{4} Fe_2O_3 \qquad (26)$$

- Reaction ① (the numbers refer to the material fluxes in Figure 15.17) corresponds to the weathering of silicates.
- Reaction ② is the most simple formulation of the photosynthesis–respiration reaction.
- Reactions ③ and ④ illustrate the oxidation of Fe(II) and sulfur compounds by photosynthetically formed O_2 to Fe(III) oxide and sulfate.

The inventories are most important for our ecosystems. O_2 and CO_2 are very small in comparison to reservoirs of the minerals in Figure 15.17. The reservoirs of O_2 and CO_2 are interconnected with numerous very large reservoirs. In the clockwork, so to speak, the rapidly turning wheels of CO_2 and O_2 are interlocked with extremely slowly turning wheels of sediment components. The interdependence of these geochemical cycles and their synchronization determine the composition of the oceans and are to a large extent responsible for the maintenance of a constant composition of the atmosphere.

The coupling of these cycles and their regulation by various feedback mechanisms impart a remarkable resilience to this interlocked system. Humans cannot change the speed of the "large wheels" but, as we have seen, the small reservoirs can be changed by civilization's activities.

The Sensitivity of Global Reservoirs to Anthropogenic Perturbation

Humans have become geochemical manipulators and agents of global change and are a major force in the transport of solid earth materials. Chemical by-products are changing the hydrosphere and atmosphere. The change on our planet involves a complex interaction between the inorganic physical processes and the biological processes. Although we depend on human intervention in the water cycle to provide adequate supplies for agriculture and people, we are often not sufficiently aware of some of the negative effects that mammoth water diversion may have on ecosystems. The consequences may not be apparent for some time, and causes and effects may be difficult to identify because of the large distances involved.

The weathering cycle is affected markedly at least locally and regionally by civilization. In local environments the proton, H^+, and the electron, e^-, balances may become upset, and significant variations in pH and redox intensity (pε) result.

Before discussing some examples of the effects of human activity on global chemical cycles, we will address some introductory considerations on the sensitivity of global reservoirs.

Example 15.3. Comparison of Global Reservoirs Establish a list of some of the more important global inventories and estimate their magnitudes and mean residence times of the molecules in these reservoirs. The simple idea behind this excercise is that, to a first approximation, the smaller the relative reservoir size and the smaller the residence time, the more sensitive is the reservoir toward perturbation.

A possible hierarchy of reservoir sizes (measured in number of molecules or atoms) is given in Figure 15.18. The mean residence time (years) is also given. Figure 15.18 gives a selection for certain elements; the list could be amplified or modified.

Obviously, the atmosphere, living biomass (mostly forests), and surface fresh waters are most sensitive to perturbation. The anthropogenic exploitation of the larger sedimentary organic carbon reservoir (fossil fuels and by-products of their combustion such as oxides and heavy metals and the synthetic chemical derived from organic carbon) can, above all, affect the small reservoirs. In recent years, we have started to recognize that biosphere processes play an important role in coupling the cycles of essential elements and in regulating the chemistry and physics of our environment. The living biomass is a relatively small reservoir and thus subject to anthropogenic interference; each species of the biosphere requires specific environmental conditions for sustenance and survival.

15.12 THE CARBON CYCLE

The carbon cycle constitutes, in terms of mass transfer between the continents, the oceans, and the atmosphere, the most active set of processes that control the behavior of many elements on the earth's surface. Because of photosynthesis, there is a continuous pump resulting in the production and respiration of organic matter. The human impact on this cycle is important; increased concentrations of CO_2 and CH_4 in the atmosphere are associated with global warming through the greenhouse effect.

We will consider here certain aspects of the combination of organic and inorganic processes where biological activity and chemical phenomena are closely interrelated. Because many of these processes are fast, most of the chemical reactions can be described by chemical equilibria. Very simple models, both on local and a global scale, have been used to describe some important

15.12 The Carbon Cycle

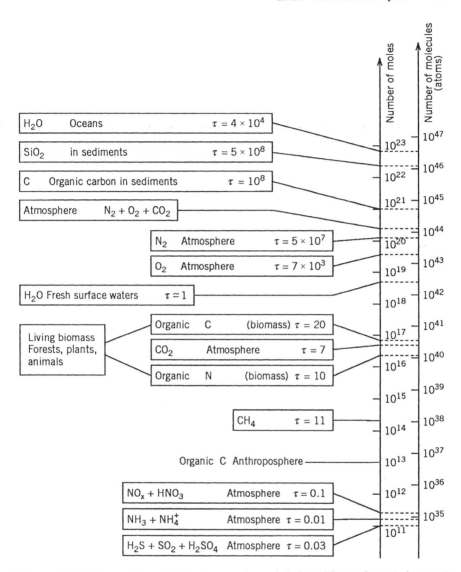

Figure 15.18. Comparison of global reservoirs and their residence times (τ in years) (Example 15.3). The reservoirs of the atmosphere, of surface fresh waters, and of living biomass are significantly smaller than the reservoirs of sediment and marine waters and are thus more susceptible to disturbance. For example, the combustion of fossil fuel (from the reservoir of organic carbon in sediments) will have an impact on the smaller reservoirs: CO_2 in the atmosphere will be markedly enlarged. This combustion also fixes some N_2 to NO and NO_2; sulfur, associated with the organic carbon, introduces CO_2 into the atmosphere. These nitrogen and sulfur compounds are washed out relatively rapidly into soil and aquatic ecosystems. The total groundwater reservoir may be twice that of surface fresh water but, however, is less accessible. (From Stumm, 1986.)

aspects of the carbon cycle. As pointed out by Wollast and Vanderborght (1994), long time predictions can also be well described in equilibrium models.

In this sense we have already described in Chapters 4 and 7 the equilibrium of the $CaCO_3(s)–H_2O–CO_2$ system. Specifically, we have used equilibrium models to characterize the concentrations of the carbonate species as a function of p_{CO_2} and of pH. We have already shown (Example 7.8) that $CaCO_3$ in surface seawater is oversaturated and we calculated in Example 4.10 how the composition of seawater changes as a result of increasing the CO_2 concentration in the atmosphere.

Example 15.4. Are Human Activities Changing the O_2 and CO_2 Content of the Atmosphere? So far we have burned $\sim 2 \times 10^{16}$ mol of fossil carbon. To what extent could this cumulative combustion affect the atmospheric contents of O_2 and CO_2?

The weight of gas above each square centimeter is about 10^3 g cm^{-2} (~ 1 atm); with an average molecular weight of air of 29 g, this corresponds to 34.5 mol. Of this, 20% is O_2 or 7.5 mol cm^{-2}. Furthermore, 1.2×10^{-2} mol of CO_2 (0.035% of air) are above each square centimeter. For the total earth surface (5.1×10^{18} cm^2), the atmospheric reservoirs of O_2 and CO_2 are 3.8×10^{19} mol O_2 and 62×10^{16} mol CO_2. Combustion of all the fossil carbon exploited thus far would lead to a production of 10^{16} mol CO_2 ($\sim 20\%$ of the CO_2 contained presently in the atmosphere) and a depletion of O_2 corresponding to 2×10^{16} mol of O_2 (0.05%) of the O_2 contained in the atmosphere. Obviously, the oxygen reservoir cannot seriously be depleted by human actions; most of the carbon released from fossil fuels will eventually end up in the ocean, where a complex cycle of circulation and other processes control its fate (Sarmiento, 1993).

Quite obviously, however, the CO_2 content of the atmosphere has increased markedly. About 50% of the fossil CO_2 added to the atmosphere has remained in the atmosphere.

This problem is taken from a lucid paper entitled, "Man's Oxygen Reserves," by Broecker (1970). Although other factors, such as the possibility of CO_2 production caused by deforestation (or increased CO_2 consumption by increased productivity due to higher p_{CO_2}), have been ignored, Broecker also showed that in the extreme case of all fossil fuel reserves being burned in one moment, the percentage change of oxygen in the atmospheric reservoir would still be trivial.

Figure 15.19 gives a description of the global carbon cycle. The inventories of the various reservoirs were already given in Table 4.1, where we noticed that the atmosphere is a relatively small reservoir with large fluxes, so that the residence time of C in the atmosphere is only a few years. The carbon system is not at steady state. Because of fossil fuel combustion and possibly also because of deforestation, the inventories of C in the atmosphere and hydrosphere are increasing. The flux related to fossil fuel combustion is nearly 1% of the total atmospheric CO_2 reservoir. The flux due to land use is more controversal but is probably $1-2 \times 10^{14}$ mol C^{-1}. As the summary at the bottom

15.12 The Carbon Cycle

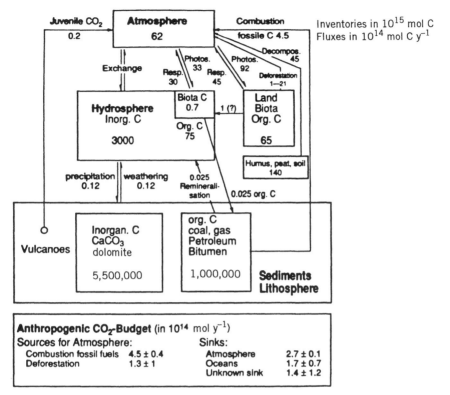

Figure 15.19. Simplified representation of the global carbon cycle. Data on photosynthesis, respiration, and decomposition from Bolin and Cook (1983). The data for the anthropogenic CO_2 budget are from Houghton et al. (1992).

of Figure 15.19 shows, there is a missing sink of $1.4 \pm 1.2 \times 10^{14}$ mol C yr^{-1}, which is not accounted for by the accumulation in the atmosphere or by transfer to the ocean. It has been suggested that this sink could be the production of extra biomass on the continents or that the ocean sink has been underestimated. It is important, however, to note that today the flux of CO_2 has been reversed between the atmosphere and the ocean.

Organic Carbon and Productivity Of the total carbon fixed annually ($\sim 120 \times 10^{14}$ mol C yr^{-1}) nearly all is respired and oxidized. Only a very small fraction of the organic matter produced in the ocean accumulates in the sediments. On a geological time scale the organic carbon preserved in the sediments is eventually remineralized to CO_2.

Rate of Removal of CO_2 from Atmosphere Depends on Ocean Circulation Ocean circulation has a major impact on the distribution of carbon between the ocean and the atmosphere. Figure 15.20 is an extension of the two-box model used in Figure 15.16. The CO_2 uptake is very rapid at first but

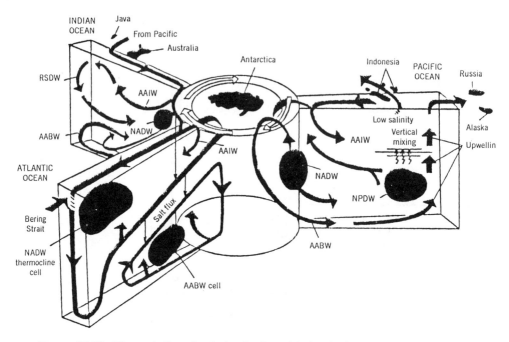

Figure 15.20. Thermohaline circulation begins with the sinking of cold, salty surface waters in the North Atlantic. The North Atlantic Deep Water (NADW) thus formed flows toward the south, where it wells up in the Antarctic, cools, mixes with other water types, and sinks again as Antarctic Bottom Water (AABW). AABW flows north along the ocean floor in all three ocean basins: the Atlantic, Pacific, and Indian. As it flows, its density is gradually reduced by mixing with waters from above. It thus moves upward, to become part of the intermediate-depth, southward-flowing "deep" bodies of water, called NADW in the Atlantic, North Pacific Deep Water (NPDW) in the Pacific, and Red Sea Deep Water (RSDW) in the Indian Ocean. The combination of the AABW flowing to the north and "deep" water flowing to the south above it forms a deep "conveyor belt" circulation. There is also shallow circulation involving the formation of intermediate-depth water at lower latitudes of the Antarctic (AAIW). This water returns to the Antarctic by a variety of pathways, many not well understood. Also necessary to close the thermohaline circulation is a shallow northward flow to supply the NADW. This probably involves transport around South Africa as well as South America. (Adapted from Sarmiento, 1993.)

gradually decreases as the added carbon dioxide enters deeper layers of the ocean, which exchange only sluggishly with the surface layer. The time history of the atmospheric carbon dioxide can be approximated as a sum of components (exponentials) with different mean lives (Sarmiento, 1993).

An Equilibrium Approach: The Uptake of CO_2 by the Ocean and the Buffer Factor

Simplifying from Figure 15.20, the following three steps must be considered. First, the relatively thin layer of well-mixed water lying above the seasonal thermocline at the ocean surface will quickly establish equilibrium with the CO_2 in the atmosphere. Second, downward mixing and the sinking of the particles generated by organisms will slowly carry this carbon to the deep sea. Finally, as the CO_2 content of the ocean rises, water masses currently supersaturated with respect to $CaCO_3$ (calcite and aragonite and their solid solutions) will become undersaturated, and the $CaCO_3$ in sediments bathed in these waters will begin to dissolve.

We will primarily consider here the first step: equilibration of surface water with excess atmospheric CO_2. The increase in C_T resulting from an increase in p_{CO_2} can be formulated by the relationship

$$\frac{\Delta p_{CO_2}}{^0p_{CO_2}} = \eta \frac{\Delta C_T}{^0C_T} \tag{27}$$

where η is the so-called buffer factor, and Δp_{CO_2} and ΔC_T are, respectively, $p_{CO_2} - {^0p_{CO_2}}$ and $C_T - {^0C_T}$; $^0p_{CO_2}$ and ^0C_T correspond to any previous CO_2 partial pressure and total inorganic carbon in seawater, respectively. For example, they might represent the preindustrial partial pressure of CO_2 and the preindustrial total inorganic carbon in seawater. If the atmospheric p_{CO_2} increases by X percent, the C_T increases by X/η percent.

We are already acquainted with the mass law equations that connect atmospheric CO_2 and surface waters. We have seen in Example 7.8 that ocean surface water is markedly supersaturated with respect to $CaCO_3$; thus the addition of CO_2 does not cause any $CaCO_3$ dissolution. Hence the buffer factor in equation 27 can be evaluated under conditions of constant alkalinity. This problem has also been examined in Example 4.10.

For a given alkalinity and an appropriate set of equilibrium constants, the relationship between p_{CO_2} and C_T can be established. With the help of the equation

$$[Alk] = \frac{K_H p_{CO_2}}{\alpha_0}(\alpha_1 + 2\alpha_2) + [OH^-] - [H^+] + B_T \alpha_{B^-}$$

the H^+ activity compatible with a given set of p_{CO_2} and [Alk] can be computed (see Example 4.10); then $C_T = K_H p_{CO_2}/\alpha_0$ is found.

The p_{CO_2}-C_T curve in Figure 15.21 can also be readily obtained by computer on the basis of the equations in Tableau 4.4.

With the results displayed in Figure 15.21, a buffer factor, $\eta = 9.7$ (15°C), is obtained for a change in p_{CO_2} from 2.7×10^{-4} to 3.0×10^{-4} atm; in other words, for an increase of 10% in p_{CO_2}, C_T increases by $\sim 1\%$. The buffer factor increases with increasing p_{CO_2}; for example, if p_{CO_2} is increased from 3.3×10^{-4} atm (approximate present global value) to 3.6×10^{-4} atm, $\eta = 15.5$; if p_{CO_2} is increased from 3.3×10^{-4} to 6.6×10^{-4} atm, $\eta = 17.4$. Thus a doubling of the present atmospheric CO_2 would increase C_T by about 5-6%. Similar effects will be observed in surface fresh waters.

Under equilibrium conditions, doubling of the CO_2 content in the atmosphere

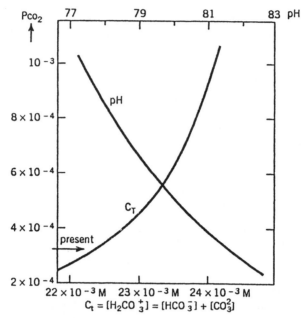

Figure 15.21. Effect of increasing p_{CO_2} upon C_T and pH of surface ocean water. The calculations have been made for the following conditions: seawater at 15°C, $p_{total} = 1$ atm, [Alk] = constant = 2.47×10^{-3} eq liter^{-1}, $[B(OH)_4^-] + [H_3BO_3] = 4.1 \times 10^{-4}$ M, $^cK_H = 4.8 \times 10^{-2}$ M atm^{-1}, $K_1' = 8.8 \times 10^{-7}$, $K_2' = 5.6 \times 10^{-10}$, and $K_{H_3BO_3}' = 1.6 \times 10^{-9}$.

will approximately double the H$^+$ concentration in these waters. The effect will be smaller in the deeper waters of the ocean because of mixing.

Production of Aqueous Organic Carbon

Gross primary production (productivity) is the total rate of photosynthesis. Net primary productivity is the rate of synthesis of organic matter in excess of its respiratory utilization during the period of measurement. The concentration of organic matter in a natural body of water results from an interplay of net productivity, exudation of organic substances by phytoplankton, and import and export (inflow, outflow, dissolution, sedimentation, etc.) of organic matter. Productive lakes have a gross primary production of organic matter on the order of a few g m^{-2} day^{-1}. Oceanic waters show gross productions of up to 1 g m^{-2} day^{-1}. Production in flowing waters is usually very high. Streams are among the most productive biological environments. The highest primary production rates occur in the recovery zones of streams polluted with sewage; this is a consequence of the supply of fertilizing elements and of additional organic nutritive sources possibly contained in sewage. High production is not limited, however, to polluted water. Salt marshes and estuaries (~0.35% of the world surface area) are estimated to produce ~2% of the net world primary production.

Example 15.5. Steady-State Distribution of Organic Matter in the Ocean Estimate from the data given in Figure 15.19 the residence time of organic C in the ocean.

The organic C reservoir is ~75 × 10^{15} mol (this corresponds to a mean concentration of 0.6 mg C liter^{-1}). The imput by photosynthesis is ~33 × 10^{14} mol C yr^{-1}. Other imputs by rain and rivers (2-3 × 10^{14} g C yr^{-1}) are negligible. Hence the residence time (= amount in reservoir per input) is ~20 years. This residence time is meaningful only with regard to the organic C in the surface waters of the oceans. As we have seen, little biodegradation of deep sea organic C occurs. We can estimate the residence time of the deep sea C by considering the output by sedimentation, which according to Figure 15.19 is ~0.025 × 10^{14} mol yr^{-1}. Since the deep ocean contains only about two-thirds of the organic C of the ocean, a residence time of 50 × 10^{15} mol ÷ 0.025 × 10^{14} mol yr^{-1} = 20,000 years for organic C is obtained. Williams et al. (1969) have determined by ^{14}C measurements the apparent "age" of organic matter at 2000 m in the northeast Pacific to be 3400 years. From this age a fraction of the primary production actually entering the deep ocean can be estimated: input into deep ocean = 50 × 10^{15} mol C ÷ 3400 ≈ 1.5 × 10^{13} mol C yr^{-1}, that is, according to this estimate only about 0.5% of primary production enters the deep ocean.

These estimations underline the fact that organic matter in the deep waters of the ocean is chronologically old and chemically or biochemically rather inert.

Concentration of Aqueous Organic Carbon

In view of the considerable variety of sources of organic material from outside the water basin, a great diversity in the concentration of organic matter must be expected, even in the absence of human contamination. In freshwater bodies, one typically encounters concentrations of a few mg C liter^{-1}; waters low in Ca^{2+} and Mg^{2+} often contain humic substances; occasionally (e.g., in bog or swamp waters), concentrations may be as high as 50 mg C liter^{-1}. In the oceans the concentrations of organic carbon range from 0.5 to 1.2 mg C liter^{-1}, with the higher values occurring in the surface waters. Particulate organic carbon, including planktonic organisms, generally accounts for about 10% of the total organic carbon in the surface waters and for about 2% in the deep waters. Over much of the ocean we have, according to Wangersky, a "dissolved" organic carbon concentration of about 500 μg C liter^{-1} and a total particulate organic carbon concentration of about 50 μg C liter^{-1}, all ultimately derived from a surface phytoplankton population equivalent to a concentration of about 5 μg C liter^{-1}. These figures are significantly higher for waters on the continental shelf, especially for regions where nutrients are returned to the surface by upwelling. Concentrations in surface layers are generally greater than in deeper waters. Below a depth of 400–600 m, the concentration variations in organic carbon become very slight. This suggests that materials present in deep waters decompose extremely slowly. Presumably much of the readily degradable organic matter has already been decomposed in the upper layers. Furthermore, life processes appear to be slower in the deep sea than in shallow water.

Organic Aggregates and Colloidal Organic Matter The nonliving particulate organic matter in waters has usually been assumed to be fecal pellets and minute remains of plankton in various stages of disintegration. However, in recent years evidence has accumulated that many of these particles are delicate plate-like aggregates a few micrometers to several millimeters in diameter. Also, as we have discussed in Chapter 14, organic colloids are present in large numbers; the distinction between dissolved and particulate carbon is operationally difficult.

Biochemicals on Their Way to Graphite

Information on the catabolic pathways of organic substances permits prediction of the type of organic substances to be found in natural waters. Table 15.7 gives a greatly condensed, simplified survey of the decomposition products of life substances and includes an abbreviated list of specific organic compounds reportedly found and identified in natural waters.

Much of the *thermodynamic stabilization* occurs in diagenesis. (Diagenesis refers to changes that take place within the sediment during and after burial.) The least stable and most reactive components *or their substituents* are gradually eliminated. This process leads with increasing age and depth of burial to a gradual stabilization, not necessarily of each individual compound but of the sedimentary organic matter as a whole. In terms of structures the transformation of open chains to saturated rings and finally to aromatic networks is favored; hydrogen becomes available for inter- or intramolecular reduction processes. Eventually, highly ordered, stable structures of graphite may be formed. The approach to thermodynamic equilibrium is very slow; unstable compounds may survive from time spans equivalent to the age of the earth.

Blumer (1975) has pointed out that the most characteristic feature of organic diagenesis is the appearance of extreme structural complexity and disorder at an intermediate stage, interposed between the high degree of biochemical order in the starting material and the even greater crystallographic order in graphite, the end product of diagenesis. Organic geochemistry is critically important to the understanding of many environmental problems. For instance, the severe effect of spilled fossil fuels is directly related to their composition and to the diagenetic processes that have produced them in the earth's interior. Our understanding of the chemistry of fossil fuels and the changes they undergo during diagenesis is directly applicable also to an understanding of their biological and chemical degradation in the environment.

Table 15.7. Naturally Occurring Organic Substances

Life Substances	Decomposition Intermediates	Intermediates and Products Typically Found in Nonpolluted Natural Waters
Proteins	Polypeptides → RCH(NH$_2$)COOH → RCOOH, RCH$_2$OHCOOH, RCH$_2$OH, RCH$_3$, RCH$_2$NH$_2$ (amino acids)	NH$_4^+$, CO$_2$, HS$^-$, CH$_4$, HPO$_4^{2-}$, peptides, amino acids, urea, phenols, indole, fatty acids, mercaptans
Polynucleotides	Nucleotides → purine and pyrimidine bases	
Lipids Fats Waxes Oils Hydrocarbons	RCH$_2$CH$_2$COOH + CH$_2$OHCHOHCH$_2$OH → RCH$_2$OH, RCOOH, shorter chain acids, RCH$_3$, RH (fatty acids) (glycerol)	CO$_2$, CH$_4$, aliphatic acids, acetic, lactic, citric, glycolic, malic, palmitic, stearic, oleic acids, carbohydrates, hydrocarbons
Carbohydrates Cellulose Starch Hemicellulose Lignin	C$_x$(H$_2$O)$_y$ → monosaccharides / oligosaccharides / chitin → hexoses / pentoses / glucosamine → polyhydroxy carboxylic acids (C$_2$H$_2$O)$_4$ → unsaturated aromatic alcohols	HPO$_4^{2-}$, CO$_2$, CH$_4$, glucose, fructose, galactose, arabinose, ribose, xylose
Porphyrins and Plant Pigments Chlorophyll Hemin Carotenes and Xantophylls	Chlorin → pheophytin → hydrocarbons	Phytane Pristane, carotenoids Isoprenoid, alcohols, ketones, acids Porphyrins
Complex Substances Formed from Breakdown Intermediates, e.g., Phenols + quinones + amino compounds Amino compounds + breakdown products of carbohydrates		→ Melanins, melanoidin, gelbstoffe → Humic acids, fulvic acids, "tannic" substances

Sediments The quantity of organic matter carried in natural waters is small compared to the total amount of organic matter found in sediments. The latter are the principal depositories of posthumous organic debris accumulated throughout the earth's history. Most of the organic matter occurs in a finely disseminated state and is associated with fine-grained sediments. In order to illustrate more vividly the ratio of inorganic to organic matter, Degens (1965) states: "From an evaluation of the total thickness of sedimentary materials that have been formed and deposited over the last 3 to 4 billion years, it can be assumed that a sediment layer of approximately 1000 m around the earth's surface has been laid down. Roughly 2 percent of this layer, namely ~ 20 m, is organic, the rest is inorganic in nature. Of this 20-m organic part, coal comprises ~ 5 cm and crude oil only a little more than 1 mm. The rest represents the finely disseminated organic matter in shales, limestones and sandstones." Surface waters and sediments are the principal sites of biodegradation and bioconversion. It can be inferred from the published literature that most of the chemical alteration of organic matter takes place in the environment of deposition and during early stages of diagenesis.

Analytical Collective Parameters Organic matter in natural waters includes a great variety of organic compounds, usually present in minute concentrations, many of which elude direct isolation and identification. *Collective parameters*, such as chemical oxygen demand (COD), biological oxygen demand (BOD), and total organic carbon (TOC) or dissolved organic carbon (DOC), are therefore often used to estimate the quantity of organic matter present. COD is obtained by measuring the equivalent quantity of an oxidizing agent, usually permanganate or dichromate in acid solution, necessary for oxidation of the organic constituents; the amount of oxidant consumed is customarily expressed in equivalents of oxygen. The BOD test measures the oxygen uptake in the microbiologically mediated oxidation of organic matter directly. In both tests not all the organic matter reacts with the oxidant. In determinations of total organic carbon, we measure the CO_2 produced in the oxidation or combustion of a water sample (from which carbonate has been removed). TOC, DOC, and COD are capacity terms; the last measures the reduction capacity of organic matter. If COD is expressed in mol O_2 liter^{-1} and TOC in mol C liter^{-1}, the "average" oxidation state of the organic carbon present can be obtained:

$$\frac{4(\text{TOC} - \text{COD})}{\text{TOC}} = \text{oxidation state}$$

The COD of a compound is very nearly proportional to its heat of formation, and a crude value of its energy of formation can often be obtained. Thus a rough estimate of the potential energy available for the metabolic needs of organisms can be made from COD data. Although measurements of TOC, DOC, BOD, and COD are conveniently obtained, these collective parameters lack physiological meaning. The rates of microbial growth and the overall use of organic matter in multisubstrate media depend in a complex way on the activities of a great variety of different enzymes and on various mechanisms by which these activities are ecologically interrelated and physiologically coordinated.

15.13 NITROGEN CYCLES: POLLUTION BY NITROGEN COMPOUNDS

Elemental nitrogen (N_2) (78% in the atmosphere) is very unreactive because of the strong bond (three electron pairs) between the nitrogen atoms. Some bacteria (associated with certain plants, leguminosae) and certain blue-green algae are able to fix nitrogen and assimilate it as organic nitrogen (Figure 15.22). Humans have also learned to fix N_2. In ^{t}the Haber–Bosch process N_2 is converted (at elevated pressure and temperature and in the presence of suitable catalysts) into NH_3:

$$2\ N_2(g) + 3\ H_2(g) = 2\ NH_3(g)$$

This NH_3 is used mainly to manufacture nigrogen fertilizers. Furthermore, N_2 is converted into nitrogen oxides in the combustion of gasoline and fossil fuel (automobile engines and thermal power plants):

$$N_2(g) + O_2(g) = 2\ NO(g)$$

$$N_2(g) + 2\ O_2(g) = 2\ NO_2(g)$$

The extent of nitrogen fixation by civilization is of the same order of magnitude as that by biological N fixation. Nitrogen fertilization in agriculture and emission of nitrogen oxides into the atmosphere have changed the distribution of nitrogen compounds between water, soil, and atmosphere. Many of the nitrogen compounds are harmful in the atmosphere and can have adverse ecological effects in terrestrial and aquatic ecosystems (Table 15.8).

The Most Important Redox Processes in the Cycle $N_2(g) \rightarrow NH_4^+ \rightleftharpoons NO_3^- \rightarrow N_2(g)$ Thermodynamically, we have already given all the important information of these transformations (Section 8.5). We can summarize it in Figure 15.23. See also Table 15.9.

$NH_4^+ \rightleftharpoons NO_2^- \rightleftharpoons NO_3^-$ The sources of NH_4^+ in the soils are synthetic fertilizers, animal wastes, biological N_2 fixation, and rain. NH_4^+ is retained in the soils by ion exchange or clays:

$$RK^+ + NH_4^+ \rightleftharpoons RNH_4^+ + K^+$$

Excess of NH_4^+ can be washed out into receiving waters. NH_4^+ is in equilibrium ($pK_a = 9.3$, 25°C) with NH_3, which is toxic to fish.

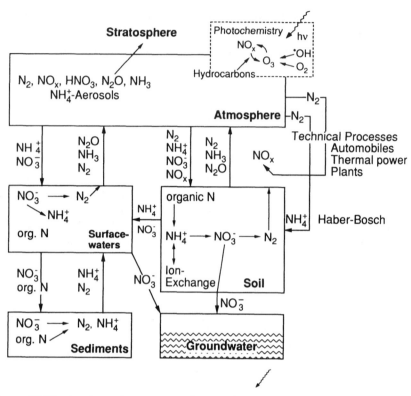

Figure 15.22. A schematic survey on the transformation and distribution of N compounds in various reservoirs of the environment. $N_2 \to$ Org N $\to NH_4^+$. As shown by reactions (4) and (5) in Table 15.9, H_2 or organic material (for our calculations we use "$\{CH_2O\}$") can—from a thermodynamic point of view—reduce nitrogen to N(-III) (organic N and NH_4^+). The free energy released in these processes are very small. As is evident from Table 8.6 and Figure 8.14, the reduction of N_2 to NH_4^+ (pH = 7) can occur only under very reducing conditions ($p\epsilon < -4$) whereas the reduction of $CO_2(g)$ to $\{CH_2O\}$ is possible under slightly less reducing ($p\epsilon < 1$) conditions. The biological fixation of $N_2(g)$ needs additional energy which is obtained from photosynthesis (bluegreen algae and bacteria in symbiosis with plants). The nitrogen fixed is initially present as organic nitrogen which upon decomposition yields NH_4^+ (and NH_3). The industrial N_2 fixation produces also NH_3.

NH_3 can volatilize [$NH_3(g) = NH_3(aq)$; $K_H = 57$ M atm^{-1}, 25°C]. The contribution of NH_3 to the atmosphere is especially large from the manure of animal feed lots. NH_3 is also used technologically to reduce pollution of NO_x in thermal energy plants [$4\ NH_3 + 6\ NO = 5\ N_2 + 6\ H_2O$]†.

---

†If an excess of NH_3 is used in this process, this ammonia is returned to the soil surface by precipitation, where it contributes to soil acidification ($NH_3 + 2\ O_2 = NO_3^- + H^+ + H_2O$).

Table 15.8. Nitrogen Compounds: Adverse Effects on the Environment

N Compound	Oxidation State	Main Origin	Systems Polluted	Effects
NO_3^-	V	Fertilizer	Groundwater, oceans	Health, drinking water eutrophication
$HNO_3(g)$	V	Combustion of fossil fuels	Atmosphere, soils	Acid rain
NO_2^-	III	Intermediate in nitrification,[a] denitrification, and NO_3^- reduction	Waters	Toxic for fish
$NO(g), NO_2(g)$	II, IV	Combustion of fossil fuels, automobiles, denitrification in soils	Atmosphere	Assists in production of ozone in troposphere,[b] toxic effects on plants
$N_2O(g)$	I	Intermediate in nitrification and NO_3^- reduction	Atmosphere	Destruction of O_3 in stratosphere[b]
$NH_3(g), NH_4^+$	−III	Fertilizer, animal feed lots	Atmosphere, soil	Nitrification of NH_4^+ (from precipitates) leads to acidification of soils[c]
			Waters	Toxicity of NH_3 to fish, increased chlorine demand in chlorination of drinking water

[a] See reactions 9 and 10 in Table 15.9.
[b] Troposphere encompasses the first ~10 km of the atmosphere; the stratosphere is above this layer.
[c] See reactions 1 and 2 in Table 15.9.

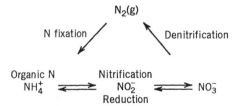

Figure 15.23. The most important biologically mediated nitrogen conversion processes (compare Table 15.9). Natural N fixation occurs through certain blue-green algae and through certain bacteria, which live in symbiosis with plants (leguminosae). Microorganisms mediate, under aerobic conditions, the nitrification of NH_4^+ (and NH_3) to NO_2^- and NO_3^-. At low pϵ (pϵ < 6 at pH = 7) the reduction to NH_4^+ is possible. Organic compounds can cause the reduction of NO_3^- to N_2 (denitrification) (see reaction 3 in Table 15.9). NO and N_2O may occur as intermediates, both in nitrification and denitrification (reactions 9 and 10 in Table 15.9).

In the presence of O_2 (pϵ > 6 at pH = 7) NH_4^+ is oxidized to NO_3^- (nitrification) (Figure 8.4); the oxidation occurs in two steps (reactions 1 and 2 in Table 15.9). Nitrification, the oxidation of NH_4^+ to NO_2^-, is mediated by the bacterium *Nitrosomanas*, and the oxidation of NO_2^- to NO_3^- by the bacterium *Nitrobacter* (Figure 15.24). NO_3^- is not retained in the soils (no ion exchange) and is readily eluted into the groundwater. Concentrations above 7×10^{-4} M in drinking water (10 mg liter^{-1} as N) are undesirable from a health point of view.

The reverse of nitrification, nitrate reduction (ammonification of NO_3^-), can also occur. At pϵ < 6 (Figure 8.4) organic material can reduce NO_3^- and NO_2^- by bacterial mediation to NH_4^+. This is in competition with denitrification.

N_2O as a By-Product Both in nitrification and in denitrification, N_2O can occur as a by-product. N_2O that is released into the stratosphere participates in the destruction of ozone.

Denitrification Organic material can usually, under anaerobic conditions, reduce microbially mediated NO_3^- to N_2 (reaction 3 in Table 15.9). Denitrification occurs probably through an indirect mechanism via NO_2^-. N_2O and other nitrogen oxides are formed as intermediates. Biologically mediated denitrification is not reversible; $N_2(g)$ cannot be oxidized to NO_3^- by biological mediation. The biological pathway of N_2 to NO_3^- goes over N_2 fixation and subsequent nitrification (Figure 15.24). Nitrogen oxides NO and NO_2 (reaction 6–8 in Table 15.9) are not only important for the formation of acid rain but also for the formation of ozone, which is catalyzed by sunlight and hydrocarbons (see Chapter 12 and reactions 6 and 8 and footnote *b* in Table 15.9).

Table 15.9. Free Energy of N Transformations at pH = 7 (25°C)

Processes		$\Delta G^{0'}$ (pH = 7) (kJ mol^{-1})
Nitrification		
(1) $NH_4^+ + 1.5\ O_2$ (0.2 atm)	$= NO_2^- + H_2O + 2\ H^+$ (10^{-7} M)	-290.4
(2) $NO_2^- + 0.5\ O_2$ (0.2 atm)	$= NO_3^-$	-72.1
Denitrification[a]		
(3) $NO_3^- + 1.25\ \{CH_2O\} + H^+$ (10^{-7} M)	$= 0.5\ N_2(g) + 1.75\ H_2O + 1.25\ CO_2(g)$	-594.6
N Fixation		
(4) $0.5\ N_2(g) + 1.5\ H_2(g) + H^+$ (10^{-7} M)	$= NH_4^+$	-39.4
(5) $0.5\ N_2(g) + 0.75\ \{CH_2O\} + 0.75\ H_2O + H^+$ (10^{-7} M)	$= 0.75\ CO_2 + NH_4^+$	-60.3
N_2 Oxidation[b] (combustion, thermal power plants, automobiles)		
(6) $0.5\ N_2(g) + 0.5\ O_2(g)$	$= NO(g)$	86.6
(7) $NO(g) + 0.5\ O_2(g)$	$= NO_2(g)$	-35.2
(8) $0.5\ N_2(g) + 1.25\ O_2$ (0.2 atm) $+ 0.5\ H_2O$	$= NO_3^- + H^+$ (10^{-7} M)	-25.7
Formation of N_2O		
(9) $NO_3^- + \{CH_2O\} + H^+$ (10^{-7} M)	$= 0.5\ N_2O(g) + 1.5\ H_2O + CO_2(g)$	-417.1
(10) $NH_4^+ + O_2$ (0.2 atm)	$= 0.5\ N_2O(g) + 1.5\ H_2O + H^+$ (10^{-7} M)	-260.2

[a] Denitrification is in competition with NO_3^- reduction (reduction of NO_3^- to NH_4^+):

$$NO_3^- + 2\ \{CH_2O\} + 2\ H^+\ (10^{-7}\ M) = NH_4^+ + 2\ CO_2(g) + H_2O, \quad \Delta G^{0'} = -655\ kJ\ mol^{-1}$$

[b] The formation of ozone in the lower atmosphere is governed by NO_2; simplified:

$$NO_2 + \text{sunlight} \rightarrow NO + O, \quad O + O_2 \rightarrow O_3$$

Figure 15.24. Typical progress in the transformation of NH_4^+ to NO_2^- and NO_3^- in a small river (23°C) (Glatt) based on a model calculation by Reichert (EAWAG).

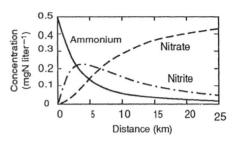

15.14 THE SULFUR CYCLE

Figure 15.25 gives a simplified representation of the S cycle and demonstrates the significant effect of civilization on this cycle. About half of the sulfur in the runoff to the oceans is from human activity. There is a substantial transfer of sulfur through the atmosphere from land to sea.

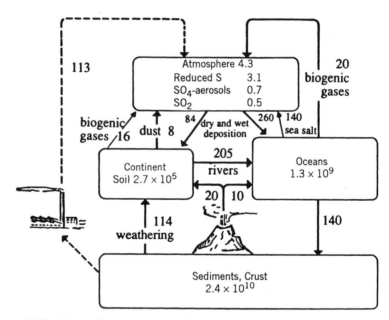

Figure 15.25. The sulfur cycle: global fluxes (million tons of S per year) and inventories (million tons of S). Weathering of S-bearing minerals [gypsum ($CaSO_4$) and pyrite (FeS_2)] and the recirculation of sea salt and biogenic gases [H_2S, dimethylsulfide (CH_3SCH_3), and carbonylsulfide (COS)] represent the natural segments of the cycle. Emissions by civiliation (especially SO_2 from combustion of fossil fuels and from smelting) are a major contribution to the atmosphere. Human activities also affect the transport of S in rivers and are responsible for a net transfer of S in the atmosphere from land to sea. (Most data are from Bolin and Cook, 1983.)

The Marine Sulfur Cycle and Global Climate Sulfate shows a highly conservative behavior in seawater and a mean residence time of ~3 million years relative to total imputs. Nevertheless, the sulfur cycle of the oceans is dynamic. The oceans are a major source of *dimethylsulfide* (DMS), CH_3SCH_3, which is produced from decomposing phytoplankton cells (decomposition of dimethylsulfoniopropionate and of carbonylsulfide, COS) (Andreae and Barnard, 1984). The flux of DMS to the atmosphere would be significantly larger if it were not for microbial degradation of DMS. In the atmosphere DMS is largely oxidized by OH^*, forming sulfate that is deposited in precipitation. The oceans emit only relatively small quantities of H_2S. In contrast the terrestrial and freshwater and wetland environments release reduced sulfur compounds mainly as H_2S with little DMS and carbonylsulfide, COS. COS is the most abundant sulfur gas in the atmosphere. The major source of COS appears to be the ocean, where it is produced by a photochemical reaction with dissolved organic matter (Ferek and Andreae, 1994).

Dimethylsulfide attains global significance for its potential effects on climate. Charlson et al. (1987) recognized that the oxidation of DMS to sulfate aerosols would increase the cloud condensation nucleii in the atmosphere, leading to greater cloudiness. Clouds over the sea reflect incoming sunlight, leading to global cooling. The production of DMS is directly related to the growth of marine phytoplankton (Andreae and Barnard, 1984; Turner et al., 1988). An increase in marine NPP (net primary production) from additions of nutrients or higher atmospheric CO_2 may increase the production of DMS. Thus DMS has the potential to act as a negative feedback on global warming that might otherwise occur by the greenhouse effect. This hypothesis for a biotic regulation on global temperature is intriguing, for it may be responsible for the moderation of global climate throughout geologic time (Schlesinger, 1991).

Given the strong arguments for global warming by increased atmospheric CO_2, the negative feedback mechanism of DMS is the subject of intense scientific scrutiny and debate. See, for example, Schwartz (1988).

As pointed out by Schlesinger (1991), there are some similarities between the global cycles of N and S. In both cases, the major annual movement of the element is through the atmosphere and a large portion of the movement is through the production of reduced gases of N and S by the biota. In contrast, the ultimate fate of *phosphorus* is incorporation of phosphate into marine sediments; its cycle is complete only as a result of long-term sedimentary uplift.

SUGGESTED READINGS

Andreae, M. O. (1986) The Oceans as a Source of Biogenic Gases, *Oceanus* **29**(4), 27–35.

Appelo, C. A. J. and Postma, D. (1993) *Geochemistry, Groundwater and Pollution*, Balkema, Rotterdam.

Bluth, G. S., and Kump, L. R. (1994) Lithological and Climatological Controls of River Chemistry, *Geochim. Cosmochim. Acta* **58**, 2341-2359.

Bolin, B., and Cook, R. B. (Eds.) (1983) *The Major Biogeochemical Cycles and Their Interactions*, Wiley-Interscience, New York.

Jannasch, H. W., and Mottl, M. J. (1985) Geomicrobiology of Deep Sea Hydrothermal Vents, *Science* **229**, 717-725.

Siegenthaler, U., and Sarmiento, J. L. (1993) Atmospheric Carbon Dioxide and the Ocean, *Nature* **365**, 119-125.

Sverdrup, H., and De Vries, W. (1994) Calculating Critical Loads for Acidity with the Simple Mass Balance Method, *Water, Air, Soil Pollut.* **72**, 143-162.

REFERENCES[†]

Abbott, M. M., and Van Ness, H. S. (1989) *Theory and Problems of Thermodynamics*, 2nd ed., McGraw-Hill, New York.

Adamson, A. W. (1990) *Physical Chemistry of Surfaces*, 5th ed., Wiley-Interscience, New York.

Ahrland, S., Clatt, S. J., and Davies, W. R. (1958) *Quart. Rev. London* **12**, 265.

Aieta, E. M., and Roberts, P. V. (1985) In *Water Chlorination*, Vol. 5, R. L. Jolley et al., Eds., Lewis, Chelsea, MI, pp. 783–794.

Aiken, G. R., McKnight, D. M., Wershaw, R. L., and MacCarthy, P. (Eds.) (1985) *Humic Substances in Soil, Sediment and Water. Geochemistry, Isolation and Characterization*, Wiley-Interscience, New York.

Al-Droubi, A., Fritz, B., Gac, J. V., and Tardy, Y. (1980) *Am. J. Sci.* **280**, 560.

Al-Ekabi, H., Serpone, N., Pelizzetti, E., Minero, C., Fox, M. A., and Draper, R. B. (1989) Kinetic Studies in Heterogeneous Photocatalysis. 2. TiO_2-Mediated Degradation of 4-Chlorophenol Alone and in a Three Component Mixture of 4-Chlorophenol, 2,4-Dichlorophenol and 2,4,5-Trichlorophenol in Air-Equilibrated Aqueous Media, *Langmuir* **51**, 251–255.

Anderson, D. M., and Morel, F. M. M. (1978) *Limnol. Oceanogr.* **23**, 283.

Anderson, M. A., and Morel, F. M. M. (1982) The Influence of Aqueous Iron Chemistry on the Uptake of Iron by the Coastal Diatom *Thalassiosira weissfloggi*, *Limnol. Oceanogr.* **27**, 789–813.

Anderson, M. A., Morel, F. M. M., and Guillard, R. R. (1978) Growth Limitation of a Coastal Diatom by Low Zinc Ion Activity, *Nature* **276**, 70–71.

Andreae, M. O. (1986a) Chemical Species in Seawater and Marine Particulates. In *The Importance of Chemical "Speciation" in Environmental Processes*, M. Bernhard, F. E. Brinkman, and P. J. Sadler, Eds., Dahlem Konferenzen, Springer-Verlag, Berlin, pp. 301–335.

Andreae, M. O. (1986b) Organoarsenic Compounds in the Environment. In *Organometallic Compounds in the Environment*, P. J. Craig, Ed., Longmans, Harlow.

Andreae, M. O. (1986c) The Oceans as a Source of Biogenic Gases, *Oceanus* **29**(4), 27–35.

Andreae, M. O., and Barnard, W. R. (1984) The Marine Chemistry of Dimethylsulfide, *Mar. Chem.* **14**, 267.

Andreae, M. O., et al. (1986) Changing Geochemical Cycles. In *Changing Metal Cycles and Human Health*, J. O. Nriagu, Ed., Springer, Berlin.

[†]Journal articles prior to 1981 are given without title.

Andrews, R. W., Biesinger, K. E., and Glass, G. E. (1977) *Water Res.* **11**, 309.

Appelo, C. A. J. and Postma, D. (1993) *Geochemistry, Groundwater and Pollution*, Balkema, Rotterdam.

Atkins, P. W. (1990) *Physical Chemistry*, 4th ed., Freeman, New York.

Atkins, P. W. (1992) *The Elements of Physical Chemistry*, Oxford University Press, Oxford.

Baes, C. F., and Mesmer, R. E. (1976) *The Hydrolysis of Cations*, Wiley-Interscience, New York.

Bahnemann, D., Cunningham, J., Fox, M. A., Pellizzetti, E., Pichat, P., and Serpone, N. (1994) Photocatalytic Treatment of Waters. In *Aquatic and Surface Photochemistry*, G. R. Helz et al., Eds., Lewis, Boca Raton, FL.

Ball, J. W., and Nordstrom, D. K. (1991) User's Manual for WATEQ4F, with Revised Thermodynamic Data Base and Test Cases for Calculating Speciation of Major, Trace, and Redox Elements in Natural Waters, *US Geol. Surv., Open-File Report*, 91-183.

Ball, W. P., and Roberts, P. V. (1991) *Organic Substances and Sediments in Water*, Vol. 2, Lewis, Chelsea, MI, pp. 273-310.

Balistrieri, L. S., and Murray, J. W. (1983) Metal-Solid Interactions in the Marine Environment: Estimating Apparent Equilibrium Binding Constants, *Geochim. Cosmochim. Acta* **47**, 1091-1098.

Barbash, J. E., and Reinhard, M. (1989) In *Biogenic Sulfur in the Environment*, E. Saltzman and W. J. Cooper, Eds., ACS Symp. **393**, American Chemical Society, Washington, DC.

Bard, A. J. (1988) Semiconductor Particles and Arrays for the Photoelectrochemical Utilization of Solar Energy, *Ber. Bunsen-Ges. Phys. Chem.* **92**, 1187-1194.

Bard, A. J., and Faulkner, L. R. (1980) *Electrochemical Methods; Fundamentals and Application*, Chap. 14, Wiley-Interscience, New York.

Bard, A. J., Parsons, R., and Jordan, J. (1985) *Standard Potentials in Aqueous Solution*. Prepared under Auspices of IUPAC, Marcel Dekker, New York.

Barry, R. C., Schnoor, J. L., Sulzberger, B., Sigg, L., and Stumm, W. (1994) Iron Oxidation Kinetics in an Acidic Alpine Lake, *Water Res.* **28**(2), 323-333.

Bartschat, B. M., Cabaniss, S. E., and Morel, F. M. M. (1992) Oligoelectrolyte Model for Cation Binding by Humic Substances, *Environ. Sci. Technol.* **26**, 284-294.

Basset, R. L., and Melchior, D. C. (1990) Chemical Modeling of Aqueous Systems: An Overview. In *Chemical Modeling of Aqueous Systems*, D. C. Melchior and R. L. Basset, Eds., ACS Symp. Ser. **416**, American Chemical Society, Washington, DC.

Bates, R. (1959) In *Treatise on Analytical Chemistry*, Part I, Vol. 1, I. M. Kolthoff and P. J. Elving, Eds., Wiley-Interscience, New York.

Bates, R. G. (1973) *Determination of pH, Theory and Practice*, 2nd ed., Wiley-Interscience, New York.

Bates, R. G. (1975) In *The Nature of Seawater*, E. Goldberg, Ed., Dahlem Conferences, Springer-Verlag, Berlin.

Bates, R., et al. (1970) *Anal. Chem.* **42**, 867.

Batina, N., Ruzić, I., and Cosović, B. (1985) An Electrochemical Study of Strongly Adsorbable Surface-Active Substances. Determinations of Adsorption Parameters for Triton-X-100 at the Mercury/Sodium Chloride Interface, *J. Electroanal. Chem. Interfacial Electrochem.* **190**, 21–32.

Batina, X., Ruzić, X., and Ćosović, X. (1985) *J. Electroanal. Chem.* **190**, 21–31.

Baur, M. E. (1978) Thermodynamics of Heterogeneous Iron–Carbon Systems: Implications for the Terrestrial Primotive Reducing Atmosphere, *Chem. Geol.* **22**, 189–206.

Beckett, R., and Hart, B. T. (1993) Use of Field-Flow Fractionation Techniques to Characterize Aquatic Particles, Colloids and Macromolecules. In *Environmental Particles*, Vol. 2, J. Buffle and H. P. van Leeuwen, Eds., Lewis, Boca Raton, FL.

Beer, J., and Sturm, M. (1992) Annual Report EAWAG.

Behra, P. (1994) Scale Effects in the Transport of Contaminants in Natural Media. In *Chemistry of Aquatic Systems, Local and Global Perspectives*, G. Bidoglio and W. Stumm, Eds., Kluwer, Dordrecht.

Behra, P., and Sigg, L. (1990) Evidence for Redox Cycling of Iron in Atmospheric Water Droplets, *Nature* **344**, 419–421.

Behra, P., Sigg, L., and Stumm, W. (1989) *Atmos. Environ.* **23**, 2691.

Behra, R., Genoni, G. P., and Sigg, L. (1993) Grundlagen für die Festlegung der Qualitätsziele für Metalle in Fliessgewässern, *Gas-Wasser-Abwasser* **73**, 942–951.

Behra, R., Genoni, G. P., and Sigg, L. (1994) Scientific Basis for Water Quality Criteria for Metals and Metalloids in Running Waters, Report No. 6 EAWAG.

Belford, R. E., and Owen, A. E. (1989) *Glasses and Glasskeramics*, M. H. Lewis, Ed., Chapman and Hall, London.

Bell, R. P. (1969) *The Proton in Chemistry*, Cornell University Press, Ithaca, NY.

Bennett, P. C., and Casey, W. H. (1994) Chemistry and Mechanism of Dissolution of Silicates by Organic Acids. In *Organic Acids in Geological Processes*, E. D. Pittman and M. D. Lewan, Eds., Springer-Verlag, Berlin.

Benson, D. (1968) *Mechanisms of Inorganic Reactions in Solution*, McGraw-Hill, New York.

Benson, S. W. (1960) *The Foundations of Chemical Kinetics*, McGraw-Hill, New York.

Berner, E. K., and Berner, R. A. (1987) *Global Water Cycle Geochemistry and Environment*, Prentice-Hall, Englewood Cliffs, NJ.

Berner, R. A. (1964) *Geochim. Cosmochim. Acta* **28**, 1497.

Berner, R. A. (1965) *Geochim. Cosmochim. Acta* **29**, 947.

Berner, R. A. (1971) *Principles of Chemical Sedimentology*, McGraw-Hill, New York.

Berner, R. A. (1976) In *The Benthic Boundary Layer*, I. M. McCave, Ed., Plenum, New York.

Berner, R. A. (1978) *Am. J. Sci.*, **278**, 1235.

Berner, R. A. (1980) *Early Diagenesis: A Theoretical Approach*, Princeton University Press, Princeton, NJ.

Berner, R. A. (1986) Kinetic Approach to Chemical Diagenesis. In *Studies in Diagenesis*, F. A. Mumpton, Ed., U.S. Geol. Surv. 1598, Washington, pp. 13–20.

Berner, R. A., and Lasaga, A. C. (1989) Modeling the Geochemical Carbon Cycle, *Sci. Am.* **260,** 74–81.

Berner, R. A., Lasaga, A. C., and Garrels, R. M. (1983) The Carbonate–Silicate Geochemical Cycle and Its Effect on Atmospheric Carbon Dioxide Over the Past 100 Million Years, *Am. J. Sci.* **283,** 641–683.

Berner, R. A., and Morse, J. W. (1974) Dissolution Kinetics of Calcium Carbonate in Seawater: IV. Theory of Calcite Dissolution, *Am. J. Sci.* **274,** 108–134.

Biber, M. V., Dos Santos Afonso, M., and Stumm, W. (1994) The Coordination Chemistry of Weathering: IV. Inhibition of the Dissolution of Oxide Minerals, *Geochim. Cosmochim. Acta* **58**(9), 1999–2010.

Bidoglio, G. (1994) Coupled Processes in Reaction-Flow Transport of Contaminants. In *Chemistry of Aquatic Systems: Local and Global Perspectives*, G. Bidoglio and W. Stumm, Eds., Kluwer Academic, Dordrecht.

Bidoglio, G., and Avogadro, A. (1989) Equilibrium and Kinetic Controls on the Subsurface Migration of Radioactive Contaminants, *Geoderma* **44,** 203–209.

Bidoglio, G., and Stumm, W. (Eds.) (1994) *Chemistry of Aquatic Systems: Local and Global Perspectives*, Kluwer Academic, Dordrecht.

Bischoff, W. D., Mackenzie, F. T., and Bishop, F. C. (1987) Stabilities of Synthetic Magnesian Calcites in Aqueous Solution: Comparison with Biogenic Materials, *Geochim. Cosmochim. Acta* **51,** 1413–1423.

Bjerrum, N. (1914) *Sammlung Chem. u Chem-Techn. Vortrage* **21,** 575.

Blandamer, M. J. (1992) *Chemical Equilibria in Solution*, Ellis Horwood, New York.

Blesa, M., Morando, B. J., and Regazzoni, A. E. (1994) *Dissolution of Metal Oxides*, CRC Press, Boca Raton, FL.

Bluth, G. S., and Kump, L. R. (1994) Lithological and Climatological Controls of River Chemistry, *Geochim. Cosmochim. Acta* **58,** 2341–2359.

Bockris, J. O. M., and Reddy, A. K. N. (1970) *Modern Electro Chemistry*, Plenum, New York.

Bodine, M. W., Holland, H. D., and Borcsik, M. (1965) Coprecipitation of Manganese and Strontium with Calcite. Symposium: Problems of Postmagmatic Ore Deposition, Vol. 2, *Prague* **2,** 401–406.

Bolin, B., and Cook, R. B. (Eds.) (1983) *The Major Biogeochemical Cycles and Their Interactions*, Wiley-Interscience, New York.

Boller, M. (1992) In IAWQ/IWSA Specialized Conference on Control of Organic Material by Coagulation and Floc Separation Processes, Sept. 1992, Geneva (Conference Reprints).

Bolt, G. H., and Van Riemsdijk, W. H. (1987) Surface Chemical Processes in Soil. In *Aquatic Surface Chemistry*, W. Stumm, Ed., Wiley-Interscience, New York, pp. 127–164.

Bondietti, G., Sinniger, J., and Stumm, W. (1993) The Reactivity of Fe(III)(hydr)oxides; Effects of Ligands in Inhibiting the Dissolution, *Colloids Surfaces A* **79,** 157–167.

Borkovec, M., Buchter, B., Sticher, H., Behra, P., and Sardin, M. (1991) Chromatographic Methods and Transport of Chemicals in Soils, *Chimia* **45,** 221–227.

Bouwer, E. J. (1992) Bioremediation of Organic Contaminants in the Subsurface, in *Environmental Microbiology*, R. Mitchell, Ed., Wiley-Liss, New York, pp. 287–318.

Boyle, E. A., et al. (1976) *Nature*, **263**, 42–44.

Boyle, E. A., Huested, S. S., and Grant, B. (1982) The Chemical Mass Balance of the Amazon Plume—II. Copper, Nickel, and Cadmium, *Deep-Sea Res.* **29**, 1355–1364.

Bradbury, M. H., and Baeyens, B. (1994) Sorption by Cation Exchange; Incorporation of a Cation Exchange Model into Geochemical Computer Codes. Bericht No. 94-07, Paul Scherrer Institute, Würenlingen.

Bradshaw, A. L., and Brewer, P. G. (1988) High Precision Measurements of Alkalinity and Total Carbon Dioxide in Seawater by Potentiometric Titration, *Mar. Chem.* **23**, 69–86.

Brady, P. V., and Walther, J. V. (1990) Kinetics of Quartz Dissolution at Low Temperatures, *Chem. Geol.* **82**, 253–264.

Brady, P. V., and Walther, J. V. (1992) Surface Chemistry and Silicate Dissolution at Elevated Temperatures, *Am. J. Sci.* **292**, 639–658.

Braun, W., Herron, J. T., and Kahaner, D. K. (1988) Acuchem: A Computer Program for Modeling Complex Chemical Reaction Systems, *Int. J. Chem. Kinet.* **20**, 51–62.

Brewer, P. G. (1975) In *Chemical Oceanography*, 2nd ed., J. P. Riley and G. Skirrow, Eds., Academic Press, New York.

Brezonik, P. L. (1990) Principles of Linear Free Energy and Structure Activity Relations and Their Application to Chemicals in Aquatic Systems. In *Aquatic Chemical Kinetics*, W. Stumm, Ed., Wiley-Interscience, New York.

Brezonik, P. L. (1994) *Chemical Kinetics and Process Dynamics in Aquatic Systems*, Lewis, Boca Raton, FL.

Bricker, O., and Rice, K. C. (1989) Acidic Deposition to Streams, *Environ. Sci. Technol.* **23**, 379–385.

Broecker, W. S. (1970) *Science* **168**, 1537.

Broecker, W. S. (1971) A Kinetic Model for the Chemical Composition of Seawater, *Quaternary Res.* **1**, 188–207.

Broecker, W. S. (1974) *Chemical Oceanography*, Harcourt, Brace Jovanovich, New York.

Broecker, W. S., and Oversby, V. M. (1971) *Chemical Equilibria in the Earth*, McGraw-Hill, New York.

Broecker, W. S., and Peng, T. H. (1974) Gas Exchange Rates Between Air and the Sea, *Tellus* **26**, 21–35.

Broecker, W. S., and Peng, T. H. (1982) Reactive Metals and the Great Particular Sweep, Chap. 4. In *Tracers in the Sea*, Eldigio Press, Palisades, New York.

Broecker, W. S., and Takahashi, T. (1985) Sources and Flow Patterns of Deep-Ocean Waters as Deduced from Potential Temperature, Salinity and Initial Phosphate Concentration, *J. Geophys. Res.* **90**, 6925–6939.

Bronowski, J. (1965) *Science and Human Values*, Harper & Row, New York.

Brown, D. S., and Allison, J. D. (1987) An Equilibrium Metal Speciation Model: Users Manual. Environ. Res. Lab, Office of Res. and Dev., U.S. Environmental Protection Agency, Athens, GA, Rept. No. EPA/600/3-87/012.

Brown, G. E. Jr., Parks, G. A., and Chisholm-Brause, C. J. (1989) In-Situ X-Ray Absorption Spectroscopic Studies of Ions at Oxide–Water Interfaces, *Chimia* **43**, 248–256.

Bruland, K. W., Knauer, G. A., and Martin, J. H. (1978) *Limnol. Oceanogr.* **23,** 618–625.

Bruland, K. W. (1983) Trace Elements in Seawater. In *Chemical Oceanography*, J. P. Riley and R. Chester, Eds., Academic Press, New York.

Bruland, K. W. (1989) Oceanic Zinc Speciation: Complexation of Zinc by Natural Organic Ligands in the Central North Pacific, *Limnol. Oceanogr.* **34,** 267–283.

Bruland, K. W. (1992) Complexation of Cadmium by Natural Organic Ligands in the Central North Pacific, *Limnol. Oceanogr.* **37,** 1008–1017.

Bruland, K. W. (1994) Reactive Trace Metals in the Stratified North Pacific, *Geochim. Cosmochim. Acta* **58,** 3171–3183.

Bruland, K. W., Donat, J. R., and Hutchins, D. A. (1991) Interactive Influences of Bioactive Trace Metals on Biological Production in Oceanic Waters, *Limnol. Oceanogr.* **36,** 1555–1577.

Bruland, K. W., and Franks, R. P. (1983) Mn, Ni, Cu, Zn, and Cd in the Western North Atlantic. In *Trace Metals in Sea Water*, C. S. Wong, E. Boyle, K. W. Bruland, J. D. Burton, and E. D. Goldberg, Eds., Plenum Press, New York.

Bruno, J. (1990) The Influence of Dissolved Carbon Dioxide on Trace Metal Speciation in Seawater, *Mar. Chem.* **30,** 231–240.

Bruno, J., Casas, I., and Puigdomènech, I. (1991) The Kinetics of Dissolution of UO_2 Under Reducing Conditions and the Influence of an Oxidized Surface Layer (UO_{2+x}): Application of a Continuous Flow-through Reactor, *Geochim. Cosmochim. Acta* **55,** 547, 647–658.

Bruno, J. I., Stumm, W. and Wersin, P. (1992) On the Influence of Carbonate in Mineral Dissolution I and II. *Geochim. Cosmochim. Acta* **56,** 1139–1155.

Buat-Menard, P. E. (1985) Fluxes of Metals Through the Atmosphere and Oceans. In *Changing Metal Cycles and Human Health*, J. O. Nriagu, Ed., Dahlem Konferenzen, Springer-Verlag, Berlin, pp. 43–69.

Buffle, J. D. (1988) *Complexation Reactions in Aquatic Systems. An Analytical Approach*, Ellis Horwood, Chichester.

Buffle, J., and Altmann, R. S. (1987) Interpretation of Metal Complexation by Heterogeneous Complexants. In *Aquatic Surface Chemistry*, W. Stumm, Ed., Wiley-Interscience, New York.

Buffle, J., and de Vitre, R. R. (Eds.) (1994) *Chemical and Biological Regulation of Aquatic Systems*, Lewis, Chelsea, MI.

Buffle, J., Perret, D., and Newman, M. (1992) The Use of Filtration and Ultrafiltration for Size Fractionation of Aquatic Particles, Colloids, and Macromolecules. In *Environmental Particles*, Vol. 1, J. Buffle and H. P. van Leeuwen, Eds., Lewis, Boca Raton, FL.

Buffle, J., and Stumm, W. (1994) General Chemistry of Aquatic Systems. In *Chemical and Biological Regulation of Aquatic Systems*, J. Buffle and R. R. de Vitre, Eds., Lewis, Chelsea, MI.

Bunker, B. C., et al. (1994) Ceramic Thin-Film Formation on Functionalized Interfaces Through Biomimeting Processing, *Science* **264,** 48–55.

Burgess, J. (1978) *Metal Ions in Solution*, Ellis Horwood, Chichester.

Burgess, J. (1988) *Ions in Solution; Basic Principles and Interactions*, Ellis Horwood, Chichester.

Bürgisser, C., Cernik, M., Borkovec, M., and Sticher, H. (1993) Determination of Nonlinear Adsorption Isotherms from Column Experiments, an Alternative to Batch Studies, *Environ. Sci. Technol.* **27**, 943-948.

Burton, W. K., Cabrera, N., and Frank, F. C. (1951) The Growth of Crystals and the Equilibrium Structure of Their Surfaces, *Philos. Trans. R. Soc.* **A243**, 299-358.

Busch, P., and Stumm, W. (1968) *Environmental Science and Technology*, **2**, 49.

Busenberg, E., and Clemency, C. V. (1976) The Dissolution Kinetics of Feldspars at 25°C at 1 Atm. CO_2 Partial Pressure, *Geochim. Cosmochim. Acta* **41**, 41-49.

Busenberg, E., and Plummer, L. N. (1986) A Comparative Study of the Dissolution and Crystal Growth Kinetics of Calcite and Aragonite. In *Studies in Diagenesis*, F. A. Mumpton, Ed., U.S. Geol. Surv. Bull. **1578**, 139-168.

Butler, J. N. (1964) *Ionic Equilibrium, A Mathematical Approach*, Addison-Wesley, Reading, MA.

Butler, J. N. (1982) *CO_2-Equilibria and Their Applications*, Lewis, Chelsea, MI.

Butler, J. N. (1992) Alkalinity Titration in Seawater: How Can the Data Be Fitted by an Equilibrium Model? *Mar. Chem.* **38**, 251-288.

Byrne, R. H., and Millero, W. L. (1985) Copper(II)Carbonate Complexation in Seawater, *Geochim. Cosmochim. Acta* **49**, 1837-1844.

Byrne, R. H., Kump, L. R., and Cantrell, K. J. (1988) The Influence of Temperature and pH on Trace Metal Speciation in Seawater, *Mar. Chem.* **25**, 163-181.

Cabaniss, S. E., and Shuman, M. S. (1988) Copper Binding by Dissolved Organic Matter. I. Suwanee River Fulvic Acid Equilibria, *Geochim. Cosmochim. Acta* **52**, 185.

Cabaniss, S. E., and Shuman, M. S. (1988) Copper Binding by Dissolved Organic Matter. II. Variation in Type and Source of Organic Matter, *Geochim. Cosmochim. Acta* **52**, 195.

Cappelos, C., and Bielski, B. H. J. (1972) *Kinetic Systems*, Wiley-Interscience, New York.

Capellos, C., and Bielski, B. H. J. (1980) *Kinetic Systems: Mathematical Description of Chemical Kinetics in Solution*, Krieger, Huntington, NY.

Capellos, C., and Bielski, B. H. J. (1980) *Kinetic Systems: Mathematical Description of Chemical Kinetics in Solution*, Krieger, Huntington, NY.

Carpenter, S. R., Cottingham, K. L., and Schnindler, D. E. (1992) Biotic Feedbacks in Lake Phosphorus Cycles, *TREE* **7**, 332-336.

Casey, W. H., Banfield, J. F., Estrich, H. R., and McLaughlin, L. (1993) What Do Dissolution Experiments Tell Us About Natural Weathering? *Chem. Geol.* **105**, 1-15.

Casey, W. H., and Sposito, G. (1992) On the Temperature Dependence of Mineral Dissolution Rates, *Geochim. Cosmochim. Acta* **56**, 3825-3830.

Casey, W. H., and Westrich, R. R. (1992) Control of Dissolution Rates of Orthosilicate Minerals by Divalent Metal-Oxygen Bonds, *Nature* **355**, 157-159.

Cederberg, G. A., Street, R. L., and Leckie, J. O. (1985) A Groundwater Mass Transport and Equilibrium Chemistry Model for Multicomponent Systems, *Water Resour. Res.* **21**, 1095-1104.

Chandar, P., Somasundaran, P., and Turro, N. J. (1987) Fluorescence Probe Studies on the Structure of the Adsorbed Layer of Dodecyl Sulfate at the Alumine Water Interface, *J. Colloid Interface Sci.* **117**, 31–46.

Charlet, L., and Manceau, A. (1993) Structure, Formation and Reactivity of Hydrous Oxide Particles: Insights from X-Ray Absorption Spectroscopy. In *Environmental Particles*, J. Buffle and H. P. van Leeuwen, Eds., Lewis, Boca Raton, FL, pp. 117–164.

Charlet, L., Schindler, P. W., Spadini, L., Furrer, G., and Zysset, M. (1993) Cation Adsorption on Oxides and Clays: The Aluminum Case, *Aquatic Sci.* **55/4**, 291–303.

Charlet, L., Wersin, P., and Stumm, W. (1990) Surface Charge of Some Carbonate Minerals, *Geochim. Cosmochim. Acta* **54**, 2329–2336.

Charlson, R. J., and Wigley, T. M. L. (1994) Sulfate Aerosol and Climate Change, *Sci. Am.* **270**(2), 28–35.

Charlson, R. J., Lovelock, J. E., Andreae, M. O., and Warren, S. G. (1987) Oceanic Phytoplankton, Atmospheric Sulfur, Cloud Albedo, and Climate, *Nature*, **326**, 655–661.

Chen, Y. S., Butler, J. N., and Stumm, W. (1973) *Environ. Sci. Technol.* **7**, 327.

Chien, S. H., and Clayton, W. R. (1980) Application of the Elovich Equation to the Kinetics of Phosphate Release and Sorption in Soils, *Soil Sci. Soc. Am. J.* **44**, 265.

Chiou, C. T., Freed, V. H., Schmedding, D. W., and Kohnert, R. L. (1977) *Environ. Sci. Technol.* **21**, 1231–1234.

Chisholm, S. W., and Morel, F. M. M. (Eds.) (1991) *What Controls Phytoplankton Production in Nutrient-Rich Areas of the Open Sea?* American Society of Limnology and Oceanography.

Chou, L., and R. Wollast (1984) Study of the Weathering of Albite at Room Temperature and Pressure with a Fluidized Bed Reactor, *Geochim. Cosmochim. Acta* **48**, 2205–2218.

Chou, L., Garrels, R. M., and Wollast, R. (1989) *Chem. Geol.* **78**, 269.

Christenson, P. G., and Gieskes, J. M. (1971) *J. Chem. Eng. Data*, **16**, 398.

Clason, R. J. (1965) *The Numerical Solution of the Chemical Equilibrium Problem*, RM-4345-PR, The Rand Corporation, Santa Monica, California.

Coale, K. H., and Bruland, K. W. (1988) Copper Complexation in the Northeast Pacific, *Limnol. Oceanogr.* **33**, 1084–1101.

Coale, K. H., and Mart, L. (1985) Analysis of Seawater for Dissolved Cadmium, Copper and Lead: An Inter-comparison of Voltammetric and Atomic Absorption Methods. *Mar. Chem.* **17**, 285–300.

Colienne, R. H. (1983) Photoreduction of Iron in the Epilimnion of Acidic Lakes, *Limnol. Oceanogr.* **28**, 83–100.

Collier, R., and Edmond, J. (1984) *Progr. Oceanogr.* **13**, 113.

Collier, R. W., and Edmond, J. M. (1983) Plankton Compositions and Trace Metal Fluxes from the Surface Ocean, in *Trace Metals in Seawater*, NATO Conf. Ser. 4, v. 9, Plenum, New York, pp. 789–809.

Comans, R. N. J. (1987) Adsorption, Desorption and Isotopic Exchange of Cadmium on Illite: Evidence for Complete Reversibility, *Water Res.* **21**, 1573–1576.

Constable, E. (1990) *Metal Ligand Interactions*, Ellis Horwood, Chichester.

Cooper, W. J., Zika, R. G., Petasne, R. G., and Plane, J. M. C. (1988) Photochemical

Formation of H_2O_2 in Natural Waters Exposed to Sunlight, *Environ. Sci. Technol.* **22,** 1156–1160.

Cornel, P. R., Summers, R. S., and Roberts, P. V. (1986) Diffusion of Humic Acid in Dilute Aqueous Solutions, *J. Colloid Interface Sci.* **110,** 149.

Ćosović, B. (1990) Adsorption Kinetics of the Complex Mixture of Organic Solutes at Model and Natural Phase Boundaries. In *Aquatic Chemical Kinetics*, W. Stumm, Ed., Wiley-Interscience, New York, pp. 291–310.

Craig, P. J. (Ed.) (1986) *Organometallic Compounds in the Environment; Principles and Reactions*, Longmans, Harlow.

Culberson, C. H. (1981) Direct Potentiometry. In *Marine Electrochemistry*, M. Whitfield and D. Jagner, Eds., Wiley-Interscience, New York.

Culberson, C. H., Latham, G., and Bates, R. G. (1978) Solubilities and Activity Coefficients of Calcium and Strontium Sulfates in Synthetic Sea Water at 0.5 and 25°C, *J. Phys. Chem.* **82,** 2693–2699.

Culberson, H. C., and Pytkovicz, R. M. (1968) *Limnol. Oceanogr.* **13,** 403.

Daly, F. P., et al. (1972) *J. Phys. Chem.* **76,** 3664.

Daskalakis, K. D., and Helz, G. R. (1992) *Environ. Sci. Technol.* **26,** 2462.

Daskalakis, K. D., and Helz, G. R. (1993) The Solubility of Sphalerite (ZnS) in Sulfidic Solutions at 25°C and 1 atm Pressure, *Geochim. Cosmochim. Acta* **57,** 4923–4932.

Dasent, W. E. (1970) *Inorganic Energetics*, Penguin Books, Baltimore, MD.

Davidson, N. (1962) *Statistical Mechanics*, McGraw-Hill, New York.

Davies, S. N., and DeWiest, R. C. M. (1966) *Hydrogeology*, Wiley-Interscience, New York.

Davis, J. A., and Leckie, J. O. (1978) Surface Ionization and Complexation at the Oxide/Water Interface. II. Surface Properties of Amorphous Iron Oxyhydroxide and Adsorption of Metal Ions, *J. Colloid Interface Sci.* **67,** 90–107.

Davis, J. A., Fuller, C. C., and Cook, A. D. (1987) A Model for Trace Metal Sorption at the Calcite Surface, *Geochim. Cosmochim. Acta* **51,** 1477–1490.

Davis, J. A., James, R. O., and Leckie, J. O. (1978) *J. Colloid Interface Sci.* **63,** 480.

Davison, W. (1978) Defining the Electroanalytically Measured Species in a Natural Water Sample, *J. Electroanal. Chem.* **87,** 395–404.

Davison, W. (1985) Conceptual Models for Transport in a Redox Boundary. In *Chemical Processes in Lakes*, W. Stumm, Ed., Wiley-Interscience, New York.

Davison, W. (1991) The Solubility of Iron Sulfides in Synthetic and Natural Waters at Ambient Temperatures, *Aquat. Sci.* **53/54,** 309–329.

Davison, W., Grime, G. W., Morgan, J. A. W., and Clarke, K. (1991) Distribution of Dissolved Iron in Sediment Pore Waters at Submillimetre Resolution, *Nature* **352,** 323–325.

Davison, W., and Zhang, H. (1994) In-Situ Speciation Measurements of Trace Components in Natural Waters Using Thin Film Gels, *Nature* **367,** 546–548.

Deffeyes, K. S. (1965) *Limnol. Oceanogr.* **10,** 412.

Degens, E. T. (1965) *Geochemistry of Sediments*, Prentice Hall, Englewood Cliffs, NJ.

Deines, P., Langmuir, D., and Harmon, R. S. (1974) *Geochim. Cosmochim. Acta* **38,** 1147.

Delaney, M., and Boyle, E. A. (1987) Cadmium/Calcium in Late Miocene Benthic Foraminifera and Changes in the Global Organic Carbon Budget, *Nature* **330**, 156–159.

Denbigh, K. G. (1971) *The Principles of Chemical Equilibrium: With Applications in Chemistry and Chemical Engineering*, 3rd ed., Cambridge University Press, Cambridge.

Denbigh, K. G., and Turner, J. C. R. (1984) *Chemical Reactor Theory*, 3rd ed., Cambridge University Press, Cambridge.

Deuser, W. G. (1967) *Nature* **215**, 1033.

Dickerson, R. E. (1969) *Molecular Thermodynamics*, Benjamin, New York.

Dickson, A. G. (1984) pH Scales and Proton Transfer Reactions in Saline Media Such as Seawater, *Geochim. Cosmochim. Acta* **48**, 2299–2308.

Dickson, A. G. (1990) Thermodynamics of the Dissociation of Boric Acid in Synthetic Seawater from 273.15 to 318.15 K, *Deep-Sea Res.* **37**, 755–766.

Dickson, A. G. (1993) pH Buffers for Sea Water Media Based on the Total Hydrogen Ion Concentration Scale, *Deep-Sea Res.* **40**, 107–118.

Dickson, A. G., Friedman, H. L., and Millero, F. J. (1988) Chemical Model of Seawater Systems, a Panel Report, *Appl. Geochem.* **3**, 27–35.

Dickson, A. G., and Riley, J. P. (1979) The Estimation of Acid Dissociation Constants in Seawater from Potentiometric Titrations with Strong Base. I. The Ion Product of Water K_W, *Mar. Chem.* **7**, 89–99.

Di Toro, D. M., Mahoney, J. D., Hansen, D. J., Scott, K. J., Carlson, A. R., and Ankley, G. T. (1992) Acid Volatile Sulfide Predicts the Acute Toxicity of Cd and Ni in Sediments, *Environ. Sci. Technol.* **26**, 96–101.

Doerner, X., and Hoskins, X. (1925) *J. Am. Chem. Soc.* **47**, 662.

Donat, J. R., and Bruland, K. W. (1990) A Comparison of Two Voltammetric Techniques for Determining Zn Speciation in Northeast Pacific Ocean Waters, *Mar. Chem.* **28**, 301–323.

Donat, J. R., Statham, P. J., and Bruland, K. W. (1986) An Evaluation of a C-18 Solid Phase Extraction Technique for Isolating Metal–Organic Complexes from Central North Pacific Ocean Waters, *Mar. Chem.* **18**, 85–99.

Donat, J. R., and Van den Berg, C. M. G. (1992) A New Cathodic Stripping Voltammetric Method for Determining Organic Copper Complexation in Seawater, *Mar. Chem.* **38**, 69–90.

Dos Santos-Afonso, M., and Stumm, W. (1992) Reductive Dissolution of Iron(III)(hydr)oxides by Hydrogen Sulfide, *Langmuir* **8**, 1671.

Drever, J. I. (1988) *The Geochemistry of Natural Waters*, 2nd ed., Prentice-Hall, New York.

Drouby, A., et al. (1976) *Sci. Geol. Bull.* **29**, 45.

Ducker, W. A., Senden, T. J., and Pashley, R. M. (1991) Direct Measurement of Colloidal Forces Using an Atomic Force Microscope, *Nature* **353**, 239–241.

Dyrssen, D. (1985) Metal Complexe Formation in Sulphidic Seawater, *Mar. Chem.* **15**, 285–293.

Dyrssen, D. (1988) Sulfide Complexation in Surface Seawater, *Mar. Chem.* **24**, 143–153.

Dyrssen, D., and Kremling, K. (1990) Increasing Hydrogen Sulfide Concentration and Trace Metal Behavior in the Anoxic Baltic Waters, *Mar. Chem.* **30**, 193–204.

Dyrssen, D., and Sillén, L. G. (1967) *Tellus* **19,** 113.

Dyrssen, D., et al. (1969) *J. Chem. Educ.* **46,** 252.

Dzombak, D. A., and Hudson, R. J. M. (1995) Ion Exchange: The Contribution of Diffuse Layer Sorption and Surface Complexation, in *Aquatic Chemistry*, C. P. Huang et al., Eds., ACS Washington, D.C.

Dzombak, D. A., and Morel, F. M. M. (1990) *Surface Complexation Modeling; Hydrous Ferric Oxide*, Wiley-Interscience, New York.

Eberhart, J. G., and Sweet, T. R. (1966) *J. Chem. Educ.* **37,** 422.

Eberson, L. (1987) *Electron Transfer Reactions in Organic Chemistry*, Springer, Berlin.

Edmond, J. M., Spivack, A., Grant, B. C., Ming-Hui, H., Zexiam, C., Sung, C., and Xiushau, Z. (1985) Chemical Dynamics of the Changjiang Estuary, *Cont. Shelf Res.* **4,** 17–36.

Edsall, J. T. (1969) In *CO_2: Chemical, Biochemical and Physiological Effects*, R. E. Foster et al., Eds., NASA SP 188, Washington, D.C.

Edwards, J. O. (1964) *Inorganic Reaction Mechanisms: An Introduction*, Benjamin, Menlo Park, CA.

Eggleston, C. M., and Stumm, W. (1993) Scanning Tunnelling Microscopy of Cr(III) Chemisorbed on α-Fe_2O_3 001 Surfaces from Aqueous Solutions, *Geochim. Cosmochim. Acta* **57,** 4843–4850.

Egozy, Y. (1980) Adsorption of Cadmium and Cobalt on Montmorillonite as a Function of Solution Composition, *Clays Clay Miner.* **28,** 311.

Eigen, M., and Tamm, K. (1962) *Z. Elektrochem.* **66,** 93.

Eisenberg, D., and Crothers, D. (1979) *Physical Chemistry*, Benjamin, Menlo Park, CA.

Eisenmann, G. (1967) *Glass Electrodes for Hydrogen and Other Cations*, Marcel Dekker, New York.

Elgquist, B. (1970) *J. Inorg. Nucl. Chem.* **32,** 437.

Emerson, E. (1975) *Limnol. Oceanogr.* **20,** 743.

Epstein, W., Thompson, P., and Yapp, C. J. (1977) *Science* **198,** 1209.

Erni, I. W. (1977) Ph.D. Thesis, Swiss Federal Institute of Technology, Zurich.

Eugster, H. P., and Hardie, L. A. (1978) In *Lakes: Chemistry, Geology, Physics*, A. Lerman, Ed., Springer-Verlag, New York.

Evans, A. (1987) *Potentiometry and Ion Selective Electrodes*, Wiley-Interscience, Chichester.

Faita, G., Longhi, P., and Mussini, T. (1967) *J. Electrochem. Soc.* **114,** 340.

Fallab, S. (1967) *Angewandte Chem. Internat. Edition*, **6,** 496.

Farley, K. J., Dzombak, D. A., and Morel, F. M. M. (1985) A Surface Precipitation Model for the Sorption of Cations on Metal Oxides, *J. Colloid Interface Sci.* **106,** 226–242.

Farrah, H., Hatton, D., and Pickering, W. (1980) The Affinity of Metals for Clay Surfaces, *Chem. Geol.* **28,** 55.

Faust, B. C. (1994a) Aqueous Phase Photochemical Reactions in Oxidant Formation, Pollutant Transformations and Atmospheric Geochemical Cycles, *Environ. Sci. Technol.* **28,** 217A–222A.

Faust, B. C. (1994b) A Review of the Photochemical Redox Reactions of Fe(III)

Species in Atmospheric, Oceanic and Surface Waters. In *Aquatic and Surface Photochemistry*, G. Helz, R. R. Zepp and D. G. Crosby, Eds., Lewis, Boca Raton, FL.

Faust, B. C., and Hoigné, J. (1987) Sensitized Photooxidation of Phenols by Fulvic Acid and in Natural Waters, *Environ. Sci. Technol.* **21**(10), 957–964.

Feitknecht, W., and Schindler, P. (1963) *Solubility Constants of Metal Oxides, Metal Hydroxides and Hydroxide Salts in Aqueous Solution*, Butterworth, London.

Ferek, R. J., and Andreae, M. O. (1994) Photochemical Production of Carbonylsulfide in Marine Surface Waters, *Nature* **307**, 148–150.

Ferri, D., Grenthe, I., Hietanen, S., Néher-Neumann, E., and Salvatore, F. (1985) Studies on Metal Carbonate Equilibria 12. Zn(II) Carbonate Complexes in Acid Solutions, *Acta Chem. Scand. A* **39**, 347–353.

Ferri, D., Grenthe, I., Hietanen, S., and Salvatore, F. (1987a) Studies on Metal Carbonate Equilibria. 17. Zinc(II) Carbonate Complexes in Alkaline Solutions, *Acta Chem. Scand.* **A41**, 190–196.

Ferri, D., Grenthe, I., Hietanen, S., and Salvatore, F. (1987b) Studies on Metal Carbonate Equilibria. 18. Lead(II) Carbonate Complexes in Alkaline Solutions, *Acta Chem. Scand.* **A41**, 349–354.

Fillela, M., and Buffle, J. (1993) *Colloids Surfaces A* **73**, 255–273.

Finlayson-Pitts, B. J., and Pitts, J. N. (1986) *Atmospheric Chemistry, Fundamentals and Experimental Techniques*, Wiley-Interscience, New York.

Fisher, F. H. (1967) *Science* **157**, 823.

Fleer, G. J., and Lyklema, J. (1983) Adsorption of Polymers, Chap. 4. In *Adsorption from Solution at the Solid/Liquid Interface*, G. D. Parfitt and C. H. Rochester, Eds., Academic Press, London.

Fletcher, P., and Sposito, G. (1989) The Chemical Modeling of Clay–Electrolyte Interactions for Montmorillonite, *Clay Miner.* **24**, 375–391.

Fouillac, C., and Criaud, A. (1984) Carbonate and Bicarbonate Trace Metal Complexes, *Geochem. J.* **18**, 297–303.

Fowkes, F. M. (1965) In *Chemistry and Physics at Interfaces*, American Chemical Society, Washington, DC.

Fowkes, F. M. (1968), Attractive Forces at Interfaces, *J. Colloid Interf. Sci.* **28**, 493.

Fraústo da Silva, J. J. R., and Williams, R. J. P. (1991) *The Biological Chemistry of the Elements; The Inorganic Chemistry of Life*, Clarendon Press, Oxford.

Freeman, J. S., and Rowell, D. L. (1981) The Adsorption and Precipitation of Phosphate onto Calcite, *J. Soil Sci.* **32**(1), 72–84.

Freeze, R. A., and Cherry, J. A. (1979) *Groundwater*, Prentice-Hall, Englewood Cliffs, NJ.

Friedlander, S. K. (1960) *J. Meteorol.* **17**, 479.

Fritz, B., and Tardy, Y. (1973) *Sci. Geol. Bull.* **26**, 339.

Furrer, G., and Stumm, W. (1986) The Coordination Chemistry of Weathering, I. Dissolution Kinetics of δ-Al_2O_3 and BeO, *Geochim. Cosmochim. Acta* **50**, 1847–1860.

Furrer, G., Westall, J. C., and Sollins, P. (1989) The Study of Soil Chemistry Through Quasi-steady State Models; I. Mathematical Definition of the Problem, *Geochim. Cosmochim. Acta* **53**, 595–601.

Furrer, G., Westall, J. C., and Sollins, P. (1990) The Study of Soil Chemistry Through Quasi-steady State Models; II. Soil Solution Acidity, *Geochim. Cosmochim. Acta* **54**(9), 2363-2374.

Fyfe, W. S. (1987) From Molecules to Planetary Environments: Understanding Global Change, In *Aquatic Surface Chemistry*, W. Stumm, Ed., Wiley-Interscience, New York, pp. 495-508.

Gardiner, W. C. (1969) *Rates and Mechanisms of Chemical Reactions*, Benjamin, Menlo Park, CA.

Garrels, R. M. (1967) In *Glass Electrodes*, G. Eisenmann, Ed., Dekker, New York.

Garrels, R. M., and Christ, C. L. (1965) *Solutions, Minerals and Equilibria*, Harper & Row, New York.

Garrels, R. M., and Mackenzie, F. T. (1967) Origin of the Chemical Composition of Some Springs and Lakes. In *Equilibrium Concepts in Natural Water Systems*, Advances in Chemistry Series, No. 67, American Chemical Society, Washington, DC.

Garrels, R. M., and Mackenzie, F. T. (1971) *Evolution of Sedimentary Rocks*, Norton, New York.

Garrels, R. M., Mackenzie, F. T., and Hunt, C. (1975) *Chemical Cycles and the Global Environment*, Kaufman, Los Altos, CA.

Garrels, R. M., and Perry, E. A., Jr. (1974) In *The Sea*, Vol. 5, E. D. Goldberg, Ed., Wiley-Interscience, New York.

Garrels, R. M., and Thompson, M. E. (1962) *Am. J. Sci.* **260**, 57.

Giovanoli, R., Schnoor, J. L., Sigg, L., Stumm, W., and Zobrist, J. (1989) Chemical Weathering of Crystalline Rocks in the Catchment Area of Acidic Ticino Lakes, Switzerland, *Clays Clay Miner.* **36**, 521-529.

Goldschmidt, V. M. (1933) *Fortschr. Miner. Kristallogr. Petrol.* **17**, 112.

Gonçalves, M. S., Sigg, L., Reutlinger, M., and Stumm, W. (1987) Metal Ion Binding by Biological Surfaces; Voltammetric Assessment in Presence of Bacteria, *The Sci. Tot. Environ.* **60**, 105-119.

Gonçalves, M. S., Sigg, L., and Stumm, W. (1985) Voltammetric Methods for Distinguishing Between Dissolved and Particulate Metal Ion Concentrations in the Presence of Hydrous Oxides. A Case Study on Lead(II). *Environ. Sci. Technol.* **19**, 141-146.

Gordon, L., Salutsky, M. L., and Willard, H. H. (1959) *Precipitation from Homogeneous Solutions*, Wiley-Interscience, New York.

Graedel, T. E., and Crutzen, P. J. (1993) *Atmospheric Change; an Earth System Perspective*, Freeman, New York.

Grahame, D. C. (1947) *Chem. Reviews* **41**, 441.

Grandstaff, D. E. (1986) The Dissolution Rate of Forsteritic Olivine from Hawaiian Beach Sand. In *Rates of Chemical Weathering of Rocks and Minerals*, S. M. Colman and D. P. Dethier, Eds., Academic Press, New York, pp. 41-59.

Gratz, A. J., Manne, S., and Hansma, P. K. (1991) Atomic Force Microscopy of Atomic-Scale Ledges and Etch Pits Formed During Dissolution of Quartz, *Science* **251**, 1343-1346.

Grätzel, M. (1989) *Heterogeneous Photochemical Electron Transfer*, CRC Press, Boca Raton, FL.

Grauer, R. (1983) Zur Koordinationschemie der Huminstoffe, *PSI-Report, Würenlingen, Switzerland* **24**, 103.

Grauer, R., and Stumm, W. (1982) Die Koordinationschemie oxidischer Grenzflächen und ihre Auswirkung auf die Auflösungskinetic oxidischer Festphasen in wässrigen Lösungen, *Colloid. Polymer. Sci.* **260**, 959–970.

Gray, H. B. (1973) *Chemical Bonds*, Benjamin-Cummings, Menlo Park, CA.

Greenland, D. J., and Hayes, M. H. B. (1978) *The Chemistry of Soil Constituents*, Wiley, Chichester.

Grenthe, I., Stumm, W., Laaksuharju, M., Nilsson, A. C., and Wikberg, P. (1992) Redox Potentials and Redox Reactions in Deep Groundwater Systems, *Chem. Geol.* **98**, 131.

Grütter, A., von Gunten, H. R., Kobler, M., and Rössler, E. (1990) Sorption, Desorption and Exchange of Cs^+ on Glaciofluvial Deposits, *Radiochim. Acta* **50**, 177–184.

Gschwend, Ph. M., and Reynolds, M. D. (1987) Monodisperse Ferrous Phosphate Colloids in an Anoxic Groundwater Plume, *J. Contam. Hydrol.* **1**(3), 309–327.

Guggenheim, E. H., and Turgeon, J. C. (1954) *Trans. Faraday Soc.* **51**, 747.

Haag, W. R., and Hoigné, J. (1983) Ozonation of Water Containing Chlorine or Chloramines. Reaction Products and Kinetics, *Water Res.* **17**(10), 1397–1402.

Haag, W. R., and Hoigné, J. (1985) Photo-sensitized Oxidation in Natural Water via OH Radicals, *Chemosphere* **14**(11/12), 1659–1671.

Haag, W. R., and Hoigné, J. (1986) Singlet Oxygen in Surface Waters. 3. Photochemical Formation and Steady-State Concentrations in Various Types of Waters. *Environ. Sci. Technol.* **20**(4), 341–348.

Hachiya, K., Sasaki, M., Saruta, Y., Mikami, N., and Yasanuga, T. (1984) Static and Kinetic Studies of Adsorption–Desorption of Metal Ions on the γ-Al_2O_3 Surface, *J. Phys. Chem.* **88**, 23–31.

Haderlein, S. B., and Schwarzenbach, R. P. (1993) Adsorption of Substituted Nitrobenzenes and Nitrophenols to Mineral Surfaces, *Environ. Sci. Technol.* **27**, 316–326.

Hahn, M. W. (1994) Unpublished doctoral dissertation, The Johns Hopkins University, Baltimore, MD.

Hamann, S. D. (1957) *Physico-Chemical Effects of Pressure*, Butterworths, London.

Hamilton-Taylor, J., Willis, M., and Reynolds, C. S. (1984) *Limnol. Oceanogr.* **29**, 695–710.

Hammet, L. P. (1940) *Physical Organic Chemistry*, McGraw-Hill, New York.

Hansson, I. (1973) *Deep-Sea Res.* **461**, 479.

Harned, H. S., and Owen, B. B. (1958) *The Physical Chemistry of Electrolytic Solutions*, 3rd ed., Van Nostrand Reinhold, New York.

Harrison, G. I., and Morel, F. M. M. (1986) Response of the Marine Diatom *Thalassiosira weissflogii* to Iron Stress, *Limnol. Oceanogr.* **31**, 989–997.

Harvie, C. E., Møller, N., and Weare, J. H. (1981) The Prediction of Mineral Solubilities in Natural Waters: the Na-K-Mg-Ca-H-Cl-SO_4-OH-HCO_3-CO_3-H_2O System to High Ionic Strength, *Geochim. Cosmochim. Acta* **48**, 723–751.

Hawker, D. W., and Connel, D. W. (1990) Simple Water/Octanol Partition System, *Environ. Sci. Technol.* **24,** 1612–1618.

Hayes, K. F. (1987) Equilibrium, Spectroscopic and Kinetic Studies of Ion Adsorption at the Oxide/Aqueous Interface, Ph.D. Thesis, Stanford University, Stanford CA.

Hayes, K. F., and Leckie, J. O. (1986) Mechanism of Lead Ion Adsorption at the Goethite–Water Interface, *Adv. Chem. Ser.* **323,** 114–141.

Hayes, K. F., Roe, A. L., Brown, G. E. Jr., Hodgson, K. O., Leckie, J. O., and Parks, G. A. (1987) In Situ X-Ray Absorption Study of Surface Complexes: Selenium Oxyanions on α-FeOOH, *Science* **238,** 783–786.

Helz, G. R., Zepp, R. G., and Crosby, D. G. (Eds.) (1994) *Aquatic and Surface Photochemistry*, Lewis, Boca Raton, FL.

Hem, J. D. (1985) Study and Interpretation of the Chemical Characteristics of Natural Water, *U.S. Geol. Surv. Water Supply Paper* 2254.

Hem, J. D., and Roberson, C. E. (1988) Aluminum Hydrolysis Reactions and Products in Mildly Acidic Aqueous Systems. In *Chemical Modeling of Aqueous Systems*, Vol. II, D. C. Melchior and R. L. Bassett, Eds., ACS Series 416, American Chemical Society, Washington, DC.

Hemond, H. F. (1990) ANC, Alk and the Acid–Base Status of Natural Waters Containing Organic Acids. *Environ. Sci. Technol.* **24,** 1486–1489.

Hemond, H. F., and Fechner, E. J. (1994) In *Chemical Fate and Transport in the Environment*, Academic Press, San Diego, CA, pp. 227–320.

Hering, J. G., and Morel, F. M. M. (1988) Kinetics of Trace Metal Complexation; Role of Alkaline Earth Metals, *Environ. Sci. Technol.* **22,** 1469–1478.

Hering, J. G., and Morel, F. M. M. (1989) Slow Coordination Reactions in Seawater, *Geochim. Cosmochim. Acta* **53,** 611–618.

Hering, J. G., and Morel, F. M. M. (1990) In *Aquatic Chemical Kinetics*, W. Stumm, Ed., Wiley-Interscience, New York.

Hering, J., and Stumm, W. (1990) Oxidative and Reductive Dissolution of Minerals. In *Reviews in Mineralogy 23: Mineral–Water Interface Geochemistry*, M. F. Hochella and A. F. White, Eds., Mineralogical Society of America, Washington, DC, pp. 427–465.

Hiemenz, P. C. (1986) *Principles of Colloid and Surface Chemistry*, 2nd ed., Marcel Dekker, New York.

Hiemstra, T., van Riemsdijk, W. H., and Bolt, H. G. (1989) Multisite Proton Adsorption Modeling at the Solid/Solution Interface of (Hydr)Oxides, I. Model Description and Intrinsic Reaction Constants, *J. Colloid Interface Sci.* **133,** 91–104.

Hiemstra, T., de Wit, J. C. M., and van Riemsdijk, W. O. (1989) Multisite Proton Adsorption Modeling at the Solid/Solution Interface of (Hydr)Oxides, II. Application to Various Important (Hydr)Oxides, *J. Colloid Interface Sci.* **133,** 105–117.

Hill, T. L. (1968) *Thermodynamics for Chemists and Biologists*, Addison-Wesley, Reading, MA.

Hoffmann, M. R. (1981) *Environ. Sci. Technol.*, **16,** 15, 345–353.

Hoffmann, M. R. (1990) Catalysis in Aquatic Environments. In *Aquatic Chemical Kinetics*, W. Stumm, Ed., Wiley-Interscience, New York.

Hoffmann, M. R., and Calvert, J. (1985) U.S. Environmental Protection Agency Report 600/3-85/017.

Hohl, H. (1982) Personal communication, unpublished.

Hohl, H., Sigg, L., and Stumm, W. (1980) Characterization of Surface Chemical Properties of Oxides in Natural Water; The Role of Specific Adsorption Determining the Specific Charge, *Particulates in Water*, American Chemical Soc., Washington, D.C., **189**, 1-31.

Hoigné, J. (1988) In *Process Technologies for Water Treatment*, S. Stucki, Ed., Plenum Press, New York.

Hoigné, J. (1990) Formulation and Calibration of Environmental Reaction Kinetics; Oxidations by Aqueous Photooxidants as an Example. In *Aquatic Chemical Kinetics*, W. Stumm, Ed., Wiley-Interscience, New York.

Hoigné, J., and Bader, H. (1978) *Int. Assoc. Water Pollut. Res. Progr. Water Technology*, **10**, 645.

Hoigné, J., and Bader, H. (1979) Ozonation of Water: "Oxidation-Competition Values" for OH Radicals of Different Types of Waters used in Switzerland, *Ozone: Sci. Eng.* **1**, 357-372.

Hoigné, J., and Bader, H. (1994) Kinetics of Reactions of Chlorine Dioxide in Water—I: Rate Constants for Inorganic and Organic Compounds, *Water Res.* **28**, 45-55.

Hoigné, J., Bader, H., Haag, W. R., and Staehelin, J. (1985) Rate Constants of Reactions of Ozone with Organic and Inorganic Compounds in Water—III: Inorganic Compounds and Radicals, *Water Res.* **19**, 993-1004.

Hoigné, J., Faust, B. C., Haag, W., Scully, F., and Zepp, R. (1989) Aquatic Humic Substances as Sources and Sinks of Photochemically Produced Transient Reactants. In *Aquatic Humic Substances*, I. H. Suffet and P. McCarthy, Eds., Adv. Chem. Ser. No. 219, American Chemical Society, Washington, DC, pp. 363-381.

Hoigné, J., Zuo, Y., and Nowell, L. (1994) Photochemical Reactions in Atmospheric Waters; Role of Dissolved Iron Species. In *Aquatic and Surface Photochemistry*, G. Helz, R. R. Zepp, and D. G. Crosby, Eds., Lewis, Boca Raton, FL.

Holland, H. D. (1978) *The Chemistry of the Atmosphere and Oceans*, Wiley-Interscience, New York.

Honeyman, B. D., and Santschi, P. H. (1992) The Role of Particles and Colloids in the Transport of Radionuclides and Trace Metals in the Ocean. In *Environmental Particles*, Vol. 1, J. Buffle and H. P. van Leeuwen, Eds., Lewis, Boca Raton, FL.

Horne, R. A. (1969) *Marine Chemistry*, Wiley, New York.

Houghton, J. T., et al. (Eds.) (1983) *Intergovernmental Panel on Climate Change (IPCC)*, Cambridge University Press, Cambridge.

Huang, C. P., and Rhoads, E. A. (1989) Adsorption of Zn(II) onto Hydrous Aluminosilicates, *J. Colloid Interface Sci.* **131**, 289-306.

Hudson, R. J. M., Covault, D. T., and Morel, F. M. M. (1992) Investigations of Iron Coordination and Redox Reactions in Seawater Using ^{59}Fe Radiometry and Ion-Pair Solvent Extraction of Amphiphilic Iron Complexes, *Mar. Chem.* **38**, 209-235.

Hudson, R. J. M., and Morel, F. M. M. (1990) Iron Transport in Marine Phytoplankton: Kinetics of Cellular and Medium Coordination Reactions, *Limnol. Oceanogr.* **35**, 1002-1020.

Hug, S. J., and Sulzberger, B. (1994) In Situ Fourier Transform Infrared Spectroscopic Evidence for the Formation of Several Different Surface Complexes of Oxalate on TiO_2 in the Aqueous Phase, *Langmuir* **10**, 3587-3597.

Hunt, J. R. (1980) In *Particulates in Water*, M. C. Kavanaugh and J. O. Leckie, Eds., Adv. Chem. Ser. No. 189, American Chemical Society, Washington, DC.

Hunter, R. J. (1989) *Foundation of Colloid Science*, Vol. 1, Clarendon Press, Oxford.

Imboden, D. M., and Lerman, A. (1979) In *Lakes*, A. Lerman, Ed., Springer-Verlag, New York, pp. 341–356.

Imboden, D. M., and Schwarzenbach, R. P. (1985) Spatial and Temporal Distribution of Chemical Substances in Lakes: Modeling Concepts, in W. Stumm, Ed., *Chemical Processes in Lakes*, Wiley-Interscience, New York, pp. 1–30.

Ingle, S. E. (1975) Solubility of Calcite in the Ocean, *Mar. Chem.* **3**, 301–319.

Ingri, N., Kakolowicz, W., Sillén, L. G., and Warnquist, B. (1967) *Talanta* **14**, 1261.

Irgolic, K. J., and Martell, A. E. (1985) In *Environmental Inorganic Chemistry*, K. J. Irgolic and A. E. Martell, Eds., VCH Verlag, Weinheim.

Irwing, H., and Williams, R. J. P. (1953) *J. Chem. Soc.* 3192.

Isaac, R. A., and Morris, J. C. (1983) In *Water Chlorination*, Vol. 4, R. L. Jolley et al., Eds., Ann Arbor Science, Ann Arbor, MI.

Israelachvili, J. N. (1991) *Intermolecular and Surface Forces*, 2nd ed., Academic Press, New York.

Israelachvili, J. N., and Adams, G. E. (1978) Measurement of Forces Between Two Mica Surfaces in Aqueous Electrolyte Solutions in the Range 0–100 nm, *J. Chem. Soc. Faraday Trans. 1* **74**(4), 975–1001.

Jackson, G. A., and Lochmann, S. (1993) Modeling Coagulation of Algae in Marine Ecosystems. In *Environmental Particles*, Vol. 2, J. Buffle and H. P. van Leeuwen, Eds., Lewis, Boca Raton, FL.

Jackson, G. A. and Morgan, J. J. (1978) *Limnol. and Oceanogr.* **23**, 268.

Jacob, D. J., Munger, J. W., Waldman, J. M., and Hoffmann, M. R. (1986) The H_2SO_4-HNO_3-NH_3 System at High Humidities and in Fogs, *J. Geophys. Res.* **91**(D1), 1073–1088, 1089–1096.

Jacobs, L. A., von Gunten, H. R., Keil, R., and Kulsys, M. (1988) Geochemical Changes Along a River–Groundwater Infiltration Flow Path: Glattfelden, Switzerland, *Geochim. Cosmochim. Acta* **52**, 2693–2706.

Jannasch, H. W., and Mottl, M. J. (1985) Geomicrobiology of Deep Sea Hydrothermal Vents, *Science* **229**, 717–725.

Jensen, W. B. (1980) *The Lewis Acid–Base Concepts*, Wiley-Interscience, New York.

Johnson, C. A. (1986) The Regulation of Trace Element Concentrations in Rivers and Estuarine Waters Contaminated with Acid Mine Drainage: The Adsorption of Cu and Zn on Amorphous Fe Oxyhydroxides, *Geochim. Cosmochim. Acta* **50**, 2433–2438.

Johnson, C. A., and Sigg, L. (1983) Acidity of Rain and Fog: Conceptual Definitions and Practical Measurements of Acidity, *Chimia* **39**, 55–61.

Johnston, C. T., Sposito, G., and Earl, W. L. (1993) Surface Spectroscopy of Environmental Particles by Fourier-Transform Infrared and Nuclear Magnetic Resonance Spectroscope. In *Environmental Particles*, Vol. II, J. Buffle and H. P. van Leeuwen, Eds., Lewis, Chelsea, MI.

Jolley, R. L., and Carpenter, J. H. (1983) In *Water Chlorination*, Vol. 4, R. L. Jolley et al., Eds., Ann Arbor Science, Ann Arbor, MI.

Kămări, J., Amann, M., Brodin, Y. W., Chadwick, M. J., Henriksen, A., Hettlingh, J. P., Kuylenstierna, J., Posch, M., and Sverdrup, H. (1992) *Ambio* **21**, 377-387.

Kester, D. R., and Pytkowicz, R. M. (1968) *Limnol. Oceanogr.* **13**, 670.

Kester, D. R., and Pytkowicz, R. M. (1969) *Limnol. Oceanogr.* **14**, 586.

Kester, D. R., Byrne, R. H., and Liang, Y. J. (1975) ACS Symposium Series No. 18, Washington D.C.

Kiefer, E. (1994) Ph.D. Thesis No. 10786, Swiss Federal Institute of Technology, Zurich, Switzerland.

Kielland, J. (1937) *J. Am. Chem. Soc.* **59**, 1675.

Kim, J. I. (1991) Actinide Colloid Generation in Groundwater, *Radiochim. Acta* **52-53**, 71-81.

King, E. J. (1965) *Acid-Base Equilibria*, Macmillan, New York.

King, E. L. (1964) *How Chemical Reactions Occur*, Benjamin, Menlo Park, CA.

Klotz, I. M. (1964) *Chemical Thermodynamics: Basic Theory and Methods*, rev. ed., Benjamin, Menlo Park, CA.

Kohler, H. P., et al. (1984) *Microbiol. Lett.* **21**, 279.

Koster, D. R., Byrne, R. H., and Liang, Y. J. (1975) In *Marine Chemistry in the Coastal Environment*, T. M. Church, Ed., Am. Chem. Soc. Symp. Ser. 18, American Chemical Society, Washington DC.

Kotronarou, A., and Sigg, L. (1993) SO_2 Oxidation in Atmospheric Water, Role of Fe(II) and Effect of Ligands, *Environ. Sci. Technol.* **27**, 2725.

Kuhn, A., Johnson, C. A., and Sigg, L. (1993) *Cycles of Trace Elements in a Lake*, Adv. Chem. Ser. No. 237, American Chemical Society, Washington, DC, pp. 473-495.

Kummert, R., and Stumm, W. (1980) The Surface Complexation of Organic Acids on Hydrous γ-Al_2O_3, *J. Colloid Interface Sci.* **75**, 373-385.

Kunz, B., and Stumm, W. (1984) Kinetik der Bildung und des Wachstums von Calcium-Carbonat, *Vom Wasser* **62**, 279-293.

Lagerstrom, G. (1959) *Acta Chem. Scand.* **13**, 722.

Laidler, K. J. (1965) *Chemical Kinetics*, 2nd ed., McGraw-Hill, New York.

Laidler, K. J. (1987) *Chemical Kinetics*, Harper, New York.

Landing, W. M., and Lewis, B. (1991) Thermodynamic Modelling of Trace Metal Speciation in the Black Sea. In *Black Sea Oceanography*, E. Izdar and J. W. Rurray, Eds., Kluwer Academic, Dordrecht.

Langmuir, D., and Whittemore, D. O. (1971) *Adv. Chem. Ser.* **106**, 209-234.

Larson, R. A., and Weber, E. J. (1994) *Reaction Mechanisms in Environmental Organic Chemistry*, Lewis, Boca Raton, FL.

Lasaga, A. C. (1981) Rate Laws of Chemical Reactions. In *Kinetics of Geochemical Processes*, A. C. Lasaga and R. J. Kirkpatrick, Eds., Society of American Reviews in Mineralogy 8, Mineralogical Society, Washington, DC, pp. 1-68.

Lasaga, A. C. (1983) Rate Laws of Chemical Reactions, Chap. 1, and Transition State Theory, Chap. 4. In *Reviews in Mineralogy 8, Kinetics of Geochemical Processes*, Mineralogical Society, Washington, DC.

Lasaga, A. C. (1990) Atomic Treatment of Mineral–Water Surface Reactions. In *Mineral–Water Interface Geochemistry*, M. Hochella and A. F. White, Eds., *Reviews in Mineralogy* 23, Mineralogical Society, Washington, DC, pp. 17–80.

Lasaga, A. C., Soler, J. M., Ganor, J., Burch, T. E., and Nagy, K. L. (1994) Chemical Weathering Rate Laws and Global Geochemical Cycles, *Geochim. Cosmochim. Acta* **58**, 2361–2386.

Latimer, W. M. (1952) *The Oxidation States of the Elements and Their Potentials in Aqueous Solutions*, 2nd ed., Prentice-Hall, Englewood Cliffs, NJ.

Lawler, D. G., O'Melia, C. R., and Tobiason, J. E. (1980) Integral Water Treatment Design. In *Particulates in Water*, M. C. Kavanaugh and J. O. Leckie, Eds. Adv. Chem. Ser. No. 189, American Chemical Society, Washington, DC.

Lee, G. F. (1967) In *Principles and Applications of Water Chemistry*, S. D. Faust and J. V. Hunter, Eds., Wiley-Interscience, New York.

Lee, T. S. (1959) In *Treatise on Analytical Chemistry*, Part I, Vol. 1, I. M. Kolthoff and P. J. Elving, Eds., Interscience, New York, pp. 185–275.

Lelieveld, J. (1994) Modeling of Heterogeneous Chemistry in the Global Environment. In *Chemistry of Aquatic Systems; Local and Global Perspectives*, G. Bidoglio and W. Stumm, Eds., Kluwer Academic, Dordrecht.

Lelieveld, J., and Crutzen, P. J. (1991) The Role of Clouds in Tropospheric Photochemistry, *J. Atmos. Chem.* **12,** 229.

Leppard, G. G. (1992) Evaluation of Electron Microscopic Techniques for the Description of Aquatic Colloids. In *Environmental Particles*, Vol. 1, J. Buffle and H. P. van Leeuwen, Eds., Lewis, Boca Raton, FL.

Lerman, A. (1979) *Geochemical Processes—Water and Sediment Environments*, Wiley-Interscience, New York.

Lewis, G. N., and Randall, M. (1923) *Thermodynamics and the Free Energy of Chemical Substances*, McGraw-Hill, New York.

Lewis, G. N., and Randall, M. (1961) *Thermodynamics*, 2nd ed., revised by K. S. Pitzer and L. Brewer, McGraw-Hill, New York.

Leyendekkers, J. V. (1975) *Mar. Chem.* **1,** 75.

Li, Y. H. (1977) *Geochim. Cosmochim. Acta* **41,** 555.

Liang, L., and Morgan, J. J. (1990) Chemical Aspects of Iron Oxide Coagulation in Water: Laboratory Studies and Implications for Natural Systems, *Aquat. Sci.* **52**(1), 32–55.

Lindberg, S. E., and Runnells, D. D. (1984) Ground Water Redox Reactions, *Science*, **225,** 925–927.

Lindberg, S. E., Lovett, G. M., Richter, D. D., and Johnson, D. W. (1986) Atmospheric Deposition and Canopy Interactions of Major Ions in a Forest, *Science* **231,** 141–145.

Lippmann, F. (1991) Aqueous Solubility of Magnesian Calcites with Different Endmembers, *Acta Mineral. Petrograph. Szeged.* **32,** 5–19.

Liss, P. S., and Slater, P. G. (1974) *Nature* **247,** 181.

Lotka, J. A. (1924) *Elements of Mathematical Biology*, reprinted by Dover, New York, 1956.

Lovelock, J. E., and Maggs, R. J. (1993) Halogenated Hydrocarbons in and over the Atlantic, *Nature* **241**, 194–196.

Lovely, D. R., Woodward, J. C., and Chapelle, F. H. (1994) Stimulated Anoxic Biodegradation of Aromatic Hydrocarbons Using Fe(III) Ligands, *Nature* **370**, 128–131.

Lowenstamm, H. A., and Weiner, S. (1985) *On Biomineralization*, Oxford University Press, New York.

Luther, G. W. III (1990) The Frontier-Molecular-Orbital Theory Approach in Geochemical Processes. In *Aquatic Chemical Kinetics*, W. Stumm, Ed., Wiley-Interscience, New York, pp. 173–198.

Luther, G. W. III, Church, T. M., Kostka, J. E., Sulzberger, B., and Stumm, W. (1992) Seasonal Iron Cycling in the Marine Environment: The Importance of Ligand Complexes with Fe(II) and Fe(III) in the Dissolution of Fe(III) Minerals and Pyrite, Respectively, *Mar. Chem.* **40**, 81–103.

Lyklema, J. (1978) Surface Chemistry of Colloids in Connection with Stability. In *The Scientific Basis of Flocculation*, K. J. Ives, Ed., Sijthoff and Noordhoff, The Netherlands, pp. 3–36.

Lyklema, J. (1985) How Polymers Adsorb and Affect Colloid Stability, Flocculation, Sedimentation, and Consolidation. In *Flocculation, Sedimentation and Consolidation: Proceedings of the Engineering Foundation Conference, Sea Island, Georgia*, B. M. Moudgil and P. Somasundaran, Eds., United Engineering Trustees, Inc., pp. 3–21.

Lyklema, J. (1991) *Fundamentals of Interface and Colloid Science*, Vol. 1, Chap. 5, Academic Press, London.

Lyman, W. J., Reehl, W. F., and Rosenblatt, D. H. (Eds.) (1982) *Handbook of Chemical Property Estimation Methods; Environmental Behavior of Organic Compounds*, McGraw-Hill, New York.

Lyman, W. J., Reehl, W. F., and Rosenblatt, D. H. (Eds.) (1990) *Handbook of Technical Property Estimation Methods*, American Chemical Society, Washington, DC.

Mabey, W., and Mill, T. (1978) Critical Review of Hydrolysis of Organic Compounds in Water Under Environmental Conditions. *J. Phys. Ref. Data* **7**(2), 383–415.

Macalady, D. L., Tratnyek, P. G., and Grundl, T. J. (1986) Abiotic Reactions of Anthropogenic Chemicals in Anaerobic Systems: A Critical Review, *J. Contam. Hydrol.* **1**, 1–28.

MacDowell-Boyer, L. M. (1992) Chemical Mobilization of Micron-Sized Particles in Saturated Porous Media Under Steady Flow Conditions, *Environ. Sci. Technol.* **26**, 586–593.

MacInnes, D. A. (1919) *J. Chem. Soc.* **41**, 1068.

Mackay, D. (1991) *Multimedia Environmental Models; the Fugacity Approach*, Lewis, Chelsea, MI.

Mackay, D., and Leinonen, P. J. (1975) *Env. Sci. Technol.* **9**, 1178.

Mackay, D., and Shiu, W. Y. (1981) A Critical Review of Henry's Law Constants for Chemicals of Enviromental Interest, *J. Phys. Chem. Ref. Data* **10,** 1175-1199.

Maity, N., and Payne, G. F. (1991) Adsorption from Aqueous Solutions Based on a Combination of Hydrogen Bonding and Hydrophobic Interaction, *Langmuir* **7,** 1247-1254.

Mangelsdorf, P. G. Jr., and Wilson, T. R. S. (1971) *J. Phys. Chem.* **75,** 1418.

Mann, S. (1988) Molecular Recognition in Biomineralization, *Nature* **332,** 119-124.

Mann, S., Webb, J., and Williams, R. J. P. (1989) *Biomineralization*, VCH Verlag, Weinheim.

Marcus, R. A. (1965) On the Theory of Electron-Transfer Reactions. VI. Unified Treatment for Homogeneous and Electrode Reactions, *J. Chem. Phys.* **43,** 679-701.

Marcus, R. A. (1975) Electron Transfer in Homogeneous and Heterogeneous Systems. In *The Nature of Seawater*, E. D. Goldberg, Ed., Dahlem Konferenzen, Berlin, pp. 477-503.

Margerum, D. W., Cayley, G. R., Weatherburn, D. C., and Pagenkopf, G. K. (1978) In *Coordination Chemistry*, Vol. 2, A. E. Martell, Ed., American Chemical Society, Washington, DC.

Martell, A. E., and Smith, R. M. (1974-1989) *Critical Stability Constants*, Vols. 1-6, Plenum Press, New York.

Martin, J. H., Bruland, K. W., and Broenkow, W. W. (1976) Cadmium Transport in the California Current, in Windom, H. and Duce, R., Eds., *Marine Pollutant* Transfer, *Heath*, Toronto.

Martin, J. H., and Fitzwater, S. E. (1988) Iron Deficiency Limits Phytoplankton Growth in the North-east Pacific Subarctic, *Nature* **331,** 341-343.

Martin, J. H., Gordon, R. M., and Fitzwater, S. E. (1991) The Case for Iron, *Limnol. Oceanogr.* **36,** 1793-1802.

Martin, J. H., Gordon, R. M., Fitzwater, S., and Broenkow, W. W. (1989) *Deep-Sea Res.* **36,** 649.

Martin, J. H., and Knauer, G. A. (1973) *GeochimiCosmochim. Acta*, **37,** 1639.

Martin, J. M., and Meybeck, M. (1979) Elemental Mass-Balance of Material Carried by Major World Rivers, *Mar. Chem.* **7,** 173.

Martin, J. M., and Whitfield, M. (1983) The Significance of the River Input of Chemical Elements to the Ocean. In *Trace Metals in Sea Water*, C. S. Wong, E. Boyle, K. W. Bruland, J. D. Burton, and E. D. Goldberg, Eds., Plenum Press, New York.

Martin, R. E., Bouwer, E. J., and Hanna, L. M. (1992) *Environ. Sci. Technol.* **26,** 1053-1058.

Mason, B. (1966) *Principles of Geochemistry*, Wiley, New York.

Mason, R. P., Fitzgerald, W. F., Hurley, J., Hanson, A. K. Jr., Donaghay, P. L., and Sieburth, J. M. (1993) Mercury Biogeochemical Cycling in a Stratified Estuary, *Limnol. Oceanogr.* **38,** 1227-1241.

Mason, R. P., Fitzgerald, W. F., and Morel, F. M. M. (1994) *Geochim. Cosmochim. Acta* **58,** 3191-3198.

Mast, M. A., and Drever, J. I. (1987) The Effect of Oxalate on the Dissolution Rates of Oligoclase and Tremolite, *Geochim. Cosmochim. Acta* **51**(9), 2559-2568.

McCarthy, J. F., and Degueldre, C. (1993) Sampling and Characterization of Colloids and Particles, in *Environmental Particles*, Vol. II, J. Buffle and H. P. van Leeuwen, Eds., Lewis, Chelsea, MI.

McCarthy, J. F., and Zachara, J. M. (1989) Subsurface Transport of Contaminants, *Environ. Sci. Technol.* **23,** 496–502.

McCarty, P. L., Reinhard, M., and Rittmann, B. E. (1981) Trace Organics in Groundwater, *Environ. Sci. Technol.* **15**(1), 40–51.

McDuff, R. E., and Morel, F. M. M. (1980) *Environ. Sci. Technol.* **14,** 1182.

McKenzie, J. A. (1985) Carbon Isotopes and Productivity in the Lacustrine and Marine Environment. In *Chemical Processes in Lakes*, W. Stumm, Ed., Wiley-Interscience, New York.

McKnight, D. M., Kimball, B. A., and Bencala, K. E. (1988) Iron Photoreduction and Oxidation in an Acidic Mountain Stream, *Science* **240,** 637–640.

McKnight, D. M., and Morel, F. M. M. (1979) Release of Weak and Strong Copper-Complexing Agents by Algae, *Limnol. Oceanogr.* **24,** 823–837.

McQuarrie, D. A. (1976) *Statistical Mechanics*, Harper Collins, New York.

Mehrbach, C., et al. (1973) *Limnol. Oceanogr.* **18,** 897.

Melchior, D. C., and Basset, R. L. (1990) *Chemical Modeling of Aqueous Systems*, American Chem. Soc. Symposium Series 416, Washington D.C.

Meybeck, M. (1979) Pathways of Major Elements from Land to Ocean Through Rivers. In *Review and Workshop on River Inputs to Ocean Systems*, FAO, Rome, pp. 18–30.

Mill, T., Mabey, W. R., Lan, B. Y., and Baraze, A. (1981) *Chemosphere* **10,** 1281.

Millero, F. J. (1969) *J. Limnol. Oceanogr.* **14,** 376.

Millero, F. J. (1979) The Thermodynamics of the Carbonic Acid System in Seawater, *Geochim. Cosmochim. Acta* **43,** 1651–1661.

Millero, F. J. (1981) The Ionization of Acids in Estuarine Waters, *Geochim. Cosmochim. Acta* **45,** 2085–2089.

Millero, F. J. (1983) Influence of Pressure on Chemical Processes in the Sea. In *Chemical Oceanography*, 2nd ed., J. P. Riley and R. Chester, Eds., Academic Press, New York.

Millero, F. J. (1984) The Activity of Metal Ions at High Ionic Strengths. In *Complexation of Trace Metals in Natural Waters*, C. J. M. Kramer and J. C. Duinker, Eds., Nijhoff/Dr. W. Junk, The Hague.

Millero, F. (1985) The Effect of Ionic Interactions on the Oxidation of Metals in Natural Waters, *Geochim. Cosmochim. Acta* **49,** 547–553.

Millero, F. J. (1992) Stability Constants for the Formation of Rare Earth Inorganic Complexes as a Function of Ionic Strength, *Geochim. Cosmochim. Acta* **56,** 3123–3132.

Millero, F. J. (1995) The Carbon Dioxide System in the Oceans, *Geochim. Cosmochim. Acta* **59**(4), 661–677.

Millero, F. J., Sotolongo, S., and Izaguirre, M. (1987) The Oxidation Kinetics of Fe(II) in Seawater, *Geochim. Cosmochim. Acta* **51,** 793–801.

Millero, F. J., Zhang, J.-Z., Fiol, S., Sototlongo, S., Roy, R. N., Lee, K., and Mane, S. (1993) The Use of Buffers to Measure the pH of Seawater, *Mar. Chem.* **44,** 143–152.

Millero, F. J., Zhang, J. Z., Lee, K., and Campbell, D. M. (1993) Titration Alkalinity of Seawater, *Mar. Chem.* **44,** 153–165.

Modi, H. J., and Fuerstenau, D. W. (1957) Streaming Potential Studies on Corundum in Aqueous Solutions of Inorganic Electrolytes, *J. Phys. Chem.* **61,** 640–643.

Moore, J. W., and Pearson, R. G. (1981) *Kinetics and Mechanism*, 3rd ed., Wiley-Interscience, New York.

Morel, F. M. M. (1983) *Principles of Aquatic Chemistry*, Wiley-Interscience, New York.

Morel, F. M. M., and Gschwend, P. M. (1987) The Role of Colloids in the Partitioning of Solutes in Natural Waters. In *Aquatic Surface Chemistry*, W. Stumm, Ed., Wiley-Interscience, New York, pp. 405–422.

Morel, F. M. M., and Hering, J. G. (1993) *Principles and Applications of Aquatic Chemistry*, Wiley-Interscience, New York, p. 374.

Morel, F. M. M. and Hudson, R. J. M. (1985) The Geobiological Cycle of Trace Elements in Aquatic Systems: Redfield Revisited. In *Chemical Processes in Lakes*, W. Stumm, Ed., Wiley-Interscience, New York.

Morel, F. M. M., Hudson, R. J. M., and Price, N. L. (1991) Limitations of Productivity by Trace Metals in the Sea, *Limnol. Oceanogr.* **36,** 1742–1755.

Morel, F. M. M., and Morgan, J. J. (1972) *Environ. Sci. Technol.* **6,** 58.

Morel, F. M. M., et al. (1994) Zinc and Carbon Co-limitation of Marine Phytoplankton *Nature* **369,** 740–742.

Morgan, J. J. (1967) In *Principles and Applications of Water Chemistry*, S. D. Faust and J. V. Hunter, Eds., Wiley-Interscience, New York.

Morgan, J. J. (1982) Factors Affecting the pH, Availability of H^+, and Oxidation Capacity of Rain. In *Atmospheric Chemistry*, E. D. Goldberg, Ed., Springer-Verlag, Berlin.

Morgan, J. J., and Stone, A. T. (1985) Kinetics of Chemical Processes of Importance in Lacustrine Environments. In *Chemical Processes in Lakes*, W. Stumm, Ed., Wiley-Interscience, New York.

Morgan, J. J., and Stumm, W. (1964) *Proceedings of The Second Conference on Water Pollution Research*, Pergamon, Elmsford, NY.

Morowitz, H. J. (1968) *Energy Flow in Biology*, Academic, New York.

Morris, J. C., and Stumm, W. (1967) *Redox Equilibria and Measurements of Potentials in Aquatic Environment*, Adv. Chem. Ser. No. 67, American Chemical Society, Washington, DC, pp. 270–285.

Morse, J. W., and Mackenzie, F. J. (1990) *Geochemistry of Sedimentary Carbonates*, Elsevier, Amsterdam.

Moser, J., Punchiheva, S., Infelta, P. P., and Graetzel, M. (1991) Surface Complexation of Colloidal Semiconductors Strongly Enhances Interfacial Electron Transfer, *Langmuir* **7,** 3012.

Mucci, A. (1983) The Solubility of Calcite and Aragonite in Seawater at Various Salinities, Temperatures and One Atmosphere Total Pressure, *Am. J. Sci.* **283,** 780–799.

Müller, B., and Sigg, L. (1990) Interaction of Trace Metals with Natural Particle Surfaces: Comparison Between Adsorption Experiments and Field Measurements, *Aquat. Sci.* **52**(1), 75–92.

Muller, F. L. L., and Kester, D. R. (1990) Kinetic Approach to Trace Metal Complexation in Seawater. Application to Zinc and Cadmium, *Environ. Sci. Technol.* **24**, 234-242.

Muller, F. L. L., and Kester, D. R. (1991) Voltammetric Determination of the Complexation Parameters of Zinc in Marine and Estuarine Waters, *Mar. Chem.* **33**, 71-90.

Nancollas, G. H. (1989) In Vitro Studies of Calcium Phosphate Crystallization, in *Biomineralization*, S. Mann et al. Eds., VCH-Verlag, Weinheim.

Nancollas, G. H., and Reddy, M. M. (1974) *Soc. Petrol. Eng. J.* **14**/2, 117.

Nancollas, G. H., and Reddy, M. M. (1974) Crystal Growth Kinetics of Minerals Encountered in Water Treatment Processes, *Aqueous Environ. Chem. Met.* 219-253.

Neihof, R. A., and Loeb, G. I. (1972) The Surface Charge of Particulate Matter in Seawater, *Limnol. Oceanogr.* **17**, 7-16.

Neilands, J. B. (1982) *Annu. Rev. Microbiol.* **36**, 285.

Nemethy, G., and Scheraga, H. A. (1962) *J. of Chemi. Physi.*, **36**, 3382.

Nieboer, E., and Richardson, D. H. S. (1980) The Replacement of the Nondescriptive Term Heavy Metals by a Biologically and Chemically Significant Classification of Metal Ions, *Environ. Pollut. Ser. B* **1**, 3-26.

Nielsen, A. E. (1964) *Kinetics of Precipitation*, Pergamon Press, Oxford.

Nielsen, A. E. (1981) Theory of Electrolyte Crystal Growth. The Parabolic Rate Law, *Pure Appl. Chem.* **53**, 2025-2039.

Nielsen, A. E. (1986) Mechanisms and Rate Laws in Electrolyte Crystal Growth from Aqueous Solution. In *Geochemical Processes at Mineral Surfaces*, J. A. Davis and K. F. Hayes, Eds., Am. Chem. Soc. Symp. Ser. 323, American Chemical Society, Washington, DC, pp. 600-614.

Nordstrom, D. K., Plummer, L. N., Langmuir, D., Busenberg, E., May, H. M., Jones, B. F., and Parkhurst, D. L. (1990) Revised Chemical Equilibrium Data for Major Water-Mineral Reactions and Their Limitations. In *Chemical Modeling of Aqueous Systems II*, D. C. Melchior and R. L. Bassett, Eds., ACS Ser. 416, American Chemical Society, Washington, DC, pp. 398-413.

Nriagu, J. O. (1986) In *The Role of the Oceans as a Waste Disposal Option*, G. Kullenberg, Ed., Reidel, Dordrecht, pp. 441-468.

Nriagu, J. O. (1989) A Global Assessment of Natural Sources of Atmospheric Trace Metals, *Nature* **338**, 47-49.

Nriagu, J. O., Kempe, A. W., Wong, H. K. T., and Harper, N. (1979) *Geochim. Cosmochim. Acta* **43**, 247-258.

Nriagu, J. O., and Pacyna, J. M. (1988) Quantitative Assessment of Worldwide Contamination of Air, Water and Soils with Trace Metals, *Nature* **333**, 134-139.

Nriagu, J. O., and Wong, H. K. (1989) *Sci. Total Environ.* **87/88**, 315-328.

O'Connor, T. P. and Kester, D. R. (1975) Adsorption of Copper and Cobalt from Freshwater and Marine Systems, *Geochim. Cosmochim. Acta* **39**, 1531.

Ochial, E.-I. (1991) Biomineralization (Part V), *J. Chem. Educ.* **68**, 827-830.

Ochs, M. (1991) *Humic Substances at Aquatic Interfaces: A Comparison of Hydro-*

phobic vs Coordinative Adsorption and Subsequent Effects on Mineral Weathering, Ph.D. Thesis, ETH Zurich, Switzerland.

Ochs, M., Ćosović, B., and Stumm, W. (1994) Coordinative and Hydrophobic Interaction of Humic Substances with Hydrophilic Al_2O_3 and Hydrophobic Mercury Surfaces, *Geochim. Cosmochim. Acta* **58**(2), 639–650.

Odum, E. P. (1969) *Science* **164**, 262.

Öhman, L. O., and Sjöberg, S. (1988) Thermodynamic Calculations with Special Reference to the Aqueous Aluminum System. In *Metal Speciation: Theory, Analysis and Application*, J. R. Kramer and H. E. Allen, Eds., Lewis, Chelsea, MI.

O'Melia, C. R. (1972) Coagulation and Flocculation. In *Processes for Water Quality Control*, W. J. Weber, Jr., Ed., Wiley-Interscience, New York, pp. 61–110.

O'Melia, C. R. (1987) Particle-Particle Interactions. In *Aquatic Surface Chemistry*, W. Stumm, Ed., Wiley-Interscience, New York, pp. 385–403.

O'Melia, C. R. (1989) Particle–Particle Interactions in Aquatic Systems, *Colloids Surf.* **39**, 255–271.

O'Melia, C. R. (1990) Kinetics of Colloid Chemical Processes in Aquatic Systems. In *Aquatic Chemical Kinetics, Reaction Rates of Processes in Natural Waters*, W. Stumm, Ed., Wiley-Interscience, New York, pp. 447–474.

O'Melia, C. R. (1991) personal communication.

O'Melia, C. R. (1994) From Algae to Aquifers: Solid–Liquid Separation in Aquatic Systems. In *Aquatic Chemistry*, C. P. Huang et al., Eds., American Chemical Society, Washington, DC.

O'Melia, C. R., and Stumm, W. (1967) *J. Colloid Interfac. Science*, **23**, 437.

O'Melia, C. R., and Tiller, C. L. (1993) Physicochemical Aggregation and Deposition in Aquatic Environments. In *Environmental Particles*, Vol. 2, J. Buffle and H. P. van Leeuwen, Eds., Lewis, Boca Raton, FL.

Overbeek, J. Th. G. (1976) Polyelectrolytes, Past, Present and Future, *Pure Appl. Chem.* **46**, 91–101.

Owen, B. B., and Brinkley, S. R. Jr. (1941) *Chem. Rev.* **29**, 461.

Pabalan, R., and Pitzer, K. S. (1988) Apparent Molar Heat Capacity and Other Thermodynamic Properties of Aqueous Potassium Chloride Solutions to High Temperatures and Pressures, *J. Chem. Eng. Data* **33**, 354–362.

Pabalan, R. T., and Pitzer, K. S. (1990) Models for Aqueous Electrolyte Mixtures for Systems Extending from Dilute Solutions to Fused Salts. In *Chemical Modeling of Aqueous Systems*, Vol. II, R. L. Basset and D. C. Melchior, Eds., American Chemical Society, Washington, DC.

Paces, T. (1983) Rate Constants of Dissolution Derived from the Measurements of Mass Balance in Hydrological Catchments, *Geochim. Cosmochim. Acta* **47**, 1855–1863.

Paces, T. (1994) Modeling the Hydrological and Biogeochemical Response of a Catchment Area to Anthropogenic Effects. In *Chemistry of Aquatic Systems, Local and Global Perspectives*, G. Bidoglio and W. Stumm, Eds., Kluwer, Dordrecht.

Palmer, D. A., and van Eldik, R. (1983) The Chemistry of Metal Carbonato and Carbon Dioxide Complexes, *Chem. Rev.* **83**, 651–731.

Pankow, J. F. (1991) *Aquatic Chemistry Concepts*, Lewis, Chelsea, MI.

Pankow, J. F., and Morgan, J. J. (1979) *Environ. Sci. Technol.* **13,** 1248.

Pankow, J. F., and Morgan, J. J. (1981) Kinetics for the Aquatic Environment, *Environ. Sci. Technol.* **15,** 1155–1164, 1306–1313.

Parkhurst, D. L., Thorstenson, D. C., and Plummer, L. N. (1990) PHREEQE—A Computer Program for Geochemical Calculations, *US Geol. Surv. Water Res. Inv.*, 80–96 (1990 revision).

Parks, G. A. (1967) Aqueous Surface Chemistry of Oxides and Complex Oxide Minerals; Isoelectric Point and Zero Point of Charge. In *Equilibrium Concepts in Natural Water Systems*, Adv. Chem. Ser. No. 67, American Chemical Society, Washington, DC.

Parks, G. A. (1984) Surface and Interfacial Free Energies of Quartz, *J. Geophys. Res.* **89,** 3997–4008.

Parks, G. A. (1990), Surface Energy and Adsorption at Mineral/Water Interfaces: An Introduction. In *Mineral-Water Interface Geochemistry*, M. F. Hochella, Jr. and A. F. White, Eds., Mineralogical Society of America, Washington, DC, pp. 133–175.

Parson, R. (1985) Standard Electrode Potentials: Units, Conventions and Methods of Determination, in *Standard Potentials in Aqueous Solutions*, A. J. Bard et al., Eds., M. Dekker, New York.

Pearson, R. G. (1963) *J. Am. Chem. Soc.* **85,** 3533.

Pennington, D. A. (1978) In *Coordination Chemistry*, Vol. 2, A. E. Martell, Ed., American Chemical Society, Washington, DC.

Perdue, E. M., and Lytle, C. R. (1983) Distribution Model for Binding of Protons and Metal Ions by Humic Substances, *Environ. Sci. Technol.* **17,** 654–660.

Pitzer, K. S. (1987) A Thermodynamic Model for Aqueous Solutions of Liquid-like Density. In *Thermodynamic Modeling of Geological Materials: Minerals, Fluids and Melts*, I. S. E. Carmichael and H. P. Eugster, Eds., *Reviews in Mineralogy* **17,** American Mineralogy Society, Washington, DC, pp. 97–142.

Pitzer, K. S., and Brewer, L. (1961) In *Thermodynamics*, G. N. Lewis and N. Randall, Eds., McGraw-Hill, New York.

Platford, R. F. (1965) *J. Mar. Res.* **23,** 55.

Platford, R. F., and Dafoe, T. (1965) *J. Mar. Res.* **23,** 68.

Plummer, L. N., Busby, J. F., Lee, R. W., and Hanshaw, B. B. (1990) Geochemical Modelling in the Madison Aquifer, *Water Resour. Res.* **26,** 1981–2014.

Plummer, L. N., and Mackenzie, F. T. (1974) Predicting Mineral Solubility from Rate Data: Application to the Dissolution of Magnesian Calcites, *Am. J. Sci.* **274,** 61–83.

Plummer, L. N., Wigley, T. M. L., and Parkhurst, D. L. (1978) The Kinetics of Calcite Dissolution in CO_2–Water Systems at 5° to 60°C and 0.0 to 1.0 atm CO_2, *Am. J. Sci.* **278,** 179–216.

Postma, D., Boesen, C., Kristiansen, H., and Larsen, F. (1991) Nitrate Reduction in an Unconfined Sandy Aquifer: Water Chemistry, Reduction Processes, and Geochemical Modeling, *Water Resourc. Res.* **27**(8), 2027–2045.

Pou, T. E., Murphy, P. J., Young, V., and Bockris, J. O. M. (1984) Passive Films on Iron: The Mechanism of Breakdown in Chloride Containing Solutions, *J. Electrochem. Soc.* **131,** 1243.

Prausnitz, J. M. (1969) *Molecular Thermodynamics of Fluid-Phase Equilibria*, Prentice-Hall, Englewood Cliffs, NJ.

Price, N. M., and Morel, F. M. M. (1990) Cadmium and Cobalt Substitution for Zinc in a Marine Diatom, *Nature* **344**, 658–660.

Prigogine, I. (1961) *Thermodynamics of Irreversible Processes*, 2nd ed., Wiley-Interscience, New York.

Pytkowicz, R. M. (Ed.) (1979) *Activity Coefficients in Electrolyte Solutions*, Vols. I and II, CRC Press, Boca Raton, FL.

Pytkowicz, R. M., and Hawley, J. E. (1974) *Limnol. Oceanogr.* **19**, 223.

Pytkowicz, R. M., and Kester, D. R. (1969) *Am. J. Sci.* **267**, 217.

Pytkowicz, R. N. (1969) *Geochem. J.*, **3**, 184.

Radke, C. J., and Prausnitz, J. M. (1972) Thermodynamics of Multi-Solute Adsorption from Dilute Liquid Solutions, *Am. Inst. Chem. Eng. J.* **18**, 761–768.

Rankama, K., and Sahama, Th. G. (1950) *Geochemistry*, University of Chicago Press, Chicago.

Rapsomanikis, S. (1986) Methyltransfer Reactions. In *Organometallic Compounds in the Environment*, P. J. Craig, Ed., Longmans, Harlow.

Redfield, A. C. (1934) On the Proportions of Organic Derivatives in Seawater and Their Relation to the Composition of Plankton. In *James Johnson Memorial Volume*, R. J. Daniel, Ed., Liverpool University Press, Liverpool.

Redfield, A. C., Ketchum, B. H., and Richards, F. A. (1963) The Influence of Organisms on the Composition of Seawater. In *The Sea*, Vol. 2, M. N. Hill, Ed., Wiley-Interscience, New York.

Reiners, W. A. (1986) Complementary Models for Ecosystems, *Am. Nat.* **127**, 59–73.

Reuss, J. D., and Johnson, D. W. (1986) *Acid Deposition and the Acidification of Soils and Waters*, Springer-Verlag, New York.

Rich, H. N., and Morel, F. M. M. (1990) Availability of Well-Defined Iron Colloids to the Marine Diatom Thalassiosira-Weissflogii, *Limnol. Oceanogr.*, **35**, 652–662.

Richards, F. A. (1965) In *Chemical Oceanogr.*, J. P. Riley and G. Skirnow, Eds., Academic, New York.

Riesen, W., Gamsjäger, H., and Schindler, P. W. (1977) *Geochim. Cosmochim. Acta* **41**, 1193.

Rimstidt, J. D., and Barnes, H. L. (1984) The Kinetics of Silica–Water Reactions, *Geochim. Cosmochim. Acta* **44**, 1683–1699.

Roberts, P. V., Goltz, M. N., Summers, R. S., Crittenden, J. C., and Nkedi-Kizza, P. (1987) The Influence of Mass Transfer on Solute Transport in Column Experiments with an Aggregated Soil, *J. Contam. Hydrol.* **1**, 375–93.

Robie, R. A., Hemingway, B. S., and Fisher, J. R. (1978) *Thermodynamic Properties of Minerals and Related Substances at 298.15 K and 1 Bar Pressure and at Higher Temperatures*, Geol. Surv. Bull. No. 1452, U.S. Government Printing Office, Washington, DC.

Robinson, R. A., and Stokes, R. H. (1959) *Electrolyte Solutions*, Butterworths, London.

Rock, P. A. (1967) *J. Chem. Educ.* **44,** 104.

Rönngren, L., Sjöberg, S., Sun, Z., Forsling, W., and Schindler, P. W. (1991) Surface Reactions in Aqueous Metal Sulfide Systems, *J. Coll. Interf. Sci.* **145,** 396–404.

Roy, R. N., Roy, L. N., Lawson, M., Vogel, K. M., Moore, C. P., Davis, W., and Millero, F. J. (1993a) Thermodynamics of the Dissociation of Boric Acid in Seawater at S = 35 from 0 to 55°C, *Mar. Chem.* **44,** 243-248.

Roy, R. N., Roy, L. N., Vogel, K. M., Moore, C. P., Pearson, T., Good, C. E., Millero, F. J., and Campbell, D. M. (1993b) Determination of the Ionization Constants of Carbonic Acid in Seawater, *Mar. Chem.* **44,** 249-258.

Ruzić, I. (1982) Theoretical Aspects of the Direct Titration of Natural Waters and Its Information Yield for Trace Metal Speciation, *Anal. Chim. Acta* **140,** 99–113.

Ruzić, I. (1984) Kinetics of Complexation and Determination of Complexation Parameters in Natural Waters. In *Complexation of Trace Metals in Natural Waters*, C. J. M. Kramer and J. C. Duinker, Eds., Nijhoff/Junk, The Hague.

Ruzić, I., and Nikolić, S. (1982) The Influence of Kinetics on the Direct Titration Curves of Natural Water Systems—Theoretical Considerations, *Anal. Chim. Acta* **140,** 131–147.

Ryan, J. N., and Gschwend, P. M. (1990) Colloid Mobilization in Two Atlantic Coastal Plain Aquifers: Field Studies, *Water Resourc. Res.* **26,** 307-322.

Sainte-Marie, J., Torna, A. E., and Gübeli, A. O. (1964) *Can. J. Chem.* **42,** 662.

Sanders, B. M., et al. (1983) Free Cupric Ion Activity in Seawater, *Science* **222,** 53.

Sarmiento, J. L. (1993) Ocean Carbon Cycle "Most of the Carbon Released from Fossil Fuels Will End up in the Oceans Where a Complex Cycle of Circulation and Other Processes Control Its Fate," *Chem. Eng. News* **72**(22) (May 31), 30-43.

Sawyer, D. T. (1991) *Oxygen Chemistry*, Oxford University Press, Oxford.

Sayles, F. L., and Mangelsdorf, P. C., Jr. (1977) *Geochim. Cosmochim. Acta* **41,** 951-960.

Sayles, F. L. (1979) The Composition and Diagenesis of Interstitial Solutions—Fluxes Across the Seawater-Sediment Interface, *Geochim. Cosmochim. Acta* **43,** 527-545.

Sayles, F. L., and Mangelsdorf, P. C. Jr. (1979) Cation-Exchange Characteristics of Amazon River Suspended Sediment and Its Reaction with Seawater, *Geochim. Cosmochim. Acta* **43,** 767-779.

Schaule, B. K., and Patterson, C. C. (1981) Lead Concentrations in the Northeast Pacific: Evidence for Global Anthropogenic Perturbations, *Earth Planet. Sci. Lett.* **54,** 97.

Schecher, W. D., and McAvoy, D. C. (1994) *MINEQL$^+$, A Chemical Equilibrium Program for Personal Computers*, Environmental Research Software, Hallowell, ME.

Scheutjens, J. M. H. M., and Fleer, G. J. (1980) Statistical Theory of the Adsorption of Interacting Chain Molecules. 2. Train, Loop, and Tail Size Distribution, *J. Phys. Chem.* **84**(2), 178–190.

Schindler, D. W. (1974) *Science* **184,** 897.

Schindler, D. W. (1977) *Science* **195,** 260.

Schindler, D. W., Mills, K. H., Mallay, D. F., Findlay, D. L., Shearer, J. A., Davies, I. J., Turner, M. A., Lindsay, G. A., and Cruikshand, D. R. (1985) Long-term Ecosystem Stress: The Effects of Experimental Acidification on a Small Lake, *Science* **228,** 1395-1401.

Schindler, P. W. (1967) Heterogeneous Equilibria Involving Oxides, Hydroxides, Carbonates and Hydroxide Carbonates. In *Equilibrium Concepts in Natural Water Systems*, Adv. Chem. Ser., No. 67, American Chemical Society, Washington DC, p. 196.

Schindler, P. W. (1984) Surface Complexation. In *Metal Ions in Biological Systems*, Vol. 18, H. Sigel, Ed., Marcel Dekker, New York.

Schindler, P. W. (1985) Grenzflächenchemie oxidischer Mineralien, *Österreichische Chem. Z.* **86**, 141-146.

Schindler, P. W. (1990) Co-Adsorption of Metal Ions and Organic Ligands: Formation of Ternary Surface Complexes. In *Mineral-Water Interface Geochemistry*, M. F. Hochella, Jr. and A. F. White, Eds., Mineralogical Society of America, Washington, DC, pp. 281-307.

Schindler, P. W., Liechti, P., and Westall, J. C. (1987) Adsorption of Copper, Cadmium and Lead from Aqueous Solutions to the Kaolinite-Water Interface, *Netherlands J. Agric. Sci.* **35**, 219.

Schindler, P. W., and Stumm, W. (1987) The Surface Chemistry of Oxides, Hydroxides and Oxide Minerals. In *Aquatic Surface Chemistry*, W. Stumm, Ed., Wiley-Interscience, New York.

Schlesinger, W. H. (1991) *Biogeochemistry, an Analysis of Global Change*, Academic Press, San Diego, pp. 40-71.

Schnoor, J. L. (1990) Kinetics of Chemical Weathering: A Comparison of Laboratory and Field Weathering Rates. In *Aquatic Chemical Kinetics*, W. Stumm, Ed., Wiley-Interscience, New York.

Schnoor, J. L., and Stumm, W. (1985) Acidification of Aquatic and Terrestrial Systems. In *Chemical Processes in Lakes*, W. Stumm, Ed., Wiley-Interscience, New York, pp. 311-338.

Schnoor, J. L., and Stumm, W. (1986) The Role of Chemical Weathering in the Neutralization of Acidic Deposition, *Schweiz. Z. Hydrol.* **48**(2), 171-195.

Schott, J., and Berner, R. A. (1985) Dissolution Mechanism of Pyroxenes and Olivines During Weathering. In *The Chemistry of Weathering*, J. I. Drever, Ed., NATO ASI Ser. C 149, pp. 35-53.

Schwartz, S. E. (1984) Gas-Aqueous Reactions of Sulfur and Nitrogen Oxides in Liquid Water Clouds. In SO_2, *NO and* NO_2 *Oxidation Mechanisms*, J. G. Calvert, Ed., Butterworth, Boston.

Schwartz, S. E. (1988) Are Global Cloud Albedo and Climate Controlled by Marine Phytoplankton? *Nature* **336**, 441-445.

Schwarzenbach, G. (1961) In *Advances in Inorganic Chemistry and Radiochemistry*, Vol. 3, H. J. Emeleus and A. G. Sharpe, Eds., Academic Press, New York.

Schwarzenbach, R. P., Giger, W., Schaffner, C., and Wanner, O. (1985) Groundwater Contamination of Volatile Halogenated Alkanes: Abiotic Formation of Volatile Sulfur Compounds Under Anaerobic Conditions, *Environ. Sci. Technol.* **19**, 322-327.

Schwarzenbach, R. P., and Gschwend, P. M. (1990) Chemical Transformations of Organic Pollutants in the Aquatic Environment. In *Aquatic Chemical Kinetics*, W. Stumm, Ed., Wiley-Interscience, New York.

Schwarzenbach, R. P., Gschwend, Ph. M., and Imboden, D. (1993) *Environmental Organic Chemistry*, Wiley-Interscience, New York.

Schwarzenbach, R. P., and Westall, J. C. (1980) *Env. Sci. Technol.* **53**, 291.

Schwarzenbach, R. P., and Westall, J. C. (1981) Transport of Nonpolar Organic Compounds from Surface Water to Groundwater. Laboratory Sorption Studies, *Environ. Sci. Technol.* **15**, 1360–1367.

Schweich, D., Sardin, M., and Gaudet, J. P. (1983) Measurement of a Cation Exchange Isotherm from Elution Curves Obtained in a Soil Column: Preliminary Results, *Soil Sci. Soc. Am. J.* **47**, 32–37.

Scott, M. J., and Morgan, J. J. (1990) In *Chemical Modeling of Aqueous Systems II*, D. C. Melchior and R. L. Bassett, Eds., ACS No. 416, American Chemical Society, Washington, DC.

Scully, F. E., and Hoigné, J. (1987) Rate Constants for Reactions of Singlet Oxygen with Phenols and other Compounds in Water, *Chemosphere* **16**, 681–694.

Scully, J. C. (1990) *The Fundamentals of Corrosion*, 3rd ed. Pergamon Press, Oxford.

Sedlak, D. L., and Hoigné, J. (1993) The Role of Copper and Oxalate in the Redox Cycling of Iron in Atmospheric Waters, *Atmosph. Environ.* **27A**, 2173–2185.

Sedlak, D. L., and Hoigné, J. (1994) Oxidation of S(IV) in Atmospheric Water by Photo-oxidants and Iron in the Presence of Copper, *Environ. Sci. Technol.* **28**, 1898–1906.

Seinfeld, J. H. (1985) *Atmospheric Chemistry and Physics of Air Pollution*, Wiley-Interscience, New York.

Servos, J. N. (1990) *Physical Chemistry from Ostwald to Pauling*, Princeton University Press, Princeton, NJ.

Settle, D., and Patterson, C. C. (1980) Lead in Albacore: Guide to Lead Pollution in Americans, *Science* **207**, 1167–1176.

Shafer, M. M., and Armstrong, D. E. (1990) Trace Elements in Lake Michigan, Abstract ACS Meeting, pp. 273–277.

Shafer, M. M., and Armstrong, D. E. (1991) in *Organic Substances and Sediments in Water*, Baker, R. A., Ed., Lewis, Chelsea, MI.

Shiao, S. Y. (1979) Ion Exchange Equilibria Between Montmorillonite and Solutions of Moderate to High Ionic Strength. In *Radioactive Water in Geological Storage*, ACS Symp. Ser. No. 100, American Chemical Society, Washington, DC.

Shiller, A. M., and Boyle, E. A. (1985) Dissolved Zinc in Rivers, *Nature* **317**, 49–52.

Sholkovitz, E. R. (1973) *Geochim. Cosmochim. Acta* **37**, 2043.

Shriver, D. F., Atkins, P. W., and Langford, C. H. (1990) *Inorganic Chemistry*, Oxford University Press, Oxford.

Siegel, D. I., and Pfannkuch, H. O. (1984) Silicate Dissolution Influence on Filson Creek Chemistry, Northeastern Minnesota, *Geol. Soc. Am. Bull.* **95**, 1446–1453.

Siegenthaler, U., and Sarmiento, J. L. (1993) Atmospheric Carbon Dioxide and the Ocean, *Nature* **365**, 119–125.

Siever, R. (1968) Sedimentological Consequences of a Steady-State Ocean–Atmosphere, *Sedimentology* **11**, 5.

Siffert, Ch., and Sulzberger, B. (1991) Light-Induced Dissolution of Hematite in the Presence of Oxalate: A Case Study, *Langmuir* **7**, 1627–1634.

Sigg, L. (1985) Metal Transfer Mechanisms in Lakes; The Role of Settling Particles. In *Chemical Processes in Lakes*, W. Stumm, Ed., Wiley-Interscience, New York, pp. 283–310.

Sigg, L. (1987) Surface Chemical Aspects of the Distribution and Fate of Metal Ions in Lakes. In *Aquatic Surface Chemistry*, W. Stumm, Ed., Wiley-Interscience, New York.

Sigg, L. (1992a) Regulation of Trace Elements in Lakes. In *Chemical and Biological Regulation of Aquatic Processes*, J. Buffle, Ed., Lewis, Boca Raton.

Sigg, L. (1992b) Regulation of Trace Elements by the Solid-Water Interface in Surface Waters. In *Chemistry of the Solid-Water Interface*, W. Stumm, Ed., Wiley-Interscience, New York, chap. 11.

Sigg, L. (1994) Regulation of Trace Elements in Lakes. In *Chemical and Biological Regulation of Aquatic Processes*, J. Buffle and R. R. de Vitre, Eds., Lewis, Chelsea, MI.

Sigg, L., Johnson, C. A., and Kuhn, A. (1991) Redox Conditions and Alkalinity Generation in an Anoxic Lake, *Mar. Chem.* **36,** 9.

Sigg, L., Kuhn, A., Xue, H. B., Kiefer, E., and Kistler, D. (1995) Cycles of Trace Elements (Copper and Zinc) in a Eutrophic Lake: Role of Speciation and Sedimentation. In *Aquatic Chemistry*, C. P. Huang et al., Eds., American Chemical Society, Washington, DC.

Sigg, L., and Stumm, W. (1981) The Interaction of Anions and Weak Acids with the Hydrous Goethite (α-FeOOH) Surface, *Colloids Surf.* **2,** 101–117.

Sigg, L., and Stumm, W. (1994) *Aquatische Chemie, Eine Einführung in die Chemie wässriger Lösungen und natürlicher Gewässer*, 3rd ed., Teubner, Stuttgart.

Sigg, L., Sturm, M., Davis, J. and Stumm, W. (1982) *Thalassia Juoslavica* **18,** 293–311.

Sigg, L., Schnoor, J., Xue, H., Kistler, D., and Stumm, W. (1995) Trace Metals in Aquatic Alpine lakes, *Aquat. Geochem.* (Submitted).

Sigg, L., Sturm, M., and Kistler, D. (1987) Vertical transport of heavy metals by settling particles in Lake Zurich, *Limnol. Oceanogr.* **32,** 112–130.

Sigg, L., Sturm, M., Stumm, W., Mart, L., and Nürnberg, H. W. (1982) Schwermetalle im Bodensee; Mechanismen der Konzentrationsregulierung. *Naturwissenschaften* **69,** 546–547.

Sigg, L., and Xue, H. B. (1994) Metal Speciation, Analysis and Effects. In *Chemistry of Aquatic Systems: Local and Global Perspectives*, G. Bidoglio and W. Stumm, Eds., Kluwer Academic, Dordrecht.

Sigg, L., Stumm, W., Zobrist, J., and Zürcher, I. (1987) The Chemistry of Fog: Factors Regulating Its Composition, *Chimia* **41,** 159–165.

Sillén, L. G. (1961) The Physical Chemistry of Seawater. In *Oceanography*, American Association for the Advancement of Science, Publ. 67, Washington, DC.

Sillén, L. G. (1965) *Ark. Kemi* **25,** Ch. 1, 159–175.

Sillén, L. G. (1967) In *Equilibrium Concepts in Natural Water Systems*, Adv. Chem. Ser. No. 67, American Chemical Society, Washington, DC.

Sillén, L. G., and Martell, A. E. (1964) *Stability Constants of Metal-Ion Complexes*, Special Publ. No. 17, Chemical Society, London.

Sillén, L. G., and Martell, A. E. (1971) *Stability Constants of Metal Ion Complexes*, Special Publ. No. 25, Chemical Society, London. (Various supplements appear in subsequent years.)

Singer, P. C., and Stumm, W. (1970) The Solubility of Ferrous Iron in Carbonate-Bearing Waters, *J. Am. Water Works Assoc.* **62,** 198–202.

Singer, P. C., and Stumm, W. (1970) *Science* **167,** 3921.

Smith, R. M., and Martell, A. E. (1974) *Critical Stability Constants*, Vols. 1–4, Plenum Press, New York. (See also Martell and Smith, 1976–1989.)

Smith, R. M., and Martell, A. E. (1975) *Critical Stability Constants*, v. 2, *Amines*, Plenum, New York.

Somasundaran, P., and Agar, G. E. (1967) Zero Point of Charge of Calcite, *J. Colloid Interface Sci.* **24/4,** 433–440.

Somasundaran, P., and Fuerstenau, D. W. (1966) Mechanisms of Alkyl Sulfonate Adsorption at the Alumina–Water Interface, *J. Phys. Chem.* **70,** 90–96.

Somasundaran, P., Healy, T. W., and Fuerstenau, D. W. (1964) Surfactant Adsorption at the Solid-Liquid Interface-Dependence, *J. of Physical Chemistry* **68,** 3562–3566.

Somorjai, G. A. (1994) *Chemistry in Two Dimensions: Surfaces*, Cornell University Press, Ithaca, NY. (An innovative treatment of modern inorganic surface chemistry.)

Sposito, G. (1981) *The Thermodynamics of Soil Solutions*, Clarendon Press, Oxford.

Sposito, G. (1983) On the Surface Complexation Model of the Oxide-Aqueous Solution Interface. *J. Colloid Interface Sci.* **91,** 329–340.

Sposito, G. (1984) *The Surface Chemistry of Soils*, Oxford University Press, New York. (This monograph gives a comprehensive and didactically valuable interpretation of surface phenomena in soils from the point of view of coordination chemistry.)

Sposito, G. (1986) The Distribution of Potentially Hazardous Trace Metals. In *Metal Ions in Biological Systems*, H. Sigel, Ed., Marcel Dekker, New York.

Sposito, G. (1989a) Surface Reactions in Natural Aqueous Colloidal Systems, *Chimia* **43,** 169–176.

Sposito, G. (1989b) *The Chemistry of Soils*, Oxford University Press, New York.

Sposito, G. (1990) Molecular Models of Ion Adsorption on Mineral Surfaces. In *Mineral-Water Interface Geochemistry*, M. F. Hochella, Jr. and A. F. White, Eds., Mineralogical Society of America, Washington, DC., pp. 261–279.

Sposito, G. (1992) Characterization of Particle Surface Charge. In *Environmental Particles*, Vol. 1, J. Buffle and H. P. van Leeuwen, Eds., Lewis, Boca Raton, FL.

Sposito, G. (1994a) *Chemical Equilibria and Kinetics in Soils*, Oxford University Press, Oxford.

Sposito, G. (1994b) Adsorption as a Problem in Coordination Chemistry: The Concept of the Surface Complex. In *Aquatic Chemistry*, C. P. Huang et al., Eds., American Chemical Society, Washington, DC.

Srinivasan, K., and Rechnitz, G. A. (1968) *Anal. Chem.* **40,** 509.

Steefel, C. I., and Van Cappellen, P. (1990) A New Kinetic Approval to Modelling Water–Rock Interaction: The Role of Nucleation, Precursors, and Ostwald Ripening, *Geochim. Cosmochim. Acta* **54,** 2657.

Stipp, S. L., and Hochella, M. F. Jr. (1991) Structure and Bonding Environments at the Calcite Surface as Observed with X-Ray Photoelectron Spectroscopy (XPS) and Low Energy Electron Diffraction (LEED), *Geochim. Cosmochim. Acta* **55**(6), 1723–1736.

Stone, A. T. (1987) *Geochim. Cosmochim. Acta* **51,** 919.

Stone, A. T. (1987) Reductive Dissolution of Mn(III,IV) Oxides by Substituted Phenols, *Environ. Sci. Technol.* **21**, 287-290.

Stone, A. T., Godtfredsen, K. L., and Deng, B. (1994) Sources and Reactivity of Reductants Encountered in Aquatic Environments. In *Chemistry of Aquatic Systems: Local and Global Perspectives*, G. Bidoglio and W. Stumm, Eds., Kluwer Academic, Dordrecht.

Stone, A. T., and Morgan, J. J. (1987) Reductive Dissolution of Metal Oxides in *Aquatic Surface Chemistry*, W. Stumm, Ed., Wiley-Interscience, New York.

Stone, A. T., and Morgan, J. J. (1990) Kinetics of Chemical Transformations in the Environment. In *Aquatic Chemical Kinetics*, W. Stumm, Ed., Wiley-Interscience, New York.

Stone, A. T., Torrents, A., Smolen, J., Vasudevan, D., and Hadley, J. (1993) Adsorption of Organic Compounds Possessing Ligand Donor Groups at the Oxide/Water Interface, *Environ. Sci. Technol.* **27**(5), 895-909.

Stumm, W. (1978) What is the pε of the Sea? *Thalassia Jugosl.* **14**, 197.

Stumm, W. (1986) Water an Endangered Ecosystem, *Ambio* **15**, 201-207.

Stumm, W. (Ed.) (1987) *Aquatic Surface Chemistry, Chemical Processes at the Particle-Water Interface*, Wiley-Interscience, New York.

Stumm, W., Ed., (1990) *Aquatic Chemical Kinetics*, Wiley-Interscience, New York.

Stumm, W. (1992) *Chemistry of the Solid-Water Interface; Processes at the Mineral-Water and Particle-Water Interface*, Wiley-Interscience, New York.

Stumm, W. (1993) Colloids as Chemical Reactants, *Colloids Surfaces A: Eng. Aspects* **73**, 1-18.

Stumm, W., and Brauner, P. A. (1975) Chemical Speciation. In *Chemical Oceanography*, J. P. Riley and G. Skirrow, Eds., Academic Press, London.

Stumm, W., Hohl, H., and Dalang, F. (1976), Interaction of Metal Ions with Hydrous Oxide Surface, *Croat. Chem. Acta* **48**, 491-504.

Stumm, W., Huang, C. P., and Jenkins, S. R. (1970) Specific Chemical Interaction Affecting the Stability of Dispersed Systems, *Croat. Chem. Acta* **42**, 223-245.

Stumm, W., and Lee, G. F. (1961) *Industrial and Engineering Chem.* **53**, 143.

Stumm, W., and Morgan, J. J. (1970) *Aquatic Chemistry*, Wiley-Interscience, New York.

Stumm, W., and Morgan, J. J. (1981) *Aquatic Chemistry*, 2nd ed., Wiley-Interscience, New York.

Stumm, W., and Morgan, J. J. (1985) On the Conceptual Significance of pε. A Comment to J. D. Hostettler's Paper, *Am. J. Sci.* **285**, 856-859.

Stumm, W., and O'Melia, C. R. (1968) Stoichiometry of Coagulation, *J. Am. Water Works Assoc.* **60**, 514-539.

Stumm, W., and Sigg, L. (1979) Kolloidchemische Grundlagen der Phosphor-Elimination in Fällung, Flockung und Filtration, *Z. Wasser Abwasser-Forschung.* **12**, 73-83.

Stumm, W., and Schnoor, J. L. (in press) Atmospheric Depositions; Impacts of Acid in Lakes. In *Lakes*, A. Lerman et al., Eds., Springer, Berlin.

Stumm, W., and Wieland, E. (1990) Dissolution of Oxide and Silicate Minerals Depend on Surface Speciation, in *Aquatic Chemical Kinetics*, W. Stumm, Ed., Wiley-Interscience, New York.

Stumm, W., and Wollast, R. (1990) Coordination Chemistry of Weathering: Kinetics of the Surface-Controlled Dissolution of Oxide Minerals, *Rev. Geophys.* **28**(1), 53–69.

Stumm, W., Furrer, F., and Kunz, B. (1983) The Role of Surface Coordination in Precipitation (Heterogeneous Nucleation) and Dissolution of Mineral Phases, *Croat. Chem. Acta* **56**, 593–611.

Suess, H. E. (1955) *Science* **122**, 415.

Sulzberger, B. (1990) Photoredox Reactions at Hydrous Metal Oxide Surfaces; a Surface Coordination Chemistry Approach. In *Aquatic Chemical Kinetics*, W. Stumm, Ed., Wiley-Interscience, New York, pp. 401–429.

Sulzberger, B., and Hug, S. I. (1994) Light Induced Processes in the Aquatic Environment. In *Chemistry of Aquatic Systems; Local and Global Perspectives*, G. Bidoglio and W. Stumm, Eds., Kluwer Academic, Dordrecht.

Sulzberger, B., Schnoor, J. L., Giovanoli, R., Hering, J. G., and Zobrist, J. (1990) Biogeochemistry of Iron in an Acidic Lake, *Aquat. Sci.* **52**, 56–74.

Sulzberger, B., Suter, D., Siffert, C., Banwart, S., and Stumm, W. (1989) Dissolution of Fe(III)(Hydr)Oxides in Natural Waters; Laboratory Assessment on the Kinetics Controlled by Surface Coordination, *Mar. Chem.* **28**, 127–144.

Sunda, W. G. (1991) Trace Metal Interaction with Marine Phytoplankton, *Biol. Oceanogr.* **6**, 411–442.

Sunda, W. G. (1994) Trace Metal/Photoplankton Interactions in the Seas. In *Chemistry of Aquatic Systems; Local and Global Perspectives*, G. Bidoglio and W. Stumm, Eds., Kluwer Academic, Dordrecht.

Sunda, W. G., and Guillard, R. L. (1976) *J. Mar. Res.* **34**, 511.

Sunda, W. G., and Hanson, P. J. (1973) In *Chemical Modeling of Aqueous Systems*, E. A. Jenne, Ed., American Chemical Society, Washington, DC.

Sunda, W. G., and Hanson, A. K. (1987) Measurement of Free Cupric Ion Concentration in Seawater by a Ligand Competition Technique Involving Copper Sorption onto C_{18} SEP-PAK Cartridges, *Limnol. Oceanogr.* **32**, 357–551.

Sunda, W. G., and Huntsman, S. A. (1985) Regulation of Cellular Manganese and Manganese Transport Rates in the Unicellular Alga, *Chlamydomonas*, *Limnol. Oceanogr.* **30**, 71–80.

Sunda, W. G., and Huntsman, S. A. (1986) Relationships Among Growth Rate, Cellular Manganese Concentrations and Manganese Transport Kinetics in Estuarine and Oceanic Species of the Diatom *Thalassiosira*, *J. Phycol.* **22**, 259–270.

Sunda, W. G., and Huntsman, S. A. (1991) The Use of Chemiluminescence and Ligand Competition with EDTA to Measure Copper Concentration and Speciation in Seawater, *Mar. Chem.* **36**, 137–163.

Sunda, W. G., and Huntsman, S. A. (1992) Mutual Feedback Interactions Between Zinc and Phytoplankton in Seawater, *Limnol. Oceanogr.* **37**, 25–40.

Sunda, W. G., and Huntsman, S. A. (1994) Photoreduction of Manganese Oxides in Seawater, *Mar. Chem.* **46**, 133–152.

Sunda, W. G., and Huntsman, S. A. (1995) Regulation of Copper Concentration in the Oceanic Nutricline by Phytoplankton Uptake and Regeneration Cycles, *Limnol. Oceanogr.* **40**, 132–137.

Sunda, W. G., and Kieber, D. J. (1994) Oxidation of Humic Substances by Manganese Oxides Yields Low-Molecular Weight Organic Substrates, *Nature* **367**, 62–64.

Sung, W., and Morgan, J. J. (1980) Kinetics and Product of Ferrous Iron Oxygenation in Aqueous Systems, *Environ. Sci. Technol.* **14,** 561–568.

Suter, D., Siffert, C., Sulzberger, B., and Stumm, W. (1988) Catalytic Dissolution of Iron(III)(Hydr)Oxides by Oxalic Acid in the Presence of Fe(II), *Naturwissenschaften* **75,** 571–573.

Sutin, N. (1986) Theory of Electron Transfer. In *Inorganic Reactions and Methods*, J. J. Zuckerman, Ed., VCH, Weinheim, pp. 16–46.

Sverdrup, H. A. (1990) *The Kinetics of Base Cation Release Due to Chemical Weathering*, Lund University Press, Lund, Sweden.

Sverdrup, H., and DeVries, W. (1994) Calculating Critical Loads for Acidity with the Simple Mass Balance Method, *Water, Air Soil Pollut.* **72,** 143–162.

Sverdrup, H., and Warfvinge, P. (1993) Calculating Field Weathering Rates Using a Mechanistic Geochemical Model PROFILE, *Appl. Geochem.* **8,** 273–283.

Sverdrup, H., Warfvinge, P., Frogner, T., Haoya, A. O., Johansson, M., and Andersen, B. (1992) Critical Loads for Forest Soils in the Nordic Countries, *Ambio* **21,** 348–355.

Swain, E. B., Engstrom, D. R., Brigham, M. F., Henning, T. A., and Brezonik, P. L. (1992) Increasing Rates of Atmospheric Hg Deposition in Midcontinental North America, *Nature* **257,** 784–787.

Sykes, P. (1986) *A Guidebook to Mechanisms in Organic Chemistry*, 6th ed., Longman Science and Technology, London.

Takahashi, T., Boecker, W. S., and Langer, S. (1985) Redfield Ratio Based on Chemical Data from Isopycnal Surfaces, *J. Geophys. Res.* **90,** 6907–6929.

Tanford, C. (1980) *The Hydrophobic Effect: Formation of Micelles and Biological Membranes*, 2nd ed., Wiley-Interscience, New York.

Tanford, C. (1991) *The Hydrophobic Effect*, 2nd ed., Krieger, Melbourne, FL.

Tejedor-Tejedor, M. I., and Anderson, M. (1986) In Situ Alternated Total Fourier Transform Studies on the Goethite Solution Interface, *Langmuir* **2,** 203.

Tessier, A. (1993) Sorption of Trace Elements on Neutral Particles in Oxic Environments. In *Environmental Particles*, Vol. 2, J. Buffle and H. P. Van Leeuwen, Eds., Lewis, Boca Raton, FL.

Thompson, M. E. (1966) *Science* **153,** 866.

Thompson, D. W., and P. G. Pownall (1989) Surface Electrical Properties of Calcite, *J. Colloid Interface Sci.* **131/1,** 74–82.

Thorstenson, D. C. (1970) *Geochim. Cosmochim. Acta* **34,** 745.

Thurman, E. M. (1985) *Organic Geochemistry of Natural Waters*, Nijhoff, Boston.

Tiller, C., and O'Melia, C. R. (1993) *Natural Organic Matter and Colloidal Stability, Colloids and Surfaces*, **73,** 89.

Tipping, E. (1981) The Adsorption of Aquatic Humic Substances by Iron Oxides, *Geochim. Cosmochim. Acta* **45,** 191–199.

Tipping, E. (1990) Interactions of Organic Acids with Inorganic and Organic Surfaces. In *Organic Acids in Aquatic Ecosystems*, E. M. Perdue and E. T. Gjessing, Eds., Wiley-Interscience, New York.

Tipping, E., Backes, C. A., and Hurley, M. A. (1988) The Complexation of Protons,

Aluminum and Calcium by Aquatic Humic Substances: A Model Incorporating Binding Site Heterogeneity and Macro-Ionic Effects, *Water Res.* **22,** 579–611.

Tipping, E., and Cooke, D. (1982) The Effect of Adsorbed Humic Substances on the Surface Charge of Goethite in Freshwater, *Geochim. Cosmochim. Acta* **56,** 75.

Tipping, E., and Hurley, M. A. (1992) A Unifying Model of Cation Binding by Humic Substances, *Geochim. Cosmochim. Acta* **56,** 3627–3642.

Tipping, E., Ready, M. M., and Hurley, M. A. (1990) Modeling Electrostatic and Heterogeneity Effects on Proton Dissociation from Humic Substances, *Environ. Sci. Technol.* **24,** 1700–1705.

Tobiason, J. E., and O'Melia, C. R. (1988) *J. Am. Water Works Assoc.* **80,** 54–64.

Tomaić, J., and Zutić, V. (1988) Humic Material Polydispersity in Adsorption at Hydrous Alumina Seawater Interface, *J. Colloid and Interface Sci.* **126,** 482–492.

Tratnyek, P. G., and Hoigné, J. (1994) Kinetics of Reactions of Chlorine Dioxide in Water–II: Quantitative Structure–Activity Relationships for Phenolic Compounds. *Water Res.* **28,** 57–66.

Tratnyek, P. G., and Hoigné, J. (1994) Photooxidation of 2,4,6-trimethylphenol in Aqueous Laboratory Solutions and Natural Water, *J. Photochem. Photobiol.* **84,** 153–160.

Troelstra, S. A., and Kruyt, H. R. (1943) Extinktiometrische Untersuchung der Koagulation. In *Kolloid-Beihefte*, W. Ostwald, Ed., Verlag Th. Steinkopf, Dresden, pp. 225–261.

Turner, D. R., Whitfield, M., and Dickson, A. G. (1981) The Equilibrium Speciation of Dissolved Components in Freshwater and Seawater at 25°C and 1 atm Pressure, *Geochim. Cosmochim. Acta* **45,** 855–882.

Turner, S. M., Malin, G., Liss, P. S., Harbour, D. S., and Holligan, P. M. (1988) The Seasonal Variation of DMS and Dimethyl Sulfonio Propionate Concentrations in Near Shore Waters, *Limnol. Oceanogr.* **33,** 364–375.

Ulrich, H.-J., Stumm, W., and Ćosović, B. (1988) Adsorption of Aliphatic Fatty Acids on Aquatic Interfaces. Comparison Between 2 Model Surfaces: The Mercury Electrode and δ-Al_2O_3 Colloids, *Environ. Sci. Technol.* **22,** 37–41.

Van Breeman, N. (1972) *Geochim. Cosmochim. Acta* **37,** 101.

Van Cappellen, P. (1991) *The Formation of Marine Apatite; A Kinetic Study*, Ph.D. Thesis, Yale University, New Haven, CT.

Van Cappellen, P., Charlet, L., Stumm, W., and Wersin, P. (1993) A Surface Complexation Model of the Carbonate Mineral–Aqueous Solution Interface, *Geochim. Cosmochim. Acta* **57,** 3505–3518.

Van den Berg, C. M. G. (1984) Determination of the Complexing Capacity and Conditional Stability Constants of Complexes of Copper(II) with Natural Organic Ligands in Seawater by Cathodic Stripping Voltammetry of Copper–Catechol Complex Ions, *Mar. Chem.* **15,** 1–18.

Van den Berg, C. M. G., Merks, A. G. A. and Duursma, E. K. (1987) Organic Complexation and Its Control of the Dissolved Concentration of Copper and Zinc in Scheld Estuary, *Estuarine Coastal Shelf Sci.* **24,** 785–797.

Van den Berg, C. M. G., Nimmo, M., Daly, P., and Turner, D. R. (1990) Effects of the Detection Window on the Determination of Organic Copper Speciation in Estuarine Waters, *Anal. Chim. Acta* **232,** 149-159.

Van der Schee, H. A., and Lyklema, J. (1984) A Lattice Theory of Polyelectrolyte Adsorption. *J. Phys. Chem.* **88,** 6661-6667.

Van Leeuwen, H. P. (1992) Dynamic Aspects of Metal Speciation in Aquatic Colloid Systems. In *Environmental Particles*, Vol. 1, Lewis, Boca Raton, FL.

Van Loosdrecht, M. C. M., Norde, W., Lyklema, J., and Zehnder, A. J. B. (1990) Hydrophobic and Electrostatic Parameters in Bacterial Adhesion, *Aquat. Sci.* **52**(1), 103-114.

Van Zeggeren, F., and Storey, S. H. (1970) *The Computation of Chemical Equilibria*, Cambridge University Press, Cambridge.

Vaslow, F., and Boyd, G. E. (1952) *J. Am. Chem. Soc.* **74,** 4691.

Velbel, M. A. (1985) Geochemical Mass Balances and Weathering Rates in Forested Watersheds of the Southern Blue Ridge, *Am. J. Sci.* **285,** 904-930.

Verwey, E. J. W., and Overbeek, Th. G. (1948) *Theory of the Stability of Lyophobic Colloids*, Elsevier, Amsterdam.

von Gunten, U., and Hoigné, J. (1992) Factors Controlling the Formation of Bromate During Ozonation of Bromide-Containing Waters, *J. Water SRT Aqua* **41,** 199-304 (Reprinted by *Ozone News*, Int. Ozone Assoc., 1993).

von Gunten, U., and Hoigné, J. (1994). Bromate Formation During Ozonation of Bromide-Containing Waters: Interaction of Ozone and Hydroxyl Radical Reactions. *Environ. Sci. Technol.* **28,** 1234-1242.

von Gunten, U., and Zobrist, J. (1993) Biogeochemical Changes in Groundwater-Infiltration Systems: Column Studies, *Geochim. Cosmochim. Acta* **57,** 3895-3906.

Waite, T. D. (1986) Photoredox Chemistry of Colloidal Metal Oxides. In *Geochemical Processes at Mineral Surfaces*, J. A. Davis and K. F. Hayes, Eds., ACS Symp. Ser. No. 323, American Chemical Society, Washington, DC.

Waite, T. D. (1990) Photo-redox Processes at the Mineral-Water Interface. In *Mineral-Water Interface Geochemistry*, M. F. Hochella, Jr. and A. F. White, Eds., Mineralogical Society of America, Washington, DC, pp. 559-603.

Waite, T. D., Wrigley, I. C., and Szymczak, R. (1988) Photoassisted Dissolution of a Colloidal Manganese Oxide in the Presence of Fulvic Acid, *Environ. Sci. Technol.* **22,** 778.

Walter, L. M., and Morse, J. W. (1984) Magnesian Calcite Solubilities: A Reevaluation, *Geochim. Cosmochim. Acta* **48,** 1059-1069.

Wardman, P. (1989) Reduction Potentials of One-Electron Couples Involving Free Radicals in Aqueous Solutions, *J. Phys. Chem. Ref. Data*, Reprint No. 372, **18,** 1637-1755.

Warfvinge, P., Holmberg, M., Posch, M., and Wright, R. F. (1992) The Use of Dynamic Models to Set Target Loads, *Ambio* **21,** 369-376.

Warfvinge, P., and Sverdrup, P. (1992) Calculating Critical Loads of Acid Deposition with PROFILE—a Steady State Soil Chemistry Model, *Water, Air Soil Pollut.* **63,** 119-143.

Weber, E. J. (1994) Abiotic Pathways of Organic Chemicals in Aquatic Ecosystems. In *Chemistry of Aquatic Systems; Local and Global Perspectives*, G. Bidoglio and W. Stumm, Eds., Kluwer Academic, Dordrecht.

Weber, W. J. Jr., McGinley, P. M., and Katz, L. E. (1991) Sorption Phenomena in Subsurface Systems: Concepts, Models and Effects on Contaminant Fate and Transport, *Water Res.* **25**(5), 499-528.

Wehrli, B. (1990) Redox Reactions of Metal Ions at Mineral Surfaces. In *Aquatic Chemical Kinetics*, W. Stumm, Ed., Wiley-Interscience, New York, pp. 311-336.

Wehrli, B., Ibrić, S., and Stumm, W. (1990) Adsorption Kinetics of Vanadyl(IV) and Chromium(III) to Aluminum Oxide: Evidence for a Two-Step Mechanism, *Colloids Surf.* **51**, 77-88.

Wehrli, B., and Stumm, W. (1989) Vanadyl in Natural Waters: Adsorption and Hydrolysis Promote Oxygenation, *Geochim. Cosmochim. Acta* **53**, 69-77.

Weidenhaupt, A. (1994) Ph.D. Thesis, ETH, Zürich.

Weilenmann, U., O'Melia, C. R., and Stumm, W. (1989) Particle Transport in Lakes: Models and Measurements, *Limnol. Oceanogr.* **34**, 1-18.

Weiss, A. (1958) Die innerkristalline Quellung als allgemeines Modell für Quellvorgänge, *Chem. Ber.* **91**, 487.

Weiss, R. (1974) Carbon Dioxide in Water and Seawater. The Solubility of a Nonideal Gas, *Mar. Chem.* **2**, 203-215.

Weissbuch, I., Addadi, L., Lahav, M., and Leiserowitz, L. (1991) Molecular Recognition at Crystal Interfaces, *Science* **253**, 637-645.

Wells, C. F., and Salam, M. A. (1968) *J. Chem. Soc.* **A1**, 24-29.

Wells, M. L., and Goldberg, E. D. (1991) Occurrence of Small Colloids in Sea Water, *Nature* **353**, 342-344.

Wells, M. L., and Goldberg, E. D. (1994) *Limnol. Oceanogr.* **39**, 286.

Wersin, P., Charlet, L., Karthein, R., and Stumm, W. (1989) From Adsorption to Precipitation; Sorption of Mn^{2+} on $FeCO_3$(s), *Geochim. Cosmochim. Acta* **53**, 2787-2796.

Wersin, P., Höhener, P., Giovanoli, R., and Stumm, W. (1991) Early Diagenetic Influences on Iron Transformations in a Freshwater Lake Sediment, *Chem. Geol.* **90**, 223-252.

Westall, J. C. (1979) MICROQL—I: A Chemical Equilibrium Program in BASIC, *EAWAG Report*, Dübendorf, Switzerland.

Westall, J. C. (1980) Chemical Equilibrium Including Adsorption on Charged Surfaces. In *Particulates in Water*, M. C. Kavanaugh and J. O. Leckie, Eds., Adv. Chem. Ser. No. 189, American Chemical Society, Washington, DC.

Westall, J. C. (1986) MICROQL, A Chemical Equilibrium Program in Basic, Report 86-02, Oregon State University, Corvallis, OR.

Westall, J. C. (1987), Adsorption Mechanisms in Aquatic Surface Chemistry. In *Aquatic Surface Chemistry*, W. Stumm, Ed., Wiley-Interscience, New York, pp. 3-32.

Westall, J. C., and Hohl, H. (1980) A Comparison of Electrostatic Models for the Oxide Solution Interface, *Adv. Colloid Interface Sci.* **12**, 265-294.

Westall, J. C., Zachara, J. L., and Morel, F. M. M. (1976) *MINEQL, a Computer Program for the Calculation of Chemical Equilibrium Composition of Aqueous Systems, Technical Note 18*, Parsons Laboratory, MIT, Cambridge.

Westall, J. C., Jones, J. D., Turner, G. D., and Zachara, J. L. (1995) Models for Association of Metal Ions with Heterogeneous Environmental Sorbents, *Environ. Sci. and Techn.* **29**, 951-959.

Weston, R. E., and Schwarz, H. A. (1972) *Chemical Kinetics*, Prentice-Hall, Englewood Cliffs, NJ.

White, A. F. (1990) Heterogeneous Electrochemical Reactions Associated with Oxidation of Ferrous Oxide and Silicate Surfaces, *Rev. Mineral.* **23**, 467–509.

White, A. F., and Yee, A. (1985) Aqueous Oxidation–Reduction Kinetics Associated with Coupled Electron Cation Transfer from Iron-Containing Silicates at 25°C, *Geochim. Cosmochim. Acta* **49**, 1263–1275.

Whitfield, M. (1973) *Mar. Chem.* **1**, 251.

Whitfield, M. (1974) *Limnol. Oceanogr.* **19**, 235.

Whitfield, M. (1975a) *Mar. Chem.* **3**, 197.

Whitfield, M. (1975b) In *Chemical Oceanography*, Vol. 1, 2nd ed., J. P. Riley and Skirrow, Eds., Academic Press, New York, pp. 44–171.

Whitfield, M. (1975c) *Geochim. Cosmochim. Acta* **39**, 1545.

Whitfield, M. (1979) The Mean Oceanic Residence Time (Mort) Concept. A Rationalization, *Mar. Chem.* **8**, 101–123.

Whitfield, M., and Jagner, D. (1981) *Marine Electrochemistry*, Wiley-Interscience, New York.

Whitfield, M., and Turner, D. R. (1987) The Role of Particles in Regulating the Composition of Seawater. In *Aquatic Surface Chemistry*, W. Stumm, Ed., Wiley-Interscience, New York.

Widmer, H. M. (1993) Ion Selective Electrodes and Ion Optodes, *Analytical Methods and Instrumentation* **1**, 60–72.

Wieland, E., Wehrli, B., and Stumm, W. (1988) The Coordination Chemistry of Weathering: III. A Generalization on the Dissolution Rates of Minerals, *Geochim. Cosmochim. Acta* **52**, 1969–1981.

Wieland, E., and Stumm, W. (1992) Dissolution Kinetics of Kaolinite in Acid Aqueous Solutions at 25°C, *Geochim. Cosmochim. Acta* **56**, 3339–3355.

Wiklander, L. (1964) *Chemistry of the Soil*, 2nd ed., F. E. Bear, Ed., Van Nostrand Reinhold, New York.

Wilkins, R. G. (1970) Mechanisms of Ligand Replacement in Octahedral Ni(II) Complexes, *Acct. Chem. Res.* **3**, 408–416.

Williams, P., Oeschger, H., and Kinney, P. (1969) *Nature* **224**, 256.

Williams, R. J. P. (1981) Physico-Chemical Aspects of Inorganic Element Transfer through Membranes, *Philos. Trans. Royal Society of London*, **B294**, 57.

Windom, H. L., Byrd, T., Smith, R. G., and Huan, F. (1991) Inadequacy of NASQUAN Data for Assessing Metal Trends in the Nation's Rivers, *Environ. Sci. Technol.* **25**, 1137–1142.

Wipf, H. K., and Simon, W. (1969) *Biochem. Biophys. Res. Commun.* **34**, 707.

Wollast, R. (1974) The Silica Problem. In *The Sea*, Vol. 5, E. D. Goldberg, Ed., Wiley-Interscience, New York, pp. 359–392.

Wollast, R. (1990), Rate and Mechanism of Dissolution of Carbonates in the System $CaCO_3$–$MgCO_3$. In *Aquatic Chemical Kinetics, Reaction Rates of Processes in Natural Waters*, W. Stumm, Ed., Wiley-Interscience, New York, pp. 431–445.

Wollast, R., and Mackenzie, F. T. (1983) Global Cycle of Silica. In *Silicon Geochemistry and Biogeochemistry*, S. R. Aston, Ed., Academic Press, New York, pp. 39–76.

Wollast, R., and Vanderborght, J. P. (1994) Aquatic Carbonate Systems, in *Chemistry of Aquatic Systems*, G. Bidoglio and W. Stumm, Eds., Kluwer Academic, Dordrecht.

Wood, J. M., and Wang, H. K. (1985) Strategies for Microbial Resistance to Heavy Metals. In *Chemical Processes in Lakes*, W. Stumm, Ed., Wiley-Interscience, New York.

Xiao, Y., and Lasaga, A. C. (1994) Ab-initio Quantum Mechanical Studies of the Kinetics and Mechanisms of Silicate Dissolution: H^+ (H_3O^+) Catalysis, *Geochim. Cosmochim. Acta* **58**, 5373–5400.

Xue, H. B., and Sigg, L. (1990) Binding of Cu(II) to Algae in a Metal Buffer, *Water Res.* **24**(9), 1129–1136.

Xue, H. B., and Sigg, L. (1993) Free Cupric Ion Concentration and Cu(II) Speciation in a Eutrophic Lake, *Limnol. Oceanogr.* **38**, 1200–1213.

Xue, H. B., and Sigg, L. (1994) Zn Speciation in Lakewaters and Its Determination by Ligand Exchange with EDTA and DPASV, *Anal. Chim. Acta* **284**, 505–515.

Xue, H. B., Sigg, L., and Kari, F. G. (1995) Speciation of EDTA in Natural Waters; Exchange Kinetics of FeEDTA in River Waters, *Environ. Sci. Technol.* **29**, 59–68.

Yates, D. E., and Healy, T. W. (1975) Mechanism of Anion Adsorption at the Ferric and Chromic Oxide/Water Interfaces, *J. Colloid Interface Sci.* **52**, 222–228.

Yates, D. E., Levine, S., and Healy, T. W. (1974) Site-binding Model of the Electrical Double Layer at the Oxide/Water Interface, *J. Chem. Soc. Faraday Trans.* **70**, 1807.

Yokoyama, A., Srinivasan, K. R., and Fogler, H. S. (1988) Stabilization Mechanism by Acidic Polysaccharides. Effects of Electrostatic Interactions on Stability and Peptization, *Langmuir* **5**(2), 534–538.

Zachara, J. M., Kittrick, J. A., and Harsh, J. B. (1988) The Mechanism of Zn^{2+} Adsorption on Calcite, *Geochim. Cosmochim. Acta* **52**, 2281–2291.

Zafiriou, O. C. (1983) Natural Water Photochemistry. In *Chemical Oceanography*, Vol. 8, Academic Press, London, pp. 339–379.

Zeleznik, F. J., and Gordon, S. (1968) *Ind. Eng. Chem.*, **60**, 27.

Zepp, R. G. (1980) In *Dynamics, Exposure and Hazard Assessment*, R. Hague, Ed., Ann Arbor Science, Ann Arbor, MI.

Zepp, R. G., and Baughman, G. L. (1978) In *Aquatic Pollutants*, O. Hutzinger et al., Eds., Pergamon, New York.

Zepp, R. G., Braun, A. M., Hoigné, J., and Leenheer, J. A. (1987a) Photoproduction of Hydrated Electrons from Natural Organic Solutes in Aquatic Environments, *Environ. Sci. Technol.* **21**, 485–490.

Zepp, R. G., Hoigné, J., and Bader, H. (1987b) Nitrate-Induced Photooxidation of Trace Organic Chemicals in Water, *Environ. Sci. Technol.* **21**(5), 443–450.

Zepp, R. G., Faust, B. C., and Hoigné, J. (1992) Hydroxyl Radical Formation in Aqueous Reactions of Fe(II) with H_2O_2: The Photo Fenton Reaction, *Environ. Sci. Technol.* **26**, 313–319.

Zhang, J. W., and Nancollas, G. H. (1990) Mechanism of Growth and Dissolution of

Sparingly Soluble Salts. In *Reviews in Mineralogy*, Vol. 23, M. F. Hochella and A. F. White, Eds., Mineralogical Society, Washington, DC, pp. 365-396.

Zhuang, G., Duce, R. A., and Kester, D. R. (1990) The Dissolution of Atmospheric Iron in Seawater of the Open Ocean, *J. Geophys. Res.* **95**, 16207-16216.

Zika, R. R., and Cooper, W. J. (1989) *Photochemistry of Environmental Aquatic Chemistry*, ACS Ser. 327, American Chemical Society, Washington, DC.

Zobrist, J., and Stumm, W. (1979) Chemical Dynamics of the Rhine Catchment Area in Switzerland; Extrapolation to the Pristine Rhine River Input into the Ocean. In *Proceedings Review and Workshop on River Inputs to Ocean Systems (RIOS)*, FAO, Rome.

Zobrist, J., Wersin, P., Jaques, C., Sigg, L., and Stumm, W. (1993) Dry Deposition Measurements Using Water as a Receptor, *Water, Air and Soil Pollution* **71**, 111-130.

Zuckerman, J. J. (Ed.) (1986) Electron Transfer and Electrochemical Reactions; Photochemical and Other Energized Reactions, Chaps. 12.1-12.2. In *Inorganic Reactions*, Vol. 15, VCH, Weinheim, pp. 3-87.

Zuo, Y., and Hoigné, J. (1992) Formation of H_2O_2 and Depletion of Oxalic Acid by Photolysis of Fe(III)hydroxo Complexes in Atmospheric Waters, *Environ. Sci. Technol.* **26**, 1014.

Zuo, Y., and Hoigné, J. (1993) Evidence for Photochemical Formation of H_2O_2 and Oxidation of SO_2 in Authentic Fog Water, *Science* **260**, 71-73.

Zutić, V., and Tomaić, J. (1988) On the Formation of Organic Coatings on Marine Particles: Interactions of Organic Matter at Hydrous Alumina/Seawater Interfaces, *Mar. Chem.* **23**, 51-67.

APPENDIXES: THERMODYNAMIC DATA

Although we refer throughout the book to thermodynamic information and equilibrium constants that, in our opinion, have been well documented in the literature, it has not been the authors' objective to critically select the "best available" data. There are various compilations available that recommend a set of equilibrium constants and/or free energy data. We mention some here.

Baes, C. F., and Mesmer, R. E. (1976) *The Hydrolysis of Cations. A Critical Review of Hydrolytic Species and Their Stability Constants in Aqueous Solution*, Wiley-Interscience, New York.

Bard, A. J., Parsons, R., and Jordan, J. (1985) *Standard Potentials in Aqueous Solution* (prepared under auspices of IUPAC), Marcel Dekker, New York.

Byrne, R. H., Kump, L. R., and Cantrell, K. J. (1988) The Influence on Temperature and pH on Trace Metal Speciation in Seawater, *Mar. Chem.* **25**, 163–181.

Nordstrom, D. K., Plummer, L. N., Langmuir, D., Busenberg, E., May, H. M., Jones, B. F., and Parkhurst, D. L. (1990) Revised Chemical Equilibrium Data from Major Mineral Reactions and Their Limitations. In *Chemical Modeling of Aqueous Systems II*, D. C. Melchior and R. L. Bassett, Eds., ACS Ser. 416, American Chemical Society, Washington, DC.

Robie, R. A., Hemingway, B. S., and Fisher, J. R. (1979) Thermodynamic Properties of Minerals and Related Substances at 298.15 K and 10^5 Pascals Pressure and at Higher Temperatures, *U.S. Geological Survey Bulletin No. 1452*, U.S. Geological Survey, Washington, DC (reprinted with corrections).

Sillén, L. G., and Martell, A. E. (1964, 1971, and later supplements) *Stability Constants of Metal Ion Complexes*, Special Publ. 17 and 25, Chemical Society, London.

Smith, R. M., and Martell, A. E. (1971-1989) *Critical Stability Constants*, six volumes, Plenum Press, New York.

Wagman, D. D., et al. *Selected Values of Chemical Thermodynamic Properties*, National Bureau of Standards, Washington, D.C. (various technical notes).

For the convenience of the reader we give three appendixes:

Appendix 1 The equilibrium data compiled by Nordstrom et al. (1990) for major water–mineral reactions.

Appendix 2 Equilibrium data compiled by Byrne et al. (1988) for trace metal speciation in seawater.

Appendix 3 A table of thermodynamic properties for common chemical species in aquatic systems.

Finally, Morel and Hering (1993) have made a very compact compilation of equilibrium constants of metal complexes. This compilation is in Appendix 6.1 of Chapter 6. The appendixes also contain information on enthalpy change, $\Delta H°$, of reactions. This permits one to calculate equilibrium constants at temperatures other than 25°C. In a first approximation, one may assume that $\Delta H°$ is independent of temperature (in the range 5–35°C). Then the following relationship is valid:

$$\ln \frac{K_{T_2}}{K_{T_1}} = \frac{\Delta H°}{R}\left(\frac{1}{T_1} - \frac{1}{T_2}\right)$$

($R = 1.987 \times 10^{-3}$ kcal mol^{-1} deg^{-1} or 8.314×10^{-3} kJ mol^{-1} deg^{-1})

APPENDIX 1: REVISED CHEMICAL EQUILIBRIUM DATA FOR MAJOR WATER–MINERAL REACTIONS[†]

Table 1A. Summary of Revised Thermodynamic Data. I: Fluoride and Chloride Species

Reaction	$\Delta H_r°$ (kcal mol^{-1})	log K	Reaction	$\Delta H_r°$ (kcal mol^{-1})	log K
$H^+ + F^- = HF^0$	3.18	3.18	$Al^{3+} + F^- = AlF^{2+}$	1.06	7.0
$H^+ + 2F^- = HF_2^-$	4.55	3.76	$Al^{3+} + 2F^- = AlF_2^+$	1.98	12.7
$Na^+ + F^- = NaF^0$	—	−0.24	$Al^{3+} + 3F^- = AlF_3^0$	2.16	16.8
$Ca^{2+} + F^- = CaF^+$	4.12	0.94	$Al^{3+} + 4F^- = AlF_4^-$	2.20	19.4
$Mg^{2+} + F^- = MgF^+$	3.2	1.82	$Al^{3+} + 5F^- = AlF_5^{2-}$	1.84	20.6
$Mn^{2+} + F^- = MnF^+$	—	0.84	$Al^{3+} + 6F^- = AlF_6^{3-}$	−1.67	20.6
$Fe^{2+} + F^- = FeF^+$	—	1.0			
$Fe^{3+} + F^- = FeF^{2+}$	2.7	6.2	$Si(OH)_4 + 4H^+ + 6F^-$ $= SiF_6^{2-} + 4H_2O$	−16.26	30.18
$Fe^{3+} + 2F^- = FeF_2^+$	4.8	10.8			
$Fe^{3+} + 3F^- = FeF_3^0$	5.4	14.0	$Fe^{2+} + Cl^- = FeCl^-$	—	0.14

[†]Adapted from Nordstrom et al. (1990). Restricted to 0–100°C and 1 atm (100 kPa); the reference state for aqueous species is infinite dilution. (Data given are for 25°C and 1 atm.) This data set is consistent with recent versions of PHREEQE (Parkhurst et al., 1990), PHREEQM, and WATEQ4F (Ball and Nordstrom, 1990). For references to the data see original publication.

Table 1A. (Continued)

Reaction	ΔH_r° (kcal mol^{-1})	log K	Reaction	ΔH_r° (kcal mol^{-1})	log K
$Mn^{2+} + Cl^- = MnCl^+$	—	0.61	$Fe^{3+} + Cl^- = FeCl^{2+}$	5.6	1.48
$Mn^{2+} + 2Cl^- = MnCl_2^0$	—	0.25	$Fe^{3+} + 2Cl^- = FeCl_2^+$	—	2.13
$Mn^{2+} + 3Cl^- = MnCl_3^-$	—	−0.31	$Fe^{3+} + 3Cl^- = FeCl_3^0$	—	1.13

Mineral	Reaction	ΔH_r° (kcal mol^{-1})	log K
Cryolite	$Na_2AlF_6 = 3Na^+ + Al^{3+} + 6F^-$	9.09	−33.84
Fluorite	$CaF_2 = Ca^{2+} + 2F^-$	4.69	−10.6

Reaction	Analytical Expressions for Temperature Dependence
$H^+ + F^- = HF^0$	$\log K_{HF} = -2.033 + 0.012645\,T + 429.01/T$
$CaF_2 = Ca^{2+} + 2F^-$	$\log K_{fluorite} = 66.348 - 4298.2/T - 25.271 \log T$

Table 1B. Summary of Revised Thermodynamic Data. II: Oxide and Hydroxide Species

Reaction	ΔH_r° (kcal mol^{-1})	log K	Reaction	ΔH_r° (kcal mol^{-1})	log K
$H_2O = H^+ + OH^-$	13.362	-14.000	$Fe^{3+} + H_2O = FeOH^{2+} + H^+$	10.4	-2.19
$Li^+ + H_2O = LiOH^0 + H^+$	0.0	-13.64	$Fe^{3+} + 2H_2O = Fe(OH)_2^+ + 2H^+$	17.1	-5.67
$Na^+ + H_2O = NaOH^0 + H^+$	0.0	-14.18	$Fe^{3+} + 3H_2O = Fe(OH)_3^0 + 3H^+$	24.8	-12.56
$K^+ + H_2O = KOH^0 + H^+$	—	-14.46	$Fe^{3+} + 4H_2O = Fe(OH)_4^- + 4H^+$	31.9	-21.6
$Ca^{2+} + H_2O = CaOH^+ + H^+$	—	-12.78	$2Fe^{3+} + 2H_2O = Fe_2(OH)_2^{4+} + 2H^+$	13.5	-2.95
$Mg^{2+} + H_2O = MgOH^+ + H^+$	—	-11.44	$3Fe^{3+} + 4H_2O = Fe_3(OH)_4^{5+} + 4H^+$	14.3	-6.3
$Sr^{2+} + H_2O = SrOH^+ + H^+$	—	-13.29	$Al^{3+} + H_2O = AlOH^{2+} + H^+$	11.49	-5.00
$Ba^{2+} + H_2O = BaOH^+ + H^+$	—	-13.47	$Al^{3+} + 2H_2O = Al(OH)_2^+ + 2H^+$	26.90	-10.1
$Ra^{2+} + H_2O = RaOH^+ + H^+$	—	-13.49	$Al^{3+} + 3H_2O = Al(OH)_3^0 + 3H^+$	39.89	-16.9
$Fe^{2+} + H_2O = FeOH^+ + H^+$	13.2	-9.5	$Al^{3+} + 4H_2O = Al(OH)_4^- + 4H^+$	42.30	-22.7
$Mn^{2+} + H_2O = MnOH^+ + H^+$	14.4	-10.59			

Mineral	Reaction	ΔH° (kcal mol^{-1})	log K
Portlandite	$Ca(OH)_2 + 2H^+ = Ca^{2+} + 2H_2O$	-31.0	22.8
Brucite	$Mg(OH)_2 + 2H^+ = Mg^{2+} + 2H_2O$	-27.1	16.84
Pyrolusite	$MnO_2 + 4H^+ + 2e^- = Mn^{2+} + 2H_2O$	-65.11	41.38
Hausmanite	$Mn_3O_4 + 8H^+ + 2e^- = 3Mn^{2+} + 4H_2O$	-100.64	61.03
Manganite	$MnOOH + 3H^+ + e^- = Mn^{2+} + 2H_2O$	—	25.34
Pyrochroite	$Mn(OH)_2 + 2H^+ = Mn^{2+} + 2H_2O$	—	15.2
Gibbsite (crystalline)	$Al(OH)_3 + 3H^+ = Al^{3+} + 3H_2O$	-22.8	8.11
Gibbsite (microcrystalline)	$Al(OH)_3 + 3H^+ = Al^{3+} + 3H_2O$	(-24.5)	9.35
$Al(OH)_3$ (amorphous)	$Al(OH)_3 + 3H^+ = Al^{3+} + 3H_2O$	(-26.5)	10.8
Goethite	$FeOOH + 3H^+ = Fe^{3+} + 2H_2O$	—	-1.0
Ferrihydrite (amorphous to microcrystalline)	$Fe(OH)_3 + 3H^+ = Fe^{3+} + 3H_2O$	—	3.0 to 5.0

Reaction	Analytical Expressions for Temperature Dependence
$H_2O = H^+ + OH^-$	log $K_w = -283.9710 + 13323.00/T - 0.05069842\,T + 102.24447 \log T - 1119669/T^2$
$Al^{3+} + H_2O = AlOH^{2+} + H^+$	log $K_1 = -38.253 - 656.27/T + 14.327 \log T$
$Al^{3+} + 2H_2O = Al(OH)_2^+ + 2H^+$	log $\beta_2 = 88.500 - 9391.6/T - 27.121 \log T$
$Al^{3+} + 3H_2O = Al(OH)_3^0 + 3H^+$	log $\beta_3 = 226.374 - 18247.8/T - 73.597 \log T$
$Al^{3+} + 4H_2O = Al(OH)_4^- + 4H^+$	log $\beta_4 = 51.578 - 11168.9/T - 14.865 \log T$

Table 1C. Summary of Revised Thermodynamic Data. III: Carbonate Species

Reaction	ΔH_r° (kcal mol^{-1})	log K	Reaction	ΔH_r° (kcal mol^{-1})	log K
$CO_2(g) = CO_2(aq)$	−4.776	−1.468	$Ca^{2+} + CO_3^{2-} = CaCO_3^0$	3.545	3.224
$CO_2(aq) + H_2O = H^+ + HCO_3^-$	2.177	−6.352	$Mg^{2+} + CO_3^{2-} = MgCO_3^0$	2.713	2.98
$HCO_3^- = H^+ + CO_3^{2-}$	3.561	−10.329	$Sr^{2+} + CO_3^{2-} = SrCO_3^0$	5.22	2.81
$Ca^{2+} + HCO_3^- = CaHCO_3^+$	2.69	1.106	$Ba^{2+} + CO_3^{2-} = BaCO_3^0$	3.55	2.71
$Mg^{2+} + HCO_3^- = MgHCO_3^+$	0.79	1.07	$Mn^{2+} + CO_3^{2-} = MnCO_3^0$	—	4.90
$Sr^{2+} + HCO_3^- = SrHCO_3^+$	6.05	1.18	$Fe^{2+} + CO_3^{2-} = FeCO_3^0$	—	4.38
$Ba^{2+} + HCO_3^- = BaHCO_3^+$	5.56	0.982	$Na^+ + CO_3^{2-} = NaCO_3^-$	8.91	1.27
$Mn^{2+} + HCO_3^- = MnHCO_3^+$	—	1.95	$Na^+ + HCO_3^- = NaHCO_3^-$	—	−0.25
$Fe^{2+} + HCO_3^- = FeHCO_3^+$	—	2.0	$Ra^{2+} + CO_3^{2-} = RaCO_3^0$	1.07	2.5

Mineral	Reaction	ΔH_r° (kcal mol^{-1})	log K
Calcite	$CaCO_3 = Ca^{2+} + CO_3^{2-}$	−2.297	−8.480
Aragonite	$CaCO_3 = Ca^{2+} + CO_3^{2-}$	−2.589	−8.336
Dolomite (ordered)	$CaMg(CO_3)_2 = Ca^{2+} + Mg^{2+} + 2CO_3^{2-}$	−9.436	−17.09
Dolomite (disordered)	$CaMg(CO_3)_2 = Ca^{2+} + Mg^{2+} + 2CO_3^{2-}$	−11.09	−16.54
Strontianite	$SrCO_3 = Sr^{2+} + CO_3^{2-}$	−0.40	−9.271
Siderite (crystalline)	$FeCO_3 = Fe^{2+} + CO_3^{2-}$	−2.48	−10.89

Siderite (precipitated)	$FeCO_3 = Fe^{2+} + CO_3^{2-}$	—	-10.45
Witherite	$BaCO_3 = Ba^{2+} + CO_3^{2-}$	0.703	-8.562
Rhodocrosite (crystalline)	$MnCO_3 = Mn^{2+} + CO_3^{2-}$	-1.43	-11.13
Rhodocrosite (synthetic)	$MnCO_3 = Mn^{2+} + CO_3^{2-}$	—	-10.39

Reaction	Analytical Expressions for Temperature Dependence
$CO_2(g) = CO_2(aq)$	$\log K_H = 108.3865 + 0.01985076\,T - 6919.53/T - 40.45154 \log T + 669365/T^2$
$CO_2(aq) + H_2O = H^+ + HCO_3^-$	$\log K_1 = -356.3094 - 0.06091964\,T + 21834.37/T + 126.8339 \log T - 1684915/T^2$
$HCO_3^- = H^+ + CO_3^{2-}$	$\log K_2 = -107.8871 - 0.03252849\,T + 5151.79/T + 38.92561 \log T - 563713.9/T^2$
$Ca^{2+} + HCO_3^- = CaHCO_3^+$	$\log K_{CaHCO_3^+} = 1209.120 + 0.31294\,T - 34765.05/T - 478.782 \log T$
$Mg^{2+} + HCO_3^- = MgHCO_3^-$	$\log K_{MgHCO_3^+} = -59.215 + 2537.455/T + 20.92298 \log T$
$Sr^{2+} + HCO_3^- = SrHCO_3^+$	$\log K_{SrHCO_3^+} = -3.248 + 0.014867\,T$
$Ba^{2+} + HCO_3^- = BaHCO_3^+$	$\log K_{BaHCO_3^+} = -3.0938 + 0.013669\,T$
$Ca^{2+} + CO_3^{2-} = CaCO_3^0$	$\log K_{CaCO_3^0} = -1228.732 - 0.299444\,T + 35512.75/T + 485.818 \log T$
$Mg^{2+} + CO_3^{2-} = MgCO_3^0$	$\log K_{MgCO_3^0} = 0.9910 + 0.00667\,T$
$Sr^{2+} + CO_3^{2-} = SrCO_3^0$	$\log K_{SrCO_3^0} = -1.019 + 0.012826\,T$
$Ba^{2+} + CO_3^{2-} = BaCO_3^0$	$\log K_{BaCO_3^0} = 0.113 + 0.008721\,T$
$CaCO_3 = Ca^{2+} + CO_3^{2-}$	$\log K_{calcite} = -171.9065 - 0.077993\,T + 2839.319/T + 71.595 \log T$
$CaCO_3 = Ca^{2+} + CO_3^{2-}$	$\log K_{aragonite} = -171.9773 - 0.077993\,T + 2903.293/T + 71.595 \log T$
$SrCO_3 = Sr^{2+} + CO_3^{2-}$	$\log K_{strontianite} = 155.0305 - 7239.594/T - 56.58638 \log T$
$BaCO_3 = Ba^{2+} + CO_3^{2-}$	$\log K_{witherite} = 607.642 + 0.121098\,T - 20011.25/T - 236.4948 \log T$

Table 1D. Summary of Revised Thermodynamic Data. IV: Silicate Species

Reaction	ΔH_r° (kcal mol^{-1})	log K
$Si(OH)_4^0 = SiO(OH)_3^- + H^+$	6.12	-9.83
$Si(OH)_4^0 = SiO_2(OH)_2^{2-} + 2H^+$	17.6	-23.0

Mineral	Reaction	ΔH_r° (kcal mol^{-1})	log K
Kaolinite	$Al_2Si_2O_5(OH)_4 + 6H^+ = 2Al^{3+} + 2Si(OH)_4^0 + H_2O$	-35.3	7.435
Chrysotile	$Mg_3Si_2O_5(OH)_4 + 6H^+ = 3Mg^{2+} + 2Si(OH)_4^0 + H_2O$	-46.8	32.20
Sepiolite	$Mg_2Si_3O_{7.5}(OH)\cdot 3H_2O + 4H^+ + 0.5 H_2O = 2Mg^{2+} + 3Si(OH)_4^0$	-10.7	15.76
Kerolite	$Mg_3Si_4O_{10}(OH)_2\cdot H_2O + 6H^+ + 3H_2O = 3Mg^{2+} + 4Si(OH)_4^0$	—	25.79
Quartz	$SiO_2 + 2H_2O = Si(OH)_4^0$	5.99	-3.98
Chalcedony	$SiO_2 + 2H_2O = Si(OH)_4^0$	4.72	-3.55
Amorphous silica	$SiO_2 + 2H_2O = Si(OH)_4^0$	3.34	-2.71

Reaction	Analytical Expressions for Temperature Dependence
$Si(OH)_4^0 = SiO(OH)_3^- + H^+$	$\log K_1 = -302.3724 - 0.050698\,T + 15669.69/T + 108.18466 \log T - 1119669/T^2$
$Si(OH)_4^0 = SiO_2(OH)_2^{2-} + 2H^+$	$\log \beta_2 = -294.0184 - 0.072650\,T + 11204.49/T + 108.18466 \log T - 1119669/T^2$
$Mg_3Si_2O_5(OH)_4 + 6H^+ = 3Mg^{2+} + 2Si(OH)_4^0 + H_2O$	$\log K_{chrysotile} = 13.248 + 10217.1/T - 6.1894 \log T$
$SiO_2 + 2H_2O = Si(OH)_4^0$	$\log K_{quartz} = 0.41 - 1309/T$
$SiO_2 + 2H_2O = Si(OH)_4^0$	$\log K_{chalcedony} = -0.09 - 1032/T$
$SiO_2 + 2H_2O = Si(OH)_4^0$	$\log K_{amorphous\ silica} = -0.26 - 731/T$

Table 1E. Summary of Revised Thermodynamic Data. V: Sulfate Species

Reaction	ΔH_r° (kcal mol^{-1})	log K	Reaction	ΔH_r° (kcal mol^{-1})	log K
$H^+ + SO_4^{2-} = HSO_4^-$	3.85	1.988	$Mn^{2+} + SO_4^{2-} = MnSO_4^0$	3.37	2.25
$Li^+ + SO_4^{2-} = LiSO_4^-$	—	0.64	$Fe^{2+} + SO_4^{2-} = FeSO_4^0$	3.23	2.25
$Na^+ + SO_4^{2-} = NaSO_4^-$	1.12	0.70	$Fe^{2+} + HSO_4^- = FeHSO_4^+$	—	1.08
$K^+ + SO_4^{2-} = KSO_4^-$	2.25	0.85	$Fe^{3+} + SO_4^{2-} = FeSO_4^+$	3.91	4.04
$Ca^{2+} + SO_4^{2-} = CaSO_4^0$	1.65	2.30	$Fe^{3+} + 2SO_4^{2-} = Fe(SO_4)_2^-$	4.60	5.38
$Mg^{2+} + SO_4^{2-} = MgSO_4^0$	4.55	2.37	$Fe^{3+} + HSO_4^- = FeHSO_4^{2+}$	—	2.48
$Sr^{2+} + SO_4^{2-} = SrSO_4^0$	2.08	2.29	$Al^{3+} + SO_4^{2-} = AlSO_4^+$	2.15	3.02
$Ba^{2+} + SO_4^{2-} = BaSO_4^0$	—	2.7	$Al^{3+} + 2SO_4^{2-} = Al(SO_4)_2^-$	2.84	4.92
$Ra^{2+} + SO_4^{2-} = RaSO_4^0$	1.3	2.75	$Al^{3+} + HSO_4^- = AlHSO_4^{2+}$	—	0.46

Mineral	Reaction	ΔH_r° (kcal mol^{-1})	log K
Gypsum	$CaSO_4 \cdot 2H_2O = Ca^{2+} + SO_4^{2-} + 2H_2O$	−0.109	−4.58
Anhydrite	$CaSO_4 = Ca^{2+} + SO_4^{2-}$	−1.71	−4.36
Celestite	$SrSO_4 = Sr^{2+} + SO_4^{2-}$	−1.037	−6.63
Barite	$BaSO_4 = Ba^{2+} + SO_4^{2-}$	6.35	−9.97
Radium sulfate	$RaSO_4 = Ra^{2+} + SO_4^{2-}$	9.40	−10.26
Melanterite	$FeSO_4 \cdot 7H_2O = Fe^{2+} + SO_4^{2-} + 7H_2O$	4.91	−2.209
Alunite	$KAl_3(SO_4)_2(OH)_6 + 6H^+ = K^+ + 3Al^{3+} + 2SO_4^{2-} + 6H_2O$	−50.25	−1.4

Reaction	Analytical Expressions for Temperature Dependence
$H^+ + SO_4^{2-} = HSO_4^-$	$\log K_2 = -56.889 + 0.006473\,T + 2307.9/T + 19.8858 \log T$
$CaSO_4 \cdot 2H_2O = Ca^{2+} + SO_4^{2-} + 2H_2O$	$\log K_{gypsum} = 68.2401 - 3221.51/T - 25.0627 \log T$
$CaSO_4 = Ca^{2+} + SO_4^{2-}$	$\log K_{anhydrite} = 197.52 - 8669.8/T - 69.835 \log T$
$SrSO_4 = Sr^{2+} + SO_4^{2-}$	$\log K_{celestite} = -14805.9622 - 2.4660924\,T + 756968.533/T - 40553604/T^2 + 5436.3588 \log T$
$BaSO_4 = Ba^{2+} + SO_4^{2-}$	$\log K_{barite} = 136.035 - 7680.41/T - 48.595 \log T$
$RaSO_4 = Ra^{2+} + SO_4^{2-}$	$\log K_{RaSO_4} = 137.98 - 8346.87/T - 48.595 \log T$
$FeSO_4 \cdot 7H_2O = Fe^{2+} + SO_4^{2-} + 7H_2O$	$\log K_{melanterite} = 1.447 - 0.004153\,T - 214949/T^2$

APPENDIX 2: THERMODYNAMIC DATA FOR TRACE METAL SPECIATION IN SEAWATER[†]

Table 2A. Thermodynamic Data for Strongly Hydrolyzed Species

Complex	$(\log \beta_n^*)^a$	$\Sigma (\Delta H_n)^b$	Complex	$(\log \beta_n^*)^a$	$\Sigma (\Delta H_n)^b$
$AlOH^{2+}$	-5.51	11.9	$InOH^{2+}$	-4.54	11.2
$Al(OH)_2^+$	-10.13	19.8	$In(OH)_2^+$	-8.65	18.4
$Al(OH)_3^0$	-15.83	25.9	$In(OH)_3^0$	-13.23	22.9
$Al(OH)_4^-$	-23.53	31.5	$In(OH)_4^-$	-22.60	30.8
$BiOH^{2+}$	-1.63	4.1	$HfOH^{3+}$	-0.79	10.3
$Bi(OH)_2^+$	-4.83	10.0	$Hf(OH)_2^{2+}$	-3.49	18.8
$Bi(OH)_3^0$	-9.69	15.0	$Hf(OH)_3^+$	-7.38	25.7
			$Hf(OH)_4^0$	-12.08	30.4
$CrOH^{2+}$	-4.54	9.7	$Hf(OH)_5^-$	-18.27	34.0
$Cr(OH)_2^+$	-10.53	19.4			
$Cr(OH)_3^0$	-18.83	29.1	$ThOH^{3+}$	-3.74	11.2
$Cr(OH)_4^-$	-27.93	36.5	$Th(OH)_2^{2+}$	-8.02	21.9
			$Th(OH)_3^+$	-13.08	30.4
$FeOH^{2+}$	-2.73	10.2	$Th(OH)_4^0$	-17.28	34.4
$Fe(OH)_2^+$	-6.50	16.9			
$Fe(OH)_3^0$	-12.83	23.9	UOH^{3+}	-1.19	11.2
$Fe(OH)_4^-$	-22.13	31.6	$U(OH)_2^{2+}$	-3.69	19.5
			$U(OH)_3^+$	-7.18	25.8
$GaOH^{2+}$	-3.14	10.6	$U(OH)_4^0$	-11.68	30.3
$Ga(OH)_2^+$	-6.73	17.1	$U(OH)_5^-$	-17.07	32.7
$Ga(OH)_3^0$	-11.3	21.4			
$Ga(OH)_4^-$	-17.13	24.7	$ZrOH^{3+}$	-0.24	9.8
			$Zr(OH)_2^{2+}$	-2.79	18.1
$TlOH^{2+}$	-1.16	7.6	$Zr(OH)_3^+$	-6.48	24.7
$Tl(OH)_2^+$	-2.40	10.9	$Zr(OH)_4^0$	-11.08	29.3
$Tl(OH)_3^0$	-4.13	11.6	$Zr(OH)_5^-$	-17.07	32.6

[a] From Baes and Mesmer (1976).
[b] From Baes and Mesmer (1981). It is assumed hydrolysis enthalpies are invariant between 0 and 0.7 M ionic strength. This assumption is supported by observations of Fe(III) and Zn(II) hydrolysis.

Hydrolysis constants appropriate to seawater ($S = 35$) were estimated using the following activity coefficients: uncharged species, $\gamma = 1$; monovalent species, $\gamma = 0.7$; divalent species, $\gamma = 0.25$; trivalent species, $\gamma = 0.05$; tetravalent species, $\gamma = 0.01$. Using these estimates, one obtains the following relationships for divalent ($z = 2$), trivalent ($z = 3$), and tetravalent ($z = 4$) metals in pure water ($S = 0$) and in seawater.

$$\log \beta_n(S = 0) - \log \beta_n \text{ (seawater)}$$

	n				
z	1	2	3	4	5
2	+0.29	+0.29	−0.02	—	—
3	+0.54	+0.83	+0.83	+0.53	—
4	+0.54	+1.09	+1.38	+1.38	+1.07

[†] From Byrne et al. (1988). For references see original paper.

Appendix 2: Thermodynamic Data for Trace Metal Speciation in Seawater

Table 2B. Thermodynamic Data for Metals Strongly Complexed with Chloride Ions

Complex	$\log \beta_n$	$\Sigma (\Delta H_n)$
$AgCl^0$	3.36^a	—
$AgCl_2^-$	5.2^a	—
$AgCl_3^{2-}$	$(5.85)^b$	-9.3^c
$AgCl_4^{3-}$	$(5.2)^b$	—
$CuCl^0$	$(3.36)^d$	—
$CuCl_2^-$	$(5.2)^d$	—
$CuCl_3^{2-}$	$(5.85)^d$	-9.3^c
$CuCl_4^{3-}$	$(5.2)^d$	—
$CdCl^+$	1.35^a	0.3^e
$CdCl_2^0$	1.70^a	0.9^e
$CdCl_3^-$	1.50^a	2.4^a
$HgCl^+$	6.72^a	-5.5^a
$HgCl_2^0$	13.23^a	-12.2^a
$HgCl_3^-$	14.2^a	-12.4^a
$HgCl_4^{2-}$	15.3^a	-14.2^a
$PdCl^+$	4.47^a	-3.0^a
$PdCl_2^0$	7.74^a	-5.6^a
$PdCl_3^-$	10.2^a	-8.2^a
$PdCl_4^{2-}$	11.5^a	-11.6^a
$TlCl^{2+}$	6.72^f	-5.5^g
$TlCl_2^+$	11.76^f	-9.9^g
$TlCl_3^0$	14.4^f	-11.0^g
$TlCl_4^-$	16.3^f	-11.3^g
$PtCl^+$	5.0^a	—
$PtCl_2^0$	9.0^a	—
$PtCl_3^-$	11.9^a	—
$PtCl_4^{2-}$	14.0^a	—

[a] Martell and Smith (1982), ionic strength = 1 M.
[b] Estimated using $\log \beta_2$ (1 M) and characterizations of stepwise $AgCl_n$ stability constant behavior at 4 M and 5 M ionic strength (Martell and Smith, 1982).
[c] Martell and Smith (1982), ionic strength = 5 M.
[d] Available data (Smith and Martell, 1976; Martell and Smith, 1982) suggest that the chloride complexation behavior of Ag^+ and Cu^+ cannot be easily distinguished. Consequently, set $\log \beta_n(AgCl_n) = \log \beta_n(CuCl_n)$.
[e] Martell and Smith (1982), ionic strength = 0.5 M.
[f] Smith and Martell (1976), ionic strength = 0.5 M.
[g] Smith and Martell (1976), ionic strength = 3 M.

Table 2C. Thermodynamic Data for Metals Strongly Complexed with Carbonate Ions

Complex	$\log \beta_n$	$\Sigma (\Delta H_n)$	Complex	$\log \beta_n$	$\Sigma (\Delta H_n)$
$LaCO_3^+$	4.85^a	0^b	$GdCO_3^+$	5.47^a	0^b
$La(CO_3)_2^-$	8.20^a	0^b	$Gd(CO_3)_2^-$	9.28^a	0^b
$LaOH^{2+}$	-9.04^c	14.1^d	$GdOH^{2+}$	-8.54^c	13.9^d
$La(OH)_2^+$	-18.23^c	28.2^d	$Gd(OH)_2^+$	-17.23^c	27.3^d
$LaCl^{2+}$	-0.1^e	0^f	$GdCl^{2+}$	-0.1^e	0^f
LaF^{2+}	2.67^g	4.0^g	GdF^{2+}	3.31^g	8.9^g
$LaSO_4^+$	1.9^h	4.4^i	$GdSO_4^+$	1.9^h	4.4^i
$CeCO_3^+$	4.96^a	0^b	$TbCO_3^+$	5.53^a	0^b
$Ce(CO_3)_2^-$	8.37^a	0^b	$Tb(CO_3)_2^-$	9.41^a	0^b
$CeOH^{2+}$	-8.84^c	13.9^d	$TbOH^{2+}$	-8.44^c	13.9^d
$Ce(OH)_2^+$	-17.93^c	27.8^d	$Tb(OH)_2^+$	-17.13^c	27.3^d
$CeCl^{2+}$	-0.1^e	0^f	$TbCl^{2+}$	-0.1^e	0^f
CeF^{2+}	2.81^g	4.8^g	TbF^{2+}	3.42^g	7.5^g
$CeSO_4^+$	1.9^h	4.4^i	$TbSO_4^+$	1.9^h	4.4^i
$PrCO_3^+$	5.07^a	0^b	$DyCO_3^+$	5.58^a	0^b
$Pr(CO_3)_2^-$	8.53^a	0^b	$Dy(CO_3)_2^-$	9.53^a	0^b
$PrOH^{2+}$	-8.64^c	13.9^d	$DyOH^{2+}$	-8.54^c	13.9^d
$Pr(OH)_2^+$	-17.83^c	28.0^d	$Dy(OH)_2^+$	-17.03^c	27.0^d
$PrCl^{2+}$	-0.1^e	0^f	$DyCl^{2+}$	-0.1^e	0^f
PrF^{2+}	3.01^g	5.7^g	DyF^{2+}	3.46^g	7.0^g
$PrSO_4^+$	1.9^h	4.4^i	$DySO_4^+$	1.9^h	4.4^i
$NdCO_3^+$	5.16^a	0^b	$HoCO_3^+$	5.62^a	0^b
$Nd(CO_3)_2^-$	8.69^a	0^b	$Ho(CO_3)_2^-$	9.66^a	0^b
$NdOH^{2+}$	-8.54^c	13.9^d	$HoOH^{2+}$	-8.54^c	13.9^d
$Nd(OH)_2^+$	-17.73^c	28.0^d	$Ho(OH)_2^+$	-16.93^c	26.9^d
$NdCl^{2+}$	-0.1^e	0^f	$HoCl^{2+}$	-0.1^e	0^f
NdF^{2+}	3.09^g	6.8^g	HoF^{2+}	3.52^g	7.3^g
$NdSO_4^+$	1.9^h	4.4^i	$HoSO_4^+$	1.9^h	4.4^i
$SmCO_3^+$	5.33^a	0^b	$ErCO_3^+$	5.66^a	0^b
$Sm(CO_3)_2^-$	8.99^a	0^b	$Er(CO_3)_2^-$	9.77^a	0^b
$SmOH^{2+}$	-8.44^c	13.8^d	$ErOH^{2+}$	-8.44^c	13.9^d
$Sm(OH)_2^+$	-17.43^c	27.6^d	$Er(OH)_2^+$	-16.73^c	26.8^d
$SmCl^{2+}$	-0.1^e	0^f	$ErCl^{2+}$	-0.1^e	0^f
SmF^{2+}	3.12^g	9.4^g	ErF^{2+}	3.54^g	7.4^g
$SmSO_4^+$	1.9^h	4.4^i	$ErSO_4^+$	1.9^h	4.4^i
$EuCO_3^+$	5.40^a	0^b	$TmCO_3^+$	5.69^a	0^b
$Eu(CO_3)_2^+$	9.14^a	0^b	$Tm(CO_3)_2^-$	9.88^a	0^b
$EuOH^{2+}$	-8.34^c	13.9^d	$TmOH^{2+}$	-8.24^c	13.9^d
$Eu(OH)_2^+$	-17.43^c	27.9^d	$Tm(OH)_2^+$	-16.73^c	27.0^d
$EuCl^{3+}$	-0.1^e	0^f	$TmCl^{2+}$	-0.1^e	0^f
EuF^{2+}	3.19^g	9.2^g	TmF^{2+}	3.56^g	8.7^g
$EuSO_4^+$	1.9^h	4.4^i	$TmSO_4^+$	1.9^h	4.4^i
$YbCO_3^+$	5.71^a	0^b	$CmCO_3^+$	5.27^j	0^b

Appendix 2: Thermodynamic Data for Trace Metal Speciation in Seawater

Table 2C. (Continued)

Complex	log β_n	$\Sigma\,(\Delta H_n)$	Complex	log β_n	$\Sigma\,(\Delta H_n)$
Yb(CO$_3$)$_2^-$	9.99^a	0^b	Cm(CO$_3$)$_2^-$	8.86^j	0^b
YbOH^{2+}	−8.24^c	13.9^d	Cm(OH)$^{2+}$	−7.84^p	13.9^p
Yb(OH)$_2^+$	−16.63^c	27.0^d	Cm(OH)$_2^+$	−16.2^p	27.0^p
YbCl^{2+}	−0.2^e	0^f	CmCl^{2+}	−0.1^p	0^p
YbF^{2+}	3.58^g	9.6^g	CmF^{2+}	3.34^h	7^o
YbSO$_4^+$	1.9^h	4.4^i	CmSO$_4^+$	1.9^h	4.4^i
LuCO$_3^+$	5.73^a	0^b	BkCO$_3^+$	5.32^j	0^b
Lu(CO$_3$)$_2^-$	10.09^a	0^b	Bk(CO$_3$)$_2^-$	8.98^j	0^b
LuOH^{2+}	−8.14^c	14.0^d	BkOH^{2+}	−7.84^p	13.9^p
Lu(OH)$_2^+$	−16.53^c	27.0^d	Bk(OH)$_2^+$	−16.2^p	27.0^p
LuCl^{2+}	−0.4^e	0^f	BkCl^{2+}	−0.1^p	0^p
LuF^{2+}	3.61^g	9.5^g	BkF^{2+}	2.89^o	7^o
LuSO$_4^+$	1.9^h	4.4^i	BkSO$_4^+$	1.9^h	4.4^i
YCO$_3^+$	5.61^j	0^b	CfCO$_3^+$	5.38^j	0^b
Y(CO$_3$)$_2^-$	9.65^j	0^b	Cf(CO$_3$)$_2^-$	9.10^j	0^b
YOH^{2+}	−8.24^c	13.9^d	CfOH^{2+}	−7.84^p	13.9^l
Y(OH)$_2^+$	−17.23^c	27.7^d	Cf(OH)$_2^+$	−16.2^p	27.0^l
YCl^{2+}	−0.1^e	0^f	CfCl^{2+}	−0.1^p	0^p
YF^{2+}	3.6^g	8.3^g	CfF^{2+}	3.03^o	7^o
YSO$_4^+$	1.9^h	4.4^i	CfSO$_4^+$	1.9^h	4.4^i
AmCO$_3^+$	5.22^j	0^b	CuCO$_3^0$	4.88^q	3^r
Am(CO$_3$)$_2^-$	8.80^j	0^b	Cu(CO$_3$)$_2^-$	7.57^s	0^r
AmOH^{2+}	−7.84^k	13.9^l	CuOH$^+$	−8.11^t	12.0^l
Am(OH)$_2^+$	−16.2^m	27.0^l	Cu(OH)$_2^0$	−16.7^t	22.0^l
AmCl^{2+}	−0.1^e	0^f	CuCl$^+$	−0.22^{q,r}	1.6^u
AmF^{2+}	3.39^n	7°	CuF$^+$	0.9^g	1^n
AmSO$_4^+$	1.9^h	4.4^i	CuSO$_4^0$	0.95^g	1.9^v

^{a}Cantrell and Byrne (1987).
^{b}Based on europium carbonate complexation behavior (Cantrell, 1986).
^{c}From Baes and Mesmer (1976); hydrolysis constants are expressed in the form of the equation $\beta_n^* = [M(OH)_n][H^+]^n[M]^{-1}$ and are appropriate to seawater ionic strength.
^{d}From Baes and Mesmer (1981).
^{e}Estimated from Smith and Martell (1976) at 1 M ionic strength and 20–25°C.
$^f\Delta H = 0$ for EuCl$^+$ (Smith and Martell, 1976); assume $\Delta H = 0$ for other trivalent lanthanides and actinides.
^{g}1 M ionic strength and 25°C (Smith and Martell, 1976).
^{h}Based on 0.5 M data for Ce, Eu, Gd, Am, and Cm (Smith and Martell, 1976).
^{i}Based on average of 0 M and 2 M data for lanthanides plus Am^{3+}, Cm^{3+}, and Cf^{3+} (Smith and Martell, 1976).
^{j}From Cantrell (1988), estimated from correlations of lanthanide ionic radii (Shannon, 1976) versus lanthanide carbonate complexation constants, plus the actinide ionic radius estimates of Shannon (1976).
^{k}From Choppin (1983).
^{l}Estimated from β_n corrected to zero ionic strength as in Table B1 (Baes and Mesmer, 1981).
^{m}Assume $\beta_2 = 0.3\,\beta_1^2$.
^{n}Smith and Martell (1976), 0.5 M ionic strength and 25°C.
^{o}From Martell and Smith (1982), 1 M and 25°C.

Table 2C. (*Continued*)

[p]Assume identical to Am^{3+}.
[q]From Byrne and Miller (1985).
[r]From Soli and Byrne (submitted).
[s]From Byrne and Miller (1985), plus the estimate $[CO_3^{2-}]/[CO_3^{2-}]_T = 0.14$ for seawater.
[t]From Paulson and Kester (1980).
[u]From Smith and Martell (1976), 2 M ionic strength and 25°C.
[v]Average of 0 M and 2 M data (Smith and Martell, 1976).

Table 2D. Thermodynamic Data for Metals that Have Transitional Seawater Chemistries

Complex	log β_n	$\Sigma (\Delta H_n)$	Complex	log β_n	$\Sigma (\Delta H_n)$
$BeOH^+$	-5.69^a	11.4^b	$ScOH^{2+}$	-4.84^a	14.4^b
$Be(OH)_2^0$	-13.94^a	21.0^b	$Sc(OH)_2^+$	-10.53^a	23.7^b
$Be(OH)_3^-$	-23.23^a	28.8^b	$Sc(OH)_3^0$	-16.93^a	30.8^b
BeF^+	4.99^c	0^c	$Sc(OH)_4^-$	-26.53^a	38.9^b
BeF_2^0	8.80^c	-1^c	$ScCO_3^+$	7.53^j	0^k
BeF_3^-	11.6^c	-2^c	$Sc(CO_3)_2^-$	12.07^j	0^k
$PbOH^+$	-8.00^a	10.7^b			
$Pb(OH)_2^0$	-17.41^a	21.8^b			
$PbCO_3^0$	5.10^d	3^e			
$Pb(CO_3)_2^{2-}$	7.26^d	—	$AcCO_3^+$	4.30^j	0^k
$PbCl^+$	0.86^f	4.4^g	$Ac(CO_3)_2^-$	7.00^j	0^k
$PbCl_2^0$	1.16^f	—	$AcOH^{2+}$	$<-10.94^a$	—
$PbCl_3^-$	1.06^f	—	$AcCl^{2+}$	-0.1^l	0^m
$PbSO_4^0$	1.29^h	2^i	AcF^{2+}	2.72^n	7^o
$Pb(SO_4)_2^{2-}$	2.48^h	2^i	$AcSO_4^+$	1.2^p	4.4^q

[a]From Baes and Mesmer (1976), expressed at seawater ionic strength.
[b]From Baes and Mesmer (1981).
[c]From Smith and Martell (1976).
[d]log β_n^0 from Turner et al. (1981) plus predictions, log β_1^0 − log β_1 (SW) = 1.90, log β_2^0 − log β_2 (SW) = 3.03, from observations of Cu behavior in pure water and seawater (Byrne and Miller, 1985). Note that $\beta_1' = [PbCO_3^0]Pb_T^{-1} [CO_3]_T^{-1} = 1.0 \times 10^4$ (Byrne, 1981) plus the estimate $Pb_T/[Pb^{2+}] = 1 + \Sigma \beta_n[Cl^-]^n = 11.7$ (Millero and Byrne, 1984) provides essentially the same result, log $\beta_1 = 5.07$.
[e]Assume similar to $CuCO_3^0$ and oxalate complexation of divalent metals.
[f]From Millero and Byrne (1984).
[g]From Smith and Martell (1976), 0 M.
[h]log β_1 at 0 M, log β_2 at 3 M plus behavior of Cd $(SO_4)_n$ at 0, 0.5, 1, 2, and 3 M (Smith and Martell, 1976).
[i]Assume similar to sulfate complexation of Cd and other divalent metals.
[j]Oxalate–carbonate correlation (Cantrell and Byrne, 1987) plus oxalate data of Smith and Martell (1977).
[k]Assume $\Delta H \simeq 0$ (similar to oxalate complexation, and carbonate complexation of europium).
[l]Estimated from Smith and Martell (1976), 1 M, 20–25°C.
[m]Assume similar to Eu.
[n]Smith and Martell (1976) 0.5 M, 25°C.
[o]Assume similar to lanthanide and actinide complexation behavior.
[p]Smith and Martell (1976), 1 M, 25°C.
[q]Assume similar to REE and actinide complexation behavior.

Appendix 2: Thermodynamic Data for Trace Metal Speciation in Seawater

Table 2E. Thermodynamic Data for Metals that Are Weakly Complexed in Seawater

Complex	$\log \beta_n$	$\Sigma (\Delta H_n)$	Complex	$\log \beta_n$	$\Sigma (\Delta H_n)$
$MnCO_3^0$	2.20^a	3^b	$ZnCO_3^0$	2.85^a	3^b
$MnOH^+$	-10.88^c	14.4^d	$ZnOH^+$	-9.25^c	13.4^d
$MnCl^+$	-0.29^e	0.5^f	$ZnCl^+$	-0.34^e	1.3^d
$MnSO_4^0$	0.9^g	1.6^h	$ZnSO_4^0$	0.90^m	1.0^n
$FeCO_3^0$	2.83^a	3^b	$LiSO_4^-$	0.27^o	0^p
$FeOH^+$	-9.79^c	13.2^d	$RbSO_4^-$	0.10^o	0^q
$FeCl^+$	-0.46^i	0.5^f	$CsSO_4^-$	-0.2^o	0^q
$FeSO_4^0$	0.9^g	1.1^h	$BaSO_4^0$	0.66^m	0.5^r
$CoCO_3^0$	3.01^a	3^b	$TlCl^0$	0.04^m	-1.5^m
$CoOH^+$	-9.94^c	14.6^d	$TlSO_4^-$	0.53^s	1.6^k
$CoCl^+$	-0.41^j	0.5^k			
$CoSO_4^0$	0.9^g	0.9^h			
$NiCO_3^0$	3.47^a	3^b			
$NiOH^+$	-10.15^c	11.9^d			
$NiCl^+$	-0.49^l	0.5^k			
$NiSO_4^0$	0.9^g	1.0^h			

[a] β_1^0 data of Turner et al. (1981) plus the relationship $\log \beta_1^0 - \log \beta_1(SW) = 1.9$ (Byrne and Miller, 1985) observed for copper in pure water and seawater.
[b] Assume $\Delta H \approx 3$ as for Cu(II).
[c] From Baes and Mesmer (1976), appropriate at seawater ionic strength.
[d] From Baes and Mesmer (1981).
[e] Derived from a refitting of the $CuCl^+$, $MnCl^+$, and $ZnCl^+$ complexation data of Short and Morris (1961) and Morris and Short (1961, 1962) plus the estimate $\beta_1 = 0.6$ for Cu^{2+}.
[f] Assume identical to Ni^{2+} and Co^{2+} estimates.
[g] Assume β_1 for $MnSO_4^0$, $FeSO_4^0$, $CoSO_4^0$, and $NiSO_4^0$ are identical to β_1 for $ZnSO_4^0$ and $CuSO_4^0$ based on β_1^0 similarities for Mn, Fe, Co, Ni, Cu, and Zn.
[h] Estimate based on ΔH behavior for $ZnSO_4^0$ formation at 0 and 2 M ionic strength plus ΔH at 0 M for Mn, Fe, Co, and Ni (Smith and Martell, 1976; Martell and Smith, 1982).
[i] From Davison (1979).
[j] From Grimaldi and Liberti (1964); β_1 results for Cu and Co at 3 M ionic strength plus the estimate $\beta_1 = 0.6$ for Cu(II) in seawater.
[k] From Martell and Smith (1982); 2 M ionic strength.
[l] From Libus and Tialowska (1975) results for Co and Ni plus β_1 result for Co^{2+} (Grimaldi and Liberti, 1964).
[m] From Smith and Martell (1976); 1 M, 25°C.
[n] Average of ΔH at 0 M and 2 M ionic strength (Martell and Smith, 1982).
[o] $\log \beta_1^0 - \log \beta_1 (0.7\ M) = 0.5$ for Na^+ and K^+ plus β_1^0 values from Martell and Smith (1982).
[p] From Martell and Smith (1982); 0 M and 25°C.
[q] Assume ΔH is identical for Li^+, Rb^+, and Cs^+.
[r] Assume ΔH is identical for $SrSO_4^0$ and $BaSO_4^0$ formation.
[s] From Martell and Smith (1982); 1 M and 0.5 M ionic strength.

APPENDIX 3: THERMODYNAMIC PROPERTIES

Table 3A. $\bar{G}_f^0$, $\bar{H}_f^0$, and $\bar{S}^0$ Values for Common Chemical Species in Aquatic Systems:[a] Valid at 25°C, 1 atm Pressure, and Standard States[b]

Species	Formation from the Elements		Entropy	Reference[c]
	$\bar{G}_f^0$ (kJ mol^{-1})	$\bar{H}_f^0$ (kJ mol^{-1})	$\bar{S}^0$ (J mol^{-1} K^{-1})	
Ag (Silver)				
Ag (Metal)	0	0	42.6	NBS
Ag$^+$(aq)	77.12	105.6	73.4	NBS
AgBr	−96.9	−100.6	107	NBS
AgCl	−109.8	−127.1	96	NBS
AgI	−66.2	−61.84	115	NBS
Ag$_2$S(α)	−40.7	−29.4	14	NBS
AgOH(aq)	−92			NBS
Ag(OH)$_2^-$ (aq)	−260.2			NBS
AgCl(aq)	−72.8	−72.8	154	NBS
AgCl$_2^-$ (aq)	−215.5	−245.2	231	NBS
Al (Aluminum)				
Al	0	0	28.3	R
Al^{3+}(aq)	−489.4	−531.0	−308	R
AlOH^{2+}(aq)	−698			S
Al(OH)$_2^+$(aq)	−911			S
Al(OH)$_3$(aq)	−1115			S
Al(OH)$_4^-$(aq)	−1325			S
Al(OH)$_3$ (amorph)	−1139			R
Al$_2$O$_3$ (Corundum)	−1582	−1676	50.9	R

AlOOH (Boehmite)	−922	−1000	17.8	R
$Al(OH)_3$ (Gibbsite)	−1155	−1293	68.4	R
$Al_2Si_2O_5(OH)_4$ (Kaolinite)	−3799	−4120	203	R
$KAl_3Si_3O_{10}(OH)_2$ (Muscovite)	−1341			R
$Mg_5Al_2Si_3O_{10}(OH)_8$ (Chlorite)	−1962			R
$CaAl_2Si_2O_8$ (Anorthite)	−4017.3	−4243.0	199	R
$NaAlSi_3O_8$ (Albite)	−3711.7	−3935.1		R
As (Arsenic)				
As (α-Metal)	0	0	35.1	NBS
H_3AsO_4(aq)	−766.0	−898.7	206	NBS
$H_2AsO_4^-$(aq)	−753.17	−904.5	117	NBS
$HAsO_4^{2-}$(aq)	−714.60	−898.7	3.8	NBS
AsO_4^{3-}(aq)	−648.41	−870.3	−145	NBS
$H_2AsO_3^-$(aq)	−587.13			NBS
Ba (Barium)				
Ba^{2+}(aq)	−560.7	−537.6	9.6	R
$BaSO_4$ (Barite)	−1362	−1473	132	R
$BaCO_3$ (Witherite)	−1132	−1211	112	R
Be (Beryllium)				
Be^{2+}(aq)	−380	−382	−130	NBS
$Be(OH)_2(\alpha)$	−815.0	−902	51.9	NBS
$Be_3(OH)_3^{3+}$	−1802			NBS
B (Boron)				
H_3BO_3(aq)	−968.7	−1072	162	NBS
$B(OH)_4^-$(aq)	−1153.3	−1344	102	NBS

Table 3A. (Continued)

	Formation from the Elements		Entropy	
Species	$\overline{G}_f^0$ (kJ mol^{-1})	$\overline{H}_f^0$ (kJ mol^{-1})	$\overline{S}^0$ (J mol^{-1} K^{-1})	Reference[c]
Br (Bromide)				
Br$_2$(l)	0	0	152	NBS
Br$_2$(aq)	3.93	−259	130.5	NBS
Br$^-$(aq)	−104.0	−121.5	82.4	NBS
HBrO(aq)	−82.2	−113.0	147	NBS
BrO$^-$(aq)	−33.5	−94.1	42	NBS
C (Carbon)				
C (Graphite)	0	0	152	NBS
C (Diamond)	3.93	−2.59	130.5	NBS
CO$_2$(g)	−394.37	−393.5	213.6	NBS
H$_2$CO$_3^*$(aq)	−623.2	−699.6	187.0	R[d]
H$_2$CO$_3$(aq) ("true")	∼ −607.1			s
HCO$_3^-$(aq)	−586.8	−692.0	91.2	s
CO$_3^{2-}$(aq)	−527.9	−677.1	−56.9	NBS
CH$_4$(g)	−50.79	−74.80	186	NBS
CH$_4$(aq)	−34.39	−89.04	83.7	NBS
CH$_3$OH(aq)	−175.4	−245.9	133	NBS
HCOOH(aq)	−372.3	−425.4	163	NBS
HCOO$^-$(aq)	−351.0	−425.6	92	NBS
CH$_2$O(aq)	−129.7			
CH$_2$O(g)	−110.0	−116.0	218.6	s
HCN(aq)	112.0	105.0	129	NBS
CN$^-$(aq)	166.0	151.0	118	NBS
COS(g)	−169.2	−137.2	234.5	NBS

CNS⁻(aq)	88.7	72.0	S
H₂C₂O₄(aq)	−697.0	−818.26	S
HC₂O₄⁻(aq)	−690.86	−818.8	S
C₂O₄²⁻(aq)	−674.04	−818.8	45.6	S

Ca (Calcium)

Ca²⁺(aq)	−553.54	−542.83	−53	R
CaOH⁺(aq)	−718.4			NBS
Ca(OH)₂(aq)	−868.1	−1003		NBS
Ca(OH)₂ (Portlandite)	−898.4	−986.0	−74.5	R
CaCO₃ (Calcite)	−1128.8	−1207.4	83	R
CaCO₃ (Aragonite)	−1127.8	−1207.4	91.7	R
CaMg(CO₃)₂ (Dolomite)	−2161.7	−2324.5	88.0	R
CaSiO₃ (Wollastonite)	−1549.9	−1635.2	155.2	R
CaSO₄ (Anhydrite)	−1321.7	−1434.1	82.0	R
CaSO₄ · 2 H₂O (Gypsum)	−1797.2	−2022.6	106.7	R
Ca₅(PO₄)₃OH (Hydroxyapatite)	−6338.4	−6721.6	194.1	R
			390.4	R

Cd (Cadmium)

Cd (γ-Metal)				
Cd²⁺(aq)	−77.58	−75.90	−73.2	R
CdOH⁺(aq)	−284.5			R
Cd(OH)₃⁻(aq)	−600.8			R
Cd(OH)₄²⁻(aq)	−758.5			R
Cd(OH)₂(aq)	−392.2			R
CdO (s)	−228.4	−258.1	54.8	R
Cd(OH)₂ (precip.)	−473.6	−560.6	96.2	R
CdCl⁺(aq)	−224.4	−240.6	43.5	R
CdCl₂(aq)	−340.1	−410.2	39.8	R

Table 3A. (Continued)

Species	Formation from the Elements		Entropy	Reference[c]
	$\overline{G}_f^0$ (kJ mol^{-1})	$\overline{H}_f^0$ (kJ mol^{-1})	$\overline{S}^0$ (J mol^{-1} K^{-1})	
CdCl$_3^-$(aq)	−487.0	−561.0	203	R
CdCO$_3$(s)	−669.4	−750.6	92.5	R
Cl (Chlorine)				
Cl$^-$(aq)	−131.3	−167.2	56.5	NBS
Cl$_2$(g)	0	0	223.0	NBS
Cl$_2$(aq)	6.90	−23.4	121	NBS
HClO(aq)	−79.9	−120.9	142	NBS
ClO$^-$(aq)	−36.8	−107.1	42	NBS
ClO$_2$(aq)	117.6	74.9	173	NBS
ClO$_2^-$(aq)	17.1	−66.5	101	NBS
ClO$_3^-$(aq)	−3.35	−99.2	162	NBS
ClO$_4^-$(aq)	−8.62	−129.3	182	NBS
Co (Cobalt)				
Co (Metal)	0	0	30.04	R
Co^{2+}(aq)	−54.4	−58.2	−113	R
Co^{3+}(aq)	134	92	−305	R
HCoO$_2^-$(aq)	−407.5			NBS
Co(OH)$_2$(aq)	−369	−518	134	NBS
Co(OH)$_2$ (blue precip.)	−450			NBS
CoO(s)	−214.2	−237.9	53.0	R
Co$_3$O$_4$ (Cobalt Spinel)	−725.5	−891.2	102.5	R

Cr (Chromium)				
Cr (Metal)	0	0	23.8	NBS
Cr²⁺ (aq)	−215.5	−143.5		NBS
Cr³⁺ (aq)	−256.0			NBS
Cr₂O₃ (Eskolaite)	−1053	−1135	308	R
HCrO₄⁻ (aq)	−764.8	−878.2	81	R
CrO₄²⁻ (aq)	−727.9	−881.1	184	R
Cr₂O₇²⁻ (aq)	−1301	−1490	50	R
Cr(OH)₃ (hydrous)	−858	−984	262	Bard et al.
Cr(OH)²⁺	−430	−495	(1051)	Bard et al.
Cr(OH)₂⁺	−653	−748	(−156)	Bard et al.
Cr(OH)₄⁻	−1013	−1169	(−27)	Bard et al.
			(238)	
Cu (Copper)				
Cu (Metal)	0	0	33.1	NBS
Cu⁺ (aq)	50.0	71.7	40.6	NBS
Cu²⁺ (aq)	65.5	64.8	−99.6	NBS
Cu(OH)₂ (aq)	−249.1	−395.2	−121	NBS
HCuO₂⁻ (aq)	−258			NBS
CuS (Covellite)	−53.6	−53.1	66.5	NBS
Cu₂S (α)	−86.2	−79.5	121	NBS
CuO (Tenorite)	−129.7	−157.3	43	NBS
CuCO₃ · Cu(OH)₂ (Malachite)	−893.7	−1051.4	186	NBS
2 CuCO₃ · Cu(OH)₂ (Azurite)		−1632		NBS
F (Fluorine)				
F₂(g)	0	0	202	NBS
F⁻ (aq)	−278.8	−332.6	−13.8	NBS

Table 3A. (*Continued*)

Species	Formation from the Elements		Entropy	Reference[c]
	$\overline{G}_f^0$ (kJ mol^{-1})	$\overline{H}_f^0$ (kJ mol^{-1})	$\overline{S}^0$ (J mol^{-1} K^{-1})	
HF(aq)	−296.8	320.0	88.7	NBS
HF$_2^-$ (aq)	−578.1	−650	92.5	NBS
Fe (Iron)				
Fe (Metal)	0	0	27.3	NBS
Fe^{2+}(aq)	−78.87	−89.10	−138	NBS
FeOH$^+$(aq)	−277.4	324.7	29	NBS
Fe(OH)$_2$(aq)	−441.0	—	—	NBS
Fe^{3+}(aq)	−4.60	−48.5	−316	NBS
FeOH^{2+}(aq)	−229.4	−324.7	−29.2	NBS
Fe(OH)$_2^+$(aq)	−438	250.8	142.0	NBS
Fe(OH)$_3$(aq)	−659.4	—	—	NBS
Fe(OH)$_4^-$(aq)	−842.2	—	34.5	NBS
Fe$_2$(OH)$_2^{4+}$(aq)	−467.27	612.1	356.0	NBS
FeS$_2$ (Pyrite)	−160.2	−171.5	52.9	R
FeS$_2$ (Marcasite)	−158.4	−169.4	53.9	R
FeO(s)	−251.1	−272.0	59.8	R
Fe(OH)$_2$ (precip.)	−486.6	−569	87.9	NBS
α-Fe$_2$O$_3$ (Hematite)[e]	−742.7	−824.6	87.4	R
Fe$_3$O$_4$ (Magnetite)	−1012.6	−1115.7	146	R
α-FeOOH (Goethite)[e]	−488.6	−559.3	60.5	R
FeOOH (amorph)[e]	−462			S
Fe(OH)$_3$ (amorph)[e]	−699(−712)			S
FeCO$_3$(Siderite)	−666.7	−737.0	105	R
Fe$_2$SiO$_4$ (Fayalite)	−1379.4	−1479.3	148	R

H (Hydrogen)			
$H_2(g)$	0	0	NBS
$H_2(aq)$	17.57	−4.18	NBS
$H^+(aq)$	0	0	NBS
$H_2O(l)$	−237.18	−285.83	NBS
$H_2O(g)$	−228.57	−241.8	NBS
$H_2O_2(aq)$	−134.1	−191.17	R
$HO_2^-(aq)$	−67.4	−160.33	NBS
Hg (Mercury)			
$Hg(l)$	0	0	NBS
$Hg_2^{2+}(aq)$	153.6	172.4	NBS
$Hg^{2+}(aq)$	164.4	171.0	NBS
Hg_2Cl_2 (Calomel)	−210.8	265.2	NBS
HgO(red)	−58.5	−90.8	NBS
HgS (Metacinnabar)	−43.3	−46.7	NBS
HgI_2 (red)	−101.7	−105.4	NBS
$HgCl^+(aq)$	−5.44	−18.8	NBS
$HgCl_2(aq)$	−173.2	−216.3	NBS
$HgCl_3^-(aq)$	−309.2	−388.7	NBS
$HgCl_4^{2-}(aq)$	−446.8	−554.0	NBS
$HgOH^+(aq)$	−52.3	−84.5	NBS
$Hg(OH)_2(aq)$	−274.9	−355.2	NBS
$HgO_2^-(aq)$	−190.3		NBS
I (Iodine)			
I_2 (Crystal)	0	0	NBS
$I_2(aq)$	16.4	22.6	NBS
$I^-(aq)$	−51.59	−55.19	NBS

Table 3A. (Continued)

Species	Formation from the Elements		Entropy	Reference[c]
	$\overline{G}_f^0$ (kJ mol^{-1})	$\overline{H}_f^0$ (kJ mol^{-1})	$\overline{S}^0$ (J mol^{-1} K^{-1})	
I_3^- (aq)	−51.5	−51.5	239	NBS
HIO(aq)	−99.2	−138	95.4	NBS
IO$^-$ (aq)	−38.5	−107.5	−5.4	NBS
HIO$_3$(aq)	−132.6	−211.3	167	NBS
IO$_3^-$	−128.0	−221.3	118	NBS
Mg (Magnesium)				
Mg (Metal)	0	0	32.7	R
Mg^{2+}(aq)	−454.8	−466.8	−138	R
MgOH$^+$(aq)	−626.8			S
Mg(OH)$_2$(aq)	−769.4	−926.8	−149	NBS
Mg(OH)$_2$ (Brucite)	−833.5	−924.5	63.2	R
Mn (Manganese)				
Mn (Metal)	0	0	32.0	R
Mn^{2+} (aq)	−228.0	−220.7	−73.6	R
Mn(OH)$_2$ (precip.)	−616			S
Mn$_3$O$_4$ (Hausmannite)	−1281			S
α-MnOOH (α-Manganite)	−557.3			S
MnO$_2$ (Manganate) (IV) (MnO$_{1.7}$–MnO$_2$)	−453.1			S
MnO$_2$ (Pyrolusite)	−465.1	−520.0	53	R
MnCO$_3$ (Rhodochrosite)	−816.0	−889.3	100	R
MnS (Albandite)	−218.1	−213.8	87	R
MnSiO$_3$ (Rhodonite)	−1243	−1319	131	R

N (Nitrogen)			
$N_2(g)$	0	191.5	NBS
$NO(g)$	86.57	210.6	S
$NO_2(g)$	51.3	240.0	S
$N_2O(g)$	104.2	220	NBS
$NH_3(g)$	−16.48	192	NBS
$NH_3(aq)$	−26.57	111	NBS
$NH_4^+(aq)$	−79.37	113.4	NBS
$HNO_2(aq)$	−55.6	153	NBS
$NO_2^-(aq)$	−37.2	140	NBS
$HNO_3(aq)$	−111.3	146	NBS
$NO_3^-(aq)$	−111.3	146.4	NBS
Ni (Nickel)			
$Ni^{2+}(aq)$	−45.6	−129	R
NiO (Bunsenite)	−211.6	38	R
NiS (Millerite)	−86.2	66	R
O (Oxygen)			
$O_2(g)$	0	205	NBS
$O_2(aq)$	16.32	111	NBS
$O_3(g)$	163.2	239	NBS
$O_3(aq)$	125.9		NBS
O_2^-	31.84		NBS
$HO_2^{\cdot}(aq)$	4.44		NBS
$H_2O_2(g)$	−105.6	232.6	NBS
$H_2O_2(aq)$	−134.1	143.9	NBS
$HO_2^-(aq)$	−67.4	23.8	NBS
$OH^{\cdot}(g)$	34.22	183.64	NBS

Table 3A. (*Continued*)

Species	Formation from the Elements		Entropy	Reference
	$\overline{G}_f^0$ (kJ mol^{-1})	$\overline{H}_f^0$ (kJ mol^{-1})	$\overline{S}^0$ (J mol^{-1} K^{-1})	
OH'(aq)	7.74			NBS
OH$^-$(aq)	−157.29	−230.0	−10.75	NBS
P (Phosphorus)				
P (α, white)	0	0	41.1	
PO$_4^{3-}$(aq)	−1018.8	−1277.4	−222	NBS
HPO$_4^{2-}$(aq)	−1089.3	−1292.1	−33.4	NBS
H$_2$PO$_4^-$(aq)	−1130.4	−1296.3	90.4	NBS
H$_3$PO$_4$(aq)	−1142.6	−1288.3	158	NBS
Pb (Lead)				
Pb (Metal)	0	0	64.8	NBS
Pb^{2+}(aq)	−24.39	−1.67	10.5	NBS
PbOH$^+$(aq)	−226.3			NBS
Pb(OH)$_3^-$(aq)	−575.7			NBS
Pb(OH)$_2$ (precip.)	−452.2			NBS
PbO (yellow)	−187.9	−217.3	68.7	NBS
PbO$_2$	−217.4	−277.4	68.6	NBS
Pb$_3$O$_4$	−601.2	−718.4	211	NBS
PbS	−98.7	−100.4	91.2	NBS
PbSO$_4$	−813.2	−920.0	149	NBS
PbCO$_3$ (Cerussite)	−625.5	−699.1	131	NBS
S (Sulfur)				
S (rhombic)	0	0	31.8	NBS
SO$_2$(g)	−300.2	−296.8	248	NBS

SO$_3$(g)	−371.1	−395.7	257	NBS
H$_2$S(g)	−33.56	−20.63	205.7	NBS
H$_2$S(aq)	−27.87	−39.75	121.3	NBS
S^{2-}(aq)	85.8^f	33.0	−14.6	NBS
HS$^-$(aq)	12.05	−17.6	62.8	NBS
SO$_3^{2-}$(aq)	−486.6	−635.5	−29	NBS
HSO$_3^-$(aq)	−527.8	−626.2	140	NBS
H$_2$SO$_3^*$	−537.9	−608.8	232	NBS^g
H$_2$SO$_3$(aq) ("true")	∼−534.5			S
SO$_4^{2-}$(aq)	−744.6	−909.2	20.1	NBS
HSO$_4^-$(aq)	−756.0	−887.3	132	NBS
Se (Selenium)				
Se (black)	0	0	42.4	NBS
SeO$_3^{2-}$(aq)	−369.9	−509.2	12.6	NBS
HSeO$_3^-$(aq)	−411.5	−514.5	135	NBS
H$_2$SeO$_3$(aq)	−426.2	−507.5	208	NBS
SeO$_4^{2-}$(aq)	−441.4	−599.1	54.0	NBS
HSeO$_4^-$(aq)	−452.3	−581.6	149	NBS
Si (Silicon)				
Si (Metal)	0	0	18.8	NBS
SiO$_2$ (α, Quartz)	−856.67	−910.94	41.8	NBS
SiO$_2$ (α, Cristobalite)	−855.88	−909.48	42.7	NBS
SiO$_2$ (α, Tridymite)	−855.29	−909.06	43.5	NBS
SiO$_2$ (amorph)	−850.73	−903.49	46.9	NBS
H$_4$SiO$_4$(aq)	−1308.0^h	−1468.6	180	NBS

Table 3A. (Continued)

Species	Formation from the Elements		Entropy	
	$\overline{G}_f^0$ (kJ mol^{-1})	$\overline{H}_f^0$ (kJ mol^{-1})	$\overline{S}^0$ (J mol^{-1} K^{-1})	Reference[c]
Sr (Strontium)				
Sr^{2+} (aq)	−559.4	−545.8	−33	R
$SrOH^+$ (aq)	−721			NBS
$SrCO_3$ (Strontianite)	−1137.6	−1218.7	97	R
$SrSO_4$ (Celestite)	−1341.0	−1453.2	118	R
Zn (Zinc)				
Zn (Metal)	0	0	29.3	NBS
Zn^{2+} (aq)	−147.0	−153.9	112	NBS
$ZnOH^+$ (aq)	−330.1			NBS
$Zn(OH)_2$ (aq)	−522.3			NBS
$Zn(OH)_3^-$ (aq)	−694.3			NBS
$Zn(OH)_4^{2-}$ (aq)	−858.7			NBS
ZnO (solid)	−318.32	−348.28	43.64	NBS
$Zn(OH)_2$ (solid β)	−553.6	−641.9	81.2	NBS
$ZnCl^+$ (aq)	−275.3			NBS
$ZnCl_2$ (aq)	−403.8			NBS
$ZnCl_3^-$ (aq)	−540.6			NBS
$ZnCl_4^{2-}$ (aq)	−666.1			S
$ZnCO_3$ (Smithsonite)	−731.6	−812.8	82.4	NBS

[a] The quality of the data is highly variable; the authors do not claim to have critically selected the "best" data. For information on precision of the data and for a more complete compendium, which includes less common substances, the reader is referred to the references. For research work, the original literature should be consulted.
[b] Thermodynamic properties taken from Robie, Hemingway, and Fisher are based on a reference state of the elements in their standard states at 1 bar (10^5 P = 0.987 atm). This change in reference pressure has a negligible effect on the tabulated values for the condensed phases. [For gas phases only data from NBS (reference state = 1 atm) are given.]

[c] NBS: D. D. Wagman et al., Selected Values of Chemical Thermodynamic Properties, U.S. National Bureau of Standards, Technical Notes 270-3 (1968), 270-4 (1969), 270-5 (1971). R: R. A. Robie, B. S. Hemingway, and J. R. Fisher, *Thermodynamic Properties of Minerals and Related Substances at 298.15 K and 1 Bar (10^5 Pascals) Pressure and at Higher Temperatures*, Geological Survey Bulletin No. 1452, Washington, DC, 1978. Bard et al.: Bard, A. J., R. Parsons and D. L. Parkhurst, *Standard Potentials in Aqueous Solution*, Marcel Dekker, New York (1985). S: Other sources (e.g., computed from data in *Stability Constants*).

[d] $[H_2CO_3^*] = [CO_2(aq)] +$ "true" $[H_2CO_3]$.

[e] The thermodynamic stability of oxides, hydroxides, or oxyhydroxides of Fe(III) depends on mode of preparation, age, and molar surface. Reported solubility products ($K_{s0} = \{Fe^{3+}\}\{OH^-\}^3$) range from $10^{-37.3}$ to $10^{-43.7}$. Correspondingly, FeOOH may have G_f° values between -452 J mol^{-1} (freshly precipitated amorphous FeOOH) and -489 J mol^{-1} (aged goethite). If the precipitate is written as Fe(OH)$_3$, its G_f° values vary from -692 to -729 J mol^{-1}.

[f] The value for this species appears too low, on the basis of recently reported pK_2 values for H$_2$S(aq).

[g] $[H_2SO_3^*] = [SO_2(aq)] +$ "true" $[H_2SO_3]$.

[h] R value yields a solubility constant for quartz more in accord with observation.

INDEX

Absorption, light, 729
Acetate, equilibrium calculation, 115
Acetic acid, rainwater, 212
Acids, 88-143
  alkalimetric titration, 180
  atmosphere, 89
  critical load, 872-879
  definitions, 90-92
  equilibrium composition, 106-110
  hard and soft, 284
  organic, 140-144
  strength, 92-94
  strong, 113
  volatile, 117
Acid deposition, 207, 210
  critical load, 237
  weathering, 879
Acid lake, 170, 235-238
  photochemical reduction of Fe(III), 752
  steady-state concentration of Fe^{2+}, 758
Acid rain, 89, 208
  genesis, 208
Acid rainwater, alkalimetric titration, 181
Acid- and base-neutralizing capacity, 138
Acid-base, titration in atmosphere, 208-211
Acid-base system:
  buffer intensity, 134-140
  diprotic, 122-125
  ionization fractions, 127-130
  titration, 130-140
Acidity, 163-169
  titration, 167-169
  capacity diagrams, 176
Acidity constant, 92-96
  composite, 95
  definition, 102
  Hammet correlation, 143
Actinide colloids, 823
Activated complex, precursor, 778
Activated complex theory, 71-76
  dissolution of a mineral, 780

  temperature and pressure effect, 74-76
  thermodynamic formulation, 72-76
Activation energy, 74, 76
  and Gibbs free energy, 706
  for an S_N2 process, 712
Activation entropy, 76
Activity:
  and activity coefficient of aqueous NaCl, 40
  and chemical potential, 35
  and concentration, 36
  infinite dilution, 336
  ionic medium scale, 337
  of the aqueous solute, 40
Activity coefficient, 40, 98-101
  and ion association, 338
  at ionic strength of seawater, 339
  ions, 102-105
  of neutral species, 104
  seawater, 340
  single-ion, 103
  specific ionic interaction model, 341
  tabulation, 104
Activity ratio diagram:
  for redox systems, 513
  solubility Fe(II), 390-393
Activity scales, 98
Acy (Acidity), titration, 183
Adhesion, 608
  and cohesion, 610
Adsorption, 519-533. *See also* Sorption
  amphipathic substances, 579
  and K_{OW}, 575
  humic acid (HM), 582
  hydrophobic, 521
  hydrophobic substances, 575-579
  isotherms, 521
  kinetics, 760-771
  microscopic reversibility, 769
  polymers, 520, 581-586
  porous media, 594

1006 Index

Adsorption (*Continued*)
  semi-infinite linear diffusion, 761
  surfactants, 519, 579, 763
  surface complex formation equilibria, 519, 534
Adsorption isotherms:
  determination column experiments, 598
  Freundlich, 524–526
  Frumkin, 526–527
  Gibbs, 530–532
  Langmuir, 521–526
Adsorption kinetics, Elovich equation, 770, 771
Adsorption rate, Co^{2+} on Al_2O_3, 766
Aerosol, 211, 233–235
Affinity, 31
Aggregation colloids, *see* Coagulation
Air–Water exchange:
  chemical enhancement, 244
  of chemicals, 242
Alk, CO_2 uptake ocean, 921
Al(III):
  as coagulant, 852–857
  hydrolysis, 273
  solubility, 273
Al_2O_3:
  dissolution, 774
  ligand- and proton-promoted dissolution, 784
Al_2O_3 surface, acidity constants, 603
Albite:
  dissolution, 148, 772
  dissolution rate, 782
Aldehyde:
  atmospheric water, 211
  reactions of SO_2, 226
Algae, composition metals, 640
Alkalinity, 163–169
  alternative definition, 165
  capacity diagrams, 176
  distribution, 149
  Gran titration, 179–186
  interstitial concentration, 903–906
  open systems, 164
  processes affecting, 173, 876
  tableau for equilibrium composition, 164
  titration, 167–169
  vs. C_T diagrams, 177
A-Metals, 284–289
  life properties, 625–628
Amino acid, microscopic constants, 97
Ammonia:
  deposition, 208–211
  in open and closed systems, 219–221

Ammonium aerosols, 233
Amphipathic substances, 521
  adsorption, 579
Ampholyte, equilibrium composition, 115
Analysis:
  carbonate system, 167–169, 179–186
  trace metals, 615–625
  voltammetric methods, 618–625
ANC (Acid Neutralizing Capacity), 138, 167–169
  buffer intensity, 140
  buffer solutions, 139
  carbonate-bearing water, 139
Anhydrite, solubility, 358
Anion, surface complexation, 544
Anode, definition, 445
Anodic stripping voltammetry, 618–625
Anorthite, dissolution, 772
Anoxic conditions, natural waters, 894
Apatite, solubility, 407
Aragonite, solubility seawater, 346
Arrhenius equation, 73
As, organic, 628–629
Ascorbate, reductive dissolution, 783
Associative mechanism, substitution reaction, 713
ASV, anodic stripping voltammetry, 618–625
Atmosphere:
  acids and bases, 88
  -water interactions, 206–213
Atmospheric CO_2, increase effect on seawater, 176
Atmospheric pollutants, 207
  washout from the atmosphere, 227–229
Avogadro's constant, 13
Azurite, 397

Ba, bioregulation seawater, 912
Bacteria, *see also* Microoganisms
  at hydrothermal vents, 903
Bacterial adhesion, hydrophobic and electrostatic parameters, 856
Balancing, detailed, 61
Band gap, (semiconductor), 755
Band gap energies, semiconductors, 757
Basalt, seawater, 899–903
Base, 88–143
  weak, 114
Base catalysis, hydrolysis reaction, 713–716
Basicity constant, 92–96
Battery, 445
Beer–Lambert law, 729
Beggiatoa, 483
Benzoate, dissolution Al_2O_3, 784

Benzoic acid, Hammet constants, 144
Bicarbonate, vs. Ca^{2+} in rivers, 189
Bioaccumulation, K_{OW}, 578
Bioavailability, metals, 632–637
Biochemical cycle, redox intensity, 467
Biochemicals, diagenesis, 924
Biogenic ligands, 306
Biogeochemical cycles, 5
  interdependence, 914
Biological regulation, of the seawater composition, 909
Biomagnification, 578
Biomimetic processing, 811
Biomineralization, 800–805, 809–811
Biotite, dissolution, 772, 782
B-metals, 284–289
  life properties, 625–628
BNC (Base Neutralizing Capacity), 138, 167–169
BOD, 926
Boltzmann constant, 14
Borate, in seawater, 156, 309
Boric acid, equilibrium composition, 111, 112
Br^-, with O_3, 694–698
Breakpoint chlorination, 700
Brines, chemical composition, 881–884
Brønsted concept, 90
Brownian diffusion, coagulation filtration, 858
Buffer factor, CO_2, ocean, 921
Buffer intensity, 116, 129–149
  carbonate system, 154
  seawater, 159
  titration curve, 137
Buffering:
  by $CaCO_3$, CO_2, 885
  natural waters, 884–886
Butylchloride, as probe in photoreaction, 736

C, geochemical cycle, 914
Ca, bioregulation seawater, 912
$Ca_{10}(PO_4)_6(OH)_2(s)$, solubility, 407
Ca^{2+}, vs. HCO_3^- in rivers, 189
$CaCO_3$:
  aragonite calcite conversion, 25
  buffering by, 885
  $CdCO_3$ solid solution, 416
  closed system, 373–380
  coprecipitation, 795
  corrosion protection, 724
  crystal growth, 791
  dissolution, 148, 375, 385
  dissolution rate, 788–791
  equilibrium constants (table), 152
  forms in lake water and in the ocean, 793
  nucleation, 811
  precipitation and dissolution, 913
  regulation composition seawater, 913
  saturation index, 357–358
  saturation profiles seawater, 795
  saturation state of lake water and seawater, 792
  size effect on solubility, 415
  solubility equilibrium, 186–192
  solubility groundwater, 383
  solubility open system, 378
  solubility seawater, 345, 346
  solubility seawater depth dependence, 382
  solubility, 374
  solubility vs. p_{CO_2}, 191
  surface charge, 564–568
  weathering, 351, 773
$CaCO_3$ system, phase rule, 412
$CaCO_3$, seawater, P and T dependence, 344, 347, 382
Calcite:
  acid lake, 237
  closed system, 373–380
  dissolution rate, 782, 789
  in seawater, 380
  point of zero charge, 539
  saturation in rivers, 189
Calomel, reference electrode, 448
$CaMg(CO_3)_2(s)$, solubility, 393–395
Capacitance, 603
  at solid-water interface, 538
  constant capacitance, 562
  Helmholtz, 562
Capacity diagram, carbonate system, 176
Carbon, distribution (table), 149
Carbon cycle, 916–927
Carbon dioxide, see also CO_2
  dissolved, 148–163
  equilibrium diagrams, 154
Carbon isotopes, 195
Carbon system, redox processes, 472
Carbon-13, 199, evolution of groundwaters, 386
Carbon-14, 195
  as a tracer for oceanic mixing, 196
Carbonate:
  ANC, 139
  equilibria, 150–163
  in a closed system, 190
  solid, equilibria, 186–192
  solid solution, 416
  solubility, 370–399
  solubility closed system, 377
  solubility open system, 378
  solubility constants (def.), 373

Carbonate (*Continued*)
  surface charge, 564–568
  surface complex formation, 563
  system closed to atmosphere, 373–380
Carbonate equilibria:
  equilibrium constants, 152, 344
  models for open and closed systems, 155
Carbonate system:
  buffering, 154
  capacity diagrams, 176
  closed, 158
  equivalence points, 156, 157
  Gran titration, 183
  open model, 160–162
  pH, 157
  seawater, 153–157, 344
  titration, 152, 156, 157
Carbonic acid:
  acidity constant, 150
  solution, 158
Carbonic anhydrase, 194
Carbonylsulfide, 932
$CaSO_4$:
  solubility, 358
  weathering, 351
Cathode, definition, 445
Cathodic protection, 721
Cathodic stripping voltammetry, 623
Cation exchange capacity, measuring, 587
Cd(II), speciation, 320
$CdCO_3$, solid solution, 417
CdS, solubility, 403
CH_3Hg^+ species, 287, 630
  stability, 631
CH_4:
  fermentation, 472
  Henry's law, 214
Charge:
  and surface potential, 835
  colloidal particles (summary), 835
  from surface complex formation, 831–839
  particle surface, 834–837
Charge reversal, by ligand, Fe_2O_3, 847–849
Chelate, 275–279
  effect, 276
Chelation, definition, 253
Chemical composition, natural waters, 872–875, 909–914
Chemical erosion, rate, 350
Chemical potential, 36, 97
  equilibrium constant, 41
  of aqueous electrolytes, 38
  of species i, 32

  pure phases and solutions, 35
  temperature and pressure, 34
Chemical reaction:
  advancement, 59
  and thermodynamics, 30
  driving force, 49
Chemical system, heterogeneous, 31
Chemical thermodynamics:
  and kinetics, 16–81
  principles, 20
Chemical weathering, *see* Weathering
Chemoautolithotrophy, (hydrothermal vents), 901
Chloramines, 699
Chlorination, 690–702
  breakpoint, 700
  phenols, 701, 716
Chlorine:
  free, definition, 699
  pε-pH diagram, 463
Chlorinity, 896
Chromate, reduction, organic, 720
Chromophore, 733
Citrate, interaction Zn(II), Cu(II), 299–301
Cl, *see also* Chlorine
  redox equilibria, 440, 463
  Cl_2, hydrolysis, 441, 442, 699
  Cl_2/Cl^-, standard potential, 450
Clay:
  binding of heavy metals, 592
  broken bonds, 592
  ion exchange, 587, 590
  ion exchange selectivity, 590
  surface complex formation, 591
ClO_2:
  oxidation by, 695
  reaction rate constants, 697
  with phenoxides (LFER), 709
Closed system:
  definition, 155
  gas-water, 216
CO_2:
  and O_2 in atmosphere, 918
  anthropogenic production, seawater, 921
  atmospheric, effect on seawater, 176
  buffer factor, 920–922
  buffering by, 885
  cycle, 149
  dissolution, 62
  equilibria, 150–163
  from atmosphere, 919
  Henry constant seawater, 345
  Henry's law, 214

kinetics of hydration, 192-195
open system, 161
seawater equilibrium, 162, 163
transfer into lakes, 195
transfer water-air, 245
CO_2 budget, anthropogenic, 919
CO_2 dissolution, thermodynamics, 46-49
Co^{2+}, adsorption rate on Al_2O_3, 766
CO_2-Acy, titration, 183
Coagulation:
 agglomeration kinetics, 859, 861
 Al(III), 845
 dosage, 854
 fatty acids, 848
 in lakes, 851
 in natural and technical systems, 862
 in the filter, 824
 kinetics, 839-842, 859
 mechanisms, 846
 polyacrylamide, 845
 Schulze-Hardy rule, 845-847
 transport, 858
 vs. filtration, 857-864
 water treatment, 852-857
Coagulation-sedimentation regimes, in particle size distribution, 829-832
Cocolithophores, 793
COD, 926
Cohesion, 608, 611
Collision efficiency factor, 839-841
 sticking coefficient, 857
Colloid stability:
 and kinetics, 839-842
 and surface charge, 837-849
 physical model, 841, 867
 surface speciation, 842-849
Colloidal, organic matter, 924
Colloidal particles, components of surface charge, 835
Colloids:
 adsorption of polymers, 849-850
 as reactants, 819
 concentrations, 818
 definition, 819
 diameter, 821
 hydrophobic and hydrophilic, 822
 in ocean, 818
 organic and biological, 820
 radionuclides, 823
 reentrainement, 864
 soil, 820
 stability, 822
 surface charge, 834-837

transport in porous media, 864-866
type present in natural systems, 820
Color constituents, removal coagulation, 854
Complex:
 definition, 255
 dissociation rate, 313
 formation rate, 313
 stability constants (table), 325-334
Complex formation:
 redox equilibria, 491
 solute vs. surface, 840
Complexation, by fulvic, humic acids, 301-304
Components, number of, 410
Computer programs, 112
 for speciation, 297
 solving equilibria, 57, 58
Concentration scales, 36, 97
 molal, 97
Condensation, of liquid water, 211
Conditional constants, 288
Conduction band, semiconductor, 753
Constant capacitance, of the oxide-electrolyte interface, 538
Constant capacitance model, 557
Constants, conditional, 288
Consumption, from increased productivity, 893
Contact angle, 608
 for solid-liquid-air interfaces, 612
Contact filtration, 862-864
Coordination chemistry, definitions, 253
Coordination number, 252-254
Copper, see Cu
Coprecipitation, $MnCO_3CaCO_3$, 417
Corrosion:
 anodic and cathodic reactions, 722
 cathodic protection, 721
 of metals, 720, 724
 passive films, 799
COS, 932, 933
Coulomb's law, 517
Coulombic Force, 13
Cr:
 equilibria, 320
 pε versus pH, 512
 speciation, 323
Cr(III), inhibition dissolution, 798
Cr(VI), 720
$Cr(OH)_3(s)$, solubility, 365-366
Critical coagulation concentration, 846, 868
Critical load, atmospheric acids, 875

Crystal growth, 812
  inhibitors, 816
  nucleation, 800–808
Cs^+, adsorption on clays, 592
Cu(II):
  ammonia, 259
  carbonate, 270
  chelate effect, 277
  complex formation various ligands, 278
  hydrolysis, 270
  phytoplankton in an eutrophic lake, 646
  solubility, 396, 399
  with citrate, 299–301
Cu^{2+}:
  analysis fresh water, 624
  in eutrophic lake, 624
CuO, solubility influence molar surface, 415

DDT, volatility, 228, 239
Debye length, 555
  and adsorbed layer thickness, 851
  and hydrodynamic thickness, 585
Debye parameter, κ, 555
Debye-Hückel limiting law, 103
Denitrification, 927–933
Deposition:
  atmospheric, 210
  coarse particles, 209
  dry, 209
  (filtration) transport, 858
  wet and dry, 207
Detergency, 611
Diagenesis:
  biochemicals, 924
  sediments, 903–908
Dialysis, 821
Diffuse layer, 556
Diffusion:
  controlling adsorption kinetics, 763
  pore water, 905
Dimethylsulfide, 932, 933
  volatility, 228, 239
Dispersion force, 518
Dissociative mechanism, substitution reaction, 713
Dissolution:
  ligand promoted, 779–783
  oxide minerals, 771–778
  proton promoted, 779–785
  transport vs surface, 775
Dissolution minerals:
  inhibition, 795–800
  rate laws, 776–786
  surface-controlled rate, 774

Dissolution rate, experimental apparatus, 787
DLVO theory:
  colloid stability, 857–871
  limitations, 871
DOC, 926
Dodecyl sulfate, adsorption, 579
Doerner and Hoskins, relation on solid solutions, 419
Dolomite:
  dissolution rate, 782, 789
  solubility, 393
Double layer:
  adsorbed polymer, 584
  diffuse, 556
  Gouy-Chapman theory, 605
  surface charge, 549
  thickness, 555
Double layer model:
  constant capacitance, 562
  diffuse, 556
DPASV:
  differential pulse anodic stripping voltammetry, 619–625
Driving force, 20

e_{aq}^-, as photoreactant, 735
Ecosystems, 4, 5
EDTA, 255, 290
  effect on solubility of $Fe(OH)_3$, 369–372
  Fe(III) exchange with Zn(II), 318
  in natural waters, 318
  regulation concentration in fresh water, 657
  thermodynamics, 49
E_H:
  electrode potential, 441–444
  from analytical information, 491
  measurement, 491
Electric double layer, *see also* Double layer
  surface charge, 549
Electrical work, maximum, 29
Electrochemical cell, 441–445
Electrochemical series, 721
  (table), 445
Electrode:
  ion selective, 498–505
  polymer membrane selective, 504
Electrode kinetics, in potential measurements, 494
Electrode potential, 441–444. *See also* Potential
  standard (table), 445
Electron:
  as a component, 438
  (photoelectron) semiconductor, 755

Electron activity, 426–432. *See also* pε
Electron charge, 13
Electron mass, 13
Electron transfer, outer-sphere, 704
Electronegativity:
  metals, 287
  of the ligand, 285
Electroneutrality, 108
Electrophile:
  def., 710
  organic, 712
Electrophoresis measurement, adsorbed polymer, 584
Electrophoresis, 558
Electrophoretic mobility:
  and zeta potential, 558
  effect of pH, 551
  γ-Al_2O_3, 836
Electrostatic correction, of constants, 568–574
Elovich equation, 770
Energy:
  conversion factors, 12
  units, 12
Enthalpy, entropy, and volume change, 34
Enthalpy change:
  and heat transferred, 27
  solid dissolution, 360–364
Entropy (S), units, 13
Entropy change, 23–26
Equilibria, basic relationships, 58
Equilibrium:
  and kinetic model, 17–20
  and rate, 60
  and thermodynamics, 16–20
  local, 81
  multicomponent, multispecies systems, 57–60
  vs. steady state, 79
Equilibrium calculations, 105–130
  graphical, 118
Equilibrium constant, 18, 41–44
  activity conventions, 335
  and pressure, 52–54
  and temperatrure, 52–54
  chemical potential, 41
  vs. rate, 703–710
Erosion, mechanical, 351
Estuary, coagulation, 862
Eutrophication, 893–895
Evaporation, isothermal, water composition, 880–884
EXAFS spectra, surface complexes, 542
Exchange current, 493

F^-, hematite surface, 573
Faraday, 13, 29, 184, 430, 443
Fatty acids, adsorption, 848
Fe, as micronutrient, 646–649
$Fe^{(VI)}$, ferrate, 510
Fe and Mn, stability relations, 513–515
Fe species:
  distribution seawater, 648
  photochemical reactions, 744, 751
Fe(aq), redox equilibria, 459
Fe(II):
  interstitial concentration, 907
  oxidation, 687
  oxygenation, 63, 683–690
  oxygenation Fe(II), role hydrolysis, 687
  oxygenation molecular orbital diagram, 687
  solubility control, 390–393
Fe(II)/Fe(III):
  as a mediator in electron transfer, 717
  at oxic-anoxic boundary, 479
  electrode potential, 448
  formal potential, 453
  role of ligands, 490
Fe(II,III)(hydr)oxides, reduction, 720
Fe(III):
  as coagulant, 852–857
  complex formation various ligands, 278
  EDTA exchange with Zn(II), 318
  hydrolysis, 264, 267–270
  phosphate colloids, 825
  solubility, 269, 273
$FeCO_3$, solubility, 374, 420
Fe(III)(hydr)oxides:
  dissolution by EDTA, 797
  dissolution by H_2S, 797
  phosphate colloid stability, 847
$Fe(OH)_3$(s):
  effect of carbonate on solubility, 422
  effect of EDTA, 369, 372
Fe_2O_3 surface:
  adsorption of F^-, 572–575
  adsorption of Pb(II), 572–575
  photoreductive dissolution, 749
  surface charge vs pH, 842–844
  zeta potential, 844
Fe_2O_3(s) → FeOOH(s), 415
FeOOH, acidity constants, 536
$Fe_3(PO_4)_2$, 906
Fe_2SiO_4(s), as a reductant, 509
Fenton Reagent, 688
FeOOH(s), solubility, 360
$FePO_4$(s), thermodynamics, 49

Ferric oxide:
  pH, surface potential, surface charge, 560
  Zn(II) adsorption, 569
Ferrichrome, 306
FeS, solubility, 400, 401, 423
Fe$-$CO$_2-$H$_2$O system, pϵ-pH diagram, 458–461
Fick's first law, 241
Film theory, for gas-water exchange, 242
Filter, pores and colloids, 821–824
Filtration:
  compared with coagulation, 857
  in natural and technical systems, 862
  operational problems in determining dissolved fraction, 825
Fine particles, solubility, effect on, 413
First order reaction, 64
Flocculation, see Coagulation
Flotation, 855
Fluorescence, 732
Fluoride, solubility, 353
Fog, 229–235
  atmospheric water, 211
  chemical composition, 231–234
Fog droplet, absorption of SO$_2$(g), NH$_3$(g), 231–234
Foraminifera, 793
Forces, at interface, 517
Formal potential, 453–455
Formaldehyde, Henry's law, 214
Free energy change, T and p dependence, 56
Free metal, see also Metals
  ecological significance, 632–637
Freon, flux into ocean, 244
Fresh water, representative, 292
Freundlich isotherm, 525
Frumkin equation, 526
FTIR spectra, surface complexes, 542
Fugacity, 35, 38
  Henry's law, 213
Fulvic acid, 141–144
  complexation by, 301–304
Furfurylalcohol:
  as probe in photoreaction, 736
  photochemistry, 741

Gas concentration, units, 217
Gas constant, 14
Gas transfer, across water-gas interface, 241–248
Gas-water:
  equilibria in closed and open systems, 216
  interface, 206–213
  partitioning, 212–216

Gibbs energy, of a system, 44
Gibbs equation:
  and adsorption, 530
  surface excess, 531
Gibbs free energy, 16, 27–29
  and extent of reaction, 45, 46
  of activation, 74
  of a system, 32
Gibbs–Duhem relationship, 34
Glass electrode, 498
Global cycling:
  anthropogenic perturbation, 915–918
  interdependence of biogeochemical cycles, 914–918
  of C, N, O, 914
  residence times, 917
Goethite, alkalimetric titration, 536
Goldschmidt, geochemical material balance, 874
Gouy–Chapman theory, 555, 560, 604, 867
Gran plots, 179–186
Greenockite, solubility, 403
Groundwater, 191
  CaCO$_3$-CO$_2$ equilibrium, 191
  calcite dissolution, 383
  chemical composition, 872–875
  dispersion, sorption and degradation, 596
  surface complex formation, 597
  transport of (ad)sorbable constituents, 594
Groundwater system, pϵ and pH response, 489
Gypsum, solubility, 358

H$^+$ balance, see Proton balance
H-Acidity, 163, 166
H-Acy, titration, 183
Hamaker constants, 518, 867–871
Hammet constants, 144
Hammet correlation, 143, 144
Hard and soft acids, 284
HCl, in atmosphere, 208
H$_2$CO$_3^*$:
  acidity constant (table), 152
  acidity constant seawater, 345
  in equilibrium with CaCO$_3$, 421
HCO$_3^-$, acidity constants (Table), 152
Heat, and work, 22
Heavy metals, see Metal
Helmholz free energy, 26
Henry's law, 206–216
  constants, 214–216
  organic compounds, 215, 239
  (table), 214
Henry's constant, and vapor pressure, 238

Heterogeneous nucleation, enhancement by specific adsorption, 811
Hg:
  biogeochemical cycling, 664
  volatility, 228, 239
Hg(II):
  activity ratio vs. pε, 665
  interaction with S(-II), 322
  methylmercury, 630
Hg(II) species, in fresh water and seawater, 665
Hg^0, equilibria, 664
HNO$_3$, equilibrium, 120
HOCl, oxidant and disinfectant, 699
Hofmeister series, 589
Hole, (Photohole) semiconductor, 755
HO$_2^{\cdot}$:
  as photoreactant, 735
  redox, 674-678
H$_2$O$_2$:
  as photoreactant, 735
  dissolution, 217
  interaction Fe(II), 688
  oxidation sulfite, 681
  photochemical production via Fe(II,III), 744
  redox, 673-678
  surface water layers, 678
H$_2$S, Henry's law, 214
H$_2$SO$_4$, production photochemical, 745
Humic acid, 141-144:
  and ^{1}O$_2$, 741
  as absorber of light, 741
  complexation, 301-304
  interaction Fe$_2$O$_3$, 848-850
  stability α-Fe$_2$O$_3$, 844
  seawater, 311
  titration curve, 142
Humic substance:
  adsorption, 581
  coagulation lakes, 851
  effect of I and pH, 585
Hydration of CO$_2$, 192-195
Hydrodynamic thickness, and Debye length, 851
Hydrogen bond, 519
  water, 7
Hydrogen electrode, standard, 446
Hydrogen peroxide, 735. *See also* H$_2$O$_2$
Hydrolysis:
  constants, 261
  metals, 260
  organic reactions, 711, 714
Hydromagnesite, solubility, 395

Hydrophobic effect, 521
Hydrophobic expulsion, 519
Hydrophobic organic pollutants, groundwater aquifer, 597
Hydrophobic substance:
  sorption, 575-579
  water solubility, 575-579
Hydrothermal vents, 901-903
  biological processes, 903
  inorganic processes, 901-903
Hydrous ferric oxide, *see* Ferric oxide
Hydroxamate, 306
Hydroxides, solubility, 359-372
Hydroxide carbonate, solubility, 389
Hydroxyl radical, 735. *See also* OH$^{\cdot}$

Illite, exchange capacity, 590
Infinite dilution scale, 98
Inhibition, mineral dissolution, 777, 795-800
Inner-Sphere, surface complexes, 540
Interception, (filtration), 858
Interfaces, *see also* Surface
  forces at, 517
  solid solution, 516-533
Interfacial energy, between two phases, 609
Interfacial tension, contact angle, 608
Intrinsic constants, oxide surface, 536
IO$_3^-$/I$^-$, seawater model, 679
Iodine, reduction by Fe(II), 491
Ion association, activity coefficients, 338
Ion exchange, 586-593
  and surface complex formation, 591
  capacity, 586
  clays seawater, 590
  concentration and selectivity, 590
  equilibria, 588
  isotherms, 589
  selectivity coefficients, 589
  sodium adsorption ratio, 591
  soils, proton balance, 879
Ion optodes, 505
Ion pair, 255-257
Ion product:
  of water, 94
  seawater, 345
Ion transport, living cells, membrane, 505
Ion-selective, electrodes, 504
Ionic interaction model, 341
Ionic medium scale, 98
Ionic strength, definition, 101
Ionization fraction, 127-130
  carbonate equilibria, 151
Iron, *see* Fe
Iron and Manganese, redox processes, 471

Iron hypothesis:
  Fe limiting in ocean, 646–649
  lakes regulation metals, 658
Irving-Williams order, 285
Isoelectric point, 561
Isomorphic substitution:
  in octahedral sheet, 565
  in tetrahedral sheet, 565
Isotope effects, 198
Isotopic fractionation, 199

K^+, valiomycin complex, 506
Kaolinite:
  dissolution rate, 782
  exchange capacity, 590
  structure, 563
  surface characteristics, 563–565, 592
Kinetics, *see also* Rate
  adsorption, 760–771
  and thermodynamics, 16–81
  complex formation, 311–319
  residence time, 5
Kinetic model, and equilibrium, 18

Lakes:
  particle size distribution, 833
  Redfield relation, 890–895
  redox sequence, 477–489
  S-transformation, 482–489
  "titrating" with biota, 485–488
Lake water, photochemical reduction of Fe(III), 752
Langmuir equation, and surface complex formation, 529
Langmuir isotherm, 521
Lanthanides, 283
Legendre transformation, 26
Lewis acid, nucleophile, electrophile, 710
Lewis concept, of acids and bases, 92
LFER:
  linear free energy relations, 702–710
  solute complex vs. surface complex, 840
Ligand:
  concentration in natural waters, 290
  definition, 253
  field effects, 282
  inorganic, 296–298
Ligand binding, by a hydrous oxide, 550, 551
Ligand exchange, 520, 534, 544–548
  analytical, metals, 623–625
Ligand-promoted dissolution, 779, 783
Light, absorption, 729
Limestone, solubility, 372

Lipophilicity, 577
London–van der Waals force, 518

Magnesian calcite, solubility, 387
Magnesite:
  dissolution rate, 789
  solubility, 395
Malachite, 397
Malonate, dissolution Al_2O_3, 784
Marcus, cross-relation, 707
Marcus theory, electron transfer, 703–710
Membrane filters, 821
  operational problems in size fractionation, 825
Membrane filtration, 823
Membranes, in living cells, 504–506
Metals:
  A and B behavior, 283, 625–628
  acidity, 260
  adsorption kinetics, 764–769
  analysis, 615–625
  anthropogenic emissions, 616
  as coagulants, 852
  bioavailability and toxicity, 632–637
  buffers for free Me, 632–635
  buffers, 279
  classification, 281–288, 625
  classification, A, B, 287, 625
  cofactors, 627
  composition phytoplankton, 640
  concentration in natural waters, 254
  concentration in rivers, lakes, sea, 618, 654, 655
  concentration regulation by interface, 648–657
  coordination (definitions), 253
  cycling in lake, 658
  distribution coefficient, 650
  ecotoxicological tolerance, 666–670
  effect of EDTA and NTA, 657
  enzymes, 627
  forms of occurrence, 258
  global cycling, 614–617, 667
  hard and soft, 283
  hydrolysis, 260, 852
  interaction particles, 256
  methylation, 288
  natural emission atmosphere, 615
  organic compounds, 287
  organometallic, 628–632
  polynuclear hydrolysis products, 261
  Redfield stoichiometry, 637–640
  regulation in oceans, 661
  removal sedimentation, 660
  rivers concentration regulation, 653

seawater speciation, 308-311
sedimentary record, 617
speciation, 258, 289, 293, 615-625
stability constants, 265
stability constants (table), 325-334
surface complex formation, 540
tableau for speciation seawater, 308
toxicity in sediments, 670
uptake by algae, 635-637
Metal binding, by a hydrous oxide, 548
Metal-Carbon, compounds, 287
Metalloids, 281
Metastability (precipitation), 356
Methane fermentation, 472
Methylation, metals, 288
Methylmercury, 287, 630
Mg(II), solubility, 394
Mg-calcite, solubility, 388
$MgNH_4PO_4(s)$, conditional solubility, 408
$MgSO_4$ system, 256
Micelle, 521
Michaelis-Menten enzyme kinetics metal uptake, 641
Microorganisms, mediation, redox processes, 468-480
Microscopic reversibility, adsorption, 769
Mineral acidity, 166
Mineral dissolution:
  enhancement and inhibition, 777, 782
  model assumptions, 778
Minerals:
  biomineralization, 800-805
  dissolution rates, 782
  surface free energies, 808
Mn and Fe, stability relations, 513-515
Mn(II), oxygenation, 684-686
Mn(III,IV)(hydr)oxides, reduction, 720
Mn^{2+}:
  cellular concentration, 643
  interstitial concentration, 903-906
$MnCO_3$, coprecipitation, 417
MnO_2, phenols (LFER), 704
$MnO_2(s)$, sediments, 906
Mn_T/Fe_T ratio, in sediments, 906-908
$Mn-CO_2-H_2O$ system, pε-pH diagram, 462
Models, 2-4
  steady state, 17
Molal concentration, 97
Molal scale, 37
Mole fraction scale, 36
Montmorillonite, 564
  exchange capacity, 590
  sorption of radionuclides, 593
  swelling, 587

N, geochemical cycle, 914
N oxides, atmosphere, 208
N_2:
  fixation, 470
  redox seawater, 679
N_2 fixation, 927-932
N_2O, nitrogen cycle, 927-933
N/P ratios, lakes, 892
Na, bioregulation seawater, 912
Na^+, and structure water, 9
Na_2CO_3:
  solution, 158
  titration, 169
$NaAlSi_3O_8$, 351
$NaHCO_3$ solution, 158
Natural water system, 2
  as open system, 17
Nernst Equation, 441-444
Nesquehonite, solubility, 395
Neutralization, $H^+ + OH^-$, 83
NH_4^+, 122:
  equilibrium, 120
  interstitial concentration, 903-906
  nitrogen cycle, 927-933
NH_3:
  dissolution, thermodynamics, 50
  equilibrium composition, 117-118
  toxicity, 121
  gas phase and water, 219-221
  Henry's law, 214
  rain, 208
NH_4Cl, equilibrium composition, 116
Nitrification, 927-933
Nitroaromatic substances, reduction, 719
Nitrogen, Redfield ratio lakes, 892
Nitrogen compounds, pollution by, 927-932
Nitrogen cycles, 914, 927-932
Nitrogen fixation, 927-932
Nitrogen system, redox processes, 469
NO, NO_2, photocatalysts for O_3, 747
NO_2^-, nitrogen cycle, 927-933, 929
NO_3^-:
  groundwater, 928
  interstitial concentration, 903-906
  photolysis, 737
  reduction, 479
  vs. PO_4, Atlantic, 890
NO_3^- reduction, 927-933
NO_x, nitrogen cycle, 927-933
NTA, 290
  regulation concentration in fresh water, 657
Nucleation:
  and crystal growth, 800-805
  homogeneous vs. heterogeneous, 805
  saturation, 803

1016 Index

Nucleophile:
  def., 710, 711
  electrophile interactions, 710–716
  sulfur, 714
Nutrients:
  limiting lakes, 891
  ocean, 889
  Redfield relations, 886–895

O, geochemical cycle, 914
O_2:
  and CO_2 in atmosphere, 918
  as oxidant, 672–679
  Henry's law, 214
  inertness interaction organic substances, 675
  one-electron steps in the reduction, 674
  oxygenation effect of hydrolysis and adsorption, 689–690
  rate vs. equilibrium constant, 689
  redox equilibrium O_3 with H_2O, 456
  dissolution, 217
  Henry's law, 214
$^1O_2$:
  as photoreactant, 735
  interaction Fe(II), 688
  redox, 674–678
O_2^-, 674, 735
O_3:
  as photoreactant, 735
  oxidation by, 691
  oxidation sulfite, 680–683
  redox, 677, 678
  troposphere and NO, NO_2, 747
  with Br^-, 694–698
Oceans:
  circulation, 920
  composition, 874, 895
  relative constancy of the composition, 895
  role of settling particles, 661–664
Octanol–water, partition, 575–579
$OH^{\cdot}$:
  interaction Fe(II), 688
  as oxidant, 676
  as photoreactant, 735
  generation in troposphere, 747
  oxidation by, 693–695
  radical production (photolysis), 737
  radical-type reactions via O_3, 691
  reaction rate constants, 695
  redox, 674–678
OH-Alk, titration, 183
Oil–water interface, 612

Open system:
  definition, 155
  gas–water, 216
  material balance, 19
Optodes, 499
Organic acids, 140–144
  rainwater, 209
Organic carbon:
  age, 923
  production, 922
Organic complexes, in seawater, 310
Organic concentration, 923
Organic ligands, effect on Zn(II) and Cu(II), 298–304
Organic reductant, and oxidant, 717
Organic substances:
  Henry's constants, 215, 239
  naturally occurring, 925
  volatility of, 238–241
  water solubility, 238–241
Organometallic compounds, 628–632
Organometalloidal compounds, 628–632
Ostwald step rule, interfacial energy, 806–808
Outer-Sphere, surface complexes, 540
Oxalate:
  adsorbed on TiO_2, 542
  dissolution Al_2O_3, 784
  ligand promoted dissolution, 775
  mineral dissolution, 783
  photoreductive dissolution of Fe_2O_3, 750
  stability $\alpha s\text{-}Fe_2O_3$, 844
Oxalato iron(III) complex, photochemistry, 744
Oxidants, photochemical, in the atmosphere, 746
Oxidation:
  pyrite, 690
  transition ions, 690
Oxidation and reduction, see Redox
Oxidation state, 426–429, 926
Oxide, surface acidity constants, 536
Oxide film, passive, 722
Oxide surface, 533
  equilibria, 534
  reactions with H^+, 533
  surface groups, 533
  titration curve, 536
Oxides, solubility, 359–372
Oxidizing capacity, in lakes, 488
Oxoanions, inhibition dissolution, 796
Oxygen, dissolution in water, 214
Oxygenation, metal ions (LFER), 710
Ozone, production photochemical, 746
Ozone, see O_3

Index 1017

P, fate in ocean, 911
p-Alk, titration, 183
Particles:
  aggregation, 822
  colloidal, 819
  concentration effect, 824
  in lakes, 833
  regulating metal ions in rivers, 653
  river-borne as colloids, 820
  role concentration regulation, 657
  settling in lakes, 659
  size spectrum, 821
  surface, 516
  surface charge components, 835
Particle size distribution:
  discrete and continuous, cumulative, 827–833
  in sewage, 832
Particle surface, metals concentration regulation, 649–657
Passivity, (corrosion), 722–724, 799
Passive film, 724
Pb(II):
  hematite surface, 572
  in lake and ocean, 662
  speciation fresh water, 289, 294–296
Pb^{2+}, distribution coefficient, 650–652
$Pb-H_2O-CO_2$, $p\varepsilon$ versus pH diagram, 511
PCB, volatility, 228, 239
$p\varepsilon$:
  and pH, 430
  as a Master Variable, 434
  def., 429–431
  effect of complex formers, 452, 489
  formal computation, 432
  from analytical information, 491
  intensity and capacity, 455
  model for the sea, 677
  nonequilibrium system, 677
Periodic table, 15
  important elements, 10
Peroxy radicals, 743
$p\varepsilon$-pH:
  diagrams, 455–464
  diagrams for biologically important elements, 461, 477
pH:
  and $p\varepsilon$, 430
  as a master variable, 118
  definitions, 101
  interstitial water, 903–906
  of CO_2, HCO_3^-, and CO_3^{2-}, 157
  seawater p_{CO_2} increase, 175
  surface charge and surface potential, 559–561
  zero point of charge, 534
pH concepts, in seawater, 343
pH conventions, 101
pH measurement:
  glass electrode, 499–503
  rain, 180
pH scales, 97
  NBS seawater, 343
  seawater (Hanson), 344
  total [H_T^+] seawater, 343, 344
Phase:
  coexistence, 409–412
  partial molar properties, 33
Phase rule, derivation, 409–412
Phenol:
  chlorination, 701
  with ClO_2 (LFER), 709
Phosphate, see also PO_4
  adsorption on Al_2O_3, 526
  elimination sewage, 863
  inhibition dissolution, 798
  in seawater, 309
  in soils, redox regulation, 485
  limitations in lakes, 892
  precipitation, 423
  solubility, 407
  solubility in presence of Fe(III), 423
Phosphorescence, 732
Photoactivation, 730
Photocatalytic:
  degradation of organochlorine compounds, 756
Photochemical, oxidation of 4-chlorophenol, 754
Photochemical processes, 726–759
Photochemical reduction, Fe(III) in acid lakes, 752
Photochemical transformations, direct and indirect, 726–733
Photolysis:
  of transition metal complexes, 743, 744
  rate, 731
Photolysis rates, organic contaminants, 734
Photons, environmental factors, 727–729
Photoreactants, 728–730, 735–743
  steady state concentration in lake, 740
Photoreductive, dissolution of Fe_2O_3, 749
Photosynthesis:
  as a photochemical process, 726
  biochemical cycle, 467–476
  effect on Alk, 173
  primary production, 922

1018   Index

Photosynthesis and respiration, Redfield relation, 886–895
Photosynthetic bacteria, 483
pH_{pzc}, 538, 835
   effect of metals and ligands, 561
   of colloidal hematite, 842
pH_{pznpc}, 538
Phthalate, 115, 125
   dissolution Al_2O_3, 784
Phyllosilicate, 564
Phytochelatin, 306
Phytoplankton, composition of metals, 640
Planck constant, 14
PO_4:
   fate ocean, 911
   stability α-Fe_2O_3, 844
PO_4 vs NO_3^-, natural waters, 890
Point of zero charge, 835
   definitions, 553
   pH dependence, 538
Polarization curve (corrosion), 722
Polyaspartic acid, stability α-Fe_2O_3, 844
Polyelectrolytes, 583–586
Polymer:
   adsorption colloid, 849, 850
   at surface, hydrodynamic thickness, 585
   sorption, 581–586
Polymer membrane, electrodes, 504
Potential, 441–445
   formal, 451
   measurements, 491–498
   mixed, 494, 495
   streaming, 558
   Zeta, 558, 603
Potential–pH diagrams, 455–464
Potentiometric determination, of solutes, 498
Potentiometric titration, 503
Power, units, 13
Precipitation:
   salts, 353–358
   types of precipitates, 356
   vs. coagulation, 852
Pressure:
   and equilibrium constant, 54
   effect on equilibrium constants seawater, 347
   units, 13
Primary production, natural waters, 922
Proton balance:
   effect vegetation, 875–879
   processes that modify, 876
Proton condition, 108
Pyrite, oxidation, 690
pzc, 553, 835
   γ-Al_2O_3, 836

pznpc, 553, 835
pzse, 553

Quality criteria, metals in fresh waters, 666
Quantum yield, 730, 734
Quartz:
   dissolution, 781
   dissolution rate, 782
   solubility, 367, 368

Rainwater, 89
   composition, 212, 230
   "pristine," 161
   stoichiometric model, 209
   washout of pollutants, 228
Rates:
   ACT elementary reaction, 76
   ACT, catalysis, 76
   adsorption, 760–771
   and equilibria, 60
   and mechanism, 61–63
   complex formation, 311–319
   complex reactions, 66
   dissociation, 78
   effect of ionic strength, 75
   effect of pressure, 76
   effect of temperature, 73–76
   elementary reactions, 64
   hydration, 78
   ligand exchange reactions, 317
   metal exchange reactions, 314
   neutralization, 78
   proton-transfer reactions, 70
   reaction order, 64–66
   redox, 78
   reversible reaction, 66–69
   solvent exchange, 78
   theory of elementary processes, 69
   vs equilibrium constant (LFER), 702–710
   water exchange, 312
Reactions:
   in series, 66
   rates of, 65–68
   reversible, 68
Reaction quotient, 42
Reaction rate, *see* Rates
Redfield stoichiometry, 886–895
   lakes, 890–895
   metals, 637–640
   rivers regulation metals, 653
Redox:
   conditions in natural waters, 464–491
   equilibrium constants (Table), 465
Redox components, in groundwater, 480
Redox equilibria, electron activity, 426–432

Redox intensity, *see also* pε
  in soil and water, 478
Redox potential:
  effect complex formers, 452–455, 490
  measured in groundwater, 481
  measuring, 491–498
  one-electron, 703
Redox process:
  free energy, 446, 477
  light-induced, 755
  sediment-water, 903–906
  sequence progressive reduction, 473–480
Redox reaction:
  balancing, 428
  in lake sediments, 906
  organic substances, 710–716, 718
  sequence of, 475–479
  tableaux for, 438–441
Redox reactivity, and thermodynamics, 701–710
Redox system, poising, 476
Reductive dissolution, minerals, 779
Reference electrode, 447
Reference state, 36, 37
Residence time, 10
  elements in fresh and seawater, 10
  global cycling, 917
  of water, 5
  seawater, 896–891, 898
Respiration, 886
Retardation, of pollutants due to adsorption, 595
Reversibility, microscopic, 61
  SiO_2 phases, 424
Reversible process, 24
River, major ion composition, 899
River water, composition, 188
ROO˙, as photoreactant, 735

S cycle, 932, 933
  bioregulation seawater, 912
S(−II), solubility (table), 400
Salicylate:
  dissolution Al_2O_3, 784
  interaction Fe_2O_3, 848–850
Salinity, 896
Saturation index, $CaCO_3$, 357–358
Schulze-Hardy rule, 845–847, 870
Seawater:
  basalt reactions, 899
  buffer intensity, 159
  calcite solubility, 380
  carbonate system, 153–157
  CO_2 equilibria, 162, 163
  composition, 909
  effect of p_{CO_2}, 176
  equilibrium constants (definition), 347
  equilibrium constants (table), 344
  equilibrium diagram, 159
  Gran titration, 186
  in equilibrium with the atmosphere, 162
  input-output balance for major ions and alkalinity, 900
  ion composition, 899
  kinetic model for the chemical major composition, 896
  residence time elements, 896–891
  speciation, 293, 308–311
  tableau for speciation, 308
  thermohaline circulation, 920
Second order reaction, 64
Sediment, metals, 617
Sediment colloids, 820
Sediment-Water, interface, 903–908
Sediments, memory record, 906
Semiconducting minerals, 753
  band gap energies, 757
Semiconductor particle, light absorption, 756
Semiconductors, 753–759
  extrinsic, 754
Sensitizers, photoreactants, 728–730
Sewage, particle size distribution, 832
Shear plane, 556
Si, bioregulation seawater, 912
Siderite, solubility, 374
Siderophores, 306, 648
Silicate:
  dissolution, 781
  in seawater, 309
  weathering, CO_2 consumption, 773
Silicate frame, dissolution, 776
Silicate mineral, "monomolecular" layer, 787
Siloxane, ditrigonal cavity, 593, 564, 653
Siloxane surface, 563, 592
Singlet oxygen, 735, 740–742
  molecular orbital diagrams, 741
Singlet states, excited, 731
SiO_2:
  aqueous, 367
  solubility, 367, 368
Sn, trialkyltin compounds, 631, 632
S_N1 reaction, 713
S_N2 reaction, 712
SO_2:
  and formaldehyde, 226
  atmosphere, 208
  cycle, 932, 933
  Henry's law, 214
  in fog, 229–233

SO_2 (*Continued*)
  oxidation, 229
  solubility water, 221–226
SO_2-water, closed system, 225
SO_4^{2-}:
  redox, 466
  sediments, 906
SO_4^{2-} production atmosphere, and solar flux, 745
SO_4–S(s)–H_2S, pε-pH diagram, 457
Soda lake, 884
Sodium adsorption ratio, 591
Sodium dodecyl sulfate, adsorption, 579
Soil, redox processes, 476–480
Soil colloids, 820
Soil system, 516
Solar radiation, water column, 728
Solid solution, 416–420
  heterogeneous, 418
Solubility:
  and hydrolysis, 272
  and saturation, 356
  effect of particle size, 413
  enhancement by complex formation, 368
  equilibria, 351, 353–358
  fine particles, 413
  metals, 272
  minerals (data), 362–369
  of oxides and hydroxides, 359–372
  product (def), 351
  salts, 353–358
  salts, pressure dependence, 349
Solubility product:
  and saturation, 357
  compilations, 355
Solvated electron, 743
Sorption:
  hydrophobic substances, 575–579
  to organic colloids, 577
Speciation:
  chemical (definition), 10, 257
  effect on transport in groundwater, 597
  elements, 10
  in fresh waters, 289, 308
  (inorganic) seawater, 289, 308–311
Spreading:
  oil-water, 613
  water on a hydrophobic solid, 611
Spreading coefficient, 609
Sr^{2+}, solid solution in $CaCO_3$, 418
Stability constants, formulation, 265
Stability ratio, colloids, 842, 868–871
Standard state, 36, 38
Steady state:
  composition, 80–82
  open system, 19
  terminology, 79
  vs. equilibrium, 79
Steady state model:
  chemical composition of natural waters, 909–914
  seawater, 897–901
Stern, surface charge density, 554, 835
Stern layer, 557
Strengite, 485
Sulfate, solubility, 353
Sulfide:
  acid-base equilibria, 126
  Black Sea, 889
  solubility, 398
  solubility transition elements, 286
Sulfite:
  oxidation by H_2O_2, 681–683
  oxidation by O_3, 680–683
Sulfur, transformations in lakes, 482–484
Sulfur cycle, 932, 933
Sulfur system:
  pε-pH diagram, 456
  redox processes, 471
Superoxide, 674, 735
Surface, molar, 414
Surface charge:
  alkalimetric and acidimetric titration curves, 535
  and the electric double layer, 549–588
  and surface potential, 555, 559–561
  as a function of pH and ionic strength, 539
  by isomorphous replacements, 552
  carbonates, silicates, sulfides, and phosphates, 562
  correction of constants for, 568–574
  density, 554
  effect of H^+, Me^{z+}, 561
  hematite, 842
  of carbonates, 564–568
  origin of, 549–554
  particle, 554
Surface chelate, 544
  weathering, 775
Surface complex:
  acidity constants, 535
  surface site density and net surface charge, 838
Surface complex formation, 520
  and surface charge, 836–838, 840
  carbonates and sulfides, 563
  intrinsic constants, 568
  Langmuir equation, 529
  pH dependence, 545, 571
  surface charge correction, 568–574

Tableau, 547
ternary, 534
Surface complexation, 519, 520
  criteria for models, 534
  outer- and inners-sphere, 540
  ternary complexes, 545
  with ligands, 544-546
  with metal ions, 540-544
Surface energy, 517
Surface free energy, minerals, 808
Surface potential:
  and net total particle charge, 554
  and surface charge, 555, 835
Surface precipitation, 812-814
  and heterogeneous nucleation, 814-816
  ferric oxide, 814
Surface protonation, 533
  and dissolution, 785
Surfactant aggregates, 580

Tableau, 110-118
Temperature, and equilibrium constant, 53
Temperature dependence, solid dissolution, 360-364
Tenorite, 397
Th colloids, 823
Thermodynamics:
  and chemical thermodynamics, 21
  and kinetics, 16-81
  CO_2 dissolution, 46-49
  equilibrium, 16-20
  first law, 22-29
  principles, 20-29
  second law, 23-29
Thermodynamic analysis, 24
Thermodynamic functions, auxiliary, 26
Time, and reaction advancement, 58-60
TiO_2:
  photochemical processes, 749
  surface complex formation, 749
Titration:
  acids, 128-140
  Alk, 179-186
  carbonate system, 152, 169
  of Na_2CO_3 solution, 169
Titration curve, buffer intensity, 137
TOC, 926
Toxicity:
  metals, 632-637
Trace elements, see also Metals
  cycling in lake, 658
  regulation in oceans, 661-664
Transition elements, 281-285
Transition metal ion, oxygenation, 690, 710

Transition-state theory, see Activated complex theory
Transport, of pollutants in groundwater, 594
Triple-Layer model, 558
Triplet, state (photoexcitation), 732

Units, international, 12
Uptake of iron, phytoplankton, 647

Valence band (semiconductor), 755
Valinomycin complex, potassium, 506
Van der Waals:
  attraction energy, 518, 841, 867
  dispersion force, 610
Vapor pressure, volatility organic substances, 238-241
Velocity gradient (coagulation), 858-860
Vents, see Hydrothermal vents
Vermiculite, 564
Vivianite, 485, 906
Voltammetric techniques:
  to distinguish between dissolved and particulate, 824

Waste technology, oxidants, 691-702
Water:
  as solvent, 6-9
  dipole moment, 7-8
  exchange, 78
  in atmosphere, 206-213
  interstitial concentration, 907
  ion-dipole moment, 9
  ion product, 55-57, 94, 345
  ion product seawater, 345
  redox stability, 456, 674
  properties, 12
  self-ionization, 94
  structure, 6-9
Water droplets, atmospheric, photochemical processes, 748
Water exchange:
  adsorption kinetics, 765
  rates, 312
Water ionization, pressure and temperature effect, 55-57, 345
Water repellency, 611
Water solubilities, octanol water partition coefficients, 575-579
Water technology and treatment:
  coagulation, filtration, flotation, 852-857
  oxidants, 691-702
  particle size distribution, 832
Weathering:
  and the proton balance, 875
  atmospheric constituents, 235-238

Weathering (*Continued*)
  dissolution minerals, 771–778
  geochemical processes, 77
Weathering rate:
  catchment area, 350
  laboratory and field, 786
Wetting, 611
Work:
  external, 28
  forms of, 22

XAD-resins, 141

Young equation, 609

Zero proton condition, and isoelectric point, 561

Zeta potential, 556, 558, 603
  adsorbed polymer, 584
  α-Fe_2O_3, 842
  Al_2O_3, 603, 604
Zn, analysis seawater, 620–623
Zn(II):
  adsorbed on hydrous iron(III) oxide, 558
  in presence of CN^-, 324
  partition, solution particles, 652
  regulation concentration in fresh waters, 656
  solubility, 273
  speciation fresh water, 294–296
  with citrate, 299–301
$Zn(OH)_2$, solubility, 275
Zn^{2+}, cellular concentration, 643–646
$ZnCO_3$, solubility, 374
ZnO(s), solubility effect of oxalate, 369

ENVIRONMENTAL SCIENCE AND TECHNOLOGY

A Wiley-Interscience Series of Texts and Monographs

Edited by JERALD L. SCHNOOR, *University of Iowa*
ALEXANDER ZEHNDER, *Swiss Federal Institute for Water Resources and Water Pollution Control*

PHYSIOCHEMICAL PROCESSES FOR WATER QUALITY CONTROL
  Walter J. Weber, Jr., Editor
pH AND pION CONTROL IN PROCESS AND WASTE STREAMS
  F. G. Shinskey
AQUATIC POLLUTION: An Introductory Text
  Edward A. Laws
INDOOR AIR POLLUTION: Characterization, Prediction, and Control
  Richard A. Wadden and Peter A. Scheff
PRINCIPLES OF ANIMAL EXTRAPOLATION
  Edward J. Calabrese
SYSTEMS ECOLOGY: An Introduction
  Howard T. Odum
INTEGRATED MANAGEMENT OF INSECT PESTS OF POME AND STONE FRUITS
  B. A. Croft and S. C. Hoyt, Editors
WATER RESOURCES: Distribution, Use and Management
  John R. Mather
ECOGENETICS: Genetic Variation in Susceptibility to Environmental Agents
  Edward J. Calabrese
GROUNDWATER POLLUTION MICROBIOLOGY
  Gabriel Bitton and Charles P. Gerba, Editors
CHEMISTRY AND ECOTOXICOLOGY OF POLLUTION
  Des W. Connell and Gregory J. Miller
SALINITY TOLERANCE IN PLANTS: Strategies for Crop Improvement
  Richard C. Staples and Gary H. Toenniessen, Editors
ECOLOGY, IMPACT ASSESSMENT, AND ENVIRONMENTAL PLANNING
  Walter E. Westman
CHEMICAL PROCESSES IN LAKES
  Werner Stumm, Editor
INTEGRATED PEST MANAGEMENT IN PINE-BARK BEETLE ECOSYSTEMS
  William E. Waters, Ronald W. Stark, and David L. Wood, Editors
PALEOCLIMATE ANALYSIS AND MODELING
  Alan D. Hecht, Editor
BLACK CARBON IN THE ENVIRONMENT: Properties and Distribution
  E. D. Goldberg
GROUND WATER QUALITY
  C. H. Ward, W. Giger, and P. L. McCarty, Editors
TOXIC SUSCEPTIBILITY: Male/Female Differences
  Edward J. Calabrese
ENERGY AND RESOURCE QUALITY: The Ecology of the Economic Process
  Charles A. S. Hall, Cutler J. Cleveland, and Robert Kaufmann
AGE AND SUSCEPTIBILITY TO TOXIC SUBSTANCES
  Edward J. Calabrese

ECOLOGICAL THEORY AND INTEGRATED PEST MANAGEMENT PRACTICE
  Marcos Kogan, Editor
AQUATIC SURFACE CHEMISTRY: Chemical Processes at the Particle Water Interface
  Werner Stumm, Editor
RADON AND ITS DECAY PRODUCTS IN INDOOR AIR
  William W. Nazaroff and Anthony V. Nero, Jr., Editors
PLANT STRESS-INSECT INTERACTIONS
  E. A. Heinrichs, Editor
INTEGRATED PEST MANAGEMENT SYSTEMS AND COTTON PRODUCTION
  Ray Frisbie, Kamal El-Zik, and L. Ted Wilson, Editors
ECOLOGICAL ENGINEERING: An Introduction to Ecotechnology
  William J. Mitsch and Sven Erik Jorgensen, Editors
ARTHROPOD BIOLOGICAL CONTROL AGENTS AND PESTICIDES
  Brian A. Croft
AQUATIC CHEMICAL KINETICS: Reaction Rates of Processes in Natural Waters
  Werner Stumm, Editor
GENERAL ENERGETICS: Energy in the Biosphere and Civilization
  Vaclav Smil
FATE OF PESTICIDES AND CHEMICALS IN THE ENVIRONMENT
  J. L. Schnoor, Editor
ENVIRONMENTAL ENGINEERING AND SANITATION, Fourth Edition
  Joseph A. Salvato
TOXIC SUBSTANCES IN THE ENVIRONMENT
  B. Magnus Francis
CLIMATE-BIOSPHERE INTERACTIONS
  Richard G. Zepp, Editor
AQUATIC CHEMISTRY: Chemical Equilibria and Rates in Natural Waters, Third Edition
  Werner Stumm and James J. Morgan
PROCESS DYNAMICS IN ENVIRONMENTAL SYSTEMS
  Walter J. Weber, Jr., and Francis A. DiGiano
ENVIRONMENTAL CHEMODYNAMICS: Movement of Chemicals in Air, Water, and Soil, Second Edition
  Louis J. Thibodeaux